# Regenerative Biology and Medicine

# Regenerative Biology and Medicine

**David L. Stocum**

AMSTERDAM • BOSTON • HEIDELBERG • LONDON
NEW YORK • OXFORD • PARIS • SAN DIEGO
SAN FRANCISCO • SINGAPORE • SYDNEY • TOKYO
Academic Press is an imprint of Elsevier

Academic Press is an imprint of Elsevier
30 Corporate Drive, Suite 400, Burlington, MA 01803, USA
525 B Street, Suite 1900, San Diego, California 92101-4495, USA
84 Theobald's Road, London WC1X 8RR, UK

This book is printed on acid-free paper. ∞



**Library of Congress Cataloging-in-Publication Data**
Application Submitted

**British Library Cataloguing-in-Publication Data**
A catalogue record for this book is available from the British Library.

ISBN 13: 978-0-12-369371-6
ISBN 10: 0-12-369371-3

For information on all Academic Press publications
visit our Web site at www.books.elsevier.com

Printed in Canada

06 07 08 09 10 9 8 7 6 5 4 3 2

# Table of Contents

# Preface

Most of my bench research career was spent at the University of Illinois Urbana-Champaign investigating the mechanisms by which the blastema of the regenerating amphibian limb self-organizes into replacement parts. Subsequently, during a relatively long decanal administrative stint at the "starship" Indianapolis campus of Indiana University (Indiana University-Purdue University Indianapolis), I became interested in regeneration from a much broader perspective, particularly when the dogma of "no new neurons" disintegrated in 1992 with the confirmation of reports from the 1960s that the mammalian brain harbored stem cells that were capable of making new neurons and glia. This new perspective led to the development of a beginning graduate course titled "Regenerative Biology and Medicine", which I have now taught in the Department of Biology for several years.

Regenerative biology and medicine is a rapidly developing field that seeks to understand the mechanisms of regeneration and apply that understanding to the development of therapeutic strategies to restore tissue structure and function compromised by injury or disease. The study of regenerative biology covers a wide range of tissues, organs and appendages in several animal models. Regenerative medicine encompasses cell transplantation, construction of bioartificial tissues ("tissue engineering") and the chemical induction of regeneration from residual tissues *in situ*. Several texts on tissue engineering are available, but these are geared primarily to engineers. No texts exist on the broader field of regenerative biology and medicine. Furthermore, in the literature on regenerative medicine, I have seen virtually no overt recognition of the fact that to realize its immense potential, we must gain a much deeper understanding of the biology of regeneration than we currently possess. Also missing is any discussion of strong non-mammalian regenerators, such as the amphibians, as model systems that are relevant to regenerative medicine. These models are important, since they regenerate tissues and complex structures that mammals do not, thus offering an opportunity to learn what is different about them and apply it to mammalian regeneration. Thus, the primary objective of the present book was to bring the many different facets of regenerative biology and medicine together in one place to give the reader an overall view of the kinds of fascinating basic and clinically-oriented research that have been, and are being done, how each informs the other, what progress has been made, and the therapeutic potential that exists.

This book is directed to a broad audience of graduate and advanced undergraduate students in biology, chemistry and bioengineering, medical students, academic and clinical physicians, and research investigators. Chapter 1 gives a general overview of the subject. Chapter 2 deals with fibrosis, the most common response to tissue damage caused by injury or disease. Chapters 3–12 deal with the regenerative biology and the corresponding regenerative medicine of specific clusters of tissues. Chapter 13 details experiments on the developmental potency of adult stem cells and Chapter 14 discusses the biology of appendage regeneration and attempts to stimulate the regeneration of appendages in adult frogs and mammals. Chapter 15 reviews the biological and bioethical issues and challenges to the development of a full-fledged regenerative medicine.

The book can be read in several ways. Reading Chapters 1 and 15, and/or all the chapter summaries will provide an abridged content. I have provided the details of regeneration and regenerative medicine for different tissues, organs and appendages between these two chapters. These details are important for understanding mechanisms of regeneration and the kinds of efforts that are being made to manipulate tissues toward regenerative pathways or faster healing by fibrosis.

The field of regenerative biology and medicine is moving incredibly fast, with many new discoveries reported each week, creating torrents of published papers. A massive amount of literature went into the synthesis of this book, yet I have sampled probably no more than 5% of the work on regenerative biology and medicine that exists. The data described here will undoubtedly already be obsolete when the book is published, but the ideas and directions of the field set forth here will, I hope, remain clear and inviting of new interpretations by current and upcoming investigators.

I am grateful to IUPUI for granting me a leave at the end of my deanship to write this book and to Eli Lilly and Company and the Indiana 21st Century Research and Technology Fund for their support of the research of the Center for Regenerative Biology and Medicine, some of which is included here. I very much appreciate the helpful critiques of the manuscript by friends and colleagues: Karen Crawford, Luisa Ann Di Pietro, E. Brady Hancock, Andrew Hawks, Ellen Heber-Katz, George Malacinski, Anton Neff, and Rosamund Smith. I am particularly indebted to my friend and colleague in New Delhi, Iqbal Niazi, who read the whole manuscript and whose insights on content and organization were invaluable. Finally, I am grateful to my editor, Luna Han, for her encouragement, guidance and patience in this endeavor. Any sins of omission or commission are mine and mine alone.

David L. Stocum
Indianapolis
August, 2006

# 1 An Overview of Regenerative Biology and Medicine

## INTRODUCTION

Perpetuation of the species requires that some minimum number of individuals survive to reproductive maturity. Individual survival in the face of environmental insults and entropic processes requires a mechanism that can maintain the functional integrity of tissues. That mechanism is regeneration. Regeneration maintains or restores the original structure and function of a tissue by recapitulating part of its embryonic development. Some tissues, such as blood and epithelia, undergo continual turnover and thus must replace themselves continually, a process called maintenance, or homeostatic regeneration. These tissues, as well as a number of others, also regenerate on a larger scale when damaged, a process called injury-induced regeneration. The relationship among regeneration, life, and death has been concisely summed up by one of the great masters of regenerative biology, Richard J. Goss, in the following words. "If there were no regeneration there could be no life. If everything regenerated there would be no death. All organisms exist between these two extremes. Other things being equal, they tend toward the latter end of the spectrum, never quite achieving immortality because this would be incompatible with reproduction" (Goss, 1969). In other words, we are in a constant battle that pits our ability to locally reverse the second law of thermodynamics (regenerate) against inexorable entropic processes, a battle that we ultimately lose as individuals, but win as a species through reproduction.

Nature has provided us with another mechanism for injury-induced repair, fibrosis, which patches a wound with scar tissue whose structure is quite different from the original tissue structure. Repair by fibrosis maintains the overall integrity of the tissue or organ, but at the expense of reducing its functional capacity. Fibrosis is the result of an inflammatory response to injury that produces a fibroblastic granulation tissue that is then remodeled into a virtually acellular collagenous scar. Mammalian tissues that do not regenerate spontaneously are repaired by fibrosis. Even tissues capable of regeneration may repair by fibrosis if they suffer wounds of a size that exceeds their regenerative capacity. Furthermore, chronic degenerative diseases can promote fibrotic repair, neutralizing inherent regenerative ability. Prominent examples of tissues that undergo repair by scar tissue when injured are the dermis of the skin, meniscus and articular cartilage, the spinal cord and most regions of the brain, the neural retina and lens of the eye, cardiac muscle, lung, and kidney glomerulus. It is not that these tissues have no ability to regenerate. Many, if not all, initiate a regenerative response to injury, but the response is overwhelmed by a competing fibrotic response.

The cost of tissue damage due to regenerative deficiency is enormous in terms of health care (estimated to exceed $400 billion/yr in the United States alone), lost economic productivity, diminished quality of life, and premature death. In the United States, the health care costs of spinal cord injuries alone exceed $8 billion per year and $1.5 million per patient over a lifetime. Diabetes, heart, liver, and renal failure, chronic obstructive pulmonary diseases such as emphysema, macular degeneration and other retinal diseases, diseases of the nervous system such as multiple sclerosis, amyotrophic lateral sclerosis, Parkinson's, Huntington's, and Alzheimer's, arthritis, burns, and traumatic injuries that damage skin, muscles, bones, ligaments, tendons, and joints are other major contributors to these fiscal and human costs. Thus, medical science seeks ways not only to prevent and cure underlying disease, but also to restore the structure and function of tissues and organs damaged by disease or traumatic injury.

There is currently great excitement over the possibility of replacing damaged body parts through the new field of **regenerative medicine**. What tends to be forgotten, however, is that regenerative medicine cannot develop its potential without a fundamental understanding of the biology of regeneration, driven by

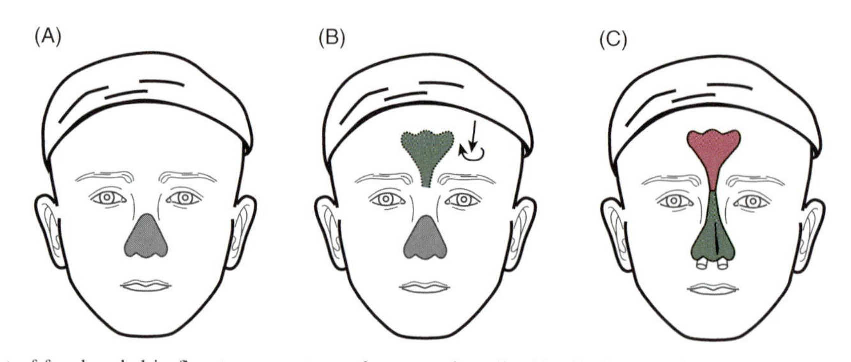


**FIGURE 1.1** Autotransplant of forehead skin flap to reconstruct the nose, described by Sushruta ~1000AD. The flap was cut to conform to the outline of a nose, peeled down and twisted 180° at the stalk (arrows). It was then positioned over the nasal area with the dermal side down and held in place with wooden rods until healed.

advances in molecular, cell and developmental biology, and information science and systems biology. This understanding is far from complete. Thus, I call this new field of tissue restoration **regenerative *biology and* medicine** to emphasize that understanding the biology of regeneration is prerequisite to establishing and practicing a regenerative medicine. The objective of regenerative biology is to understand the cellular and molecular mechanisms of regeneration where it occurs naturally and how these mechanisms differ from the mechanism of fibrosis. Regenerative medicine then seeks to use this understanding to devise therapies that will stimulate the functional regeneration of damaged human tissues that do not regenerate spontaneously, or whose regenerative capacity has been compromised.

## A BRIEF HISTORY OF REGENERATIVE BIOLOGY AND MEDICINE

Our current knowledge of repair by fibrosis and regeneration has a long and fascinating history. Imprints of hands with missing fingers on the walls of Paleolithic caves have been interpreted as examples of amputations (Goss, 1991). Severe injuries such as penetrating wounds, multiple fractures, spinal cord compression, or damage to the eyes must have occurred frequently in prehistoric times, often with high morbidity and mortality. Methods to facilitate wound healing and surgical intervention were central foci in the medicine of ancient Sumerian, Egyptian, Chinese, Indian, and Incan civilizations (Majno, 1975; Brown, 1992; Falabella, 1998). The cleansing and debridement of wounds was a common practice in ancient cultures, and many different vegetable and mineral concoctions were applied topically to treat wounds. Honey and wine were used by the ancient Chinese and Egyptians as antiseptics. For 2000 years the Chinese have used a bread mold to treat minor burns (Majno, 1975; Fu et al., 2001). Trephination to relieve intracranial pressure was performed by the Incas in the treatment of head wounds (Majno, 1975). A thousand years ago, the Indian physician, Sushruta, described the use of autogeneic skin transplants to reconstruct severed noses and ears (Majno, 1975). The method is illustrated in **FIGURE 1.1**.

The Greek and Roman physicians Hippocrates (460–370 BC), Celsus (25 BC–50 AD), and Galen (130–201 AD) contributed greatly to the development of medicine, including the treatment of wounds (Brown, 1992). Celsus described the four cardinal signs of wound inflammation, "rubor et tumor cum calore et dolore"(redness and swelling with heat and pain). Part of Galen's career was spent caring for injured gladiators, which gave him extensive experience in treating wounds to many parts of the body. Galen compiled virtually all that was known about anatomy, physiology, and medical treatment at the time (much of it erroneous) into at least 35 volumes. When the Western Roman Empire collapsed near the end of the 5th century AD, Galen's texts were translated into Arabic, which was the language of medicine in the Eastern Roman Empire, and later from Arabic into Latin. These texts were the dominant guides for medical practice to the end of the Middle Ages.

A creative resurgence in biology and medicine took place from the 14th to the 17th centuries, based on more accurate and detailed observations of form and function. Major contributions were the anatomical descriptions of embryonic and adult structure made by Da Vinci (1452–1519), Vesalius (1514–1564), Paracelsus (1493–1541), Fallopius (1523–1562), and Fabricius (1537–1614). Descartes (1596–1650) and Borelli (1608–1679) wrote important texts on physiology, and Harvey (1578–1657) produced his famous treatises on the circulation of the blood and the reproduction of animals.

The surgeon Guy de Chruliac (1300–1370) published *Wounds and Fractures* in 1363, a book that details many kinds of wounds and how to treat them. Wilhelm Fabry (1560–1634) described nearly 70 topically applied formulations for treatment of wounds, many of which have been recently re-examined and found to have true therapeutic value (Kirkpatrick et al., 1996).

In the 17th century, the scientific method, formalized in Newton's four rules of reason, provided a powerful way of answering questions about how things work, changing our view of the world forever (Van Doren, 1991). Galileo (1564–1627), Kepler (1571–1630), and Newton (1642–1727) discovered the physical laws relating forces and motion. These discoveries, and the mathematical ideas of Descartes (1596–1650), established the philosophy of mechanism, which viewed all natural objects, including living organisms, as machines that operated according to strict mathematical laws of physics.

In biology, the invention of the compound microscope in the early 1600s made it possible to view biological structure in greater detail than ever before and thus better understand the nature of biological phenomena. This technological leap forward led to the development of the science of microscopic anatomy in the 18th century. The comparative anatomist John Hunter (1728–1793) studied healing skin wounds and discovered granulation tissue and the transitional role it played in scar tissue formation (Brown, 1992). The dogma of organismic preformation, which held that the mechanism of ontogeny was growth of a tiny predelineated adult within the egg, was toppled by the studies of C.F. Wolff on chick embryos, which showed that the embryo developed in a continuum of epigenetic steps that built form out of amorphous substance.

The 19th century saw numerous medical and surgical advances that improved the prospects of recovery from serious injury and disease (Allen, 1977). The development of ether anesthesia made pain-free surgery possible and thus increased the types of surgical interventions that could be made in the human body to treat different conditions. Surgery, however, also increased the possibility of death by systemic bacterial infection. Joseph Lister, a student of Pasteur, introduced the use of dressings soaked in carbolic acid and strict hygenic measures in hospitals to combat sepsis after surgical operations (Brown, 1992). Nineteenth-century surgeons revived the techniques of skin grafting described by Sushruta many centuries earlier (Majno, 1975).

The most significant development of the 19th century for the future of biology and medicine, however, was the rise of materialistic biology. Until the mid-18th century, the motive power of biological organisms was believed to be an immaterial vital force (Coleman, 1977). However, experiments begun by Lavoisier (1743–1794) in the second half of the 18th century, and carried on by others into the 19th century, showed that life depended on chemical reactions that were reproducible in the laboratory. A key conceptual advance was the formulation of the cell theory by Schleiden and Schwann in 1838–1839 and the later microscopic observations of Virchow, Remak, and others, which led to the idea that cells are the fundamental units that carry out the chemical reactions of life and that new cells are created by the division of existing cells.

Regeneration has always been a topic of interest in the history of science and medicine. Primitive people were likely aware of the ability of certain food animals, such as male deer and elk, to shed and regenerate antlers, and crayfish or lobsters to regenerate limbs. They were undoubtedly cognizant of the fact that hair and nails grow continually. The regeneration of organs and appendages is a theme found in the ancient Greek mythologies of the Hydra and of Prometheus, recounted by Homer and Hesiod (Dinsmore, 1998). The Second Labor of Hercules was to slay the nine-headed Hydra, feared for its remarkable power of regenerating two heads for every one sliced off. The Titan god Prometheus made the mistake of offending the supreme god, Zeus, by stealing fire from Olympus for the benefit of humankind. As punishment for this deed, Zeus had him chained him to a rock, where each day his liver was eaten by an eagle, only to regenerate at night. This cycle of hepatectomy and regeneration would have been eternal had not Prometheus been rescued by Hercules. Accounts of regeneration of severed human arms and legs are also woven into the superstitions and descriptions of miracles in the Middle Ages (Goss, 1991).

Regeneration became a focus of systematic scientific investigation in the 18th century. Abraham Trembly performed detailed experiments on the regeneration of hydra that made a deep impression on the biologists of the time (Lenhoff and Lenhoff, 1991), while Reaumer and Spallanzani reported observations on the regeneration of limbs in crustaceans and newts, respectively (Skinner and Cook, 1991; Dinsmore, 1991).

Studies on limb development and regeneration in the latter part of the 19th and the early part of the 20th centuries made major contributions to the understanding of development. Prior to the 20th century, limb regeneration in amphibians and crustaceans was explained in the context of preformation. The limbs of these animals were presumed to contain multiple copies of preformed appendages, the growth of which was stimulated by amputation. In the early 20th century, regeneration was recognized as a regulative process that restored the whole from the remaining part. A major research aim of Thomas Hunt Morgan (1866–1945), before he turned to genetics, was to explain regeneration in terms of chemical and physical principles. Over a century later, we are still in the process of formulating this explanation.

The 20th century saw an unparalleled explosion of biological and medical knowledge. Major advances were the discovery and production of antibiotics, the development of molecular replacement therapies for diseases such as diabetes, an understanding of the immune system that revealed the antigenic differences between "self" and "nonself," and the development of highly sophisticated imaging and surgical technologies. These advances, coupled with advances in engineering and materials science and the development of immunosuppressive drugs, have given us the ability to transfuse blood and replace damaged and dysfunctional tissues and organs through tissue and organ transplants and implants of bionic devices.

Without question, however, the most fundamental and far-reaching event of 20th-century biology was the discovery, in mid-century, that DNA is the hereditary material (Avery et al., 1944; Hershey and Chase, 1952) and that it has a helical structure consisting of two deoxyribose sugar-phosphate backbones held together by complementary base pairs, adenine to thymine and guanine to cytosine (Watson and Crick, 1953). The enormous power of this structure to explain how the genetic material is replicated and mutates, and how information for protein structure is encoded in it and expressed, led to exponential advances in our knowledge of cell, developmental, and evolutionary biology. In the process, the age-old dream of being able to regenerate tissues and organs that do not regenerate spontaneously has re-emerged. Now, in the first decade of the 21st century, that goal seems attainable.

## THE BIOLOGY OF REGENERATION

### 1. Regeneration Takes Place at All Levels of Biological Organization

All organisms regenerate, though the degree of regenerative ability varies among species and with level of biological organization within the individual organism (Goss, 1969). For example, a single carrot cell can regenerate a whole carrot (Steward et al., 1964). Some species, such as planaria and hydra, can regenerate whole organisms from fragments of the body (Goss, 1969; Baguna, 1998; Sanchez-Alvarado, 2000). Certain amphibians can regenerate complex structures such as limbs and tails, as well as many other tissues. Compared to these life forms, the regenerative capacity of mammals, including humans, is limited, but no less vital. Within individual organisms, regeneration takes place from the molecular to the tissue levels of biological organization.

#### a. Molecular Level

On the molecular level, regeneration is ubiquitous. All cells can adjust the balance of protein synthesis and degradation in response to biochemical or mechanical load. For example, cardiomyocytes replace most of their molecules over the course of two weeks and adjust their rate of protein synthesis upwards under a sustained increase in blood pressure, becoming hypertrophied (Gevers, 1984).

#### b. Cell Level

On the single-cell level, regenerative capacity is more restricted. The free-living unicellular protozoans can regenerate complete cells after removal of large fragments as long as nuclear material is present in the remaining part (Goss, 1969). For example, as little as 1/80 of an amoeba is capable of reconstituting a complete amoeba (Vorontsova and Liosner, 1960). In vertebrates, the axons of sensory and motor nerves are capable of regeneration *in vivo* after crush or transection, provided that the endoneurial tubes encasing them remain intact and in register at the site of injury (Yannis, 2001). The ends of the axons on the proximal side of the injury are sealed off and their distal part degenerates. The sealed proximal segment of the axon then sprouts an advancing growth cone, behind which the axon extends through the endoneurial tube to make new synapses with target skin and muscle (Griffin and Hoffman, 1993; Yannas 2001).

#### c. Tissue Level

There are three prerequisites for regeneration at the tissue level. First, tissues must contain mitotically competent cells; that is, cells that have the receptors and signal transduction pathways to respond to a regeneration-permissive environment. Second, the injury environment of the tissue must contain the necessary signals to promote the proliferation and differentiation of these cells in an organized way. Third, factors inhibitory to regeneration must be absent from the injury environment, suppressed, or neutralized. Blood, epithelia, bone, skeletal muscle, liver, pancreas, small blood vessels, and kidney epithelium are examples of mammalian tissues containing mitotically competent cells that engage in maintenance and injury-induced regeneration.

### 2. Mechanisms of Regeneration at the Tissue Level

There are three mechanisms of tissue regeneration in vertebrates: compensatory hyperplasia, activation of

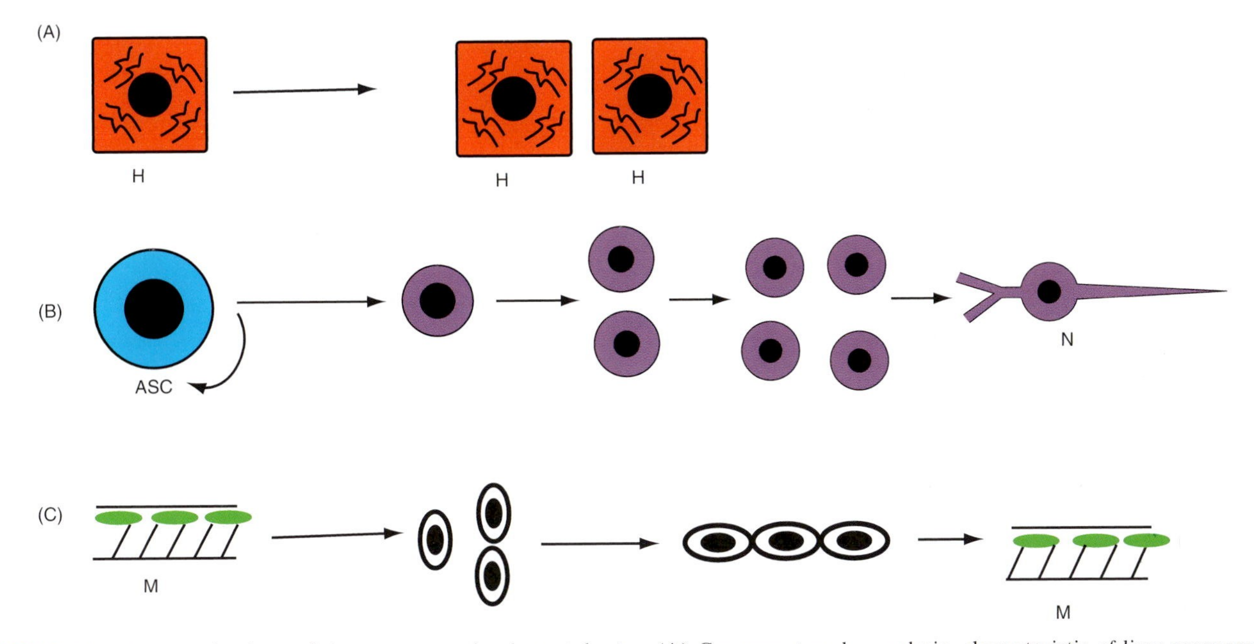


**FIGURE 1.2** The three mechanisms of tissue regeneration in vertebrates. **(A)** Compensatory hyperplasia, characteristic of liver regeneration. New hepatocytes (H) are produced by mitosis of pre-existing hepatocytes in the differentiated state. **(B)** Activation of resident undifferentiated adult stem cells (ASC). The stem cell divides to produce one daughter committed to a specific lineage differentiation, such as a neuron (straight arrows), while the other daughter is renewed as a stem cell (curved arrow). **(C)** Dedifferentiation of a mature cell. A myofiber (M) is illustrated. The myofiber fragments and degrades its contractile proteins to produce single stem-like cells that divide and redifferentiate to form new myofibers.

reserve adult stem cells, and dedifferentiation of mature cells (**FIGURE 1.2**). **TABLE 1.1** lists some regenerating tissues in vertebrates according to the mechanism by which they regenerate.

### a. Compensatory Hyperplasia

Compensatory hyperplasia is the proliferation of differentiated cells to regenerate new tissue. The differentiated cells are released from their ECM or syncytial complex and divide while maintaining all or some of their differentiated functions. The classic example of regeneration by compensatory hyperplasia is the liver (Michalopoulos and DeFrances, 1997; Trembly and Steer, 1998). After partial hepatectomy, the hepatocytes of the liver, as well as its nonparencymal cell types (Kupffer, Ito, bile duct epithelial, and fenestrated epithelial cells) divide while performing their functions of glucose regulation, synthesis of blood proteins, secretion of bile, and drug metabolism, until the original mass of the liver is restored.

### b. Activation of Resident or Circulating Adult Stem Cells

Late in embryonic or fetal life, subsets of lineage-restricted, but not fully differentiated, cells are set aside as reserve adult stem cells (ASCs). These cells may

**TABLE 1.1** Regenerating Adult Wild-Type Vertebrate Tissues According to Mechanism

| Compensatory Hyperplasia | Adult Stem Cell | Dedifferentiation |
|---|---|---|
| Liver | Hematopoietic | *Lens |
| Pancreas | Blood vessels | **Neural retina |
| Tendon | Epithelia | **Spinal cord |
| Ligament | epidermis | **Heart |
| | nails | **Limbs/fins |
| | g.i. tract | *#Tails |
| | airway | *Jaws |
| | alveoli | *Gonad |
| | kidney tubule | Blood vessels |
| | urinary tract | |
| | Muscle | |
| | Bone | |
| | Auditory hair cells | |
| | Olfactory bulb | |
| | Olfactory nerve | |
| | Hippocampal neurons | |
| | Periodontal ligament | |
| | Prostate | |
| | Antlers | |
| | Digit tips | |
| | Rabbit ear hole? | |

*Denotes amphibian or **amphibian and fish. #Denotes lizard.

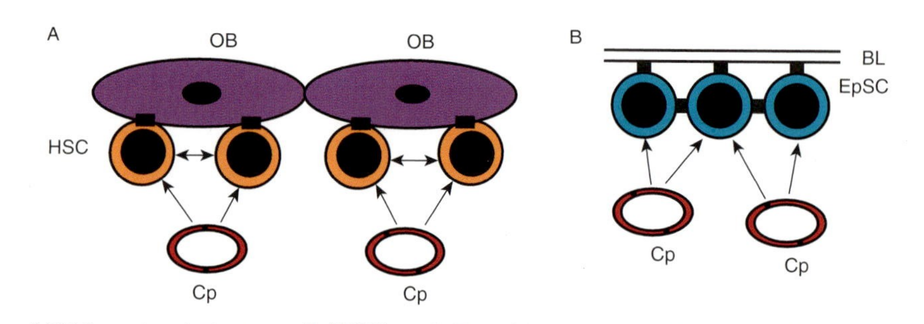


**FIGURE 1.3** Maintenance of **(A)** hematopoietic stem cell (HSC), and **(B)** epithelial stem cell (EpSC) niches. The HSC niche is maintained by paracrine signals between HSCs (double arrows) and from capillaries (Cp, single arrows) and by juxtacrine contact (bars) of HSCs with stromal cells such as osteoblasts (OB). Epithelial cells are maintained in their niche by juxtacrine contact (bars) between themselves and with their basement membrane (BM) and by paracrine signals between the EpSCs and capillaries.

reside within a tissue, or circulate in the blood. Some are used for juvenile growth after birth or hatching and others for regeneration throughout life (Fuchs and Segre, 2000; Weissman, 2000). ASCs have several characteristics. They are self-renewing and typically divide asymmetrically to give rise to a cell with a more restricted lineage and another stem cell. They have varying degrees of developmental potential, depending on the tissue they serve and/or where they are in the developmental stages of a given lineage. Regeneration via ASCs is the most common avenue of tissue regeneration in multicellular organisms, including planarians, which regenerate via stem cells called neoblasts (Baguna, 1998; Sanchez-Alvarado, 2000).

Epithelia, which make up 60% of the differentiated tissue types in the body (Slack, 2000), regenerate from ASCs that reside in the epithelium. Examples are skin epidermis, hair follicles, the ventricular walls of the brain, the acoustic epithelium, and the epithelium of the respiratory and digestive tracts (Jensen et al., 1999; Potten et al., 1997; Slack, 2000; Gage, 2000; Roberson et al., 1995). The liver, which is also an epithelial organ, contains a population of stem cells that are activated when the liver is damaged beyond the capacity for regeneration by differentiated liver cells (Michalopolous and DeFrances, 1997).

Mesodermal derivatives, such as blood, skeletal muscle, and bone, regenerate from non-epithelial hematopoietic stem cells (HSCs), mesenchymal stem cells (MSCs), and muscle stem cells (satellite cells, SCs) associated with the tissue or residing in the bone marrow (Ham and Cormack, 1979; Hansen-Smith and Carlson, 1979).

ASCs normally differentiate into a range of phenotypes within their lineage, called their prospective significance, or fate. For example, HSCs, which reside in the bone marrow, are multipotential and give rise to erythrocytes, myeloid cells, and the several cell types of the immune system. Others, such as liver stem cells, are bipotent and give rise only to hepatocytes and bile duct cells. Still others, such as epidermal stem cells, appear to be unipotent, giving rise only to one cell type, in this case keratinocytes.

All ASCs reside in environmental microniches that maintain their stem cell character (**FIGURE 1.3**). Studies from both *Drosophila* germ cells and vertebrate hematopoetic and epithelial stem cells indicate that maintaining stem cells in a quiescent state requires interactions among the stem cells and between the stem cells and surrounding cells through soluble and insoluble signals (Watt and Hogan, 2000; Fuchs et al., 2004). In the *Drosophila* ovary, the adherens junction proteins, DE-cadherin and β-catenin, are concentrated at the border of the germ cells and cap cells. Mutations in these molecules result in a failure to recruit and maintain germ cells (Song et al., 2002). In the bone marrow, maintenance of HSCs relies on their adherence to osteoblasts of the stroma through N-cadherin of adherens junctions and fibronectin-binding integrins ($\alpha4\beta1$, $\alpha5\beta1$) (Zhang et al., 2003). Blocking this adhesion inhibits hematopoiesis in long-term bone marrow cultures (Whetton and Graham, 1999).

Whether or not an epithelial stem cell divides symmetrically or asymmetrically appears to depend on whether the intercellular contacts between stem cells or the contacts between stem cells and basement membrane of the epithelium are stronger (**FIGURE 1.4**). In the former case, the mitotic spindle lines up parallel to the basement membrane, cell fate determinants are equally segregated into the halves of the cell, and two identical stem cells are produced. In the latter case, the spindle lines up perpendicular to the basement membrane, cell fate determinants are distributed unequally, and the result is a stem cell adherent to the basement membrane and a free lineage-committed daughter (Yamashita et al., 2003). An asymmetric division moves

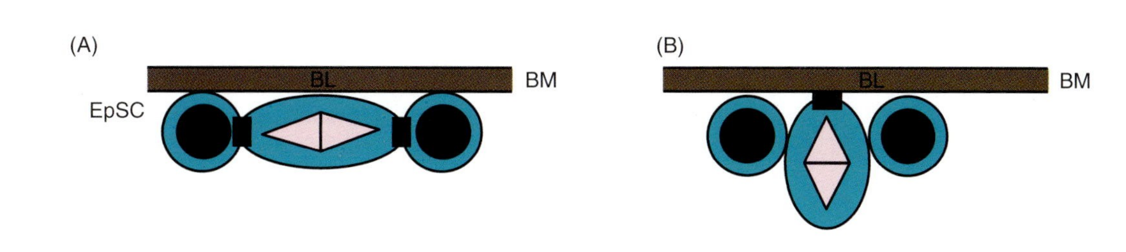


**FIGURE 1.4** Symmetrical vs. asymmetrical division of an epithelial stem cell (EpSC). **(A)** Symmetrical division. The contacts (bars) between EpSCs are stronger than the contacts between EpSCs and the basement membrane (BM) leading to orientation of the mitotic spindle in parallel with the BM to produce two stem cells. **(B)** Asymmetrical division. The contacts between the the EpSCs and their BM are stronger than the contacts between EpSCs and the spindle aligns perpendicular to the BM to produce a stem cell and a daughter committed to differentiate.

the nonrenewing daughter to another part of the niche that promotes proliferation and differentiation.

Activation of stem cells to divide may take place differently in maintenance versus injury-induced regeneration. In maintenance regeneration, the stem cells undergo continual but slow division in response to environmental signals, feeding a constant excess of progeny into a new environment that induces them to differentiate. In injured tissues that normally undergo slow turnover for maintenance, stem cells remain relatively dormant until injury-induced signals mobilize them.

### c. Creation of Stem Cells via Dedifferention

Dedifferentiation is a loss of phenotypic specialization that converts differentiated cells into adult stem cells, which then proliferate and differentiate into replacement tissue (Brockes, 1998; Carlson, 1998; Geraudie et al., 1998; Stocum, 2004a). Dedifferentiation is a relatively common mechanism of regeneration in lower vertebrates. Fish regenerate fins and barbels (teleosts) by dedifferentiation, and certain species of lizards can regenerate tails by this mechanism. The divas of dedifferentiation in the vertebrate world, however, are the anuran tadpoles and the larval and adult urodele amphibians. These animals can regenerate the same tissues as mammals via compensatory hyperplasia and resident ASCs, but use dedifferentiation to regenerate a wide variety of tissues and complex structures that mammals cannot regenerate, including limbs, tails, jaws, lens, spinal cord, neural retina, and intestine.

### d. Signal Transduction Pathways in Regeneration-Competent Cells

Resident stem cells interact with other cells in the microniche through signaling pathways that regulate self-renewal and proliferation during asymmetric division. **FIGURE 1.5** illustrates the six major signaling/receptor pathways implicated in both embryonic development and the behavior of ASCs.

#### 1. Notch Pathway

The Notch receptor is a major mediator of stem cell self-renewal during embryonic development in *Drosophila* (Go et al., 1998). Vertebrates contain four Notch receptors that control fate decisions during embryonic development (Lundkvist and Lendahl, 2001). Notch is a transmembrane protein that is signaled (activated) by the membrane-bound ligands Delta, Jagged, and Serrate on neighboring cells. Ligand binding causes a conformational change in Notch that results in the release of its intracellular domain (NICD) by the presenilin enzyme (Lecourtois and Schweisguth, 1998; Schroeter et al., 1998). The NICD translocates to the nucleus, where it interacts with the DNA-binding protein RBP-Jκ and the histone acetyltransferases p300 and PCAF (Wallberg et al., 2002) to activate target genes whose products act as transcriptional repressors of genes whose activity leads to cell differentiation (Nakamura et al., 2000).

Notch ICD activity is inhibited by the intracellular membrane-associated protein, Numb (Jan and Jan, 1998). Asymmetric expression of Numb in one daughter cell during the division of an ASC thus leads to differentiation of that cell, whereas the other daughter self-renews. Notch activity in the self-renewing daughter cell is regulated at the translational level by Nrp-1 (Musashi-1 in *Drosophila*), a RNA-binding protein (Sakakibara et al., 1996). NRP-1 binds *numb* mRNA, preventing it from being translated (Imai et al., 2001).

#### 2. Canonical Wnt Pathway

An important set of transcriptional factors that maintains ASCs in a quiescent state is the Tcf/Lef family,

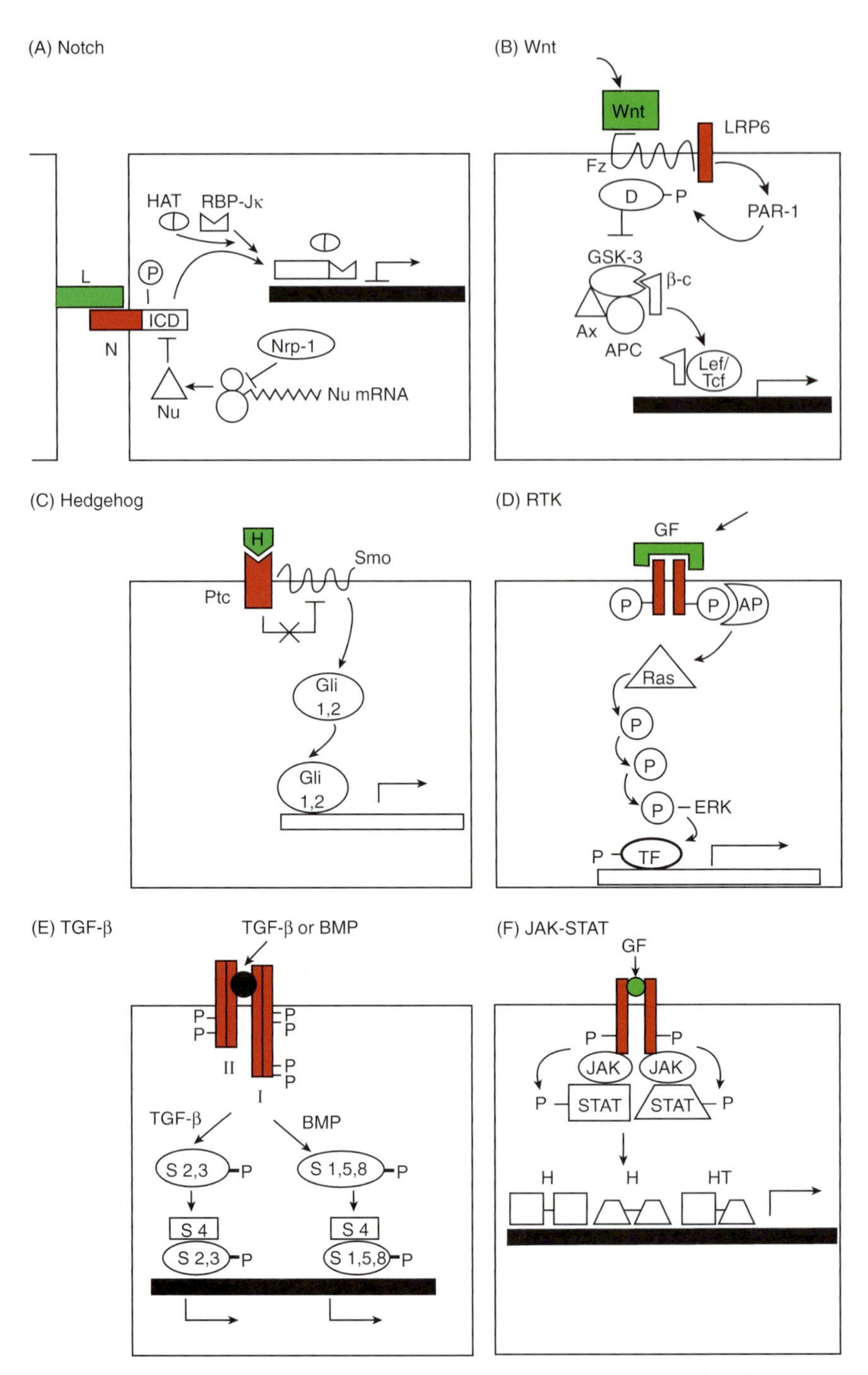


**FIGURE 1.5** Simplified sketches of six major signaling pathways used for development and regeneration. Descriptions in text. The signal in the Notch pathway actively maintains cells as stem cells whereas the signals in the other pathways activate cells to divide asymmetrically. **(A) Notch pathway**. L = ligand, N = notch protein, ICD = Notch intracellular domain, P = presenelin, HAT = histone acetyltransferase, RBK-Jκ = DNA binding protein. Presenelin cleaves off the ICD. The ICD, HAT and RBP-Jκ form a complex that binds to the regulatory regions of differentiation genes to block their transcription. The blockade is maintained by the protein NRP-1, which prevents the translation of Numb (Nu) mRNA. If translated, Numb inhibits the removal of the ICD from Notch and the transcription-inhibiting complex is not formed, activating the cell. **(B) Wnt pathway**. Fz and LRP5/6 = co-receptors for Wnt. PAR-1 = protein kinase that phosphorylates (P) the Disheveled protein (D). GSK-3 = glycogen synthase kinase-3, Ax = axin, APC = adenomatous polyposis coli protein, β-C = beta catenin. Lef and Tcf = transcription factors that in the absence of Wnt signaling inhibit transcription of genes that activate stem cells, but stimulate transcription of these genes when complexed with β-catenin. **(C) Hedgehog pathway**. Ptc = patched protein, Smo = smoothened protein, Gli = transcription factor that is the counterpart of cubitus interruptus in *Drosophila*. **(D) RTK pathway**. GF = growth factor, P = phosphate, AP = adaptor protein that recognizes phosphorylated receptor and activates the Ras protein, leading to a chain of phosphorylation reactions that ends in ERK, a mitogen activated protein kinase (MAPK) that phosphorylates transcription factors (TF). **(E) TGF-β (Smad) pathway**. P = phosphate, S = Smad proteins. The type I and II receptors are each dimers. **(F) JAK-STAT pathway**. P = phosphate. The JAK and STAT proteins form transcriptional complexes that are homodimers (H) or heterodimers (HT).

downstream effectors of the canonical Wnt signaling pathway. The family consists of four members, Tcf-1–3 and Lef-1. These transcription factors are widely expressed during embryonic development, where in combination with Groucho-related proteins, they serve as transcriptional repressors in the absence of a Wnt signal (Cavallo et al., 1998; Roose et al., 1998; Brantjes et al., 2001). However, when complexed with Wnt-induced stable β-catenin, the Tcf/Lef transcription factors become activators of transcription (Roel et al., 2002; Bejsovec, 2005).

Stem cells can be maintained in a quiescent state by destruction of β-catenin through inhibition of Wnt signaling. In the absence of Wnt signaling, β-catenin is continually degraded in the cytoplasm by a proteasome complex consisting of glycogen synthase kinase-3 (GSK-3), axin and adenomatous polyposis coli (APC) protein. GSK-3 holds β-catenin in the complex, where it is broken down by APC. GSK-3 is inhibited by Wnt signaling, allowing β-catenin to dissociate from the complex and accumulate in the nucleus, where it forms a transcriptional complex with Lef and Tcf transcription factors. Wnt proteins, of which 15 have been identified, bind to the Frizzled (Fz) receptor and its co-receptor LRP5/6 (Wherli et al., 2000). This binding activates a protein kinase called PAR-1, which phosphorylates the protein Disheveled (Sun et al., 2001). In turn, Disheveled inhibits the activity of GSK-3. Wnt signaling is inhibited by a number of antagonists that mimic the molecular structure of Fz and also by the protein Dickopf-1, which binds to LRP-5/6 (Bafico et al., 2001).

### 3. Hedgehog Pathway

The hedgehog family of signaling molecules has three members, sonic hedgehog (Shh), Indian hedgehog (Ihh), and desert hedgehog (Dsh). These proteins have been implicated in the self-renewal and proliferation of stem cells in a wide variety of tissues (Zhang et al., 2001; Reya et al., 2001; Lai et al., 2003). Unlike the Notch ligands or the Wnts, the hedgehog signal acts to inhibit the action of its receptor, Patched. Patched is bound to, and inhibits, the actual signal transducer for hedgehog, Smoothened. When hedgehog binds to Patched, Smoothened is activated. To be active, hedgehog must be cleaved to form an amino-terminal peptide that is esterified at its C-terminus to a cholesterol molecule (Porter et al., 1996; Taipale and Beachy, 2001). Smoothened is phosphorylated and, in *Drosophila*, releases the Cubitus interruptis (Ci) protein that is tethered to microtubules. Ci is a transcription factor that acts as a repressor or activator, depending on whether it is cleaved. In the absence of hedgehog signal, the carboxy-terminal domain of Ci is cleaved off and moves to the nucleus, where it acts as a transcriptional repressor. In the presence of hedgehog signal, the entire Ci molecule is released and translocated to the nucleus, where it acts as a transcriptional activator (van den Heuvel, 2001; Lum and Beachy, 2004). Different sets of genes are activated depending on the concentration of Ci. In vertebrates, three different proteins have evolved to effect transcriptional repression or activation in the hedgehog pathway (Stecca and Ruiz i Altaba, 2002). In the absence of hedgehog signal, Gli 2 and 3 repress transcription by removal of their carboxy-terminal domain. In the presence of hedgehog signal, Gli 1 and 2 both act as transcriptional activators (Wang et al., 2000; Aza-Blanc et al., 2000). The hedgehog and Wnt pathways share certain similarities, suggesting that they may be sister pathways that synergize to affect stem cell proliferation (Lum and Beachy, 2004).

### 4. Receptor Tyrosine Kinase (RTK) Pathway

The receptor tyrosine kinase (RTK) pathway is used by growth factors such as fibroblast growth factors (FGF), platelet derived growth factors (PDGF), epidermal growth factors (EGF), vascular endothelial growth factor (VEGF), and stem cell factor (SCF) (Schlessinger, 2000). These signaling ligands bind to specific RTKs. The RTK is a transmembrane protein that is dimerized by the ligand and undergoes a conformational change that results in autophosphorylation of specific tyrosines on the cytoplasmic domain of the receptor. One of these tyrosine residues is recognized by an adaptor protein. This leads to the activation of G proteins such as Ras, which activates a cascade of phosphorylation reactions. The last members of this cascade, phosphorylated extracellular signal-regulated kinases [ERK proteins (also called mitogen-activated protein (MAP) kinases], enter the nucleus where they phosphorylate and activate transcription factors.

### 5. The Transforming Growth Factor Beta (TGF-β) Pathway

The TGF-β family of growth factors consists of two subfamilies, the TGF-β/Activin/Nodal subfamily and the bone morphogenetic protein (BMP)/growth and differentiation factor (GDF)/Muellerian inhibiting substance (MIS) subfamily (Shi and Massague, 2003). These molecules signal through serine-threonine kinase receptors that consist of two transmembrane proteins known as the type I and II receptors (Attisano and Wrana, 2002). In vertebrates, there are seven different type I receptors and five type II receptors. Different type I and II receptors form heterodimers, depending on which ligand they bind. Upon binding ligand, the type II receptor phosphorylates the type I receptor, activating its kinase domain. Depending on the ligand, the activated receptor then phosphorylates different classes of Smad proteins. There are eight Smads that fall into three classes. The receptor Smads (R-Smads, 1, 2, 3, 5, 8) are the only ones directly phosphorylated by the receptors. They accumulate in the nucleus where they associate with a Co-Smad (Smad 4) and other proteins into complexes that activate or repress transcription. Smads 6 and 7 compete for binding sites with the other Smads and are inhibitory. Three of the type I receptors transduce TGF-β-like signals by phosphorylating Smads 2 and 3, whereas the other four type I receptors activate Smads 1, 5, and 8 to mediate BMP signals.

### 6. The JAK-STAT Pathway

The JAK-STAT pathway (JAK = janus kinases; STAT = signal transducers and activators of transcription) is activated by many cytokines and growth factors that bind to receptors lacking intrinsic tyrosine kinase activity (Aaronson and Horvath, 2002). The JAK proteins bind to the intracellular domains and phosphorylate tyrosine residues, converting the receptor into a tyrosine kinase. This conversion creates a docking site for STAT proteins, where they are activated by phosphorylation. The STAT proteins form homo- and heterodimers and move rapidly into the nucleus, where they associate with other proteins to form transcriptional complexes. In mammals, there are four JAK genes and seven STAT genes, which provide diversity of receptor activation and transcriptional binding.

### 7. Other Pathways

These pathways are able to cross-talk with one another, thereby conferring additional flexibility in the regulation of gene activity. There are additional signal transduction pathways that operate through other soluble factors and ECM components. One of the most important is the apoptotic (programmed cell death) pathway. Embryonic development overproduces cells that must be "pruned" to arrive at the correct number and shape for optimum function of an organ system. In the adult body, tens of billions of cells die and are replaced each day. Embryonic and regenerating adult tissues rely on apoptosis to regulate cell number and turnover and to sculpt the shape of developing or regenerating organs (Nicholson and Thornberry, 2003; Gilbert, 2003). Some cell types such as erythrocytes are programmed to die in the absence of signals (in the case of red blood cells, erythropoietin) that prevent apoptosis, whereas others die as the result of specific signals that activate apoptosis. Cells have two opposing sets of genes that regulate apoptosis, one that preserves the cell and the other that destroys it by enzymatic action. The life-savers are members of the Bcl-2 gene family, most prominently *Bcl-2 and Bcl-X*. The killers are apoptotic protease activating factor 1 (Apaf1), which in the presence of cytochrome-*c* activates caspase-9 and 3. These caspases are proteases that digest the contents of the cell. The life-and-death genes are remarkably conserved across vast phylogenetic distances and in fact were first discovered in the nematode worm *C. elegans* (Ellis and Horvitz, 1986).

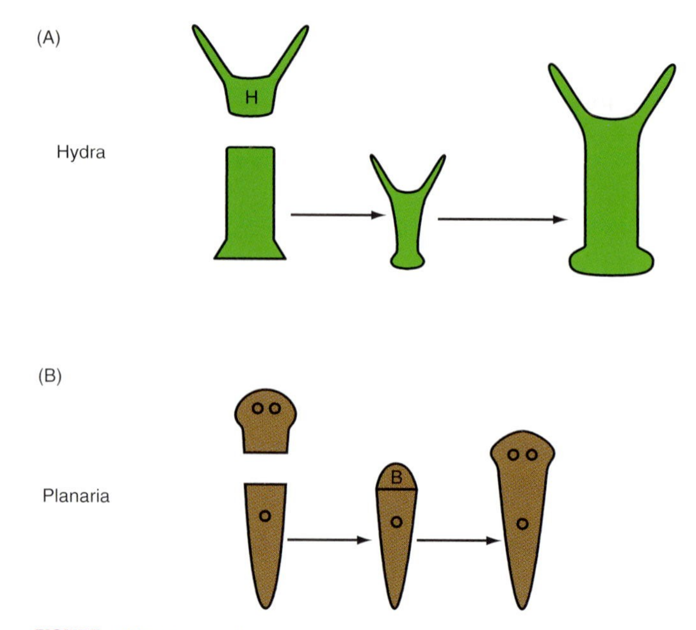


**FIGURE 1.6** Morphallactic vs. epimorphic regeneration. **(A)** Morphallactic regeneration of hydra. After amputating the head region (H) the remaining cell mass is reorganized into a smaller version of the original hydra, which then grows to final size. **(B)** Epimorphic regeneration of planaria. Amputation of the head region is followed by the formation of a regeneration blastema (B) at the amputation surface without any reorganization of the remaining cell mass. The blastema grows and at the same time is organized into a new head to restore the structure and size of the worm.

## 3. Epimorphosis and Morphallaxis

Morgan (1901) distinguished two modes of regeneration, morphallaxis and epimorphosis, that are independent of the types of regeneration-competent cells used for regeneration (**FIGURE 1.6**). Morphallaxis is the regeneration of missing parts in the absence of growth, by repatterning the remaining tissue into a normally proportioned but smaller whole, followed by growth to the original size. It is typical of regeneration in multicellular animals with relatively simple tissue organizations, such as *Hydra*. Epimorphic regeneration restores body parts by the proliferation of stem or progenitor cells, followed by their differentiation into the tissues that were lost. Differentiation is thus linked to growth in epimorphic regeneration, whereas it is not in morphallactic regeneration. Epimorphic regeneration is characteristic of flatworms and annelid worms, as well as amphibian appendages and jaws. Mammals are also capable of the epimorphic regeneration of some structures, such as deer antlers, rabbit ear tissue, and fetal digits.

### 4. Evolutionary Significance of Regeneration and Fibrosis

Wound repair and tissue regeneration are ubiquitous to multicellular organisms. These processes are universally adaptive, in that they are obligatory for the survival of multicellular organisms; thus, natural selection will always favor them. Epimorphic regeneration, however, is restricted to only a relatively few species within each phylum. This fact is sometimes taken to mean that epimorphic regeneration arose in a few species from nonregenerating ancestors by selection for favorable mutations conferring this ability. However, Goss (1987, 1992) has pointed out that the *de novo* origin of epimorphic regeneration by positive selection is improbable, because it would have had to have been an all or nothing event involving at least several simultaneous mutations.

Many tissues harbor regeneration-competent cells (see below), yet regenerative responses of these tissues when injured have been suppressed in favor of patching the injury with scar tissue. Why does fibrosis prevail after injury in so many adult tissues and appendages of most vertebrate species? One argument is that regeneration confers no adaptive advantage, even in those species that possess it (Goss, 1987; 1992). Regeneration may require considerably more energy expenditure over a longer period of time than fibrosis, particularly with regard to large and complex structures such as limbs. In mammals, which are warm-blooded and whose injured tissues are particularly susceptible to bacterial infection and water loss, rapid wound closure and scar formation would be particularly advantageous because it prevents fluid loss, suppresses bacterial proliferation, and provides a relatively quick patch to the wound.

The most logical conclusion, therefore, is that regeneration is suppressed in most adults because they exist under conditions where fibrosis is more energy-efficient and provides more assurance of survival. This suppression has been related to the maturity of the immune system in both fetal versus adult mammalian skin and limb regeneration in the frog, *Xenopus* (McCallion and Ferguson, 1996; Harty et al., 2003).

## STRATEGIES OF REGENERATIVE MEDICINE

Three main strategies are being used to develop clinical regenerative therapies (**FIGURE 1.7**). These are cell transplantation, implantation of bioartificial tissue constructs ("tissue engineering"), and the chemical (pharmaceutical) induction of regeneration from tissue at the site of injury or recruited from elsewhere in the body. Which strategy is used depends on which is more appropriate for the nature of the tissue and the extent of the damage to be repaired. In general, cell transplantation and pharmaceutical induction of regeneration have been used to correct smaller tissue deficiencies, whereas bioartificial tissue constructs have been used to correct larger tissue deficiencies.

### 1. Cell Transplantation

Transplanted cells can be autogeneic, allogeneic, or xenogeneic, differentiated cells, fetal cells, derivatives of embryonic stem cells (ESCs), or ASCs. They can be normal or genetically modified to boost production of important signaling or ECM molecules, or molecules that neutralize factors inhibitory to cell survival and proliferation. The advantage of allogeneic or xenogeneic cells is that they can be expanded and banked in advance for use in cases where there is not enough time to expand a patient's own cells for use. Their disadvantage is that they elicit rejection if seen by the immune system of the recipient. Differentiated cells can be harvested directly from the patient's body, but with few exceptions, their expansion *in vitro* is difficult. Fetal cells are derived from spontaneously aborted fetuses; thus, the supply is quite limited. By far, the current research focus is on the transplantation of ESC derivatives, and ASCs.

#### *a. Embryonic Stem Cells*

All of the trillions of cells in the adult human body, including ASCs, are derived from a few embryonic stem cells that constitute the inner cell mass of the preimplantation blastocyst (Rossant, 2001) (**FIGURE 1.8**). These cells express species and stage-specific embryonic antigens (SSEAs), alkaline phosphatase, and high levels of telomerase (Rippon and Bishop, 2004). In mammals, all the cells of the inner cell mass are pluripotent. Prior to implantation, the pluripotent cells become restricted to the epiblast of the embryo. Epiblast cells are self-renewing and pluripotent up to gastrulation, when they begin progressive differentiation into the three germ layers and their derivatives (Smith,

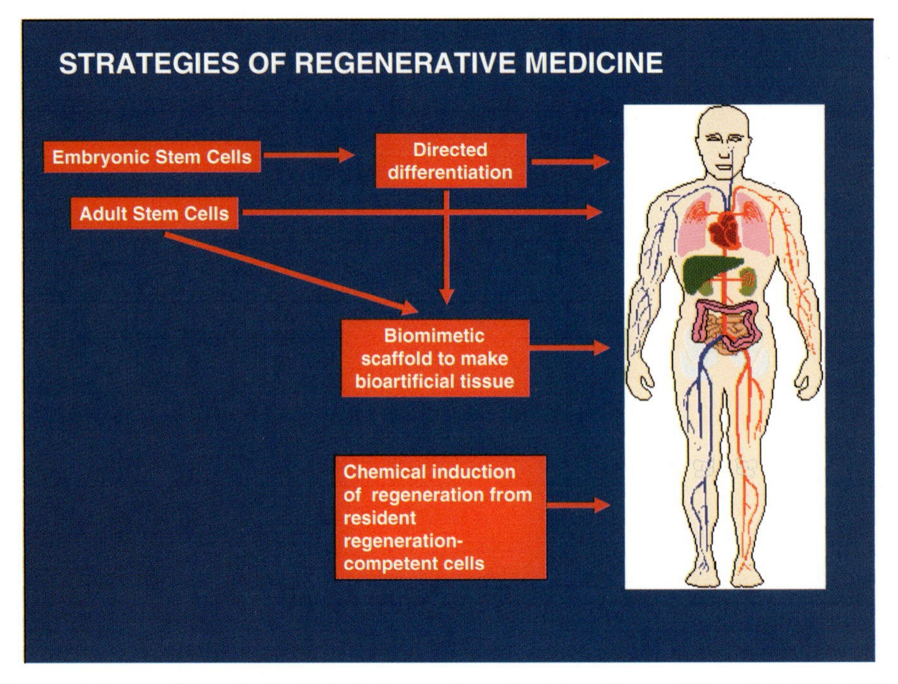


**FIGURE 1.7** Strategies of regenerative medicine. In the first strategy, embryonic stem cells are differentiated to adult stem or progenitor cells or stem or progenitor cells are harvested from adult tissues and transplanted into the lesion as suspensions or aggregates. In the second strategy, embryonic stem cell derivatives or adult stem cells are seeded into a biomimetic scaffold to make a bioartificial tissue, which is then implanted into the body. In the third strategy, regeneration-permissive molecules, as well as neutralizers of regeneration-inhibiting molecules, are applied to the lesion.

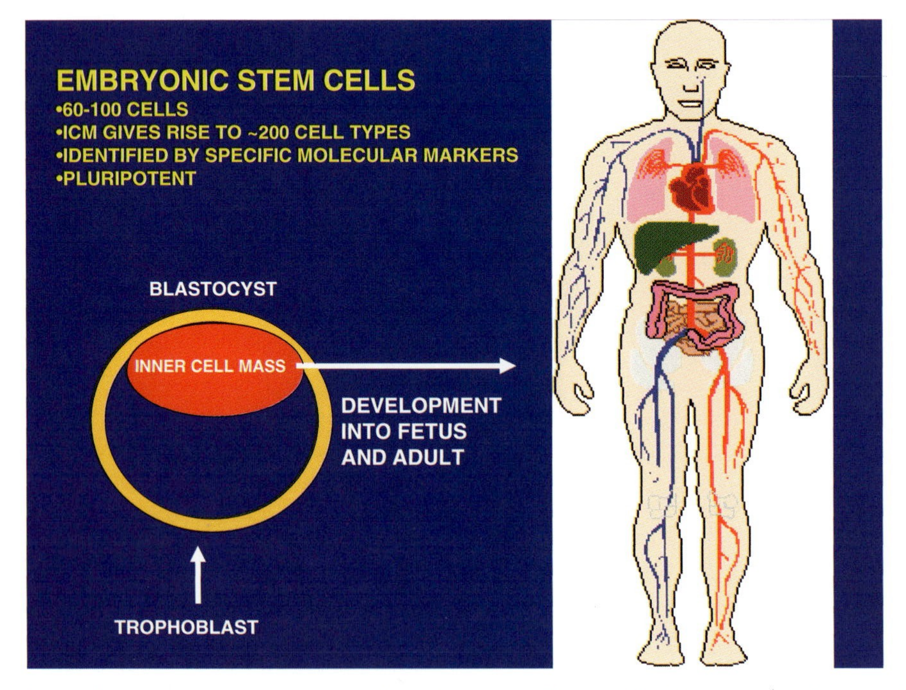


**FIGURE 1.8** Origin of the human body from embryonic stem cells of the inner cell mass. The blastocyst (~5 days post-fertilization) consists of 60–100 cells, of which fewer than 25 constitute the inner cell mass. The inner cell mass gives rise to the fetus and adult, while the remaining cells (trophoblast) are incorporated into the placenta.

2001). This pluripotency was demonstrated *in vivo* for murine ESCs by injecting them into host blastocysts, where they made contributions to all tissues to form a chimeric embryo (Smith, 2001). The pluripotency of human ESCs has been shown by demonstrating that they can differentiate as teratomas containing representatives of all three germ layers when implanted into immunodeficient mice (Thomson et al., 1998; Amit et al., 2000; Reubinoff et al., 2000).

The acquisition and maintenance of mouse ESC pluripotency requires the cooperation of leukemia inhibitory factor (LIF) and BMPs. BMPs induce the expression of inhibitor of differentiation (Id) genes via Smad transcription factors (Ying et al., 2003). LIF activates STAT-3. In turn, STAT-3 activates the homeodomain transcription factors OCT4 (Smith, 2001), SOX2 (Avilion et al., 2003), and Fox D3 (Hanna et al., 2002). Recently, a LIF/STAT3-independent transcription factor, Nanog, has been discovered that is also essential for mouse and human ESC pluripotency and self-renewal (Mitsui et al., 2003; Chambers et al., 2003). Nanog is down-regulated as cells exit from the primitive streak to become mesoderm, while OCT4 is still expressed. As development proceeds, however, all the pluripotency genes are down-regulated except in the germ cells (Hart et al., 2004).

ESCs give rise to the ectodermal, endodermal, and mesodermal germ layers of the embryo. All the differentiated adult tissues of the body are derived from these layers. As ontogenesis proceeds, the cells of the germ layers become progressively more diversified in precise spatial patterns to block out the emerging organ systems. Ultimately, diversification proceeds to the determination of progenitor cells for each terminal cell phenotype in every tissue. The progenitor cells divide to make more progenitor cells (amplification), which then differentiate into the functional cells. The nervous system is primarily an ectodermal derivative, and the musculoskeletal, cardiovascular, and urogenital systems are primarily mesodermal derivatives. The skin, digestive, and respiratory systems are composites of ectodermal or endodermal and mesodermal derivatives.

ESC cultures have been established from the early embryos of fish, birds, and a variety of mammals (Smith, 2001; Rippon and Bishop, 2004). Human ESC cultures have been established from unused frozen blastocysts produced by *in vitro* fertilization in assisted reproduction facilities (Thomson et al., 1998; Amit et al., 2000; Reubinoff et al., 2000; Richards et al., 2002) and from primordial germ cells of 5–9-week embryos (Shamblott et al., 1998). The procedure for creating mammalian embryonic stem cell lines is diagrammed in **FIGURE 1.9**. The inner cell mass is removed from the blastocyst, dissociated, and the cells grown on a layer of feeder fibroblasts or on an adhesive substrate in defined medium without feeder cells. The cells are grown in the presence of factors (LIF for mouse; FGF-2 for human) that maintain them in an undifferentiated state. ESCs cultured for long periods of time retain their pluripotency and are capable of giving rise to any of the more than 200 differentiated cell types of the body. Multipotent germline stem cells (mGSCs) have also been isolated and cultured from neonatal mouse testis (Kanatsu-Shinohara et al., 2004). These cells are similar to ESCs in their molecular phenotype, including the transcription factors associated with pluripotency, and exhibited pluripotency *in vivo* and *in vitro*.

Upon withdrawal of LIF or FGF-2 and being placed in high-density suspension culture, hanging drop culture, or medium containing methylcellulose, ESCs differentiate into embryoid bodies containing precursors of the derivatives of each germ layer (Rippon and Bishop, 2004). The embryoid bodies are then transferred to culture conditions that promote adhesion to a substrate, whereupon cells grow out and differentiate randomly into specific cell types. Protocols have been developed to direct the differentiation of both murine and human embryoid body cells specifically into neural cells, cardiomyocytes, chondrocytes, osteoblasts and lung alveolar epithelium (Rippon and Bishop, 2004), but directed differentiation of human ESCs remains a major technical difficulty.

ESCs are viewed as a prime source of transplantable cells for regenerative medicine therapies because they

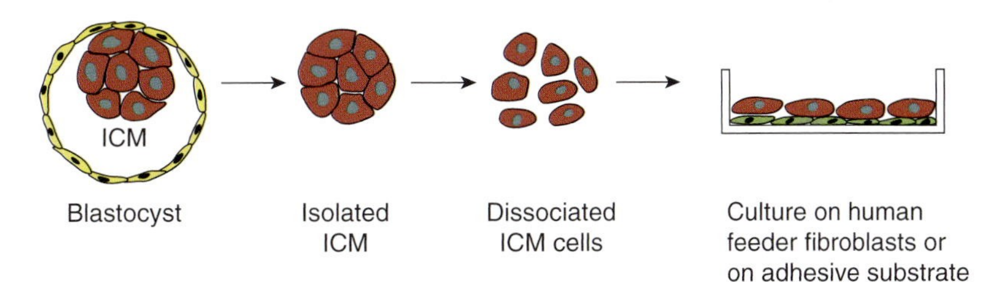


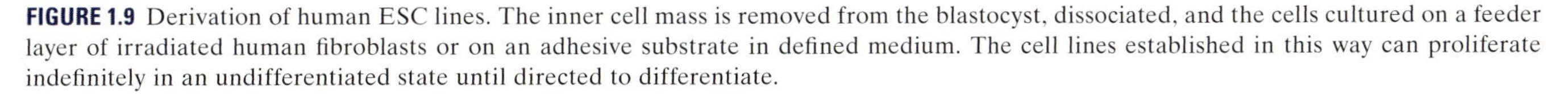

**FIGURE 1.9** Derivation of human ESC lines. The inner cell mass is removed from the blastocyst, dissociated, and the cells cultured on a feeder layer of irradiated human fibroblasts or on an adhesive substrate in defined medium. The cell lines established in this way can proliferate indefinitely in an undifferentiated state until directed to differentiate.

can be expanded indefinitely in culture to provide the large numbers of cells required to produce derivatives while retaining their potential to differentiate into virtually every cell type in the body. Major technical hurdles to their use besides directed differentiation are the work required to maintain them in an undifferentiated state, their potential to form teratomas when transplanted, and the fact that their differentiated derivatives, like other allogeneic cells, are subject to immune rejection (Rippon and Bishop, 2004). Different hESC lines have distinct HLA profiles, suggesting that they would elicit different intensities of immune response (Carpenter et al., 2004). The use of hESC derivatives for transplantation is also the subject of intense bioethical debate, because their production requires the destruction of embryos (Chapter 15).

### b. Adult Stem Cells

Research on human ESCs has been going on for only seven years, compared to over 20 years of research on animal ESCs and many decades of research on ASCs. The use of human ASCs for cell transplant therapy gets around all of the bioethical issues raised by ESCs and most of the biological issues as well. Autogeneic ASCs are not immunorejected and at least some of them, such as bone marrow, umbilical cord blood, and fat stem cells, are easy to harvest and expand *in vitro*. However, ASCs have their own problems. Most, such as neural stem cells, are difficult to harvest and expand *in vitro*. In some tissues, their numbers decline with age and they are less robust in their proliferation and differentiation potential. Allogeneic ASC transplants elicit an immunorejection response, requiring the use of immunosuppressants.

One of the major reasons cited by proponents of using ASCs over ESCs or their derivatives for cell therapy are numerous reports that ASCs have a developmental potential that is greater than their prospective significance. That is, they are developmentally plastic and can be induced to differentiate into cells other than their prospective significance (Fuchs and Segre, 2000; Weissmann, 2000; Stocum, 2004b). This feature would potentially allow us to use just one type of autogeneic stem cell, say from the easily accessible bone marrow, for differentiation into any desired cell type. Whether or not ASCs have this range of developmental potential, however, is a subject of intense debate (Chapter 13). On the other hand, there is some evidence that the bone marrow and connective tissue compartments of multiple organs might harbor pluripotent stem cells that share a number of qualities with ESCs, though they are different from ESCs, or that such cells can be derived by extensive culturing and deprogramming [Jiang et al., 2002; Young and Black, 2004 (Chapter 13)]. Either way, such cells would obviate two of the major problems of ESCs, the need to combat immunorejection and the bioethical issues associated with the production of ESCs (Chapter 15).

A second reason often cited by opponents of ESC research is that ASC therapy has been successful in curing many diseases. However, this is true only for hematopoietic disorders. Transplantation of adult bone marrow cells or umbilical cord blood stem cells has been used for several decades to reconstitute the hematopoietic system after high-dose ablation chemotherapy or irradiation for genetic and malignant hematopoietic diseases (Chapter 12). Transplantation of ASCs for other types of diseases and injuries in experimental animals and humans has had much less success so far.

### c. The Chemical and Physical Environment of Transplanted Cells Is Critical to Their Survival, Proliferation, and Differentiation

A major requirement of transplanted cell suspensions or aggregates is that they differentiate and self-assemble into the three-dimensional architecture of the original tissue and become integrated with surrounding host tissue. This depends on the presence of the appropriate signals and adhesion molecules in the injury environment, as well as the three-dimensional architecture of the surrounding healthy tissue. We know that tissues undergoing maintenance regeneration provide the signals necessary for maintaining the organization of the tissue, and the injury environments of spontaneously regenerating tissues must also have the requisite signals. Some of these signals must also be present in the injury environments of tissues that fail to regenerate, because such tissues often initiate a regenerative response that is then suppressed by fibrosis. To regenerate these tissues, it will be necessary to provide additional regeneration-permissive signals, as well as neutralizers of fibrosis-inducing molecules, to the injury site.

## 2. Bioartificial Tissues

Bioartificial tissue construction aims to set up an appropriate replacement tissue organization and morphology prior to implantation, through the use of scaffolds shaped like the damaged tissue or organ (Nerem and Sambanis, 1995; Ogle et al., 1998). Ideally, the scaffolds of bioartificial tissues would mimic the ECM *in vivo*, providing not only the geometry and physical/chemical properties to maximize the migration of cells throughout the scaffold, but also being capable of sequestering and releasing biological signals essential for cell proliferation and differentiation (Langer and Tirell, 2004). Natural biomaterials such as type I collagen either by itself or in combination with other ECM molecules, or decellularized connective tissue matrix are the most widely used scaffolds for bioartificial tissue construction or as templates to induce regeneration from host tissues (Badylak, 2002). Type I collagen is easy to extract and mold into a variety of shapes, and is biodegradable. A great deal of attention is also being paid to the development of synthetic biomimetic materials as scaffolds. The advantage of synthetic biomaterials is that they can be manufactured according to strict specifications and in potentially unlimited quantities. Current synthetic biomaterials frequently used in bioartificial tissue construction or as regeneration templates are polydioxanone, poly (epsilon-caprolactone), poly (glycolic acid), poly (lactic acid), and ceramics (Langer and Tirell, 2004). All are biodegradable, although ceramic materials degrade very slowly.

Bioartificial tissues can be open (vascularized by the host) or closed (cells encapsulated in a biomaterial and dependent on diffusion for survival). If the cells of an open construct are allogeneic or xenogeneic, they will be subject to immunorejection; those of a closed construct will be immunoprotected. The matrix supporting a closed construct must be resistant to degradation. By contrast, scaffolds of open constructs should be biodegradable so that they are replaced over weeks and months with the natural matrix made by the cells. In closed bioartificial tissues, cells are placed into porous microspheres or microcapsules $<0.5$-mm diameter, porous macrocapsules shaped as rods, sheaths or discs 0.5–1-mm in diameter, or vascular devices in which cells are placed in an extracellular sheath surrounding a tubular membrane ~1 mm in diameter that can be attached to blood vessels. Microcapsules are made of alginate hydrogels, and macrocapsules and vascular devices are often made of acrylonitrile/vinyl chloride copolymers. Open systems are constructed of cells attached to biodegradable polymer (natural or artificial) matrices. The matrix supports the proliferation, vascularization, and differentiation of the cells into tissue that becomes integrated into the natural tissues as the matrix degrades.

Bioartificial tissues have been constructed for the replacement of several types of tissues in experimental animals and humans. Bioartificial skin equivalents are in wide human clinical use to treat burns, large acute excisional wounds, and chronic venous and diabetic ulcers. They use primarily allogeneic cells and are thus temporary living dressings that cover the wound while being rejected and replaced by host cells. There has been moderate success in regenerating urinary conduit tissue through acellular regeneration templates, and some progress toward creating bioartificial bone, blood vessels, and urinary conduit tissue, but overall, the quest to replace damaged tissues with bioartificial equivalents has been largely elusive.

## 3. Induction of Regeneration *In Situ*

Although the initial objective of cell transplantation was the differentiation of the transplanted cells into new tissue that becomes integrated into the host tissue, there is substantial evidence that transplanted cells also secrete paracrine factors that are protective to host cells, promote their survival, proliferation, and differentiation, and suppress scarring. Thus, a third strategy of regenerative medicine is to identify these factors, or small molecules that mimic their action or stimulate their signaling pathways, and apply them topically by injection or in regeneration templates at the site of injury, where they will induce the regeneration of new tissues from local or circulating regeneration-competent cells. The advantage of this approach is that it eliminates problems associated with the logistics of cell transplantation (culturing, implantation), immunorejection, and bioethical issues in one stroke and would be relatively low-cost compared to other strategies.

The success of this strategy depends on two things. First, regeneration-competent cells must be present in nonregenerating tissues or be circulating in the blood, or we must be able to create them by inducing the dedifferentiation or compensatory hyperplasia of mature cells. Second, our understanding of the differences in molecular pathways that distinguish regeneration from scarring is limited and inadequate. Thus, we need to look at experimental systems that enable us to identify these pathways and the different combinations of factors that constitute a regeneration-permissive environment. There is evidence that mammals have much more inherent regenerative potential than they actually display, and that it is their tissue environment that determines whether they will engage in regeneration or fibrosis. Furthermore, several experimental systems can be exploited to analyze the molecular pathways that distinguish regeneration from repair by scar tissue formation.

### a. Evidence for Inherent Regenerative Potential of Nonregenerating Tissues

First, there is substantial evidence that many mammalian tissues house ASCs that normally are not acti-

**TABLE 1.2** Mammalian Tissues That Do Not Regenerate *in Vivo*, but Harbor Stem Cells That Can Proliferate and Differentiate *in Vitro*

| Tissue | Differentiation *in vitro* |
|---|---|
| Spinal cord | Neurons, glia |
| *Hippocampus | Neurons, glia |
| Striatum | Neurons, glia |
| Cerebral cortex | Neurons, glia |
| Neural retina | Retinal neurons, glia |
| Dermis | Neurons, glia |
| Myocardium | Cardiomyocytes, endothelium |

*Indicates lack of regeneration after injury as opposed to maintenance regeneration.

vated or participate in scar tissue formation after injury, indicating that mammals have considerable latent capacity for regeneration that is suppressed (Stocum, 2004b). **TABLE 1.2** summarizes the nonregenerating tissues in which stem cells have been found so far. Both the spinal cord and the heart initiate regeneration that is then aborted by inhibitory factors in the injury environment leading to the formation of scar tissue (Chapters 5, 11). Thus, by changing a nonpermissive injury environment to a regeneration-permissive one, we may be able to initiate and/or complete the regenerative process.

Second, regenerative responses have been induced or enhanced in a number of tissues of experimental animals. Biodegradable, cell-free artificial regeneration templates have been used to induce dermal regeneration in excisional skin wounds and improve regeneration across gaps in peripheral nerves, though the results have been far from perfect (Yannas, 2001). A variety of neuroprotective agents, as well as agents that neutralize molecules inhibitory to axon regeneration, and enzymes that degrade glial scar, have been used to improve spinal cord regeneration and slow the loss of neurons in Parkinson's disease and amyotrophic lateral sclerosis (ALS). Cell-free ceramic templates can induce bone regeneration across large gaps (Constanz et al., 1995; Yaszemski et al., 1995). Attempts to induce epimorphic limb regeneration from the non- or poorly regenerating limbs of adult frogs have also elicited or enhanced regenerative responses.

Third, mammalian muscle cells, which do not normally cellularize in response to injury, were induced to do so when treated *in vitro* with a protein extract of regenerating newt limb (McGann et al., 2001) (**FIGURE 1.10**). The resulting mononucleate cells were reported to dedifferentiate on a morphological and molecular level and become capable of differentiating into muscle, cartilage, and adipocytes, like mesenchymal stem cells of the bone marrow. This result suggests that mammalian cells have the inherent ability to dedifferentiate, but that the signal(s) and/or receptor(s) to initiate the process of dedifferentiation when injured are normally lacking.

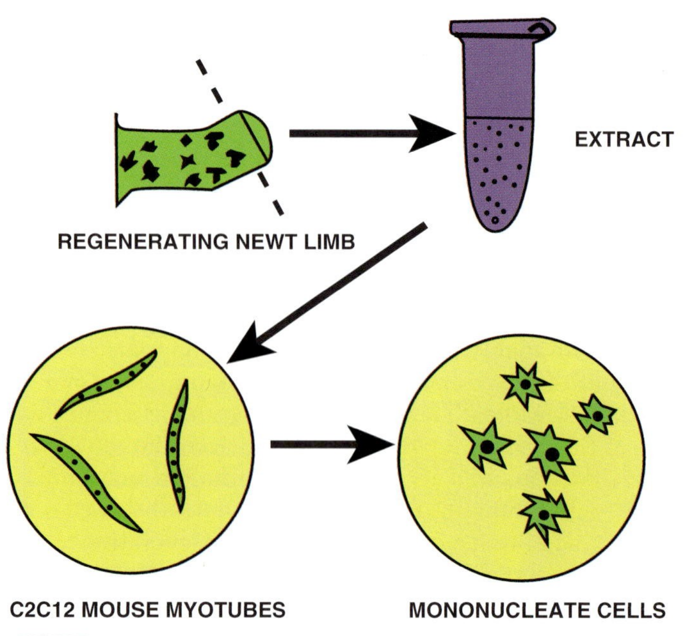


**FIGURE 1.10** Cellularization and dedifferentiation of cultured mammalian muscle cells (C2C12 mouse myofibers) by application of protein extract (PE) from regenerating newt limbs. The multinucleate myofibers cellularize into mononucleate cells that lose their phenotypic specializations.

### b. Models to Analyze the Molecular Differences That Distinguish Regeneration from Fibrosis

The chemical induction of regeneration from the body's own tissues, like cell transplantation and bioartificial tissue implants, requires that we understand the molecular differences between fibrosis and regeneration. We need to know what signals are present in the regeneration-competent environment versus those in the regeneration-deficient environment, and what inhibitory factors might be present in the regeneration-deficient environment that need to be suppressed or neutralized.

The standard approach to understanding why regeneration fails in mammalian tissues that do not regenerate spontaneously is to lesion these tissues, attempts to identify molecules in the injury environment that are inhibitory to regeneration, and devise ways to neutralize them. While this approach is useful, a more direct strategy is to compare and contrast the transcriptomes and proteomes of regenerating versus nonregenerating tissues and define the stimulatory and inhibitory molecules that differentiate regeneration from fibrosis. Three experimental models can be used for these comparisons. The first model compares wild-type tissues to genetic variations that confer a gain or loss of regenerative capacity. For example, there are several strains of mice that can regenerate ear and heart tissue (Heber-Katz et al., 2004), whereas the *short-toe* mutant of the axolotl causes deficiencies in limb regeneration (Humphrey, 1967). The second model is to compare the same tissues at developmental stages when they are capable of regeneration versus stages when they are not. For example, fetal skin in many mammalian species regenerates perfectly, but late in gestation the injury response switches to scar tissue formation characteristic of the adult (McCallion and Ferguson, 1996). The third model compares the same tissue between two species, one of which regenerates the tissue and the other does not. For example, the ability of cultured myotubes to re-enter the cell cycle in response to serum factors has been compared in the newt and mouse (Tanaka et al., 1999).

The amphibians are particularly useful for the latter two comparative strategies. Frogs regenerate many tissues in early tadpole stages, but lose the capacity to regenerate at late tadpole stages or by the time metamorphosis is complete (Stocum, 1995). Thus, cellular and molecular comparisons can be made between tissues at regeneration-competent and regeneration-deficient stages of development. Or, regeneration-competent juvenile or adult urodele tissues can be compared with regeneration-deficient anuran tissues. Genomic studies have repeatedly shown that developmentally important genes are strongly conserved across vast evolutionary distances. Much of what we know of vertebrate developmental and molecular genetics has been derived from studies on *Drosophila* and other invertebrates. Thus, it is not unreasonable to expect that the same genes involved in the regeneration of amphibians are conserved in humans and other mammals, though their activity is either suppressed or neutralized by inhibitory factors.

## SUMMARY

Regeneration is a repair process that maintains or restores the original architecture of a tissue by recapitulating part of its original embryonic development, as opposed to fibrosis (wound healing), which repairs tissues with scar tissue. Advances in the biological sciences and in our understanding of regeneration over the 18th and 19th centuries have, in the 21st century, culminated in a serious effort to develop a regenerative medicine by which we can restore damaged tissues and organs. Regeneration takes place from the molecular to the tissue levels of biological organization. There are three mechanisms of regeneration at the tissue level, compensatory hyperplasia of differentiated cells, activation of resident adult stem cells that self-renew within a specific microniche, and the creation of regeneration-competent cells by dedifferentiation, seen mainly in amphibians and fish. The activation and proliferation of these cells is regulated through a number of major signal transduction pathways. While regeneration is ubiquitous throughout the animal kingdom, only a few species within each phylum are able to regenerate complex structures such as appendages. Such regeneration appears to confer no selective advantage, so why some animals have this ability and others do not is a mystery.

Three strategies are being pursued in the development of a regenerative medicine. These are (1) the replacement of damaged tissue by transplanting cell suspensions or aggregates, (2) the implantation of bioartificial tissues or organs constructed in the laboratory to replace the natural ones, and (3) the induction of regeneration from tissues at the site of injury by pharmaceutical means. None of these strategies has achieved broad success as yet. There are two major problems with cell transplants. The first is to identify cell sources by which to provide sufficient cells to an injury site that will survive and either differentiate into new cells or secrete paracrine survival and proliferation factors that allow host cells to regenerate new tissue. The second is to avoid immunorejection. Adult stem cells and derivatives of embryonic stem cells are the two main cell sources that are being explored. The advantage of adult stem cells is they can be autogeneic, thus eliminating immunorejection. However, they are not easy to expand to large numbers. Embryonic stem cells are easy to expand, but are allogeneic and thus subject to immune rejection. Furthermore, their differentiation must be directed to a much greater extent than adult stem cells. The construction of bioartificial tissues faces all these challenges plus the challenge of providing a biodegradable scaffold incorporating the chemical and physical cues and signals necessary for the development of the tissue *in vitro*. The pharmaceutical induction of regeneration *in situ* would be the ultimate in regenerative therapy, but faces two major challenges. The first is identifying the combination of signals that promote regeneration over fibrosis. Genomic and proteomic approaches to comparing regeneration-competent versus deficient tissues will help in this endeavor. The second is the question of whether or not regeneration-competent cells are widely distributed in the body or whether they can be created on site by dedifferentiation. Numerous studies have now shown that many non-regenerating adult mammalian tissues actually harbor regeneration-competent cells that initiate regeneration, but are overwhelmed by fibrosis initiated by an inflammatory response. Furthermore, the application of extracts from regenerating amphibian limbs to cultured mammalian muscle cells results in their cellularization and dedifferentiation, suggesting that mammalian cells have the inherent capacity to dedifferentiate.

The conclusion is inescapable that the nonregenerating tissues of mammals, including humans, have a latent regenerative capacity that can potentially be evoked in the presence of the appropriate signaling environment. Thus, with continuing research into the biology of regeneration, using a variety of animal models, along with translating this research into therapeutic approaches, we should be able to achieve our goal of developing a regenerative medicine during the first half of the 21st century.

## REFERENCES


Aaronson DS, Horvath CM (2002) A road map for those who don't know JAK-STAT. Science 296:1653–1655.

Allen G (1977) Life Science in the Twentieth Century. Cambridge University Press, Cambridge.

Amit M, Shariki C, Margulets V, Itskovitz-Eldor J (2004) Feeder-layer and serum-free culture of human embryonic stem cells. Biol Reprod 70:837–845.

Attisano L, Wrana JL (2002) Signal transduction by the TGF-β superfamily. Science 296:1646–1647.

Avery OT, McCleod CM, McCarty M (1944) Studies on the chemical nature of the substance inducing transformation of pneumococcal types. J Exp Med 79:139–158.

Avilion AA, Nicolis S, Pevny LH, Perez L, Vivian N, Lovell-Badge R (2003) Multipotent cell lineages in early mouse development depend on SOX2 function. Genes Dev 17:126–140.

Aza-Blanc P, Lin H-Y, Ruiz i Altaba A, Kornberg TB (2000) Expression of the vertebrate Gli proteins in *Drosophila* reveals a distribution of activator and repressor activities. Development 127:4293–4301.

Badylak SF (2002) The extracellular matrix as a scaffold for tissue reconstruction. Sem Cell Dev Biol 13:377–383.

Bafico A, Liu G, Yaniv A, Gazit A, Aaronson SA (2001) Novel mechanism of Wnt signaling inhibition mediated by Dickopf-1 interaction with LRP6/Arrow. Nature Cell Biol 3:683–686.

Baguna J (1998) Planarians. In: Ferretti P, Geraudie J (eds.). Cellular and Molecular Basis for Regeneration. New York: John Wiley & Sons Ltd, pp 135–166.

Bejsovec A (2005) Wnt pathway activation: New relations and locations. Cell 120:11–14.

Brantjes H, Roose J, van de Wetering M, Clevers H (2001) All Tcf HMG box transcription factors interact with Groucho-related co-repressors. Nuc Acids Res 29:1410–1419.

Brockes JP (1998) Progenitor cells for regeneration: origin by reversal of the differentiated state. In: Ferretti P, Geraudie J (eds.). Cellular and Molecular Basis of Regeneration. New York: John Wiley & Sons Ltd, pp 63–78.

Brown H. (1992) Wound healing research through the ages. In: Cohen IK, Diegelmann RF, Lindblad WJ (eds.). Wound Healing: Biochemical and Clinical Aspects. Philadelphia: WB Saunders, pp 5–18.

Carlson BM (1998) Development and regeneration with special emphasis on the amphibian limb. In: Ferretti P, Geraudie J (eds.). Cellular and Molecular Basis of Regeneration. New York: John Wiley & Sons Ltd, pp 45–62.

Carpenter MK, Rosler ES, Fisk GJ, Brandenberger R, Ares X, Miura T, Lucero M, Rao MS (2004) Properties of four human embryonic stem cell lines maintained in a feeder-free culture system. Dev Dynam 229:243–258.

Cavallo RA, Cox RT, Moline MM, Roose J, Polevoy GA, Clevers H, Peifer M, Bejsovec A (1998) *Drosophila* Tcf and Groucho interact to repress Wingless signalling activity. Nature 395: 604–608.

Chambers I, Colby D, Robertson M, Nichols J, Lee S, Tweedie S, Smith A (2003) Functional expression cloning of Nanog, a pluripotency sustaining factor in embryonic stem cells. Cell 113:643–655.

Coleman W (1977) Biology in the Nineteenth Century: Problems of Form, Function and Transformation. Cambridge: Cambridge University Press, pp 118–159.

Constanz BR, Ison IC, Fulmer MT, Poser RD, Smith ST, VanWagoner M, Ross J, Goldstein SA, Jupiter JB, Rosenthal DI (1995) Skeletal repair by in situ formation of the mineral phase of bone. Science 267:1796–1799.

Dinsmore CE (1991) Lazzaro Spallanzani: Concepts of generation and regeneration. In: Dinsmore CE (ed.). A History of Regeneration Research. Cambridge: Cambridge University Press, pp 67–90.

Dinsmore CE (1998) Conceptual foundations of metamorphosis and regeneration: from historical links to common mechanisms. Wound Repair Reg 6:291–301.

Ellis HM, Horvitz HR (1986) Genetic control of programmed cell death in the nematode C. elegans. Cell 44:817–829.

Falabella A (1998) Debridement of wounds. Wounds 10 (Suppl):1C–9C.

Fu X, Wang Z, Sheng Z (2001) Advances in wound healing research in China: From antiquity to the present. Wound Rep Reg 9:2–10.

Fuchs E, Segre JA (2000) Stem cells: a new lease on life. Cell 100:143–156.

Fuchs E, Tumbar T, Guasch G (2004) Socializing with the neighbors: Stem cells and their niche. Cell 116:769–778.

Gage FH (2000) Mammalian neural stem cells. Science 287: 1433–1438.

Geraudie J, Akimenko M-A, Smith MM (1998) The dermal skeleton. In: Ferretti P, Geraudie J (eds.). Cellular and Molecular Basis for Regeneration. New York: John Wiley & Sons Ltd, pp 167–186.

Gilbert SF (2003) Developmental Biology, 7th ed. Sunderland: Sinauer Associates Inc, pp 164–166.

Gevers W (1984) Protein metabolism in the heart. J Mol Cell Cardiol 16:3–32.

Go MJ, Eastman DS, Artavanis-Tsakonas S (1998) Cell proliferation control by Notch signaling in *Drosophila* development. Development 125:2031–2040.

Goss RJ (1969) Principles of Regeneration. New York: Academic Press.

Goss RJ (1987) Why mammals don't regenerate—or do they? News in Physiol Sci 2:112–115.

Goss RJ (1991) The natural history (and mystery) of regeneration. In: Dinsmore CE (ed.). A History of Regeneration Research. Cambridge: Cambridge University Press, pp 7–23.

Goss RJ (1992) Regeneration vs. repair. In: Cohen IK, Diegelmann RF, Lindblad WJ (eds.). Wound Healing: Biochemical and Clinical Aspects. Philadelphia: WB Saunders, pp 20–39.

Griffin JW, Hoffman PN (1993) Degeneration and regeneration in the peripheral nervous system. In: Dyck PJ, Thomas PK, Griffin JW, Low PA, Poduslo J (eds.). Peripheral Neuropathy, 3rd ed, Philadelphia: WB Saunders Co, pp 361–376.

Ham AW, Cormack DH (1979) Histology, 8th ed. Philadelphia: JB Lippincott, pp 377–462.

Hanna LA, Foreman R, Tarasenko IA, Kessler DS, Labosky PA (2002) Requirement for Foxd3 in maintaining pluripotent cells of the early mouse embryo. Genes Dev 16:2650–2661.

Hansen-Smith FM, Carlson BM (1979) Cellular responses to free grafting of the extensor digitorum longus muscle of the rat. J Neurol Sci 41:149–173.

Hart AH, Hartley L, Ibrahaim M, Robb L (2004) Identification, cloning and expression analysis of the pluripotency promoting nanog genes in mouse and human. Dev Dynam 230:187–198.

Harty M, Neff AW, King MW, Mescher AL (2003) Regeneration or scarring: An immunologic perspective. Dev Dynam 226:268–279.

Heber-Katz E, Leferovich JM, Bedelbaeva K, Gourevitch D (2004) Spallanzani's mouse: a model of restoration and regeneration. In: Heber-Katz E (ed.). Regeneration: Stem Cells and Beyond. Curr Top in Microbiol and Immunol 280:165–190.

Hershey AD, Chase M (1952) Independent functions of viral protein and nucleic acid in growth of bacteriophage. J Gen Physiol 36: 39–56.

Humphrey RR (1967) Genetic and experimental studies on a lethal trait ("Short Toes") in the Mexican axolotl (*Ambystoma mexicanum*). J Exp Zool 164:281–296.

Imai T, Tokunaga A, Yoshida T, Hashimoto M, Mikoshiba K, Weinmaster G, Nakafuku M, Okano H (2001) The neural RNA-binding protein Musashi1 translationally regulates mammalian *numb* gene expression by interacting with its mRNA. Mol Cell Biol 21:3888–3900.

Jan YN, Jan LY (1998) Asymmetric cell division. Nature 392: 775–778.

Jensen UB, Lowell S, Watt F (1999) The spatial relationship between stem cells and their progeny in the basal layer of human epidermis: A new view based on whole mount labeling and lineage analysis. Development 126:2409–2418.

Jiang Y, Jahagirdar BN, Reinhardt R, Schwarts RE, Keene CD, Ortiz-Gonzalez XR, Reyes M, Lenvik T, Lund T, Blackstad M, Du J, Aldrich S, Lisberg A, Low WC, Largaespada DA, Verfaille CM (2002) Pluripotency of mesenchymal stem cells derived from adult marrow. Nature 418:41–49.

Kanatsu-Shinohara M, Inoue K, Lee J, Yoshimoto M, Ogonuki N, Miki H, Baba S, Kato T, Kazuki Y, Toyokuni S, Toyoshima M, Niwa O, Oshimura M, Heike T, Nakahata T, Ishino F, Ogura A, Shinohara T (2004) Generation of pluripotent stem cells from neonatal mouse testis. Cell 119:1001–1012.

Kirkpatrick JJR, Curtis B, Naylor IL (1996) Back to the future for wound care? The influences of Padua on wound management in Renaissance Europe. Wound Repair Reg 4:326–334.

Lai K, Kaspar BK, Gage FH, Schaffer DV (2003) Sonic hedgehog regulates adult neural progenitor proliferation *in vitro* and *in vivo*. Nature Neurosci 6:21–27.

Langer R, Tirell DA (2004) Designing materials for biology and medicine. Nature 428:487–492.

Lecourtis M, Schweisguth F (1998) Indirect evidence for Delta-dependent intercellular processing of Notch in *Drosophila* embryos. Curr Biol 8:771–774.

Lenhoff HM, Lenhoff SG (1991) Abraham Trembly and the origins of research on regeneration in animals. In: A History of Regeneration Research, Dinsmore CE, ed. Cambridge: Cambridge University Press, pp 47–66.

Lum L, Beachy PA (2004) The hedgehog network: Sensors, switches, and routers. Science 304:1755–1759.

Lundkvist J, Lendhal U (2001) Notch and the birth of glial cells. Trends Neurosci 9:492–494.

Majno G (1975) The Healing Hand: Man and Wound in the Ancient World. Cambridge: Harvard University Press.

McCallion RL, Ferguson MWJ (1996) Fetal wound healing and the development of antiscarring therapies for adult wound healing. In: Clark RAF (ed.). The Molecular and Cellular Biology of Wound Repair, 2nd ed. New York: Plenum Press, pp 561–600.

McGann C, Odelberg SJ, Keating MT (2001) Mammalian myotube dedifferentiation induced by newt regeneration extract. Proc Natl Acad Sci (USA) 98:13699–13703.

Michalopoulos GK, DeFrances MC (1997) Liver regeneration. Science 276:60–66.

Mitsui K, Tokuzawa Y, Itoh H, Segawa K, Murakami M, Takahashi K, Maruyama M, Maeda M, Yamanaka (2003) The homeoprotein Nanog is required for maintenance of pluripotency in mouse epiblast and ES cells. Cell 113:631–642.

Morgan TH (1901) Regeneration. New York: Macmillan.

Nakamura Y, Sakakibara S-I, Miyata T, Ogawa M, Shimazaki T, Weiss S, Kageyama R, Okano H (2000) The bHLH gene *Hes1* as a repressor of the neuronal commitment of CNS stem cells. J Neurosci 20:283–293.

Nerem RM, Sambanis A (1995) Tissue engineering: From biology to biological substitutes. Tiss Eng 1:3–13.

Nicholson DW, Thornberry NA (2003) Life and death decisions. Science 299:214–215.

Ogle B, Gairola P, Balik J, Mooradian DL (1998) Tissue engineering: At the interface between engineering, biology and medicine. J Minn Acad Sci 63:47–57.

Porter JA, Young KE, Beachy PA (1996) Cholesterol modification of Hedgehog signaling proteins in animal development. Science 274:255–259.

Potten CS, Booth C, Pritchard DM (1997) The intestinal epithelial stem cell: the mucosal governor. Int J Path 78:219–243.

Reubinoff BE, Itsykson P, Turetsky T, Pera MF, Reinhartz E, Itzik A, Ben-Hur T (2001) Nerual progenitors from human embryonic stem cells. Nature Biotech 19:1134–1140.

Reya T, Morrison SJ, Clarke MF, Weissman IL (2001) Stem cells, cancer, and cancer stem cells. Nature 414:105–111.

Richards M, Fong C, Chan WK, Wong PC, Bongso A (2002) Human feeders support prolonged undifferentiated growth of human inner cell masses and embryonic stem cells. Nature Biotech 20:933–936.

Rippon HJ, Bishop AE (2004) Embryonic stem cells. Cell Prolif 37:23–34.

Roberson DW, Rubel EW (1995) Hair cell regeneration. Curr Opin Otolaryngol Head Neck Surg 3:302–307.

Roel G, Hamilton FS, Gent Y, Bain AA, Destree O, Hoppler S (2002) Lef-1 and Tcf-3 transcription factors mediate tissue-specific Wnt signaling during *Xenopus* development. Curr Biol 12:1941–1945.

Roose J, Molenaar M, Peterson J, Hurenkamp J, Brantjes H, Moerer P, van de Wetering M, Destree O, Clevers H (1998) The *Xenopus* Wnt effector XTcf-3 interacts with Groucho-related transcriptional repressors. Nature 395:608–612.

Rossant J (2001) Stem cells from the mammalian blastocyst. Stem Cells 19:477–482.

Sakakibara S-I, Imai T, Hamaguchi K, Okabe M, Aruga J, Nakajima K, Yasutomi D, Nagata T, Kurihara Y, Uesugi S, Miyata T, Ogawa M, Mikoshiba K, Okano H (1996) Mouse Musashi-1, a neural RNA-binding protein highly enriched in the mammalian CNS stem cell. Dev Biol 176:230–242.

Sanchez-Alvarado A (2000) Regeneration in the metazoans: Why does it happen? Bioessays 22:578–590.

Schlessinger J (2000) Cell signaling by receptor tyrosine kinases. Cell 103:211–225.

Schroeter EH, Kisslinger JA, Kopan R (1998) Notch-1 signaling requires ligand-induced proteolytic release of intracellular domain. Nature 393:382–386.

Shamblott MJ, Axelman J, Wang S, Bugg EM, Littlefield JW, Donovan PJ, Blumenthal D, Huggins GR, Gearhart JD (1998)

Derivation of pluripotent stem cells from cultured human primordial germ cells. Proc Natl Acad Sci USA 95:13726–13731.

Shi Y, Massague J (2003) Mechanisms of TGF-β signaling from cell membrane to the nucleus. Cell 113:685–700.

Skinner DM, Cook JS (1991) New limbs for old: Some highlights in the history of regeneration in Crustacea. In: Dinsmore CE (ed.). A History of Regeneration Research. Cambridge: Cambridge University Press, pp 25–46.

Slack JMW (2000) Stem cells in epithelial tissues. Science 287: 1431–1433.

Smith A (2001) Embryonic stem cells. In: Marshak DR, Gardner R, Gottlieb D (eds.). Stem Cell Biology. Cold Spring Harbor: Cold Spring Harbor Laboratory Press, pp 205–230.

Song X, Zhu CH, Doan C, Xie T (2002) Germline stem cells anchored by adherens junctions in the *Drosophila* ovary niches. Science 296:1855–1857.

Stecca B, Ruiz i Altaba A (2002) The therapeutic potential of modulators of the hedgehog-Gli signaling pathway. J Biol 1:9.1–9.4.

Steward FC, Mapes MO, Kent AE, Holsten RD (1964) Growth and development of cultured plant cells. Science 143:20–27.

Stocum DL (2004a) Amphibian regeneration and stem cells. In: Heber-Katz E (ed.). Regeneration: Stem Cells and Beyond. Curr Topics in Microbiol and Immunol 280:1–70.

Stocum DL (2004b) Tissue restoration through regenerative biology and medicine. Adv in Anat Embryol Cell Biol 176:1–101.

Sun T-Q, Lu B, Feng J-J, Reinhardt C, Jan YN, Fanti WJ, Williams LT (2001) PAR-1 is a disheveled-associated kinase and a positive regulator of Wnt signaling. Nature Cell Biol 3:628–635.

Taipale J, Beachy PA (2001) The hedgehog and Wnt signaling pathways in cancer. Nature 411:349–354.

Thomson JA, Itskovitz-Eldor J, Shapiro SS, Waknitz MA, Swiergiel JJ, Marshall VS, Jones JM (1998) Embryonic stem cell lines derived from human blastocysts. Science 282:1145–1147.

Trembly JH, Steer CJ (1998) Perspectives on liver regeneration. J Minn Acad Sci 63:37–46.

Van den Heuvel M (2001) Straight or split: Signals to transcription. Nature Cell Biol 3:E155–E156.

Van Doren C (1991) A History of Knowledge. Ballantine Books, New York.

Vorontsova MA, Liosner LD (1960) Asexual Propagation and Regeneration. Permagon Press, New York, pp 143–162.

Wallberg AE, Pedersen K, Lendahl U, Roeder RG (2002) p300 and CAF act cooperatively to mediate transcriptional activation from chromatin templates by Notch intracellular domains in vitro. Mol Cell Biol 22:7812–7819.

Wang B, Fallon JF, Beachy PA (2000) Hedgehog-regulated processing of Gli3 produces an anterior/posterior repressor gradient in the developing vertebrate limb. Cell 100:423–434.

Watson JD, Crick FHC (1953) A structure for deoxyribose nucleic acid. Nature 171:737–738.

Watt FM, Hogan BL (2000) Out of Eden: Stem cells and their niches. Science 287:1427–1430.

Weissman IL (2000) Stem cells: units of development, units of regeneration, and units in evolution. Cell 100:157–168.

Wehrli M, Dougan ST, Caldwell K, O'Keefe L, Schwartz S, Vaizel-Ohayon D, Schejter E, Tomlinson A, DiNardo S (2000) *arrow* encodes an LDL-receptor-related protein essential for Wingless signaling. Nature 407:527–530.

Whetton AD, Graham GJ (1999) Homing and mobilization in the stem cell niche. Trends Cell Biol 9:233–238.

Yamashita YM, Jones DL, Fuller MT (2003) Orientation of asymmetric stem cell division by the AC tumor suppressor and centrosome. Science 301:1547–1550.

Yannas IV (2001) Tissue and Organ Regeneration in Adults. New York: Springer.

Yaszemski MJ, Payne RG, Hayes WC, Langer RS, Aufdemorte TB, Mikos AG (1995) The ingrowth of new bone tissue and initial mechanical properties of a degrading polymeric composite scaffold. Tiss Eng 1:41–52.

Ying Q-L, Nichols J, Chambers I, Smith A (2003) BMP induction of Id proteins suppresses differentiation and sustains embryonic stem cell self-renewal in collaboration with STAT3. Cell 115: 281–292.

Young H, Black AC Jr (2004) Adult stem cells. Anat Rec 276A: 75–102.

Zhang Y, Kalderon D (2001) Hedgehog acts as a somatic stem cell factor in the *Drosophila* ovary. Nature 410:599–604.

Zhang J, Niu, Ye L, Huang H, He X, Tong WE, Ross J, Haug J, Johnson T, Feng JQ, et al. (2003) Identification of the hematopoietic stem cell niche and control of the niche size. Nature 425: 836–841.

# 2 Repair of Skin Wounds by Fibrosis

## INTRODUCTION

The skin is the largest organ of the vertebrate body. Skin has many functions, including thermoregulation, sensory transduction, and acting as a mechanical barrier to protect the body from dessication, invasion by microorganisms, and environmental insults such as UV irradiation, mechanical trauma, and chemical or thermal burns. The preservation of normal skin structure, especially facial skin, has a high social value, since facial appearance and expression are such important determinants of image and communication. Thus cosmetic treatments and plastic surgeries for the skin constitute a multibillion dollar industry worldwide.

The dermis of the skin does not regenerate, but repairs by fibrosis. Pig skin most closely resembles human skin anatomically and physiologically, and thus is the best model for the study of wound repair by fibrosis. Although pigs are becoming the model of choice, most work has been done on rodents (rats and mice) or lagomorphs (rabbits) because of their ease of breeding and handling, and relatively low maintenance costs (Sullivan, 2001). The results of wound healing studies in these animals, however, cannot be generalized *in toto* to human skin, because the skin of these animals differs from human skin in a major way. Human skin is tight and contraction does not play a large role in healing. Rodent and rabbit skin is loose and undergoes significant contraction to reduce the size of the wound to be repaired. A general problem of wound healing studies is that they are often hard to compare, due to variability in the type of wound model used, strain of animal, sex, age, and location and size of wound (Dorsett-Martin, 2004). Thus, a caveat to interpreting skin repair research, regardless of model used, is that the parameters on which observations and measurements are made have not been well-standardized. Nevertheless, we have established a general picture of skin repair by fibrosis at the tissue, cellular, and molecular levels.

## STRUCTURE OF ADULT MAMMALIAN SKIN

The skin is composed of two layers, the epidermis and dermis (**FIGURE 2.1**). The epidermis is a stratified squamous epithelium consisting primarily of keratinocytes in various stages of differentiation, from mitotically active basal cells (stratum basale or stratum germinatiuum) to the heavily keratinized superficial cells (stratum corneum) that are continually sloughed. Keratinocytes are held together laterally by adhesion belts, desmosomes, and tight junctions to form a water-impermeable sheet. The ECM of the stratum basale contains hyaluronic acid (HA), a large polyionic, nonsulfated glycosaminoglycan (GAG) that binds water, and the basal cells express the CD44 receptor for HA (Chen and Abatangelo, 1999). Stem cells in the stratum basale are constantly dividing to self-renew and give rise to more differentiated keratinocytes that migrate (or are pushed) upward to replace the cells of the stratum corneum as they slough off. Three other nonepithelial cell types are found in lesser numbers in the epidermis: melanocytes, which give the skin its color; Langerhans cells, antigen-presenting dendritic cells of the immune system; and Merkel cells, which are thought to function as mechanoreceptors (Ham and Cormack, 1979). A number of epidermal appendages project downward into the dermis: hair follicles, sweat glands, and sebaceous glands.

The dermis consists of two layers of fibroblasts embedded in ECM, the papillary layer next to the basal layer of the epidermis and the deeper reticular layer (Ham and Cormack, 1979). The papillary layer derives its name from the fact that it is thrown into papillae that project up into the epidermis. This layer is pervaded by a capillary network that provides nourishment to the epidermis and acts as a heat exchanger. The ECM of the papillary layer contains a network of thin collagen and elastic fibers. Mast cells, immune cells that release histamine in allergic reactions, are present in the dermis, as well as in other connective tissue compart-

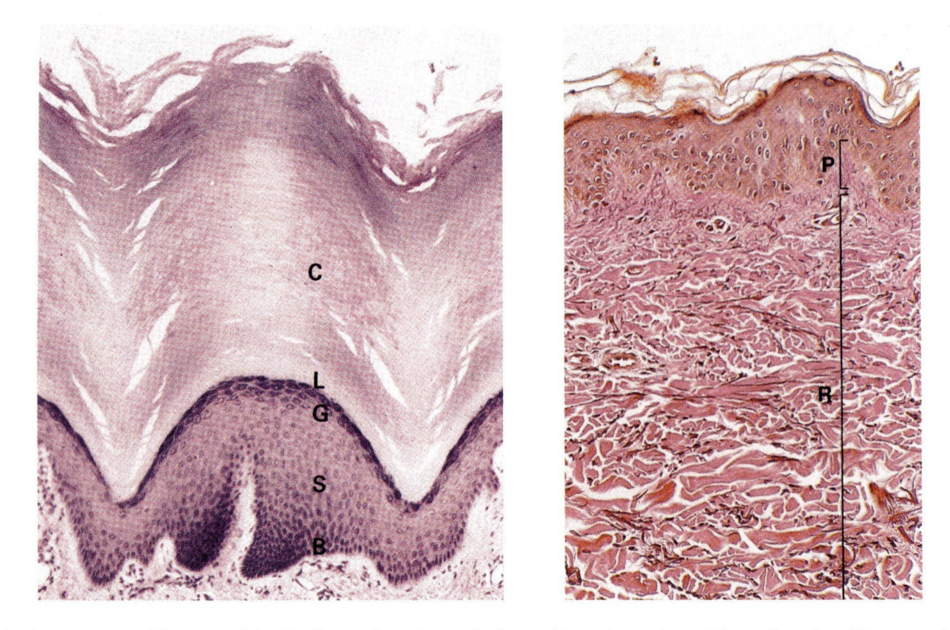


**FIGURE 2.1** Histological structure of human skin. **Left**, section through the epidermis and papillary dermis of human thick fingertip skin (H & E stain). The epidermis is composed of multiple layers of cells that differentiate into progressively flattened keratinocytes from the basal layer (stratum germanitivum or stratum basale, B, to the outer highly keratinized stratum corneum C). Note the intense staining of the stratum basale that is due to its active stem cell and transit amplifying population. S and G indicate the stratum spinosum and stratum granulosum, respectively, which represent successive stages of decreased mitosis and increased keratinization. L = the stratum lucidum, a homogeneous layer of keratinocytes present only in thick skin. **Right**, section through the epidermis and dermis of thin skin (elastic van Gieson stain). The papillary dermis (P) is thrown into folds of highly vascularized loose connective tissue. The reticular layer (R) of the dermis is composed of intertwined coarse collagen bundles (red) that are cut in many planes. This layer contains larger blood vessels and fewer capillaries. Elastin fibers stain black. They form a fine network in the papillary dermis and follow the course of the thick collagen fibers in the reticular dermis. (Reproduced with permission from Wheater et al., Wheater's Functional Histology, 3rd ed. Copyright 1997, Elsevier.

ments of the body. Tissue macrophages (phagocytic cells) patrol the dermis, and fat cells may be present.

The reticular layer of the dermis is thicker than the papillary layer and is characterized by an ECM containing a network of coarse collagen fibers and elastin. The collagen and elastin fibers in both the papillary and reticular layers are organized in a reticular (basket weave) pattern. Fewer capillaries course through the reticular layer. The reticular layer rests on a superficial fascia, or hypodermis, that is not part of the skin. Spaced bundles of collagen fibers extending from the reticular layer anchor it into the hypodermis. These anchors have a range of motion that permits a variable planar movement of the skin with respect to the hypodermis, depending on species.

The proteins of the dermal ECM fall into three classes: proteoglycans, fibrous proteins, and adhesive proteins (Alberts et al., 1994; Clark, 1996). Proteoglycans (PGs) are proteins linked to sulfated GAGs, of which dermatan sulfate, heparan sulfate, and chondroitin sulfate are prominent in dermal ECM. Significant PGs in dermal ECM are the large PG versican, and the small PG decorin. Multiple PGs are linked to molecules of HA to create hyaluronic acid-PG complexes. These complexes, which are among the largest biological molecules known, bind water avidly through versican, causing the dermal matrix to swell. The binding of water gives the uninjured dermal ECM its property of resisting compressive force and creates space for cell migration in injured skin. The major fibrous proteins in dermal ECM are the collagens, with smaller amounts of elastins. Collagens give tensile strength to the ECM and elastins confer resiliency, allowing the skin to be stretched and then assume its original shape. The major collagens in the dermal matrix are type I collagen (80%) and type III collagen. Smaller amounts of other collagens are also present. Type VI collagen forms a highly branched network of filaments surrounding the type I collagen fibrils. Type IV collagen is part of the basement membrane of blood vessels, and type VIII collagen is located around hair follicles and small blood vessels. The predominant dermal adhesive proteins are fibronectin (Fn), vitronectin (Vn), and tenascin-C (Tn-C). Fibronectin and Vn serve as a substrate to which cells can adhere when they are either migrating or stationary (Clark, 1996). Tenascin-C is an anti-adhesive protein that, with Fn, helps control the degree of cellular adhesion to the ECM substrate.

The proteins that make up dermal ECM, as well as the ECM of other tissues and organs, have short amino acid recognition sequences that allow them to bind to cells through a variety of receptors. The major cell receptor family for ECM molecules such as collagens I, III, and Fn is the integrins, low-affinity, heterodimeric linker proteins consisting of two noncovalently associated transmembrane glycoprotein subunits, $\alpha$ and $\beta$. The integrins are linked on the inside of the plasma membrane to the cytoskeleton. Integrins function in three capacities: adhesion of cells to ECM; migration of cells on ECM; and maintenance and modulation of gene expression by the transmission of signals to the nucleus (Adams and Watt, 1993). Tenascin C and type I collagen have epidermal growth factor (EGF) repeat domains that bind to the epidermal growth factor receptor (EGFR), a tyrosine kinase (Tran et al., 2004). Other recognition domains allow proteins to bind to each other, thus regulating the organization of the ECM.

A basement membrane approximately 100-nm thick is synthesized by the epidermis and connects the epidermis to the papillary layer of the dermis (Yannas, 2001). It consists of two layers, the lamina lucida directly beneath the epidermal cells, and the lamina densa next to the papillary dermis. The lamina lucida consists mainly of the glycoprotein laminin (Ln), whereas the lamina densa is composed of type IV collagen, perlecan, and another protein, entactin, that is believed to bind type IV collagen and Ln together. Perlecan is a heparan sulfate PG that has both a structural and filtering function. The epidermal cells are anchored into the lamina lucida by hemidesmosomes, through the $\alpha 6\beta 4$ integrin, while type VII collagen fibrils anchor the lamina densa into the ECM of the dermal papillary layer (Alberts et al., 1994; Yannas, 2001).

The dermal ECM serves as a reservoir for growth factors, binding them in latent form, and releasing them upon injury. Growth factors are signaling molecules that stimulate or inhibit proliferation, migration, and differentiation, depending on the cell type. For example, transforming growth factor beta (TGF-$\beta$), which exists in three isoforms, binds to decorin, type IV collagen, Fn, and thrombospondin (Roberts and Sporn, 1996). Most growth factors signal cells through receptor tyrosine kinase (RTK) pathways that initiate intracellular phosphorylation cascades by other kinases, resulting in the activation of transcription factors that up- or down-regulate gene activity. The exception is the TGF-$\beta$ family, which signals target genes through the serine-threonine kinase/Smad pathways (Roberts and Sporn, 1996; see Chapter 1).

The ECM, growth factors, and cell adhesion molecule (CAM) and growth factor receptors play important roles in orchestrating the healing of a dermal wound that are only partially understood.

## THE EFFECT OF WOUND TYPE AND EXTENT ON DERMAL REPAIR

Injury to the epidermis alone results in regeneration without scar, whereas wounds that penetrate the dermis result in repair by fibrosis (Yannas, 2001). The degree of dermal fibrosis depends on the type of tissue and the extent of the wound. The original tissue architecture is most closely restored in wounds made by shallow surgical incisions, where there is no loss of tissue and the wound edges either do not pull apart or are sutured together. These wounds heal primarily by simple fibroblast proliferation with minimal scar formation because there is little wound space that needs to be filled in. Deep incisions in which the wound edges pull apart, as well as excisional wounds and burns that destroy substantial amounts of tissue, repair by the formation of extensive scar tissue.

An interesting and important fact is that both regeneration and scarring occur simultaneously in excisional wounds. While the dermal connective tissue heals by fibrosis, the epidermis and the vasculature regenerate, though the epidermis may lack structures such as hair follicles and sebaceous glands. Regeneration of the basic epidermal structure and the vasculature are constant features associated with repair by regeneration or scarring and reflect selective pressure for re-forming a barrier to the external environment and nourishment and oxygenation of underlying tissue. Why epidermal and endothelial cells respond to the wound environment by regenerating and dermal fibroblasts do not is a question of considerable interest.

## PHASES OF REPAIR IN EXCISIONAL WOUNDS

Repair of an excisional skin wound can be divided into three tightly integrated phases that occupy variable time frames, depending on the size of the wound. These phases are hemostasis, inflammation, and structural repair (Kirsner and Eaglstein, 1993; Linares, 1996; Sicard et al., 1998; Yannas, 2001; Schultz et al., 2003). Each phase initiates and overlaps the next. The phase of structural repair can be divided into the formation of granulation tissue and the remodeling of granulation tissue to form scar, which may last as long as six months to two years.

Nine cell types play major roles in the epidermal and dermal repair process. Platelets, neutrophils, macrophages, T-cells, mast cells, and injured axons of sensory and sympathetic post-ganglionic nerves, are involved in hemostasis and inflammation. Epidermal cells, dermal fibroblasts, and endothelial cells provide the means for structural repair. These cell types orchestrate the process of repair through the production of ECM molecules, proteases, growth factors, cytokines and chemokines that stimulate or inhibit specific cell activities. **FIGURE 2.2** illustrates the stages of repair. **FIGURE 2.3** and **TABLE 2.1** summarize the cells and molecular factors that regulate each phase of repair.

(A)

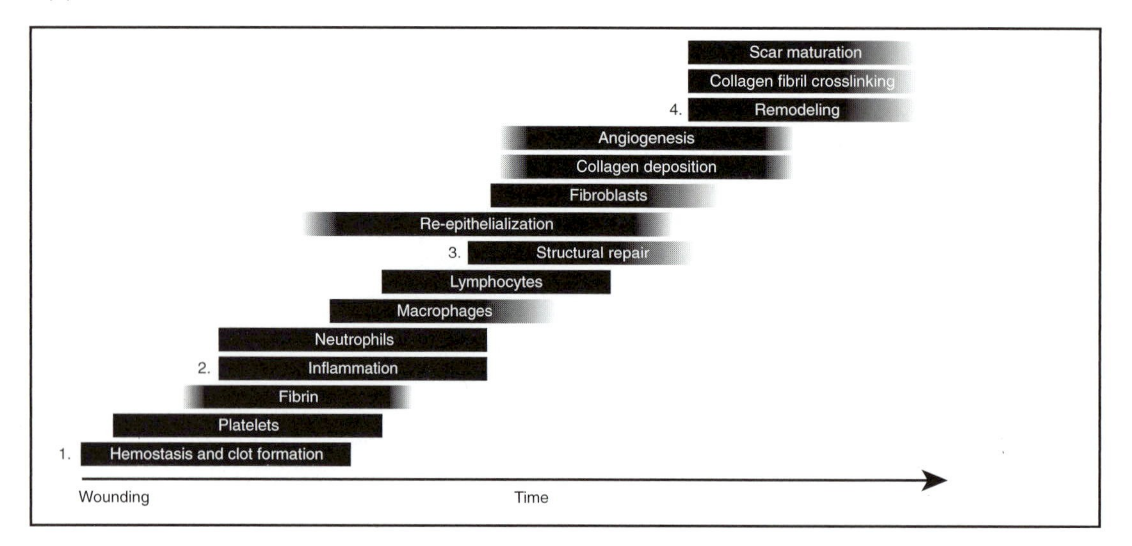


(B)

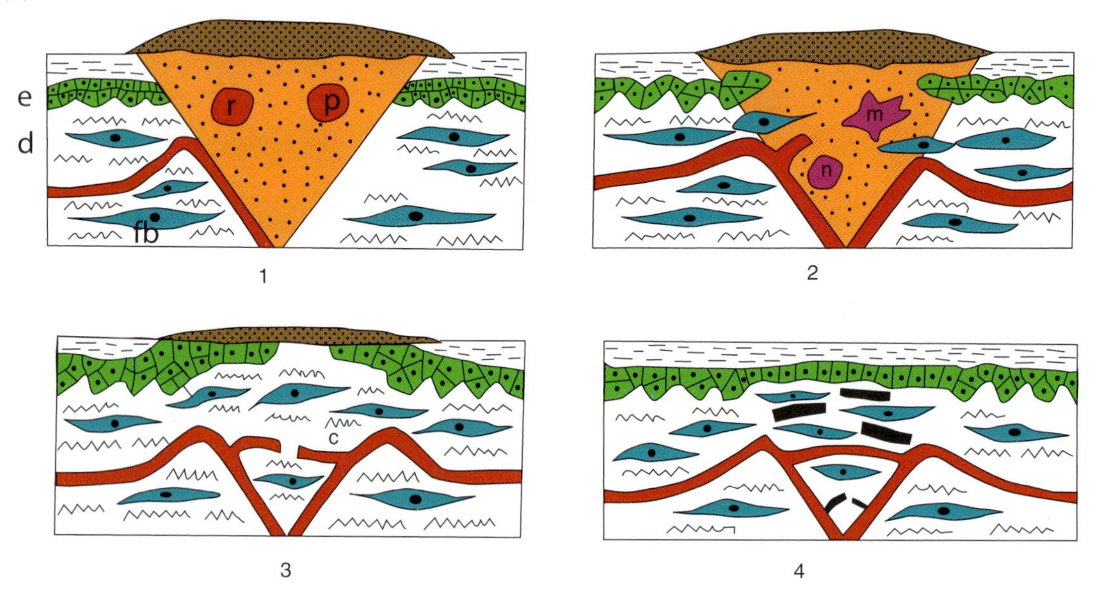

**FIGURE 2.2** Overlapping stages of excisional wound repair in skin. **(A)** Progression of the phases with time. **(B)** Diagrams of clot formation, inflammation, formation of granulation tissue and cross-linking of collagen fibers. (1) Clot formation. The fibrin clot (light stipple) traps platelets (p) and red blood cells (r). The clot has dried at the surface to form a scab (heavy stipple). e = epidermis, d = dermis, fb = fibroblasts. (2) Inflammation. The clot has been invaded by macrophages (m) and neutrophils (n). Basal epidermal cells have begun migrating into the clot under the scab. Fibroblasts begin entering the clot toward the end of the inflammation phase. (3) Fibroblasts populate the wound and replace the fibrin clot with new ECM, forming granulation tissue penetrated by new capillaries (c). Re-epithelialization continues. (4) Re-epithelialization and angiogenesis are complete and the collagen fibers of the granulation tissue are being cross-linked into thick bundles parallel to the wound surface.

## 1. Hemostasis and Clot Formation

Wounding severs blood vessels and damages epidermal and connective tissue cells and ECM. The first response to wounding is hemostasis, which occurs within a matter of minutes to stop bleeding and seal off the wound by three mechanisms: formation of platelet clumps, vasoconstriction, and formation of a fibrin clot (Clark, 1996).

Platelets are anucleate fragments of megakaryocytes, a blood cell lineage of hematopoietic stem cells in the bone marrow. Immediately upon wounding, blood suffuses the wound. The cells in the wall of the blood vessel at the site of injury release ADP,

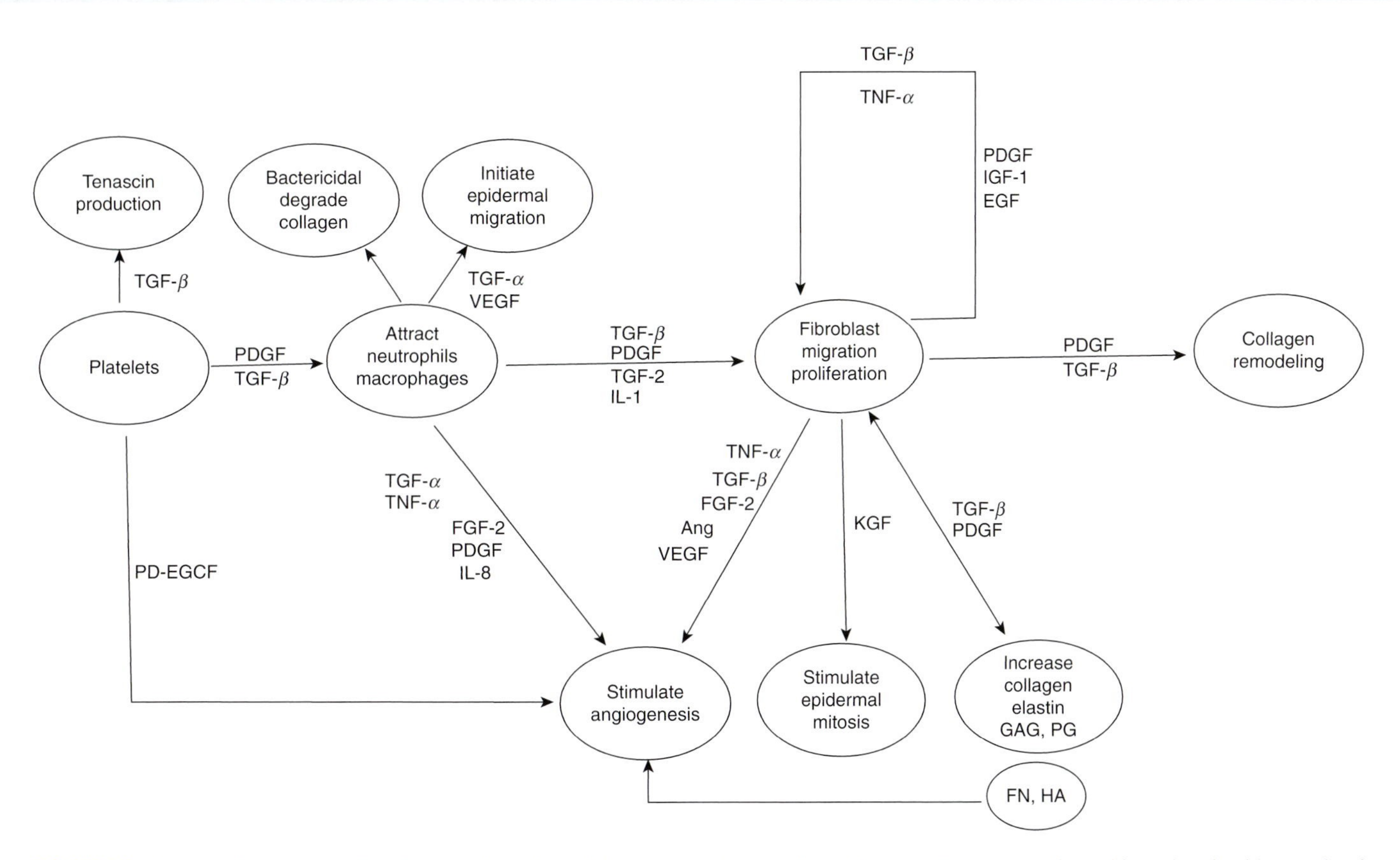


**FIGURE 2.3** Diagram showing simplified network of the cellular interactions, mediated by growth factors and cytokines, involved in repair of an excisional skin wound.

which causes clumps of platelets to adhere to the injury site. The clumped platelets degranulate upon contact with exposed collagen in the walls of the injured vessels, releasing more ADP (which attracts more platelets), as well as arachadonic acid, fibrinogen, fibronectin, thrombospondin, and von Willebrand factor VIII. Fibrinogen, fibronectin, and thrombospondin act as ligands for platelet aggregation. Von Willebrand factor VIII helps mediate the adhesion of platelets to collagen in the injured blood vessel wall. The net result is a sticky mass of platelets, which plugs the vessel.

The arachodonic acid (AA) released by the platelets is converted to thromboxane A2, which belongs to one of four classes of eicosanoids formed by a cyclooxygenase-dependent pathway. The eicosanoids play a prominent role in pain, fever, and inflammation and are thus major targets for drugs developed to inhibit their activities. Thromboxane A2 and serotonin from injured nerve axons are powerful vasoconstrictors that slow blood flow into the wound by constricting blood vessels. The injured nerves also release substance P, a neuropeptide that causes mast cells in the dermis to degranulate, releasing histamine that increases the permeability of vessel walls, allowing further leakage of plasma into the wound.

The blood plasma present in the wound space contains coagulation factors that induce clot formation (Clark, 1996). Hagemann Factor VII, a plasma protein, initiates a cascade of reactions involving about 12 clotting factors (I–XII) and requiring $Ca^{++}$ as an essential co-factor (Lansdown, 2002). Tissue factor is produced at the end of the cascade and converts prothrombin to its active form, the enzyme thrombin. Thrombin catalyzes the conversion of plasma fibrinogen to fibrin, the major structural protein of the clot (Yamada and Clark, 1996). Fibrin molecules are organized into fibers that intertwine to form a meshwork that traps erythrocytes, leukocytes, and platelets. This meshwork also contains plasma Fn and Vn, as well as thrombospondin from platelets (Yamada and Clark, 1996) and collagen types I, III, and IV (Lindblad, 2003). Several studies have indicated that these collagens are synthesized by invading blood-borne cells, probably monocytes (Lindblad, 2003; Fathke et al., 2004). Once formed, the provisional matrix contracts and extrudes serum by the action of a contractile platelet factor, thrombosthenin, which helps prevent gaping of the wound. The surface of the clot dehydrates to form a scab, preventing further fluid loss. Cells of the immune system now invade the deeper part of the clot to begin the phase of inflammation.

**TABLE 2.1** Phases of Skin Wound Repair, Summarizing Cell Types and Functions, and the ECM Molecules, Growth Factors, Enzymes, and Bactericides Produced by the Cells That Mediate Their Function

| Cells | Functions and Activities | Bactericides, Enzymes and Enzyme Inhibitors | ECM Molecules | Growth Factors |
|---|---|---|---|---|
| **A. HEMOSTATSIS AND INFLAMMATION** | | | | |
| Platelets | 1. Attraction of neutrophils, monocytes | — | — | PDGF,TGF-β |
| | 2. Stimulate tenascin production | — | — | TGF-β |
| | 3. Stimulate angiogenesis | — | — | PD-EGCF |
| Neutrophils | 1. Kill bacteria | Bactericidal peptides and proteins | — | — |
| | 2. Phagocytosis | | — | — |
| | | $H_2O_2$ | — | — |
| | | Hypochlorous acid | — | — |
| Macrophages | 1. Kill bacteria | $H_2O_2$ | — | — |
| | 2. Phagocytosis | Hypochlorous acid | | |
| | 3. Degrade collagens | Collagenases | — | — |
| | 4. Promote migration/ proliferation of fibroblasts | — | — | TGF-β, PDGF, IL-1, FGF-2 |
| | 5. Initiate re-epithelization | | | |
| | 6. Initiate angiogenesis | — | — | TGF-α |
| | | — | — | TGF-α, TNF-α FGF-2, VEGF |
| Fibroblasts | 1. Synthesize substratum for epithelial migration | — | Tenascin Fibronectin | KGF |
| Epidermis | 1. Cover wound | — | — | VEGF |
| **B. FORMATION OF GRANULATION TISSUE** | | | | |
| Fibroblasts | 1. Migration/Proliferation | — | | PDGF, IGF-1, EGR activity, TGF-β, TNF-α |
| | 2. Production of new ECM | TIMPS | Fibronectin Hyaluronate Sulfated PGs Collagens I, II, III, IV, V, VII Elastin | PDGF, TGF-β |
| | 3. Stimulated epidermal cell mitosis | — | — | KGF |
| | 4. Stimulate angiogenesis | — | Fibronectin Hyaluronate | TNF-α, TGF-β AFGF, bFGF VEGF Angiogenin E-Prostaglandins IL-8 |
| | 5. Traction in wound contraction | — | — | — |
| Myofibroblasts | 1. Anchoring wound contraction | — | — | — |
| Endothelial | 1. Form new capillaries | — | — | — |
| **C. COLLAGEN REMODELING** | | | | |
| Fibroblasts | 1. Collagen degradation and x-linking into bundles | Collagenases Lysyl oxidase TIMPs | — | PDGF TGF-β |
| Epidermis | 1. Collagen degradation | Collagenases | — | — |

## 2. Inflammation

The phase of inflammation lasts for approximately 5–7 days. Within a day after injury, chemotactic factors released during clot formation increase the permeability of surrounding capillaries, attracting neutrophils, monocytes, and T-lymphocytes, which crawl out between endothelial cells (diapedesis) into the provisional fibrin matrix (Pierce, 1991; Wahl and Wahl, 1992; Grotendorst, 1992). Chemotactic factors bind to receptors on the immune cells that trigger intracellular signaling leading to directional cell movement and capability for phagocytosis. The leukocytes insert themselves into the space between endothelial cells via binding of platelet endothelial cell adhesion molecule (PECAM) on the surface of the leukocytes to PECAM on extrusions of the endothelial cell surface (Mamdouh et al., 2003). Neutrophils and monocytes migrate into the clot simultaneously, but neutrophils migrate in greater numbers. Both cell types use Fn in the clot as an adhesive substrate and express the appropriate integrin receptors to bind to Fn. After entering the wound and adhering to Fn, the monocytes differentiate into macrophages (Pierce, 1991; Grotendorst, 1992; Martin, 1997).

Chemotactic factors for neutrophils and monocytes include fibrinopeptides cleaved from fibrinogen by thrombin, fibrin degradation products produced by the action of plasmin, complement fragment C5A in the plasma, and collagen and elastin fragments. Plasmin is derived from plasminogen by the action of tissue plasminogen activator (tPA); it degrades fibrin both by activating matrix metalloproteinases (MMPs) and by acting directly on fibrin.

The most important chemoattractants for the inflammatory phase are TGF-β and platelet-derived growth factor (PDGF) supplied by degranulating platelets (Clark, 1996; Riches, 1996). TGF-β has three isoforms, of which TGF-β1 is the most abundant in all tissues, followed by isoforms two and three. TGF-β1 is the only isoform in human platelets and comprises 85% of the TGF-β in wound fluid (Roberts and Sporn, 1996). Activated mast cells also attract neutrophils and monocytes by the release of these TGF-β isoforms (Gottwald et al., 1998). In rat skin wounds, both PDGF and TGF-β1 are present in the clot by one hr after injury. TGF-β1 persists for 12 hr and PDGF for 72 hr, after which they disappear (Whitby and Ferguson, 1991). PDGF and TGF-β1 both are chemotactic for monocytes at femtomolar concentrations, and PDGF is chemotactic for neutrophils as well (Duel et al., 1982; Ross et al., 1986; Heldin and Westermark, 1990).

Within the clot, macrophages secrete TGF-β, PDGF, and other chemoattractants that amplify the number of neutrophils and macrophages entering the wound, such as leukotriene B4, monocyte-chemoattractant proteins 1,2,3 (MCPs), macrophage inflammatory protein 1α and β, and the chemotactic protein CAP37 (Clark, 1996; Haslett and Hansen, 1996; Riches, 1996). Neutrophil influx diminishes within 3–4 days after injury. T-cell lymphocyte infiltration peaks by the end of the first week or later (Swift et al., 2001). T-cells, particularly the $T_H1$ subset of $CD4^+$ cells and $CD8^+$ cells, may play a role in regulating macrophage-induced activities (Linares, 1996; Park and Barbul, 2004).

An important task of inflammatory neutrophils and macrophages is to kill bacteria and debride the wound (Clark, 1996). Both cell types kill bacteria through oxygen-dependent mechanisms that generate hydrogen peroxide and hypochlorous acid. Neutrophils also produce bactericidal peptides and proteins, including the enzyme cathepsin G, which kills by an action separate from its catalytic action. Neutrophils and macrophages also play a central role in degrading collagen within the wound via the production of MMPs, particularly the collagenases MMP-1 and MMP-8 (Jeffrey, 1992). MMP-1 degrades type I and III collagens and MMP-8 degrades type I collagen (Wahl and Wahl, 1992; Barbul, 1992; Nwomeh et al., 1998). Synthesis of MMP-1 (interstitial collagenase) is augmented more than 25-fold by contact of macrophages with denatured type I and III collagens, but not by contact with other ECM components (Shapiro et al., 1993). The neutrophils and macrophages clean up the wound site by phagocytizing dead bacteria, as well as cellular and molecular debris. Interestingly, neutrophils appear to prevent re-epithelialization from occurring too rapidly by accelerating keratinocyte differentiation at the wound edges, thus slowing proliferation and migration. This effect may insure that the wound, particularly a heavily contaminated wound, remains open to the air long enough for the oxygen-dependent neutrophils to exert their bactericidal effects and to allow the sloughing of cellular debris (Dovi et al., 2004).

Neutrophils and macrophages normally have a limited lifetime in the wound area. Continual influx of these cells, which occurs in some pathological conditions, leads to chronic tissue injury, necrosis, and excessive scarring. Thus, under normal conditions, there must be mechanisms to limit the immigration of neutrophils and macrophages, control the secretory activities of these cells, and ultimately remove them. Little is known about the mechanisms that inhibit neutrophil migration into the wound. There is evidence, however, that neutrophils undergo apoptosis within a few hours after entering the wound and are ingested intact by the macrophages so that their contents are not released into the wound (Haslett and Henson, 1996). Macrophages are thought to recognize changes in apoptotic cell surfaces by specific integrin, lectin, and phosphatidylserine receptors. The macrophages themselves must be cleared from the wound after they have performed their functions. Macrophages can undergo apoptosis *in vitro*. Assuming that they meet a similar fate *in vivo*, the mechanism by which apoptotic macrophages are removed is unknown.

## 3. Structural Repair

The phase of structural repair involves several subphases: re-epithelialization; formation of granulation tissue; and remodeling of granulation tissue into scar.

### a. *Re-epithelialization*

Within a few hours after injury, cells in the basal layer of the epidermis at the edge of the wound loosen their attachments to one another and begin migrating as a sheet through the fibrin clot to cover the wound surface. The migrating cells do not divide, but the sheet expands because basal cells just interior to the wound edge divide and feed progeny into the migrating sheet. Later, after the migrating epidermis has expanded to cover the wound surface, its cells divide vertically to thicken the epidermis and synthesize a new basement membrane. In wounds with extensive dermal damage, the epidermis does not regenerate appendages such as hair and glands. The mechanisms of epidermal regeneration will be discussed in more detail in Chapter 3.

### b. Development of Granulation Tissue

This phase of wound healing is characterized by replacement of the fibrin clot by a capillary-rich fibroblastic tissue. Within two days after injury, well before the inflammatory phase is over, fibroblasts begin migrating from the surrounding dermis into the clot. Simultaneously, new capillaries regenerate into the nascent fibroblastic tissue, which is called granulation tissue because the capillaries give it a reddish, granular appearance.

The formation of granulation tissue is initiated by growth factors secreted by macrophages, primarily PDGF and TGF-β. Experiments involving the depletion of neutrophils and macrophages showed that wound healing could take place in the absence of neutrophils, but not in the absence of macrophages (Simpson and Ross, 1972). Leibovich and Ross (1975) reported that healing of skin wounds was impaired in guinea pigs treated with antimacrophage serum and steroids (hydrocortisone) to suppress production of circulating monocytes. Macrophages, fibroblasts, and new capillaries invade the wound space as a unit (Clark, 1996), which correlates well with the biological interdependence of these cells during tissue repair. The macrophages secrete growth factors that stimulate fibroblast migration and proliferation, the fibroblasts synthesize a new transitional matrix to replace the fibrin matrix, and the new capillaries carry oxygen and nutrients to sustain the metabolism required for these cellular processes.

#### Fibroblast Migration and Proliferation

The source of fibroblasts for structural repair appears to be two-fold, resident dermal fibroblasts and fibroblasts differentiating from circulating mesenchymal stem cells from the bone marrow that enter the wound from the vasculature (Fathke et al., 2004). The marrow-derived fibroblasts synthesize both collagens I and III, whereas the resident fibroblasts synthesize only collagen I. The resident fibroblasts come primarily from the deep layers of the reticular dermis and the hypodermal (subcutaneous) tissue (Ham and Cormack, 1979). There is evidence for significant heterogeneity in fibroblasts cloned from human skin (Bayreuther et al., 1988; Sempowski et al., 1995; Gross, 1996; Lindblad, 1998). Subpopulations of skin fibroblasts show variation in morphology, the amount of collagen expressed per cell, and in their response to cytokines derived from macrophages. Fibroblasts from the papillary layer, though morphologically similar to reticular fibroblasts, exhibit higher metabolic and proliferative activity and cell density at confluence (Sempowski et al., 1995). Differences in cell surface markers between these layers have also been noted (Phipps et al., 1989). The expression of type I and III procollagen mRNAs is highest in the papillary layer, whereas expression of collagenase mRNA is lower in fibroblasts of the reticular layer (Ali Bahar et al., 2004). Similar heterogeneity has been noted in the fibroblasts derived from healing intra-oral wounds and fetal dermis (Sempowski et al., 1995). What these differences mean to potential differential functions of fibroblasts in the formation of granulation tissue is unknown.

The movement of fibroblasts into the fibrin matrix of the wound is mediated by their integrin receptors. Fibronectin and HA are the major components of the early granulation tissue matrix in rat skin wounds (Alexander and Donoff, 1980; Grinnell, 1984; Chen and Abatangelo, 1999). Hyaluronic acid-PG complexes bind water, which expands extracellular space, thereby facilitating fibroblast migration. The fibroblasts express the CD44 receptor, which mediates cell attachment to and locomotion on HA substrates, and RHAMM (receptor for hyaluronan-mediated motility), which mediates cell locomotion in response to soluble HA (Clark, 1996). Fibronectin acts as a substrate for the migration of fibroblasts and vascular endothelial cells into the wound (Grinnell, 1984; Risau and Lemmon, 1988).

Fibroblast migration, proliferation, and ECM synthesis is stimulated *in vitro* by macrophages through their production of several growth factors during the inflammatory phase, of which TGF-β and PDGF play central roles (Pierce, 1991; Grotendorst, 1992; Lawrence and Diegelmann, 1994; Gottwald et al., 1998). *In vivo* studies on rats using wound chambers implanted subcutaneously showed that animals in which wound healing is impaired by destroying the precursors of macrophages have greatly reduced levels of PDGF, TGF-β, and EGF activity (Grotendorst et al., 1988). PDGF appears to be the major growth factor that renders fibroblasts competent to leave $G_0$ and enter the $G_1$ phase of the cell cycle, although FGF-2 can also act as a competence factor (Morgan and Pledger, 1992). PDGF receptors are expressed by dermal fibroblasts and the induction of competence requires the expression of a number of proteins, the genes for which are activated by PDGF (Morgan and Pledger, 1992). PDGF interacts synergistically with IGF-1 and a protein with EGF-like activity to reach the so-called V point in $G_1$, 6hr after attaining competence. Only IGF-1 is required to reach the $G_1/S$ transition. Once this point is reached, the fibroblast cell cycle proceeds without the need for growth factors (Wharton, 1983).

Paracrine-acting PDGF is provided by macrophages, but macrophage-secreted IL-1 stimulates the fibroblasts themselves to produce PDGF, which then acts through an autocrine loop to induce competence (Wahl and Wahl, 1992). Interferon-β (Ifn-β) may normally inhibit the entry of fibroblasts into $G_1$ (Taylor-Papadimitriou and Rosengurt, 1985). Addition of Ifn-β to cultured fibroblasts within 6hr after PDGF stimulation inhibits proliferation and Inf-β inhibits PDGF-induced protein synthesis, including proteins encoded by genes activated by PDGF (Morgan and Pledger, 1992).

TGF-β and TNF-α also stimulate fibroblast proliferation, but how this stimulation is effected is unclear. TGF-β may exert its effect indirectly, by stimulating PDGF synthesis (Leof et al., 1986). TNF-α stimulates DNA synthesis only in some fibroblast cell lines; in most it is cytotoxic, making its role *in vivo* enigmatic (Morgan and Pledger, 1992). Activin A, another member of the TGF-β family, is also strongly upregulated during the inflammation phase, reaching a peak as the wound fills with granulation tissue (Hubner et al., 1996). The function of activin A, however, is unknown.

As the number of fibroblasts invading the wound space increases and the number of neutrophils decreases, the fibrin provisional matrix is degraded into soluble fragments through the conversion of plasminogen to plasmin by tPA secreted by the endothelial cells of regenerating capillaries, and MMPs secreted by fibroblasts. As the fibrin degrades, tPA is inhibited and the expression of plasminogen activator inhibitor (PAI) is stimulated, thus reducing the conversion of plasminogen to plasmin. In this way, the tPA/PAI system slows the influx of neutrophils and macrophages (Li et al., 2003).

### Extracellular Matrix Synthesis by Fibroblasts

As fibroblasts invade the wound space, they replace the provisional fibrin matrix with an ECM consisting of Fn, HA, sulfated PGs and type I and III collagens (Miller and Gay, 1992; Weitzhandler and Bernfield, 1992). Hyaluronic acid predominates over Fn in early granulation tissue, opening up migration space for fibroblasts. Type III is the major collagen synthesized at this time (Whitby and Ferguson, 1991; Miller and Gay, 1992). The early appearance of type III collagen is associated with the deposition of Fn, and in fact Fn may act as a template for the deposition of type III collagen fibrils (McDonald, 1988). The collagens are at first organized in the normal reticular network.

Dermal repair then begins to show visible signs of following a fibrotic rather than a regenerative pathway. HA synthesis is replaced by synthesis of chondroitin sulfate and dermatan sulfate-PGs. The fibroblasts synthesize predominantly Type I collagen and more collagen is synthesized than in uninjured skin.

These shifts in the pattern of ECM synthesis appear to be orchestrated by PDGF and TGF-β. Whereas these growth factors are initially produced by platelets and macrophages during the inflammatory phase, they are produced by the fibroblasts themselves during the formation of granulation tissue. PDGF promotes the early synthesis of HA and the later synthesis of sulfated GAGs (Pierce, 1991). TGF-β stimulates the early synthesis of Fn and the later synthesis of type I collagen, elastin, and sulfated PGs, and inhibits collagen degradation (Pierce, 1991; Roberts and Sporn, 1996). TGF-β reduces collagen degradation by two complementary mechanisms, the reduction of collagenase gene transcription, and an increase in the synthesis of tissue inhibitors of metalloproteinases (TIMPs) (Jeffrey, 1992).

### Angiogenesis in Granulation Tissue

Angiogenesis, the regeneration of new blood vessels from existing vessels, is a crucial regenerative response in all injured tissues, whether the end result is scar tissue formation or restoration of normal tissue architecture. Thus, both regenerative and fibrotic environments both have common elements that promote blood vessel regeneration, insuring an adequate supply of oxygen and nutrients in each case. The initial fibrin matrix is normally hypoxic, thus, low-oxygen tension may be a signal for initiating angiogenesis in any kind of wound (Hunt and Hussain, 1992). The new capillaries of granulation tissue are sprouted from pre-existing vessels. Initially, the granulation tissue is hypervascularized, giving it a characteristic reddish appearance. As the granulation tissue is remodeled into scar tissue, the extra blood vessels are resorbed by apoptosis that may be induced by macrophages. Studies on the pupillary membrane, a vascular membrane covering the pupil of newborn rats, have shown that macrophages induce the apoptosis of both endothelial cells and their supporting cells, the pericytes (Diez-Roux et al., 1999). Blood vessel regeneration is described in more detail in Chapter 11.

### Reinnervation of Granulation Tissue

Peripheral sensory and sympathetic postganglionic nerves also regenerate in healing wounds and play a significant role in the formation of granulation tissue

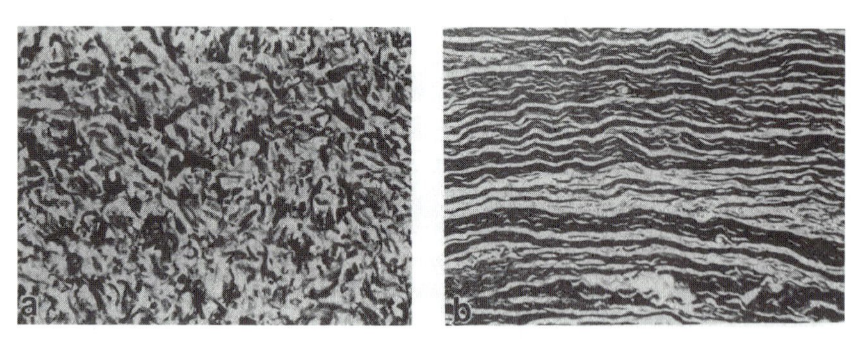

**FIGURE 2.4** Comparison of collagen organization in normal uninjured dermal architecture **(a)** vs collagen organization in dermal scar tissue **(b)**. The collagen of the uninjured dermis has a reticular ("basket-weave") organization, whereas the collagen in a repaired dermis is cross-linked into thick bundles parallel to the wound surface. Reproduced with permission from Linares, 1996, From wound to scar. Burns 22:339–352. Copyright 1996, Elsevier.

because of their effects on inflammation (Gottwald et al., 1998; Kim et al., 1998). After wounding, the portion of the nerves distal to the injury degenerates within one to two days, then regenerates over the next two weeks, resulting in hyperinnervation of the granulation tissue. Subsequently, many of the nerve fibers regress. Thus reinnervation follows the same pattern of advance and recession as angiogenesis; in fact, it is possible that blood vessel and nerve regeneration are mechanistically coupled.

Denervation of wounded rat skin significantly delays wound contraction and re-epithelialization (Fukai, 2005). Injured sensory and sympathetic postganglionic nerves appear to help mediate the inflammation phase ("neurogenic inflammation") via antidromic (conduction in reversed direction) stimulation of release of neuropeptides such as substance P (Gottwald et al., 1998). There may be a relationship between the release of these neuropeptides and mast cell activation. There is an intimate microanatomic association of mast cells and nerves in the dermis. Nerve injury causes degranulation of mast cells, releasing histamine and cytokines that effect vasodilation, increased vascular permeability, attraction of neutrophils and macrophages, and increased fibroblast activity. Denervation of sensory or sympathetic postganglionic fibers decreases these inflammatory effects (Gottwald et al., 1998; Kim et al., 1998; Egozi et al., 2003). However, the absence of mast cells from the skin has no effect on any aspect of dermal repair, so their effects appear to be redundant (Egozi et al., 2003).

### c. Remodeling of Granulation Tissue into Scar

In the final phase of structural repair, the granulation tissue is remodeled into a relatively acellular fibrous scar tissue (**FIGURE 2.4**). The scar differs from normal dermis in several ways (Miller and Gay, 1992; Davidson et al., 1992; Linares, 1996). Fibronectin and HA levels return to normal, but the level of decorin PG is lower than in normal skin, and the level of chondroitin-4-sulfate PG is much higher. The organization of the ECM is also different. The number of elastin fibers is reduced in scar tissue. Instead of the random basket-weave organization of normal dermis, type I collagen fibers in scar are broken down by MMPs and cross-linked by the enzyme lysyl oxidase into thick bundles oriented parallel to the surface of the wound. The MMPs appear to be produced by both the epidermis and fibroblasts of the granulation tissue. MMP synthesis by the fibroblasts appears to require an interaction with the epidermis, since synthesis is much reduced *in vitro* in the absence of epidermis (Grillo and Gross, 1967). Growth factors do not influence the cross-linking process itself, only the amounts of collagen available to be cross-linked (Mast, 1992). As the scar matures, the density of the vascular and neural networks in the granulation tissue returns to normal and there is a reduction in the number of fibroblasts by apoptosis (Mast, 1992; Miller and Gay, 1992; Davidson et al., 1992; Grinnell, 1994; Desmouliere et al., 1995; Lorena et al., 2002). Establishment of a mature, stable scar takes ~80 days in rodents, but at least six months in humans. Although the tensile strength of the scar tissue increases in proportion to the degree of cross-linking, it achieves no more than 70%–80% of the tensile strength of normal dermal tissue (Mast, 1992).

There must be a variety of feedback loops involving inhibitory factors that bring the repair process to an end, but these are not well understood. The factors that play a role in terminating the remodeling of granulation tissue matrix and which cells produce them remain largely to be worked out.

## 4. The Role of Wound Contraction in Dermal Repair

In mammals that have loose skins, closure of excisional skin wounds is aided by contraction of the dermis

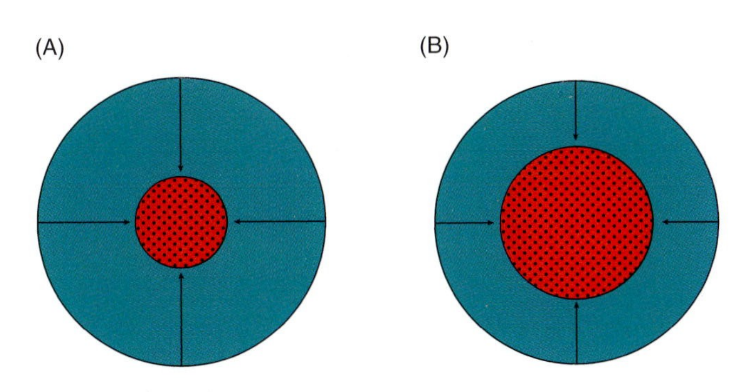


**FIGURE 2.5** Diagram illustrating contraction of the skin in rodent **(A)** vs human **(B)** circular excisional wounds. Outer circle = initial perimeter of the wound. Inner circle = position to which the skin is drawn by contraction. Arrows indicate distance and direction moved by the skin. Stipple indicates area of granulation tissue. Human skin contracts much less, leaving a greater relative area to be filled in by granulation tissue and scar.

to decrease the area that needs to be covered by epidermis and filled in by scar tissue. Contraction is characterized by the sliding and stretching of perilesional skin over the defect and should not be confused with contracture, which is the shortening of scar tissue, leading to deformity and loss of function. Dermal contraction accounts for a much greater percentage of wound closure in rodents than in pigs or humans (**FIGURE 2.5**). *In vivo*, contraction accounts for up to ~90% of wound closure in mice (Yannas, 2001). In humans, less than 50% of excisional wound closure is due to contraction; the majority is due to scar tissue formation. In addition to dermis, contraction has been shown to help close wounds in peripheral nerve, ligaments, ureter, esophagus, and duodenum (Yannas, 2001).

The skin of developing frog tadpoles and fetal mammals regenerates in the absence of dermal contraction (see below), but as development proceeds, there is an increase in contraction accompanied by a decrease in the capacity for regeneration (Yannas, 1996, 2001). This implies an inverse relationship between the degree of contraction and the degree of regeneration. However, inhibiting wound contraction in mammals by pharmacological agents such as steroids does not lead to regeneration and the wound fills with granulation tissue (Yannas, 2001). Thus, although contraction reduces the area to be covered in adult wounds, it does not appear to be a causal factor in determining whether regeneration occurs.

Dermal contraction in mammals is initiated early in the phase of structural repair by fibroblasts that differentiate into myofibroblasts under the influence of TGF-β1 and ECM (Gabbiani, 1998; Lorena et al., 2002; Desmouliere et al., 2005). Myofibroblasts have characteristics of both fibroblasts and smooth muscle cells. Their contractile apparatus is like that of smooth muscle, consisting of large bundles of the α-smooth muscle form of actin microfilaments running along the inner surface of the plasma membrane (Rudolph et al., 1992; Clark, 1996). Myofibroblasts assemble Fn into fibrils at their surfaces and form fibronexi, transmembrane connections linking the actin microfilaments with the extracellular Fn fibrils (Newman and Tomasek, 1996; Tomasek et al., 1999). Treatment of fibroblasts *in vitro* with TGF-β1 increased the expression of αv β1 integrins. Blocking the function of αv and β1-containing integrins with antibodies inhibited TGF-β1-induced smooth muscle actin expression and the ability of myofibroblasts to contract collagen gel, implicating the involvement of these integrins in the differentiation of myofibroblasts (Lygoe et al., 2004).

Interestingly, contraction takes place normally after excision of the central granulation tissue in back and flank skin wounds of guinea pigs (Gross, 1996). This suggests that the mechanism of contraction is one in which the coordinated action of myofibroblasts around the edge of the wound pulls the dermis inward.

## MOLECULAR COMPARISON OF WOUNDED VERSUS UNWOUNDED SKIN

We do not yet know all the molecular elements involved in fibrotic wound repair. To obtain a comprehensive picture of gene activity in wounded adult skin, transcriptional profiles of fibroblasts or whole skin have been examined. Iyer et al. (1999) analyzed the response of human neonatal foreskin fibroblasts to serum stimulation *in vitro*. DNA microarray hybridization was used to measure the temporal changes in mRNA levels of over 8,600 genes. Bioinformatic analysis revealed clusters of genes that shared specific expression profiles. The clusters included known genes involved in transduction of serum signals, entrance and progression through the cell cycle, and in wound repair, including 10 genes associated with clotting and hemostasis, 8 with inflammation, 6 with re-epithelialization, 12 with angiogenesis, and 19 with remodeling of granulation tissue. The response to serum was rapid, with the most rapid response (within 15 min) exhibited by genes encoding transcription factors and other proteins involved in signal transduction. Over 200 genes in this study were novel, with unidentified function.

In wounded whole human skin, microarray analysis revealed that at 30 min postinjury, 124/4,000 (3%) of the genes in the array were upregulated by two-fold or more (**FIGURE 2.6**); these genes again were involved primarily in signaling and signal transduction (Cole et al., 2001). By one hour, 46 genes (1.15%) were upregulated and 264 (6.6%) were downregulated two-fold or more. Analysis of a cDNA library made by subtraction between unwounded and wounded mouse back skin

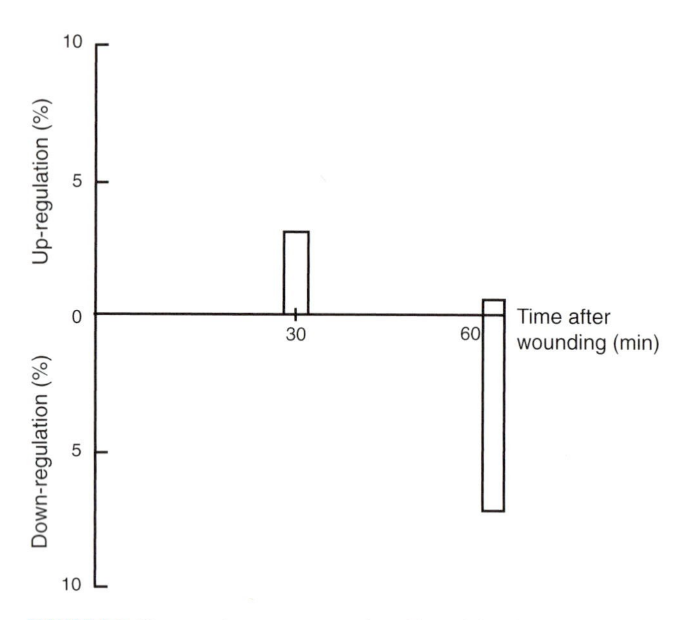


**FIGURE 2.6** Changes in gene expression 20 and 60 min after wounding of human skin. At 30 min, 3% of the total number of active genes were up-regulated and none were down-regulated, but by one hour nearly 7% were down-regulated and slightly over 1% were up-regulated. Data from Cole et al., (2001) Wound Rep Reg 9:360–370.

identified several wound-regulated genes (Kaesler et al., 2004). One of these encodes the chemokine receptor CCR1, which binds several chemokines. CCR1 was barely detectable in unwounded skin, but was strongly expressed after injury in macrophages and neutrophils. However, CCR1 knockout mice healed wounds normally, indicating that there is redundancy in chemokine/receptor signaling in skin wounds.

While still in their infancy, these kinds of studies, coupled with more powerful methods of bioinformatic analysis and systems biology approaches, may lead to a much more complete molecular description of each phase of wound repair by scarring and, even more importantly, how the synthesis or activation of these molecules is coordinated to bring about the repair.

## FETAL SKIN HEALS WITHOUT SCARRING

Fetal mammalian skin repairs without scarring after excisional wounding (Olutoye and Cohen, 1996). Re-epithelialization and fibroblast migration is more rapid in fetal wounds and there is less contraction. The granulation tissue lays down the normal architecture of ECM. Late in gestation, fetal skin changes its response to injury from regeneration to the adult response of fibrosis. In the rat and mouse, this transition takes place at 16–18 days of gestation, 3–5 days prior to birth. Comparison of fetal wound healing with adult wound healing is a valuable approach to elucidating the differences between regeneration and fibrosis and how we might intervene to prevent scarring in adults.

### 1. Cellular and ECM Differences Between Fetal and Adult Wound Healing

A number of differences have been noted between the repair of fetal and adult wounds. Scanning electron and confocal microscopy studies on amputated embryonic mouse (11.5 days of gestation) hindlimb buds *in vitro* indicated that re-epithelialization is complete by 24 hr (McCluskey and Martin, 1995). Re-epithelialization occurs not by lamellipodal crawling (migration) of cells as in the adult, but by a purse-string contraction of a filamentous actin cable assembled in the basal layer of cells at the edges of the wound. Observations on DiI-labeled mesenchymal cells revealed that the mesenchymal edge of the wound moves inward by 50% within 24 hr. This movement is not due to the conversion of mesenchymal cells to myofibroblasts and contraction as in adults, however, but to the active movement of the cells over the ECM substrate (Mast et al., 1997). Ihara and Motobayashi (1992) reported similar findings for wounded fetal rat skin.

Collagen and noncollagen protein synthesis are both elevated above normal in fetal and adult rat wounds. However, the ratio of collagen/total protein in wounded adult skin is higher than in unwounded adult skin, whereas there is no significant difference in the collagen/total protein ratio between wounded and unwounded fetal skin (Houghton et al., 1996). The fibroblasts of the fetal wound synthesize the same collagens as fibroblasts in the adult wound but there is no excessive deposition of type I collagen fibrils in fetal wounds and the fibrils are organized in the basket-weave pattern of normal dermis (Frantz et al., 1992; Mast et al., 1997).

This normal pattern of collagen synthesis and architecture is associated with three other differences from adult wounds in ECM synthesis. First, fetal wound fibroblasts synthesize higher levels of HA and HA receptor (DePalma et al., 1989; Longaker et al., 1989; Alaish et al., 1993), giving them more opportunity to bind HA and thus for cell movement. HA has been shown to inhibit fetal fibroblast proliferation (Mast et al., 1993) and to decrease scar formation in wounds of adult tympanic membranes (Hellstrom and Laurent, 1987). Conversely, treatment of fetal rabbit skin wounds with hyaluronidase or HA degradation products alters the regenerative response toward fibrosis (Mast et al., 1991, 1995). Second, sulfated PG synthesis does not accompany collagen synthesis in fetal wounds (Whitby and Ferguson, 1992). Third, fetal skin has a higher ratio of type III to type I collagen (Epstein, 1974; Merkel et al., 1988).

## 2. Fetal Wounds Have a Minimal Inflammatory Response

A major difference between fetal and adult wounds is that fetal wounds exhibit a minimal inflammatory response (McCallion and Ferguson, 1996; Yang et al., 2003; Ferguson and O'Kane, 2004), suggesting that as it matures during development, the immune system response to injury suppresses regeneration in favor of scar tissue formation. Platelets are few in number in the fetal wound. Only small numbers of platelets, neutrophils, and macrophages are present up to 18 hr after wounding (Cowin et al., 1998). All types of macrophages were eventually recruited to the fetal wound, but their numbers and persistence were much lower than in adult wounds. These observations suggest that the types and proportions of growth factors and cytokines associated with the inflammatory environment are different in fetal wounds compared to adult wounds.

Consistent with this idea, there is a much lower level of PDGF, TGF-β1 and 2 and their receptors in unwounded and wounded fetal rat skin (Whitby and Ferguson, 1991; Sullivan et al., 1993; Durham et al., 1989; Ferguson and O'Kane, 2004; Chen et al., 2005). Conversely, TGF-β3, which is synthesized by keratinocytes and fibroblasts, is low in adult wounds, but is high in fetal wounds (Ferguson and O'Kane, 2004; Chen et al., 2005). In adult wounds, PDGF induces the persistent expression of IL-6 by fibroblasts, which helps to maintain an environment that promotes production and deposition of fibrotic matrix. Fetal wounds also express IL-6, but this expression rapidly disappears (Liechty et al., 2000).

The negative influence of the inflammatory response on regeneration has been further established by experiments in which the molecular composition of the wound environment is manipulated. Placing pellets of bacteria in PVA sponges subcutaneously in rabbits elicited an acute inflammatory response, including recruitment of neutrophils and large numbers of leukocytes that initiated an adultlike fibrosis (Frantz et al., 1993). Adding IL-6 to fetal wounds induced scarring, whereas an adenoviral construct overexpressing IL-10, which decreases the production of IL-6, reduced inflammation and scarring when applied to adult mouse wounds (Leichty et al., 1998, 2000). Treating fetal rat wounds with exogenous TGF-β1 induced scarring (Krummel et al., 1988). TGF-β1 decreased endogenous MMPs and increased TIMPs, which would increase collagen accumulation and scarring (Soo et al., 2001). At high concentrations, PDGF also caused a shift to a more adultlike pattern of repair in fetal skin wounds (Haynes et al., 1990). Conversely, reducing the levels of TGF-β1 and 2, but not PDGF or FGF, *in vivo* or *in vitro* by application of neutralizing antibodies reduced scarring in adult rat wounds (Shah et al., 1994, 1995; Houghton et al., 1995). Reduction in scarring was enhanced only by applying the antibodies immediately after injury. This suggests that TGF-β1 and 2 exert early inflammatory effects in adult wounds that carry through the whole molecular cascade of wound repair events and that intervention must be done prior to the point where this cascade is easily able to withstand perturbation (O'Kane and Ferguson, 1997; Shah, 2000; Ferguson and O'Kane, 2004).

Addition of exogenous TGF-β3 to adult wounds at levels found in fetal wounds resulted in reduced or absent scarring (Shah et al., 1995) (**FIGURE 2.7**). Wounds in the skin of fetal TGF-β3 null mice showed the adult repair response. Fetal fibroblasts of these mice exhibited slower migration in collagen gels, an effect that could be rescued by treatment with exogenous TGF-β3, but not by TGF-β1 or 2 (Ferguson and O'Kane, 2004). These results indicate that the proportion of TGF-β3/β1/β2 is critical to determining whether dermal tissue will regenerate or scar in response to injury. Finally, fetal wounds may contain inhibitors of TGF-β, such as decorin and α2-macroglobulin (Yamaguchi et al., 1990; Danielpour and Sporn, 1990).

There may be species and locational differences in the composition of the fetal versus adult wound environment and in the environments of adult wounds. For example, Longaker et al. (1994) reported that macrophages were recruited into wounds in fetal sheep skin and that the concentration TGF-β2 was higher in the fetal wound than in the adult wound. Interestingly, adult mammalian saliva contains high levels of TGF-β2, yet oral lesions in adults heal with little scarring, suggesting that inhibitors of TGF-β could play a role in minimizing scarring.

Most studies comparing adult and fetal wounds have been on incisional wounds. However, studies on fetal sheep indicate that deep dermal burn wounds heal without scarring as well, whereas they heal with scarring in lambs (Fraser et al., 2005). This difference is associated with only a slight increase in TGF-β1 after wounding in the fetus, but a massive increase in the lamb that does not begin to fall off until 21 days postinjury and has not reached the control value even by day 60.

Some studies suggest that, in addition to differences in the inflammatory response, there are intrinsic differences between fetal and adult fibroblasts that are the result of differentiation independent of the maturation of the immune system. Secondary fibroblasts derived from primary cultures of human dermal fibroblasts fall into seven types, based on morphology and patterns of protein synthesis, that are a function of population doubling number and time kept in secondary culture

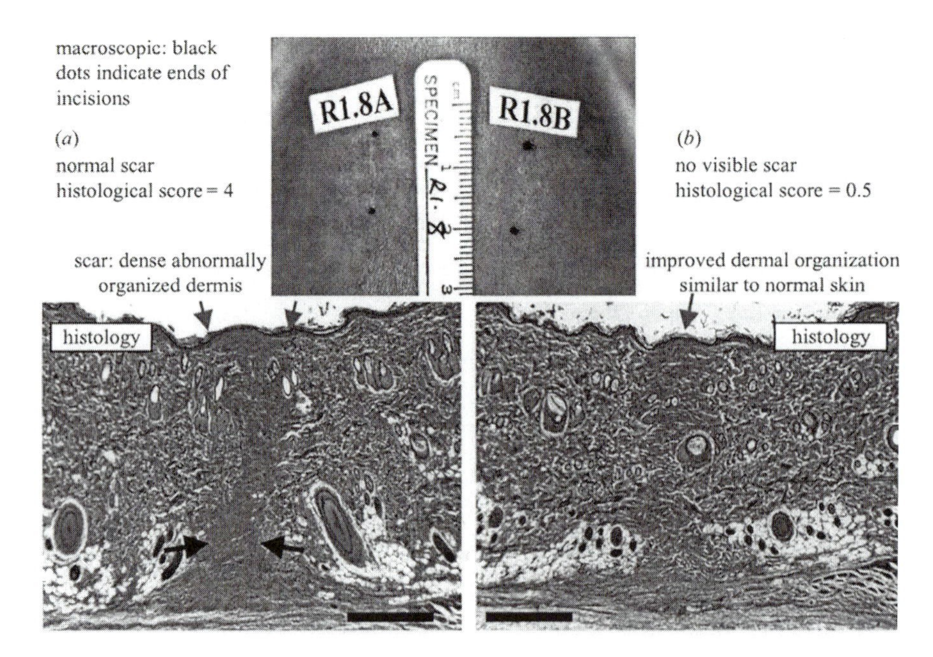


**FIGURE 2.7** Effect of treating an adult rat deep (1 cm) incisional wound with TGF-β3, viewed 84 days post-wounding. **(a)** surface view (top) and histology (bottom) of placebo-treated wound. Scar tissue is the result (arrows). **(b)** Surface view (top) and histology (bottom) of TGF-β3 treated wound. No scar is visible in the surface view and dermal histological organization is similar to that of normal skin. Reproduced with permission from Ferguson and O'Kane, Scar-free healing: from embryonic mechanisms to adult therapeutic intervention. Phil Trans R Soc Lond B 359:839–850. Copyright 2004, The Royal Society.

(Bayreuther et al., 1988). This heterogeneity suggests the existence of a fibroblastic stem cell that gives rise to a seven-stage terminal cell lineage. It is possible that earlier parts of the lineage are capable of responding to injury with a regenerative response. Fetal fibroblasts may be those that comprise the earlier part of the lineage. The transition from fetal to the adult pattern of wound healing might then be the result of differentiation of the majority of these cells to a terminal cell type that is unable to respond in a regenerative way.

Fetal fibroblasts were reported to have a unique phenotype that differs from that of the adult fibroblast, including production of and response to growth factors, synthesis of matrix molecules, pericellular HA coats and antigen determinants (Moriarty et al., 1996; Ellis et al., 1997; Gosiewska et al., 2001). They maintained their regenerative response when grafted subcutaneously into adult athymic mice, even though these mice heal by scarring (Lorenz et al., 1992; Lin et al., 1994). The transition from regeneration to scarring in incisional wounds of cultured 14-day mouse limb buds takes place autonomously, in the absence of immune cells and systemic factors (Chopra et al., 1997), and this presumably would involve transition of fibroblast molecular phenotype.

Nevertheless, the fact that changing the concentrations of growth factors and cytokines associated with inflammation can shift the nature of the injury response in both fetal and adult dermis strongly suggests that the wound environment is the major determining factor in the response. There may indeed be intrinsic differences between fetal and adult skin fibroblasts, but it appears that these differences can either be largely eliminated in either direction by the appropriate signaling molecules, or by the recruitment of fibroblasts that give a regenerative response. One type of experiment that could be done to investigate these possibilities is to examine the molecular changes in phenotype that might occur following treatment of fetal or adult fibroblasts *in vivo* or *in vitro* with the appropriate growth factor and cytokine combinations.

Other data lend support to these conclusions. Observations on tattoos show that the adult human dermis can respond to injury either by regeneration or scar formation, depending on the extent of the injury (Ferguson and O'Kane, 2004). Tatoo artists use needles to deposit patterns of ink particles in the dermis, which are then phagocytosed by fibroblasts. Despite the large total area injured, tattooed dermis heals without scarring, because each individual injury from the needle is small and does not evoke an inflammatory response. Tatoos can be removed by lysing the fibroblasts containing the pigment particles, which again creates numerous microwounds in the skin. These wounds, however, exhibit a minimal inflammatory response and heal without scarring. The microwounds of tattooing

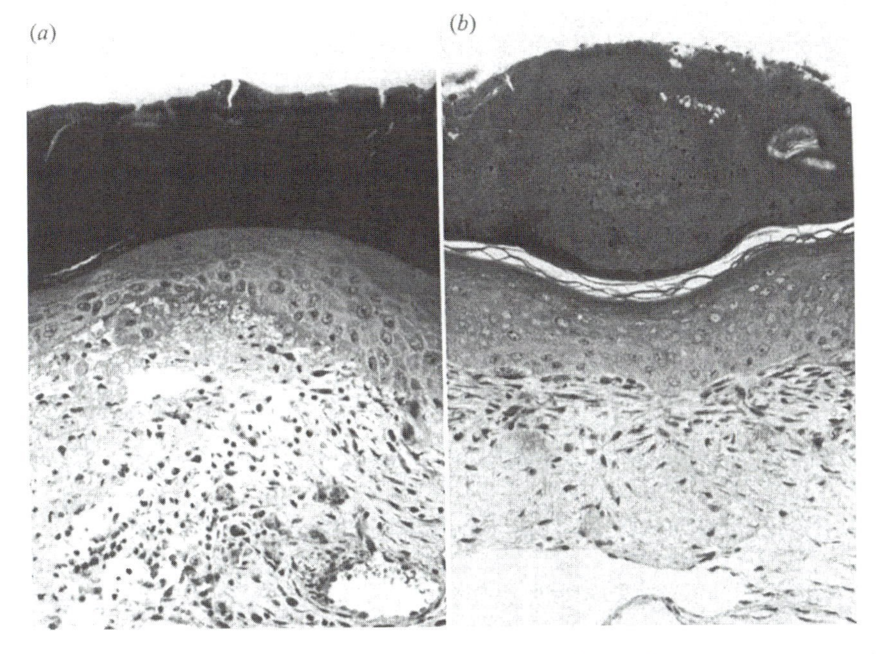


**FIGURE 2.8** Healing of incisional skin wounds in wild-type **(a)** and PU.1 null neonatal mice **(b)**, 2–3 days post-wounding. The wild-type wound is covered by epidermis migrating beneath the scab. Below the epidermis is disorganized connective tissue with many inflammatory cells. This wound will form a scar. The PU.1 wound has a much better formed epidermis beneath the scab, and dermal tissue that appears similar to that of unwounded skin. Reproduced with permission from Redd et al., Wound healing and inflammation: embryos reveal the way to perfect repair. Phil Trans R Soc Lond B 359:777–784. Copyright 2004, The Royal Society.

regenerate, whereas incisional or excisional wounds of equal total volume heal by fibrosis! Tatoo wounds have been shown to have low levels of TGF-β1 and higher levels of TGF-β3 (Ferguson and O'Kane, 2004).

PU.1 null mice lack a hematopoietic lineage transcription factor that results in the absence of both macrophages and neutrophils. These mice die within 24 hr after birth unless they are given a wild-type bone marrow transplant because they are prone to infection, showing the importance of these cells for phagocytic and bactericidal functions (Dovi et al., 2004). Antibiotics can be used to prolong the lives of neonate PU.1 mice. Martin et al. (2003) investigated excisional wound repair in PU.1 mouse neonates maintained on antibiotics. Not only were excisional wounds repaired in antibiotic-maintained PU.1-null mice at the same rate as wounds in wild-type mice, they were repaired by regeneration, not scar (Martin et al., 2003; Redd et al., 2004 **FIGURE 2.8**.) Regeneration was associated with greatly reduced expression of IL-6 and TGF-β mRNA (Redd et al., 2004). These results suggest that PU.1-null mice do not undergo the transition from the fetal to adult wound healing response! Whether the absence of macrophages in these mice is causal to the failure to shift to the adult healing response is unknown. It would be interesting to compare wounded versus unwounded PU.1 null mice to learn how gene activity differs between the two in the absence of macrophages (Martin et al., 2003). We will encounter differences between fetal and adult wound healing in other chapters as well, some of which have been correlated with the absence or presence of an inflammatory response.

## SUMMARY

Excisional wounds of skin exhibit both regenerative and fibrotic responses to injury. The dermis repairs by scar formation, but within this context the epidermis and vasculature regenerate, although the epidermis may lack structures such as hair follicles and sebaceous and sweat glands. Repair of an excisional wound can be divided into three tightly integrated and overlapping phases: hemostasis, inflammation, and structural repair. During hemostasis, one set of serum factors causes adhesion of platelets to injured blood vessel walls, sealing them off, while another set induces formation of a fibrin clot that forms a provisional matrix for the migration of immune cells, epidermal cells, and fibroblasts into the wound. The phase of inflammation is initiated by factors in the fibrin clot, as well as PDGF and TGF-β from degranulating platelets, that attract neutrophils, monocytes and T-cells. The neutrophils and macrophages kill bacteria, degrade collagen, and debride the wound. Neutrophils die by apoptosis within a few hours after entering the wound and are removed by macrophage phagocytosis. The macrophages stay longer and initiate the phase of structural repair. Re-epithelialization begins and PDGF and TGF-β secreted by the macrophages promotes fibroblast proliferation and migration into the fibrin matrix to form granulation tissue. In mammals with loose skins, myofibroblasts appear at the edges of the wound in the early phase of structural repair and contract the skin. Granulation tissue is at first hypervascularized and hyperinnervated by the regeneration of capillaries and nerves, but these later regress. The fibroblasts of the granulation tissue are stimulated to secrete ECM by PDGF and TGF-β that replaces the fibrin clot, which is degraded. In the final stages of structural repair, the number of fibroblasts is reduced by apoptosis and the organization of the ECM is modified. The number of elastin fibers is reduced and collagen I fibers are broken down by MMPs and cross-linked into thick bundles parallel to the skin surface by lysyl oxidase. These dense, relatively acellular bundles of collagen fibers constitute the visible scar.

Fetal mammalian skin does not scar after wounding and regenerates perfectly until late in gestation, when the response shifts to the adult scarring pattern. The fibroblasts of the fetal wound synthesize the same collagens as fibroblasts in adult wounds, but lay down the normal architecture of the ECM. This regenerative response is associated with several differences between fetal and adult skin. Fetal skin has a higher ratio of type III to type I collagen than does adult skin, sulfated PG synthesis does not accompany collagen synthesis in fetal wounds, and fetal wound fibroblasts synthesize higher levels of HA and its receptor. This in turn is associated with a minimal inflammatory response in fetal wounds. Platelets, neutrophils, and macrophages are present in very low numbers if at all, and the adult levels of the growth factors and cytokines produced by these cells are correspondingly lower, with the exception of TGF-β3, which is higher in fetal wounds. Adding adult levels of growth factors (TGF-β1 and 2) and cytokines to fetal wounds induces their scarring, whereas reducing the levels of these molecules in, or adding TGF-β3 to, adult wounds reduces scarring. These data suggest that it is possible, by manipulating the wound environment, to confer regenerative capacity on the adult dermis after excisional wounding. This notion is backed up by studies on tattooing, which injures a large area of tissue with thousands of small wounds that do not evoke an inflammatory response, and on neonatal PU.1 null mice, which lack macrophages and neutrophils. These mice repair skin wounds by regeneration, not scarring, suggesting that they fail to undergo the transition from the fetal to the adult wound healing response.

Microarray analysis of wounded versus unwounded skin is now being used in an attempt to identify all the molecular elements that are upregulated and downregulated during wound repair by fibrosis. These analyses have shown that genes that are up-regulated early in wound repair are primarily genes involved in signal transduction pathways. Such comparisons could also be made between the wounded skin of wild-type and PU.1 neonatal mice, or the wounded skins of fetal and adult wild-type mice to identify what differences in gene activity characterize regenerating vs scarring skin. Coupled with bioinformatics and systems biology approaches, these data will be invaluable in providing a complete molecular description of scarring vs regenerative pathways in wounded skin.

## REFERENCES


Adams JC, Watt FM (1993) Regulation of development and differentiation by the extracellular matrix. Development 1993; 117: 1183–1198.

Alaish SM, Yager D, Diegelmann RF, et al. (1993) Hyaluronate receptor expression in fetal fibroblasts. Surg Forum 44:733–735.

Alberts B, Bray D, Lewis J, Raff M, Roberts K, Watson J (1994) Molecular Biology of the Cell, 3rd ed. New York: Garland Press, pp 960–1009.

Alexander SA, Donoff RB (1980) The glycosaminoglycans of open wounds. J Surg Res 29:422–429.

Ali Bahar M, Bauer B, Tredget EE, Ghahary A (2004) Dermal fibroblasts from different layers of human skin are heterogeneous in expression of collagenase and types I and III procollagen mRNA. Wound Rep Reg 12:175–182.

Barbul A (1992) Role of the immune system. In: Cohen IK, Diegelmann RF, Lindblad WJ (eds.). Wound Healing: Biochemical and Clinical Aspects. Philadelphia: WB Saunders, pp 282–291.

Bayreuther K, Rodemann H, Francz PI, Maier K (1988) Differentiation of fibroblast stem cells. In: Lord BI, Dexter T (eds.). Stem Cells (J Cell Sci Suppl 10) Cambridge: Company of Biologists Ltd, pp 115–130.

Chen WY, Abatangelo G (1999) Functions of hyaluronan in wound repair. Wound Rep Reg 7:79–89.

Chen W, Fu X, Ge S, Sun T, Zhou G, Jiang D, Sheng Z (2005) Ontogeny of expression of transforming growth factor-β and its receptors and their possible relationship with scarless healing in human fetal skin. Wound Rep Reg 13:68–75.

Chopra V, Blewett CJ, Ehrlich HP, Krummel TM (1997) Transition from fetal to adult repair occurring in mouse forelimbs maintained in organ culture. Wound Rep Reg 5:47–51.

Clark RAF (1996) Wound repair: overview and general considerations. In: Clark RAF (ed.). Molecular and Cellular Biology of Wound Repair. New York, Plenum Press, pp 3–50.

Cole J, Tsou R, Wallace K, Gibran N, Isik F (2001) Early gene expression profile of human skin to injury using high-density cDNA microarrays. Wound Rep Reg 9:360–370.

Cowin AJ, Brosnan MP, Holmes TM, Ferguson MWJ (1998) Endogenous inflammatory response to dermal wound healing in the fetal and adult mouse. Dev Dynam 21:385–393.

Danielpour D, Sporn MB (1990) Differential inhibition of transforming growth factor β1 and β2 activity by $\alpha_2$-macroglobulin. J Biol Chem 265:6973–6977.

Davidson JM, Giro MG, Quaglino D (1992) Elastin repair. In: Cohen IK, Diegelmann RF, Lindblad WJ (eds.). Wound Healing:

Biochemical and Clinical Aspects. Philadelphia, WB Saunders, pp 223–236.
DePalma RL, Krummel TM, Durham LA, Michna BA, Thomas BL, Nelson JM, Diegelmann RF (1989) Characterization and quantitation of wound matrix in the fetal rabbit. Matrix 9: 224–231.
Desmouliere A, Redard M, Darby I, Gabbiani G (1995) Apoptosis mediates the decrease in cellularity during the transition between granulation tissue and scar. Am J Pathol 146:56–66.
Desmouliere A, Chaponnier C, Gabbiani G (2005) Tissue repair, contraction, and the myofibroblast. Wound Rep Reg 13:7–12.
Diegelman RF, Cohen IK, McCoy BJ (1979) Growth kinetics and collagen synthesis of normal skin , normal scar and keloid fibroblasts in vitro. J Cell Physiol 98:341–346.
Diez-Roux G, Argilla M, Makarenkova H, Ko K, Lang R (1999) Macrophages kill capillary cells in $G_1$ phase of the cell cycle during programmed vascular regression. Development 126:2141–2147.
Dorsett-Martin WA (2004) Rat models of skin wound healing. Wound Rep Reg 12:591–599.
Dovi JV, Szpaderska AM, Di Pietro LA (2004) Neutrophil function in the healing wound: adding insult to injury. Thromb Haemost 92:275–280.
Duel TF, Senior RM, Huang JS, Griffin GL (1982) Chemotaxis of monocytes and neutrophils to platelet-derived growth factor. J Clin Invest 69:1046–1049.
Durham LA, Krummel TM, Cawthorn JW, Thomas B, Diegelmann RF (1989) Analysis of transforming growth factor beta receptor binding in embryonic, fetal, and adult rabbit fibroblasts. J Pediatr Surg 24:784–788.
Egozi E, Ferreira AM, Burns AL, Gamelli RL, DiPietro LA (2003) Mast cells modulate the inflammatory but not the proliferative response in healing wounds. Wound Rep Reg 11:46–54.
Ehrlich HP (1988) The modulation of contraction of fibroblast populated collagen lattices by types I, II and III collagen. Tiss Cell 20:47–50.
Ellis I, Banyard J, Schor S (1997) Differential response of fetal and adult fibroblasts to cytokines: cell migration and hyaluronan synthesis. Development 124:1593–1600.
Epstein EH (1974) {alpha 1 (III)} human skin collagen. J Biol Chem 249:3225–3231.
Fathke C, Wilson L, Hutter J, Kapoor V, Smith A, Hocking A, Isik F (2004) Contribution of bone marrow-derived cells to skin: collagen deposition and wound repair. Stem Cells 22:812–822.
Ferguson MWJ, O'Kane S (2004) Scar-free healing: from embryonic mechanisms to adult therapeutic intervention. Phil Trans R Soc Lond B 359:839–850.
Frantz FW, Diegelmann RF, Mast BA, Cohen IK (1992) Biology of fetal wound healing: collagen biosynthesis during dermal repair. J Ped Surg 27:945–949.
Frantz FW, Bettinger DA, Haynes JH, Johnson DE, Harvey KM, Dalton HP, Yager DR, Diegelmann RF, Cohen IK (1993) Biology of fetal repair: The presence of bacteria in fetal wounds induces an adult-like healing response. J Pediatr Surg 28:428–434.
Fraser JF, Cuttle L, Kempf M, Phillips GE, O'Rourke PK, Choo K, Hayes MT, Kimble RM (2005) Deep dermal burn injury results in scarless wound healing in the ovine fetus. Wound Rep Reg 13:189–197.
Fukai T, Takeda A, Uchinuma E (2005) Wound healing in denervated rat skin. Wound Rep Reg 13:175–180.
Gabbiani G (1998) Evolution and clinical implications of the myofibroblast concept. Cardiovasc Res 38:545–548.
Ghahary A, Shen YJ, Nedelec B, Scott PG, Tredget EE (1995) Interferons gamma and alpha-2b differentially regulate the expression of collagease and tissue inhibitor of metalloproteinase-1 messenger RNA in human hypertrophic and normal dermal fibroblasts. Wound Rep Reg 3:176–184.
Gosiewska A, Yi C-F, Brown LJ, Cullen B, Silcock D, Geesin JC (2001) Differential expression and regulation of extracellular matrix-associated genes in fetal and neonatal fibroblasts. Wound Rep Reg 9:213–222.
Gottwald T, Coerper S, Schaffer M, Koveker G, Stead RH (1998) The mast cell-nerve axis in wound healing: a hypothesis. Wound Rep Reg 6:8–20.
Greenalgh DG (1996) The role of growth factors in wound healing. J Trauma, Injury, Infection, Critical Care 41:159–167.
Grillo HC, Gross J (1967) Collagenolytic activity during mammalian wound repair. Dev Biol 15:300–317.
Grinnell F (1984) Fibronectin and wound healing. J Cell Biochem 26:107–116.
Grinnell F (1992) Cell adhesion. In: Cohen IK, Diegelmann RF, Lindblad WJ, (eds.). Wound Healing: Biochemical and Clinical Aspects. Philadelphia, WB Saunders, pp 209–222.
Gross J (1996) Getting to mammalian wound repair and amphibian limb regeneration: a mechanistic link in the early events. Wound Rep Reg 4:190–202.
Grotendorst GR (1992). Chemoattractants and growth factors. In: Cohen IK, Diegelmann RF, Lindblad WJ (eds.). Wound Healing: Biochemical and Clinical Aspects. Philadelphia, WB Saunders, pp 237–247.
Grotendorst GR, Grotendorst CA, Gilman T (1988) Production of growth factors (PDGF and TGF-beta) at the site of tissue repair. In: Hunt TK, Pines E, Barbul A, et al. (eds.). Biological and Clinical Aspects of Tissue Repair. New York, Alan R Liss, pp 47–54.
Ham AW, Cormack DH (1979) Histology, 8th ed. Philadelphia, JB Lippincott, pp 614–644.
Haslett C, Henson P (1996) Resolution of inflammation. In: The Molecular and Cellular Biology of Wound Repair. Clark RAF (ed.). New York, Plenum Press, pp 143–170.
Haynes JH, Johnson DE, Mast BA, Diegelmann RF, Salzberg DA, Cohen IK, Krummel TM (1994) Platelet-derived growth factor induces fetal wound fibrosis. J Pediatr Surg 29:1405–1408.
Heldin C-H, Westermark B (1990) Platelet-derived growth factor: Mechanism of action and possible *in vivo* function. Cell Regul 1:555–556.
Hellstrom S, Laurent C (1987) Hyaluronan and healing of typanic membrane perforations. An experimental study. Acta Otolaryngol (Stockholm) 42 (Suppl): 54–61.
Houghton PE, Keefer KA, Krummel TM (1995) The role of transforming growth factor-beta in the conversion from "scarless" healing to healing with scar formation. Wound Rep Reg 3: 229–236.
Houghton PE, Keefer KA, Diegelmann RF, Krummel TM (1996) A simple method to assess the relative amount of collagen deposition in wounded fetal mouse limbs. Wound Rep Reg 4:489–495.
Hubner G, Hu Q, Smola H, Werner S (1996) Strong induction of activin expression after injury suggests an important role of activin in wound repair. Dev Biol 173:490–498.
Hudson-Goodman P, Girard N, Jones MB (1990) Wound repair and the potential use of growth factors. Heart and Lung 18:379–384.
Hunt TK, Hussain Z. Wound microenvironment (1992) Cohen IK, Diegelmann RF, Lindblad W (eds.). Wound Healing: Biochemical and Clinical Aspects. Philadelphia, WB Saunders, pp 274–281.
Ihara S, Motobuyashi Y, Nagao E, Kistler A (1990) Ontogenetic transition of wound healing pattern in rat skin occurring at the fetal stage. Development 110:671–680.
Ihara S, Motobayashi Y (1992) Wound closure in foetal rat skin. Development 114:573–582.

Ishai-Michaeli R, Eldor A, Vlodasky I (1990) Heparanase activity expressed by platelets, neutrophils, and lymphoma cells releases active fibroblast growth factor from extracellular matrix. Cell Reg 1:833–842.

Iyer VR, Eisen MB, Ross DT, Schuler G, Moore T, Lee JCF, Trent JM, Staudt LM, Hudson J Jr, Boguski MS, Lashkari D, Shalon D, Botstein D, Brown PO (1999) The transcriptional program in the response of human fibroblasts to serum. Science 283:83–87.

Jeffrey JJ (1992) Collagen degradation. In: Cohen IK, Diegelmann RF, Lindblad WJ (eds.). Wound Healing: Biochemical and Clinical Aspects. Philadelphia, WB Saunders, pp 177–194.

Kaesler S, Bugnon P, Gao J-L, Murphy PM, Goppelt A, Werner S (2004) The chemokine receptor CCR1 is strongly up-regulated after skin injury but dispensable for wound-healing. Wound Rep Reg 12:193–204.

Kim LR, Whelpdale K, Zurowski M, Pomerantz B (1998) Sympathetic denervation impairs epidermal healing in cutaneous wounds. Wound Rep Reg 6:194–201.

Kirsner RS, Eaglstein WH (1993) The wound healing process. Dermatol Clinics 11:629–639.

Kratz G, Compton CC (1997) Tissue expression of transforming growth factor-b1 and transforming growth factor-a during wound healing in human skin explants. Wound Rep Reg 5:222–228.

Krummel TM, Michna BA, Thomas BL, Sporn MB, Nelson JM, Salzberg AM, Cohen IK, Diegelmann RF (1988) Transforming growth factor beta induces fibrosis in a fetal wound model. J Pediatr Surg 23:747–752.

Ladin DA, Garner WL, Smith DJ (1995) Excessive scarring as a consequence of healing. Wound Rep Reg 3:6–14.

Lansdown ABG (2002) Calcium: A potential central regulator in wound healing in the skin. Wound Rep Reg 10:271–285.

Lawrence WT, Diegelman R (1994) Growth factors in wound healing Clin in Dermatol 12:157–169.

Leibovich SB, Ross R (1975) The role of the macrophage in wound repair: A study with hydrocortisone and anti-macrophage serum. Am J Pathol 1978:71–91.

Leof EB, Proper JA, Govstin AS, et al. (1986) Induction of c-sis mRNA and activity similar to platelet-derived growth factor by transforming growth factor beta. A proposed model for indirect mitogenesis involving autocrine activity. Proc Natl Acad Sci USA 83:2453–2457.

Li W-Y, Chong SSN, Huang EY, Tuan T-L (2003) Plasminogen activator/plasmin system: A major player in wound healing? Wound Rep Reg 11:239–247.

Liechty KW, Adzick NS, Cromblehome TM (2000) Diminished interleukin 6 (IL-6) production during scarless human fetal wound repair. Cytokine 12:671–676.

Lin RY, Sullivan KM, Argenta PA, Lorenz HP, Adzick NS (1994) Scarless human fetal skin repair is intrinsic to the fetal fibroblast and occurs in the absence of an inflammatory response. Wound Rep Reg 2:297–305.

Linares HA (1996) From wound to scar. Burns 22:339–352.

Lindblad WJ (1998) Collagen expression by novel cell populations in the dermal wound environment. Wound Rep Reg. 6:186–193.

Lindblad WJ (2003) Stem cells in dermal wound healing. In: Sell S (ed.). Stem Cells Handbook. Towota, Humana Press, pp 101–106.

Longaker MT, Chiu ES, Harrison MR, Crombleholme M, Langer JC, Duncan BW, Adzick NS, Verrier ED, Stern R (1989) Studies in fetal wound healing: IV. Hyaluronic acid stimulating activity distinguishes fetal from adult wound fluid. Ann Surg 219:667–672.

Longaker MT, Bouhana KS, Harrison MR, Danielpour D, Roberts AB, Banda MJ (1992) Wound healing in the fetus: Possible role for inflammatory macrophages and transforming growth factor-beta isoforms. Wound Rep Reg 2:104–112.

Lorena D, Uchio K, Monte Alto Costa A, Desmouliere A (2002) Normal scarring: Importance of myofibroblasts. Wound Rep Reg 10:86–92.

Lorenz HP, Longaker MT, Perkocha LA, Jennings RW, Harrison MR, Adzick NS (1992) Scarless wound repair: A human fetal skin model. Development 114:253–259.

Lygoe KA, Norman JT, Marshall JF, Lewis MP (2004) $\alpha_v$ integrins play an important role in myofibroblast differentiation. Wound Rep Reg 12:461–470.

Mamdouh Z, Chen X, Pierini LM, Maxfield FR, Muller WA (2003) Targeted recycling of PECAM from endothelial surface-connected compartments during diapedesis. Nature 421:748–753.

Martin P (1997) Wound healing—aiming for perfect skin regeneration. Science 276:75–81.

Martin P, D'Souza D, Martin J, Grose R, Cooper L, Maki R, McKercher SR (2003) Wound healing in the PU.1 mouse—tissue repair is not dependent on inflammatory cells. Curr Biol 13: 1122–1128.

Mast BA (1992) The skin. In: Cohen IK, Diegelmann RF, Lindblad WJ (eds.). Wound Healing: Biochemical and Clinical Aspects. Philadelphia, WB Saunders, pp 344–355.

Mast BA, Flood LC, Haynes JH, DePalma RL, Cohen IK, Diegelmann RF, Krummel TK (1991) Hyaluronic acid is a major component of the matrix of fetal rabbit skin and wounds:implications for healing by regeneration. Matrix 11:63–68.

Mast BA, Diegelmann RF, Krummel TM, Cohen IK (1993) Hyaluronic acid modulates proliferation, collagen and protein synthesis of cultured fetal fibroblasts. Matrix 13:441–446.

Mast BA, Frantz FW, Diegelmann RF, Krummel TM, Cohen IK (1995) Hyaluronic acid degradation products induce neovascularization and fibroplasias in fetal rabbit wounds. Wound Rep Reg 3:66–72.

Mast BA, Haynes JH, Krummel TM, Cohen IK, Diegelmann RF (1997) Ultrastructural analysis of fetal rabbit wounds. Wound Rep Reg 6:243–248.

McCallion RL, Ferguson MWJ (1995) Fetal wound healing and the development of anti-scarring therapies for adult wound healing. In: Clark RAF (ed.). The Molecular and Cellular Biology of Wound Repair, 2nd ed, New York, Plenum Press, pp 561–600.

McDonald JA (1988) Extracellular matrix assembly. Ann Rev Cell Biol 4:183–207.

McCluskey J, Martin P (1995) Analysis of the tissue movements of embryonic wound healing—DiI studies in the limb bud stage mouse embryo. Dev Biol 170:102–114.

Merkel JR, DiPaolo BR, Hallok GG, Rice DC (1988) Type I and type III collagen content of healing wounds in fetal and adult rats. Proc Soc Exp Biol Med 187:493–497.

Messadi DV, Berg S, Sun-Cho K, Lesavoy M, Bertolami C (1994) Autocrine transforming growth factor-b1 activity and glysoaminoglycan synthesis by human cutaneous scar fibroblasts. Wound Rep Reg 2:284–291.

Miller EJ, Gay S (1992) Collagen structure and function. In: Cohen IK, Diegelmann RF, Lindblad WJ (eds.). Wound Healing: Biochemical and Clinical Aspects. Philadelphia, WB Saunders, pp 130–151.

Morgan CJ, Pledger WJ (1992) Fibroblast proliferation. In: Cohen IK, Diegelmann RF, Lindblad WJ (eds.). Wound Healing: Biochemical and Clinical Aspects. Philadelphia, WB Saunders, pp 63–76.

Moriarty KP, Cromblehome TM, Gallivan EK, O'Donnell C (1996) Hyaluronic acid-dependent pericellular matrices in fetal fibroblasts: Implication for scar-free wound repair. Wound Rep Reg 4:346–352.

Newman SA, Tomasek JJ (1996) Morphogenesis of connective tissues. In: Comper WD (ed.). Extracellular Matrix, Vol. 2, Molecular Components and Interactions. Amsterdam: Harwood Acadacemic Pub GmBh, pp 335–369.

Nwomeh C, Liang H-X, Diegelmann R, Cohen K, Yager D, et al. (1998) Dynamics of the matrix metalloproteinases MMP-1 and MMP-8 in acute open human dermal wounds. Wound Rep Reg 6:127–134.

O'Kane S, Ferguson MWJ (1997) Transforming growth factor betas and wound healing. Int J Biochem Cell Biol 29:63–78.

Olutoye O, Cohen IK (1996) Fetal wound healing: An overview. Wound Rep Reg 4:66–74.

Park JE, Barbul A (2004) Understanding the role of immune regulation in wound healing. Am J Surg 187:11S–16S.

Pearson CA, Pearson D, Shibahara D, Hofsteenge J, Chiquet-Ehrismann R (1988) Tenascin: cDNA cloning and induction by TGF-beta. EMBO J 7:2977–2981.

Phipps RP, Penney DP, Keng P, Quill H, Paxhia A, Derdak S, Felch ME (1989) Characterization of two major populations of lung fibroblasts: Distinguishing morphology and discordant display of Thy 1 and class II MHC. Am J Respir Cell Mol Biol 1:65–74.

Pierce GF (1991) Tissue repair and growth factors. In: Dulbecco R (ed.). Encyclopedia of Human Biology. New York, Academic Press, pp 499–509.

Rappolee DA, Mark D, Banda MJ, et al. (1988) Wound macrophages express TGF-alpha and other growth factors *in vivo*: Analysis by mRNA phenotyping. Science 241:708–712.

Redd MJ, Cooper L, Wood W, Stramer B, Martin P (2004) Wound healing and inflammation: Embryos reveal the way to perfect repair. Phil Trans R Soc Lond B 359:777–784.

Repesh LA, Fitzgerald TJ, Furcht LT (1982) Fibronectin involvement in granulation tissue and wound healing in rabbits. J Histochem Cytochem 30:351–358.

Riches DWH (1996) Macrophage involvement in wound repair, remodeling, and fibrosis. In: Clark RAF (ed.). The Molecular and Cellular Biology of Wound Repair, 2nd ed. New York, Plenum Press, pp 95–142.

Risau W, Lemmon V (1988) Changes in the vascular extracelular matrix during embryonic vasculogenesis and angiogenesis. Dev Biol 125:441–450.

Roberts AB, Sporn MB (1996) Transforming growth factor-beta. In: Clark RAF (ed.). The Molecular and Cellular Biology of Wound Repair, 2nd ed. New York, Plenum Press, pp 275–310.

Ross R, Raines EW, Bowen-Pope DF (1986) The biology of platelet-derived growth factor. Cell 46:155–169.

Rudolph R, VandeBerg J, Ehrlich HP (1992) Wound contraction and scar contracture. In: Cohen IK, Diegelmann RF, Lindblad WJ (eds.). Wound Healing: Biochemical and Clinical Aspects. Philadelphia, WB Saunders, pp 177–194.

Schultz GS, Sibbald RG, Falanga V, Ayello EA, Dowsett C, Harding K, Romanelli M, Stacey MC, Teot L, Vanscheidt W (2003) Wound bed preparation: A systematic approach to wound management. Wound Rep Reg 11:1–28.

Sempowski GD, Borrello MA, Blieden T, Barth RK, Phipps R (1995) Fibroblast heterogeneity in the healing wound. Wound Rep Reg 3:120–131.

Shah M, Foreman DM, Ferguson MW (1994) Neutralizing antibody to TGF-beta 1,2 reduces cutaneous scarring in adult rodents. J. Cell Sci 107:1137–1157.

Shah M, Foreman D, Ferguson MW (1995) Neutralization of TGF-beta 1 and TGF beta 2 or exogenous addition of TGF beta 3 to cutaneous rat wounds reduces scarring. J Cell Sci 108:985–1002.

Shapiro JD, Kabayashi DK, Pentland AP, et al. (1993) Induction of macrophage metalloproteinases by extracellar matrix. J Biol Chem 268:8170–8175.

Sicard RE, Shearer JD, Caldwell MD (1998) Wound repair. J Minn Acad Sci 63:31–36.

Simpson DM, Ross R (1972) The neutrophilic leukocyte in wound repair, a study with anti-neutrophil serum. J Clin Invest 51: 2009–2023.

Stiles CD, Capone GT, Scher CD, et al. (1979). Dual control of cell growth by somatomedins and platelet-derived growth factor. Proc Natl Acad Sci USA 76:1279–1283.

Sullivan K, Lorenz HP, Adzick NS (1993) The role of transforming growth factor-β in human fetal wound healing. Surg Forum 44:625–627.

Sullivan TP, Eaglstein W, Davis SC, Mertz P (2001) The pig as a model for human wound healing. Wound Rep Reg 9:66–76.

Swift ME, Burns A, Gray KL, Di Pietro LA (2001) Age-related alterations in the inflammatory response to dermal injury. J Invest Dermatol 117:1027–1035.

Taylor-Papadimitriou J, Rosengurt E (1985) Interferons as regulators of cell growth and differentiation. In: Taylor-Papadimitriou J (ed.). Interferons: Their Impact in Biology and Medicine. Oxford, Oxford University Press, pp 81–98.

Tomasek JJ, Vaughn B, Haaksma CJ (1999) Cellular structure and biology of Dupuytren's disease. Hand Clinics 15:1–15.

Tran KT, Griffity L, Wells A (2004) Extracellular matrix signaling through growth factor receptors during wound healing. Wound Rep Reg 12:262–268.

Van Beurden H, Snoek PAM, Von Den Hoff JW, Torensma R, Kiijpers-Jagtman A-M (2003) Fibroblast subpopulations in intra-oral wound healing. Wound Rep Reg 11:55–63.

Wahl LM, Wahl SM Inflammation (1992) In: Cohen IK, Diegelmann RF, Lindblad WJ (eds.). Wound Healing: Biochemical and Clinical Aspects. Philadelphia, WB Saunders, pp. 40–62.

Wanaka A, Milbrandt J, Johnson EM (1991) Expression of FGF receptor gene in rat development. Development 111:455–468.

Weitzhandler M, Bernfield MR (1992) Protoglycan glyconjugates. In: Cohen IK, Diegelmann RF, Lindblad WJ (eds.). Wound Healing: Biochemical and Clinical Aspects. Philadelphia, WB Saunders, pp 195–208.

Wharton W (1983) Hormonal regulation of discrete portions of the cell cycle: Commitment to DNA synthesis is commitment to cellular division. J Cell Physiol 117:423–429.

Whitby DJ, Ferguson MWJ (1991) The extracellular matrix of lip wounds in fetal, neonatal and adult mice. Development 112:651–668.

Whitby DJ, Ferguson MWJ (1992) Immunohistochemical studies of the extracellular matrix and soluble growth factors in fetal and adult wound healing. In: Adzick NS, Longaker (eds.). Fetal Wound Healing. New York, Elsevier, pp 161–179.

Yamada KM, Clark RAF (1996) Provisional matrix. In: Clark RAF (ed.). The Molecular and Cellular Biology of Wound Repair, 2nd ed. New York, Plenum Press, pp 51–94.

Yamaguchi Y, Mann DM, Ruoslahti E (1990) Negative regulation of transforming growth factor-β by the proteoglycan decorin. Nature 346:281–284.

Yang GP, Lim IJ, Phan T-T, Lorenz HP, Longaker MT (2003) From scarless fetal wounds to keloids: Molecular studies in wound healing. Wound Rep Reg 11:411–418.

Yannas IV, Colt J, Wai YC (1996) Wound contraction and scar synthesis during development of the amphibian Rana catesbeiana. Wound Rep Reg 4:29–39.

Yannas IV (2001) Tissue and Organ Regeneration in Adults. Springer, New York.

# 3 Regeneration of Epidermal and Dental Tissues, Lens and Cornea

## INTRODUCTION

The epidermis of the skin and its appendages, hair follicles (or their counterparts, feathers and scales), sebaceous glands, sweat glands, and nails are derived embryonically from the prospective epidermal ectoderm, with contributions from the neural crest. In addition, the epidermal ectoderm gives rise to the dentition and to the lens and cornea of the eye. The development of all epidermal ectoderm derivatives requires interactions with underlying or surrounding mesenchyme. All vertebrates regenerate the epidermis, as well as periodontal tissues and corneal epithelium, but only a few species can regenerate teeth and lens. The epidermis, hair, nails, periodontium, and cornea turn over continually and maintain their structure by maintenance regeneration, as well as regenerating in response to injury. Because they are nonneural structures derived from the ectoderm, these tissues are treated here as a cluster.

## REGENERATION OF EPIDERMIS AND HAIR

### 1. Maintenance Regeneration

#### a. *Interfollicular Epidermis*

The epidermis of the mammalian skin is a continuous, multilayered epithelium interspersed with several appendages: hair follicles, sebaceous glands, and sweat glands. The regions of epidermis between the hair follicles and their associated sebaceous and sweat glands are termed the interfollicular epidermis (IFE).

The cells of the IFE are arranged in columnar structural-proliferative units consisting of a single layer of basal cells at the bottom of each column (the stratum basale), two intermediate layers of differentiating keratinocytes, the stratum spinosum and stratum granulosum, and the top layers of flattened keratnized cells, the stratum corneum. The stratum corneum cells have the geometry of flattened orthic tetrakaidecahedrons, 14-sided polygons that are the only geometric forms able to stack into regular columns and interdigitate with adjacent cells in six surrounding columns, thus completely eliminating space between them (Ham and Cormack, 1979). The architecture of the epidermis constitutes a mechanical and physiological barrier that prevents loss of water, intake of noxious chemicals, and protects against mechanical, chemical, and thermal insults.

The basal layers of the IFE are a mixture of epidermal stem cells (EpSCs), transit amplifying cells, and postmitotic cells committed to undergo terminal differentiation (Potten and Morris, 1988). Epidermal regeneration is maintained by the EpSCs. Transit amplifying progeny of the EpSCs detach from their niche on the basement membrane, proliferate and migrate upward, undergoing progressive differentiation into the dead, keratinized squames of the stratum corneum (**FIGURE 3.1**). Basal cells *in vitro* express three β1 integrins that mediate adhesion to the basement membrane and respond to microniche factors that regulate the onset of commitment to terminal differentiation and upward migration (Adams and Watt, 1990; Jones and Watt, 1993). These are $\alpha_5\beta_1$ (fibronectin receptor), $\alpha_2\beta_1$ (binds to collagen and laminin), and $\alpha_3\beta_1$ (receptor for laminin). Commitment to terminal differentiation and upward movement of basal cells are associated with the downregulation of the β1 integrins, resulting in a decreased ability to adhere to the basement membrane (Adams and Watt, 1990; Hotchin et al., 1995). The microniche factors that insure a steady stream of detachment of basal cells are poorly understood, but undoubtedly involve some form of communication between the top (exiting) layers of the epidermis and entering basal cells.

No unique surface antigens or combinations of transcription factors have been identified that specifically

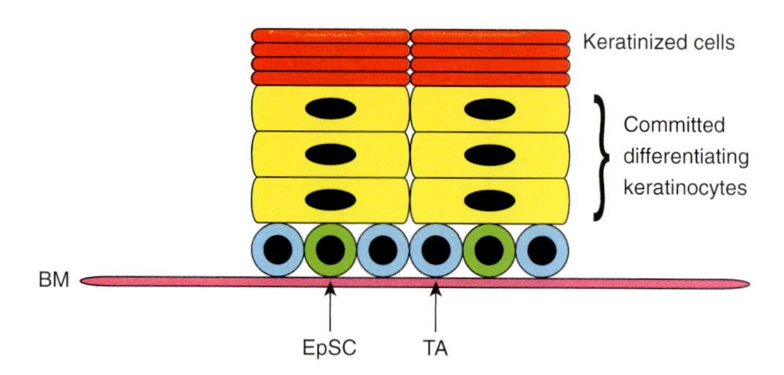


**FIGURE 3.1** Production of transit amplifying (TA) cells from epidermal stem cells (EpSCs) and their progressive differentiation into keratinized cells of the stratum corneum. EpSCs (green) express β1 integrins that firmly attach them to the basement membrane (BM), whereas TAs down-regulate β1 integrins to detach from the BM and migrate upward.

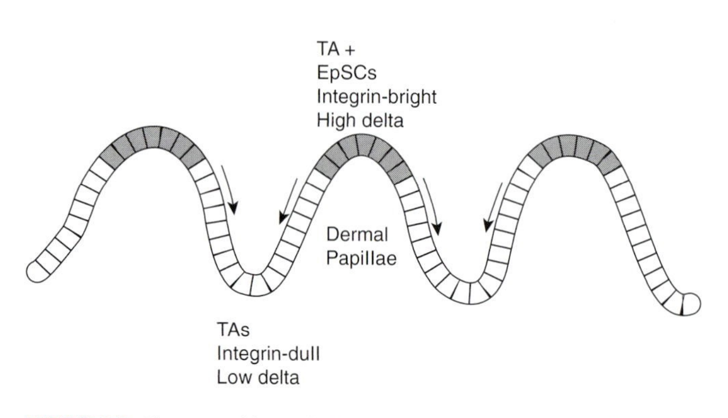


**FIGURE 3.2** Clusters of integrin-bright, Delta-bright stem cells mixed with integrin-dull, Delta-dull transient amplifying (TA) cells reside in the basal epidermis atop the dermal papillae. TA cells migrate laterally to fill in the valleys between the papillae.

define EpSCs of the IFE. However, EpSCs have been tentatively identified by their expression pattern of the notch ligand, Delta 1 and integrins. Mouse basal epidermis lying over the apex of the dermal papillae contains clusters of cells that express high levels of the transmembrane Notch ligand, Delta 1 (Lowell et al., 2000) and 2–3 times the level of $\alpha_2\beta_1$ and $\alpha_3\beta_1$ integrins ("integrin-bright" cells) than surrounding "integrin-dull" cells (Jones and Watt, 1993; Jones et al., 1995) (**FIGURE 3.2**). High Delta and integrin-bright cells constitute about 40% of the cells in the basal layer. This population expresses higher levels of noncadherin-associated β-catenin than other keratinocytes (Zhu and Watt, 1999), is highly adhesive, and can generate large numbers of keratinocytes *in vitro*, suggesting that it is rich in EpSCs. The low Delta and integrin-dull cells are believed to be transit amplifying cells and cells committed to becoming keratinocytes (Barrandon and Green, 1987; Jones et al., 1995). Consistent with this idea, integrin-bright EpSCs have been shown to divide infrequently *in vivo* to give rise to a more profligate transit amplifying cell population that undergoes 3–5 rounds of division within the basal layer (Jones et al., 1995; Jensen et al., 1999). The transit amplifying cells spread out over the basement membrane to fill any intermediate spaces (Jensen et al., 1999) and commit to keratinocyte differentiation. They become integrin-dull and move upward toward the epidermal surface, where β1 integrin expression is ultimately extinguished as the cells become terminally differentiated.

EpSCs plated *in vitro* reconstruct integrin-bright and dull patches of epidermis. This self-organizational tendency suggests that stem cell patterning in the epidermis is an intrinsic property of keratinocytes, rather than depending on environmental cues (Jones et al., 1995). Computer simulations based on the assumption that committed basal keratinocytes migrating upward occupy less crowded regions showed that the squamous epithelial architecture of orthic tetrakaidecahedrons is spontaneously organized from basal cells supplied at random (Honda et al., 1996), consistent with the *in vitro* evidence for self-organization of epidermal patterning.

Interestingly, the basal epithelium of the esophagus (an endodermal derivative) also contains patches of β1 integrin-bright cells surrounded by β1 integrin-dull cells. However, clonogenic analyses *in vitro* showed that the proportion of integrin-dull cells forming keratinocyte clones was 2.5 times higher that the integrin-bright cells, suggesting that the integrin-dull cells are the stem cells of the esophageal basal layer and the integrin-bright cells are the transit-amplifying population (Seery and Watt, 2000).

### b. Hair

The hair follicles are epidermal tubes that project into the dermis (**FIGURES 3.3, 3.4**). The proximal tip of the follicle is invaginated to form a cap over the dermal papilla, a condensation of dermal fibroblasts associated with the follicle (Hardy, 1992). This cap is called the matrix, and its cells are the transit amplifying stem cells that differentiate into the hair shaft. Together, the dermal papilla and the matrix constitute the bulb of the hair follicle. The hair shaft is formed by the proliferation and differentiation of the matrix cells upward through the lumen of the follicle. The differentiated hair consists of keratinocytes that form a medullary core surrounded by a cortex. The walls of the hair follicle above the matrix differentiate into the inner and outer root sheaths. The IFE is continuous with the epidermis of the outer root sheath of the hair follicles, but hair follicle matrix cells differentiate into specialized keratinocytes that form hair instead of squames of keratinized epidermis. Two epidermal thickenings are present in the upper third of the outer root sheath. The

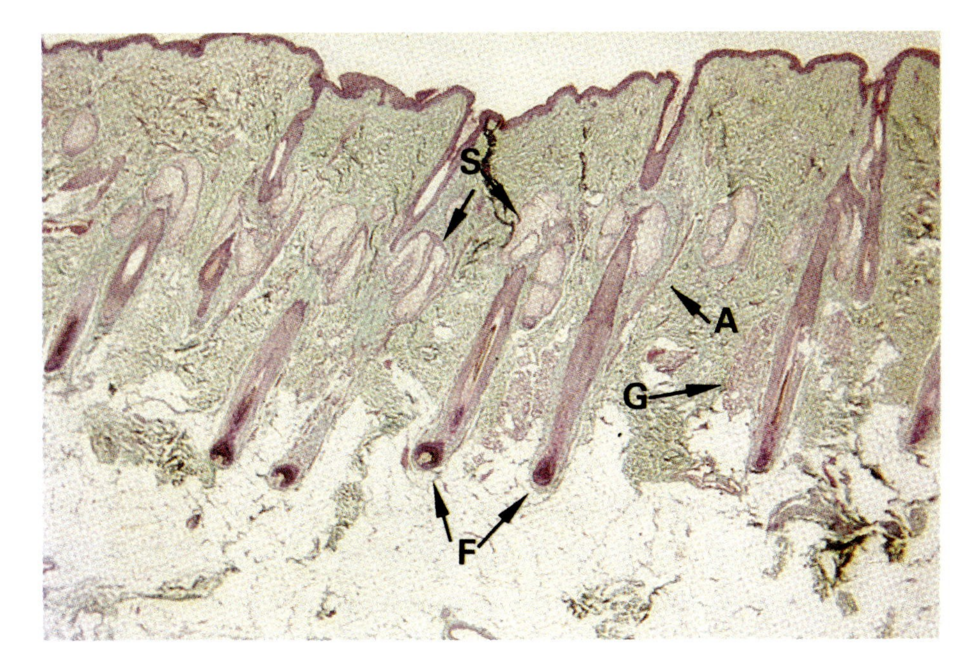


**FIGURE 3.3** Section of human scalp skin (Masson's green trichrome) showing hair follicles (F), sebaceous glands (S), sweat glands (G) and arrector pili muscles (A). Reproduced with permission from Wheater et al., Wheater's Functional Histology (3rd ed). Copyright 1997, Elsevier.

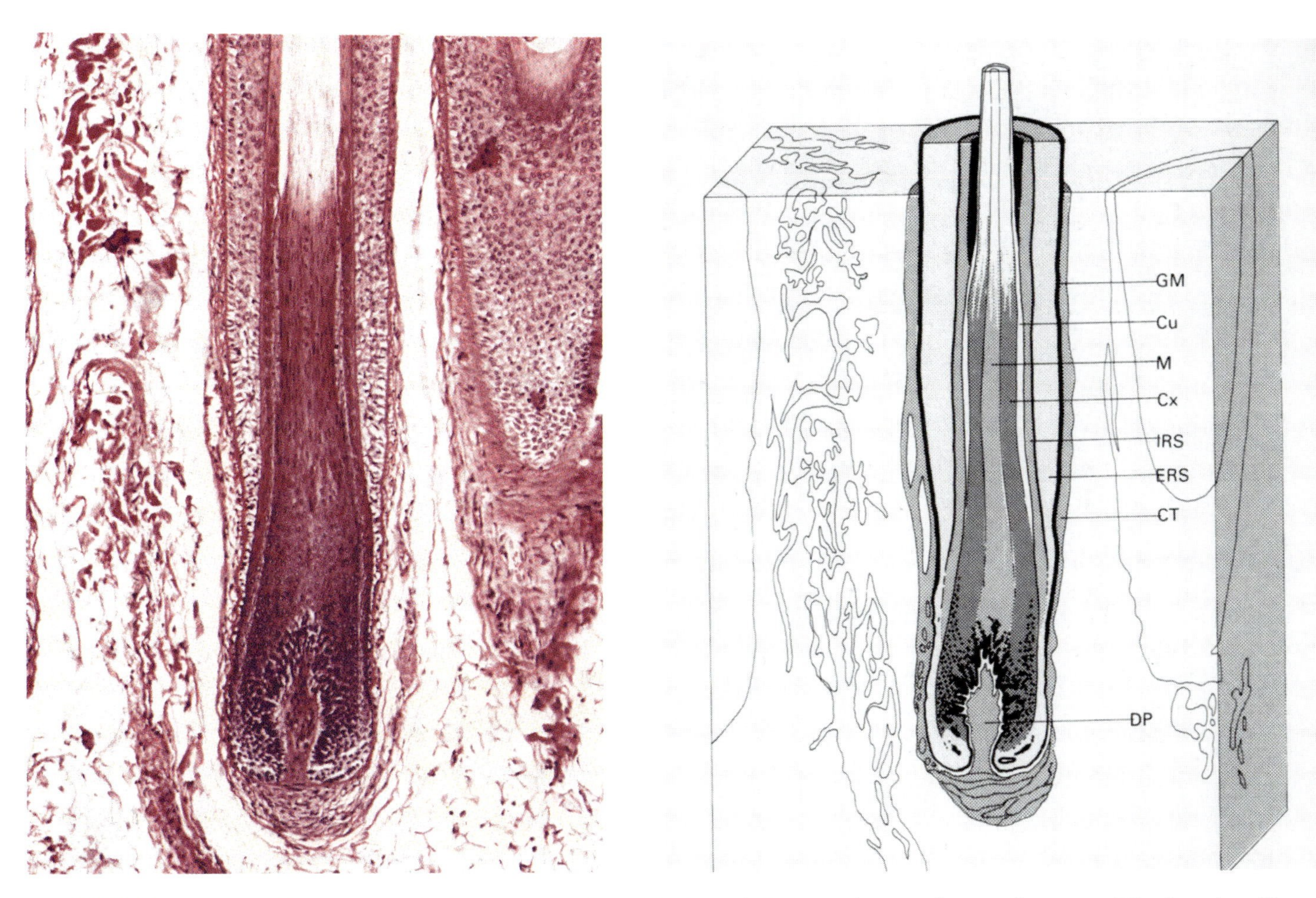


**FIGURE 3.4** Structure of the hair follicle. **Left**, section of a follicle (H & E stain). **Right**, explanatory diagram of structure. DP = dermal papilla, CT = connective tissue sheath surrounding the follicle, ERS = external root sheath (outer root sheath), GM = the glassy membrane that separates the external root sheath from the outer connective tissue layer, IRS = internal root sheath, Cx = the highly keratinized cortex of the hair shaft, M = the medulla or core of the hair shaft, Cu = the thin cuticle that that covers the surface of the hair shaft. The irregular black color is the matrix, where transit amplifying cells migrate upward to differentiate into the parts of the hair. Reproduced with permission from Wheater et al., Wheater's Functional Histology (3rd ed). Copyright 1997, Elsevier.

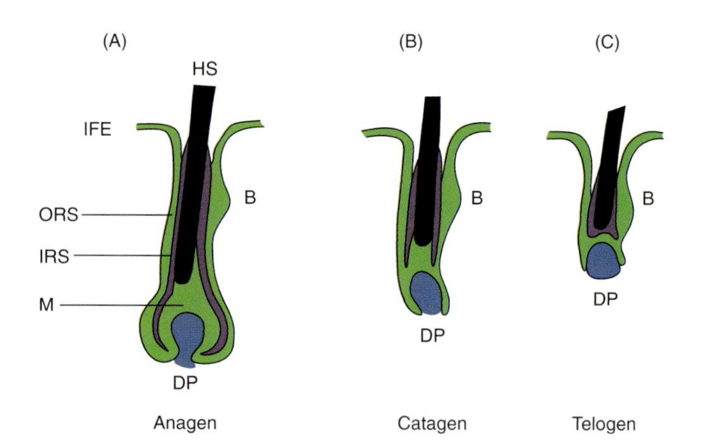


**FIGURE 3.5** Stages in the adult hair regeneration cycle. **(C),** Telogen. Apoptosis during catagen has shrunk the hair follicle upward to its maximum extent. **(A),** The anagen stage, where signals from the dermal papilla induce EpScs in the bulge to proliferate, extending the hair follicle downward and regenerating a new hair. **(B),** Catagen. The matrix and inner and outer root sheath cells are undergoing apoptosis, causing the follicle to regress.

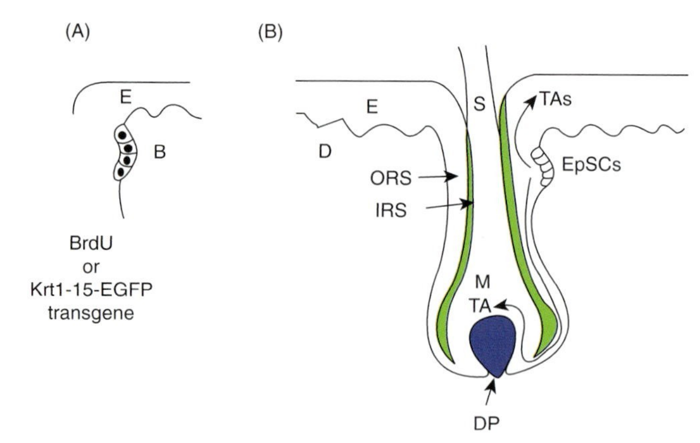


**FIGURE 3.6** Cell flow in the regenerating hair follicle. **(A)** EpSCs of the bulge (B) were marked in mice by labeling with BrdU at anagen or making mice transgenic for the enhanced green fluorescent protein (EGFP) under the control of the EnSC-specific *Krt1-15* promoter. **(B)** EpSCs in the basal epithelium of the bulge divide to form transient amplifying (TA) cells that migrate down through the outer root sheath (ORS) into the matrix (M), where they move upward and differentiate into keratinocytes of the hair shaft (S). After injury to the interfollicular epidermis, bulge TA cells also migrate upward to repair the epidermis. E = interfollicular epidermis, D = dermis, IRS = inner root sheath, D = dermal papilla. Arrows = path of migration.

upper thickening forms the sebaceous gland. The lower thickening is called the "bulge" and its basal layer is the site of stem cells for renewal of the hair follicle and attachment of the arrector pili muscle. The epidermal follicle is surrounded by a sheath of dermal cells.

Mammalian hairs are constantly falling out and regenerating. Hairs are lost and regenerated in three phases (Hardy, 1992; Messenger, 1993): (1) catagen, the cessation of hair growth and regression of the hair follicle; (2) telogen, or follicular rest; and (3) anagen, the regeneration of the follicle and a new hair (**FIGURE 3.5**). During catagen, cells in the follicle wall undergo apoptosis and the hair follicle regresses upward so that the dermal papilla contacts the bulge. In telogen, the epithelial cells are in a resting state and the hair shaft remains within the shortened follicle. The dermal papilla shrinks in volume during catagen and telogen, but it is not known if this is due to cell loss or compaction (Matsuzaki and Yoshizato, 1998). At anagen, the dermal papilla induces stem cells of the bulge to proliferate. The progeny of bulge stem cells flow downward, extending the hair follicle, and pushing the dermal papilla downward to reconstitute the hair bulb and matrix. A new hair shaft and inner root sheath are then differentiated from matrix transit amplifying cells. As it grows, the new hair pushes the old one out of the hair follicle. The hair follicle is a highly useful system in which to study the microniche factors that control the quiescence and activation of stem cells in a cyclic manner (see below).

There is strong evidence that the EpSCs responsible for hair follicle growth during anagen reside in the bulge region of the outer root sheath. $^{3}$H-thymidine or BrdU label a population of slowly cycling (i.e., label-retaining) bulge cells in rodent and human hair follicles (Cotsarelis et al., 1990; Kobayashi et al., 1993; Rochat et al., 1994; Lavker et al., 1999; Morris and Potten, 1999; Taylor et al., 2000). The proliferative capacity of these cells and their ability to form keratinocyte colonies *in vitro* is much higher than cells from other regions of the follicle. Further evidence that the bulge cells are hair follicle stem cells was obtained by long-term BrdU labeling studies in mice (Taylor et al., 2000). Eight weeks after labeling, all the follicles were in telogen and only bulge cells contained label. By 10 weeks, however, as the follicles entered anagen, many of the matrix cells below the bulge contained label, indicating that they were derived from bulge cells.

Stem cells have been isolated directly from the basal epidermal layer of the outer root sheath of the bulge by the generation of ROSA26 mice transgenic for the reporter genes β-galactosidase or green fluorescent protein (GFP) under the control of keratin promoters specific to bulge cells (Morris et al., 2004; Tumbar et al., 2004; Khavari, 2004). The reporters were activated at anagen by transcription factors targeting these promoters, thus marking the bulge cells (**FIGURE 3.6**). The cells could be followed *in vivo* or they could be isolated for *in vitro* studies by FACS. *In vivo*, the marked cells contributed primarily to the regenerated hair. Only a few labeled cells were found in the IFE and sebaceous glands, confirming that maintenance regeneration of

IFE is accomplished primarily through the activity of IFE basal stem cells. The kinetics of cell migration in the bulge is similar to those in the IFE. The basal EpSCs of the bulge are attached to a basement membrane. The suprabasal cells are a transit amplifying population that feeds into the matrix.

Blanpain et al. (2004) showed, by clonal analysis *in vitro*, that although the basal and suprabasal cells of the bulge express different cell surface markers, both are multipotent and can form all the cell types of the hair follicle. This means that, even though the suprabasal cells of the bulge are committed to differentiating into keratinocytes of the hair shaft, this commitment is reversible. Another indication of multipotency is the fact that isolated bulge stem cells could regenerate skin epithelium with hair follicles, sebaceous glands, and interfollicular epidermis when combined *in vitro* with neonatal mouse dermal cells and implanted subcutaneously in immunodeficient mice (Morris et al., 2004).

The molecular phenotype of the bulge cells is also in line with their function as stem cells (Morris et al., 2004; Tumbar et al., 2004). Comparison of gene arrays of bulge basal cells versus nonbulge basal cells (upper outer root sheath, sebaceous gland, interfollicular epidermis) revealed over 150 genes that were differentially expressed in the basal bulge cells. The upregulated genes were those associated with quiescent stem cells, such as inhibitors of the Wnt signaling pathway (see below), the stem cell marker CD34, Tn-C and genes associated with growth arrest or inhibition of differentiation. Genes associated with cell cycling were downregulated. Genes encoding FGF-18 and BMP-6, which are inhibitory to proliferation and terminal differentiation, are upregulated in basal bulge cells. These growth factors slowed proliferation without inducing terminal differentiation of bulge cells *in vitro*, suggesting that they are part of the mechanism that regulates the rate of feed into the transit amplifying population in the suprabasal layers and from there into the matrix (Blanpain et al., 2004).

The outer root sheath, including the bulge and matrix, contains neural crest-derived cells. Neural crest cells normally differentiate into the pigmented cells of the hair, but in clonal culture they form colonies containing neurons and smooth muscle cells, as well as melanocytes, and can be induced by neuregulin-1 (glial growth factor-1) and BMP-2 to differentiate Schwann cells and chondrocytes, respectively (Sieber-Blum et al., 2004).

We do not know what cells other than the EpSCs themselves and their basement membrane might contribute to the microniche of the bulge that maintains EpSCs in an undifferentiated state (Christiano, 2004). Melanocytes and Merkel cells reside in the bulge and may contribute factors (Nishimura et al., 2002; Narisawa et al., 1994), as could arrector pili muscle cells that attach to the bulge (Akiyama et al., 1995).

The growth and regression cycle of the hair follicle is regulated by signals from the dermal papilla (Inamatsu et al., 1998). When the lower part of the hair follicle regresses during catagen, the dermal papilla is brought into proximity to the bulge. Papilla cells are different from surrounding dermal fibroblasts in that the *Hoxc1* gene is active and they express more laminin, fibronectin, and versican (Messenger et al., 1991; du Cros et al., 1995), as well as secreting hepatocyte growth factor (HGF) and vascular endothelial growth factor (VEGF) (Shimaoka et al., 1994; Lachgar et al., 1996). These growth factors may be the initial signals required to activate EpSCs in the bulge at the start of anagen, since HGF stimulates the growth of mouse vibrissal follicles *in vitro* (Jindo et al., 1994). *Fgf5* is expressed in the outer root sheath of mouse hair follicles during anagen (Hebert et al., 1994). Mice with null alleles of this gene have abnormally long hair, suggesting that *Fgf5* functions to inhibit hair growth.

The dermal papilla can be regenerated if it is lost. When the lower third or half of the follicle is amputated, new hair bulbs regenerate from the remaining part of the follicle (Horne et al., 1986). This regeneration is associated with the regeneration of a new dermal papilla from dermal sheath cells (Matsuzaki and Yoshizato, 1998). Tissue recombination experiments have shown that matrix cells can induce dermal sheath cells to become dermal papilla cells and can also rejuvenate passaged noninductive dermal papilla cells to become inductive again (Reynolds and Jahoda, 1996).

#### c. Role of the Wnt Pathway in Regulating EpSC Quiescence and Activation

The choice between quiescence and activation of EpSCs, whether in the interfollicular epidermis or the hair follicle bulge, involves the Wnt signal transduction pathway ending in Tcf/Lef1, which also operates in the embryonic development of epidermal structures (Gat et al., 1998; Zhu and Watt, 1999; Das Gupta and Fuchs, 1999). The *Wnt3* gene is expressed in developing and mature hair follicles and its effector, Disheveled 2 (Dvl2) is expressed in a subset of cells in the outer root sheath and in precursor cells of the hair shaft (Millar et al., 1999). Overexpression of both *Wnt3* and *Dvl2* give a short-hair and cyclical balding phenotype in mice resulting from structural defects and an abnormal profile of protein expression in the hair shaft. In quiescent telogen hair follicles, Lef-1 represses the activity of EpSCs, because active GSK-3β prevents β-catenin from dissociating from the APC protein, which degrades β-catenin. At anagen, Wnt proteins activate EpSCs by inhibiting the enzyme GSK-3β through Dvl2, allowing

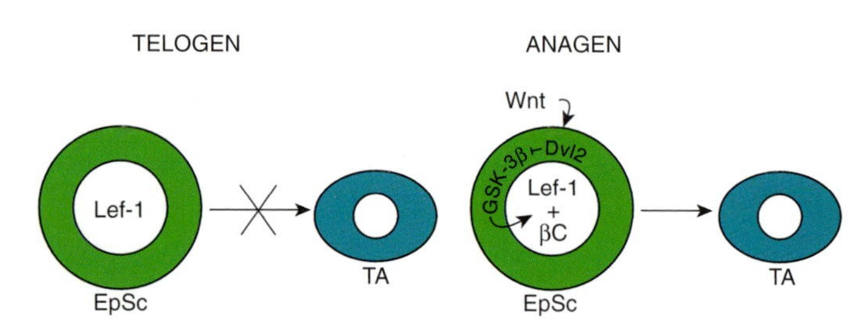


**FIGURE 3.7** Diagram showing how Wnt proteins activate quiescent EpSCs of the hair follicle to become transit amplifying (TA) cells. At telogen, Lef-1 acts as a transcriptional repressor in the absence of β-catenin (β-C), which is degraded by GSK-3β. At anagen, Wnt activates the Disheveled 2 (Dvl-2) protein, which inactivates GSK-3β. β-C is stabilized and translocates to the nucleus, where it forms a complex with Lef-1 that stimulates gene transcription characteristic of TA cells.

β-catenin to dissociate from APC and complex with Lef-1 to activate genes that release the EpSCs from a quiescent state (**FIGURE 3.7**). Expression of stabilized β-catenin by deletion or mutation of GSK-3β phosphorylation sites at the N-terminus increased the proportion of β-1 integrin-expressing EpSCs to nearly 90% of the proliferative population *in vitro*, whereas unstable β-catenin mutants inhibited proliferation of these cells (Zhu and Watt, 1999). Co-transfection of N-terminally truncated (stable) β-catenin and Lef-1 into differentiating mouse keratinocytes caused them to revert to a proliferative state from which they could differentiate into either hair follicles or interfollicular epidermis (Gat et al., 1998).

Mice lacking the retinoblastoma (Rb) proteins p107/p130, which are required for cell cycling, have fewer hair follicles and display defects in the terminal differentiation of the epidermis. The basal epidermal cells of these mice were found to have increased nuclear accumulation of β-catenin (Ruiz et al., 2004). The increase was due to the increased expression of Frat, a GSK-3β binding protein that prevents degradation of β-catenin. Thus, components of the Wnt pathway in hair follicle bulge cells appear to be linked to cell cycle regulation through the Rb family.

## 2. Injury-Induced Regeneration of the IFE

In excisional wounds of adult skin, the epidermis recovers the wound via the centripetal migration of basal cells. Basal cells of the interfollicular epidermis detach from the basement membrane and begin migrating into the fibrin matrix of the wound within a day or two after injury. The initial signal for migration may be a "free edge" effect, in which a lack of neighbors on one side stimulates the cells to alter their morphology and internal structure to accommodate movement (Woodley, 1996). The migrating cells lose their apical-basal polarity and dissolve the desmosomes that hold them together laterally, as well as the hemidesmosomes that anchor them to the basement membrane. Simultaneously, the epidermal cells form the peripheral actin locomotory apparatus that gives them motility through the active protrusion of lamellapodia and filopodia, a process that is mediated by the small GTPases Rac and Cdc42 (Fenteany et al., 2000). The cells may move as a sheet, or in "leapfrog" fashion, with cells just behind the leading edge moving over the lead cells to establish a new edge. The migrating cells do not divide to increase epidermal area. Instead, the area of re-epithelialization increases by the division of basal cells at the edge of the wound to provide more cells for migration. To facilitate their passage through the fibrin matrix the migrating cells produce collagenase and tPA to activate it (Woodley, 1996). This process continues during the formation of granulation tissue until the epidermal sheet covers the wound and migration stops due to contact inhibition. The regenerated wound epidermis resynthesizes a new basement membrane as well as Type VII collagen anchoring fibrils (Miller and Gay, 1992).

The migrating epidermal cells express the α5β1 and $\alpha_v\beta6$ Fn/Tn-C and $\alpha_v\beta5$ vitronectin integrin receptors, as well as type IV collagen receptors, as they move through the provisional fibrin matrix. The initiation of migration is correlated with the synthesis of Tn-C by dermal fibroblasts at the wound edge (Repesh et al., 1982; Whitby and Ferguson, 1991). Tenascin-C is an anti-adhesive molecule and thus may help in the detachment of basal cells from the basement membrane. TGF-β1 appears in rat skin wounds prior to the appearance of Tn-C (Whitby and Ferguson, 1991) and is strongly upregulated in keratinocytes migrating into a full-thickness punch wound in human foreskin *in vitro* (Kratz et al., 1997). The synthesis of Tn-C after the appearance of TGF-β1 in the wound might reflect an inductive effect of TGF-β on tenascin synthesis. In any event, a balance of adhesive and de-adhesive forces is likely to play a role in regulating the rate of epithelial migration across the fibrin clot and granulation tissue.

Macrophages are thought to play a crucial role in initiating epidermal cell migration after injury via their production of EGF and TGF-α (**see FIGURE 2.4**). These growth factors promote the spreading of epithelial sheets *in vitro* (Woodley, 1996). However, since antibiotic-maintained PU.1 mice, which lack macrophages, re-epithelialize normally (Martin et al., 2003), the factors produced by macrophages may be redundant to the same or different factors produced by other cell types. The epidermal cells themselves are also induced to produce EGF and TGF-α, thus sustaining their own migration through an autocrine mechanism. Activin B is strongly upregulated in migrating and proliferating keratinocytes, suggesting that this member of the TGF-β family also plays an autocrine role in these processes (Hubner et al., 1996).

Granulation tissue fibroblasts produce keratinocyte growth factor (KGF, or FGF-7), which stimulates mitosis in the peripheral epidermis, allowing it to increase in area, and later in the re-established epidermis, allowing it to increase in thickness (Pierce, 1991). Expression of a dominant-negative transgene for KGF receptor in mice reduces the rate of proliferation of keratinocytes at the wound edge, thus substantially delaying re-epithelialization of the wound (Werner et al., 1994). Granulocyte macrophage colony stimulating factor (GM-CSF) is also produced by fibroblasts of granulation tissue. KGF, EGF, and GM-CSF are sufficient to promote full epidermal differentiation *in vitro* (Ghalbzouri and Ponec, 2004). The quantitative perfection of epidermal and basement membrane structure, however, requires the production of additional (unknown) factors mediated by keratinocytes, since a completely normal epidermis is produced only when keratinocytes and fibroblasts of granulation tissue can interact.

Epiregulin is a member of the EGF family that promotes epidermal proliferation and thickening and has been shown to enhance the repair of mouse excisional skin wounds (Draper et al., 2003). All members of the EGF family are anchored into the plasma membrane of cells and are cleaved to produce active fragments. The active fragments of epiregulin exert their effects by binding to the tyrosine kinase EGF receptors erbB1 and erbB4. Respiratory epithelial cells produce heregulin, which is anchored to the plasma membrane on the apical side of the cells and binds to erbB2. When injured, respiratory epithelium regenerates from basal stem cells, as in the epidermis. In the epidermal and respiratory epithelia, tight junctions separate the apical and basal extracellular domains of epithelial cells, preventing any diffusible molecular interactions between the domains. The spatial locations of the epiregulin receptors have not been determined in epidermis, but in respiratory epithelium the erbB2 receptor is on the basolateral surface of the cells. *In vitro* experiments in which a break was made in the epithelium showed that injury allowed access of heregulin to erbB2, followed by proliferation (**FIGURE 3.8**). The same result was achieved by the removal of calcium, which is required to maintain tight junctions (Vermeer et al., 2003; Mostov and Zegers, 2003). These results suggest that the injury-promoted access of active fragments of ligands belonging to the EGF family to their receptors on the opposite side of the cell may be a common trigger mechanism for initiation of basal cell proliferation at the edges of a wound.

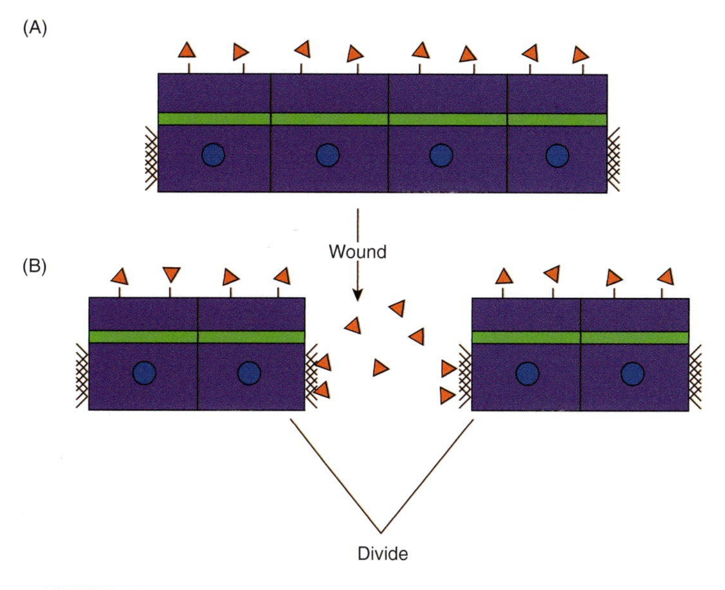


**FIGURE 3.8** Proposed mechanism to trigger proliferation of basal epidermal cells at the edge of a wound. **(a)** Epiregulin (orange triangles) cannot interact with its basolaterally located receptors erbB1 and erbB4 because the domains of the two are separated by tight junctions. **(b)** Injury separates the apical and basolateral domains so that epiregulin can bind to erbB1 and erbB4, triggering proliferation.

Bulge cells of the hair follicle make a significant contribution to the regeneration of injured IFE. Experiments with pigs over 60 years ago had suggested that the external root sheath of hair follicles made a substantial contribution to the epidermis recovering an extensive excisional wound (Ham, 1944; Argyris, 1976; Ham and Cormack, 1979). A contribution of bulge EpSCs to the regeneration of injured IFE was confirmed by double-labeling bulge cells with BrdU and $^3$H-thymidine after making a full-thickness wound in the dorsal skin of mice. Subsequent examination revealed an increase of double-labeled cells in the regenerating IFE and a corresponding decrease in double-labeled cells in the follicular epithelium (Taylor et al., 2000). Thus bulge stem cells feed downward into the hair follicle to regenerate hair, as well as upward to aid in the regeneration of injured IFE.

## REGENERATION OF NAILS

Nails are hardened epidermal variants of hair. They grow continuously from the nail organ, an epidermal invagination on the dorsal side of the digit tips that is the equivalent of a large hair follicle. **FIGURE 3.9** illustrates the anatomy and function of the nail organ

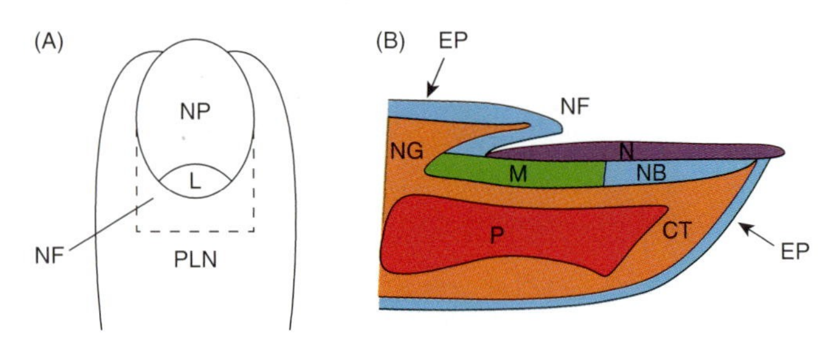


**FIGURE 3.9** Regeneration (growth) of nails. **(A)** External view of nail. NP = nail plate (the nail), PLN = proximal limit of the nail, L = lunula, the half-moon shaped region of the nail where the matrix extends from under the nail fold (NF). **(B)** Longitudinal section through the distal phalange (P). The NF is the rim of an epidermal invagination that forms the nail groove (NG). The epidermal cells of the dorsal wall of the nail groove constitute the matrix (M), which produces the nail (N). The nail bed (NB) consists of epidermal cells under the nail distal to the lunula. CT = connective tissue.

(Fleckman and Allan, 2001). The dorsal wall of the invagination is short and extends to the cuticle, while the ventral wall extends nearly to the tip of the digit. The ventral wall forms the nail bed. At the proximal end of the nail bed is the matrix, a mass of stem cells that is equivalent to the transit amplifying cells in the matrix of the hair follicle. The matrix is the immediate source of the cells that form the nail.

Matrix cells divide and differentiate to form a hard keratin as they migrate distally to differentiate into the nail. The growing nail slides along the surface of the nail bed, while remaining firmly attached to it. Nails lost by infection or by amputation of the terminal phalanx distal to the proximal end of the nail bed will be regenerated from the matrix cells.

## REGENERATION OF DENTAL TISSUES

The adult human dentition consists of 32 teeth and associated periodontium, 16 in the upper and 16 in the lower jaw. **FIGURE 3.10** illustrates a section of an adult cat incisor. The bulk of the tooth consists of a central, highly vascularized and innervated pulp of mesenchymal cells surrounded by dentin, a thick, calcified collagenous matrix secreted by odontoblasts, highly differentiated postmitotic mesenchymal cells just under the dentin. Two-thirds of the dentin is embedded in a socket of alveolar jaw bone; this part of the tooth is called the root. Each tooth has multiple roots. Externally, the alveolar sockets are covered by epidermis and underlying connective tissue to form the gum. A thin layer of calcified matrix, the cementum, covers the roots of the teeth. The periodontal ligament connects the alveolar bone and the cementum of the tooth, thus anchoring the tooth into the alveolar socket. The part of the dentin above the gum line is capped with enamel, the hardest substance in the human body, forming the crown of the tooth. Enamel is an organic matrix containing protein, carbohydrate, and calcium phosphate in the form of apatite.

Teeth, like nails, are variants of hair follicles that develop by a combination of epithelial-mesenchymal interactions. During fetal development, the ectoderm over the developing jaw forms a thickening called the dental lamina. In discrete areas of the dental lamina, the epithelium buds downward to form tooth buds (**FIGURE 3.11**). As each bud penetrates into the neural crest-derived mesenchyme below, it forms a bell-shaped enamel organ that encloses a mesenchymal dental papilla. The outer cells of the enamel organ differentiate into the ameloblasts that secrete the enamel of the tooth. The inner cells of the enamel organ differentiate into the odontoblasts that form the dentin of the tooth. The mesenchyme cells of the dental papilla remaining after tooth development constitute the pulp of the tooth.

Mammalian tooth bud formation requires sequential and reciprocal interactions between the dental epithelium and underlying cranial neural crest mesenchyme (Chai and Slavkin, 2003; Zhang et al., 2005). At the prebud stage of development, the dental epithelium is capable of inducing cranial neural crest mesenchyme to form teeth, but the mesenchyme has no ability to induce tooth bud formation from the epithelium. As the tooth bud forms, the dental epithelium loses its inducing power, but the dental papilla mesenchyme gains the ability to instruct nondental epithelium to form tooth buds. The epithelium exerts its inducing activity through production of Fgf8 and 9, Bmp 2, 4, and 7, Shh, and Wnt 10a and b. These signaling molecules activate the production of the transcription factors Pitx 1 and 2 in the epithelium and Pax9, Msx1 and 2, Lef1, Dlx1 and 2, and Gli 1 in the mesenchyme. In turn, these mesenchymal transcription factors activate the genes for signaling molecules, such as Bmp4, Fgf3, activin βA and Wnt 5a, which act back on the epithelium. At the stage of the bell-shaped enamel organ, when the mesenchyme gains its inducing power, a nonproliferating group of epithelial cells called the enamel knot expresses Bmp 2, 4, and 7. The enamel knot acts as a signaling center that controls proliferation and apoptosis and determines the number and positions of tooth cusps. Tooth development is coordinated with the differentiation of alveolar bone, through ligands and receptors of the TNF family (Ohazama et al., 2003). These same factors regulate bone remodeling in general and will be taken up in Chapter 9.

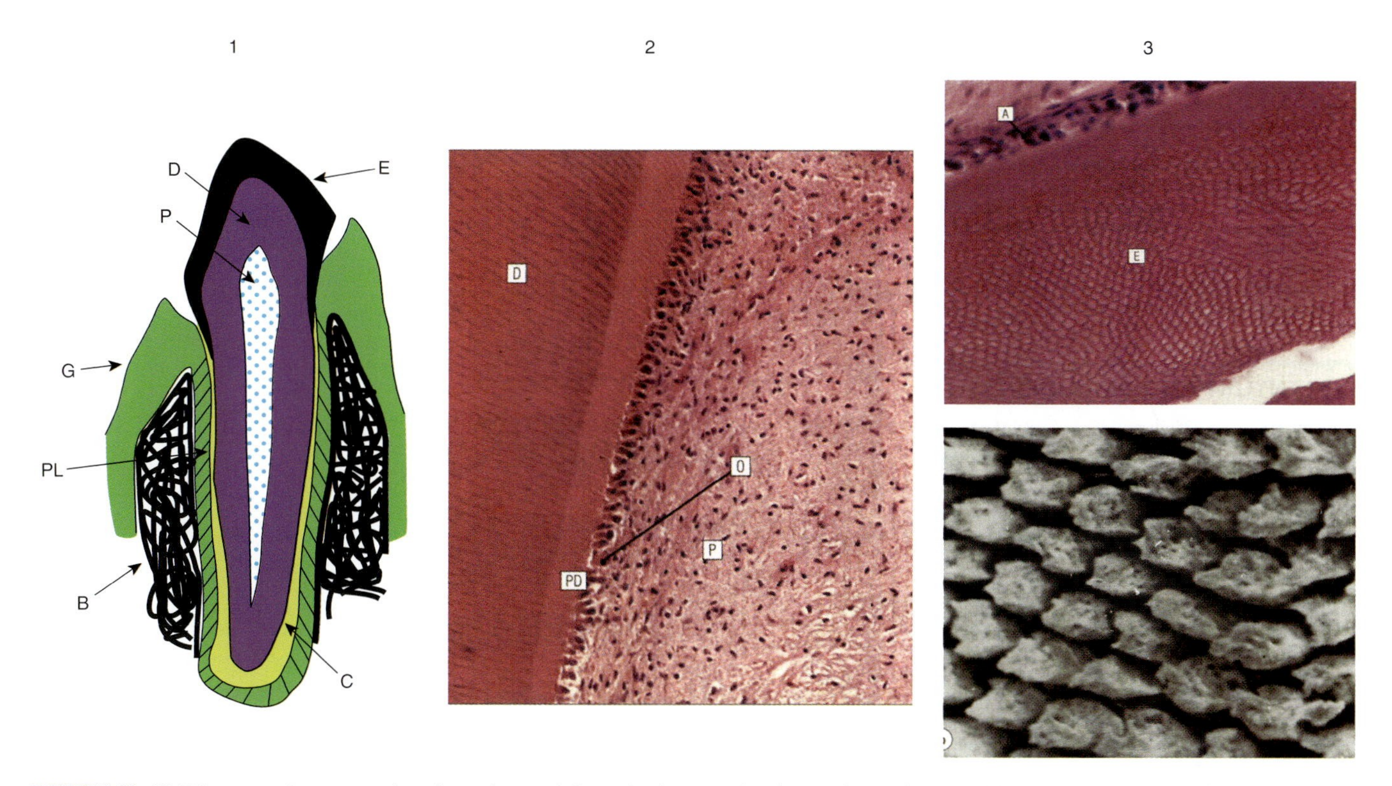


**FIGURE 3.10** **(1)** Diagram of cross-section through an adult cat incisor tooth. The tooth consists of a highly vascularized and innervated pulp (P, blue stipple) surrounded by dentin (D, purple) with a crown capped by enamel (E, black). The part of the tooth below the crown is the root. The tooth is embedded in a socket of alveolar bone (B, black). Cells in the periodontal ligament secrete a coating of cementum (C) onto the tooth root. The PL anchors the tooth into the bone socket by its attachments to the bone and cementum. Oral epidermis and underlying connective tissue constitute the gingiva, or gum of the tooth. **(2)** Section through tooth showing pulp (P) lined externally by a layer of odontoblasts (O). The odontoblasts secrete predentin (PD), which becomes mineralized dentin (D). **(3) Top**: Ameloblasts (A) secreting enamel (E), which is arranged in tightly packed rods. **Bottom**: Scanning electron micrograph showing the packing of enamel rods. **2**, **3** reproduced with permission from Stevens and Lowe, Human Histology (3rd ed), Mosby/Elsevier. Copyright 2005, Elsevier.

In humans, two sets of tooth buds develop during fetal life, representing the primary or deciduous ("baby") teeth and the adult teeth, respectively. The set of adult tooth buds forms later than, but in close proximity to, the primary tooth buds. The primary teeth are shed over a period extending from about age 6 to 12 as they are pushed out by the growth of their corresponding adult teeth.

Adult teeth have some ability to regenerate certain components. Dentin is produced throughout life by the odontoblasts of the pulp to compensate for the erosion of enameled biting surfaces (Linde and Goldberg, 1993). As the enamel thins and the dentin thickens, the pulp cavity gets smaller. The odontoblasts also increase secretion of dentin in response to tooth injury, such as cavity formation and restorative dental procedures, that leaves some dentin intact over the pulp (Linde and Goldberg, 1993; Murray and Garcia-Godoy, 2004). More serious injury or bacterial erosion that completely destroys the dentin and exposes the pulp, destroying the odontoblasts, evokes the regeneration of a "reparative dentin" bridge to cover the pulp. Unless the cavity is successfully decontaminated and filled, reparative dentin regeneration loses out to bacterial erosion, necessitating root canal therapy or extraction of the tooth. Although the source of the cells that produce reparative dentin is controversial (Yamamura, 1985; Goldberg and Lasfargues, 1995), $^{3}$H-thymdine labeling studies (Feit et al., 1970; Fitzgerald, 1979, 1990) suggest that it is formed by mesenchymal stem cells below the odontoblast layer of the pulp. These stem cells, called dental pulp stem cells (DPSCs), have been isolated from adult human teeth (Gronthos et al., 2000). The transcriptional patterns of these cells are very similar to the mesenchymal stem cells of bone marrow (Shi et al., 2001). DPSCs can reconstitute a complete dentin-pulp structure when transplanted subcutaneously into immunocompromised mice.

Another type of stem cell has been isolated from the pulp of normally exfoliated deciduous human incisors (Miura et al., 2003). These cells are called SHED (stem cells from human exfoliated deciduous teeth). They

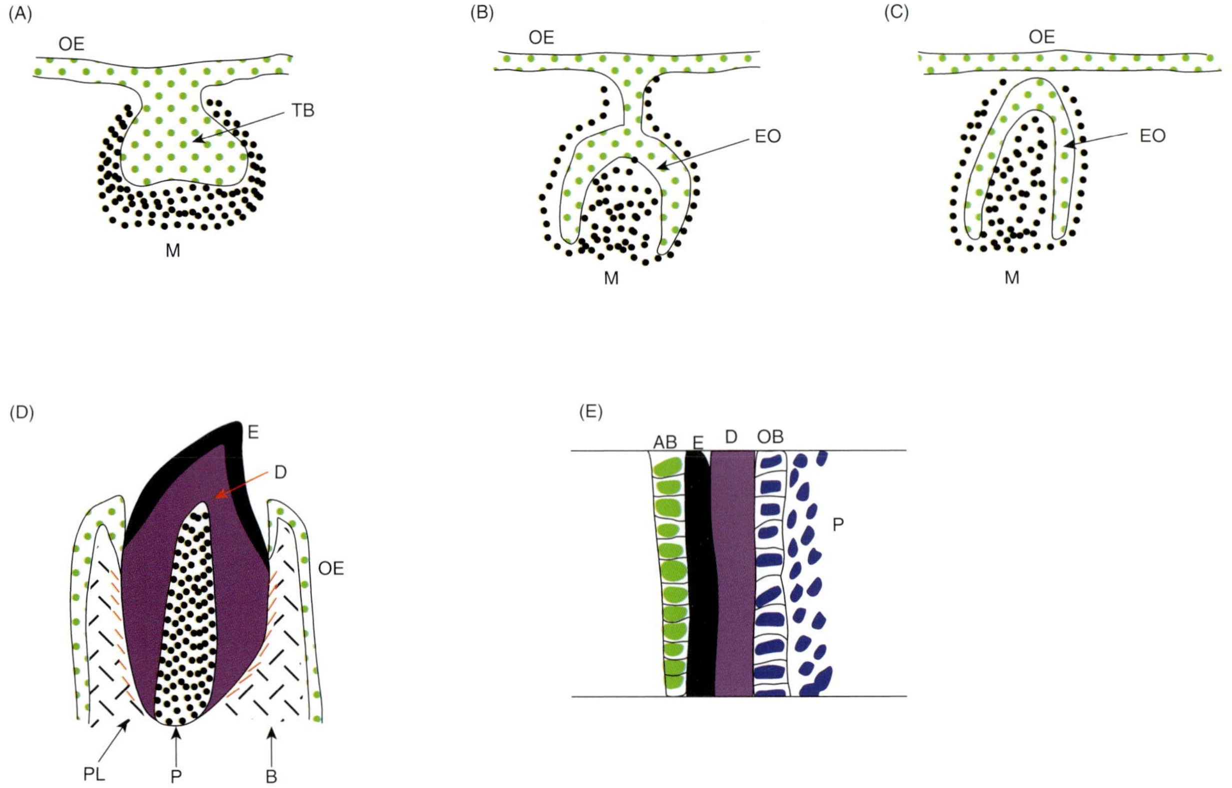

**FIGURE 3.11** Tooth development. **(A)** the oral epthelium (OE) thickens to form a dental lamina, which is induced by underlying mesenchyme (M) to invaginate and form a tooth bud (TB) on a stalk. **(B)** The tooth bud forms a bell-shaped enamel organ (EO) enclosing the inducing mesenchyme and **(C)** detaches from the oral epithelium. **(D)** The enamel organ forms dentin (D) and enamel (E) as the tooth develops attachments to the alveolar bone (B) by the periodontal ligament (PL, red). **(E)** Formation of dentin and enamel. The outer epithelium of the enamel organ contains ameloblasts (AB) that secrete enamel inward. The inner epithelium contains odontoblasts (OB) that secrete dentin outward. As a result, dentin and enamel interface with each other.

express the mesenchymal stem cell markers STRO-1 and CD146, as well as other stromal and vascular-related markers (alkaline phosphatase, MEPE, FGF-2, and endostatin) and are associated with pulp capillaries *in vivo*, suggesting that factors secreted by endothelial cells may play a role in defining the stem cell niche. SHED are distinct from DPSCs in that they have a higher proliferation rate, have osteoconductive capacity *in vivo*, and generate only dentin-like structures in subcutaneous transplant experiments.

Still another stem cell type has been isolated from human periodontal ligament, called periodontal ligament stem cells (PDLDCs; Seo et al., 2004). Like SHED, PDLSCs express STRO-1 and CD146 but differ in their developmental potential. These cells are able to regenerate the periodontal ligament itself, and also cementum and alveolar bone (Kletsas et al., 1998; Sakallioglu et al., 2004).

Adult mammals cannot regenerate lost teeth. Adult urodele amphibians, however, can regenerate lost teeth from rows of reserve tooth buds in the dental lamina (epidermis) medial to the functional teeth (Orban, 1953; Kerr, 1958). Furthermore, adult urodeles can regenerate amputated jaws via dedifferentiation and proliferation of cells at the amputation surface (Goss and Stagg, 1958; Ghosh et al., 1994, see Chapter 14). Functional teeth and multiple rows of reserve tooth buds are regenerated as part of this process. Crocodilians and sharks also replace their teeth throughout life from reserve tooth buds.

## REGENERATION OF THE LENS

### 1. Structure of the Eye and Lens

The general structure of the eye is depicted in **FIGURE 3.12**. The eye consists of three layers, the outer sclera, a tough connective tissue coat that becomes the transparent cornea on the anterior surface of the eye. The middle layer is the heavily pigmented vascular layer, the blood vessels of which nourish the eye. The inner layer

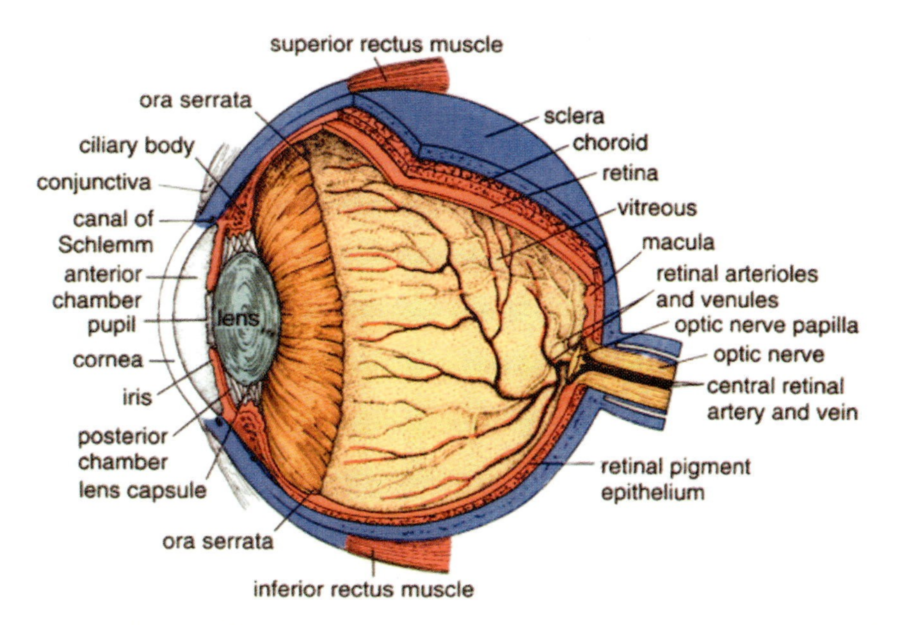


**FIGURE 3.12** Structure of the mammalian eye. The outer connective tissue capsule of the eye is the sclera, which grades into the cornea on the anterior side of the eye. The conjunctiva is a mucus membrane that covers the sclera. The iris (which includes its unseen part, the ora serrata) and lens form a partition between the anterior and posterior chambers of the eye. Muscles in the ciliary body control the configuration of the lens for focus. The anterior chamber is filled with an aqueous solution akin to blood plasma; the posterior chamber is filled with a gel-like vitreous body of hyaluronic acid and a complex protein called vitrein. The iris is a continuation of the pigmented (outer) layer of the retina. The inner layers of the retina are neural layers. The axons of the optic nerve originate in the ganglion layer of the neural retina and converge at the back of the eye, where they exit on their way to the brain. Reproduced with permission, courtesy of Dr. K. W. Condon, http://anatomy.iupui.edu/courses/histo_D502/D502f04/lecture.f04/Eyef04/1.jpg.

is the retina, which is composed of two layers, a thinner pigmented retina, and a thicker neural retina of several cell layers that transduces light photons into electrical signals that pass over the optic nerve to the brain. Toward the anterior part of the eye, the vascular layer thickens to form a muscular ring around the eye called the ciliary body. The vascular layer extends further anteriorly to form the iris with its pupillary opening. The iris contains smooth muscle that controls the diameter of the pupil. Lining the iris posteriorly are two layers of pigmented epithelial cells (PECS) that extend from the pigmented and neural layers of the retina. The lens is a transparent, flexible oblate body suspended just behind the PEC layers in the pupillary opening by ligaments attached to processes of the ciliary body. The suspensory ligaments hold the lens under tension; contraction of smooth muscle in the ciliary body eases this tension, allowing the lens to focus light.

The structure of the adult lens is shown in **FIGURE 3.13**. The lens consists of three components, anucleate transparent lens fibers characterized by their synthesis of α, β, and γ crystallin proteins, an anterior lens epithelium, and a covering capsule that is a thick basement membrane laid down during eye development by lens epithelial cells (LECs). During development, the LECs of the posterior half of the epithelium are the first to differentiate into lens fibers. The LECs of the anterior

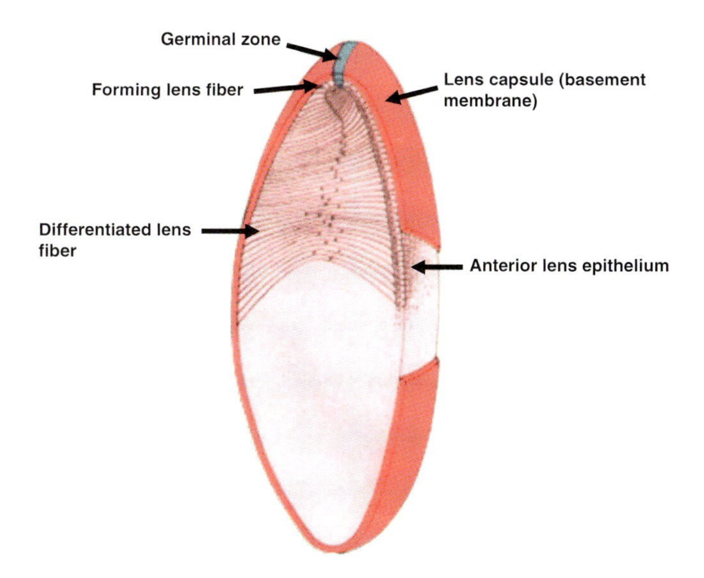


**FIGURE 3.13** Diagram of a slice of lens. The anterior lens epithelium is the source of new lens fibers at the equator in the germinal zone. The epithelial cells of the germinal zone elongate and differentiate into lens fibers that express α, β and γ crystallins. The lens is surrounded by a basement membrane, the lens capsule, which is secreted by the epithelium early in development. Reproduced with permission, courtesy of Dr. K. W. Condon, http://anatomy.iupui.edu/courses/histo_D502/D502f04/lecture.f04/Eyef04/1.jpg.

half of the epithelium persist into adult life. Their proliferation is responsible for the growth of the lens. LECs proliferate throughout the anterior epithelium of the growing lens, but differentiate into new lens fibers only at the equator of the eye.

The embryos of several vertebrate species, including fish, frogs, birds, and rats, as well as the tadpoles of the anuran, *Xenopus laevis*, and the adult newt, can regenerate the lens (Reyer, 1977; Stroeva and Mitashov, 1983; Mitashov, 1996). Lens regeneration is an epimorphic process involving formation of a blastema of dedifferentiated cells. The source of the lens blastema in *Xenopus* tadpoles is the inner corneal epithelium (Freeman, 1963; Carinato et al., 2000). The newt lens blastema is derived by the dedifferentiation of pigmented epithelial cells (PECs) of the dorsal iris (Reyer, 1977; Stroeva and Mitashov, 1983; Mitashov, 1996; Eguchi, 1998; Del-Rio Tsonis et al., 2003; Del-Rio Tsonis and Eguchi, 2004). The capacity for lens regeneration by the newt iris declines from dorsal to ventral. Pieces of ventral iris fail to form lens under experimental conditions in which the dorsal iris readily regenerates lens (Reyer, 1977). The cells of the ventral iris, however, have not lost the capacity for lens regeneration, because they can form lens when dissociated and cultured *in vitro* (Eguchi, 1998).

## 2. Cellular Events of Lens Regeneration in the Newt Eye

**FIGURE 3.14** illustrates the stages of lens regeneration in the newt eye. Two days after lentectomy, there is an increase in the synthesis of ribosomal RNA by the amplification of rRNA genes and their increased rate of transcription (Reese et al., 1969; Collins, 1972). Dedifferentiation and proliferation begin by four days (Yamada and Roesel, 1969, 1971; Reyer, 1971). Expression of the *c-myc* proto-oncogene is enhanced in dedifferentiating PECs and proliferative activity is increased (Agata et al., 1993). The cells take on an irregular shape and eliminate melanosomes, which are phagocytized by macrophages invading the iris epithelium (Reyer, 1982, 1990a, b). The numbers of mitochondria, free ribosomes, and microfilaments increase (Reyer, 1977). Proteoglycans are lost from the surface of the dedifferentiating cells (Zalik and Scott, 1972, 1973).

Eguchi (1988) identified a specific cell surface glycoprotein, 2NI-36, that disappears from the surface of dorsal but not ventral, iris cells. Ventral iris treated with antibody to this antigen and implanted into a lentectomized eye can regenerate a lens. These observations suggest that the disappearance or inhibition of 2NI-36 is necessary and sufficient to trigger dedifferentiation of pigmented epithelial cells. 2NI-36 is also found in many other tissues of the newt and has been postulated to exert a general stabilizing effect on differentiation (Imokawa et al., 1992; Imokawa and Eguchi, 1992).

With continued proliferation, the dedifferentiated dorsal iris cells differentiate into α- and β-crystallin-expressing lens epithelial cells that form a small lens vesicle at the edge of the dorsal iris (Yamada and McDevitt, 1974). The lens epithelial cells facing the retina subsequently withdraw from the cell cycle, express γ-crystallins and differentiate into lens fibers (Takata et al., 1966; Eguchi, 1967). Newts (*Cynops pyrrhogaster)* transgenic for the EGFP gene under the regulation of the βB1-crystallin promoter express EGFP as the lens epithelial cells begin to differentiate (Ueda et al., 2005). The new lens enlarges to normal size by the continuing division of lens epithelial cells and their differentiation into lens fibers.

## 3. Transcription Factors Regulating Lens Regeneration

A number of transcription factors are required for eye development. Among the most important is Pax-6, which is considered a master gene for eye development and is also required for development of the nervous

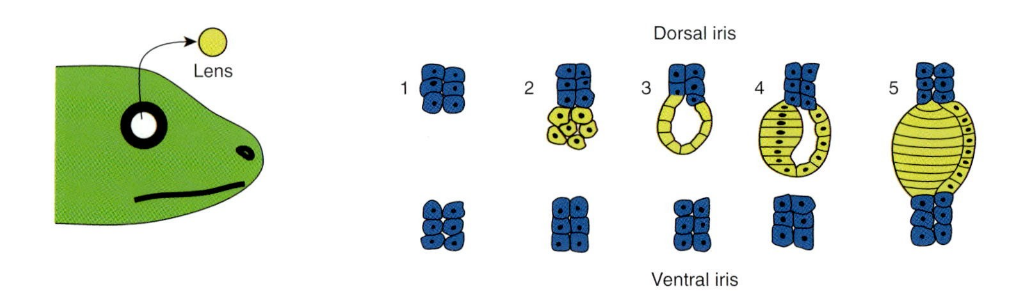


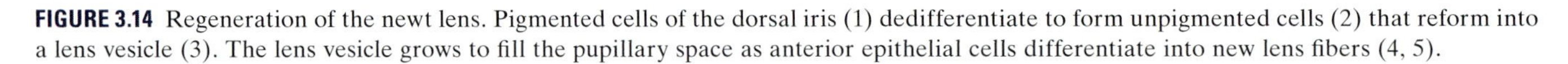
**FIGURE 3.14** Regeneration of the newt lens. Pigmented cells of the dorsal iris (1) dedifferentiate to form unpigmented cells (2) that reform into a lens vesicle (3). The lens vesicle grows to fill the pupillary space as anterior epithelial cells differentiate into new lens fibers (4, 5).

system and the pancreas. Only cells that express Pax-6, Sox-2, and Six-3 can express lens crystallin genes (Cveki et al., 1995; Richardson et al., 1995; Del-Rio Tsonis et al., 1995; Oliver et al., 1995), suggesting that Pax-6 must act in combination with Sox-2 and Six-3 to activate these genes. The *Pax-6* gene is expressed in the intact retina of the adult newt and in the cells of the regenerating lens vesicle at all stages of its development. *Pax-6* expression is also observed in the developing retina and lens of the embryonic and larval axolotl, but declines as the axolotl gets older and becomes unable to regenerate its lens (Del-Rio Tsonis et al., 1995). A number of *Hox* genes appear to be downregulated in both the dorsal and ventral iris during newt lens regeneration (Jung et al., 1998). However, one *Hox* gene, *prox-1*, is upregulated specifically in the dorsal iris after lentectomy (Del-Rio Tsonis et al., 1999).

Since the dorsal iris or cornea does not regenerate lens unless the lens is removed, it has been proposed that the lens produces a signal or signals that normally inhibit iris or corneal cells from dedifferentiating (Reyer, 1977). Such signals have not yet been identified, but they might act through micropthalmic transcription factor (Mitf). *Mitf* encodes a basic helix-loop-helix (HLH)-leucine zipper transcription factor and is active in PECs, the ear and pigmented cells of the skin. The expression patterns of *Pax-6* and *mitf* are complementary in chick embryo pigmented epithelial cells, both *in vivo* and *in vitro*. Overexpression of *Mitf* inhibits Fgf-2-induced transformation of cultured chick pigmented epithelial cells to lens cells and simultaneously inhibits expression of *Pax-6* (Mochii et al., 1998). Although not yet replicated in amphibians, these results suggest that the negative regulation of *Pax-6* by Mitf may be a crucial event in preventing dorsal iris cells from dedifferentiating and participating in lens regeneration.

## 4. Signals Involved in Lens Regeneration

### a. *Thrombin Activated Protein*

As in other wounded tissues, serum clotting factors promote hemostasis when the pigmented epithelium is wounded by lentectomy. A major discovery was that thrombin is required to induce PECs of the newt dorsal iris cells to re-enter the cell cycle (Imokawa and Brockes, 2003). One of the first events in lens regeneration is the selective transient upregulation of active thrombin in the dorsal iris within 20–30 minutes (**FIGURE 3.15**). Inactivation of thrombin by the inhibitors PPACK or antithrombin III resulted in failure of PECs to re-enter the cell cycle. The adult newt and the axolotl both regenerate their limbs, but the axolotl cannot regenerate its lens. Thrombin is upregulated after limb amputation in both newt and axolotl (see Chapter 14), but not after lentectomy in the axolotl, further strengthening the evidence that thrombin plays a key role in initiating newt lens regeneration (Imokawa and Brockes, 2003).

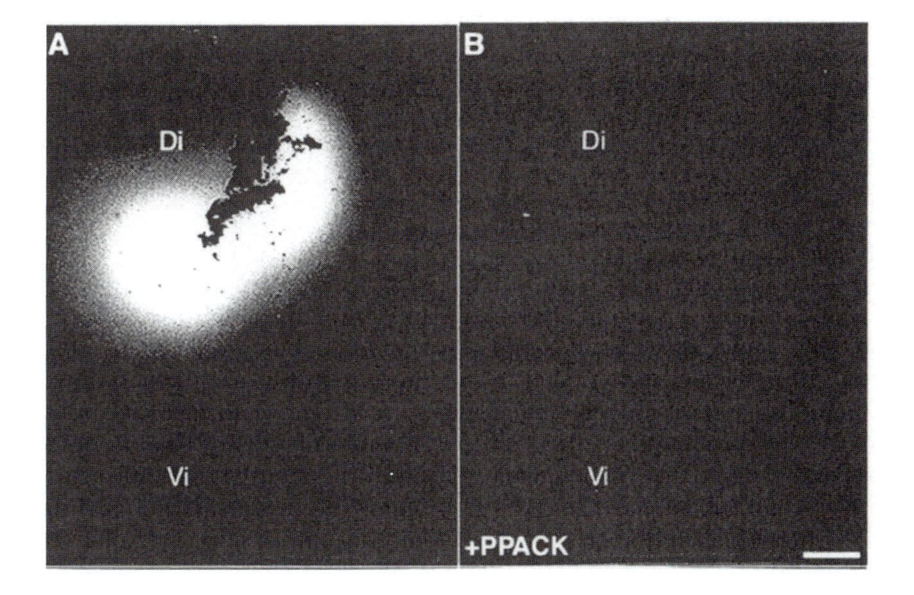


**FIGURE 3.15** **(A)** Expression of thrombin activity in the dorsal iris (Di) but not the ventral iris (Vi) after lentectomy, as visualized by overlaying a section with a fluorogenic thrombin substrate. **(B)** A similar section treated with the thrombin inhibitor PPACK. No reaction is now visible in the dorsal iris. Pigmented epithelial cells fail to re-enter the cell cycle in PPACK-treated lentectomized eyes. Scale bar = 100 μm. Reproduced with permission from Imokawa and Brockes, Selective activation of thrombin is a critical determinant for vertebrate lens regeneration. Curr Biol 13:877–881. Copyright 2003, Elsevier.

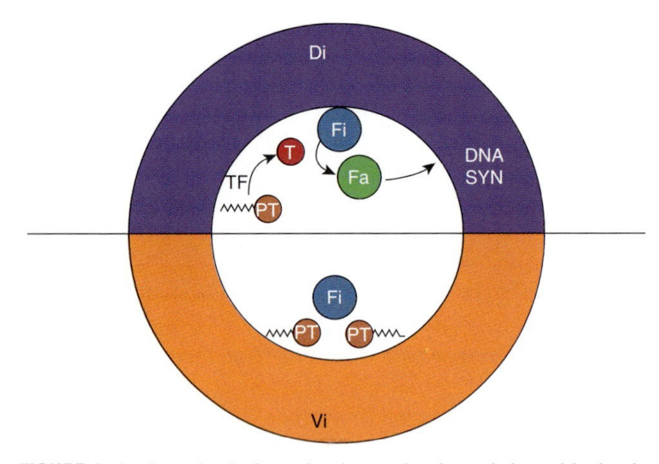


**FIGURE 3.16** Hypothesis for selective activation of thrombin in the newt dorsal iris following lentectomy. Lentectomy releases prothrombin (PT) and an as yet unidentified inactive plasma protein (Fi) in both the dorsal (Di) and ventral (Vi) iris. Tissue factor (TF) is expressed only dorsally, where it acts with clotting factors to convert prothrombin to thrombin (T, red). Thrombin then converts Fi to its active form (Fa). Fa induces DNA synthesis of pigmented epithelial cells in the dorsal iris.

Thrombin itself is produced through the activation of prothrombin by tissue factor. Prothrombin is released from damaged vasculature in the lentectomized newt eye, but there is no marked dorsoventral difference in the vascular density of the iris that could account for the selective activation of prothrombin dorsally. Imokawa and Brockes (2003) have therefore hypothesized that tissue factor is selectively released dorsally; the resulting thrombin then activates a protein that triggers PEC re-entry into the cell cycle (**FIGURE 3.16**), as has been shown for the regenerating newt limb (Tanaka and Brockes, 1999; Chapter 14). This hypothesis remains to be tested.

Cell cycle re-entry and proliferation of dorsal iris PECs appears to involve downregulation of the cell-cycle inhibiting retinoblastoma (Rb) protein, not by hypophosphorylating it as in other cells (Thitoff et al., 2003). As the new lens forms, Rb expression is upregulated again in the differentiating lens fibers, but not in the lens epithelium. The PECs of lentectomized newts immersed in solutions of the cyclin-dependent kinase 2-selective inhibitor, SU9516, were able to dedifferentiate and form a small lens vesicle, but the cells of the vesicle were unable to proliferate to form a new lens (Tsonis et al., 2004). This result suggests that dedifferentiation and cell cycle re-entry by PECs are not mechanistically linked. Two members of the complement system, C3 and C5, have been shown to stimulate mitosis of monocytes and B-cells at low concentration *in vitro* and may play a similar role in the proliferation of cells in the regenerating lens vesicle of the newt (Kimura et al., 2003). C3 is strongly expressed in the stroma and the PECs of the newt iris after lentectomy, and C5 is expressed in the regenerating lens vesicle.

### b. FGFs

Lens regeneration requires a nonspecies-specific interaction of the dorsal iris or cornea with the neural retina (NR). Fragments of newt dorsal iris cultured in the dorsal fin or brain ventricle (Reyer et al., 1973) or isolated *in vitro* (Yamada et al., 1973) do not regenerate lens, although the cells sometimes undergo depigmentation. Newt dorsal iris cultured *in vitro* with frog NR or implanted into limb regeneration blastemas regenerates a lens, whereas newt ventral iris does not (Yamada et al., 1973; Reyer et al., 1973), implying that the same factor(s) required for lens regeneration that are made by the NR are also made by the limb blastema. *Xenopus* corneal tissue will not regenerate lens when isolated *in vitro* but will do so when cultured with neural retina (Bosco et al., 1993). During embryonic development, the optic vesicle exerts a late inductive effect on presumptive lens ectoderm. Both lens and nonlens presumptive ectoderm implanted into the anterior chamber of mature lentectomized *Xenopus* tadpole eyes is stimulated to form lens cells in a fashion resembling the embryonic induction by the optic vesicle (Henry and Mittleman, 1995).

The NR signaling molecule(s) appears to be a member or members of the FGF family. FGF-1 transcripts are expressed in the intact eye and are upregulated in regenerating lens cells of newts (Del-Rio Tsonis et al., 1997). FGF-1 stimulates newt dorsal iris PECs and *Xenopus* corneal cells to transdifferentiate into lens cells *in vitro*, although this transformation is not associated with the formation of a normally organized lens (Bosco et al., 1993, 1997). Exogenous FGF-1, which binds to FGFR-1 and to the KGFR variant of FGFR-2, and FGF-4, which binds to the *bek* variant of FGFR-2, cause structural abnormalities in regenerating newt lenses when applied via implanted beads, a result similar to that seen in the developing lenses of FGF transgenic mice (Del-Rio Tsonis, 1997). *In situ* hybridization experiments showed that transcripts for FGFR-1 are expressed in the neural retina and lens epithelium of the intact adult newt eye (**FIGURE 3.17**). After lentectomy, expression of the FGFR-1 gene remained the same in the retina, but was upregulated during dedifferentiation of PECs and throughout their transdifferentiation into lens cells (Del-Rio Tsonis et al., 1998). Transcripts were also expressed in the ciliary body and the ventral iris, though at a lower level than in the dorsal iris. A similar pattern of expression of FGFR-2 and FGFR-3 was observed in the intact eye and regenerating lens (Del-Rio Tsonis et al., 1997). Despite the widespread expression of FGFR-1 transcripts, immu-

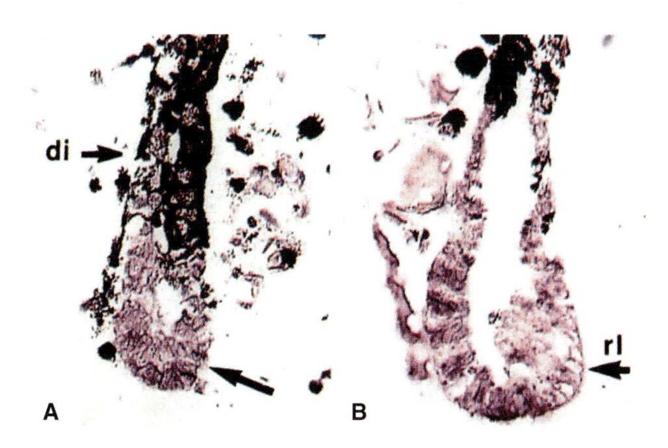


**FIGURE 3.17** Expression of FGFR-1 mRNA (purple) in the regenerating newt lens. **(A)** expression in the lens vesicle (long arrow) derived from the edge of the dorsal iris (di) 10 days post-lentectomy. **(B)**, expression at 15 days post-lentectomy. Lens fibers are differentiating from the anterior epithelium of the regenerating lens (rl). Reproduced with permission from Del-Rio Tsonis et al., Regulation of lens regeneration by fibroblast growth factor receptor 1. Dev Dynam 213:140–146. Copyright 1998, Wiley-Liss.

nostaining showed that the FGFR-1 protein was confined to the dedifferentiating cells of the newt dorsal iris that form the lens, indicating that expression of the gene is regulated at the level of translation, as well as transcription (Del-Rio Tsonis et al., 1998). A further indication of the central importance of FGFR-1 (and FGF-1) was the inhibition of lens regeneration by SU5402, a 3-substituted indolin-2-one that specifically inhibits the autophosphorylation, and thus activation, of this receptor (Del-Rio Tsonis et al., 1998). The protein expression patterns of FGFR-2 and 3, as well as FGFR-1, were also examined by immunoblotting (McDevitt et al., 1997). They reported no regional, temporal, qualitative, or quantitative differences for FGFR-1 and -2 expression. A higher level of FGFR-3, however, was observed in intact dorsal iris and in dorsal iris cells as they engaged in lens regeneration.

FGF-2 was reported to be unable to effect transdifferentiation of newt PECs *in vivo* (Del-Rio Tsonis et al., 1995), although it could induce the transdifferentiation of chick PECs to lens cells *in vitro* (Mochii et al., 1998). However, Hayashi et al. (2004) reported that a single injection of FGF-2 into the intact newt eye induced the expression of *Pax6* and *Sox2* genes in both dorsal and ventral PECs, but development of a second lens only by PECs of the dorsal iris. Furthermore, FGF-2 protein accumulated in the iris after lentectomy and injection of a soluble form of FGF receptor to titrate FGF2 inhibited all molecular and morphological changes in PECs that are associated with lens regeneration. These results suggest that FGF-2 plays an important role in triggering and maintaining lens regeneration, and that the failure of ventral iris to regenerate lens is due to a block in a later phase of the lens developmental pathway. FGF-2 and FGF receptors are normally expressed in the retina of birds and mammals (Noji et al., 1990; Park and Hollenberg, 1993; Tsonis and Del-Rio Tsonis, 2004), but neither the expression pattern of FGF-1 or FGF-2 has been reported in the newt eye after lentectomy.

### c. Hedgehog Proteins

Other signaling molecules that may be important in lens regeneration are the hedgehog proteins and their receptors, Patched 1 and 2. In the developing eye, Sonic hedgehog (Shh) is expressed in the ganglion cell layer of the retina. Indian hedgehog (Ihh) is expressed exclusively in PECS, and desert hedgehog (Dhh) is expressed in PECs and the neural retina. The patched receptors are also expressed in these tissues. Perturbing the Shh signal transduction pathway in various ways (mutations, inhibitors, overexpression) results in eye defects (Sasagawa et al., 2002; Perron et al., 2003; Tsonis et al., 2004). The patched receptors are found in the lens and in the dorsal and ventral iris of the intact eye in the adult newt. Neither Ihh nor Shh are found in the adult lens, nor is Shh found in the dorsal or ventral iris of the intact eye. However, Ihh is expressed in both dorsal and ventral iris. During lens regeneration, *shh* and *ihh* transcripts are found in both the dorsal and ventral iris and in the lens vesicle (Tsonis et al., 2004). Inhibition of the hedgehog pathway by implanting beads impregnated with KAAD (3-keto, N-amino-ethyl aminocaproyl dihydrocinnamoyl) or pellets of mammalian cells transiently expressing hedgehog interacting protein (HIP) prevent lens regeneration completely or allow the formation of only a small vesicle that does not undergo fiber differentiation. Currently, the exact function of hedgehog proteins in lens regeneration is unknown, but given their role in stem cell renewal and proliferation (Chapter 1), it is likely that they play a similar role in lens regeneration.

### d. Retinoic Acid

Retinoic acid (RA), the acid derivative of vitamin A (MW = ~300) is a small molecule critical to eye development and lens regeneration. RA binds to two sets of nuclear receptors, the retinoic acid receptors (RARs) and the retinoid X receptors (RXRs). The RARs have three isoforms, α, β and γ (δ in the urodele amphibian). This binding activates the receptors, which are transcription factors that bind to retinoic acid response elements (RAREs) on the 5′ side of target genes. Eye development is highly dependent on the activation by RA of the αB crystallin gene through its nuclear receptors, specifically the RAR-δ isoform of RAR (Tsonis et al., 2000). *RAR*-δ transcripts are expressed at low levels in the ganglion cell layer of the retina, but not elsewhere in the intact eye. After lentectomy, the *RAR*-δ gene is expressed in the dedifferentiated PECs that form the new lens vesicle, and expression increases to its highest level during new lens fiber differentiation (Tsonis et al., 2000).

The activation of the α B crystalline gene by RA signaling is also dependent on Pax-6, as shown by the fact that in mice mutated for *Pax-6*, RA signaling in the eye is decreased and the developing lens does not respond to exogenous RA (Enwright and Grainger, 2000). Inhibition of RA synthesis by disulfram, or inhibition of RAR function by the RAR antagonist 193109, results in the inhibition, retardation, or abnormal morphogenesis of the regenerating lens, although in some cases ectopic lenses regenerate from the ventral iris or the cornea (Tsonis et al., 2000). Collectively, these observations suggest that RARs and Pax-6 act together in promoting the differentiation of dedifferentiated dorsal iris cells to lens cells. However, elsewhere in the eye (ventral iris) RARs may function to inhibit ectopic lens regeneration.

### 5. Mammals Have Some Capacity for Lens Regeneration

Rabbits, cats, and mice can regenerate an imperfect lens if the lens is removed from the lens capsule, leaving the capsule behind (Gwon et al., 1990; Call et al., 2004). A new lens is formed by the proliferation and differentiation of residual lens epithelium cells that remain adherent to the lens capsule. Interestingly, in human cataract surgery to replace the lens with a plastic lens, the posterior, and part of the anterior, lens capsule is left behind to hold the artificial lens. Residual lens epithelial cells adhering to the anterior part of the capsule sometimes undergo an epithelial to mesenchymal transformation and proliferate across the whole posterior lens surface, where they essentially form new cataracts, a process called posterior capsule opacification (PCO) (Wormstone et al., 2001). In these cases, the surgeon uses a laser to ablate the proliferating cells and the posterior capsule. PCO happens in other mammals as well, but is diminished considerably by 20 days postlentectomy (Call et al., 2004).

If residual lens epithelial cells in human lens capsules could be directed to proliferate and differentiate into proper lens fibers, it might be possible to simply regenerate a lens, rather than implant an artificial one. A culture system to explore this possibility has been devised in which residual lens epithelial cells proliferate within the capsules (Wormstone et al., 2001). FGF-2, FGFR1, Pax-6, and Six-3 have been shown to be expressed in these cultures. Inhibition of FGFR1 by the specific antagonist SU5402 retarded proliferation of the epithelial cells.

Mammals do not regenerate a lens from PECs of the dorsal iris. Neverthless, mammalian PECs have the ability to form lenses under the right conditions. Cells of the human H80HrPE-6 dedifferentiated PEC line, derived from an 80-year-old man, formed clear lentoid-like aggregates that expressed crystallins after four days of culture on Matrigel (Eguchi, 1998; Tsonis et al., 2001). Thus, if all the factors involved in lens regeneration from the dorsal iris in newts become known, it may be possible to intervene to regenerate a lens from the dorsal iris in humans.

## REGENERATION OF THE CORNEA

The cornea is continuous with the tough connective tissue sheath of the eye, the sclera. The cornea consists of a transparent, avascular stroma of flattened fibroblasts sandwiched between layers of collagen that are covered anteriorly by an epithelium several layers thick and posteriorly by a single layer of endothelial cells, Descemet's endothelium (**FIGURE 3.18**). The anterior epithelium produces a basement membrane that lies on another acellular layer produced by the stromal cells called Bowman's membrane. Bowman's membrane constitutes a protective barrier that is resistant to

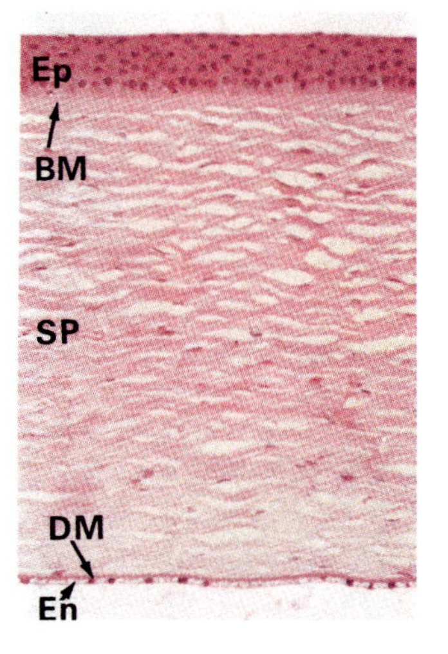


**FIGURE 3.18** Histological structure of the human cornea (H&E stain). The clear (non-keratinized), outer corneal epithelium rests on its basement membrane (BM, Bowman's membrane). Transparent connective tissue layers constitute the substantia propria (SP), which forms the bulk of the cornea. The posterior side of the cornea is lined with an endothelium that synthesizes another basement membrane, Descemet's membrane. The cornea is avascular and nourishment of its cells is via diffusion from the aqueous humor and limbal blood vessels. Reproduced with permission from Wheater et al., Wheater's Functional Histology (3rd ed), Copyright 1997, Elsevier.

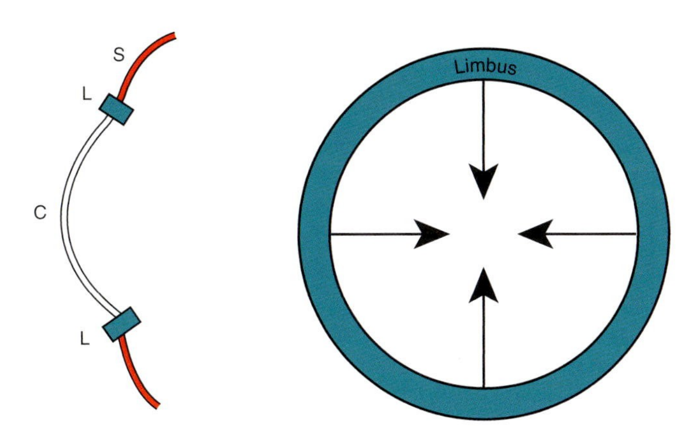


**FIGURE 3.19** Diagrams of corneal regeneration. **Left**, saggital section of the eye showing the location of the limbus (L, boxes) at the junction of the cornea (C) and sclera (S). **Right**, external view of limbus surrounding the cornea and feed of TA cells toward the center of the corneal epithelium.

trauma and bacterial invasion. The corneal epithelium is very important to vision and must be kept wet by the blinking of the eyelids and tear flow. The layer of endothelium produces a basement membrane termed Descemet's membrane (Ham and Cormack, 1979). Corneas can be allografted with considerable success because their avascularity and large amounts of ECM hides their antigenicity from the immune system.

Corneal epithelium undergoes both maintenance and injury-induced regeneration (FIGURE 3.19). Long-term labeling studies with $^3$H-thymidine showed that regeneration is accomplished by slow-cycling basal epithelial stem cells located in the limbus of the cornea, the region where the cornea undergoes transition into the sclera (Cotsarelis et al., 1989). These limbal stem cells have extensive proliferative capacity *in vitro* (Lavker et al., 1998; Pellegrini et al., 1999). They give rise to proliferating transit amplifying cells that migrate toward the center of the cornea to replace cells that turn over or are damaged by injury (Buck, 1985; Auran et al., 1995; Tseng and Sun, 1999). Corneal epithelium will not regenerate in the absence of the limbic region (Huang and Tseng, 1991). Transplants of limbic stem cells will rescue severely damaged or depleted corneal epithelium (Kenyon and Tseng, 1989; Tsai et al., 1990; Tsubota et al., 1995; Holland and Schwartz, 1996; Pellegrini et al., 1997; see Chapter 4).

## SUMMARY

The epidermis, hair, nails, periodontium, and cornea are all epithelial structures that undergo maintenance and injury-induced regeneration in mammals through the activation of resident adult stem cells. The lens is an epithelial structure that regenerates imperfectly in some adult mammals from residual lens epithelial cells after lentectomy. The adult newt regenerates the lens perfectly from dedifferentiated cells of the dorsal iris.

The stem cells that regenerate the epidermis are located in the stratum basale. Integrin-bright, Delta-expressing stem cells self renew while giving rise to transit amplifying progeny that detach from the basal lamina anddifferentiate into keratinocytes as they move upward to the stratum corneum. The basal epidermal cells are continuous with the hair follicle basal cells of the outer root sheath. The stem cells of the hair follicle are located in a special region called the bulge, which feeds transit-amplifying cells upward toward the surface of the epidermis where they differentiate into epidermal keratinocytes, and downward to the matrix of the hair follicle where they proliferate and differentiate as the hair shaft. Hairs go through a three-stage maintenance cycle of catagen (follicle regression), telogen (follicular rest), and anagen (regeneration of the follicle and new hair growth). This cycle is regulated by growth factor signals from the dermal papilla at the base of the hair follicle. Transcriptional repression by Lef1 maintains the stem cells of the epidermis and hair follicles in a quiescent state. The cells are activated to proliferate by Wnt signaling, which stabilizes β-catenin, allowing it to complex with Lef1 to form a transcriptional activating complex.

Injury makes a gap in the epidermis that is filled in by the division of basal stem cells at the edges of the wound that form a sheet of cells migrating laterally through the provisional fibrin matrix of the wound. Adherence to the substrate is mediated by Fn, Tn, and integrin receptors. Migration is initiated by TGF-α and EGF produced by macrophages and division at the edges of the wound is promoted by KGF and GM-CSF by fibroblasts of the granulation tissue. Once the wound is recovered, these same factors promote vertical division to thicken the epidermis. The actual trigger for basal cell proliferation may be the binding of epiregulin fragments on the apical cell surface to EGF receptors on the basolateral surface. Normally, tight junctions separate the apical and basolateral domains, but injury allows them to interact. Stem cells of the hair follicle bulge have also been shown to make a significant contribution to the injury-induced regeneration of interfollicular epidermis.

Nails and dental tissues are variants of hair follicles. The nail regenerates directly from a matrix of transient-amplifying cells located at the proximal end of the nail bed, but the actual stem cells may be located further distally in a region under the nail bed that is equivalent to the bulge of the hair follicle. Teeth contain two types of stem cells in the pulp. One type is the odontoblasts located just under the dentin, which produce dentin throughout life to compensate for the erosion of enameled biting surfaces. The other is a fibroblastic stem cell below the ontoblast layer that is activated to produce a "reparative dentin" when the pulp is exposed. In addition, stem cells are located in the periodontal ligament. These cells continually maintain the ligament, which is under constant stress, but they can also regenerate injured alveolar bone. Although adult mammals cannot regenerate lost teeth, adult urodele amphibians and crocodilians can do so, and therefore may be good research models to understand what they have and we don't, and how we might induce mammalian teeth to regenerate.

Lens regeneration has been studied extensively in the adult newt. After lentectomy, PECs of the dorsal iris appear to dedifferentiate, proliferate, and differentiate into lens epithelial cells and then into lens fibers. The transcription factor Mitf has been implicated in the production of a signal by the lens that normally inhibits PECs from dedifferentiation. Removal of the lens releases the PECs from this inhibition. The initial event of lens regeneration is the activation of prothrombin, perhaps by selective release of tissue factor dorsally. The resulting thrombin then activates an unidentified protein that initiates cell-cycle re-entry. Re-entry of PECs into the cell cycle is associated with downregulation of the cell cycle-inhibiting Rb protein and the complement factors C3 and C5 may promote proliferation of PECs. Several signaling proteins regulate the proliferation and differentiation of PECs into lens cells. Lens regeneration requires an interaction of the dorsal iris PECs with the neural retina through the production of FGF1 by the neural retina. *Indian hedgehog* is expressed exclusively in PECs of the uninjured eye and *Ihh* and *Shh* are expressed in dedifferentiated PECs, along with the *RAR*-δ gene. Interference with the signal transduction pathways for FGF1 or Hedgehog prevents lens regeneration completely or allows only the formation of small, undifferentiated lens vesicle. RA is a very important signaling molecule for lens fiber differentiation that, through the RAR-δ receptor, acts together with the transcription factor Pax-6 to activate the αB crystallin gene.

The corneal epithelium is regenerated continuously or after injury by adult stem cells located in the limbus, the region where the cornea undergoes a transition into the sclera of the eye.

## REFERENCES


Adams C, Watt FM (1990) Changes in keratinocyte adhesion during terminal differentiation reduction in fibronectin binding precedes $alpha_5beta_1$ integrin loss from the cell surface. Cell 63: 425–435.

Akiyama M, Dale BA, Sun TT, Holbrook KA (1995) Characterization of hair follicle bulge in human fetal skin: The human fetal bulge is a pool of undifferentiated keratinocytes. J Invest Dermatol 105:844–850.

Argyris TS (1976) Kinetics of epidermal production during epidermal regeneration following abrasion in mice. Am J Path 83: 329–337.

Auran JD, Koester CJ, Kleiman NJ, Rapaport R, Bomann JS, Wirotsko BM, Florakis GJ, Koniarek JP (1995) Scanning slit confocal microscopic observations of cell morphology and movement within the normal human anterior cornea. Opthalmology 102: 33–41.

Barrandon Y, Green H (1987) Three clonal types of Keratinocyte with different capacities for multielication. Proc Natl Acad Sci USA 84:2302–2306.

Blanpain C, Lowry WE, Geoghegan A, Polal L, Fuchs E (2004) Self-renewal, multipotency and the existence of two cell populations within an epithelial stem cell niche. Cell 118:635–648.

Bosco L, Valle C, Willems D (1993) *In vivo* and *in vitro* experimental analysis of lens regeneration in larval *Xenopus laevis*. Dev Growth Diff 35:257–270.

Bosco L, Venturini G, Willems D (1997) *In vitro* lens transdifferentiation of *Xenopus laevis* outer cornea induced by Fibroblast Growth Factor (FGF). Development 124:421–428.

Brock J, Midwinter K, Lewis J, Martin P (1996) Healing of incisional wounds in the embryonic chick wing bud: Characterization of the actin purse-string and demo nstration of a requirement for Rho activation. J Cell Biol 135:1097–1107.

Buck RC (1985) Measurement of centripetal migration of normal corneal epithelial cells in the mouse. Invest Opthalmol Visual Sci 26:1296–1299.

Call MK, Grogg MW, Del-Rio Tsonis K, Tsonis PA (2004) Lens regeneration in mice: Implications in cataracts. Exp Eye Res 78:297–299.

Carinato M, Walter BE, Henry JJ (2000) *Xenopus laevis* gelatinase B (Xmmp-9): development, regeneration, and wound healing. Dev Dynam 217:377–387.

Chai Y, Slavkin HC (2003) Prospects for tooth regeneration in the 21st century: A perspective. Microsc Res 60:469–479.

Christiano AM (2004) Epithelial stem cells: Stepping out of their niche. Cell 118:530–532.

Collins JM (1972) Amplification of ribosomal ribonucleic acid cistrons in the regenerating lens of Triturus. Biochemistry 11:1259–1263.

Cotsarelis G, Cheng SZ, Dong G, Sun TT, Lavker RM (1989) Existence of slow-cycling limbal epithelial basal cels that can be preferentially stimulated to proliferate: implications on epithelial stem cells. Cell 57:201–209.

Cotsarelis G, Sun TT, Lavker RM (1990) Label-retaining cells reside in the bulge area of the pilosebaceous unit: Implications for follicular stem cells, hair cycle, and skin carcinogenesis. Cell 61:1329–1337.

Cvekl A, Sax C, Li X, McDermott J, Piatogorski J (1995) Pax-6 and lens-specific transcription of the chicken delta1-crystallin gene. Proc Natl Acad Sci USA 92:4681–4685.

Das Gupta R, Fuchs E (1999) Multiple roles for activated LEF/TCF transcription complexes during hair follicle development and differentiation. Development 126:4557–4568.

Del-Rio Tsonis K, Eguchi G (2004) Lens Regeneration. In: Lovicu FJ, Robinson ML (eds.). Cambridge, Cambridge University Press, pp 290–311.

Del-Rio Tsonis K, Washabaugh CH, Tsonis PA (1995) Expression of *pax-6* during urodele eye development and lens regeneration. Proc Natl Acad Sci USA 92:5092–5096.

Del-Rio Tsonis K, Jung JC, Chiu I-M, Tsonis PA (1997) Conservation of fibroblast growth factor function in lens regeneration. Proc Natl Acad Sci USA 94:13701–13706.

Del-Rio Tsonis K, Trombley MT, McMahon G, Tsonis PA (1998) Regulation of lens regeneration by fibroblast growth factor receptor 1. Dev Dynam 213:140–146.

Del-Rio Tsonis K, Tomarev SI, Tsonis PA (1999) Regulation of Prox 1 during lens regeneration. Invest Opthalmol Visual Sci 40: 2039–2045.

Del-Rio Tsonis K, Tsonis PA (2003) Eye regeneration at the molecular age. Dev Dynam 226:211–224.

Draper BK, Komurasaki T, Davidson MK, Naney B (2003) Topical epiregulin enhances repair of murine excisional wounds. Wound Rep Reg 11:188–197.

DuCros DL, LeBaron RG, Couchman JR (1995) Association of versican with dermal matrices and its potential role in hair follicle development and cycling. J Invest Dermatol 105:426–431.

Eguchi G (1967) *In vitro* analyses of of Wolffian lens regeneration: Differentiation of the regenerating lens rudiment of the newt, *Triturus pyrrhogaster*. Embryologia 9:246–266.

Eguchi G (1988) Cellular and molecular background of Wolffian lens regeneration. In: Eguchi G, Okada TS, Saxen L (eds.). Regulatory Mechanisms in Developmental Processes. Cell Diff Dev 25 (Suppl):147–158.

Eguchi G (1998) Transdifferentiation as the basis of eye lens regeneration. In: Cellular and molecular basis of regeneration. Ferretti P, Geraudie J (eds.). John Wiley & Sons, New York, pp 207–229.

Elias A, Zheng T, Einarsson O, Landry M, Trow T, Rebert N, Panuska J (1994) Epithelial interleukin-11. Regulation by cytokines, respiratory syncytial virus and retinoic acid. J Biol Chem 269:22261–22268.

Enwright JF, Grainger RM (2000) Altered retinoid signaling in the heads of small eye mouse embryos. Dev Biol. 221:10–22.

Fenteany G, Janmey PA, Stossel TP (2000) Signaling pathways and cell mechanics involved in wound closure by epithelial cell sheets. Curr Biol 10:831–838.

Fleckman P, Allan C (2001) Surgical anatomy of the nail unit. Dermatol Surg 27:257–260.

Freeman G (1963) Lens regeneration from cornea in *Xenopus laevis*. J Exp Zool. 154:39–65.

Gat U, Dasgupta R, Degenstein L, Fuchs E (1998) *De novo* hair follicle morphogenesis and hair tumors in mice expressing a truncated beta-catenin in skin. Cell 95:605–614.

Ghalbzouri A, Ponec M (2004) Diffusable factors released by fibroblasts support epidermal morphogenesis and deposition of basement membrane components. Wound Rep Reg 12:359–367.

Ghosh S, Thorogood P, Ferretti P (1994) Regenerative capability of upper and lower jaws in urodele amphibians. Int Dev Biol 38: 479–490.

Goss RJ, Stagg MW (1958) Regeneration of lower jaws in adult newts. J Morph 102:289–310.

Gronthos S, Mankani M, Brahim J, Robey P, Shi S (2000) Postnatal human dental pulp stem cells (DPSCs) *in vitro* and *in vivo*. Proc Natl Acad Sci USA 97:13625–13630.

Gwon A, Gruber LJ, Mantras CJ (1993) Restoring lens capsule integrity enhances lens regeneration in New Zealand albino rabbits and cats. Cataract Refract Surg 19:735–746.

Ham AW (1944) Experimental study of histopathology of burns, with particular reference to sites of fluid loss in burns of different depths. Ann Surg 120:689–692.

Ham AW, Cormack DH (1979) Histology, 8th ed, Philadelphia, JB Lippincott, pp 614–644.

Hammersen F (1980) Histology, 2nd ed. Baltimore, Urban and Schwarzenberg.

Hardy MH (1992) The secret life of the hair follicle. Trends Genet 8:55–61.

Hayashi T, Mizuno N, Ueda Y, Okamoto M, Kondoh H (2004) FGF2 triggers iris-derived lens regeneration in newt eye. Mech Dev 121:519–526.

Hebert JM, Rosenquist T, Gotz J, Martin GR (1994) FGF5 as a regulator of the hair growth cycle: Evidence from targeted and spontaneous mutations. Cell 78:1017–1025.

Henry JJ, Mittleman JM (1995) The matured eye of *Xenopus laevis* tadpoles produces factors that elicit a lens-forming response in embryonic ectoderm. Dev Biol 171:39–50.

Holland E, Schwartz G (1996) The evolution of epithelial transplantation for severe ocular surface disease and a proposed classification system. Cornea 15:549–556.

Honda H, Tanemura M, Imayama S (1996) Spontaneous architectural organization of mammalian epidermis from random cell packing. J Invest Dermatol 106:312–315.

Horne KA, Jahoda CAB, Oliver RF (1986) Whisker growth induced by implantation of cultured vibrissa dermal papilla cells in the adult rat. J Embryol Exp Morph 97:111–124.

Hotchin NA, Gandarillas A, Watt FM (1995) Regulation of cell surface beta$_1$ Integrin levels during keratinocyte terminal differentiation. J Cell Biol 128:1209–1219.

Huang AJW, Tseng SCG (1991) Corneal epithelial wound healing in the absence of limbal epithelium. Invest Oththalmol Visual Sci 32:96–105.

Hubner G, Hu Q, Smola H, Werner S (1996) Strong induction of activin expression after injury suggests an important role of activin in wound repair. Dev Biol 173:490–498.

Imokawa Y, Eguchi G (1992) Expression and distribution of regeneration-responsive molecule during normal development of the newt, *Cynops pyrrhogaster*. Intl J Dev Biol 36:407–412.

Imokawa Y, Ono S-I, Takeuchi T, Eguchi G (1992) Analysis of a unique molecule responsible for regeneration and stabilization of differentiated state of tissue cells. Intl J Dev Biol 36:399–405.

Imokawa Y, Brockes JP (2003) Selective activation of thrombin is a critical determinant for vertebrate lens regeneration. Curr Biol 13:877–881.

Inamatsu M, Matsuzaki T, Iwanari H, Yoshizato K (1998) Establishment of rat dermal papilla cell lines that sustain the potency to induce hair follicles from afollicular skin. J Invest Dermatol 111:767–775.

Jacinto A, Woolner S, Martin P (2002) Dynamic analysis of dorsal closure in *Drosophila*: from genetics to biology. Dev Cell 3: 9–19.

Jensen AM, Wallace VA (1997) Expression of Sonic hedgehog and its putative role as a precursor cell mitogen in the developing mouse retina. Development 124:363–371.

Jensen UB, Lowell S, Watt F (1999) The spatial relationship between stem cells and their progeny in the basal layer of human epidermis: a new view based on whole mount labeling and lineage analysis. Development 126:2409–2418.

Jindo T, Tsuboi R, Imai R, Takamori K, Rubin JS, Ogawa H (1994) Hepatocyte growth factor/scatter factor stimulates hair growth of mouse vibrissae in organ culture. J Invest Dermatol 103: 306–309.

Jones PH, Watt FM (1993) Separation of human epidermal stem cells from transit amplifying cells on the basis of differences in integrin function and expression. Cell 73:713–724.

Jones PH, Harper S, Watt FM (1995) Stem cell patterning and fate in human epidermis. Cell 80:83–93.

Jung JC, Del-Rio Tsonis K, Tsonis PA (1998) Regulation of homeobox-containing genes during lens regeneration. Exp Eye Res 66:361–370.

Keefe JR (1973) An analysis of urodelan retinal regeneration: I. Studies of the cellular source of retinal regeneration in *Notopthalmus viridescens* utilizing $^3$H-thymidine and colchicines. J Exp Zool 184:185–206.

Keefe JR (1973) An analysis of urodelan retinal regeneration: II. Ultrastructural features of retinal regeneration in *Notophthalmus viridescens*. J Exp Zool 184:207–232.

Keefe, JR (1973) An analysis of urodelan retinal regeneration: III. Degradation of extruded melanin granules in *Notopthalmus viridescens*. J Exp Zool 184:233–238.

Kenyon KR, Tseng SC (1989) Limbal autograft transplantation for ocular surface disorders. Opthalmology 96:709–722.

Kerr T (1958) Development and structure of some actinopterygian and urodele teeth. Proc Zool Soc Lond 133:401–424.

Khavari P (2004) Profiling epithelial stem cells. Nature Biotech 22:393–394.

Kimura Y, Madhavan M, Call MK, Santiago W, Tsonis PA, Lambris JD, Del-Rio Tsonis K (2003) Expression of complement 3 and complement 5 in newt limb and lens regeneration. J Immunol 170:2331–2339.

Kletsas D, Basda E, Papavassiliou A (1998) Mechanical stress induces DNA synthesis in PDL fibroblasts by a mechanism unrelated to autocrine growth factor action. FEBS Lett 430:358–362.

Kobayashi K, Nishimura E (1989) Ectopic growth of mouse whiskers from implanted lengths of plucked vibrissa follicles. J Invest Dermatol 92:278–282.

Kobayashi K, Rochat A, Barrandon Y (1993) Segregation of keratinocyte colony-forming cells in the bulge of the rat vibrissa. Proc Natl Acad Sci USA 90:7391–7395.

Kratz G, Compton CC (1997) Tissue expression of transforming growth factor-β1 and transforming growth factor-α during wound healing in human skin explants. Wound Rep Reg 5:222–228.

Lachgar, Moukadiri, Jonca F, Charveron M, Bouhaddioui N, Gall Y, Bonafe JL, Plouet J (1996) Vascular endothelial growth factor is an autocrine growth factor for hair dermal papilla cells. J Invest Dermatol 106:17–23.

Lavker RM, Wei Z-G, Sun T-T (1998) Phorbol ester preferentially stimulates mouse fornical conjunctival and limbal epithelial stem cells to proliferate in vivo. Invest Opthalmol Visual Sci 39:101–107.

Lavker RM, Bertolino AP, Freedberg IM, Sun T-T (1999) Biology of hair follicles. In: Freedberg IM, Eisen Z, Wolff K (eds.). Fitzpatrick's Dermatology in General Medicine, 5th Ed. New York, McGraw Hill, pp 230–238.

Lowell S, Jones P, Le Roux I, Dunne J, Watt FM (2000) Stimulation of human epidermal differentiation by Delta-Notch signaling at the boundaries of stem-cell clusters. Curr Biol 10:491–500.

Martin P (1997) Wound healing—aiming for perfect skin regeneration. Science 276:75–81.

Martin P, D'Souza D, Martin J, Grose R, Cooper L, Maki R, McKercher SR (2003) Wound healing in the PU.1 mouse—tissue repair is not dependent on inflammatory cells. Curr Biol 13:1122–1128.

Matsuzaki T, Inamatsu M, Yoshizato K (1996) The upper dermal sheath has a potential to regenerate the hair in the rat follicular epidermis. Differentiation 60:287–297.

Matsuzaki T, Yoshizato K (1998) Role of hair papilla cells on induction and regeneration processes of hair follicles. Wound Rep Reg 6:524–530.

McCluskey J, Martin P (1995) Analysis of the tissue movements of embryonic wound healing—DiI studies in the limb bud stage mouse embryo. Dev Biol 170:102–114.

McDevitt DS, Brahma SK, Courtois Y, Jeanny J-C (1997) Fibroblast growth factor receptors and regeneration of the eye lens. Dev Dynam 208:220–226.

Messenger AG, Ellit K, Westgate GE, Gibson WT (1991) Distribution of extracellular matrix molecules in human hair follicles. Ann NY Acad Sci 642:253–262.

Messenger AG (1993) The control of hair growth: an overview. J Invest Dermatol (Suppl) 101:4S–9S.

Millar SE, Willert K, Salinas PC, Roelink H, Nusse R, Sussman DJ, Barsh GS (1999) WNT signaling in the control of hair growth and structure. Dev Biol 207:133–149.

Miller EJ, Gay S (1992) Collagen structure and function. In: Cohen IK, Diegelmann RF, Lindblad WJ (eds.). Wound Healing: Biochemical and Clinical Aspects. Philadelphia, WB Saunders, pp 195–208.

Miura M, Gronthos S, Zhao M, Lu B, Fisher LW, Robey PG, Shi S (2002) SHED: Stem cells from human exfoliated deciduous teeth. Proc Natl Acad Sci USA 100:5807–5812.

Mochii M, Mazaki Y, Mizuno N, Hayashi H, Eguchi G (1998) Role of Mitf in differentiation and transdifferentiation of chicken pigmented epithelial cell. Dev Biol 193:47–62.

Morris RJ, Potten CS (1999) Highly persistent label-retaining cells in the hair follicles of mice and their fate following induction of anagen. J Invest Dermatol 112:470–475.

Morris RJ, Liu Y, Marles L, Yang Z, Trempus C, Li S, Lin JS, Sawicki JA, Cotsarelis G (2004) Capturing and profiling adult hair follicle stem cells. Nature Biotechnol 22: 411–417.

Mostov K, Zegers M (2003) Just mix and match. Nature 422: 267–268.

Narisawa Y, Hashimoto K, Kohda H (1994) Merkel cells of the terminal hair follicle of the adult human scalp. J Invest Dermatol 102:506–510.

Nishimura EK, Jordan SA, Oshima H, Yoshida H, Osawa M, Moriyama M, Jackson IJ, Barrandon Y, Miyachi Y, Nishikawa S-I (2002) Dominant role of the niche in melanocyte stem-cell fate determination. Nature 416:854–860.

Ohazama A, Courtney J-M, Sharpe PT (2003) *Opg*, *Rank*, and *Rankl* in tooth development: Co-ordination of odontogenesis and osteogenesis. J Dent Res 83:241–244.

Oliver G, Mailhos A, Wehr R, Copeland NG, Jenkins NA, Gruss P (1995) six-3, a murine homologue of the sine oculis gene, demarcates the most anterior border of the developing neural plate and is expressed during eye development. Development 121:3959–3967.

Orban B (1953) Oral Histology and Embryology. St Louis, CV Mosby.

Pellegrini G, Golisano O, Paterna P, Lambaise A, Bonini S, Rama P, De Luca M (1999) Location and clonal analysis of stem cells and their differentiated progeny in the human ocular surface. J Cell Biol 145:769–782.

Perron M, Boy S, Amato MA, Viczian A, Koebernick K, Pieler WA, Harris WA (2003) A novel function for hedgehog signaling in retinal pigment epithelial differentiation. Development 130: 1565–1577.

Pierce GF (1991) Tissue repair and growth factors. In: Encyclopedia of Human Biology. Dulbecco R (ed.). New York, Academic Press, pp 499–509.

Potten CS (1974) The epidermal prolferative unit: The possible role of the central basal cell. Cell Tiss Kinet 7:77–88.

Potten CS, Morris RJ (1988) Epithelial stem cells *in vivo*. J Cell Sci (Suppl) 10:45–62.

Reese DH, Puccia E, Yamada T (1969) Activation of ribosomal RNA synthesis in initiation of Wolffian lens regeneration. J Exp Zool 170:259–268.

Repesh LA, Fitzgerald TJ, Furcht LT (1982) Fibronectin involvement in granulation tissue and wound healing in rabbits. J Histochem Cytochem 30:351–358.

Reyer RW (1971) DNA synthesis and the incorporation of labeled iris cells into the lens during lens regeneration in adult newts. Dev Biol 24:533–558.

Reyer RV, Woolfitt RA, Withersty LT (1973) Stimulation of lens regeneration from the newt dorsal iris when implanted into the blastema of the regenerating limb. Dev Biol 32:258–281.

Reyer RW (1977) The amphibian eye: development and regeneration. In: Crescitelli F (ed.). Handbook of Sensory Physiology, Vol. VII/5. Berlin, Springer-Verlag pp 309–390.

Reyer RW (1982) Dedifferentiation of iris epithelium during lens regeneration in newt larvae. Am J Anat 163:1–23.

Reyer RW (1990) Macrophage invasion and phagocytic activity during lens regeneration from the iris epithelium in newts. Am J Anat 188:329–344.

Reyer RW (1990) Macrophage mobilization and morphology during lens regeneration from the iris epithelium in newts: studies with correlated scanning and transmission electron microscopy. Am J Anat 188:345–365.

Reynolds AJ, Jahoda CAB (1996) Hair matrix germinative epidermal cels confer follicle-inducing capabilities on dermal sheath and high passage papilla cells. Development 122:3085–3094.

Richardson J, Cvekl A, Wistow G (1995) Pax-6 is essential for lens-specific expression of zeta-crystallin. Proc Natl Acad Sci USA 92:4676–4680.

Rochat A, Kobayashi K, Barrandon Y (1994) Location of stem cells of human hair follicles by clonal analysis. Cell 76:1063–1073.

Ruiz S, Segrelles C, Santos M, Lara MF, Paramio JM (2004) Functional link between retinoblastoma family of proteins and the Wnt signaling pathway in mouse epidermis. Dev Dynam 230:410–418.

Sakallioglu U, Acikgoz G, Ayas B, Kirtiloglu T, Sakalliuglu E (2004) Healing of periodontal defects treated with enamel matrix proteins and root-surface conditioning—an experimental study in dogs. Biomaterials 25:1831–1840.

Sasagawa S, Takabatake, Takabatake Y, Muramatsu T, Takeshima K (2002) Axes establishment during eye morphogenesis in *Xenopus* by coordinate and antagonistic actions of BMP4, Shh and RA. Genesis 33:86–96.

Seery JP, Watt FM (2000) Asymmetric stem-cell divisions define the architecture of human oesophageal epithelium. Curr Biol 10: 1447–1450.

Seiber-Blum M, Grim M, Hu YF, Szeder V (2004) Pluripotent neural crest stem cells in the adult hair follicle. Dev Dynam 231:258–269.

Seiberg M, Marthinuss J (1995) Clusterin expression within skin correlates with hair growth. Dev Dynam 202:294–301.

Seo B-M, Miura M, Gronthos S, Bartold PM, Batouli S, Brahim J, Young M, Robey PG, Wang C-U, Shi S (2004) Investigation of multipotent postnatal stem cells from human periodontal ligament. The Lancet 364:149–155.

Shi S, Robey PG, Grnthos S (2001) Comparison of human dental pulp and bone marrow stromal cells by cDN microarray analysis. Bone 29:532–539.

Shimaoka S, Imai R, Ogawa H (1994) Dermal papilla cells express hepatocyte growth factor. J Dermatol Sci 7 (Suppl):S79–83.

Strocva OG, Mitashov VI (1983) Retinal pigment epithelium: proliferation and differentiation during development and regeneration. Intl Rev Cytol 83:221–293.

Takata C, Albright JF, Yamada T (1964) Lens antigens in a lens-regenerating system studied by the immunofluorescent technique. Dev Biol 9:385–397.

Tanaka EM, Dreschel N, Brockes JP (1999) Thrombin regulates S phase re-entry by cultured newt myoblasts. Curr Biol 9:792–799.

Taylor G, Lehrer MS, Jensen PJ, Sun T-T, Lavker RM (2000) Involvement of follicular stem cells in forming not only the follicle but also the epidermis. Cell 102:451–461.

Thitoff AR, Call MK, Del-Rio Tsonis, Tsonis PA (2003) Unique expression patterns of the retinoblastoma (*Rb*) gene in intact and lens regeneration—undergoing newt eyes. Anat Rec 271A:185–188.

Tsai RJ, Li LM, Chen JK (2000) Reconstruction of damaged corneas by transplantation of autologous limbal epithelial cells. New Eng J Med 343:86–93.

Tseng SCG, Sun T-T (1999) Stem cells: Ocular surface maintenance. In: Brightbill FS (ed.). Corneal Surgery: Theory, Technique, Tissue. St Louis, Mosby, pp 9–18.

Tsonis PA, Trombley MT, Rowland T, Chandraratna RAS, Del-Rio Tsonis K (2000) Role of retinoic acid in lens regeneration. Dev Dynam 219:588–593.

Tsonis PA, Jang W, Del-Rio Tsonis K, Eguchi G (2001) A unique aged human retinal pigmented epithelial cell line useful for studying lens differentiation *in vitro*. Intl J Dev Biol 45:753–758.

Tsonis PA, Tsavaris M, Call MK, Chandraratna RAS, Del-Rio Tsonis K (2002) Expression and role of retinoic acid receptor alpha in lens regeneration. Dev Growth Diff 44:391–394.

Tsonis PA, Vergara MN, Spence JR, Madhavan M, Kramer EL, Call MK, Santiago WG, Vallance JE, Robbins DJ, De-Rio Tsonis K (2004) A novel role of the hedgehog pathway in lens regeneration. Dev Biol 267:450–461.

Tsonis PA, Madhavan M, Call MK, Gainer S, Tice A, Del-Rio Tsonis K (2004) Effect of a CDK inhibitor on lens regeneration. Wound Rep Reg 12:24–29.

Tsubota K, Toda, Saito H, Shinozaki N, Shimazaki J (1995) Reconstruction of the corneal epithelium by limbal autograft transplantation for corneal surface reconstruction. Opthalmology 102:1486–1496.

Tumbar T, Guasch G, Greco V, Blanpain C, Lowry WE, Rendl M, Fuchs E (2004) Defining the epithelial stem cell niche in skin. Science 303:359–363.

Ueda Y, Kondoh H, Mizuno N (2005) Generation of transgenic newt *Cynops pyrrhogaster* for regeneration study. Genesis 41:87–98.

Vermeer PD, Einwalter LA, Moninger TO, Rokhlina T, Kern JA, Zabner J, Welsh MJ (2003) Segregation of receptor and ligand regulates activation of epithelial growth factor receptor. Nature 422:322–326.

Werner S, Smola H, Liao X, Longaker MT, Krieg T, Hofschneider PH, Williams LT (1994) The function of KGF in morphogenesis of epithelium and reepithelialization of wounds. Science 266: 819–822.

Whitby DJ, Ferguson MWJ (1991) The extracellular matrix of lip wounds in fetal, neonatal and adult mice. Development: 112: 651–668.

Woodley DT (1996) Reepithelialization. In: Clark, RAF (ed.). The Molecular and Cellular Biology of Wound Repair, 2nd (ed.). New York, Plenum Press, pp 339–354.

Wormstone IM, Del-Rio Tsonis K, McMahon G, Tamiya S, Davies PD, Marcantonio JM, Duncan G (2001) FGF: An autocrine regulator of human lens cell growth independent of added stimuli. Invest Opthalmol Visual Sci 42:1305–1311.

Yamada T, Roesel M (1969) Activation of DNA replication in the iris epithelium by lens removal. J Exp Zool 171:425–432.

Yamada Y, Roesel M (1971) Control of mitotic activity in Wolffian lens regeneration. J Exp Zool 177:119–128.

Yamada T, Dumont JN (1972) Macrophage activity in Wolffian lens regeneration. J Morph 136:367–384.

Yamada T, Reese DH, McDevitt DS (1973) Transformation of iris into lens in vitro and its dependency on neural retina. Differentiation 1:65–82.

Yamada T, McDevitt DS (1974) Direct evidence for transformation of differentiated iris epithelial cells into lens cells. Dev Biol 38:104–118.

Zalik SE, Scott V (1972) Cell surface changes during dedifferentiation in the metaplastic transformation or iris into lens. J Cell Biol 55:134–146.

Zalik SE, Scott V (1973) Sequential disappearance of cell surface components during dedifferentiation in lens regeneraion. Nature (New Biol) 244:212–214.

Zhang YD, Chen Z, Song YQ, Liu C, Chen YP (2005) Making a tooth: Growth factors, transcription factors, and stem cells. Cell Res 15:301–316.

Zhu AJ, Watt FM (1999) Beta catenin signaling modulates proliferative potential of human epidermal keratinocytes independently of intercellular adhesion. Development 127:2285–2298.

# 4 Regenerative Medicine of Skin, Hair, Dental Tissues, and Cornea

## INTRODUCTION

Clinical and/or experimental regenerative therapies exist for mammalian skin, hair, dental tissues, and cornea, but not for lens.

Acute (normally healing) incisional and excisional skin wounds account for over 4.5 million medical procedures per year in the United States (Langer and Vacanti, 1993). Among the most severe excisional wounds are burns, which account for nearly 10,000 deaths each year in the United States. Burn survivors can suffer physical disability and depression due to scar tissue contracture and the poor cosmetic appearance of the healed tissue. Chronic (nonhealing) wounds are a major health problem the world over. They are primarily ulcers resulting from compromised circulation due to pressure, diabetes, and venous incompetence. Chronic wounds are characterized by a failure to re-epithelialize and by defective production and remodeling of ECM (Mast and Schultz, 1996; Mustoe, 2004). Therapies that accelerate normal skin repair and production of stronger scar tissue are desirable because they would lead to quicker recovery from burns and surgical operations, particularly in older individuals. Therapies that heal chronic wounds would save tissue that might otherwise be lost to necrosis and infection and, in some cases, save the life of the patient.

Although skin has received the most attention, hair follicles, teeth, and cornea are also tissues for which regenerative therapies would be invaluable. Although not life-threatening or even compromising function, male pattern baldness is a source of great distress to many men. Thus, ways are being sought to stimulate the regeneration of hair in men with this condition. Loss of teeth to periodontal disease or injury is treated with bridges or implants. Useful as they are, these substitutes for teeth do not have the structural strength of natural teeth and are expensive to make and install. Thus, dental researchers envision the ability to regenerate missing teeth. Maintaining the transparency of the cornea is absolutely essential to vision. Injuries to or diseases of the cornea that compromise the capacity for maintenance regeneration lead to corneal opaqueness and loss of vision. Thus, therapies have been developed to restore the capacity for maintenance regeneration.

## REPAIR OF SKIN

Deep incisions are repaired by stitching the wound edges together, resulting in a thin line of scar tissue on the surface of the skin. Deep excisional wounds require more extensive intervention to avoid major scarring. The gold standard treatment for a deep excisional skin wound of limited area is a full-thickness skin autograft (epidermis plus both layers of dermis). Full-thickness grafts give the best structural and cosmetic results because there is minimal contraction and scarring at the edges of graft and host skin. Furthermore, they take well because they contain a vascular plexus that can quickly establish connections with the vasculature at the wound site. The large area of harvested skin required to cover extensive excisional wounds, however, makes full-thickness skin grafts impractical because healing of the donor area is compromised.

To increase the area that can be covered by a skin autograft, surgeons use meshed split-thickness skin grafts (MSTSGs). A small area of partial-thickness donor skin consisting of epidermis and the papillary layer of the dermis is shaved off with a dermatome and the skin is meshed so it can be stretched out to cover a much wider wound area. The results obtained with MSTSGs, while saving lives, are often not satisfactory. Their vascular plexus is not as extensive as that of a full-thickness graft. Thus, in wounds where there has been extensive destruction of the reticular dermis, autografted MSTSGs can suffer weak attachment and epidermal blistering, reducing the percentage of grafts that take. Even in successful grafts, the healed skin is cosmetically unattractive due to excessive scar formation and contraction in the interstices of the mesh.

A major goal in treating normal and chronic skin wounds is to either accelerate their healing by fibrosis or to actually regenerate skin with the original architecture and function, using interventions based on what is known about the biology of skin repair. These interventions are the topical application of growth factors, cytokines, inhibitors of inflammation, and other compounds to accelerate fibrosis or regenerate skin, the use of keratinocyte transplants or bioartificial skin equivalents as living wound dressings until the host skin is repaired, or the use of acellular regeneration templates to induce skin repair from the sides and bottom of the wound.

## 1. Acceleration of Acute Wound Repair by Topically Applied Agents

A wide variety of topical agents have been employed to accelerate the normal repair process and produce stronger scar tissue in acute wounds of normal skin. Application of healing accelerants as early as possible after wounding is desirable, because of the cascade effects that carry through to later stages of healing. **TABLE 4.1** summarizes some of the many topical agents that have been used to accelerate the repair of wounded normal skin. What is usually measured in animal experiments is how rapidly the diameter or area of a wound decreases with respect to controls. This measure is correlated to biochemical and structural parameters of repair such as collagen synthesis and accumulation and histological appearance of repair tissue.

**TABLE 4.1** Agents That Have Been Found, Singly or in Combination, to Accelerate the Repair of Normal Skin Wounds When Applied Topically

| Growth Factors |
|---|
| EGF |
| FGF-2 |
| TGF-β |
| Human GH |
| **Other Accelerants** |
| Extract of *C. argentia* leaf |
| Vanadate |
| Oxandrolone |
| Fentanyl |
| Pig enamel matrix derivative |
| Ketanserin |
| Angiotensin |
| Oleic n-9 fatty acids |
| HB-107 cecropin B peptide |
| **Antiscarring Agents** |
| Chitosan |
| N-O carboxylmethyl chitosan |
| Celicoxib |
| Anti-TGF-β1, 2 antibodies |
| Hyaluronic acid/chondroitin sulfate hydrogel |
| **Debriding Agents** |
| Vibriolysin |
| Papain/urea |

### a. Growth Factors

Growth factors known to have a role in the early phases of wound repair have been applied topically to excisional wounds of experimental animals and patients in order to accelerate their healing (reviewed by Fu et al., 2005). Recombinant human EGF and FGF-2 (rhEGF, rhFGF-2) were reported to accelerate the healing of burn wounds in human patients by 1–4 days (Brown et al., 1989; Fu et al., 1998). TGF-β and human growth factor (hGF) were reported to accelerate normal wound repair in both young and old rats (Puolakkainen et al., 1995; Roberts, 1995). Topical application of TGF-β increased the rate of re-epithelialization and contraction in rat and pig incisional wounds and in guinea pig and pig punch wounds (Franzen et al., 1995). Collagen synthesis was increased, leading to increased strength of scar tissue (Roberts, 1995). Intravenous delivery of 100–500 μg/kg of TGF-β to old rats prior to wounding or 4 hr after wounding increased the tensile strength of scar tissue to the level seen in the untreated wounds of young rats. This effect was considerably diminished if the TGF-β was delivered 48 hr after wounding, suggesting that the beneficial effect of the growth factor is through the attraction of neutrophils, macrophages, and fibroblasts early in the repair process. Estrogen levels affect the levels of TGF-β in females and thus wound repair (Ashcroft et al., 1997). Older women have reduced rates of healing and lower levels of scarring, both of which are associated with reduced levels of TGF-β and are reversed by hormone replacement therapy. Ovariectomized young female rats exhibit delayed wound repair that is reversed by estrogen administration, which results in an increase of TGF-β secretion by dermal fibroblasts.

Recombinant human growth hormone (rhGH) has also been shown to increase the mechanical strength of granulation tissue in incisional skin wounds of rats up to 94% that of unwounded skin (Jorgensen and Oxlund, 1996). The increase was dose-dependent and was associated with a nearly 150% increase in the deposition rate of collagen in the wound. Local application of rhGH in subcutaneous wound chambers, or systemic administration of the hormone, stimulated the formation of granulation tissue (Steenfos and Jansson, 1992). Human growth hormone was reported to accelerate healing of donor skin sites in burned children (Herndon et al., 1990; Gilpin et al., 1994). The effects of rhGH may be both local and systemic. Growth hormone stimulates the production of IGF-I by the liver, elevating the serum level of IGF-I, which as a cell cycle competence factor would then stimulate fibroblast proliferation (Jorgensen and Oxlund, 1996). rhGH may also have a direct effect on fibroblast proliferation, since receptors for the hormone are present in skin fibroblasts (Lobie et al., 1990).

Some studies have obtained less positive results (reviewed by Greenalgh, 1996). FGF-2 was reported to have no effect on the rate of re-epithelialization, wound closure, or scarring at donor sites of partial-thickness skin grafts in burned children (Greenalgh and Rieman, 1994) and Cohen et al. (1995) found no effect of EGF on the rate of epithelial closure. On balance, the evidence would suggest that TGF-β, FGF-2, EGF, GH, and IGF-I are able to accelerate the repair of acute wounds in normal skin (Fu et al., 2005).

### b. Other Agents

Agents not known to be part of the normal wound repair pathways have also been found to enhance skin repair in a variety of ways.

Several of these agents appear to work by enhancing the formation of granulation tissue. Extract of the *Celosia argentea* leaf, which has antimicrobial properties and is used in traditional medicine to treat skin conditions, has also been reported to accelerate normal wound repair in a rat burn model (Priya et al., 2004). Wound closure took place in half the time for controls, and there was a more rapid increase in collagen and hexosamine content of the granulation tissue. Studies *in vitro* indicated that the enhanced wound healing was due to enhancement of motility and proliferation of the fibroblasts. Vanadate was reported to optimize the organization of collagen fibers in developing granulation tissue (Mackay et al., 2003). The anabolic steroid, oxandrolone, decreased time to wound closure, and produced a more cellular granulation tissue with more mature and densely packed collagen fibers (Demling, 2000).

Increased angiogenesis has been associated with accelerated wound repair by several agents. Applied for one week, opioids, particularly fentanyl, accelerated the healing of excisional wounds by 3–4-fold (Poonawala et al., 2005). There was a 1.5–2.5-fold increase in cellular density of granulation tissue and a 45%–87% increase in angiogenesis that was associated with upregulation of nitric oxide synthetase (NOS) and the VEGF receptor Flk1. Acceleration was inhibited by the opioid receptor antagonist naloxone, suggesting that the effect was mediated by peripheral opioid receptors. Pig enamel matrix derivative (EMD) is a composite of amelogenin proteins and is used topically to promote regeneration of connective tissues damaged in periodontal surgery (Heijl et al., 1997; Yukna and Mellonig, 2000; Wennstrom and Lindhe, 2002). EMD significantly accelerated wound closure in full-thickness excisional wounds of rabbits. The effect was postulated to be on angiogenesis because the production of VEGF and MMP-2 is enhanced in cultured fibroblasts treated with EMD (Mirastschijski et al., 2004). Ketanserin, a serotonin receptor blocker used in the treatment of essential hypertension, accelerated angiogenesis and re-epithelialization of full-thickness skin wounds in the ears of nude mice in a dose-dependent fashion (Kim et al., 1995) (**TABLE 4.2**). The effect was considered to be directly on endothelial cells rather than indirectly through change in the amount of blood flow, since there was no effect on terminal arteriole diameter. Angiotensin (1-7), a nonhypertensive fragment of angiotensin II, accelerated dermal healing in guinea pig burn wounds by increasing both angiogenesis and keratinocyte proliferation in hair follicles (Rodgers et al., 2001).

**TABLE 4.2** 20% ketanserin (K) accelerates epithelialization and angiogenesis in circular wounds (2.25 mm diameter, 0.125 mm depth) made in the dorsal ear skin of male nude mice. Numbers are the surface area of the wound within the advancing epithelial and vascular regenerative fronts, expressed as percent of the initial wound area. Data from Kim et al. (1995)

| | *Epithelial Front* | | | *Vascular Front* | | |
|---|---|---|---|---|---|---|
| | 8 days | 14 days | 16 days | 11 days | 20 days | 28 days |
| Control | 42 | 3 | 0 | 75 | 35 | 5 |
| 20% K | 8 | 0 | 0 | 60 | 8 | 0 |

Oleic n-9 fatty acids induced more rapid skin wound closure in mice (Cardoso et al., 2004), as did HB-107 (Lee et al., 2004), a peptide lacking bactericidal activity derived from cecropin B, an antimicrobial (gram negative bacteria) peptide found in the moth *Hyalophora cecropia* (Steiner et al., 1981). The fact that HB-107 promotes wound repair suggests that antimicrobial peptides can enhance wound repair not only by bactericidal action, but also through other pathways (Lee et al., 2004).

Debriding agents are also beneficial to normal wound repair. Pig partial-thickness burn wounds exhibited a significant dose-dependent stimulation of granulation tissue formation when vibriolysin was used to digest the dessicated eschar of the wounds (Nanney et al., 1995). Repair of pig full-thickness incisional wounds and partial-thickness burn wounds was improved by the topical application of a papain/urea debriding combination (Hebda et al., 1998). Papain/urea proved substantially better in this regard than other commonly used debriding agents such as papain/urea/chlorophyllin copper complex, bacterial collagenase, and fibrolysis/DNAse.

### c. Reduction of Scarring by Modulating the Inflammatory Response

A number of topical interventions that reduce the inflammatory response have been tried in an attempt to reduce scarring in adult skin wounds, particularly to reduce the presence of TGF-β1 and 2, or create a more fetal-like ECM.

Hepatocyte growth factor (HGF) is an important growth factor in liver regeneration (Chapter 7). HGF has angiogenic, angioprotective, anti-inflammatory and antifibrotic activities (Matsumoto and Nakamura, 1996), but its expression has not been studied in healing skin wounds. However, injecting rat incisional skin wounds with a combination of rhFGF-2 protein (Fiblast™) and a plasmid expressing the HGF gene resulted in elevated fibroblast apoptosis in the granulation tissue and less extensive scarring than observed with either agent alone (Ono et al., 2004). The mechanism by which the antifibrotic effect was achieved is not clear, but the rhFGF-2 may have had an apoptotic effect and the HGF an anti-inflammatory effect at early stages of repair.

Chitosan, a high-molecular-weight, positively charged polysaccharide extracted from the chitin of crab shells, has a significant positive effect on the repair of subcutaneous wounds in rats (Diegelman et al., 1996). Chitosan prolongs the presence of neutrophils and delays the appearance of macrophages, thus reducing TGF-β1 and 2 production, capillary ingrowth, fibroblast migration, and collagen deposition. The collagen produced was in the form of fine reticulin-like fibrils rather than the mature bands of dense collagen seen in controls. N-O carboxymethylchitosan (N-O-CMC, a GAG hydrogel derivative of chitin) has ECM-like properties that prevent or minimize fibrosis and adhesions when applied topically to the injured caecum of rats (Krause et al., 1996). The mechanisms of action of chitosan and N-O-CMC are unclear. Chitosan has blood clotting activity that is independent of the normal platelet-dependent cascade (Malette et al., 1983) and might modulate platelet function, thus altering the inflammatory phase of repair. N-O-CMC is hydrophilic and would have low affinity for fibronectin, thus preventing hydrophilic interactions between the molecules involved in the formation of adhesions.

Reduction of TGF-β1 in wounds reduces scarring. The enzyme cyclooxygenase-2 (COX-2) catalyzes the conversion of arachadonic acid to prostaglandins during the inflammatory response. Prostaglandins induce collagen production in adult wounds (Talwar et al., 1996), and $PGE_2$ induces scar formation in fetal wounds (Wilgus et al., 2003). Daily treatment of excisional wounds in mice with the COX-2 inhibitor, celicoxib, decreased $PGE_2$ by 50% and TGF-β1 by one-third at 48-hr postwounding (Wilgus et al., 2003). These decreases were associated with a later reduction in scar tissue formation without disrupting re-epithelialization or decreasing tensile strength of the repair tissue. The mean collagen content and scar width was only half that of controls.

Studies *in vivo* and *in vitro* have shown that reducing the levels of TGF-β1 by topical application of neutralizing antibodies, or the application of TGF-β3 immediately after wounding reduces scarring in skin wounds (**FIGURE 4.1**) (Shah et al., 1994; Houghton et al., 1995). Anti-TGF-β1 antibodies are currently being developed commercially as agents to reduce scarring during skin wound repair (Ferguson and O'Kane, 2004).

Biointeractive hydrogel films composed of cross-linked hyaluronic acid and chondroitin sulfate were tested on full-thickness skin wounds of mice to determine their efficacy in reducing scarring (Kirker et al., 2002). HA is present in much higher quantities in fetal skin than in adult skin and is associated with the ability of fetal skin to regenerate. The films did not effect any changes in the inflammatory response or degree of wound contraction, but there was a significant increase in re-epithelialization and in the amount of dermal collagen, as well as organization of the collagen.

## 2. Acceleration of Repair in Chronic Wounds by Topically Applied Agents

Chronic wounds exhibit a decrease/delay in the proliferation of inflammatory cells or a persistent inflammatory phase that prevents transition to the formation of granulation tissue (Mast and Schultz, 1996). Hypoxia due to compromised circulation is hypothesized to be a major factor in the failure of chronic wounds to heal where there is a persistent inflammatory phase. Poor circulation in diabetic patients is correlated with impaired proliferation, adhesion and incorporation of circulating endothelial precursor cells into vascular structures (Tepper et al., 2002), and diabetic ulcers are reported to have significantly decreased numbers of these cells (Keswani et al., 2004). **TABLE 4.3** lists a variety of agents that have been used to accelerate the repair of chronic wounds.

### a. Growth factors

Consistent with evidence that their production is impaired in chronic wounds, topically applied growth factors have been shown to improve the repair of these wounds in animal and human studies. Hyperbaric

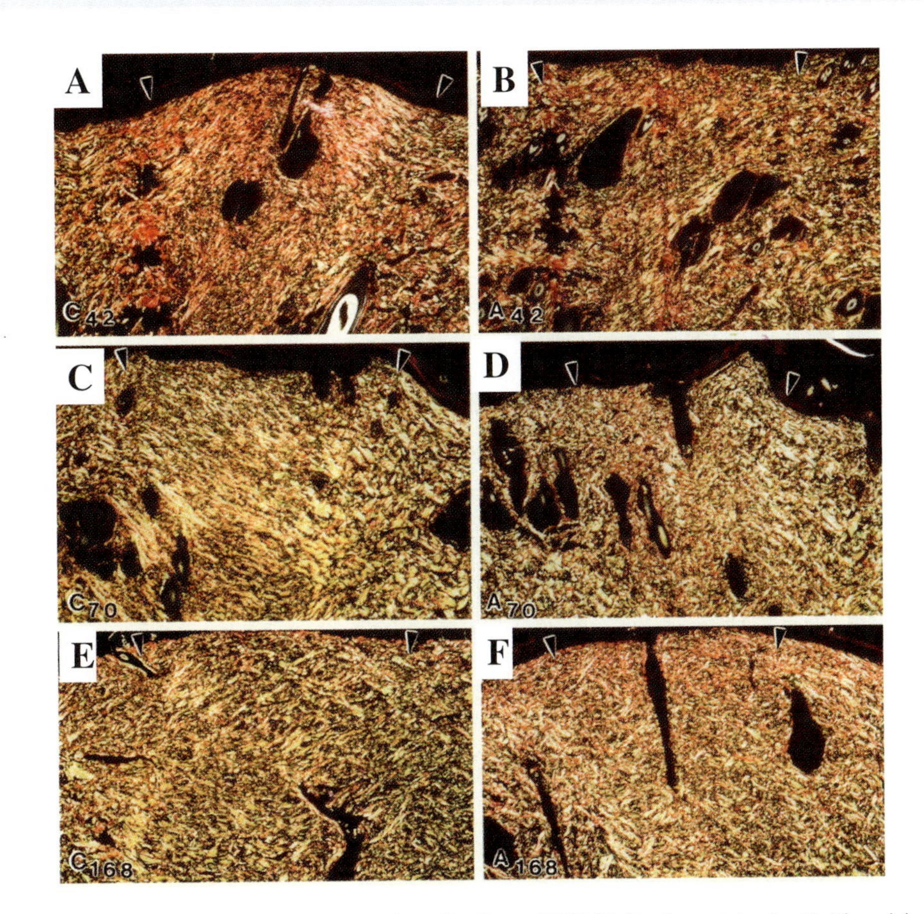


**FIGURE 4.1** Effect of treating incisional skin wounds of rats with antibodies to TGF-β1. Sections stained with Picrosirius red and viewed under a polarizing microscope to visualize collagen fibers. C = control, A = antibody-treated. Numbers indicate days after wounding. Arrowheads indicate the junction between unwounded dermis and neodermis. **(A, C, E)** Control wounds. The collagen fibers are compactly arranged in a parallel pattern to form distinct scars. **(B, D, E)** Wounds treated with TGF-β1 antibody. The collagen fibers form a less compact reticular pattern that resembles the dermal architecture of normal skin. Reproduced with permission from Shah et al., Neutralization of TGF-beta 1 and TGF-beta 2 or exogenous addition of TGF-beta 3 to cutaneous rat wounds reduces scarring. J Cell Sci 108:985–1002. Copyright 1994, Company of Biologists Ltd.

**TABLE 4.3** Agents That Have Been Found, Singly or in Combination, to Accelerate the Repair of Chronic Wounds When Applied Topically

| Growth Factors and Hormones |
|---|
| β-NGF |
| FGF-2 |
| PDGF-BB |
| IGFI, II |
| EGF |
| TGF-α |
| TGF-β |
| VEGF |
| KGF-2 |
| Insulin |

| Other Accelerants |
|---|
| Angiotensin 1–7 |
| Thymosin β-4 |
| L-arginine |
| Pentoxifylline |
| Infrared and near-infrared light |

oxygen in combination with growth factors may have an additive effect on the healing of chronic wounds (Greenalgh, 1996).

Beta-NGF (1 μg/day) was reported to significantly decrease healing times in diabetic mice by promoting re-epithelialization (Muangman et al., 2004). Recombinant FGF-2 and PDGF-BB each stimulated marked neovascularization of ischemic rabbit ear wounds, but only PDGF-BB accelerated and augmented granulation tissue formation and re-epithelialization (Mustoe et al., 1994). Combinations of PDGF and IGF II, and insulin plus EGF, were more effective than the individual growth factors (Greenalgh, 1996). In a rabbit ear dermal ulcer model, topical application of 1–4 μg of IGF-I plus IGF-I binding protein at a molar ratio of 5:1 or 11:1 resulted in a 52% increase in granulation tissue compared to untreated controls or ears treated with the growth factor or binding protein alone. This dose of IGF-I/IGF-I binding protein did not augment healing

**TABLE 4.4** Epithelial gap (EG) size and granulation tissue area (GTA) in healing 8 $mm^2$ wounds made in the back skin of three different types of diabetic mouse, 7 days after administering phospho-buffered saline (PBS), adenoviral *LacZ*, or adenoviral *PDGFB*. Measurements of EG in mm and GTA in $mm^3$. Data from Keswani et al. (2004)

| Mouse | PBS | | *LacZ* | | *PDGFB* | |
|---|---|---|---|---|---|---|
| Model | EG | GTA | EG | GTA | EG | GTA |
| *db/db* | 4.6 | 7 | 5.2 | 8.0 | 3.0 | 24.0 |
| STZ | 2.9 | 23.0 | 3.0 | 23.0 | 1.5 | 37.0 |
| NOD | 5.2 | 9.0 | 5.0 | 14.0 | 2.5 | 28.0 |
| Control | 0.6 | 35.0 | 0.6 | 35.0 | 0.6 | 35.0 |

**TABLE 4.5** Effect of KGF2 administered topically twice per week for 12 weeks on healing of chronic venous ulcers in a subset of patients with wounds less than or equal to 15 $cm^2$ and less than or equal to 18 months duration. Data expressed as the percentage of patients that showed 50–100% healing of wounds. KGF2 at 60 μg/$cm^2$ was the most effective. Data from Robson et al. (2001)

| Percent of Wound | *Percentage of Subjects* | | |
|---|---|---|---|
| Healed | Placebo | 20 μg/$cm^2$ KGF2 | 60 μg/$cm^2$ KGF2 |
| 50 | 70 | 84 | 85 |
| 75 | 50 | 84 | 85 |
| 90 | 45 | 60 | 85 |
| 100 | 40 | 40 | 60 |

in ischemic ear wounds, but a much higher dose of 43 μg at a molar ratio of 10:1 significantly enhanced repair of ischemic wounds (Zhao et al., 1995).

The healing of full-thickness skin wounds in older diabetic mice was accelerated by a combination of PDGF-BB and IGF-I applied topically in methylcellulose at a dose of 4 μg/$cm^2$ by an average of 30% faster than PDGF alone applied at 4 or 40 μg/$cm^2$ (Kiritsy et al., 1995). PDGF plus TGF-α also stimulated re-epithelialization more than either growth factor alone. Growth factor treatments may directly or indirectly affect angiogenesis. Injection of an adenoviral PDGF-B construct into flank wounds of diabetic rats increased the number of endothelial precursor cells in the wound to normal, enhancing neovascularization, formation of granulation tissue and epithelialization (Keswani et al., 2004) (**TABLE 4.4**). Furthermore, topically applied VEGF improved time to wound closure by 25% in genetically diabetic rats (Kirchner et al., 2003).

Topical treatment of chronic pressure or diabetic ulcers with PDGF-B, FGF-2, EGF, and TGF-β in patients resulted in significant improvement in the rate of wound closure, but neither IL-1 nor 8 had any effect (Greenalgh, 1996; Robson et al., 1995; Smiell et al., 1999; Fu et al., 2005). TGF-β2 in a collagen sponge accelerated the healing of venous ulcers, decreasing the wound size by 57% compared to 30% for collagen sponge alone and 9% for standard wound dressings (Robson, 1997). hrFGF-2 is marketed in the United States as Trafermin and in Japan as Fiblast™ spray. This spray has been successfully used to heal a variety of chronic skin ulcers and chronic wounds with deep soft-tissue defects and exposed bone (Inadomi, 2004; Ichioka et al., 2005). PDGF-B and recombinant human keratinocyte growth factor-2 (rhKGF-2) are marketed in the USA as topical treatments for diabetic and venous ulcers. PDGF-B is sold under the brand name of Regranex® (Ethicon, Inc). rhKGF-2 was developed by Human Genome Sciences from a screen of unknown proteins for the selective promotion of proliferation and migration of keratinocytes *in vitro* and is known as Repifermin. In a Phase II clinical trial on venous ulcers, topical application of Repifermin resulted in a significantly higher proportion of patients achieving wound closure (Robson et al., 2001) (**TABLE 4.5**).

Other observations suggest that the impairment of healing in chronic wounds may also be due to the presence of inflammatory mediators in addition to a deficiency of growth factors in the wound. The addition of fluid from healing wounds to fibroblasts *in vitro* induced greater mitotic activity compared to fluid from chronic wounds. The profiles of growth factors and inflammatory cytokines are different in acute and chronic wounds of human patients (Mast and Schultz, 1996). The fluids of acute pressure wounds contain TGF-α and β, EGF and IGF-I, all of which are mitogenic for fibroblasts and all but TGF-β are mitogenic for epidermal cells as well. By contrast, the fluids of chronic pressure wounds contain PDGF-AB, FGF-2, IGF-I, and EGF, but not TGF-β, and the level of IGF-I is substantially higher than in acute wound fluids. The activity of pro-inflammatory cytokines appears to be much higher in the chronic pressure wounds. The ratio of TNF-α and IL-1b to their inhibitors (p55 and IL-1ra) in acute pressure wounds is 1:6 and 1:320, respectively. These ratios drop to 1:3 and 2.1:1, respectively, in chronic wounds (Mast and Schultz, 1996). Furthermore, growth factors and receptors get degraded faster in chronic wounds and MMP levels are higher, while TIMP levels are lower. Trengrove et al. (2000) reported that the wound fluid of human leg ulcers had significantly higher concentrations of the pro-inflammatory cytokines IL-1, IL-6, and TNF-α, but there were no significant differences between healing and chronic wounds in the levels of PDGF, EGF, FGF-2, or TGF-β.

As outlined in Chapter 2, wound healing is a very complex process. We do not know all of the molecular elements required for repair, nor the details of all the regulatory pathways of which these elements are a part. It is likely that much better therapeutic success will be achieved by the topical use of synergistic combinations of growth or other factors. In the future, gene and proteomic array studies will provide a more complete picture of wound repair in normal and chronic wounds that will better enable us to determine what combinations of topically applied agents will be most effective in enhancing the healing of chronic wounds (Tomic-Canic and Brem, 2004). These agents will include delivering growth factors by gene therapy. A system of wet wound healing has been devised that heals wounds with less scarring and allows delivery of antibiotics, growth factors, cells or plasmids to the wound tissue (Eriksson and Vranckx, 2004). This system should be highly useful in testing the therapeutic value of various combinations of molecules known to be involved in wound repair, as well as other agents that may accelerate healing or reduce fibrosis.

#### b. Other Chemical Agents

Several other agents have been shown in animals to accelerate the repair of chronic wounds. Angiotensin (1–7) accelerates the healing of full-thickness excisional wounds in diabetic mice (Rodgers et al., 2001). Thymosin β4 is an anti-inflammatory and angiogenic factor that promotes endothelial cell migration and tube formation. This protein binds to cytoskeletal actin filaments and prevents the addition of actin monomers to the ends of the polymers. Thymosin β4 accelerates wound repair in both young and old diabetic mice by significantly increasing wound contraction and collagen deposition. The wounds of old mice also exhibited increased keratinocyte migration. A synthetic peptide that duplicated the actin-binding domain of thymosin β4 promoted wound repair in aged mice to a degree comparable to that of the whole molecule (Philp et al., 2003). Supplemental L-arginine also enhances wound repair in diabetic rats, most likely because it is the substrate for nitric oxide (NO) synthesis, which is critical for normal healing (Shi et al., 2003). Lastly, pentoxifylline, a modified methyl xanthine that is effective in treating peripheral occlusive arterial disease, significantly accelerated the healing of venous ulcers in human patients in a large randomized, double-blinded study (Falanga et al., 1998).

#### c. Infrared or Near-Infrared Light

Infrared (700 nm–1200 nm wavelength) and near-infrared (600–700 nm) light delivered through lasers or light-emitting diodes (LEDs) has been reported to significantly accelerate the repair of chronic skin wounds (Rochkind et al., 1989; Conlan, 1996; Schindl et al., 2000; Enwemeka, 2004), an effect first observed by Mester et al. (1968). Spectroscopic measurements indicate that photons at wavelengths of 630–800 nm penetrate through the skin and muscles of the forearm and lower leg (Chance et al., 1988; Beauvoit et al., 1994, 1995). The effect of the light may be to stimulate cytochrome *c* oxidase in the mitochondria, resulting in increased oxygen consumption and production of ATP (Karu, 1999).

Studies on diabetic rats indicated significant increases in the amount of collagen and in tensile strength of light-treated wounds over controls (Stadler et al., 2001; Reddy et al., 2001). In combination with hyperbaric oxygen, light-treated skin wounds in rats closed faster (Yu et al., 1997), an effect that was associated with a more uniform rise and fall in VEGF and FGF-2 instead of the sharp peaks at day four and subsequent rapid drop-off observed in control wounds (Whelan et al., 2001). *In vitro,* proliferation of mouse fibroblasts was increased by over 150% and that of human epithelial cells by 155%–171% (Whelan et al., 2001). Whelan et al. (2001) also reported that wound-healing time was decreased by 50% aboard a submarine, where the atmosphere is lower in oxygen and higher in carbon dioxide, and that children suffering from oral mucositis as a result of chemotherapy experienced a 47% reduction in pain.

Despite these interesting findings, infrared and near-infrared light does not appear to have been tested as yet on wounds in large randomized, double-blind clinical trials.

### 3. Keratinocyte Transplants

Lack of re-epithelialization is a common reason for failure of wound repair. Keratinocyte transplantation is a way to enhance re-epithelialization of an extensive wound area. A 1 $cm^2$ piece of epidermis can yield enough cells *in vitro* to cover the whole surface of the human body (Green et al., 1979). Sheets of autogeneic keratinocytes have been grown *in vitro* from epidermal stem cells to cover large wounds (O'Connor et al., 1981; Gallico et al., 1984), but there are several drawbacks to these transplants (Boyce, 2001; Balasubramani et al., 2001; Shakespeare, 2001). The time required (usually 2–3 weeks) to culture a sufficient number of autologous keratinocytes from a small biopsy necessitates covering the wound with temporary dressings, increasing the chances of infection.

Allogeneic keratinocytes can also be used to cover large wounds. Their advantage is that epidermal sheets can be cultured in advance and banked for use off the shelf, but the down side is that they are immunorejected. However, it has been shown that mixtures of allogeneic and autogeneic keratinocytes, containing from 5%–50% autogeneic cells, elicit only a weak immunogenic response in mice (Rouabhia et al., 1995; Suzuki et al., 1996; Larochelle et al., 1998). The allogeneic cells are slowly rejected and replaced by host epidermis. These chimeras seem to induce partial tolerance of the allogeneic cells. The use of keratinocyte mixtures with low numbers of autogeneic cells would significantly reduce the time required to grow enough of these cells for autografting.

Adherence of the epidermal sheets to the wound bed is particularly dependent on binding of keratinocytes to basement membrane laminin-5, through $\alpha3$ integrin (Yancy, 1995). Resynthesis of a damaged or lost basement membrane by keratinocytes requires interactions with the dermis and is highly variable when the dermis is badly damaged, leading to low take rates. Keratinocytes require factors such as KGF (FGF-7) for mitosis that are produced by dermal fibroblasts; thus, epidermal regeneration is poor in the absence of dermis. Restoration of anchoring fibrils is slow, leading to blistering of the epidermis (Woodley et al., 1990).

A further problem is that epidermal sheets grown *in vitro* are fragile and difficult to transfer to the wound. The fragility issue has been overcome by the use of carriers that allow epidermal cells to be transferred to the wound bed, for example, sheets of collagen (Horch et al., 2001), polyethylene glycol terephthalate (PEGT)/polybutylene terephthalate (PBT) copolymer (Van Den Bogaerdt et al., 2004), and fibrin (Mis et al., 2004). More rapid proliferation of epidermal stem cells, allowing transfer of keratinocytes to the wound within 10 days, was obtained with porcine gelatin beads because of their high surface area for growth (Liu et al., 2004). Two applications three days apart of porcine gelatin microbeads harboring proliferating keratinocytes on their surface were shown to be more effective than cells grown on collagen pads in completely healing recalcitrant venous ulcers in a group of five patients (**FIGURE 4.2**).

## 4. Bioartificial Skin Equivalents

Bioartificial skin equivalents have been developed to cover extensive excisional wounds such as burns (Boyce, 2001; Yannas, 2001; Balasubramani et al., 2001; Kearney, 2001). These constructs consist of a collagen-based or biomimetic matrix seeded with allogeneic fibroblasts and topped with a MSTSG or cultured autogeneic or allogeneic keratinocytes (**FIGURE 4.3**). Because allogeneic cells are rejected and replaced by host cells, bioartificial skin equivalents do not replace host skin, but rather are temporary living dressings that accelerate wound closure by host cells through their secretion of growth factors and cytokines (Ehrlich, 2004; Jimenez and Jimenez, 2004). While saving lives, the cost of skin substitutes is a limiting factor in their use. The cost of keratinocyte sheets alone ranges from $1,000 to $1,300 for every one percent of body surface area covered, and these costs are approximately doubled when keratinocytes are combined with a bioartificial dermal component (Boyce, 2001).

The first skin-equivalent was developed by Bell et al. (1981). They constructed a bioartificial dermis by seeding autogeneic rat dermal fibroblasts in a collagen lattice. A few days later, autogeneic epidermal cells were seeded onto the dermis and the construct was autografted onto an open skin wound made on the back of the animal. The grafts became vascularized, most retained their original size and shape, and wound contraction was minimized. Histological examination 9–10 weeks after grafting revealed a well-developed but abnormal skin, in which the epidermis was hyperplastic and the dermis was more compact and half the thickness of adjacent dermis.

Many types of bioartificial dermal scaffolds have been tested in an attempt to find the optimum scaffold that best supports fibroblasts and epidermal cells alike and results in the highest quality skin repair, but success has been elusive. Collagen gels and polyester mesh are the basis of several dermal substitutes that are now in clinical use (Boyce, 2001; Yannas, 2001). Collagen gels are reported to have two major drawbacks. When hydrated they tend to depress both epidermal and fibroblast proliferation (Kono et al., 1990), and epidermal cells on collagen gels secrete an unidentified factor(s) that stimulates intense collagenolytic activity by fibroblasts (Yoshizato et al., 1986; Shimizu-Nishikawa and Yoshizato, 1990). Investigators have tried to minimize these problems by using gelatin scaffolds, scaffolds made by mixing collagen with gelatin (Koide et al., 1993; Yoshizato and Yoshikawa, 1994) or elastin (DeVries et al., 1994) and/or glycosaminoglycans, and decellularized natural ECM.

A further problem in the clinical failure of bioartificial skin is that it lacks blood vessels. Thus, vascularization of the graft from the host is slow, unless the underlying wound bed is well-vascularized (Sahota et al., 2003). Seeding endothelial cells into the construct might solve this problem. To test the feasibility of this approach, human dermal microvascular endothelial cells were seeded onto decellularized human dermal ECM *in vitro* (Sahota et al., 2004). The endothelial cells penetrated into the ECM, although no capillary formation was reported. Adding angiogenic growth factors (VEGF, FGF-2) or epidermal keratinocytes and fibroblasts to the endothelial cells had no effect on the number of endothelial cells that migrated into the ECM. However, hypoxia and introducing the endothelial cells on the papillary surface of the ECM significantly enhanced endothelial cell penetration.

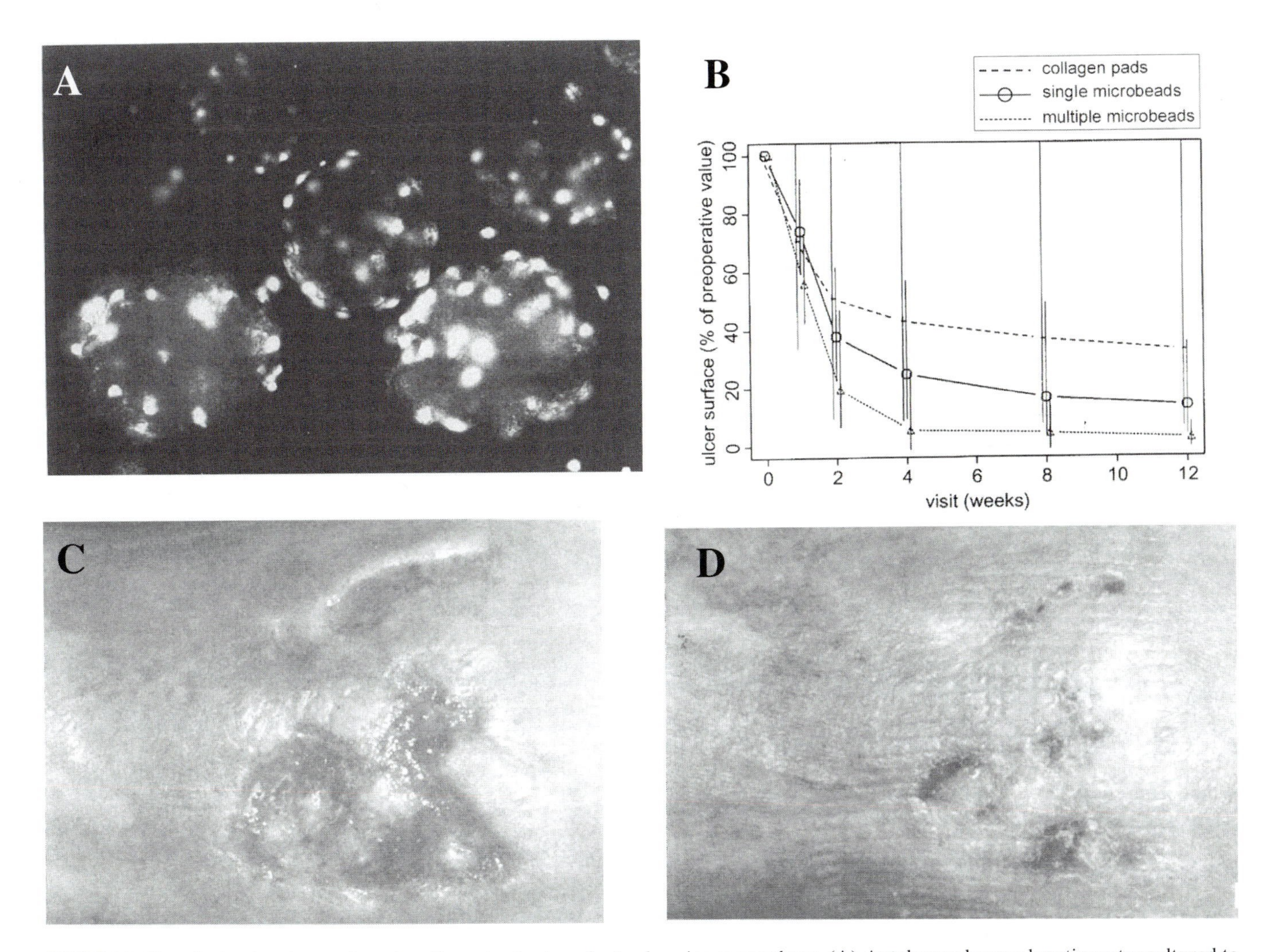


**FIGURE 4.2** Use of microbeads to culture keratinocytes for transfer to chronic venous ulcers. **(A)** Autologous human keratinocytes cultured to confluence on porcine gelatin microbeads (x100). **(B)** Rate of closure of ulcer surface after transferring keratinocytes on collagen pads, after a single application of microbeads, and after two successive applications of microbeads. Ulcers closed significantly faster with one or two microbead treatments. **(C, D)** Photographs of venous ulcer closure after two applications of microbeads 3 days apart. **C**, one week after last application, **D**, two weeks after last application, showing virtually complete wound closure. Reproduced with permission from Lu et al., Autologous cultured keratinocytes on porcine gelatin microbeads effectively heal chronic venous leg ulcers. Wound Rep Reg 12:148–156. Copyright 2004, Blackwell Pub Co.

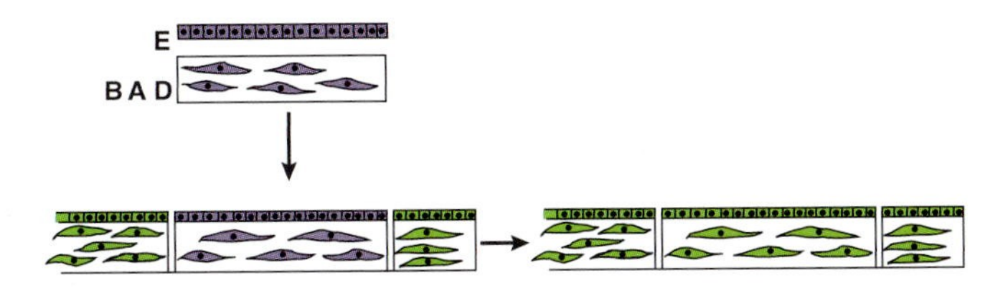


**FIGURE 4.3** Design of bioartificial skin equivalents. A bioartificial dermal scaffold (BAD) is made *in vitro* from polymer mesh, or from collagen mixed with other ECM components such as elastin and chondroitin 6-sulfate. The scaffold is seeded with autogeneic or allogeneic fibroblasts and covered with either autogeneic or allogeneic keratinocytes or a meshed split-thickness skin graft to provide an epidermis (E). The construct (purple) is fitted into the recipient wound (green) and acts as a living dressing that is subsequently replaced by host cells if the construct contains allogeneic cells.

Other studies in which acellular cadaver dermis was seeded with keratinocytes alone or with keratinocytes and fibroblasts showed that the presence of fibroblasts greatly reduces contraction and promotes thicker epidermis and vascularization after grafting to full-thickness wounds of athymic mice (Erdag and Sheridan, 2004).

The collagen-based Apligraf® (Organogenesis) and the polyester-based Dermagraft® (Advanced Cell Science) are two allogeneic skin equivalents that have been approved by the FDA for use on chronic skin wounds. Apligraf® is a modification of the collagen dermal equivalent originally described by Bell et al. (1981). Human neonatal foreskin fibroblasts are mixed with bovine Type I collagen. The fibroblasts contract the collagen fibrils in the gel, reducing the volume of the matrix and forming a dense meshwork of fibrils. Keratinocytes or a MSTSG are then layered onto the gel and induced to differentiate into an epidermis with stratum corneum by raising the construct to the air–liquid interface. When placed on full-thickness skin wounds of athymic mice, the graft was rapidly incorporated into the host tissue and the bovine collagen was progressively replaced by human and mouse collagens (Guerret et al., 2003). Clinical trials of Apligraf® were reported to result in a higher frequency of healed wounds and a reduced median time for complete wound closure of venous ulcers (Sabolinski et al., 1996; Falanga et al., 1998) and diabetic ulcers (Veves et al., 2001; Marston et al., 2003; Boulton et al., 2004).

Dermagraft® is constructed by growing human neonatal foreskin fibroblasts to confluence on a mesh of polyglactin-910 (Vicryl) that is degraded in the wound by hydrolysis. After placement on the wound bed, the Dermagraft® is overlaid with a MSTSG or culture-expanded keratinocytes (Hansbrough et al., 1992a, b). The fibroblasts have been shown to synthesize a matrix containing dermal type I, III, and VI collagens, elastin, tenascin, fibronectin, hyaluronic acid, chondroitin sulfates, and the major dermal proteoglycan core protein, decorin, as well as mRNAs for IGF-1 and 2, FGF-2, PDGF, HGF and VEGF (Landeen et al., 1992), all of which are involved in skin repair *in vivo*. The construct has high tensile strength (Cohen et al., 1991). Re-epithelialization and vascularization of the wound is rapid, and there is minimal inflammatory response. The epidermis formed by the MSTSG exhibits a mesh pattern, but the mesh outline is less pronounced that that observed with a MSTSG alone. Immunostaining for laminin and type IV collagen revealed the presence of a continuous basement membrane at the epidermal-dermal junction. Clinical studies indicate that Dermagraft® significantly enhances healing in diabetic foot ulcers (Gentzkow et al., 1996). One case has been reported in which Dermagraft® was used to help close a postsurgical defect resulting from a multivisceral transplant in a pediatric patient (Charles et al., 2004). Five applications of this artificial dermis were applied over the course of eight months posttransplant. The Dermagraft® stimulated re-epithelialization and closure of the incision.

Other skin equivalents are in experimental stages. Van Dorp et al. (1998) seeded fibroblasts into BISKIN-M, an elastomeric poly(ethylene glycol terephthalate)/poly(butylene terephthalate) with gradually changing pore size. This construct was tested on full-thickness skin wounds of Yucatan pigs. Wound contraction was significantly reduced and new blood vessels and fibrous tissue infiltrated the graft within two weeks. The polymer degraded over a period of 24 months and was replaced by a mature connective tissue with the same distribution pattern observed in normal dermis.

Although bioartificial skin constructs obviate the need to cover wounds with full-thickness autografts and promote the healing of chronic wounds and burn excisional wounds, they have not proven to be any better than MSTSGs in many respects (Boyce, 2001). Neither MSTSGs nor skin equivalents restore normal cosmetic appearance of the skin, nor do patients experience full restoration of sensation. A major deficiency of bioartificial skin equivalents is that hair follicles, sebaceous glands, and sweat glands are not restored. The lack of sweat glands means loss of thermal regulation, which is a serious problem when patients are burned over large areas of the body. L'Heureux et al. (1999) have described the construction of an experimental bioartificial skin equivalent with hair. They formed a dermal equivalent of stacked fibroblast sheets grown *in vitro* and covered it with keratinocytes. After four weeks, this construct formed a continuous and structurally organized basement membrane containing laminin and collagen IV and VII. Hair follicles were inserted into the construct, where they significantly increased the rate of penetration of hydrocortisone in comparison with control skin equivalent without hair follicles.

## 5. Acellular Dermal Regeneration Templates

Bioartificial skin equivalents are cryopreserved until use, a procedure that may seriously reduce the viability of fibroblasts in the dermal component (Mansbridge et al., 1998). Thus, the dermal matrix of cryopreserved complete skin equivalents may actually be much less cellular when used than when they were constructed, or may even be acellular. The trend today is, in fact, toward the use of much less expensive acellular dermal templates overlaid with MSTSGs or cultured keratinocytes to induce the regeneration of host dermal tissue as the scaffold degrades (Yannas, 2001). Epidermis and dermal regeneration template can be applied in either a one-step or two-step procedure. In the former, the dermal regeneration template and a MSTSG or keratinocytes are placed on the wound together (**FIGURE 4.4**). However, the epidermal component does not survive well because of the initial lack of vascularization in the template. This problem is overcome in a two-step procedure, where the template is placed on the wound first and allowed to be invaded by fibroblasts and vascularize before adding the MSTSG or keratinocytes.

One of the effects shared by dermal regeneration templates and bioartificial skin equivalents is the reduction of wound contraction, which results in reduced

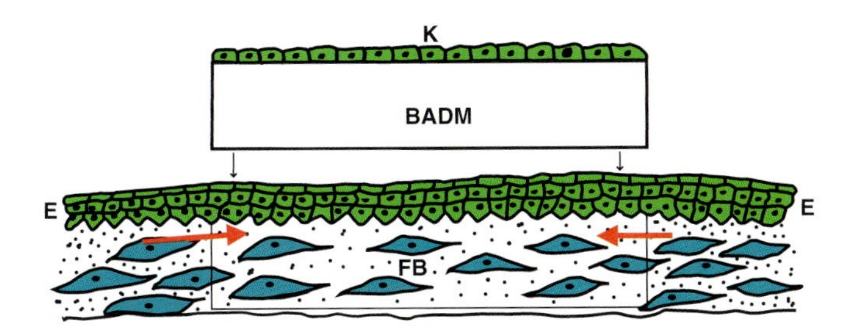


**FIGURE 4.4** Use of a biomimetic regeneration template to promote regeneration of host skin. A bioartificial dermal matrix (BADM) is placed in the wound bed and overlaid with keratinocytes (K) or a meshed split-thickness skin graft. Host fibroblasts grow into the matrix (arrows), which degrades over time.

scarring and the induction of partial regeneration (Yannas, 2001). The extent of partial regeneration is determined by the chemical composition, average pore size, and degradation rate of the templates (Yannas, 2001). These structural features control the following cell-matrix interaction features, respectively: matrix ligand identity, ligand density, and ligand duration on the surface of the matrix. Collagen-GAG mixtures are often used as templates, and it has been found that the GAG component is critical for the activity of the template in inhibiting contraction and promoting regeneration (Shafritz et al., 1994). Chondroitin-6-sulfate is the commonly used GAG, but dermatan sulfate or decorin were more effective in reducing contraction and promoting regeneration, perhaps due to their ability to neutralize TGF-β and mitigate the inflammatory response. One of the best results, on full-thickness porcine wounds, was obtained with a type I collagen/elastin mixture (Devries et al., 1994) (**FIGURE 4.5**).

Several dermal regeneration templates are in clinical use. Alloderm® (Life Cell Corporation) aims to regenerate dermis by providing a normal dermal ECM derived from cadaver skin, including a basement membrane. The skin is processed to remove the epidermis and fibroblasts and is then freeze-dried. After placement on the wound bed, it is overlaid with an MSTSG or cultured keratinocytes. Alloderm® is reported to induce the regeneration of dermis with the normal orientation of collagen and elastin fibrils instead of the cross-linked collagen and lack of elastin seen in scar tissue (Wainwright et al., 1995). The fibroblasts that repopulate the Alloderm® lack the myofibroblastic phenotype of granulation tissue and the grafts exhibit minimal contraction. The basement membrane of the dermis interacts with overlying keratinocytes to induce the formation of hemidesmosomes and Type VII collagen anchoring fibrils. Alloderm® has been evaluated clinically for burns and is reported to provide good to excellent cosmetic appearance and functional performance (Wainwright et al., 1996; Lattari et al., 1997; Sheridan and Choucair, 1997). However, in a study of burn wounds in children grafted with a different regeneration template derived from just the papillary layer of human split thickness skin and covered with autogeneic keratinocytes, vascularization was successful in less than half of the cases compared to the MSTSG controls, and no differences were found in the quality of the regenerated skin over a 12-month period compared to MSTSGs (Sheridan et al., 2001).

Other biodegradable biological acellular dermal templates are derived from animal tissues. SIS (Cook Biotech) is an infection-resistant porcine small intestine submucosa matrix (Badylak et al., 1994). It consists mainly of collagen (types I, III, and V) with smaller amounts of hyaluronic acid, chondroitin sulfate A, heparan sulfate, and TGF-β (Hodde et al., 1996; Voytik-Harbin et al., 1997, 1998; McPherson and Badylak, 1998). A variant of SIS, Oasis™, is effective in accelerating the healing of diabetic ulcers. SIS and Permacol™ (Tissue Science Laboratories), a porcine dermal matrix, have been tested for their ability to support the viability of MSTSGs and reduce the contraction of MSTSGs (MacLeod et al., 2004). Neither one supported the viability of the MSTSG or reduced contraction when applied to wounds in rat skin as a one-stage procedure. When a two-stage procedure was used, the MSTSG still showed necrosis on Permacol™ but maintained viability on SIS. Contraction of the MSTSG, however, was less on Permacol™ than on SIS, but was no better than that observed for MSTSGs alone. PriMatrix™ (TEI Biosciences), which is fetal bovine dermis, promotes the healing of chronic wounds and acute incisional and excisional wounds, incuding burns.

Burke et al. (1981) designed a bioartificial dermal matrix consisting of co-precipitated and cross-linked bovine hide collagen and chondroitin 6-sulfate, known today as Integra®. This biodegradable matrix is overlaid with a thin layer of Silastin that functions to prevent water loss. Wounds are treated in a two-step procedure. The first step is to place Integra™ on the wound. When the dermal substitute has been invaded by host fibroblasts and blood vessels (10–14 days), the second step is performed. The Silastin sheet is removed and replaced by a MSTSG or autogeneic keratinocytes that are expanded *in vitro* during this time period. Growing the keratinocytes on a fibrin sheet greatly improved their adhesion to the Integra® and epithelial development (Mis et al., 2004). The Integra® matrix is degraded by 30 days, and the invading fibroblasts synthesize a matrix reported to be histologically similar to that of normal skin.

Assessments of Integra® in full-thickness skin wounds of minipigs and in clinical trials for burn wounds have reported results superior to those of other constructs

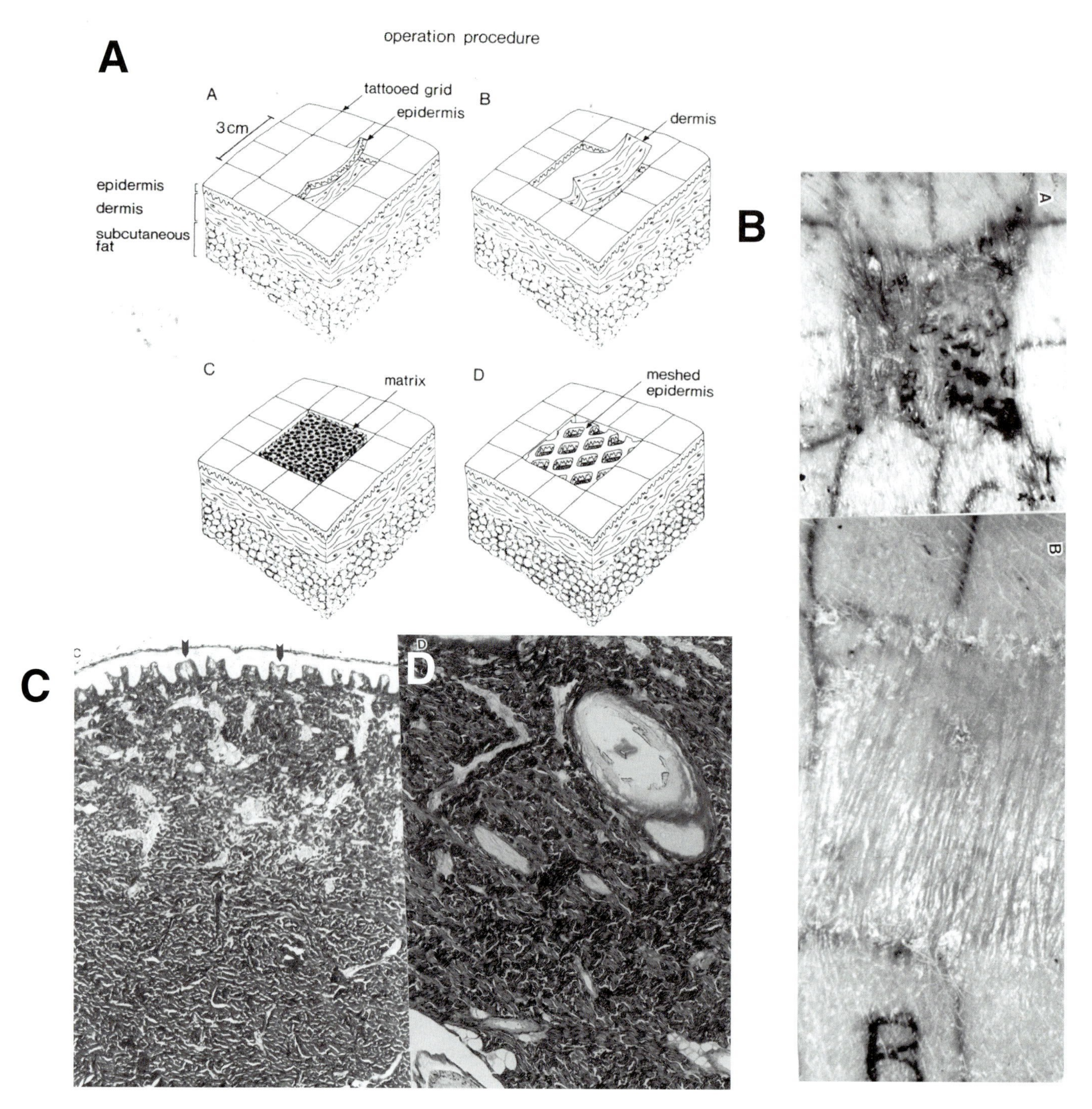


**FIGURE 4.5** Regeneration of skin in a pig excisional wound using a regeneration template consisting of non-denatured bovine collagen I plus elastin. **(A)** Diagram of operation. A grid is tattooed on the surface of the skin and a split-thickness skin graft is harvested. The remainder of the dermis is then removed to make a full-thickness defect. The regeneration template is inserted in the wound and covered by the meshed split-thickness skin graft. **(B) Top**, new skin regenerated using a native collagen/fibronectin template exhibited an uneven and scarred surface with severe wound contraction, as evidenced by distortion of the grid. **Bottom**, new skin regenerated using a native collagen/elastin template has good cosmetic appearance with little contraction. **(C)** Histological appearance of the latter skin. The collagen fibers have the reticular, basket-weave architecture characteristic of normal skin **(D)**. Reproduced with permission from Devries et al., Dermal regeneration in native non-cross-linked collagen sponges with different extracellular matrix molecules. Wound Rep Reg 2:37–47. Copyright 1994, Blackwell Pub Co.

(Druecke et al., 2004; Heimbach et al., 1988; Sheridan et al., 1994; Pandya et al., 1998; Heitland et al., 2004), as well as for the reconstruction of breast skin in young women who had suffered significant burns on the anterior chest wall (Palao et al., 2003). The take rate of Integra® is comparable to that of allogeneic MSTSGs, but is about 15% lower than autogeneic MSTSGs. The performance of Integra® is considered by patients and surgeons to be equivalent or superior to MSTSGs alone. Very thin autogeneic MSTSGs can be used with Integra®; thus, donor skin graft sites heal more quickly. The most serious problem reported with Integra® is with the Silastin layer, which acts as a nonpermeable occlusive wound dressing that cannot be changed easily during the period between the time Integra® is placed on the wound and the time an MSTSG or cultured keratinocytes are applied (Heitland et al., 2004). Seeding keratinocytes onto the underside of the Integra® may allow a single-step solution. The keratinocytes migrate upward after application of the template to pig burn wounds and form a confluent epidermal layer (Jones et al., 2003).

## USE OF ACELLULAR REGENERATION TEMPLATES FOR ABDOMINAL WALL REPAIR

The use of regeneration templates to repair skin has a counterpart in the repair of fascia during the surgical closure of abdominal wall hernias. The templates are used to reinforce the repair and encourage new tissue ingrowth. Polypropylene or polytetrafluorethylene (PTFE) mesh has been widely used in both animal and human clinical studies because they have good tensile strength and support fibroblast ingrowth. However, they do not biodegrade, cause an inflammatory response, and are prone to infection and fistula formation (Voyles et al., 1981). Polyglactin and polyglycolic acid meshes, which are biodegradable, have also been used, but have low tensile strength and a high number of patients suffer recurrent hernia at the site of mesh placement (Tyrell et al., 1989).

Recently, surgeons have turned to collagen-based biological templates for hernia repair. Permacol™ (Tissue Science Laboratories) has been found to be effective in hernia repair. SIS (Surgisis™) has been evaluated for hernia repair. Comparison of untreated or sterilized SIS to gluteraldehyde-treated porcine heart valve after subcutaneous implantation into rats showed that whereas the gluteraldehyde-treated tissue suffered significant calcification, the SIS had virtually no calcium content after eight weeks (Owen et al., 1997). SIS and two other collagen-based porcine scaffolds, urinary bladder submucosa (UBS) and renal capsule matrix, have been compared with polypropylene mesh in the repair of hernias in dogs and in surgically created abdominal wall defects in rats (Clarke et al., 1996; Soiderer et al., 2004). The results showed that these materials were superior to polypropylene mesh in having reduced inflammation, adhesion and contraction, and maintenance of tensile strength, as well as being completely replaced by organized collagenous tissue. In similar experiments on surgically created abdominal wall defects in rats, SIS placement in combination with fascial repair significantly improved the strength of the repaired abdominal wall over SIS placement alone or fascial closure alone (Zhang et al., 2003).

SIS was used to make 58 successful hernia repairs in 53 patients with only one recorded case of infection, even though the operative field in 20 procedures was potentially contaminated and was grossly contaminated in 13 of the procedures (Franklin et al., 2002, 2004). In another study, SIS was compared with polypropylene mesh in the laparoscopic repair of hernias, with comparable results in both groups of patients (Edelman, 2002). SIS has also been found to be useful and safe in reinforcing the gastrojejunostomy in Roux-en-Y gastric bypass (Kini et al., 2001) and in the repair of hiatal hernias (Oelschlager et al., 2003).

## STIMULATION OF HAIR REGENERATION

Androgenetic alopecia, or male pattern baldness, is a genetic predisposition in which there is thinning and loss of the hair on the temples and crown of the scalp in an "M"-shaped pattern (Springer et al., 2003). The cause is sensitivity of hair follicles in these regions to the hormone dihydrotestosterone (DHT), which is formed in scalp cells from testosterone by the action of the enzyme 5-α-reductase. DHT acts to progressively shorten the anagen stage of hair growth while lengthening telogen, resulting in smaller hair follicles, thinner and shorter hairs and ultimately, many empty and involuted hair follicles. The follicles on the sides and back of the scalp are not responsive to DHT and so are spared.

The standard treatment for pattern baldness is the transplantation of clusters of 1–4 follicles from the sides and back of the scalp to the temples and crown, very close together, a time-consuming and expensive process. Two over-the-counter medications, minoxidil (marketed as Rogaine and applied topically) and finasteride (marketed as Propecia and taken orally) are available to combat hair loss. Minoxidil is converted to minoxidil sulfite, which activates potassium channels in follicle

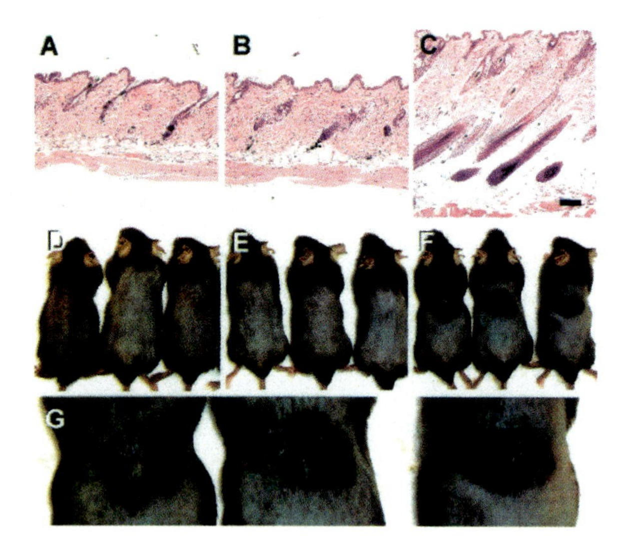


**FIGURE 4.6** Regrowth of hair in cyclophosphamide-treated mice after intradermal injection of the *Shh* gene in an adenoviral vector. **Top**, histological appearance of cyclophosphamide-treated control skin **(A)** and skin treated with adenoviral vector alone **(B)**, showing dystrophic hair follicles. **(C)** AdShh-treated skin showing large growing hair follicles. **Middle and bottom**, gross appearance of control skin **(D)** and skin treated with adenoviral vector alone **(E)**, showing lack of hair growth. **(F, G)** Mice treated with Shh, showing robust growth of hair. Reproduced with permission from Sato et al., Effect of adenovirus-mediated expression of sonic hedgehog gene on hair regrowth in mice with chemotherapy-induced alopecia. J Natl Cancer Inst 93:1858–1864. Copyright 2001, Oxford University Press.

cells, slowing hair loss by a mechanism that is unclear (Baker et al., 1994). Finasteride blocks the conversion of testosterone to DHT by inhibiting 5-$\alpha$-reductase, restoring the length of anagen and promoting hair growth in 66% of men after two years of treatment (McClellan and Markham, 1999).

Transient expression of *shh* induces anagen in the hair follicles of postnatal mice (Sato et al., 1999) and is essential for hair development and regeneration (Wang et al., 2000). Hair loss in mice can be induced by administration of cyclophosphamide, which is cytotoxic for cycling cells because it binds to and cross-links DNA strands. *Shh*, delivered by intradermal injection in an adenoviral vector to the skin of mice suffering from cyclophosphamide-induced alopecia, stimulated rapid hair regrowth (Sato et al., 2001), suggesting that Shh or its agonists may be useful in the treatment of alopecias (**FIGURE 4.6**).

While research continues to better understand the molecular elements and pathways involved in hair regeneration in order to identify targets for drug intervention, the transplantation of stem cells from hair follicles is viewed as another promising therapy to regenerate hair (Costarelis and Millar, 2001). Intact dermal papilla or cultured dermal papilla cells (Horne et al., 1989; Jahoda et al., 1984), as well as pieces of lower dermal sheath or cultured dermal sheath cells (Horne and Jahoda, 1992; Jahoda et al., 1992) can form new dermal papillae in association with follicle epidermis. Current research aims to culture follicular bulge stem cells with dermal papilla cells from DHT-insensitive follicles *in vitro* to create unlimited numbers of cell pellets that can be implanted into the scalp, where they would form normal follicles (**FIGURE 4.7**).

## REPAIR OF TEETH AND PERIODONTIUM

There are no dental restorative materials that have the same physical and chemical characteristics of natural tooth tissues. The success rate of using such materials to repair teeth in which the pulp has been exposed is very low, which means that dentists will commonly recommend either a root canal (pulpotomy) or extraction of the tooth. To fill even tiny early cavities, a disproportionate amount of enamel must be drilled out in order to insure adhesion of the resins or metal alloys currently used as restorative materials. However, a paste of modified hydroxyapatite (the inorganic component of bone matrix) has recently been developed that can be applied to microcavities without drilling and which integrates perfectly into the enamel (Yamagishi et al., 2005). This material has high durability and acid tolerance.

A quarter of adults in the United States aged 65–74 have had all their teeth extracted. Furthermore, 25% of these patients who have been fitted with either dentures or osseointegrated implants are not satisfied with them (Murray and Garcia-Godoy, 2004). A high percentage of tooth loss is due to periodontal disease. Periodontal disease is caused by bacterial infection below the gum line that destroys the supporting structure, or periodontium of the tooth, which consists of the periodontal ligament (PDL), cementum, alveolar bone, and gingiva. The inflammation associated with periodontal disease has been implicated in cardiovascular disease as well (Desvarieux et al., 2003; Elter et al., 2003).

Periodontal defects are currently treated by guided tissue regeneration, which involves scraping the roots of the teeth and inserting a membrane between the gum and tooth, or by allografts of ground cadaver bone matrix to induce regeneration of alveolar bone. Various growth factors, including PDGF, IGF, BMP-2, TGF-$\beta$, and FGF-2 have been found to have a positive effect on the regeneration of alveolar bone in experimental animals (Takayama et al., 2001). PDGF-BB, derived from the patient's platelets, is used to enhance guided tissue regeneration (Cochran and Wozney, 2000; Nevins et al., 2003).

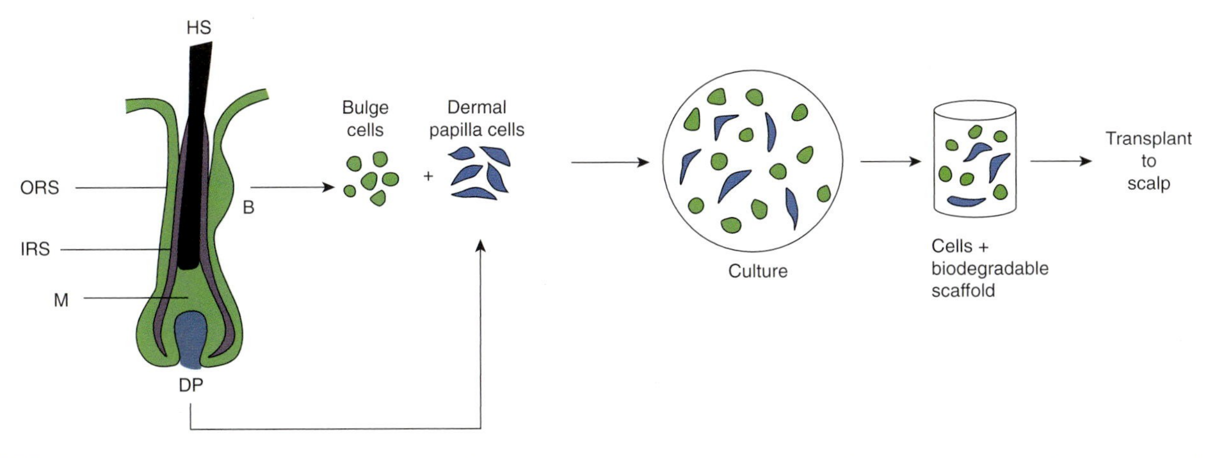


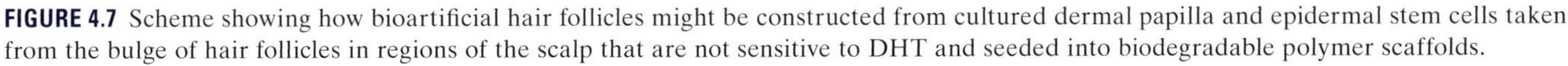
**FIGURE 4.7** Scheme showing how bioartificial hair follicles might be constructed from cultured dermal papilla and epidermal stem cells taken from the bulge of hair follicles in regions of the scalp that are not sensitive to DHT and seeded into biodegradable polymer scaffolds.

Infrared laser light was reported to induce the regeneration of periodontal ligament, cementum, and bone significantly better than guided tissue regeneration in surgically inflicted periodontal defects of beagles (Crespi et al., 1997). While these results are intriguing, clinical trials on the use of infrared or near-infrared light have not been reported. Presumably, the regeneration induced by infrared or near-infrared light would be achieved by enhanced activation of stem cells in the PDL.

Experiments have been done to examine the feasibility of using PDL stem cells to regenerate periodontal tissues and tooth components. Nakahara et al. (2004) performed experiments on dogs in which cultured autogeneic PDL cells were seeded into freeze–dried biodegradable sponges made of porcine Type I and Type III collagen and the constructs implanted into fenestration defects made in the alveolar bone and cementum of the canine tooth roots. The only difference between the cell-seeded and control teeth was the regeneration of more cementum in the former. Human PDL stem cells have been tested for their ability to regenerate cementum and PDL (Seo et al., 2004). STRO-$1^+$ cells were isolated by magnetic bead separation from the PDL of extracted third molars and single colonies were expanded *in vitro*. STRO-1 is an antigen that may be specific for mesenchymal stem cells. Cells derived from a single colony were then mixed with particles of hydroxyapatite/tricalcium phosphate carrier and transplanted subcutaneously into immunocompromised mice. The transplants formed typical cementum/PDL-like structures in which the collagen fibers mimicked the connection of Sharpey's fibers to the cementum (FIGURE 4.8). PDL stem cells were also implanted into periodontal defects made in the cementum of rat mandibular molars. The cells showed occasional attachment to the surface of alveolar bone and teeth, but there was no evidence that they regenerated periodontal tissues.

Whole teeth can be successfully allotransplanted into the empty sockets of missing teeth (Andreasen et al., 1990), but donor shortages restrict the flexibility of this technique. There is great interest in the possibility of regenerating dentin in damaged teeth by the transplantation of stem cells from dental pulp (DPSCs or SHED) and even of creating new tooth buds *in vitro* from these cells that could be transplanted in place of missing teeth (Kim and Vacanti, 1999; Murray and Garcia-Godoy, 2004). To regenerate dentin, pulp stem cells expanded *in vitro* could be injected into the lesion site where they would differentiate into odontoblasts (FIGURE 4.9). Because the cells would be present in high concentration, they would more quickly begin the process of dentin production and completely recover the pulp.

No attempts have yet been reported to make tooth buds using DPSCs and SHED, but rat and pig tooth buds have been created *in vitro* by seeding dissociated molar tooth bud cells into polymer scaffolds. When these constructs were implanted into the omentum they were able to form dentin, enamel, and pulp tissues (Young et al., 2002; Duailibi et al., 2004). Ohazama et al. (2004) combined mandibular oral epithelium from E10 mouse embryos transgenic for GFP, with either ESCs, NSCs, or bone marrow cells and implanted the constructs under the renal capsules of host mice. The oral epithelium induced an odontogenic response in the stem cells and the constructs formed tooth tissues and bone. In addition, tooth buds from E14.5 embryos were able to form an ectopic tooth when transplanted into the maxillar soft tissue between the normal teeth of adults, indicating the potential feasibility for the similar development of tooth buds created *in vitro*. The results of these experiments suggest the feasibility of

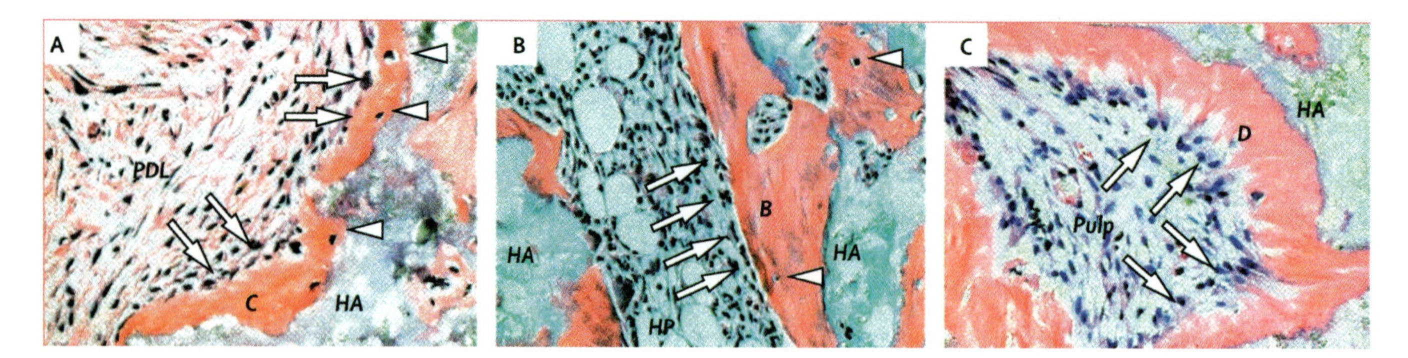


**FIGURE 4.8** Generation of cementum-like and periodontal ligament-like structures in vivo by human periodontal ligament stem cells (PDLSCs) derived from third molars. The cells were transplanted in hydroxyapatite carriers into 2 mm surgically created periodontal defects in the mandibular molar of immunocompromised rats. **(A)** After 8 weeks, the PDLSCs differentiated into cementoblast-like cells (arrows) that formed a cementum-like structure (C) on the surface of the hydroxyapatite (HA). Cementocyte-like cells (triangles) and PDL-like tissue were also regenerated. **(B)** Control transplant of bone marrow stem cells to show the differentiation of osteoblasts (arrows), osteocytes (triangles), elements of bone matrix (B) and hematopoietic marrow (HP). **(C)** Control pulp stem cell transplant that differentiated a dentin/pulp-like structure containing odontoblasts (arrows) and dentin-like (D) and pulp-like tissue. Reproduced with permission from Seo et al., Investigation of multipotent postnatal stem cells from human periodontal ligament. The Lancet 364:149–155. Copyright 2004, Elsevier.

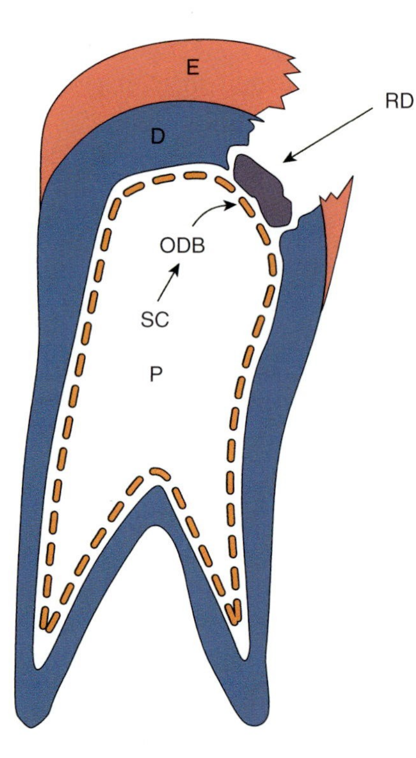


**FIGURE 4.9** Model for stem cell involvement in dental repair. Tooth enamel (E) and dentin (D) are eroded by the action of bacteria to expose the pulp (P), destroying underlying odontoblasts (ODB). Pulp stem cells (SC) would be injected into the lesion site, where they would become odontoblasts that secrete reparative dentin (RD) and minerals required for repair.

making tooth buds *in vitro* from embryonic oral epithelium and bone marrow cells that can develop into mature teeth in the adult jaw. This strategy would be even more useful if adult oral epithelium could be used, as has been done in the regeneration of cornea (see below).

The ultimate goal is to create "tooth farms" using human cells. The National Institute of Dental and Craniofacial Research, in its workshop on Strategies for Tooth Structure Regeneration (2000), has stated its vision of being able to create single-rooted teeth using dental tissue precursors by the year 2010 through interdisciplinary research goals that include using animal models in addition to mice. These models could include urodele amphibians, sharks, and a smaller relative of the crocodile, the cayman, all of which continually replace lost teeth, with the aim of understanding the biology and chemistry of tooth development and biomineralization of tooth structures.

## CORNEAL REGENERATION

The standard therapy for treatment of corneas rendered opaque by trauma or disease is a donor corneal transplant. However, the corneal epithelium is continually sloughed off and renewed by limbal epithelial stem cells. If the host limbus is damaged in addition to the cornea, as in chemical burns or diseases such as Stevens-Johnson syndrome or ocular pemphigoid, the epithelium of the transplant will not be renewed. The surface of the cornea will be invaded by fibroblasts from the conjunctiva, causing the graft to scar and fail.

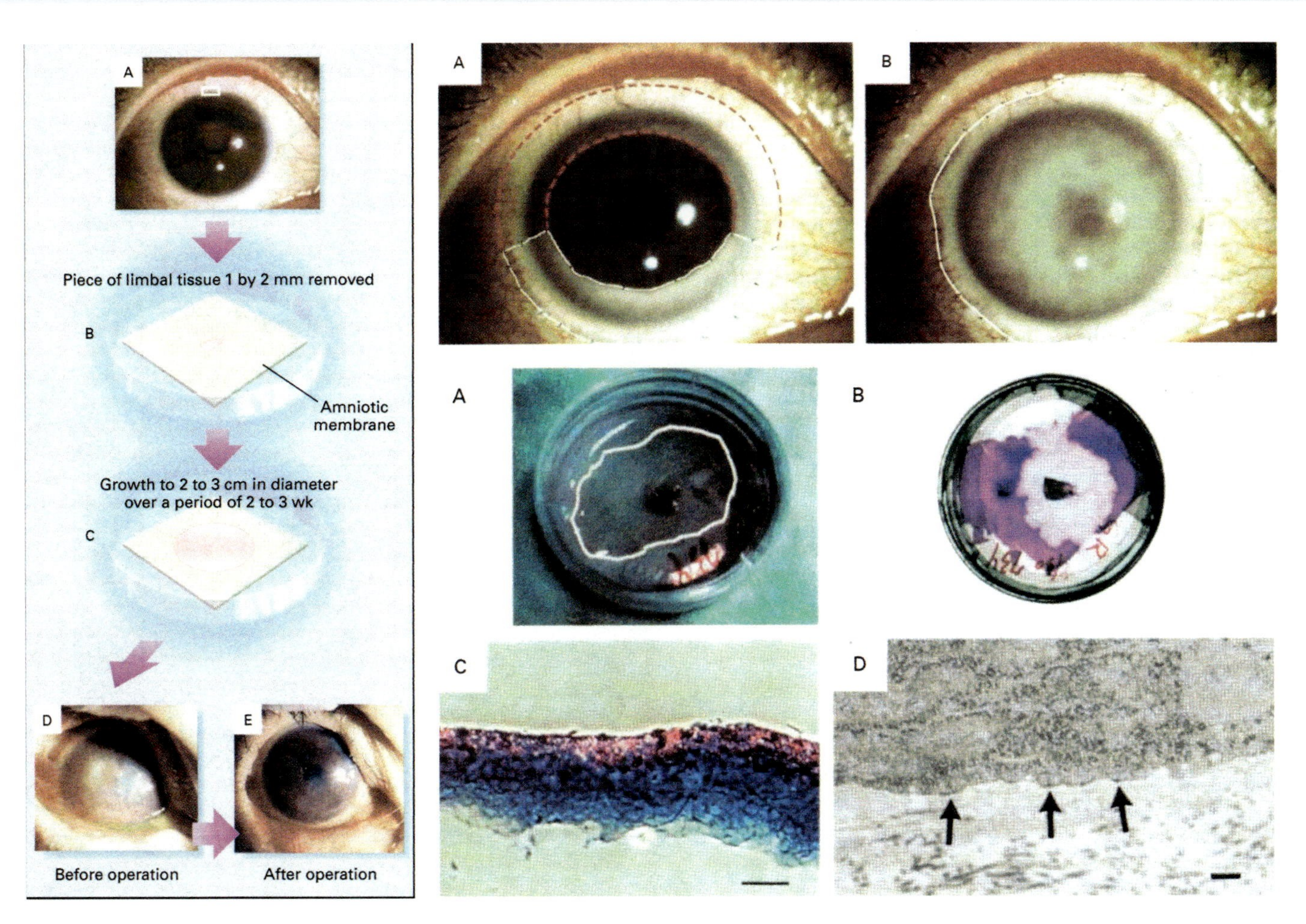


**FIGURE 4.10** Transplants of limbal tissue derived from cultured autogeneic limbal cells. **Left**, **A–E**. Explants of limbal tissue are grown *in vitro* on amniotic membrane and grafted to the edge of the cornea. **Right**, **Top (A)** For eyes with limbal and corneal damage but with a normal central cornea the limbal tissue was transplanted as a half (white lines) or whole circumference (red + white lines). **(B)** For eyes with damage to the entire limbal and corneal surface, a complete ring of limbal tissue with amniotic membrane was transplanted. **Bottom**, **(A)** Cultured limbal tissue (outlined by white line) on amniotic membrane in 35 mm dish with 1.5 ml culture medium. **(B)** Limbal tissue does not stain with Periodic acid-Schiff reagent and Alcian blue, but amniotic membrane stained purple. **(C)** Histological examination shows that the limbal epithelium was composed of 4–5 layers of cells at its margin. **(D)** Electron micrograph shows development of basement membrane with electron-dense focal condensations at the junction between basal cells and amniotic membrane. Scale bars = 200 μm. Reproduced with permission from Tsai et al., Reconstruction of damaged corneas by transplantation of autologous limbal epithelial cells. New Eng J Med 343:86–93. Copyright 2000, Massachusetts Medical Society.

Grafts of limbal tissue have been successful in restoring the cornea in patients who have suffered limbal damage. For eyes in which only one eye, or part of the limbus in both eyes is damaged, autogeneic transplants are possible (Tsai et al., 2000) (**FIGURE 4.10**). In this procedure, a 1 × 2-mm piece of healthy limbal tissue was removed from the patient and cultured for 2–3 weeks on human amniotic membrane, by which time it formed a layer 2–3 cm in diameter. After clearing the cornea of fibrovascular tissue, the amnion/limbal epithelium construct was used as a sectorial or ring graft (in cases where the central cornea was still clear) or as a graft that completely covered the surface of the cornea (in cases with damage to the entire limbal and corneal surface). Vision was improved in five of the six patients autografted in this way.

Autografting of limbal tissue is not possible when patients suffer bilateral damage or disease that results in total corneal stem cell deficiency. In these cases, allografts of limbal tissue can be used. Tsubota et al. (1999) allografted rings of cadaver limbal tissue to patients with severe ocular surface disorders and limbal dysfunction (**FIGURE 4.11**). If the stroma of the cornea was not involved, the diseased epithelium was removed from the cornea. Human amniotic membrane was then applied to the surface of the eye to facilitate epithelialzation and reduce inflammation, and the limbal allograft was sutured in place. If the stroma was involved, the

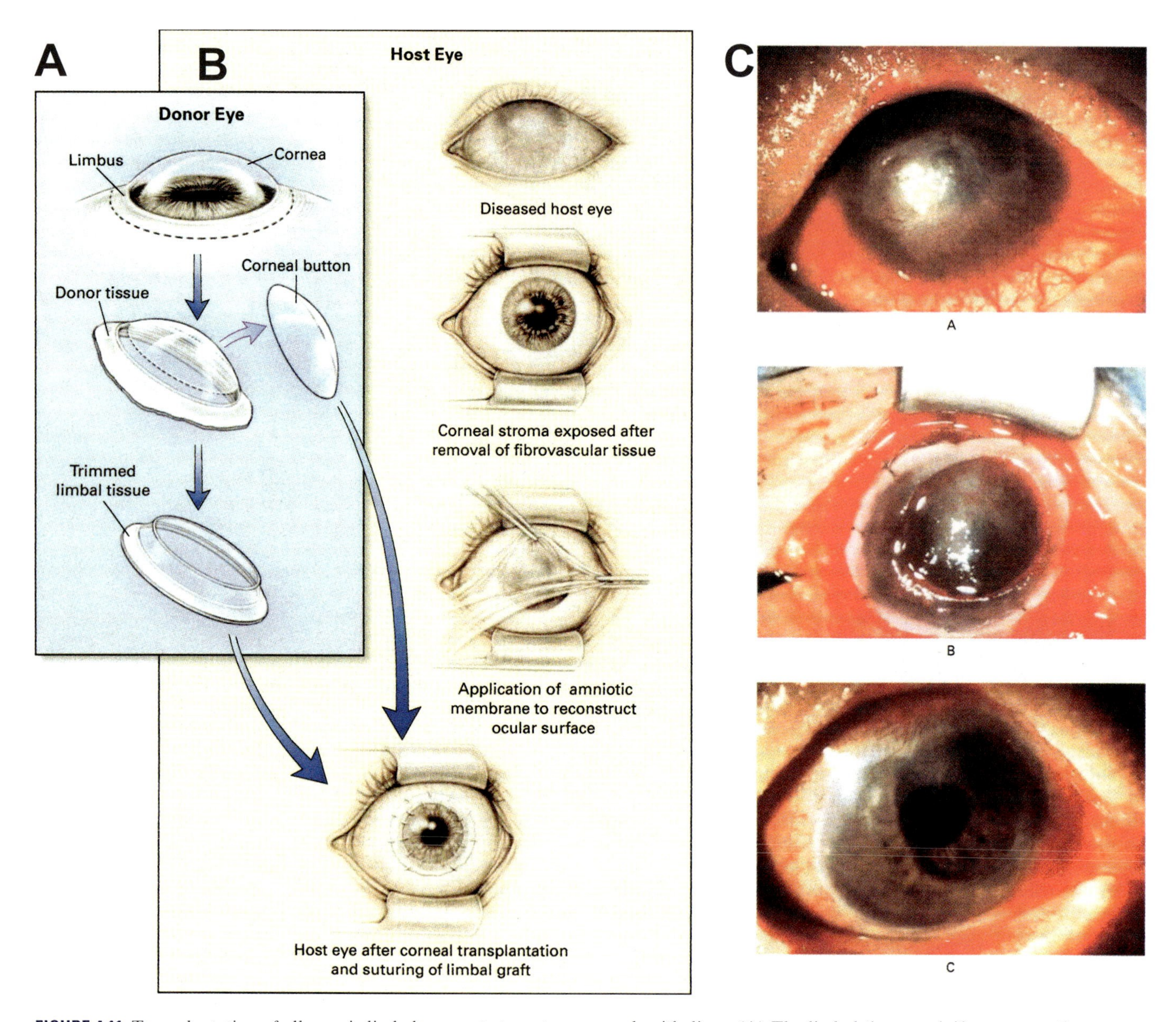


**FIGURE 4.11** Transplantation of allogeneic limbal segments to restore corneal epithelium. **(A)** The limbal tissue and, if necessary, the cornea, were removed from the donor eye. **(B)** The diseased fibrovascular tissue was removed from the host eye to expose the corneal stroma and a sheet of amniotic membrane was applied as a substrate for re-epithelialization and to reduce inflammation. The limbal graft was then sutured in around the rim of the eye. If the host stroma was diseased, it was removed and the donor corneal button transplanted as well. **(C)** Results of the operation in a 32 year old woman with Stevens-Johnson syndrome. **Top**, the opaque ocular surface. **Middle**, donor limbal tissue (arrow) sutured in place. **Bottom**, Seventeen months after surgery a large portion of the cornea is clear. Reproduced with permission from Tsubota et al., Treatment of severe ocular-surface disorders with corneal epithelial stem-cell transplantation. New Eng J Med 340:1697–1703. Copyright 1999, Massachusetts Medical Society.

recipient cornea was removed and a donor cornea was allografted in addition to the limbal tissue. A clear cornea was established and vision improved in 35% of the 43 eyes transplanted.

These allogeneic transplants required immunosuppression and the rejection rate was high. Therefore, Nishida et al. (2004) used autologous oral epithelium as a source of epithelial cells for cases of total limbal insufficiency FIGURE 4.12. A small piece of full-thickness oral epithelium was excised and enzymatically treated to prepare a single cell suspension. The epithelial cells were cultured on one side of a temperature-sensitive polymer, poly(N-isopropylacrylamide), with mitomycin C-treated NIH 3T3 feeder cells on the other side. The polymer is a thin film at 37°C, but reducing the temperature below 30°C causes the

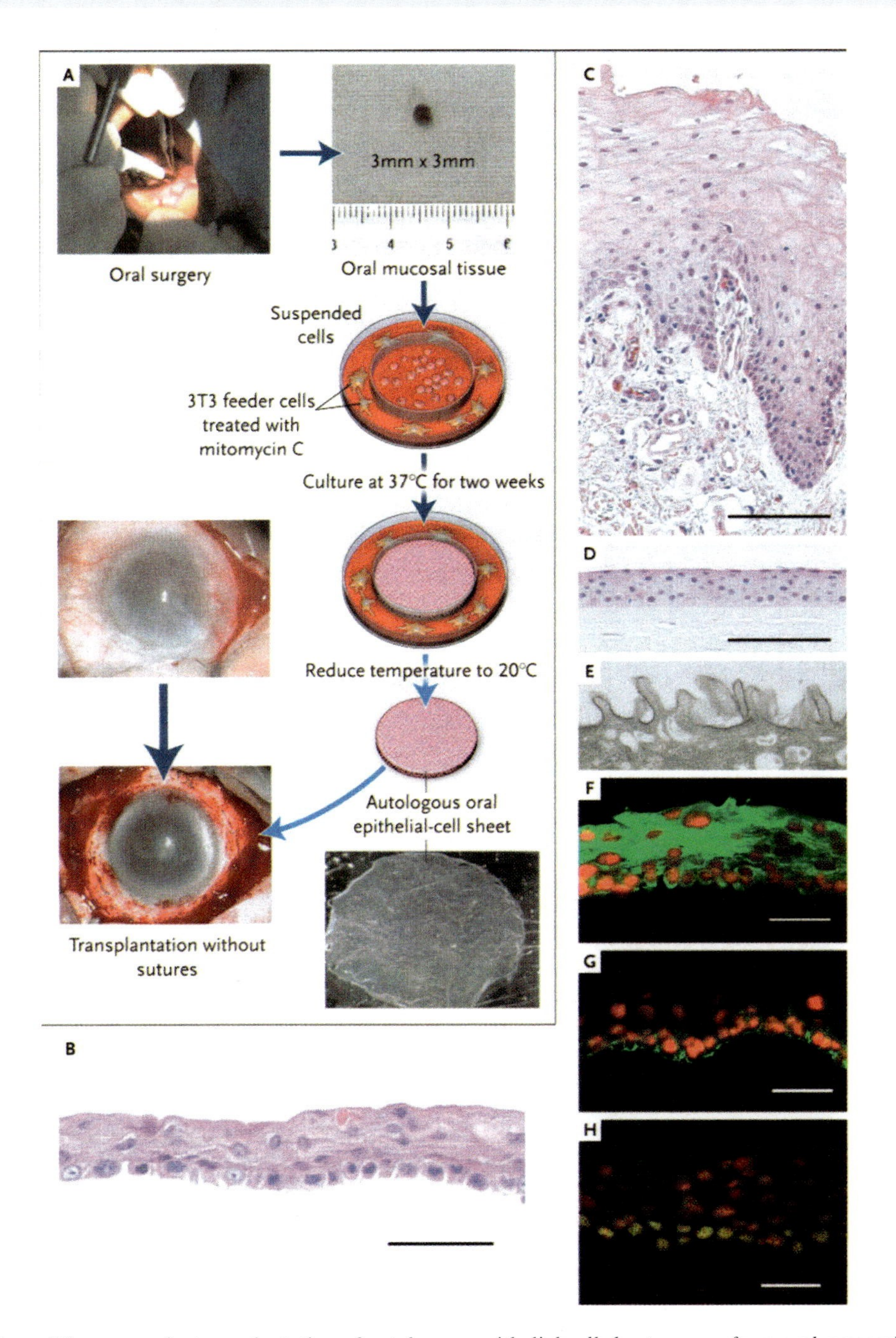


**FIGURE 4.12** Regeneration of the cornea by transplantation of autologous epithelial cell sheets grown from oral mucosal epithelium. **(A)** Oral epithelial cells were grown on temperature-sensitive poly(N-isopropylacrylamide) film at 37°C. The sheet of cells deadheres from the film when the temperature is reduced to 20°C. The sheet is transplanted without sutures to the cornea after removal of scar tissue. The cultured cell sheet **(B)** resembles corneal epithelium **(D)** more closely than oral epithelium **(C)**. **(E)** Electron micrograph of microvilli that developed on the apical surface of the cultured cells. **(F–H)** Cultured epithelial cell sheets stained with antibodies to keratin 3 **(F)**, integrin β1 **(G)**, and the epithelial cell marker p-63 **(H)**. Scale bars in B, F, G and H = 50 μm; scale bars in C, = 100 μm. Reproduced with permission from Nishida et al. Corneal reconstruction with tissue-engineered cell sheets composed of autologous oral epithelium. New Eng. J. Med. 351:1187–1196.

film to hydrate rapidly, allowing detachment of the sheets formed by the epithelial cells. The epithelial cell sheets were applied to the corneal surface without sutures after removal of scar tissue. Four patients were autografted in this way, and all had significantly improved vision. The corneas remained clear over 15 months of follow-up. The corneal epithelium is regenerated by stem cells in the mucosal epithelium, the presence of which was demonstrated by colony-forming assays.

There have been numerous attempts to construct bioartificial corneas (Trinkaus-Randall, 2000). Perhaps the most successful have been corneal equivalents constructed from immortalized cell lines derived from each layer of the human cornea (Griffith et al., 1999). These cells were seeded into a collagen-chondroitin sulfate scaffold crosslinked with glutaraldehyde and the construct was cultured for two weeks. The differentiated constructs were transparent and responded to irritation in the same way that natural human and rabbit corneas do. While these bioartificial corneas have not yet been used to replace diseased and injured corneas, they have applications as alternatives to animal models to evaluate chemicals as irritants, and to study wound healing and cell-matrix interactions.

## SUMMARY

Regenerative therapies are available for acute and chronic skin wounds, hair loss, periodontal injuries and disease, and diseases and injuries of the cornea.

A wide variety of topical agents have been tested for their efficacy in accelerating repair of acute wounds in normal skin. The growth factors TGF-β, FGF-2, EGF, GH, and IGF-1 can accelerate the repair of acute wounds in experimental animals. FGF-2 and GH have this effect in human patients. Other agents reported to accelerate the repair of skin wounds are extract of the *Celosia argentea* leaf, vanadate, oxandrolone, the opoid fentanyl, ketanserin, oleic fatty acids, pig enamel matrix, and the peptide HB-107. These agents increase the rate and extent of re-epithelialization, angiogenesis, and cellularity of granulation tissue. Removal of eschar from burn wounds by debriding agents such as papain/urea improves repair. Still other topical agents act to reduce scarring by decreasing levels of TGF-β, thus mimicking a fetal wound environment more closely. Chitosan, the COX-2 inhibitor celicoxib, HGF, and anti-TGF-β antibodies all reduce TGF-β in wounds promote healing with less scarring, as do hydrogels composed of cross-linked hyaluronic acid and chondroitin sulfate.

The inability of chronic skin wounds to heal on their own is by far the biggest clinical problem in wound repair. A wide variety of regenerative therapies have been devised to treat chronic wounds, including topically applied agents, infrared and near-infrared light, keratinocyte transplants, bioartificial skin equivalents, and acellular dermal regeneration templates.

The topically applied growth factors β-NGF, FGF-2, PDGF-B, EGF, TGF-β, IGF-1 have been reported to improve healing of chronic wounds in animals. Combinations of PDGF-B plus IGF-1, EGF plus insulin, and TGF-β plus PDGF-B were more effective than the individual growth factors alone. In human patients, PDGF-B, FGF-2, EGF, TGF-β, and rhKGF-2 all accelerated the closure of chronic wounds. Currently, only FGF-2, PDGF-B and rhKGF-2 are approved for clinical use. FGF-2 is sold in Japan as Fiblast, and PDGF-B and rhKGF-2 are sold in the USA as Regranex® and repifermin, respectively. Growth factors may accelerate chronic wound repair by decreasing inflammation and enhancing angiogenesis. Other topically applied agents that accelerate the repair of chronic wounds are angiotensin (1–7), thymosin β4, L-arginine, and pentoxifylline. These agents, too, may exert their effect through anti-inflammatory and angiogenesis-promoting activities. Infrared laser light and near-infrared light emitted by LEDs has also been reported to accelerate the repair of wounds in both animals and human patients. If confirmed by large randomized, double-blind clinical trials, acceleration of wound repair by LEDs could become the treatment of choice because of simplicity of treatment, minimal invasiveness, and low cost.

Sheets of allogeneic keratinocytes are used to enhance re-epithelialization of extensive wound areas. They are only slowly rejected by recipients and are replaced by host keratinocytes. The sheets are very fragile, so they are grown on polymer carriers that enable easier transfer to the wound surface. They suffer low take rates when applied to wounds in which the dermis has been badly damaged, because they depend on dermal fibroblasts for basement membrane resynthesis and mitotic factors such as KGF. Thus, complete skin equivalents constructed of a biodegradable polymer scaffold seeded with allogeneic fibroblasts from human foreskin and covered with keratinocytes or an MSTSG have been designed to cover wounds with extensive dermal damage.

The purpose of bioartificial skin is to obviate the need to take large areas of donor skin for full-thickness skin grafts. Bioartificial skin equivalents are temporary living dressings that prevent wound contraction and stimulate repair of the wound by host cells. There are a number of FDA-approved skin equivalents on the market that are similar in nature. The first two of these were Apligraf® and Dermagraft®, which are used for the treatment of the most recalcitrant chronic wounds. Apligraf® consists of a bioartificial dermis of human fibroblasts in bovine type I collagen covered with allogeneic keratinocytes or an autogeneic MSTSG. The bioartificial dermis of Dermagraft® is constructed by growing human fibroblasts on a mesh of Vicryl. Keratinocytes or an MSTSG are added after placement of the construct on the wound bed. Apligraf® has been shown to accelerate the healing of venous ulcers and increase the percentage of ulcers that are closed, while Dermagraft® has been shown to do the same for diabetic ulcers. Skin equivalents have proven their worth in avoiding the need to harvest autogeneic full-thickness skin grafts. They suffer several problems, however. First, lacking blood vessels, they are slow to vascularize if the underlying wound bed is severely damaged; thus, ways are being sought to seed endothelial cells into the construct that will form blood vessels. Second, they do no better than MSTSGs alone in restoring cosmetic appearance, sensation, or skin functions such as thermal regulation, because the repaired skin has no hair follicles, sebaceous glands, or sweat glands. Third, they are cryopreserved until use, which reduces their cellularity. Thus, it is cheaper to use an acellular dermal regeneration matrix covered with keratinocytes or MSTSGs to treat wounds with deep dermal damage. The matrix slowly degrades as it is invaded by host fibroblasts, which repair the dermis.

Dermal regeneration templates can be either natural (derived from tissue) or bioartificial. Several processed collagenous matrices have been approved for clinical use. One of these, Alloderm®, is cadaver dermal matrix. It has been evaluated clinically for burns and was reported to provide good cosmetic appearance and function when used in full thickness. Permacol® is porcine dermal matrix and is effective in hernia repair. SIS is porcine small intestine submucosa. One variant of SIS, Surgisis™, has been shown to be superior to polypropylene mesh in hernia repair, and another, Oasis™, accelerates the healing of diabetic ulcers. Primatrix™ is fetal bovine dermal matrix that is approved for chronic wounds and acute incisional and excisional wounds, including burns. The most widely used bioartificial dermal matrix is Integra®, which consists of bovine dermal collagen and chondroitin 6-sulfate. Clinical assessments of Integra® have reported results superior to those of other constructs for excisional wounds, including burns. Epidermal coverings do not take well on dermal regeneration templates when the dermis is badly damaged, due to slow vascularization. Thus, they are often applied in a two-step procedure in which the dermal template is put on the wound first and allowed to revascularize, after which keratinocytes or an MSTSG are added.

Male pattern baldness is a genetic deficiency leading to hair loss on the temples and crown of the scalp due to sensitivity of the hair follicles in these regions to dihydrotestosterone (DHT). This hormone shortens the anagen stage and lengthens the telogen stage of hair regeneration, leaving empty and involuted hair follicles. Two drugs, minoxidil (Rogaine) and finasteride (Propecia) are available to combat hair loss. Rogaine slows hair loss by an unknown mechanism, whereas Propecia promotes hair growth by blocking the formation of DHT. Shh stimulates hair regrowth in mice suffering from cyclophosphamide-induced alopecia, suggesting that it may be useful in the treatment of hair loss. Other approaches to restoring hair involve the culture of bulge stem cells with dermal papilla cells from DHT-insensitive follicles to create new hair follicles that can be implanted into the scalp.

The current technique to restore periodontal tissue is guided tissue regeneration, which does not produce predictably good outcomes. Other experimental approaches in animals that have reported some success are irradiation with infrared laser light, and the use of stem cells isolated from the periodontal ligament and expanded in culture. When mixed with hydroxyapatite/tricalcium phosphate carrier and transplanted subcutaneously in mice, the cells formed cementum/PDL-like structures. Also envisioned are the regeneration of dentin from pulp stem cells and the creation of tooth buds by combining alveolar epithelium and pulp stem cells. Porcine tooth buds have been created in this way and can grow and develop after implantation into the abdomen of host rats. Nonmammalian research models such as urodele amphibians, sharks, and crocodilians, all of which naturally replace tooth buds, will be useful in understanding the biology and chemistry of tooth development.

The cornea is unable to regenerate if limbal tissue, which provides the stem cells for regeneration, is compromised. If part of the limbus in either eye remains undamaged, pieces of this tissue can be removed and cultured on human amniotic membrane to form a larger sheet. Pieces of this epithelial sheet can then be transplanted as an autogeneic sectorial or ring graft to regenerate and maintain the damaged cornea. In the absence of autogeneic limbal epithelium, oral epithelium has been used as a source of stem cells. These cells form corneal epithelium when transplanted to the corneal stroma. These techniques have been reported to improve vision in human patients.

## REFERENCES


Ashcroft GS, Dodsworth J, van Boxtel E, Tarnuzzer RW, Horan MA, Schultz GS, Ferguson WJ (1997) Estrogen accelerates cutaneous wound healing associated with an increase in TGF-β1 levels. Nature Med 3:1209–1215.

Badylak SF, Coffey AC, Lantz GC, Tacker WA, Geddes LA (1994) Comparison of the resistance to infection of intestinal submucosa arterial autografts versus polytetrafluoroethylene arterial prostheses in a dog model. J Vasc Surg 19:465–472.

Baker CA, Uno H, Johnson GA (1994) Minoxidil sulfation in the hair follicle. Skin Pharmacol 7:335–339.

Balasubramani M, Kumar TR, Babu M (2001) Skin substitutes: A review. Burns 27:534–544.

Beauvoit B, Evans SM, Jenkins TW, Miller EE, Chance B (1995) Contribution of the mitochondrial compartment to the optical properties of the rat liver: A theoretical and practical approach. Analyt Biochem 226:167–174.

Beauvoit B, Kitai T, Chance B (1994) Correlation between the light scattering and the mitochondrial content of normal tissues and transplantable rodent tumors. Biophys J 67:2501–2510.

Bell E, Ehrlich H, Buttle DJ, Nakatsuji T (1981) Living tissue formed in vitro and accepted as skin-equivalent tissue of full thickness. Science 211:1052–1054.

Boulton AJM, Kirsner RS, Vileikyte L (2004) Neuropathic diabetic foot ulcers. New Eng J Med 351:48–55.

Boyce ST (2001) Design principles for composition and performance of cultured skin substitutes. Burns 27:523–533.

Brown G, Nanney LB, Griffin J, Gramer AB, Yancey JM, Curtsinger LJ, Holtzin L, Schultz GS, Jurkiewicz MJ, Lynch JB (1989) Enhancement of wound healing by topical treatment with epidermal growth factor. New Eng J Med 321:76–79.

Burke JF, Yannas IV, Quinbu WC, Bondoc CC, Jung WK (1981) Successful use of a physiologically acceptable artificial skin in the treatment of extensive burn injury. Ann Surg 194:413–428.

Cardoso CRB, Souza MA, Ferro EAV, Favoreto S Jr, Pena JDO (2004) Influence of topical administration of n-3 and n-6 essential and n-9 nonessential fatty acids on the healing of cutaneous wounds. Wound Rep Reg 12:235–243.

Chance B, Nioka S, Kent J, McCully K, Fountain M, Greenfield R, Holtom G (1988) Time-resolved spectroscopy of hemoglobin and myoglobin in resting and ischemic muscle. Analyt Biochem 174:698–707.

Charles CA, Kato T, Tzakis AG, Miller BN, Kirsner RS (2004) Use of a living dermal equivalent for a refractory abdominal defect after pediatric multivisceral transplantation. Dermatol Surg 30: 1236–1240.

Clarke KM, Lantz GC, Salisbury SK, Badylak SF, Hiles MC, Voytik SL (1996) Intestine submucosa and polypropylene mesh for abdominal wall repair in dogs. J Surg Res 60:107–114.

Cochran DL, Wozney JM (1999) Biological mediators for periodontal regeneraton. Periodontol 2000 19:40–58.

Cohen R, Zimber M, Hansbrough JF, Fung YC, Debes J, Skalak R (1991) Tear strength properties of a novel cultured dermal tissue model. Ann Biomed Eng 19A:600 (abstract).

Cohen IK, Crossland MC, Garrett A, Diegelmann RF (1995) Topical application of epidermal growth factor onto partial-thickness wounds in human volunteers does not enhance reepithelialization. Plast Reconstruct Surg 96:251–254.

Conlan MJ, Rapley JW, Cobb CM (1996) Biostimulation of wound healing by low-energy laser irradiation. J Clin Periodont 23:492–496.

Costarelis G, Millar SE (2001) Trends Mol Med 7:293–301.

Crespi R, Covani U, Margone E, Andreana S (1997) Periodontal tissue regeneration in beagle dogs after laser therapy. Lasers in Surg Med 21:395–402.

Demling RH (2000) Oxandrolone, an anabolic steroid, enhances the healing of a cutaneous wound in the rat. Wound Rep Reg 8:97–102.

Desvarieux M, Demmer RT, Rundek T, Rundek T, Boden-Albala B, Jacobs DR, Papapanou PN, Sacco RL (2003) Relationship between periodontal disease, tooth loss, and carotid artery plaque: The Oral Infections and Vascular Disease Epidemiology Study (INVEST). Stroke 34:2120–2125.

DeVries HJC, Middelkoop E, Mekkes JR, Dutrieux RP, Eildevurr CHR, Westerhof W (1994) Dermal regeneration in native non-cross-linked collagen sponges with different extracellular matrix molecules. Wound Rep Reg 2:37–47.

Diegelmann RF, Dunn JD, Lindblad WJ, Cohen IK (1996) Analysis of the effects of chitosan on inflammation, angiogenesis, fibroplasias, and collagen deposition in polyvinyl alcohol sponge implants in rat wounds. Wound Rep Reg 4:48–52.

Druecke D, Lamme EN, Hermann S, Pieper J, May PS, Steinau H-U, Steinstraesser L (2004) Modulation of scar tissue formation using different dermal regeneration templates in the treatment of experimental full-thickness wounds. Wound Rep Reg 12:518–527.

Duailibi MT, Duailibi SE, Young CS, Bartlett JD, Vacanti JP, Yelick PC (2004) Bioengineered teeth from cultured rat tooth bud cells. J Dent Res 83:523–528.

Edelman DS (2002) Laparoscopic herniorraphy with porcine small intestinal submucosa: A preliminary study. J Soc Laparoendoscopic Surgeons 6:203–205.

Ehrlich H (2004) Understanding experimental biology of skin equivalent: From laboratory to clinical use in patients with burns and chronic wounds. Am J Surg 187S:29S–33S.

Elter JR, Offenbacher S, Toole JF, Beck JD (2003) Relationship of periodontal disease and edentulism to stroke/TIA. J Dent Res 82:998–1001.

Enwemeka CS (2004) Therapeutic light. Rehab Mgmt, Jan/Feb.

Erdag G, Sheridan RL (2004) Fibroblasts improve performance of cultured composite skin substitutes on athymic mice. Burns 30:322–328.

Eriksson E, Vranckx J (2004) Wet wound healing: From laboratory to patients to gene therapy. Am J Surg 188S:36S–41S.

Falanga V, Margolis D, Alvarez O, Auletta M, Maggiacomo F, Altman M, Jensen J, Sabolinski M, Hardin-Young J (1998) Rapid healing of venous ulcers and lack of clinical rejection with an allogeneic cultured human skin equivalent. Arch Dermatol 134: 293–299.

Falanga V, Sabolinski M (1999) A bilayered living skin construct (APLIGRAF®) accelerates complete closure of hard-to-heal venous ulcers. Wound Rep Reg 7:201–207.

Falanga V, Fujitani RM, Diaz C, Hunter G, Jorizzo J, Lawrence PF, Lee BY, Menzoian JO, Tretbar LL, Holloway GA, Hoballah J, Seabrook GR, Mcmillan DE, Wolf W (1999) Systemic treatment of venous leg ulcers with high doses of pentoxifylline: Efficacy in a randomized, placebo-controlled trial. Wound Rep Reg 7:208–213.

Ferguson MWJ, O'Kane S (2004) Scar-free healing: From embryonic mechanisms to adult therapeutic intervention. Phil Trans R Soc Lond B 359:839–850.

Franklin ME, Gonzalez JJ, Michaelson R, Glass JL, Chock DA (2002) Preliminary experience with new bioactive prosthetic material for repair of hernias in infected fields. Hernia 6:171–174.

Franklin ME, Gonzalez JJ, Glass JL (2004) Use of porcine small intestinal submucosa as a prosthetic device for laparoscopic repair of hernias in contaminated fields: 2-year follow-up. Hernia 8:186–189.

Franzen LE, Ghassemifar N, Nordman J, Schultz G, Skogman R (1995) Mechanisms of TGF-β action in connective tissue repair of rat mesenteric wounds. Wound Rep Reg 3:322–329.

Fu X, Shen Z, Chen Y, Xie J, Guo Z, Zhang M, Sheng Z (1998) Randomized placebo-controlled trial use of topical recombinant bovine basic fibroblast growth factor for second-degree burns. Lancet 352:1661–1664.

Fu X, Li X, Chen W, Sheng Z (2005) Engineered growth factors and cutaneous wound healing: Success and possible questions in the past 10 years. Wound Rep Reg 13:122–130.

Gallico GG III, O'Connor NE, Compton CC, Kehinde O, Green H (1984) Permanent coverage of large burn wounds with autologous cultured human epithelium. New Eng J Med 34: 448.

Gentzkow GD, Iwasaki SD, Hershon KS, Mengel M, Prendergast JJ, Ricotta JJ, Steed DP, Lipkin S (1996) Use of Dermagraft, a cultured human dermis, to treat diabetic foot ulcers. Diabetes Care 19:350–354.

Gilpin DA, Barrow RE, Rutan RL, Broemling L, Herndon DN (1994) Recombinant human growth hormone accelerates wound healing in children with large cutaneous burns. Ann Surg 220: 19–24.

Green H, Kehinde O, Thomas J (1979) Growth of human epidermal cells into multiple epithelia suitable for grafting. Proc Natl Acad Sci USA 76:5665–5668.

Greenalgh DG (1996) The role of growth factors in wound healing. J Trauma Injury Infection 41:159–167.

Greenalgh DG, Rieman M (1994) Effects of basic fibroblast growth factor on the healing of partial-thickness donor sites. A prospective, randomized double-blind trial. Wound Rep Reg 2:113–121.

Griffith M, Osborne R, Munger R, Xiong X, Doillon CJ, Laycock NLC, Hakim M, Song Y, Watsky MA (1999) Functional human corneal equivalents constructed from cell lines. Science 286: 2169–2172.

Guerret S, Govignon E, Hartmann D, Ronfard V (2003) Long-term remodeling of a bilayered living human skin equivalent (Apligraf®) grafted onto nude mice: Immunolocalization of human cells and characterization of extracellular matrix. Wound Rep Reg 11: 35–45.

Hansbrough JF, Cooper ML, Cohen R, Spielvogel R, Greenleaf G, Bartel R, Naughton G (1992a) Evaluation of a biodegradable matrix containing cultured human fibroblasts as a dermal replacement beneath meshed skin grafts on athymic mice. Surgery 111: 438–446.

Hansbrough JF, Dore C, Hansbrough WB (1992b) Clinical trials of a living dermal tissue replacement placed beneath meshed, split-thickness skin grafts on excised burn wounds. J Burn Care Rehabil 13:519–529.

Hebda PA, Flynn KJ, Dohar JE (1998) Evaluation of the efficacy of enzymatic debriding agents for removal of necrotic tissue and promotion of healing in porcine skin wounds. Wounds 10:83–96.

Heijl L, Heden G, Svardstrom G, Ostgren A (1997) Enamel matrix derivative (EMDOGAIN) in the treatment of intrabony periodontal defects. J Clin Periodontol 24:705–714.

Heimbach D, Luterman A, Burke J, Cram A, Herndon D, Hunt J, Jordan M, McManus W, Solem L, Warden G, Zawacki B (1988) Artificial dermis for major burns. Ann Surg 208:313–320.

Heitland A, Piatkowski A, Noah EM, Pallua N (2004) Update on the use of collagen/glycosaminoglycan skin substitute—six years of experiences with artificial skin in 15 German burn centers. Burns 30:471–475.

Herndon DN, Barrow RE, Kunkel KR, Broemeling L, Rutan RL (1990) Effects of human growth hormone on donor-site healing in severely burned children. Ann Surg 212:424–431.

Hodde JP, Badylak SF, Brightman AO, Voytik-Harbin SL (1996) Glycosaminoglycan content of small intestinal submucosa: Bioscaffold for tissue replacement. Tiss Eng 229–217.

Horch RE, Bannasch H, Stark GB (2001) Transplantation of cultured autologous keratinocytes in fibrin sealant biomatrix to resurface chronic wounds. Transplant Proc 33:642–644.

Horne KA, Forrester JC, Jahoda CAB (1989) Isolated human hair follicle dermal papillae induce hair growth in in athymic mice. British J Dermatol 122:267.

Horne KA, Jahoda CAB (1992) Restoration of hair growth by surgical implantation of follicular dermal sheath. Development 116: 563–571.

Houghton PE, Keefer KA, Krummel TM (1995) The role of transforming growth factor-beta in the conversion from "scarless" healing to healing with scar formation. Wound Rep Reg 3:229–236.

Ichioka S, Ohura N, Nakatsuka T (2005) The positive experience of using a growth-factor product on deep wounds with exposed bone. J Wound Care 14:105–109.

Inadomi T (2004) Basic fibroblast growth factor: A new approach to treat skin ulcers. Prog Med 24:451–453.

Jahoda CAB, Horne KA, Oliver RF (1984) Induction of hair growth by implantation of cultured dermal papilla cells. Nature 311:560–562.

Jahoda CAB, Horne KA, Mauger A, Bard S, Sengel P (1992) Cellular and extracellular involvement in the regeneration of the rat lower vibrissa follicle. Development 114:887–897.

Jimenez PA, Jimenez SE (2004) Tissue and cellular approaches to wound repair. Am J Surg 187S:56S–64S.

Jones I, James SE, Rubin P, Martin R (2003) Upward migration of cultured autologous keratinocytes in Integra™ artificial skin: A preliminary report. Wound Rep Reg 11:132–138.

Jorgensen PH, Oxlund H (1996) Growth hormone increases the biomechanical strength and collagen deposition rate during the early phase of skin wound healing. Wound Rep Reg 4:40–47.

Karu T (1999) Primary and secondary mechanisms of action of visible to near-IR radiation on cells. J Photochem Photobiol B: Biology 49:1–17.

Kearney JN (2001) Clinical evaluation of skin substitutes. Burns 27:545–551.

Keswani SG, Katz AB, Lim F-Y, Zoltick P, Radi A, Alaee D, Heryln M, Crombleholme TM (2004) Adenoviral mediated gene transfer of PDF-B enhances wound healing in type I and type II diabetic ulcers. Wound Rep Reg 12:497–504.

Kim SS, Vacanti JP (1999) The current status of tissue engineering as potential therapy. Sem Pediatr Surg 8:119–123.

Kim M, Ustuner ET, Schuschke D, Morsing A, Kjolseth D, Fingar V, Wieman J, Kamler M, Tobin GR, Bond S, Barker JH (1995) Ketanserin accelerates wound epithelialization and neovascularization. Wound Rep Reg 3:506–511.

Kini S, Gagner M, de Csepel J, Gentileschi P, Dakin G (2001) A biodegradable membrane from porcine intestinal submucosa to reinforce the gastrojejunostomy in laparascopic Roux-en-Y gastric bypass: preliminary report. Obesity Surg 11:469–474.

Kirchner LM, Merbaum SO, Gruber BS, Knoll AK, Bulgrin J, Taylor RAJ, Schmidt SP (2003) Effects of vascular endothelial growth factor on wound closure rates in the genetically diabetic mouse model. Wound Rep Reg 11:127–131.

Kiritsy CP, Antoniades HN, Carlson MR, Beaulieu MT, D'Andrea M, Lynch SE (1995) Combination of platelet-derived growth factor-BB and insulin-like growth factor-1 is more effective than platelet-derived growth factor-BB alone in stimulating complete healing of full-thickness wounds in "older" diabetic mice. Wound Rep Reg 3:340–350.

Kirker KR, Luo Y, Nielson JH, Shelby J, Prestwich GD (2002) Glycosaminoglycan hydrogel films as bio-interactive dressings for wound healing. Biomaterials 23:3661–3671.

Koide M, Osaki K, Konishi J, Oyamada K, Katakura T, Takahashi A (1993) A new type of biomaterial for artificial skin: Dehydrothermally cross-linked composites of fibrillar and denatured collagens. Biomed Mater Res 27:79–87.

Kono T, Tanii T, Furukawa M, Mizuno N, Kitajima J, Ishii M, Hamada T, Yoshizato K (1990) Cell cycle analysis of human dermal fibroblasts cultured on or in hydrated type I collagen lattices. Arch Dermatol Res 282:258–262.

Krause TJ, Zazanis GA, McKinnon RD (1996) Prevention of postoperative adhesions with the chitin derivative N-O-carboxymethylchitosan. Wound Rep Reg 4:53–57.

Landeen LK, Zeigler FC, Halberstadt C, Cohen R, Slivka SR (1992) Characterization of a human dermal replacement. Wounds 4:167–175.

Langer R, Vacanti JP (1993) Tissue engineering. Science 260:920–926.

Larochelle F, Ross G, Rouhabhia M (1998) Permanent skin replacement using engineered epidermis containing fewer than 5% syngeneic keratinocytes. Lab Invest 78:1089–1099.

Lattari V, Jones LM, Varcelotti JR, Latenser BA, Sherman HF, Barrette RR (1997) The use of a permanent dermal allograft in full-thickness burns of the hand and foot: A report of three cases. J Burn Care Rehabil 18:147–155.

Lee PHA, Rudisill A, Lin KH, Zhang L, Harris SM, Falla TJ, Gallo RL (2004) HB-107, a nonbacteriostatic fragment of the antimicrobial peptide cecropin B, accelerates murine wound repair. Wound Rep Reg 12:351–358.

L'Heureux MM, Pouliot R, Xu W, Auger F, Germain L (1999) Characterization of a new tissue-engineered human skin equivalent with hair. In Vitro Cell Dev Biol Anim 35:318–326.

Liu JY, Hafner J, Dragieva G, Seifert B, Burg G (2004) Autologous cultured keratinocytes on porcine gelatin microbeads effectively heal chronic venous leg ulcers. Wound Rep Reg 12:148–156.

Lobie PE, Breipohl W, Lincoln DT, Garcia-Aragon J, Waters MJ (1990) Localization of growth hormone receptor/binding protein in skin. J Endocrin 126:467–471.

Mackay DJ, Moyer KE, Saggers GC, Myers RL, Mackay DR, Ehrlich HP (2003) Topical vanadate optimizes collagen organization within granulation tissue. Wound Rep Reg 11:204–212.

MacLeod TM, Sarathchandra P, Williams G, Sanders R, Green C (2004) Evaluation of a porcine origin acellular dermal matrix and small intestinal submucosa as dermal replacements in preventing secondary skin graft contraction. Burns 30:431–437.

Malette WG, Quigley HJ, Gaines RD, Johnson ND, Rainer WG (1983) Chitosan: A new hemostatic. Ann Thoracic Surg 36:55–58.

Mansbridge J, Liu K, Patch R, Symons K, Pinney E (1998) Three-dimensional fibroblast culture implant for the treatment of diabetic foot ulcers: Metabolic activity and therapeutic range. Tiss Eng 4:403–414.

Marston WA, Hanft J, Norwood P, Pollak R (2003) The efficacy and safety of Dermagraft in improving the healing of chronic diabetic foot ulcers results of a prospective randomized trial. Diabetes Care 26:1701–1705.

Mast BA, Schultz GS (1996) Interactions of cytokines, growth factors and proteases in acute and chronic wounds. Wound Rep Reg 4:411–420.

Matsumoto K, Nakamura T (1996) Emerging multipotent aspects of hepatocyte growth factor. J Biochem 119:591–600.

McClellan KJ, Markham A (1999) Finasteride: A review of its use in male pattern hair loss. Drugs 57:111–126.

McPherson TB, Badylak SF (1998) Characterization of fibronectin derived from porcine small intestinal submucosa. Tiss Eng 4:75–83.

Mester E, Ludany M, Seller M (1968) The stimulating effect of low power laser ray on biological systems. Laser Rev 1:3.

Mirastschijski U, Konrad D, Lundberg E, Lyngstadaas S, Jorgensen LN, Agren MS (2004) Effects of a topical enamel matrix derivative on skin wound healing. Wound Rep Reg 12:100–108.

Mis B, Rolland E, Ronfard V (2004) Combined use of a collagen-based dermal substitute and a fibrin-based cultured epithelium: A step toward a total skin replacement for acute wounds. Burns 30:713–719.

Muangman P, Muffley LA, Anthony JP, Spenny ML, Underwood RA, Olerud JE, Gibran NS (2004) Nerve growth factor accelerates wound healing in diabetic mice. Wound Rep Reg 12:44–52.

Murray PE, Garcia-Godoy F (2004) Stem cell responses in tooth regeneration. Stem Cells Dev 13:255–262.

Mustoe T (2004) Understanding chronic wounds: A unifying hypothesis on their pathogenesis and implications for therapy. Am J Surg 187S:65S–70S.

Mustoe TA, Cutler NR, Allman RM, Goode PS, Deuel TF, Prause JA, Bear M, Serdar CM, Pierce GF (1994) A phase II study to evaluate recombinant platelet-derived growth factor BB in the treatment of stage 3 and 4 pressure ulcers. Arch Surg 129:213–219.

Nakahara T, Nakamura T, Kobayashi E, Kuremoto K-I, Matsuno T, Tabata Y, Eto K, Shimizu Y (2004) *In situ* tissue engineering of periodontal tissues by seeding with periodontal ligament-derived cells. Tiss Eng 10:537–544.

Nanney L, Fortney DZ, Durham DR (1995) Effect of vibriolysin, an enzymatic debriding agent, on healing of partial-thickness burn wounds. Wound Rep Reg 3:442–448.

National Institute of Dental and Craniofacial Research: Strategies for Tooth Structure Regeneration (2000) http://rarediseases.info.nih.gov/news-reports/workshops/tooth010227.html.

Nevins M, Camelo M, Nevins ML, Schenk RK, Lynch SE (2003) Periodontal regeneration in humans using recombinant human platelet-derived growth factor BB (rhPDGF-BB) and allogeneic bone. J Periodontol 74:1282–1292.

Nishida K, Yamato M, Hayashida Y, Watanabe K, Yamamoto K, Adachi E, Nagai S, Maeda N, Watanabe H, Okano T, Tano Y (2004) Corneal reconstruction with tissue-engineered cell sheets composed of autologous oral mucosal epithelium. New Eng J Med 351:1187–1196.

O'Connor NE, Mulliken JB, Banks-Schlegel S, Kehinde O, Green H (1981) Grafting of burns with cultured epithelium prepared from autologous epidermal cells. Lancet 175–178.

Oelschlager BK, Barreca M, Chang L, Pellegrini CA (2003) The use of small intestine submucosa in the repair of paraesophageal hernias: Initial observations of a new technique. Am J Surg 186:4–8.

Ohazama A, Modino SAC, Miletich I, Sharpe PT (2004) Stem-cell-based tissue engineering of murine teeth. J Dent Res 83:518–522.

Ono I, Yamashita T, Hida T, Jin H-Y, Ito Y, Hamada H, Akasaka Y, Ishii T, Jimbow K (2004) Combined administration of basic fibroblast growth factor protein and the hepatocyte growth factor gene enhances the regeneration of dermis in acute insisional wounds. Wound Rep Reg 12:67–79.

Owen T, Lantz GC, Hiles MC, VanFleet J, Martin BR, Geddes LA (1997) Calcification potential of small intestinal submucosa in a rat subcutaneous model. J Surg Res 71:179–186.

Palao R, Gomez P, Huguet P (2003) Burned breast reconstructive surgery with integra dermal regeneration template. British J Plastic Surg 56:252–259.

Pandya AN, Woodward B, Parkhouse N (1998) The use of cultured autologous keratinocytes with Integra in the resurfacing of acute burns. Plastic and Reconstruct Surg 102:825–827.

Philp D, Badamchian M, Scheremeta B, Nguyen M, Goldstein A, Kleinman HK (2003) Thymosin β4 and a synthetic peptide containing its actin-binding domain promote dermal wound repair in db/db diabetic mice and in aged mice. Wound Rep Reg 11:19–24.

Poonawala T, Levay-Young BK, Hebbel R, Gupta K (2005) Opioids heal ischemic wounds in the rat. Wound Rep Reg 13:165–174.

Priya KS, Arumugam G, Rathinam B, Wells A, Babu M (2004) Celosia argentea Linn leaf extract improves wound healing in a rat burn model. Wound Rep Reg 12:618–625.

Puolakkainen A, Reed MJ, Gombotz WR, Twardzik DR, Abrass IB, Sage EH (1995) Acceleration of wound healing in aged rats by topical application of transforming growth factor-$\beta_1$. Wound Rep Reg 3:330–339.

Reddy G, Stehno-Bittel L, Enwemeka CS (2001) Laser photostimulation accelerates wound healing in diabetic rats. Wound Rep Reg 9:248–255.

Roberts AB (1995) Transforming growth factor-β: Activity and efficacy in animal models of wound healing. Wound Rep Reg 3:408–418.

Robson MC (1997) The role of growth factors in the healing of chronic wounds. Wound Rep Reg 5:12–17.
Robson MC, Phillip LG, Cooper DM, Lyle WG, Robson LE, Odom L, Hill DP, Hanham AF, Sander GA (1995) Safety and effect of transforming growth factor—beta$_2$ for treatment of venous stasis ulcers. Wound Rep Reg 3:157–167.
Robson C, Phillips TJ, Falanga V, Odenheimer DJ, Parish LC, Jensen JL, Steed DL (2001) Randomized trial of topically applied repifermin (recombinant human keratinocyte growth factor-2) to accelerate wound healing in venous ulcers. Wound Rep Reg 9:347–352.
Rochkind S, Rousso M, Nissan M, Villarrea M, Barr-Nea L, Rees DG (1989) Systemic effects of low-power laser irradiation on the peripheral and central nervous system, cutaneous wounds and burns. Lasers in Surg Med 9:174–182.
Rodgers K, Xiong S, Felix J, Roda N, Espinoza T, Maldonado S, Dizerega G (2001) Development of angiotensin (1–7) as an agent to accelerate dermal repair. Wound Rep Reg 9:238–247.
Rouabhia M (1996) Permanent skin replacement using chimeric epithelial cultured sheets comprising xenogeneic and syngeneic keratinocytes. Transplant 61:1290–1300.
Sabolinski ML, Alvarez O, Auletta M, Mulder G, Parentau NL (1996) Cultured skin as a "smart material" for healing wounds: Experience in venous ulcers. Biomaterials 17:311–320.
Sahota PS, Burn JL, Heaton M, Freelander E, Suvarna SK, Brown NJ, Mac Neil S (2003) Development of a reconstructed human skin model for angiogenesis. Wound Rep Reg 11:275–284.
Sahota PS, Burn JL, Brown N, Macneil S (2004) Approaches to improve angiogenesis in tissue-engineered skin. Wound Rep Reg 12:635–642.
Sato N, Leopold P, Crystal RG (1999) Induction of the hair growth phase in postnatal mice by localized transient expression of sonic hedgehog. J Clin Invest 104:855–864.
Sato N, Leopold PL, Crystal RG (2001) Effect of adenovirus-mediated expression of sonic hedgehog gene on hair regrowth in mice with chemotherapy-induced alopecia. J Natl Cancer Inst 93:1858–1864.
Schindl A, Schindl M, Pernerstorfer-Schon H, Schindl L (2000) Low-intensity laser therapy: A review. J Invest Med 48:312–326.
Seo B-M, Miura M, Gronthos S, Bartold PM, Batouli S, Brahim J, Young M, Robey PG, Wang C-U, Shi S (2004) Investigation of multipotent postnatal stem cells from human periodontal ligament. The Lancet 364:149–155.
Shafritz TA, Rosenberg LC, Yannas IV (1994) Specific effects of glycosaminoglycans in an analog of extracellular matrix that delays wound contraction and induces regeneration. Wound Rep Reg 2:270–276.
Shah M, Foreman D, Ferguson MW (1994) Neutralization of TGF-beta 1 and TGF-beta 2 or exogenous addition of TGF-beta 3 to cutaneous rat wounds reduces scarring. J Cell Sci 108:985–1002.
Shakespeare P (2001) Burn wound healing and skin substitutes. Burns 27:517–522.
Sheridan RL, Tompkins R, Burke J (1994) Management of burn wounds with prompt excision and immediate closure. J Intensive Care Med 9:6–19.
Sheridan RL, Morgan JR, Cusik JL, Petras LM, Lydon MM, Tompkins RG (2001) Initial experience with a composite autologous skin substitute. Burns 27:421–424.
Sheridan RL, Choucair RJ (1997) Acellular allogeneic dermis does not hinder initial engraftment in burn wound resurfacing and reconstruction. J Burn Care Rehabil 18:496–499.
Shi HP, Most D, Efron DT, Witte MB, Barbul A (2003) Supplemental L-arginine enhances wound healing in diabetic rats. Wound Rep Reg 11:198–203.
Shimizu-Nishikawa K, Yoshizato K (1990) An epidermal factor which stimulates the synthesis of collagenase in fibroblasts in a reconstituted dermal model. Biomed Res 11:231–241.
Smiell JM, Wieman TJ, Steed DL, Perry B, Sampson A, Schwab BH (1999) Efficacy and safety of becaplermin (recombinant human platelet-derived growth factor-BB) in patients with nonhealing, lower extremity diabetic ulcers: A combined analysis of four randomized cases. Wound Rep Reg 7:335–346.
Soiderer EE, Lantz GC, Kazacos EA, Hodde JP, Wiegand RE (2004) Morphologic study of three collagen materials for body wall repair. J Surg Res 118:161–175.
Springer K, Brown M, Stulberg DL (2003) Common hair loss disorders. Am Family Physician 68:93–102.
Stadler I, Lanzafame RJ, Evans R, Narayan V, Dailey B, Buehner N, Naim JO (2001) 830-nm irradiation increases the wound tensile strength in a diabetic murine model. Lasers in Surg and Med 28:220–226.
Steenfos HH, Jansson J-O (1992) Growth hormone stimulates granulation tissue formation and insulin-like growth factor I gene expression in wound chambers in the rat. J Endocrin 132:293–298.
Steiner H, Hultmark D, Engstrom A, Bennich H, Boman HG (1981) Sequence and specificity of two antibacterial proteins involved in insect immunity. Nature 292:246–248.
Suzuki S, Matsuda K, Nishimura Y, Maruguchi Y, Maruguchi T, Ikada Y, Morita S-I, Morota K (1996) Review of acellular and cellular artificial skins. Tiss Eng 2:267–275.
Takayama S, Murakami S, Shimabukuro Y, Kitamura M, Okada H (2001) Periodontal regeneration by FGF-2 (bFGF) in primate models. J Dent Res 80:2075–2079.
Talwar M, Moyana TN, Bharadwaj B, Tan LK (1996) The effect of a synthetic analogue of prostaglandin E2 on wound healing in rats. Ann Clin Lab Sci 26:451–457.
Tepper OM, Galiano RD, Capla JM, Kalka C, Gagne PJ, Jacobowitz GR, Levine JP, Gurtner GC (2002) Human endothelial progenitor cells from type II diabetics exhibit impaired proliferstion, adhesion, and incorporation into vascular structures. Circulation 106:2781–2786.
Tomic-Canic M, Brem H (2004) Gene array technology and pathogenesis of chronic wounds. Am J Surg 188S:67S–72S.
Trengrove NJ, Bielfeldt-Ohmann H, Stacey MC (2000) Mitogenic activity and cytokine levels in non-healing and healing chronic leg ulcers. Wound Rep Reg 8:13–25.
Trinkaus-Randall V (2000) Cornea. In: Lanza RP, Langer R, Vacanti J (eds.). Principles of Tissue Engineering, 2nd ed. New York, Academic Press, pp 471–491.
Tsai R J-F, Li L-M, Chen J-K (2000) Reconstruction of damaged corneas by transplantation of autologous limbal epithelial cells. New Eng J Med 343:86–93.
Tsubota K, Satake Y, Kaido M, Shinozaki N, Shimmura S, Bissen-Miyajima H, Shimazaki J (1999) Treatment of severe ocular-surface disorders with corneal epithelial stem-cell transplantation. New Eng J Med 340:1697–1703.
Tyrell J, Silberman H, Chandrasoma P, Niland J, Shull J (1989) Absorbable versus permanent mesh in abdominal operations. Surg Gynecol Obstet 188:227–232.
Van Den Bogaerdt AJ, Ulrich MMW, Van Galen MJM, Reijnen L, Verkerk M, Pieper J, Lamme EN, Middelkoop E (2004) Upside-down transfer of porcine keratinocytes from a porous, synthetic dressing to experimental full-thickness wounds. Wound Rep Reg 12:225–234.
Van Dorp AGM, Verhoeven MCH, Koerten HK, Van Der Nat-Van Der Meij TH, Van Blitterswijk CA, Ponec M (1998) Dermal regeneration in full-thickness wounds in Yucatan miniature pigs using a biodegradable copolymer. Wound Rep Reg 6:556–568.

Veves A, Falanga V, Armstrong DG, Sabolinski ML (2001) Graftskin, a human skin equivalent, is effective in the management of non-infected neuropathic diabetic foot ulcers: A prospective randomized multicenter clinical trial. Diabetes Care 24: 290–295.

Voyles CR, Richardson JD, Bland KI, Tobin GR, Flint LM, Polk HC Jr (1981) Emergency abdominal wall reconstruction with polypropylene mesh. Short-term benefits versus long-term complications. Ann Surg 194:219–223.

Voytik-Harbin SL, Brightman AO, Waisner BZ, Robinson JP, Lamar CH (1997) Small intestinal submucosa: A tissue-derived extracellular matrix that promotes tissue-specific growth and differentiation of cells *in vitro*. Tiss Eng 4:157–174.

Wainwright D (1995) Use of an acellular allograft dermal matrix (Alloderm) in the management of full-thickness burns. Burns 21:243–248.

Wang LC, Liu ZY, Gambardella L, Delacour A, Shapiro R, Yang J, Sizing I, Rayhorn P, Garber EA, Benjamin CD, Williams KP, Taylor FR, Barrandon Y, Ling L, Burkly LC (2000) Conditional disruption of hedgehog signaling pathway defines its critical role in hair development and regneration. J Invest Dermatol 114:901–908.

Wennstrom J, Lindhe J (2002) Some effects of enamel matrix proteins on wound healing in the dento-gingival region. J Clin Periodontol 29:9–14.

Whelan HT, Smits L, Buchmann EV, Whelan NT, Turner SG, Margolis DA, Cevenini V, Stinson H, Ignatius R, Martin T, Cwiklinski J, Philippi AF, Graf WR, Hodgson B, Gould L, Kane M, Chen G, Caviness J (2001) Effect of NASA light-emitting diode (LED) irradiation on wound healing. J Clin Laser Med Surg 19:305–314.

Wilgus R, Vodovotz Y, ittadini E, Clubbs EA, Oberyszyn TM (2003) Reduction of scar formation in full-thickness wounds with topical celecoxib treatment. Wound Rep Reg 11:15–34.

Woodley DT, Briggaman A, Herzog SR, Meyers AA, Peterson HD, O' Keefe EJ (1990) Characterization of "neodermis" formation beneath cultured human epidermal autografts transplanted on muscle fascia. J Invest Dermatol 95:20–26.

Wysocki AB, Staiano-Coico L, Grinnell F (1993) Wound fluid from chronic leg ulcers contans elevated levels of metalloproteinases MMP-2 and MMP-9. J Invest Dermatol 101:64–68.

Yamagishi K, Onuma K, Suzuki T, Okada F, Tagami J, Otsuki M, Senawangse P (2005) A synthetic enamel for rapid tooth repair. Nature 433:819.

Yancy KB (1995) Adhesion molecules II. Interactions of keratinocytes with epidermal basement membrane. J Invest Dermatol 104:1008–1014.

Yannas IV (2001) Tissue and Organ Regeneration in Adults. New York, Springer.

Yoshizato K, Yoshikawa E (1994) Development of bilayered gelatin substrate for bioskin: a new structural framework of the skin composed of porous dermal matrix and thin basement membrane. Mater Sci Eng C1:95–105.

Young CS, Terada S, Vacanti JP, Honda M, Bartlett JD, Yelick PC (2002) Tissue engineering of complex tooth structures on biodegradable polymer scaffolds. J Dent Res 81:695–700.

Yu W, Naim JO, Lanzafame RJ (1997) Effects of photostimulation on wound healing in diabetic mice. Lasers Surg Med 20:56–63.

Yukna RA, Mellonig JT (2000) Histologic evaluation of periodontal healing in humans following regenerative therapy with enamel matrix derivative. A 10-case series. J Periodontol 71: 752–759.

Zhang F, Zhang J, Lin S, Oswald T, Sones W, Cai Z, Dorsett-Martin W, Lineaweaver WC (2003) Small intestinal submucosa in abdominal wall repair after TRAM flap harvesting in a rat model. Plast Reconstruct Surg 112:565–569.

Zhao LL, Galiano RD, Cox GN, Roth SI, Mustoe TA (1995) Effects of insulin-like growth factor-1 and insulin-like growth factor binding protein-1 on wound healing in a dermal ulcer model. Wound Rep Reg 3:316–321.

# 5 Regeneration of Neural Tissues

## INTRODUCTION

The nervous system is derived from the dorsal portion of the embryonic ectoderm, which rolls into an epithelial neural tube, and from the neural crest. It has three major divisions: central, peripheral, and autonomic (Ham and Cormack, 1979). The central nervous system (CNS) is derived from the neural tube and consists of the brain and spinal cord. The CNS controls the voluntary actions of the body, as well as some involuntary actions, such as reflexes. The retina and optic nerve (cranial nerve II) are considered part of the central nervous system, since they develop from, and project to, the diencephalon of the brain. The peripheral division includes the spinal nerves, as well as cranial nerves I (olfactory) and III–XII. The motor component of the spinal nerves develops from the neural tube and the sensory component from the neural crest. Cranial nerves I and III–XII develop from ectodermal placodes and neural crest. The autonomic system, a complex subset of the peripheral nervous system, is derived from the neural crest and controls involuntary activities, such as heart rate, temperature, and the smooth muscle activity of the vascular and digestive systems.

Neural tissue in all three divisions is made up of neurons and associated glial cells (**FIGURE 5.1**). Neurons consist of a cell body that sends electrical signals over axons and receives signals from the axons of other neurons through shorter dendrites. Neurons are linked at synapses, junctions where axons from one neuron meet the dendrites or cell body of another neuron. All nerve cell bodies and axons are associated with glial cells. In the brain and spinal cord, the major glial cell types are oligodendrocytes, which form insulating myelin, and astrocytes, of which several types reside in both the gray and white matter. Astrocytes are crucial for neuron survival, clearing of the neurotransmitter glutamate, maintaining acid-base balance, degrading and forming synapses, and modulating neuron responses (Travis, 1994). The glial cells associated with the optic nerve are primarily astrocytes. In the peripheral nervous system, Schwann cells are the glial counterparts of CNS oligodendrocytes and form the myelin sheath of the axons.

Mammals are able to regenerate peripheral nerve axons, neurons of the olfactory epithelium and bulb, and neurons of the hippocampus, but cannot regenerate axons or neurons of the spinal cord, optic nerve, or acoustic sensory epithelium. However, fetal stages of mammals and the adults of other vertebrates have the capacity to regenerate the neurons and axons of these tissues and are thus valuable research models to learn why regeneration fails in the mammal.

## AXON REGENERATION

Axons of both central and peripheral neurons in adult mammals have the intrinsic potential to regenerate after crush or transection, as indicated by the fact that they initiate sprouting after such injuries. In most vertebrates, spinal nerves and olfactory nerves are able to regenerate across a lesion and reinnervate their target tissues. The optic nerve and the ascending and descending tracts of the spinal cord in fish and larval and adult urodele amphibians are also able to regenerate. By contrast, axons of the reptilian, avian and mammalian central nervous system fail to regenerate further after initial sprouting.

A large body of evidence suggests that whether or not axons regenerate depends in large part on their associated glial cells (Yannas, 2001). Differences in the ability of peripheral and central glial cell populations to support regeneration have been well documented by experiments in which the regeneration of central axons was promoted by peripheral nerve sheaths grafted into the central nervous system, whereas central nerve sheaths inhibit the regeneration of peripheral axons (Aguayo, 1985). These differences appear to reside largely in the adhesion molecules and soluble signals synthesized by glial cells. Glial cells that support regeneration provide most or all of the molecules that are

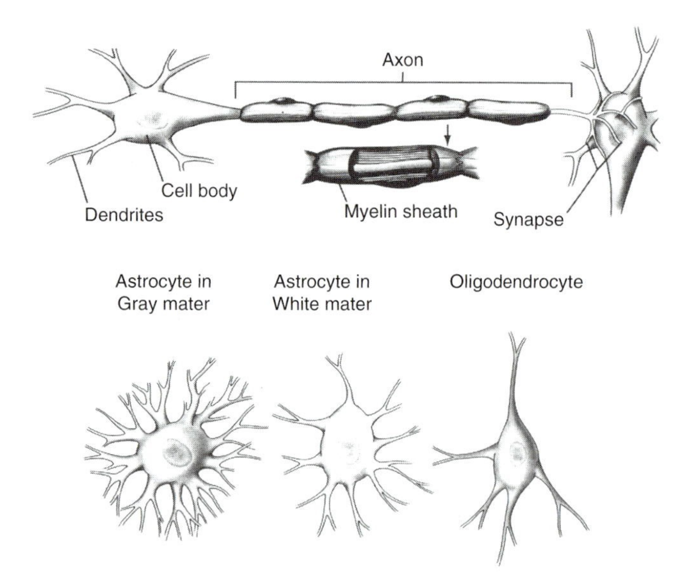


**FIGURE 5.1** Neurons and glial cells. Neurons have a large metabolically active cell body with short dendrites on one side and a long axon on the other. Axons synapse with the cell bodies and/or dendrites of adjacent or more distant neurons. Axons are sheathed in myelin made by oligodendrocytes in the CNS and Schwann cells in the PNS. Astrocytes in both the gray and white matter interact with neurons in a variety of dynamic functions.

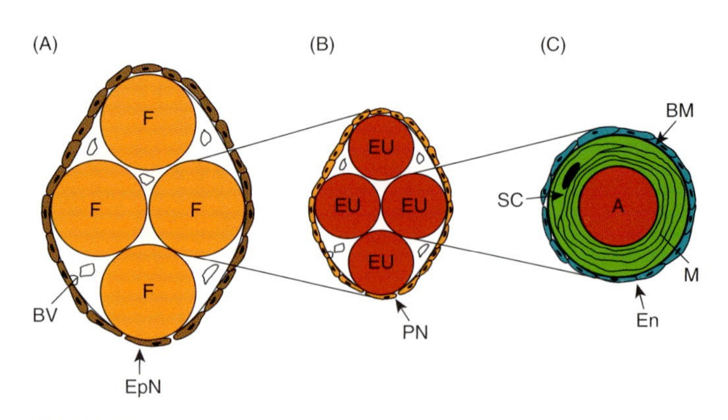


**FIGURE 5.2** Cross sectional organization of an individual spinal nerve. **(A)** The whole nerve is made up of fasicles (F) composed of endoneurial units (EU). An epineurium (EpN) surrounds the nerve. **(B)** Structure of an individual fascicle, surrounded by a perineurium (Pn). **(C)** Endoneurial unit. The axon (A) is myelinated (M) by wraps of Schwann cell (SC) plasma membrane. The endoneurium (En) synthesizes a basement membrane (BM).

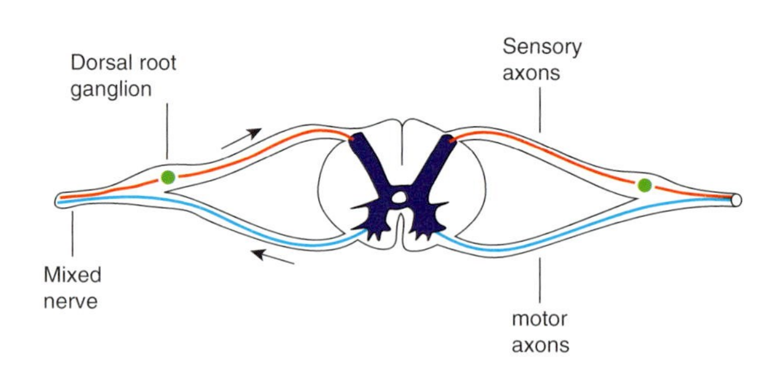


**FIGURE 5.3** Entry of mixed spinal nerves into the spinal cord. Sensory nerves are red, motor nerves are blue. The dorsal (posterior) root ganglion houses sensory neuron cell bodies (green dot). These neurons have a second axon that enters the white matter on the dorsal (posterior) side of the cord. Motor neuron cell bodies are located in the ventrolateral (anterolateral) gray matter of the cord. Their axons exit ventrally and join the incoming sensory fibers distal to the DRG. Within the gray matter (purple), interneurons connect sensory to motor neurons.

important for neuron survival and axon outgrowth, whereas regeneration fails where glial cells do not make these molecules and/or synthesize molecules inhibitory to regeneration.

## 1. Axons of Mammalian Spinal (Peripheral) Nerves

Spinal nerves are organized into three structural levels (FIGURE 5.2). The fundamental unit is the endoneurial unit, composed of an axon, its associated Schwann cell sheath, and surrounding connective tissue sheath, the endoneurium. The endoneurium and Schwann cells synthesize a basement membrane between them. Endoneurial units are organized into fascicles, each of which is surrounded by a perineurium. The fascicles are bundled into the nerve trunk, which is encased by the epineurium. The nerve trunks are richly vascularized by epineurial, intrafascicular, and perineurial arteries and arterioles, and a capillary network in the endoneurium. Each nerve trunk is a mixture of afferent (sensory) and efferent (motor) nerves that branch distally to innervate target tissues and organs (Ham and Cormack, 1979).

In humans, there are 31 pairs of spinal nerves. Just before it enters the spinal cord, each nerve splits into sensory (dorsal) and motor (ventral) roots (FIGURE 5.3). The cell bodies of the incoming sensory axons are located outside the spinal cord in the dorsal root ganglia. These cell bodies extend a second major axon into the dorsal side of the white matter of the cord to form the ascending tracts to sensory centers in the brain. The cell bodies of the motor axons are located in the ventral horn of the cord and synapse with descending tracts of motor axons from the motor centers of the brain. The motor neurons send axons peripherally to join the sensory axons distal to the dorsal root ganglia. Some of these motor axons belong to the autonomic nervous system. These autonomic efferents (preganglionc fibers) synapse with a second motor neuron in a series of autonomic paravertebral ganglia outside the vertebral column. The axons of the second neuron (postganglionic fibers) innervate the viscera and glands. Interneurons within the cord synapse with both a branch of the sensory root axons and the motor cell bodies to integrate sensory and motor information and mediate reflexes (Ham and Cormack, 1979).

Distal to the dorsal root ganglia, mammalian spinal nerves regenerate relatively well, provided that the endoneurial tubes remain intact and in register at the site of injury (Yannas, 2001). The ends of the axons on the proximal side of the injury seal off and their distal parts degenerate along with their myelin, a process called Wallerian degeneration. The proximal axon stumps then regenerate through the endoneurial tubes to reinnervate target tissues. Regeneration is good following a crush injury because continuity of the endoneurial tubes is maintained. Regeneration after nerve transection is more problematic. Axons from the proximal stump initiate regeneration but often form tangles called neuromas, due to obstruction and misdirection by nerve sheath fibroblasts proliferating in the lesion space (Yannas, 2001). In this case, surgeons attempt to direct sprouting axons into the distal endoneurial tubes by suturing the cut ends of the nerve together through the epineurium. However, many endoneurial tubes still may not be in register and the quality of the regeneration is not as good as after a crush injury.

Axon regeneration has three phases: sprouting, elongation, and maturation (McQuarrie, 1983). As Schwann cells dedifferentiate and proliferate, the proximal stumps of the axons sprout by the actin-driven formation of growth cones (Sinicropi and McIlwain, 1983). As the growth cone advances, it elongates the axon by pulling out a thin daughter cylinder of the proximal axon stump. The elongating part of the axon is stabilized by the polymerization of microtubules behind the growth cone. Growth cone sprouting is inhibited by cytochalasin B, which destabilizes actin microfilaments, and axon elongation is inhibited by colchicine, which destabilizes microtubules. The growth cones of the elongating axons are guided toward their targets by the Schwann cells and basement membranes of the endoneurial tubes. When a growth cone contacts its target, the axon enlarges radially to its mature diameter.

### a. RNA and Protein Synthesis

During regeneration, axotomized neurons switch from a transmitting mode to a growth mode and express growth-associated proteins (Fu and Gordon, 1997). Labeling, electron microscope, and immunocytochemical studies have shown that the materials for building new cell membrane and cytoskeleton in the elongating axon are supplied by an upregulation of RNA and protein synthesis in the neuron cell body (Grafstein, 1983) and the anterograde transport of the materials. The growth cone signal transduction proteins, GAP-43 and CAP-23, the cytoskeletal proteins actin and tubulin, and cell adhesion molecules (CAMs) such as L1 and neural cell adhesion molecule (NCAM) are strongly upregulated, while choline acetyltransferase (CAT), the neurotransmitter substance P and neurofilament proteins are downregulated (Bisby, 1995; Fu and Gordon, 1997). The different proteins are transported distally at different rates. GAP-43 is transported rapidly, whereas G-actin and tubulin subunits are transported slowly (McQuarrie, 1983). There is a close correlation between the rate of transport of tubulin and actin and the rate of axonal regeneration (McQuarrie and Lasek, 1989). The radial maturation of regenerated axons is associated with the re-elaboration of neurofilaments (McQuarrie, 1983).

To regenerate successfully to their targets, sprouting neurons must survive and the endoneurial tubes must be able to provide optimal growth support. Neuron survival and axon elongation require both soluble and insoluble factors produced by immune cells and fibroblasts, target tissues, and the neurons themselves. The most important source of survival and elongation factors, however, is the Schwann cells of the endoneurial tubes, which dedifferentiate to a nonmyelinating support function during Wallerian degeneration.

### b. Wallerian Degeneration and Schwann Cell Dedifferentiation and Proliferation

The distal part of an injured axon is degraded by a process called Wallerian degeneration. The mechanism of Wallerian degeneration is distinct from pathological neurodegenerative processes such as dying back from the target tissue or the pruning process observed during embryonic neurogenesis that matches the proper number of axons to their targets (Raff et al., 2002).

During Wallerian degeneration, axons and their myelin layers distal to the injury disintegrate over a period of several days. Myelin proteins are inhibitory to axon regeneration and must be removed from the endoneurial tubes (Fu and Gordon, 1997). Electron microscopic, metabolic labeling, and cell surface marker expression studies have shown that the Schwann cells degrade small bits of myelin, but that the bulk of the myelin is phagocytosed and degraded by macrophages that enter the nerve tubes (Goodrum and Bouldin, 1996) (**FIGURE 5.4**). The resulting cholesterol and free fatty acids are complexed to apolipoprotein E within the macrophages to form lipoprotein particles. These particles are then released and taken up by Schwann cells via low-density lipoprotein (LDL) receptors, to be reused in myelin synthesis during axonal regeneration.

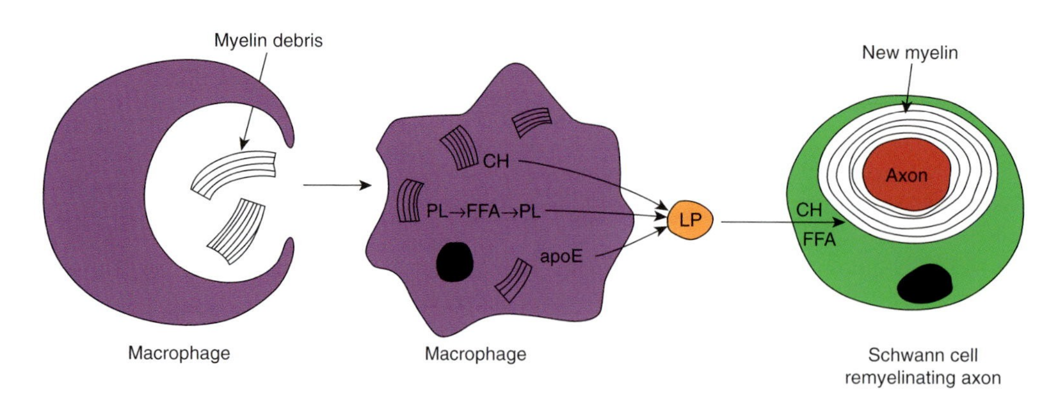


**FIGURE 5.4** Wallerian degeneration of axons and salvage pathway for re-myelination. Macrophages penetrate the basement membrane to phagocytose myelin debris. Within the macrophage, myelin phospholipids (PL) are hydrolyzed to free fatty acids (FFA). Up to half the FFA are re-incorporated into phospholipids. These phospholipids and all the cholesterol (CH) are complexed to apolipoprotein E (apoE) to form lipoprotein particles (LP) that are then taken up by Schwann cells via low-density lipoprotein receptors to be re-used for myelin synthesis.

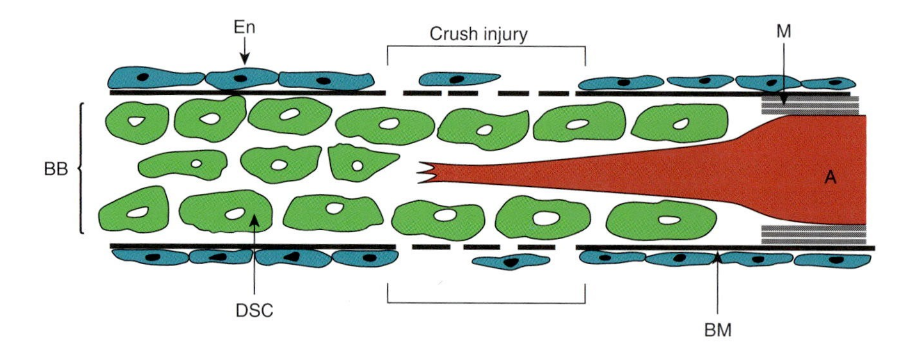


**FIGURE 5.5** Dedifferentiated Schwan cells (DSC) proliferate on both sides of a crush injury to bridge the lesion. The cords of DSCs are called bands of Bungner (BB). Axons (A) sprout into the bands of Bungner of the individual endoneurial tubes. En = endoneurium, BM = basement membrane, M = myelin.

As their myelin is degraded, the Schwann cells dedifferentiate within the basement membrane to form cords of cells called the bands of Bungner (**FIGURE 5.5**). They proliferate and migrate to form a continuous bridge across the lesion and down the length of the distal endoneurial tubes. Axotomized neurons, along with macrophages and platelets that have invaded the injury site, produce growth factors and cytokines that are mitogenic for Schwann cells (**TABLE 5.1**). Many of these molecules are the same ones responsible for the fibrosis of skin repair, such as FGF, PDGF, IL 1,2, and 6, TGF-β and IFN-γ, as well as nerve-specific factors such as glial growth factor (GGF) (Fu and Gordon, 1997). When the gap in the nerve is too large to be bridged by dedifferentiating Schwann cells, fibroblasts enter the wound space and these same factors induce scar formation, preventing regeneration.

### c. Neuron Survival Factors

The survival of axotomized neurons is promoted by neurotrophic factors, neurotransmitters, and neuropeptides, and their receptors (**TABLE 5.1**). These factors are made primarily by dedifferentiated Schwann cells (Fu and Gordon, 1997; Yannas, 2001). Reduction or denial of access to these factors results in neuronal cell death. The sensory neurons of the dorsal root ganglion are particularly sensitive to lack of survival factors, whereas motor neurons survive for periods of more than one year (Fu and Gordon, 1997).

The neurotrophins are an important family of molecules secreted by Schwann cells that are essential for the survival of axotomized neurons. This family includes nerve growth factor (NGF), brain-derived neurotrophic factor (BDNF), and neurotrophins 3 and 4 (NT-3, NT-4). NGF supports the survival of sensory and sympathetic axons, whereas BDNF, NT-3, and NT-4 promote survival of regenerating motor neurons (Raivich and Kreutzberg, 1993; Birling and Price, 1995; Fu and Gordon, 1997). The neurotrophins act through specific receptor tyrosine kinases (Trks). NGF acts through TrkA, BDNF and NT-4 through TrkB, and NT-3 through TrkC. In addition, all the neurotrophins bind to a common low-affinity receptor, $p75^{NTR}$, which selectively modulates the Trk-mediated survival response of neurons to the various neurotrophins (Davies, 2000).

Ciliary neurotrophic factor (CNTF) and IL-6, which share common structural motifs and use JAK/STAT signaling pathways, and glial derived neurotrophic factor (GDNF), are also important to the survival of axotomized neurons (Fu and Gordon, 1997). CNTF is released from Schwann cells after injury; macrophages and fibroblasts are the primary producers of IL-6. GDNF enhances survival of axotomized motor neurons and is produced primarily by Schwann cells (Fu and Gordon, 1997).

### d. Axon Elongation Factors

The same neurotrophins that enhance survival of axotomized neurons also enhance axon elongation (**TABLE 5.1**), through their interaction with the $p75^{NTR}$ receptor (Davies, 2000). In the absence of neurotrophins this receptor activates RhoA, one of three Rho isoforms belonging to the Ras superfamily of GTP-binding proteins that when active, prevent axon extension via the actin cytoskeleton. Ciliary ganglion cells extended neurites when RhoA was inactivated by the binding of NGF, BDNF, and NT-3 to the receptor. Conversely, introduction of a constitutively active mutant of RhoA inhibited neurite outgrowth from these neurons (Yamashita et al., 1999).

**TABLE 5.1** Soluble factors that are mitogenic for dedifferentiated Schwann cells and that are required for neuron survival and axon elongation after axotomy. These factors are produced by macrophages, platelets and the axons and Schwann cells themselves.

| *Factors Mitogenic for Schwann Cells* | | *Factors required for Neuron Survival and Elongation* | |
|---|---|---|---|
| M-phages and Platelets | Axons | M-phages | Schwann Cells |
| FGF | GGF | IL-6 | NGF[a] |
| PDGF | | | BDNF[b] |
| TGF-β | | | NT 3,4[b] |
| IL-1,2,6 | | | CNTF[b] |
| Ifn-γ | | | GDNF[b] |

[a] Sensory and sympathetic neurons.
[b] Motor neurons.

The dedifferentiated Schwann cells within the endoneurial tubes play a major role in promoting axon outgrowth (Bunge, 1987; Fu and Gordon, 1997; Yannas, 2001). In addition to their synthesis of neurotrophins, Schwann cells synthesize cell adhesion molecules and basement membrane that are substrates for the growth cones of regenerating axons (**FIGURE 5.6**). *In vitro*, both Schwann cells and the endoneurial basement membrane are effective in promoting axon outgrowth. Important cell adhesion molecules for neurite extension are L1 of the axon growth cones, NCAM and N-cadherin on Schwann cells, and Fn and Ln of the basement membrane. L1 and NCAM mediate homophilic calcium-independent cell adhesion, and N-cadherin mediates calcium-dependent adhesion between Schwann cells and axons and between axons. Antibodies to L1 inhibit at least 80% of sensory, ciliary, and retinal ganglion cell neurite outgrowth on Schwann cell surfaces *in vitro*, whereas antibodies to N-CAM are less effective (Fu and Gordon, 1997). N-cadherin stimulates neurite outgrowth *in vitro* that is partially blocked by antibodies to N-cadherin (Doherty and Walsh, 1991).

The Fn and Ln of the basement membrane promote neurite extension (Rogers et al., 1983; Westerfield, 1987). Merosin is the predominant isoform of laminin present in the basement membrane of the endoneurial tube. Antibodies to merosin inhibit neurite outgrowth and Schwann cell migration *in vitro* and block the

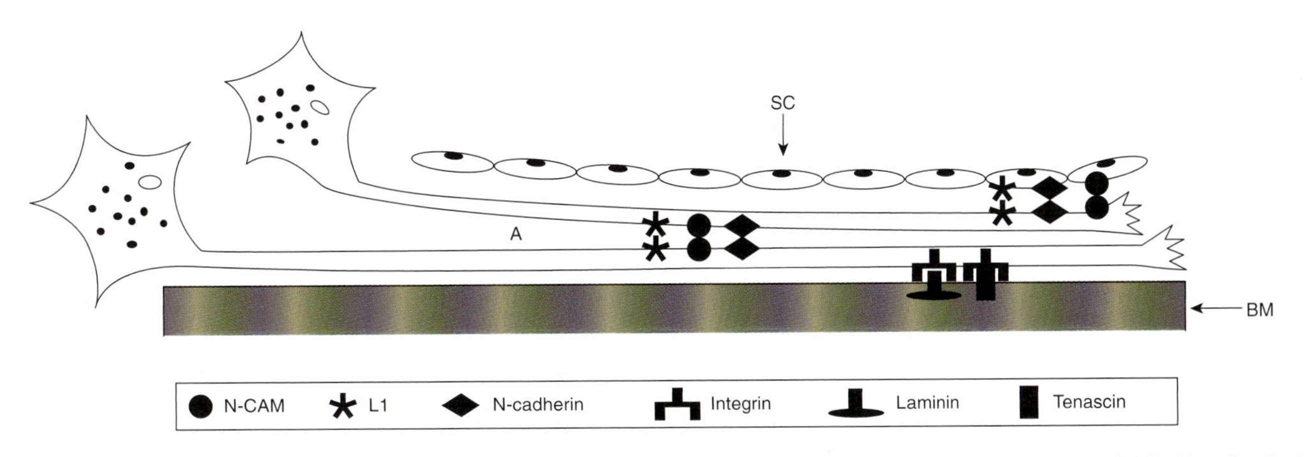


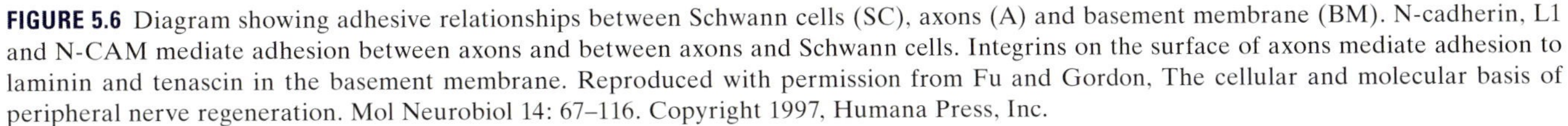


**FIGURE 5.6** Diagram showing adhesive relationships between Schwann cells (SC), axons (A) and basement membrane (BM). N-cadherin, L1 and N-CAM mediate adhesion between axons and between axons and Schwann cells. Integrins on the surface of axons mediate adhesion to laminin and tenascin in the basement membrane. Reproduced with permission from Fu and Gordon, The cellular and molecular basis of peripheral nerve regeneration. Mol Neurobiol 14: 67–116. Copyright 1997, Humana Press, Inc.

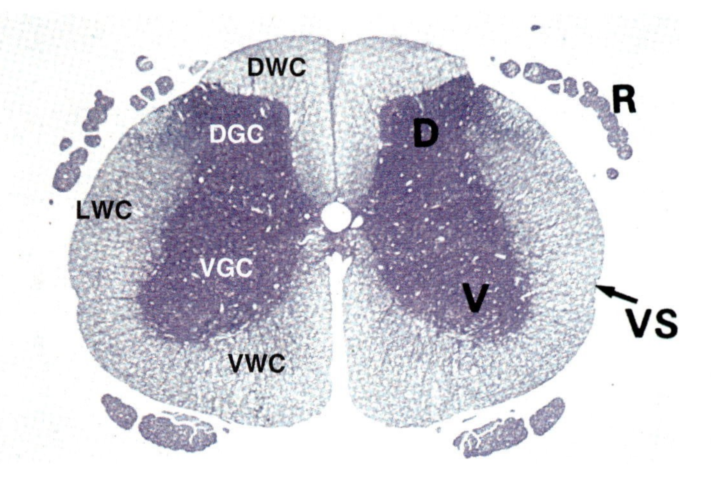


**FIGURE 5.7** Cross section of cat lumbar spinal cord (Weigart-Pal stain). Gray matter stains purple, white matter (myelinated axons) stains mottled blue. D = dorsal, V = ventral. VS = ventrolateral sulcus where ventral nerve roots exit. DWC = dorsal white column, DGC = dorsal gray column. LWC = lateral white column. VWC = ventral white column, VGC = ventral gray column. R = dorsal nerve root. Reproduced with permission from Wheater et al., Wheater's Functional Histology (3rd ed). Copyright 1997, Elsevier.

regeneration of sympathetic fibers *in vivo* (Anton et al., 1994). The integrin fibronectin receptor α5β1 is synthesized by both Schwann cells and regenerating axons. Thus, Fn is important both for Schwann cell migration and for neurite extension on Schwann cells during regeneration (Lefcourt et al., 1992). Tenascin C also promotes neurite outgrowth *in vitro* (Wehrle and Chiquet, 1990) and is expressed by Schwann cells for incorporation into the basement membrane of the endoneurial tube. The upregulation of TnC in Schwann cells may be due to IL-1 and TGF-β released by invading macrophages; these molecules enhance TN-C expression in other cell types (Pearson et al., 1988; McCachren and Lightner, 1992).

## 2. Axons of Mammalian Spinal Cord

A cross-section of cat spinal cord is illustrated in **FIGURE 5.7**. In mammals, spinal cord injury that destroys substantial numbers of axons and neurons causes sensory deprivation and paralysis below the level of injury, followed by muscle atrophy, spasticity, and bone loss (McDonald and Sadowsky, 2002; Eser et al., 2004). Cervical injuries result in disturbances in autonomic functions such as regulation of blood pressure, heart rate, and temperature. The spinal cord of mammals contains NSCs in the ependyma and subventricular zone (Weiss et al., 1996; Kehl et al., 1997; Johansson et al., 1999; Doetsch et al., 1999), but these cells respond to injury by participating in the formation of scar tissue.

Although spinal cord axons do not regenerate after crush injury or transection that damages the pia mater, they transiently upregulate genes involved in growth cone formation and neurite extension (GAP-43, CAP-23, L1, NCAM), sprout briefly, and then undergo growth arrest and retraction (Lu and Waite, 1999). Failure to continue regeneration is due to the lack of expression, by associated glial (oligodendrocyte and/or astrocyte) cells, of the appropriate neuronal survival and elongation-promoting molecules. Instead, these cells contribute to the formation of a "glial scar" that is inimical to regeneration through the following series of events, illustrated in **FIGURE 5.8** (Barron, 1983; Steward et al., 1999; McDonald and Sadowsky, 2002; Chernoff et al., 2003).

First, blood flow is interrupted, depriving the injured area of oxygen and glucose. Excess plasma leaking from damaged vessels causes the cord to swell, further compressing the tissue and killing many neurons and glial cells outright.

Second, undamaged neurons become overexcited, releasing excess amounts of the neurotransmitter, glutamate. Glutamate kills undamaged neurons by opening membrane channels that allow the influx of toxic amounts of calcium, a phenomenon called "glutamate toxicity." A further source of toxicity is the production of free radicals by lipid peroxidation.

Third, the insulating myelin from both dead and surviving axons breaks down. Within a few hours after injury, the breakdown products of myelin and dead cells begin to spread the initial damage to neighboring uninjured regions, a process that can go on for weeks, yielding secondary damage even greater than the primary damage caused by the trauma and causing many intact neurons to undergo apoptosis.

Fourth, myelin proteins resulting from the breakdown of myelin have been shown to inhibit axon regeneration *in vitro* or *in vivo* (Kapfhammer and Schwab, 1992; Filbin, 2000). Several such proteins have been identified, including "Nogo," oligodendrocyte myelin glycoprotein (Omgp), myelin-associated glycoprotein (MAG), and Ephrin-B3 (Filbin, 2000; Chen et al., 2000; GrandPre et al., 2000; Prinjha et al., 2000; Yiu and He, 2003; Benson et al., 2005). Chondroitin sulfate proteoglycans (CSPGs) synthesized by astrocytes also inhibit the growth of spinal cord axons in culture (Niederost et al., 1999). The normal developmental roles of Nogo, Omgp, and MAG are unknown, but Ephrin-B3 has been shown to be a midline axon repellant during CNS development for axons of the cortico-

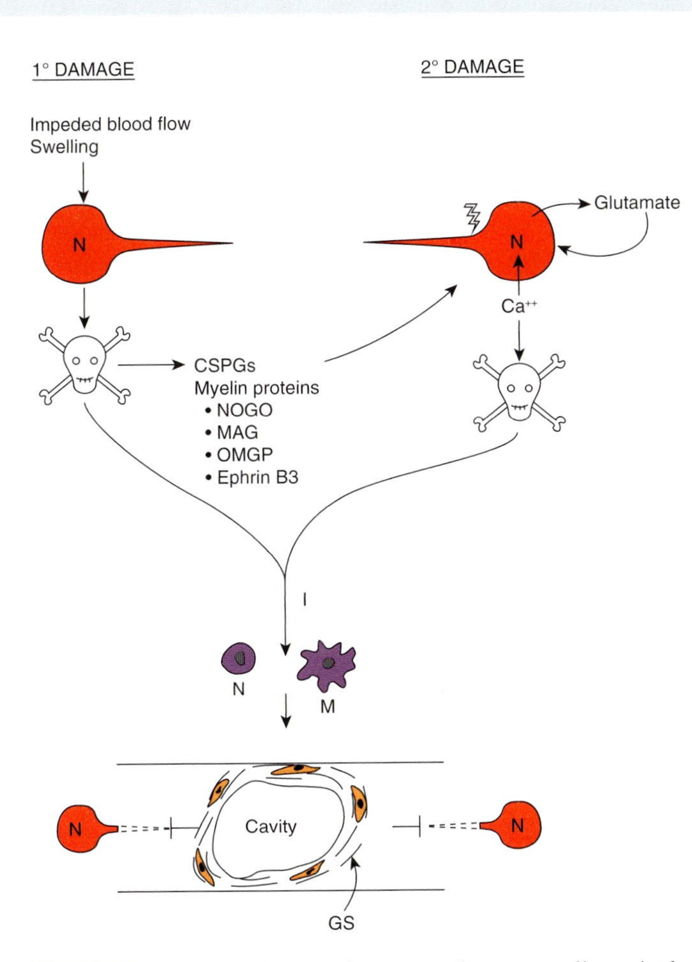


**FIGURE 5.8** Summary of events that occur after mammalian spinal cord injury. The primary damage is caused by interrupted blood flow and compression due to swelling of the cord, leading to neuron (N) death. Chondroitin sulfate proteoglycans (CSPGs) and several myelin breakdown proteins are produced as a result. Secondary death of surviving neurons is caused by these molecules and by glutamate excitotoxicity (lightning bolt), which results in flooding of cells with extracellular calcium. An inflammatory response (I) is initiated in which bacteria and cellular debris is cleared by neutrophils (N) and macrophages (M), creating cavities in the cord tissue. The inflammatory response also leads to the formation of a glial scar (GS) that impedes axon regeneration (dashed lines).

spinal tract (Dickson, 2002). There are undoubtedly many more inhibitory proteins awaiting identification.

Fifth, as the damage spreads, an inflammatory response is mounted. PDGF and TGF-β from platelets attract neutrophils and macrophages (microglia) into the lesion to ingest bacteria and debris. The cleanup of debris continues over the whole area of damage, creating cavities in the cord tissue that vary in size, depending on species. Astrocytes and fibroblasts from the injured pia mater proliferate around the periphery of the cavities to form a glial scar that is both a physiological and mechanical barrier to axon regeneration. The astrocytes of the scar produce the same types of inhibitory CSPGs as oligodendrocytes (Filbin, 2000).

The major role that this barrier plays in regenerative failure is highlighted by the fact that the spinal cords of mice, transected in a way that minimizes displacement of the ends of the cord and infiltration of fibroblasts, regenerate accompanied by a high degree of functional recovery (Seitz et al., 2002). This suggests that upregulation of all the genes necessary for neurite extension is maintained after a clean transection that does not induce an inflammatory response leading to scar tissue formation. Furthermore, it would appear that under these conditions regenerating axons make functional synapses on either side of the lesion. The regenerative events that take place in this transection model deserve further exploration. Is there axon degeneration after transection? Does excitatory neurotoxicity take place? Are myelin inhibitory proteins and proteoglycans synthesized? What consequences of cord injury are missing that permit this regeneration?

The fetal spinal cord of birds and mammals can regenerate (Shimizu et al., 1990; Hasan et al., 1991; Nicholls and Saunders, 1996; Iwashita et al., 1996). Furthermore, the spinal cord of the newborn South American opossum, *Monodelphus domestica*, is able to regenerate completely with functional recovery (Treherne et al., 1992; Saunders et al., 1995). The mammalian fetal versus adult spinal cord and the newborn opossum are thus models that can be used to compare the cellular features and molecular activities of injured cord at regeneration-competent versus regeneration-deficient stages of development. In the injured spinal cord of the newborn opossum, microarray analysis showed changes in the expression patterns of 129 genes, of which 88% were upregulated (Farlow et al., 2000). Kierstead et al. (1995) showed that delaying the myelination of spinal cord axons in the chick embryo extends the period of development over which regeneration can occur, suggesting that regenerative inhibition of the spinal cord coincides with oligodendrocyte differentiation. Consistent with this notion, Nogo is present in axons, and its receptor (NgR) is present on chick nerve cell bodies during these permissive developmental stages (O'Neill et al., 2004). Nogo only becomes inhibitory when the locus of its expression shifts from neurons to oligodendrocytes later in development.

Regeneration of the mouse spinal cord in the absence of an inflammatory response suggests that the loss of ability to regenerate spinal cord and transition to scarring is correlated with maturation of the immune system, as in skin (Chapter 2). Thus, it might be instructive to ask whether PU.1 mice, in which macrophages and neutrophils are absent, might be able to regenerate their spinal cord, and whether suppression of macrophages or inhibitors of TGF-β improve spinal cord regeneration in wild-type animals, as is the case in adult skin?

### 3. Axons of Amphibian Spinal Cord

Unlike mammals, larval and adult urodeles regenerate the axons of ascending and descending nerve tracts after transection or ablation of the spinal cord at the level of the trunk, with recovery of function (Piatt, 1955; Butler and Ward, 1965; Egar and Singer, 1972, 1981; Nordlander and Singer, 1978; Holder and Clarke, 1988; Chernoff et al., 2003; Ferretti et al., 2003). The ependymal (NSC) layer of the cord plays a central role in this process (Holder and Clarke, 1988; Ferretti et al., 2003; Chernoff et al., 2003).

The ependymal cells in amphibians span the width of the cord wall, branching within the white matter to terminate in expanded end-feet that form the glia limitans under the pia mater (Holder et al., 1990). Morphologically, these cells resemble the radial glial cells of the embryonic spinal cord of birds and mammals (Clarke and Ferretti, 1998; Ferretti et al., 2003). However, they are different from the radial glial cells of birds and mammals in that they express GFAP, but not nestin or the intermediate cytoskeletal filament vimentin (Zamora and Mutin, 1988). They also express the epithelial keratins 8 and 18, which are not expressed in the ependymal cells of mammals, but are expressed in the ependymal cells of adult fish and lampreys, which can also regenerate the spinal cord (Holder et al., 1990; O'Hara et al., 1992; Bodega et al., 1994; Clarke and Ferretti, 1998; Chernoff et al., 2003). These cytoskeletal features have sometimes been cited as possible markers of regenerative potential. This is unlikely, however, given that adult frog ependymal cells share these features but do not regenerate the spinal cord (Chernoff et al., 2003). Chernoff et al. (2003) have identified axolotl and *Xenopus* ependymal NSCs that express Nrp-1, the vertebrate homolog of Musashi-1 in *Drosophila* that suppresses the translation of *numb* mRNA, thereby keeping Notch active and neural stem cells in a state of quiescence or self-renewal (Sakakibara et al., 2002).

The ependymal response to injury appears to provide an environment that protects axon stumps from degeneration and promotes their growth to make new synaptic connections that allow recovery of function (**FIGURE 5.9**). After transection, the ependymal cells on either side of the lesion undergo an epithelial to mesenchymal transformation (EMT) in which they suppress production of Ln, GFAP, and epithelial cytokeratins and express Fn and vimentin (Vn) (O'Hara et al., 1992; Chernoff et al., 2003). The mesenchymal cells proliferate to form blastemas that subsequently fuse to bridge the gap between the cut ends of the cord. The proliferating mesenchyme then undergoes a mesenchymal to epithelial transformation (MET) to become an ependyma again. The new ependymal cells become radially arranged, and form end-feet with channels through which axons regenerate (O'Hara et al., 1992; Chernoff et al., 2003). In newts, the regenerated cord is thinner than the intact cord (**FIGURE 5.10**), there are fewer axons and not all of the connections made by regenerating axons are correct even after weeks or months (Stensaas, 1983; Davis et al., 1989), but there is functional recovery of swimming (Davis et al., 1990). In axolotls, innervation from the brain reached control levels in long-term studies of up to 23 months (Clarke et al., 1988). Substantial numbers of new neurons are formed by ependymal NSCs in larval urodeles that are not formed in adults. This difference is most likely related to the fact that the larvae are growing rapidly and are therefore generating new neurons as part of this growth.

The factors involved in the transformation of ependymal cells to mesenchyme and back again are incompletely known. Matrix metalloproteinases (MMPs) may play a role by digesting ependymal ECM. In the uninjured axolotl cord, MMP activity is not detectable and TIMP-1 is expressed. The ependymal mesenchyme of the injured cord expresses MMP-1, 2, and 9 and TIMP-1 expression disappears (Chernoff et al., 2000, 2003). TGF-β1 and PDGF in combination cause ependymal mesenchyme blastemas *in vitro* to break apart and migrate as cords of cells, suggesting that these growth factors play a role in mediating the organization of the cells *in vivo* (**FIGURE 5.11**) (O'Hara and Chernoff, 1992, 1994). Experiments *in vitro* suggest that migration and proliferation of mesenchymal cells from cultured ependymal blastemas *in vitro* are dependent on EGF and are inhibited by TGF-β1 (Chernoff et al., 2003). The EMT/MET response of injured amphibian spinal cord deserves closer study to see how it compares with similar responses in other systems, such as the proximal tubules of the mammalian kidney, where EMT is regulated by TGF-β1 signaling that can lead to fibrosis and MET is regulated by BMP-7 signaling (see Chapter 7).

How the ependymal cells protect and encourage extension of regenerating axons is a major research issue. One possibility is that they may be able to buffer the effect of glutamate excitotoxicity by the uptake and sequestration of calcium, which in turn might trigger epithelial to mesenchymal transformation through second messenger pathways (Chernoff et al., 2003). Retinoic acid promotes axolotl neurite outgrowth *in vitro*. The ependymal cells can take up retinol, convert it to retinaldehyde and then to RA. The secreted RA is then taken up by neurons (Hunter et al., 1991). Endogenous retinol and RA have been detected in the urodele spinal cord, implying that RA might be a crucial molecule for axon extension *in vivo*. Most or all of the other survival and guidance factors that have

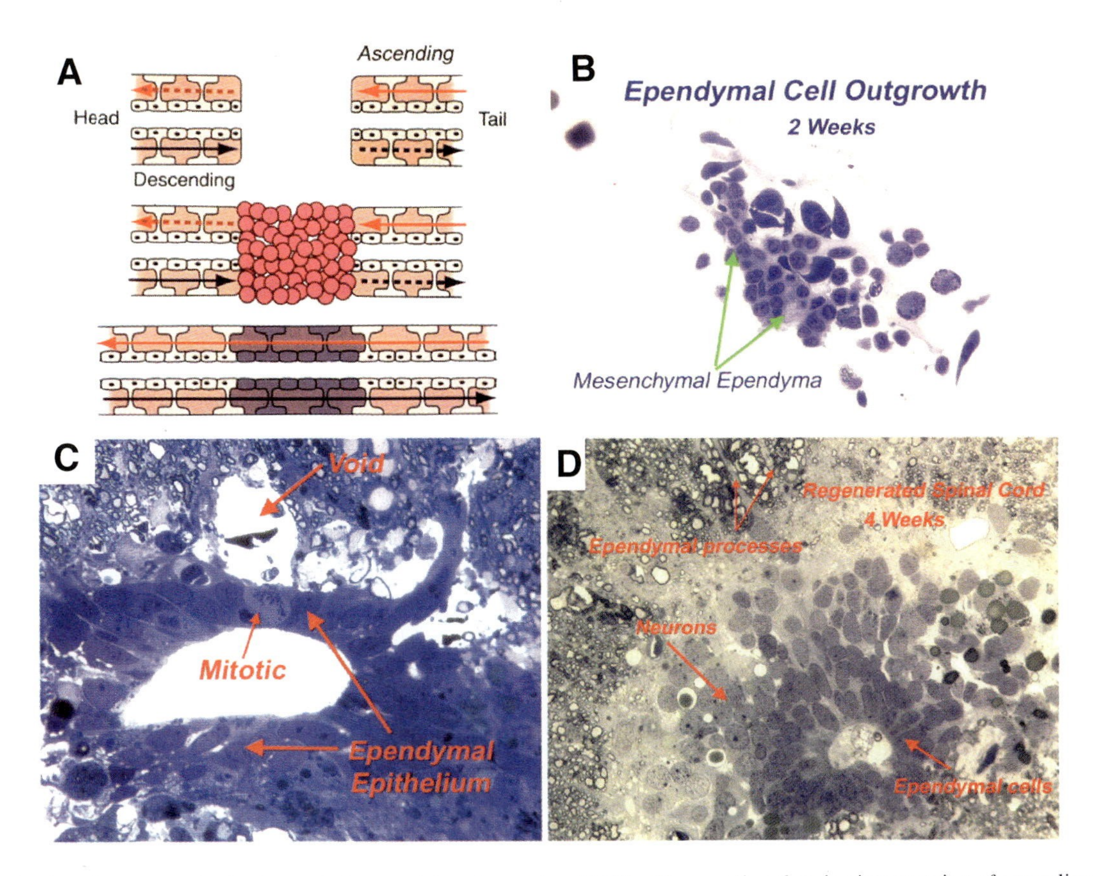


**FIGURE 5.9** Spinal cord regeneration in the axolotl, a urodele salamander. **(A)** Cord transection showing interruption of ascending and descending axon tracts (red and black arrows). The spinal canal is lined by ependymal epithelial cells. These cells transform into mesenchymal cells (red) that proliferate to bridge the gap. The mesenchymal cells then transform back to epithelial cells to re-establish the ependyma and provide paths for axon regeneration. **(B)** Section showing mesenchymal cells two weeks after transection. **(C)** Anterior to the lesion the ependymal cells, some of which are in mitosis, remain in epithelial form. Void spaces can be seen in the damaged tissue. **(D)** Four weeks after lesioning, the cord has regenerated new neurons and axons. The ependymal epithelium is arranged around the lumen of the cord with processes and end feet on the periphery. Reproduced with permission from Chernoff et al., Urodele spinal cord regeneration and related processes. Dev Dynam 226:280–294. Copyright 2003, Wiley-Liss.

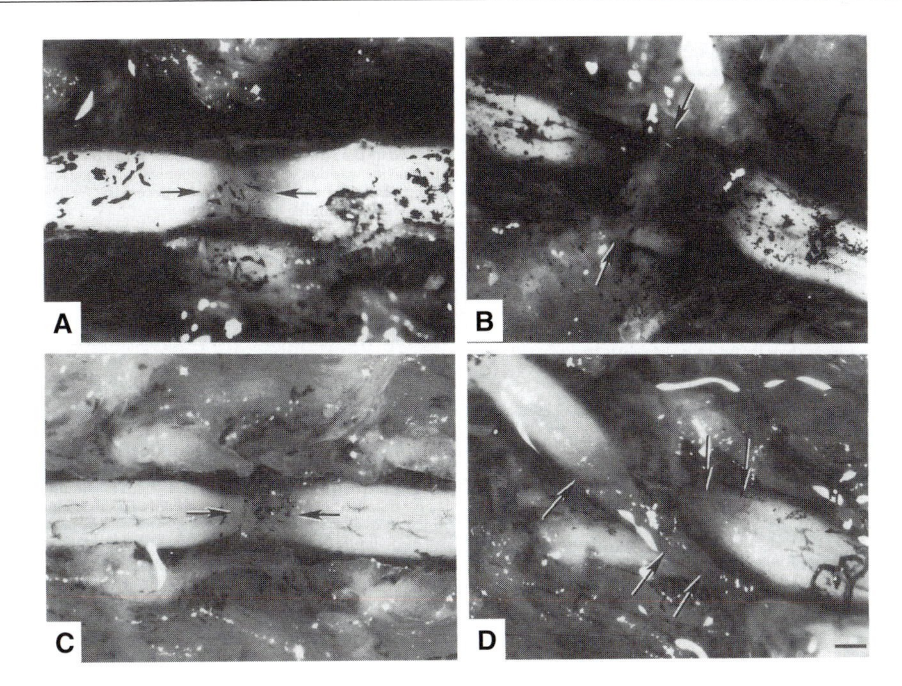


**FIGURE 5.10** Spinal cord regeneration after transection in the adult newt *Notophthalmus viridescens*. **(A)** and **(B)** are six weeks after transection; **(C)** and **(D)** are 12 weeks after transection. **(A)** and **(C)** are examples of successful regeneration where a new ependyma has bridged the gap (arrows) and new axons have re-grown. **(B)** and **(D)** are examples of unsuccessful regeneration where the gap was not bridged (**B**, arrows) or the ends of the cord grew past each other (**D**, arrows). Reproduced with permission from Davis et al., Time course of salamander spinal cord regeneration and recovery of swimming: HRP retrograde pathway tracing and kinematic analysis. Exp Neurol 108:198–213. Copyright 1990, Elsevier.

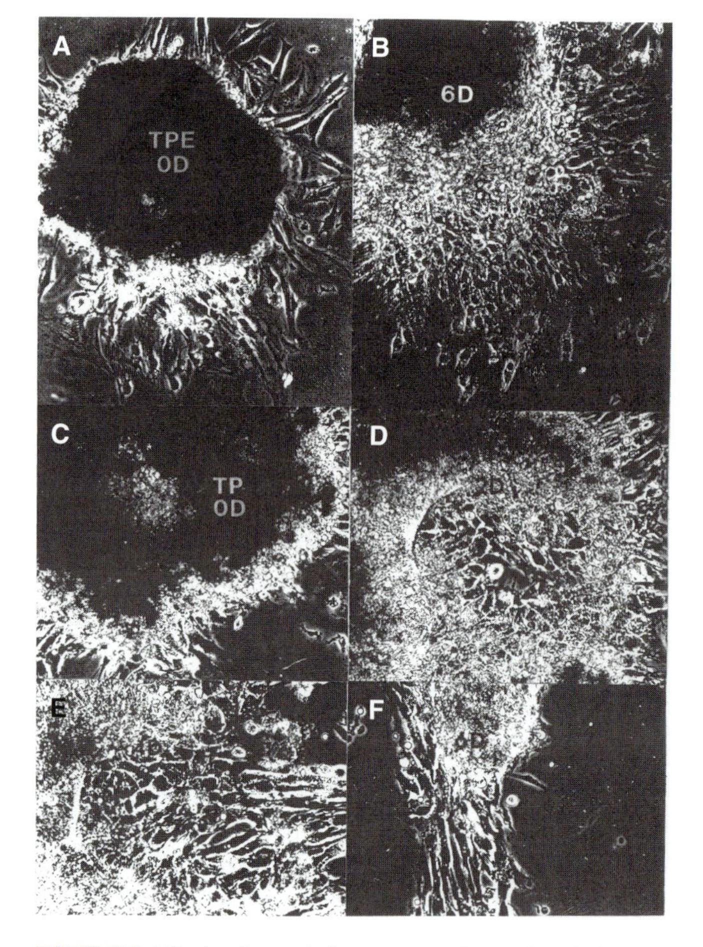


**FIGURE 5.11** Effects of growth factors on explants of axolotl spinal cord tissue. **(A, B)** Medium containing TGF-β and EGF (TPE) supports ependymal outgrowth at 0 days (0D) and 6 days. **(C–F)** TGF-β (T) and PDGF (P) cause ependymal explants to break apart and disperse on the dish as cords of cells over a six-day period in culture. Courtesy of Dr. Ellen Chernoff.

been shown to be important for peripheral nerve regeneration would presumably also be involved in urodele spinal cord regeneration, but this question has not yet been addressed.

An important feature of cord injury in amphibians is that myelin proteins inhibitory to spinal cord regeneration in mammals appear not to be present or to be removed. Nogo is not present in the regenerating cord of fish and larval *Xenopus*, although other molecules such as MAG, CSPGs and tenascin-R are present but do not inhibit regeneration (Wanner et al., 1995; Lang et al., 1995; Becker et al., 1999). Tenascin-R and MAG are rapidly removed in injured adult newt spinal cord (Becker et al., 1999), probably by macrophages ingesting fragments of myelin. The presence and disposition of these molecules have not yet been studied in the regenerating cord of other urodeles. In metamorphosing *Xenopus*, spinal cord myelin becomes nonpermissive to axon outgrowth and reacts with the anti-Nogo antibody (Lang et al., 1995), consistent with the studies on the chick outlined above showing that loss of the ability to regenerate the cord coincides with oligodendrocyte differentiation and the shift of Nogo expression from neurons to oligodendrocytes as development progresses (Kierstead et al., 1995; O'Neill et al., 2004).

## 4. Axons of Amphibian and Fish Optic Nerve

The optic nerve is a cranial nerve that forms by extension of axons from the ganglion cell layer of the neural retina to the optic tectum of the brain. Mammals cannot regenerate the transected or crushed optic nerve, a failure that is related to astrocytic reactive gliosis and early postnatal downregulation of the apoptosis inhibitor Bcl-2 (Cho et al., 2005). Adult fish and larval and adult amphibians, however, can regenerate their optic nerves and are therefore good models with which to identify regeneration-permissive conditions (Matthey, 1925; Sperry 1944; Gaze, 1959; Attardi and Sperry, 1963). The original connections of the optic nerve to the tectum are re-established in these animals and vision is recovered.

In contrast to the mammalian optic nerve, spinal cord or brain, where astrocytes help form a glial scar that impedes axon extension, the astrocytes of the optic tract promote axon extension in the adult newt and in *Xenopus* tadpoles (Reier and Webster, 1974; Turner and Singer, 1974; Stensaas and Feringa, 1977; Reier, 1979; Scott and Foote, 1981; Bohn et al., 1982). The astrocytes in the degenerated distal portion of the cut optic nerve hypertrophy and form a longitudinal band within the basement membrane synthesized by the pia mater (Turner and Singer, 1974; Bohn et al., 1982). These astrocytes are different from those of mammals in that they do not express GFAP, but do express nestin and cytokeratin intermediate filaments and desmosomal proteins (Rungger-Brandle et al., 1989).

The growth cones of the regenerating optic nerve axons associate preferentially with end-feet of the astrocytes that project toward the pia, so all the regenerating axons are found just under the pia (Gaze and Grant, 1978; Bohn et al., 1982). Eliminating the astrocyte band by resection of a segment of the optic nerve results in a decrease in the number of regenerating axons crossing the lesion into the nerve stump and an increase in the number of axons deviated into inappropriate locations (Bohn et al., 1982). Thus, the astrocytes of the regenerating amphibian optic nerve appear to support axon regeneration in the same way that ependymal cells support the regeneration of spinal cord axons and Schwann cells support the regeneration of spinal nerves in mammals.

Studies on goldfish optic nerve regeneration suggest that the regenerating axons and glial cells of the optic tract in fish and amphibians require at least some of the same molecules required for axon sprouting and elongation as do spinal nerve axons in mammals. The optic nerves of goldfish have good capacity for regeneration (Attardi and Sperry, 1963), with visual recovery occurring by 5–6 weeks (Edwards et al., 1981). RNA and protein synthesis are significantly elevated in goldfish retinal ganglion neurons during regeneration, and provide growth-associated proteins that are transported down the length of successfully regenerating axons (Grafstein, 1991; Stuermer et al., 1992). Cell adhesion and basement membrane molecules are associated with optic nerve regeneration, but these have not yet been shown to be made by optic nerve astrocytes. The adhesive substrate preferred by regenerating optic nerve axons of stage 47–50 *Xenopus* tadpoles is laminin, followed by collagen I > polylysine = polyornithine > fibronectin (Grant and Tseng, 1986). The retinal ganglion cells of regenerating goldfish optic nerve express GAP-43 and NCAMs (Stuermer et al., 1992) and two other surface proteins called reggie-1 and 2 (Schulte et al., 1997) that probably interact with the surface of the optic nerve astrocytes.

The ability to regenerate the optic nerve in fish and amphibians is also associated with the absence of inhibitory myelin proteins in the lesion. Tenascin-R and MAG are removed from the degenerated optic nerve in the newt, probably by macrophages, as in regenerating spinal nerves of mammals (Becker et al., 1999). Cohen et al. (1990) reported that goldfish optic nerves produce a factor that is cytotoxic to myelin-producing oligodendrocytes. The factor is recognized by antibodies to the growth factor IL-2, and its size indicates that it might be a dimer of IL-2 (Eitan and Schwartz, 1993). Consistent with this hypothesis, an enzyme identified as a nerve transglutaminase was purified from regenerating optic nerves of fish, and was shown to dimerize human IL-2. The dimerized IL-2 was cytotoxic to cultured rat brain oligodendrocytes. Thus the transglutaminase-catalyzed dimerization of IL-2 might be a mechanism in the fish and amphibian optic nerve that prevents oligodendrocyte inhibition of regeneration.

## MAINTENANCE REGENERATION OF NEURONS IN THE MAMMALIAN CNS

### 1. Discovery of Neural Stem Cells

Fifty years ago it was dogma that neurogenesis ceased after birth in the CNS of mammals. Then, in the early 1960s, $^{3}$H-thymidine labeling studies in adult rats revealed cells actively synthesizing DNA in the hippocampus, leading to the hypothesis that maintenance regeneration is taking place in this region of the brain (Messier et al., 1958; Smart, 1961; Altman, 1962, 1963). The idea of maintenance regeneration in the CNS of any vertebrate was not accepted, however, until Nottebohm and colleagues showed in the 1980s that in male canaries there is a tremendous increase each spring in the number of neurons in their vocal control brain nuclei. These neurons are recruited into song-learning circuits and die when the mating season is over (Goldman and Nottebohm, 1983; Paton and Nottebohm, 1984). Thus, it was shown conclusively that there is neuron turnover and replacement in the adult canary brain, but the dogma of "no new neurons" remained in force for the mammalian brain.

This dogma was shattered for good in the 1990s by the discovery of neural stem cells (NSCs) in different parts of the mammalian brain. **FIGURE 5.12** illustrates how NSCs (and stem cells in general) are identified.

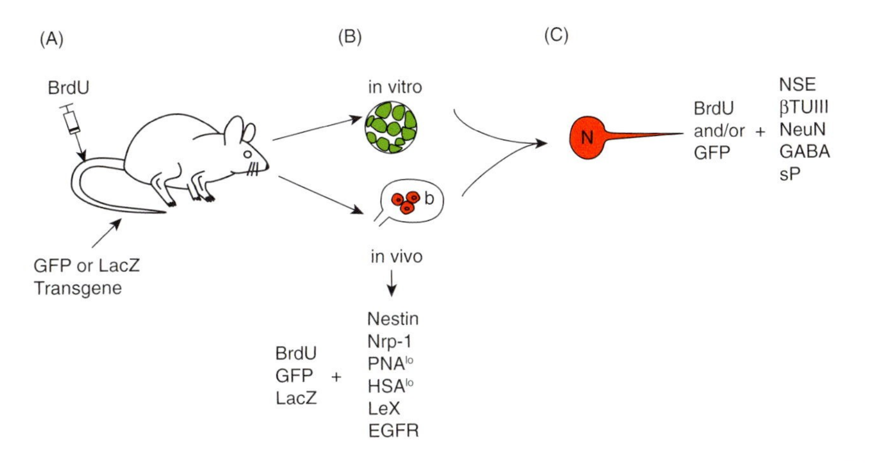


**FIGURE 5.12** Identification of NSCs and production of new neurons in the mouse brain. **(A)** Animals transgenic for GFP or β-galactosidase are injected with BrdU. **(B)** Labeled cells that divide and make neurospheres *in vitro* or labeled cells that divide in the brain (b) *in vivo* are potential NSCs. Such immature cells express a variety of identifying markers in addition to the initial label. **(C)** Proof that the labeled cells are NSCs is their differentiation into neurons that express markers of mature neurons.

First, animals are made transgenic for *LacZ* or the green fluorescent protein (GFP) gene. Second, BrdU is incorporated into the DNA of the labeled cells to show DNA synthesis. Labeled, dividing cells are then shown to express stem cell morphology and molecular markers of stemness, such as nestin, Nrp-1, EGF receptor, peanut agglutinin low ($PNA^{lo}$), heat-stable antigen low ($HSA^{lo}$), and LeX (SSEA-1), and then differentiate to express markers for immature neurons (doublecortin, Dc) and finally mature neurons, such as neuron-specific enolase (NSE), class III β-tubulin (β-TuIII), neuron nuclear transcription factor (NeuN) and the neurotransmitters γ-aminobutyric acid (GABA) and substance P (McKay, 1997; Rao, 1999; Guena et al., 2001; Goh et al., 2003; Bottai et al., 2003). To show that the labeled cells differentiate into neurons, it is also important to show that they exhibit the morphology and electrophysiology of functional neurons. Labeled, FACS-purified NSCs have been subjected to clonal analysis *in vitro* and shown to both self-renew and differentiate into neurons and glia.

Reynolds and Weiss (1992) showed conclusively that the midbrain of adult mice harbors NSCs. Lois and Alvarez-Buylla (1994) and Luskin (1993) demonstrated neurogenesis by NSCs in the walls of the lateral ventricles of the adult mouse. Mouse NSCs cultured *in vitro* proliferated in serum-free medium in the presence of EGF or FGF-2 and formed neurospheres reactive to antinestin antibodies. The cells of the neurospheres clonally differentiated into new neurons and glia, as well as forming new neurospheres, indicating their ability to self-renew (Reynolds and Weiss, 1992).

Similar studies have led to the conclusion that NSCs reside in several regions of the adult mammalian CNS. However, in the uninjured nervous system *in vivo*, maintenance regeneration of neural tissue is known to occur in only two regions, the ventricular walls of the hippocampus, where NSCs give rise to granule cells, and the lateral ventricle walls of the forebrain, where they give rise to olfactory bulb neurons. Proliferating NSCs have been observed in other regions of the uninjured CNS, but they normally give rise only to new glial cells (Gage, 2000; Temple, 2001).

## 2. Regeneration of Olfactory Nerve and Bulb Neurons

The olfactory nerve is a cranial nerve arising from nerve cell bodies in the nasal epithelium that are specialized for discriminating between different odorants. The axons of olfactory sensory nerves (OSNs) extend into the olfactory bulb (OB) of the forebrain, where they synapse on OB neurons. The OB is particularly well developed in nonhuman animals that depend heavily on the sense of smell to assess their environment. In these animals, the olfactory nerve and bulb are continually turning over and regenerating via NSCs.

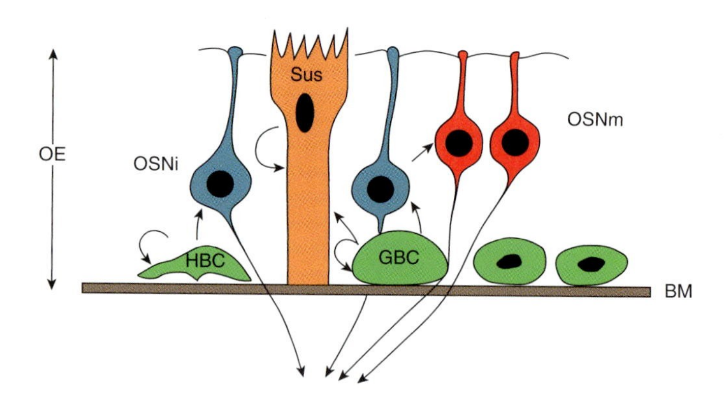


**FIGURE 5.13** Regeneration of olfactory sensory neurons (OSN). The basal layer of the olfactory epithelium (OE) consists of several cell types. The main type is the self-renewing globose basal stem cell (GBC). The GBC gives rise to immature olfactory sensory neurons (OSNi) which are closest to the surface of the epithelium. Another basal cell type is the self-renewing horizontal basal cell (HBC) that may also give rise to OSNi. OSNm = mature OSN. Large sustentacular (Sus) cells, which phagocytose dead neurons and may also act as detoxifiers, span the epithelium. Axons from the OSNs pierce the basement membrane (BM) on their way to the olfactory bulb.

Olfactory receptor neurons have a life span of 4–8 months, depending on species, and are replaced continuously throughout the life of the animal by neurons regenerated from stem cells located in the basal layer of the olfactory epithelium (Schwob, 2002), as illustrated in **FIGURE 5.13**. When the olfactory nerve is transected, the receptor neurons and their axons degenerate and are replaced by these stem cells in the same way (Ramon-Cuerto and Valverde, 1995). In both cases, the axons of the replacement neurons must extend into the OB to synapse with neurons there.

The ability of olfactory axons to extend into the OB, in contrast to the inability of other CNS mammalian axons to regenerate, is due to their association with glial olfactory ensheathing cells (OECs) derived from the olfactory placode during development (Mendoza et al., 1982; Chuah and Au, 1991). These cells are distinct from other mammalian glial cell populations in terms of their ultrastructure and cytoskeletal and cell surface molecules expressed, but they express soluble and insoluble factors that promote neuron survival and axon extension (Ramon-Cueto and Valverde, 1995).

No other mammalian cranial nerves are known to regenerate, but like the optic nerve, the acoustic nerve (cranial nerve VIII) of frogs regenerates well, restoring the specificity of central connectivity that characterizes the auditory system in normal animals (Zakon and Capranica, 1981). It would be interesting to know whether this regeneration, too, is dependent on an associated glial cell population and what the characteristics of these cells are.

The olfactory bulb interneurons of most nonhuman mammals are constantly turned over and replaced by NSCs in the walls of the anterior lateral ventricles (Schwob, 2002; Bottai et al., 2003). These NSCs have some of the characteristics of astrocytes. They divide asymmetrically to self-renew and give rise to a transit amplifying population (Doetsch et al., 1999; Rietze et al., 2001). In mice, approximately 15% of these cells are actively dividing with a cell cycle time of approximately 14 hr (Smith and Luskin, 1998). Chains of transit amplifying cells migrate into the olfactory bulb along a restricted pathway called the rostral migratory stream, where they differentiate into new neurons and glia (Lois and Alvarez-Buylla, 1994; Lois et al., 1996).

The exact location within the lateral ventricle wall of the NSCs that have long-term self-renewal potential has been the subject of controversy (Bottai et al., 2003). BrdU incorporation experiments on adult mice suggested that the ependymal cells of the lateral ventricles are labeled first and are therefore the actual long-term stem cells (Johansson et al., 1999; Rietze et al., 2001). Mouse ependymal cells from the lateral ventricles were fractionated by FACS followed by testing the fractions for their ability to form neurospheres under clonal conditions (Rietze et al., 2001). This procedure revealed a discrete population of cells that was nestin$^+$, had low binding affinity for PNA and HSA, and did not express differentiated neuron or glial cell markers. Clonally derived neurospheres from this population differentiated into astrocytes (GFAP$^+$), oligodendrocytes (O4$^+$), and neurons ($\beta$-Tu type III$^+$). Further evidence that the PNA$^{lo}$ HSA$^{lo}$ cells were NSCs was a six-fold reduction in the percentage of these cells in the mouse mutant *querkopf*, which has a greatly reduced number of olfactory neurons (Rietze et al., 2001). The ependymal cells were negative for the cell surface marker LeX1 (SSEA-1), a complex carbohydrate expressed on the surface of ESCs and in the germinal zones of the embryonic CNS. However, cells in the subventricular zone do express LeX1. Furthermore, FACS-isolated LeX1$^+$ cells of the lateral ventricle wall make neurospheres, but the LeX$^-$ cells of the ependyma do not, suggesting that long-term renewing NSCs reside in the subventricular zone (Capela and Temple, 2002).

Astrocyte-like NSCs are also found in the lateral ventricle walls of the human brain, but unlike the brains of other mammals, the cells are arranged in a single ribbon, rather than in multiple chains (Sanai et al., 2004). Approximately 0.7% of these NSCs are in division. Single NSCs isolated from the ribbon can form neurospheres *in vitro* that differentiate into astrocytes, oligodendrocytes, and neurons. *In vivo*, however, the NSCs are not observed to migrate along a rostral pathway to replace neurons in the olfactory bulb (Sanai et al., 2004). Unless the NSCs migrate as single cells, these observations suggest that human olfactory bulb neurons may not undergo maintenance regeneration. Their normal fate *in vivo* is unknown.

## 3. Regeneration of Hippocampal Neurons

The hippocampus is an area of the brain that is crucial for cognitive activities, learning, and memory. Proliferating NSCs with long-term renewal potential have been identified in the hippocampus of mice and rats (Rao, 1999; Gage, 2000; Momma et al., 2000; Guena et al., 2001; Rietze et al., 2001), marmoset monkeys (Gould et al., 1998), and deceased human patients given BrdU as part of a cancer study (Eriksson et al., 1998).

The precise location of hippocampal NSCs is still unclear. The dentate gyrus, located within the hippocampus itself, is one possibility. Human dentate gyrus cells purified by FACS and transfected with the GFP gene under the control of the nestin enhancer or the T$\alpha$1 tubulin promoter proliferated *in vitro* and differentiated morphologically, antigenically, and electrophysiologically as neurons (Roy et al., 2000). However, these cells have limited self-renewal capacity, suggesting that they are a transit-amplifying population of progenitors, whereas the lateral ventricular wall adjacent to the hippocampus has been shown to contain multipotent NSCs with long-term self-renewal capacity (Seaberg and Van der Kooy, 2002).

Regardless of their location, the presence of NSCs in the hippocampus suggests that maintenance regeneration might be crucial for the retention or formation of memory and for learning new information and tasks. Several experimental results support this idea. Studies *in vivo* strongly suggest that the number of new neurons born in the hippocampus of mice is influenced by both physical and cognitive activity. Running on a treadmill, or placement in an enriched environment (more mice per cage to increase social interactions, mouse toys and treats, and re-arrangable sets of tunnels) increased stem cell proliferation and neurogenesis in the dentate gyrus of mice above the level of controls (Kempermann et al., 1998; van Praag et al., 1999a, b). Enriched-environment mice also learned a maze faster than controls. Control mice exhibited a decline in the number of BrdU-labeled new neurons in the hippocampus with age that was correlated with a reduction in the speed at which the maze was learned. The reduction in number of BrdU-labeled differentiated neurons in aged animals was reduced by more than half when they were placed in an enriched environment (Kempermann et al., 1998).

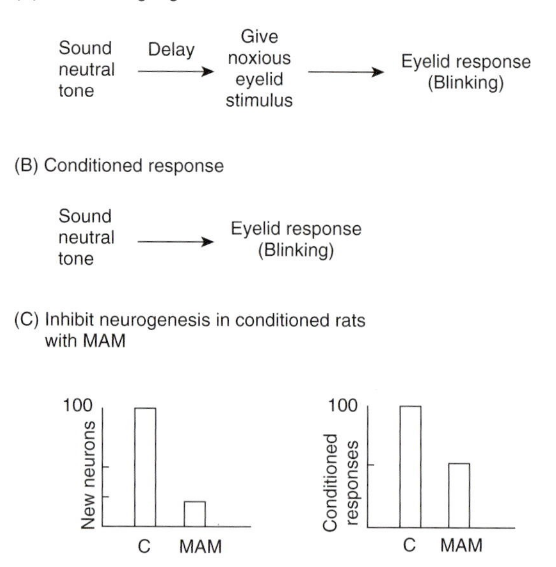


**FIGURE 5.14** Behavioral experiment to demonstrate the integration of NSC-derived neurons into hippocampal circuitry. **(A, B)** Rats were conditioned by a noxious eyelid stimulus to respond to a neutral tone by blinking. Hippocampal neurogenesis was then inhibited by methylazoxymethane (MAM). This drug drastically reduced the number of new neurons produced by NSCs and simultaneously reduced the frequency at which rats exhibited the conditioned response to the tone.

Other studies have shown that new neurons born in the hippocampus are functionally integrated into hippocampal neural circuitry. Van Praag et al. (2002) transfected proliferating cells in the mouse hippocampus with a retroviral GFP transgene. GFP-labeled cells differentiated into mature neurons that received synaptic input, as shown by ultrastructural analysis and immunostaining with synaptophysin. Electrophysiological measurements indicated that the neurons functioned normally and received normal input from the perforant pathway, the main excitatory input to dentate granule cells. Shors et al. (2001) used behavioral studies to demonstrate integration into the normal hippocampal circuitry (**FIGURE 5.14**). Rats were "trace conditioned" to associate a neutral tone with delayed noxious eyelid stimulation. Neurogenesis was then inhibited by the DNA methylating agent methylazoxymethane (MAM), which kills proliferating cells. The number of new neurons generated was reduced by 80%, which in turn reduced the frequency of conditioned responses by 50%.

## 4. Regeneration of Cortical Neurons

Whether or not there is maintenance regeneration of the neocortex by NSCs in the lateral ventricles is controversial. Gould et al. (1999) injected macaque monkeys with BrdU and assessed incorporation into cells in the prefrontal, posterior, parietal, and inferior temporal cortex at two hours and at 1–2 weeks postinjection. One–two weeks later, labeled cells were observed streaming from the lateral ventricles through the subcortical white matter into the cortex. Most of the labeled neurons within the cortex reacted for markers associated with mature neurons (MAP-2, NeuN), whereas cells in the streams outside the cortex reacted with a marker for immature neurons, TOAD-64. These observations suggested that NSCs in the walls of the lateral ventricles replenish cortical neurons. Injection of dyes into cortical areas known to be projection targets of the newly differentiated neurons resulted in their retrograde filling, indicating that they had extended axons that became part of the cortical circuitry.

Another study of BrdU-injected adult macaques also revealed labeled ventricular cells that moved to the cortex. However, these cells failed to express markers of mature neurons but did express GFAP, indicating that they had differentiated as glia (Kornack and Rakic, 2001). Detailed examination of sections with confocal optics suggested that the close apposition of BrdU-labeled glial cells and neurons could give the illusion of BrdU-labeled cells that also expressed neuronal markers. Magavi et al. (2000) reached a similar conclusion with regard to the mouse cortex. They, too, observed BrdU-labeled cells in the walls of the lateral ventricles that entered the cortex, but these cells differentiated as glia, remained undifferentiated, or died.

A novel approach that has been called "archeo-cell biology" (Croce and Calin, 2005) appears to have settled this question in favor of the view that the NSCs do not maintain the cerebral cortex and that the complement of neurons present at birth or shortly thereafter is all there will be thereafter (Spalding et al., 2005). The method was to measure the amount of $^{14}C$ in the nuclei of occipital cortex neurons and other tissues known to be maintained by stem cells, taking advantage of the fact that atomic bomb testing between the years 1955 and 1963 generated large amounts of atmospheric $^{14}C$ that peaked in 1963 and fell off by 50% every 11 years thereafter. The measurements were made in deceased persons born prior to, during, and after this period and were compared to the amounts of $^{14}C$ in tree rings to establish the birth date of the neurons. The result was that neurons born during a time of high atmospheric $^{14}C$ maintained this level until they died, whereas other tissues known to turn over rapidly and be maintained by stem cells had undetectable levels of $^{14}C$, indicating that they had been replaced.

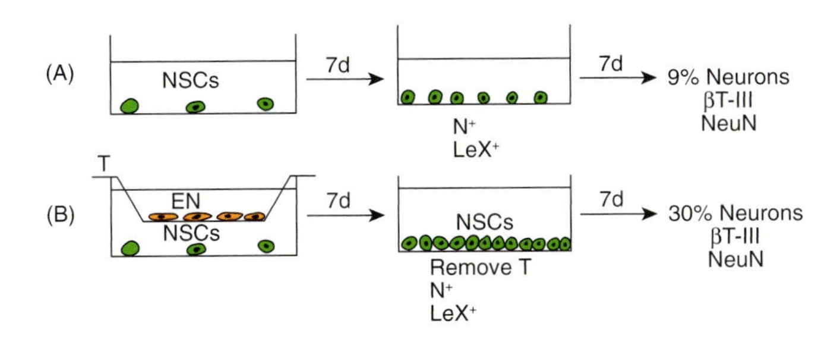


**FIGURE 5.15** Experiment showing that soluble factors from endothelial cells (EN) promote expansion and commitment to neural differentiation of NSCs. **(A)** NSCs cultured alone for 7 days proliferated and 9% formed neurons after an additional 7 days in culture. **(B)** By contrast, NSCs incubated for 7 days with endothelial cells, but separated from them by a transwell insert, proliferated to a much greater degree and 30% differentiated into neurons after removal of the insert. N, Le = markers for NSCs; βT-III, NeuN = markers of mature neurons.

### 5. The CNS Neural Stem Cell Microniche

In the adult vertebrate brain, NSCs of the hippocampus, lateral ventricles, and songbird higher vocal center are in close proximity to blood vessels, like the stem cells of the dental pulp (Palmer et al., 2000; Louissaint et al., 2002). Factors secreted by blood vessel cells are suggested to help create the microniche that controls the balance between maintaining NSCs in an undifferentiated state and their asymmetric division. The relationship of NSCs to blood vessels begins during embryonic development, when cells of the ventricular walls of the neural tube produce VEGF, directing the growth of capillaries toward them. Shen et al. (2004) established a direct functional relationship between endothelial cells of blood vessels and NSCs from a variety of regions of the embryonic CNS, as well as adult lateral ventricular wall NSCs (**FIGURE 5.15**). NSCs co-cultured in wells with endothelial cells residing on the other side of transwell filter inserts were much more strongly expanded than NSCs cultured with blood vessel smooth muscle or nonvascular cells, while simultaneously being primed to differentiate a significantly higher percentage of projection neurons and interneurons after removal of the inserts. These results suggest that the endothelial cells secrete diffusible factors that affect NSC proliferation and differentiation. The proliferation and differentiation of hippocampal NSCs is also enhanced by astrocytes (see below), so each NSC niche is likely to be determined by a variety of cell types, as well as by the NSCs themselves (Goh et al., 2003). The niche signals that regulate NSC quiescence versus proliferation and differentiation remain largely unidentified, but a number of growth factors such as EGF, FGF-2, and TGF-α, as well as sonic hedgehog, are known to elevate the proliferation of NSCs after infusion into the brain or *in vitro*, and BMP can promote the differentiation of NSCs into glial cells (Goh et al., 2003). It would not be surprising to find involvement of the Wnt pathway in the activation of NSCs as well.

Niche signals regulate the activity of transcription factors that maintain stemness and self-renewal of NSCs. Delta/Notch signaling upregulates the production of Hes 1, a transcriptional repressor of neuron differentiation genes (Nakamura et al., 2000; Yamamoto et al., 2001; Bottai et al., 2003). Neurosphere formation is abolished in NSCs lacking Nrp 1 and 2. Another key transcriptional factor that maintains NSCs in an undifferentiated state is neuronal restricted silencing factor (NRSF/REST), a zinc-finger protein of the kruppel family that represses neuron-specific genes (Chong et al., 1995). A small, noncoding double-stranded (ds) RNA, NRSE dsRNA, plays a key role in the differentiation of NSCs into neurons by binding to the NRSF/REST transcriptional complex of proteins, allowing neuron-specific genes such as Mash 1, neurogenin 1 and 2, and Neuro D to be activated (Kuwabara et al., 2004; Bottai et al., 2003).

## INJURY-INDUCED CNS REGENERATION

Regeneration of new neurons takes place in several CNS tissues following injury: auditory sensory cells in birds, the mammalian hippocampus and cortex of the brain, the retina of embryonic birds and mammals, the retina of adult fish and amphibians, and the regenerating tails of larval anurans and urodeles, as well as the regenerating tails of adult urodeles and lizards. The regenerative response is by NSCs resident to the tissue or created by dedifferentiation. In some cases, the response is robust and the outcome is functional recovery, but in others, primarily adult mammals, it is too minimal to result in any meaningful recovery. Nevertheless, the fact that there is a response at all in mammals suggests that if a more regeneration-permissive environment could be provided, functional recovery might be possible.

### 1. Avian and Mammalian Auditory Sensory Neurons

The vertebrate inner ear is an organ of hearing and balance consisting of the cochlea and the vestibular apparatus (Ham and Cormack, 1979; Ashmore and Gale, 2000). The cochlea senses sound vibrations from the tympanic membrane. The vestibular apparatus, consisting of the utricle, saccule, and semicircular canals, is a sensor of gravity and linear and angular acceleration (Ham and Cormack, 1979). The energy of

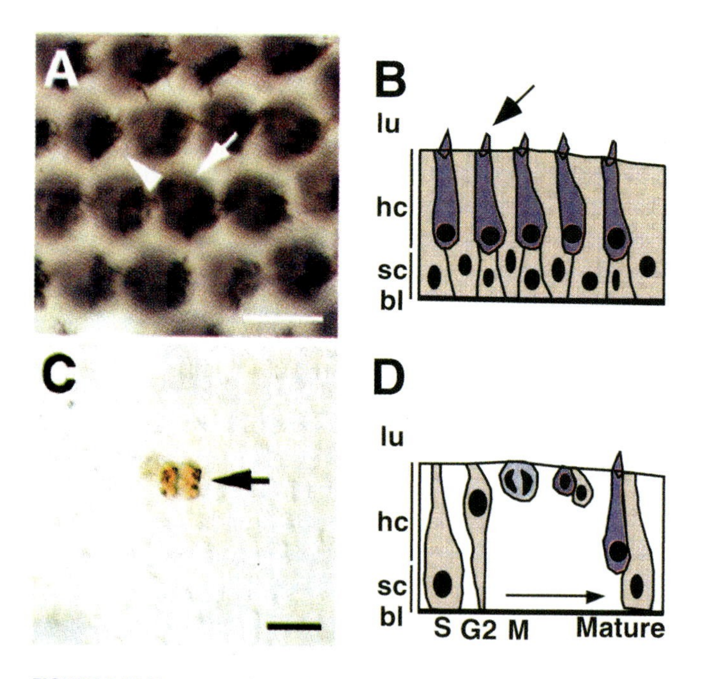


**FIGURE 5.16** Regeneration of hair cells in the chick basilar epithelium after antibiotic-induced injury. **(A)** A whole mount of untreated basilar epithelium immunostained for the hair cell marker calmodulin, showing the alternating pattern of hair cells (purple, arrow) and support cells (arrowhead). **(B)** Diagram showing arrangement of hair cells and supporting cells in transverse section. Hair cells (hc, purple) reside closer to the lumen (lu) and do not contact the basement membrane (bl), whereas support cells rest on the bl and span the epithelium. Arrow points to stereocilia. **(C)** Support cells divide near the lumen of the injured epithelium as shown by whole-mount BrdU staining. **(D)** Support cell nuclei migrate toward the lumen and the cell detaches from the basal lamina and undergoes mitosis into a pair of daughter cells, one of which differentiates into a hair cell, while the other remains a stem (support) cell. Reproduced with permission from Stone and Rubel, *Delta 1* expression during avian hair cell regeneration. Development 126:961–973. Copyright 1999, The Company of Biologists.

sound and motion is transduced to electrical impulses by auditory sensory cells called hair cells, which reside in the sensory epithelium lining these structures (**FIGURE 5.16**). Each hair cell is flask-shaped and is surrounded by support cells. The hair cells extend from the luminal surface part way toward the basement membrane of the epithelium, but do not contact it, whereas the support cells are columnar and span the whole distance from the basement membrane to the lumen. The mechanosensitive organelles of the hair cells are the stereocilia, or "hair bundles," which are supported by a dense core of actin filaments that are completely replaced via treadmilling over a 48-hr period (Schneider et al., 2002).

A major cause of hearing loss and balance problems is dysfunction of the inner ear caused by loss of hair cells due to noise trauma, toxic drug damage (for example, aminoglycoside antibiotics such as gentamycin), infection, or age-related degeneration. Up to 120 million persons worldwide are estimated to suffer from hearing impairment (Lowenheim, 2000).

Adult birds are able to regenerate sensory cells in both the sensory epithelium of the cochlea and the utricle of the vestibular apparatus. In mature birds, the epithelial cells of the cochlea are mitotically quiescent. Sensory hair cells in the vestibular sensory epithelium, however, normally turn over by apoptosis and are replenished by maintenance regeneration (Kil et al., 1997; Stone and Rubel, 2000). Hair cells of both cochlea and utricle are able to regenerate after injury (Roberson and Rubel, 1995; Oesterle and Rubel, 1996; Kil et al., 1997; Stone and Rubel, 2000), with virtually complete recovery of hearing.

The stem cells responsible for both maintenance and injury-induced hair cell regeneration in birds are most likely the support cells, based on the size, shape, and location of the nuclei of cells that divide and differentiate into hair cells in the epithelium (**FIGURE 5.16**). Support cell nuclei are small, are oval-shaped and lie near the basement membrane of the epithelium, whereas hair cell nuclei are large, are round, and lie closer to the lumen (Fekete, 1996; Fekete and Wu, 2002). The nuclei of the hair cell progenitors migrate from the basal lamina toward the lumen as they progress from S phase to M (Cotanche et al., 1994; Tsue et al., 1994; Kil et al., 1997), similar to the migration of dividing cells observed in the differentiation of the neural tube (Sauer, 1935), suggesting that they are derived from support cells. At the lumen, one member of the pair of cells generated by mitosis differentiates into a hair cell, the other into a support cell (Stone and Cotanche, 1994). A similar asymmetric division has also been reported in the neuromasts of the lateral line system in the salamander, which detects motion of the water (Jones and Corwin, 1996).

A number of markers are expressed by all support cells within the epithelium of the cochlea, but only the homeodomain transcription factor, c(hicken)Prox1 is differentially regulated in a way that suggests it might have a key role in determining hair cell fate and differentiation (Stone et al., 2004). Prox1 is required for cell cycle withdrawal in a number of vertebrate and invertebrate embryonic tissues. In the damaged auditory epithelium of the cochlea, some quiescent support cells do not express cProx1, but others exhibit a low level of expression. CProx1 is absent in half the activated support cells three days after gentamycin treatment, but present in the other half. Sibling pairs of cells following division exhibit three patterns of cProx1 expression, in approximately equivalent numbers: symmetrical low, symmetrical high, and asymmetric (**FIGURE 5.17**). In the undamaged vestibular epithelium of the utricle, none of the quiescent support cells

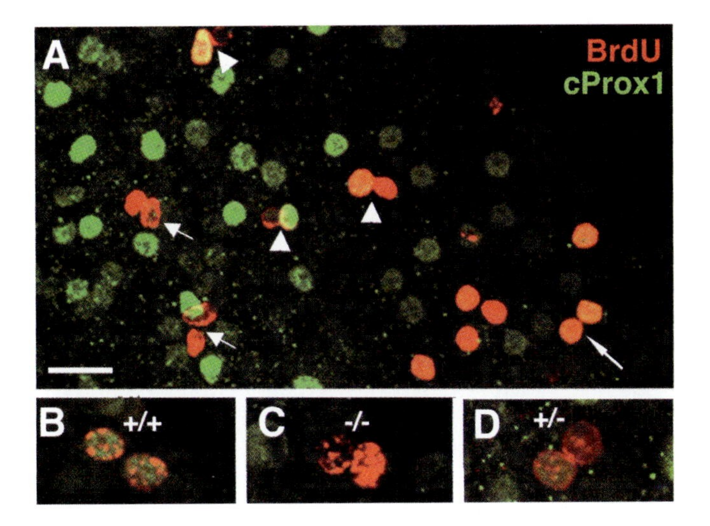


**FIGURE 5.17** Expression of cProx1 (green) in sibling pairs of BrdU-labeled (red) cells derived by the division of support cells in the damaged basilar epithelium. Whole-mount immunostaining. **(A)** Arrowheads = asymmetric expression of cProx1. Short arrows = asymmetric low levels of cProx1 expression. Long arrow = symmetric high levels of cProx1 expression. **(B–D)** Higher magnification of symmetrical high (+/+), symmetrical low (–/–) and asymmetric (+/–) expression of cProx1 at 10, 8 and 17 hr post-BrdU, respectively. Reproduced with permission from Stone et al., cProx1 immunoreactivity distinguishes progenitor cells and predicts hair cell fate during avian hair cell regeneration. Dev Dynam 230:597–614. Copyright 2004, Wiley-Liss.

expresses cProx. During maintenance regeneration, sibling pairs of cells are at first negative for cProx1, but then the majority of them exhibit asymmetric expression. The cell of the pair that expresses cProx1 differentiates into a hair cell in both the cichlea and the utricle. **FIGURE 5.18** illustrates all the patterns of cProx1 expression and their relationships to patterns of differentiation.

The pattern of one hair cell surrounded by several support cells is believed to be generated by lateral inhibition from one member of a mitotic pair to the other (Cotanche, 1987). The Delta/Notch signaling system has been implicated in the lateral inhibition of sensory cell differentiation of the zebrafish inner ear (Haddon et al., 1998a, b). Delta and Notch are at first equally expressed in both members of support cell mitotic pairs in the gentamycin-damaged chick vestibular epithelium; Delta is then downregulated in the cell that will become a support cell and upregulated in the hair cell (Stone and Rubel, 1999). The factors regulating this asymmetric regulation of Delta expression are unknown. In the chick cochlea, the FGF/FGFR3 signaling pathway has been implicated in maintaining the stemness of support cells in the normal basal epithelium (Bermingham-McDonogh et al., 2001). FGFR3 is highly expressed in the support cells of the epithelium.

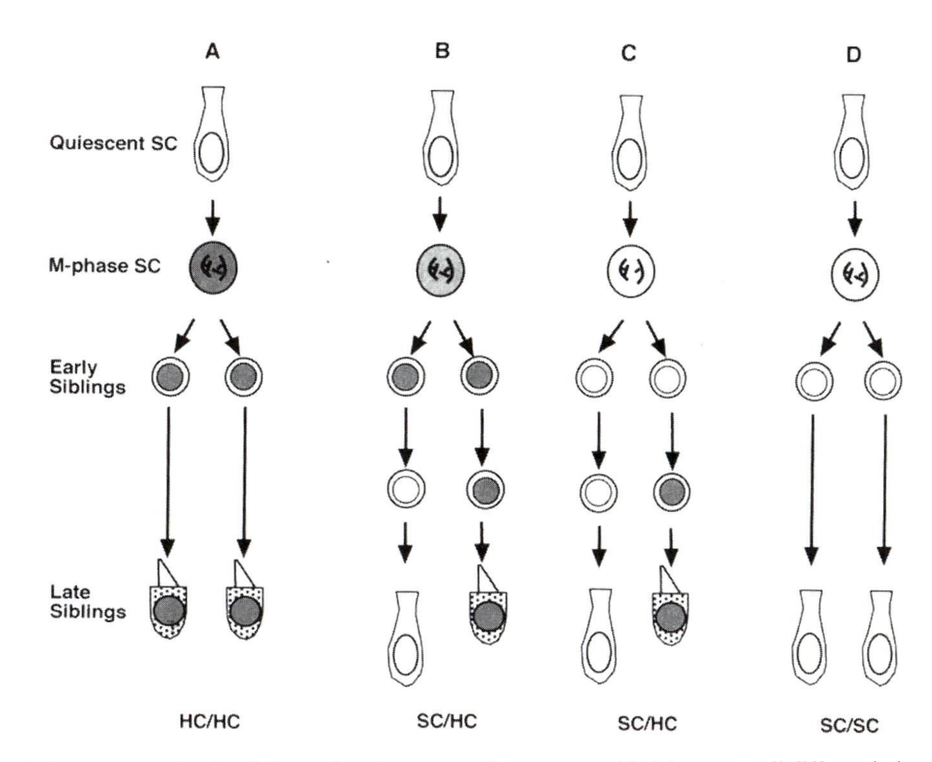


**FIGURE 5.18** Patterns of cProx1 expression in sibling pairs of support cell progeny and hair/support cell differentiation during hair cell regeneration. Gray is high expression, white is cProx1-negative. **(A)** Production of two hair cells (HC/HC). **(B, C)** Production of a support cell and a hair cell (SC/HC). **(D)** Production of two support cells. **A, B, D** are all observed in the damaged basilar epithelium. **C** is the rule in the undamaged utricle. Reproduced with permission from Stone et al., cProx1 immunoreactivity distinguishes progenitor cells and predicts hair cell fate during avian hair cell regeneration. Dev Dynam 230:597–614. Copyright 2004, Wiley-Liss.

The hypothesis is that adjacent hair cells secrete an FGF ligand that binds to a Trk on the support cells and prevents their differentiation. When hair cells are damaged, the ligand is no longer available, FGFR3 is downregulated in the support cells, and they enter the cell cycle. Once new support cells differentiate and production of ligand resumes, their expression of FGFR3 is upregulated and their proliferation is inhibited.

The hair cells of mammals normally do not regenerate and permanent hearing and balance deficits are the usual result of injury to the sensory epithelium. Nevertheless, there is *in vivo* and *in vitro* evidence that hair cells can regenerate to some extent in the vestibular apparatus of guinea pigs and humans. Scanning and transmission electron microscopy of gentamycin-treated utricular sensory epithelium in guinea pigs showed complete hair cell loss within a week after treatment. By four weeks, however, cells with organized bundles of stereocilia had reappeared at the luminal surface (Forge et al., 1993). Proliferating support cells were observed throughout guinea pig and human utricular sensory epithelium cultured *in vitro* with gentamycin and labeled with BrdU or $^3$H-thymidine. Some of the labeled cells developed into immature hair cells with small apical tufts of stereocilia (Warchol et al., 1993).

The developing mouse organ of Corti synthesizes retinoic acid (RA). Exogenous RA exerts a concentration-dependent stimulation of hair cell development in cultured mouse organs of Corti (Kelley et al., 1993). Lefebvre et al. (1993) have reported that RA and TGF-$\alpha$ stimulate the regeneration of hair cells in neomycin-treated rat organ of Corti sensory epithelium.

In mice, mitosis is normally completed in the sensory epithelial cells of the organ of Corti of mice by embryonic day 14. The negative regulator of cell cycle progression, the p27$^{Kip1}$ protein, mediates their exit from the cell cycle, and they differentiate into hair cells and support cells (Parker et al., 1995; Nakayama et al., 1998). Disruption of the p27$^{Kip1}$ gene allows cell division to continue in the sensory epithelium of the organ of Corti in postnatal and adult mice. However, p27$^{Kip1}$ expression is downregulated in differentiating hair cells and is unlikely to maintain mitotic arrest in these cells (Lowenheim et al., 1999; Chen and Segil, 1999). Sage et al. (2005) identified the Rb gene in a microarray analysis as the regulator of mitotic arrest in hair cells (see also Taylor and Forge, 2005).

The effect of knocking out specific genes can be tested by transgenic Cre/lox techniques (Branda and Dymecki, 2004). Transgenic animals are created in which genes or portions of genes are flanked by *loxP* sequences that are the recognition sites for the bacteriophage Cre recombinase. Such genes are said to be "floxed." The animals are also made transgenic for *Cre* under the control of a cell-specific promoter, or *Cre* or a *Cre*-promoter construct can be introduced into tissues or cells *in vitro* by a viral vector. *Cre* will be activated in those cells with transcription factors for the cell-specific promoter. Cre-recombinase binds to the *loxP* sites and excises the gene flanked by these sites.

Sage et al. (2005) tested the effect of knocking out Rb in utricles by culturing utricles from transgenic mice carrying *Rb* flanked with loxP sites (**FIGURE 5.19**). The utricles were transfected with adenovirus constitutively making Cre recombinase, which removed the *Rb* gene. The result was that hair cells re-entered the cell cycle and divided while still differentiated. That is, they produced more hair cells by compensatory hyperplasia. This means that it might be possible to treat deafness by transient inactivation of *Rb*; more broadly, it suggests that it might be possible to induce compensatory hyperplasia in other differentiated cell types as a mechanism of restoring normal numbers of such cells instead of relying only on resident stem cells or stem cells created by dedifferentiation.

## 2. Regeneration of Neurons in the Mammalian Brain

There is evidence that injured mammalian brain tissue can respond to injury by a low-level increase in neurogenesis. Although Magavi et al. (2000) found no evidence of maintenance neurogenesis in the mouse cortex, they did observe neurogenesis after destruction of a subset of pyramidal neurons in the lower layer (VI) of the cortex that projects to the thalamus. Thalamic neurons were injected with chromophore-conjugated chlorin $e_6$ nanospheres, which were retrogradely transported to layer VI cortical neurons. The layer VI neurons were then killed by activating the chromophore with laser light at a wavelength of 674 nm. Subsequent proliferation of BrdU-labeled lateral ventricle cells was no greater than in uninjured animals, but cells labeled with BrdU and expressing doublecortin were observed along a path to the cortex. One to two percent of the BrdU-labeled cells within the cortex expressed the mature neuron marker NeuN. Retrograde labeling showed that the regenerated neurons were pyramidal neurons that projected to the thalamus; i.e., the neurons that were destroyed were selectively regenerated and functionally integrated into the normal circuitry of the brain. Other studies have shown increased production of granule cells in the dentate gyrus of the hippocampus in rats and gerbils after neuronal degeneration induced by focal or global ischemia (Gould and Tanapat, 1997; Jin et al., 2001; Liu et al., 1998). Thus, it would appear that injury signals can diffuse (or be carried by the circulation) far enough to activate and "home" NSCs to the injury site, and are sufficient to increase

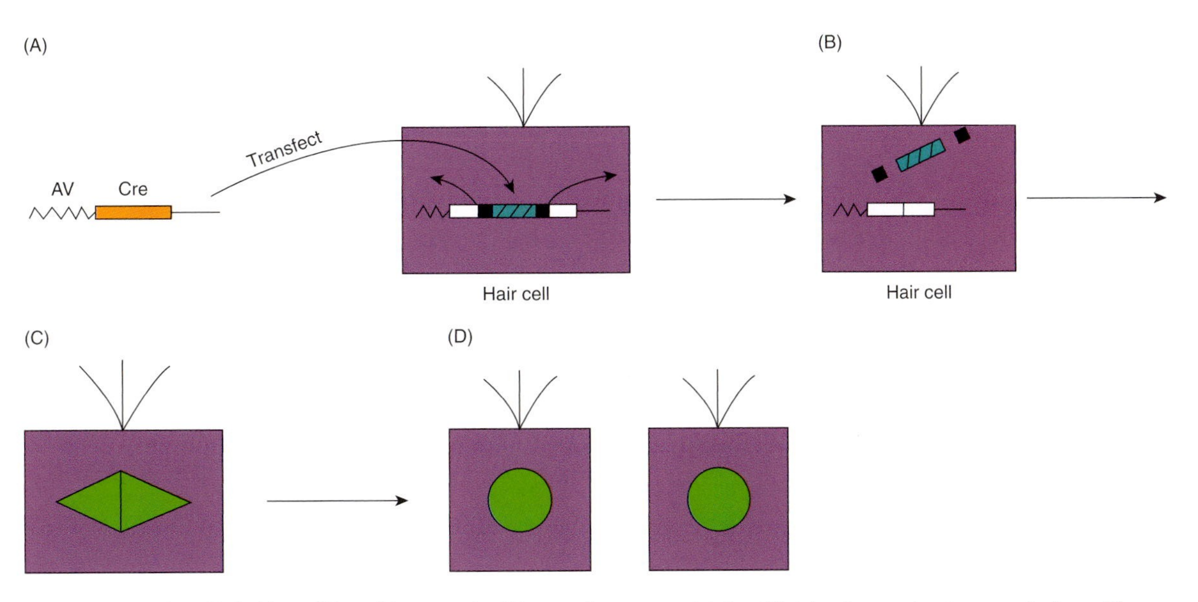


**FIGURE 5.19** Effect on hair cell division of knocking out the Rb gene in mouse utricles. Utricles from mice transgenic for a *Rb* gene carrying lox P (P) sites flanking exon 19 (hatched) were cultured *in vitro*. **(A)** The utricles were transfected with an adenovirus (AV) constitutively expressing the *Cre* gene. **(B)** The Cre enzyme excised exon 19 at the lox P sites, creating a defective Rb gene whose product could no longer block entry into the cell cycle. **(C, D)** The result was mitosis of the hair cells.

the number of NSCs that differentiate into neurons. However, only a small fraction of the degenerated neurons are replaced, not enough for functional recovery.

The small increase in the frequency of ischemia-induced differentiation of hippocampal NSCs into pyramidal neurons can be markedly enhanced by exogenous FGF-2 and EGF (Nakatomi et al., 2002). Intraventricular injection of FGF-2 and EGF together boosted the number of regenerated neurons to 40% of the number that were lost. The neurons were functionally integrated into the normal hippocampal circuitry as determined by microscopy, the electrophysiological properties of synapses, and the performance of the rats on behavioral tasks. These findings are consistent with the fact that FGF-2 expression is upregulated after CNS injury in mammals by fibrous astrocytes (Fawcett and Asher, 1999) and helps to keep NSCs dividing and undifferentiated *in vitro* (Temple, 2001).

Neonatal astroglia induce the proliferation and differentiation of adult hippocampal NSCs into neurons *in vitro*, whereas the effect of adult astrocytes is only half that of neonatal cells (Song et al., 2002). Furthermore, the effect seems specific to hippocampal astrocytes, since spinal cord astrocytes do not support hippocampal neurogenesis. These results suggest that astrocytes are involved in generation of the hippocampal NSC niche, but that the low level of increase in neurogenesis in the injured adult hippocampus is due to deficient FGF-2 and EGF signaling from astrocytes to NSCs.

## 3. Regeneration of the Retina

### a. *Structure of the Retina*

**FIGURE 5.20** illustrates the structure of the retina. The retina receives photons focused by the lens and converts them to electrical impulses that travel to the optic tectum of the brain. The retina is multilayered in all vertebrates (Ham and Cormack, 1979; Litzinger and Del-Rio Tsonis, 2002). The inner layer consists of the ganglion cells, which give rise to the optic nerve. The photoreceptor layer is composed of the light-transducing segments of the rods and cones that constitute the outer layer of the retina. Between these two layers are several others that vary in number depending on the species, but that always include an inner nuclear layer next to the ganglion cells and an outer nuclear layer closer to the photoreceptors. The outer nuclear layer is actually that portion of the rods and cones where their nuclei reside. The inner nuclear layer contains horizontal, bipolar, and amacrine cells that relay impulses from the photoreceptors to the ganglion cells. The synapses made between the photoreceptors in the outer nuclear layer and the bipolar cells of the inner nuclear layer show up in sections as the outer plexiform layer. A similar inner plexiform layer is created by the synapses between the neurons of the inner nuclear layer and the ganglion cell layer. The ends of the photoreceptors are embedded in the epithelial cells of the pigmented layer of the retina.

Outside the retina is the retinal pigmented epithelium (RPE) and outside the RPE is the choroid, a layer

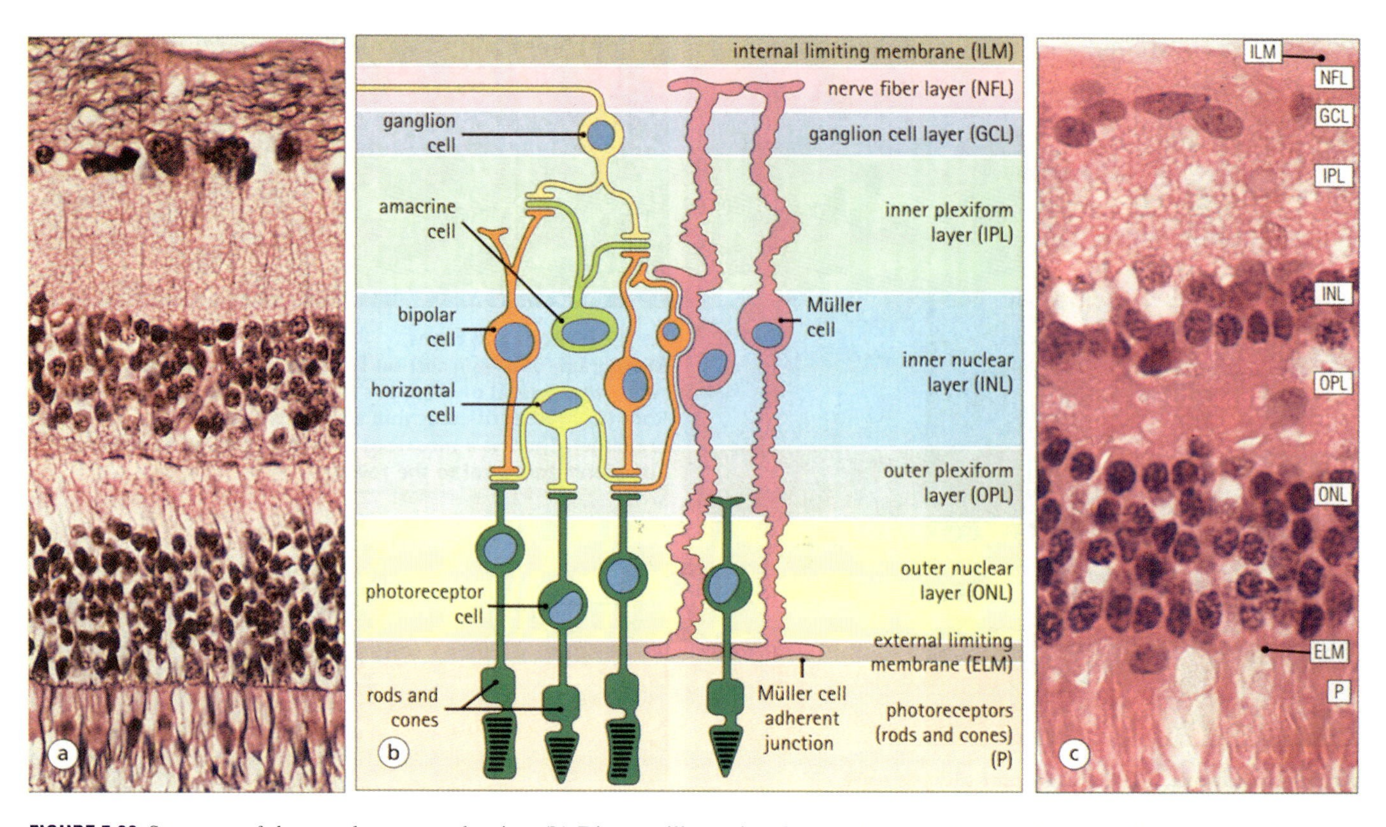


**FIGURE 5.20** Structure of the vertebrate neural retina. **(b)** Diagram illustrating the nine layers of the neural retina. The internal and external limiting membranes are basement membranes. Muller cells are glia spanning the neural retina. **(a)** The nine layers of the retina as outlined by silver staining. **(c)** Histological section stained with hematoxylin and eosin. Reproduced with permission from Stevens and Lowe, Human Histology (3rd ed), Mosby/Elsevier. Copyright 2005, Elsevier.

of heavily vascularized connective tissue. The RPE functions to provide support factors for the photoreceptors. It transports and regulates the concentrations of ions, water, growth factors and nutrients to the photoreceptors. RPE cells also take up, store and re-isomerize retinol, which is used to synthesize the photopigment rhodopsin.

The retina can be surgically removed from the eye or destroyed by cutting off its blood supply, which results in the degeneration of all retinal neurons. The retina of adult birds and mammals is unable to regenerate after injury with any degree of functional recovery. However, the retina of adult teleost fish and the newt, as well as the retina of some vertebrate embryos, does regenerate.

### *b. Regeneration in Teleost Fish*

The retina in adult teleost fish grows continually by increase in size of neurons and by addition of new neurons appositionally and interstitally (Raymond and Hitchcock, 2000; Haynes and Del-Rio Tsonis, 2004). Appositional growth occurs from a ring of neuroepithelial cells at the margin of the retina, called the circumferential germinal zone (CVG). Interstitial growth occurs by the addition of new rods into the layer of photoreceptors from rod precursor cells scattered throughout the outer nuclear layer. If the CVG is removed from the fish eye, it is regenerated from stem cells residing in both the peripheral neural retina and the anterior ciliary body (Jimeno et al., 2003).

Surgical removal of a piece of retina induces regeneration of the removed piece by formation of a blastema of proliferating cells around the margin of the wound that have the characteristics of germinal zone stem cells (Hitchcock et al., 1992). Destruction of the retina with oubain results in more extensive regeneration (Maier and Wolberg, 1979). **FIGURE 5.21** illustrates regeneration of the fish retina. The source of the cells that regenerate the retina appears to be a population of stem cells in the inner nuclear layer (INL stem cells) that express the transcription factor Pax-6 (Raymond and Hitchcock, 1997; Julian et al., 1998; Otteson et al., 2001, 2003). During larval stages of development, INL stem cells give rise to rod precursor cells, which differentiate into

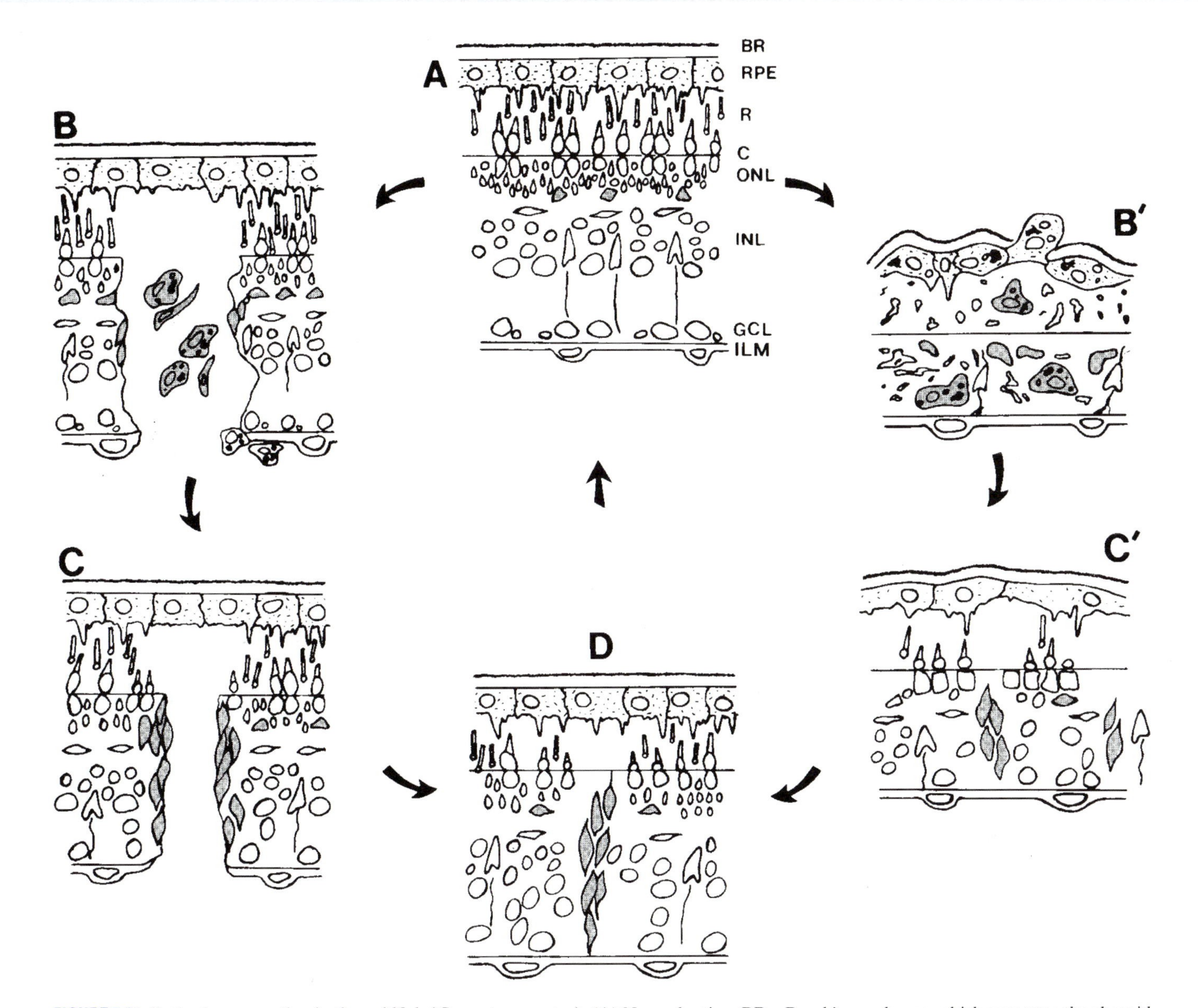


**FIGURE 5.21** Retinal regeneration in the goldfish (*Carassius auratus*). **(A)** Normal retina. BR = Bruch's membrane, which separates the choroid layer of the eye from the pigmented retina, RPE = retinal pigmented epithelium, R = rods, C = cones, ONL = outer nuclear layer, INL = inner nuclear layer, GCL = ganglion cell layer, ILM = inner limiting membrane. Whether injured locally by surgical means **(B, C)** or globally by oubain **(B', C')**, NSCs (slender gray cells), as well as macrophages (larger gray cells) proliferate **(B, C; B', C')**. The NSCs redifferentiate into all the retinal layers **(D)**. Reproduced with permission from Hitchcock and Raymond, Retinal regeneration. Trends in Neurosci. 15:103–108. Copyright 1992, Elsevier.

rods, the last retinal cell type to differentiate in the eye. In the injured adult retina, a blastema of proliferating INL stem cells is established that can regenerate all the cell types of the retina.

To initiate retinal regeneration in fish, the damage must include destruction of photoreceptor cells. Selective destruction of the retinal layers interior to the photoreceptor layers by low doses of oubain or the selective destruction of ganglion cells by injecting propidium iodide into the optic nerve does not lead to regeneration of any of the destroyed retinal layers. Selective destruction of photoreceptors by tunicamycin or laser light, however, leads to the regeneration of new rods and cones (Raymond and Hitchcock, 1997).

A number of hormones and growth factors have been tested for their effects on adult goldfish eyecups *in vitro*

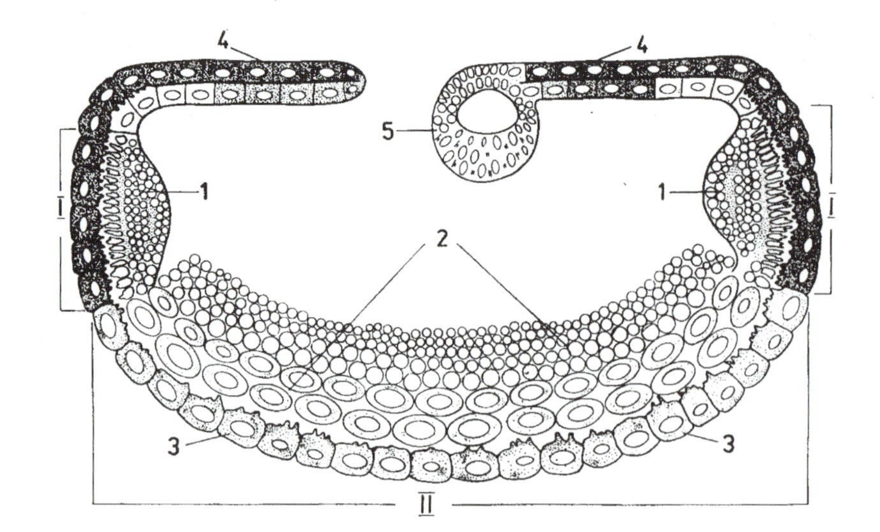


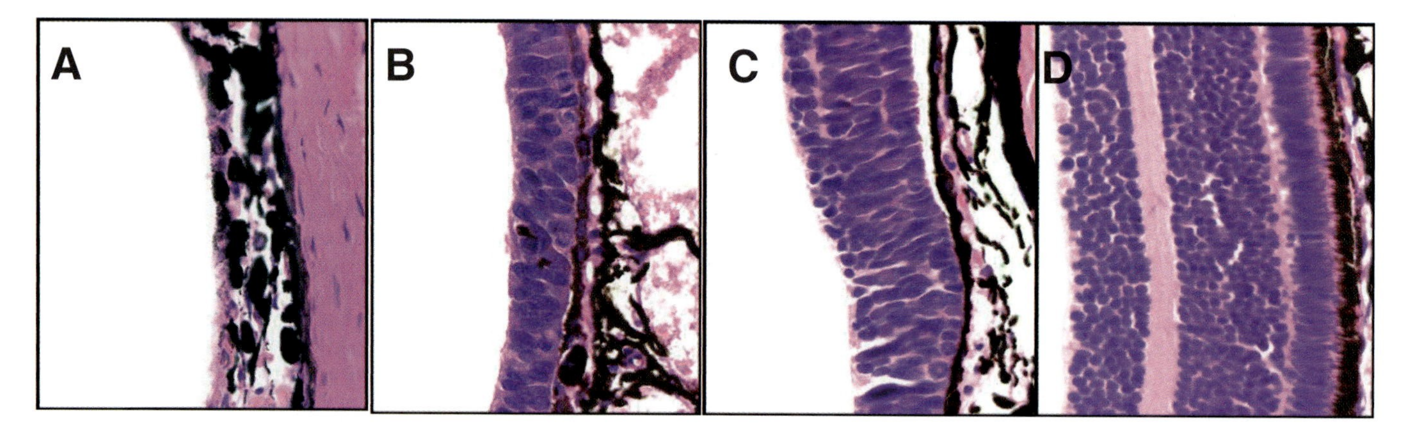


**FIGURE 5.22** Regeneration of the newt retina. **Top**, diagram of regenerating retina. I = retina periphery, II = central portion of retina. When surgically removed or destroyed by cutting off the vascular supply, the retina regenerates from retinal pigmented epithelial cells (3) that undergo dedifferentiation to form a neuroepithelial blastema (2) that redifferentiates into all the neural layers of the retina. Dedifferentiated cells are also contributed from the ora serrata (1), the part of the iris (4) not visible from the exterior of the eye. The diagram also shows lens (5) regenerating from PECs of the dorsal iris. Reproduced with permission from Mitashov, Mechanisms of retina regeneration in urodeles. Int J Dev Biol 40:833–844. Copyright 1996, University of the Basque Country Press. **Bottom**, histological views of the regenerating retina. **(A)** Pigmented epithelial layer after retinectomy. **(B)** Neuroepithelial blastema formed by dedifferentiation of pigmented epithelial cells. **(C)** Stratification of the neuroepithelial blastema into the different layers of the retina. **(D)** Fully regenerated retina. Images courtesy of Ms. Natalia Vergara and Dr. Katia Del-Rio Tsonis.

in an attempt to identify molecules important in the retinal response to injury (Boucher and Hitchcock, 1995). Factors tested included insulin, IGF-I and I, NGF, EGF and FGF 1 and 2. Insulin and IGF-I stimulated the mitosis of cells in the germinal neuroepithelium, but none of the factors tested had any effect on the mitotic activity of rod precursor cells or cells of the injury-induced blastema.

### c. Regeneration in the Adult Newt

Cells in the ciliary marginal zone (CMZ) of the newt eye proliferate throughout life to add new neurons of all types to the retina as the eye grows (Perron et al., 1998). These cells can be stimulated to proliferate after injury. However, the adult newt regenerates the retina primarily by dedifferentiation and transdifferentiation of RPE cells into a neuroepithelium that differentiates into the layers of the new retina (**FIGURE 5.22**), in the same way that PECs of the dorsal iris dedifferentiate and transdifferentiate to regenerate the lens (Mitashov, 1996; Raymond and Hitchcock, 1997; Tsonis and Del-Rio Tsonis, 2004; Haynes and Del-Rio Tsonis, 2004). The RPE cells also act as scavengers to clean up cell debris (Klein et al., 1990; Mitashov, 1996). The dedifferentiated cells facing the vitreous chamber proliferate and differentiate into the different layers of the neural retina: retinal ganglion cells, inner and outer nuclear cells, and photoreceptor cells (Stone, 1950; Keefe, 1973 a, b, c; Levine, 1977; Reyer, 1977). The dedifferentiated cells next to the choroid layer of the eye redifferentiate as RPE cells to regenerate the RPE. The new retinal ganglion cells then extend their axons into the brain and restore normal retinal-tectal connections, with full visual recovery (Sperry, 1944; Gaze, 1959; Mitashov, 1996; Raymond and Hitchcock, 1997).

Dedifferentiation of the RPE cells and their re-entry into the cell cycle involves changes in their shape, internal structure, and gene activity that are regulated by the ECM and soluble signals (Reyer, 1977). RPE cells change shape from cuboidal to columnar within two days after destruction of the retina, followed over the next week by cell enlargement and depigmentation by extrusion of melanosomes. Though not well studied, it is possible that the change in shape of the RPE cells is related to enzymatic breakdown of the ECM and the loss of epithelial cell–cell contacts and that these shape changes lead to melanosome extrusion and altered patterns of gene activity via reorganization of the cytoskeleton. Dedifferentiated and proliferating newt RPE cells express an antigen that binds the mouse monoclonal antibody RPE-1. Expression of RPE-1 is extinguished only when the cells express retinal molecular markers (Negeshi et al., 1992; Mitashov, 1996), suggesting that at least part of the expression pattern of dedifferentiated RPE cells overlaps with that of retina cells during proliferation and transdifferentiation. It would be of interest to know whether thrombin activity is an early event in newt (or fish) retinal regeneration that precedes re-entry into the cell cycle, as it is in lens regeneration from the dorsal iris (Imokawa and Brockes, 2003).

Laminin appears to be an important regulator of retina regeneration. Dedifferentiation and transdifferentiation of *Rana* tadpole RPE cells is promoted by growing them on a laminin substrate *in vitro* (Reh et al., 1987). During the *in vivo* regeneration of the retina in *Rana* tadpoles, the first new retina neurons arise in association with the vitreal vascular membrane (Reh and Nagy, 1987), which contains a high concentration of laminin (Reh et al., 1987). Regulation of regeneration by hormones and growth factors has not yet been studied in the newt.

### d. Regeneration in the Bird

During optic cup formation in stage 9–10 embryos, the prospective RPE (outer layer of the optic cup) differentiates as neural retina if its contact with the prospective retina (inner layer of the optic cup) is broken (Orts-Lorca and Genis-Galvez, 1960), suggesting that RPE cells can transdifferentiate to retinal neurons and that the retina normally inhibits the RPE from differentiating as retina. By stage 22–24 (~96 hr), the RPE and retina are fully differentiated and the RPE will not regenerate retinal neurons when the retina is completely removed.

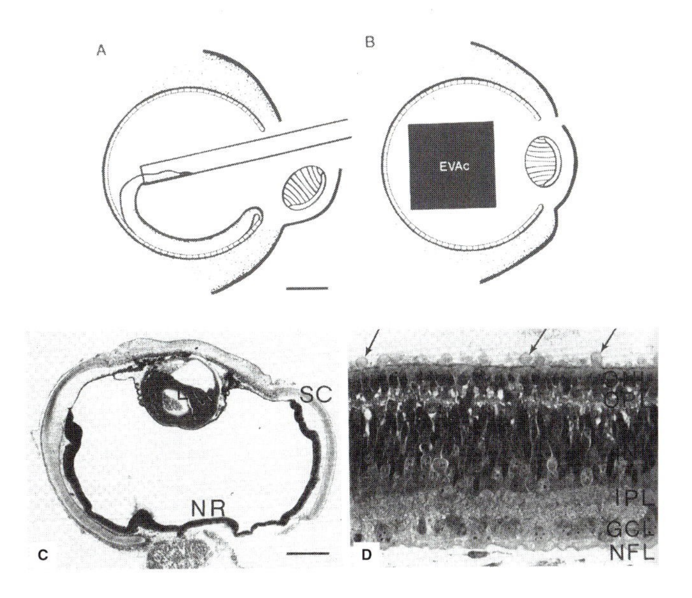


**FIGURE 5.23** Induction of neural retina regeneration in chick embryos. **(A)** The neural retina is suctioned from a stage 22–24 eye by micropipette, leaving the RPE. **(B)** Insertion of FGF-2 impregnated EVAc copolymer implants into the vitreous humor. **(C)** Low magnification view of neural retina (NR) regenerated from the RPE in the presence of FGF-2. SC = sclera. **(D)** Higher magnification of the regenerated neural retina in (C). The polarity of the regenerated retina is reversed, with developing photoreceptors (arrows) protruding into the vitreous cavity toward the lens instead of facing the RPE. **A** and **B** reproduced with permission from Park and Hollenberg, Growth factor-induced retinal regeneration *in vivo*. Int Rev Cytol 146:49–71. Copyright 1993, Elsevier. **C** and **D** reproduced with permission from Park and Hollenberg, Induction of retinal regeneration *in vivo* by growth factors. Dev Biol 148:322–333. Copyright 1991, Elsevier.

However, the RPE can be artificially stimulated at these stages to regenerate the retina by implanting a piece of retina back into the vitreous chamber (Coulombre and Coulombre, 1965, 1970). Implanted blocks of ethylene/vinyl coacetate (EVAc) polymer impregnated with 10 ng of FGF-2, or, at higher concentrations, FGF-1, can substitute for the piece of retina (Park and Hollenberg, 1991). In both cases, the polarity of the regenerated retina is reversed, with the photoreceptor layer facing the implant, and the RPE is not regenerated (**FIGURE 5.23**). Incubation of stage 9–10 chick optic vesicles in medium containing FGF-2 also induced the prospective RPE to differentiate as retina, producing an eye with two retinas of normal polarity (Pittack et al., 1997). FGF-2 has also been shown to induce RPE cells to form retinal neurons *in vitro* (Pittack et al., 1991). FGF-2 and FGF receptors are normally expressed in the retina of birds and mammals

(Noji et al., 1990; Park and Hollenberg, 1993; Tsonis and Del-Rio Tsonis, 2004). The embryonic chick retina can also regenerate from stem cells in the CMZ/ciliary body region. In this case, the regenerated retina has the normal polarity (Willbold and Layer, 1992).

Collectively, these results suggest that, during chick embryogenesis, the RPE and the retina exert a reciprocal inhibitory effect on one another. The retina prevents the RPE from becoming retina through as yet unidentified signals, while the RPE prevents the retina from producing FGF-2. If the retina is separated from the RPE, or if it is removed and a piece put back, it produces FGF-2, which stimulates RPE cells to regenerate the retina. The RPE cells experiencing the highest concentration of FGF (next to the implant) become photoreceptors. Since in amphibians the RPE regenerates the retina after complete destruction of the original retina, either no FGF signal is required for regeneration, or enough FGF is released from the degenerating retina, or the RPE is released from an inhibitory influence of the retina and itself produces FGF, which acts in an autocrine fashion.

The role of other signaling molecules and transcription factors involved in retina regeneration has also been investigated in the chick. Overexpression of Mitf has the same inhibitory effects on FGF-2-induced dedifferentiation and transdifferentiation of cultured chick RPE cells into retina cells as it does on the transdifferentiation of dorsal iris PECs into lens cells (Mochii et al., 1998). Conversely, induction of dedifferentiation and transdifferentiation of RPE cells by FGF-2 inhibits Mitf expression. These observations suggest that downregulation of *Mitf* might be essential for retina regeneration in the same way that it appears to be for lens regeneration.

The adult avian retina has a minimal regenerative response to injury. Different cell populations of the adult avian retina can be selectively destroyed by different agents (Fischer and Reh, 2002). *N*-methyl-D-aspartate (NMDA) destroys amacrine and bipolar neurons, kainate eliminates these neurons plus ganglion cells, and colchicine destroys only ganglion cells. Studies applying these agents to chick eyes suggest that Muller glia cells can act as the equivalent of RPE cells in the newt or embryonic chick retina. In response to NMDA-induced excitotoxic damage or destruction by kainate, numerous Muller glia in the retina dedifferentiate, proliferate, and express markers (including Pax-6) characteristic of embryonic retinal progenitors (Fischer and Reh, 2001). Most of the dedifferentiated cells remain undifferentiated, but a few differentiate into Muller glia cells and amacrine or bipolar neurons (Fischer and Reh, 2002a). Destruction of ganglion cells by colchicine results in the regeneration of ganglion cells from dedifferentiated Muller glia cells that is stimulated by a combination of insulin and FGF-2 (Fischer and Reh, 2002b, 2003a, b; Fischer et al., 2002). Whether the axons of the ganglion cells form an optic nerve that makes functional connections with the optic tectum is not known. Regeneration of photoreceptors has never been observed in the adult avian retina.

#### e. Regeneration in Mammals

The mammalian retina does not regenerate. However, cells of the pigmented ciliary margin of adult rats and mice have proliferative potential *in vitro* and can differentiate into retinal neurons after dedifferentiation (Ahmad et al., 2000; Tropepe et al., 2000; Shatos et al., 2001; Zhao et al., 2005). The dedifferentiated cells self-renew and express retinal progenitor markers such as Chx10. Cultured limbal epithelial cells express neural markers (Zhao et al., 2002) and cultured iris cells transfected with crx (a crucial developmental gene for genesis of photoreceptors) have been induced to differentiate into retinal cells, including photoreceptors (Haruta et al., 2001).

Sonic hedgehog (shh) and its receptor, Patched (ptc), are expressed in the developing mouse retina. Treatment of cultured mouse NR cells with the amino-terminal fragment of the Shh protein promotes their proliferation (Jensen and Wallace, 1997). $Ptc^{+/-}$ mice have fewer Ptc receptors. Since the function of Ptc is to inhibit Smoothened (Smo), this means that the hedgehog signal transduction pathway is more active. These mice thus exhibit an increased number of proliferating cells in the nonpigmented region of the ciliary body up to three months of age. Cells of the nonpigmented ciliary epithelium in the progeny of crosses between these mice and mice transgenic for retinal degeneration respond to retinal injury by proliferation and express markers for retinal neurons (Moshiri and Reh, 2004).

Retinal stem cells have also been identified in the ciliary margin of the human eye (Haynes and Del-Rio Tsonis, 2004; Coles et al., 2004). Coles et al. (2004) isolated these RSCs from early postnatal to 7th-decade human eyes by dissociating cells from different parts of the eye and culturing them in suspension. Single cells capable of forming floating sphere colonies were considered to be stem cells. Single cells from these primary spheres were able to form secondary spheres, indicating their ability for self-renewal. Cells migrating from spheres plated in differentiation medium gave rise to all the cell types of the retina, although the vast majority of these were photoreceptors. Spheres derived from EGFP-labeled RSCs were dissociated and transplanted into the eyes of 3.5-day chick embryos, when host ganglion and inner nuclear layer cells were differentiating, and into the eyes of 1-day postnatal mice, when host photoreceptors and Muller glia cells are differentiating.

The donor RSCs differentiated into all these retinal cell types in response to host differentiation signals, indicating their multipotential nature. In addition, human scleral and choroid cells were shown to have the potential to differentiate toward a neural lineage (Arsenijevic et al., 2003).

## 4. Regeneration of the Spinal Cord in Amputated Amphibian and Lizard Tails

Histological and labeling studies have shown that new spinal cord neurons are produced relatively frequently in uninjured juvenile axolotls up to 6–7 months of age, but infrequently in older animals (Holder et al., 1991). Although axons of surviving neurons regenerate in larval and adult urodeles, fewer new neurons are regenerated after transection or ablation of trunk cord (Butler and Ward, 1965; Nordlander and Singer, 1978; Davis et al., 1989). By contrast, the spinal cord of larval and adult urodeles, including new neurons, can be regenerated after amputation of the tail (Clarke and Ferretti, 1998). Urodele spinal cord contains nine groups of neurons, but the fate of each type during either gap replacement regeneration or caudal regeneration from the amputated tail has not been studied (Chernoff et al., 2003).

Tail regeneration in larval urodeles is accomplished by the dedifferentiation of cartilage, muscle, and dermal fibroblasts to mesenchymal stem cells (**FIGURE 5.24**). These cells form a blastema of proliferating cells at the cut surface of the tail. The mesenchymal cells differentiate into new cartilage, muscle, and dermis of the regenerated tail. A separate tube of dividing ependymal, or radial glial, NSCs extends from the cut end of the spinal cord into the blastema. These ependymal cells maintain their epithelial organization during tail regeneration, as opposed to the transformation into mesenchyme characteristic of gap replacement regeneration. The central canal of the spinal cord is sealed off at the amputation plane by the formation of a terminal vesicle of nonproliferating ependymal cells. Ependymal cells proximal to the vesicle proliferate to extend the ependymal tube (Holtzer, 1952; Egar and Singer, 1972; Nordlander and Singer, 1978). Nestin and vimentin are expressed by the proliferating ependymal NSCs (Ferretti, 2000; Chernoff et al., 2003). ECM and cell adhesion molecules expressed during spinal cord development are re-expressed by ependymal cells during regeneration (Caubit et al., 1994; Clarke and Ferretti, 1998). The distribution of tenascin is like that in developing animals. The polysialated embryonic form of N-CAM (PSA-N-CAM) is expressed at a low level in the ependymal cells of uninjured tail cord, but is strongly upregulated in the regenerating cord (Caubit et al., 1993). Studies in which ependymal cells in the regenerating tail were pulse-labeled with BrdU have shown, using confocal microscopy, that they give rise to new glia (GFAP$^+$) and neurons (NSE$^+$) *in vivo* and *in vitro* (Benraiss et al., 1999).

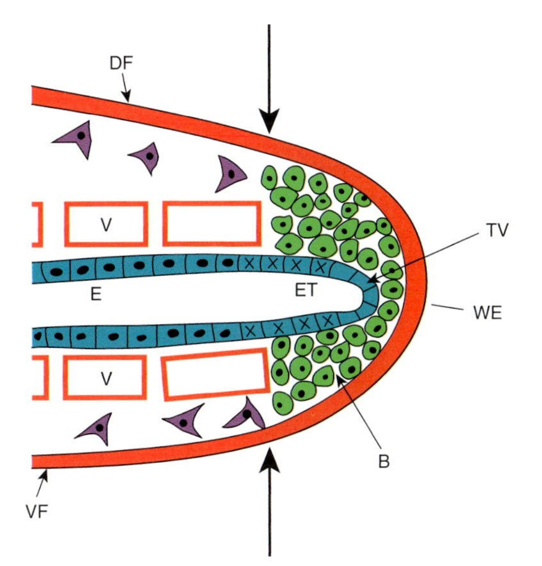


**FIGURE 5.24** Tail tip and spinal cord regeneration in the larval salamander. Thick arrows indicate amputation plane. A blastema (B) forms, covered by a wound epithelium (WE). The blastema forms by the dedifferentiation of vertebral cartilage (V), dermal fibroblasts, connective tissue of the dorsal and ventral fins (DF, VF), and muscle (not shown). The spinal cord extends an ependymal tube (ET) into the blastema. The cells of the terminal vesicle (TV) of the ependymal tube do not divide, but the tube lengthens by mitosis of cells proximal to the terminal vesicle (x).

The regenerating ependymal tube resembles the neural tube of the embryo. As the tube grows distally, the proliferating cells closest to the amputation plane become radially arranged and extend end-feet to form a glia limitans. The end-feet overlap to form channels that guide the outgrowth of regenerating axons from nerve cell bodies above the level of transection, as in the gap regeneration of the urodele spinal cord (**FIGURE 5.9**) or optic nerve (Simpson, 1983; Clarke and Ferretti, 1998). Some of the proliferating ependymal cells differentiate into new motor neurons, interneurons, and glia (Piatt, 1955; Geraudie et al., 1988; Holder et al., 1991; Arsanto et al., 1992; Echeverri and Tanaka, 2002). The ependymal NSCs also regenerate neural crest derivatives, such as sensory ganglia, Schwann cells, fin mesenchyme, and pigment cells (Benraiss et al., 1996). In adult newts, however, no new neurons are formed, and the spinal cord of the regenerated tail consists only of axons (Duffy et al., 1992; Stensaas, 1983; Benraiss, 1997).

The fate of ependymal NSCs has been followed live in the thin, transparent amputated tails of newly hatched axolotl larvae (Echeverri and Tanaka, 2002). Individual ependymal cells were marked by microinjection of the cDNA for green fluorescent protein (GFP), under the control of the glial fibrillary acidic protein (GFAP) promoter. The labeled cells were observed to divide and, as expected, many of their progeny differentiated into neurons and glia in the regenerating spinal cord, as well as neural crest derivatives, such as fin mesenchyme. Some NSCs, however, migrated out of the ependymal tube and transdifferentiated into muscle, which was identified with antibody to muscle-specific myosin heavy chain, and into chondrocytes, at frequencies of 24% and 12%, respectively.

Little is known about the signaling molecules and transcription factors that control proliferation of ependymal NSCs and their patterns of neural differentiation and synaptogenesis in the regenerating urodele tail. FGF-2 may play a central role in proliferation. It is not expressed in the ependymal cells of the uninjured cord, though it is expressed in a subset of neurons. In the proliferating ependymal cells of the amputated *Pleurodeles* tail, there is strong expression of FGF-2 and FGF receptors that gradually decreases as differentiation of the new tail progresses, until expression can no longer be detected in the ependyma. Exgenous FGF-2 injected into the regenerating tail increases the number of proliferating ependymal cells (Zhang et al., 2000). These observations suggest that FGF-2 is important for ependymal cell proliferation after tail amputation. This idea is further supported by the fact that FGF-2 is an important factor in the proliferation of blastema cells during urodele limb regeneration (see Chapter 14) and it fits with the fact that FGF-2 can enhance the regeneration of pyramidal neurons of the rat hippocampus (Nakatomi et al., 2002). Thus, in both the spinal cord of mammals and urodele amphibians, FGF-2 appears to be a survival and proliferation factor, but in mammals these functions of FGF are thwarted by factors inhibitory to regeneration that are not present in urodeles.

*Wnt-10* is expressed in the uninjured cord of the adult newt *Pleurodeles waltlii* (but not in the cord of other species) and is upregulated weakly in the regenerating *Pleurodeles* cord (Caubit et al., 1997a, b). Since the *Wnt* proteins play crucial roles in the activation and proliferation of epithelial and other adult stem cells, as well as in the embryonic development of the nervous system, it is possible that maintenance of *Wnt-10* expression in the adult *Plurodeles* cord and regenerative ability are causally related (Clarke and Ferretti, 1998).

In *Xenopus* embryos and tadpoles, the tail can regenerate after amputation from stage 42 until metamorphosis, except for a refractory period between stages 45 and 47 (Beck et al., 2003). BMP signaling has been shown to be a requirement for the formation of tail structures during ontogenesis (Beck et al., 2001). BMP and its likely downstream targets, *Msx1* and *2*, as well as the Notch signaling pathway components Notch-1, X-delta-1, and lunatic fringe, are all expressed in regenerating tails during regeneration-competent stages of tail development, but not in regeneration-incompetent stages. Inhibition of the BMP pathway in *Xenopus* transgenic for the BMP antagonist, Noggin, or for the dominant-negative BMP receptor under the control of a heat shock promoter, blocked tail regeneration under conditions of heat shock. Conversely, tadpoles transgenic for the BMP or Notch receptor under the control of a heat shock promoter were able to regenerate during the refractory period (Beck et al., 2003). The BMP pathway suffices for the regeneration of all tail tissues, whereas the Notch pathway activates regeneration of spinal cord and notochord, but not muscle.

How the outgrowth and synaptic patterns of axons and dendrites in the regenerating spinal cord are regulated is not known. *PwDlx-3*, a homeobox transcription factor related to the *distal-less* gene, which is essential for appendage development in *Drosophila* (Cohen and Jurgens, 1989), is upregulated in the ventrolateral region of the cord in the regenerating *Pleurodeles* tail (Nicolas et al., 1996). This is the region from which cells migrate to form the spinal ganglia of the regenerated cord (Geraudie and Singer, 1988). Expression of the gene is not detected in the mammalian central nervous system; thus, it has been suggested that in urodeles, its expression might be related to regenerative capacity (Nicolas, 1996). Another important gene involved in the patterning of motor neurons in the developing spinal cord is *shh* (Altmann and Brivanlou, 2001), but patterns of *shh* expression and other components of the hedgehog pathway have not yet been reported in regenerating amphibian spinal cord.

Lizards regenerate their tails and are the highest class of vertebrate to display any degree of spinal cord regeneration during tail regeneration. The regeneration blastema of the lizard tail does not faithfully restore the original tail architecture. The myotomic patterns of the tail muscles are regenerated, but the vertebrae are replaced by a thin, unsegmented cartilage tube (Simpson, 1965; Alibardi and Sala, 1988). The cartilage tube encloses the regenerating spinal cord, and in fact is induced to form by the cord. The cord regenerates in the same way as described for the salamander, except that it is not as highly differentiated. Most of the cells in the regenerated cord are ependymal cells and only a few are neurons and glia. A pale, round cell differentiates that is probably a precursor to neural and glial cells (Alibardi et al., 1992). A few axons regenerate into the regenerated tail from neurons above the level of amputation (Alibardi and Miolo, 1990).

The amphibian and lizard spinal cord, whether regenerating axons across a gap, or neurons after tail amputation, are underused model systems with significant potential for identifying regeneration-permissive factors that can be applied to the nonregenerating mammalian spinal cord. For example, molecular comparisons could be made between the regenerating axolotl cord and the regeneration-deficient adult frog cord, or between axolotl and adult mouse.

## SUMMARY

Different parts of the nervous system exhibit differential ability to regenerate, within and between species. The axons of the PNS and CNS in all vertebrate species have the innate capacity for regeneration after injury and undergo initial sprouting, but only PNS axons can complete their regeneration in all vertebrates. In mammals, transected spinal cord and optic nerve axons fail to complete regeneration, but these axons regenerate in fish and larval and adult urodeles, effectively reversing paralysis and restoring vision.

Whether or not axons regenerate depends on whether or not they are associated with regeneration-permissive or regeneration-inhibitory glial cells. Axons of the mammalian spinal cord are associated with oligodendrocytes, which make the myelin sheath. Upon injury, neurons are killed by trauma, glutamate excitiotxicity, and free radical production. The myelin sheath breaks down and its products spread the initial damage to uninjured regions. Furthermore, oligodendrocytes and other cells synthesize inhibitory proteins such as Nogo, MAG, OMgp, Ephrin-B3, and CSPGs. An inflammatory response is mounted in which neutrophils and microglia enter the lesion, kill bacteria and ingest debris, creating cavities in the cord. The microglia also secrete PDGF and TGF-β, which attract astrocytes and fibroblasts from the injured pia mater into the lesion, where they proliferate and form a glial scar. The scar is a mechanical barrier to axon regeneration, as well as a chemical barrier that produces the same kind of CSPGs as oligodendrocytes.

PNS axons of all vertebrates are associated with Schwann cells, and the optic nerve axons of fish and amphibians are associated with astrocytes. These cells are regeneration-permissive for axons. In the transected amphibian spinal cord, ependymal cells play the same role as Schwann cells and optic nerve astrocytes. The ependymal cells dedifferentiate into mesenchymal cells that proliferate to bridge the gap. The mesenchymal cells then differentiate back to epithelial cells that extend end-feet to the pia mater. The end-feet form channels that support axon regeneration both ways across the gap to make synapses with neurons on the other side. The epithelial to mesenchymal transformation involves downregulation of epithelial markers such as Ln, GFAP, and epithelial chtokeratins and upregulation of the mesenchymal markers Fn and Vn. The production of MMPs appears to promote ependymal transformation and TGF-β and PDGF play a role in the migration of the mesenchymal cells. However, our understanding of this whole process is rudimentary at best.

Regeneration-permissive glial and epyndemal cells permit regeneration in several ways. First, they produce soluble signals such as neurotrophins and neuropeptides that are survival factors for neuron cell bodies. Second, they and the basement membrane of endoneurial tubes provide substrates on which sprouting axons can extend. Third, they phagocytize and digest myelin breakdown fragments, thus neutralizing their ability to extend damage. Fourth, they either fail to produce some of the inhibitory myelin proteins made by CNS ologodendrocytes, or these proteins are removed.

Interestingly, the fetal avian and mammalian spinal cord, like fetal skin, can regenerate, as can the spinal cord of the newborn South American opossum. The shift to regeneration-deficiency late in development is correlated with myelination of spinal cord axons and a shift in the locus of Nogo from nerve cell bodies to oligodendrocytes. Whether or not the maturation of an inflammatory response also contributes to this shift is unknown, but might be tested in PU.1 mice, which retain the ability to regenerate dermis after birth.

Whether or not the CNS can regenerate new neurons for structural maintenance or in response to injury depends on the region of the CNS and the environment associated with that region. In the 1990s, NSCs were identified in several regions of the uninjured mammalian brain. In most of these locations, these NSCs engaged in maintenance regeneration of glial cells, but in two locations they were found to give rise to new neurons. NSCs in the lateral ventricle walls of the forebrain were shown to replace neurons turning over in the olfactory bulb. In nonhuman animals, OB and olfactory nerve neurons turn over and are replaced by NSCs. NSCs of the lateral ventricle walls proliferate and migrate rostrally into the OB, where they differentiate into interneurons and glia. At the same time, NSCs in the basal layer of the olfactory epithelium in the nose replace dying olfactory receptor neurons and extend axons into the OB to maintain the olfactory nerve. The glial cells of the olfactory nerve are called olfactory ensheathing cells (OECs) and they provide the essential signals and substrates for axon extension. NSCs in the ventricular walls of the hippocampus, which plays a major role in memory formation and in learning new information and tasks, have been shown to continuously regenerate new neurons at a low level.

Animals whose physical and cognitive activities were enriched exhibited increased NSC proliferation and neurogenesis. Some studies have suggested that maintenance regeneration of cortical neurons takes place via NSCs in the lateral ventricle walls, but other studies have failed to confirm this and the idea remains controversial.

The nature of the microniche that maintains NSCs is of great interest. A direct functional relationship has been established between NSCs and the endothelial cells of capillaries. NSCs cultured on the opposite side of a filter with endothelial cells proliferated to a higher degree than NSCs cultured with blood vessel smooth muscle or nonvascular cells and were primed to differentiate a significantly higher percentage of projection neurons and interneurons once away from the influence of the endothelial cells. Factors that might be secreted by the endothelial cells to affect these results are largely unidentified, but might include EGF, FGF-2, TGF-α, and Shh, which are known to enhance the proliferation of NSCs when infused into the brain or administered *in vitro*. The Notch pathway is an important mediator of stemness. Notch maintains the activity of transcription factors involved in the repression of neuronal differentiation genes, such as Hes1 and the NRSF/REST protein complex.

New neurons can regenerate after injury to several structures of the adult CNS. These include the auditory hair cells of the cochlea and utricle of the vestibular apparatus in birds, hippocampal neurons of the mammalian brain, the several cell types of the neural retina of fish and amphibians, and spinal cord neurons in the regenerating tail of amphibians and lizards.

Auditory hair cells, which reside in the sensory epithelium of the cochlea and utricle, have bundles of cilia that transduce the energy of sound and motion to electrical impulses that travel to the brain over the acoustic nerve. Loss of these cells results in hearing impairment. The support cells of the epithelium are the stem cells that give rise to new hair cells by asymmetric division. The cell that differentiates as a hair cell expresses the transcription factor cprox1, Delta, and FGF-2, whereas the cell that differentiates as a support (stem) cell expresses Notch and FGFR3. Delta and FGF-2 expressed by the hair cell maintains the support cell as a stem cell by lateral inhibition of differentiation. The damaged sensory epithelium of mammals exhibits a low level of hair cell regeneration that is enhanced by administration of RA and TGF-α, but not enough for functional recovery of hearing. The Rb protein appears to keep cells in mitotic arrest, preventing regeneration. Knocking out Rb in mouse utricles *in vitro* by cre recombinase allowed hair cells themselves to re-enter the cell cycle and divide. This result suggests that cells other than hepatocytes might be induced to undergo compensatory hyperplasia (see Chapter 5) and that deafness might be reversible by the controlled inactivation of the *Rb* gene.

Killing thalamic projection neurons in the cortex of the mammalian brain or granule neurons in the dentate gyrus of the hippocampus by focal or global ischemia results in the proliferation of NSCs in the lateral ventricle walls and hippocampal ventricle walls, respectively. Only a small fraction of the degenerated neurons is replaced, not enough for functional recovery. However, intraventricular injection of a combination of FGF-2 and EGF can elevate the number of regenerated hippocampal neurons to 40% of the number lost and these neurons have been shown to be functionally integrated into the hippocampal circuitry. The small number of regenerated neurons in the absence of these growth factors may be due to inadequate output of growth factors by astrocytes, since neonatal hippocampal astrocytes induce hippocampal NSCs to differentiate into neurons *in vitro*, whereas adult astrocytes have only half the effect.

The retina can be surgically removed from the eye of teleost fish or the adult newt, or it can be destroyed by cutting off its blood supply. Consistent with the ability to regenerate transected optic nerve, these animals can regenerate the retina and recover their vision. A population of stem cells that reside in the inner nuclear layer and express the transcription factor Pax-6 is responsible for regeneration of the fish retina. These cells form a blastema that can regenerate all the retinal cell types. The adult newt regenerates the retina by the dedifferentiation and proliferation of RPE into a neuroepithelium that differentiates into the layers of the new retina. The adult avian and mammalian retina does not regenerate, but the RPE of stage 22–24 chick embryos can be stimulated to regenerate the retina by implanting a piece of retina or polymer impregnated with FGF-2 into the vitreous chamber. This suggests that FGF-2 is an important signal that stimulates retinal regeneration from the RPE. Laminin promotes the dedifferentiation of frog tadpole RPE cells and may be an important regulator of retinal regeneration. Although it does not regenerate, retinal stem cells reside in the ciliary margin of the mammalian retina, including that of humans. These cells can form neurospheres *in vitro*, exhibit enhanced proliferation in the presence of Shh, and differentiate into retinal neurons.

New neurons are not regenerated by the urodele spinal cord after simple transection, although the axons of surviving neurons are. However, the tail of larval urodeles regenerates after amputation by dedifferentiation of muscle, cartilage, and connective tissue to form a tail blastema. A separate tube of dividing ependymal cells extends from the cut end of the cord into this blastema. This tube never loses its epithelial nature, although its cells express vimentin, nestin, and ECM and cell adhesion molecules that are expressed during embryonic development. As the ependymal tube grows distally, the cells closest to the amputation plane extend end-feet that form channels to promote the regeneration of axons from above the level of transection. Some of the ependymal tube cells differentiate into new motor neurons, interneurons, and glia. Others give rise to neural crest derivatives (sensory ganglia, Schwann cells, fin mesenchyme, and pigment cells), and still others transdifferentiate into cartilage and muscle cells. Interestingly, no new neurons are formed in the regenerating tail of the adult newt. This cannot be attributed to a more mature immune system in which an inflammatory response prevents regeneration, since such a response should also inhibit regeneration of the other components of the tail. This difference between larval and adult urodele remains to be explained. Furthermore, adult lizards regenerate their tails, although the original vertebral architecture is replace by a thin cartilaginous tube. The spinal cord, however, regenerates the same way as in the urodele, except that it does not regenerate as many neurons and glia.

While some work has been done, little is known about the signal transduction pathways and transcriptional repressors and effectors in the regenerating urodele tail. FGF-2, Wnt, BMP, and Notch signaling pathways all appear to be involved, but details of their roles are lacking.

## REFERENCES


Aguayo AJ (1985) Axonal regeneration from injured neurons in the adult mammalian central nervous system. In: CW Cotman (ed.). Synaptic Plasticity. New York, Guilford Press, pp 457–483.

Ahmad I, Tang L, Pham H (2000) Identification of neural progenitors in the adult mammalian eye. Biochem Biophys Res Comm 270:517–521.

Alibardi L, Sala M (1988) Fine structure of the blastema in the regenerating tail of the lizard *Podarcis sicula*. Boll Zool 55:307–313.

Alibardi L, Miolo V (1990) Fine observation on nerves colonizing the regenerating tail of the lizard *Podarcis sicula*. Histol Histopath 5:387–396.

Alibardi L, Gibbons J, Simpson SB (1992) Fine structure of cells in the young regenerating spinal cord of the lizard *Anolis carolinensis* after $H^3$-thymidine administration. Biol Struct Morph 4:45–52.

Altman J (1962) Are new neurons formed in the brains of adult mammals? Science 135:1127–1128.

Altman J (1963) Autoradiographic investigation of cell proliferation in the brains of rats and cats. Anat Rec 145:573–591.

Altmann CR, Brivanlou AH (2001) Neural patterning in the vertebrate embryo. Int Rev Cytol 203:447–482.

Anton ES, Sandrock AW, Matthew WD (1994) Merosin promotes neurite outgrowth and Schwann cell migration *in vitro*: Evidence using an antibody to merosin, ARM-1. Dev Biol 164: 133–146.

Arsanto J-P, Komorowski TE, Dupin F, Caubit X, Diano M, Geraudie J, Carlson BM, Thouveny Y (1992) Formation of the peripheral nervous system during tail regeneration in urodele amphibians: Ultrastructure and immunohistochemical studies of the origin of the cells. J Exp Zool 264:273–292.

Arsenijevic Y, Taverney N, Kostic C, Tekaya M, Riva F, Zografos L, Schorderet D, Munier F (2003) Non-neural regions of the adult human eye: A potential source of neurons? Invest Opthalmol Vis Sci 44:799–807.

Ashmore J, Gale J (2000) The cochlea. Curr Biol 10:R325–R327.

Attardi DG, Sperry RW (1963) Preferential selection of central pathways by regenerating optic nerve fibers. Exp Neurol 7:46–64.

Barron KD (1983) Axon reaction and central nervous system regeneration. In: Seil FJ (ed.). Nerve, Organ and Tissue Regeneration: Research Perspectives. New York, Academic Press, pp 3–36.

Beck CW, Slack JMW (1998) Analysis of the developing *Xenopus* tail bud reveals separate phases of gene expression during determination and outgrowth. Mech Dev 72:41–52.

Beck CW, Christen B, Slack JMW (2003) Molecular pathways needed for regeneration of spinal cord and muscle in a vertebrate. Dev Cell 5:429–439.

Becker CG, Becker T, Meyer RL, Schachner M (1999) Tenascin-R inhibits the growth of optic fibers in vitro but is rapidly eliminated during nerve regeneration in the salamander *Pleurodeles waltl*. J Neurosci 19:813–827.

Benraiss A, Caubit X, Coulon J, Nicolas S, Le Parco Y, Thouveny Y (1996) Clonal cell cultures from adult spinal cord of the amphibian urodele *Pleurodeles waltl* to study the identity and potentialities of cells during tail regeneration. Dev Dynam 205: 135–149.

Benraiss A, Arsanto JP, Coulon J, Thouveny Y (1997) Neural crest-like cells originate from the spinal cord during tail regeneration in adult amphibian urodeles. Dev Dynam 209:15–28.

Benraiss A, Arsanto JP, Coulon J, Thouveny Y (1999) Neurogenesis during caudal spinal cord regeneration in adult newts. Dev Genes Evol 209:363–369.

Bermingham-McDonough O, Stone JS, Reh TA, Rubel EW (2001) FGFR3 expression during development and regeneration of the chick inner ear sensory epithelia. Dev Biol 238:247–259.

Benson MD, Romero MI, Lush ME, Lu QR, Henkemeyer M, Parada LF (2005) Ephrin-B3 is a myelin-based inhibitor of neurite outgrowth. Proc Natl Acad Sci USA 102:10694–10699.

Birling M-C, Price J (1995) Infuence of growth factors on neuronal differentiation. Curr Opinion Cell Biol 7:878–884.

Bisby MA (1995) Regeneration of peripheral nervous system axons. In: Waxman SG, Kocsis J, Stys PK (eds.). The Axon: Structure, Function and Pathophysiology. New York, Oxford University Press, pp 553–578.

Bodega G, Suarez I, Rubio M, Fernandez B (1994) Ependyma: Phylogenetic evolution of glial fibrillary acidic protein (GFAP) and vimentin expression in vertebrate spinal cord. Histochem 102:113–122.

Bohn R, Reier PJ, Sourbeer EB (1982) Axonal interactions with connective tissue and glial substrata during optic nerve regeneration in *Xenopus* larvae and adults. Am J Anat 165:397–419.

Bottai D, Fiocco R, Gelain F, Defilippis L, Galli R, Gritti A, Vescovi AL (2003) Neural stem cells in the adult nervous system. J Hematother Stem Cell Res 12:655–670.

Boucher S-EM, Hitchcock PF (1995) Insulin-related factors stimulate *in vitro* proliferation of pluripotent neuroepithelial cells, but not rod precursors or injury-induced progenitors, in adult goldfish retina. Soc Neurosci 21:1557 (abstract).

Branda CS, Dymecki SM (2004) Talking about a revolution: The impact of site-specific recombinases on genetic analyses in mice. Dev Cell 6:7–28.

Briscoe J, Ericson J (2001) Specification of neuronal fates in the ventral neural tube. Curr Opin Neurobiol 11:43–49.

Bunge R (1987) Tissue culture observations relevant to the study of axon-Schwann cell interactions during peripheral nerve development and repair. J Exp Biol 132:21–34.

Butler EG, Ward MB (1965) Reconstitution of the spinal cord following ablation in urodele larvae. J Exp Zool 160:47–66.

Butler EG, Ward MB (1967) Reconstitution of the spinal cord following ablation in adult *Triturus*. Dev Biol 15:464–486.

Capela A, Temple S (2002) LeX/SSEA-1 is expressed by adult mouse CS stem cells, identifying them as nonependymal. Neuron 35:865–875.

Caroni P, Schwab ME (1988) Two membrane protein fractions from rat central myelin with inhibitory properties for neurite growth and fibroblast spreading. J Cell Biol 106:1281–1288.

Caubit X, Arsanto J-P, Figarella-Branger D, Thouveny Y (1993) Expression of polysialylated neural cell adhesion molecule (PSA-N-CAM) in developing, adult and regenerating caudal spinal cord of the urodele amphibians. Intl J Dev Biol 37:327–336.

Caubit X, Riou JF, Coulon J, Arsanto JP, Benraiss A, Boucaut JC, Thouveny Y (1994) Tenascin expression in developing, adult and regenerating caudal spinal chord in the urodele amphibians. Intl J Develop Biol 38:661–672.

Caubit X, Nicolas S, Shi DL, Le Parco Y (1997a) Reactivation and graded axial expression pattern of *Wnt-10a* gene during early regeneration stages of adult tail in amphibian urodele *Pleurodeles waltl*. Dev Dyn 208:139–148.

Caubit X, Nicolas S, Le Parco Y (1997b) Possible roles for *Wnt* genes in growth and axial patterning during regeneration of the tail in urodele amphibians. Dev Dyn 210:1–10.

Chen P, Segil N (1999) p27(kip1) links cell proliferation to morphogenesis in the developing organ of Corti. Development 126:1581–1590.

Chen MS, Huber AB, Can der Haar ME, Fran M, Schnell L, Spillmann AA, Christ F, Schwab ME (2000) Nogo-A is a myelin-associated neurite outgrowth inhibitor and an antigen for monoclonal antibody IN-1. Nature 403:434–439.

Chernoff EAG, O'Hara CM, Bauerle D, Bowling M (2000) Matrix metalloproteinase production in regenerating axolotl spinal cord. Wound Rep Reg 8:282–291.

Chernoff EAG, Stocum DL, Nye HLD, Cameron JA (2003) Urodele spinal cord regeneration and related processes. Dev Dynam 226:280–294.

Cho K-S, Yang L, Lu B, Ma HF, Huang X, Pekny M, Chen DF (2005) Re-establishing the regenerative potential of central nervous system axons in postnatal mice. J Cell Sci 118:863–872.

Chong JA, Tapia-Ramirez J, Kim S, Toledo-Aral JJ, Zheng Y, Boutros MC, Aktshuller YM, Frohman MA, Kraner SD, Mandel G (1995) REST: A mammalian silencer protein that restricts sodium channel gene expression to neurons. Cell 80:949–957.

Chuah MI, Au C (1991) Olfactory Schwann cells are derived from precursor cells in the olfactory epithelium. J Neuro Sci Res 29: 172–180.

Clarke JDW, Alexander R, Holder N (1988) Regeneration of descending axons in the spinal cord of the axolotl. Neruosci Lett 89:1–6.

Clarke JDW, Ferretti P (1998) CNS regeneration in lower vertebrtates. In: Ferretti P, Geraudie J (eds.). Cellular and Molecular Basis of Regeneration. New York, John Wiley and Sons, pp 255–272.

Cohen SM, Jurgens G (1989) Proximal-distal pattern formation in *Drosophila*: Cell autonomous requirement for distal-less gene activity in limb development. EMBO J 8:2045–2055.

Cohen A, Sivron T, Duvdedani R, Schwartz M (1990) Oligodendrocyte cytotoxic factor associated with fish optic nerve regeneration: implications for mammalian CNS regeneration. Brain Res 537:24–32.

Coles BLK, Angenieux B, Inoue T, Del Rio-Tsonis K, Spence JR, McInnes RR, Arsenijevic Y, van der Kooy D (2004) Facile isolation and the characterization of human retinal stem cells. Proc Natl Acad Sci USA 101:15772–15777.

Cotanche DA, Lee KH, Stone JS, Picard DA (1994) Hair cell regeneration in the bird cochlea following noise damage or ototoxic drug damage. Anat Embryol Berlin 189:1–18.

Coulombre JL, Columbre AJ (1965) Regeneration of neural retina from the pigmented epithelium in the chick embryo. Dev Biol 12:79–92.

Coulombre JL, Coulombre AJ (1970) Influence of mouse neural retina on regeneration of chick neural retina from chick embryonic pigmented epithelium. Nature 228:559–560.

Cotanche DA (1987) Regeneration of hair cell stereociliary bundles in the chick cochlea following severe acoustic trauma. Ear Res 30:181–194.

Croce CM, Calin GA (2005) Archeo-cell biology: Carbon dating is not just for pots and dinosaurs. Cell 122:4–7.

Davies AM (2000) Neurotrophins: neurotrophic modulation of neurite growth. Curr Biol 10:R198–R200.

Davis BM, Duffy MT, Simpson, SB Jr (1989) Bulbospinal and intraspinal connection in normal and regenerated salamander spinal cord. Exp Neurol 103:41–51.

Davis BM, Ayers JL, Koran L, Carlson J, Anderson MC, Simpson SB (1990) Time course of salamander spinal cord regeneration and recovery of swimming: HRP retrograde pathway tracing and kinematic analysis. Exp Neurol 108:198–213.

Dickson BJ (2002) Molecular mechanisms of axon guidance. Science 298:1959–1964.

Doetsch F, Caille I, Lim DA, Garcia-Verdugo JM, Alvarez-Bullya A (1999) Subventricular zone astrocytes are neural stem cells in the adult mammalian brain. Cell 97:703–716.

Doherty P, Walsh FS (1991) The contrasting roles of N-CAM and N-cadherin as neurite outgrowth-promoting molecules. J Cell Sci 15 (Suppl):13–21.

Duffy MT, Liebich DR, Garner LK, Hawrych A, Simpson SB Jr, Davis BM (1992) Axonal sprouting and frank regeneration in the lizard tail spinal cord:correlation between changes in synaptic circuitry and axonal growth. J Comp Neurol 316:363–374.

Echeverri K, Tanaka EM (2002) Ectoderm to mesoderm lineage switching during axolotl tail regeneration. Science 298:1993–1996.

Edwards DL, Alpert R, Grafstein B (1981) Recovery of vision in regeneration of goldfish optic axons: enhancement of axonal outgrowth by a conditioning lesion. Exp Neurol 72:672–686.

Egar M, Singer M (1972) The role of ependyma in spinal cord regeneration in the urodele, Triturus. Exp Neurol 37:422–430.

Egar M, Singer M (1981) The role of ependyma in spinal cord regrowth. In: Becker RO (ed.). Mechanisms of Growth Control. Springfield, Charles Thomas, pp 93–106.

Eitan S, Schwartz M (1993) A transglutaminase that converts interleulin-2 into a factor cytotoxic to oligodendrocytes. Science 261:106–108.

Eng LF, Reier PJ, Houle JD (1987) Astrocyte activation and fibrous gliosis: Glial fibrillary acidic protein immunostaining of astrocytes following intraspinal cord grafting of fetal CNS tissue. Prog Brain Res 71:439–455.

Eriksson PS, Perfilieva E, Bjork-Eriksson T, Alborn AM, Nordberg C, Peterson DA, Gage FH (1998) Neurogenesis in the adult hippocampus. Nature Med 4:1313–1317.

Eser P, Schiessl H, Willnecker J (2004) Bone loss and steady state after spinal cord injury: A cross-sectional study using pQCT. J Musculoskel Neuron Interact 4:197–198.

Farlow DN, Vansant G, Cameron A, Chang J, Khoh-Reiter S, Pham N-L, Wu W, Sagara Y, Nicholls JG, Carlo DJ, Ill, CR (2000) Gene expression monitoring for gene discovery in models of peripheral and central nervous system differentiation, regeneration, and trauma. J Cellular Biochem 80:171–180.

Fawcett JW, Asher RA (1999) The glial scar and central nervous system repair. Brain Res Bull 49:377–391.

Fekete DM (1996) Cell fate specification in the inner ear. Curr Opin Neurobiol 6:533–541.

Fekete DM, Wu DK (2002) Revisiting cell fate specification in the inner ear. Curr Opin Neurobiol 12:35–42.

Ferretti P, Zhang F, O'Neill P (2003) Changes in spinal cord regenerative ability through phylogenesis and development. Dev Dynam 226:245–256.

Filbin MT (2000) Axon regeneration: Vaccinating against spinal cord injury. Curr Biol 10:R100–103.

Fischer AJ, Seltner RLP, Poon J, Stell WK (2001) Muller glia are a potential source of neural regeneration in the post-natal chicken retina. Nat Neurosci 4:247–252.

Fischer AJ, Reh TA (2002) Exogenous growth factors stimulate the regeneration of ganglion cells in the chicken retina. Dev Biol 251:367–379.

Fischer AJ, Dierks BD, McGuire C, Reh TA (2002) Insulin and FGF2 activate a neurogenic program in Muller glia of the chicken retina. J Neurosci 22:9387–9398.

Fischer AJ, Reh TA (2003a) Potential of Muller glia to become neurogenic retinal progenitor cells. Glia 43:70–76.

Fischer AJ, Reh TA (2003b) Growth factors induce neurogenesis in the ciliary body. Dev Biol 259:225–240.

Forge A, Li L, Corwin JT, Nevill G (1993) Ultrastructural evidence for hair cell regeneraton in the mammalian inner ear. Science 25:1616–1619.

Fu SY, Gordon T (1997) The cellular and molecular basis of peripheral nerve regeneration. Mol Neurobiol 14:67–116.

Gage FH (2000) Mammalian neural stem cells. Science 287:1433–1438.

Garrity PA, Zipursky SL (1995) Neuronal target recognition. Cell 83:177–185.

Gaze RM (1959) Regeneration of the optic nerve in *Xenopus laevis*. Quart J Exp Physiol 44: 209–308.

Gaze M, Grant P (1978) The diencephalic course of regenerating retinotectal fibers in *Xenopus* tadpoles. J Embryol Exp Morph 44:201–216.

Geraudie J, Nordlander R, Singer M, Singer J (1988) Early stages of spinal ganglion formation during tail regeneration in the newt, *Notophthalmus viridescens*. Am J Anat 183:359–370.

Go MJ, Eastman DS, Artavanis-Tsakonas S (1998) Cell proliferation control by Notch signaling in *Drosophila* development. Development 125:2031–2040.

Goh ELK, Ma D, Ming G-L, Song H (2003) Adult neural stem cells and repair of the adult central nervous system. J Hematother Stem Cell Res 12671–679.

Goldman SA, Nottebohm F (1983) Neuronal production, migration, and differentiation in a vocal control nucleus of the adult female canary brain. Proc Natl Acad Sci USA 80:2390–2394.

Goodrum JF, Bouldin TW (1996) The cell biology of myelin degeneration and regeneration in the peripheral nervous system. J Neuropath Exp Neurol 55:943–953.

Gould E, Tanapat P (1997) Lesion-induced proliferaton of neuronal progenitors in the dentate gyrus of the adult rat. Neurosci 80: 427–436.

Gould E, Tanapat P, McEwen BS, Flugge G, Fuchs E (1998) Proliferation of granule cell precursors in the dentate gyrus of adult monkeys is diminished by stress. Proc Natl Acad Sci USA 95:3168–3171.

Gould E, Reeves AJ, Graziano MSA, Gross CG (1999) Neurogenesis in the neocortex of adult primates. Science 286:548–552.

Grafstein B (1983) Chromatolysis reconsidered: A new view of the reaction of the nerve cell body to axon injury. In: Seil FC (ed.). Nerve, Organ and Tissue Regeneration: Research Perspectives. New York, Academic Press, pp 37–50.

Grafstein B (1991) The goldfish visual system as a model for the stdy of regeneraton in the central nervous system. Development and plasticity of the visual system. In: Cronley-Dillon JR (ed.). Vision and Visual Dysfunction Vol II. London, Macmillan, pp 190–205.

GrandPre T, Nakamura F, Vartanian T, Strittmatter SM (2000) Identification of the nogo inhibitor of axon regeneration as a reticulon protein. Nature 403:439–443.

GrandPre T, Li S, Strittmatter SM (2002) Nogo-66 receptor antagonist peptide promotes axonal regeneration. Nature 417:547–555.

Grant P, Tseng Y (1986) Embryonic and regenerating *Xenopus* retinal fibers are intrinsically different. Dev Biol 114:475–491.

Guena S, Borrione P, Fornaro M, Giacobini-Robecchi MG (2001) Adult stem cells and neurogenesis: historical roots and state of the art. Anat Rec 265:132–141.

Haddon C, Smithers L, Schneider-Maunoury S, Coche T, Henrique D, Lewis J (1998a) Multiple Delta genes and lateral inhibition in zebrafish primary neurogenesis. Development 125:359–370.

Haddon C, Jiang Y-J, Smithers L, Lewis L (1998b) Delta-Notch signaling and the patterning of sensory cell differentiation in the zebrafish ear: Evidence from the mind bomb mutant. Development 125:4645–4654.

Ham AW, Cormack DH (1979) Histology, 8th ed. Philadelphia, JB Lippincott, pp 614–644.

Haruta M, Kosaka M, Kanagae Y, Saito I, Inoue T, Kageyama R, Nishida A, Honda Y, Takahashi M (2001) Induction of photoreceptor-specific phenotypes in adult mammalian iris tissue. Nature Neurosci 4:1163–1164.

Hasan SJ, Nelson BH, Valenzuela JL, Kierstead HS, Shull SE, Ethell DW, Steeves JD (1991) Functional repair of transected spinal cord in embryonic chick. Restor Neurol Neurosci 2:137–154.

Haynes T, Del-Rio Tsonis K (2004) Retina repair, stem cells and beyond. Curr Neurovasc Res 1:1–8.

Hitchcock PF, Raymond PA (1992) Retinal regeneration. Trends in Neurosci 15:103–108.

Hitchcock PF, Lindsey KJ, Easter SS Jr, Mangione-Smith R, Jones DD (1992) Local regeneration in the retina of the goldfish. J Neurobiol 23:187–203.

Holder N, Clarke JDW (1988) Is there a correlation between continuous neurogenesis and directed axon regeneration in the vertebrate nervous system? Trends Neurosci 11:94–99.

Holder N, Clarke JDW, Wilson S, Hunter K, Tonge DA (1989) Mechanisms controlling directed axon regeneration in the peripheral and central nervous systems of amphibians. In: Kiorstsis V, Koussoulakos S, Wallace H (eds.). Recent Trends in Regeneration Research, NATO ASI Series 172:179–190.

Holder N, Clarke JDW, Kamalati T, Lane EB (1990) Heterogeneity in spinal radial glia demonstrated by intermediate filament expression and HRP labeling. J Neurocytol 19:915–928.

Holder N, Clarke JDW, Stephens,N, Wilson SW, Orsi C, Bloomer T, Tonge DA (1991) Continuous growth of the motor system in the axolotl. J Comp Neurol 303:534–550.

Holtzer H (1952) Reconstitution of the urodele spinal cord following unilateral ablation. J Exp Zool 119:263–301.

Horner PJ, Gage FH (2000) Regenerating the damaged central nervous system. Nature 407:963–969.

Huang DW, McKerracher L, Braun PE, David M (1999) A therapeutic vaccine approach to stimulate axon regeneration in the adult mammalian spinal cord. Neuron 24:639–647.

Hunter K, Tonge DA, Holder N (1988) In vitro neurite outgrowth from axolotl neurons. Neurosci Lett (Suppl) 32:S68.

Hunter K, Maden M, Summerbell D, Eriksson U, Holder N (1991) Retinoic acid stimulates neurite outgrowth in the amphibian spinal cord. Proc Natl Acad Sci USA 88:3666–3670.

Hynes M, Rosenthal A (1999) Specification of dopaminergic and serotoergic neurons in the vertebrate CNS. Curr Opinion Neurobiol 9:26–36.

Imokawa Y, Brockes JP (2003) Selective activation of thrombin is a critical determinant for vertebrate lens regeneration. Curr Biol 13:877–881.

Iwashita Y, Kawaguchi S, Murata ME (1994) Restoration of function by replacement of spinal cord segments in the rat. Nature 367167–170.

Jensen AM, Wallace VA (1997) Expression of Sonic hedgehog and its putative role as a precursor cell mitogen in the developing mouse retina. Development 124:363–371.

Jimeno D, Lillo C, Cid E, Aijon J, Velasco A, Lara JM (2003) The degenerative and and regenerative processes after elimination of the proliferative peripheral retina in fish. Exp Neurol 179: 210–218.

Jin K, Minami M, Lan JQ, Mao XO, Batteur S, Simon RP, Greenberg DA (2001) Neurogenesis in dentate subgranular zone and rostral subventricular zone after focal cerebral ischemia in the rat Proc Natl Acad Sci USA 98:4710–4715.

Johansson C, Momma S, Clarke D, Risling M, Lendahl U, Frisen J (1999) Identification of a neural stem cell in the adult mammalian central nervous system. Cell 96:25–34.

Jones JE, Corwin JT (1996) Regeneration of sensory cells after laser ablation in the lateral line system: Hair cell lineage and macrophage behavior revealed by time-lapse video microscopy. J Neurosci 16:649–662.

Julian D, Ennis K, Korenbrot JL (1998) Birth and fate of proliferative cells in the inner nuclear layer of the mature fish retina. J Comp Neurol 394:271–282.

Kapfhammer JP, Schwab ME (1992) Modulators of neuronal migration and neurite growth. Curr Opinion Cell Biol 4:863–868.

Keefe JR (1973a) An analysis of urodelan retinal regeneration: I. Studies of the cellular source of retinal regeneration in *Notopthalmus viridescens* utilizing $^3$H-thymidine and colchicines. J Exp Zool 184:185–206.

Keefe JR (1973b) An analysis of urodelan retinal regeneration: II. Ultrastructural features of retinal regeneration in *Notophthalmus viridescens*. J Exp Zool 184:207–232.

Keefe JR (1973c) An analysis of urodelan retinal regeneration: III. Degradation of extruded melanin granules in *Notopthalmus viridescens*. J Exp Zool 184:233–238.

Kehl LJ, Fairbanks CA, Laughlin TM, Wilcox GL (1997) Neurogenesis in postnatal rat spinal cord: A study in primary culture. Science 276:586–589.

Kelley MW, Xu X-M, Wagner MA, Warchol ME, Corwin JT (1993) The developing organ of Corti contains retinoic acid and forms supernumerary hair cells in response to exogenous retinoic acid in culture. Development 119:1041–1053.

Kempermann G, Kuhn HG, Gage FH (1998) Experience-induced neurogenesis in the senescent dentate gyrus. J Neurosci 18: 3206–3212.

Kierstead HS, Dyer JK, Sholomenko G, McGraw J, Delaney KR, Steeves JD (1995) Axonal regeneration and physiological activity following transection and immunological disruption of myelin within the hatchling chick spinal cord. J Neurosci 15:6963–6974.

Kil J, Warchol ME, Corwin JT (1997) Cell death, cell proliferation, and estimates of hair cell life spans in the vestibular organs of chicks. Hear Res 114:117–126.

Klein LR, MacLeish PR, Wiesel TN (1990) Immunolabeling by a newt retinal pigment epithelium antibody during retinal development and regeneration. J Comp Neurol 293:331–339.

Kornack D, Rakic P (2001) Cell proliferation without neurogenesis in adult primate neocortex. Science 294:2127–2130.

Kuwabara T, Hsieh J, Nakashima K, Taita K, Gage FH (2004) A small modulatory dsRNA specifies the fate of adult neural stem cells. Cell 116:779–793.

Lang DM, Rubin BP, Schwab ME, Stuermer CA (1995) CNS myelin and oligodendrocytes of the *Xenopus* spinal cord—but not optic nerve—are nonpermissive for axon growth. Neurosci 15:99–109.

Lefcourt F, Venstrom K, McDonald JA, Reichardt LK (1992) Regulation of expression of fibronectin and its receptor, $\alpha_5\beta_1$, during development and regeneration of peripheral nerve. Development 116:767–782.

Lefebvre P, Malgrange B, Staecker H, Moonen G, Van De Water R (1993) Retinoic acid stimulates regeneration of mammalian auditory hair cells. Science 260:692–695.

Levine R (1977) Regeneration of the retina in the adult newt, *Triturus cristatus*, following surgical division of the eye by a post-limbal incision. J Exp Zool 200:41–54.

Litzinger TC, Del-Rio Tsonis K (2002) Eye anatomy. Encyclopedia of Life Sciences, MacMillan Publishers Ltd, Nature Publishing Group, pp 1–7.

Liu J, Solway K, Messing RO, Sharp FR (1998) Increased neurogenesis in the dentate gyrus after transient global ischemia in gerbils. J Neurosci 18:7768–7778.

Liu S, Qu Y, Stewart J, Howard MJ, Chakraborty S, Holekamp F, McDonald JW (2000) Embryonic stem cells differentiate into oligodendrocytes and myelinate in culture and after spinal cord transplantation. Proc Natl Acad Sci USA 97:6126–6131.

Livesey FJ, O'Brien A, Li M, Smith AG, Murphy LJ, Hunt SP (1997) A Schwann cell mitogen accompanying regeneration of motor neurons. Nature 390:614–618.

Lois C, Alvarez-Buylla A (1994) Long-distance neuronal migration in the adult mammalian brain. Science 264:1145–1148.

Lois C, Garcia-Verdugo JM, Alvarez-Buylla A (1996) Chain migration of neuronal precursors. Science 271:978–981.

Louissaint A Jr, Rao S, Leventhal C, Goldman SA (2002) Coordinated interaction of neurogenesis and angiogenesis in the adult songbird brain. Neuron 34:945–960.

Lowenheim H (2000) Regenerative biology of hearing: Taking the brakes off the cell cycle engine. J Reg Med (e-biomed) 1:21–24.

Lowenheim H, Furness DN, Kil J, Zinn C, Gultig K, Fero ML, Frost D, Gummer AW, Roberts JM, Rubel EW, Hackney C, Zenner HP (1999) Gene disruption of a p27 (Kip 1) allows cell proliferation in the postnatal and adult organ of Corti. Proc Nat Acad Sci USA 96:4084–4088.

Lu J, Waite P (1999) Advances in spinal cord regeneration. Spine 24:926–930.

Luskin M, Parnavelas JG, Barfield JA (1993) Neurons, astrocytes, and oligodendrocytes of the rat cerebral cortex originate from separate progenitor cels: an ultrastructural analysis of clonally related cells. J Neurosci 13:1730–1750.

Magavi SS, Leavitt BR, Macklis JD (2000) Induction of neurogenesis in the neocortex of adult mice. Nature 405:951–955.

Maier W, Wolburg H (1979) Regeneration of the goldfish retina after exposure to different doses of oubain. Cell Tiss Res 202:99–118.

Matthey R (1925) Recuperation de la vue après resection des nerfs optiques, chez le *Triton*. Comp Rend Soc Biol 93:904–906.

McCachren SS, Lightner V (1992) Expression of human tenascin in synovitis and its regulation by interleukin-1. Arthritis Rheum 35:1185–1196.

McDonald JW, Sadowsky C (2002) Spinal cord injury. The Lancet 359:417–425.

McKay R (1997) Stem cells in the central nervous system. Science 276:66–71.

McQuarrie IG (1983) Role of the axonal cytoskeleton in the regenerating nervous system. In: Seil FJ (ed.). Nerve, Organ and Tissue Regeneration: Research Perspectives. New York, Academic Press, pp 51–88.

McQuarrie IG, Lasek RJ (1989) Transport of cytoskeletal elements from parent axons into regenerating daughter axons. J Neurosci 9:436–446.

Mendoza AS, Breipohl W, Miragall F (1982) Cell migration from the chick olfactory placode: A light and electron microscopic study. J Embryol Exp Morph 69:47–59.

Messier B, LeBlond CP, Smart I (1958) Presence of DNA synthesis and mitosis in the brain of young adult mice. Exp Cell Res 14:224–226.

Miller RH, Liuzzi FJ (1986) Regional specialization of the radial glial cells of the adult frog spinal cord. J Neurocytol 15:187–196.

Mitashov VI (1996) Mechanisms of retina regeneration in urodeles. Int J Dev Biol 40:833–844.

Mocchetti I, Rabin S, Colangelo AM, Whittemore SR, Wrathall JR (1996) Increased basic fibroblast growth factor expression following contusive spinal cord injury. Exp Neurol 141:154–164.

Momma S, Johansson SB, Frisen J (2000) Get to know your stem cells. Curr Opinion Neurobiol 10:45–49.

Moshiri A, Reh TA (2004) Persistent progenitors at the retinal margin of ptc$^{+/-}$ mice. J Neurosci 24:229–237.

Nakamura Y, Sakakibara S, Miyata T, Ogawa M, Shimazaki T, Weiss S, Kageyama R, Okano H (2000) The bHLH gene hes 1 as a repressor of the neuronal commitment of CNS stem cells. J Neurosci 20:283–293.

Nakatomi H, Toshihiko Kuriu, Okabe S, Yamamoto S, Hatano O, Kawahara N, Tamura A, Kirino T, Nakafuku M (2002) Regeneration of hippocampal pyramidal neurons after ischemic brain injury by recruitment of endogenous neural progenitors. Cell 110:429–441.

Nakayama K, Nakayama K (1998) Cip/Kip cyclin-dependent kinase inhibitors: brakes of the cell cycle engine during development. Bioessays 20:1020–1029.

Negeshi K, Shinagawa S, Ushijima M, Kaneko Y, Saito T (1992) An immunohistochemical study of regenerating newt retinas. Dev Brain Res 68:255–264.

Nicholls J, Saunders N (1996) Regeneration of immature mammalian spinal cord after injury. Trends Neurosci 19:229–234.

Nicolas S, Caubit X, Massacrier A, Cau P, Le Parco, Y (1999) Two Nkx-3-related genes are expressed in the adult and regenerating central nervous system of the urodele Pleurodeles waltl. Dev Genet 24:319–328.

Nicolas S, Massacrier A, Caubit X, Cau P, Le Parco Y (1996) A distal-less-like gene is induced in the regenerating central nervous system of the urodele Pleurodeles waltl. Mech Dev 56:209–220.

Niederost B, Zimmerman DR, Schwab ME, Bandtlow CE (1999) Bovine CNS myelin contains neurite growth-inhibitory activity associated with chondroitin sulfate proteoglycans. J Neurosci 19:8979–9889.

Noji S, Matsuo T, Koyama E, Yamaai T, Nohno T, Matsuo N, Tanaguchi S (1990) Expression pattern of acidic and basic fibroblast growth factor genes in adult rat eyes. Biochem Biophys Res Comm 168:343–349.

Nordlander R, Singer M (1978) The role of ependyma in regeneration of the spinal cord in the urodele amphibian tail. J Comp Neurol 180:349–374.

Oesterle EC, Rubel EW (1996) Hair cell regeneration in vestibular sensory epithelia. Ann NY Acad Sci 781:34–70.

O'Hara CM, Egar MW, Chernoff EAG (1992) Reorganization of the ependyma during axolotl spinal cord regeneration: Changes in intermediate filament and fibronectin expression. Dev Dynam 193:103–115.

O'Hara CM, Chernoff EAG (1994) Growth factor modulation of injury-reactive ependymal cell proliferation and migration. Tissue and Cell 26: 599–611.

O'Neill P, Whalley K, Ferretti P (2004) Nogo and Nogo-66 receptor in human and chick: implications for development and regeneration. Dev Dynam 231:109–121.

Orts-Llorca F, Genis-Galvez JM (1960) Experimental production of retinal septa in the chick embryo: Differentiation of pigment epithelium into neural retina. Acta Anat 42:31–70.

Otteson DC, D'Costa AR, Hitchcock PF (2001) Putative stem cells and the lineage of rod photoreceptors in the mature retina of the goldfish. Dev Biol 232:62–76.

Otteson DC, Hitchcock PF (2003) Stem cells in the teleost retina: Persistent neurogenesis and injury-induced regeneration. Vision Res 43:927–936.

Palmer TD, Willhoite AR, Gage FH (2000) Vascular niche for adult hippocampal neurogenesis. J Comp Neurol 425:479–494.

Park C, Hollenberg MJ (1991) Induction of retinal regeneration *in vivo* by growth factors. Dev Biol 148:322–333.

Park C, Hollenberg MJ (1993) Growth factor-induced retinal regeneration *in vivo*. Intl Rev Cytol 146:49–71.

Parker SB, Eichele G, Zhang P, Rawls A, Sands AT, Bradley A, Olson EN, Harper JW, Elledge SJ (1995) p53-independent expression of p21Cip1 in muscle and other terminally differentiating cells. Science 267:1024–1027.

Paton A, Nottebohm FN (1984) Neurons generated in the adult brain are recruited into functional circuits. Science 225:1046–1048.

Pearson CAD, Shibabhara S, Hofsteenge J, Chiquet-Ehrismann R (1988) Tenascin: cDNA cloning and induction by TGF-beta. EMBO J 7:2977–2982.

Perron M, Kanekar S, Vetter ML, Harris WA (1998) The genetic sequence of retinal development in the ciliary margin of the *Xenopus* eye. Dev Biol 99:185–200.

Piatt J (1955) Regeneration of the spinal cord in the salamander. J Exp Zool 129:177–208.

Pittack C, Jones M, Reh TA (1991) Basic fibroblast growth factor induces retinal pigment epithelium to generate neural retina in vitro. Development 113:577–588.

Pittack C, Grunwald GB, Reh TA (1997) Fibroblast growth factors are necessary for neural retina but not pigmented epithelium differentiation in chick embryos. Development 124:805–816.

Prinjha R, Moore SE, Vinson M, Blake S, Morrow R, Christie G, Michalovich D, Simmons DL, Walsh FS (2000) Inhibitor of neurite outgrowth in humans. Nature 403:383–384.

Raff MC, Whitmore AV, Finn JT (2002) Axonal self-destruction and neurodegeneration. Science 296:868–871.

Raivich G, Kreutzberg GW (1993) Peripheral nerve regeneration: Role of growth factors and their receptors. Int J Dev Neurosci 11:311–324.

Ramon-Cueto A, Valverde F (1995) Olfactory bulb ensheathing glia: A unique cell type with axonal growth-promoting properties. Glia 14:163–173.

Rao MS (1999) Multipotent and restricted precursors in the central nervous system. Anat Rec 257:137–148.

Raymond PA, Hitchcock PF (1997) Retinal regeneration: Common principles but a diversity of mechanisms. Adv in Neurol 72:171–184.

Raymond PA, Hitchcock PF (2000) How the neural retina regenerates. Results Problems Cell Diff 31:197–218.

Reh TA, Nagy Y (1987) A possible role for the vascular membrane in retinal regeneration in Rana catesbiena tadpoles. Dev Biol 122:460–482.

Reh TA, Nagy T, Gretton H (1987) Retinal pigment epithelial cells induced to transdifferentiate to neurons by laminin. Nature 330:68–71.

Reier PJ (1979) Penetration of grafted astrocytic scars by regenerating optic nerve axons in *Xenopus* tadpoles. Brain Res 164:61–68.

Reier PJ, Webster H deF (1974) Regeneration and remyelination of *Xenopus* tadpole optic nerve fibers following transection or crush. J Neurocytol 3:591–618.

Reyer RW (1977) The amphibian eye: Development and regeneration. In: Crescitelli F (ed.), Handbook of Sensory Physiology, Vol. VII/5. Berlin, Springer-Verlag, pp 309–390.

Reynolds B, Weiss S (1992) Generation of neurons and astrocytes from isolated cells of the adult mammalian central nervous system. Science 255:1707–1710.

Reynolds BA, Weiss S (1996) Clonal and population analyses demonstrate that an EGF-responsive mammalian embryonic CNS precursor is a stem cell. Dev Biol 175:1–13.

Rietze RL, Valcanis H, Brooker G, Thomas T, Voss AK, Bartlett PF (2001) Purification of a pluripotent neural stem cell from the adult mouse brain. Nature 412:736–739.

Roberson DF, Rubel EW (1995) Hair cell regeneration. Curr Opin Otolaryngol Head Neck Surg 3:302–307.

Rogers SL, Letourneau PC, Palm SL, McCarthy J, Furcht LT (1983) Neurite extension by peripheral and central nervous system neurons in response to substratum-bound fibronectin and laminin. Dev Biol 98:212–220.

Roy NS, Wang, Jiang L, Kang J, Benraiss A, Harrison P, Restelli C, Fraser R, Couldwell WT, Kawaguchi A, Okano H, Nedergaard M, Goldman SA (2000) *In vitro* neurogenesis by progenitor cells isolated from the adult human hippocampus. Nat Med 6:271–277.

Rungger-Brandle E, Achtstatter T, Franke W (1989) An epithelium-type cytoskeleton in a glial cell: Astrocytes of amphibian optic nerves contain cytokeratin filaments and are connected by desmosomes. J Cell Biol 109:705–716.

Sage C, Huang M, Karimi K, Gutierrez G, Vollrath MA, Zhang D-S, Garcia-Anoveros J, Hinds PW, Corwin JT, Corey DP, Chen Z-Y (2005) Proliferation of functional hair cells *in vivo* in the absence of the retinoblastoma protein. Science 307:1114–1118.

Sakakibara S, Nakamura Y, Yoshida T, Shibata S, Koike M, Takano H, Ueda S, Uchiyama Y, Noda T, Okano H (2002) RNA-binding protein Musashi family: Roles for CNS stem cells and a subpopulation of ependymal cells revealed by targeted disruption and antisense ablation. Proc Natl Acad Sci USA 99:15194–15199.

Sanai N, Tramontin AD, Quinones-Hinojosa A, Barbaro NM, Gupta N, Kunwar S, Lawton MT, McDermott MW, Parsa AT, Verdugo J M-G, Berger MS, Alvarez-Buylla A (2004) Unique astrocyte ribbon in adult humans brain contains neural stem cells but lacks chain migration. Nature 427:740–744.

Sauer F (1935) Mitosis in the neural tube. J Comp Neurol 62: 377–405.

Saunders NR, Deal A, Knott GW, Varga ZM, Nicholls JG (1995) Repair and recovery following spinal cord injury in a neonatal marsupial (monodelphus domestica). Clin Exp Pharmacol Physiol 22:518–526.

Schneider ME, Belyantseva IA, Azevedo RB, Kachar B (2002) Rapid renewal of auditory hair bundles. Nature 418:837–838.

Schulte T, Paschke KA, Laessing U, Lottspeich F, Stuermer CAO (1997) Reggie-1 and reggie-2, two cell surface proteins expressed by retinal ganglion cells during axon regeneration. Development 124:577–587.

Schwob JE (2002) Neural regeneration and the peripheral olfactory system. Anat Rec 269:33–49.

Scott TM, Foote J (1981) A study of degeneration, scar formation and regeneration after section of the optic nerve in the frog, *Rana pipiens*. J Anat 133:213–225.

Seaberg RM, Van der Kooy D (2002) Adult rodent neurogenic regions: the ventricular subependyma contains neural stem cells, but the dentate gyrus contains restricted progenitors. J Neurosci 22:1784–1793.

Seitz A, Aglow E, Heber-Katz E (2002) Recovery from spinal cord injury: A new transection model in the C57B1/6 mouse. J Neurosci Res 67:337–345.

Shatos MA, Mizumoto K, Mizumoto H, Kurimoto Y, Klassen H, Young MJ (2001) Multipotent stem cells from the brain and retina of green mice. J Reg Med 2:13–15 (E-biomed).

Shen Q, Goderie SK, Jin L, Karanth N, Sun Y, Abramova N, Vincent P, Pumiglia K, Temple S (2004) Endothelial cells stimulate self-renewal and expand neurogenesis of neural stem cells. Science 304:1338–1340.

Shimizu I, Oppenheim RW, Obrien M, Shneiderman A (1990) Anatomical and functional recovery following spinal cord transection in the chick embryo. J Neurobiol 21:918–937.

Shors T, Miesegaes G, Beylin A, Zhao M, Rydel T, Gould E (2001) Neurogenesis in the adult is involved in the formation of trace memories. Nature 410:372–375.

Simpson SB Jr (1965) Regeneration of the lizard tail. In: Kiortisis V, Trampusch HA (eds.). Regeneration in Animals and Related Problems. Amsterdam: North-Holland Pub Co, pp 431–443.

Sinicropi DV, McIlwain DL (1983) Changes in the amounts of cytoskeletal proteins within the perikarya and axons of regenerating frog motoneurons. J Cell Biol 96:240–247.

Smart I (1961) The subependymal layer of the mouse brain and its cell production as shown by radioautography after thymidine-$H^3$ injection. J Comp Neurol 116:325–348.

Smith CM, Luskin MB (1998) Cell cycle length of olfactory bulb neuronal progenitors in the rostral migratory system. Dev Dynam 213:220–227.

Song H, Stevens CS, Gage FH (2002) Astroglia induce neurogenesis from adult neural stem cells. Nature 417:39–44.

Spalding KL, Bhardwaj RD, Buchholtz BA, Druid H, Frisen J (2005) Retrospective birth dating of cells in humans. Cell 122: 133–143.

Sperry RW (1944) Optic nerve regeneration with return of vision in anurans. J Neurophysiol 7:57–69.

Stensaas LJ (1983) Regeneration in the spinal cord of the newt *Notophthalmus (Triturus) pyrrhogaster*. In: Kao CC, Bunge RP, Reier PJ (eds.). Spinal Cord Reconstruction. New York, Raven Press, pp 121–149.

Stensaas LJ, Feringa ER (1977) Axon regeneration across the site of injury in the optic nerve of the newt *Triturus pyrrhogaster*. Cell Tiss Res 179:501–516.

Steward O, Schauwecker PE, Guth L, Zhang Z, Fujiki M, Inman D, Wrathall J, Kempermann G, Gage FH, Saatman KE, Raghupathi R, McIntosh T (1999) Genetic approaches to neurotrauma research: Opportunities and potential pitfalls of murine models. Exp Neurol 157:19–42.

Stocum DL (2002) A tail of transdifferentiation. Science 298: 1901–1903.

Stokes BT, Fox P, Hollinden G (1983) Extracellular calcium activity in the injured spinal cord. Exp Neurol 80:561–572.

Stone LS (1950) The role of retinal pigment cells in regenerating neural retinae of adult salamander eyes. J Exp Zool 113:9–31.

Stone JS, Cotanche DA (1994) Identification of the timing of S phase and the patterns of proliferation during hair cell regeneration in the chick cochlea. J Comp Neurol 341:50–67.

Stone JS, Rubel EW (1999) *Delta 1* expression during avian hair cell regeneration. Development 126:961–973.

Stone JS, Rubel EW (2000) Avian auditory hair cell regeneration. Proc Natl Acad Sci USA 97:11714–11721.

Stone JS, Shang L, Tomarev S (2004) cProx1 immunoreactivity distinguishes progenitor cells and predicts hair cell fate during avian hair cell regeneration. Dev Dynam 230:597–614.

Stuermer CO, Bastmeyer M, Bahr M, Strobel G, Paschke K (1992) Trying to understand axonal regeneration in the CNS of fish. J Neurobiol 23:537–559.

Taylor R, Forge A (2005) Life after deaf for hair cells? Science 307:1056–1058.

Temple S (2001) The development of neural stem cells. Nature 414:112–117.

Tsonis PA, Del-Rio Tsonis K (2004) Lens and retina regeneraton: transdifferentiation, stem cells and clinical applications. Exp Eye Res 78:161–172.

Travis J (1994) Glia: The brain's other cells. Science 266:970–972.

Treherne JM, Woodward SKA, Varga ZM, Ritchie JM, Nicholls JG (1992) Restoration of conduction and growth of axons through injured spinal cord of the neonatal opossum in culture. Proc Natl Acad Sci U 89:431–434.

Tropepe V, Coles BLK, Chiasson BJ, Horsford D, Elia J, McInnes RR, van der Kooy D (2000) Retinal stem cells in the adult mammalian eye. Science 287:2032–2036.

Tsue TT, Watling DL, Weisleder P, Coltrera MD, Rubel EW (1994) Identification of hair cell progenitors and intermitotic migration of their nuclei in the normal and regenerating avian inner ear. J Neurosci 14:140–152.

Turner JT, Singer M (1974) The ultrastructure of regeneration in the severed newt optic nerve. J Exp Zool 190:249–288.

Van Praag H, Kempermann G, Gage FH (1999a) Running increases cell proliferation and and neurogenesis in the adult mouse dentate gyrus. Nature Neurosci 2:266–270.

Van Praag H, Christie BR, Sejnowski TJ, Gage FH (1999b) Running enhances neurogenesis, learning and long-term potentiation in mice. Proc Natl Acad Sci USA 96:13427–13431.

Van Praag H, Schinder AF, Christie BR, Toni N, Palmer T, Gage FH (2002) Fuctional neurogenesis in the adult hippocampus. Nature 415:1030–1034.

Wakabayashi Y, Komori H, Kawa UT, Mochida K, Takahashi M, Qi M, Otake K, Shinomiya K (2001) Functional recovery and regeneration of descending tracts in rats after spinal cord transection in infancy. Spine 26:1215–1222.

Wanner M, Lang DM, Bandtlow CE, Schwab ME, Bastmeyer M, Steuermer CA (1995) Reevaluation of the growth-permissive substrate properties of goldfish optic nerve myelin and myelin proteins. J Neurosci 15:7500–7508.

Warchol ME, Lambert PR, Goldstein BJ, Forge A, Corwin JT (1993) Regenerative proliferation in inner ear sensory epithelia from adult guinea pigs and humans. Science 259:1619–1622.

Wehrle B, Chiquet M (1990) Tenascin is accumulated along developing peripheral nerves and allows neurite outgrowth in vitro. Development 110:401–415.

Weiss SM, Dunne C, Hewson J, Wohl C, Wheatley M, Peterson AC, Reynolds BR (1996) Multipotent CNS stem cells are present in the adult mammalian spinal cord and ventricular neuraxis. J Neurosci 16:7599–7609.

Westerfield M (1987) Substrate interqactions affecting motor growth cone guidance during development and regeneration. J Exp Biol 132:161–176.

Willbold E, Layer PG (1992) A hidden retina regenerative capacity from the chick ciliary margin is reactivated *in vitro*, that is accompanied by downregulation of butyrylcholinesterase. Eur J Neurosci 4:210–220.

Yamamoto S, Nagao M, Ugimori M, Kosako H, Nakatomi H, Yamamoto N, Takebayashi H, Nabeshima Y, Kitamura T, Weinmaster G, Nakamura K, Nakafuku M (2001) Transcription factor expression and Notch-dependent regulation of of neural progenitors in the adult rat spinal cord. J Neurosci 21:9814–9823.

Yamashita T, Tucker KL, Barde YA (1999) Neurotrophin binding to the p75 receptor modulates Rho activity and axonal outgrowth. Neuron 24:585–593.

Yannas IV (2001) Tissue and Organ Regeneration in Adults. New York, Springer.

Yiu G, He Z (2003) Signaling mechanisms of the myelin inhibitors of axon regeneration. Curr Opin Neurobiol 13:545–551.

Zakon H, Capranica RR (1981) Reformation of organized connections in the auditory system after regeneration of the eighth nerve. Science 213:242–244.

Zamora AJ, Mutin M (1988) Vimentin and glial fibrillary acidic protein in radial glia of the adult urodele spinal cord. Neurosci 27:279–288.

Zhang FC, Clarke JDW, Ferretti P (2000) FGF-2 up-regulation and proliferation of neural progenitors in the regenerating amphibian spinal cord in vivo. Dev Biol 225:381–391.

Zhao X, Das AV, Thoreson WB, James J, Wattnem TE, Rodriguez-Sierra J, Ahmad I (2002) Adult corneal limbal epithelium: A model for studying neural potential of non-neural stem cells/progenitors. Dev Biol 250:317–331.

Zhao X, Das A, Soto-Leon F, Ahmad I (2005) Growth factor-responsive progenitors in the postnatal mammalian retina. Dev Dynam 232:349–358.

# 6 Regenerative Medicine of Neural Tissues

## INTRODUCTION

Injuries to, and degenerative diseases of, the nervous system are among the most difficult to deal with medically and surgically. The healthcare costs of neurological disorders are enormous, to say nothing of the loss in economic productivity and quality of life they inflict. In addition, many of these injuries and diseases place a large emotional and physical burden on family members and other caregivers. As the planet's human population ages, the incidence of neurodegenerative diseases will continue to rise, putting a further physical and financial strain on healthcare systems, insurance companies, and families. Finding ways to combat these injuries and diseases, and repair the tissue damage they cause, is one of the major goals of modern medicine. In this chapter, progress toward regenerative therapies for damage to the peripheral and central nervous systems is reviewed.

## THERAPIES FOR INJURED PERIPHERAL NERVE

Crush injuries of peripheral nerves regenerate relatively well because the proximal and distal segments of the nerve remain aligned so that proximal axon stumps can regenerate into distal endoneurial tubes. Transections, or injuries that necessitate removing segments of nerve, are more difficult to repair. The cut ends of the nerves retract, and it is difficult to suture them back together. Fibroblasts invade the wound space, resulting in the formation of scar tissue that leads sprouting proximal axons to form neuromas. Continuous longitudinal sutures have been used successfully to bridge small gaps in the rat sciatic nerve (Scherman et al., 2004), but this does not suffice for larger gaps.

The strategy to regenerate nerves across large gaps is to bridge the gap with tissue autografts or biomimetic materials that promote the survival, elongation, and guidance of regenerating axons. Autografts of nerve trunks from elsewhere in the body are currently the treatment of choice to bridge gaps in nerves because they contain endoneurial tubes to guide regenerating axons and blood vessels that can quickly connect to the local circulation. The current gold standard is a sensory nerve autograft, such as the sural nerve or saphenous branch of the femoral nerve. However, endoneurial tubes appear to have some degree of specificity in the type of nerve axon they will support. Sensory nerve autografts have been found to support the regeneration of mixed nerves poorly compared to motor and mixed nerve autografts (Nichols et al., 2004). Most peripheral nerves are mixed nerves, suggesting that motor or mixed nerve allografts might be a better choice for inducing regeneration across gaps. The main limitation to this idea is that there are few expendable motor or mixed nerves. Thus, surgeons have turned to autografts of other tissues such as freeze-dried muscle, blood vessels, and tendons, but these have not worked as well as nerve autografts (Hall, 1997; Hems and Glasby, 1993), even when cultured Schwann cells are added to them (Nishiura et al., 2004). Furthermore, any autograft taken from one nerve to repair another nerve requires two surgeries and compromises the site innervated by the donor nerve, a general limitation of autografts.

The limitations of autografted tissues have prompted a search for biomimetic materials to serve as bridges for gap regeneration of peripheral nerves (**FIGURE 6.1**). Nonbiodegradable silicon tubes are the reference standard for comparison with bridges made from other biomaterials (Lundborg et al., 1982). The metric for comparison is the critical axon elongation length ($L_c$), defined as the injury gap length at which the frequency of reinnervation drops below 50% (Yannas, 2001; Yannas and Hill, 2004). The rat sciatic nerve is a commonly used model of peripheral nerve regeneration. The $L_c$ for the regeneration of the rat sciatic nerve in a silicon tube is $9.7 \pm 1.8$ mm (Yannas, 2001). This is equivalent to the performance of a nerve autograft, but to be more clinically useful than nerve autografts,

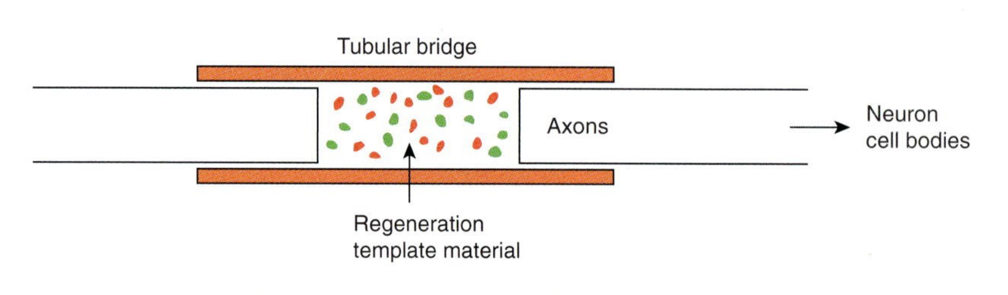


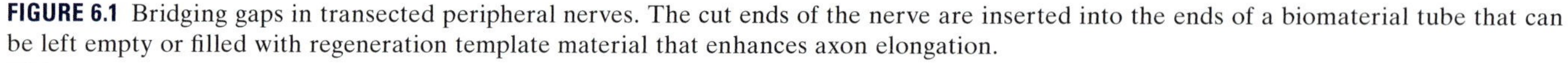

**FIGURE 6.1** Bridging gaps in transected peripheral nerves. The cut ends of the nerve are inserted into the ends of a biomaterial tube that can be left empty or filled with regeneration template material that enhances axon elongation.

bridge materials need to support regeneration across gaps of 25–80 mm.

The course of regeneration after insertion of the distal and proximal ends of a transected rat sciatic nerve into a silicon tube is as follows (Yannas, 2001). The tube first fills with endoneurial-derived hypertonic wound fluid containing PDGF, FGF-1, and NGF. By 7 days, the fluid coagulates into a cable of longitudinally oriented fibrin fibers containing fibronectin and trapped red blood cells. Over the course of the next month, Schwann cells, fibroblasts, and blood vessels grow from the proximal and distal cut ends of the nerve to meet in the middle of the tube, and nonmyelinated and myelinated axons grow through the tube from the proximal nerve stump, using the dedifferentiated Schwann cells as adhesive substrates. A connective tissue sheath of myofibroblasts is formed around the regenerating nerve. Axons that fail to re-enter endoneurial tubes do not synthesize an endoneurium, although they are bundled together in minifascicles by a perineurium. The regenerated nerve never reaches the diameter of the normal nerve and does not synthesize an epineurium, resulting in a lower conduction velocity.

A wide variety of biomaterial tubes, including silicon, ethylene-vinyl acetate (EVAc), poly (lactic-co-glycolic acid) (PLGA), polyhydroxybutyrate (PHB), and type I collagen, with or without supplementation with adhesion, growth and neurotrophic factors, have been tested for their effectiveness as regeneration templates (Bellamkonda and Aebischer, 1995; Yannis, 2001; Constans, 2004). The degree of successful regeneration depends on four parameters: chemical composition of the tube and its supplements; orientation of the structural materials of the tube wall surface; porosity of the tube; and the degradation rate of the biomaterial (Yannas, 2001). Tubes composed of longitudinally oriented collagen and related ECM materials that are permeable to cells and degrade relatively rapidly are superior to synthetic polymers (Yannas, 2001). Longitudinally oriented molecules or microgrooves in the tube walls promote the straight growth of neurites better than molecules oriented perpendicularly. The porosity of the walls has to be large enough to allow gas and nutrient exchange between the inside of the guide and the external environment. Biodegradability of nerve guide tubes is a desirable feature; otherwise the tubes must be removed from the regenerated nerve by a second surgical operation. Biodegradable guides need to survive for ~4–12 weeks, and their degradation products should not have an adverse effect on regeneration.

Many different kinds of tubes and tube fills have been tested for their efficacy in promoting peripheral nerve regeneration across variable-sized gaps (see **TABLE 6.1**, Yannas, 2001). Silicon tubes filled with a collagen/chondroitin 6-sulphate matrix or an agarose matrix containing the laminin recognition peptide CDPGYIGSR or FGF promote sciatic nerve regeneration beyond the $L_c$, but the increase in length does not approach the desirable 25 mm (Yannas, 2001; Aebischer et al., 1989; Bellamkonda and Aebischer, 1995). Regeneration is enhanced to a $L_c$ of over 21 in the presence of a Schwann cell suspension (Ansselin et al., 1997). Porous collagen, poly-L-lactic acid, and polyglycolic acid promote substantial regeneration of the sciatic nerve as well (Henry et al., 1985; Molander et al., 1983; Yannas, 2001). The best regeneration was obtained with a collagen tube filled with a copolymer of type I collagen and chondroitin 6-sulfate with longitudinally oriented pores. This construct produced an $L_c$ of 25, more than three times that of a silicon tube (Spilker, 2000; cited in Yannas, 2001). This template also promoted an increase in average axon diameter from 30–60 weeks when used to bridge a 10-mm gap in rat sciatic nerve and significantly increased the conduction velocity and amplitude of the regenerated nerve over a control phosphobuffered saline-filled collagen tube (Chamberlain et al., 1998).

Functional assessment of template-guided sciatic nerve regeneration showed that it does not restore normal walking patterns in rats (Ijkema-Paassen et al., 2004). There are long-term abnormalities in walking and electromyographic patterns, as well as abnormalities in neuromuscular contacts and shifts in the histochemical properties of target muscles. These deficits are likely due to the lack of specificity of the biomaterial templates for guiding regenerating axons into their previous endoneurial tubes distally, resulting in random

**TABLE 6.1** Damage and Regeneration-Inhibitory Factors Involved in SCI, Their Effects, and the Strategies and Therapies That Have Been Devised to Mitigate Them

| Factor | Effect | Strategy/Therapy |
|---|---|---|
| Compression | Kills neurons by trauma | Relieve compression |
| Glutamate toxicity, free radical toxicity | Secondary damage leading to apoptosis of healthy neurons | Neuroprotective molecules and cell transplants |
| Myelin inhibitory proteins | Growth cone collapse | Neutralize proteins or the pathways they activate<br>Regeneration templates, with or without glial cells |
| Glial scar components | Collapse growth cones, mechanical barrier | Inhibit inflammatory cells, enzymatic digestion |
| All | Loss of neural circuitry | NSC transplants<br>Bioartificial spinal cord<br>Rehabilitation programs |

innervation of target muscles, and failure to synthesize endoneurium and epineurium. Higher-quality outcomes of peripheral nerve regeneration might be achieved by making bridges of many parallel nanotubes that replicate the features of sensory and motor endoneurial tubes and give specific guidance cues to sensory and motor targets.

Near-infrared laser or LED light has been reported to accelerate the rate of peripheral nerve regeneration. Daily transcutaneous irradiation of the facial nerve of rats after crush injury, as assessed by retrograde transport of HRP, resulted in a statistically significant increase in the number of labeled neurons in the facial motor nucleus as compared to nonirradiated controls (Anders et al., 1993). Crush injury to the sciatic nerve drastically lowers the electrophysiological activity of the nerve, but irradiation of either the crush injury or the neuron cell bodies in the spinal cord at the sciatic nerve level maintains the activity at near-normal levels (Rochkind et al., 1987, 2001). Again, cytochrome *c* oxidase is implicated in absorption of the light energy and increasing ATP levels. Irradiation of cultured primary neurons with a 670-nm LED upregulates the activity of this enzyme in cultured primary neurons. Potassium cyanide (KCN) irreversibly inhibits cytochrome *c* oxidase. LED irradiation at 10–100-μM KCN partially restores the inhibited enzyme activity and significantly reduces neuron death, but at 1–100-mM KCN, the neuroprotective effect of the LED is abolished (Wong-Riley et al., 2005).

## THERAPIES FOR INJURED SPINAL CORD

The annual incidence of spinal cord injuries (SCI) in developed countries varies from 2.1 to 123.6 per million persons; in the United States the injury rate has been estimated at 43–55 per million (Harkey et al., 2003). SCIs result primarily from automobile accidents, recreational accidents, sports injuries, and gunshot wounds and are suffered predominantly by males. Harkey et al. (2003) have summarized the classification of SCIs and their current clinical treatment. Most injuries involve vertebral fracture and are incomplete, with varied functional outcomes in terms of walking. A large study of surviving SCI patients found that 37% could not walk, 24% could walk but used a wheelchair to get around, and 39% were functional walkers (Curt et al., 2004). The total direct cost of spinal cord injury in the United States (data in constant 1995 dollars) was nearly $8B/yr a few years ago (De Vivo, 1997). Maddox (1993) estimated that the regeneration of only two to five percent of the normal number of axons in a transected human spinal cord would be required to regain a reasonable measure of function.

To achieve successful regeneration, neurons must survive and their axons must be able to enter the lesion, traverse it, exit the other side, and continue on to re-innervate targets. Several cell transplantation and pharmaceutical therapies have been developed to promote mammalian spinal cord regeneration, based on our knowledge of the events that occur after a spinal cord injury. Pharmaceutical (chemical induction) approaches include the use of neuroprotective agents, neutralization of myelin proteins, inhibition of gliosis, and degradation of glial scar ECM, and implanting biomaterial bridges to bypass glial scar and promote axon extension (Horner and Gage, 2000; Bjorklund and Lindvall, 2000; Talac et al., 2004). Glial cell transplants have been used to provide factors essential for axon extension, and NSCs and other cell types appear to have paracrine neuroprotective effects (Mansergh et al., 2004). Finally, rehabilitation programs have been designed that are aimed at "rewiring" spared circuitry through synaptic plasticity.

Ramer et al. (2005) have published an outstanding comprehensive review of the strategies used to regenerate spinal cord axons and neurons that should be consulted for details not covered here. **TABLE 6.1** summarizes the effects of various damage or regeneration-

**TABLE 6.2** Pharmaceutical Neuroprotection Therapies for SCI

| Agent | Effect | Target |
|---|---|---|
| Methylprednisolone | Stabilization | Cell membranes |
| Gacyclidine | Inhibit toxic calcium influx | NMDA, the glutamate receptor |
| NMDA | Inhibit toxic calcium influx | Glutamate |
| P2X7R antagonists (OXATP, PPADS) | Inhibit toxic calcium influx | P2X7R (ATP receptor) |
| Carboxy buckyballs | Neutralize | Free radicals |
| 4-aminopyridine | Improve conduction in spared demyelinated axons | Potassium channels |
| Carbamylated EPO | Reduce inflammation | Unknown receptor |

**TABLE 6.3** Pharmaceutical Therapies to Enhance Axon Elongation in Mammals with SCI (gc = growth cone)

| Agent | Effect | Target |
|---|---|---|
| cAMP | Restore normal calcium responses | Calcium channels |
| Rolipram | Restore normal calcium responses | Phospodiesterase IV |
| Bacterial C3 transferase | Prevent gc collapse | Rho pathway |
| Inhibitors of Rho kinase | Prevent gc collapse | Rho pathway |
| BDNF, GDNF, NGF, NT-3 | Prevent gc collapse | Rho pathway |
| Nogo 66 | Prevent gc collapse | NgR |
| Anti-Nogo Ab | Prevent gc collapse | Nogo |
| Antimyelin protein Abs (via vaccination) | Prevent gc collapse | Myelin proteins |
| Minocycline | Limit cavitation and glial scar | Microglia, immune cells |
| CM101 | Limit glial scar | Inflammatory cells |
| Chondroitinase ABC | Increase adhesivity of glial scar PGs | CS GAGs of PGs |

inhibiting factors and the strategies/therapies that have been used to counter them.

## 1. Pharmaceutical Therapies

**TABLES 6.2** and **6.3** summarize the types of molecular interventions that have been tested to enhance neuron survival and axon elongation after SCI.

### a. *Neuroprotective Agents*

A large number of neuroprotective agents have been tested in animals and humans for their ability to mitigate neuron damage after spinal cord injury. The most widely used neuroprotective agent for human spinal cord injury has been high-dose methylprednisolone (MP). This steroid stabilizes cell membranes, and its use doubles the chances for functional recovery if administered within the first eight hours after injury (Bracken et al., 1990, 1992). However, MP is potentially toxic (Kwon et al., 2004) and other countries, including Canada, have abandoned it as a standard for treatment (Hugenholtz, 2003). GM-1 gangliosides were reported to accelerate the rate of recovery from SCI in humans, but did not significantly improve ultimate functional outcome (Geisler et al., 2001). Gacyclidine, a noncompetitive antagonist of the glutamate receptor N-methyl-D-aspartate (NMDA), reduced toxic glutamate release and improved recovery from spinal cord contusion injury in rats, as did N-methyl-D-aspartate itself (Gaviria et al., 2000; Feldblum et al., 2000). Polyethylene glycol (PEG), in conjunction with fampridine (4-aminopyridine), induces the recovery of action potentials in guinea pig spinal cord after compression injury to 72% of pre-injury levels (Luo et al., 2002). Fampridine is a potassium channel blocker that improves conduction in axons that have been preserved after SCI or that have suffered demyelination (Targ and Kocsis, 1985).

In France, clinical trials of gacyclidine on patients within three hr of SCI, with a second dose delivered within the next four hr, showed no statistically significant improvement in functional recovery over placebo-treated control patients, though there was a trend suggesting some improvement at the highest doses and in a subgroup of patients who had suffered cervical injury (Steeves et al., 2004). In the United States, ongoing trials of fampridine by Acorda Therapeutics in New York suggest that it may reduce spasticity and stiffness and improve bladder, bowel, and sexual functions (Steeves et al., 2004).

Other agents reported to have neuroprotective effects in animal models are α-amino-3-hydroxy-5-methyl-4-isoazolepropionate keinate (Kaku et al., 1993) and water-soluble carboxy "buckyballs," spheres of carbon to which are attached six pairs of carboxylic acid molecules (Dugan, 1997). Buckyballs are proficient at soaking up toxic free radicals generated by injury. When tested on oxygen and glucose-starved neurons *in vitro* (conditions that mimic the effects of stroke), neuron death was reduced by 75%. Carbamylated erythropoietin (CEPO) lacks erythropoietic activity, but is neuroprotective and significantly improved motor function in rat spinal cord injury, even when administered as late as 72hr after injury (Leist et al., 2004). Spinal cord injury in the rat causes sustained release of ATP in the peritraumatic zone and sustained expression of P2X7 purine receptors that bind ATP, leading to calcium influx and apoptosis (Wang et al., 2004). Injection of the P2X7R antagonists OxATP or PPADS rostral and caudal to the lesion immediately after injury rescued neurons and improved functional recovery. Similarly, intracerebroventricular injection of inhibitors of acid-sensing ion channels (ASICs) protect the brain by reducing glutamate excitotoxicity (Xiong et al., 2004) and might be able to do the same in the injured spinal cord.

Growth and neurotrophic factors enhance the survival of axotomized spinal cord neurons and promote axon regeneration. Schnell et al. (1994) reported that injection of NGF or NT-3 increased the initial sprouting of injured axons, though the distance of axon extension was limited. NGF, NT-3, and GDNF also promoted the selective regrowth of damaged axons across the dorsal root entry zone and into the spinal cord, making functional connections with dorsal horn neurons (Ramer et al., 2000). Transduction of spinal motor neurons in the L3–L6 region with an adenoviral NT-3 gene construct resulted in a significant increase in the concentration of NT-3 that enhanced the ability of axons to grow from the intact corticospinal tract after hemisection of the opposite tract (Zhou et al., 2003). The level of cAMP in the cell bodies of transected central axons drops significantly (Pearse et al., 2004). The augmentation of intracellular cAMP in dorsal root ganglion neurons *in vivo* enhances the sprouting of central axons following lesions to the dorsal columns and allows the axons of dorsal root ganglion neurons to elongate on normally inhibitory substrates, such as myelin associated glycoprotein (Cai et al., 2001, 2002). The level of cAMP is controlled by phosphodiesterases that degrade it. The phosphodiesterase IV inhibitor, rolipram, causes cAMP to accumulate by preventing its degradation. Rolipram administered systemically after SCI was reported to stimulate axon regeneration and functional recovery in rats (Nikulina et al., 2004).

Administration of cAMP to nonregenerating axons of zebrafish promotes their regeneration. The axon of the Mauthner neuron, which is part of the escape behavior circuit of the fish, dies back after a spinal cord lesion and either does not regenerate at all or regenerates poorly. This poor regenerative ability appears to be related to intrinsic limitations of the neuron, because the axons of other neurons in the zebrafish cord can regenerate (Becker et al., 1998). Mauthner axons can be induced to regenerate by direct application of dibutryl cyclic AMP (FIGURE 6.2). This regeneration is correlated with restoration of normal calcium responses of Mauthner neurons and escape behavior after stimulation (Bhatt et al., 2004).

One of the consequences of SCI is the spread of damage and disappearance of neurons beyond the immediate area of injury. The assumption has been that the missing neurons died, but there is evidence that they only atrophy and can be revived. One year after a cervical SCI, the cell bodies of severed descending rubrospinal axons, which were thought to have died, reappeared after injection of BDNF in the vicinity of the red nucleus (Kwon et al., 2002). Direct application of a variety of neurotrophic factors following SCI can prevent the atrophy of rubrospinal, corticospinal, and ascending propriospinal neurons, stimulate the expression of axonal growth-associated genes, and prevent corticospinal and ascending sensory axonal dieback (Ramer et al., 2005).

Anti-inflammatory agents that interefere with activation of microglia and macrophages may be useful as neuroprotective agents. For example, systemic administration of the tetracycline derivative minocycline inhibits microglial activation, promotes oligodendrocyte survival, reduces cavity formation, and prevents retraction or dieback of injured axons in rats, which exhibit improved functional recovery (Ramer et al., 2005).

While many different chemical agents have been tested for their neuroprotective ability, it is likely that different types of neurons may respond differently to damage and therefore may have different requirements for neuroprotection. Thus, improvements in neuroprotection may come from the use of combinations of drugs that address different aspects of neurotoxicity.

### b. Agents That Promote Axon Elongation

#### Neutralizers of Myelin Inhibitory Proteins

Chondroitin sulfate PGs produced by astrocytes (Niederost et al., 1999; Ramer et al., 2005) and myelin proteins produced by oligodendrocytes (Filbin, 2000; Ramer et al., 2005) inhibit the regeneration of injured axons after SCI and are constituents of the glial scar (Chapter 5). The myelin proteins Nogo, MAG, and Omgp act through the Nogo-66 receptor (Ng-66R) and its co-receptors, p75/TROY and LINGO-1a, to activate the GTPase RhoA, which leads to remodeling of the actin cytoskeleton and the collapse of axon growth

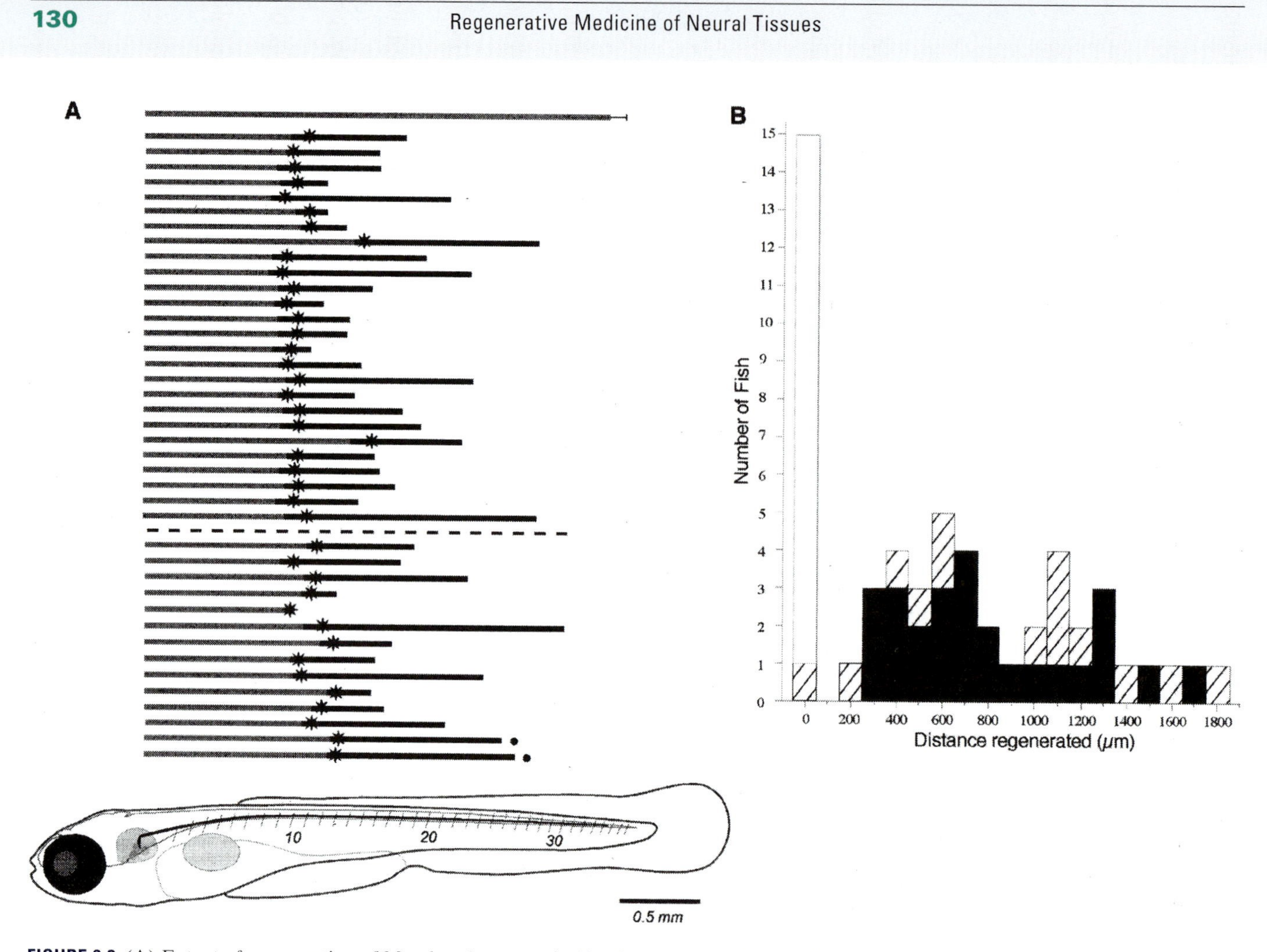


**FIGURE 6.2** **(A)** Extent of regeneration of Mauthner's neuron in 40 zebrafish treated with either dibutryl-cAMP (above the dashed line) or 8-OH-cAMP (below the dashed line). Top gray bar = length of unlesioned neuron. Asterisks indicate lesion site. Remaining gray bars represent the length of the axons after lesioning and immediately preceding treatment, and black bars represent axon regeneration 24–32 hr after treatment. The bars are aligned with a diagram of the fish to show axon growth in relation to body segments. Treatment was done 3–5 days after lesioning, except for the bottom two fish (marked by dots), which were treated 14 days after lesioning. **(B)** Histogram showing distance of axon regeneration for control fish (white bar), db-cAMP-treated (black bars) and 8-OH-cAMP-treated (hatched bars) fish. Reproduced with permission from Bhatt et al., Cyclic AMP-induced repair of zebrafish spinal circuits. Science 305:254–258. Copyright 2004, AAAS.

cones (Ramer et al., 2005). Other inhibitory ECM molecules also appear to signal through the Rho pathway, suggesting that this pathway is a convergence point for a number of inhibitors of axon outgrowth. Interventions have thus been targeted to the inhibiting proteins, the Nogo receptor, and the Rho pathway itself.

Dergham et al. (2002) and Fournier et al. (2003) reported that targeting the Rho pathway with bacterial C3 transferase or with specific inhibitors of Rho kinase after SCI in rats resulted in corticospinal axon regeneration and some functional recovery. Synthetic amino-terminal peptide fragments of the 66-residue domain of Nogo are competitive antagonists of Nogo for the Nogo receptor and promoted axon regeneration in rats when administered to lesions created by severing the dorsal halves of the corticospinal tracts (GrandPre et al., 2002). Growth factors such as BDNF, GDNF, and NGF block the inhibitory effect of MAG (Cai et al., 1999).

Myelin inhibitory proteins also cause growth cone collapse by increasing cytosolic calcium levels in neurons (Bandtlow et al., 1993). The effect is through the calcium-dependent phosphorylation of the EGF receptor (EGFR), which has kinase activity (Koprivica et al., 2005). EGFR is expressed in most parts of the mature nervous system. Inhibitors of EGFR kinase activity (AG1478 and PD 168393) promoted neurite outgrowth from cerebellar granule cells (CGCs) grown on a myelin substrate and from retinal explants treated with CSPG preparations derived from chicken brain. Nogo 66 and Omgp triggered rapid phosphorylation of EGFR in cultured CGCs, and calcium chelators significantly reduced this effect. Furthermore, the introduction of kinase inhibitors into the lesioned optic nerve

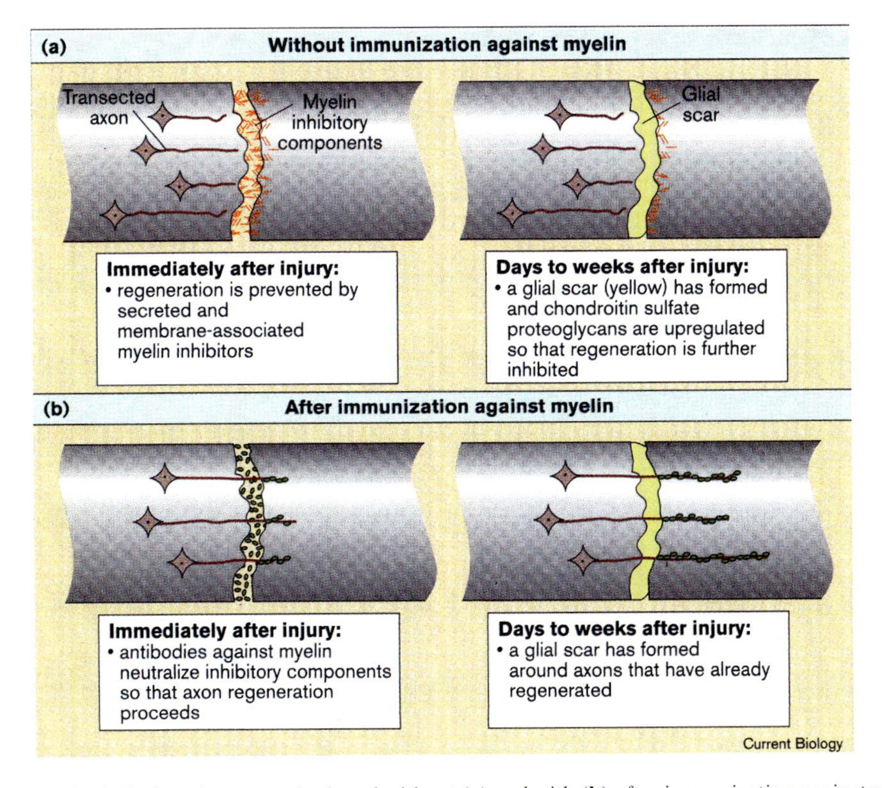


**FIGURE 6.3** Axon regeneration in the lesioned mouse spinal cord without **(a)** and with **(b)** after immunization against myelin. Reproduced with permission from Filbin, Axon regeneration: vaccinating against spinal cord injury. Curr Biol 10:R100–103. Copyright 2000, Elsevier.

of mice resulted in a significant incease in axon outgrowth over controls. Koprivica et al. (2005) found that the EGFR inhibitor erlotinib (Tarceva), which is already FDA-approved for the treatment of cancer, was effective in neutralizing the inhibition of neurite outgrowth by myelin inhibitory proteins. Thus, EGFR kinase inhibitors may be another way to treat SCI.

Treatment of lesioned rat spinal cord with monoclonal antibodies to Nogo resulted in axon regeneration; however, the number of axons that regenerated was low and few or no functional synapses were made (Schnell and Schwab, 1990). Combining anti-Nogo antibody treatment with injection of NT-3 enhanced corticospinal axon regeneration, so that 5%–10% of the corticospinal fibers were regenerated in successful cases (Schnell et al., 1994). The cord was only partially transected in these experiments, however, and the regenerated axons went through the intact part of the cord, not through the lesion itself.

Targeting several inhibitory proteins at once is more effective than targeting them singly. This has been done in two ways. First, mice were vaccinated with preparations of spinal cord enriched in myelin, followed by dorsal hemisection of corticospinal tracts (Huang et al., 1999; Filbin, 2000) (**FIGURE 6.3**). Retrograde and anterograde labeling indicated that numerous axons regenerated across the lesion, many through the glial scar. At least some of these axons appeared to make functional synapses, since 58% of the vaccinated animals showed some functional recovery, defined as the ability to lift their foot and place it on a support surface when the dorsal surface of the hindlimb was touched. Coordinated locomotion, however, was not reported. Whether antibodies to inhibitory proteins are required to get regenerating axons through the initial stages only, or need to be present for the whole course of regeneration, is unknown. Second, dorsal hemisection was performed on the rat spinal cord and Nogo-66, MAG, and Omgp were inhibited by administration of soluble Ng-66R to the lesion site via a catheter connected to an osmotic minipump implanted subcutaneously (Li et al., 2004). This treatment resulted in axonal regeneration that was correlated with improved electrical conduction within the cord and improved locomotion.

## Limiting or Digesting the Glial Scar

Glial scar is a prominent physical and chemical barrier to axon regeneration. Much of the glial scar is composed of ECM molecules laid down by astrocytes. Microarray analysis of rapidly upregulated astrocytic mRNAs has identified a large number of potential inhibitors of axon elongation (Ramer et al., 2005). These include chondroitin sulfate PGs (neurocan,

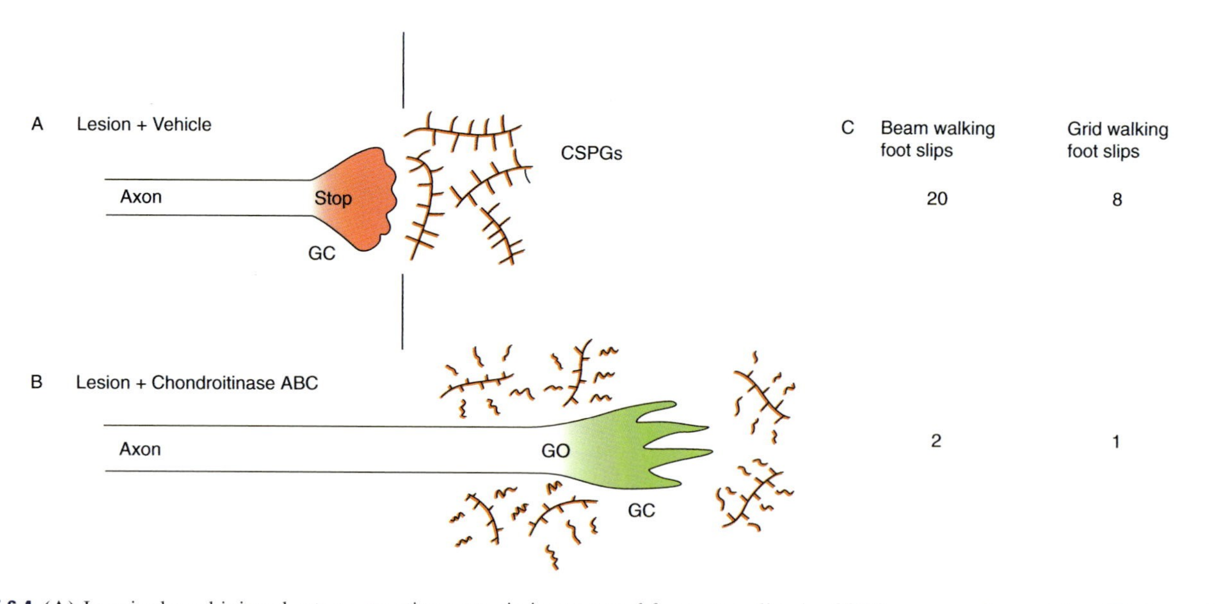


**FIGURE 6.4** **(A)** In spinal cord injured rats a sprouting axon tip is prevented from extending by CSPGs and myelin proteins. **(B)** Degradation of the side chains of the CSPG molecules by chondroitinase ABC (ChABC) clears pathways for axon regeneration through the glial scar. This enzyme and antibodies to myelin proteins could potentially be used together, along with neurotrophic factors, to stimulate regeneration. **(C)** Functional recovery of spinal cord injured rats after treatment with chondroitinase ABC. Rats treated with ChABC exhibited a much lower number of foot slips when walking a beam or a grid, compared to rats treated with vehicle, but they did not recover the ability to sense and remove pieces of adhesive tape with their forepaws.

versican, NG2), heparan sulfate PGs (syndecan-1), and keratan sulfate PGs (lumican), as well as a number of proteins involved in ECM remodeling and stabilization, such as lysyl oxidase, thrombomodulin, osteopontin, and TIMPs.

Wamil et al. (1998) reported that application of CM101, a polysaccharide derived from group B streptococcus, to the lesioned spinal cords of rats prevents angiogenesis and the infiltration of inflammatory immune cells, thus limiting glial scar formation. The number of surviving animals was high and they regained some locomotory function. Glial scar components have also been cleared by enzymatic treatment (**FIGURE 6.4**). Treatment of lesions with chondroitinase ABC in rats paralyzed by crushing the dorsal columns at the level of the fourth vertebra resulted in improvement in walking pattern and the ability to walk a beam without foot slipping (Bradbury et al., 2002). This enzyme removes antiadhesive chondroitin sulfate GAGs from the core protein of CS proteoglycans, increasing the adhesivity of the PGs as a substrate for elongating axons. The significant degree of functional motor recovery suggests that axons regenerating from neurons in the motor cortex were able to exit the lesion and make a significant number of functional synapses on the other side of the lesion. HSPGs and KSPGs have not yet been targets for intervention.

Lysyl oxidase, osteopontin, thrombomodulin and TIMPs stabilize the glial scar. On the other hand, some proteins that inhibit glial scar formation, such as decorin and SC1, an antiadhesive protein, are upregulated by astrocytes (Ramer et al., 2005). The counteractive effects of these two sets of molecules probably reflect the fact that, as in other wounds, both regenerative and scarring processes are initiated after SCI, but the scarring pathway dominates.

## 2. Regeneration Templates

**FIGURE 6.5** summarizes the variety of biological and biomimetic regeneration templates that have been used as bridges to promote axon regeneration through or around glial scar, and into healthy tissue. Such templates must satisfy the requirements of three potentially different critical regions: the entry or "on-ramp"; the surface of the template itself; and the exit or "off-ramp", each of which may have specific design requirements (Geller and Fawcett, 2002). Region-specific adhesive molecules and soluble survival and axon extension-promoting molecules can be incorporated into the templates to help meet design requirements. No biomimetic templates have yet been designed that meet all the requirements, particularly those of exiting the lesion and continuing on to make synapses with targets, but there has been considerable success in inducing axons to enter and cross lesions by implants of biomimetic materials, neural tissue, and glial and other cells.

Goldsmith and de la Torre (1992) implanted a collagen matrix containing laminin or 4-aminopyridine into

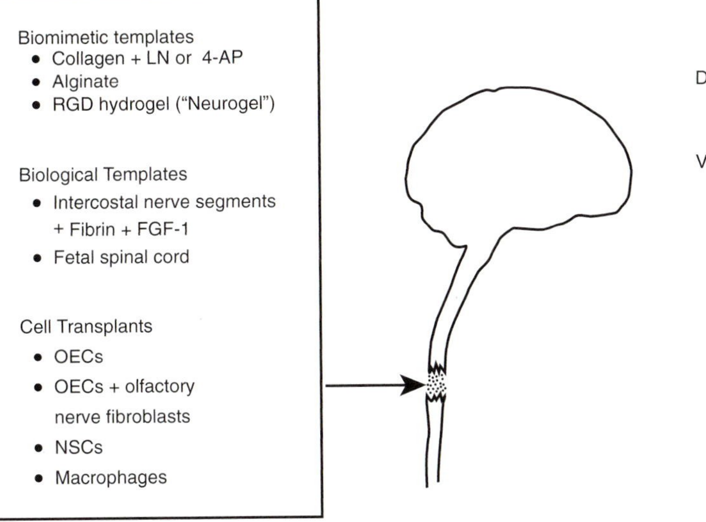


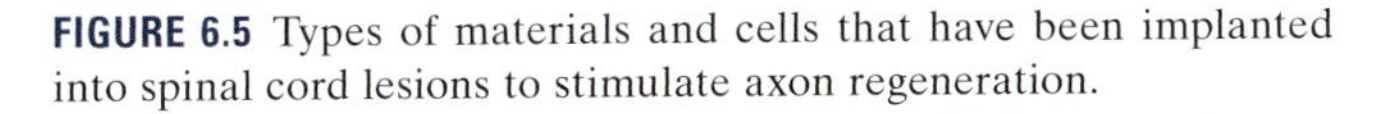

**FIGURE 6.5** Types of materials and cells that have been implanted into spinal cord lesions to stimulate axon regeneration.

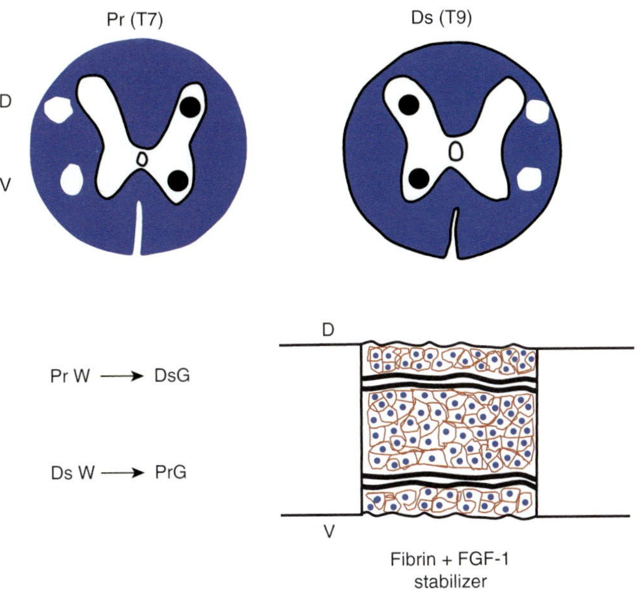


**FIGURE 6.6** White and gray matter were bridged in the rat spinal cord by implanting intercostal nerve segments in a stabilizer of fibrin containing FGF-1. The grafts connected the proximal (T7) white matter to the distal (T9) gray matter and *vice versa*, allowing regenerating axons to bypass glial scar. D = dorsal, V = ventral.

cat spinal cords (six animals in each group) following complete transection (Goldsmith and de la Torre, 1992). Regeneration of supraspinal fibers was reported for a distance of up to 90 mm below the transection site. The synaptogenic marker, synaptophysin, was detected on preganglionic sympathetic neurons in association with dopaminergic and noradrenergic-containing varicosities below the collagen bridge, suggesting that synaptic connections had been made with these sympathetic neurons. One cat in each group showed recovery of coordinated locomotion, the remainder did not.

Other biomimetic templates that have been reported to promote axon regeneration into and across lesions are alginate gel (Suzuki et al., 2000; Kataoka et al., 2004), and "Neurogel," a hydrogel incorporating the RGD adhesive sequence (Woerly et al., 2001). Functional recovery was not measured or recovery was minimal in these studies. Studies are underway to construct injectable spinal cord regeneration templates that gel in the lesion and deliver agents that are neuroprotective, and promote regeneration (Constans, 2004).

Iwashita et al. (1994) excised a 1.5–2-mm section of the lower thoracic cord from 1–2-day-old rats and replaced it with an equal length of fetal spinal cord. The transplanted fetal cord united with the cut ends of the host cord and promoted the regeneration of host axons across the lesion. Normal neural connections were established, as evidenced by anterograde and retrograde labeling with fluorescent dyes, and by the recovery of walking ability and righting reflexes. Iwashita et al. (1994) observed that the lumen of the transplanted fetal cord became closed, forming a mass of cells reminiscent of the mesenchymal blastema formed by the epithelial to mesenchymal transformation during regeneration of the urodele spinal cord (Chernoff et al., 2003). It would be interesting to know whether anything resembling an epithelial-mesenchymal tranformation takes place in such transplants, and how this might be related to the success of axonal outgrowth. Animals implanted with a section of sciatic nerve regenerated a few short neural connections across the graft, but these were not functional, consistent with the findings of others (Richardson et al., 1980). Nor did fetal spinal cord implants induce the regeneration of adult spinal cord (Iwashita et al., 1994), suggesting that the 1–2-day neonatal cord still had a high degree of plasticity.

Adult rat spinal cord axons were induced to regenerate across lesions by implanting segments of intercostal nerves embedded in a fibrin matrix impregnated with FGF-1 (Cheng et al., 1996) (**FIGURE 6.6**). The endoneurial tubes of the nerve segments bridged white and gray matter, allowing host axons to regenerate through them, protected from inhibitory proteins and bypassing glial scar. Anterograde and retrograde labeling showed that the axons regenerated from the white matter through the endoneurial tubes to the gray matter. There was some functional recovery in that the rats were able to support their weight on the paralyzed hindlimbs and displayed movement in all leg joints, but they did not recover coordinated locomotion. Inclusion of FGF-1 in the fibrin matrix was essential for regeneration. No improvement was noted in the absence of FGF-1, indicating that this growth factor is essential for axon extension.

A small number of human SCI patients have been treated with this experimental protocol. Cheng et al. (2004) reported that a patient who had suffered a thoracic SCI four years previously experienced motor and sensory improvements after receiving a graft of a sural nerve along with FGF-1 treatment. However, similar treatment of eight patients in Brazil (Tarcisio Barros, University of Sao Paulo) who had suffered functionally complete SCIs due to gunshot wounds failed to improve sensory or motor function over a five-year recovery period (Steeves et al., 2004).

## 3. Cell Transplant Therapies

A variety of cell types have been transplanted into SCIs of experimental animals and human patients to encourage spinal cord regeneration (**TABLE 6.4** and **FIGURE 6.5**). Glial cells derived from glial precursors provide neuron survival and axon elongation factors that enable regeneration and remyelination. NSCs may differentiate into new neurons, but their main effect is differentiation into glial cells that promote axon elongation, and host neuroprotection and reduction of scarring by paracrine factors. The idea that neural cell transplants promote regeneneration by stimulatory paracrine factors rather than differentiating into new neurons is strengthened by the fact that a number of nonneuronal cell types improve regenerative capacity as well.

### a. *Glial Cells*

Oligodendrocyte precursors derived from human ESCs enhanced remyelination and improved motor function when injected into adult rat spinal cord lesions 7 days after injury, but not 10 months after injury (Kierstead et al., 2005). Glial-restricted NSCs from E13.5 rat embryos injected into lesioned adult rat spinal cord were reported to differentiate into oligodendrocytes and astrocytes and to reduce scarring and expression of inhibitory proteoglycans by host cells. The transplanted cells also altered the morphology of host corticospinal tract axons toward that of growth cones, although they did not induce long-distance regeneration of the axons (Hill et al., 2004).

Suspensions of Schwann cells and olfactory ensheathing cells have been used to induce axon regeneration after SCI. Rat and human Schwann cells have been isolated and expanded *in vitro* and implanted into SCI rats (Guest et al., 1997; Bunge, 2002). A common model has been to fill a PVC tube with a Schwann cell suspension and implant this construct into the spinal cord lesions of adult rats (reviewed by Bunge, 2002). These implants alone have supported the regeneration of axons across the lesion with modest recovery of function that is enhanced somewhat when combined with other treatment, such as MP (Chen et al., 1996). The major problem noted with these grafts is that supraspinal axons typically fail to re-enter the cord after crossing the bridge, perhaps because CSPGs are present at a higher concentration at the caudal graft/host interface (Plant et al., 2001). Significant axon regeneration was reported in rat SCI after transplanting collagen-imbedded Schwann cells genetically modified to hypersecrete NGF (Weidner et al., 1999).

Olfactory ensheathing cells are receiving increasing attention as therapeutic agents for spinal cord injury because of their ability to promote long distance growth of regenerating axons and to remyelinate CNS axons (Boyd et al., 2003; Franklin, 2003). Culture-expanded OECs injected on either side of a Schwann cell-containing Matrigel bridge induced axon extension

**TABLE 6.4** Cell Transplant Therapies That Promote Axon Elongation Across SCI Lesions (Rodent models used except where human patients noted)

| Cells Transplanted | Effect |
|---|---|
| Schwann cell suspension in PVC tube | Entry into and traverse lesion, no exit |
| Schwann cells in Matrigel tube, OECs injected on either side of tube | Entry, traverse lesion, and exit; partial functional recovery |
| OECs injected into SCI of human patients | Improved control of sweating and reduction of spasticity; improvement of sensory and motor scores in some patients |
| Macrophages injected into SCI of human patients | Modest functional improvement |
| Allogeneic NSCs | Regained weight-bearing and partly coordinated stepping movements; differentiation into new interneurons and glia |
| Splenic dendritic cells *in vivo* | Axon sprouting and proliferation; proliferation of ***host*** NSCs into new neurons; partial functional recovery |
| Splenic dendritic cells *in vitro* | Enhanced survival and proliferation of NSCs |
| Human umbilical cord blood cells | Partial recovery of locomotion |
| Fibroblasts transgenic for NT-3 | Partial recovery of locomotion |

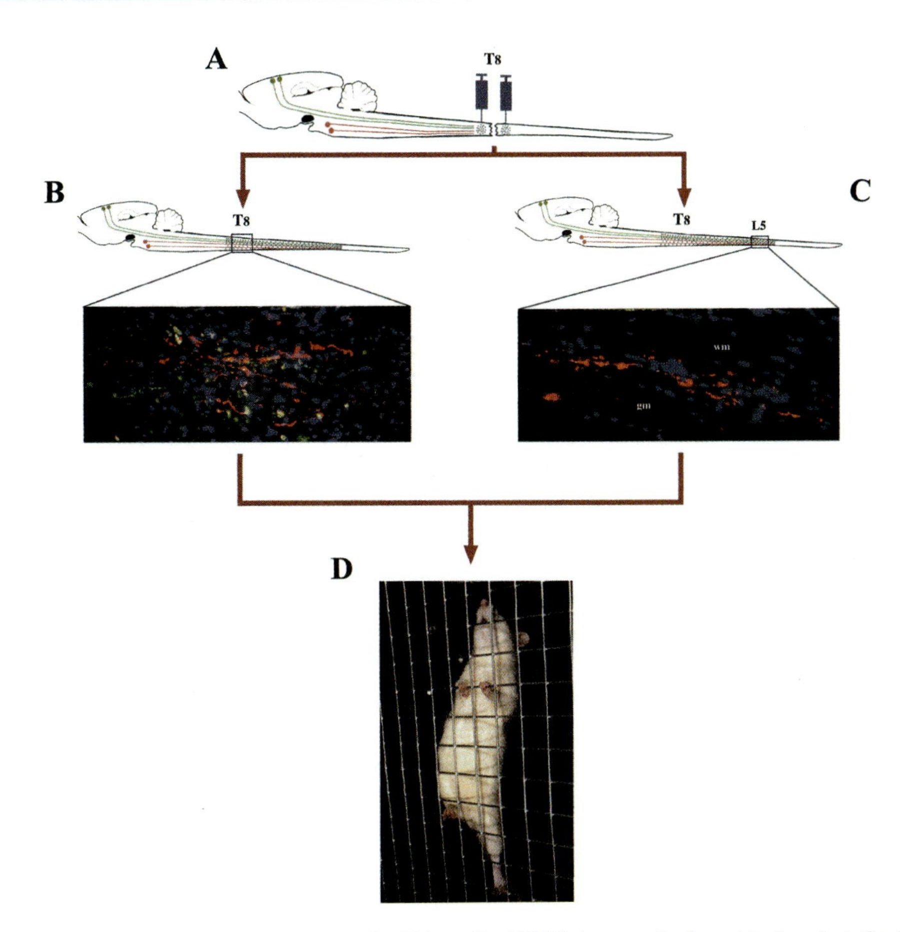


**FIGURE 6.7** Promotion of axon regeneration by olfactory ensheathing glia (OEG) in rat spinal cord lesioned at the T8 level. **(A)** OEG were injected into both stumps of the completely transected cord. Green = corticospinal axons, red = brainstem axons. **(B)** OEG promoted regeneration of corticospinal (green label) and brainstem axons (red label) through the glial scar. **(C)** Section showing corticospinal axons (red) that regenerated to the L5 level (3 cm, the longest distance analyzed). **(D)** Photograph of an OEG-transplanted paraplegic rat 7 months after spinal cord transection, showing that the rat can properly place its paws on the rungs of the climbing grid and support its body weight with its hindlimbs. Reproduced with permission from Santos-Benito and Ramon-Cueto, Olfactory ensheathing glia transplantation: A therapy to promote repair in the mammalian central nervous system. Anat Rec (Part B: New Anat) 271B:77–85. Copyright 2003, Wiley-Liss.

onto the bridge and the Schwann cells promoted regeneration across it (Ramon-Cueto et al., 1998). Glial scar formation was not inhibited, but the regenerating axons were able to penetrate the scar. Many investigators have reported that OECs alone, injected directly into or rostral and caudal to SCI lesions in rodents promote axon regeneration across the lesions, accompanied by partial functional recovery (Li et al., 1998; Ramon-Cueto et al., 2000; Lu et al., 2001; Nash et al., 2002; Santos-Benito and Ramon-Cueto, 2003; Andrews and Stelzner, 2004) (**FIGURE 6.7**).

One problem in translating animal studies to the clinic is providing enough cells for transplant. OECs can be expanded *in vitro* as primary cultures, but even when initiated from an OEC-enriched preparation primary cultures contain contaminating cells; furthermore, they survive only a few weeks. To circumvent these problems, DeLucia et al. (2003) have tested a homogeneous OEC clonal cell line immortalized with by the SV40 large T cell antigen for its ability to survive and support functional recovery in rats after SCI. These cells were found to produce nearly all of the same neurotrophic factors *in vitro* as primary cell cultures. Immortalized OECs transplanted into thoracic hemisection lesions supported functional recovery of tactile sense and proprioceptive functions required to walk a grid without missing the wires.

How do OECs facilitate functional recovery? A number of studies suggest that OECs are neuroprotective through their secretion of neurotrophic factors, promote axon extension through their expression of adhesion molecules, prevent cavitation, enhance vascularization, and promote branching of neighboring intact axons (Verdu et al., 2003; Ramer et al., 2004; Chua et al., 2004). The effectiveness of OEC transplants is further increased by the inclusion of olfactory nerve fibroblasts, suggesting that interactions with fibroblasts are important for OEC function (Barnett and Chang, 2004).

In China and Portugal, human OECs have been harvested, expanded *in vitro* and transplanted into SCI patients (Steeves et al., 2004). Honyun Huang (Beijing Chaoyang Hospital) has reported the results of about 300 fetal OEC transplants, in which ~$1 \times 10^6$ cells were injected from 6 months to 31 years after injury (Huang, 2003). Patients experienced improved control of sweating and reduction of spasticity within 2–3 days after injection, and some patients showed improvement in sensory and motor test scores. In Portugal, Carlos Lima (Lisbon Egaz Moniz Hospital) has transplanted autologous whole olfactory epithelium into seven patients who had sustained SCI at least 6 months prior to operation. In some cases, glial scar tissue was removed before transplant. The patients were reported to show some improvement in autonomic and bladder functions and reduction in spasticity, but no improvement in sensory or motor function. A third set of trials by Alan Mackay-Sim (Griffith University, Brisbane), using adult OECs expanded in culture (Bianco et al., 2004), is scheduled to be undertaken in Australia.

### b. NSCs

Teng et al. (2002) reported significant axon regeneration from host neurons in lesioned adult rat spinal cords after implanting NSCs in a PGA-based matrix. Partial restoration of function in contused rat spinal cords was reported after injecting neural/glial precursors differentiated from mouse ESCs *in vitro* into the lesion nine days after injury (McDonald et al., 1999). Staining with antibodies specific for mouse proteins and for glial and neuronal markers showed that many of the implanted cells survived, migrated throughout the injured area, and differentiated into new interneurons, oligodendrocytes, and astrocytes. These findings were correlated with regaining the ability to bear weight on their hind legs and by restoration of partly coordinated stepping movements, suggesting that signal transmission was partially restored between brain and hind legs.

Transduction of NSCs with the gene for neurogenin-2 (Ngn2), a transcription factor involved in the determination of neural lineages during ontogenesis, promoted neuron differentiation *in vitro*, but not after transplantation (Hofstetter et al., 2005; see also Klein and Svendsen, 2005). However, transplanted Ngn2 NSCs differentiated primarily into oligodendrocytes as opposed to nontransduced NSCs, which differentiated mainly into astrocytes, and significantly improved hindlimb motor and sensation recovery over that observed with nontransduced NSCs. This improvement was associated with significantly greater areas of white matter, suggesting that increased remyelination played a role in functional recovery.

Neural stem cells may be nonimmunogenic in an allogeneic setting. Allogeneic NSCs survive when transplanted into the brain or spinal cord because they are in an immune-privileged site. NSCs also survive when grafted into nonprivileged sites in mice such as beneath the kidney capsule (Hori et al., 2003). This lack of immunogenicity, if it extends across species, might make it possible to use xenogeneic NSCs for transplant to humans.

### c. Nonneural Cells

Macrophages have been reported to promote axon regeneration and functional recovery after implantation into SCI lesions of adult rats (Rapalino et al., 1998). The effect of the macrophages was explained by their phagocytosis of inhibitory myelin fragments (Lazarov-Spiegler et al., 1998). In a small Phase I trial (Proneuron Biotechnologies, Israel), five SCI patients were transplanted within two weeks of injury with autologous macrophages activated by preincubation with skin tissue (Steeves et al., 2004). The patients showed modest functional improvement.

Splenic dendritic cells transplanted into mouse spinal cords lesioned by hemisection at the T8 level were reported to activate the proliferation and differentiation of host NSCs into new neurons and induce axon sprouting through a trophic effect, accompanied by partial recovery from hindlimb paralysis (Mikami et al., 2004) (**FIGURE 6.8**). Co-culture of spinal cord NSCs with dendritic cells significantly enhanced the survival and proliferation of the NSCs. By contrast, medium conditioned by dendritic cells had only one-tenth the enhancing activity seen in co-cultures, indicating that the major effect of the dendritic cells on neurons is mediated by cell contact. Dendritic cells secrete NT-3 *in vitro* and *in vivo*, so the minor enhancing activity of conditioned medium may be exerted through this trophic molecule. It is also possible that the dendritic cells had an indirect effect, by activating resident microglia and macrophages to clear myelin debris, or secreting NT-3.

Fibroblasts transgenic for NT-3 promoted partial functional recovery when transplanted into lesions of the rat spinal cord created by dorsal hemisection (Grill et al., 1997). Rats injected intravenously with human fluroscein isothiocyanate-labeled umbilical cord blood (which contains HSCs) five days after lesioning the spinal cord were reported to achieve partial recovery of locomotory behavior (Saporta et al., 2003). Histological examination indicated that the labeled cells, of which fewer than a thousand survived, were neither neurons nor glia, suggesting that cord blood cells provided

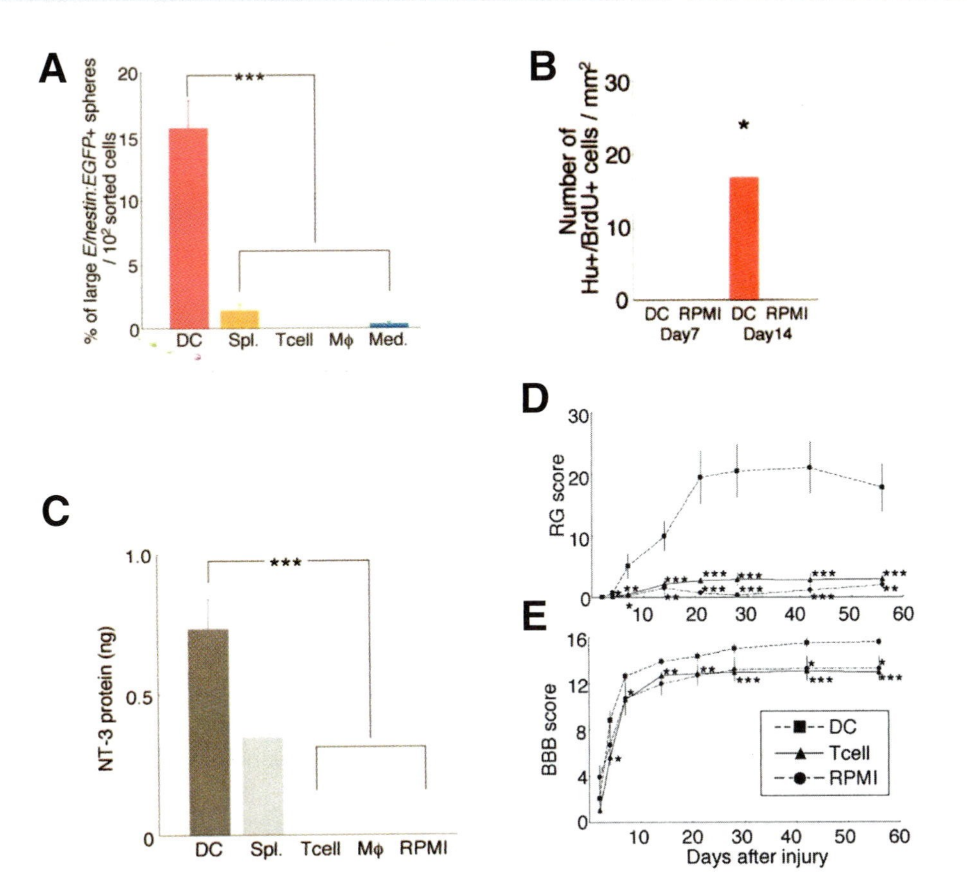


**FIGURE 6.8** Effects of dendritic cells on mouse NSCs. **(A)** Enhancement of neurosphere formation *in vitro* by dendritic cells (DC) is significantly greater than by splenocytes (Spl), T-cells, macrophages (Mϕ) or neurosphere culture medium (Med). Asterisks indicate statistical significance. **(B)** Number of BrdU-labeled NSCs in the vicinity of the lesion that stain for Hu, a marker of early neuron differentiation, after treatment with culture medium (RPMI, control) or dendritic cells. No such cells appeared in control lesion sites, but a significant number had appeared by 14 days in lesions treated with dendritic cells. **(C)** In contrast to other cell types, dendritic cells secrete large amounts of neurotrophin-3 (NT-3) into the culture medium. Asterisks indicate statistical significance. **(D, E)** RG scores (frequency of vertical movements during a 10 min period) and Basso-Beattie-Bresnahan (BBB) locomotor ratings two months after spinal cord lesioning. Values are significantly higher for mice treated with dendritic cells. Reproduced with permission from Mikami et al., Implantation of dendritic cells in injured adult spinal cord results in activation of endogenous neural stem/progenitor cells leading to de novo neurogenesis and functional recovery. J Neurosci Res 76:453–465. Copyright 2004, Wiley-Liss.

paracrine factors that rescued host neurons and glia. Infusions of bone marrow cells into a group of 32 patients 2–12 years after complete SCI (Tarcisio Barros, University of Sao Paulo) were reported to modestly improve lower extremity function in 15 patients (Steeves et al., 2004)

## 4. Bioartificial Spinal Cord

Another approach to spinal cord regeneration is to build bioartificial cord from biomimetic materials seeded with NSCs that would integrate into the host neural tissue. The scaffold material would contain factors that direct the differentiation of the NSCs primarily into neurons (Constans, 2004). Vacanti et al. (2001) constructed bioartificial spinal cord tissue by seeding culture expanded, GFP-labeled rat spinal cord NSCs into PGA mesh. The construct was implanted into surgically created gaps of 3–4 mm in the thoracic cord of host rats. The implanted rats were reported to recover coordinated locomotion after four weeks. Histological examination of implanted cords after four months showed GFP-labeled cells with neuronal morphology and markers, while anterograde labeling with DiI revealed that neurons had sent processes from the implant into the lumbar region of the cord. In another study, embryonic rat neural stem and progenitor cells from cortical and subcortical neuroepithelium were embedded in rat tail tendon collagen and the construct was cultured in a serum-free medium containing FGF-2 (Ma et al., 2004). The cells proliferated and differentiated into neurons that formed a three-dimensional neuronal network and developed functional synapses. The next step would be to see whether this type of construct can replace a segment of injured cord tissue and make functional connections with axons regenerating from host tissue.

## 5. Rehabilitation Programs and Synaptic Plasticity

Newborn rats regain the ability to walk without intervention after complete spinal cord transection. This recovery is due to stereotypical locomotor patterns produced by central pattern generators within the cord itself (Burke, 2001). Thus, the rats regenerate those nerve tracts that control posture, balance, and coordination, but do not regenerate the corticospinal tract, which is the primary conduit for signals controlling voluntary movements (Wakabayashi et al., 2001). This means that there is a tremendous amount of synaptic plasticity in the spinal tracts that control nonvoluntary motor functions, allowing them to compensate for the inability of the corticospinal tract to regenerate.

Synaptic plasticity involves creating and eliminating synapses among the neurons still present after an injury, as well as strengthening or weakening of existing synapses. There is substantial evidence that this plasticity takes place where dendritic spines make excitatory synapses with axons and involves changes in spine morphology based on the dynamic reorganization of actin filaments in the spines and influenced by trophic factors, hormones, and astrocyte activity (Matus, 2000). Electron microscopic studies of dendritic segments of pyramidal neurons in the mouse barrel cortex showed that, while dendritic structure is stable, dendritic spines appear and disappear, depending on changing sensory experience (Trachtenberg et al., 2002). This remodeling of synaptic organization is similar to morphallaxis and can be called "synaptallaxis." Young granule cells of the rat hippocampus, recently born from NSCs, have mechanisms to facilitate synaptic plasticity that more mature granule cells do not have, which may contribute to the ability to learn new tasks (Schmidt-Heiber et al., 2004).

Synaptic plasticity involved in locomotion can be induced by muscle use, through either exercise regimens or functional electrical stimulation. Recovery from spinal cord injury is possible in younger animals by using these means, particularly in cases of incomplete injury (Peckham and Creasy, 1992), but becomes poorer with age. For example, Muir (1999) showed that chicks learned to walk on their own after hemisectional thoracic injury, but could not relearn swimming movements. By stimulating their feet so that their legs moved in a swimming pattern, the chicks were able to make leg movements that approximated the swimming motions made prior to hemisection.

In humans, motor behaviors can be learned even late in life (Wolpaw and Tennissen, 2001). This implies retention by the cord of a significant amount of plasticity in whatever spared circuits remain that can be tapped to restore motor circuits after SCI. Treadmill exercises involving repetitive coordinated motion have been of great help in retraining the spinal cord for walking motions in persons with incomplete SCI, but are of minimal use in complete SCI patients (Wernig et al., 1995, 1998; Wirz et al., 2001; Ramer et al., 2005). Of 44 incomplete SCI wheelchair-bound patients, treadmill training enabled two-thirds to walk independently over short distances and about half of these could ascend stairs. Electrical stimulation of muscles in patterns that mimic coordinated movements also helps regain some motor function and improve respiratory, bladder, bowel, and sexual function (Peckham and Creasy, 1992).

## 6. Combinatorial Approaches May Bring Better Success

None of the interventions used for spinal cord injury in experimental animals or humans have by themselves resulted in complete recovery from paralysis. Protocols for experiments are highly variable, and many studies have not been comparable or reproducible. Results are inconsistent and tests for functional recovery often are not part of studies or the tests used vary widely from one study to another. Furthermore, there are differences among mice, rats, and humans, as well as differences among strains of animals, in their response to neural trauma that affect results (Steward et al., 1999). Standardization of protocols is essential if we are to be able to reproduce and compare results.

Nevertheless, a major accomplishment has been the induction of axon regeneration across lesions. What is still missing is the growth of substantial numbers of axons into endoneurial sheaths on either side of the lesion, and the establishment of new synapses that results in the restoration of a functional circuitry. What inhibits further growth is not clear. Myelin fragments in the empty endoneurial tubes may not be phagocytosed, inhibitory myelin proteins may be present, and appropriate adhesion molecules for axon extension may be lacking. On the other hand, the axons of completely transected adult mouse spinal cords regenerated perfectly, with full functional recovery, if the transection was done in a way that avoided dural damage and thus scarring in the vicinity of the lesion (Seitz et al., 2001), suggesting that the endoneurial tubes above and below a lesion are capable of supporting axon regeneration. This observation is worthy of further study.

Combinatorial approaches may result in better success. Ramer et al. (2005) have summarized the combination therapies that have been tried so far. Cell transplants and neurotrophic or growth factors have been combined many times, but have not produced significant recovery of function. Combinations of biological agents plus rehabilitative treatments are rare.

One promising combination is augmentation of the intracellular concentration of cAMP by rolipram administration, combined with Schwann cell transplant (Pearse et al., 2004). This combination treatment in rats prevents the decrease in cAMP caused by SCI and promotes significant sparing and myelination of supraspinal and proprioceptive axons (**TABLE 6.5**, **FIGURE 6.9**). Furthermore, when rolipram is combined with an injection of dibutryl cAMP near the Schwann cell graft, cAMP levels are increased above control levels, further enhancing axonal sparing and myelination, promoting regeneration of serotonergic fibers

**TABLE 6.5** Effects of rolipram (RP) and RP in combination with dibutryl cAMP (db-cAMP) and Schwann cells (SC) on cAMP levels (pg/ml), 14 days after moderate contusion injury to different parts of the CNS. The parietal cortex, which does not project directly to the spinal cord, was the only brain area that had no significant decrease in cAMP injured controls. RP alone elevates the cAMP level above that of the injured controls, but not to the level of the uninjured controls. RP in combination with db-cAMP and Schwann cells elevates cAMP levels above uninjured control levels. Data from Pearse et al. (2004)

| CNS Area | Uninjured Control | Injured Control | RP | RP/db-cAMP/SC |
|---|---|---|---|---|
| Spinal cord | 100 | 50 | 80* | 140* |
| Brainstem | 90 | 40 | 60 | 110* |
| Motor cortex | 100 | 40 | 60 | 90* |
| Parietal cortex | 160 | 150 | 150 | 160 |

*Statistical significance relative to injured control

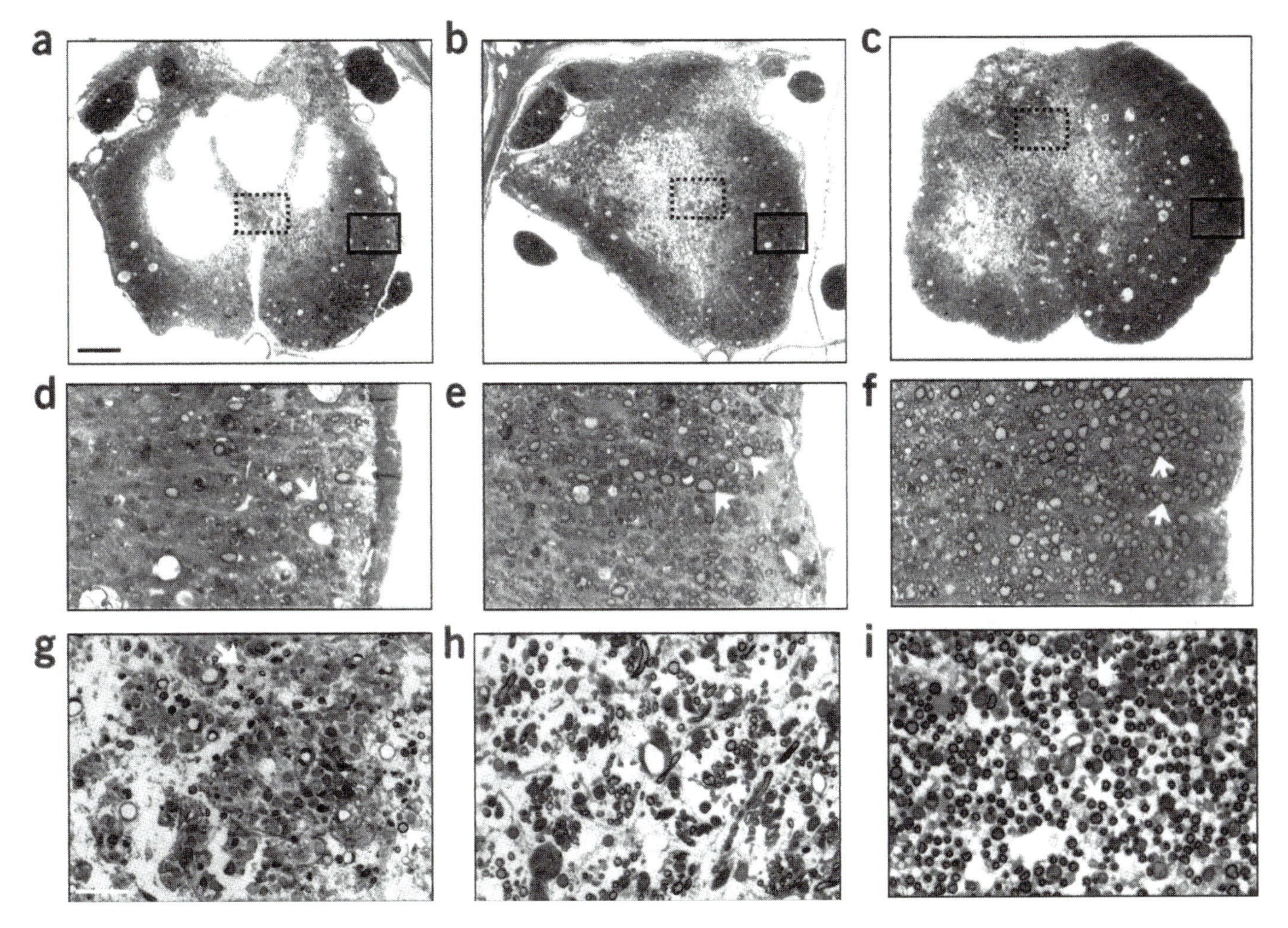


**FIGURE 6.9** Transverse sections showing that elevation of cAMP promotes sparing of central myelinated axons and myelination of axons by Schwann cell grafts. **(a)** Injured control cord with cavitation. **(b)** Cord with Schwann cell graft. **(c)** Lesion treated with Schwann cell graft in combination with rolipram and db-cAMP. **(d, e, f)** Higher magnification of the closed rectangle areas in **a–c** shows the differences in density of spared central myelinated axons (arrows) between the treatment groups. **(g, h, i)** Higher magnification of the dotted rectangle areas illustrates the differences in density of peripheral axons that were remyelinated (black) by grafted Schwann cells (arrows). Reproduced with permission from Pearse et al., cAMP and Schwann cells promote axonal growth and functional recovery after spinal cord injury. Nature Med 10:610–616. Copyright 2004, Nature Publishing Group.

into and beyond the graft, and significantly improving locomotion.

Useful approaches may come from other directions as well. Comparative studies on spinal cord regeneration in regeneration-competent versus regeneration-deficient situations may reveal the differences between the two. For example, microarray and proteomic analysis of lesioned axolotl spinal cord (regeneration-competent) versus adult frog cord (regeneration-deficient) would be helpful in identifying molecules that are regeneration-permissive, as well as molecules inhibitory to regeneration. Cho et al. (2005) have correlated changes in regenerative capacity of the mouse optic nerve at different developmental stages with Bcl-2 expression. In mice, the optic nerve loses its ability to regenerate as fetal development advances. Regenerative incompetence is associated with the maturation of astrocytes and the loss of Bcl-2 expression by ganglion cells. Prior to postnatal day 4, when astrocytes are immature, optic nerve regeneration to tectal targets can be induced by the overexpression of Bcl-2. After day 4, however, Bcl-2 overexpression will induce regeneration only if reactive gliosis by maturing astrocytes is downregulated.

Another interesting approach would be to apply near-infrared (NIR) light to SCI lesions. NIR from an LED was reported to enhance the survival and regeneration of the rat optic nerve and retina, after damage by formic acid produced by methanol intoxication, and to enhance healing of the laser-burned retina in rabbits and monkeys (Whelan et al., 2004). Microarray studies revealed that several groups of genes were upregulated in methanol-treated versus untreated rats, including genes of the cytochrome family, consistent with other evidence indicating that cytochrome oxidase is a target for NIR in skin wound repair studies (see Chapter 4).

## THERAPIES FOR NEURODEGENERATIVE DISEASES

Demyelinating disorders, Parkinson's disease, Huntington's disease, Alzheimer's disease, and amyotrophic lateral sclerosis (ALS) are neurodegenerative diseases that strike humans at different ages and show variable courses of progression. Huntington's disease is clearly due to a single gene mutation, as are a small percentage of ALS and Parkinson's cases. The vast majority of ALS cases, demyelinating disorders, and Parkinson's disease, however, are multifactorial in origin. Currently, about 10% of the population suffers from neurodegenerative diseases. The over-60 population will double by 2050, greatly increasing this percentage and placing a tremendous strain on healthcare systems.

Although each neurodegenerative disorder differs in the kinds of neurons affected, etiology, symptoms, and progression, a common feature is the presence of intraneuronal neurofibrillary tangles (NFTs) and intra- or interneuronal plaques in the affected areas of the CNS. These pathological inclusions are due to protein misfolding caused by genetic mutation or undefined causes. Current thinking is that the misfolded proteins themselves are neurotoxic and that sequestering them in plaques and NFTs is a neuroprotective mechanism (Mattson, 2004; Arrasate et al., 2004). However, some results with agents that prevent aggregation of the misfolded proteins in Alzheimer's disease suggest that the plaques and NFTs themselves may also be neurotoxic. NFTs in Alzheimer's disease are composed of the hyperphosphorylated microtubule-associated tau protein. Interestingly, transgenic mice expressing doxycycline-suppressible tau accumulated NFTs, and exhibited neuronal loss, and cognitive defects that were reversed by suppressing tau expression (SantaCruz et al., 2005). However, NFTs continued to accumulate, suggesting that it is not the NFTs themselves that are neurotoxic in Alzheimer's disease.

Animal experiments indicate that fetal neural cells and adult NSCs have a remarkable ability to migrate throughout, and integrate into, the CNS of fetal and adult hosts (Tate et al., 2001). Mouse neural precursors derived from ESCs integrate into the hippocampal circuitry and differentiate as neurons when placed on slice cultures of the hippocampal region of young rats (Benninger et al., 2003). Human fetal brain cells (53–74 days postconception), marked with a retroviral *lacZ* construct, were transplanted into the brain ventricles of embryonic (E17–18) or neonatal rats and tracked by several markers: β-galactosidase expression; a human-specific probe to the alu repeat element; and a human-specific antibody to glutathione-S-transferase (Flax et al., 1998; Brustle et al., 1998). The transplanted cells migrated into every region of the host brain, where they differentiated into site-specific neurons and glia in the organization characteristic of each region. Human NSCs behaved likewise when injected into adult rat brains, though their response was less marked than in host neonatal brain (Fricker et al., 1999). Human NSCs differentiated *in vitro* from ESCs by FGF-2 or EGF plus FGF-2, localized to subventricular areas after grafting into the cerebral ventricles of neonatal mice (Zhang et al., 2001; Reubinoff et al., 2001). Freeze damage to the adult mouse brain induces the production of stem cell factor in neurons within the injury zone and the migration of NSCs by their activation of the SCF receptor, c-kit. Furthermore, damage to brain neurons by ischemia and neurodegenerative disease results in increased NSC proliferation and differentiation into new neurons (Raber et al., 2004; Curtis et al., 2003; Jin et al., 2004), indicating that the brain initiates regeneration by NSCs in response to injury. These observations, among many others, have justified the hope that pharmaceutical agents and/or cell therapy will be potentially useful in treating neurodegenerative diseases (Zhang et al., 2003; Rice and Scolding, 2004; Goldman, 2005).

## 1. Demyelinating Disorders

Brustle et al. (1999) injected glial precursor cells differentiated *in vitro* from mouse ESCs into the spinal cord and brain of rats that have the equivalent of a human demyelinating disorder, Pelizaeus–Merzbacher disease. This disease is caused by a mutation in the X-linked gene for myelin proteolipid protein (PLP). The mouse ESCs were placed in defined medium with FGF-2 and PDGF to produce proliferating glial precursors. These precursor cells were then injected into the spinal cord and brain of seven-day-old myelin-deficient rats. Two weeks after injection, electron microscopy and staining of the tissues with a probe to mouse satellite DNA and antibody to PLP showed that the injected cells had differentiated into oligodendrocytes that formed myelin sheaths around the axons (**FIGURE 6.10**).

Nistor et al. (2005) developed a reliable protocol for the directed differentiation of oligodendrocyte precursors from human ESC cultures over 42 days, based on signaling molecules known to regulate the survival, proliferation, migration, and differentiation of cells at various steps of the oligodendrocyte lineage (Grinspan, 2002). The principal components restricting differentiation to the oligodendroglial lineage were insulin, triiodothyroidin hormone, FGF, and EGF. Commitment of the hESCs to the oligo lineage was confirmed by their expression of a panel of molecular markers including Olig1, SC10, A2B5, NG2, and PDGFRα. When these cells were transplanted into the spinal cord of *shiverer* mice, which suffer from dysmyelination due to homozygosity for a mutation in the myelin basic protein gene (Chernoff, 1981), they integrated into the cord tissue, differentiated into oligodendrocytes and produced compact myelin. Likewise, clonal NSCs

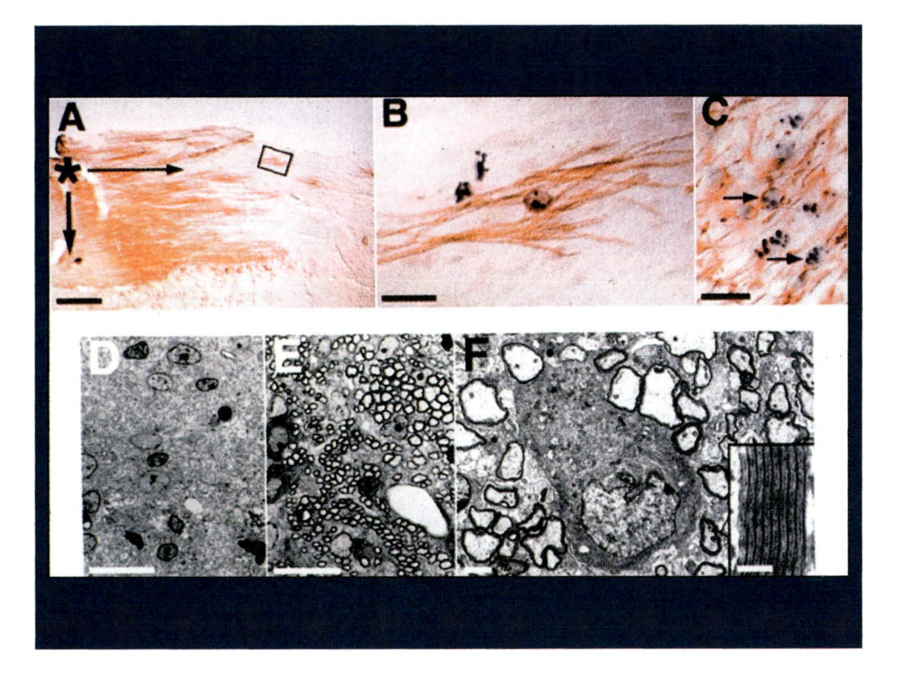


**FIGURE 6.10** Differentiation of ESC-derived mouse glial precursors into oligodendrocytes after injecting them into the spinal cord (dorsal columns) of myelin-deficient rats. **(A)** Saggital section stained with antibody to myelin proteolipid protein (PLP). The asterisk marks the injection site and the arrows indicate the directions of migration of the injected cells. Abundant PLP is evident. **(B)** *In situ* hybridization of the boxed area in **(A)** with a probe to mouse satellite DNA shows that the myelin is generated by donor cells. **(C)** Double staining with a probe to mouse satellite DNA and an antibody to GFAP shows that the ESC derivatives also differentiated into astrocytes. Toluidine blue-stained semithin cross sections show the myelin deficiency in an untreated rat **(D)** and **(E)** new myelin derived from ESC glial derivatives two weeks after injection. **(F)** Electron micrograph of an oligodendrocyte contacing numerous myelinated axons. Inset, high magnification shows the layered structure of the myelin. Reproduced with permission from Brustle et al., Embryonic stem cell-derived glial precursors. A source of myelinating transplants. Science 285-754-756. Copyright 1999, AAAS.

transplanted into the cerebral ventricles of *shiverer* mice at birth migrated throughout the brain and remyelinated up to 52% of axons, reducing the tremor of a number of recipient animals (Yandava et al., 1999).

Human olfactory ensheathing cells were used to remyelinate rat dorsal spinal cord axons that were focally demyelinated by X-irradiation and injection of ethidium bromide (Kato et al., 2000). The OECs, prepared from adult olfactory nerves removed during surgery for nasal melanoma, were injected into the lesioned areas. Examination of tissue sections three weeks later by electron microscopy and *in situ* hybridization with the COT-1 human DNA probe revealed remyelination of the axons by the human cells.

Multiple sclerosis (MS) is an autoimmune CNS disorder in which myelin is damaged and astrocytes proliferate to form scar, resulting in the blockade of electrical impulses along nerve axons, loss of sensation and coordination, and in severe cases, paralysis and blindness (Steinman, 1996). It affects ~2.5M people worldwide and is one of the most common neurological diseases of young adults (Zamvil and Steinman, 2003). Typically, the disease follows a relapsing and remitting pattern in which acute demyelinating episodes are followed by the generation of new oligodendrocytes and remyelination, but with increasing numbers of lesions that fail to remyelinate.

Arresting the progression of the disease has been a major focus of research on MS. Currently, the interferon-based drugs Avonex (INFβ-Ia), Betaseron (INFβ-Ib), and Rebif (INFβ-Ia), and the synthetic peptide co-polymer glatiramer acetate (Copaxone) are FDA-approved to treat relapsing forms of MS. Experimental work on a mouse model of MS, autoimmune encephalomyelitis (EAE), and clinical trials in humans suggest that attacks of MS can be aborted or diminished by blocking the $\alpha_4$ integrin subunit of $\alpha_4\beta_1$ and $\alpha_4$ $\beta_7$ integrins on the surface of the attacking immune cells (Yednock et al., 1992; Steinman, 2001; Miller et al., 2003) by the use of drugs such as Tysabri (natalizumab), interfering with immune cell interactions with anti-CD154 antibody (Couzin, 2005), or by high-dose immunosuppression and reconstitution of the immune system by autologous transplantation of HSCs (Fassas and Kazis, 2003).

Regardless of how far the disease has progressed at the point of arrest, we want to be able to repair whatever damage has occurred to that point. Such repair has been accomplished in EAE mice by the intravenous injection of neurospheres derived from NSCs of the lateral ventricles of the brain (Pluchino et al., 2003). The NSCs express the same α4 integrin expressed by attacking immune cells; they homed to sites of demyelination where they differentiated into oligodendrocytes and new neurons. Astrogliosis and axon damage were markedly reduced, and this was associated with a reduction in the expression of TGF-β and FGF-2, factors known to promote astrogliosis, while expression of other factors such as CNTF, NT3, NGF, GDNF, BDNF, and LIF were unchanged (**FIGURE 6.11**). Nearly 27% of the mice experienced remyelination with complete functional recovery from paralysis, whereas controls showed no sign of recovery. Interestingly, donor NSCs provided only 20% of the new oligodendrocyte precursors in sites of remyelination, the remainder being host cells. This observation suggests that the injected cells can induce host NSCs to differentiate into oligodendrocytes and astrocytes. The cells induce apoptosis of inflammatory T-cells that infiltrate the CNS, conferring long-lasting neuroprotection (Pluchino et al., 2005).

A chronic form of EAE can be induced by injecting wild-type mice with myelin oligodendrocyte glycoprotein (MOG). Mice deficient in Nogo A showed a significant delay in onset of EAE and significantly milder symptoms, suggesting that Nogo A plays a role in the development of EAE. MOG-induced development of EAE is suppressed in mice vaccinated against a Nogo A peptide (Nogo 623-640) or passively immunized with anti-Nogo 623-640 IgGs. This peptide thus appears able to induce an immune response that blunts the development of EAE and may be an appropriate target for the development of therapies for MS (Karnezis et al., 2004).

Recent work suggests that the bHLH transcription factor Olig1 is required to repair demyelinated lesions. Olig1 and 2 are closely related transcription factors expressed in myelinating oligodendrocytes and their progenitors in the developing CNS. Olig2 is required for specification of both motor neurons and oligodendrocytes (Lu et al., 2002). Olig1 is not required for this process, since the CNS of Olig1-null mice develops normally, with full myelination (Arnett et al., 2004). However, Olig1-null mice treated with cuprizone or lysolecithin, which demyelinate the brain and spinal cord, respectively, are unable to remyelinate. Remyelination in wild-type mice after these demyelinating treatments showed that Olig1 is located in the nucleus in oligodendrocyte precursors, but translocates to the cytoplasm in the differentiating oligodendrocytes, suggesting that this translocation plays an essential role in the repair process (Arnett et al., 2004).

## 2. Parkinson's Disease

The striatum of the brain consists of several basal ganglia: the caudate nucleus; the putamen; and the globus pallidus. The electrical output of these ganglia is an important regulator of movement through a

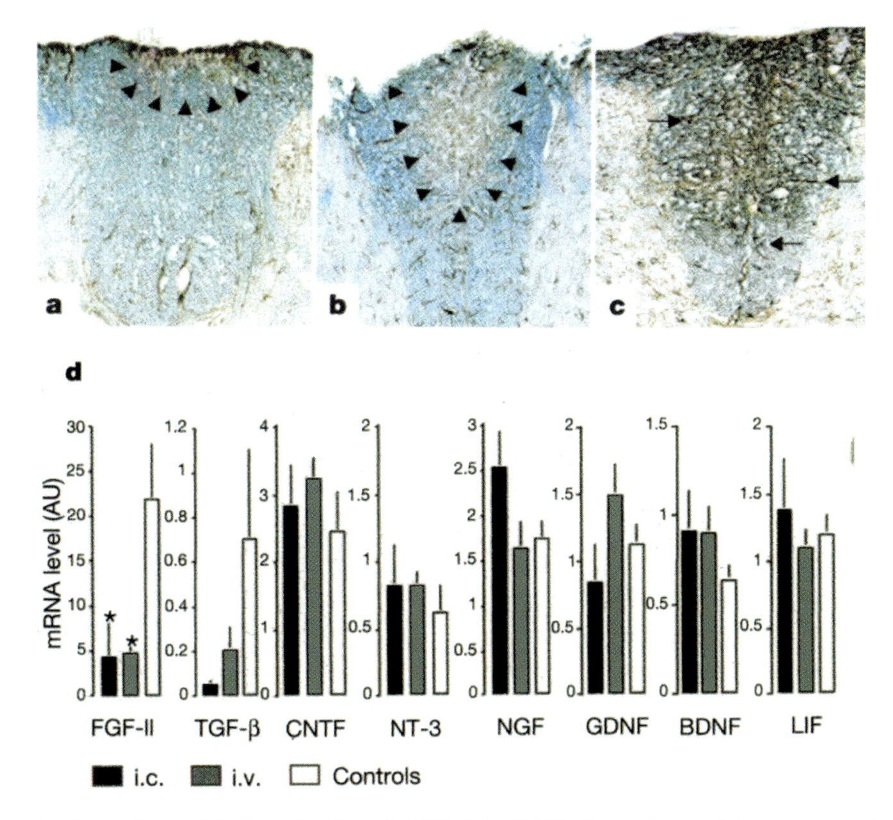


**FIGURE 6.11** Intravenous or intracerebroventricular injection of NSCs into EAE mice reduces glial scarring and modulates expression of neurotrophic mRNAs. **a**, **b**, reduction of area of reactive astrogliosis within demyelinating areas (arrowheads) as evidenced by Luxol fast blue staining, compared to **(c)** sham controls. Arrows indicate astrocyte processes. **d**, injection of NSCs intravenously (grey bars) or intracerebroventricularly (black bars) caused a significant drop in factors known to drive reactive gliosis (FGF-2 and TGF-β), whereas expression level of neurotrophic factors remained essentially unchanged compared to sham controls (white bars). Reproduced with permission from Pluchino et al., Injection of adult neurospheres induces recovery in a chronic model of multiple sclerosis. Nature 422:688–695. Copyright 2003, Nature Publishing Group.

pathway known as the striatopallidothalamic output pathway (SPTOP) (**FIGURE 6.12**). The SPTOP begins with dopaminergic neurons (DANS) in the substantia nigra that project to the basal ganglia, The DANs secrete dopamine, which stimulates electrical output by the basal ganglia to the thalamus and from there to the motor cortex.

Parkinson's disease is characterized by the apoptosis of DANs in the substantia nigra, which in turn results in lower dopamine output by the striatum, hyperactivity of the SPTOP, and impaired motor function (Bjorkland and Lindvall, 2000). Parkinson's is invariably progressive, characterized by tremors at rest, akinesia and bradykinesia, muscle rigidity, postural instability, and lack of facial expression (Rosenthal, 1998). Neuronal death is thought to be the result of genetic mutations and environmental toxins acting through common mechanisms that result in mitochondrial impairment, oxidative stress, and neurotoxic aberrant α-synuclein production (Valente et al., 2004; Greenamyre and Hastings, 2004). The genes *Engrailed*

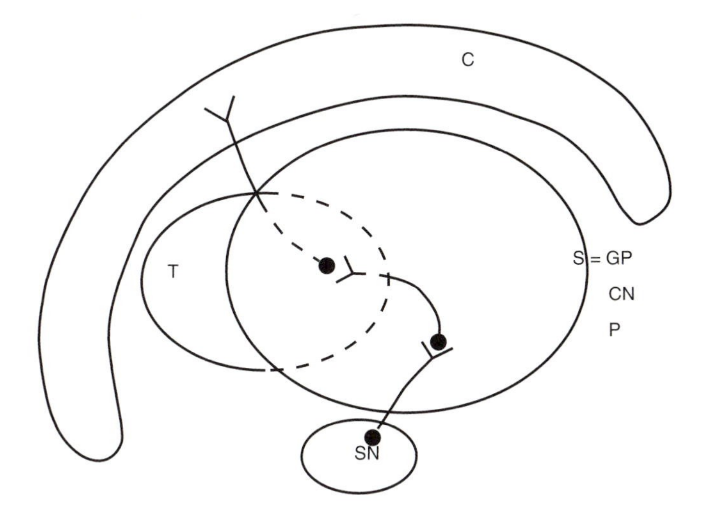


**FIGURE 6.12** Pathway from the substantia nigra (SN) to the motor cortex (C). Dopamine is secreted at the synapses of dopaminergic neurons in the SN with neurons in the striatum (S), which consists of the globus pallidus (GP), caudate nucleus (CN) and putamen (P) basal ganglia. Axons of neurons from the basal ganglia synapse with neurons in the thalamus (T), which then project to the cortex.

*1* and *2* may also be involved in the apoptosis of DANS. These genes are active in DANS shortly after their formation and throughout adulthood. In *En1* and *2* null mice, however, apoptosis of these neurons occurs about two-thirds of the way through embryonic development, leading to the speculation that changes in the expression levels of *En 1* and *2* during adulthood might be involved in the apoptosis of DANS in Parkinson's disease (Alberi et al., 2004).

The three primary therapies for Parkinson's are (1) to increase the dopamine output of the remaining viable DANs by administration of L-dopa, which is taken up and converted to extra dopamine that compensates for the lost output of dying cells, (2) pallidotomy, and (3) inhibitory electrical stimulation of the subthalamic nucleus, which normally stimulates the globus pallidus (Rosenthal, 1998; Bjorklund and Lindvall, 2000). These therapies reverse the akinesia, bradykinesia, and rigidity of the disease, but not the tremors. L-dopa treatment produces severe side effects after several years and eventually has no effect on disease symptoms when the number of viable DANs becomes too low. Many symptoms recur a few years after pallidotomy or electrical stimulation of the subthalamic nucleus. None of the therapies slows the disease progression.

Investigators have turned to transplants of NSCs in the hope of a cure. These transplants have been performed on both human patients and a rat model of Parkinson's in which the neurotoxin 6-hydroxydopamine is injected unilaterally into the striatum (6-OHDA model). The rats suffer muscle rigidity on the contralateral side and exhibit ipsilateral turning movements in response to amphetamine, as well as akinesia and bradykinesia. The uninjected side serves as a control.

Fetal human (6–8 weeks old) mesencephalic cells (which include dopaminergic NSCs) were the first cells to be transplanted in an attempt to cure Parkinson's. When injected into the striatum of patients, they differentiated into DANs and made synaptic connections with host neurons, restoring the activity of the SPTOP toward normal. The results of such transplants, however, have been highly variable. In the best cases there have been dramatic clinical improvements that have lasted 5–10 years (Rosenthal, 1998; Bjorklund and Lindvall, 2000; Lindvall and Hagell, 2001; Bjorklund et al., 2003). The improvements are correlated with an increased output of dopamine, as visualized by increased uptake of $^{18}$fluoro-dopa in PET scans. In other cases, improvements have been minimal, or patients have continued to deteriorate. Autopsies of two patients who died, as well as transplant experiments on Parkinsonian rats suggest that this variation is due to the differential survival of transplanted cells (Rosenthal, 1998). It is thought that a minimum of 80,000 DANs (~20% of the normal number of DANs in the human substantia nigra) are required to obtain a beneficial effect. Clouding the picture is the fact that several double-blind studies of patients receiving transplants of fetal neural cells have suggested a large placebo effect on Parkinson's symptoms, meaning that sham control patients also showed improvement of symptoms (Lazic and Barker, 2003). This result makes it difficult to ascertain the true efficacy of the transplants. Furthermore, transplantation does not seem to benefit older patients (Freed et al., 2001).

Assuming that Parkinson's symptoms can be ameliorated by dopamine output from the transplanted cells, practical and ethical considerations dictate that fetal tissues cannot be a reliable source of cells for transplant. Thus, investigators have turned to other potential sources of DANS. Carotid body cells, which regulate respiratory rate through the medulla by monitoring changes in blood levels of $O_2$, $CO_2$, and $H^+$, are rich in dopamine and thrive in low $O_2$. Since only one carotid body is necessary, the other could be used as a source of DANS for autotransplants. When injected into the striatum of rats with Parkinson's, over 90% of the carotid body cells survived and dopamine levels increased to 65% of unlesioned controls (Espejo et al., 1998). The rat striatum is normally innervated by ~30,000 DANs, but only 400–600 carotid body cells were needed to achieve this result. Amphetamine-induced turning movements were eliminated, but there was less improvement in akinesia and bradykinesia.

Transplantation of ESCs or their derivatives is also being explored as a treatment for Parkinson's. Undifferentiated ESCs have been injected at low concentrations into the striatum of 6-OHDA Parkinsonian rats (Bjorklund et al., 2002). Low concentrations were used to allow the cells to respond to their surroundings rather than self-generated cues. In 14 of 25 rats, the cells survived and differentiated into DANS, and this was associated with recovery from amphetamine-induced turning movements. This is an exciting result that indicates that the striatal environment contains the signals necessary for site-specific differentiation of ESCs. However, the grafts did not survive in six of the animals and developed into teratomas in five others.

Kim et al. (2002) used a five-stage protocol (Lee et al., 2000) to differentiate mouse ESCs transfected with the nuclear receptor related-1 (nurr-1) gene into DANS. Nurr-1 is a transcription factor that promotes the differentiation of mesencephalic precursor cells into tyrosine hydroxylase $(TH)^+$ dopaminergic cells in the presence of FGF-8 and Shh (Hynes and Rosenthal, 1999; Wurst and Bally-Cuif, 2001). Grafts of $5 \times 10^5$ cells were injected into the striatum of rats lesioned unilaterally by 6-OHDA. The ESC-derived $TH^+$ cells were functional DANS as assessed by morphological, neurochemical, electrophysiological, and behavioral criteria. They released dopamine and extended axons

**TABLE 6.6** Behavioral effects of Nurrl transfected ESCs grafted into the 6OH-DA lesioned rat striatum. WTES = wild-type ES cells. Amphetamine-induced rotation was evaluated for 70 min, numbers are given for 8 wks post-grafting. The other measurements were made starting 9 wks post-grafting. Spontaneous ipsilateral rotations were counted over a 5 min period. The adjusting step test measures the number of adjusting steps by the paw on the lesioned side vs the paw on the non-lesioned side when the animals were moved backward by the experimenter. For the paw-reaching test, rats were maintained on a restricted diet until they lost 10–15% of their initial weight. The number of food pellets eaten with the lesioned vs non-lesioned paw was then determined over a 7d period. The cylinder test measures the number of contacts with the cylinder wall the rats make with the lesioned paw when rearing and landing. Minus numbers indicate a change from ipsilateral turning to contralateral turning. $^{*}P < 0.05$; $^{**}P < 0.001$ compared with the sham group. Data from Kim et al (2004).

| | Amphetamine-induced rotations | Spontaneou ipsilateral rotations | % adjusting steps taken by lesioned paw | Fraction of pellets eaten with lesioned paw | # of times lesioned paw used in cylinder test |
|---|---|---|---|---|---|
| Sham | 900 | 8 | 46 | 0.26 | 30 |
| WTES | 400* | –1 | 54 | 0.34 | 53 |
| Nurrl ES | –300** | –4* | 76* | 0.45* | 70* |

into the host striatum. The axons formed functional synapses as indicated by significant recovery from amphetamine-induced rotation and improvement in step-adjusting, cylinder, and paw-reaching tests (**TABLE 6.6**). The next step in both these sets of experiments would be to determine whether DANs derived from human ESCs would improve motor function in the rat Parkinson's model.

Barberi et al. (2003) developed a protocol to direct the differentiation of EGFP-expressing mouse ESCs, derived from blastocysts of normally fertilized or somatic nuclear transfer eggs, into a variety of neuronal phenotypes, including DANs. Transplantation of $1 \times 10^5$ DANs derived from either source into the ipsilateral striatum reduced amphetamine-induced turning in 6-OHDA-treated mice by over 70%. Histological analyses showed that 2%–22% of the cells survived and extended over large areas of the striatum, with the highest density of $TH^+$ cells at the interface of host and grafted cells.

The potential of xenogeneic neural precursor cells from pig fetus to alleviate symptoms of Parkinson's has been tested in the 6-OHDA rat model after expansion in culture (Armstrong et al., 2001). When implanted into the striatum of immunosuppressed rats, many of the cells differentiated into neurons that extended axons to the normal targets of DANS and made synapses at distant sites. However, only a few cells were $TH^+$, and these cells were unable to effect any functional recovery.

In a very interesting set of experiments, Dezawa et al. (2004) reported that transducing cultured rat and human MSCs with the portion of the Notch gene that encodes the intracellular domain (NICD) and treatment of the cells with several neurotrophic factors induced them to become neurons. When GDNF was included in the trophic factor treatment, $TH^+$ DANS were induced with an efficiency of 41%. Transplantation of either rat or human GFP-labeled cells into unilateral 6-OHDA treated rats led to substantial recovery from apomorphine-induced rotational behavior and significant improvement in step adjustment and paw-reaching. Immunohistochemistry of brain sections indicated that grafted cells had migrated and extended beyond the injection site. Concentrations of dopamine were measured in the medium of cultured tissue slices from the lesioned and unlesioned sides of control animals and from the transplanted and unlesioned sides of experimental animals. The ratio of lesioned/engrafted to unlesioned concentration was 0.57 in control animals and 0.67 in experimental animals. These data strongly suggest that not only can MSCs be converted in high numbers to neurons, including DANS, the DANS reconstruct new tissue that is integrated into the neural circuitry of the recipients.

There is also substantial evidence that cell transplants provide host DANS of animal Parkinson's models with neuroprotective factors, thus reducing their loss. Ourednik et al. (2002) induced symptoms of Parkinson's in aged mice via systemic administration of 1-methyl-4-phenyl-1,2,3,6 tetrahydropyridine (MPTP) and transplanted marked NSCs unilaterally into the substantia nigra one to four weeks after treatment. The cells migrated extensively into both hemispheres. Tyrosine hydroxylase and dopamine expression was increased throughout the midstriatal system. While some of the transplanted NSCs differentiated into $TH^+$ cells, 90% of the $TH^+$ cells were host cells. This result strongly suggests that the transplanted cells secreted neurotrophic factors that prevented further death of host DANs or induced neural regeneration (Marconi et al., 2003).

Neuroprotective factors can be delivered directly to the brain tissue of Parkinsonian animals via gene therapy or infusion of protein. Glial-derived neurotrophic factor (GDNF) has been shown to improve the symptoms of Parkinson's disease in experimental animals. Injection of rat brains near the substantia nigra with a replication-defective adenoviral vector encoding human GDNF reduced the loss of DANS after 6-OHDA treatment by approximately threefold when the brains were examined six weeks after injec-

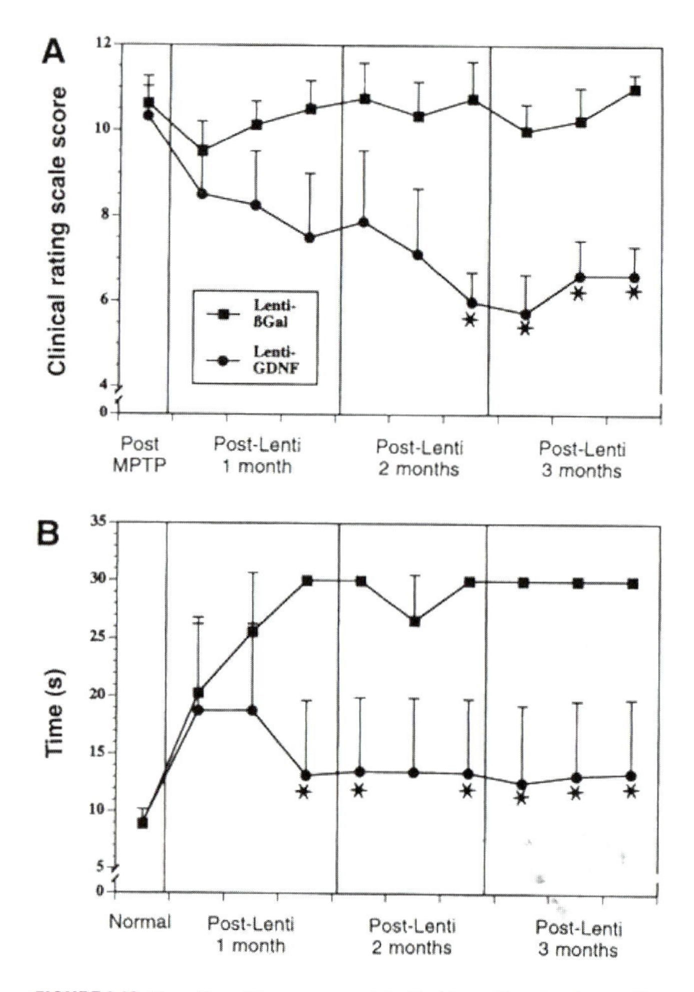


**FIGURE 6.13** Functional improvement in Parkinsonian monkeys after treatment with a GDNF-lentiviral construct. Monkeys receiving the construct showed statistically significant improvement over controls receiving a β-gal-lentiviral construct in clinical ratings **(A)** and **(B)** a hand reach task (seconds required to retrieve food treats from a recessed well). Reproduced with permission from Kordower et al., Neurodegeneration prevented by lentiviral vector delivery of GDNF in primate models of Parkinson's disease. Science 290:767–773. Copyright 2000, AAAS.

tion (Choi-Lundberg et al., 1997). Kordower et al. (2000) induced Parkinsonian symptoms in monkeys by intracarotid injections of 1-methyl-4-phenyl-1,2,3,6 tetrahydropyridine (MPTP), followed by injections into the substantia nigra and striatum of lentiviral constructs encoding GDNF. Three months later, motor performance of the monkeys was vastly improved (**FIGURES 6.13, 6.14**) and PET scans with $^{18}$fluorodopa showed an uptake of the tracer over 300% greater than in control animals injected with constructs containing the β-galactosidase gene. Histological studies revealed abundant GDNF$^+$/TH$^+$ neurons in the substantia nigra, suggesting the rescue of DANS, although the activation of stem cells in the striatum and their differentiation into DANS could not be ruled out. In another study, primary astrocytes of mice transduced with a retroviral GDNF/lac-Z construct were injected into the striatum or substantia nigra of mice six days prior to unilateral 6-OHDA lesioning (Cunningham and Su, 2002). GDNF levels were elevated compared to controls and TH$^+$ immunostaining was 89% of the unlesioned side, indicating that the GDNF conferred neuroprotection on host DANS. Dopamine depletion in the substantia nigra was completely prevented, as was the acquisition of amphetamine-induced rotational behavior. These studies indicate the feasibility of treating Parkinson's disease by gene therapy.

Cells of the subventricular zone proliferate *in vitro* in response to TGF-α, suggesting that this growth factor might be used to activate NSCs in the striatum or lateral ventricle walls that would then migrate to the degenerating substantia nigra and differentiate into DANs. To test this idea, a total of 50 µg of TGF-α was administered to the striatum of unilateral Parkinsonian rats via a shoulder-implanted minipump and cannula at the rate of 0.5 µl/hr for two weeks (Fallon et al., 2000). The infused rats showed a 31.5% improvement in rotational behavior over controls. Studies of BrdU incorporation and immunostaining for nestin and differentiated neuron markers revealed that NSCs proliferated, migrated to the striatum and differentiated into new neurons, some of which were presumably DANs.

Finally, Shh, which is crucial for the development of the nervous system, reduced amphetamine-induced rotation and forelimb akinesia when injected into the striatum of 6-OHDA rats (Tsuboi and Shults, 2002). This effect was associated with partial preservation of striatal dopaminergic axons, but only a modest preservation of nigral TH$^+$ neurons, suggesting that Shh may induce behavioral improvement by mechanisms other than action on nigral DANS.

## 3. Huntington's Disease

Huntington's disease is a progressive disorder of movement accompanied by severe cognitive deterioration. The symptoms are associated with the death of multiple populations of striatal neurons, particularly medium spiny neurons, resulting in the loss of inhibitory connections from the striatum to the globus pallidus. In normal cells, the huntingin protein, complexed with huntingtin-associated protein-1 (HAP-1) and a dynactin subunit, p150$^{Glued}$, mediates vesicular transport of brain-derived neurotrophic factor (BDNF) along neuron microtubules (**FIGURE 6.15**). Huntington's disease is associated with a mutation in the N-terminal region of the huntingtin gene that results in a polyglutamine (CAG) repeat, creating a misfolded protein that is neurotoxic (Huntington's Disease Collaborative

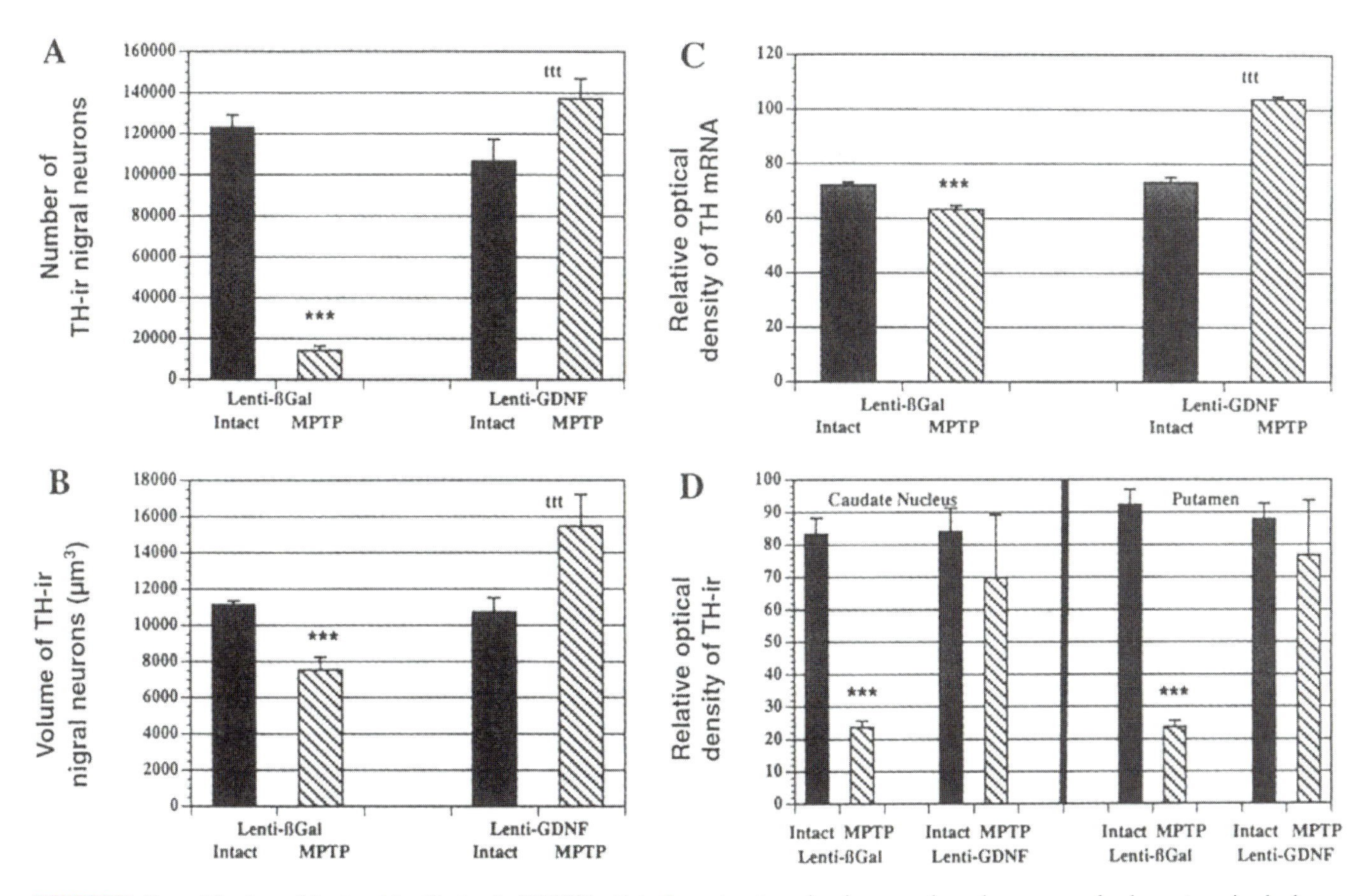


**FIGURE 6.14** Quantification of the trophic effects of a GDNF lentiviral construct on nigral neuronal number, neuronal volume, tyrosine hydroxylase (TH) mRNA, and striatal TH immunoreactivity (ir). Asterisks indicate statistically significant decreases relative to the intact side; ttt denotes significant increases relative to the intact side. Injection of the GDNF construct clearly rescues TH$^+$ neurons. Reproduced with permission from Kordower et al., Neurodegeneration prevented by lentiviral vector delivery of GDNF in primate models of Parkinson's disease. Science 290:767–773. Copyright 2000, AAAS.

Research Group, 1993). The mutation in the protein reduces the association of molecular motor proteins with microtubles, reducing the transport of BDNF (Gauthier et al., 2004). Huntington's disease can be mimicked in primates by injecting quinolinic acid into the striatum or by intramuscular injection of nitropropionic acid (NPA).

Fetal striatal tissue grafted to the striatum of marmoset or macaque monkeys with NPA-induced Huntington's reversed the symptoms of the disease (Kendall et al., 1998; Palfi et al., 1998). Immunohistochemical studies indicated good survival and differentiation of the grafted neurons with establishment of functional connections with host tissue. Preliminary clinical trials in human patients given grafts of human fetal striatal tissue have shown that the tissue survives and that the symptoms of the disease are alleviated to some extent, with persistent benefits to some patients three years postgrafting (Freeman et al., 2000; Bachoud-Levy et al., 2002; Hauser et al., 2002; Rosser et al., 2002; Liu et al., 2003).

Evidence for a possible neuroprotective effect of cell transplants on striatal neurons was obtained in experiments on the quinolinic acid model of Huntington's disease in monkeys (Emerich et al., 1997). The monkeys were first given unilateral intrastriatal implants of polymer-encapsulated baby hamster kidney (BHK) fibroblasts alone or BHK fibroblasts transgenic for the human ciliary neurotrophic factor (CNTF). A week later, the implanted region was injected with quinolinic acid. The loss of striatal neurons was significantly attenuated in the CTNF implant region, compared to control implant regions with BHK cells lacking the CTNF construct. Human umbilical cord stem cells or cultured human neural precursor cells injected into a mouse model of Huntington's also improved the condition (Cova et al., 2004). While it is possible that the transplanted cells differentiated into new neurons, it is likely that their major effect was to provide paracrine factors that stabilized host neuron survival.

A conditional model of Huntington's disease has been developed that suggests that it might be possible

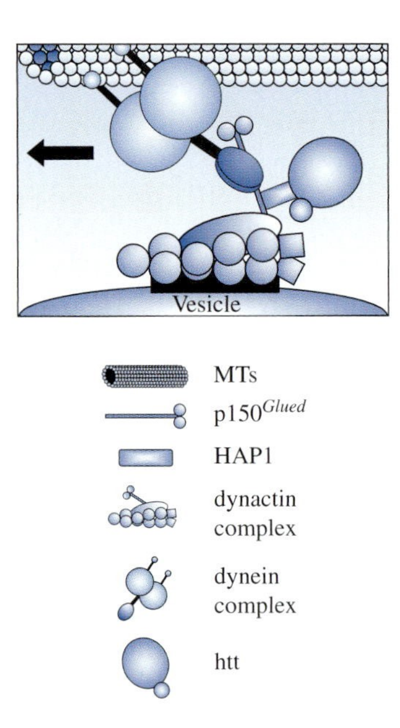


**FIGURE 6.15** Molecular motor that transports brain derived growth factor (BDNF) along mictotubules (MTs). p150$^{Glued}$ = subunit of dynactin, HAP1 = huntingtin associated protein, htt = huntingtin. Reproduced with permission from Gauthier et al., Huntingtin controls neurotrophic support and survival of neurons by enhancing BDNF vesicular transport along microtubules. Cell 118:127–138. Copyright 2004, Elsevier.

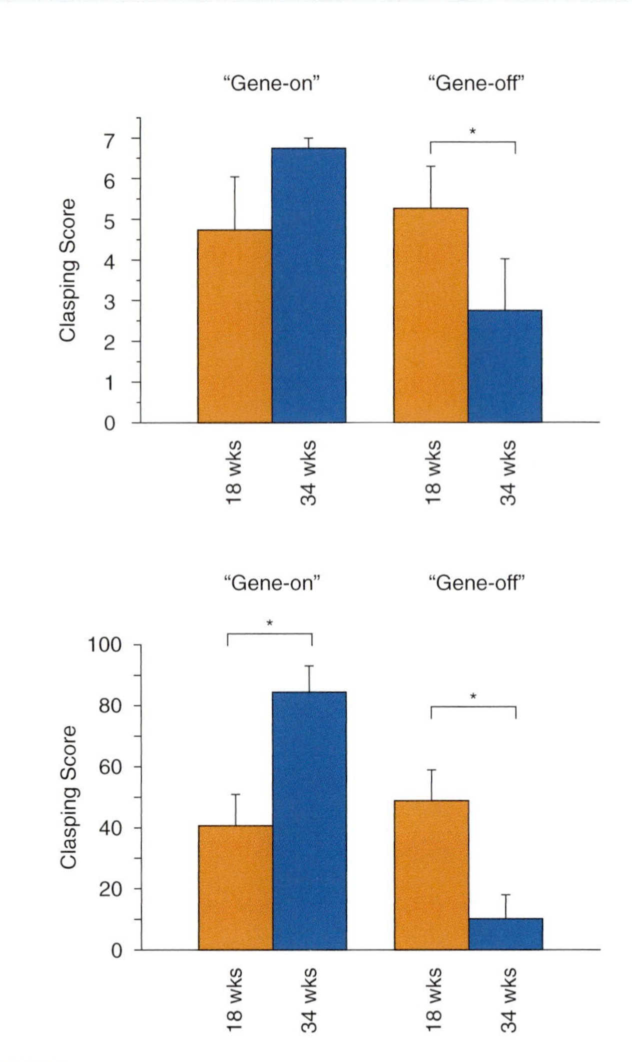


**FIGURE 6.16** Effect of inhibiting the expression of the polyQ-htt gene on the clasping behavior of HD94 mice at 18 and 34 weeks. In the "gene-off" experiment, 2 mg/ml doxycycline was given for 16 weeks in a 5% sucrose solution that was provided to the mice to drink. Both clasping score and duration in the tail suspension test was significantly increased in controls (left) from 18–34 weeks, but significantly decreased in the gene-off group. Asterisk indicates statistical significance. Reproduced with permission from Yamamoto et al., Reversal of neuropathology and motor dysfunction in a conditional model of Huntington's disease. Cell 101:57–66. Copyright 2000, Elsevier.

to reverse the pathological changes that lead to neurodegeneration (Yamamoto et al., 2000). Mice were made doubly transgenic for the tetracycline-regulated transactivator (tTA) and a mutant 94 CAG repeat of the huntingtin gene under the direction of the tetO bidirectional promoter. Both transgenes were constitutively expressed. Doxycycline, a tetracycline derivative, inhibits expression of the mutant Huntington gene by binding to tTA and decreasing its affinity for the tetO promoter. These transgenic mice exhibited aggregates of the mutant Huntington protein, decrease in striatal size, reactive astrocytosis in the striatum, and a behavior defect typical of Huntington's in mice, the clasping of limbs when suspended by the tail. Yamamoto et al. (2000) showed that continuous expression of the mutant huntingtin was required for progression of the disease. Administration of doxycycline for 16 weeks, beginning at 16 weeks of age, reversed aggregate formation, as well as the other pathological symptoms of the disease, to the level of eight-week-old transgenic mice (**FIGURE 6.16**). This result suggests that abnormal protein deposits can be cleared from dysfunctional cells. Orr and Zoghbi (2000) have suggested that this might also be done by an RNA interference strategy (see ALS, ahead).

Environmental enrichment has also been shown to delay the onset of symptoms in a transgenic mouse Huntington's model (van Dellen et al., 2000). Loss of motor coordination, clasping of limbs, and onset of seizures were all significantly delayed. Striatal volume was 13% larger in the enriched HD mice than the nonenriched mice. These improvements were not accompanied, however, by reductions in protein aggregates in striatal cells, suggesting that the effect of enrichment may be to overcome deficiencies in synaptic plasticity.

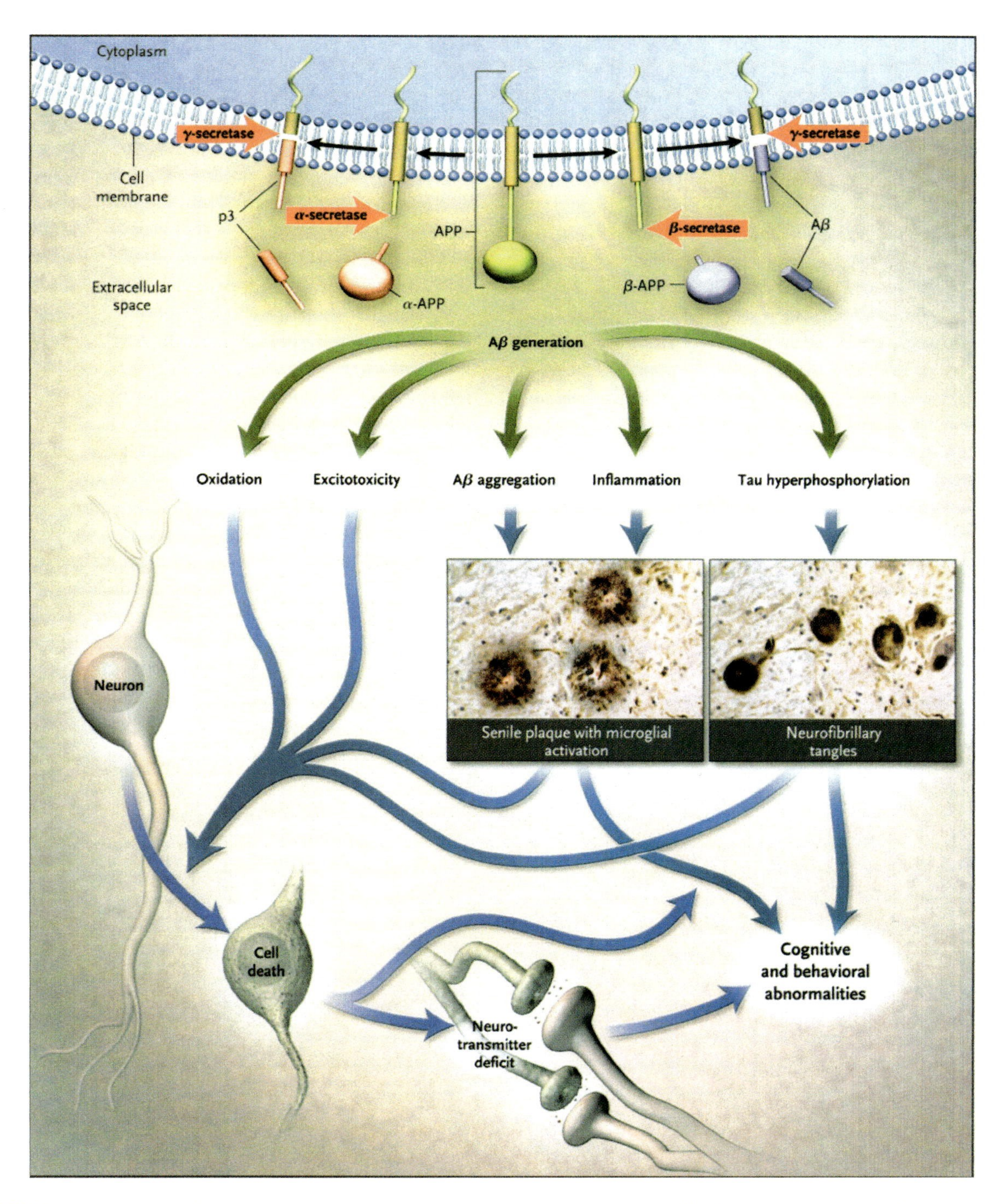


**FIGURE 6.17** Hypothesis for development of Alzheimer's disease. β and γ secretases sequentially cleave the cytoplasmic domain of the amyloid precursor protein (APP) to yield β-APP and amyloidβ (Aβ). Aβ leads to neurodegeneration through multiple pathways, including the formation of plaques and neurofibrillary tangles. Reproduced with permission from Cummings, Alzheimer's disease. New Eng J Med 351:56–67. Copyright 2004, Massachusetts Medical Society.

## 4. Alzheimer's Disease

Alzheimer's disease places one of the highest economic and social burdens on caregivers of any neurodegenerative disease. The incidence of Alzheimer's increases to 30%–40% in persons over the age of 85 and has an economic impact of $80–$100 B/year (Mattson, 2004). The causes of the disease are believed to be both genetic and environmental. Proteolytic processing of β-amyloid precursor protein into amyloid-β peptides, and hyperphosphorylation of the microtubule-associated tau protein lead to the plaques and neurofibrillary tangles that characterize Alzheimer's (FIGURE 6.17). Initial symptoms of Alzheimer's are subtle, but as the disease advances, there is clear-cut progressive loss of memory and cognition until the persona is destroyed and all control over physical function is lost (Cummings, 2004).

Various rat models of Alzheimer's are available for experimentation. Cells from a variety of cholinergic tissues (fetal basal forebrain, nodosal ganglion, intracortical adrenal chromaffin cells, and chemically differentiated neuroblastoma cell lines) survived when injected ectopically into the terminal areas of the forebrain cholinergic projection system of these models, and significantly reversed deficits in cognitive behavior (Hodges et al., 1991, a, b, c). These transplants contained both neurons and glial cells. Transplants of noncholinergic cells from the hippocampus had no effect. If glial cells and neurons from cholinergic areas were transplanted separately, the glial cells promoted the recovery of cognitive function, but the cholinergic neurons did not (Bruckner and Arendt, 1992; Sinden et al., 1995). Thus, when cells from a cholinergic area were transplanted, the glial cells of this tissue may have been increasing acetylcholine (Ach) secretion by host cholinergic neurons, as well as fostering axonal outgrowth and Ach secretion by donor cholinergic neurons. These results suggest that the transplanted cells secrete paracrine factors that exert a neuroprotective effect on host neurons, opening the possibility for pharmaceutical treatments that can halt the disease and permit the regeneration of new neurons from resident NSCs.

Pharmaceutical companies are testing more than two-dozen experimental treatments for Alzheimer's. Among these are several drugs that may control or reduce symptoms of Alzheimer's, although none can halt its progression (Cummings, 2004; Pallarito, 2004). Cholinesterase-inhibiting drugs, which work by slowing the rate of breakdown of Ach, are Aricept (donepezil), Exelon (rivastigmine), and Reminyl (galantamine). According to the Alzheimer's Association, these drugs slow cognitive decline in at least half of patients who have mild cognitive impairment. Namenda (memantine) is a new drug that is thought to work by protecting brain cells from excess glutamate. Two drugs, Alzhemed and Clioquinol (PBT-1), target the aggregation of β amyloid peptides into plaques. A small Phase II trial of Clioquinol indicated that it was able to reduce the rate of decline of cognitive function in patients with moderate to severe Alzheimer's (Ritchie et al., 2003). Still another drug in Phase III trials is Axonyx, which can inhibit both cholinesterases and prevent plaque formation.

One of the problems with drugs directed at inhibiting plaque formation is that they are too small to effectively block the interaction between β-amyloid peptides. A new strategy to prevent the aggregation of β-amyloid peptides is to sterically "bulk up" a small bifunctional molecule that binds on one end to β-amyloid peptide and on the other end to a large molecule that also sticks to β-amyloid peptide. Gestwicki et al. (2004) made such a small molecule by coupling the dye Congo Red (CR) to a synthetic ligand (FKBPL) that binds to the FK506 binding protein (FKBP) chaperone family of peptidyl prolyl cis-trans isomerases (**FIGURE 6.18**). When added to β-amyloid along with FKBP, the CR-FKBPL molecule delayed or completely prevented aggregation of β-amyloid in test-tube assays and prevented neuronal cell death *in vitro* caused by β-amyloid added to the medium. CR, however, does not enter cells (where the β-amyloid is being produced), so an alternative to CR needs to be found before this approach can be developed into a clinical drug (Wickelgren, 2004). These results suggest that aggregates of β-amyloid peptides are neurotoxic.

Provision of an enriched environment (including exercise) to mice suffering from a familial form of AD (Lazarov et al., 2002) reduces the levels of β-amyloid peptides and amyloid deposition in the hippocampus and cortex (Lazarov et al., 2005). They showed by microarray analysis that a number of genes involved with learning and memory, vasculogenesis, neurogenesis, cell survival pathways, amyloid-β sequestration, and prostaglandin synthesis were upregulated in the hippocampus of enriched mice. Neprilysin (NEP), an enzyme that degrades amyloid, was significantly elevated, suggesting enhanced clearing activity. These results are consistent with previous results demonstrating the effects of enrichment and exercise on learning and memory in young and old mice (Chapter 5), the effects of physical therapy regimens on spinal cord injury, and in the delay of onset of neurological signs in Huntington's disease. Although many questions remain about the degree to which environmental enrichment for mice corresponds to the exercise of mental capabilities in humans, and what may be cause or effect, Karsten and Geschwind (2005) point out that other work shows that higher degrees of educational attainment, exercise, occupational status and a strong social network all have positive effects on maintaining cognitive powers with age.

## 5. Amyotrophic Lateral Sclerosis

ALS is a progressive, fatal disease characterized by the death of spinal cord large motor neurons. The cause of the disease is unknown, but may include genetic susceptibility, cellular aging, and as yet unidentified environmental factors (Rowland and Scheider, 2001). Ninety to ninety-five percent of all cases are sporadic.

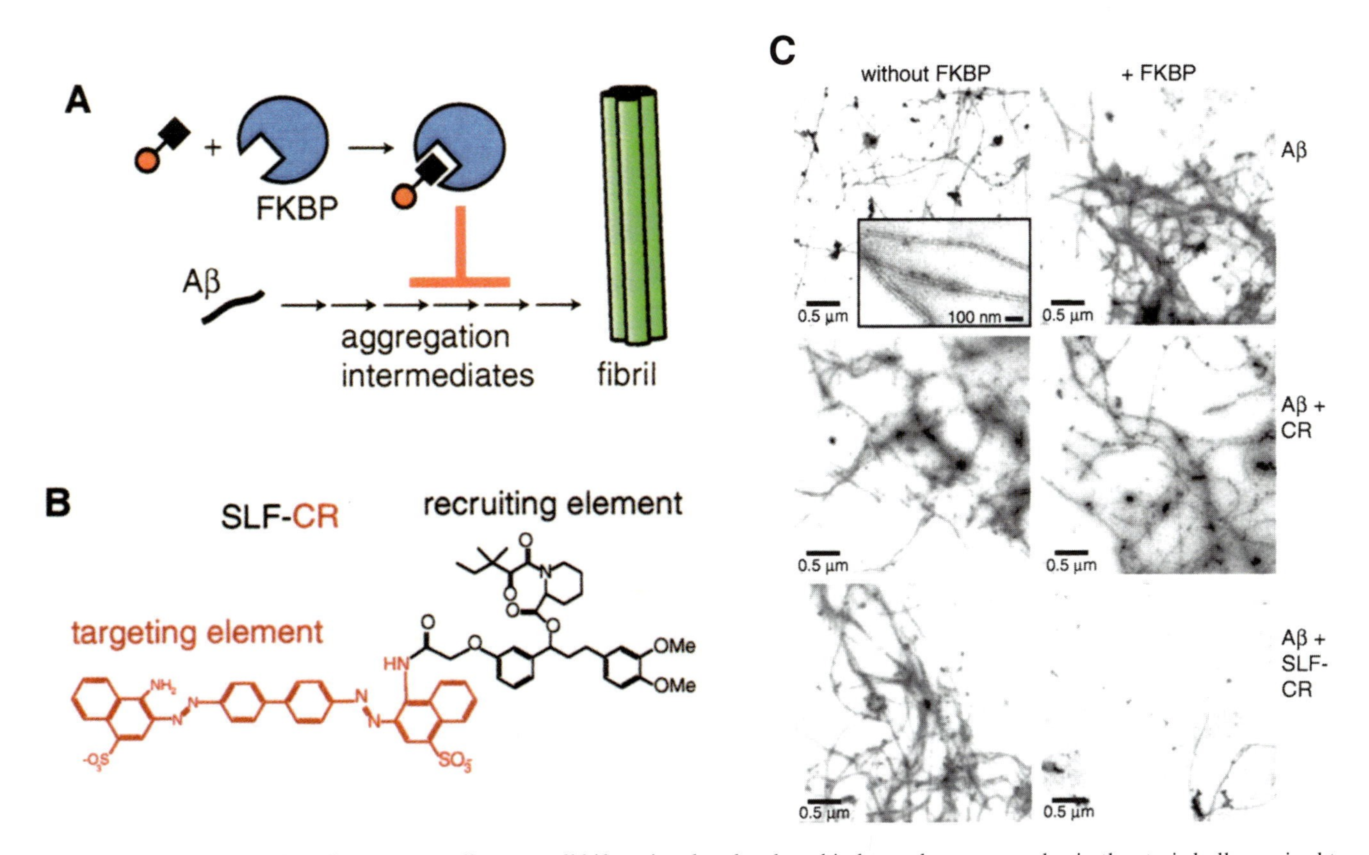


**FIGURE 6.18** **(A)** A "Trojan Horse" strategy to allow a small bifunctional molecule to bind to a chaperone and gain the steric bulk required to disrupt amyloid β aggregation. One end of the bifunctional molecules binds to the FK506 binding protein, a member of the FKBP chaperone family of peptidyl prolyl cis-trans isomerases that are highly expressed in all mammalian cells. The other end of the bifunctional molecule interacts with Aβ to block aggregation. **(B)** One such bifunctional molecule is SLF (synthetic ligand for FKBP) coupled to CR (Congo Red). CR interacts with Aβ. **(C)** Electron micrographs of Aβ fibrils formed in solution when incubated with or without FKBP in the presence of CR or SLF-CR. SLF-CR + FKBP clearly inhibits the formation of Aβ fibrils. Reproduced with permission from Gestwicki et al., Harnessing chaperones to generate small-molecule inhibitors of amyloid β aggregation. Science 306:865–869. Copyright 2004, AAAS.

The remainder show familial inherited forms linked to mutations in the Cu/Zn-dependent superoxide dismutase-1 (SOD-1) gene. SOD-1 is a mitochondrial enzyme that converts the highly reactive and destructive superoxide anions produced by the one-electron reduction of $O_2$ in metabolic reactions into hydrogen peroxide and $O_2$. This mutation may thus cause cellular destruction because it does not function properly and may also cause the protein to fold abnormally and have a cytotoxic effect on motor neurons (Cova et al., 2004).

Chronic glutamate-toxicity is a major factor in the progressive death of motor neurons (Shaw and Ince, 1997). In the spinal cord and brain, astrocytes prevent glutamate toxicity by extending processes into the vicinity where two motor neurons synapse. These processes have glutamate transporters (GLT1) that recover glutamate molecules, thus turning down the glutamate signal after it has been received. In mouse models of ALS, and in the human disease, GLT1 expression is downregulated in astrocytes and ALS patients have elevated levels of glutamate in their cerebrospinal fluid (Plaitakis et al., 1988; Rothstein et al., 1990, 1992, 1995, 2005; Lin et al., 1998). There is also a defect in the editing of the mRNA encoding the GluR2 subunit of glutamate α-amino-3-hydroxy-5-methyl-4-isoxazole-propionate (AMPA) receptors on spinal motor neurons (Kawahara et al., 2004). These features suggest that key elements of progressive neuronal cell death in ALS are astrocyte dysfunction that results in reduction in glutamate clearance and compromised functioning of glutamate receptors on spinal motor neurons. Another feature of ALS is the accumulation of neurofilaments. Mice transgenic for the mouse neurofilament light gene and the human neurofilament heavy gene display large motor neuron degeneration associated with neurofilament accumulation and altered neurofilament phosphorylation (Xu et al., 1993; Cote et al., 1993); these features have been observed in human ALS as well (Manetto et al., 1988; Hirano and Kato, 1992). Neurons die by apoptosis, in which the caspase family plays a prominent role (Namura et al., 1998).

Several mouse models of ALS are useful for developing therapeutic strategies. The *pmn/pmn* mouse exhibits progressive motor neuron degeneration detectable as hindlimb weakness during the third week of life, and dies at ~6 weeks of age (Schmalbruch et al., 1991). The G(lysine)93A(alanine) mouse is transgenic for a mutated human SOD-1 gene associated with human familial ALS (Wong et al., 2002). This mouse also exhibits a progressive motor neuron degeneration that manifests itself in locomotor deficits at 13 weeks of age, and dies at ~4–5 months of age. A third model is the organotypic culture of mouse spinal cord in the presence of the glutamate transport inhibitor, threo-hydroxyaspartate (THA), which causes glutamate-mediated motor neuron cell death (Corse and Rothstein, 1995). These models have been used to test the efficacy of pharmaceutical and cell therapies on ALS.

### a. Neuroprotective Cell Transplants

Wild-type astrocytes delay the degeneration and significantly extend the survival of mutant SOD1-expressing motoneurons in ALS mice (Clement et al., 2003; Klein et al., 2005), suggesting that they may counter the glutamate toxicity characteristic of ALS.

Transplants of stem cells into the spinal cord have been reported to have positive effects in the G93A mutant mouse model. Human umbilical cord blood cells were reported to delay the progression of ALS in these mice (Chen and Ende, 2000; Ende et al., 2000). The use of these cells was predicated on the idea that NSCs are "migrating HSCs," based on the fact that NSCs are in close contact with capillaries and on the reports that HSCs could be induced to differentiate into neurons (Chapter 13). The cells migrated toward sites of motor neuron degeneration and expressed neural and glial markers (Garbuzova-Davis et al., 2003), suggesting that they had undergone lineage conversion and replaced host motor neurons. However, the number of injected cells expressing neural markers was low. More likely, their effect was due largely to a paracrine protective action on host motor neurons through secreted factors. Dental pulp stem cells were reported to promote the survival of sensory neurons of the trigeminal ganglion and of injured motor neurons of mouse spinal cord through their production of GDNF (Nosrat et al., 2004).

Injection of neurons from a human neuron cell line (hNT) into the ventral horn of the spinal cord in the G93A mutant mouse was reported to delay the onset of motor dysfunction and extend the average lifespan of the mice (Garbuzova-Davis et al., 2002). The mechanism of action of the injected cells was not clear, however, and could have been through replacement of dying motor neurons, protection of host motor neurons through paracrine factors, or both.

Autologous peripheral blood stem cells and bone marrow cells have been injected into the spinal cords of several ALS patients in Phase I clinical cell therapy trials (Janson et al., 2001; Mazzini et al., 2003; Cova et al., 2004; Silani et al., 2004). These trials indicated that cells could be transplanted into the spinal cord with minor and reversible side effects. In the bone marrow cell trial (Mazzini et al., 2003), the linear decline in strength was slowed somewhat in some of the muscles in four of seven patients, and two of the seven actually experienced a mild increase in strength.

Fetal OECs have also been used by Huang Hongyen to treat ALS patients in Bejing's Chaoyang Hospital (Watts, 2005; Cyranoski, 2005; Curt and Dietz, 2005). Injection of over 1–2 × $10^6$ fetal OECs into the frontal lobes of ALS patients under local anesthetic was reported to stabilize and even improve the patient's condition in about half the cases through paracrine effects. About 100 ALS patients from around the world have been treated in this way.

### b. Neuroprotective Pharmaceutical Agents

Pharmaceutical neuroprotection is an area of intense focus for the treatment of ALS. Two neurotrophins known to promote the survival of motor neurons, CNTF and BDNF, were tested in clinical trials for their efficacy in halting the progression of ALS. Both drugs were failures (ALS CNTF Study Group, 1996; Cederbaum et al., 1998). Some positive effect of IGF-1 was reported in a North American clinical trial (Lai et al., 1997), but a similar European trial did not show any effect (Borasio et al., 1998). In mice, IGF-1, GDNF, and NT-4/5 were protective of motor neurons in spinal cord cultures treated with THA, whereas BDNF, CNTF, NT-3, FGF-2, and LIF were not protective (Corse et al., 1999), consistent with the results of the human clinical trials (**FIGURE 6.19**).

GDNF is of particular interest as a possible therapeutic agent, given its effectiveness in protecting dopaminergic neurons and motor neurons after axotomy-induced cell death in adults (Oppenheim et al., 1995) and during the embryonic period of programmed neural cell death (Zhao et al., 2004).

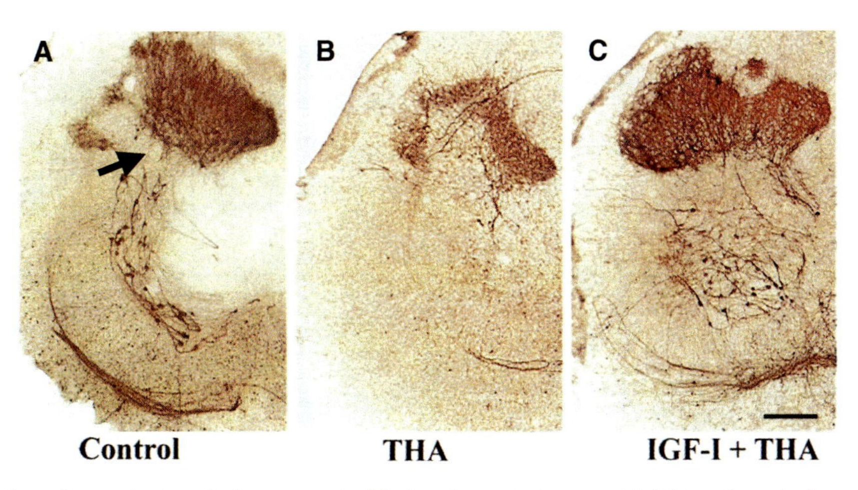


**FIGURE 6.19** Sections of cultured rat spinal cord slices treated with the glutamate transport inhibitor threo-hydroxyaspartate (THA). Arrow indicates motor neurons. The sections were stained with a monoclonal antibody against neurofilament-H (THA). **(A)** Control cord. **(B)** THA-treated, showing severe loss of motoneurons. **(C)** IGF-I protects motor neurons from TA. Reproduced with permission from Corse et al., Preclinical testing of neuroprotective neurotrophic factors in a model of chronic motor neuron degeneration. Neurobiol Dis 6:335–346. Copyright 1999, Elsevier.

GDNF administered to *pmn/pmn* mice by subcutaneously implanting encapsulated BHK fibroblasts transfected with the GDNF gene reduced the loss of facial motor neurons by 50%, but did not extend the lifespan of the mice because, although it protected neuron cell bodies, it did not prevent axon degeneration (Sagot et al., 1996). GDNF retroviral constructs, delivered in transfected myoblasts to the gastrocnemius muscles of G93A mutant mice, resulted in a shift in the size distribution of motor neurons toward larger neurons compared with untreated G93A controls, although the average size and number of large motor neurons were less than in wild-type mice (Mohajeri et al., 2004). Quantitative RT-PCR showed that the GDNF DNA and mRNA persisted for up to 12 weeks postgrafting. Furthermore, GDNF prolonged the onset phase of the disease, slowed muscle atrophy, and delayed the deterioration of performance in tests of motor behavior.

Klein et al. (2005) demonstrated the feasibility of using human cells expressing GDNF as therapy for ALS. Human neural progenitor cells isolated from fetal cortex were expanded *in vitro* and infected with a GDNF lentiviral construct. When transplanted into the lumbar spinal cord of a G93A rat model of ALS, the cells differentiated into astrocytes, survived for up to 11 weeks and were integrated into the gray and white matter of the cord without observable side effects. The transplanted cells were shown to secrete GDNF in the area of transplant and to stimulate the upregulation of cholinergic markers in nerve fibers. Although counts of ChAT positive neurons revealed no differences in motor neuron number between treated rats and controls, there was a significant difference between the size of the cell body in the treated rats versus controls. However, the transplanted cells did not improve measures of limb strength and coordination.

$VEGF^{\partial\partial}$ mice, in which the hypoxia-response element in the VEGF promoter is deleted, have an impaired ability to upregulate VEGF levels under conditions of stress and develop ALS-like neuropathology, suggesting that motor neurons are sensitive to lack of VEGF (Oosthuyse et al., 2001). Injection of a lentiviral VEGF construct bilaterally into the gastrocnemius, diaphragm, intercostal, facial, and tongue muscles of G93A mutant mice prior to ALS onset slowed progression of the disease and extended life expectancy by as much as 30% (**TABLE 6.7**). There was a 15% increase in life expectancy when the construct was injected after disease symptoms were manifest (Azzouz et al., 2004). Interestingly, injection of the GDNF gene in a similar construct resulted in only a 5% increase in life span.

Molecules other than neurotrophic and growth factors may hold promise in treating ALS. Overexpression of the gene for the apoptosis inhibitor, Bcl-2, delayed the onset of ALS symptoms by 20%, increased survival time by 15%, and preserved significantly more motor neurons and myelinated axons in the phrenic nerve (Kostic et al., 1997). Caspase-3 is activated in mouse and human ALS (Li et al., 2000). Inhibiting caspase activity in G93A mice with N-benzyloxycarbonyl-Val-Asp-fluoromethylketone (zVAD-fmk) delayed the onset of ALS and prolonged survival time by 22%. At 110 days of age, treated mice exhibited a significantly higher number of motor neurons in the cervical region of the spinal cord. The antibiotic minocycline, which also inhibits the activity

**TABLE 6.7** Effects of equine infectious anemia virus-vascular endothelial growth factor (EIAV-VEGF) gene construct on ALS age of onset, age at death and number of surviving facial motor neurons at the end stage in G93A mice. EIAV-LacZ served as a control construct. VEGF improves all parameters. Data from Azzouz et al. (2004)

| | EIAV Vector | |
|---|---|---|
| | LacZ | VEGF |
| Age at onset | 95 ± 4 | 123 ± 11 |
| Age at death | | |
| Given at 3 wks | 125 ± 6 | 163 ± 15 |
| Given at 3 months | 127 ± 6 | 146 ± 10 |
| Number of facial motor neurons (control = 3500) | 1800 | 2700 |

of caspase-1 and caspase-3, delayed the onset of impaired motor function in G93A mice by 21% and extended survival time by 9% (Zhu et al., 2002). The action of the drug was through the inhibition of cytochrome *c* release, which strongly stimulates the activation of caspase-1 and -3.

Corcoran et al. (2002) found that mice deprived of vitamin A developed symptoms of motor neuron disease. Neurofilaments accumulated in the cell bodies of lumbar and cervical motor neurons and the lumbar cord had 34% fewer motor neurons than normally fed rats. In a rotarod test (ability to maintain footing on a rotating rod), the performance of the retinoid-deficient rats was half that of normal rats. *In situ* hybridization for transcripts of *RARs* and three retinaldehyde dehydrogenases (*raldhs*, the enzymes that convert retinaldehyde to RA) showed that motor neurons expressed only *RARα* and *raldh2*. Retinoid-deficient rats were deficient in *RARα* but not *raldh2*, suggesting that receptor expression, but not the enzyme, is regulated by retinoids.

Corcoran et al. (2002) also found significant deficits in the number of motor neurons expressing *RARα* and *raldh2*, and in the levels of *islet-1, RARα*, and *raldh2* compared to the housekeeping enzyme *gapdh* in spinal cord sections of patients who died from ALS (**FIGURE 6.20, TABLE 6.8**). Thus, ALS in humans is different from the motor neuron syndrome in retinoid-deficient rats in that the factors that maintain *raldh2* are absent. Thus, one aspect of the etiology of ALS may be downregulation of the *raldh2* gene, leading to loss of cellular RA, and then the loss of *RARα*. Identification of the factors that regulate *raldh2* synthesis could possibly lead to treatments that have a positive effect on the course of ALS by bringing levels of *raldh2* back to normal.

Recently, the results of screening 1,040 FDA-approved drugs and nutritionals for their effect on the expression of GLT1, using organotypic spinal cord slice cultures prepared from postnatal day 9 rats, were reported (Rothstein et al., 2005). The screen revealed that 15 different β-lactam antibiotics, including penicillin and its derivatives, as well as cephalosporin antibiotics, were highly active in stimulating the expression of GLT1 protein in the slices and in the brains of whole animals and that the mechanism in human astrocytes was by upregulating transcription of the *GLT1* gene. The antibiotics were neuroprotective in organotypic spinal cord cultures treated with THA. Furthermore, G93A mutant mice treated with one of the antibiotics, ceftriaxone, at the time of ALS onset significantly delayed loss of muscle strength and body weight, and increased overall survival of the mice by 10 days. GLT1 expression was increased and there was significant prevention of motor neuron loss (**TABLE 6.9**). Ceftriaxone previously had been reported to provide some improvement in a third of 21 patients treated with the drug for 5–8 weeks (Italian ALS Study Group, 1996).

Since none of these treatments alone cures ALS in mouse models, but only prolongs survival and delays loss of muscle strength (and motor neuron loss), therapies using combinations of several or all of these molecules to affect multiple targets involved in the onset and progression of ALS might be worth pursuing to see if they might have synergistic effects on ameliorating the disease.

**TABLE 6.8** Reduction of transcripts of RARα, RALDH2, and Islet-1 in motor neurons of ALS patients. Data from Corcoran et al. (2002)

| | Normal | ALS |
|---|---|---|
| % motor neurons expressing | | |
| RARα | 62 | 25 |
| RALDH2 | 56 | 16 |
| Islet-1 | 61 | 48 |
| % reduction of transcripts in surviving motor neurons | | |
| RARα | — | 31 |
| RALDH2 | — | 49 |
| Islet-1 | — | 56 |

**TABLE 6.9** Effects of ceftriaxone vs saline treatment on G93A mice. Numbers shown for grip strength and body weight are given for 17 and 18 wks of age, respectively. Drug therapy was begun at 12 wks of age. GLT1 protein is in arbitrary units. Lumbar motor neurons were counted after two wks of drug treatment starting at 70 days of age. Data from Rothstein et al. (2005).

| | G93A | | Wild type |
|---|---|---|---|
| | Ceftriaxone | Saline | |
| Grip strength (g) | 35 | 20 | — |
| Body wt (g) | 24.5 | 21 | — |
| Overall survival time (d) | 132 | 122 | — |
| GLT1 protein | 55 | 6 | 20 |
| # lumbar motor neurons | 600 | 400 | 700 |

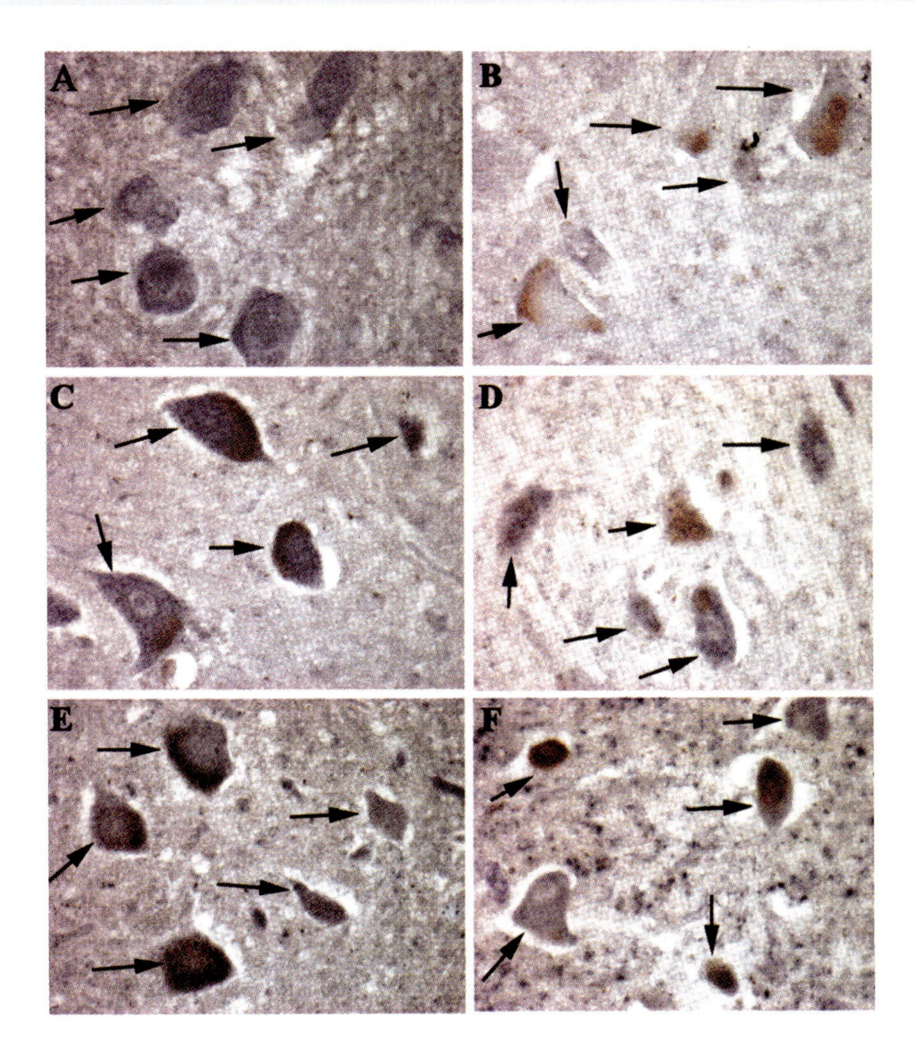


**FIGURE 6.20** **Left column**, sections of normal human lumbar spinal cord. **Right column**, sections of lumbar cord from patients with ALS. Sections subjected to *in situ* hybridization. **(A, B)** Expression of *islet-1*. **(C, D)** *RARα* expression. **(E, F)** *raldh-2* expression. Arrows indicate motor neurons. Expression of all three genes was decreased in the human ALS cord. Reproduced with permission from Corcoran et al., Absence of retinoids can induce motoneuron disease in the adult rat and a retinoid defect is present in motoneuron disease patients. J Cell Sci 115:4735–4741. Copyright 2002, The Company of Biologists.

### c. RNA Interference Therapy

A different form of gene therapy is RNA interference (RNAi) (Novina and Sharpe, 2004). RNAi is a natural biological defense mechanism against double-stranded RNA viruses and is also involved in gene regulation during development (Mello and Conte, 2004; Meister and Tuschl, 2004; Ambrose, 2004). Short interfering RNAs (siRNAs) are duplexes of RNA 21 to 23 nucleotides long. These duplexes can be synthesized so that the antisense strand is complementary to the mRNA transcribed by a specific gene. The antisense strand becomes complexed with a protein complex called the RNA-induced silencing complex (RISC). When the antisense strand of the siRNA binds to its complementary mRNA, the RISC either cleaves or prevents translation of the mRNA, thus effectively blocking the activity of the gene. SiRNAs are thus potentially useful in silencing disease-causing genes.

Two sets of experiments have demonstrated the potential usefulness of siRNAs in treating familial ALS by knocking down mutant SOD. SiRNAs designed to silence SOD1 mRNA were delivered in lentiviral vectors either into the lumbar spinal cord of G93A mice at sites that targeted the hindlimb muscles (Raoul et al., 2005), or into the hindlimb muscles themselves (Ralph et al., 2005). When injected into the spinal cord, the mutant protein was reduced by over 50% in mice younger than 90 days that had not yet displayed the symptoms of ALS. The neuromuscular function of the mice was improved by 40% and motoneuron survival increased by 60%. When injected into the hindlimbs, the lentivirus transduced over 50% of spinal neurons in the ventral lumbar cord and mutant protein expres-

**TABLE 6.10** Effect of silencing mutant *SOD1* expression in G93A mice by an EIAV-SOD1 RNA interference construct on ALS age of onset, age of end stage, and number of surviving lumbar cord motor neurons. Data from Ralph et al. (2005)

| | Uninjected control | EIAV control | EIAV-SOD1 |
|---|---|---|---|
| Age of onset (days) | 94.3 ± 2.9 | 94.1 ± 5 | 202.1 ± 7.5 |
| Age of end stage (days) | 128.3 ± 2.9 | 129.3 ± 1.6 | 227.8 ± 11.3 |
| Number of surviving motor neurons per section at end stage | 25 | 5 | 15 |

sion was reduced by 40%. **TABLE 6.10** shows that the age at which the disease was first manifested increased by a factor of two, life span was increased by 77%, and weight loss was not observed until late in the course of the disease. Motor neuron survival at the end stage of the disease was more than three times that of untreated mutant mice and about 60% that of wild-type mice. These results suggest that complete knockdown of the mutant SOD-1 gene theoretically might effect a cure for familial ALS in humans.

## SUMMARY

Although peripheral nerves regenerate well after a crush injury, they form scar tissue in the wound space after transection and develop neuromas. Biomimetic regeneration templates that bridge proximal and distal segments promote the regeneration of transected nerves. Autogeneic sensory nerve grafts are currently used as bridges, but the critical axon elongation length ($L_c$) promoted by these autografts and by silicon tubes is only 10 mm, well below the 25–80 mm deemed desirable. Thus, a variety of biomaterials, with or without supplementation with adhesion, growth and neurotrophic factors, have been tested for the ability to promote nerve regeneration. The degree to which a biomaterial supports nerve regeneration depends on its chemical composition, orientation of its structural materials, porosity, and degradation rate. Many biomaterials and supplements have been tried, but the best regeneration so far ($L_c = 25$) have been obtained with a copolymer bridge of type I collagen and chondroitin 6-sulfate with longitudinally oriented pores. The regenerated nerves do not synthesize endoneurium or epineurium within the bridge and do not reach the normal diameter of the nerve. In rats, normal walking patterns are not restored in experiments involving sciatic nerve regeneration, probably because the bridge material lacks the specificity to guide regenerating axons into their previous endoneurial tubes. Exposure to near infrared light accelerates the rate of facial nerve regeneration after crush injury. Whether or not a similar exposure of transected nerves to NIR elicits a similar effect is unknown.

In the central nervous system, regenerative therapies are being directed at the injured spinal cord and neurodegenerative diseases. Most of these are under development in experimental animals, although trials with human patients have begun in some cases. Therapies for spinal cord injury include treatment with neuroprotective agents, and inducing axon regeneration across the lesion by neutralizing inhibitory proteins, limiting the extent of glial scar, bridging the lesion with scaffolds and cell transplants, and inducing remodeling of synaptic organization by rehabilitation programs.

Neuroprotection aims to increase the survival of neurons. Neuroprotective agents work in a variety of ways, including stabilizing cell membranes (methylprednisolone), reducing calcium influx and glutamate excititoxicity (antagonists of ATP receptors, gacyclidine, NMDA, inhibitors of acid-sensing ion channels), neutralizing toxic free radicals (buckyballs), improving conduction in demyelinated axons (the potassium channel blocker, 4-aminopyridine) inhibiting inflammation (minocycline), promoting axon sprouting (neurotrophic and growth factors such as NT-3 and FGF-2), and augmenting cAMP levels (phosphodiesterase IV inhibitor, rolipram). High-dose methylprednisolone doubles the chances for functional recovery in human patients if administered within the first eight hr after injury, but is toxic. Clinical trials of gacycidine have not been very successful, but clinical trials of 4-aminopyridine suggest some improvement in bladder, bowel, and sexual functions.

Inhibiting the action of the myelin proteins Nogo, myelin associated glycoprotein and oligodendrocyte myelin protein with antibodies (Nogo), neurotrophic factors (MAG), or inhibitors of Rho kinase (all) promotes axon regeneration, as does vaccinating mice with myelin preparations prior to lesioning. Limiting the buildup of glial scar with CM101, or digesting glial scar with chondroitinase ABC resulted in significant functional recovery from paralysis in rats.

SCI lesions have been bridged by scaffolds of biological or biomimetic materials, or cell transplants. Scaffolds reported to promote axon regeneration across spinal cord lesions include a collagen matrix containing Ln or 4-aminopyridine, alginate gel, a hydrogel incorporating the RGD adhesive sequence, and segments of peripheral nerves embedded in a fibrin matrix impregnated with FGF-1. Peripheral nerve grafts connect the two sides of the lesion and allow regenerating axons to bypass glial scar. Nerve graft bridges have been used to treat SCI in nine human patients, but only one experienced any sensory or motor imorovement. Suspensions of Schwann cells and olfactory ensheathing cells also successfully promote CNS axon regeneration, as they do in peripheral nerves. Clinical trials using olfactory ensheathing cells or epithelium to treat human cases of SCI indicated that patients experience improved autonomic functions, but improvement in sensory and motor function was variable. Attempts to replace killed neurons by transplantation of NSCs have led to the discovery that that the NSCs, as well as several other types of transplanted cells including macrophages, splenic dendritic cells, and fibroblasts genetically modified to secrete neurotrophic factors, promote host neuron survival and axon regeneration, and reduce scarring.

The spinal cord retains synaptic plasticity into late life. In human SCI that spares circuitry, this plasticity can be tapped to restore motor circuits, either by treadmill exercises involving repetitive corrdinated motion, or electrical stimulation of muscles in patterns that mimic coordinated movements.

Complete functional recovery in cases of severe spinal cord trauma has not been achieved by any of these approaches, either in experimental animals or in humans. Combinations of different therapeutic approaches may be beneficial. For example, combining administration of rolipram and dibutryl cAMP with Schwann cell transplantation significantly improved locomotor function in SCI rats. Near-infrared light irradiation combined with other treatments may prove beneficial. Finally, a deeper understanding of spinal cord regeneration by the study of animal models that regenerate the cord well, such as amphibians, can potentially be translated to mammals.

The destruction of neurons by neurodegenerative diseases is another challenge for regenerative medicine. We do not understand the origin of most of these diseases, so we do not know whether regenerative therapy can "cure" them by replacing the lost cells. In cases where there is a genetic component to the disease, we may be able to replace the defective cells or the defective gene or missing molecules. Regenerative therapies, or therapies that slow cell loss, have been attempted for demyelinating disorders, Parkinson's disease, Huntington's disease, Alzheimer's disease, and amyotrophic lateral sclerosis. These therapies have been primarily cell or chemical therapies.

Glial precursors differentiated from ESCs have been shown to differentiate into oligodendrocytes and remyelinate axons when injected into the spinal cords and brains of mice suffering from demyelination. Human OECs can also remyelinate focally demyelinated spinal cord axons. Neurospheres derived from lateral ventricle NSCs differentiated into oligodendrocytes and remyelinated axons in a mouse model of multiple sclerosis, but they also induced host NSCs to differentiate into oligodendrocytes. Multiple sclerosis in humans is treated with several types of drugs, but clinical trials of remyelination therapies have not yet been conducted.

Parkinson's disease is the result of apoptosis of dopaminergic neurons (DANs) in the substantia nigra of the brain. Cell therapies have been sought to replace the apoptotic DANs. Parkinson's patients receiving striatal transplants of fetal mesencephalic cells showed variable degrees of improvement that was correlated with differential survival rates of the transplanted cells. Since fetal cells cannot be a standard source of transplanted cells, other sources of DANs have been sought. The unilateral 6-OHDA rat model of Parkinson's has been used to test the effect of transplanting carotid body cells, NSCs, or DANs derived by the directed differentiation of ESCs. Survival of the cells is variable, but dopamine production is increased. One aspect of symptoms in this model, amphetamine-induced turning, is significantly reduced, but other symptoms such as rigidity, bradykinesia, and akinesia are not improved. The effects of cell transplants are duplicated in Parkinsonian mice by injection of viral constructs encoding glial-derived neurotrophic factor or by administration of TGF-$\alpha$ to the striatum. These factors promoted the survival of host DANs or induced NSCs to differentiate into DANs, indicating the feasibility of pharmaceutical therapies that rescue DANs or stimulate their local regeneration.

Fetal striatal tissue grafted to the striatum of a monkey model of Huntington's disease differentiated new striatal neurons and reversed disease symptoms. Preliminary clinical trials have shown that grafts of fetal striatal tissue survive and alleviate the symptoms of Huntington's to some degree in human patients. Huntington's symptoms are also improved in monkey and mouse models by transplants of human umbilical cord stem cells or baby hamster kidney cells transgenic for ciliary neurotrophic factor, again suggesting the feasibility of a pharmaceutical approach to slowing neuron loss and regenerating new neurons.

Transplants of glial cells from cholinergic tissue, but not cholinergic neurons themselves, promoted the recovery of cognitive function when transplanted into the terminal area of the forebrain cholinergic projecton system of rat models of Alzheimer's disease. This suggests that the glial cells stimulated Ach secretion by host cholinergic neurons, as well as fostering axonal outgrowth and Ach secretion by donor cholinergic neurons. Several drugs used to reduce symptoms of

Alzheimer's and slow cognitive decline in human patients work by slowing the rate of breakdown of Ach. Other drugs target the aggregation of β-amyloid peptides into plaques and another drug in phase II trials does both. People who exercise physically and cognitively, and who have a stong social network, are at lower risk for Alzheimer's, a fact that is reflected in mouse models of the disease, where enriched environment reduces the levels of β-amyloid peptide and increases the levels of neprilysin, an amyloid-degrading enzyme.

Amyotrophic lateral sclerosis is a disease in which motor neurons undergo apoptosis. Chronic glutamate toxicity due to downregulation of glutamate transporters (GLT1) plays a prominent role in progression of the disease. In most cases, the cause of ALS is unknown, but 5%–10% of cases are due to familial inheritance of mutations in the superoxide dismutase-1 (SOD-1) gene. A great deal of effort is being put into finding therapies that will slow the progress of the disease. The G93A mouse model, which is transgenic for a mutated human SOD-1 gene associated with ALS, is the most common test-bed for potential therapies. So far, the therapeutic approaches have been cell transplants, administration of neuroprotective agents, and RNA interference. Cells that have been transplanted into the spinal cord of the G93A mouse are human umbilical cord blood cells, dental pulp stem cells, and neurons from a human neuron cell line (hNT). The transplants delayed the onset of ALS and extended the life span of the mouse but did not effect a cure. ALS patients have been treated by injections of autologous peripheral blood stem cells or bone marrow cells into the spinal cord, or autologous OECs into the frontal lobes. Of 100 patients treated with OECs, about half reported stabilization or even improvement of their condition.

The effect of the cell transplants on ALS is believed to be due to neuroprotection. Several neuroprotective agents have been identified that delay the onset and slow the progress of ALS in the G93A mouse model. These include the genes for GDNF, VEGF, and Bcl-2 delivered in viral vectors. Bcl-2 inhibits apoptosis; administration of agents that inhibit caspase activity had the same effect. Fifteen different β-lactam antibiotics were found to be neuroprotective through upregulation of the *GLT1* gene. One of these antibiotics, ceftriaxone, was reported to delay loss of muscle strength and increase survival time of G93A mice. Retinoic acid deficiency may also play a role in loss of motor neurons, since the surviving motor neurons of ALS patients are found to be deficient in RALDH2, the enzyme that converts retinaldehyde to RA.

Finally, injection of siRNAs that silence mutant SOD1 mRNA were effective in delaying the onset of ALS in G93A mice and extended life span significantly. Motor neuron survival at the end stage of the disease was more than three times that of untreated G93A mice and about 60% that of wild-type mice.

Combinations of these treatments may potentially be even more effective in combating ALS.

## REFERENCES


Aebischer P, Salessiotis AN, Winn SR (1989) Basic fibroblast growth factor released from synthetic guidance channels facilitates peripheral nerve regeneration across long nerve gaps. J Neurosci Res 23:282–289.

Alberi L, Sgado P, Simon HH (2004) Engrailed genes are cell-autonomously required to prevent apoptosis in mesencephalic dopaminergic neurons. Development 131:3229–3236.

ALS CNTF Study Group (1996) A double-blind placebo-controlled clinical trial of subcutaneous recombinant human ciliary neurotrophic factor (rHCNTF) in amyotrophic lateral sclerosis. Neurology 46:1244–1249.

Ambrose V (2004) The functions of animal microRNAs. Nature 431:350–355.

Anders JJ, Borke RC, Woolery SK, Van de Merwe WP (1993) Low power laser irradiation alters the rate of regeneration of the rat facial nerve. Lasers in Surg and Med 13:72–82.

Andrews MR, Stelzner DJ (2004) Modification of the regenerative response of dorsal column axons by olfactory ensheathing cells or peripheral axotomy in adult rat. Exp Neurol 190:311–327.

Ansselin AD, Fink , Davey DF (1997) Peripheral nerve regeneration through nerve guides seeded with adult Schwann cells. Neuropathol Appl Neurobiol 23:387–398.

Armstrong RJ, Barker RA (2001) Neurodegeneration: A failure of neuroregeneration? Lancet 358:1174–1176.

Armstrong RJE, Hurelbrink CB, Tyera P, Ratcliffe EL, Richards A, Dunnett SB, Rosser AE, Barker RA (2002) The potential for circuit reconstruction by expanded neural precursor cells explored through porcine xenografts in a rat model of Parkinson's disease. Exp Neurol 175:98–111.

Arnett HA, Fancy SPJ, Alberta JA, Zhao C, Plant SR, Kaing S, Raine CS, Rowitch DH, Franklin RJM, Stiles CD (2004) bHLH transcription factor Olig1 is required to repair demyelinated lesions in the CNS. Science 306:2111–2115.

Arrasate M, Mitra S, Schweitzer ES, Segal MR, Finkbeiner S (2004) Inclusion body formation reduces levels of mutant huntingtin and the risk of neuronal death. Nature 431:805–810.

Azzouz M, Ralph GS, Storkebaum E, Walmsley LE, Mitrophanous K, Kingsman SM, Carmeliet P, Mazarakis ND (2004) VEGF delivery with retrogradely transported lentivector prolongs survival in a mouse ALS model. Nature 429:413–417.

Bachoud-Levy AC, Hantraye P, Peschanski M(2002) Fetal neural grafts for Huntington's disease: A prospective view. Movement Disorders 17:439–444.

Bandtlow CE, Schmidt MF, Hassinger TD, Schwab ME, Kater SB (1993) Role of intracellular calcium in NI-35-evoked collapse of neuronal growth cones. Science 259:80–83.

Barberi T, Klivenyi P, Calingasan NY, Lee H, Kawamata H, Loonam K, Perrier AL, Bruses J, Rubio ME, Topf N, Tabar V, Harrison NL, Beal MF, Moore MAS, Studer L (2003) Neural subtype specification of fertilization and nuclear transfer embryonic stem cells and application in parkinsonian mice. Nature Biotech 21:1200–1207.

Barnett SC, Chang L (2004) Olfactory ensheathing cells and CNS repair: Going solo or in need of a friend? Trends Neurosci 27:54–60.

Becker T, Bernhardt RR, Reinhard E, Wullimann MF, Tongiorgi E, Schachner M (1998) Readiness of zebrafish brain neurons to

regenerate spinal axon correlates with differential expression of specific cell recognition molecules. J Neurosci 18:5789–5803.
Bellamkonda R, Aebischer P (1995) Tissue engineering in the nervous system. In: Bronzino JD (ed.). The Biomedical Engineering Handbook. Boca Raton, FL, CRC Press, pp 1754–1773.
Bhatt DH, Otto SJ, Depoister B, Fetcho JR (2004) Cyclic AMP-induced repair of zebrafish spinal circuits. Science 305:254–258.
Bianco JI, Perry C, Harkin DG, Mackay-Sim A, Feron F (2004) Neurotrophin-3 promotes purification and proliferation of olfactory ensheathing cells from human nose. Glia 45:111–123.
Bjorklund A, Lindvall O (2000) Cell replacement therapies for central nervous system disorders. Nature Neurosci 3:537–544.
Bjorklund LM, Sanchez-Pernaute R, Chung S, Anderson T, Chen IYC, McNaught KP, Brownell A-L, Jenkins BG, Wahlstedt C, Kim K-S, Isacson O (2002) Embryonic stem cells develop into functional dopaminergic neurons after transplantation in a Parkinson rat model. Proc Natl Acad Sci USA 99:2344–2349.
Bjorklund A, Dunnett SB, Brundin P, Stoessl AJ, Freed CR, Breeze RE, Levivier M, Peschanski M, Studer L, Barker R (2003) Neural transplantation for the treatment of Parkinson's disease. Lancet Neurol 2:437–445.
Borasio GD, Robberecht W, Leigh PN, Emile J, Guiloff RJ, Jerusalem F, Silani V, Vos PE, Wokke JHJ, Dobbins T, European ALS/IGF-I Study Group (1998) A placebo-controlled trial of insulin-like growth factor I in amyotrophic lateral sclerosis. Neurology 51:585–586.
Boyd JG, Skihar V, Kawaja M, Doucette R (2003) Olfactory ensheathing cells: Historical perspective and therapeutic potential. Anat Rec (Part B: New Anat) 271B:49–60.
Bradbury EJ, Moon LDF, Popat RJ, King VR, Bennet GS, Patel PN, Fawcett JW, McMahon SB (2002) Chondroitinase ABC promotes functional recovery after spinal cord injury. Nature 416:636–640.
Bracken MB, Shepard MJ, Collins WF, Holford TR, Young W, Baskin DS, Eisenberg HM, Flamm E, Leo-Summers L, Maroon JC, Marshall LF, Perot DL, Piepmeier J, Sonntag E, Wagner FC, Wilberger JL, Winn HR, Young W (1990) A randomized controlled trial of methylprednisolone or nalaxone in the treatment of acute spinal cord injury. New Eng J Med 322:1405–1461.
Bracken MB, Shepard MJ, Collins WF, Holford TR, Young W, Baskin DS, Eisenberg HM, Flamm E, Leo-Summers L, Maroon JC, Marshall LF, Perot DL, Piepmeier J, Sonntag E, Wagner FC, Wilberger JL, Winn HR, Young W (1992) Methylprednisolone or nalaxone treatment after acute spinal cord injury: 1-year follow-up data. J Neurosurg 76:23–31.
Bruckner MK, Arendt T (1992) Intracortical grafts of purified astrocytes ameliorate memory deficits in rat induced by chronic treatment with alcohol. Neurosci Lett 141:251–254.
Brustle O, Choudhary K, Karram K, Huttner A, Murray K, Dubois-Dalcq M, McKay RDG (1998) Chimeric brains generated by intraventricular transplantation of fetal human brain cells into embryonic rats. Nature Biotech 16:1040–1044.
Brustle O, Jones N, Learish R, Karram K, Choudhary K, Wiestler OD, Duncan ID, McKay RDG (1999) Embryonic stem cell-derived glial precursors. A source of myelinating transplants. Science 285:754–756.
Bunge MB (2002) Bridging the transected or contused adult rat spinal cord with Schwann cell and olfactory ensheathing glia transplants. Prog Brain Res 137:275–282, 2002.
Burke RE (2001) The central pattern generator for locomotion in mammals. Adv Neurol 87:11–24.
Cai D, Shen Y, DeBellard M, Tang S, Filbin MT (1999) Prior exposure to neurotrophins blocks inhibition of axonal regeneration by MAG and myelin via a cAMP-dependent mechanism. Neuron 22:89–101.
Cai D, Qiu J, Cao Z, McAtee M, Bregman BS, Filbin MT (2001) Neuronal cycle cAMP controls the developmental loss in ability of axons to regenerate. J Neurosci 21:4731–4739.
Cai D, Deng K, Mellado W, Lee J, Ratan RR, Filbin MT (2002) Arginase I and polyamines act downstream from cyclic AMP in overcoming inhibition of axonal growth by MAG and myelin *in vitro*. Neuron 35:711–719.
Cedarbaum JM, Stambler N, Brooks BR, Bradley W (1998) Brain-derived neurotrophic factor in amyotrophic lateral sclerosis: A failed drug or a failed trial. Ann Neurol 44:506–507.
Chamberlain LJ, Yannas IV, Hsu H-P, Strichartz G, Spector M (1998) Collagen-GAG substrate enhances the quality of nerve regeneration through collagen tubes up to level of autograft. Exp Neurol 154:315-329.
Chen A, Xu X, Kleitman B, Bunge MB (1996) Methylprednisolone administration improves axonal regeneration into Schwann cell grafts in transected adult rat thorasic spinal cord. Exp Neurol 138:261–276.
Chen MS, Huber AB, Can der Haar ME, Fran M, Schnell L, Spillmann AA, Christ F, Schwab ME (2000) Nogo-A is a myelin-associated neurite outgrowth inhibitor and an antigen for monoclonal antibody IN-1. Nature 403:434–439.
Chen R, Ende N (2000) The potential for the use of mononuclear cells from human umbilical cord blood in the treatment of amyotrophic lateral sclerosis in SOD1 mice. J Med 31:21–30.
Cheng H, Cao Y, Olson L (1996) Spinal cord repair in adult paraplegic rats: Partial restoration of hind limb function. Science 273:510–513.
Cheng H, Liao KK, Liao SF, Chuang TY, Shih YH (2004) Spinal cord repair with acidic fibroblast growth factor as a treatment for a patient with chronic paraplegia. Spine 29:E284–E288.
Chernoff EAG (1996) Spinal cord regeneration: A phenomenon unique to urodeles? Int J Dev Biol 40:823–832.
Chernoff GF (1981) Shiverer: An atosomal recessive mutant mouse with myelin deficiency. J Hered 72:128–130.
Cho K-S, Yang L, Lu B, Ma HF, Huang Z, Pekny M, Chen DF (2005) Re-establishing the regenerative potential of central nervous system axons in postnatal mice. J Cell Sci 118:863–872.
Choi-Lundberg DL, Lin Q, Chang Y-N, Chiang YL, Hay CM, Mohajeri H, Davidson BL, Bohn MC (1997) Dopaminergic neurons protected from degeneration by GDNF gene therapy. Science 275:838–841.
Chua MI, Choi-Lundberg D, Weston S, Vincent AJ, Chung RS, Vickers JC, West AK (2004) Olfactory ensheathing cells promote collateral axonal branching in the injured adult rat spinal cord. Exp Neurol 185:15–25.
Clement A, Nguyen D, Roberts EA, Garcia ML, Boille S, Rule M, McMahon AP, Doucette W, Siwek D, Ferrante RJ, Brown RH, Julien J-P, Goldstein LSB, Cleveland DW (2003) Wild-type non-neuronal cells extend survival of SOD1 mutant motor neurons in ALS mice. Science 302:113–116.
Constans A (2004) Neural tissue engineering. The Scientist June 21, pp 40–42.
Corcoran S, So PL, Maden M (2002) Absence of retinoids can induce motoneuron disease in the adult rat and a retinoid defect is present in motoneuron disease patients. J Cell Sci 115:4735–4741.
Corse AM, Rothstein JD (1995) Organotypic spinal cord cultures and a model of chronic glutamate-mediated motot neuron degeneration. In: Ohnishi ST, Ohnishi T (eds.). Central Nervous System Trauma Research Techniques. New York, CRC Press, pp 341–351.
Corse AM, Bilak MM, Bilak SR, Lehar M, Rothstein JD, Kuncl RW (1999) Preclinical testing of neuroprotective neurotrophic factors in a model of chronic motor neuron degeneration. Neurobiol Dis 6:335–346.

Cote F, Collard J-F, Julien J-P (1993) Progressive neuronopathy in transgenic mice expressing the human neurofilament heavy gene: A mouse model of amyotrophic lateral sclerosis. Cell 73:35–46.
Couzin J (2005) Magnificient obsession. Science 307:1713–1715.
Cova L, Ratti A, Volta M, Fogh I, Cardin V, Corbo M, Silani V (2004) Stem cell therapy for neurodegenerative diseases: The issue of transdifferentiation. Stem Cells Dev 13:121–131.
Cunningham LA, Su C (2002) Astrocyte delivery of glial cell line-derived neurotrophic factor in a mouse model of Parkinson's disease. Exp Neurol 174:230–242.
Cunnings JL (2004) Alzheimer's disease. New Eng J Med 35:56–67.
Curt A, Schwab ME, Dietz V (2004) Providing the clinical basis for new interventional therapies: refined diagnosis and assessment of recovery after spinal cord injury. Spinal Cord 42:1–6.
Curt A, Dietz V (2005) Controversial treatments for spinal-cord injuries. Lancet 365:841.
Curtis MA, Penney EB, Pearson AG, van Roon-Mom WMC, Butterworth NJ, Dragunow M, Connor B, Faull RLM (2003) Increased cell proliferation and neurogenesis in the adult human Huntington's disease brain. Proc Natl Acad Sci 100:9023–9027.
Cyranoski D (2005) Paper chase. Nature 437:810–811.
DeLucia TA, Connors JJ, Brown TJ, Cronin CM, Khan T, Jones KJ (2003) Use of a cell line to investigate olfactory ensheathing cell-enhanced axonal regeneration. Anat Rec (Part B: New Anat) 271B:61–70.
De Vivo MJ (1997) Causes and costs of spinal cord injury in the United States. Spinal Cord 35:809–813.
Dergham P, Ellezam B, Essagian C, Avedissian H, Lubell WD, McKerracher L (2002) Rho signaling pathway targeted to promote spinal cord repair. J Neurosci 22:6570–6577.
Dugan L, Turetsky DM, Du C, Lobner D, Wheeler M, Almli CR, Clifton K-FS, Luh T-Y, Choi DW, Lin T-S (1997) Carboxyfullerenes as neuroprotective agents. Proc Natl Acad Sci USA 94:9434–9439.
Emerich DF, Winn SR, Hantraye PM, Peshanski M, Chen E-Y, Chu Y, McDermott P, Baetge EE, Kordower JH (1997) Protective effect of encapsulated cells producing neurotrophic factor CNTF in a monkey model of Huntington's disease. Nature 386:395–398.
Ende N, Weistein F, Chen R, Ende M (2000) Human umbilical cord blood effect on SOD mice (amyotrophic lateral sclerosis). Life Sci 67:53–59.
Espejo EF, Montoro RJ, Armengol JA, Lopez-Barneo J (1998) Cellular and functional recovery of Parkinsonian rats after intrastriatal transplantation of carotid body cell aggregates. Neuron 20:197–206.
Fallon J, Reid S, Kinyamy R, Opole I, Opole R, Baratta J, Korc M, Endo TL, Duong A, Nguyen G, Karkehabadhi M, Twardzik D, Loughlin S (2000) *In vivo* induction of massive proliferation, directed migration, and differentiation of neural cells in the adult mammalian brain. Proc Natl Acad Sci USA 97:14686–14691.
Fassas A, Kazis A (2003) High-dose immunosuppression and autologous hematopoietic stem cell rescue for severe multiple sclerosis. J Hematother Stem Cell Res 12:701–711.
Feldblum S, Arnaud M, Simon M, Rabin O, d'Arbigny P (2000) Efficacy of a new neuroprotective agent, gacyclidine, in a model of rat spinal cord injury. J Neurotraum 17:1079–1093.
Filbin MT (2000) Axon regeneration: vaccinating against spinal cord injury. Curr Biol 10:R100–R103.
Flax JD, Aurora S, Yang C, Simonin C, Wills AM, Billinghurst LL, Jendoubi M, Sidman RL, Wolfe JH, Kim SU, Snyder EY (1998) Engraftable human neural stem cells respond to developmental cues, replace neurons and express foreign genes. Nature Biotech 16:1033–1039.
Fournier AE, Takizawa BT, Strittmatter SM (2003) Rho kinase inhibition enhances axonal regeneration in the injured CNS. J Neurosci 23:1416–1423.
Franklin RJM (2003) Remyelination by transplanted olfactory ensheathing cells. Anat Rec (Part B: New Anat) 271B:71–76.
Freed CR, Greene PE, Breeze RE, Tsai W-Y, Du Mouchel W, Kao R, Dillon S, Winfield H, Culver S, Trojanowski Q, Eidelberg D, Fahn S (2001) Transplantation of embryonic dopamine neurons for severe Parkinson's disease. New Eng J Med 344:710–719.
Freeman TB, Cicchetti F, Hauser RA, Deacon TW, LI X-J, Hersch SM, Nauert GM, Sanberg PR, Kordower JH, Saporta S, Isacson O (2002) Transplanted fetal stratum in Huntington's disease: Phenotypic development and lack of pathology. Proc Natl Acad USA 97:13877–13882.
Fricker RA, Carpenter MK, Winkler C, Greco C, Gates MA, Bjorklund A (1999) Site-specific migration and neuronal differentoation of human neural progenitor cells after transplantation in the adult rat brain. J Neurosci 19:5990–6005.
Fournier AE, Takizawa T, Strittmatter SM (2003) Rho kinase inhibition enhances axonal regeneration in the injured CNS. J Neurosci 23:1416–1423.
Garbuzova-Davis, Willing AE, Milliken M, Saporta S, Zigova T, Cahill DW, Sanberg PR (2002) Positive effect of transplantation of hNT neurons (Ntera 2/D1 cell-line) in a model of familial amyotrophic lateral sclerosis. Exp Neurol 174:169–180.
Garbuzova-Davis S, Willing AE, Zighova T, Saporta S, Justen EB, Lane JC, Hudson JE, Chen N, Davis CD, Sanberg PR (2003) Intravenous administraton of human umbilical cord blood cells in mouse model of amyotrophic lateral sclerosis: distribution, migration, and differentiation. J Hematother Stem Cell Res 12:255–270.
Gauthier LR, Vharrin BC, Borrell-Pages M, Dompierre JP, Rangone H, Cordelieres FP, De Mey J, MacDonald ME, Lebmann V, Humbert S, Saudou F (2004) Huntingtin controls neurotrophic support and survival of neurons by enhancing BDNF vesicular transport along microtubules. Cell 118:127–138.
Gaviria M, Privat A, d'Arbigny P, Kamenka J, Haton H, Ohanna F (2000) Neuroprotective effects of a novel NMDA antagonist, gacyclidine, after experimental contusive spinal cord injury in adult rats. Brain Research 874:200–209.
Geisler FH, Coleman WP, Grieco G, Poonian D (2001) The Sygen multicenter acute spinal cord injury study. Spine 26:S87–S98.
Geller HM, Fawcett JW (2002) Building a bridge: Engineering spinal cord repair. Exp Neurol 174:125–136.
Gestwicki JE, Crabtree GR, Graef I (2004) Harnessing chaperones to generate small-molecule inhibitors of amyloid β aggregation. Science 306:865–869.
Goldman S (2005) Stem and progenitor cell-based therapy of the human central nervous system. Nature Biotech 23:862–866.
Goldsmith HS, de la Torre JC (1992) Axonal regeneration after spinal cord transection and reconstruction. Brain Res 589: 217–224.
GrandPre T, Li S, Strittmatter SM (2002) Nogo-66 receptor antagonist peptide promotes axonal regeneration. Nature 417:547–555.
Greenamyre JT, Hastings TG (2004) Parkinson's—divergent causes, convergent mechanisms. Science 304:1120–1122.
Grill R, Murai K, Blesch A, Gage FH, Tuszynski MH (1997) Cellular delivery of neurotrophin-3 promotes corticospinal axonal growth and partial functional recovery after spinal cord injury. J Neurosci 17:5560–5572.
Grinspan J (2002) Cells and sinaling in oligodendrocyte development. J Neuropathol Exp Neurol 61:297–306.
Guest JD, Rao A, Olson L, Bunge MB, Bunge RP (1997) The ability of human Schwann cell grafts to promote regeneration in the transected nude rat spinal cord. Exp Neurol 148:502–522.

Hall S (1997) Axonal regeneration through acellular muscle grafts. J Anat 190:57–71.

Harkey HL III, White EA IV, Tibbs RE, Haines DE (2003) A clinician's view of spinal cord injury. Anat Rec (Part B: New Anat) 271B:41–48.

Hauser RA, Furtado S, Cimino CR, Delgado H, Eichler S, Schwartz S, Scott D, Nauert GM, Soety E, Sossi V, Holt DA, Sanberg PR, Stoessl AJ, Freeman TB (2002) Bilateral human fetal striatal transplantation in Huntington's disease. Neurology 58:687–695.

Hems TEJ, Glasby MA (1993) The limit of graft length in the experimental use of muscle grafts for nerve repair. J Hand Surg (British and European) Volume 18:165–170.

Henry EW, Chiu T-H, Nyilas E, Brushart TM, Dikkes P, Sidman RL (1985) Nerve regeneration through biodegradable polyester tubes. Exp Neurol 90:652–676.

Hill CE, Proschel C, Noble M, Mayer-Proschel M, Gensel JC, Beattie MS, Bresnahan JC (2004) Acute transplantation of glial-restricted precursor cells into spinal cord contusion injuries: survival, differentiation, and effects on lesion environment and axonal regeneration. Exp Neurol 190:289–310.

Hirano A, Kato S (1992) Fine structural study of sporadic and familial amyotrophic lateral sclerosis. In: Handbook of of Amyotrophic Lateral Sclerosis, Smith RA (ed.). New York, Marcel Dekker Inc, pp 193–192.

Hodges H, Allen Y, Sinden J, Lantos PL, Gray JA (1991a) Effects of cholinergic-rich neural grafts on radial maze performance of rats after excitotoxic lesions of the forebrain cholinergic projection system. I. Amelioration of cognitive defects by transplants into cortex and hippocampus but not into basal forebrain. Neurosci 45:587–607.

Hodges H, Allen Y, Sinden J, Lantos PL, Gray JA (1991b) Effects of cholinergic-rich neural grafts on radial maze performance of rats after excitotoxic lesions of the forebrain cholinergic projection system. II. Cholinergic drugs as probes to investigate lesion induced deficits and transplant-induced functional recovery. Neurosci 45:609–623.

Hodges H, Allen Y, Sinden J, Mitchell SN, Arendt T, Lantos PL, Gray JA (1991c) The effects of cholinergic drugs and cholinergic-rich fetal neural transplants on alcohol-induced deficits in radial-maze performance in rats. Behav Brain Res 43:7–28.

Hofstetter CP, Holmstrom NAV, Lilja JA, Schweinhardt P, Hao J, Spenger C, Wiesenfeld-Hallin Z, Kurpad SN, Frisen J, Olson L (2005) Allodynia limits the usefulness of intraspinal neural stem cell grafts; directed differentiation improves outcome. Nature Neurosci 8:346–353.

Hori J, NgTF, Shatos M, Klassen H, Streilein JW, Young MJ (2003) Neural progenitor cells lack immunogenecity and resist destruction as allografts. Stem Cells 21:405–416.

Horner PJ, Gage FH (2000) Regenerating the damaged central nervous system. Nature 407:963–969.

Huang DW, McKerracher L, Braun D (1999) A therapeutic vaccine approach to stimulate axon regeneration in the adult mammalian spinal cord. Neuron 24:639–647.

Huang H (2003) Influence of patient's age on functional recovery after transplantation of olfactory ensheathing cells into injured spinal cord injury. Chin Med J (Engl) 116:1488–1491.

Hugenholtz H (2003) Methylprednisolone for acute spinal cord injury: not a standard of care. Can Med Assoc J 168:1145–1146.

Hynes M, Rosenthal A (1999) Specification of dopaminergic and serotoergic neurons in the vertebrate CNS. Curr Opinion Neurobiol 9:26–36.

Huntington's Disease Collaborative Research Group (1993) A novel gene containing a trinucleotide repeat that is expanded and unstable on Huntington's disease chromosomes. Cell 72:971–983.

Ijkema-Paassen J, Jansen K, Gramsbergen A, Meek MF (2004) Transection of peripheral nerves, bridging strategies and effect evaluation. Biomats 25:1583–1592.

Italian ALS Study Group (1996) Eur J Neurol 3:295.

Iwashita, Kawaguchi S, Murata M (1994) Restoration of function by replacement of spinal cord segments in the rat. Nature 367:167–170.

Janson CG, Ramesh TM, During MJ, Leone P, Heywood J (2001) Human intrathecal transplantation of peripheral blood stem cells in amyotrophic lateral sclerosis. J Hematother Stem Cell Res 10:913–915.

Jin K, Peel AL, Mao XO, Xie L, Cottrell BA, Henshall DC, Greenberg DA (2004) Increased hippocampal neurogenesis in Alzheimer's disease. Proc Natl Acad Sci 101:343–347.

Kaku DA, Giffard RG, Choi DW (1993) Neuroprotective effects of glutamate antagonists and extracellular acidity. Science 260: 1516–1518.

Karnezis T, Mandemakers W, McQualter JL, Zheng B, Ho PP, Jordan KA, Murray BM, Barres B, Tessier-Lavigne M, Bernard CCA (2004) The neurite outgrowth inhibitor Nogo A is involved in autoimmune-mediated demyelination. Nature Neurosci 7:736–744.

Karsten SL, Geschwind DH (2005) Exercise your amyloid. Cell 120:572–574.

Kataoka K, Suzuki Y, Kitada M, Hashimoto T, Chou H, Bai H, Ohta M, Wu S, Suzuki K, Ide C (2004) Alginate enhances elongation of early regenerating axons in spinal cord of young rats. Tiss Eng 10:493–504.

Kato T, Honmou O, Uede T, Hashi K, Kocsis JD (2000) Transplantation of human olfactory ensheathing cells elicits remyelination of demyelinated rat spinal cord. Glia 30:209–218.

Kawahara Y, Ito K, Sun H, Aizawa H, Kanazawa I, Kwak S (2004) RNA editing and death of motor neurons. Nature 427:801.

Kendall AL, Rayment FD, Torres EM, Baker HF, Ridley RM, Dunnett SB (1998) Functional integration of striatal allografts in a primate model of Huntington's disease. Nature Med 4:727–729.

Kierstead HS, Nistor G, Bernal G, Totoiu M, Cloutier F, Sharp K, Steward O (2005) Human embryonic stem cell-derived oligodendrocyte progenitor cell transplants remyelinate and restore locomotion after spinal cord injury. J Neurosci 25:4694–4705.

Kim J-H, Auerbach J, Rodriguez-Gomez JA, Velasco I, Gavin, Lumelsky N, Lee S-H, Nguyen J, Sanchez-Pernaute R, Bankiewicz K, McKay R (2002) Dopamine neurons derived from embryonic stem cells function in an animal model of Parkinson's disease. Nature 418:50–56.

Klein S, Svendsen CN (2005) Stem cells in the injured spinal cord: reducing the pain and increasing the gain. Nature Neurosci 8:259–260.

Klein SM, Behrstock S, McHugh J, Hoffmann K, Wallace K, Suzuki M, Aebischer P, Svendsen CN (2005) GDNF delivery using human neural progenitor cells in a rat model of ALS. Human Gene Ther 16:509–521.

Koprivica V, Cho K-S, Park JB, Yiu G, Atwal J, Gore B, Kim JA, Lin E, Tessier-Lavigne M, Chen DF, He Z (2005) EGFR activation mediates inhibition of axon regeneration by myelin and chondroitin sulfate proteoglycans. Science 310:106–110.

Kordower JH, Emborg ME, Bloch J, Ma SY, Chu Y, Leventhal L, McBride J, Chen E-Y, Palfi S, Roitberg BZ, Brown WD, Holden JE, Pyzalski R, Taylor MD, Carvey P, Ling ZD, Trono D, Hantraye P, Deglon N, Aebischer P (2000) Neurodegeneration prevented by lentiviral vector delivery of GDNF in primate models of Parkinson's disease. Science 290:767–773.

Kostic V, Jackson-Lewis V, de Bilbao F, Dubois-Dauphin M, Przedborski S (1997) Bcl-2: Prolonging life in a transgenic mouse model of familial amyotrophic lateral sclerosis. Science 277:559–562.

Kwon BK, Liu J, Messerer C, Kobayashi NR, McGraw J, Oshchipok L, Tetzlaff W (2002) Survival and regeneration of rubrospinal neurons 1 year after spinal cord injury. Proc Natl Acad Sci USA 99:3246–3251.

Kwon BK, Tetzlaff W, Grauer JN, Beiner J, Vaccaro AR (2004) Pathophysiology and pharmacologic treatment of acute spinal cord injury. Spine J 4:451–464.

Lai EC, Felice KJ, Festoff BW, Gawel MJ, Gelinas DF, Kratz R, Murphy MF, Natter HM, Norris FH, Rudnicki SA (1997) Effect of recombinant human insulin-like growth factor-I on progression of ALS. A placebo-controlled study. Neurology 49:1621–1630.

Lazarov O, Lee M, Peterson DA, Sisodia SS (2002) Evidence that synaptically released beta-amyloid accumulates as extracellular deposits in the hippocampus of transgenic mice. J Neurosci 22:9785–9793.

Lazarov O, Robinson J, Tang Y-P, Hairston I, Korade-Mirnics Z, Lee V M-Y, Hersh LB, Sapolsky RM, Mirnics K, Sisodia SS (2005) Environmental enrichment reduces Aβ levels and amyloid deposition in transgenic mice. Cell 120:701–713.

Lazarov-Spiegler O, Solomon AS, Zeev-Brann AB, Hirschberg DL, Lavie V, Schwartz M (1998) Transplantation of activated macrophages overcomes central nervous system regrowth failure. FASEB J 10:1296–1302.

Lazic SE, Barker RA (2003) The future of cell-based transplantation therapies for neurodegenerative disorders. J Hematother Stem Cell Res 12:635–642.

Lee SH, Lumelsky N, Auerbach JM, McKay RD (2000) Efficient generation of midbrain and hindbrain neurons from mouse embryonic stem cells. Nature Biotech 18:675–679.

Leist M, Ghezzi P, Grasso G, Bianchi R, Villa P, Fratelli M, Savino C, Bianchi M, Nielsen J, Gerwien J, Kallunki P, Larsen AK, Helboe L, Christensen S, Pedersen LO, Nielsen M, Torup L, Sager T, Sefacteria A, Erbayraktar S, Erbayrakter Z, Gokmen N, Yilmaz O, Cerami-Hand C, Xie Q-W, Coleman T, Cerami A, Brines M (2004) Derivatives of erythropoietin that are tissue protective but not erythropoietic. Science 305:239–242.

Li Y, Field PM, Raisman G (1998) Regeneration of adult rat corticospinal neurons induced by transplanted olfactory ensheathing cells. J Neurosci 18:10514–10524.

Li M, Ona VO, Guegan C, Chen M, Jackson-Lewis V, Andrews LJ, Olzewski AJ, Steig PE, Lee J-P, Przedborski S, Friedlander RM (2000) Functional role of caspase-1 and caspase-3 in an ALS transgenic mouse model. Science 288:335–339.

Li S, Liu BP, Budel S, Li M, Ji B, Walus L, Li W, Jirik A, Rabacchi S, Choi E, Worley D, Sah DWY, Pepinsky B, Lee D, Relton J, Strittmatter SM (2004) Blockade of Nogo-66, myelin-associated glycoprotein, and oligodendrocyte myelin glycoprotein by soluble Nogo-66 receptor promotes axonal sprouting and recovery after spinal injury. J Neurosci 24:10511–10520.

Lin C-L, Bristol LA, Jin L, Dykes-Hoberg M, Crawford T, Rothstein JD (1998) Aberrant RNA processing in a neurodegenerative disease: A common cause for loss of glutamate trans[ort EAAT2 protein in sporadic amyotrophic lateral sclerosis. Neuron 20:589–602.

Lindvall O, Hagell P (2001) Cell therapy and transplantation in Parkinson's disease. Clin Chem Lab Med 39:356–361.

Liu CY, Westerlund U, Svensson M, Moe MC, Varghese M, Berg-Johnsen J, Apuzzo MLJ, Tirrell DA, Langmoen IA (2003) Artificial niches for human adult neural stem cells: Possibility for autologous transplantation therapy. J Hematother Stem Cell Res 12:689–699.

Lu QR, Sun T, Zhu Z, Ma N, Garcia M, Stiles CD, Rowitch DH (2002) Common developmental requirement for *Olig* function indicates a motor neuron/oligodendrocyte connection. Cell 109:75–86.

Lu J, Feron F, Ho SM, Mackay-Sim A, Waite PM (2001) Transplantation of nasal olfactory tissue promotes partial recovery in paraplegic adult rats. Brain Res 889:344–357.

Lundborg G, Dahlin LB, Danielsen N, Gelberman RH, Longo FM, Powell HC, Varon S (1982) Nerve regeneration in silicone model chambers: Influence of gap length and of distal stump components. Exp Neurol 76:361–375.

Luo J, Borgens R, Shi R (2002) Polyethylene glycol immediately repairs neuronal membranes and inhibits free radical production after acute spinal cord injury. J Neurochem 83:471–480.

Ma W, Fitzgerald W, Liu Q-Y, O'Shaughnessy TJ, Maric D, Lin HJ, Alkon DL, Barker JL (2004) CNS stem and progenitor cell differentiation into functional neuronal circuits in three-dimensional collagen gels. Exp Neurol 190:276–288.

Maddox S (1993) The Quest for Cure. Restoring Function After Spinal Cord Injury. Washington, DC, Paralyzed Veterans of America.

Manetto V, Sternberger NH, Gambetti P (1988) Phosphorylation of neurofilaments is altered in amyotrophic lateral sclerosis. J Neurol Sci 47:642–653.

Mansergh FC, Wride MA, Rancourt DE (2004) Neurons, stem cells and potential therapies. In: Sell S (ed.), Stem Cells Handbook. Towota, Humana Press, pp 177–190.

Marconi MA, Park I, Teng D, Ourednik J, Ourednik V, Taylor RM, Marciniak AE, Daadi MM, Rose HL, Lavik EB, Langer R, Auguste KI, Lachyankar M, Freed CR, Redmond DE, Sidman L, Snyder E (2003) Neural stem cells. In Sell S (ed.), Stem Cells Handbook, Towota, Humana Press, pp 191–208.

Mattson MP (2004) Pathways towards and away from Alzheimer's disease. Nature 430:631–639.

Matus A (2000) Actin-based plasticity in dendritic spines. Science 290:754–758.

Mazzini L, Fagioli F, Boccaletti R, Mareschi K, Oliveri G, Oliveri C, Pastore I, Marasso R, Madon E (2003) Stem cell therapy in in amyotrophic lateral sclerosis: a methodological approach in humans. Amyotroph Lateral Scler Other Motor Neuron Disord 4:158–161.

McDonald JW, Liu XZ, Qu Y, Liu S, Mickey SK, Turetsky D, Gottleib DL, Choi DW (1999) Transplanted embryonic stem cells survive, differentiate, and promote recovery in injured rat spinal cord. Nature Med 12:1410–1412.

Meister G, Tuschl T (2004) Mechanisms of gene silencing by double-stranded RNA. Nature 431:343–349.

Mello CC, Conte D Jr (2004) Revealing the world of RNA interference. Nature 431:338–342.

Mikami Y, Okano H, Sakaguchi M, Nakamura M, Shimazaki T, Okano HJ, Kawakami Y, Toyama Y, Toda M (2004) Implantation of dendritic cells in injured adult spinal cord results in activation of endogenous neural stem/progenitor cells leading to de novo neurogenesis and functional recovery. J Neurosci Res 76:453–465.

Miller DH, Khan OA, Sheremat WA, Blumhardt LD, Rice GPA, Libonati MA, Willmer-Hulme A, Dalton CM, Miszkiel KA, O'Connor PW (2003) A controlled trial of natalizumab for relapsing multiple sclerosis. New Eng J Med 348:15–23.

Mohajeri MH, Figlewicz DA, Bohn MC (2004) Intramuscular grafts of myoblasts genetically modified to secrete glial cell line-derived neurotrophic factor prevent motoneuron loss and disease progression in a mouse model of familial amyotrophic lateral sclerosis. Human Gene Ther 10:1853–1866.

Molander H, Engkvist O, Hagglund J, Olsson, Torebjork E (1983) Nerve repair using a polyglactin tube and nerve graft: An experimental study in the rabbit. Biomaterials 4:276–280.

Muir GD (1999) Locomotor plasticity after spinal injury in the chick. Neurotraum 16:705–711.

Namura S, Zhu J, Fink K, Endres M, Srinivasan A, Tomaselli KJ, Yuan J, Moskowitz MA (1998) Activation and cleavage of caspase-3 in apoptosis induced by experimental cerebral ischemia. J Neurosci 18:3659–3668.

Nash HH, Borke RC, Anders JJ (2002) Ensheathing cells and methylprednisolone promote axonal regeneration and functional recovery in the lesioned adult rat spinal cord. J Neurosci 22:425–435.

Nichols CM, Brenner MJ, Fox IK, Tung TH, Hunter DA, Rickman SR, Mackinnon SE (2004) Effect of motor versus sensory nerve grafts on peripheral nerve regeneration. Exp Neurol 190:347–355.

Niederost B, Zimmerman DR, Schwab ME, Bandtlow CE (1999) Bovine CNS myelin contains neurite growth-inhibitory activity associated with chondroitin sulfate proteoglycans. J Neurosci 19:8979–9889.

Nikulina E, Tidwell JL, Dai HN, Bregman BS, Filbin MT (2004) The phosphodiesterase inhibitor rolipram delivered after a spinal cord lesion promotes axonal regeneration and functional recovery. Proc Natl Acad Sci USA 101:8786–8790.

Nishiura Y, Brandt J, Nilsson A, Kanje M, Dahlin LB (2004) Addition of cultured Schwann cells to tendon autografts and freeze-thawed muscle grafts improves peripheral nerve regeneration. Tiss Eng 10:157–164.

Nistor GI, Totoiu MO, Haque N, Carpenter MK, Keirstead HS (2005) Human embryonic stem cells differentiate into oligodendrocytes in high purity and myelinate after spinal cord transplantation. Glia 49:385–396.

Nosrat IV, Smith CA, Mullally P, Olson L, Nosrat CA (2004) Dental pulp cells provide neurotrophic support for dopaminergic neurons and differentiate into neurons *in vitro*; implications for tissue engineering and repair in the nervous system. Eur J Neurosci 19:2388–2401.

Oosthuyse B, Moons L, Storkebaum E, Beck H, Nuyens D, Brusselmana K, Van Dorpe J, Hellings P, Gorselink M, Heymans S, Theilmeier G, Dewerchin M, Laudenbach V, Vermylen P, Raat H, Acker T, Vleminckx V, Van Den Bosch L, Cashman N, Fujisawa H, Drost MR, Sciot R, Bruyninckx F, Hicklin DJ, Ince C, Gressens P, Lupu F, Plate KH, Robberecht W, Herbert J-M, Collen D, Carmielet P (2001) Deletion of the hypoxia-response element in the vascular endothelial growth factor promoter causes motor neuron degeneration. Nature Genet 28:131–138.

Oppenheim RW, Houneou LJ, Johnson JE, Lin LF, Lin Lo AC, Newsome AL, Prevette DM, Wang S (1995) Developing motor neurons rescued from programmed and axotomy-induced cell death by GDNF. Nature 373:344–346.

Orr HT, Zoghbi HY (2000) Reversing neurodegeneration: A promise unfolds. Cell 101:1–4.

Ourednik V, Ourednik J, Lynch WP, Snyder EY, Schachner M (2002) Neural stem cells display an inherent mechanism for rescuing dysfunctional neurons. Nature Biotech 20:1103–1110.

Palfi S, Conde F, Riche D, Brouillet E, Dautry C, Mittoux V, Chibois A, Peschanski M, Hantraye P (1998) Fetal striatal allografts reverse cognitive deficits in a promate model of Huntington's disease. Nature Med 4:963–966.

Pallarito K (2004) The ailing brain: A pressing need for new treatments. The Scientist, October 25, pp 48–50.

Pearse D, Pereira FC, Marcillo AE, Bates ML, Berrocal YA, Filbin MT, Bunge MB (2004) cAMP and Schwann cells promote axonal growth and functional recovery after spinal cord injury. Nature Med 10:610–616.

Peckham PH, Creasy GH (1992) Neural prostheses: Clinical applications of functional electrical stimulation in spinal cord injury. Paraplegia 30:96–101.

Plant GW, Bates ML, Bunge MB (2001) Inhibitory proteoglycan immunoreactivity is higher at the caudal than the rostral Schwann cell graft-transected spinal cord interface. Mol Cell Neurosci 17:471–487.

Plaitakis A, Constantakakis E, Smith J (1988) The neuroexcitotoxic amino acids glutamate and aspartate are altered in the spinal cord and brain in amyotrophic lateral sclerosis. Ann Neurol 24:446–449.

Pluchino S, Quattrini A, Brambilla E, Gritti A, Salani G, Dina G, Galli R, Del Carro U, Amadio S, Bergami A, Furlan R, Comi G, Vescovi A, Martino G (2003) Injection of adult neurospheres induces recovery in a chronic model of multiple sclerosis. Nature 422:688–695.

Pluchino S, Zanotti L, Rossi B, Brambilla E, Ottoboni L, Salani G, Martinello M, Cattalini A, Bergami A, Furlan R, Comi G, Constantin G, Martino G (2005) Neurosphere-derived multipotent precursors promote neuroprotection by an immunomodulatory mechanism. Nature 436:266–271.

Raber J, Fan Y, MatsumoriY, Liu Z, Weinstein PR, Fike JR, Liu J (2004) Irradiation attenuates neurogenesis and exacerbates ischemia-induced defects. Ann Neurol 55:381–389.

Ralph GS, Radcliffe A, Day D, Carthy JM, Leroux MA, Lee DCP, Wong L-F, Bilsland LG, Greensmith L, Kingsman SM, Mitrophanous KA, Mazarakis ND, Azzouz M (2005) Silencing mutant SOD1 using RNAi protects against neurodegeneration and extends survival in an ALS model. Nature Med 11:429–433.

Ramer MS, Priestly JV, McMahon SB (2000) Functional regeneration of sensory axons into the adult spinal cord. Nature 403: 312–316.

Ramer LM, Ramer MS, Steeves JD (2005) Setting the stage for functional repair of spinal cord injuries: A cast of thousands. Spinal Cord 43:134–161.

Ramon-Cueto A, Plant GW, Avila J, Bunge MB (1998) Long-distance axonal regeneration in the transected adult rat spinal cord is promoted by olfactory ensheathing glia transplants. J Neurosci 18:3803–3815.

Raoul C, Abbas-Terki T, Bensadoun J-C, Guillot S, Haase G, Szulc J, Henderson C, Aebischer P (2005) Lentiviral-mediated silencing of SOD1 through RNA interference retards disease onset and progression in a mouse model of ALS. Nature Med 11:423–428.

Rapalino O, Lazarov-Spiegler O, Agranov E, Velan GJ, Yoles E, Fraidakis M, Solomon A, Gepstein R, Katz A, Belkin M, Hadani M, Schwartz M (1998) Implantation of stimulated homologous macrophages results in partial recovery of paraplegic rats. Nature Med 4:814–821.

Reubinoff BE, Itsykson P, Turetsky T, Pera MF, Reinhartz E, Itzik A, Ben-Hur T (2001) Neural progenitors from human embryonic stem cells. Nature Biotech 19:1134–1140.

Rice CM, Scolding NJ (2004) Adult stem cells—reprogramming neurological repair? Lancet 364:193–199.

Richardson PM, McGuiness UM, Aguayo AJ (1980) Axons from CNS neurons regenerate into PNS grafts. Nature 284:264–265.

Ritchie CW, Bush AI, Mackinnon A, Macfarlane S, Mastwyk M, MacGregor L, Kiers L, Cherny R, Li Q-X, Tammer A, Carrington D, Mavros C, Volitakis I, Xilinas M, Ames D, Davis S, Beyreuther K, Tanzi RE, Masters CL (2003) Metal-protein attenuation with iodochlorhydroxyquin (Clioquinol) targeting A beta amyloid deposition and toxicity in AD: a pilot Phase 2 clinical trial. Arch Neurol 60:1685–1691.

Rochkind S, Barr-Nea L, Razon N, Bartal A, Schwartz M (1987) Stimulatory effect of He-NE low dose laser on injured sciatic nerves of rats. Neurosurg 20:843–847.

Rochkind S, Nissan M, Alon M, Shamir M, Salame K (2001) Effects of laser irradiation on the spinal cord for the regeneration of crushed peripheral nerve in rats. Lasers in Surg and Med 28:216–219.

Rosenthal A (1998) Autotransplants for Parkinson's disease? Neuron 20:169–172.

Rosser AE, Barker RA, Harrower T, Watts C, Farrington M, Ho AK, Burnstein RM, Menon DK, Gillard JH, Pickard J, Dunnett SB (2002) Unilateral transplantation of human primary fetal tissue in four patients with Huntington's disease: NEST-UK safety report ISRCTN #36485475. J Neurol Neurosurg Psychiatr 73678–685.

Rothstein JD, Jin L, Dykes-Hoberg M, Kuncl RW (1992) Decrerased glutamate transport by the brain and spinal cord in amyotrophic lateral sclerosis. New Eng J Med 236:1464–1468.

Rothstein JD, Tsai G, Kuncl RW, Clawson L, Cornblath DR, Drachman DB, Pestronk A, Stauch BL, Coyle JT (1990) Abnormal excitatory amino acid metabolism in amyotrophic lateral sclerosis. Ann Neurol 28:18–25.

Rothstein JD, Van Kammen M, Levey A, Martin LJ, Kuncl RW (1995) Selective loss of glial glutamate transporter GLT-1 in amyotrophic lateral sclerosis. Ann Neurol 38:73–84.

Rothstein JD, Patel S, Regan MR, Haenggeli C, Huang YH, Bergles DE, Jin L, Hoberg MD, Vidensky S, Chung DS, Toan SV, Bruijin LI, Su Z-z, Gupta P, Fisher PB (2005) β-lactam antibiotics offer neuroprotection by increasing glutamate transporter expression. Nature 433:73–77.

Rowland LP, Schneider LA (2001) Amyotrophic lateral sclerosis. New Eng J Med 344:1688–1700.

Sagot Y, Tan SA, Hamang JP, Aebischer P, Kato AC (1996) GDNF slows loss of motoneurons but not axonal degeneration or premature death of *pmn/pmn* mice. J Neurosci 16:2335–2341.

SantaCruz K, Lewis J, Spires T, Paulson J, Kotilinek L, Ingelsson M, Guimaraes A, DeTure M, Ramsden M, McGowan E, Forster C, Yue M, Orne J, Janus C, Mariash A, Kuskowski M, Hyman B, Hutton M, Ashe KH (2005) Tau suppression in a neurodegenerative mouse model improves memory function. Science 309: 476–480.

Santos-Benito FF, Ramon-Cueto A (2003) Olfactory ensheathing glia transplantation: A therapy to promote repair in the mammalian central nervous system. Anat Rec (Part B: New Anat) 271B:77–85.

Saporta S, Kim J-J, Willing AE, Fu ES, Davus CD, Sanberg PR (2003) Human umbilical cord blood stem cells infusion in spinal cord injury: Engraftment and beneficial influence on behavior. J Hematother Stem Cell Res 12:271–278.

Scherman P, Kanje M, Dahlin LB (2004) Local effects of triiodothyronine-treated polyglactin sutures on regeneration across peripheral nerve defects. Tiss Eng 10:455–464.

Schmalbruch H, Jensen H-JS, Bjaerg M, Kamieniecka Z, Kurland L (1991) A new mouse mutant with progressive motor neuropathy. J Neuropathol Exp Neurol 50:192–204.

Schmidt-Heiber C, Jonas P, Bischofberger J (2004) Enhanced synaptic plasticity in newly generated granule cells of the adult hippocampus. Nature 429:184–187.

Schnell L, Schwab ME (1990) Axonal regeneration in the rat spinal cord produced by an antibody against myelin-associated neurite growth inhibitors. Nature 343:269–272.

Schnell L, Schneider R, Kolbeck, Barde Y-A, Schwab ME (1994) Neurotrophin-3 enhances sprouting of corticospinal tract during development and after adult spinal cord lesion. Nature 367: 170–173.

Seitz A, Aglow E, Heber-Katz E (2002) Recovery from spinal cord injury: A new transection model in the C57B1/6 mouse. J Neurosci Res 67:337–345.

Shaw PJ, Ince PG (1997) Glutamate excitotoxicity and amyotrophic lateral sclerosis. J Neurol 244(S2):S3–S14.

Silani V, Cova L, Corbo M, Ciammola A (2004) Stem-cell therapy for amyotrophic lateral sclerosis. Lancet 364:200–202.

Sinden JD, Hodges H, Gray JA (1995) Neural transplantation and recovery of cognitive function. Behavioral Brain Sci 18:10–35.

Steeves J, Fawcett J, Tuszynski M (2004) Report of international clinical trials workshop on spinal cord injury. Spinal Cord 42: 591–597.

Steinman L (1996) Multiple sclerosis: A coordinated immunological attack against myelin in the central nervous system. Cell 85:299–302.

Steinman L (2001) Multiple sclerosis: A two-stage disease. Nature Immunol 2:762–765.

Steinman L (2003) Collateral damage repaired. Nature 422:671–672.

Steward O, Schauwecker PE, Guth L, Zhang Z, Fujiki M, Inman D, Wrathall J, Kempermann G, Gage FH, Saatman KE, Raghupathi R, McIntosh T (1999) Genetic approaches to neurotrauma research: Opportunities and potential pitfalls of murine models. Exp Neurol 157:19–42.

Suzuki K, Suzuki Y, Tanihara M, Ohnishi, Hashimoto T, Endo K, Nishimura Y (2000) Reconstruction of rat peripheral nerve gap without sutures using freeze-dried alginate gel. J Biomed Mater Res 49:528–533.

Talac R, Friedman JA, Moore MJ, Lu L, Jabbari E, Windebank A, Currier B, Yaszemski MJ (2004) Animal models of spinal cord injury for evaluation of tissue engineering treatment strategies. Biomats 25:1505–1510.

Targ EF, Kocsis JD (1985) 4-aminopyridine leads to restoration of conduction in demyelinated rat sciatic nerve. Brain Res 328: 358–361.

Tate BA, Bower KA, Snyder EY (2001) Transplant therapy. In Tao MS (ed.) Stem Cells and CNS Development. Towota, Humana Press, pp 291–112.

Teng Y, Lavik EB, Qu X, Park KI, Ourednik J, Zurakowski D, Langer R, Snyder EY (2002) Functional recovery following traumatic spinal cord injury mediated by a unique polymer scaffold seeded with neural stem cells. Proc Natl Acad Sci USA 99: 3024–3029.

Tobias DA, Dhoot NO, Wheatley MA, Tessler A, Murray M, Fischer I (2001) Grafting of encapsulated BDNF-producing fibroblasts into the injured spinal cord without immune suppression in adult rats. J Neurotraum 18:287–301.

Trachtenberg JT, Chen BE, Knott GW, Feng G, Sanes JR, Welker E, Svoboda K (2002) Long-term *in vivo* imaging of experience-dependent synaptic plasticity in adult cortex. Nature 420:788–794.

Tsuboi K, Shults CW (2002) Intrastriatal injection of sonic hedgehog reduces behavioral impairment in a rat model of Parkinson's disease. Exp Neurol 173:95–104.

Vacanti MP, Leonard JL, Dore B, Bonassar LJ, Cao Y, Stachelek S, Vacanti JP, O'Connell F, Yu CS, Farwell AP, Vacanti CA (2001) Tissue-engineered spinal cord. Transpl Proc 33:592–598.

Valente EM, Abou-Sleiman PM, Caputo V, Muqit MMK, Harvey K, Gispert S, Ali Z, Del Turco D, Bentivoglio AR, Healy DG, Albanese A, Nussbaum R, Gonzalez-Maldonado R, Deller T, Salvi S, Cortelli P, Gilks WP, Latchman DS, Harvey RJ, Dallapiccola B, Auberger G, Wood NW (2004) Science 304: 1158–1160.

Van Dellen A, Blakemore C, Deacon R, York D, Hannan AJ (2000) Delaying the onset of Huntington's in mice. Nature 404:721–722.

Verdu E, Garcia-Alias G, Fores J, Lopez-Vales R, Navarro X (2003) Olfactory ensheathing cells transplanted in lesioned spinal cord prevent loss of spinal cord parenchyma and promote functional recovery. Glia 42:275–286.

Wakabayashi Y, Komori H, Kawa UT, Mochida K, Takahashi M, Qi M, Otake K, Shinomiya K (2001) Functional recovery and regeneration of descending tracts in rats after spinal cord transection in infancy. Spine 26:1215–1222.

Wamil AW, Wamil BD, Hellerqvist CG (1998) CM101-mediated recovery of walking ability in adult mice paralyzed by spinal cord injury. Proc Natl Acad Sci USA 95:13188–13193.

Wang X, Arcuino G, Takano T, Lin J, Peng WG, Wan P, Li P, Xu QX, Liu QS, Goldman SA, Nedergaard M (2004) P2X7 receptor inhibition improves recovery after spinal cord injury. Nature Med 10:821–827.

Watts J (2005) Controversy in China. Lancet 365:109–110.

Weidner N, Blesch A, Grill RJ, Tuszynski MH (1999) Nerve growth factor-hypersecreting Schwann cell grafts augment and guide spinal cord axonal growth and remyelinate central nervous system axons in a phenotypically appropriate manner that correlates with expression of L1. J Comp Neurol 413:495–506.

Wernig A, Muller S, Nanassy A, Cagol E (1995) Laufband therapy based on "rules of spinal locomotion" is effective in spinal cord injured persons. Eur J Neurosci 7:823–829.

Wernig A, Nanassy A, Muller S (1998) Maintenance of locomotor abilities following Laufband (treadmill) therapy in para- and tetraplegic persons: Follow-up studies. Spinal Cord 36:744–749.

Whelan HT, Wong-Riley MTT, Eells JT, VerHoeve JN, Das R, Jett M (2004) DARPA soldier self care: Rapid healing of laser eye injuries with light emitting diode technology. Paper presented at the RTO HFM Symposium "Combat Casualty Care in Ground Based Tactical Situations: Trauma Technology and Emergency Medical Procedures," St Pete Beach, FL, August 2004. Published in RTO-MP-HFM-109.

Wickelgren I (2004) A wily recruiter in the battle against toxic β amyloid aggregation. Science 306:791–792.

Wirz M, Colombo G, Dietz V (2001) Long term effects of locomotor training in spinal humans. J Neurol Neurosurg Psychiatr 71: 93–96.

Woerly S, Pinet E, deRobertis L, VanDeip D, Bousmina M (2001) Spinal cord repair with PHPMA hydrogel containing RGD peptides (NeuroGel®). Biomats 22:1095–1111.

Wolpaw JR, Tennissen AM (2001) Activity-dependent spinal cord plasticity in health and disease. Ann Rev Neurosci 24:807–843.

Wong PC, Cai H, Borchelt DR, Price DL (2002) Genetically engineered mouse models of neurodegenerative diseases. Neurosci 5:633–639.

Wong-Riley MTT, Liang HL, Eells JT, Chance B, Henry MM, Buchmann E, Kane M, Whelan HT (2005) Photobiomodulation directly benefits primary neurons functionally inactivated by toxins: Role of cytochrome oxidase. J Biol Chem 280:4761–4771.

Wurst W, Bally-Cuif L (2001) Neural plate patterning upstream and downstream of the isthmic organizer. Nature Rev Neurosci 2:99–108.

Xiong Z-G, Zhy X-M, Chu X-, Minami M, Hey J, Wei W-L, MacDonald JF, Wemmie JA, Proce MP, Welsh MJ, Simon RP (2004) Neuroprotection in ischemia: Blocking calcium-permeable acid-sensing ion channels. Cell 118:687–698.

Xu Z, Cork LC, Griffin JW, Cleveland DW (1993) Increased expression of neurofilament subunit NF-L produces morphological alterations that resemble the pathology of human motot disease. Cell 73:23–33.

Yamamoto A, Lucas JJ, Hen R (2000) Reversal of neuropathology and motor dysfunction in a conditional model of Huntington's disease. Cell 101:57–66.

Yandava BD, Billinghurst LL, Snyder EY (1999) "Global" cell replacement is feasible via neural stem cell transplantation. Evidence from the dysmyelinated *shiverer* mouse brain. Proc Natl Acad Sci UA 96:7029–7034.

Yannas IV (2001) Tissue and Organ Regeneration in Adults. New York, Springer.

Yannas IV, Hill BJ (2004) Selection of biomaterials for peripheral nerve regeneration using data from the nerve chamber model. Biomats 25:1593–1600.

Yednock TA, Cannon C, Fritz LC, Sanchez-Madrid F, Steinman L, Karin N (1992) Prevention of experimental autoimmune encephalomyelitis by antibodies against α4β1 integrin. Nature 356:63–66.

Zamvil SS, Steinman L (2003) Diverse targets for intervention during inflammatory and neurodegenerative phases of multiple sclerosis. Neuron 38:685–688.

Zhao Z, Alam S, Oppenheim RW, Prevette DM, Evenson A, Parsadanian A (2004) Overexpression of glial cell line-derived neurotrophic factor in the CNS rescues motoneurons from programmed cell death and promotes their long-term survival following axotomy. Exp Neurol 190:356–372.

Zhang S-U, Wernig M, Duncan ID, Brustle O, Thomson JA (2001) *In vitro* differentiation of transplantable neural precursors from human embryonic stem cells. Nature Biotech 19:1129–1133.

Zhou L, Baumgartner BJ, Hill-Felberg SJ, McGowen LR, Shine HD (2003) Neurotrophin-3 expressed *in situ* induces axonal plasticity in the adult injured spinal cord. J Neurosci 23:1424–1439.

Zhu S, Stavrovskaya IG, Drozda M, Kim BYS, Ona V, Li M, Sarang S, Liu AS, Hartley DM, Wu DC, Gullans S, Ferrante RJ, Przedborski S, Kristal BS, Friedlander RM (2002) Minocycline inhibits cytochrome c release and delays progression of amyotrophic lateral sclerosis in mice. Nature 417:74–78.

# 7 Regeneration of Digestive, Respiratory, and Urogenital Tissues

## INTRODUCTION

The digestive, respiratory, and urogenital systems are derived from the endodermal and mesodermal germ layers of the embryo. The digestive system consists of the pharynx, esophagus, stomach, small and large intestines, gall bladder, liver, and pancreas. The respiratory system includes the trachea, bronchial tubes, and lungs. The urogenital system is made up of the kidneys, ureters, bladder, and urethra, plus the reproductive organs. All the epithelia of the tubular portions of these systems have a relatively high rate of turnover and undergo both maintenance and injury-induced regeneration. The liver and pancreas are solid epithelial organs with low turnover and correspondingly slow maintenance regeneration. The liver has extensive powers of injury-induced regeneration, and the pancreas has the ability to regenerate in response to injury as well. This chapter will focus on regeneration of the intestinal epithelium, liver, and pancreas of the digestive system, and the alveolar epithelium of the respiratory system.

## INTESTINAL EPITHELIUM

### 1. Structure of the Intestinal Tract

The intestinal wall, and in fact that of the whole gastrointestinal tract, is composed of four major layers, going from the inside out (**FIGURE 7.1**). The first layer is the mucous membrane, which is made up of three sublayers. The inner sublayer is a simple columnar epithelium that faces the lumen. Directly underneath the epithelium is the lamina propria, a loose connective tissue sublayer that secretes the basement membrane on which the epithelium lies and which binds to the third sublayer, the muscularis mucosa. The muscularis mucosa consists of two thin layers of smooth muscle fibers. The inner layer of fibers has a circular orientation and the outer layer a longitudinal orientation. The lamina propria and overlying columnar epithelium are thrown into numerous villi that project into the intestinal lumen to give very large absorptive areas. The villous epithelium is differentiated into epithelial columnar cells, goblet cells, and enteroendocrine cells. The second major layer is the submucosa, consisting of a collagenous loose connective tissue permeated with blood vessels and a plexus of nerve fibers. The third layer consists of two sublayers of smooth muscle, the inner circular fibers and the outer longitudinal fibers, which are responsible for peristaltic contraction. The final layer is the serosa, a connective tissue layer covered with a layer of flat mesothelial cells.

### 2. Regeneration of the Villous Epithelium

The villous epithelium undergoes maintenance regeneration by the migration and differentiation of intestinal stem cells (ISCs) residing in the crypts of Lieberkuhn, which are microniches for ISCs at the bottom of the villous folds containing ~250 cells each. **FIGURE 7.2** illustrates the process of epithelial regeneration. There are many more crypts than villi, so the epithelium of each villus is replenished by multiple crypts. Crypt stem cells also regenerate the epithelium after chemical or radiation injury (Potten, 1997; Brittan and Wright, 2004).

Good *in vitro* culture methods are not available for the growth and differentiation of ISCs, although a promising *in vivo* subcutaneous culture method in mice has been devised (Booth et al., 1999). Most of the data on the location, number, and kinetics of proliferation of ISCs have been provided by *in vivo* DNA labeling experiments on mice (Potten et al., 1997; Brittan and Wright, 2004). These data indicate that ISCs of the small intestine reside in a ring lying 3–5 cells from the base of the crypt, with their average position four cells above the base. In the large intestine, ISCs appear to reside at the bottom of the crypt. Above the stem cell zone is a larger region consisting of transit amplifying

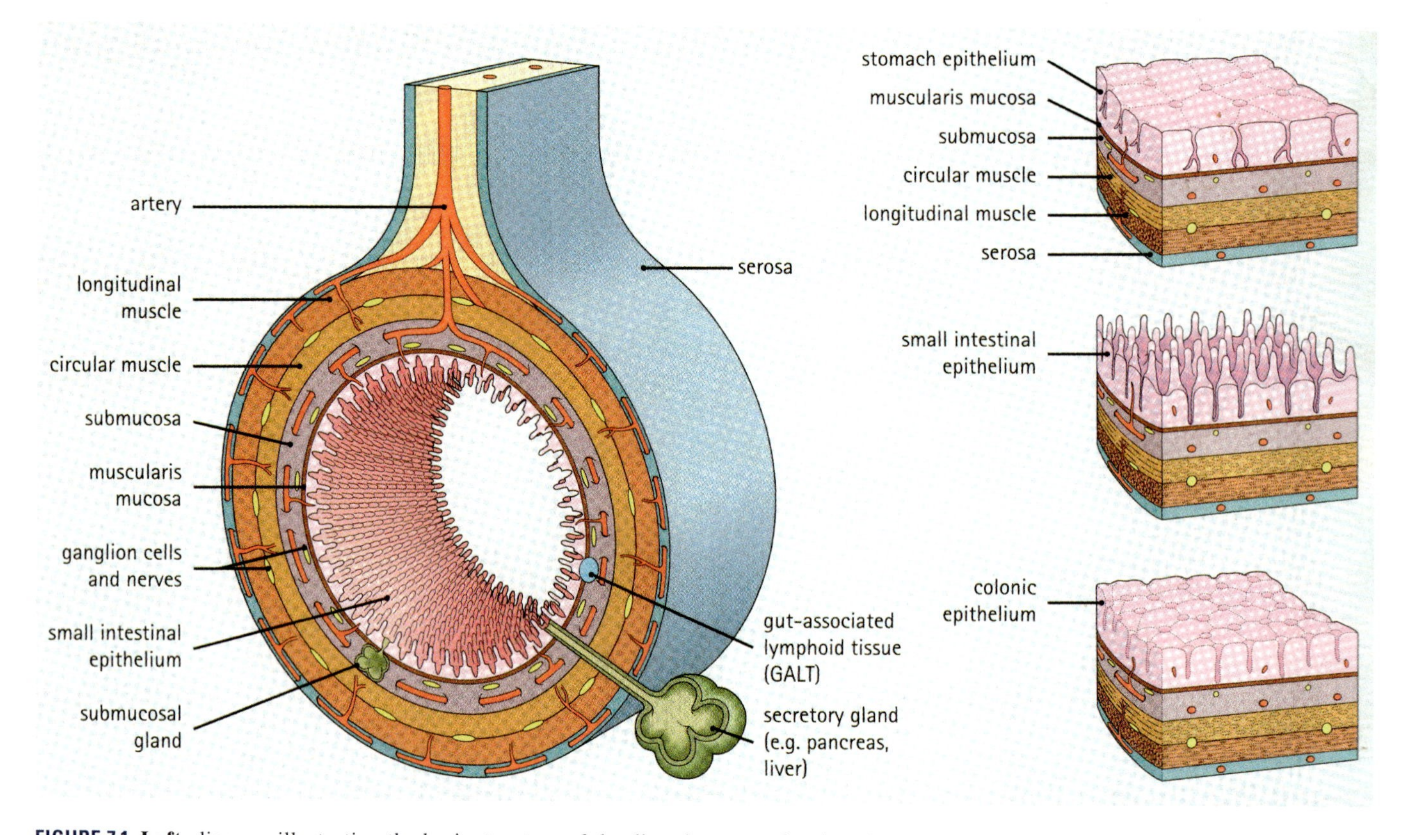


**FIGURE 7.1 Left,** diagram illustrating the basic structure of the digestive tract, showing the epithelium, muscularis mucosa, submucosa, layers of circular and longitudinal smooth muscle, and serosa. The epithelium of the small intersine is thrown into villi and exocrine secretion products of the liver and pancreas empty into the duodenum via the bile and pancreatic ducts. **Right,** differences in the epithelia of different parts of the digestive tract. Reproduced with permission from Stevens and Lowe, Human Histology (3rd ed), Mosby/Elsevier. Copyright 2005, Elsevier.

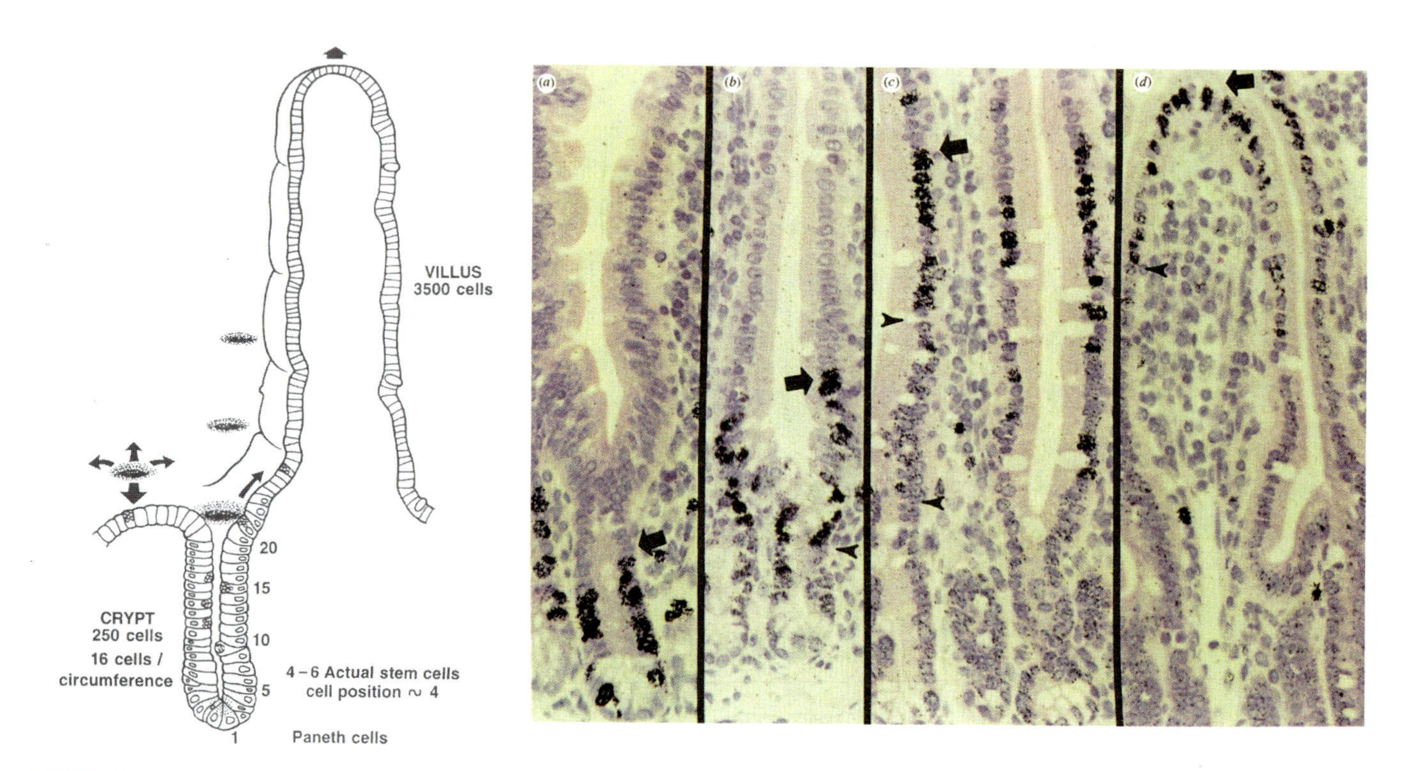


**FIGURE 7.2 Left,** diagram of crypt and villus of the epithelium of the mucus membrane of the small intestine. Numbers in the crypt indicate cell position. Arrows indicate that crypt cells feed multiple villi. **Right,** autoradiograms of $^{3}$H-thymidine labeled crypts/villi at (a) 40 min, (b) 24 hr, (c) 48 hr, (d) 72 hr. Movement of labeled cells from deep in the crypt (a) to the tip of the villus (d) is evident. Reproduced with permission from Potten, Stem cells in gastrointestinal epithelium: Numbers, characteristics and death. Phil Trans R Soc London B 353:821–830. Copyright 1998, The Royal Society.

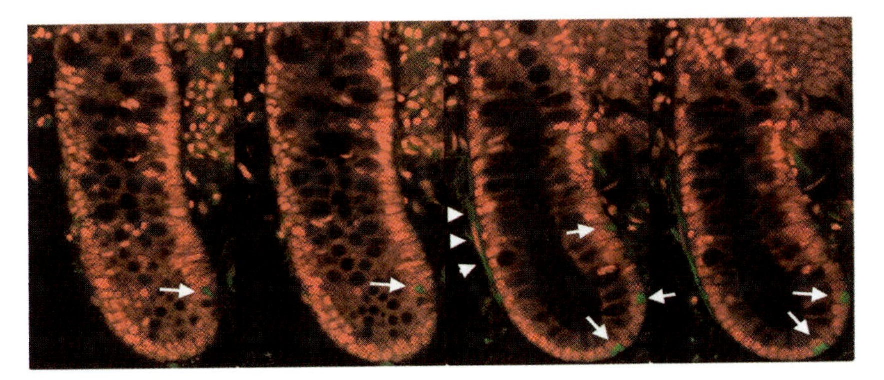

**FIGURE 7.3** Confocal images of Musashi (Nrp)-1 expression in the crypt epithelium of the human colon. Red, propidium iodide stain; green, Mushashi-1 expression. Musashi-1 positive cells were predominantly locted in the lower parts of crypts (arrows) but expression was also seen in pericryptal fibroblasts (arrowheads). Reproduced with permission from Nishimura et al., Expression of Musashi-1 in human normal colon crypt cells: a possible stem cell marker of human colon epithelium. Dig Dis Sci 48:1523–1529. Copyright 2003, Springer-Verlag.

cells that grades into maturing cells at the rim of the crypt. The transit amplifying cells have the potential for up to six divisions each. Each crypt has an output of 200–300 differentiated cells per day.

Clonal analyses *in vivo* following irradiation and repopulation of the crypts by surviving stem cells suggest that the crypt could be maintained under steady-state conditions by between four and six stem cells (Potten et al., 1997). Studies marking ISCs *in vivo* by chemical mutagenesis suggest that there is a mechanism by which only one of these cells per crypt contributes progeny to the villous epithelium (Winton and Ponder, 1990; Williams et al., 1992; Winton, 2001; Brittan and Wright, 2004). Chemical mutation of the *Dlb-1*$^{b}$ allele of *Dlb-1*$^{b}$/*Dlb-1*$^{a}$ heterozygous mice results in loss of lectin binding to the dolichos biflorus agglutinin on intestinal epithelium. The mutation results in monoclonal crypts composed entirely of cells unable to bind DLB lectin. A ribbon of monoclonal cells from each crypt can then be visualized on the villi. Using the number of cells in these ribbons, backcalculation of the number of stem cells/crypt required to produce each ribbon gives a value of one (Cosentino et al., 1996). The number of ISCs in the crypt niche is tightly regulated by apoptosis (Potten, 1997). Light and electron microscopy, and TUNEL (Tdt-mediated dUTP-biotin nick end labeling) analysis, which reveals apoptotic cells, showed that spontaneous apoptosis occurs in 5%–10% of the stem cells in the crypt.

Markers specific for ISCs have not yet been identified, but Kayahara et al. (2003), Potten et al. (2003), and Nishimura et al. (2003) found that, like neural stem cells, ISCs express Nrp-1 and Hes-1 (**FIGURE 7.3**). Like the EpSCs of the epidermis, integrins mediate the adhesion of ISCs to the basement membrane. Epithelial cells in the base of the crypts of the small intestine express α2β1 integrin (Beaulieu et al., 1992). ISCs express the transcription factor Tcf-4, which represses transcription in combination with TLE-1 or CREB-binding protein (CBP) and maintains them in a quiescent state (Brittan and Wright, 2004). Tcf-4 null mice are devoid of small intestinal ISCs, indicating that Tcf-4 is required to lay down the adult pattern of stem cells in the small intestine during embryogenesis (Korinek et al., 1998). The activation and proliferation of ISCs are regulated similarly to the EpSCs of epidermis and hair follicles. In the absence of a Wnt signal, the APC protein continually degrades β-catenin. In the presence of a Wnt signal, β-catenin is stabilized and interacts with Tcf-4 to activate transcription of downstream genes (Winton, 2001; Van de Wetering et al., 2002). One downstream target of the Wnt pathway is the hyaluronic acid receptor, CD44. CD44 mRNA is expressed throughout the crypt in a pattern similar to that of Tcf-4 (Wielenga et al., 1999).

Interactions between the villous epithelium and the underlying mesenchymal cells of the lamina propria positively and negatively regulate the transition of crypt cells to proliferating and differentiating cell populations (Potten et al., 1997; Winton, 2001). The mesenchymal cells of the lamina propria produce signals, including Wnt, that lead to ISC activation, migration out of the crypt, and proliferation. A number of heterogeneous sublines of lamina propria fibroblasts have been cloned that differ in their ability to induce proliferation and differentiation of intestinal endoderm from 14-day fetal rats (Winton, 2001), suggesting that different types of fibroblasts emit different signals.

Several growth factors have been shown to have effects on the proliferation and differentiation of ISCs. IGF-1 and 2 are expressed by intestinal epithelial cells and have been shown to increase proliferation of ISCs *in vivo* and *in vitro* in mice and humans (Conteas et al., 1987; Park et al., 1992; Booth et al., 1995; Potten et al., 1997). IL-4 stimulates proliferation of ISCs *in vitro* (McGee and Vitkus, 1996), while TGF-β, IL-6, and IL-11 inhibit proliferation *in vivo* (Migdalska et al., 1991; Potten et al., 1995) and *in vitro* (Barnard et al., 1989; Booth et al., 1995; Booth and Potten, 1995; Peterson et al., 1996). TGF-β induces the production of IL-6 and 11 *in vitro*, suggesting an interaction between these growth factors in regulating ISC proliferation (McGee et al., 1993; Elias et al., 1994). The FGFR-3 receptor, which binds FGF 1, 2, and 9, is strongly expressed in injured adult intestinal epithelium, suggesting that these growth factors are involved in regulating epithelial regeneration (Vidrich et al., 2004). FGF-2 and FGF-7 have been shown to enhance ISC survival after irradiation (Khan et al., 1997; Houchen et al., 1999) and FGF-10 enhances the healing of experimentally induced ulcers in the small intestine of rats (Han et al., 2000). Collectively, these factors might act to regulate the output of cells from the crypt stem cell compartment, as well as regulate exit from the cell cycle and entry into terminal differentiation (Potten, 1997). Commitment of ISCs to enteroendocrine precursors appears to require the transcription factor BETA 2, because BETA 2 null mice lack enteroendocrine cells expressing secretin and cholecystokinin (Mutoh et al., 1998). The homeobox transcription factor, Cdx-2, is involved in the final maturation of columnar epithelial cells (Lorentz et al., 1997).

Cell adhesion molecules and the ECM also play a role in regulating the rate of migration and differentiation from the crypt microniche. The basement membrane of the intestinal epithelium contains a number of adhesion molecules, including E-cadherin, Fn, Tn, Ln, and collagen IV that regulate ISC migration (Potten, 1997; Winton, 2001). Fibronectin (adhesive) expression is more abundant in the crypt, whereas tenascin (anti-adhesive) expression is more abundant in the villus (Potten, 1997), which is in accord with faster rates of cell migration with increasing distance from the stem cell positions (Kaur and Potten, 1986). Expression of laminin isoforms and α integrin chains also changes along the length of the crypt and villous epithelium (Beaulieu and Vachon, 1994; Beaulieu, 1992). Cadherins appear to play an important role in regulating the differentiation of intestinal epithelium. In mice transgenic for a dominant negative mutant N-cadherin construct regulated by a promoter expressed only in enterocytes, there was an increased rate of migration along the crypt-villus axis and loss of differentiation. When a variation of the mutation was expressed along the entire crypt-villus axis, the animals developed lesions in the transfected cells that resembled Crohn's disease (Hermiston and Gordon, 1995a, b).

### 3. Regeneration of Transected Intestine

The intestine of frogs, newts, and rats is able to regenerate epimorphically after complete transection. Histological and $^3$H-thymidine labeling studies after transection of the adult frog intestine (Goodchild, 1956) and newt (O'Steen and Walker, 1962; Grubb, 1975) indicate that serosal and smooth muscle cells at each cut end undergo dedifferentiation and proliferate to form a mesenchymal blastema that regenerates the intestinal walls. The intestinal epithelium was not observed to contribute to the blastema. The two blastemas grow by proliferation and fuse to bridge the gap, much like gap regeneration in the urodele spinal cord. Cords of intestinal epithelium then spread into the mesenchymal mass, join together, and hollow out to reform the intestinal lumen, reminiscent of the secondary lumen formation observed during ontogenesis of the amphibian intestine (Ballard, 1970). The mesenchymal cells differentiate into submucosal, smooth muscle, and serosal cell layers to reconstitute the original intestinal architecture.

The small and large intestine of the rat regenerate after simple transection (Dumont et al., 1980). The colon regenerates even after separating the transected ends with a 2–3-cm piece of silastic tubing (Dumont et al., 1984). These results suggest that humans might have the ability to regenerate segments of the colon that have been surgically removed as a result of disease. Presumably, the regeneration of the rat intestine is via a blastema, similar to the newt intestine. No histological studies of rat intestinal regeneration, however, have been done, so nothing is actually known about the mechanism of this regeneration.

## LIVER

### 1. Structure and Function of the Mammalian Liver

The mammalian liver is the classic example of regeneration by compensatory hyperplasia. The hepatocytes of the liver function in both an endocrine and exocrine secretory capacity, reflected in their tremendous numbers of mitochondria and volume of rough endoplasmic reticulum and Golgi stacks. The endocrine activity involves the conversion of glycogen to glucose and the secretion of a large number of serum factors, including albumin, prothrombin, fibrinogen, and the protein component of lipoproteins. The exocrine secretion of hepatocytes is bile, which contains both waste products, such as bilirubin, and bile salts required for intestinal absorption. Hepatocytes also have a major role in carbohydrate, ammonia, and triglyceride metabolism, as well as detoxifying metabolic byproducts and toxic substances.

The structural organization of the liver reflects these secretory and metabolic functions (**FIGURES 7.4, 7.5**). The hepatocytes, which constitute 80% of the liver, are arranged as trabeculae two cells thick (Ham and Cormack, 1979). Spaces between the hepatocytes of the trabeculae constitute biliary canaliculi, which convey bile to the bile ductules via short canals of Hering, which are lined by a mixture of hepatocytes and bile duct cells. The trabeculae are separated by vascular sinusoids lined by fenestrated endothelial cells and macrophages called Kupffer cells. Between the fenestrated endothelium of the sinusoids and the hepatic trabeculae is the space of Disse. The fenestrated endothelium and the space of Disse provide the hepatocytes with maximum exposure to hepatic blood flow. The hepatocytes have numerous microvilli projecting from their surface into the space of Disse, providing them with a large surface area for absorption of molecules from the blood. Interspersed between hepatocytes and also abutting on the space of Disse are Ito cells, lipocytes that store vitamin A.

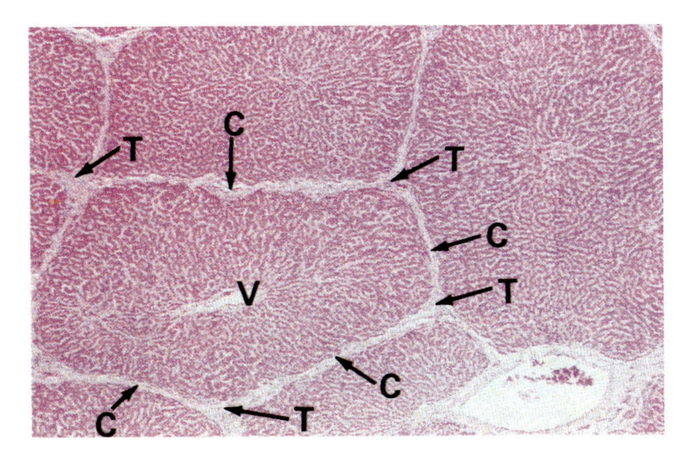


**FIGURE 7.4** Section through a pig liver (H&E stain). The pig liver is organized into roughly hexagonal lobules separated by connective tissue septa (C). V = centrilobular venule, T = portal tracts containing the terminal branches of the hepatic artery and vein. Blood from the terminal branch of the hepatic artery passes into the sinusoids (narrow light spaces), into the centrolobular venule and out again through the terminal branch of the hepatic vein. The human liver is also lobular in organization, but the interlobular septa are absent. Reproduced with permission from Wheater et al., Wheater's Functional Histology (3rd ed). Copyright 1987, Elsevier.

The hepatic trabeculae in mammals such as the rat and pig are organized into lobules. The lobules are defined by "portal areas" connected by connective tissue septa to roughly form a hexagon. Hepatic trabeculae radiate from the periphery of the lobule to a central vein. In humans, the connective tissue septae outlining lobules are not present, but the arrangement of the portal areas is the same and lobules can be "seen" by drawing imaginary lines between the portal areas. The portal areas themselves consist of a branch of the portal vein, hepatic artery, and bile duct, together with a lymphatic vessel (Ham and Cormack, 1979).

The ECM of the liver is concentrated mainly in its outer connective tissue capsule, blood vessels, and bile

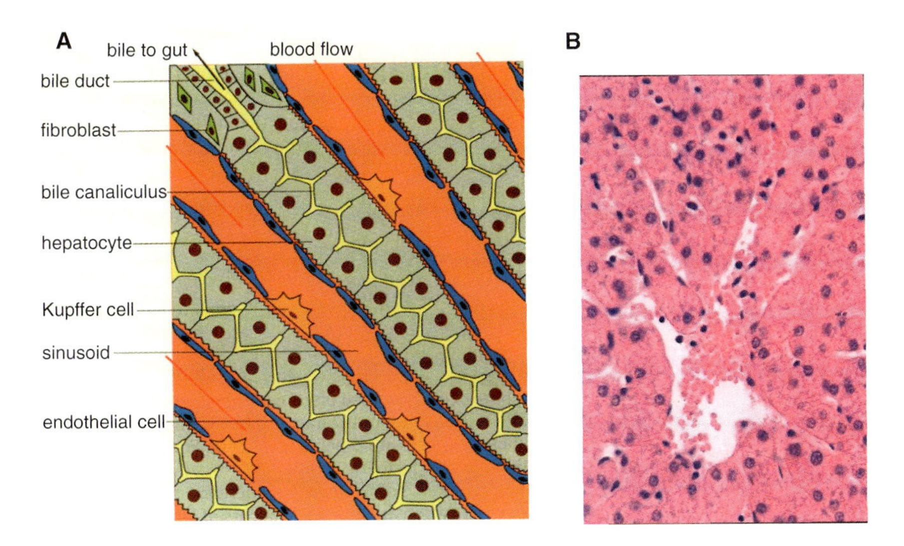


**FIGURE 7.5** **(A)** Diagram showing the two-layered hepatic trabeculae, bile cabaliculi and sinusoids lined with fenestrated endothelium. Reproduced with permission from Alberts et al., Molecular Biology of the Cell (3rd ed), Garland Press. Copyright 1994, Taylor and Francis Group LLC. **(B)** Histological section of pig liver showing hepatocyte plates and sinusoids draining into the terminal branch of the centrolobular venule. Reproduced with permission, courtesy of Dr. R. A. Bowen, http://arbl.cvmbs.colostate.edu/hbooks/pathphys/digestion/liver/histo_sinusoids.html.

ducts, with only small amounts associated with hepatocytes. Collagens I, II, V, VI, VII, fibronectin, and tenascin are found in the outer capsule and blood vessels. Blood vessels and bile ducts all have basement membranes containing laminin, entactin, collagen IV, and perlecan. The sinusoids, however, lack basement membrane. Fibronectin is abundant in the space of Disse, and free collagen IV is also found there (Martinez-Hernandez and Amenta, 1995).

## 2. Regeneration by Compensatory Hyperplasia

The mammalian liver turns over slowly, with the average life span of hepatocytes estimated at ~200–300 days (Bucher and Malt, 1971). Cell marking studies indicate that during normal liver turnover, hepatocytes are replaced by compensatory hyperplasia of existing hepatocytes, not by stem cells (Ponder, 1996; Grompe and Feingold, 2001). All vertebrates exhibit the capacity to regenerate the liver following tissue loss due to trauma, chemical insults, and viral infections.

The most widely used, and best studied, model for injury-induced liver regeneration is partial hepatectomy (PH) in the rat, in which two-thirds of the liver is surgically removed (Higgins and Anderson, 1931). The rat liver has four lobes, two large and two small. Excision of the two large lobes removes two-thirds of the mass. The excised lobes do not grow back, but the remaining two small lobes grow rapidly to attain the original mass of the liver within 14 days after operation. There is no cell damage when discrete lobes of the liver are removed; consequently, cell proliferation occurs in the absence of cell death or inflammatory events (Fausto and Webber, 1994). Other kinds of damage do involve an inflammatory phase, but the regenerative outcome is the same. How the inflammatory response is held in check to avoid fibrosis and allow regeneration to proceed is unknown but is an important question.

Hepatocyte nuclear transcription factors (HNFs), which control liver and pancreas development during ontogenesis, are also expressed in adult hepatocytes and islet cells of the pancreas (Odom et al., 2004). HNF-1α, HNF4-α, and HNF-6 are at the center of the transcriptional network of both hepatocytes and pancreatic islet cells. All three are required for the normal function of these cells. Gene location experiments indicate that HNF-1α and HNF-6 bind to approximately 1.6% and 1.7% of the promoters on a microarray containing portions of the promoters of 13,000 human genes, respectively, while HNF-4α binds to 12%. HNF-4α is a highly abundant and constitutively active transcription factor. The transcriptional network regulated by these factors during liver ontogenesis remains largely the same during liver regeneration. Partial hepatectomy triggers a series of regulatory events that superimpose a gene activity pattern characteristic of proliferation onto the hepatocyte-specific transcriptional pattern (Fausto and Webber, 1994; Michaelopoulos and De Frances, 1997; Trembly and Steer, 1998).

### a. Activation and Proliferation of Liver Cells

#### Capacity for and Kinetics of Proliferation

The Greek legend of Prometheus describes a prodigious capacity of the liver for regeneration. In this myth, the Titan god, Prometheus, breaks a taboo and gives fire to humans. He is punished for this act by the supreme god, Zeus, who has Prometheus chained to a rock. By day his liver is eaten by an eagle, only to regenerate at night, insuring eternal torture that is escaped only because Hercules rescues Prometheus.

Whether the ancient Greeks really knew that the liver regenerates is doubtful, but the myth of Prometheus has been validated over 2000 years later by experimental studies (Fausto, 2004). The hepatocyte has enormous clongenic potential. Experiments in which $1–2 \times 10^5$ adult mouse hepatocytes transgenic for *lacZ* were transplanted into mice with a functional liver deficit showed that individual hepatocytes have the capacity to divide at least 12 times (Rhim et al., 1994). Mice with hereditary tyrosinemia type I, a fatal recessive liver disease caused by a deficiency of fumarylacetoacetate hydrolase (FAH) leading to the accumulation of a hepatotoxic metabolite of tyrosine, were rescued by injecting as few as 1,000 normal hepatocytes into their livers. Serial transplant experiments showed that these mice could be rescued through four generations, suggesting that a single hepatocyte can divide at least 70 times (Overturf et al., 1997). This capacity for proliferation is remarkable, given the fact that most mature hepatocytes have a 4C ploidy, with some having even higher ploidy numbers.

All the cell types of the liver proliferate after PH, though the kinetics of DNA synthesis for each cell type are different (Michaelopoulos and De Frances, 1997) (**FIGURE 7.6**). Hepatocyte DNA synthesis is initiated 10–12 hr after partial hepatectomy and peaks at 24 hr after PH, whereas DNA synthesis in biliary ductular cells, Kupffer cells, and Ito cells begins later and peaks later. The endothelial cells of the sinusoids begin proliferating last, reaching a peak of DNA synthesis at four days. In a young adult rat, up to 95% of hepatocytes divide at least once in restoring the original liver size. In older animals, liver regeneration is slower, is less complete, and involves the proliferation of fewer hepatocytes.

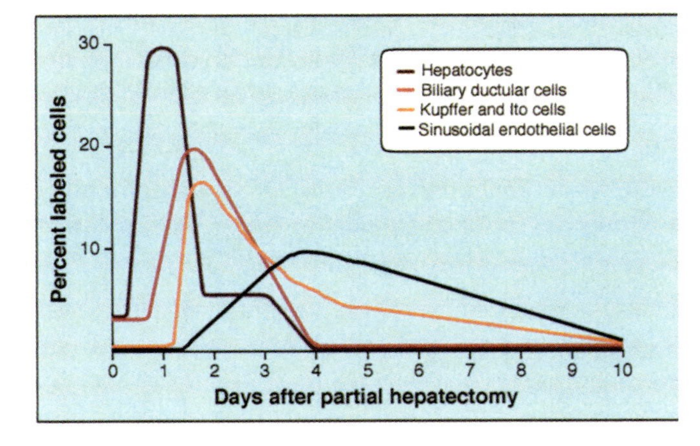


**FIGURE 7.6** Kinetics of DNA synthesis for various cell types in the liver after partial hepatectomy. Reproduced with permission from Michalopoulos and DeFrances, Liver regeneration. Science 276:60–66. Copyright 1997, AAAS.

## Priming Phase for Entry into the Cell Cycle

The cells of the liver are maintained in a mitotically quiescent state by the transcription factor CAAT/enhancer binding protein alpha (C/EBPα), which is expressed at high levels in the intact liver and exerts its effect by direct interaction with cyclin dependent kinases (cdks), preventing entry into the cell cycle (Wang et al., 2001, 2002). After PH, the liver cells are "primed," or acquire competence, to exit $G_0$ into $G_1$ (**FIGURE 7.7**). The levels of C/EBPα are reduced and the activity of over 70 "immediate-early" genes is induced over a period of 30 min–3 hr (Taub, 1996). Induction of this group of immediate early genes is independent of new protein synthesis; i.e., it requires only the activation of existing transcription factors through post-translational modifications. The most important of these transcription factors are STAT3, partial hepatectomy factor (PHF/NF-κB, a liver-specific form of NF-κB), AP-1 (active complex of the proto-oncogene proteins, c-Jun and c-Fos), and C/EBPβ (Diehl, 1998; Fausto, 2000). Many of the immediate-early genes contain binding sequences for these transcription factors in their promoters and code for transcription factors involved in initiating the early $G_1$ phase of the cell cycle, such as the protooncogene c-Myc and liver regeneration factor-1 (LRF-1), which also forms DNA-binding complexes of the AP-1 type. High levels of these complexes are detected for several hours after the $G_0$ to $G_1$ transition (Taub, 1996).

The pattern of liver-specific protein synthesis shifts slightly during priming. A number of proteins identical to those produced in the fetal liver appear in regenerating liver as the result of activation of immediate early genes. These include α-fetoprotein, hexokinase, and

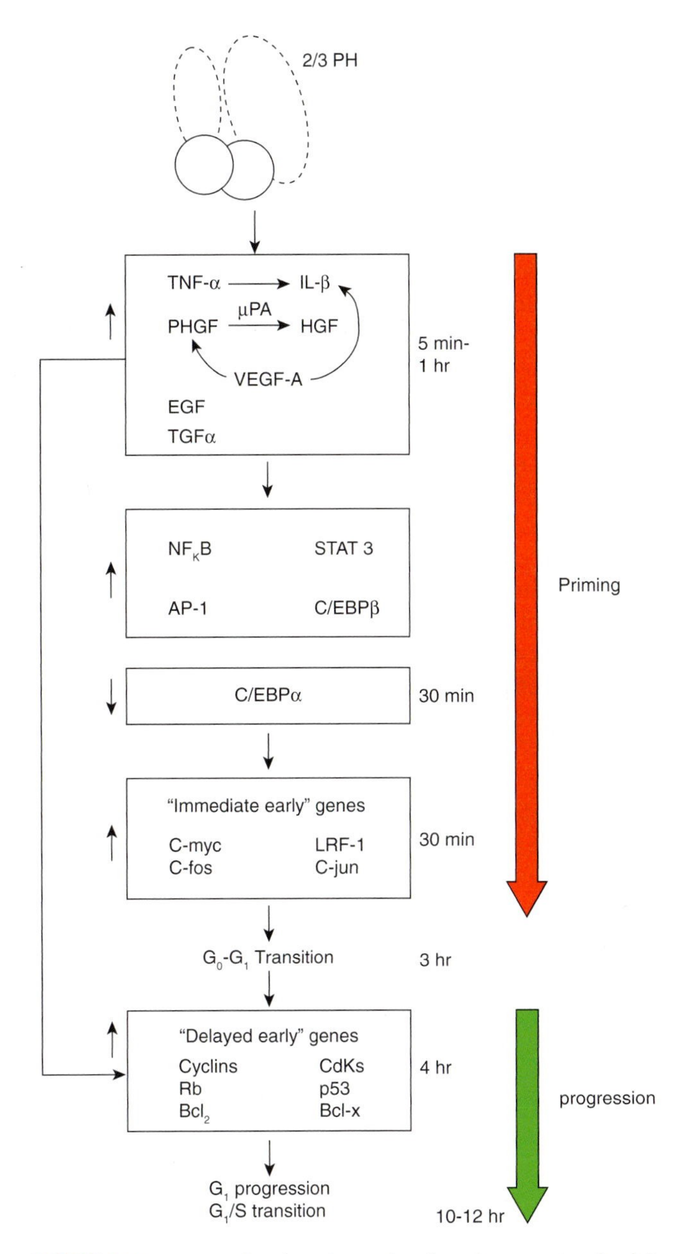


**FIGURE 7.7** Diagram showing the molecular elements involved in priming and progression of hepatocytes through the cell cycle after partial hepatectomy. Short arrows on left side indicate up-regulation or down-regulation of molecular activity in the boxes. Details in the text.

fetal isozymes of aldolase and pyruvate kinase. In addition, the expression of several liver function proteins is upregulated to compensate for tissue loss, including albumin and several genes that encode proteins involved in glucose regulation and metabolism, such as glucose-6-phosphatase, insulin-like growth factor binding protein-1, and phosphoenolpyruvate carboxykinase (Taub, 1996).

A major question is how the immediate-early gene response is initiated. Mitogenic signals appear in the blood after PH, as shown by the fact that liver tissue or hepatocytes transplanted to ectopic locations replicate DNA following partial hepatectomy of the host liver. Furthermore, hepatectomy of one member of a parabiosed pair of rats induces growth of the intact liver of the other member of the pair (Michalopoulos and De Frances, 1997).

The mitogenic signals are growth factors produced by cells of the liver other than hepatocytes, as well as by nonliver cells elsewhere in the body (Fausto and Webber, 1995; Fausto et al., 1994; Michaelopoulos and De Frances, 1997; Trembly and Steer, 1998). The TNF/TNFR1 signaling pathway is important for the initiation of liver regeneration, as is hepatocyte growth factor (HGF), EGF, and TGF-α. TNF-α and IL-6 play an essential role in the transition of liver cells from $G_0$ to $G_1$. TNF-α exerts its effects by binding to its receptor, TNFR1. Knockout mice lacking either TNFR1 or IL-6 are deficient in their ability to regenerate liver. DNA synthesis is severely impaired in mice with a TNF-α receptor deficiency and in normal mice treated with antibodies to TNF-α. In both cases, there is a failure to activate STAT3 and NF-κB and to increase the production of c-jun and AP1.

These deficiencies can be corrected by the injection of IL-6. IL-6 is secreted by Kupffer cells and plasma concentrations reach high levels by 24 hr after PH. In mice homozygous for deletion of the IL-6 gene, STAT3 activation, AP1 activity, Myc, and cyclin D1 are all markedly reduced, and DNA synthesis is suppressed. Secretion of IL-6 is stimulated by TNF-α. Collectively, these observations suggest that IL-6 is a mitogen essential for priming that is regulated by TNF-α. The results suggest a signaling pathway in which partial hepatectomy induces expression of TNF-α followed by activation of NF-κB, which induces IL-6, causing activation of STAT3. Activation of STAT3 and NF-κB initiates immediate-early gene expression. However, liver regeneration proceeds to completion in IL-6 deficient mice, indicating that other pathways can compensate for the loss of IL-6 (Michaelopoulos and De Frances, 1997). Thus, although TNF-α and IL-6 are required for the *normal* pace of liver regeneration, they are not indispensable for initiating regeneration.

HGF and EGF/TGFα are also able to initiate liver regeneration (Michaelopoulos and De Frances, 1997). The inactive precursor of HGF (pro-HGF) is found in many tissues, including the liver, where it is produced primarily by endothelial cells. Pro-HGF is activated by urokinase plasminogen activator (uPA). Activated pro-HGF forms a heterodimer that binds to its receptor, c-met. HGF is a potent mitogen for hepatocytes cultured in the absence of serum (i.e., in the absence of other growth factors) and is thus a "complete mitogen." Within one hour after PH the plasma concentration of active HGF rises over 20-fold (Mars et al., 1995). The mechanism of this rise may be related to release of pro-HGF from the ECM by matrix degradation. Large amounts of pro-HGF are bound to the liver ECM (Michaelopoulos and De Frances, 1997). Injecting HGF into uninjured liver results in a weak proliferative response, but infusing collagenase into the liver prior to injection of HGF greatly magnifies the response, suggesting that degradation of the matrix around hepatocytes and other cells plays a role in initiating regeneration by releasing pro-HGF. Furthermore, within 1–5 min after partial hepatectomy, urokinase receptor activity rises due to translocation of the receptor to the hepatocyte plasma membrane (Mars et al., 1995). This results in increased uPA activity, which in addition to activating pro-HGF, converts plasminogen to plasmin, activating MMPs. Thus, the release of pro-HGF by matrix degradation and its cleavage by uPA could explain the rapid rise in active plasma HGF after PH.

The angiogenic growth factor VEGF-A plays a role in elevating the production of HGF and IL-6 (LeCouter et al., 2003; see also Davidson and Zon, 2003). VEGF-A stimulates hepatocyte division when injected *in vivo* but stimulates hepatocyte mitosis *in vitro* only in the presence of sinusoidal endothelial cells (**FIGURE 7.8**). Injury activates two distinct pathways in liver endothelial cells that lead to elevated levels of HGF and IL-6 by upregulating hepatocyte VEGF-A production. VEGF-A binds to both the VEGFR-1 and 2 receptors on endothelial cells. Binding to VEGR-1 results in enhanced secretion of HGF and IL-6. Binding to VEGF-2 enhances endothelial cell proliferation, thus increasing the number of cells producing higher levels of the growth factors.

EGF is also a complete mitogen for hepatocyte proliferation *in vitro*. EGF mRNA increases by 10-fold in the liver within 15 min after PH and is expressed in hepatocytes, endothelial cells, Kupffer cells, and Ito cells (Michaelopoulos and De Frances, 1997; Trembly and Steer, 1998). Plasma EGF levels rise somewhat later than HGF levels after partial hepatectomy, but the rise is less than 30%. The number of EGF receptors doubles in the first three hr. EGF protein accumulates inside hepatocytes but is also synthesized by the salivary glands and released into the bloodstream, suggesting that it acts through both autocrine and endocrine mechanisms. The latter mechanism may be the more important. Removal of the salivary glands two weeks prior to partial hepatectomy reduced plasma EGF concentration by 50%, and abolished the increase in EGF level after hepatectomy (Lambotte et al., 1997). DNA synthesis and mitosis were also reduced by 50% when the salivary glands were removed at the time of, or within three hr after, partial hepatectomy. Salivary gland removal six hr or more after hepatectomy had no

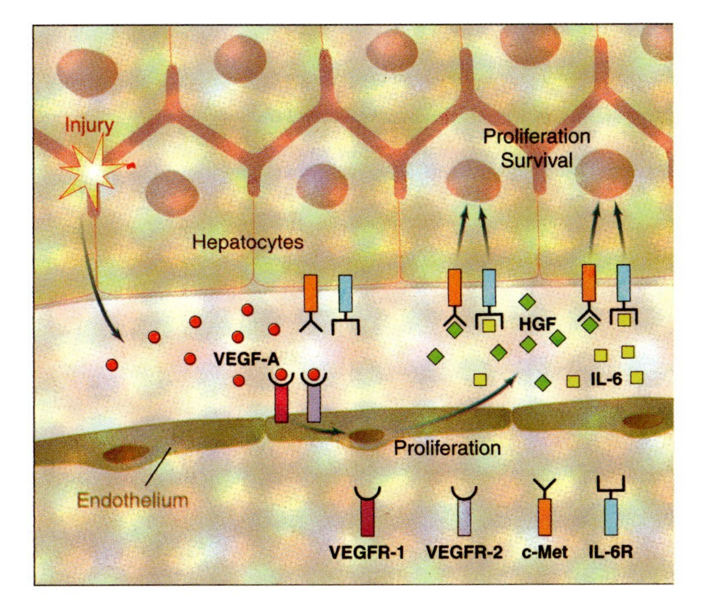


**FIGURE 7.8** Diagram showing how VEGF-A promotes hepatocyte survival and proliferation after liver injury. Injury induces hepatocytes and non-hepatocytes to secrete VEGF-A (red dots). If VEGF-A binds to VEGFR-1 (red rectangle) on the endothelial cells of the sinusoids, it induces a higher output of HGF (green diamonds) and IL-6 (yellow squares) by the endothelial cells. If VEGF-A binds to VEGFR-2 (purple rectangle) it enhances endothelial cell proliferation, which also increases the amount of HGF and IL-6 available to hepatocytes. HGF and IL-6 act by binding to their receptors on the hepatocytes, cMet (orange) and IL-6R (blue). Reproduced with permission from Davidson and Zon, Love, honor, and protect (your liver). Science 299:835–837. Copyright 2003, AAAS.

effect. Administration of exogenous EGF to sialoadenectomized rats restored normal regenerative activity. In general, diminished EGF levels delayed the regeneration response by 24 hr, but the liver nevertheless was fully regenerated by seven days. These observations suggest that EGF affects events immediately following HGF-promoted entry into $G_1$. Insulin and norephinephrine amplify the mitogenic signals of HGF and EGF by binding to the α1 adrenergic receptor (Michaelopoulos and De Frances, 1997).

TGF-α also binds to the EGF receptor and is induced in hepatocytes within 2–3 hr after PH, rising to a peak between 12 and 24 hr (Michaelopoulos and De Frances, 1997; Trembly and Steer, 1998). However, there is only a small increase in plasma TGF-α protein after PH, despite a large increase in TGF-α mRNA. Furthermore, liver regeneration proceeds normally in mice carrying a homozygous deletion of the TGF-α gene. Thus, it is unclear whether TGF-α has an important role to play in liver regeneration insofar as it affects hepatocyte proliferation. However, along with other growth factors produced by hepatocytes of regenerating liver such as FGF-1 and VEGF-A, TGF-α may be important for inducing the rise in proliferation of other liver cell types.

Pleiotrophin, another cytokine that is expressed in a variety of tissues, has been implicated in the regulation of the early immediate gene response. Rat hepatocytes will proliferate successfully *in vitro* if co-cultured with Swiss 3T3 fibroblasts or in a medium conditioned by these cells. The active molecule in the conditioned medium is pleiotrophin and it is expressed at twice the level in the regenerating liver as in control intact livers (Sato et al., 1999).

### Progression Phase of Proliferation

A second set of genes, called "delayed-early" genes, is induced independently of new transcription factor synthesis starting at about 4 hr after PH (Fausto and Webber, 1994; Trembly and Steer, 1998). Several of the same growth factors involved in priming hepatocytes, such as HGF, EGF, and TGF-α also regulate their progression through $G_1$ to the S phase. Enhanced expression of TGF-α in hepatocytes under the influence of the albumin promoter leads to sustained high levels of DNA synthesis and to tumor formation, implying that it plays a role in progression of hepatocytes through the $G_1$/S transition and beyond. Delayed-early genes encode proteins that take the cells through $G_1$ and the $G_1$/S transition (**FIGURE 7.7**). Important proteins synthesized are (1) the cyclins and cyclin-dependent kinases (cdks), which phosphorylate proteins such as the Rb protein, allowing passage from $G_1$ into S, (2) p53, a transcription factor that activates *p21*, which encodes a protein that inhibits cyclin/cdk activity, and (3) Bcl-X and Bcl-2, which, along with Rb, protect cells against apoptosis while going through the cell cycle.

Rb exhibits peaks in expression at 12, 30, and 72 hr, with expression at 30 hr representing a greater than 100-fold increase over expression in nonregenerating liver. DNA synthesis in the livers of transgenic mice overexpressing *p21* is less than 15% of normal after PH and the mice have correspondingly decreased liver mass. Expression of *Bcl-2* and *Bcl-X* occurs at 6 hr following PH. *Bcl-2* transcripts are expressed by nonhepatocyte cells at a 2-fold level above normal, whereas *Bcl-X* transcripts are expressed by hepatocytes at a 20-fold level, but the proteins encoded by these genes do not fluctuate significantly during regeneration (Michaelopoulos and DeFrances 1997; Trembly and Steer, 1998).

The broad picture that has thus emerged is one in which a variety of signals are necessary for normal liver regeneration. TNF-α, IL-6, HGF, and EGF appear to play definitive roles in initiating regeneration, with other growth factors playing facultative roles. How events as rapid as the increase in uPA activity within

5 min and the activation of immediate early genes within 30 min are triggered is not yet clear.

### Cessation of Proliferation

DNA synthesis in regenerating rat liver is complete by 72 hr. Little is known, however, about the mechanism by which proliferation is shut down. TGF-β1, which is normally made by Ito cells, may play a role in preventing and cessation of proliferation (Michaelopoulos and De Frances, 1997). *In vitro*, TGF-β1 inhibits the mitosis of hepatocytes. However, liver regeneration, though slowed, proceeds to completion in mice transgenic for increased TGF-β1 gene expression, suggesting that other factors act synergistically with TGF-β1 to terminate hepatocyte proliferation (Michaelopoulos and DeFrancis, 1997).

### b. Remodeling

When proliferation ceases, hepatocytes exist as clusters of 10–14 cells lacking sinusoids and ECM (Martinez-Hernandez and Amenta, 1995). The normal organization of the liver is then re-established by regulation of cell number by apoptosis, cell movements, and ECM synthesis. The pro-apoptotic protein Bax is most abundant in regenerating liver during the phase of reorganization and thus may be involved in regulation of cell numbers. Three liver cell types synthesize and secrete ECM molecules: hepatocytes, endothelial cells, and Ito cells. Ito cells invade the hepatocyte clusters and synthesize laminin. The laminin is not organized into a basement membrane, perhaps because of a lack of entactin. It may serve as a stimulus for the subsequent invasion of endothelial cells that separate the hepatocytes into trabeculae two cells wide. In this manner, the sinusoidal spaces and space of Disse are re-established. Simultaneously, the other ECM molecules characteristic of normal liver are synthesized and deposited (Martinez-Hernandez and Amenta, 1995).

## 3. Injury-Induced Regeneration via Stem Cells

Several models of liver injury in rats show that the liver contains a population of stem cells that enable it to regenerate when the ability of hepatocytes to proliferate is compromised (Shafritz and Dabeva, 2002; Dabeva and Shafritz, 2003). Compounds that destroy the ability of hepatocytes to regenerate are D-galactosamine (D-galN), which induces extensive hepatocyte necrosis, 2-acetylaminofluorine (2-AAF), which causes extensive hepatocyte DNA damage, and retrosine, a

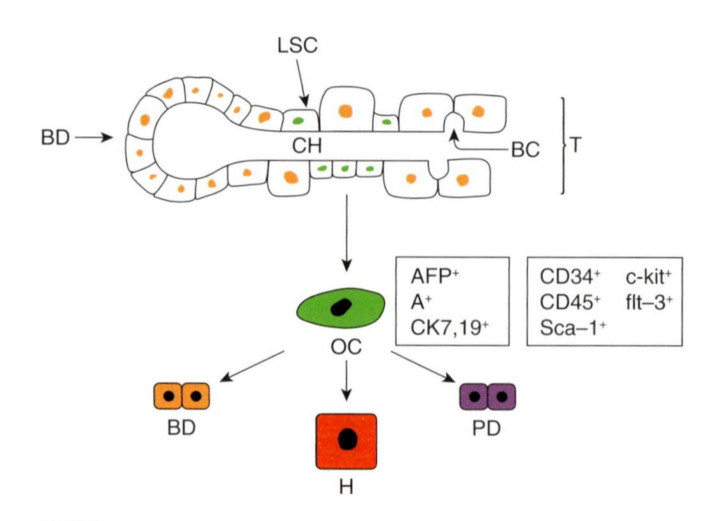


**FIGURE 7.9** Origin of oval cells for liver regeneration. T = liver trabeculum. Bile canaliculi (BC) connect to bile ductules (BD) by a short canal of Hering (CH). The walls of the canal of Hering are composed of hepatocytes and duct-like epithelial cells, some or all of which are liver stem cells (LSC). Under conditions of liver injury where the hepatocytes cannot proliferate, the LSCs proliferate and become oval cells (OC), which express α-fetoprotein (AFP), albumin (A), the cytokeratins 7 and 19 (CK 7, 19) and some cell surface antigens that are also expressed by hematopoietic stem cells. Oval cells can differentiate into bile duct cells, hepatocytes (H) or pancreatic duct (PD) cells.

DNA alkylating agent that inhibits hepatocyte proliferation. Administration of the latter two compounds is followed by partial hepatectomy to elicit a regeneration-promoting environment, whereas this is automatically accomplished with D-galN. Liver regeneration under these circumstances is accomplished by the proliferation and differentiation of small cells with oval nuclei that are unaffected by the treatments (**FIGURE 7.9**). These oval cells arise from a stem cell microniche in the epithelium of the canals of Hering (Thorgeirsson, 1996; Alison et al., 1996; Thiese et al., 1999; Grompe and Feingold, 2001; Sell, 2001; Shafritz and Dabeva, 2002; Dabeva and Shafritz, 2003; Fausto, 2004). The canals of Hering are enclosed by a basement membrane. The oval cells are thought to enzymatically digest the basement membrane and pass through it into the periportal tissue. Thereafter, they proliferate and differentiate.

Oval cells are tripotential and can differentiate into hepatocytes, bile duct epithelium, or pancreatic duct epithelium (Sirica, 1995; Thorgeirsson, 1996). They express both embryonic hepatoblast markers, such as α-fetoprotein (AFP) and albumin, and the bile ductule cell markers cytokeratins 7 and 19, as well as γ glutamyl transferase and others, suggesting that they can recapitulate the embryonic differentiation of hepatocytes (Thorgeirsson, 1996; Hixson et al., 1997, 2003; Petersen et al., 1998; Fausto, 2004). In addition to hepatoblast and bile ductule markers, these cells also express CD34, CD 45, Sca-1, Thy-1, c-Kit, and flt-3 receptor, suggesting a potential overlap in the oval cell and hematopoietic stem cell phenotypes (Fujio et al., 1994; Petersen et al., 1998, 2003). Using antibodies against Thy-1.1, a 95%–97% pure population of oval cells was isolated by FACS from rat liver treated with 2-AAF and injured by $CCl_4$ or partial hepatectomy (Petersen et al., 1998). Based on these and other data, Sell (2001) proposed that the stem cells that give rise to oval cells have their origin in the bone marrow. If so, it would be expected that transplanted bone marrow cells could repopulate the liver after injury. However, marked bone marrow cells did not repopulate the liver in various injury models that evoke oval cell proliferation (Dabeva and Shafritz, 2003; Kanazawa and Verma, 2003; Wang et al., 2003).

## 4. Heterogeneity of Hepatocyte Size and Growth Potential

There is heterogeneity in the size and growth potential of hepatocytes. Tateno isolated two hepatocyte populations, small and large, by FACS (Tateno and Yoshizato, 1999; Tateno et al., 2000). The small hepatocytes have low granularity and autofluorescence, whereas the large hepatocytes are the opposite. Both populations require nonparenchymal cells to proliferate, but the small hepatocytes have a much higher growth potential (**FIGURE 7.10**). Small hepatocytes are bipotent, being able to differentiate as hepatocytes or bile duct cells. This would suggest that they are early stages of oval cell differentiation. Consistent with this idea, small hepatocytes are observed during liver regeneration after chemical damage that suppresses compensatory hyperplasia of hepatocytes and evokes proliferation of oval cells (Fausto, 1994). However, the small hepatocytes of Tateno et al. (2000) were derived from uninjured liver, so it would seem unlikely that they represent a stage of hepatocyte differentiation from oval cells in the uninjured liver or after PH, given that maintenance regeneration of the liver or its regeneration after PH is due solely to the proliferation of fully differentiated hepatocytes. Another possibility is that small hepatocytes represent a unique population of progenitor cells that are not derived from oval cells. Consistent with this idea, oval cell proliferation in the retrorsine/PH model of liver regeneration is modest, but the liver is fully regenerated by the rapid expansion of small hepatocytes that express phenotypic characteristics of fetal hepatoblasts, oval cells, and fully differentiated hepatocytes (Gordon et al., 2000a, b).

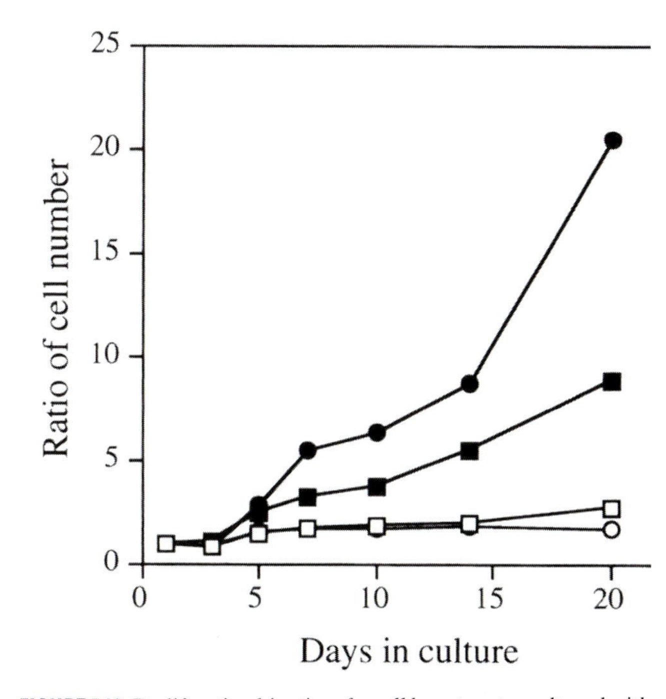


**FIGURE 7.10** Proliferation kinetics of small hepatocytes cultured with (closed circle) and without (open circle) non-parenchymal cells and parenchymal hepatocytes cultured with (closed square) and without (open square) non-parenchymal cells. The ratio of the number of hepatocytes on each day of culture to the number of hepatocytes on day 1 of culture was calculated. Both small hepatocytes and parenchymal hepatocytes proliferated in the presence of non-parenchymal cells, but the small hepatocytes proliferated to a much greater extent. Reproduced with permission from Tateno et al., Heterogeneity of growth potential of adult rat hepatocytes *in vitro*. Hepatol 31:65–74. Copyright 2000, Wiley-Liss.

# PANCREAS

## 1. Structure and Function of the Pancreas

The pancreas is made up of lobules that contain two types of cell clusters, tubules of exocrine cells called acini, and clusters of endocrine cells called islets of Langerhans (**FIGURES 7.11, 7.12**). Both cell types develop during ontogenesis from a network of epithelial duct-like structures that form from the endodermal epithelium of the pancreatic bud. The acinar cells secrete a large number of digestive enzymes into the lumens of small intercalary ducts that connect the acini to larger intralobular ducts that connect to still larger interlobular ducts. The interlobular ducts empty into a single pancreatic duct that joins with the duodenum. The epithelium of the intercalated ducts secretes fluid with a high concentration of sodium bicarbonate that serves to buffer acid arriving in the duodenum from the stomach.

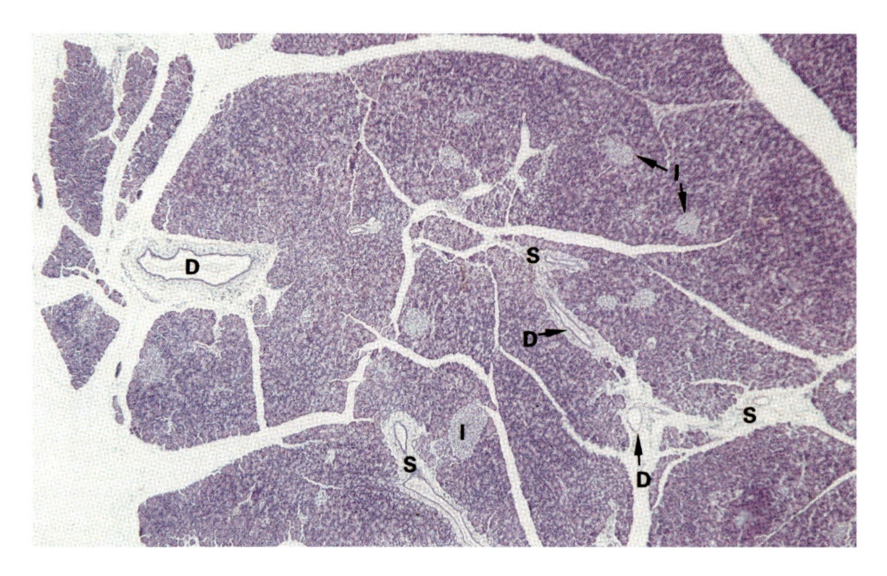


**FIGURE 7.11** Section through the pancreas (H&E stain). The pancreas is organized into lobules with thin septa (S) between the lobules. Each lobule is composed of a highly branched network of ducts surrounded by masses of exocrine tissue (acini, purple) in which are imbedded islands of endocrine tissue (islets of Langerhans, I). DF = interlobular ducts of various sizes. Reproduced with permission from Wheater et al., Wheater's Functional Histology (3rd ed), Copyright 1987, Elsevier.

The islets of Langerhans are interspersed among the acini and constitute only 1%–2% of the mass of the pancreas. They consist of cords and irregular clumps of cells and capillaries. The average human pancreas has approximately one million islets. The islets are supplied with extensive vasculature into which they secrete their hormones directly. Islets consist of four functional cell types, each of which secretes a different hormone. Insulin and glucagon, the main hormones involved in the regulation of blood glucose, are produced by the β-cells and α-cells, respectively. Insulin lowers blood glucose concentration and glucagon raises it. The δ-cells secrete somatostatin, which regulates the secretion of insulin and glucagon. The F-cells secrete pancreatic polypeptide, which affects the exocrine secretion of pancreatic enzymes, water and electrolytes, but the exact physiological role of this molecule has not been defined.

A number of transcription factors are necessary for normal pancreatic development from the endoderm, such as PDX-1, HNF1, HNF3β, HNF4, HNF6, Isl-1, and Pax-6 (Kritzik et al., 2000; Murtaugh and Melton, 2003). PDX-1 is expressed in adult β-cells and regulates the expression of a number of molecules characteristic of islet cell phenotype and function, such as insulin, glucokinase, and glucose transporter-2 (GLUT-2). In turn, PDX-1 activity is regulated by extracellular signal-related kinase (ERK-1/2) and by the survival kinase Akt, which is the downstream effector of another kinase, phosphatidylinositol 3′-kinase (P13-K).

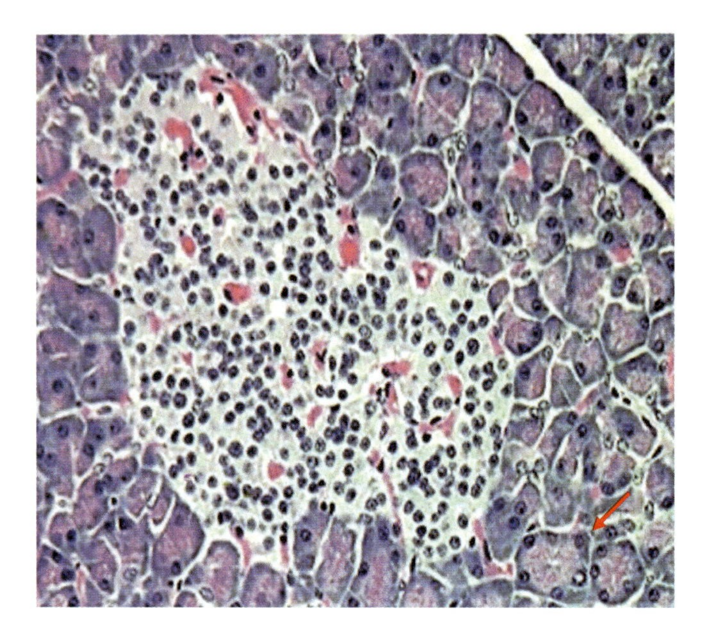

**FIGURE 7.12** Higher magnification of a section of monkey pancreas, showing a paler-staining islet surrounded by acinar tissue. Red arrow points to an acinus. Reproduced with permission, courtesy of Dr. R.A. Bowen, http:/arbl.cvmbs.colostate.edu/hbooks/pathphys/digestion/pancreas/histo_endo.html.

## 2. Regeneration of the Pancreas

The human pancreas is not known to regenerate. In the absence of diabetes, the mouse pancreas exhibits maintenance regeneration and is able to regenerate after injury (Ham and Cormack, 1979; Slack, 1995; Kritzik and Sarvetnick, 2001; Dor et al., 2004). Acinar tissue regenerates rapidly after its selective destruction by temporary ligation of the pancreatic duct or ethionine treatment that spares the islet cells and pancreatic ducts. Surgical removal of 60%–90% of the pancreas leads to the regeneration of acinar and islet tissue, an observation that dates back to the 1920s (Slack, 1995). Treatment with streptozotocin (STZ) selectively ablates the β-cell population, but there is significant regeneration of β-cells if partial pancreatectomy (PP) is performed.

## 3. Origin of Regenerated β-Cells

What is the origin of regenerated β-cells? Are they derived from a stem cell population in the pancreas, by the division of pre-existing β-cells, or by neogenesis from the non-β islet cells, acinar, or ductal cells? The ability to derive and expand β-cells by any of these routes might provide a potential source of transplantable β-cells to replace those lost to diabetes.

Since the liver and pancreas are derived from the same endodermal embryonic rudiment and express many of the same early transcription factors, it would be logical to assume that the pancreas might regenerate like the liver, through compensatory hyperplasia of existing islet cells, as well as through resident stem cells.

To determine the origin of regenerated β-cells under both noninjury and injury conditions, Dor et al. (2004) selectively marked β-cells with an inducible Cre gene and a reporter gene (**FIGURE 7.13**). Mice were generated that were transgenic for a construct encoding a Cre-estrogen receptor fusion protein (*CreER*) driven by the insulin promoter and a reporter construct consisting of the human placental alkaline phosphatase gene (HPAP) driven by a ubiquitous CMV/β-actin promoter with a *lac Z* floxed stop cassette between the promoter and the reporter. The CreER fusion protein is expressed only in β-cells, but is excluded from the nucleus. A pulse of injected tamoxifen allows the CreER protein to translocate into the nucleus and excise the *lac Z* stop cassette, activating the HPAP reporter. The marked β-cells can then be tracked during normal growth and turnover or after PP. If new β-cells arise from noninsulin-producing stem cells, they would be unmarked and dilute the percentage of marked cells over time, but if they arose from pre-existing β-cells, they would express HPAP.

All islets of the intact pancreas contained HPAP$^+$ β-cells after a tamoxifen pulse. The percentage of marked cells did not change significantly over time, but their absolute number increased 6.5-fold between 3 and 12 months of age. This means that growth and maintenance regeneration of the pancreas occurs by the generation of new β-cells from existing β-cells. In another set of experiments, a 70% PP was performed 14 days after a tamoxifen pulse. BrdU was then administered for two weeks after operation to label dividing HPAP$^+$ cells. Any new β-cells that arose from stem cells should have been labeled with BrdU, but not HPAP. Instead, the BrdU-labeled cells were HPAP$^+$, showing that they were derived from pre-existing β-cells.

These data suggest that regenerated β-cells are derived from pre-existing β-cells via compensatory hyperplasia or by dedifferentiation and redifferentiation. Gershengorn et al. (2004) reported that adult

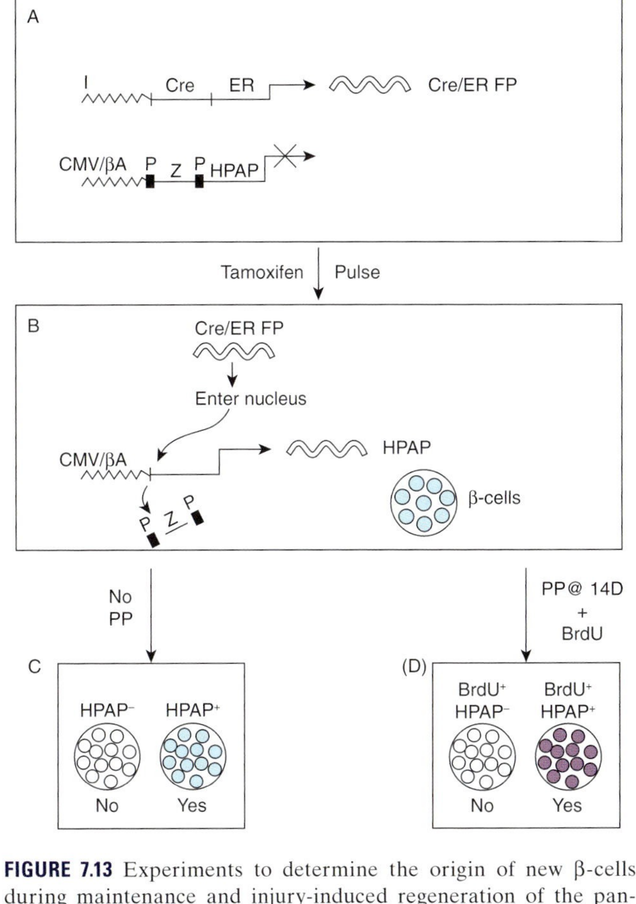


**FIGURE 7.13** Experiments to determine the origin of new β-cells during maintenance and injury-induced regeneration of the pancreas. **(A)** Mice were made transgenic for a Cre/estrogen receptor gene (*Cre/ER*) driven by the insulin promoter (I) and a human placental alkaline phosphatase (HPAP) gene driven by a *CMV/β-actin* promoter. The Cre/ER fusion protein (FP) is expressed, but cannot enter the nucleus in the absence of a hormone pulse. The HPAP protein is not expressed because a stop cassette consisting of a floxed *LacZ* sequence is inserted between the promoter and the HPAP gene. **(B)** tamoxifen pulse induces nuclear localization of the Cre/E fusion protein. The floxed stop cassette is excised selectively in the β-cells, allowing expression of HPAP. Marked cells were tracked in the intact, growing pancreas **(C)** or partial pancreatectomy was performed, followed by labeling with BrdU **(D)**. HPAP$^-$ β-cells were not observed in either case, indicating that regenerated β-cells were derived from pre-existing β-cells.

human β-cells *in vitro* undergo a dedifferentiative transformation to mesenchymal cells in serum-containing medium. During this transformation, both epithelial and endocrine markers decrease and mesenchymal markers increase. The mesenchymal cells, called hIPCs (human islet precursor cells), could be expanded by almost $10^{12}$ over three months of culture. Upon withdrawal of serum, hIPCs differentiated into islet-like cell aggregates (ICAs) that expressed proinsulin mRNA 100–1000-fold, and proglucagon mRNA over 100-fold over hIPC levels. This level of proinsulin expression, however, was less than 0.02% of that in human islets *in vivo*. Nevertheless, measurements of human C-peptide in the blood of SCID mice with ICSs implanted under the kidney capsule indicated that levels of the protein were similar to levels found to reverse hyperglycemia in mice transplanted with insulin-expressing cells differentiated from human fetal liver progenitor cells (Zalzman et al., 2003).

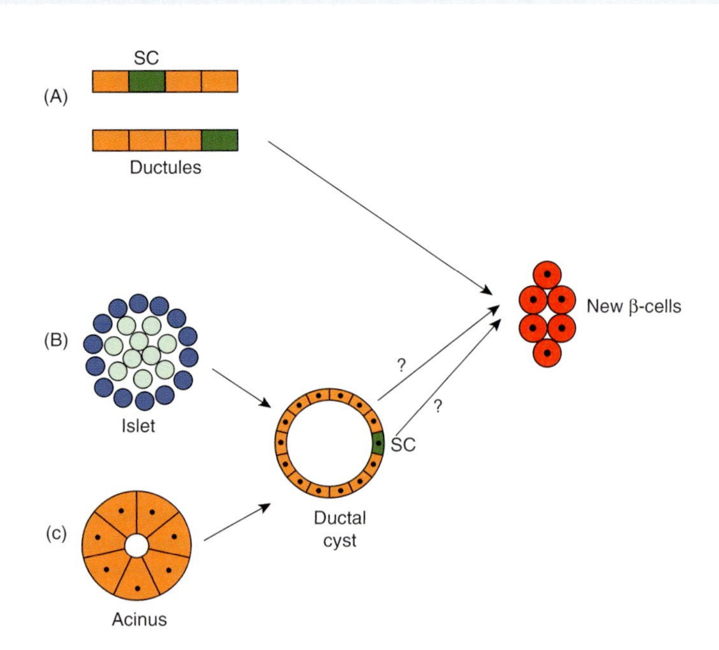


**FIGURE 7.14** Potential sources of new β-cells other than pre-existing β-cells. **(A)** Stem cells in pancreatic ductules can regenerate β-cells. **(B)** β-cells (green) in cultured islets undergo apoptosis and the other endocrine cells (blue) transform into an epithelial cyst. These cysts can be induced to give rise to β-cells. **(C)** Cultured acinar cells also give rise to ductal cysts that regenerate β-cells. Whether the islets and acini contain a stem cell that is incorporated into the ductal cyst, or whether the cyst epithelial cells take on stem cell properties is unknown.

Other studies suggest that stem cells in the pancreatic ductules or the islets themselves regenerate new β-cells or that non-β endocrine cells of the islets and acinar cells can transform into ductal epithelium, which then gives rise to new β-cells (Bonner-Weir and Weir, 2005) (**FIGURE 7.14**). Stem cells have been reported to exist in adult mouse, rat, and human pancreatic ductules that regenerate β-cells after PP and are able to differentiate into β-cells *in vitro* (Bonner-Weir et al., 1993; 2000; Ramiya et al., 2000; Holland et al., 2004). In mice transgenic for the interferon-γ (IFN-γ) gene linked to an insulin promoter, the IFN-γ protein induces a severe inflammatory response in the pancreas that destroys islet cells and stimulates their regeneration from duct cells. BrdU labeling studies showed that the duct cells have a high proliferative activity. Immunohistochemical staining for insulin, carbonic anhydrase II (a duct cell antigen), and amylase (an exocrine enzyme) indicated that these cells were insulin-positive and co-expressed carbonic anhydrase II and amylase, suggesting their derivation from a ductal cell population (Gu and Sarvetnick, 1993; Gu et al., 1994). Histological analysis of the pancreas of STZ-treated mice that had undergone 50% PP suggested that new islets regenerated from pancreatic ductules (Hardikar et al., 1999). Ductule cells involved in regeneration expressed the stem cell marker nestin, but not cytokeratin-20 or vimentin, markers for duct epithelial and mesenchymal cells, respectively (Kim et al., 2004). Ductules isolated *in vitro* after PP displayed many nestin-positive proliferating cells that differentiated into insulin-secreting β-cells that responded to glucose stimulation.

Islet neogenesis was also stimulated from ductal cells of diabetic hamsters (STZ-treated) by administration of islet neogenesis associated protein ($INGAP^{104-118}$), a 15-amino acid segment of a β-cell regeneration-inducing protein called Reg that is a marker for injured pancreas and is expressed *in vitro* and *in vivo* by mouse and human islets. Both the mouse and human Reg proteins are capable of activating β-cell proliferation in non-obese diabetic (NOD) mice (Zhang et al., 2003). $INGAP^{104-118}$ also stimulated the uptake of $^{3}$H-thymidine in duct cells, as well as the differentiation of these cells into insulin-secreting β-cells (Rafaeloff et al., 1997). Islet number increased by 75% and circulating levels of blood glucose and insulin returned to normal, reversing the diabetic state (Rosenberg et al., 2004). $INGAP^{104-118}$-induced β-cell regeneration was associated with increased expression of PDX-1 in duct cells.

Ligation of exocrine ducts draining the splenic half of the pancreas in rats resulted in the conversion of acinar cells to ductal cells in the ligated part (Rooman et al., 2002). Intravenous administration of gastrin for three days induced β-cell neogenesis in the ligated part of the pancreas, resulting in a doubling of the β-cell mass. The new β-cells appeared to be derived from the converted ductal cells. Pancreatic acinar cells cultured *in vitro* were reported to convert to spheroids of cells with a ductal phenotype (Song et al., 2004). Cells at the periphery of these spheroids were observed to differentiate into insulin-positive cells.

Jamal et al. (2005) reported the *in vitro* transformation of quiescent human pancreatic islets into highly proliferative epithelial cysts by apoptosis of the β-cells and conversion of remaining cells into epithelium, driven by the apoptotic enzymes c-Jun N-terminal kinase-1/2 (JNK-1/2) and caspase 3. Expression of islet-specific hormones and transcription factors such as insulin, ISL-1, and NKX-2.2 was downregulated in the epithelial cysts, whereas the expression of pancreatic precursor ductal markers, such as PDX-1, CK-19, and carbonic anhydrase II, and stemness markers such as α-fetoprotein, nestin, and NGN-3, was upregulated. Short-term treatment of epithelial cysts with $INGAP^{104-118}$ induced them to differentiate into functional islets expressing islet transcription factors (PDX-1, NKX 2.2, and ISL-1) essential to produce a mature islet phenotype, and hormones (insulin, glucagon, somatostatin, and pancreatic polypeptide) at levels comparable to freshly isolated human islets. The new islets arose directly from $CK\text{-}19^{+}$ epithelial cells. Studies inhibiting P13-K showed that the signaling pathway for $INGAP^{104-118}$ induction was the same as that used in islet development, through P13-K activation of Akt. Akt regulates the expression of NGN-3, which in turn participates in the induction of ISL-1, NeuroD/β2, and NKX-2.2.

The mechanism by which the phenotypic switch from islet to epithelial cyst is unclear. Jamal et al. (2005) have proposed four possible mechanisms. The first is transdifferentiation of islet cells to epithelial cells. The second postulates an islet stem or progenitor cell that proliferates and differentiates into ductal epithelium. The presence of PDX-1-expressing resident islet stem cells in injured pancreas *in vivo* and *in vitro* that are capable of differentiating into new β-cells has been reported by a number of investigators (Fernandes et al., 1997; Guz et al., 2001; Zulewski et al., 2001; Petropavlovskaia and Rosenberg, 2002; Wang et al., 2004; Seaberg et al., 2004). Choi et al. (2004) reported that fractions of mouse pancreas enriched for islets lost their three-dimensional architecture and generated morphologically heterogeneous cell populations that no longer produced insulin when cultured on fibronectin in serum-free medium in the presence of FGF-2 and LIF. The loss of insulin production was associated with extensive apoptosis. The remaining cells of the islets proliferated rapidly and expressed a number of stem cell markers, including the ESC transcription factors Nanog and Oct-4, and the embryonic pancreas transcription factor Sox 10. In addition, they expressed a number of NSC markers. When replated at high density on Ln-1, they formed islet-like cell clusters that produced insulin, and also cells with embryonic CNS and neural crest phenotypes. These data were interpreted to mean that the β-cells of the islets died during culture and were replaced by the differentiation of multipotential stem cells in the islet tissue that also have the potential to form neuronal cells. Third, the epithelial cells might be derived by a combination of endocrine cell transdifferentiation and stem cells. Fourth, the non-β endocrine cells might be the sole source of the epithelial cyst, but associated with them is at least one stem cell that can be stimulated to give rise to islets.

A number of growth factors, hormones and their receptors stimulate the regeneration of β-cells by inducing the expression of PDX-1 (Risbud and Bhonde, 2002; Bonner-Weir and Weir, 2005). Betacellulin (BTC, a member of the EGF family), glucagon-like peptide-1 (GLP-1), and its receptor exendin-4 (Ex-4) promote β-cell regeneration in STZ mice after 90% pancreatectomy (Li et al., 2003; Xu et al., 1999; Tourrel et al., 2001) and cholecystokinin (CCK) is effective in stimulating mitosis of β-cells (Miyasaka et al., 1999).

## ALVEOLAR EPITHELIUM OF THE LUNG

### 1. Structure of the Respiratory System

The structure of the respiratory system is shown in **FIGURE 7.15**. The respiratory system consists of the air conducting system (mouth, nasopharynx, larynx, trachea, bronchi, and bronchioles) and the gas exchange system of the lung, the alveoli. The trachea, bronchi, bronchioles, and alveoli are all derived by branching morphogenesis from a ventral outgrowth of the embryonic pharyngeal endoderm, the lung bud. The lung bud forms the epithelium of the respiratory system and is surrounded by mesenchyme that differentiates into smooth muscle and connective tissue. The upper part of the lung bud grows straight to form the trachea, whereas the lower end bifurcates into two tubes that will form the paired bronchi. The bronchi branch repeatedly into the bronchioles and then into the thin alveolar air sacs of the lung.

The lining of the air conducting system is a pseudostratified ciliated epithelium interspersed with nonciliated goblet cells that secrete a protective mucous coat onto its surface. Basal cells lie beneath the differentiated cells of the epithelium, attached to a basement membrane. The ciliated and goblet cells of the epithelium turn over and are regenerated by stem cells in the basal layer that can form both ciliated cells and

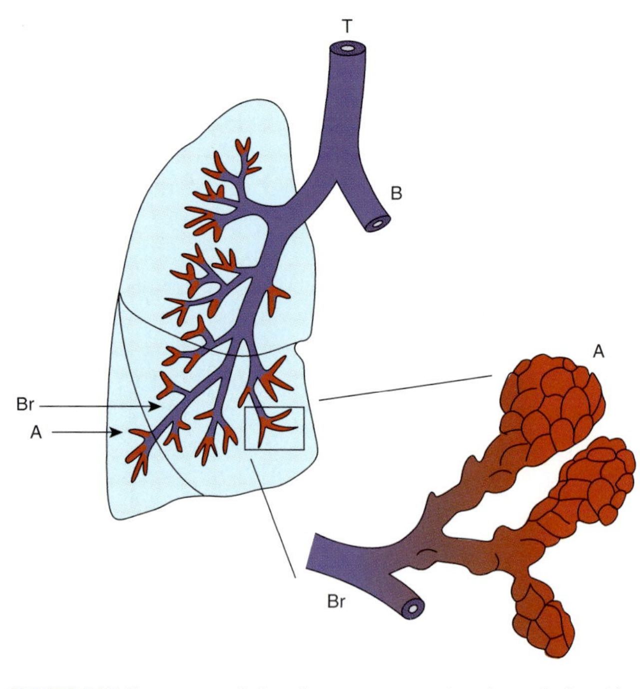


FIGURE 7.15 Structure of the airway system, showing relationships between trachea (T), bronchial tubes (B), bronchioles (Br) and alveoli (A). Boxed area is enlarged to the right.

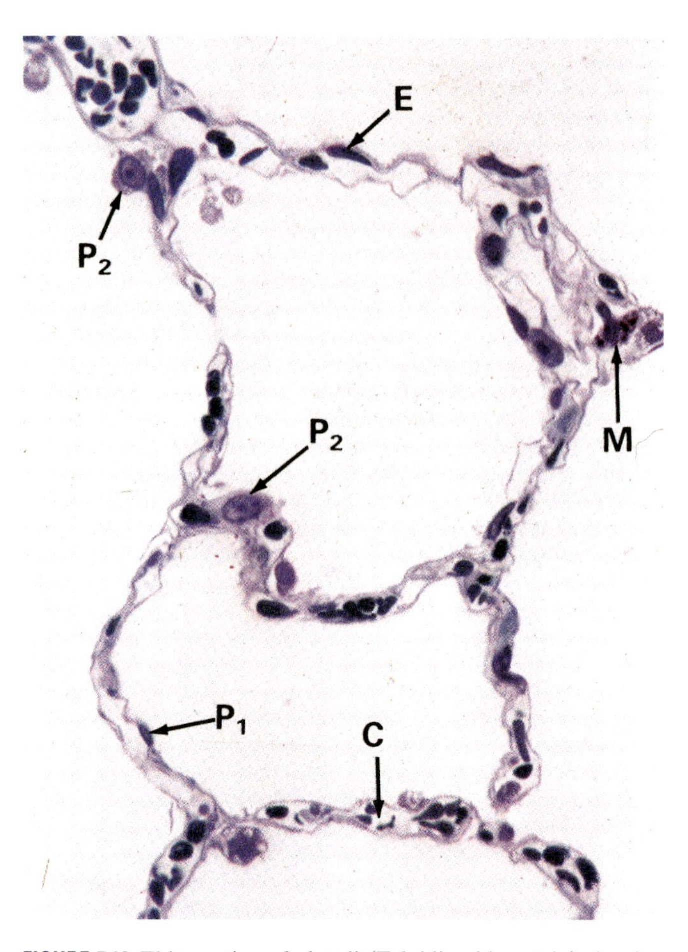


FIGURE 7.16 Thin section of alveoli (Toluidine blue stain) showing Type I (P1) and type II (P2) pneumocytes. Note that type II pneumocytes are larger than type I; they contain surfactant globules. C = capillaries, E = capillary endothelial cells, M = alveolar macrophages. Reproduced with permission from Wheater et al., Wheater's Functional Histology (3rd ed). Copyright 1997, Elsevier.

goblet cells (Ford and Terzaghi-Howe, 1992; Johnson and Hubbs, 1990; Liu et al., 1994; Hong, 2004).

The smallest bronchioles, the respiratory bronchioles, open into alveolar ducts, which bud off multiple alveolar saccules. Each saccule consists of several alveoli. The epithelium of the alveoli is very thin and the interalveolar walls are highly vascularized with capillaries, thus allowing gas exchange between the air in the lungs and the blood. It is estimated that adult human lungs contain around 300 million alveoli, with a surface area of 70–80 $ft^2$ for gas exchange (Ham and Cormack, 1979).

The alveolar epithelium consists of two types of cells, Type 1 and 2 pneumocytes (**FIGURE 7.16**). Type 1 pneumocytes are the more numerous. They are squamous cells and it is through them that gas exchange takes place. Type 2 pneumocytes are interspersed among the Type 1 pneumocytes. These cells are larger and have the important function of secreting a pulmonary surfactant. The role of the surfactant is to reduce surface tension of the tissue fluid by diminishing the intermolecular forces between water molecules, facilitating the inflation of the alveoli on inspiration (Ham and Cormack, 1979). Macrophages also patrol the alveoli from the blood to phagocytize incoming dust and debris. They move out of the alveoli with their cargo into the air conducting system, where they are swept along by cilia to be coughed up or swallowed.

## 2. Regeneration of Alveolar Epithelium

Pulmonary alveoli can be injured in a number of ways, such as hyperoxia and inhalation of toxic substances, that collectively are called acute lung injury (ALI). Mists of acid, the antibiotic bleomycin, elastase, or lipopolysaccharide can be used to induce ALI in experimental animals. Alveolar injury is believed to contribute to the development of chronic obstructive pulmonary diseases (COPD) such as emphysema and pulmonary fibrosis (Adamson et al., 1988). Type 1 pneumocytes are the cells that are predominantly damaged in ALI (Voelker and Mason, 1990). Type 1 pneumocytes are regenerated by the proliferation and differentiation of type 2 pneumocytes that act as stem cells (Adamson and Bowden, 1975; Evans et al., 1975; Witschi, 1976; Voelker and Mason, 1990; Bishop, 2004). Not all type 2 pneumocytes are stem cells, however.

Reddy et al. (2004) separated cultured type 2 pneumocytes, derived from hyperoxia-treated alveoli of rats, into E-cadherin negative and positive subpopulations. The E-cadherin$^+$ subpopulation contained the majority of damaged cells, was quiescent, and expressed low levels of telomerase. The E-cadherin$^-$ subpopulation was resistant to hyperoxia, proliferated, and expressed high levels of telomerase, suggesting that this is the stem cell subpopulation. The molecular phenotypes of these subpopulations of cells, however, have not been characterized.

Several growth factors are important in regulating injury-induced alveolar regeneration. FGF-1 (Mason et al., 1992), EGF (Voelker and Mason, 1990), and HGF (Mason et al., 1994) were shown to stimulate the mitosis of type 2 pneumocytes *in vitro*. Of these, HGF appears to play a central role in initiating alveolar repair, as it does in liver regeneration. HGF synthesis is induced in endothelial cells and alveolar macrophages within 3–6 hr after ALI, accompanied by a rapid disappearance of the HGF receptor, c-Met, indicating HGF:receptor binding and endocytosis (Yanagita et al., 1992, 1993). Alveolar epithelial cells express u-PA, which cleaves pro-HGF, and urokinase receptor (Marshall et al., 1990) and HGF upregulates u-PA activity in cultured mouse alveolar epithelial cells (Dohi et al., 2000). As in the liver, the C/EBP$\alpha$, $\beta$, and $\delta$ transcription factors play a role in the early events of alveolar regeneration. These factors are all found in type 2 alveolar cells and the first two are also found in alveolar macrophages. C/EBP$\alpha$ and $\beta$ mRNAs are significantly upregulated in lipopolysaccharide or bleomycin-induced ALI (Sugahara et al., 1999), suggesting that HGF-induction of the activity of these two transcription factors may play a role in activating a set of early immediate genes involved in the proliferation of type 2 pneumocytes.

Further evidence for the importance of HGF in alveolar regeneration is that intravenous injection of hrHGF into mice elicits proliferation of type 2 pneumocytes after HCl-induced injury *in vivo* (Omichi et al., 1996) and continuous intravenous administration of rhHGF to mice with bleomycin-induced pulmonary fibrosis suppresses the fibrotic changes (Yaekashiwa et al., 1997) by inhibiting collagen accumulation (Dohi et al., 2000). High concentrations of HGF are detected in the pulmonary edema fluid of patients with ALI (Verghese et al., 1998). Partially purified edema fluids increased DNA synthesis in cultured rat type 2 pneumocytes and anti-HGF antibody attenuated the increase by 66%.

KGF (FGF-7) also appears to play an important role in restoration of the alveolar epithelium. KGF is a critical factor in lung development (Shiratori et al., 1996) and ameliorates the effects of hyperoxic and bleomycin-induced injury when given intratracheally prior to the injury (Panos et al., 1995; Yi et al., 1996). Not only does it stimulate proliferation of type 2 pneumocytes, it also stimulates expression of surfactant proteins by these cells *in vitro* and *in vivo* (Sugahara et al., 1995; Ulich et al., 1994). Low concentrations of KGF are found in the pulmonary edema fluids of ALI patients (Verghese et al., 1998). The KGF in these fluids has a stimulatory effect on proliferation of type 2 pneumocytes. Anti-KGF antibody attenuated the increase in their DNA synthesis by 53%, almost as much as anti-HGF antibody.

The lung as an organ does not regenerate after removal. However, the contralateral lung experiences increased proliferation of airway basal cells and type II pneumocytes (Brody et al., 1978; Cagle et al., 1990). This proliferation appears to be stimulated by a post-operative increase in HGF levels in the lung, liver and kidney that is associated with an increase in plasma HGF level and a transitory increase in c-Met receptors on type 2 pneumocytes. Antibody neutralization of endogenous HGF suppressed the compensatory growth of the lung, whereas administration of hrHGF to pneumonectomized mice stimulated DNA synthesis (Sakamaki et al., 2002).

## KIDNEY AND URINARY SYSTEM

### 1. Structure of the Kidney and Urinary System

The mammalian kidney develops from the intermediate mesoderm between the lateral plate and the somites. Paired nephric ducts form on the outer edge of the intermediate mesoderm. The metanephric mesoderm medial to the ducts induces them to form an evagination called the ureteric bud, which grows into the nephrogenic mesoderm, branches, and reciprocally induces it to form the nephrons, the units of the kidney that eliminate metabolic waste products and regulate water, electrolyte, and acid-base balance. The human kidney contains approximately one million nephrons.

The structure of the mammalian kidney is illustrated in **FIGURE 7.17**. Most of the nephrons are located toward the outer surface of the cortex, but about 15% are located deep in the cortex. Each nephron consists of a renal corpuscle attached to a renal tubule. The renal tubule is differentiated into a proximal convoluted tubule, closest to the renal corpuscle, an intermediate looped segment, the loop of Henle, and a distal convoluted tubule that flows into a collecting duct. The different parts of the renal tubule have specialized roles in water resorption and ion pumping. The loops of Henle and the collecting ducts extend into the medulla of the kidney. The collecting ducts flow into larger ducts called calyces, which eventually merge to flow into the

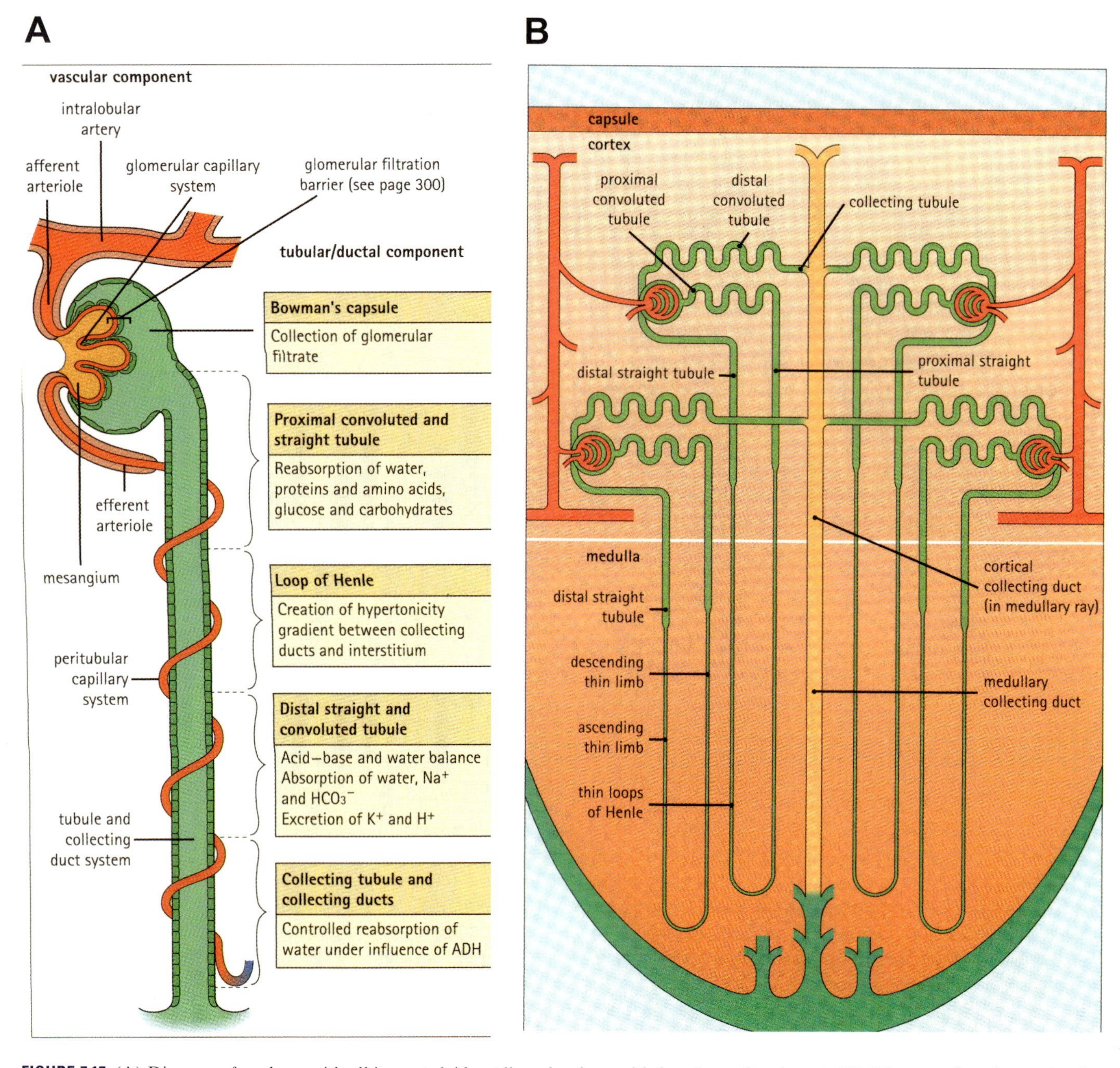


**FIGURE 7.17** **(A)** Diagram of nephron with all its parts laid out linearly, along with functions of each part. **(B)** Diagram of nephrons showing the actual spatial relationships of the different parts of the nephron. Reproduced with permission from Stevens and Lowe, Human Histology (3rd ed), Mosby/Elsevier. Copyright 2005, Elsevier.

ureters. Urine is collected in the bladder and excreted through the ureters and urethra.

The renal corpuscle of the nephrons lies in the outer portion of the kidney, called the cortex. The renal corpuscle functions to filter waste from the blood. The corpuscle consists of a double-walled epithelial cup, Bowman's capsule, enclosing a tuft of capillaries called the glomerulus (**FIGURE 7.18**). The outer wall of the capsule is the parietal epithelial layer, and the inner wall is the visceral layer. The space between the parietal and visceral layers of the capsule is continuous with the lumen of the renal tubule. The space enclosed by the visceral layer is occupied by the glomerulus. The epithelial cells of the visceral layer are called podocytes because they send long processes laterally that extend short "feet" onto the basement membrane of the capillary endothelium (**FIGURE 7.19**). The basement membrane is ~threefold thicker than other basement

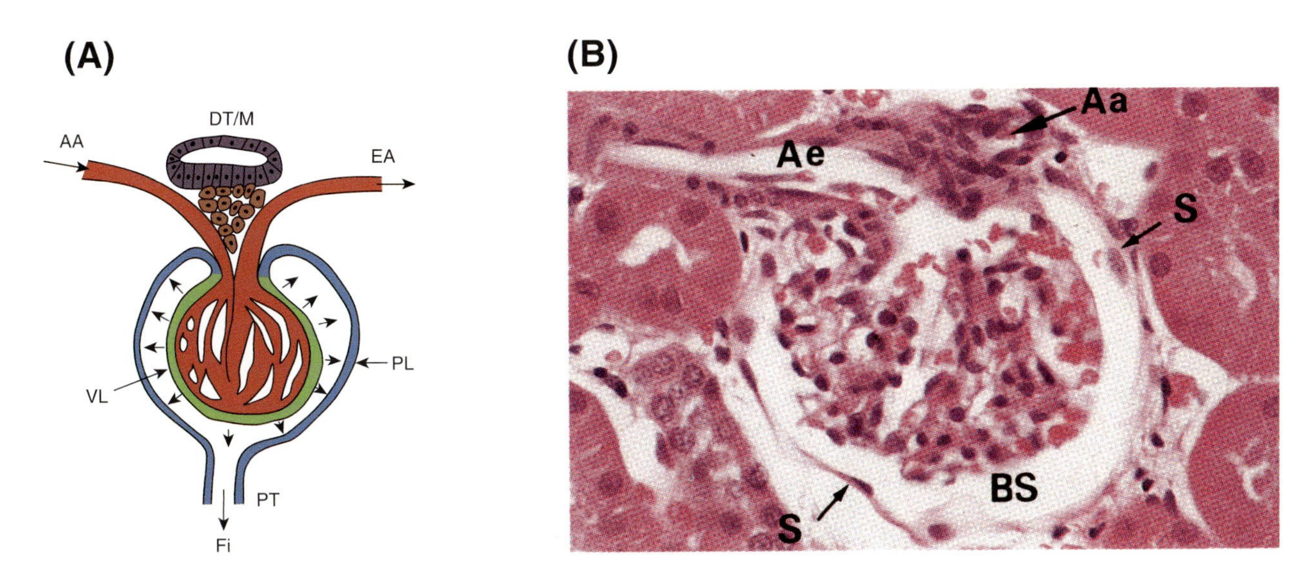


**FIGURE 7.18** **(A)** Diagram of renal corpuscle showing how urinary filtrate (Fi) is made. The corpuscle consists of the parietal layer (PL) and visceral layer (VL) of Bowman's capsule, and a glomerular tuft of capillaries (red) connecting afferent (AA) and efferent (EA) arterioles. The distal tubule and mesangial cells (DT/M) of the nephron are closely associated with the arterioles. The wall of the distal tubule next to the mesangial cells (macula densa) is thicker and has a higher cell density. Blood flows via the afferent arteriole into the capillary tuft and is filtered through the visceral epithelium into the space of Bowman's capsule (arrows). The filtrate then enters the proximal tubule (PT) where water and ions are reabsorbed. **(B)** Section through a renal corpuscle showing the afferent arteriole (Aa) cut transversely and the efferent arteriole (Ae) cut longitudinally. S = nuclei of flattened squamous cells of Bowman's capsule. BS = Bowman's space (size is artifactual. **B** reproduced with permission from Wheater et al., Wheater's Functional Histology (3rd ed). Copyright 1997, Elsevier.

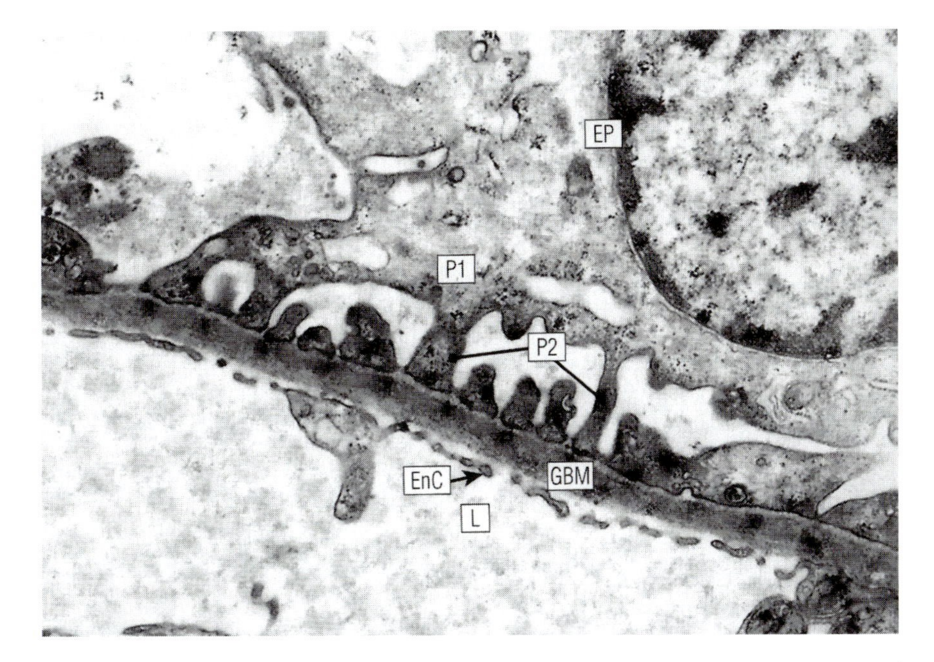


**FIGURE 7.19** Electron micrograph of the interface between the glomerular capillary epithelium and the visceral epithelium of Bowman's capsule, which contains podocytes (EP). The interface is a thick glomerular basement membrane (GBM). The cells of the visceral epithelium (podocytes) extend numerous cytoplasmic processes ("feet") that rest on the basement membrane (P1, P2). The narrow spaces between the feet are the slit pores. The plasma is first filtered through the fenestrated capillaries (EnC), which are permeable to water and small molecules, but not macromolecules. The filtrate then is sieved through the slit pores of the basement membrane and enters the space of Bowman's capsule. L = capillary lumen. Reproduced with permission from Stevens and Lowe, Human Histology (3rd ed). Elsevier/Mosby, Copyright 2005, Elsevier.

membranes and serves to filter proteins from the plasma. The filtered plasma enters the internal space of Bowman's capsule and moves into the renal tubule. Water is resorbed and electrolyte balance maintained by diffusion and ion pumps as it moves through the renal tubule.

Nephrons as a whole do not regenerate, but the epithelium of the kidney tubules undergoes maintenance and injury-induced regeneration via stem cells in the basal layer of the epithelium, as does the epithelium of the bladder and ureters.

## 2. Regeneration of Kidney Tubule Epithelium

The tubule epithelial cells of the uninjured adult mammalian kidney have a low rate of turnover and replacement, as evaluated by PCNA and Ki-67 staining (Nadasdy et al., 1994). This implies that the kidney epithelium harbors a stem cell population for maintenance regeneration (Al-Awqati and Oliver, 2002). To locate these cells, mice were given a seven-day pulse of BrdU, followed by a two-week chase (Maeshima et al., 2003). Cells that retained the BrdU label were considered to represent stem cells. Small numbers of label-retaining cells (LRCs) were found scattered throughout the proximal tubules, loops of Henle, distal tubules and collecting ducts, with the proximal tubules exhibiting the greatest frequency of LRCs. Most of the LRCs were adjacent to intertubular capillaries, suggesting that the capillary endothelium might provide factors that constitute a stem cell microniche. Interestingly, the LRCs also stained with agglutinins specific for the different segments of the renal tubule in which they resided, implying that like the hepatocytes of liver or the β-cells of the pancreas, they have differentiated functions.

A standard injury model to demonstrate the regeneration of tubule epithelium is ischemia/reperfusion, in which the blood supply to the kidney is temporarily clamped off and the kidney then allowed to reperfuse. Under these circumstances, epithelial cells die and are sloughed off their basement membrane, but are rapidly replaced (Bonventre, 2003). BrdU pulse-chase experiments have shown that replacement is accomplished by the division of LRCs to give rise to a transit-amplifying population of cells (Maeshima et al., 2003).

FIGURE 7.20 depicts the replacement process. The LRCs undergo an epithelial to mesenchymal transformation (EMT) (Witzgall et al., 1994; Bonventre, 2003; El Nahas, 2003) and express the mesenchymal markers vimentin and α-smooth muscle actin, as well as Pax-2, a transcription factor critical for kidney ontogenesis (Imgrund et al., 1999; Maeshima et al., 2002). Maeshima et al. (2003) observed pairs of BrdU-labeled LRCs 18 hr after reperfusion in pulse-chase experiments. Only one member of the pair expressed vimentin, suggesting that after division of a LRC, one daughter self-renews as a LRC and the other dedifferentiates to become a dividing mesenchymal cell. The number of vimentin-expressing transient amplifying cells increases in proportion to dilution of the BrdU label. It is not known whether maintenance regeneration of the uninjured renal tubule takes place by a similar EMT of LRCs. Transient amplifying cells show relocation of basal integrins to their lateral borders, increased expression of NCAM, and upregulation of cellular Fn, HA, uPA, and MMP-2 and 9 (Pawar et al., 1995; Bonventre, 2003), all consistent with dedifferentiation and preparation for migration. Another molecule that may be involved in the migration of tubule cells is kidney injury molecule-1 (Kim-1), a transmembrane protein of the Ig superfamily that is strongly upregulated in the post-ischemic rat kidney (Ichimura et al., 1998).

Kidney tubule EMT is regulated by changes in the expression of a number of growth factors, of which TGF-β plays the most prominent role (Bonventre, 2003; El Nahas, 2003; Kalluri and Neilson, 2003; Zeisberg and Kalluri, 2004; Neilson, 2005). TGF-β signals through Smads 2, 3 and 4, which complex with Wnt-stabilized β-catenin and LEF1/TCF in the nucleus to activate a gene program that transforms the epithelial cells to fibroblastic cells (Kim et al., 2002; Kalluri and Neilson, 2003; Zeisberg and Kalluri, 2004).

After migrating to cover the denuded areas of the basement membrane, the mesenchymal cells undergo a mesenchymal to epithelial transformation (MET). This transformation is regulated by BMP-7 (Zeisberg et al., 2003). BMP-7 stabilizes differentiation and branching during development of renal tubules (Zeisberg and Kalluri, 2004). BMP-7 expression is highest in the kidney and it reverses TGF-β1-induced epithelial to mesenchymal transformation by re-induction of the key epithelial cell adhesion molecule, E-cadherin (Zeisberg et al., 2003).

Important modulators of the EMT and MET are "trap" proteins (FIGURE 7.20), which are either attached to cell membranes, where they act as co-receptors to enhance ligand binding to receptors (positive traps), or are soluble entities that can block receptor activation by ligand (negative traps) (Shi and Massague, 2003; Neilson, 2005). Negative trap proteins for TGF-β include decorin, latency-associated polypeptide (LAP), and α-macroglobulin. Noggin, chordin, and follistatin are negative trap proteins for BMP-7. Connective tissue growth factor (CTGF) is a positive trap for TGF-β and a negative trap for BMP-7. The first protein that has been identified as a positive trap for BMP-7 is keilin/chordin-like protein (KCP). KCP is different in that it

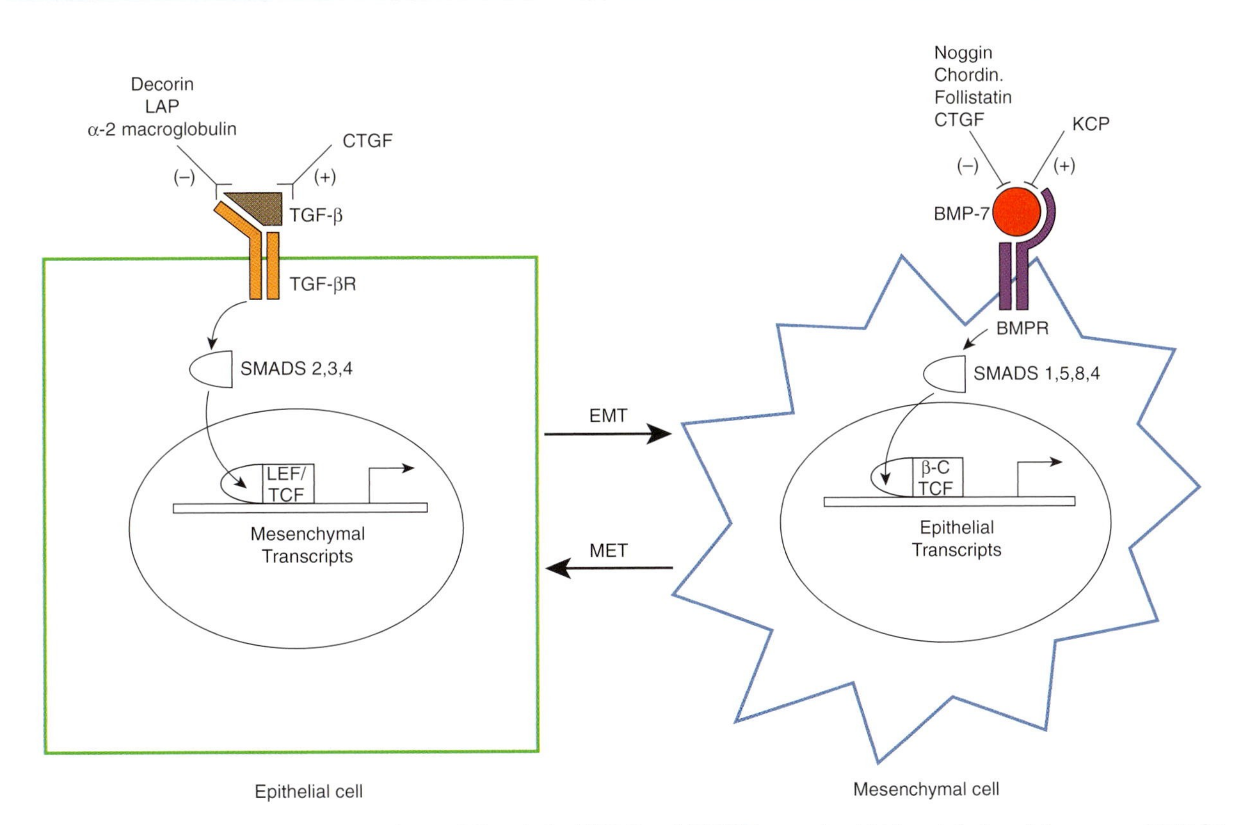


**FIGURE 7.20** Regulation of epithelial regeneration and fibrosis by TGF-β and BMP7 in proximal kidney tubules of the mouse. TGF-β induces an epithelial to mesenchymal transformation (EMT) via the Smad 2,3,4 pathway. The EMT involves the production of matrix metalloproteinases by the epithelial cells, degradation of the basement membrane, detachment of the cells from the basement membrane and the production of fibroblast (mesenchymal)-specific transcripts. BMP7 induces a mesenchymal to epithelial transformation (MET) via the Smad 1,5,8,4 pathway that results in the production of epithelial-specific transcripts. The binding of TGF-β and BMP7 ligands to their receptors is positively and negatively modulated by ligand-trap proteins. When regulation is balanced, the mesenchymal cells created by EMT migrate and divide to cover the denuded basement membrane, followed by MET and regeneration of tubule epithelial cells is the result. However, if the EMT predominates over the MET, interstitial fibrosis of kidney tubules results.

is a soluble enhancer of BMP-7 signaling (Lin et al., 2005).

While a MET regenerates epithelial cells, the TGF-β-induced inflammatory process associated with kidney injury and disease that results in EMT can prevent MET, leading to permanent scarring of nephric tissue through the accumulation of collagens and/or cross-linking of ECM molecules, rendering them resistant to breakdown by MMPs (El Nahas, 2003). Whether EMT is followed by MET or fibrosis therefore depends on the balance between the TGF-β and BMP-7 signaling pathways, as shown by the fact that pharmacological doses of BMP-7 reverse the fibrosis associated with EMT (Zeisberg et al., 2003) and $Kcp^{-/-}$ mice are more susceptible to developing renal fibrosis (Lin et al., 2005).

In contrast to a model of tubule epithelial cell regeneration by EMT and MET, Kale et al. (2003) have proposed that epithelial-forming stem cells are recruited from the bone marrow to repair tubules after ischemic injury. Irradiated C57BL/6J mice were engrafted with ROSA26 $Lin^-$ $Sca^+$ c-$kit^+$ HSCs (which express β-galactosidase) and their kidneys subjected to transient ischemic injury. One week later, ~20% of the renal tubules contained epithelial cells that co-expressed β-galactosidase and the tubule epithelial-specific protein megalin. The blood urinary nitrogen (BUN) of these animals was reduced to control levels, whereas irradiated animals that did not receive stem cells had significantly elevated BUN. Kale et al. (2003) reported that the majority of the cells present in the previously necrotic tubules of the transplanted animals were bone marrow-derived.

The mammalian kidney as an organ responds to growth-enhancing treatments and to partial nephrectomy by compensatory hypertrophy, or nephron enlargement, but does not regenerate new nephrons. Removal of one kidney or treatment of the animal with folic acid or growth hormone causes the remaining kidney to undergo compensatory hypertrophy. In the case of nephrectomy, the individual nephrons enlarge by an increase in epithelial cell size that involves increases in RNA and protein synthesis (Halliburton, 1969). Folic acid and growth hormone treatment, on the other hand, cause nephron enlargement by increased proliferation of epithelial cells (Nicholls et al., 1975; Norman et al., 1988). However, teleost fish and the skate, an elasmobranch fish, respond to partial nephrectomy or chemical injury by regenerating new nephrons (Salice et al., 2001; Elger et al., 2003; Drummond, 2003). The source of the new nephrons appears to be a population of stem cells resident to a special nephrogenic zone. In the skate, injury to one kidney also stimulates new nephron formation in the contralateral, uninjured kidney. Because the skate kidney shows similarities to the mammalian kidney in its development, anatomy, and physiology, it may be an excellent model to study how we might induce nephron regeneration in mammals.

## GONADS

The gonads of many species, including mammals, respond to partial gonadectomy by compensatory hypertrophy (Goss, 1964), but regeneration of the original structure does not take place. Complete regeneration has been reported from gonad remnants in salmonid fish (Robertson, 1961; Kersten et al., 2001), carp (Clippinger and Osborne, 1984; Underwood et al., 1986), and blue gourami (Johns and Liley, 1970). It is not known whether regeneration takes place by proliferation of somatic and germinal cells, by resident stem cells, or by dedifferentiation of somatic and germinal cells. In *Drosophila*, spermatogonia can dedifferentiate to repopulate the germline stem cell (GSC) niche (Brawley and Matunas, 2004). GSCs null for the JAK-STAT signaling pathway differentiate into spermatogonia without self-renewal. A temperature-sensitive mutant of this pathway was used to shut off JAK-STAT signaling and deplete germline stem cells in adult flies, then restore signaling. Spermatogonia that had initiated differentiation were able to dedifferentiate to become functional germline stem cells.

The effect of hormone-induced sex reversal in females or treatment of mature males with estradiol on gonadal regeneration in rainbow trout has been investigated by Kersten et al. (2001). They found that estradiol had no effect on gonadal regeneration in males, which always regenerated testes, and that the testes of masculinized females always regenerated as testes. These results indicate that regenerating testes cannot be sex-reversed and that sex reversal during embryogenesis is maintained during gonadal regeneration in mature fish.

Amphibians also can regenerate gonadal tissue. The testes of male adult newts, *Notophthalmus viridescens*, were restored to normal size after partial gonadectomy (Scadding, 1977). The regrowth was due to regeneration, not compensatory hypertrophy, since the restored testis was histologically the same as the unoperated control. There is no data, however, on whether regeneration is due to resident stem cells, dedifferentiation, or compensatory hyperplasia. Ovaries of the newt were not able to regenerate after partial gonadectomy, nor did the remnant ovarian tissue undergo compensatory hypertrophy. In the toad *Discoglossus pictus*, the testis has been reported to regenerate after its total removal (Furnori-Savoca, 1967), but again, the source of the regenerated tissue is unknown.

## PROSTATE TISSUE

The prostate is an epithelial gland consisting of numerous tubules comprised of three major cell types, luminal cells that produce secretory proteins, basal cells between the luminal cells and the basement membrane, and neuroendocrine cells, which are present in the luminal layer and are thought to regulate the secretory activities of the luminal cells by paracrine factors (Abate-Shen and Shen, 2000). The luminal cells are dependent on androgen for their viability. The number of these cells is drastically reduced upon androgen withdrawal, leaving the basal layer, but the luminal layer regenerates upon re-addition of androgen. In some male mammals that are seasonal breeders, such as the ram, the prostate regresses and regenerates in response to seasonal changes in androgen levels (Kurita et al., 2001). Furthermore, mouse prostate regenerates from dissociated epithelial cells when they are cultured together with urogenital sinus mesenchyme (Xin et al., 2003). Regeneration requires signaling through the hedgehog pathway (Karhadkar et al., 2004).

The regenerative ability of the prostate implies that it harbors a population of stem cells. These cells were believed to reside in the basal layer of the epithelium, because the *p63* gene, a homologue of the *p53* tumor suppressor gene, is essential for the differentiation of epithelia and is expressed in prostate basal epithelial cells (Yang et al., 1999) and the prostate is not formed in *p63* null mice (Signoretti et al., 2000). However, urogenital sinus tissue isolated from *p63* null mice is able to form prostatic tissue with luminal and neuroendocrine, but not basal, cells when grafted under the kidney capsule of nude mice (Kurita et al., 2004). The *p63* null prostate tissue regressed in response to castration, yet was able to regenerate in response to androgen administration, suggesting that stem cells may reside in the luminal layer of the prostate epithelium. This does not rule out the existence of stem cells in the basal layer, but shows that basal cells are not essential for prostate regeneration.

## SUMMARY

The villous epithelium of the intestines is regenerated by stem cells deep in the crypts of Lieberkuhn. These cells share some of the characteristics of other epithelial basal stem cells in that they express components of the Notch signal transduction pathway, the transcription factor Tcf-4, and the α2β1 integrin, and they are activated to proliferate by Wnt signaling. ISC proliferation and differentiation are regulated by growth factors and cytokines, cell adhesion molecules and ECM synthesized by the fibroblasts of the lamina propria. Cytokines and growth factors that promote ISC proliferation are IGF-1 and 2, IL-4, and perhaps FGF-1, 2, and 9, whereas TGF-β, IL-6, and IL-11 inhibit proliferation. Alpha integrins, E-cadherin, Fn, Ln, and collagen IV control the rate of migration out of the crypt and N-cadherin is essential for the differentiation of migrating ISCs.

Transected amphibian intestine regenerates epimorphically by the dedifferentiation of serosal and smooth muscle cells to form two blastemas that grow together, followed by the migration of cords of intestinal epithelium into the mesenchyme. The cords hollow out to reform an intestinal lumen while the mesenchymal cells differentiate into the other layers of the intestine. The rat also regenerates the transected intestine, but the mechanism of this regeneration has not been defined.

The classic example of regeneration by compensatory hyperplasia is the mammalian liver. Hepatocytes have an enormous capacity for replication, up to at least 70 times. They are maintained in a nonproliferative state by C/EBPα inhibition of cdks. Partial hepatectomy triggers the appearance of TNF-α, IL-6, HGF, and EGF, which are mitogenic signals that prime the hepatocytes for entry into the cell cycle by activation of the transcription factors STAT3, PHF/NF-κB, AP-1 and C/EBPβ. These transcription factors induce the activity of sets of "early-immediate" and "delayed-immediate" genes that encode other transcription factors involved in entering and progressing through the $G_1$ phase of the cell cycle. HGF appears to play a central role in this process. Pro-HGF is released by matrix degradation and its synthesis is elevated by VEGF stimulation of sinusoidal endothelial cells. Pro-HGF is activated by uPA and binds to the c-met receptor, triggering entry into and progression through the cell cycle. Once the original mass of the liver is attained, proliferation ceases and the original tissue architecture of the liver is restored by processes that are not well understood.

The liver can also regenerate by stem cell proliferation when the ability of hepatocytes to proliferate is compromised by chemical injury. The current consensus is that these cells reside in the terminal bile ducts and give rise to a transit amplifying population of small oval cells that differentiate into hepatocytes. Small hepatocytes with a high proliferative potential and the ability to differentiate either as hepatocytes or bile duct cells have been isolated from the uninjured liver. These cells may represent another population of progenitor cells with the capacity for regenerating the liver. The relationships between oval cells, small hepatocytes and fully differentiated hepatocytes are not well understood.

Beta-cells of the pancreas can also regenerate. Genetic marking experiments have revealed that during either maintenance or injury-induced regeneration of the pancreas, new β-cells are derived from pre-existing β-cells. This regeneration can be initiated by a number proteins: β-cell regeneration protein (Reg), islet neogenesis associated protein (INGAP, a 15 amino acid fragment of Reg), betacellulin (a member of the EGF family), and GLP-1. However, the mechanisms leading to proliferation of β-cells are not as well understood as those of hepatocyte proliferation. Regeneration of β-cells from epithelial stem cells in the pancreatic ductules after partial pancreatectomy or administration of INGAP has also been reported. Furthermore, although not proven to take place *in vivo*, a number of investigators have shown that acinar cells or other endocrine cells of the islets are able to dedifferentiate and proliferate and then differentiate into new β-cells. It is also possible that the islets harbor a population of stem cells that can produce new β-cells. These studies all suggest that in cases of diabetes, if the progress of the disease can be arrested, it may be possible to provide enough new β-cells to restore endocrine function.

Injured type I pneumocytes of the alveolar epithelium are regenerated by a subset of type II pneumocytes that do not express E-cadherin, are resistant to hyperoxia, and express high levels of telomerase. As in the liver, HGF plays a central role in initiating entry into the cell cycle by activating C/EBPα, β, and δ transcription factors through binding to the c-met receptor. EGF and FGF-1 have also been shown to stimulate the mitosis of type 2 pneumocytes *in vitro*, and KGF does the same *in vitro* and *in vivo*. Interestingly, continuous administration of rhHGF to mice with bleomycin-induced pulmonary fibrosis suppresses the fibrotic changes.

The epithelial cells of the kidney tubules have a low rate of turnover. Stem cells have been demonstrated in the epithelium of the kidney tubules that do maintenance regeneration and that respond to ischemia/reperfusion injury by rapidly replacing dead cells. This rapid replacement is accomplished by a TGF-β1 mediated transformation of epithelial cells to mesenchymal cells that show upregulation of HA, FN, NCAM, Kim-1, MMP2, and 9, uPA, and PAI, all consistent with dedifferentiation, degradation of basement membrane, and preparation for migration. After covering the denuded areas of the epithelium, the mesenchymal cells reacquire the epithelial phenotype, through BMP-7 signaling. A shift toward predominance of the TGF-β pathway in this process by renal injury or disease leads to fibrosis that compromises kidney function.

The mammalian kidney does not regenerate new nephrons, but can enlarge by compensatory hypertrophy of nephrons. However, the skate, an elasmobranch fish, is able to regenerate new nephrons from a special nephrogenic zone containing a population of stem cells. The skate kidney is similar to the mammalian kidney in its metanephric development, anatomy and physiology and thus may be an excellent model to determine how we might induce nephron regeneration in mammals.

The only part of the mammalian reproductive system known to regenerate is the prostate. Prostate tissue regenerates from a population of stem cells that may reside in both the basal and luminal layers of the organs. The testes and ovaries of some teleost fish can regenerate, and the testes of adult newts, but not the ovaries, can also regenerate. The mechanism of regeneration, whether by compensatory hyperplasia, activation of stem cells, or dedifferentiation, is unknown.

## REFERENCES


Abate-Shen C, Shen MM (2000) Molecular genetics of prostate cancer. Genes Dev 14:2410–2434.

Adamson IYR, Bowden DH (1975) Derivation of type 1 epithelium from type 2 cells in the developing rat lung. Lab Invest 32: 736–745.

Adamson IYR, Young L, Bowden DH (1988) Relationship of alveolar epithelial injury and repair to the induction of pulmonary fibrosis. Am J Pathol 130:377–383.

Al-Awqati Q, Oliver JA (2002) Stem cells in the kidney. Kidney Int 61:387–395.

Alison MR, Golding MH, Sarraf CE (1996) Pluripotential liver stem cells. Facultative stem cells located in the biliary tree. Cell Prolif 29:373–402.

Alison MR, Poulsom R, Jeffery R, Dhillon P, Quaglia A, Jacob J, Novelli M, Prentice G, Williamson J, Wright NA (2000) Hepatocytes from non-hepatic adult stem cells. Nature 406:257.

Ballard WW (1970) Archenteric origin of midgut lumen in amphibia. Dev Biol 21:424–439.

Barnard JA, Beauchamp RD, Coffey RJ, Moses HL (1989) Regulation of intestinal epithelial cell growth by transforming growth factor. Proc Natl Acad Sci USA 86:1578–1582.

Beaulieu J-F (1992) Differential expression of the VLA family of integrins along the crypt-villus axis in the human small intestine. J Cell Sci 102:427–436.

Beaulieu J-F, Vachon PH (1994) Reciprocal expression of laminin A-chain isoforms along the crypt-villus axis in the human small intestine. Gasteroenterology 106:829–839.

Bishop AE (2004) Pulmonary epithelial stem cells. Cell Prolif 37: 89–96.

Bonner-Weir S, Baxter LA, Schuppin GT, Smith FE (1993) A second pathway for regeneration of adult exocrine and endocrine pancreas: A possible recapitulaton of embryonic development. Diabetes 42:1715–1720.

Bonner-Weirs, Weir GC (2005) New sources of pancreatic β-cells. Nature Biotech 23:857–861.

Bonventre JV (2003) Dedifferentiation and proliferation of surviving epithelial cells in acute renal failure. J Am Soc Nephrol 14: S55–S61.

Booth C, Potten CS (1995) Effects of IL-11 on the growth of intestinal epithelial cells *in vitro*. Cell Prolif 28:581–594.

Booth C, Evans GS, Potten CS (1995) Growth factor regulation of of proliferation in primary cultures of small intestinal epithelium. In Vitro Cell Dev Biol 31:234–243.

Booth C, O'Shea JA, Potten CS (1999) Maintenance of functional stem cells in isolated and cultured adult intestinal epithelium. Exp Cell Res 249:359–366.

Brawley C, Matunis E (2004) Regeneration of male germline stem cells by spermatogonial dedifferentiation in vivo. Science 304: 1331–1334.

Brittan M, Wright NA (2004) The gastrointestinal stem cell. Cell Prolif 37:35–53.

Brody JS, Burki R, Kaplan N (1978) Deoxyribonucleic acid synthesis in lung cells during compensatory lung growth after pneumonectomy. Am Rev Respir Dis 117:307–316.

Bucher NLR, Malt RA (1971) Regeneration of the Liver and Kidney. Boston, Little, Brown and Co.

Cagle PT, Langston C, Goodman JC, Thurlbeck WM (1990) Autoradiograhic assessment of the sequence of cellular proliferation in post-pneumonectomy lung growth. Am J Respir Cell Mol Biol 3:153–158.

Cardoso WV (1995) Transcription factors and pattern formation in the developing lung. Am J Physiol 269:L429–L442.

Choi Y, Ta M, Atouf F, Lumelsky N (2004) Adult pancreas generates multipotent stem cells and pancreatic and non-pancreatic progeny. Stem Cells 22:1070–1084.

Clippinger DH, Osborne JA (1984) Surgical sterilization of grass carp, a nice idea. Aquatics 6:9–10.

Conteas CN, Adhip P, Majumdar N (1987) The effects of gastrin, epidermal growth factor and somatostatin on DNA synthesis in a small intestinal cell line (IEC6). Proc Soc Exp Biol Med 184: 307–311.

Cosentino L, Shaver-Walker P, Heddle JA (1996) The relationships among stem cells, crypts, and villi in the small intestine of mice as determined by mutation tagging. Dev Dynam 207:420–428.

Dabeva MD, Shafritz DA (2003) Hepatic stem cells and liver repopulation. In: Berk PD, Thorgeirsson S, Grisham JW (eds.). Seminars in Liver Disease. New York, Thieme Medical Pub Inc, pp 349–361.

Davidson AJ, Zon LI (2003) Love, honor, and protect (your liver). Science 299:835–837.

Diehl AM (1998) Roles of CCAAT/enhancer-binding proteins in regulation of liver regenerative growth. J Biol Chem 273:30843–30846.

Dohi M, Hasegawa T, Yamamoto K, Marshall B (2000) Hepatocyte growth factor attenuates collagen accumulation in a murine model of pulmonary fibrosis. Am J Respir Crit Care Med 162:2302–2307.

Dor Y, Brown J, Martinez OI, Melton DA (2004) Adult pancreatic β-cells are formed by self-duplication rather than stem cell differentiation. Nature 429:41–46.

Drummond I (2003) The skate weighs in on kidney regeneration. J Am Soc Nephrol 14:1704–1705.

Dumont AE, Martelli AB, Illiesceu H (1980) Spontaneous reconstitution of the mammalian intestinal tract following complete transection. Proc Soc Exp Biol Med 164:545–549.

Dumont AE, Martelli AB, Schinella RA (1984) Regeneration-induced lengthening of the cut ends of the rat colon. Brit J Exp Pathol 65:155–163.

Elger M, Hentschel H, Litteral J, Wellner M, Kirsch T, Luft FC, Haller H (2003) Nephrogenesis is induced by partial nephrectomy in the elasmobranch Leucoraja erinacei. J Am Soc Nephrol 14: 1506–1518.

Elias JA, Zheng T, Einarsson O, Landry M, Trow T, Rebert N, Panuska J (1994) Epithelial interleukin-11. Regulation by cytokines, respiratory syncytial virus and retinoic acid. J Biol Chem 269:22261–22268.

El Nahas AM (2003) Plasticity of kidney cells: Role in kidney remodeling and scarring. Kidney Internatl 64:1553–1563.

Evans MJ, Cabral LJ, Stephens RJ, Greeman G (1975) Renewal of alveolar type 2 cells to type 1 cells following exposure to $NO_2$. Am J Pathol 22:142–150.

Fausto N (1994) Liver stem cells. In: Arias IM, Boyer JL, Fausto N, Jakoby WB, Schacter DA, Shafritz DA (eds.). New York, Raven Press, Ltd, pp 1501–1518.

Fausto N (2000) Liver regeneration. J Hepatol 32:19–31.

Fausto N (2004) Liver regeneration and repair: Hepatocytes, progenitor cells and stem cells. Hepatol 39:1477–1487.

Fausto N, Webber EM (1994) Liver regeneration. In: Arias IM, Boyer JL, Fausto N, Jakoby WB, Schachter DA, Shafritz DA (eds.). The Liver: Biology and Pathobiology, 3rd ed. New York, Raven Press, pp 1059–1084.

Fausto N, Laird AD, Webber EM (1995) Role of growth factors and cytokines in hepatic regeneration. FASEB J 9:1527–1536.

Fernandes A, King LC, Guz Y, Stein R, Wright CVE, Teitelman G (1997) Differentiation of new insulin-producing cells is induced by injury in adult pancreatic islets. Endocrin 138:1750–1762.

Ford J, Terzaghi-Howe M (1992) Basal cells are the progenitors of primary tracheal cell cultures. Exp Cell Res 198:69–77.

Fujio K, Evarts RP, Hu Z, Marsden R, Thorgeirsson SS (1994) Expression of stem cell factor and its receptor, c-kit, during liver regeneration from putative stem cells in adult rat. J Lab Invest 70:511–516.

Furnari-Savoca G (1967) Regeneration of the testes in the clawed toad (*Discoglossus pictus* Otth.) after complete surgical removal. Experentia 23:218–219.

Gershengorn M, Hardikar A, Wei C, Geras-Raaka E, Marcus-Samuels B, Raaka BM (2004) Epithelial-to-mesenchymal transition generates proliferative human islet precursor cells. Science 306:2261–2264.

Goodchild CG (1956) Reconstitution of the intestinal tract in the adult leopard frog *Rana pipiens Schreber.* J Exp Zool 131:301–327.

Gordon GJ, Coleman WB, Hixson DC, Grisham JW (2000a) Liver regeneration in rats with retrorsine-induced hepatocellular injury proceeds through a novel cellular response. Am J Pathol 156:607–619.

Gordon GJ, Coleman WB, Grisham JW (2000b) Temporal analysis of hepatocyte differentiation by small hepatocyte-like progenitor cells during liver regeneration in retrorsine-exposed rats. Am J Pathol 157:771–786.

Grompe M, Finegold M (2001) Liver stem cells. In: Marshak DR, Gardner RL, Gottlieb D (eds.). Stem Cell Biology. Cold Spring Harbor, Cold Spring Harbor Laboratory Press, pp 455–497.

Grubb RB (1975) An autoradiographic study of the origin of intestinal blastema cels in the newt *Notophthalmus viridescens.* Dev Biol 47:185–195.

Gu D-L, Sarvetnick N (1993) Epithelial cell proliferation and islet neogenesis in IFN-gamma transgenic mice. Development 118: 33–46.

Gu D-L, Lee MS, Krahl T, Sarvetnick N (1994) Transitional cells in the regenerating pancreas. Development 120:1873–1881.

Guz Y, Nasir I, Teitleman G (2001) Regeneration of pancreatic β-cells from intra-islet precursor cells in an experimental model of diabetes. Endocrin 142:4956–4968.

Halliburton IW (1969) Effect of unilateral nephrectomy and of diet on composition of kidney. In: Nowinski WW, Goss RJ (eds.). Compensatory Renal Hypertrophy. New York, Academic Press, pp 101–130.

Ham AW, Cormack DH (1979) Histology, 8th ed. Philadelphia, JB Lippincott.

Han DS, Li F, Holt L, Connolly K, Hubert M, Miceli R, Okoye Z, Santiago G, Windle K, Wong E, Sartor RB (2000) Keratinocyte growth factor-2 (FGF-10) promotes healing of experimental small intestinal ulceration in rats. Am J Physiol 279:G1011–G1022.

Hardikar AA, Karandikar MS, Bhonde RR (1999) Effect of partial pancreatectomy on diabetic status in BALB/c mice. J Endocrinol 162:189–195.

Hermiston ML, Gordon JI (1995a) Organization of the crypt-villus axis and evolution of its stem cell hierarchy during intestinal development. Am J Physiol 268:G813–G822.

Hermiston ML, Gordon JI (1995b) In vivo analysis of cadherin function in the mouse intestinal epithelium: Essential roles in adhesion, maintenance of differentiation, and regulation of programmed cell death. J Cell Biol 129:489–506.

Higgins GM, Anderson RM (1931) Experimental pathology of the liver: I. Restoration of the liver of the white rat following partial surgical removal. Arch Pathol 12:186–202.

Hixson D, Chapman L, McBride A, Faris, R, Yang L (1997) Antigenic phenotypes common to rat oval cells, primary hepatocellular carcinomas, and developing bile ducts. Carcinogenesis 18:1169–1175.

Hixson D (2003) Animal models for assessing the contribution of stem cells to liver development. In: Sell S (ed.). Stem Cell Handbook. Towota Humana Press, pp 353–366.

Holland AM, Gonez LJ, Harrison LC (2004) Progenitor cells in the adult pancreas. Diabetes/Metabolism Res Rev 20:13–27.

Hong KU, Reynolds SD, Watkins S, Fuchs E, Stripp BR (2004) Basal cells are a multipotent progenitor capable of renewing the bronchial epithelium. Am J Pathol 164:577–588.

Houchen CW, George RJ, Sturmoski MA, Cohn SM (1999) FGF-2 enhances intestinal stem cell survival and its expression is induced after radiation injury. Am J Physiol 276:G249–G258.

Imgrund M, Grone E, Grone HJ, Kretzler M, Holzman L, Schlondorff LD, Rothenpieler UW (1999) Re-expression of the developmental gene Pax-2 during experimental acute tubular necrosis in mice. Kidney Int 56:1423–1431.

Jamal A-M, Lipsett M, Sladek R, Laganiere S, Hanley S, Rosenberg L (2005) Morphogenetic plasticity of adult human pancreatic islets of Langerhans. Cell Death Diff 12:702–712.

Johns LS, Liley NR (1970) The effects of gonadectomy and testosterone treatment on the reproductive behavior of the male blue gourami *Trichogaster trichopterus*. Canad J Zool 48:977–987.

Johnson NF, Hubbs AF (1990) Epithelial progenitor cells in the rat trachea. Am J Respir Cell Mol Biol 3:579–585.

Kale S, Karihaloo A, Clark PK, Kashgarian M, Krause DS, Cantley LG (2003) Bone marrow stem cells contribute to repair of the ischemically injured renal tubule. J Clin Invest 112:42–49.

Kalluri R, Neilson EG (2003) Epithelial-mesenchymal transition and its implications for fibrosis. J Clin Invest 112:1776–1784.

Kanazawa Y, Verma I (2003) Little evidence of bone marrow-derived hepatocytes in the replacement of injured liver. Proc Natl Acad Sci USA 100:11850–11853.

Karhadkar SS, Bova GS, Abdallah N, Dhara S, Gardner D, Maitra A, Isaacs JT, Berman DM, Beachy A (2004) Hedgehog signaling in prostate regeneration, neoplasia and metastasis. Nature 431: 707–712.

Kayahara T, Sawada M, Takaishi S, Fukui H, Seno H, Fukuzawa H, Suzuki K, Hiai H, Kageyama R, Okano H, Chiba T (2003) Candidate markers for stem and early progenitor cells, Mushashi-1 and Hes 1, are expressed in crypt base columnar cells of mouse small intestine. FEBS Lett 535:131–135.

Kaur P, Potten S (1986) Cell migration velocities in the crypts of the small intestine after cytotoxic insult are not dependent on mitotic activity. Cell Tiss Kinet 19:601–610.

Kersten CA, Krisfalusi M, Parsons JE, Cloud JG (2001) Gonadal regeneration in masculinized female or steroid-treated rainbow trout (*Oncorhynchus mykiss*). J Exp Zool 290:396–401.

Khan B, Shui CX, Ning SC, Knox SJ (1997) Enhancement of murine intestinal stem cell survival after irradiation by keratinocyte growth factor. Radiat Res 148:248–253.

Kim K, Lu Z, Hay ED (2002) Direct evidence for a role of β-catenin/LEF-1 signaling pathway in induction of EMT. Cell Biol Int 26:463–476.

Kim S-Y, Lee S-H, Kim B-M, Kim E-H, Min B-H, Bendayan M, Park I-S (2004) Activation of nestin-positive duct stem (NPDS) cells in pancreas upon neogenic motivation and possible cytodifferentiation into insulin-secreting cells from NPDS cells. Dev Dynam 230:1–11.

Kobayashi M, Aida K, Stacey NE (1991) Induction of testis development by implantation of 11-ketotestosterone in female goldfish. Zool Sci 8:389–393.

Korinek V, Barker N, Moerer P, van Donselaar E, Huls G, Peters PJ, Clevers H (1998) Deletion of epithelial stem-cell compartments in the small intestine of mice lacking Tcf-4. Nat Genet 19:379–383.

Kritzik MR, Krahl T, Good A, Krakowski M, St-Onge L, Gruss P, Wright C, Sarvetnik N (2000) Transcription factor expression during pancreatic islet regeneration. Mol Cell Endocrinol 164: 99–107.

Kritzik MR, Sarvetnick N (2001) Pancreatic stem cells. In: Marshak DR, Gardner RL, Gottlieb D (eds.). Stem Cell Biology. Cold Spring Harbor, Cold Spring Harbor Laboratory Press, pp 499–513.

Kurita T, Wang Z, Donjacour AA, Zhao C, Lydon J, O'Mammey BW, Isaacs JT, Dahiya R, Cunha GR (2001) Paracrine regulation of apoptosis by steroid hormones in the male and female reproductive system. Cell Death Diff 8:192-200.

Kurita T, Medina RT, Mills AA, Cunha GR (2004) Role of p63 and basal cells in the prostate. Development 131:4955–4964.

Lambotte L, Saliez A, Triest S, Maiter D, Baranski A, Barker A, Li B (1997) Effect of sialoadenectomy and epidermal growth factor administration on liver regeneration after partial hepatectomy. Hepatology 25:607–612.

LeCouter J, Moritz DR, Li B, Phillips G, Liang XH, Gerber H-P, Hillan KJ, Ferrara N (2003) Angiogenesis-independent endothelial protection of the liver: Role of VEGFR-1 Science 299:890–893.

Li F, Rosenberg E, Smith CI, Notarfrancesco K, Reisher SR, Shuman H, Feinstein SI (1995) Correlation of expression of transcription factor C/EBPα and surfactant protein genes in lung cells. Am J Physiol 269:L241–L247.

Li L, Seno M, Yamada H, Kojima I (2003) Betacellulin improves glucose metabolism by promoting conversion of intraislet precursor cells to beta cells in streptozotocin-treated mice. Am J Physiol Endocrinol Metab 285:E577–E583.

Lin J, Patel SR, Cheng X, Cho EA, Levitan I, Ullenbruch M, Phan SH, Park JM, Dressler GR (2005) Keilin/chordin-like protein, a novel enhancer of BMP signaling, attenuates renal fibrotic disease. Nature Med 11:387–393.

Liu JY, Nettesheim P, Randell SH (1994) Growth and differentiation of tracheal epithelial progenitor cells. Am J Physiol 266: L296–L307.

Lorentz O, Duluc I, DeArcangelis A, Simon-Assmann P, Kedinger M, Freund JN (1997) Key role of the cgx-2 homeobox gene in extracellular matrix-mediated intestinal cell differentiation. J Cell Biol 139:1553–1565.

Maeshima A, Yamashita S, Nojima Y (2003) Identification of renal progenitor-like tubular cells that participate in the regeneration processes of the kidney. J Am Soc Nephrol 14:3138–3146.

Mars M, Liu M-L, Kitson RP, Goldfarb RH, Gabauer MK, Michalopoulos GK (1995) Immediate early detection of urokinase receptor after partial hepatectomy and its implications for initiation of liver regeneration. Hepatology 21:1695–1701.

Martinez-Hernandez A, Amenta P (1995) The extracellular matrix in hepatic regeneration. FASEB J 9:1401–1410.

Mason R, Leslie CC, McCormick-Shannon K, Deterding R, Nakamura T, Rubin JS, Shannon JM (1994) Hepatocyte growth factor is a growth factor for rat alvolear type II cells. Am J Respir Cell Mol Biol 11:561–567.

McGee DW, Beagley KW, Aicher WK, McGee JR (1993) Transforming growth factor beta and IL-1 act in synergy to enhance IL-6 secretion by the intestinal epithelial cell line IEC6. J Immunol 151:970–980.

McGee DW, Vitkus SJD (1996) IL-4 enhances IEC6 intestinal epithelial cell proliferation yet has no effect on IL-6 secretion. Clin Exp Immunol 105:274–277.

Michaelopoulos GK, DeFrances MC (1997) Liver regeneration. Science 276:60–66.

Migdalska A, Molineaux G, Demuynck H, Evans GS, Ruscetti F, Dexter TM (1991) Growth inhibitory effects of transforming growth factor β *in vivo*. Growth Factors 4:239–245.

Miyasaka K, Ohta M, Masuda M, Funakoshi A (1997) Retardation of pancreatic regeneration after partial pancreatectomy in a strain of rats without CCK-A receptor gene expression. Pancreas 14: 391–399.

Miyazawa K, Shimomura T, Kitamura N (1996) Activation of hepatocyte growth factor in injured tissues is mediated by hepatocyte growth factor activator. J Biol Chem 271:3615–3618.

Murtaugh LC, Melton DA (2003) Genes, signals, and lineages in pancreas development. Ann Rev Cell Dev Biol 19:71–89.

Mutoh H, Naya F, Tsai M-J, Leiter AB (1998) The basic helix-loop-helix protein BETA2 interacts with p300 to coordinate differentiaton of secretin-expressing enteroendocrine cells. Genes Dev 12:820–830.

Neilson EG (2005) Setting a trap for tissue fibrosis. Nature Med 11: 373–374.

Nicholls DM, Chan YPM, Girgis GR (1975) Aminoacyl-tRNA binding activity in regenerating kidney following contralateral nephrectomy or administration of folic acid. Dev Biol 47: 1–11.

Nishimura S, Wakabayashi N, Toyoda K, Kasshima K, Mitsufuji S (2003) Expression of Musashi-1 in human normal colon crypt cells: A possible stem cell marker of human colon epithelium. Dig Dis Sci 48:1523–1529.

Norman JT, Bohman RE, Fischmann G, Bowen JW, McDonough A, Slamon D, Finne LG (1988) Patterns of mRNA expression during early cell growth differ in kidney epithelial cells destined to undergo compensatory hypertrophy versus regenerative hyperplasia. Proc Natl Acad Sci USA 85:6768–6772.

Odom DT, Zizlsperger N, Gordon DB, Bell GW, Rinaldi NJ, Murray HL, Volkert TL, Schreiber J, Rolfe PA, Gifford DK, Fraenkl E, Bell GI, Young RA (2004) Control of pancreas and liver gene expression by HNF transcription factors. Science 303:1378–1381.

Ohmichi H, Matsumoto K, Nalamura T (1996) In vivo mitogenic action of HGF on lung epithelial cells: Pulmotrophic role in lung regeneration. Am J Physiol 270 (Lung Cell Mol Physiol 14): L1031–L1039.

O'Steen WK, Walker BE (1962) Radioautographic studies of regeneration in the common newt. III. Regeneration and repair of the intestine. Anat Rec 142:179–188.

Overturf K, al-Dhalimy M, Ou CN, Finegold M, Grompe M (1997) Serial transplantation reveals the stem-cell-like regenerative potential of adult mouse hepatocytes. Am J Pathol 151:1273–1280.

Panos RJ, Bak PM, Simonet WS, Rubin JS, Smith LJ (1995) Intratracheal instillation of keratinocyte growth factor decreases hyperoxia-induced mortality in rats. J Clin Invest 96:2026–2033.

Park H, McCusker RH, Van Derhoof JA, Mohammadpour H, Harty RF, Macdonald RG (1992) Secretion of insulin-like growth factor II (IGF II) and IGF-binding protein-2 by intestinal epithelial (IEC6) cells: Implications for autocrine growth regulation. Endocrinology 131:1359–1368.

Peterson RL, Bozza MM, Dorner AJ (1996) Interleukin-11 induces intestinal epithelial cell growth arrest through effects on retinoblastoma protein phosphorylation. Am J Pathol 149:895–902.

Petersen BE, Goff J, Greenberger JS, Michalopoulos GK (1998) Hepatic oval cells express the hematopoietic stem cell marker Thy-1 in the rat. Hepatology 27:433–445.

Petersen BE, Zajac V, Michalopoulos GK (1998) Hepatic oval cell activation in response to injury following chemically induced periportal or pericentral damage in rats. Hepatology 27:1030–1038.

Petersen BE, Bowen WC, Patrene KD, Mars M, Sullvan AK, Murase N, Boggs SS, Greenberger JS, Goff JP (1999) Bone marrow as a potential source of hepatic oval cells. Science 284:1168–1170.

Peterson L, Bozza MM, Dorner AJ (1996) Interleukin-11 induces intestinal epithelial cell growth arrest through effects of retinoblastoma protein phosphorylation. Am J Pathol 149:895–902.

Petropavlovskaia M, Rosenberg L (2002) Identification and characterization of small cells in the adult pancreas: Potential progenitor cells? Cell Tiss Res 310:51–58.

Ponder KP (1996) Analysis of liver development, regeneration and carcinogenesis by genetic marking studies. FASEB J 10:673–682.

Potten CS (1974) The epidermal prolferative unit: The possible role of the central basal cell. Cell Tiss Kinet 7:77–88.

Potten CS, Morris RJ (1988) Epithelial stem cells *in vivo*. J Cell Sci (Suppl) 10:45–62.

Potten CS, Booth C, Pitchard DM (1997) The intestinal epithelial stem cell: The mucosal governor. Int J Exp Pathol 78:219–243.

Potten CS, Booth C, Tudor GL, Booth D, Brady G, Hurley P, Ashton G, Clarke R, Sakakibara S, Okano H (2003) Identification of a putative intestinal stem cell and early lineage marker; musashi-1. Differentiation 71:28–38.

Rafaeloff R, Pittenger GL, Barlow SW, Quin XF, Yan B, Rosenberg L, Duguid WP, Vinik AI (1997) Cloning and sequencing of the pancreatic islet neogenesis associated protein (INGAP) gene and its expression in islet neogenesis in hamsters. J Clin Invest 99: 2100–2109.

Ramiya V, Maraist M, Arfors KE, Schatz DA, Peck B, Cornelius JG (2000) Reversal of insulin-dependent diabetes using islets generated *in vitro* from pancreatic stem cells. Nature Med 6:278–282.

Reddy R, Buckley S, Doerken M, Barsky L, Weinberg K, Anderson KD, Warburton D, Driscoll B (2004) Isolation of a putative progenitor subpopulation of alveolar epithelial type 2 cells. Am J Physiol Lung Cell Mol Physiol 286:L658–L667.

Rhim JA, Sandgren EP, Degen JL, Palmiter RD, Brinster RL (1994) Replacement of diseased mouse liver by hepatic cell transplantation. Science 263:1149–1152.

Risbud MV, Bhonde RR (2002) Models of pancreatic regeneration in diabetes. Diabet Res Clin Practice 58:155–165.

Rooman I, Lardon J, Bouwens L (2002) Gastrin stimulates β-cell regeneration and increases islet cell mass from transdifferentiated but not from normal exocrine pacreas tissue. Diabetes 51:686–690.

Rosenberg L, Lipsett M, Yoon J-W, Prentki M, Wang R, Jun H-S, Pittenger GL, Taylor-Fishwick D, Vinik AI (2004) A pentadecapeptide fragment of islet neogenesis-associated protein increases beta-cell mass and reverses diabetes in C57BL/6J mice. Ann Surg 240:875–884.

Ryu S, Kodama S, Ryu K, Schoenfeld DA, Faustman DL (2001) Reversal of established autoimmune diabetes by restoration of endogenous β cell function. J Clin Invest 108:63–72.

Sakamaki Y, Matsumoto K, Mizuno S, Miyoshi S, Matsuda H, Nakamura T (2002) Hepatocyte growth factor stimulates proliferation of respiratory epithelial cells during postpneumonectomy compensatory lung growth in mice. Am J Respir Cell Mol Biol 26:525–533.

Salice CJ, Rokous JS, Kane AS, Reinschuessel R (2003) New nephron development in goldfish (*Carassius auratus*) kidneys following repeated gentamicin-induced nephrotoxicosis. Comp Med 51:56–59.

Sato H, Funahashi M, Kristensen DB, Tateno C, Yoshizato K (1999) Pleiotrophin as a Swiss 3T3 cell-derived potent mitogen for adult rat hepatocytes. Exp Cell Res 246:152–164.

Scadding SR (1977) Response of the adult newt, *Notophthalmus viridescens*, to partial or complete gonadectomy. Anat Rec 189: 641–647.

Seaberg RM, Smukler SR, Kieffer TJ, Enikolopov G, Asghar Z, Wheeler MB, Korbutt G, van der Kooy D (2004) Clonal identification of multipotent precursors from adult mouse pancreas that generate neural and pancreatic lineages. Nature Biotech 22:1115–1124.

Sell S (2001) Heterogeneity and plasticity of hepatocyte lineage cells. Hepatology 2001; 33:738–750.

Shafritz DA, Dabeva MD (2002) Liver stem cells and model systems for liver repopulation. J Hepatol 36:552–564.

Shi Y, Massague J (2003) Mechanisms of TGF-β signaling from cell membrane to the nucleus. Cell 113:685–700.

Shiratori M, Oshika E, Ung LP, Singh G, Shinozukaa H, Warburton D, Michaelopoulos G, Katyal SL (1996) Keratinocyte growth factor and embryonic rat lung morphogenesis. Am J Respir Cell Mol Biol 15:328–338.

Signoretti S, Waltregny D, Dilks J, Isaac B, Lin D, Garraway L, Yang A, Montoroni R, McKeon F, Loda M (2000) p63 is a prostate basal cell marker and is required for prostate development. Am J Pathol 157:1769–1775.

Sirica AE (1995) Ductular hepatocytes. Histol Histopathol 10: 433–456.

Slack M (1995) Developmental biology of the pancreas. Development 121:1569–1580.

Song K-H, Ko SH, Ahn Y-B, Yoo S-J, Chin H-M, Kaneto H, Yoon K-H, Cha B-Y, Lee K-W, Son H-Y (2004) In vitro transdifferentiation of adult pancreatic acinar cells into insulin-expressing cells. Biochem Bloghys Res Comm 316:1094–1100.

Sugahara K, Rubin JS, Mason RJ, Aronsen EL, Shannon FM (1995) Keratinocyte growth factor increases mRNAs for SP-A and SP-B in adult rat alveolar type II cells in culture. Am J Physiol 269: L344–L350.

Sugahara K, Iyama K, Sano K, Kuroki Y, Akino T, Matsumoto M (1996) Overexpression of surfactant protein SP-A, SP-B, and SP-C mRNA in rat lungs with lipopolysaccharide-induced injury. Lab Invest 74:209–220.

Sugahara K, Iyama K, Kuroda MJ, Sano K (1998) Double intratracheal instillation of keratinocyte growth factor prevents bleomycin-induced lung fibrosis in rats. J Pathol 186:90–98.

Sugahara K, Sadohara T, Sugita M, Iyama K, Takiguchi M (1999) Differential expression of CCAAT enhancer binding protein family in rat alveolar epithelial cell proliferation and in acute lung injury. Cell Tiss Res 297:261–270.

Tanaguchi H, Yamato E, Tashiro F, Ikegami H, Ogihara T, Miyazaki J (2003) Beta-cell neogenesis induced by adenovirus-mediated gene delivery of transcription factor pdx-1 into mouse pancreas. Gene Therapy 10:15–23.

Tateno C, Yoshizato K (1999) Growth potential and differentiation capacity of adult rat hepatocytes in vitro. Wound Rep Reg 7:36–44.

Tateno C, Takai-Kajihara K, Yamasaki C, Sato H, Yoshizato K (2000) Heterogeneity of growth potential of adult rat hepatocytes *in vitro*. Hepatol 31:65–74.

Taub H (1996) Transcriptional control of liver regeneration FASEB J 10:413–427.

Taylor SI (1999) Deconstructing type 2 diabetes. Cell 97:9–12.

Thiese ND, Saxena R, Portmann BC, Thung SN, Yee H, Chiriboga L, Kumar A, Crawford JM (1999) The canals of Hering and hepatic stem cells in humans. Hepatology 30:1425–1433.

Thorgeirsson S (1996) Hepatic stem cells in liver regeneration. FASEB J 10:1249–1256.

Tourrel C, Bailbe D, Meile M, Kergoat M, Portha B (2001) Glucagon-like peptide-1 and exendin-4 stimulate β-cell neogenesis in streptozotocin-treated newborn rats resulting in persistently improved glucose homeostasis at adult age. Diabetes 50:1562–1570.

Trembly JH, Steer CJ (1998) Perspectives on liver regeneration. J Minn Acad Sci 63:37–46.

Ulich TR, Yi ES, Longmuir K, Yin S, Blitz R, Morris CF, Housley RM, Pierce GF (1994) Keratinocyte growth factor is a growth factor for type II pneumocytes in vivo. J Clin Invest 93:1098–1306.

Underwood JL, Hestad RS III, Thompson BZ (1986) Gonad regeneration in grass carp following bilateral gonadectomy. Prog Fish Cult 48:54–56.

Van de Wetering M, Sancho E, Verweij C, de Lau W, Oving I, Hurlstone A, van der Horn K, Batlle E, Coudreuse D, Haramis A-P, Tjon-Pon-Fong M, Moerer P, van den Born M, Soete G, Pals S, Eilers M, Medema R, Clevers H (2002) The β-catenin/TCF-4 complex imposes a crypt progenitor phenotype on colorectal cancer cells. Cell 111:241–250.

Verghese GM, McCormick-Shanon K, Mason RJ, Matthay MA (1998) Hepatocyte growth factor and keratinocyte growth factor in the pulmonary edema fluid of patients with acute lung injury: Biologic and clinical significance. Am J Respir Crit Care Med 158:386–394.

Vidrich A, Buzan JM, Ilo C, Bradley L, Skaar K, Cohn SM (2004) Fibroblast growth factor receptor-3 is expressed in undifferentiated intestinal epithelial cells during murine crypt morphogenesis. Dev Dynam 230:114–123.

Voelker DR, Mason RJ (1990) Alveolar type II epithelial cells. In: Massaro DJ (ed.). Lung Biology in Health and Disease Vol 41. New York, Dekker, pp 487–538.

Wang H, Iakova P, Wilde M, Welm A, Goode T, Roesler WJ, Timchenko NA (2001) C/EPBα arrests cell proliferation through direct inhibition of cdk2 and cdk4. Mol Cell 8:817–828.

Wang H, Goode T, Iakova P, Albrecht J, Timchenko NA (2002) C/EBPα triggers proteasome-dependent degradation of cdk4 during growth arrest. EMBO J 21:930–941.

Wang X, Foster M, Al-Dhalimy M, Lagasse E, Finegold M, Grompe M (2003) The origin and liver repopulating capacity of murine oval cells. Proc Natl Acad Sci USA 100:11881–11888.

Wang R, Li J, Yashpal N (2004) Phenotypic analysis of c-Kit expression in epithelial monolayers derived from postnatal rat pancreatic islets. J Endocrinol 182:113–122.

Wielenga VJM, Smits R, Korinek V, Smits L, Kielman M, Fodde R, Clevers H, Pals ST (1999) Expression of CD44 in Apc and Tcf mutant mice implies regulation by the Wnt pathway. Am J Pathol 154:515–523.

Williams ED, Lowes AP, Williams D, Williams GT (1992) A stem cell niche theory of intestinal crypt maintenance based on somatic mutation in colonic mucosa. Am J Pathol 141:773–776.

Winton DJ (2001) Stem cells in the epithelium of the small intestine and colon. In: Marshak DR, Gardner RL, Gottleib D (eds.). Stem Cell Biology. Cold Spring Harbor, Cold Spring Harbor Laboratory Press, pp 515–536.

Winton DJ, Ponder BAJ (1990) Stem-cell organization in mouse small intestine. Proc Royal Soc London B Biol Sci 241:13–18.

Witschi HP (1976) Proliferation of type II alveolar cells: A review of common responses in toxic lung injury. Toxicology 5:267–277.

Witzgall R, Brown D, Schwarz C, Bonventre JV (1994) Localization of proliferating cell nuclear antigen, vimentin, c-Fos, and clusterin in the postischemic kidney. Evidence for a heterogeneous genetic response among nephron segments, and a large pool of mitotically active and dedifferentiated cells. J Clin Invest 93: 2175–2188.

Xin L, Ide H, Kim Y, Dubey P, Witte ON (2003) *In vivo* regeneration of murine prostate from dissociated cell populations of postnatal epithelia and urogenital sinus mesenchyme. Proc Natl Acad Sci USA 100:11896–11903.

Xu G, Stoffers DA, Habener JF, Bonner-Weis S (1999) Exendin-4 stimulates both beta cell replication and neogenesis, resulting in increased beta cell mass and improved glucose tolerance in diabetic rats. Diabetes 48:2270–2276.

Yaekashiwa M, Nakayama S, Ohnuma K, Sakai T, Abe T, Satoh K, Matsumoto K, Nakamura T, Takahashi T, Nukiwa T (1997) Simultaneous or delayed administration of hepatocyte growth factor equally represses the fibrotic changes in murine lung injury induced by bleomycin: A morphologic study. Am J Respir Crit Care Med 156:1937–1944.

Yanagita K. Matsumoto K, Sekiguchi K, Ishibashi H, Niho Y, Nakamura T (1993) Hepatocyte growth factor may act as a pulmotrophic factor on lung regeneration after acute lung injury. J Biol Chem 268:21212–21217.

Yang A, Schweitzer R, Sun D, Kaghad M, Walker B, Bronson RT, Tabin C, Sharpe A, Caput D, Crum C, McKeon F (1999) p63 is essential for regenerative proliferaton in limb, craniofacial and epithelial development. Nature 398:714–718.

Yi ES, Williams ST, Lee H, Malicki DM, Chin EM, Yin S, Tarpley J, Ulich TR (1996) Keratinocyte growth factor ameliorates radiation- and bleomycin-induced lung injury and mortality. Am J Pathol 149:1963–1970.

Zalzman M, Gupta S, Giri RK, Berkovich I, Sappal BS, Karnieli O, Zern MA, Fleischer N, Efrat S (2003) Proc Natl Acad Sci USA 100:7253–7258.

Zeisberg M, Hanai J-I, Sugimoto H, Mammato T, Charytan D, Strutz F, Kalluri R (2003) BMP-7 counteracts TGF-β1-induced epithelial-to-mesenchymal transition and reverses chronic renal injury. Nature Med 9:964–968.

Zeisberg M, Kalluri R (2004) The role of epithelial-to-mesenchymal transition in renal fibrosis. J Mol Med 82:175–181.

Zhang Y, Ding L, Lai M (2003) Reg gene family and human diseases. World J Gastroenterol 9:2635–2641.

Zulewski H, Abraham EJ, Gerlach MJ, Daniel PB, Moritz W, Muller B, Vallejo M, Thomas MK, Habener JF (2001) Multipotential nestin-positive stem cells isolated from adult pancreatic islets differentiate ex vivo into pancreatic endocrine, exocrine, and hepatic phenotypes. Diabetes 50:521–533.

# 8 Regenerative Medicine of Digestive, Respiratory, and Urinary Tissues

## INTRODUCTION

The epithelia of the intestinal tract and respiratory and urogenital systems turn over at a relatively high rate compared to the hepatocytes of the liver and β-cells of the pancreas, so maintenance regeneration is more prominent in the former than the latter. These epithelia and the liver and pancreas regenerate after injury as well. However, with the exception of the intestine in the rat, the full-thickness walls of the intestinal, respiratory, and urogenital tracts do not regenerate after injury, nor do lungs, kidneys, or glomeruli and nephrons of the kidney. The regenerative capacity for tissues of the digestive, respiratory, and urogenital systems may become insufficient to compensate for advancing fibrosis of a disease state or for cells lost to disease. Currently, dysfunctional parts of the digestive tract are replaced by allogeneic organ transplants, which suffer from the problems of donor shortage and immunorejection. The strategies of cell transplantation, bioartificial tissue implants, and chemical induction via regeneration templates, small molecules, and gene therapy, have enabled the regeneration or replacement of some of these tissues in experimental animals and in some cases, human patients.

## REGENERATIVE THERAPIES FOR THE LIVER

Acute or chronic liver failure and genetic deficiency liver diseases are treatable today only by an allogeneic whole or partial liver transplant. To counter the organ donor shortage, infusions of hepatocyte suspensions are seen as a way to repopulate the failing liver, or extracorporeal liver assist devices (LADs) representing a fraction of the liver are envisioned to bridge liver failure patients to transplant or, better yet, take over liver function until the host liver can regenerate over a period of months. The liver of a 70-kg adult is estimated to contain ~$2.8 \times 10^{11}$ hepatocytes. Humans can survive up to 90% hepatectomy, so a cell transplant or LAD needs to provide at least 10% of the normal liver mass, which amounts to ~120 g of liver tissue, or ~$25 \times 10^{9}$ cells (Asonuma et al., 1992; Jauregui, 2000).

Autogeneic cells from a patient's own liver are ideal for transplant, if the nature of the liver problem allows harvesting of a small sample of remaining healthy tissue. Allogeneic hepatocytes can be transplanted under immunosuppression regimens and xenogeneic pig hepatocytes can be used in LADs, although there is still debate about the risk of xenogeneic viral transmission from pig cells. A major obstacle is provision of enough cells to rescue liver function. This problem could be overcome by expanding autogeneic or allogeneic hepatocytes *in vitro*, or harvesting the requisite number of fresh xenogeneic hepatocytes from multiple livers of pigs. Primary hepatocytes are difficult to culture, but rapid clonal growth has been reported in medium supplemented with HGF, EGF, and TGFα, which drive hepatocyte proliferation *in vivo* (Block et al., 1996). The cultured hepatocytes dedifferentiated, but could be induced to redifferentiate by Matrigel. Lines of allogeneic human hepatocytes are available, but there is debate about the efficacy of their function (Strain and Neuberger, 2002).

### 1. Hepatocyte Transplants

Adult hepatocytes have been injected as suspensions, or incorporated into polymer scaffolds and transplanted in a wide variety of animal experiments (Kim and Vacanti, 1995; Jauregui, 2000; Xu et al., 2000). Allogeneic hepatocytes injected into the liver or portal vein, or delivered intraperitoneally on microcarrier beads, were reported to reduce bilirubin levels in Gunn rats, which are hyperbilirubinemic due to a functional deficiency of the uridine diphosphate glucuronyltransferase enzyme (Matas, 1976; Demetriou et al., 1986). Allogeneic hepatocytes encapsulated in alginate microcapsules to protect them from the immune system

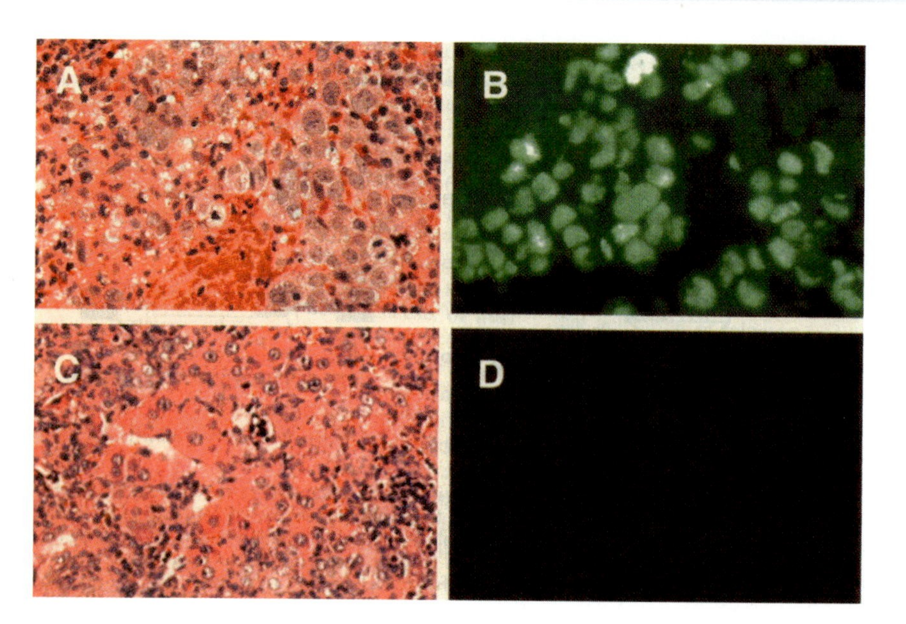


**FIGURE 8.1** Sections of rat spleens 4 days after 90% hepatectomy and intrasplenic implantation of human immortalized and reverted NKNT-3 cells. **A** and **C** were stained with hematoxylin and eosin, **B** and **D** with monoclonal antibody to SV40T. **(A, B)** Non-reverted NKNT-3 cells. **(C)** Cre-reverted NKNT-3 cells with hepatocellular morphology, organized in a trabecular pattern. **(D)** No SV40T staining is seen in reverted cells, indicating excision of the immortalizing gene. Reproduced with permission from Kobayashi et al., Prevention of acute liver failure in rats with reversibly immortalized human hepatocytes. Science 287:1258–1261. Copyright 2000, AAAS.

expressed characteristic molecular markers such as albumin, and cleared the metabolic breakdown products of bilirubin and urea when transplanted into Gunn rats (Demetriou et al., 1986; Dixit et al., 1990).

To insure that adequate vascularization and hepatotrophic factors were available to transplanted hepatocytes, empty polyvinyl sponges were first prevascularized by implanting them subcutaneously or into the mesenteric folds of immunosuppressd Gunn rats (Uyama et al., 1993). The rats were then partially hepatectomized to initiate the production of hepatotrophic factors involved in regeneration and a postcaval shunt was introduced to route the factors to the implant site. A few days later, a liver-equivalent of allogeneic hepatocytes ($5 \times 10^8$ cells) was injected into the sponges. The hepatocytes synthesized DNA, formed plates of cells, and serum bilirubin level decreased 30% compared to controls.

Kobayashi et al. (2000) generated large numbers of partially dedifferentiated human hepatocytes *in vitro* by transfecting the cells with the immortalizing SV40T gene flanked by LoxP recombination targets (NKNT-3 cell line). When transfected with *Cre*, the cells reverted to a fully differentiated form. Xenotransplantation of $5 \times 10^7$ immortalized and reverted human hepatocytes (5% of the total number of hepatocytes in an adult rat) into the spleen rescued immunosuppressed host rats that had undergone 90% hepatectomy. The host rats would have died within 36 hours without treatment. Bilirubin, prothrombin time, and blood ammonia were significantly decreased during the first week post-transplantation, and histological examination showed islets of hepatocytes in the spleen and ongoing regeneration of the remaining host liver (**FIGURES 8.1, 8.2**).

Insuring the post-transplant survival of enough hepatocytes for the length of time required to obtain a transplant or to recover host liver function is the most crucial problem facing hepatocyte transplant therapy. Most experiments have reported only a transient increase in liver function after cell transplant, suggesting that post-transplant hepatocyte survival is low. In fact, the number of cells remaining at one week post-injection into implanted and prevascularized polyvinyl sponges was found to be only 11%–18% of the initial number (Uyama et al., 1993). The problem is not a loss of intrinsic survival and proliferation capacity after transplantation, because *Lac*-Z marked wild-type hepatocytes injected into the spleen of the Alb-uPA transgenic mouse, which exhibits a chronic liver growth stimulus, migrated to the liver and underwent up to 12 doublings (Rhim et al., 1994).

Lack of space for the infused cells may be the limiting survival factor. The number of infused cells that can be accommodated by a normal host liver in a single infusion is 1%–2% of a liver-equivalent (Gupta et al., 1991; Ponder et al., 1991). This 1%–2%, however, exhibits long-term survival. Wild-type hepatocytes infused into the spleen of DPPIV$^-$ rats migrated into the liver sinusoids and intercalated into the plates of host liver cells, where they could be found eight months

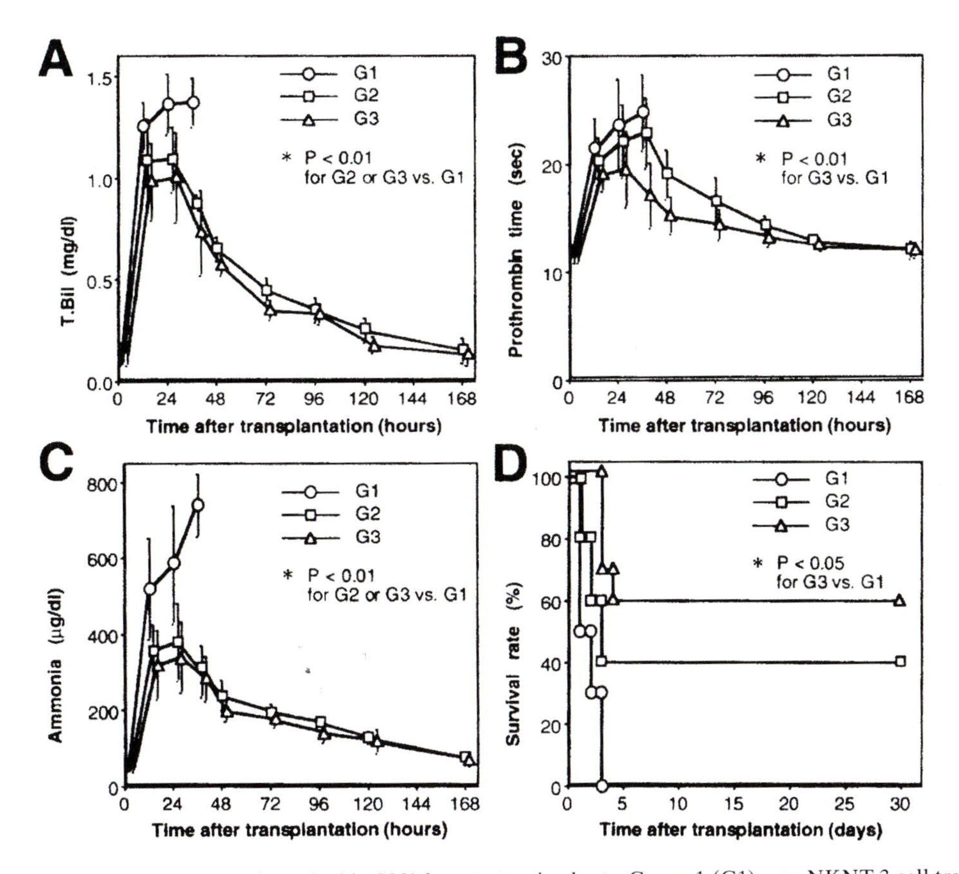


**FIGURE 8.2** Measurements of liver function and survival in 90% hepatectomized rats. Group 1 (G1) = no NKNT-3 cell transplant, G2 = intrasplenic transplant of non-reverted NKNT-3 cells, G3 = intrasplenic transplant of reverted NKNT-3 cells. Asterisk = statistical significance. Both non-reverted and reverted cells significantly improved liver functions and survival, but survival was higher with reverted cells. Reproduced with permission from Kobayashi et al., Prevention of acute liver failure in rats with reversibly immortalized human hepatocytes. Science 287:1258–1261. Copyright 2000, AAAS.

later (Gupta et al., 1995). The intercalation of donor cells into host liver plates suggests that sequential infusions of hepatocytes could be carried out to achieve the level of seeding required to significantly restore functional liver mass.

The results of hepatocyte transplantation in human patients suffering from genetic liver dysfunction or nongenetic liver failure have been encouraging. Autogeneic hepatocytes transfected with a gene for the LDL receptor reduced cholesterol levels in a familial hypercholesterolemia patient for 18 months post-transplant, although the reduction was not subsequently sustained (Grossman et al., 1994). Crigler-Najar syndrome type I, a recessive genetic disorder characterized by the absence of hepatic uridine diphosphosphoglucuronate glucuronosyltransferase activity, results in a high level of unconjugated bilirubin. An 11-year-old female patient was treated for this disorder by the infusion into her liver of $7.5 \times 10^9$ hepatocytes derived from the liver of a 5-year-old boy who could not be matched for transplantation (Fox et al., 1998). This number of hepatocytes attempted to achieve a 2.5% reconstitution of the liver, assuming 50% engraftment. The patient was immunosuppressed to avoid graft rejection. Eleven months later, total serum bilirubin was half the starting value of ~26 mg/dl, although the patient was still receiving 6–7 hr of adjunct phototherapy daily. In another trial, cryopreserved allogeneic hepatocytes, at numbers ranging from $7.5 \times 10^6$ to $1.9 \times 10^8$ were infused into the spleens of five patients suffering from liver failure (Strom et al., 1997). The transplanted hepatocytes improved liver function values immediately and bridged the patients to a standard liver transplant. Subsequently, two patients died. Serial sections of the spleen in one of them revealed nodules of hepatocytes with normal structure.

## 2. Extracorporeal Liver Assist Devices

Extracorporeal liver assist devices (LADs) have been developed to temporarily take over liver function until an organ transplant can be secured or the patient's liver has regenerated. The different large animal models

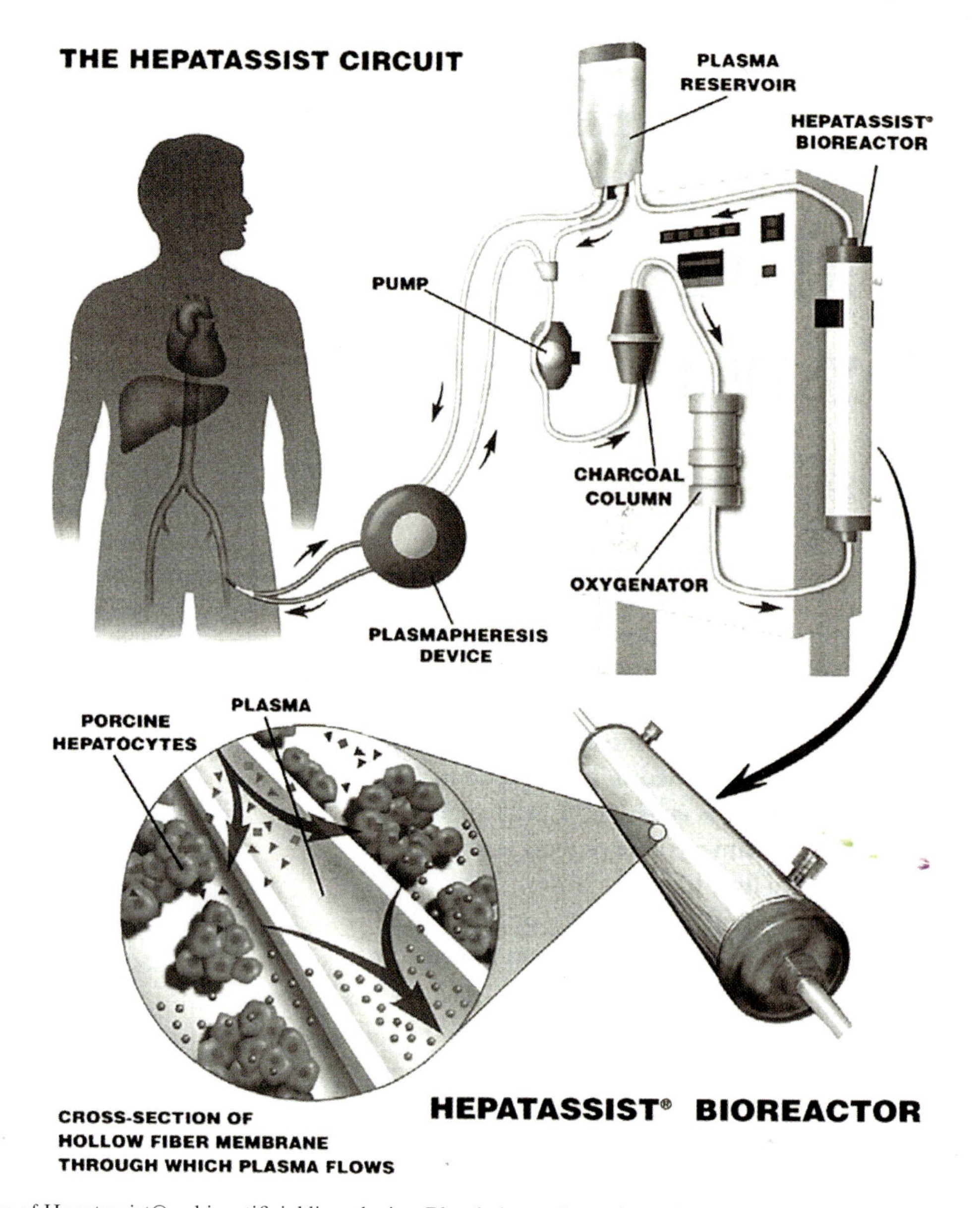


**FIGURE 8.3** Diagram of Hepatassist®, a bioartificial liver device. Blood plasma flows through a plasmapheresis device, then through a bioreactor consisting of hollow fiber membranes surrounded by porcine hepatocytes, before being returned through the plasmapheresis device to the bloodstream. Reproduced with permission from Mullon and Solomon, HepatAssist liver support system. In: Lanza RP, Langer R, Vacanti J (eds). Principles of Tissue Engineering, 2nd ed. New York: Academic Press, pp 553–558. Copyright 2000, Elsevier.

of liver failure used to test LADs and their advantages and drawbacks have been reviewed by van de Kerkhove et al. (2004).

The typical LAD is a hollow fiber bioreactor containing hepatocytes (Jauregui, 2000; Xu et al., 2000). Liver assist devices can use either allogeneic or xenogeneic hepatocytes. To be maximally effective, it is estimated that the number of hepatocytes in the bioreactor should be at least $10^{10}$ (Strain and Neuberger, 2002) and preferably the full human liver-equivalent of $2.8 \times 10^{11}$. This number has not yet been achieved, but cultured and cryopreserved human or pig hepatocytes would make it feasible.

FIGURE 8.3 illustrates an example of an LAD, the HepatAssist™ liver support system (Circe Biomedical, Inc.) (Mullon and Solomon, 2000). The bioreactor in this system consists of polysulfone hollow fibers with a 0.15-μm pore size. The spaces between the fibers contain $7 \times 10^9$ pig hepatocytes attached to collagen-coated dextran beads. Blood flows from the femoral artery and is first passed through a plasmapheresis machine. Plasma then goes through two charcoal columns and through the hollow fibers of the bioreactor, where it bathes the hepatocytes through the fiber pores. The plasma is then returned to the circulation via the femoral vein.

A Phase I/II clinical study of this system on 39 patients found that the procedure was well tolerated by patients and that the bioreactor performed well. Six patients survived without transplantation, indicating that their liver had regenerated, and the remainder received liver transplants. The overall $30^+$ days survival rate for all patients was 90%, compared to 50%–60% for acute liver failure patients, including those who received transplants that failed (Mullon and Solomon, 2000).

Hepatocyte function in bioreactors is better if they are in the form of spheroids rather than monolayers. The hepatocytes reacquire normal polarization (asymmetric localization of surface proteins that confer different functions on different parts of the cell) when aggregated into spheroids (Wu et al., 1995; Strain and Neuberger, 2002). Rat hepatocyte spheroids embedded in collagen gel within the spaces between the polysulfone tubules of a bioreactor provided a 4-fold improvement in albumin synthesis and P-450 enzyme activity over that of hepatocytes dispersed in the collagen (Wu et al., 1995). Even more improvement in function might be achieved by seeding bioreactors with tiny organoids composed of more than one liver cell type. Spherical organoids have been made with hepatocytes and fibroblasts that exhibited the histological and functional characteristics of liver, with structures resembling bile canaliculi and the secretion of albumin (Takezawa et al., 1992).

## REGENERATIVE THERAPIES FOR THE PANCREAS

Type I and type II diabetes is a major worldwide health problem. Together, these two types of diabetes afflict 5% of the population, with type 2 accounting for over 90% of the cases. Diabetes is the third leading cause of death in the United States and accounts for approximately 6% of total personal healthcare costs.

Type II (adult-onset) diabetes is attaining epidemic proportions and is associated primarily with obesity, which has effects on underlying genetic susceptibility (Bell and Polonsky, 2001). Adult-onset diabetes is caused by insulin resistance, which leads to deficiency of insulin secretion (Taylor, 1999; Rhodes, 2005). Most research is directed at Type 1 (juvenile) diabetes, which is a much more serious disease caused by mutations in a number of β-cell transcription factors leading to cell apoptosis and autoimmune attack by T-cells (Bell and Polonsky, 2001; Mathis et al., 2001). The autoimmune response is primarily against insulin (Nakayama et al., 2005; Kent et al., 2005), as well as other antigens such as glutamic acid decarboxylase (GAD), an enzyme produced by the β-cells (Yoon et al., 1999), leading to β-cell destruction, pancreatic fibrosis, and near-total insulin deficiency. The damage done to cardiovascular, renal, and neural systems is severe, and wound healing is impaired. Environmental factors appear to have a strong influence in accelerating the processes involved in Type I diabetes, although the identity of these factors is largely unknown (Zimmet et al., 2001). One candidate that has been proposed is a *Streptomyces* toxin (Myers et al., 2001). Interestingly, a *Streptomyces* species is the source of streptozotocin, which is used to induce diabetes in experimental rats and mice.

The rate of β-cell destruction and fibrosis caused by diabetes is greater than the capacity of the β-cells to regenerate, so the output of insulin steadily decreases to zero. Since the discovery of insulin and its actions in the 1920s, patients with Type 1 diabetes have controlled their blood glucose levels by daily injections of insulin. The relief from monitoring blood glucose concentration and insulin administration, and the restoration of feedback control over insulin concentration that would be afforded by the ability to halt the disease process and replace the damaged β-cells, would be enormous.

### 1. Cell Transplants

Allotransplants of cadaver pancreatic islets to treat Type I diabetes were begun in 1990, but the harsh dissociation procedures and immunosuppressive regimes led to such poor cell survival that only 12.4% of recipients showed insulin-independence for more than one week and only 8.2% for more than one year. In a protocol now known as the Edmonton Procedure, Shapiro et al. (2000) introduced a gentler dissociation protocol and a glucocorticoid-free immunosuppression regimen (sirolimus, tacrolimus, and daclizumab) that greatly enhanced donor cell survival. Islet preparations that had more than 4,000 islet equivalents/kg recipient body weight in a packed-tissue volume of less than 10 ml were transplanted into the liver of seven patients through a portal vein catheter. The procedure was done under local anesthetic and sedation and took only 20 min. Two transplants several weeks apart, totaling between 65 and $200 \times 10^6$ β-cells, were required to attain normoglycemia, which occurred rapidly. Patients recovered control of glucose fluctuation. There were no clinically evident symptoms of graft rejection, and the immunosuppressive regimen appeared to be effective in preventing autoimmune recurrence of diabetes. Ryan et al. (2001) reported results from 12 additional patients, 11 of whom achieved exogenous insulin independence.

In 2004, the Collaborative Islet Transplant Registry of the National Institute for Digestive and Kidney Disorders reported the results of the Edmonton Procedure from a total of 86 patients who were transplanted in 13 centers in Edmonton and throughout the United States and who had suffered from diabetes for an average of 30 years. Twenty-eight patients received one islet infusion (8,665 total islet equivalents/kg, about 30%–50% of the number in a nondiabetic pancreas), 44 received two (14,102 total equivalents/kg), and 14 received three (22,922 total equivalents/kg). Six months after the last infusion, 61% of the recipients were insulin-independent, and at one year, 58% were insulin-independent. Improvements in blood glucose control were greater in the two- and three-transplant patients.

A major drawback to this procedure, aside from the need for immunosuppression, is the shortage of donor pancreati as a cell source. This problem may be alleviated somewhat by new techniques that increase the efficacy of transplantation. A clinical trial involving eight female diabetics found that transplantation of an average of 7,271 islets/kg body wt from a single donor pancreas, preceded by induction immunosuppression and followed by maintenance immunosuppression, sufficed to achieve insulin independence for over a year (Hering et al., 2005). Another approach would be to harvest a portion of the healthy pancreas from a live donor and use it to supply islets for transplant. One such transplant has been performed from a mother to her diabetic daughter (Matsumoto et al., 2005). Pancreatitis is a serious inflammatory disease that could be alleviated by removal of the organ, but this would produce severe diabetes. The disease does not affect the endocrine functions of the pancreas, however. Thus, a transplant procedure might be developed to harvest the islets from the excised pancreas and infuse them back into the patient's liver. The digestive enzymes of the exocrine cells would have to be supplied exogenously.

One alternative is to derive large numbers of β-cells *in vitro* (Bonner-Weir and Weir, 2005). Differentiated β-cells cannot yet be expanded significantly *in vitro* (Halvorsen et al., 2000), thus attention has turned to stem cells as an expandable source of β-cells (Hussain and Theise, 2004). As outlined in Chapter 7, cells with stem cell phenotypes have been isolated from pancreatic ductules, and β-cells and non-β-cells of islets have been reported to dedifferentiate/redifferentiate *in vitro* into β-cells. If a small number of these cells were isolated from the remaining healthy tissue of a patient's pancreas, or from an allogeneic donor, large numbers of new β-cells might be derived from them for transplant. Immunosuppressive drugs would be required in the case of allogeneic transplants unless tolerance could be induced. Ramiya et al. (2000) induced pancreatic ductal cells isolated from prediabetic NOD (nonobese diabetic) mice to produce functional α, β, and δ cells *in vitro* (**FIGURE 8.4**). When masses of these cells were implanted subcutaneously or under the kidney capsule of diabetic mice that were being maintained on insulin, they became vascularized and reversed the diabetic state for up to three months.

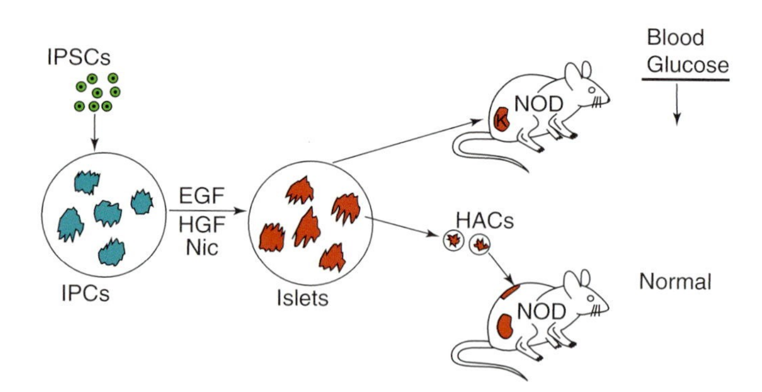


**FIGURE 8.4** Reversal of diabetes in non-obese diabetic (NOD) mice by transplantation of islets derived from pre-diabetic islet pancreatic stem cells (IPSCs) of ducts. The IPSCs were induced to differentiate into islet progenitor cells (IPCs). Further differentiation to immature islets was achieved with epidermal growth factor (EGF), hepatocyte growth factor (HGF) and nicotinamide (Nic). The immature islets were then transplanted under the capsule of one kidney, where they differentiated further and reduced blood glucose level (arrow), or were implanted subcutaneously encapsulated in hyaluronic acid capsules (HACs), in which case blood glucose level was normalized.

Another potential source of allogeneic β-cells is ESCs. Lumelsky et al. (2001) reported the differentiation of mouse ESCs into insulin-secreting islets, but other experiments showed that the differentiated cells did not express insulin transcripts and that the insulin detected was taken up from serum in the culture medium (Rajagopal et al., 2003). Almost certainly, protocols will be developed to direct the differentiation of human ESCs into *bonafide* β-cells.

Bone marrow cells might be still another source of expandable insulin-producing cells. The nonadherent fraction of adult rat bone marrow cells (the HSC-containing fraction) was reported to undergo lineage conversion into β-cells when cultured for three days in a defined medium and then switched to a serum-supplemented medium with high glucose concentration (Oh et al., 2004). Surviving cells were reported to differentiate into organized clusters of cells that expressed insulin transcripts and protein. Immunogold electron microscopy revealed insulin-containing secretory vesicles resembling those of the native β-cell. Transplantation of these cell clusters under the kidney capsule of NOD/*scid* diabetic mice rescued them from the disease and maintained comparatively normal blood glucose levels for up to 90 days post-transplantation (Oh et al., 2004). Whether or not sufficient numbers of converted cells are maintained over the long term is unknown.

Still another potential way to provide large numbers of β-cells is to immortalize them, expand them in the dedifferentiated state, and then revert them to the differentiated state, as has been done with hepatocytes.

Narushima et al. (2005) have established an immortalized clonal human β-cell line (NAKT-15) by transfecting β-cells with retroviral vectors containing *SV40T* and *TERT*, both flanked by loxP recombination sites to allow removal of these genes and reversion to differentiated β-cells. When reverted by transfection with a *Cre* adenoviral vector, this cell line expressed the β-cell transcription factors Isl-1, Pax 6, Nkx 6.1, and Pdx-1, hormone processing enzymes, secretory granule proteins, and secreted insulin in response to glucose. Transplantation of $3 \times 10^6$ reverted NAKT-15 cells under the kidney capsule of *scid* mice with STZ-induced diabetes resulted within two weeks in normoglycemia that was maintained longer than 30 weeks. However, the insulin content of reverted NAKT-15 cells was only 40% that of fresh β-cells, and the level of transcription factors, insulin processing enzymes and secretory granule proteins was only 20%–40%, so the number of NAKT-15 cells needed for transplant may be from 2–5 times the normal number of β-cells. In addition, NAKT-15 cells are allogeneic, so patients would still have to be on immunosuppressive regimens.

## 2. Suppression of Autoimmunity and Regeneration from Remaining β-Cells

A number of experiments on mice suggest that if the autoimmune attack on β-cells can be halted, the pancreas can recover its function by the rescue of surviving β-cells or the regeneration of new β-cells from preexisting ones or from stem cells in the pancreas. The immune mechanisms of foreign islet graft rejection and recurrent autoimmunity appear to be distinct. Although allogeneic or xenogeneic cell grafts are capable of permanent engraftment in nonautoimmune mice after transient ablation of MHC I expression, they do not survive recurrent β-cell autoimmunity in diabetic mice (Faustman and Coe, 1991; Markmann et al., 1992; Osorio et al., 1994). Autoreactive T-cells of NOD mice are miseducated by a defect in the display of endogenous peptides by MHC class I molecules, rendering them intolerant to β-cell self-antigens and vulnerable to TNF-α-induced apoptosis (Faustman et al., 1991). However, peripheral tolerance to self-antigens can be achieved in these mice by injection of complete Freund's adjuvant (CFA) and normal splenocytes.

Ryu et al. (2001) eliminated autoreactive T-cells in hyperglycemic NOD mice by administering CFA to induce TNF-α expression and the apoptosis of autoreactive T-cells. Tolerance of newly emerging naive T-cells to self-antigens was achieved by transplanting, under the capsule of one kidney, islets or splenocytes (spleen cells, which include lymphocytes and other circulating cells) from nonautoimmune allogeneic donors temporarily rendered incapable of expressing MHC I antigens on their cell surface. The mice attained normoglycemia that was maintained even after the kidney with the transplant was removed at 120 days. The transplant itself was destroyed, but β-cells reappeared in the host pancreas, suggesting the rescue of surviving β-cells or the regeneration of new β-cells from surviving cells, thus allowing recovery of the animals.

Zorina et al. (2003) induced tolerance to autogeneic β-cells in prediabetic NOD mice by sublethal whole-body irradiation, followed by transplantation of T-cell depleted GFP-marked allogeneic bone marrow to create lymphocyte chimerism. As little as 1% chimerism was able to abolish the genesis of diabetes. GFP-marked syngeneic islets were then transplanted under the capsule of one kidney. The transplanted islet cells returned the mice to normoglycemia within 24 hr. Immunochemical staining for insulin on sections of the pancreas three weeks post-transplantation showed an almost total lack of β-cells in the host pancreas, indicating its destruction by diabetes. The islet transplant continued to maintain normoglycemia and was removed 15 weeks post-transplantation. Six weeks later, the state of normoglycemia continued to be maintained, indicating the reappearance of β-cells in the host pancreas, which was confirmed by histological examination. None of the β-cells was $GFP^+$, indicating that they were not derived from bone marrow cells or from the islet graft. Whether the regenerated cells originated from surviving β-cells, from stem cells within the islets or pancreatic ductules, or from some combination of these, is unknown. Preliminary results on two mice indicated that, rather than transplanting islets to maintain the mice until the pancreas had regenerated, the mice could be maintained on injected insulin.

In another experiment (**FIGURE 8.5**), Kodama et al. (2003) showed that treatment of prediabetic or diabetic

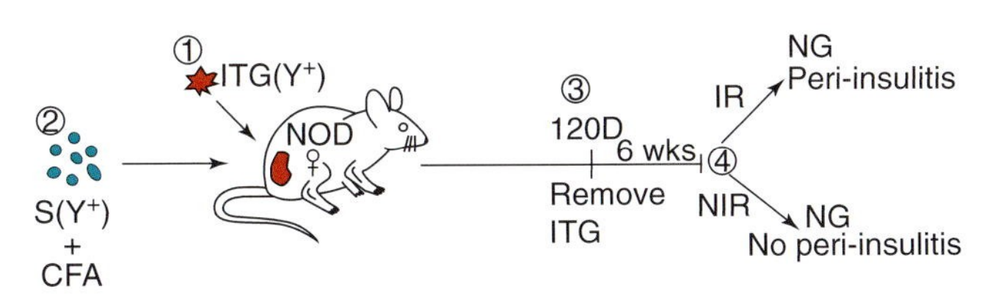


**FIGURE 8.5** Experimental scheme for induction of T-cell chimerism and reversal of diabetes in NOD mice given bi-weekly injections of live or irradiated male ($Y^+$) splenocytes (S) plus complete Freund's adjuvant (CFA) for 40 days. (1) A wild-type male temporary islet graft (ITG) was placed under the kidney capsule of severely diabetic female NOD mice. (2) Male splenocytes + CFA were injected intravenously. (3) The ITG was removed 120 days later. (4) After a further 6 weeks, pancreati were examined. Mice receiving irradiated splenocytes were normoglycemic (NG) but had peri-insulitis. Mice receiving live splenocytes were normoglycemic and had no peri-insulitis.

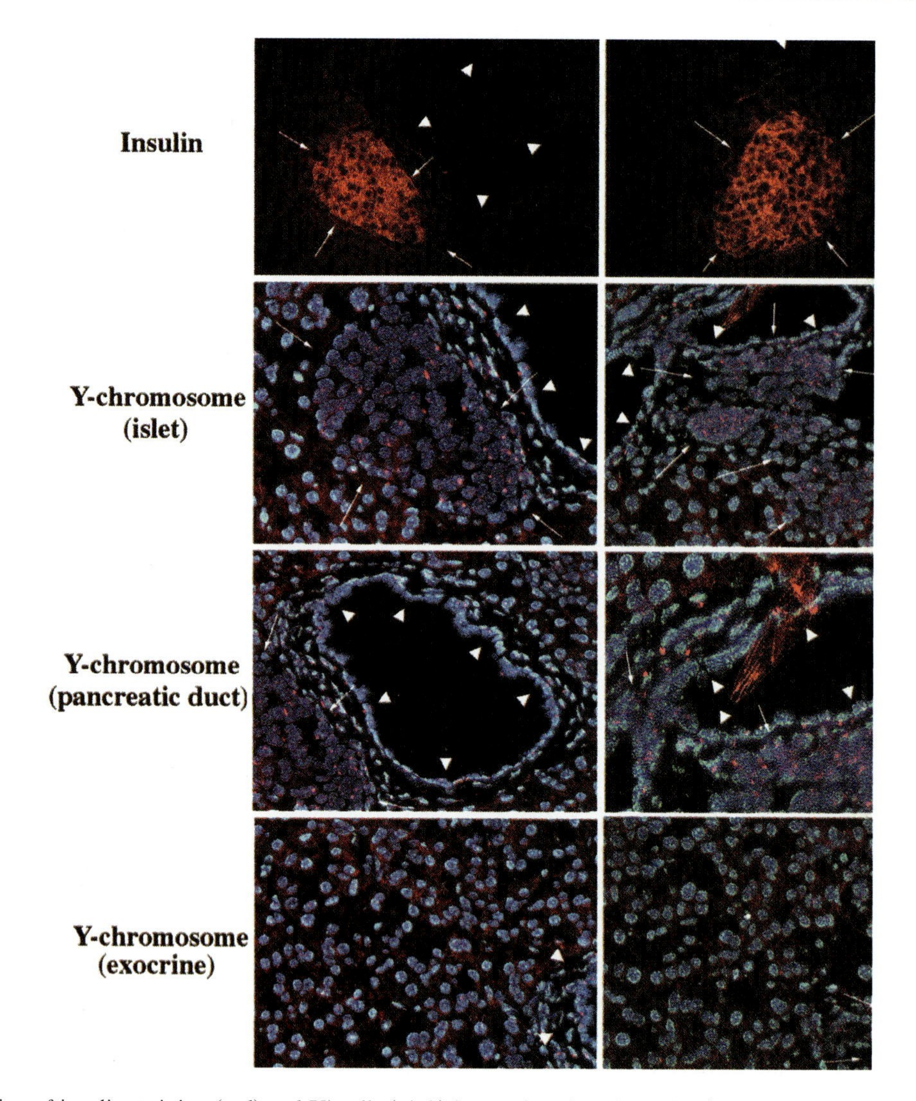


**FIGURE 8.6** Visualization of insulin staining (red) and $Y^+$ cells (pink) in two female NOD mice (left and right columns) treated successfully with live male splenocytes. Arrows outline the islets, arrowheads outline pancreatic ducts. Nuclei (blue) are stained with DAPI. $Y^+$ cells contributed to regenerated islets and, to a lesser extent, pancreatic ducts, but not to exocrine (acinar) tissue. Reproduced with permission from Kodama et al., Islet regeneration during the reversal of autoimmune diabetes in NOD mice. Science 302:1223–1227. Copyright 2003, AAAS.

female NOD mice with CFA and either live or irradiated splenocytes from normal male mice is followed by islet regeneration. Normoglycemia of the hosts was maintained by a temporary implant of syngeneic islets under the capsule of one kidney, which was subsequently removed. If the temporary islet graft was left in place for 120 days, the vast majority of the animals treated with either live or irradiated splenocytes maintained normoglycemia, although those mice treated with irradiated splenocytes experienced insulitis and eventually died because they were not rendered tolerant by the dead donor cells. This result indicates that β-cells were regenerated. Regeneration was confirmed by probing histological sections of the pancreas for the X and Y chromosomes (**FIGURES 8.6, 8.7**). In mice treated with live splenocytes, the presence of both $Y^+$ and $X^+$ islet cells in the recovered pancreas indicated that the β-cells were derived both from cells in the host pancreas and from the transplanted splenocytes. In mice treated with irradiated splenocytes, the new β-cells could only have come from pre-existing β-cells or pancreatic stem cells. This was verified by the lack of $Y^+$ islets and was consistent with the fact that normoglycemia could be maintained in animals treated with irradiated splenocytes only if the temporary islet graft remained in place for 120 days, whereas in animals treated with live splenocytes, the graft could be removed at 40 days, indicating faster islet regeneration. To determine what cells in the live splenocyte transplant contributed to the regeneration of β-cells, the experiment

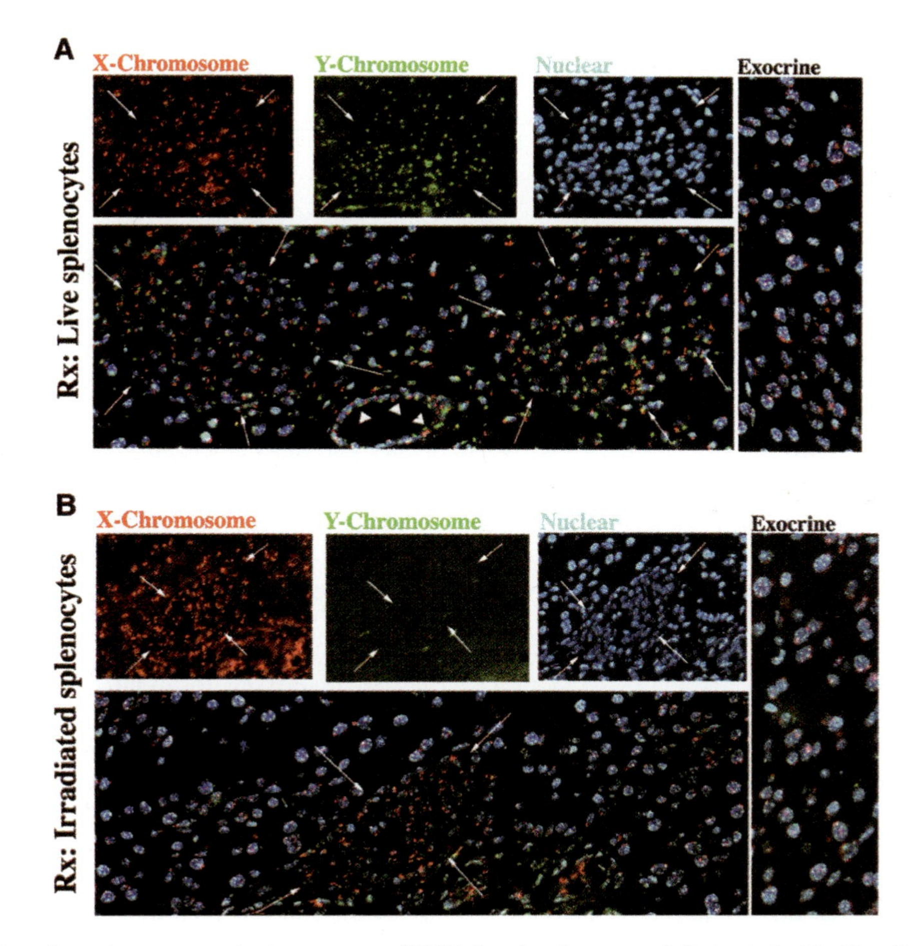


**FIGURE 8.7** Visualization of sex chromosomes in the pancreas of NOD female mice successfully treated with either live **(A)** or irradiated **(B)** splenocytes from male donors. Arrows outline islets. Live splenocyte treatment results in islets that stain for both $Y^+$ (donor) cells and X chromosomes, suggesting regeneration of islets from donor and host cells. The mouse treated with irradiated splenocytes stains only for the X chromosome, indicating that neoislets were regenerated from host cells only. Reproduced with permission from Kodama et al., Islet regeneration during the reversal of autoimmune diabetes in NOD mice. Science 302:1223–1227. Copyright 2003, AAAS.

was repeated by transplanting GFP-labeled $CD45^+$ (lymphocytic), $CD45^-$ (mesenchymal), or unfractionated splenocytes into diabetic mice. The mesenchymal splenocytes were found to give rise to the β-cells, suggesting that MSCs in the spleen have the ability to differentiate into β-cells.

We are not yet able to reverse the autoimmunity of human diabetes. The key to doing so will be to understand the pathways by which tolerance to self-antigens is normally generated and then to devise minimally complex ways of correcting those pathways when they have gone wrong. Furthermore, assuming that we learn how to induce tolerance, we must learn how to induce regeneration from residual cells of the human pancreas.

## 3. Bioartificial Pancreas

Allogeneic or xenogeneic islets can be immunoprotected by encapsulating them prior to transplantation, either in microcapsules or in perfusion-based vascular devices (Wang and Lanza, 2000). Allogeneic β-cells generated *in vitro* from pancreatic ductal stem cells and encapsulated in hyaluronic acid reversed insulin-dependent diabetes when injected into diabetic NOD mice under the kidney capsule or subcutaneously (Ramiya et al., 2000). Alginate microcapsules containing allogeneic islets were injected into the intraperitoneal cavity of Type I spontaneously diabetic dogs (Soon-Shiong et al., 1993). The blood glucose level of the dogs was reduced to normal and they survived for six months without insulin (the length of the experiment). The alginate capsules must have a high guluronic acid content to work in large animal models; alginate that is high in mannuronic acid induces fibrosis of the capsule via growth factor release.

A clinical trial of this system was carried out on a 38-year-old patient who was a Type I diabetic for 30 years, had peripheral neuropathies and foot ulcers, and end-stage renal failure that required a kidney trans-

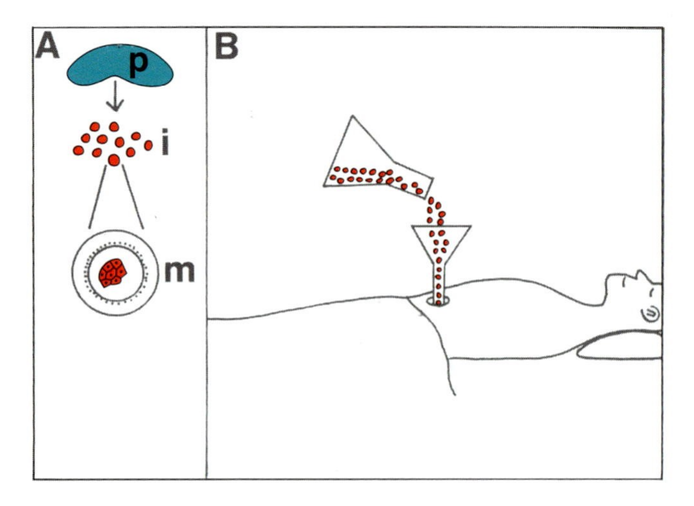


**FIGURE 8.8** Reversal of human diabetes by infusion of islets encapsulated in high-guluronic acid alginate microcapsules. **(A)** The microcapsules were prepared by digesting pancreatic tissue (P) to release the islets (i). The islets were then mixed with alginate and the mixture was sprayed from an atomizer into a $CaCl_2$ bath, cross-linking the alginate. **(B)** Delivery of the microcapsules into the peritoneal cavity through a funnel.

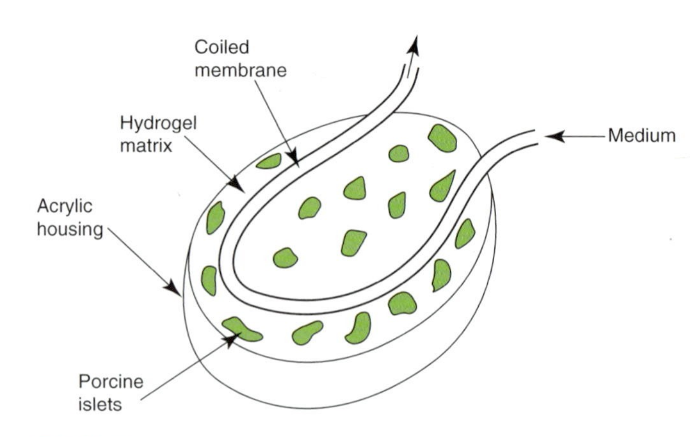


**FIGURE 8.9** Design of an extracorporeal bioartificial pancreas. A coiled tubular membrane is placed in an acrylic housing and the spaces between the coils filled with a hydrogel containing porcine islets. Medium is pumped into the tubular membrane through an inlet port, circulates to nourish the islets by diffusion and exits with metabolic waste products through an exit port.

plant (Soon-Shiong et al., 1994). **FIGURE 8.8** illustrates the procedure. The patient received an initial dose of 10,000 islets in alginate microcapsules/kg body weight, harvested from cadavers by collagenase digestion and purified by gradient separation, through a 2-cm abdominal incision into the intraperitoneal cavity. Six months later, he received another dose of 5,000 islets/kg. Nine months after the initial implantation, exogenous insulin was discontinued, peripheral neuropathy had abated, foot ulcers had healed at a normal pace, and metabolic indices were markedly improved. Measurements of pro-insulin levels, however, indicated that the total dose of 15,000 islets/kg was subcritical and that the islets were being stressed to maintain insulin independence. Studies on diabetic dogs indicated that 20,000 encapsulated islets are required to achieve prolonged insulin independence, but the optimum human intraperitoneal dose is not yet known.

The use of xenogeneic islets would overcome the donor shortage problem. Diabetic mice implanted with rat islet-containing microspheres maintained normal blood sugar concentrations for over two years in some cases (Lum et al., 1992). Canine, porcine, and bovine islets seeded into alginate and put into semipermeable acrylic tubular membrane chambers (200–400 islets per chamber) normalized the fasting blood sugar of streptozotocin-induced diabetic rats for over 10 weeks after intraperitoneal implantation (Lanza et al., 1991). Long-term histological examination (30–130 days) revealed normal-appearing α, β, and δ cells. Diabetic monkeys treated by intraperitoneal transplant of pig islets in alginate-polylysine-alginate microcapsules were rendered insulin-independent for 120 days to over two years with fasting blood glucose levels and glucose clearance in the normal range (Sun et al., 1996). Although hyperglycemia reappeared, probably because of lysis of encapsulated islets, it could be reversed again by transplantation of another set of microcapsules. Microcapsules recovered from two recipients after three months were found to be physically intact. The islets within were able to secrete insulin in response to a glucose challenge.

Microcapsules eventually break down or fracture and must be surgically removed and replaced by another set. To avoid the surgical removal procedure, composite microcapsules made from alginate and poly (L-lysine) have been engineered to biodegrade over variable lengths of time corresponding to the functional longevity of the encapsulated cells (Lanza et al., 1999). By changing the alginate concentration, stability for varying lengths of time was achieved.

Perfusion-based pancreatic devices have been tested on dogs. These devices consist of a plastic housing with a coiled permselective membrane (usually acrylic copolymers) that is connected at each end to vascular grafts anastomosed to the iliac vessels as an arteriovenous shunt (**FIGURE 8.9**). The space between the housing and coiled membrane is filled with islets embedded in a hydrogel such as alginate (Maki et al., 1996; Wang and Lanza, 2000). The device is small enough to be implanted intraperitoneally. Allogeneic and xenogeneic (pig) islets implanted in such a device maintained normal blood glucose concentrations and other metabolic indicators for several months in Type I diabetic dogs (Sullivan et al., 1991; Maki et al., 1996). The device remained patent for up to 3.5 years in normal dogs, demonstrating its excellent biocompatibility. However, late vascular thrombosis was encountered in the majority of dogs used in the study (Maki et al., 1996).

### 4. Gene Therapy to Induce Islet Neogenesis

A common gene therapy strategy to reverse diabetes in diabetic animals is the use of a viral vector to deliver an insulin gene coupled to a glucose-responsive promoter to the liver (Dong and Woo, 2001). There are problems with this approach, however. Since the liver does not contain pro-insulin processing enzymes, the construct must be modified to produce a single-chain insulin or one that contains new cleavage sites so that liver proteases can activate the proinsulin. Furthermore, although the liver has glucose-sensing mechanisms, they are different from those of β-cells, leading to suboptimal blood glucose control. A different approach is to deliver genes for transcription factors or signaling molecules to the liver that are important for islet cell development. Kojima et al. (2003) reported that a combination of the genes for the pancreatic transcription factor NeuroD and the β-cell stimulating hormone betacellulin induced islet neogenesis when delivered to STZ diabetic mice in a helper-dependent adenoviral (HDAD) vector. The islet cells were found primarily under the capsule of the liver, but their origin, whether from hepatocytes or liver stem cells, is not clear. Three months after treatment, the mice exhibited normal glucose and insulin levels during a glucose tolerance test.

## REGENERATIVE THERAPIES FOR THE ESOPHAGUS AND INTESTINE

### 1. Esophagus

Attempts to regenerate full-circumference defects in the esophagus by implanting synthetic scaffolds have generally been disappointing, with stenosis as the usual outcome. Silicon rubber tubes surrounded with a Dacron mesh were used by Fukushiwa et al. (1983) to regenerate 5–7-cm segments of the esophagus in 16 dogs. Seven dogs survived more than 12 months, and four survived more than six years. An esophageal epithelium regenerated into the tubes from the adjacent esophageal ends, but the center of the tube was covered only with fibrous connective tissue, causing stenosis. The tubes had no skeletal muscle and no motility, so they served only as a conduit. Similar outcomes have been reported after bridging 5-cm esophageal defects in dogs with biodegradable lactic and glycolic acid copolymers (Grower et al., 1989) or SIS (Badylak et al., 2000). Endothelium and connective tissue grew over the scaffolds to reunite the cut ends of the esophagus, but the circumference in the defect region was reduced by 50% (Badylak et al., 2000). The stenosis may be due to inadequate stiffness of the scaffold combined with lack of intraluminal pressure, leading to collapse of the scaffold.

Greater success has been achieved using SIS and Alloderm™ scaffolds to regenerate esophageal tissue in dogs after creating noncircumferential defects in the wall of the esophagus (Badylak et al., 2000; Isch et al., 2001). No strictures developed, and squamous epithelium and skeletal muscle regenerated over the scaffold from the edges of the defect. The scaffolds degraded, leaving only regenerated tissue filling the defect.

### 2. Intestine

Short bowel syndrome, intestinal ischemia, tumors, and inflammatory bowel disease are all conditions that can result in a deficiency in the amount of intestinal area required for proper electrolyte balance and nutrition. The only current treatment available in many instances is allogeneic transplantation of whole intestine, which is limited by donor organ shortage and the requirement for long-term immunosuppression. Successful patching of defects in the wall of the small or large intestine in experimental animal models has been accomplished by transplanting autologous serosa, peritoneum, or muscle flaps from other parts of the intestine or abdominal wall (Chen and Bierele, 2004). New mucosa with the qualities of the original regenerated over the grafts. These methods, however, are not readily applicable to human patients because of the desire to avoid the morbidity associated with harvesting tissue from donor sites. Thus, investigators have turned to biomimetic templates to induce regeneration of intestinal tissue across defects made in the intestinal wall or to seed with cells to create whole segments of neointestine (Chen and Bierele, 2004). **FIGURE 8.10** illustrates these strategies.

Nonresorbable materials such as Dacron and polytetrafluorethylene did not effectively promote the regeneration of intestinal wall tissue across patch defects. However, biodegradable materials such as SIS and polylactic or polyglycolic acid mesh were successful in inducing patch regeneration and in constructing full-circumference segments of intestine. Chen et al. (2001) made 7 × 3-cm defects (50%–60% of the circumference) in the wall of the small intestine of dogs and patched it with a multilayered SIS sheet. Thirteen of 15 dogs were healthy until the end of the experiment at one year, with no evidence of intestinal dysfunction. The grafts were found to have contracted to a maxi-

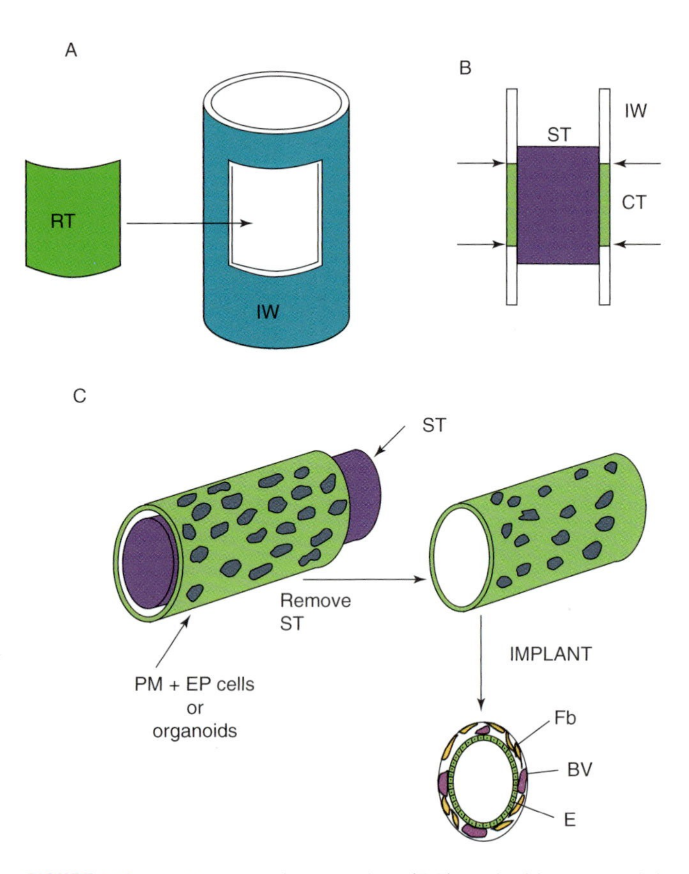


**FIGURE 8.10** **(A)** Regeneration template (RT) grafted into a partial circumferential defect of intestinal wall (IW). **(B)** Replacement of a full-circumference segment of IW with a tubular regeneration template, in this case a collagen tube (CT). A support tube (ST) was used to stabilize the CT and was later removed. **(C)** Construction of a segment of bioartificial intestine. A polymer mesh (PM) seeded with intestinal epithelial cells (EP) or intestinal organoids was wrapped around a support tube (ST) and briefly cultured *in vitro*. The support tube was removed and the seeded polymer mesh was implanted in place of the segment of intestine that was removed. The organoids reconstituted the intestinal wall. Fb = fibroblasts, BV = blood vessels, E = epithelium.

mum of 80% at one year, but exhibited no stenosis. Histological evaluation revealed that normal intestinal wall composed of a mucosal epithelium, varying amounts of smooth muscle, sheets of collagen, and a serosal covering had regenerated.

Hori et al. (2001) resected a 5-cm-long segment of the jejunem in dogs and replaced it with a silicone tube wrapped in a porcine collagen sponge. One month later the silicone tube was removed endoscopically. By four months, intestinal wall had regenerated over the luminal surface of the sponge and the sponge itself had resorbed.

The common bile duct empties from the gall bladder and liver into the duodenum of the small intestine. Bile duct strictures often occur due to injury during surgery and are difficult to repair by surgical means. Autografted pieces of jejenum, artery, vein, ureter, appendix, fascia, or split-thickness skin, or grafts of a wide variety of synthetic materials, have also failed due to scarring and recurrent stricture formation (see Rosen et al., 2002). SIS (Surgisis$^{TM}$), however, was successfuly used to replace the anterior two-thirds of the bile duct in dogs (Rosen et al., 2002). The SIS sheet was formed into a tube by wrapping it around a stent and stitching the edges together. Cholangiography performed three months after transplant indicated patent ducts. Histological evaluation of the grafts at two weeks revealed the expected inflammatory infiltrate, but also infiltration by blood vessels and biliary epithelium and fibroblasts invading the ends of the graft. By five months, the SIS was completely replaced by biliary epithelium lining a fibrous wall of native collagen.

Whole segments of bioartificial neointestine have been constructed from intestinal epithelial cells, intestinal "organoids," and polymer tubes (Organ and Vacanti, 2000). Organ et al. (1992) seeded intestinal epithelial cells of rats onto degradable polyglycolic acid mesh. The seeded mesh was wrapped around a silastic tube for structural support and implanted into the mesentery. The silastic tubes were removed a week later. Fibrous tissue with accompanying blood vessels grew into the polymer mesh and formed a tubular structure lined with stratified intestinal epithelium 5–7-cell layers thick. However, no goblet cell differentiation or villus formation was observed by 28 days.

Kim et al. (1999) seeded polyglycolic acid/polylactic acid tubes with syngeneic intestinal organoids prepared by enzymatic digestion of fragments of rat small intestine and implanted the constructs into the omentum to become vascularized. Three weeks later, a 75% small bowel resection was performed and the ends of the remaining bowel were anastomosed for continuity. The construct and native small intestine were anastomosed side-to-side three weeks later. Histological evaluation of the constructs at 10 weeks revealed patent lumens and the development of a vascularized tissue with a neomucosa supported in some places by smooth muscle. The neomucosa exhibited invaginations resembling crypt-villus structures. Interestingly, the neotissue developed by constructs implanted without anastomosis to the bowel was significantly less differentiated than the neotissue developed in anastomosed constructs, indicating that the resected small intestine provides factors favorable to the regeneration of intestinal wall. In another study, the anastomosed constructs were observed to grow significantly in length and diameter over a 36-week period (Kaihara et al., 2000). Similar experiments with bioartificial colon constructs using organoids derived from colon showed that rats with the constructs anastomosed to the ilium after an end ileostomy could recapitulate some of the major physiologi-

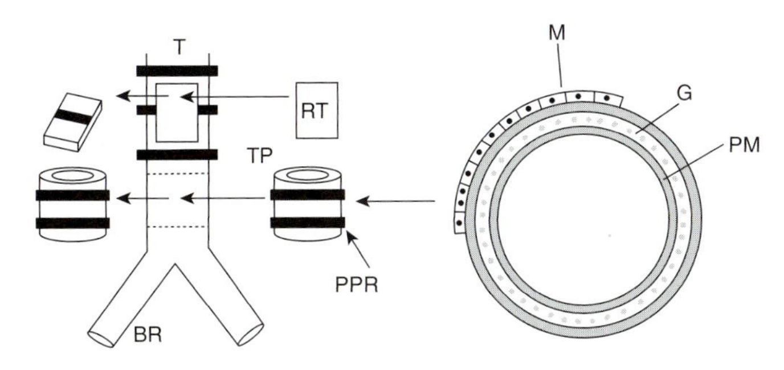


**FIGURE 8.11 Left,** use of a regeneration template (RT) to patch a portion of the tracheal (T) circumference, or replacement of a tracheal segment with a bioengineered tracheal prosthesis (TP). BR = bronchial tube, black bars = cartilage rings of the trachea. PPR = polypropylene rings to substitute for cartilage rings. **Right,** cross-section of the prosthesis. PM = polymer mesh, G = gelatin, M = mucosal epithelial cells on polymer mesh. The mucosal cells migrate into the gel and line the lumen.

cal functions of the native colon (Grikscheit et al., 2002).

While animal experiments have provided proof of concept that acellular or cell-seeded biomimetic templates can promote regeneration of gastrointestinal tissue, how well they would work in human patients is unknown. More detailed information from animal experiments on the long-term structure and function of the tissue regenerated by such constructs is required before conducting clinical trials.

## REGENERATIVE THERAPIES FOR THE RESPIRATORY SYSTEM

### 1. Trachea

Reconstruction of large circumferential defects in the trachea is a surgical challenge. Grafts of autologous pericardium and costal cartilage have been used in reconstructive procedures, but their use is associated with donor site morbidity and complications due to scar tissue formation (Wanamaker and Eliacher, 1995). Tracheal prostheses to patch defects or replace segments of the trachea have been constructed using scaffolds made of silicone, polytetrafluoroethylene, collagen-conjugated mesh, poly-L-lactic acid, polyglycolic acid, and pig SIS (Surgisis™). The aim is to develop a prosthesis that is airtight, causes minimal inflammatory reaction, is readily incorporated into surrounding tissue, permits the in-growth of respiratory epithelium over the walls of the lumen while resisting fibroblastic invasion, and resists bacterial invasion of the lumen. FIGURE 8.11 illustrates these techniques.

Surgisis™ was used to repair a 2-mm-width (about 30% of the tracheal circumference) × 6-mm-length surgical defect in adult rat trachea (De Urgarte et al., 2003). Over the four-week period of the experiment, the rats exhibited no respiratory distress, whereas controls with no patch died. The lumen in the defect region remained patent, and the wall area was 70% of normal. Histological studies indicated that the patch had become vascularized, had not degraded, and was completely covered on its internal surface by ciliated columnar epithelium from the host tracheal epithelium. Thus, the scaffold alone was sufficient to promote epithelialization, but there was no regeneration of cartilage rings. One requisite for success in this experiment was the use of 8-ply Surgisis™. This thickness provided the rigidity necessary to prevent collapse of the patch during inspiration; thinner ply material did not work.

Kim et al. (2004) reported the development, for dogs, of a 5-cm-long complete tracheal prosthesis consisting of two polymer mesh cylinders with gelatin in between. Keratinocytes from abdominal skin patches were seeded into the outer polymer mesh cylinder. Ten polypropylene rings were attached to the exterior surface of the construct and it was enveloped in omentum to encourage vascularization. After culture in the peritoneal cavity for a week, the construct was sutured into the trachea to bridge a 5-cm surgical defect. Bronchoscopic examination showed that by two months, the grafts had completely integrated into the native tracheal tissue at their ends and that the lumen was lined with ciliated epithelium. Histological findings were not ready to be reported, but previous histological evaluation of mucosa derived from oral keratinocytes had indicated that after six months, ciliated columnar epithelium was confluent from one end of the graft to the other (Suh et al., 2000). The extent to which the ciliated columnar epithelium was derived from transformed donor keratinocytes versus replacement by host stem cells differentiating into ciliated epithelium was not clear.

An artificial trachea for sheep was constructed from cartilage and epithelial cells derived from nasal septum biopsy (Kojima et al., 2003). To exploit a more universal source of cells to make tracheal cartilage, sheep bone marrow cells were cultured on a nonwoven polyglycolic acid fiber mesh in the presence of TGF-β2 and IGF-1 for one week (Kojima et al., 2004). The cell-seeded mesh was wrapped around a helical silicone template, and this construct was coated with microspheres impregnated with TGF-β2. The construct was implanted subcutaneously into nude mouse hosts and harvested six weeks later. Significant angiogenesis was observed on the surface of the construct and cartilage differentiated that showed great similarity to native tracheal cartilage in terms of appearance, texture, and flexibility. The cartilage resembled native cartilage histologically, and no significant differences were found between the bioartificial cartilage and native tracheal cartilage in hydroxyproline and GAG content.

There are several problems with the current state of the art for tracheal patches or prostheses. First, whether or not scar formation and stenosis occur over the long term after a patch repair is not yet known. Second, prostheses have not included the smooth muscle layer of the trachea. And third, replacement of segments greater than 5 cm in length has not been accomplished because of the limits of revascularization. A 5-cm segment defect is currently repairable by surgically anastomosing the ends of the trachea, so segmental bioartificial constructs have so far added little of clinical value to current surgical procedures.

Nevertheless, the first use of a bioartificial patch to repair a 1.5 × 1.5 cm defect on the ventrolateral aspect of the tracheobronchial anastomosis after removal of a lung was successful in a human patient (MacChiarini et al., 2004). Thoracic muscle cells and fibroblasts were harvested by biopsy, expanded *in vitro*, and seeded onto jejunal pig SIS (5% muscle cells, 95% fibroblasts). This construct was incubated for a further three weeks to allow ECM turnover. The construct was folded to double its thickness, placed over the defect, and anchored with sutures to neighboring mediastinal tissues to form an airtight seal. Periodic endoscopies showed an airtight graft that was surfaced with ciliated respiratory epithelium and vascularized by regeneration from the host airway. By 12 weeks, the patch was completely integrated functionally and morphologically into the host airway tissue. The patient's quality of life returned to normal.

### 2. Lung

Chronic obstructive pulmonary disease (COPD) is a rapidly growing worldwide problem. The disease involves the proliferation of fibroblasts and their deposition of ECM that progressively obliterates the alveolar air spaces. The fibroblasts are presumed to originate within the lung, but Hashimoto et al. (2004) reported that some of the fibroblasts seen in bleomycin-induced COPD mice engrafted with $GFP^+$ bone marrow come from the bone marrow. Treatment of 2-day-old postnatal mice with dexamethasone for 10 days results in permanent alveolar abnormalities that mimic COPD. There are fewer and larger alveoli, which is consistent with a failure of normal postnatal alveolar septation, reducing the surface area for gas exchange. Among the factors that are important for normal alveolar septation are retinoids and their receptors (Cardoso, 1995). Retinoic acid, the RA-synthesizing enzymes, RARs, and cytoplasmic RA-binding proteins (CRABPs) are all expressed during postnatal alveolar development in mouse and rat lungs (Geevarghese and Chytil, 1994; McGowan et al., 1995; Hind et al., 2002a, b; Maden and Hind, 2003). Intraperitoneal administration of disulfram, an inhibitor of RA synthesis, inhibits postnatal alveolar development in mice. Retinoic acid may be a potential agent to treat COPD (**FIGURE 8.12**). Rat alveoli destroyed by dexamethasone were regenerated after treatment with RA (Massaro and Massaro, 2000), and dexamethasone-treated mice treated with RA from postnatal days 30–42 regenerated alveoli to control levels (Hind and Maden, 2004). Furthermore, systemic RA treatment reversed the pathology of emphysema induced experimentally by elastase in adult rats (Massaro and Massaro, 1997; Belloni et al., 2000; Tepper et al., 2000).

## REGENERATIVE THERAPIES FOR THE URINARY SYSTEM

### 1. Nephrons

Human nephrons do not regenerate, so the research focus has been on constructing a bioartificial kidney consisting of a hemofilter (glomerulus) connected in tandem to a bioartificial renal tubule (Humes, 2000). A functional hemofilter has been successfully achieved by seeding endothelial cells onto the internal surface of acrylonitrile/vinyl chloride copolymer tubes (Humes et al., 1997a). The endothelial cells were transfected with the hirudin gene as a means of preventing thrombosis (Humes et al., 1997b). A functional proximal tubule has been constructed by seeding cultured allogeneic tubule epithelial cells on the inside of polysulfone tubes coated with ECM. The endothelial cells are protected from the immune system by the tubes. They were able to transport insulin, glucose, and tetraethylammonium from the blood of a uremic patient (Humes and Cieslinski, 1992) and to transport salt and water along osmotic and oncotic gradients (MacKay et al., 1998).

The ultimate goal is to develop this single-fiber device as an implantable bioartificial kidney (**FIGURE 8.13**). As presently conceived, the glomerulus would be housed inside a cylinder that collects the filtrate from the endothelial cells. The glomerulus would be connected at its ends to the iliac artery and vein. The cylinder would be connected in series to one end of the proximal tubule, which would be embedded in the peritoneal membrane and connected at its other end to the ureter. However, before an implantable device is fully developed, it will be used as a renal assist device (RAD). The scale-up

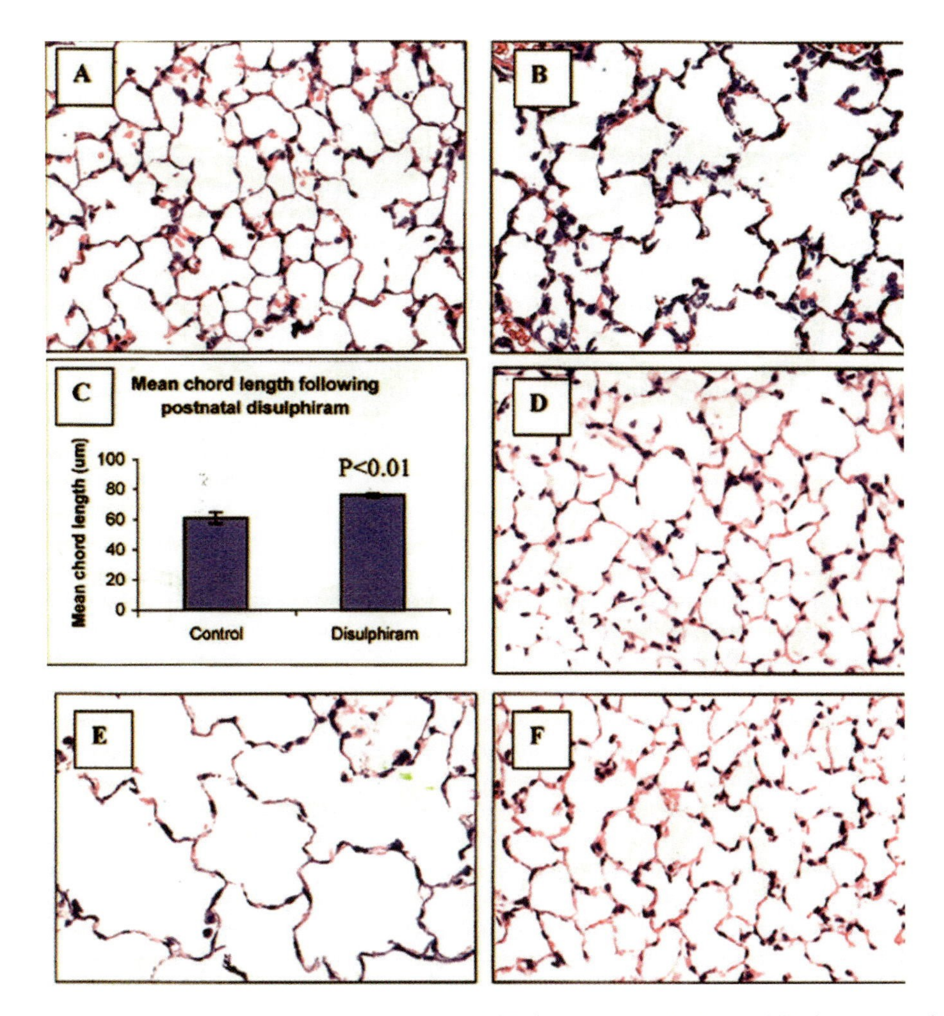


**FIGURE 8.12** Regeneration of alveoli in mouse lungs after retinoic acid (RA) treatment. **(A)** Normal 24 day postnatal mouse lung, showing typical arrangement of alveoli. **(B)** Lung of same age treated with disulfiram from postnatal days 2–14, showing the larger than normal air spaces caused by inhibition of alveolar development. **(C)** Average diameter of alveoli in control and disulfiram-treated lungs. **(D)** Lung of a postnatal day 90 mouse. **(E)** Postnatal day 90 lung from mouse treated with dexamethasone from postnatal days 2–14, showing abnormally enlarged air spaces. **(F)** Dexamethasone-treated (postnatal days 2–14) mouse that was treated with RA from postnatal days 30–42. The alveoli have regenerated to control status. Reproduced with permission from Maden and Hind, Retinoic acid, a regeneration-inducing molecule. Dev Dynam 226:237–244. Copyright 2003, Wiley-Liss.

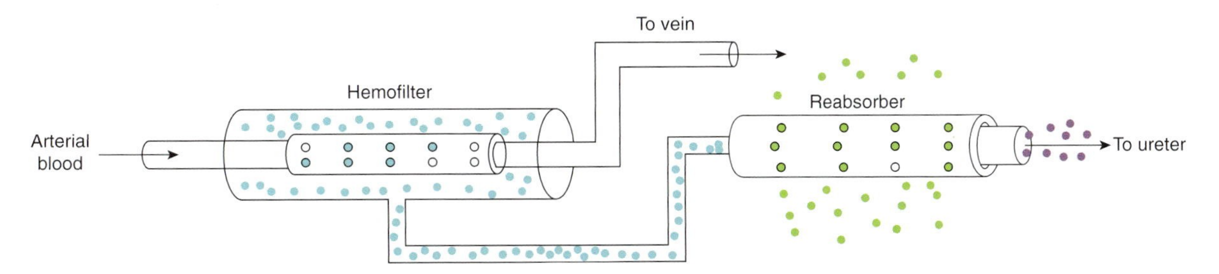


**FIGURE 8.13** Diagram of an implantable bioartificial kidney. Tubes carrying arterial blood pass through a hemofilter unit (HU) lined with endothelial cells and a reabsorbing unit (RU) lined with proximal tubule epithelial cells. The hemofilter mimics the renal capsule of the nephron and the reabsorber mimics the proximal tubule. Water and electrolytes (green dots) are reabsorbed from the filtrate created by the hemofilter (blue dots). The remainder of the filtrate (urine, purple) is carried to the ureter. Redrawn after Humes, 2000.

from a single-fiber device to a multifiber bioreactor has been successfully accomplished (Humes et al., 1999a, b). The RAD proved effective in controlling fluid and electrolyte balance in uremic dogs, excreted ammonia, reclaimed filtered glutathione, and produced 1,25-$(OH)_2$ D3.

## 2. Urinary Conduit Issue

The urinary conduit consists of the renal pelvis of the kidney, ureters, bladder, and urethra. The wall of the conduit consists of three layers, an inner uroepithelium, an intermediate connective tissue layer, and an outer smooth muscle layer. Surgical reconstruction of parts of the conduit from other tissues, or diversion of urine into the ileum or colon, has been discouraging because of stenosis, adhesions, kidney infection and the morbidity of the donor sites. Therefore, attention has turned to the use of regeneration templates or construction of bioartificial urinary wall tissue.

### *a. Bladder*

The standard material for use in augmentation cystoplasty is autogeneic ilium wall, though other tissues such as peritoneum, fascia, omentum, pericardium, and dura have been used (Kim et al., 2000). These autografts rarely reproduce the function of the bladder wall, and complications include metabolic abnormalities, infection, perforation, and malignancy.

Acellular regeneration templates have been used to patch the bladder wall in animal models where 40% of the bladder wall is routinely excised (**FIGURE 8.14**). Sutherland et al. (1996) and Piechota et al. (1998) described the use of allogeneic bladder matrix to regenerate the bladder wall in rats. Regeneration was excellent, with complete epithelialization of the graft by 4 days, evidence of smooth muscle and vascular regeneration by two weeks, and near-normal contractile response to electrical, muscarinic, purinergic, and $\alpha$ and $\beta$-adrenergic drug stimulation, but there was a high incidence of bladder stone formation. Reddy et al. (2000) used allogeneic bladder matrix to repair pig bladders. By 4 weeks, all the layers of the bladder wall were regenerated, with the only abnormality noted in the organization of the smooth muscle.

The drawback of allogeneic bladder matrix for human use is donor shortage. Xenogeneic matrix such as pig SIS could be prepared for use off the shelf. Bladder augmentation in dogs by an SIS patch resulted in regeneration of the bladder wall into the patch (Kropp et al., 1996). Histological evaluation showed that all three layers of the bladder were regenerated from the edges of the defect. The thickness of the regenerated tissue was the same as that of normal canine bladder, but the smooth muscle was haphazardly organized. Regenerating nerves re-innervate the smooth muscle as it differentiates (Pope et al., 1997). Over the course of the experiment, the dogs maintained normal bladder capacity and blood chemistry. The SIS was completely degraded by 8–12 weeks after transplantation (Badylak et al., 1998).

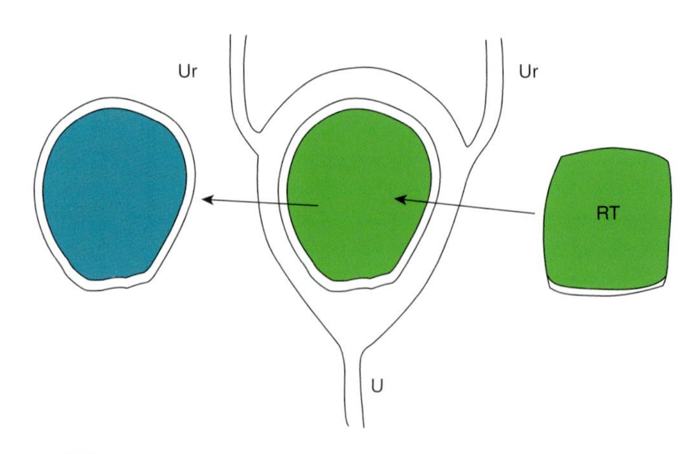


**FIGURE 8.14** Patching an area of bladder wall with a regeneration template (RT). U = urethra, Ur = ureter.

A study in Yucatan minipigs, however, indicated that pig SIS gave inferior results compared to complete ilial wall transplants in augmentation cystoplasty (Paterson et al., 2002). The SIS grafts shrank, there was minimal regeneration of bladder wall tissue, and there were osseous deposits and scar formation in the graft. The authors did not describe the origin of the SIS used or the age of the animals from which it was derived. Kropp et al. (2004) found in a comparative study that only SIS from the distal ilium of sows greater than three years of age produced consistent bladder regeneration in dogs without osseous deposits or severe graft shrinkage.

A comparative study of commercially available SIS versus a porous cross-linked type I collagen scaffold in a rabbit model indicated that both were equally effective in inducing full-thickness regeneration of the bladder wall over a two-month period, although the density of smooth muscle was lower in the center of the grafts than toward the edges (Nuininga et al., 2004). Bladder stones were found in 5 of 56 animals, but the authors did not report their distribution between the SIS and collagen-grafted animals.

Complete replacement of the bladder requires prefabricating a bioartificial bladder wall *in vitro*, fashioning it into the shape of the normal bladder, and transplanting it in place of the excised bladder. Atala and colleagues (Atala, 1998; Kim and Atala, 2000) have pioneered the techniques required to make bioartificial bladder wall by seeding cells into polyester scaf-

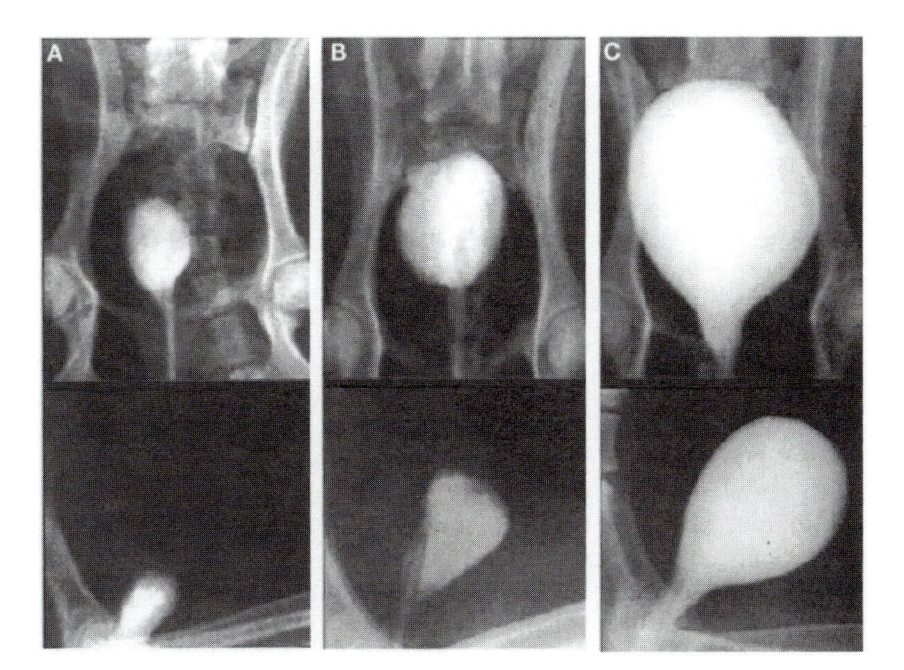


**FIGURE 8.15** Radiographic cystograms of dog bladders after subtotal cystectomy. **(A)** Without reconstruction. **(B)** Implant of PLA:PLGA scaffold only. **(C)** Implant of bioartificial bladder tissue. Reproduced with permission from Oberpenning et al., De novo reconstitution of a functional mammalian urinary bladder by tissue engineering. Nature Biotech 17:149–155. Copyright 1999, Nature Publishing Group.

folds. Initial proof of concept involved seeding rat urothelial cells onto degradable polyglycolic acid mesh, which was then wrapped around a silastic tube for structural support. These constructs were implanted into the rat mesentery, and the silastic tubes removed a week later. Fibrous tissue with accompanying blood vessels grew into the polymer mesh and formed a tubular structure of urothelium and connective tissue (Atala et al., 1992).

Oberpenning et al. (1999) constructed dog neobladders by molding a biodegradable polyglycolic acid mesh coated with poly-dl-lactide-co-glycolide 50:50 into the shape of a bladder. The luminal side of the scaffold was seeded with biopsy-derived autologous cultured urothelial cells and the opposite side was seeded with autologous smooth muscle cells. The bioartificial bladders were transplanted to each donor dog after partial cystectomy that spared the top (ureteral side) of the bladder. Control dogs received no construct or acellular scaffold. Urodynamic studies, radiographic cystograms, gross anatomical inspection, and histological and immunocytochemical evaluation showed that the controls that received no construct did not regenerate new bladder wall and had low bladder capacity (**FIGURE 8.15A**). Controls that received polymer scaffold only showed drastic shrinkage of the graft, scar tissue formation and adhesions, and minimal tissue regeneration (**FIGURE 8.15B**). Animals that received the bioartificial bladder, however, had excellent bladder capacity and a histologically normal trilayered structure with ingrowth of neural tissue (**FIGURE 8.15C**). A similar experiment was performed using allogeneic bladder submucosa (UBS) as the scaffold (Yoo et al., 1998) and autogeneic urothelial cells. Cruciate partial cystectomies were performed and the defects closed by grafting either seeded or unseeded UBS. Histologically, the grafts in both groups had a normal trilayered cellular organization with nerve fibers, but bladders augmented with seeded UBS showed a 99% increase in capacity compared to only a 30% increase in the unseeded group.

Bioartificial bladder wall has also been fabricated *in vitro* from cultured pig urothelial and bladder smooth muscle cells seeded onto the opposite sides of either pig SIS or bladder wall matrix (Gabouev et al., 2003). Both matrices supported development of a double layer of urothelial and smooth muscle cells. In another experiment, human bladder smooth muscle cells were first seeded onto pig SIS, followed by seeding urothelial cells on top of the smooth muscle cells (Zhang et al., 2004). The constructs were incubated for 14 days, folded with the cells facing the lumen, and implanted subcutaneously into the flanks of nude mice, with unseeded SIS implanted as controls. Grafts harvested at 12 weeks showed an open lumen lined with several layers of uroepithelium with early smooth muscle formation peripherally. Muscle-derived stem cells may also be useful in constructing the smooth muscle layer of bioartificial bladder wall. These cells are able to develop contractile activity and increase pliability when seeded into an SIS scaffold (Lu et al., 2003, 2005).

A full-thickness urinary wall has also been made *in vitro* without using any scaffold material (Fossum et al., 2004). Human penile urothelial cells, fibroblasts, and smooth muscle cells were expanded separately in culture for a week. The urothelial cells, fibroblasts, and smooth muscle cells were then resuspended and plated on top of one another, in that order. After three weeks of co-culture, a well-organized three-layered tissue was produced that could be mechanically handled, although the tissue was thin and the urothelial component was only a monolayer. No tests of this bio-artificial tissue to replace native bladder wall have yet been made.

### b. Ureter and Urethra

Stress urinary incontinence develops in a significant fraction of older individuals, due to deficiency in the urethral sphincter. A minimally invasive way to treat incontinence is to inject a bulking agent into the sphincter. The most common and successful material used for injection is an injectable paste of bovine collagen cross-linked by glutaraldehyde. While short-term results are good, long-term success is lacking, primarily because the injected collagen is reabsorbed (Hershorn et al., 1996).

In an attempt to get a better long-term outcome, muscle cells have been explored as living bulking agents. Cells of an immortalized myoblast cell line transduced with the *lac-Z* gene and incubated with fluorescent latex microspheres were injected into the bladder wall and dorsal proximal urethra of rats to determine whether the cells would integrate into surrounding tissue and differentiate into myofibers (Chancellor et al., 2000). The injected myoblasts fused and differentiated into myotubes and myofibers. Muscle-derived stem cells also differentiated into myotubes after injection into the bladder wall (Yokoyama et al., 2001). These cells are easy to prepare from a biopsy of the patient's own muscle and may soon be tested as bulking agents in human clinical trials.

Ureteral stricture is another problem that is difficult to treat surgically. Pig SIS has been used as a scaffold to patch defects in the pig ureter by inducing regeneration of host tissues (Liatsikos et al., 2001a, b; Smith et al., 2002). After inserting a stent in the left ureter, one-half or two-thirds of the circumference of the ureteral wall was excised over a 2–7-cm length, and a tubularized or elliptical patch of SIS was stitched in to cover the defect. The stent was removed after one week. Histological evaluation showed that normal urothelium regenerated on the luminal surface. Grafts of tubular pig SIS have been used to bridge 11-mm gaps in rabbit ureters (Jaffe et al., 2001). By postoperative day 35, all three layers of the ureter had regenerated and there was no evidence of obstruction. Sofer et al. (2004) bridged 2-cm gaps in the pig ureter with tubular SIS. They reported, however, that although urothelium and smooth muscle cells regenerated into the graft, they were embedded in fibrotic tissue that completely occluded the lumen. Control ureters, in which the cut ends were anastomosed end to end, remained patent. Shalav et al. (1999) obtained a similar result.

One patient has been treated for a stricture of the distal ureter by replacing 5 cm of the ureter with a single sheet of 4-ply SIS rolled into a tube (O'Connor et al., 2002). Ureteroscopy performed at 12 weeks postsurgery revealed that new tissue had regenerated across the gap. Peristalsis was not observed, so it is likely that the new tissue was serving only as a conduit.

Urethral strictures are commonly treated by urethrotomy and placement of a vascularized flap of penile or preputial skin. If the graft does not take, however, there is the danger of scarring and further stricture. As an alternative, biomimetic scaffolds have been used to promote regeneration of urethral tissue. Pig SIS has been shown to promote the regeneration of all three layers of the urethral wall in male rabbits (Kropp et al., 1998; Rotariu et al., 2002). Cadaver UBS was used to treat 28 human patients ages 22–61 with a diagnosis of fibrotic ventral urethral stricture (El-Kassaby et al., 2003). The stricture region was opened and the fibrotic ventral tissues were excised. Single graft strips were stitched to the healthy dorsal tissues. The outcome was successful in 24 of the 26 patients. Retrograde urethrography showed a wide patent urethra. Uroflowmetry showed that the average and maximum urine flow rates were doubled by this procedure. New urethral wall tissue, indistinguishable from normal, had regenerated across the grafts.

## SUMMARY

Cell transplants and bioartificial tissues have been developed as regenerative therapies for both liver and pancreas. Animal studies have shown the feasibility of infusing adult hepatocytes into the liver or spleen to rescue liver function. When transplanted into humans, autogeneic hepatocytes transfected with the gene for the LDL receptor temporarily reduced cholesterol levels in a familial hypercholesterolemia patient, and cryopreserved allogeneic hepatocytes have bridged patients to whole organ transplant. An extracorporeal liver assist device in which blood flows across hepatocytes contained in a bioreactor successfully bridged patients to whole organ transplant and even allowed several of them to survive while their liver regenerated. Research suggests that the function of the bioreactor can be improved by seeding them with tiny organoids composed of more than one liver cell type. Providing enough cells for transplant or bioreactor is a major problem. Large numbers of human and rat hepatocytes can be generated by immortalization with the SV40T gene, and liver assist devices will be able to use xenogenic hepatocytes from pigs, thus potentially solving this problem.

Type 1 diabetes is an autoimmune disease that has been successfully treated by the Edmonton protocol, in which islets from multiple cadavers were transplanted into the liver. The success of the procedure is due to gentle harvest of islets and a milder immunosuppression regimen. Donor shortage is a major problem, however, even though new techniques now require only one donor pancreas. The hope is thus to derive large numbers of β-cells by the directed differentiation of ESCs, by the proliferation and differentiation of regeneration-competent ductule or islet cells, or by the use of allogeneic immortalized β-cells that can be expanded and reverted. Another strategy would be to tolerize the patient so that the progression of diabetes is halted and stimulate regeneration from residual healthy cells. Tolerance to autogeneic islets has been achieved in diabetic mice, allowing regeneration of new islets. Transplants of allogeneic or xenogeneic islets immunoprotected by microcapsules is another approach that has been tried on one patient. Alginate microcapsules containing islets were infused into the intraperitoneal cavity, allowing the patient to go off insulin. Microcapsules eventually break down, however, and need to be replaced. Gene therapy with genes encoding factors that promote islet development have shown promise in reversing STZ-induced diabetes in mice.

Patch defects in the esophagus of dogs have been bridged with regeneration templates of polylactic and polyglycolic copolymers, SIS, and Alloderm™. The synthetic polymers induced the regeneration of epithelium and fibrous tissue, which created a stenosis, whereas the biological templates regenerated epithelium, smooth muscle, and fibrous tissue with no stricture. Polylactic or polyglycolic mesh and SIS all promoted the regeneration of normal, trilayered intestinal wall across patch defects of dog intestine. Full-circumference segments of rat intestine have been constructed by seeding polyglycolic/polylactic acid tubes with intestinal "organoids" prepared by enzymatic digestion of intestinal wall. When anastomosed to a host intestine, these constructs developed mucosa and smooth muscle, though the amount of smooth muscle was small. These approaches have not yet been successful enough to try on human patients, however.

Replacement of tracheal tissue by either regeneration templates or bioartificial trachea has not been particularly successful so far. Surgisis™ induced the regeneration of tracheal ciliary epithelium across a gap in the rat trachea, but there was no regeneration of cartilage rings. Bioartificial tracheal segments have been made by seeding keratinocytes onto a scaffold of polylatic and polyglycolic acid, which was then wrapped around a polypropylene mesh cylinder with 10 polypropylene rings to mimic tracheal cartilage rings. After culture in the peritoneal cavity for a week, the construct was used to replace a 5-cm segment of trachea in dogs. The grafts integrated into the host tracheal tissue and became lined with ciliated epithelium. A major problem with these grafts is that gaps of more than 5 cm cannot be bridged due to the limits of revascularization. Thus, they are not clinically useful, since a 5-cm gap can be eliminated by surgical anastomosis. However, a bioartificial tissue patch constructed by seeding pig SIS with autogeneic thoracic muscle cells and fibroblasts has been used to successfully repair a defect in the tracheobronchial anastomosis of a human patient. Another promising research result in respiratory tissue regeneration is the finding that retinoic acid induces the regeneration of rat and mouse alveoli destroyed by dexamethasone and elastase.

An implantable bioartificial kidney is under development, consisting of a hemofilter connected in tandem to a proximal tubule. The hemofilter is made by seeding endothelial cells onto the inner surface of polymer tubes and the proximal tubule by seeding tubule epithelial cells onto the inside of polysulfone tubes. This unit has been shown to have some of the functional capabilities of nephrons. Acellular regeneration templates or bioartificial tissue have been tested for patching defects in the walls of the urinary conduit system, or replacing segments of the system. Regeneration templates of allogeneic bladder matrix, pig SIS, or collagen I promoted the regeneration of normal three-layered bladder wall in rats, pigs, dogs, or rabbits. SIS rolled into tubes has been successful in promoting the regeneration of segments of ureter in experimental animals and in one patient. The major deficiency in regeneration is in smooth muscle, which regenerates in subnormal amounts and organization. Whole bladders have been fabricated by seeding the inner side of a polyglycolic acid mesh, molded into the shape of a bladder, with autologous urothelial cells and the outer side with autologous smooth muscle cells. When implanted in dogs, the bioartificial bladders had excellent capacity and were found to have a histologically normal trilayered structure with ingrowth of neural tissue. Further advances in bioartificial bladder wall construction without the necessity of a scaffold are being made *in vitro* by plating urothelial cells, fibroblasts, and smooth muscle cells on top of one another in succession.

## REFERENCES


Asonuma K, Gilbert TG, Stein JE, Takeda T, Vacanti JP (1992) Quantitation of transplanted hepatic mass necessary to cure the Gunn rat model of hyperbilirubinemia. J Pediatr Surg 27: 298–301.

Atala A (1998) Autologous cell transplantation for urologic reconstruction. J Urol 159:2–3.

Atala A, Vacanti JP, Peters CA, Mandell J, Retik AB, Freeman MR (1992) Formation of urothelial structures *in vivo* from dissociated cells attached to biodegradable polymer scaffolds *in vitro*. J Urol 148:658–662.

Badylak SF, Kropp B, McPherson T, Liang H, Snyder PW (1998) Small intestinal submucosa: A rapidly resorbed bioscaffold for augmentation cystoplasty in a dog model. Tiss Eng 4:379–387.

Badylak SF, Meurling S, Chen M, Spievack A, Simmons-Buryd A (2000) Resorbable bioscaffold for esophageal repair in a dog model. J Pediatr Surg 35:1097–1103.

Bell GI, Polonsky KS (2001) Diabetes mellitus and genetically programmed defects in β-cell function. Nature 414:788–791.

Belloni PN, Garvin L, Mao CP, Bailey-Healy I, Leaffer D (2000) Effects of all-trans retinoic acid in promoting alveolar repair. Chest 117:235S–241S.

Block GD, Locker J, Bowen WC, Petersen BE, Katyal S, Strom SC, Riley T, Howard TA, Michaelopoulos GK (1996) Population expansion, clonal growth, and specific differentiation patterns in primary cultures of hepatocytes induced by HGF/SF, EGF and TGFα in a chemically defined (HGM) medium. J Cell Biol 132:1133–1149.

Bonner-Weir S, Weir GC (2005) New sources of pancreatic β-cells. Nature Biotech 23:857–860.

Cardoso WV (1995) Transcription factors and pattern formation in the developing lung. Am J Physiol 269:L429–L442.

Chancellor MB, Yokoyama, Tirney S, Mattes CE, Ozawa H, Yoshimura N, de Groat WC, Huard J (2000) Preliminary results of myoblast injection into the urethra and bladder wall: A possible method for the treatment of stress urinary incontinence and impaired detrusor contractility. Neurourol Urodynam 19:279–287.

Chen MF, Badylak SF (2001) Small bowel tissue engineering using small intestinal submucosa as a scaffold. J Surg Res 99:352–358.

Chen MK, Bierele EA (2004) Animal models for intestinal tissue engineering. Biomats 25:1675–1681.

Collaborative Islet Transplant Registry (July 1, 2004) National Institute of Diabetes and Digestive and Kidney Diseases.

Demetriou AA, Whiting JF, Feldman D, Levenson SM, Chowdhury NR, Moscioni AD, Kram M, Chowdhury JR (1986) Replacement of liver function in rats by transplantation of microcarrier-attached hepatocytes. Science 233:1190–1192.

De Urgarte DA, Puapong D, Roostaeian J, Gillis N, Fonkalsrud EW, Atkinson JB, Dunn JCY (2003) Surgisis patch tracheoplasty in a rodent model for tracheal stenosis. J Surg Res 112:65–69.

Dixit V, Darvasi R, Arthur M, Brezina M, Lewin K, Gitnick G (1990) Restoration of liver function in Gunn rats without immunosuppression using transplanted microencapsulated hepatocytes. Hepatol 12:1342–1349.

Dong H, Woo SL (2001) Hepatic insulin production for type I diabetes. Trends Endocrinol Metab 12:441–446.

El-Kassaby AW, Retik AB, Yoo JJ, Atala A (2003) Urethral stricture repair with an off-the-shelf collagen matrix. Urol 169: 170–173.

Faustman D, Coe C (1991) Prevention of xenograft rejection by masking donor HLA class I antigens. Science 252:1700–1702.

Fossum M, Nordenskjold A, Kratz G (2004) Engineering of multilayered urinary tissue *in vitro*. Tiss Eng 10:175–180.

Fox IJ, Chowdhury JR, Kaufman SS, Goertzen TC, Chowdhury NR, Warkentin PI, Dorko, Sauter BV, Strom SC (1998) Treatment of the Crigler-Najjar syndrome type I with hepatocyte transplantation. New Eng J Med 338:1422–1427.

Fukushima M, Kako N, Chiba K, Kawaguchi T, Kimura Y, Sato M, Yamauchi M, Koie H (1983) Seven-year follow-up study after the replacement of the esophagus with an artificial esophagus in the dog. Surgery 93:70–77.

Gabouev AI, Schultheiss D, Mertsching H, Koeppe M, Schloter N, Wefer J, Jonas U, Stief CG (2003) In vitro construction of urinary bladder wall using porcine primary cells reseeded on acellularized bladder matrix and small intestinal submucosa. Int J Artif Organs 26:935–942.

Geevarghese SK, Chytil F (1994) Depletion of retinyl esters in the lungs coincides with lung prenatal morphological maturation. Biochem Biophys Res Comm 200:529–535.

Grikscheit TC, Ogilvie JB, Ochoa ER, Alsberg E, Mooney D, Vacanti JP (2002) Tissue-engineered colon exhibits function in vivo. Surgery 132:200–204.

Grossman M, Raper SE, Kozarsky K, Stein EA, Negelhardt JF, Muller D, Lupien PJ, Wilson M (1994) Successful ex vivo gene therapy directed to liver in a patient with familial hypercholesterolaemia. Nat Genet 6:335–341.

Grower MF, Russell EA, Cutright DE (1989) Segmental neogenesis of the dog esophagus utilizing a biodegradable polymer frame-work. Biomater Artif Cells Artif Organs 17: 291–314.

Gupta S, Aragona E, Vemuru RP, Bhargava K, Burk RD, Chowdhury R (1991) Hepatology 14:144–149.

Gupta S, Rajvanshi P, Lee C-D (1995) Integration of transplanted hepatocytes into host liver plates demonstrated with dipeptidyl peptidase IV-deficient rats. Proc Natl Acad Sci USA 92: 5860–5864.

Hashimoto N, Jin H, Liu T, Chensue SW, Phan H (2004) Bone marrow-derived progenitor cells in pulmonary fibrosis. Am Soc Clin Invest 113:243–252.

Hering BJ, Kandaswamy R, Ansite JD, Eckman P, Nakano M, Sawada T, Matsumoto I, Ihm S-H, Zhang H-J, Parkey J, Hunter DW, Sutherland DER (2005) Single-donor, marginal-dose islet transplantation in patients with type I diabetes. J Am Med Assoc 293:830–835.

Hershorn S, Steele DJ, Radomski SB (1996) Followup of intraurethral collagen for female stress urinary incontinence. J Urol 156:1305–1309.

Hind M, Maden M (2004) Retinoic acid induces alveolar regeneration in the adult mouse lung. Eur Respir J 23:20–27.

Hind M, Corcoran J, Maden M (2002a) Alveolar proliferation, retinoid synthesizing enzymes and endogenous retinoids in the postnatal mouse lung: Different roles for Raldh-1 and Raldh-2. Am J Respir Cell Mol Biol 26:67–73.

Hind M, Corcoran J, Maden M (2002b) The temporal/spatial distribution of retinoid binding proteins and RAR isoforms in the postnatal lung. Am J Physiol 282:L1–L9.

Hori Y, Nakamura T, Matsumoto K, Kurokawa Y, Satomi S, Shimizu Y (2001) Tissue engineering of the small intestine by acellular collagen sponge scaffold grafting. Intl J Artif Organs 24:50–54.

Humes HD (2000) Renal replacement devices. In: Lanza RP, Langer R, Vacanti J (eds.). Principles of Tissue Engineering, 2nd ed. New York, Academic Press, pp 645–653.

Humes HD, Cieslinski DA (1992) Interaction between growth factors and retinoic acid in the induction of kidney tubulogenesis inb tissue culture. Exp Cell Res 201:8–15.

Humes HD, MacKay SM, Funke AJ, Buffington DA (1997a) Acute renal failure: Growth factors, cell therapy, gene therapy. Proc Assoc Am Physicians 109:547–557.

Humes HD, Mackay SM, Funke AJ, Buffington DA (1997b) The bioartificial renal tubule assist device to enhance CRRT in acute renal failure. Am J Kidney Dis 30:S28–S31.

Humes HD, Buffington DA, MacKay SM, Funke AJ, Weitzel WF (1999a) Replacement of renal function in uremic animals with a tissue-engineered kidney. Nature Biotech 17:451–455.

Humes HD, MacKay SM, Funke AJ, Buffington DA (1999b) Tissue engineering of a bioartificial renal tubule assist device: *In vitro* transport and metabolic characteristics. Kidney Intl 55:2502–2514.

Hussain MA, Thiese ND (2004) Stem-cell therapy for diabetes mellitus. Lancet 364:203–205.

Isch JA, Engum SA, Ruble CA, Davis MM, Grosfeld JL (2001) Patch esophagoplasty using Alloderm as a tissue scaffold. J Pediatr Surg 36:266–268.

Jaffe JS, Ginsberg PC, Yanoshak SJ, Costa LE Jr, Ogbolu FN, Moyer CP, Greene CH, Finkelstein LH, Harkaway RC (2001) Ureteral segment replacement using a circumferential small-intestine submucosa xenogeneic graft. J Invest Surg 14:259–265.

Jauregui HO (2000) Liver. In: Lanza R, Langer R, Vacanti J (eds.). Principles of Tissue Engineering, 2nd ed. New York, Academic Press, pp 541–552.

Kaihara S, Kim SS, Kim BS, Mooney D, Tanaka K, Vacanti JP (2000) Long-term follow-up of tissue-engineered intestine after anastomosis to native small bowel. Transplantation 69:1927–1932.

Kent SC, Chen Y, Bregoli L, Clemmings SM, Kenyon NS, Ricordi C, Hering BJ, Hafler DA (2005) Expanded T cells from pancreatic lymph nodes of type I diabetic subjects recognize an insulin epitope. Nature 435:224–228.

Kim SS, Kaiara S, Benvenuto MS, Choi RS, Kim B-S, Mooney DJ, Vacanti JP (1999) Effects of anastomosis of tissue-engineered neointestine to native small bowel. J Surg Res 87:6–13.

Kim TH, Vacanti JP (1995) Tissue engineering of the liver. In: Bronzino JD (ed.). The Biomedical Engineering Handbook. Boca Raton, FL, CRC Press, pp 1745–1753.

Kim J, Suh SW, Shin JY, Kim JH, Choi Y, Kim H (2004) Replacement of a tracheal defect with a tissue-engineered prosthesis: Early results from animal experiments. J Thor Cardiovasc Surg 128:124–129.

Kim B-S, Mooney DJ, Atala A (2000) Genitourinary system. In: Lanza RP, Langer R, Vacanti J (eds.). Principles of Tissue Engineering, 2nd ed. New York, Academic Press, pp 655–667.

Kobayashi N, Fujiwara T, Westerman KA, Inoue Y, Sakaguchi M, Noguchi H, Miyazaki M, Cai J, Tanaka N, Fox IJ, Leboulch P (2000) Prevention of acute liver failure in rats with reversibly immortalized human hepatocytes. Science 287:1258–1261.

Kodama S, Kuhtreiber W, Fujimura S, Dale EA, Faustman DL (2003) Islet regeneration during the reversal of autoimmune diabetes in NOD mice. Science 302:1223–1227.

Kojima H, Fujimiya M, Matsumura K, Younan P, Imaeda H, Maeda M, Chan L (2003) NeuroD-betacellulin gene therapy induces islet cell neogenesis in the liver and reverses diabetes in mice. Nature Med 9:596–603.

Kojima K, Bonassar LJ, Roy AK, Mizuno H, Cortiella J, Vacanti CA (2003) A composite tissue-engineered trachea using sheep nasal chondrocyte and epithelial cells. FASEB J 17:823–828.

Kojima K, Ignotz RA, Kushibiki T, Tinsley KW, Tabata Y, Vacanti CA (2004) Tissue-engineered trachea from sheep marrow stromal cells with transforming growth factor β2 released from biodegradable microspheres in a nude rat recipient. J Thor Cardiovasc Surg 128:147–153.

Kropp BP, Rippy MK, Badylak SF, Adams MC, Keating MA, Rink RC, Thor KB (1996) Regenerative urinary bladder augmentation using small intestinal submucosa: Urodynamic and histopathologic assessment in long-term canine bladder augmentations. J Urol 155:2098–2104.

Kropp BP, Ludlow JK, Spicer D, Rippy MK, Badylak SF, Adams MC, Keating MA, Rink RC, Birhle R, Thor KB (1998) Rabbit urethral regeneration using small intestinal submucosa onlay grafts. Urology 52:138–142.

Kropp BP, Cheng EY, Lin H-K, Zhang Y (2004) Reliable and reproducible bladder regeneration using unseeded distal small intestinal submucosa. J Urol 172:1710–1713.

Lanza RP, Butler DH, Borland KM, Staruk JE, Faustman DL, Solomon BA, Muller TE, Rupp RG, Maki T, Monaco AP, Chick WL (1991) Xenotransplantation of canine, bovine and porcine islets in diabetic rats without immunosuppression. Proc Natl Acad Sci USA 88:11100–11104.

Lanza RP, Jackson R, Sullivan A, Ringeling J, McGrath C, Kuhtreiber WM, Chick WL (1999) Xenotransplantation of cells using biodegradable microcapsules. Transplantation 67:1105–1111.

Liatsikos EN, Dinlenc CZ, Kapoor R, Bernardo NO, Pikashov D, Anderson AE, Smith AD (2001a) Ureteral reconstruction: Small intestine submucosa for the management of strictures and defects of the upper third of the ureter. J Urol 165:1719–1723.

Liatsikos EN, Dinlenc CZ, Kapoor R, Alexianu M, Yohannes P, Anderson AE, Smith AD (2001b) Laparoscopic ureteral reconstruction with small intestine submucosa. J Endourol 15:217–220.

Lu S-H, Cannon TW, Chermanski C, Pruchnic R, Somogyi G, Sacks M, de Groat WC, Huard J, Chancellor MB (2003) Muscle-derived stem cells seeded into acellular scaffolds develop calcium-dependent contractile activity that is modulated by nicotinic receptors. Urology 61:1285–1291.

Lu S-H, Sacks MS, Chung SY, Gloeckner C, Prichnic R, Huard J, de Groat WC, Chancellor MB (2005) Biomats 26:443–449.

Lum ZP, Tai P, Krestow I, Tai T, Sun AM (1992) Xenografts of microencapsulated rat islets into diabetic mice. Transplantation 53:1180–1183.

Lumelsky N, Blondel O, Laeng P, Velasco I, Ravin R, McKay R (2001) Differentiation of embryonic stem cells to insulin-secreting structures similar to pancreatic islets. Science 292:1389–1394.

MacChiarini P, Walles T, Biancosino C, Mertsching H (2004) First human transplantation of a bioengineered airway tissue. J Thor Cardiovasc Surg 128:638–641.

Maden M, Hind M (2003) Retinoic acid, a regeneration-inducing molecule. Dev Dynam 226:237–244.

Maki T, Monaco AP, Mullon CJP, Solomon BA (1996) Early treatment of diabetes with porcine islets in a bioartificial pancreas. Tiss Eng 2:299–306.

Markmann JF, Bassiri H, Desai NM, Odorico JS, Kim JI, Koller BH, Smithies O, Barker CF (1992) Indefinite survival of MHC class I-deficient murine pancreatic islet allografts. Transplantation 54:1085–1089.

Massaro GD, Massaro D (1997) Retinoic acid treatment abrogates elastase-induced pulmonary emphysema in rats. Nat Med 3:675–677.

Massaro GD, Massaro D (2000) Retinoic acid treatment partially rescues failed septation in rats and in mice. Am J Physiol Lung Cell Mol Physiol 278:L955–L960.

Matas AJ, Sutherland DER, Steffes MW, Maurer SM, Lowe A, Simmons RL, Najarian JS (1976) Hepatocellular transplantation for metabolic deficiencies: Decrease of plasma bilirubin in Gunn rats. Science 192:892–894.

Mathis D, Vence L, Benoist C (2001) β-cell death during progression to diabetes. Nature 414:792–798.

Matsumoto S, Okitsu T, Iwanaga Y, Noguchi H, Nagata H, Yonekawa Y, Yamada Y, Fukuda K, Tsukiyama K, Suzuki H, Kawasaki Y, Shimodaira M, Matsuoka K, Shibata T, Kasai Y, Maekawa T,

Shapiro AMJ, Tanaka K (2005) Insulin independence after living-donor distal pancreatectomy and islet allotransplantation. Lancet 365:1642–1644.

McGowan SE, Harvey CS, Jackson SK (1995) Retinoids, retinoic acid receptors and cytoplasmic retinoid binding proteins in perinatal rat lung fibroblasts. Am J Physiol 269:L463–L472.

MacKay SM, Funke AJ, Buffington DA, Humes HD (1998) Tissue engineering of a bioartificial renal tubule. ASAIO J 44:179–183.

Mullon C, Solomon BA (2000) HepatAssist liver support system. In: Lanza RP, Langer R, Vacanti J (eds.). Principles of Tissue Engineering, 2nd ed. New York, Academic Press, pp 553–558.

Myers M, Mackay I, Rowley M, Zimmet P (2001) Dietary microbial toxins and Type I diabetes—a new meaning for seed and soil. Diabetologia 44:1199–1200.

Nakayama M, Abiru N, Moriyama H, Babaya N, Liu E, Miao D, Yu L, Wegmann DR, Hutton JC, Elliot JF, Eisenbarth GS (2005) Prime role for an insulin epitope in the development of type 1 diabetes in NOD mice. Nature 435:220–223.

Narushima M, Kobayashi N, Okitsu T, Tanaka Y, Li S-A, Chen Y, Miki A, Tanaka K, Nakaji S, Takei K, Gutierrez AS, Rivas-Carrillo D, Navarro-Alvarez N, Jun H-S, Westerman KA, Noguchi H, Akey JRT, Leboulch P, Tanaka N, Yoon J-W (2005) A human β-cell line for transplantation therapy to control type 1 diabetes. Nature Biotech 23:1274–1282.

Nuininga JE, van Moerkerk H, Hanssen A, Hulsbergen CA, Oosterwijk-Wakka J, Oosterwijk E, de Gier RPE, Schalken JA, van Kuppevelt H, Feitz WFJ (2004) A rabbit model to tissue engineer the bladder. Biomats 25:1657–1661.

Oberpenning F, Meng J, Yoo JJ, Atala A (1999) De novo reconstitution of a functional mammalian urinary bladder by tissue engineering. Nature Biotech 17:149–155.

O'Connor RC, Hollowell CMP, Steinberg GD (2002) Distal ureteral replacement with tabularized porcine small intestine submucosa. Urology 60:697–701.

Oh SH, Muzzonigro TM, Bae S-H, LaPlante JM, Hatch H, Petersen BE (2004) Adult bone marrow-derived cells trans-differentiating into insulin-producing cells for the treatment of type 1 diabetes. Lab Invest 84:607–617.

Organ GM, Mooney DJ, Hansen LK, Schloo B, Vacanti JP (1992) Transplantation of enterocytes utilizing polymer-cell constructs to produce a neointestine. Transplant Proc 24:3009–3011.

Organ GM, Mooney DJ, Hansen LK, Schloo B, Vacanti JP (1993) Enterocyte transplantation using cell-polymer devices to create intestinal epithelial-lined tubes. Transpl Proc 25:998–1001.

Organ GM, Vacanti JP (2000) Alimantary tract. In: Lanza RP, Langer R, Vacanti J (eds.). Principles of Tissue Engineering, 2nd ed. New York, Academic Press, pp 525–540.

Osorio RW, Asher NL, Melzer JS, Stock PG (1994) Beta-2 microglobulin gene disruption prolongs murine islet allograft survival in NOD mice. Transplant Proc 26:752.

Paterson RF, Lifshitz DA, Beck SD, Siqueira TM, Jr, Cheng L, Lingeman JE, Shalav AL (2002) Multilayered small intestinal submucosa is inferior to autologous bowel for laparascopic bladder augmentation. J Urol 168:2253–2257.

Piechota HJ, Dahms SE, Nunes LS, Dahiya R, Lue TF, Tanagho EA (1998) In vitro functional properties of the rat bladder regenerated by the bladder acellular matrix graft. J Urol 159:1717–1724.

Ponder KP, Gupta S, Leland F, Darlington G, Finegold M, DeMayo J, Ledley FD, Roy Chowdhury J, Woo SLC (1991) Mouse hepatocytes migrate to liver parenchyma and function indefinitely after intrasplenic transplantation. Proc Natl Acad Sci USA 88:1217–1221.

Pope JC, David MM, Smith ER Jr, Walsh MJ, Ellison PK, Rink RC, Kropp BP (1997) The ontogeny of canine small intestinal submucosa regenerated bladder. J Urol 158:1105–1110.

Rajagopal J, Anderson W, Kume S, Martinez OI, Melton DA (2003) Insulin staining of ES cell progeny from insulin uptake. Science 299:363.

Ramiya VK, Maraist M, Arfors KE, Schatz DA, Peck AB, Cornelius JG (2000) Reversal of insulin-dependent diabetes using islets generated *in vitro* from pancreatic stem cells. Nature Med 6:278–282.

Reddy PP, Barrieras DJ, Wilson G, Bagli DJ, McLorie GA, Khoury AE, Merguerian PA (2000) Regeneration of functional bladder substitutes using large segment acellular matrix allografts in a porcne model. J Urol 164:936–941.

Rhim J, Sangren EP, Degan JL, Palmiter RD, Brinster RL (1994) Replacement of diseased mouse liver by hepatic cell transplantation. Science 263:1149–1152.

Rhodes CJ (2005) Type 2 diabetes—a matter of β-cell life and death? Science 307:380–384.

Rosen M, Ponsky J, Petras R, Fanning A, Brody F, Duperier F (2002) Small intestinal submucosa as a bioscaffold for biliary tract regeneration. Surgery 132:480–486.

Rotariu P, Yohannes P, Alexianu M, Gershbaum D, Pinkashov D, Morgenstern N, Smith AD (2002) Reconstruction of rabbit urethra with Surgisis® small intestinal submucosa. J Endourol 16:617–620.

Ryan EA, Lakey JRT, Rajotte RV, Korbutt GS, Kin T, Imes S, Rabinovitch A, Elliot JF, Bigham D, Kneteman NM, Warnock GL, Larsen I, Shapiro AMJ (2001) Clinical outcomes and insulin secretion after islet transplantation with the Edmonton protocol. Diabetes 50:710–719.

Ryu S, Kodama S, Ryu K, Schoenfeld A, Faustman DL (2001) Reversal of established autoimmune diabetes by restoration of endogenous β cell function. J Clin Invest 108:63–72.

Shalav AL, Elbahnasy AM, Bercowsky E, Kovacs G, Brewer A, Maxwell KL, McDougall EM, Clayman RV (1999) Laparoscopic replacement of urinary tract segments using biodegradable materials in a large animal model. J Endourol 13:241–244.

Shapiro AMJ, Lakey JRT, Ryan EA, Korbutt GS, Toth E, Warnock GL, Netman NM, Rajotte RV (2000) Islet transplantation in seven patients with type I diabetes mellitus using a glucocorticoid-free immunosuppressive regimen. New Eng J Med 343:230–238.

Smith TG, Gettman M, Lindberg G, Napper C, Pearle MS, Cadeddu JA (2002) Ureteral replacement using porcine small intestine submucosa in a porcine model. Urology 60:931–934.

Sofer M, Rowe E, Forder DM, Denstedt JD (2004) Ureteral segmental replacement using multilayer porcine small-intestinal submucosa. J Endourol 16:27–31.

Soon-Shiong P, Feldman E, Nelson R, Heintz R, Yao Q, Yao Z, Zheng T, Merideth N, Skjak-Braek G, Espevik T, Smidsrod O, Sandford P (1993) Long-term reversal of diabetes by the injection of immunoprotected islets. Proc Natl Acad Sci USA 90:5843–5847.

Soon-Shiong P, Heintz RE, Merideth N, Yao QX, Zheng T, Murphy M, Moloney MK, Schmehl M, Harris M, Mendez R, Sandford PA (1994) Insulin independence in a type 1 diabetic patient after encapsulated islet transplantation. Lancet 343:950–951.

Strain A, Neuberger JM (2002) A bioartificial liver—state of the art. Science 295:1005–1009.

Strom SC, Fisher RA, Thompson MT, Sanyal AJ, Cole PE, Ham JM, Posner MP (1997) Hepatocyte transplantation as a bridge to orthotopic liver transplantation in terminal liver failure. Transplantation 63:559–569.

Suh SW, Kim J, Beak CH, Kim H (2000) Development of new tracheal prosthesis: Autogenous mucosa-lined prosthesis made from polypropylene mesh. Intl Artif Organs 23:261–267.

Sullivan SJ, Maki T, Borland KM, Mahoney MD, Solomon BA, Muller TE, Monaco A, Chick WL (1991) The biohybrid artificial pancreas: Long-term implantation studies in diabetic, pancreatctomized dogs. Science 252:718–722.

Sun Y, Ma X, Zhou D, Vacek I, Sun AM (1996) Normalization of diabetes in spontaneously diabetic cynomolgus monkeys by xenografts of microencapsulated porcine islets without immunosuppression. J Clin Invest 98:1417–1422.

Sutherland RS, Baskin LS, Hayward SW, Simon W, Cunha GR (1996) Regeneration of bladder urothelium, smooth muscle, blood vessels and nerves into an acellular tissue matrix. J Urol 156: 571–577.

Takezawa T, Kamazaki M, Mori Y, Yonaha T, Yoshizato K (1992) Morphological and immuno-cytochemical characterization of a hetero-spheroid composed of fibroblasts and hepatocytes. J Cell Sci 101:495–501.

Taylor SI (1999) Deconstructing type 2 diabetes. Cell 97:9–12.

Tepper J, Pfeiffer J, Aldrich M, Tumas D, Kern J, Hoffman E, McLennan G, Hyde D (2000) Can retinoic acid ameliorate the physiologic and morphologic effects of elasatase instillation in the rat? Chest 117:242S–244S.

Uyama S, Kaufman P-M, Takeda T, Vacanti JP (1993) Delivery of whole liver-equivalent hepatocyte mass using polymer devices and hepatotrophic stimulation. Transplantation 55: 932–935.

van de Kerkhove M-P, Hoekstra R, van Gulik TM, Chamuleau RAFM (2004) Large animal models of fulminant hepatic failure in artificial and bioartificial liver support research. Biomats 25:1613–1625.

Wanamaker JR, Eliacher I (1995) An overview of treatment options for lower airway obstructions. Otolaryngol Clin North Am 28: 751.

Wang TG, Lanza RP (2000) Bioartificial pancreas. In: Lanza RP, Langer R, Vacanti J (eds.). Principles of Tissue Engineering, 2nd ed. New York, Academic Press, pp 495–507.

Wu FJ, Peshwa MV, Cerra FB, Hu W-S (1995) Entrapment of hepatocyte spheroids in a hollow fiber bioreactor as a potential bioartifical liver. Tiss Eng 1:29–40.

Xu ASL, Luntz TL, Macdonald JM, Kubota H, Hsu E, London RE, Reid LM (2000) Lineage biology and liver. In: Lanza RP, Langer R, Vacanti J (eds.). Principles of Tissue Engineering, 2nd ed, New York, Academic Press, pp 559–598.

Yokoyama T, Huard J, Pruchnic R, Yoshimura N, Qu Z, Cao B, de Groat WC, Kumon H, Chancellor MB (2001) Muscle-derived cell transplantation and differentiation into lower urinary tract smooth muscle. Urol 57:826–831.

Yoo JJ, Meng J, Oberpenning F, Atala A (1998) Bladder augmentation using allogeneic bladder submucosa seeded with cells. Urology 51:221–225.

Yoon J-W, Yoon C-S, Lim H-W, Huang QQ, Kang Y, Pyun KH, Hirasawa K, Sherwin RS, Jun H-S (1999) Control of autoimmune diabetes in NOD mice by GAD expression or suppression in β cells. Science 284:1183–1187.

Zhang Y, Kropp BP, Lin H-K, Cowan R, Cheng EY (2004) Bladder regeneration with cell-seeded small intestinal submucosa. Tiss Eng 10:181–187.

Zimmet P, lberti KGMM, Shaw J (2001) Global and societal implications of the diabetes epidemic. Nature 414:782–787.

Zorina TD, Subbotin VM, Bertera S, Alexander AM, Haluszczak C, Gambrell B, Bottino R, Styche AJ, Trucco M (2003) Recovery of endogenous β cells function in the NOD model of autoimmune diabetes. Stem Cells 21:377–388.

# 9 Regeneration of Musculoskeletal Tissues

## INTRODUCTION

The musculoskeletal system is derived from the mesodermal germ layer of the embryo and consists of the muscles, bones, joints, tendons, and ligaments. Musculoskeletal injuries and diseases are among the most common disorders suffered by the human population. Bones and muscles regenerate well when torn or fractured, but nonunion fractures or surgically created gaps in bone, and extensive injuries to or surgically created gaps in muscle are filled in by scar tissue. The articular cartilages and menisci of joints regenerate poorly and heal with fibrocartilaginous scar tissue. Tendons and ligaments repair essentially by forming a scar that is similar to the original tissue, but lacking the original strength. In this chapter, I discuss the regenerative mechanisms of these tissues.

## REGENERATION OF SKELETAL MUSCLE

### 1. Structure of Skeletal Muscle

Skeletal muscles are composed of multinucleated myofibers organized into fascicles that are grouped together to form individual muscles (**FIGURE 9.1**). Individual myofibers develop by end-to-end fusion of mononucleate myoblasts to form a multinucleate syncytium (**FIGURE 9.2**). Each mononucleate cell is a contractile unit called a sarcomere bounded by Z lines, structures that anchor the actin filaments into each end of the sarcomere. The sarcomere shortens when the actin filaments slide along the thick myosin filaments toward the middle of the sarcomere (Ham and Cormack, 1979; Alberts et al., 1994). Myofibers are organized into fascicles surrounded by perimysia, and the fascicles collectively constitute the muscle, which is surrounded by an epimysium. Skeletal muscles are highly vascularized within and between the mysial sheaths and are heavily innervated at specialized contacts called neuromuscular junctions. At their ends, muscles grade into fascia or tendons, which attach them to bones.

The fibroblasts of the endomysium synthesize an ECM consisting of a typical basement membrane of Fn, Ln, and collagen IV that surrounds each myofiber, as well as collagen I and sulfated PGs outside the basement membrane, including a muscle-specific sulfated PG (Caplan, 1991). In addition, the endomysium synthesizes tetranectin, a protein that serves as an interactive agent with other ECM proteins, cell surface receptors, cytokines, and proteases (Wewer et al., 1998). Tetranectin binds sulfated polysaccharides, suggesting that it might interact with the GAG chains of PGs. It is particularly prominent at myotendinous junctions, which are considered to be the equivalent of focal adhesion sites in muscle (Wewer et al., 1998). The actin cytoskeleton of the myofibers is linked to the ECM by a dystrophin-glycoprotein complex (DGC), a multisubunit complex comprised of intracellular dystrophin and syntrophins and three types of sarcolemmal proteins: dystroglycans, sarcoglycans, and sarcospan. Disruption of this linkage by mutations in dystrophin or the sarcoglycans results in sarcolemmal damage during contraction, rendering the myofibers susceptible to the necrosis that is a feature of muscular dystrophy (Cohn et al., 2002).

### 2. Satellite Cells Are the Source of Regenerated Muscle

Vertebrate skeletal muscle in the neonates and adults of virtually every species examined contains a population of stem cells called "satellite cells" (SCs) located between the sarcolemma and the overlying basement membrane (**FIGURE 9.3**). Satellite cells were first identified by electron microscopy in frog skeletal muscle (Mauro, 1961). DNA labeling studies subsequently showed them to be the source of regenerated muscle after injury in several species (Hinterberger and Cameron, 1990). Satellite cell nuclei constitute about

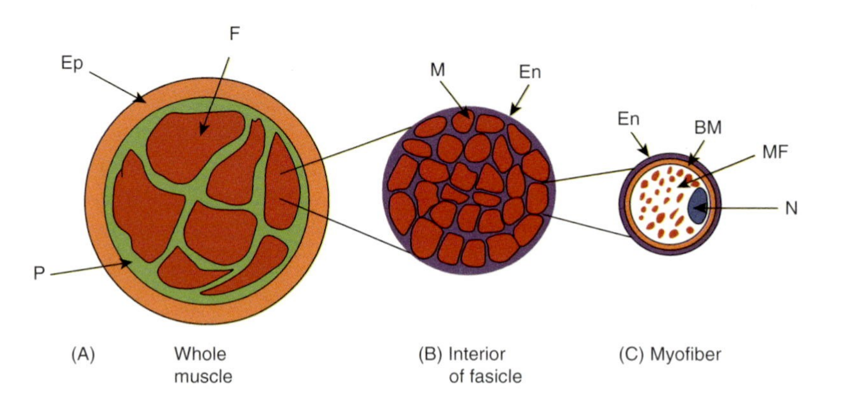


**FIGURE 9.1** Diagram of a cross-section of a skeletal muscle showing **(A)** divisions into fascicles (F) surrounded by perimysium (P), **(B)** the interior of a fascicle composed of myofibers (M) surrounded by endomysium (En). **(C)** Structure of an individual myofiber (endomysial unit). MF = myofilaments, BM = basement membrane, N = nucleus. The whole muscle is surrounded by an epimysium (Ep).

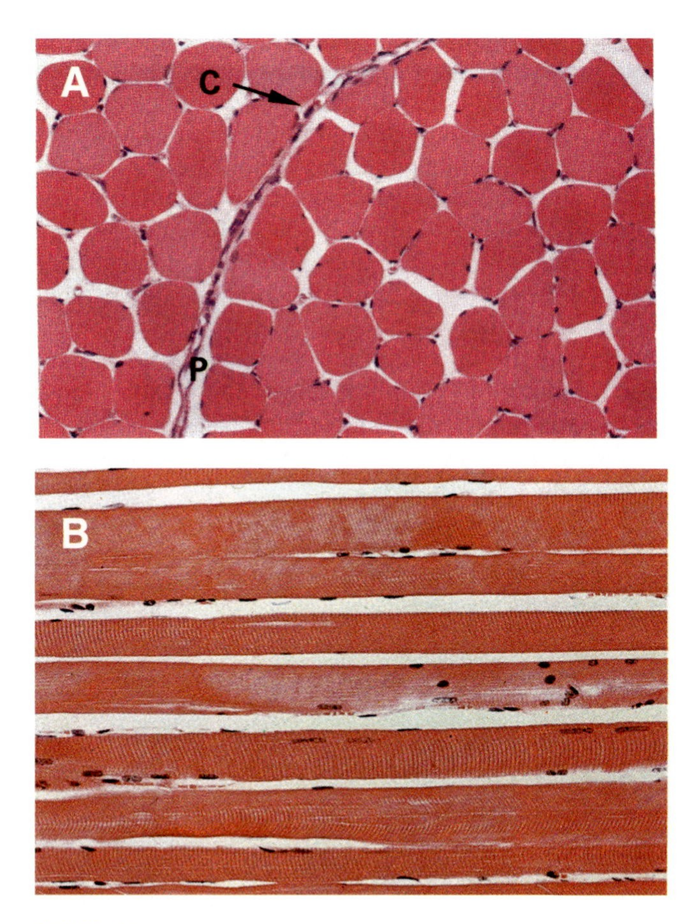


**FIGURE 9.2 (A)** Cross-section of skeletal muscle (H&E stain) showing individual myofibers. P = perimysium that surrounds a fasicle. C = capillary. Note the peripheral location of myonuclei (dark dots). **(B)** Longitudinal section of skeletal muscle (H&E stain). The cross striations of the actin-myosin contractile protein complex are clearly visible in the individual myofibers. Reproduced with permission from Wheater et al., Wheater's Functional Histology (3rd ed.). Copyright 1997, Elsevier.

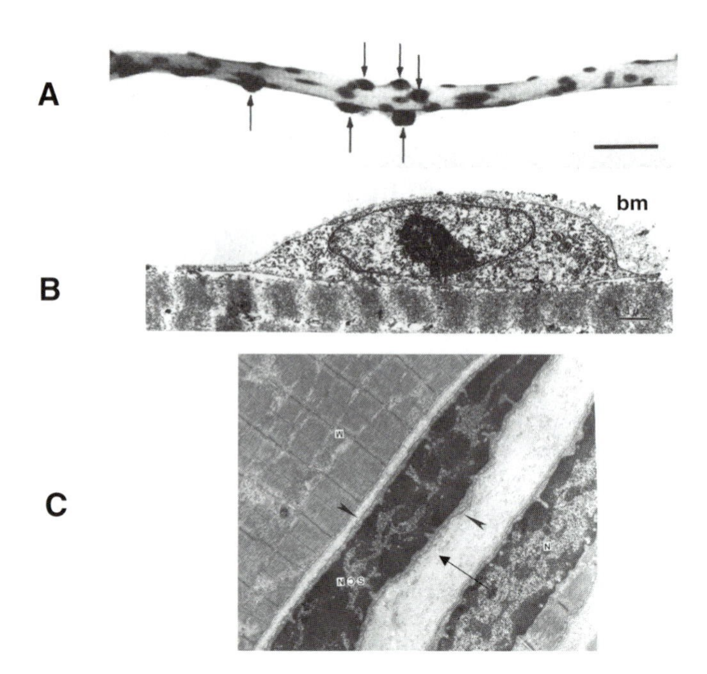


**FIGURE 9.3 (A)** Autoradiogram of a single chicken myofiber labeled *in vitro* with $^3$H-thymidine. Arrows point to labeled satellite cell nuclei. Reproduced with permission from Bischoff, Interaction between satellite cells and skeletal muscle fibers. Development 109:943–952. Copyright 1990, Company of Biologists. **(B)** Electron micrograph of a satellite cell lying under the basement membrane (bm) of a cultured chick myofiber. Reproduced with permission from Konigsberg et al., The regenerative response of single mature muscle fibers isolated *in vitro*. Dev Biol 45:260–275. Copyright 1975, Elsevier. **(C)** Electron micrograph of a satellite cell nucleus (SCN) of a newt limb myofiber (M). In newt skeletal muscle, the satellite cells are encased in a basement membrane (arrowheads) that is separate from the basement membrane of the myofiber (arrow). N = nucleus of adjacent myofiber. Courtesy of Dr. Jo Ann Cameron.

30% of myofiber nuclei in newborns. This percentage decreases with age; satellite cell nuclei represent between 1%–5% of the total nuclei in adult mammalian muscles (Allbrook, 1981). Satellite cells separated from chicken pectoral muscle by Percoll density centrifugation and cultured *in vitro* give rise to large clones of cells that form myofibers (Yablonka-Reuveni et al., 1987). Experiments implanting SCs into dystrophic muscles showed that not only do the implanted cells contribute to the regeneration of myofibers, but they also give rise to more SCs, indicating that they are self-renewing (Blaveri et al., 1999).

Quiescent SCs are defined by the collective expression of several markers: (1) myocyte nuclear factor (Garry et al., 1997), which may prevent the synthesis of muscle transcription factors that promote myogenic differentiation; (2) the c-met receptor (Cornelison and Wold, 1997), which plays a key role in the activation of SCs (see ahead); (3) p130, a protein that blocks cell cycle progression by binding to E2F transcription factors and also inhibits the differentiation of cultured mouse C2 myoblasts by inhibiting the expression of MyoD (Carnac et al., 2000); and (4) syndecans 3 and 4, which are transmembrane heparan sulfate proteoglycans (HSPGs) (Cornelison et al., 2001). In addition, SCs have the surface antigen phenotype CD34$^+$ Sca1$^-$ (Zammit and Beauchamp, 2001) and express the transcription factor *Pax 7*, which is also expressed in early myoblasts (Seale et al., 2000; Halevy et al., 2004). *Pax-7* is required for the formation of SCs during development, as shown by the fact that *Pax-7* null mice exhibit normal muscle development, but lack SCs (Seale et al., 2000). Another cell type associated with the microvessels in the endomysium is CD34$^+$ Sca1$^+$, but these cells are not stem cells (Zammit and Beauchamp, 2001). Possibly, they may be pericytes.

Studies of the kinetics of SC proliferation have indicated that resident SCs are fully capable of regenerating skeletal muscle (Zammit et al., 2002). Nevertheless, there are reports that skeletal muscle harbors bone marrow-derived CD45$^+$ Sca1$^+$ side population muscle stem cells that have both hematopoietic and myogenic developmental potential, contribute to muscle regeneration, and perhaps continually replenish the supply of resident SCs (Seale et al., 2000; Seale and Rudnicki., 2000; Seale et al., 2004; McKinney-Freeman et al., 2002).

To test the myogenic potential of the various cell types associated with muscle fibers, Sherwood et al. (2004) used a two-step enzymatic dissociation procedure to separate muscle-resident cells into myofiber-associated and interstitial compartments, followed by FACS into subsets of cells expressing different combinations of surface antigens. They found that the only cell type that had intrinsic clonal myogenic capability

**TABLE 9.1** Myogenic capacity of the four types of cells associated with myofibers. SC = satellite cell; HSC = hematopoietic stem cell. Data from Sherwood et al. (2004)

| | True SC | Fibroblastic | Inducible myogenic | HSC |
|---|---|---|---|---|
| Phenotype | CD45$^-$ SCA1$^-$ CD34$^+$ | CD45$^-$ Sca1$^+$ CD34$^+$ | CD45$^-$ Sca1$^+$ CD34$^-$ | CD45$^+$ |
| Myogenic? | | | | |
| *In vitro* | Yes | No | Yes (co-culture) | No |
| *In vivo* | Yes | No | No | No |

*in vitro* corresponded to the SC, with phenotype [CD45 Sca1 Mac-1]$^-$ [CXCR β1-integrin CD34 c-met]$^+$ (**TABLE 9.1**). Another cell population of phenotype CD45$^-$ CD34$^-$ Sca1$^+$ cells could generate myosin heavy chain-expressing cells at low frequency (~1% of input cells), but only when co-cultured with endogenous myogenic cells. Two other populations were sorted that displayed no myogenic activity, a CD45$^-$ CD34$^+$ group that formed fibroblastic colonies and a CD45$^+$ group of myeloid lineage cells derived from bone marrow hematopoietic cells. The roles of the latter three cell types in muscle function are not clear. They may perform support functions for myogenic cells, act as precursors for muscle-associated mesenchymal lineages, or be tissue resident immune cells (Sherwood et al., 2004). Thus, self-renewing SCs appear to be the only cell type that regenerates injured muscle.

Collins et al. (2005) have provided further proof for the sufficiency of satellite cells in regenerating muscle. Single intact myofibers transgenic for *lacZ* were transplanted into radiation-ablated tibialis anterior muscles of *mdx* nude mice. *Myf5* is expressed in satellite cells and newly formed myofibers, but not in mature myofibers. The satellite cells from one myofiber (average, 22 SCs) were sufficient to completely regenerate the irradiated muscle, while self-renewing, as indicated by the co-expression of *Myf-5* and *Pax-7* in SC nuclei on the periphery of the new myofibers and *Pax-7*$^-$ central nuclei with weak β-galactose staining, indicating their recent derivation from graft SCs.

Skeletal muscles and SCs have a common origin from Pax3/Pax 7-expressing cells of the dermamyotome (Relaix et al., 2005; Gros et al., 2005). Satellite cells contribute to the growth and regeneration of muscle in rapidly growing juvenile animals, but these cells appear to be a subpopulation that differs from the SCs that contribute to regeneration in the adult (Pastoret and Sebille, 1995). Regeneration of skeletal muscle in young postnatal mice produces myofibers with the peripherally located nuclei characteristic of muscle formed during embryogenesis, whereas the nuclei of myofibers regenerated from older mice are located centrally within the fiber, suggesting differences in the cells involved in regeneration at these points in the life cycle. Consistent with this idea, continuous labeling of mitotically active cells with BrdU in immature rats revealed two distinctly different sets of labeling kinetics in SCs (Schultz, 1996). One population, consisting of 80% of the cells, labeled to saturation within five days. The other 20% had a much slower kinetics of labeling and was not saturated even after 14 days of continuous labeling, leaving about 10% of the cells unlabeled. This result suggests that the SCs used for regeneration in the adult are derived from this slow labeling population, whereas the rapidly labeling population is used for juvenile growth.

In addition, different subpopulations of the adult SC compartment appear to reside in different myofiber types. A superfast myosin in cat jaw muscles is expressed only in myofibers regenerated from SCs of the jaw muscle, and not by other muscle fibers (Hoh and Hughes, 1988). Satellite cells from slow myofibers form myotubes *in vitro* that express the slow type I myosin heavy chain, whereas myotubes derived from the SCs of fast myofibers do not (Dusterhoft and Pette, 1993). It is not known whether these fiber type-specific subpopulations of SCs are the remains of distinct subpopulations of myogenic cells giving rise to slow and fast myofibers during embryogenesis, or whether they reflect an influence of the myofiber with which they become associated.

Different muscles also exhibit a different regenerative response to injury that is correlated with differences in their SCs. In rats, the masseter muscle of the jaw does not regenerate as well as the tibialis anterior muscle of the leg. The biochemical properties of these muscles are different and have different embryonic origins. The SCs of the masseter muscle were lower in number and rate of proliferation than the satellite cells of the tibialis anterior (Pavlath et al., 1998). These features are likely the result of intrinsic differences in SCs, but might also reflect differences in extrinsic factors, such as growth factor production by inflammatory cells during regeneration of the two muscles.

## 3. Cellular and Molecular Events of Skeletal Muscle Regeneration

Skeletal muscle regeneration is triggered by weight bearing and exercise, which may be the equivalent of the maintenance regeneration and remodeling that is seen in bone (see ahead) and by injury (Grounds, 1991). It has also been noted in astronauts after space flight, probably due to a return from weightlessness to 1g conditions (Pastoret and Partridge, 1998). Muscle does not regenerate across a gap, and excised segments of muscle are not regenerated. Incisions made through myofibers by small transverse cuts or laser that do not remove tissue and leave the cut ends of the myofibers apposed evoke regeneration by cytoplasmic budding and rejoining of the cut fibers. Most muscle injuries, however, are more extensive and result in tissue death. There are several experimental animal models of such injury, each of which has its advantages (Carlson, 2003). Genetic myopathies, such as muscular dystrophy or administration of myotoxins damage only myofibers, while leaving vascular and neural fibers intact. Relatively localized ischemic injuries can be created by crushing, cutting, freezing, and vascular clamp. Extensive ischemic degeneration can be achieved by removing the muscle and grafting it back to the muscle bed (free graft), or mincing the muscle and grafting the mince back to the muscle bed (Carlson, 1983, 2003; Pastoret and Partridge, 1998). There are two major, overlapping phases in the regeneration of a free-grafted muscle, a phase of myofiber breakdown and inflammation, and a developmental phase in which SCs partially recapitulate embryonic muscle development. These phases have been well-described in the regeneration of the free-grafted extensor digitorum longus muscle, in both amphibians and mammals (Hansen-Smith and Carlson, 1979; Carlson, 2003). **FIGURE 9.4** illustrates the phases of regeneration of a free-grafted muscle.

### a. Myofiber Breakdown and Inflammation

The first event to occur after free-grafting a rat muscle is a wave of necrosis that sweeps from the periphery of the graft to the center over a 7-day period. Myofibers are destroyed by complement activation and by calcium influx that activates calcium-dependent neutral proteases (Pastoret and Partridge, 1998). As the wave of necrosis sweeps through the muscle, a typical inflammatory response follows, in which soluble chemoattractants attract neutrophils and macrophages into the degenerating muscle (Grounds and Yablonka-Reuvini, 1993; Pastoret and Partridge, 1998). The macrophages and neutrophils kill bacteria and remove the cellular debris resulting from myofiber breakdown.

Zymographic studies indicate that MMP-9 and MMP-2 are upregulated during the inflammatory phase. *In situ* hybridization showed that the mRNA for MMP-9 is localized to satellite cells (Kherif et al., 1999). MMP-9 breaks down ECM components of degenerating myofibers for digestion by macrophages. This degradation may also serve to release ECM-bound growth factors that play a role in SC proliferation. Proliferation and fusion of SCs into new myofibers takes place within the basement membranes and endomysial sheaths of the degenerated myofibers, much like axon regeneration takes place within endoneurial tubes. The basement

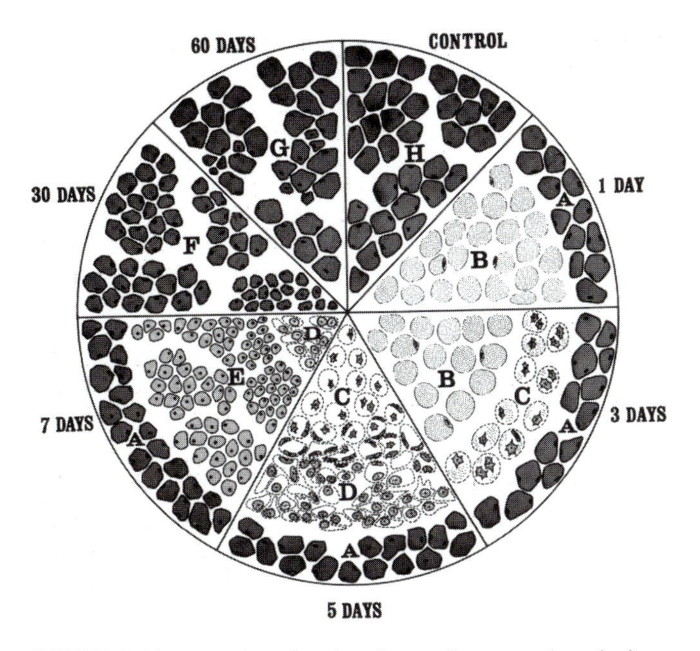


**FIGURE 9.4** Diagram of overlapping phases of regeneration of a free-grafted rat gastrocnemius muscle. Myofibers at the periphery of the grafted muscle (Zone A) survive ischemia. The majority of the interior myofibers degenerate (Zone B), but their satellite cells survive (small dark ovals). By three days, macrophages begin to ingest the debris from the necrotic myofibers (Zone C), leaving only the proliferating satellite cells inside the basement membranes. By the fifth day, the satellite cells begin fusing to form new myofibers with the old basement membranes (Zone D). These changes occur in centripetal waves. By 7 days, various stages in the differentiation of the new myofibers are visible, with a centripetal gradient of differentiation (Zone E). Reproduced with permission from Hansen and Carlson, Cellular responses to free grafting of the extensor digitorum longus muscle of the rat. J Neurol Sci 41:149–173. Copyright 1979, Elsevier.

membranes of the myofibers are "tattered" due to a partial breakdown that allows detachment of the SCs from the membrane (Gulati et al., 1983). MMP-2, which degrades the basement membrane components Fn, Ln, heparan sulfate proteoglycan (HSPG), and collagen IV, may be involved in this breakdown. Cultured C2C12 myoblasts and normal skeletal muscle synthesize MMP-2, the expression of which is upregulated during the period of maximal SC proliferation and fusion into myotubes (Kherif et al., 1999). As the new myofibers differentiate, continuity of their basement membranes is restored by fibroblasts of the endomysial sheaths (Hansen-Smith and Carlson, 1979).

Carlson (1978) has reviewed the histochemical and biochemical data on muscle regeneration. The activity of some enzymes involved in the generation of ATP through glycolysis (phosphofructokinase, α-glycerophosphate) and glycogen metabolism (phosphorylase) falls to very low values soon after muscle transplantation, and remains lower than normal throughout muscle regeneration. The activities of other glycolytic enzymes such as pyruvate kinase and lactate dehydrogenase, and the activities of adenylate kinase and creatine kinase show a less precipitous decline, and then rise and approach normal levels as regeneration proceeds. The glycolytic enzyme hexokinase and glucose-6-phosphate dehydrogenase, the first enzyme in the pentose phosphate pathway, exhibits activities much higher than normal during the first few days after muscle transplantation (Wagner et al., 1976; 1977). In the pentose phosphate pathway, glucose-6-phosphate is oxidized to ribose-5-phosphate, generating NADPH, which provides the energy for reductive biosynthesis. Ribose-5-phosphate provides the starting point for synthesis of ATP, CoA, FAD, and the units of the sugar-phosphate chains of RNA and DNA. Collectively, these data suggest a shift from aerobic to anaerobic generation of ATP and biosynthetic activities during the early phases of regeneration, and back again, consistent with the initial degeneration of the transplanted muscle, followed by proliferation, revascularization, and differentiation of SCs (Wagner et al., 1977).

### b. Developmental Phase

Myoblast determination, proliferation, and differentiation during ontogenesis are regulated by the bHLH muscle regulatory factors (MRFs) Myf5, myogenin, and MyoD. These factors are activated first in the somite to determine the myogenic fate of myotome cells (**FIGURE 9.5**). The myotome has a lateral hypaxial component that produces the muscles of the body wall and limbs, and a medial epaxial component that produces the back muscles. In the hypaxial myotome, *Wnt 7a* from the dorsal ectoderm activates the expression of *Pax3*, the counterpart of the adult *Pax7* of SCs, and *MyoD* (Hinterberger et al., 1991; Tajbakhsh et al., 1997, 1998; Zhao and Hoffman, 2004). Wnt 1, 3a, and Shh are expressed in the neural tube and in the notochord and floor plate of the neural tube, respectively, and activate *Myf5* in the epaxial myotome. *MyoD* is not activated in splotch (*Pax3/Myf5* double-mutant) mice, suggesting that *Pax3* acts to turn on *MyoD* in the hyapxial myotome and *Myf5* acts to turn on *MyoD* in the epaxial myotome. BMP 4 from the lateral plate mesoderm promotes the expression of *Pax3* (Amthor et al., 2002). In general, cells lacking either *Myf5* or *MyoD* can become muscle cells, indicating redundancy of function, but if both are lacking, cells cannot form muscle.

These transcription factors continue to be expressed throughout the proliferation and migration of myoblasts into their definitive positions within the musculoskeletal system (Zhao and Hoffman, 2004). Fusion of myoblasts into multinucleated myotubes requires the expression of myogenin. In addition, a number of

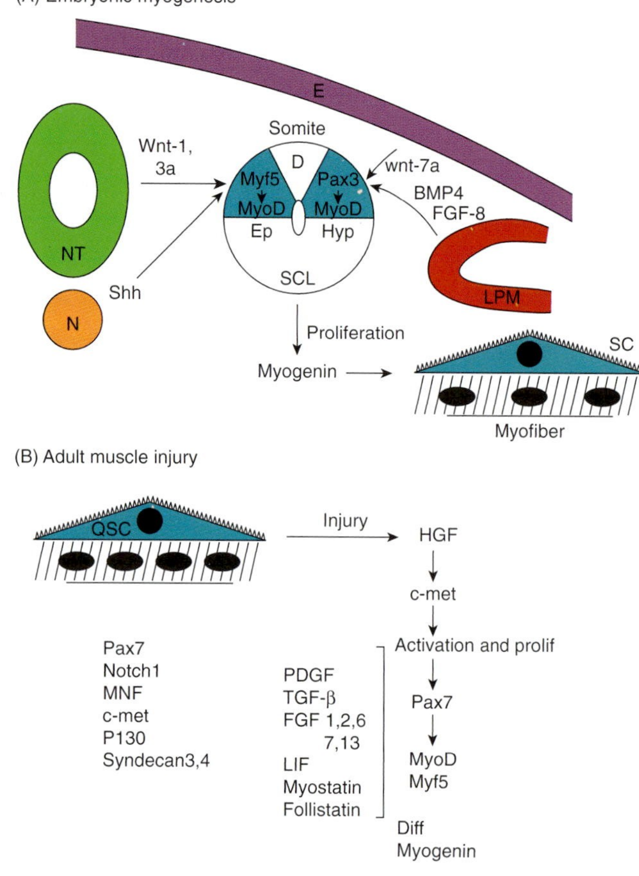


**FIGURE 9.5** Diagram showing molecular signaling pathways during embryonic myogenesis vs muscle regeneration. **(A)** Embryonic myogenesis. The somite has four subdivisions, the dermatome (D), the sclerotome (SCL), the epaxial myotome (Ep) that gives rise to the back muscles and the hypaxial myotome (Hyp) that gives rise to the appendicular muscles. Wnt-1 and 3a signals from the neural tube (NT) and Shh from the notochord (N) and floor plate of the neural tube elicit the expression of myf5, which activates MyoD to determine the epaxial myoblasts. Wnt 7a from the ectoderm (E), along with BMP-4 and FGF-8 from the lateral plate mesoderm (LPM), elicits expression of Pax3, which activates MyoD to determine the hypaxial myoblasts. After proliferating, the myoblasts in either case express myogenin, which leads to fusion and expression of myofiber proteins. Some myoblasts remain undifferentiated as satellite cells (SC). **(B)** Muscle regeneration. The quiescent satellite cell (QSC) expresses Pax 7, Notch 1, myocyte nuclear factor (MNF), the c-met receptor, the p130 protein, and syndecans 3 and 4. Injury releases HGF from the ECM. HGF binds to c-met and activates the expression of Pax7, which leads to the expresson of MyoD and myf5 and the proliferation of the satellite cells under the modulating influence of a variety of other growth factors. As in embryonic myogenesis, fusing and differentiating myoblasts express myogenin.

growth factors are involved in the proliferation and differentiation of embryonic skeletal muscle. The most important of these appears to be FGF8, which binds to FGFR4 (Marics et al., 2002).

There are both similarities and differences in the developmental programs of skeketal myogenesis during ontogenesis and during regeneration in adult muscles (**FIGURE 9.5**). Pax7 is strongly upregulated after injury in SCs, consistent with the fact that SCs are determined already as myogenic cells (Zhao and Hoffman, 2004). Myf5 may be expressed in quiescent SCs of adult rats at a very low level (Beauchamp et al., 2000), but SCs do not express any of the other bHLH muscle factors active in the determination of myotome cells (Koishi et al., 1995; Megeny et al., 1996). The expression of all the MRFs is upregulated by SCs *in vitro* or after injury *in vivo* (Smith et al., 1994; Launay et al., 2001; Zhao and Hoffman, 2004), consistent with their roles in proliferation and differentiation. However, the sequence of bHLH factor expression is different from the sequence seen in embryonic development, at least *in vitro* (Smith et al., 1994). MyoD was expressed first, at 12 hr after plating. Upregulation of myf5 and MRF 4 expression began at 48 hr and was correlated with the first division cycle. Myogenin expression was detected sporadically at 48 hr, and in all cultures at 72 hr, coincident with the ability to immunostain cultures for sarcomeric myosin heavy chain and 24 hr before myoblast fusion, consistent with the role ascribed to myogenin in myogenic differentiation. Regeneration was impaired in MyoD-null mice (Megeny et al., 1996).

A slightly different expression pattern of MRFs was observed when their mRNAs were assayed in single activated satellite cells of cultured individual myofibers via single-cell polymerase chain reaction (Cornelison and Wold, 1997). Either MyoD or myf5 were expressed first. Most cells then expressed myf5 and MyoD simultaneously, followed by myogenin expression in cells expressing both myf5 and MyoD. Ultimately, many of the cells expressed myf5, MyoD, and myogenin simultaneously. Thus, there may be a great deal of heterogeneity of expression of MRFs among activated satellite cells that may be masked by the "average" pattern.

The *in vitro* growth rates and fusion characteristics of embryonic myoblasts and SCs are similar. Both exhibit the same increase in creatine kinase activity and shift in creatine kinase isozyme profile as myofiber formation proceeds (Jones, 1982). Like myoblasts of embryonic muscle, SCs of regenerating mouse muscle express M-cadherin (Moore and Walsh, 1993). Normal adult chicken muscle synthesizes only the small heparan sulfate and dermatan sulfate proteoglycans, whereas regenerating muscle reverts to the embryonic pattern of abundant synthesis of large chondroitin sulfate proteoglycans (Carrino et al., 1988).

Embryonic and regenerating myofibers of chicken muscle both synthesize the embryonic myosin heavy chain, the α fast light myosin chain, and β-tropomyosin (Gorza et al., 1983; Matsuda et al., 1983; Dix and Eisenberg, 1991). This is true of myofibers derived from SCs of either fast (pectoralis major) or slow (anterior latissmus dorsi) muscle. Muscle cells also contain the S1 protein, a variant of EF-1α, which is ubiquitous to all cells (Khalyfa et al., 1999). The ratio of EF-1α to S1 is very high in embryonic muscle, but is low in adult muscle. Regenerating muscle reverts to the embryonic ratio. However, increased expression of EF-1α is also observed in cells undergoing apoptosis (Duttaroy et al., 1998); thus, the increase in EF-1α in regenerating muscle may reflect myofiber death as well as the events of regeneration. Tetranectin is expressed by the satellite cells, myotubes, and the stumps of damaged myofibers (Wewer et al., 1998).

The synthesis of light myosin chains differs in fast muscle embryonic versus regenerating myofibers. During embryonic development, fast and slow myofibers synthesize both fast and slow myosin light chains. During regeneration, the SCs of the slow myofibers repeat this pattern, but the SCs of the fast fibers synthesize only fast light chains (Matsuda et al., 1983). Expression of the slow myosin light chain mRNA isoform is nerve-dependent, whereas expression of the fast isoform is not (Esser et al., 1993). The difference in myosin light chain synthesis by regenerating fast and slow fibers probably represents a divergence in the commitment of SCs during development (Matsuda et al., 1983).

Other differences between embryonic and regenerating muscle are the presence of a cell surface glycoprotein marker, OX-2, on regenerating myofibers that is absent during fetal and postnatal myogenesis (Grim and Bukoversusky, 1986) and the prominence of the pentose phosphate pathway in regenerating muscle (Wagner et al., 1977). The use of the pentose phosphate pathway is clearly related to the anaerobic environment in which a transplanted muscle initially begins regeneration.

## 4. The Activation and Proliferation of SCs Are Regulated by a Variety of Signaling Molecules

The Notch pathway is active in keeping SCs in an undifferentiated state. Notch1 expression inhibits myoblast differentiation. Notch1 is strongly upregulated in regenerating mouse muscles (Conboy and Rando, 2002; Zhao and Hoffman, 2004). Numb is asymmetrically expressed in the daughter cells derived from a single SC, driving one into the myogenic differentiation pathway by inhibiting Notch, while the other is renewed as a stem cell (Conboy and Rando, 2002).

Polesskaya et al. (2003) reported that injured muscle produces the Wnt5a, 5b, 7a, and 7b isoforms as early as 24hr post-injury, suggesting that the Wnt signaling pathway is involved in SC activation. By contrast, Zhao and Hoffman (2004) found no differential expression of Wnt5a, b, or 7a in regenerating versus intact muscles over 27 time points after muscle injury. The same was true for elements of the Shh and BMP signaling pathways, with the exception of BMP1, which has not been implicated in embryonic myogenesis. BMP1 was upregulated 2–4-fold from 2–16 days after muscle injury. What role BMP1 might play in muscle regeneration, if any, is not known. These results suggest that the Wnt, Shh, and BMP pathways are not involved in SC activation. These conflicting results should be investigated further, since one or more of the Wnt, Shh, and BMP pathways have been shown to be important for activation of other stem cell types, for example, EpSCs and ISCs, and MSCs in bone development (see ahead).

Several growth factors are involved in the activation and proliferation of SCs (**FIGURE 9.5**). In a search for the factors that activate and promote the proliferation of SCs, Bischoff (1986) found that a fraction of crushed muscle extract induces early entry of SCs into the cell cycle. PDGF-BB, FGF-2, and transferrin are present in crushed muscle extract (Chen et al., 1994), but these factors, nor IGF-1, IGF-2, TGF-β1, or TGF-β2 can activate quiescent SCs (Johnson and Allen, 1995). Several lines of evidence suggest that, like hepatocytes, SC activation and proliferation is induced by HGF. First, HGF is the only growth factor that can mimic the active fraction of extract (Allen et al., 1995). Second, the ability of crushed muscle extract to stimulate the early entry of SCs into the cell cycle is abolished by anti-HGF antibodies (Tatsumi et al., 1998). Third, although HGF is not expressed in uninjured adult myofibers, it is present in the ECM surrounding the muscle fibers, is expressed by proliferating SCs, and is downregulated upon their fusion and differentiation into myotubes (Jennische et al., 1993; Tatsumi et al., 1998). Fourth, the receptor for HGF, c-met, is present in the plasma membrane of quiescent SCs (Allen et al., 1995; Cornelison and Wold, 1997; Tatsumi et al., 1998) and is co-localized with HGF in the activated SCs of regenerating muscles in *mdx* dystrophic mice (Tatsumi et al., 1998). Lastly, SC activation is stimulated by the injection of HGF into the uninjured tibialis anterior muscle of 12-month-old rats (Tatsumi et al., 1998). These observations suggest that injury releases HGF from the ECM of the muscle cells, triggering SC activation and proliferation. As part of the activation process, the SCs themselves express HGF, which acts in an autocrine fashion. It is likely that the HGF bound to muscle ECM is activated in the same way as it is in regenerating liver, by the uPA → plasminogen activator → plasmin cascade (Miyazawa et al., 1996). Presumably, some of the HGF produced by proliferating SCs would become bound to newly synthesized ECM molecules of the regenerating muscle, to be available for release in any subsequent round of regeneration.

Although not activators of SCs, the PDGF, TGF-β, FGF-2, and leukemia inhibitory factor (LIF) found in crushed muscle extract have a modulating or cooperative effect on the proliferative activity of SCs *in vitro* (Johnson and Allen, 1995; Pastoret and Partridge, 1998). In uninjured muscle, these factors are bound to ECM components and their receptors are not expressed by quiescent SCs (DiMario et al., 1989). Following injury, they are released from the ECM and are also produced by macrophages to interact with upregulated receptors on activated SCs (Pastoret and Partridge, 1998). The factors produced by macrophages, however, do not appear to be essential for activation of SCs. Abolishing the inflammatory response in rats by total body irradiation reduces the efficiency of regeneration in muscles injured by incision or crush, but does not abolish regeneration (Robertson et al., 1992). Preventing the proliferation of SCs by heavy local irradiation, however, which does not inhibit the inflammatory response, results in failure of the muscles to regenerate.

Members of the FGF family play an important collaborative role in SC proliferation. High affinity FGF-2 receptors are not present in quiescent SCs, but are upregulated by HGF *in vitro* and *in vivo* following their activation, coincident with the expression of FGF-2 (Garrett and Anderson, 1995). FGF-1, 6, 7, and 13 are also differentially upregulated in regenerating mouse muscle, as is the FGF receptor, FGFR4 (Zhao and Hoffman, 2004). The regenerating muscles of *mdx* dystrophic mice exhibit abnormally high levels of FGF in the endomysium (Anderson et al., 1991), perhaps because of the high levels of macrophage MMPs that are degrading the ECM (Kherif et al., 1999). The muscles of these mice also express LIF and IL-6. LIF binds to ECM components and stimulates the formation of larger myotubes when infused continuously into *mdx* muscle, presumably by increasing the proliferation of activated SCs (Kurek et al., 1996). Thus, it appears that HGF and a variety of FGFs together promote SC proliferation. Conversely, IGF I and II and TGF-β2 suppress SC proliferation and promote differentiation into myofibers.

Heparan sulfate is required for the activation of SCs by HGF or FGF, perhaps by promoting dimerization of their receptors (Rapraeger, 2000). Inhibition of HSPG sulfation by treatment of intact myofibers with chlorate results in delayed proliferation and altered MyoD expression of SCs in injured muscle (Cornelison et al., 2001). This suggests that syndecan 3 and 4 expression by SCs may provide the heparan sulfate required for dimerization of HGF and FGF receptors during SC activation.

Two other proteins secreted by SCs and differentiating myofibers appear to have a role in muscle regeneration. These are the activin-binding protein, follistatin, and a member of the TGF-β growth factor family, myostatin (GDF-8). During embryogenesis, myostatin negatively regulates myoblast proliferation, while follistatin positively regulates myoblast differentiation (Lee and McPherson, 2001). The same appears to be true in regenerating mouse soleus muscle (Armand et al., 2003). Myostatin mRNA progressively accumulates throughout regeneration to slow proliferation and follistatin is most strongly expressed in forming myotubes, promoting differentiation.

## 5. Tension and Innervation Are Requirements for Normal Muscle Regeneration

Muscle atrophies from disuse, but function is not required for myofiber regeneration per se, since regeneration takes place normally in salamander larvae immobilized under continuous anesthesia (Carlson, 1970). The shape of the regenerated muscle is at least partly determined by the surrounding tissues, as shown by experiments in which minced Gelfoam, soaked in supernatants of muscle homogenates, was substituted for the gastrocnemius muscle in rats (Carlson, 1970). Muscle was not regenerated under these conditions, but tendinous connections were made with the Gelfoam, and it was molded into a structure approximating the shape of a regenerated muscle.

Tension is important for the normal orientation of myofibers in regenerating muscle. A whole muscle can regenerate from a remaining short muscle stump in some vertebrates like the chicken and rat. In the rat, elongation of the stump of the gastrocnemius muscle has been correlated with the successful regeneration of a functional Achilles tendon, which connects to the muscle (Carlson, 1970, 1974). The tension exerted on the muscle stump by the tendon is thought to be an important factor in the ability of the muscle to regenerate.

Regenerating myofibers exhibit slow spontaneous contractions toward the end of the first post-transplantation week. Reinnervation of the regenerating muscle begins during the second week post-transplantation, and the speed of contraction subsequently increases until it approaches normal at 30–40 days after transplantation (Carlson and Gutmann, 1972). Denervation of intact muscle induces the proliferation of SCs (Weis et al., 2000). Muscle regeneration proceeds under conditions of denervation, but denervation prevents or retards the full structural and functional differentiation of the regenerated myofibers (Carlson and Gutmann, 1975). The regenerated fibers are small and atrophied

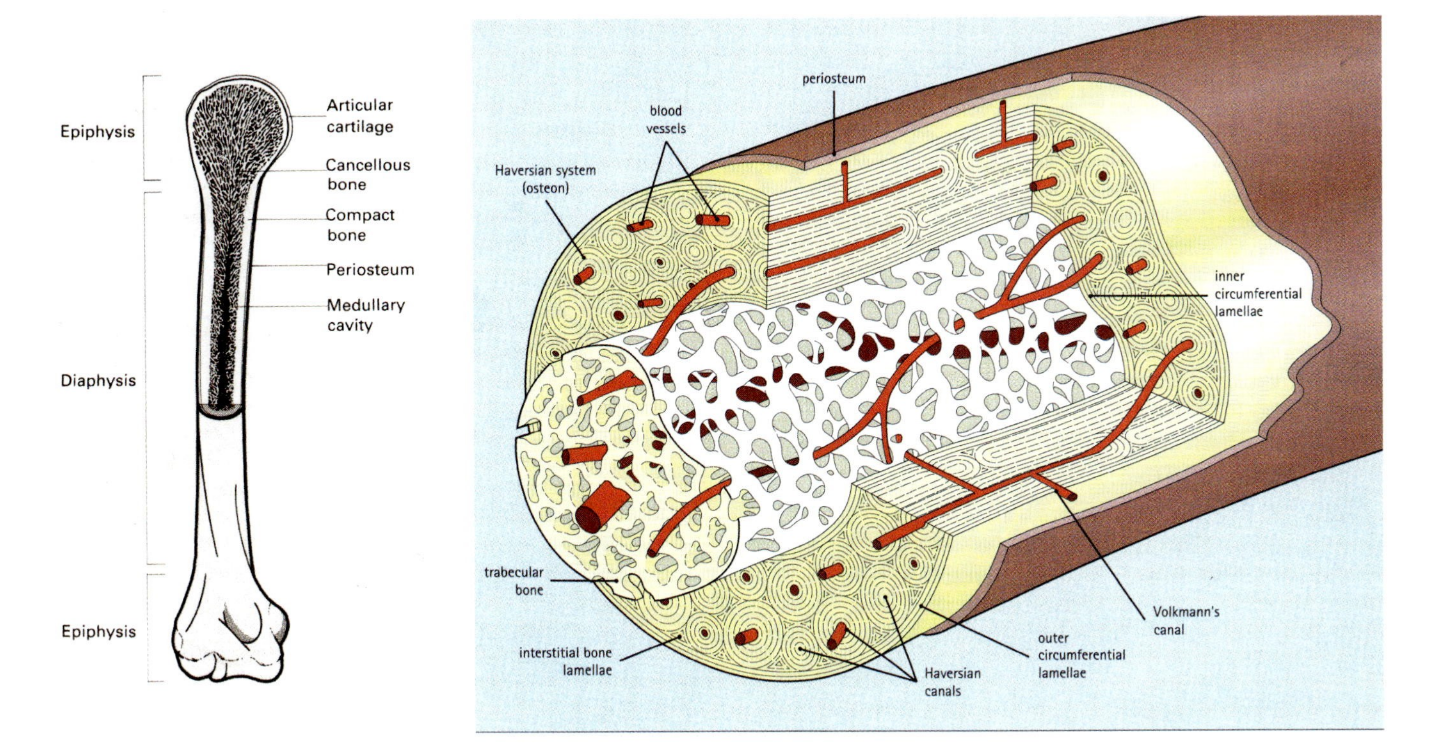


**FIGURE 9.6** Diagrams of endochondral bone structure. **(A)** Cut-away anatomy of a long bone. The bone consists of a shaft, the diaphysis, and two heads, the epiphyses, that are covered with articular cartilage. The medullary (marrow) cavity is lined by cancellous (spongy, or trabecular) bone. Spongy bone is enclosed by compact bone, which is covered with a periosteum. Reproduced with permission from Wheater et al., Wheater's Functional Histology (3rd ed). Copyright 1997, Elsevier. **(B)** Detailed internal structure of the diaphysis. Compact bone is composed of inner and outer circumferential lamellae between which lie Haversian systems, or osteons, consisting of concentric rings of bone around a central Haversian canal. Blood vessels course longitudinally through the Haversian canals and transversely through Volkmann's canals. Between the osteons are interstitial lamellae of bone matrix. Reproduced with permission from Stevens and Lowe, Human Histology (3rd ed), Elsevier/Mosby. Copyright 2005, Elsevier.

(d'Albis, 1988) and differentiation of the specialized intrafusal fibers of the muscle spindles, sensory receptors that function in the stretch reflex and in the regulation of muscle tone, is poor (Rogers, 1982). In *Xenopus*, denervation reduces MRF4 levels in regenerating muscle (Nicolas et al., 2000). In regenerating mouse soleus muscle, denervation upregulates follistatin and downregulates myostatin, suggesting that differentiation may occur prematurely, contributing to the small size of the muscles (Armand et al., 2003). It is likely that the nerves provide survival and proliferation factors to the regenerating muscle that maintain myofiber number and size.

## REGENERATION OF BONE

### 1. Structure and Development of Bone

Bone is composed of osteocytes surrounded by a rigid, highly calcified organic matrix. A long bone is organized into several parts (Baron, 1999). The greater length of the bone is occupied by a cylindrical shaft or diaphysis. On either end of the diaphysis is a flared metaphysis, followed by a disc-shaped epiphysis capped with articular cartilage. **FIGURE 9.6** illustrates bone structure in the diaphyseal region. The external part of the bone cylinder is composed of dense cortical or compact bone, which becomes progressively thinner in the metaphysis and epiphysis. The osteocytes of this bone are embedded in small lacunae. They form layers organized concentrically around blood vessels into units called Haversian systems, or osteons. The osteocytes communicate with one another and with endosteal and periosteal cells by long processes that ramify throughout a network of canaliculi in the bone matrix. Thin trabeculae of bone are formed interior to the compact bone. This trabecular bone has the appearance of a sponge and is often called spongy bone. The spaces within the trabeculae are continuous with the medullary cavity of the diaphysis.

The medullary cavity is lined with an endosteal connective tissue and is filled with bone marrow. The endosteum consists of fibroblasts, mesenchymal stem cells (MSCs), pre-osteoblasts and osteoblasts, while the bone marrow consists of a mixture of MSCs, fibroblasts, adipocytes, macrophages and endothelial cells of sinusoids. These cells, along with the endosteal connective tissue layers, constitute the stroma of the bone marrow. Embedded in the stroma, and dependent on it for survival, are hematopoietic stem cells (HSCs) and endothelial stem cells (EnSCs). The outer surface of the bone is covered with another connective tissue sheath, the periosteum. The periosteum and linings of the Haversian canals also contain MSCs, pre-osteoblasts and osteoblasts. Ninety percent of the organic material in the bone matrix is collagen I, with the remaining 10% consisting of various glycoproteins and proteoglycans. Interspersed throughout the organic matrix are crystals of hydroxyapatite $[3Ca_3(PO_4)_2]\cdot(OH)_2$, which give the bone matrix its stiffness.

Flat bones, such as the bones of the skull, develop during embryogenesis by the direct differentiation of MSCs into osteoblasts, a process called intramembranous bone formation. Long bones, however, exhibit endochondral development; i.e., a calcified cartilage template of the bone is formed first, which is then replaced by bone (Olsen, 1999). The developmental stages of an endochondral bone are illustrated in **FIGURE 9.7**. Mesenchymal stem cells condense and differentiate into chondrocytes, starting at the center of the condensation. Chondrocyte differentiation spreads toward both ends of the bone, forming a cartilage template. Periosteal cells encasing the template differentiate into osteoblasts that form a shell of bone around the template that also advances toward the ends of the bone. As this process continues, the chondrocytes of the template hypertrophy and die, releasing angiogenic signals that trigger the sprouting of capillaries in the periosteum, while osteoclasts excavate much of the calcified matrix (Alini et al., 1996; Carlevaro et al., 1997). The periosteal capillaries invade the matrix, accompanied by perivascular MSCs. Some of these MSCs differentiate into osteoblasts that replace the cartilage matrix with bone matrix, others form the endosteum and stroma of the bone marrow, and still others remain within the endosteum and stroma as MSCs. The epiphyseal ends of the bone are the last to undergo cartilage differentation and ossify. In juvenile animals, a cartilaginous growth plate is maintained between the metaphysis and the epiphysis that maintains growth of the bone until it is replaced by bone at the end of the growth period. The growth plate is illustrated in **FIGURE 9.7**.

## 2. Maintenance Regeneration of Bone

Bone is a highly dynamic tissue that functions in several ways, including supporting and protecting soft tissues, serving as levers for muscle action, and supporting blood cell regeneration. One of its most important functions is to act as a storehouse that can sequester and release calcium to maintain normal blood calcium levels. These functions require that bone be continuously degraded and regenerated throughout life. Ten percent of the skeletal bone mass is replaced every year in adult vertebrates, so the equivalent of the whole skeleton is replaced each decade (Alliston and Derynck, 2002). At any given time, this process is taking place at about two million microscopic sites throughout the skeleton (Harada and Rodan, 2003). At these sites, bone is removed by large multinucleated cells called osteoclasts and is regenerated directly by osteoblasts, the bone-matrix forming cells. This maintenance regeneration is often referred to as bone remodeling and is illustrated in **FIGURE 9.8**. As outlined ahead, both bone resorption and bone regeneration are driven primarily by systemic and local factors acting through the osteoblast.

Biomechanical stress (bending of bone, or strain) is an osteogenic stimulus that leads to increased bone formation (Burr et al., 2002). Available data suggest that deformation of bone by bending strains causes fluid within the spaces surrounding osteocytes and canicular processes of the bone to flow from more concave to more convex bone surfaces, as originally proposed by Frost (1964). This triggers an osteogenic response by periosteal osteoblasts, although how these mechanical signals are converted to cellular signals is still unknown. The osteogenic response is maximized, at least in rats, by short periods of high-rate loading alternating with 4–8-hr rest periods (Burr et al., 2002).

Imbalances in the rate of bone removal and replacement lead to skeletal abnormalities. When the rate of removal exceeds the rate of replacement, the result is low bone mass, or osteoporosis. When the rate of replacement exceeds the rate of removal, the result is high bone mass, or osteopetrosis. There are many genetic abnormalities associated with syndromes and disorders in the development of the bone remodeling system (Ducy et al., 2000; Teitelbaum, 2000; Harada and Rodan, 2003; Boyle et al., 2003; Zelzer and Olsen, 2003; Hofbauer et al., 2004; Whyte and Mumm, 2004).

### a. Osteoclast Origin and Function

Osteoclasts are giant cells with 4–20 nuclei, derived by the fusion of macrophages, and differentiated for the specialized function of bone matrix removal (Teitelbaum, 2000; Harada and Rodan, 2003) (**FIGURE 9.9**). Osteoclasts form on the connective tissue surfaces that line the bone: periosteum and endosteum. When stimulated by agents such as parathyroid hormone (PTH),

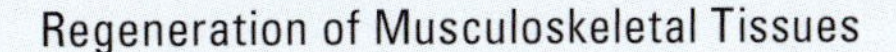

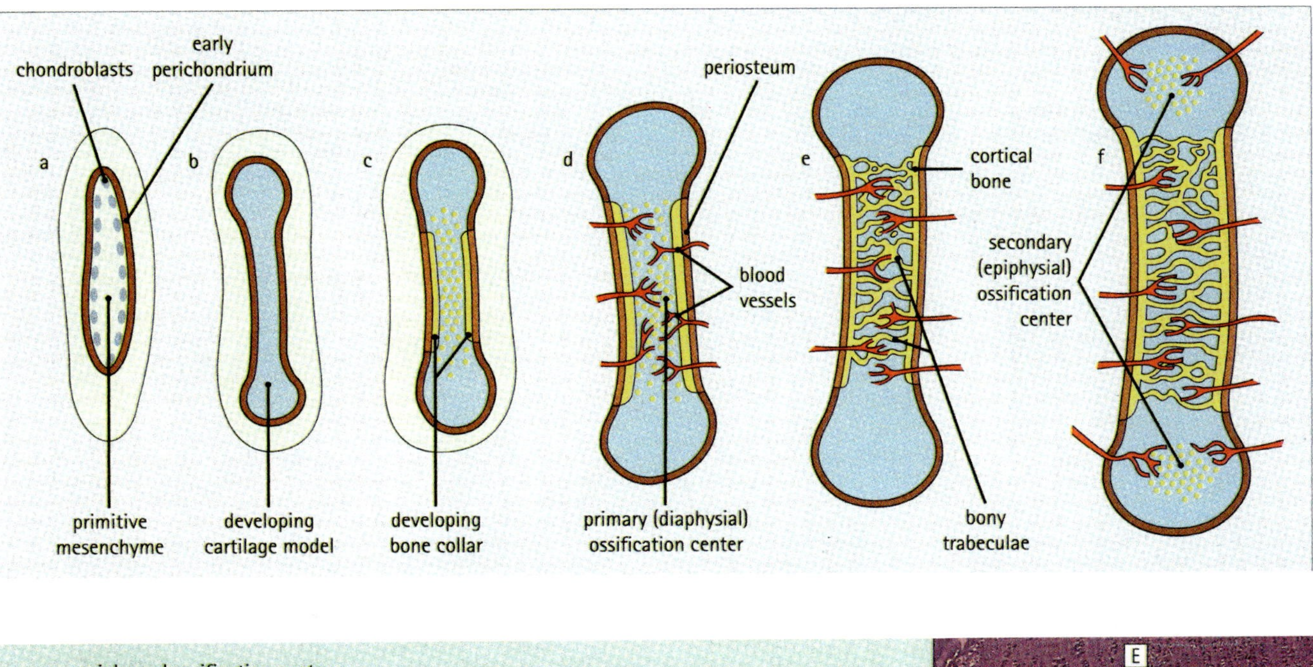


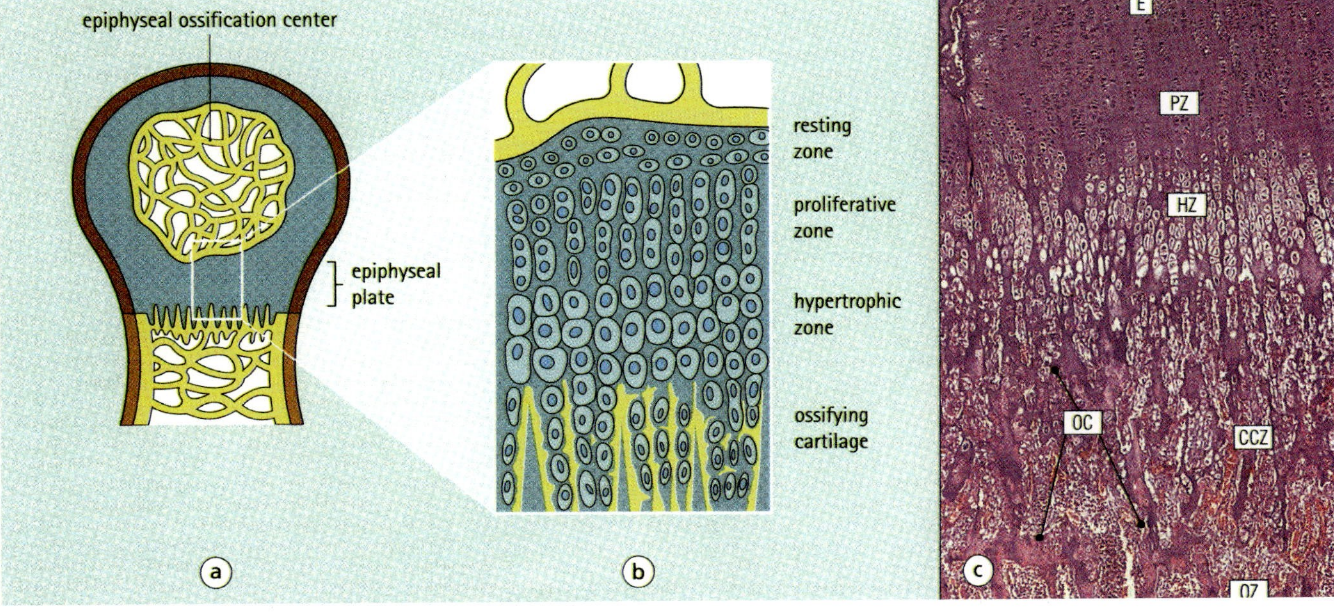


**FIGURE 9.7** Endochondral bone development. **Top: (a)** Mesenchymal cells condense and differentiate as chondroblasts surrounded by an early perichondrium, which becomes the periosteum. **(b, c)** A cartilage template is formed that develops from the center of the bone to the ends. A periosteal shell of compact bone (yellow) is formed around the cartilage template by the periosteum. **(d, e, f)** The chondrocytes of the template hypertrophy, the cartilage matrix calcifies and the periosteal bone shell and calcified cartilage matrix are penetrated by blood vessels with associated mesenchymal and hematopoietic stem cells that differentiate into osteoblasts that form trabecular bone and stroma of the marrow. In late stages of bone development, the epiphyses go through the same process of forming a cartilage template that is then replaced by trabecular bone. **Bottom:** Development of the growth plate between the metaphysis and diaphysis. **(a, b)** Ossifying cartilage of the epiphysis and diaphysis with growth plate (epiphyseal plate) between them. The layers of chondrocytes in the growth plate reflect the ongoing process of cartilage replacement by bone as the bone grows. The youngest chondrocytes are in the resting zone and below them are proliferating chondrocytes that feed the hypertrophic zone. Below the hypertrophic zone, osteoblasts are taking the place of chondrocytes. **(c)** Low power micrograph of histological section through the decalcified growth plate. E = resting chondrocytes, PZ = proliferation zone, HZ = hypertrophic zone, CCZ = zone of calcified cartilage with ossifying plates of cartilage (OC) extending from the ossification zone (OZ). Reproduced with permission from Stevens and Lowe, Human Histology (3rd ed), Elsevier/Mosby. Copyright 2005, Elsevier.

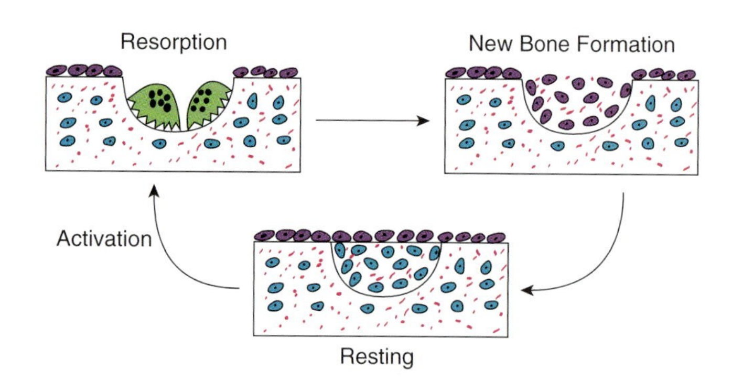


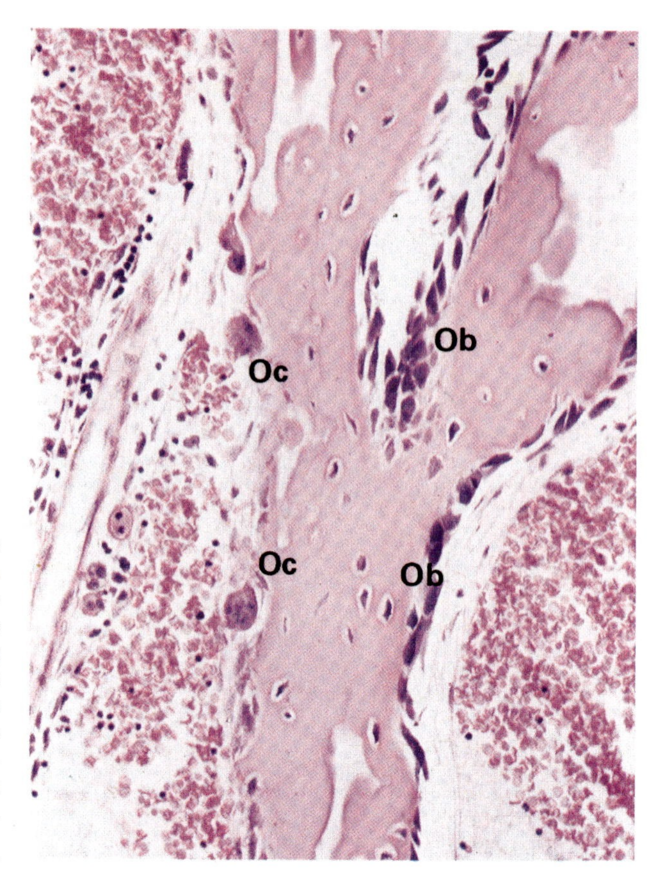


**FIGURE 9.8** Bone remodeling. **Left,** bone remodeling cycle. Osteoclasts (green cells) are activated to resorb the bone, creating a resorption lacuna. The osteoclasts then disappear and MSCs (purple cells) enter the resorption lacuna and become osteoblasts, which secrete new bone matrix. Once filled with new bone (blue cells), the area of remodeling enters a resting phase before the process is repeated. The initial matrix deposited by the osteoblasts is organic matrix, or osteoid, followed by the deposition of the mineral component of the matrix. **Right,** H&E-stained section of fetal bone showing erosion of bone by osteoclasts (Oc) in one location and simultaneousbone regeneration in another by osteoblasts (Ob). Reproduced with permission from Wheater et al., Wheater's Functional Histology (3rd ed). Copyright 1997, Elsevier.

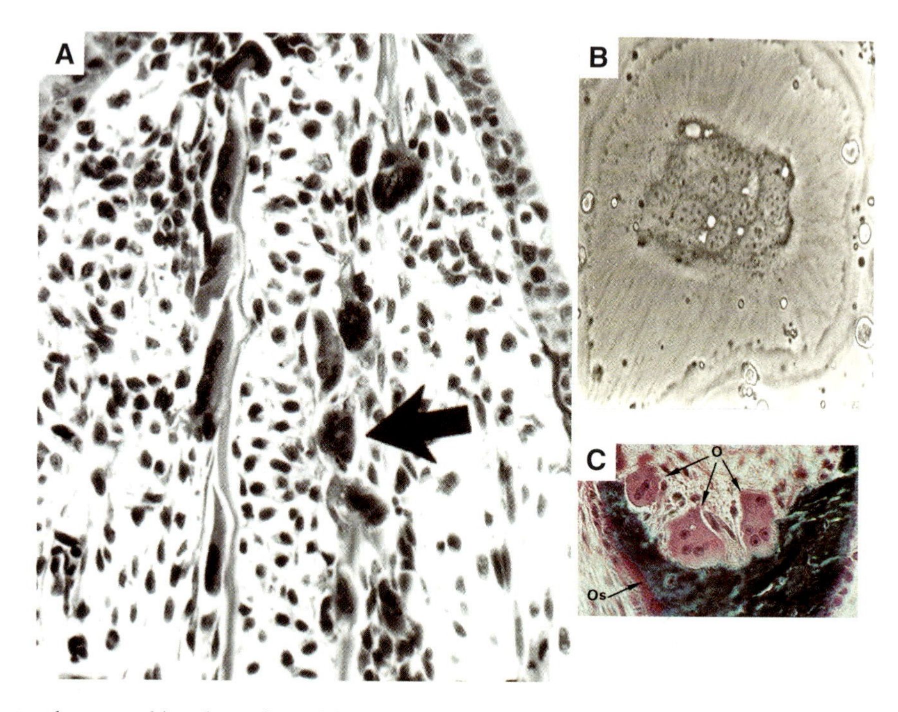


**FIGURE 9.9** **(A)** Multiple osteoclasts resorbing the periosteal bone shell of the humerus in an amputated larval salamander limb (see Chapter 14). Arrow points to one of the osteoclasts. **(B)** Living osteoclast that migrated out of an explanted blastema of a regenerating larval salamander limb. The osteoclast has approximately 14 nuclei. **(C)** Section through bone of a woman with low calcium level and excessive bone resorption. O = osteoclasts, Os = osteoid of newly deposited bone matrix. **A, B,** Photos by D. L. Stocum; **C,** reproduced with permission from Wheater et al., Wheater's Functional Histology (3rd ed). Copyright 1997, Elsevier.

the osteoblasts in these tissues produce factors that induce the differentiation of macrophages into osteoclasts (Teitelbaum, 2000; Boyle et al., 2003; Martin, 2004).

Two factors that are necessary and sufficient to induce osteoclastogenesis are macrophage colony-stimulating factor (M-CSF) and the ligand of receptor for activation of nuclear factor kappa B (RANKL) (**FIGURE 9.10**). M-CSF binds to its receptor, c-Fms, on macrophages. RANKL is a stromal cell surface molecule that binds to its receptor (RANK) on the macrophage cell surface; thus, macrophage differentiation into osteoclasts requires contact between macrophages and stromal cells. RANKL and RANK are members of the TNF and TNF receptor superfamilies of growth factors, respectively. Osteoblasts make another soluble protein, osteoprotegerin (OPG) that competes with RANK for RANKL and inhibits macrophage differentiation into osteoclasts. Osteoclast differentiation is thus regulated by a balance between the concentrations of M-CSF and RANKL versus OPG (Karsenty, 2003; Martin, 2004). The transcription factor, c-Fos, is a key intracellular effector that drives the expression of the osteoclast molecular phenotype (Grigoriadis et al., 1994).

Osteoclast differentiation is also regulated by a negative feedback mechanism activated within the osteoclasts themselves (Takayanagi et al., 2002; Alliston and Derynck, 2002; Boyle et al., 2003). One of the genes activated by c-Fos is the interferon-β gene. Takayanagi et al. (2002) proposed that interferon-β is released from the osteoclasts, binds to its receptors on the cell surface, and downregulates c-Fos, bringing osteoclast differentiation to a halt. They provided several kinds of experimental evidence that support this model. First, mice lacking either interferon-β expression or the interferon-β receptor have low bone mass and exhibit increased bone resorption, suggesting that interferon-β is required to inhibit osteoclast differentiation. Second, in mice lacking interferon-β expression, bone loss could be reversed with exogenous interferon-β. Third, exogenous RANKL induced interferon-β expression in cultured macrophages. Fourth, by adding exogenous interferon-β to the cultures, the macrophages were prevented from differentiating into osteoclasts.

Activated osteoclasts cause the stromal cells on which they lie to retract, exposing the bone matrix (Mundy, 1999). The osteoclasts resorb the matrix by first demineralizing it and then degrading its organic components (Teitlebaum, 2000) (**FIGURE 9.10**). The osteoclast becomes polarized, forming a ruffled membrane on one side. A "sealing ring" around the ruffled membrane attaches the osteoclast to the bone matrix. Through ion transport pumps, HCl is secreted into the space circumscribed by the sealing ring to achieve a pH of ~4.5, which dissolves the hydroxyapatite from the matrix. Subsequently, the organic components are solubilized by a lysosomal protease, cathepsin K, and MMPs to form a "resorption lacuna" in the bone. The osteoclasts then disappear by apoptosis and the resorption lacuna is occupied by osteoblasts that synthesize new bone matrix (Mundy, 1999). How osteoclasts recognize the places where bone is to be resorbed is unknown, as are the factors that regulate their apoptosis and the subsequent appearance of osteoblasts on the scene. As Teitlebaum (2000) has stated, "The factors that orchestrate the sequential appearance of osteoclasts and osteoblasts at bone remodeling sites are one of the great enigmas of bone biology."

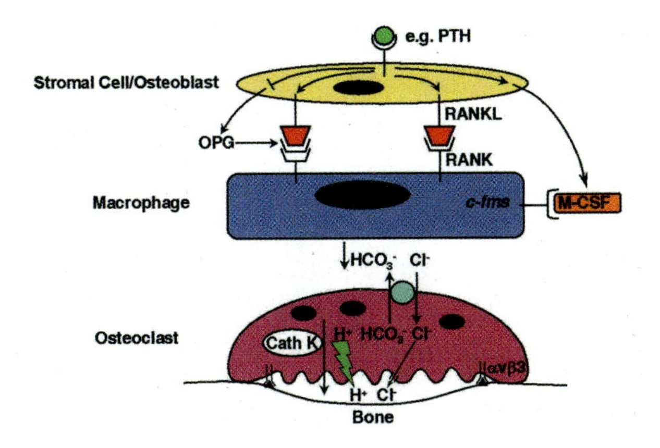


**FIGURE 9.10** Factors mediating the differentiation of macrophages into osteoclasts. Signals such as parathyroid hormone (PTH) upregulate RANKL and M-CSF, while depressing the expression of the RANKL decoy molecule, OPG. RANKL binds to its receptor on the macrophage, RANK, and M-CSF to its receptor, *c-fms*. Reproduced with permission from Teitelbaum, Bone resorption by osteoclasts. Science 289:1504–1508. Copyright 2000, AAAS.

### b. Osteoblast Origin and Function

Osteoblasts are postmitotic cells derived from MSCs of the endosteum, periosteum, and bone marrow (**FIGURE 9.11**). The existence of MSCs in adult bone marrow was first indicated by the ability of marrow cells to form bone when transplanted to ectopic sites (Urist, 1965; Urist and McLean, 1962; Tavassoli and Crosby, 1968). *In vitro*, MSCs are distinguished from hematopoietic stem cells (HSCs) of the marrow by the fact that they are adherent to culture dishes, whereas HSCs are not. Clonally derived MSCs can differentiate into bone, cartilage, and adipocytes *in vitro* or when implanted subcutaneously in diffusion chambers or ceramic blocks (Friedenstein et al., 1970; Owen, 1987; Lian et al., 1999; Pittinger et al., 1999). Human MSCs are positive for the antigens SH2, SH3, CD29, CD44, CD71, CD90, CD106, CD120a, and CD124, negative for the hematopoietic markers CD14, CD344, and

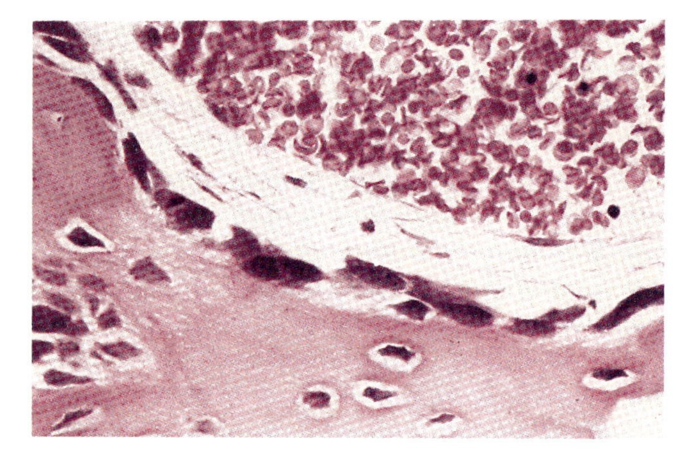

FIGURE 9.11 Higher magnification of section showing osteoblasts laying down osteoid (organic bone matrix). The high rate of protein and proteoglycan synthesis of these cells is reflected in their large size and abundant basophilic cytoplasm. Reproduced with permission from Wheater et al., Wheater's Functional Histology (3rd ed). Copyright 1997, Elsevier.

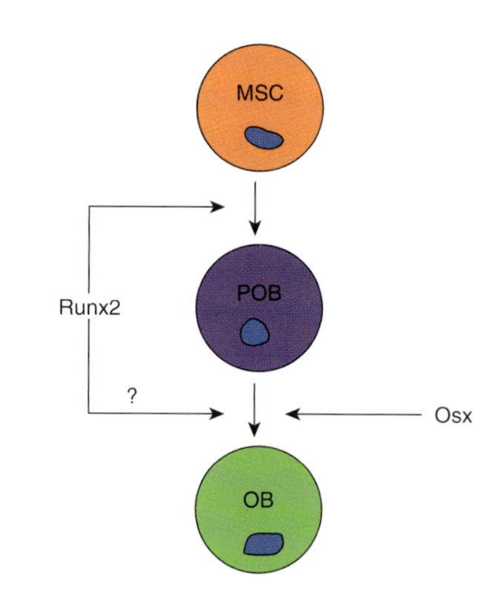


FIGURE 9.12 Diagram illustrating the differentiation of a MSC to a collagen I pre-osteoblast (POB) progenitor under the influence of Runx2 and the further step to a functional osteoblast (OB) expressing collagen I, bone sialoprotein, and osteocalcin, mediated by Osx and perhaps also Runx2.

CD45 and negative for endothelial surface antigens (P-selectin and von Willebrand factor VIII; Pittinger et al., 1999; Pittinger and Marshak, 2001). FAC-sorted bone marrow cells expressing the cell surface antigen STRO-1 have been isolated that differentiate into bone, cartilage, and adipocytes; this antigen may uniquely identify MSCs (Aubin and Liu, 1996). MSCs in micromass pellet culture behave like mesenchymal cells of the limb bud in micromass culture, differentiating as cartilage in high-glucose medium (Mackay et al., 1998) and expressing the BMP-1B receptor (Chen et al., 1998).

Two transcription factors that play key roles in the differentiaton of MSCs into osteoblasts are Runx2 (Cbfa1), which is specific for the chondrocyte/osteoblast lineage, and Osterix (Osx), which is specific for pre-osteoblasts (Ducy et al., 2000; Harada and Rodan, 2003). Runx2 is expressed in the mesenchymal cell condensations of the embryonic endochondral skeleton and induces an osteoblast-specific pattern of gene expression. Both Runx2 and Osx are expressed in the periosteum, indicating that the periosteum harbors MSCs equivalent to those forming cartilage condensations in the embryo (Mundlos and Olsen, 1997; Nakashima et al., 2002; Harada and Rodan, 2003). Together, Runx2 and Osx specify osteoblast differentiation (FIGURE 9.12) with Osx appearing to act downstream of Runx2 (Harada and Rodan, 2003). Runx2 also appears to control the rate of bone matrix formation by differentiated osteoblasts. One of the downstream targets of Runx2 is the osteocalcin (*Bgp*) gene, which is activated only in differentiated osteoblasts and is a negative regulator of matrix production (Ducy et al., 2000). Mice with mutations that affect the *Runx2* and *Osx* genes have a perfectly patterned cartilaginous skeleton, but lack osteoblasts and exhibit deficiencies in ossification.

## c. Systemic and Local Control of Bone Remodeling

Bone resorption and regeneration is under both endocrine (systemic) and growth factor (local) control. Most of the factors regulating bone mass act on osteoblasts to either directly affect osteoblast differentiation, or regulate the production of M-CSF and RANKL by osteoblasts, thus indirectly affecting the differentiation of osteoclasts.

### Systemic Control

Circulating hormones have regulatory effects on both osteoclasts and osteoblasts (FIGURE 9.13). Continuous exposure to PTH, parathyroid hormone-related protein (PTHrP), and low doses of 1, 25-dihydroxyvitamin $D_3$ stimulates stromal cells, including osteoblasts, to express M-CSF and RANKL, leading to osteoclast production and increased bone resorption (Teitelbaum, 2000; Erben, 2001). Thyroid hormone (T3) and glucocorticoids also stimulate bone resorption (Canalis, 1988; Mundy, 1999). Glucocorticoids inhibit calcium resorption from the gut, stimulating production of PTH from the parathyroid gland (Mundy, 1999). T3 acts through thyroid hormone receptors on osteoblasts to upregulate their production of RANKL, thus stimulating osteoclast differentiation. Thyroid stimulating hormone (TSH) also has a role in bone

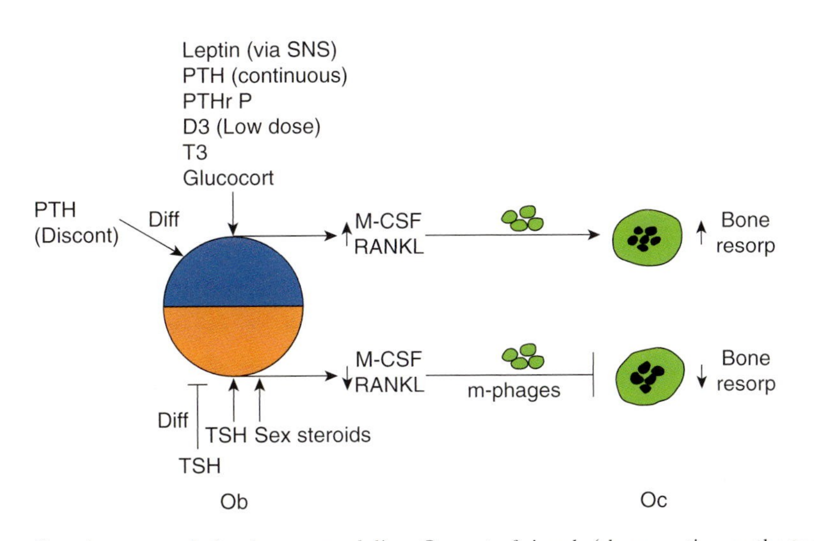


**FIGURE 9.13** Systemic (hormonal) pathways regulating bone remodeling. One set of signals (shown acting on the top half of the osteoblast, Ob) stimulates osteoclast (Oc) differentiation from macrophages (m-phages), and thus increased bone resorption, by increased output of RANKL and M-CSF. Discontinuous doses of parathyroid hormone (PTH) increase production of RANKL and M-CSF by stimulating the differentiation of more osteoblasts, whereas the other signals act to directly increase the output of M-CSF and RANKL by osteoblasts. Leptin upregulates RANKL production by osteoblasts via the hypothalamus and sympathetic nervous system (SNS). The other set of signals (shown acting on the lower half of the osteoblast) leads to the inhibition of osteoclast differentiation from macrophages and decreased bone resorption. Thyroid stimulating hormone (TSH) inhibits osteoblast differentiation, thus indirectly down-regulating M-CSF and RANKL concentration. At the same time TSH and sex steroids directly down-regulate the production of M-CSF and RANKL.

remodeling. In normal cells, TSH inhibits the differentiation of osteoblasts and osteoclasts, both of which express TSH receptors (TSHRs) on their surfaces. The inhibition of osteoclast differentiation is via negative regulation of RANKL and TNFα in osteoblasts. Osteoblast differentiation is inhibited via the suppression of LRP-5 expression in the Wnt pathway (see ahead). In TSHR-null mice, RANKL and TNFα, as well as LRP-5, are overexpressed in osteoblasts, leading to increased osteoblast and osteoclast differentiation simultaneously. However, the rate of osteoclast formation, and thus bone resorption, outpaces the rate of osteoblast formation and bone regeneration, and TSHR-null mice have severe osteoporosis (Abe et al., 2003).

Sex steroids, such as estradiol and testosterone, inhibit osteoclast differentiation by reducing expression of RANKL or increasing the expression of OPG by osteoblasts, thus reducing resorption (Mundy, 1999; Ducy et al., 2000; Teitelbaum, 2000; Boyle et al., 2003). The osteoporosis experienced by aging men and women is due to lower production of sex steroids. The onset of osteoporosis is earlier in women because of menopause, and the extent of bone loss is greater at first, but bone mass becomes more equal in men and women of the same age as time goes on (Harada and Rodan, 2003; Zelzer and Olsen, 2003). PTH administered intermittently stimulates bone regeneration by inducing precursor cells to differentiate as osteoblasts (Hock, 2001) and vitamin $D_3$ does the same at high doses, by increasing the number of osteoblast precursors (Erben, 2001). Insulin enhances osteoblast function by stimulating amino acid transport, the synthesis of RNA, collagen and noncollagen protein synthesis, and proteoglycan synthesis.

An important systemic regulator of bone mass is leptin, which functions through an inhibitory action of the hypothalamus on bone formation (Ducy et al., 2000; Harada and Rodan, 2003). Leptin is made by adipocytes and functions to suppress appetite and inhibit bone formation by binding to receptors in the hypothalamus. Mice and humans deficient in leptin or its hypothalamic receptor are obese, but have a higher than normal bone mass. Infusion of leptin into the cerebral ventricles of *ob/ob* mice rescues them from obesity and restores normal bone mass (Ducy et al., 2000).

Leptin does not affect osteoblasts directly, because no transcripts for leptin receptors can be detected in osteoblasts (Ducy, 2000). The effector pathway from hypothalamus to bone is the sympathetic nervous system (Elmquist and Strewler, 2005). Bone is heavily innervated by sensory and sympathetic fibers that directly contact bone cells, and a variety of neuromediators and their receptors have been detected in bone (Chenu, 2004). Sympathetic neurons produce noradrenaline, which binds to β2-adrenergic receptors (β2-AR) on osteoblasts (Takeda et al., 2002; Harada and Rodan, 2003; Chenu, 2004). Mutant mice lacking β2-AR have increased bone mass like *ob/ob* mice, but unlike *ob/ob* mice, they do not respond to leptin by

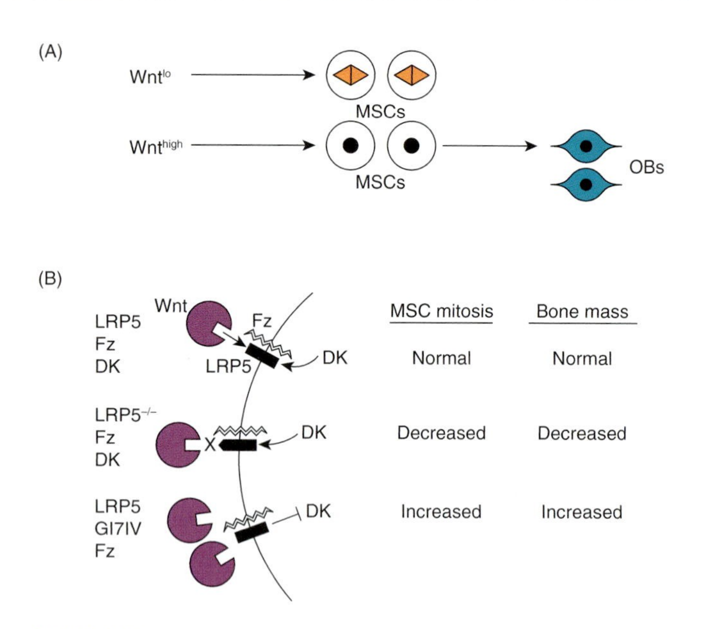


**FIGURE 9.14** Local control of bone remodeling by the Wnt pathway. **(A)** Low concentrations of Wnt stimulate proliferation of mesenchymal stem cells (MSCs), whereas high concentrations promote the differentiation of SCs into osteoblasts (Obs). **(B)** Interactions of Dickopf-1 (DK) with the Wnt receptor Frizzled (Fz) and co-receptor LRP5 regulate mitosis of MSCs and bone mass. Top, DK binding to LRP5 (arrow) internalizes LRP5, balancing activation by Wnt, leading to normal mitosis and bone mass. Middle, loss of function mutation in LRP5 blocks binding of Wnt to the co-receptor, leading to decreased mitosis and bone mass. Bottom, gain of function mutation in LRP5 inhibits binding of DK, leading to increased Wnt binding and increasing MSC mitosis and bone mass.

reduction in bone mass (Elfteriou et al., 2005). Furthermore, ovariectomized β2-AR null mice fail to lose bone mass, in contrast to ovariectomized wild-type mice. The latter mice become osteoporotic and exhibit a significant reduction in the number of nerves that penetrate bone (Burt-Pichat et al., 2005). Thus, maintenance of the sympathetic nervous system in bone may require estrogen. Adrenergic signals activate protein kinase A (PKA) in osteoblasts, leading to the phosphorylation of the ATF4 transcription factor and the upregulation of RANKL production. RANKL, in turn, induces macrophages to differentiate as osteoclasts, which reduce bone mass (Elefteriou et al., 2005).

## Local Control

The Wnt signaling pathway is a major component in the local control of bone regeneration (**FIGURE 9.14**). Low levels of Wnt signaling components promote the proliferation of MSCs (De Boer et al., 2004). Higher levels of Wnt signaling components (Wnt 3a, constitutitively active β-catenin, or LiCl) promote the differentiation of MSCs to osteoblasts (Harada and Rodan, 2003; De Boer et al., 2004). The low-density lipoprotein (LDL)-receptor-related protein 5 (LRP5) on MSCs functions as a co-receptor with the Wnt receptor Frizzled to bind Wnt 1 and Wnt 3a (Wehrli et al., 2000; Tamai et al., 2000). Null mutation of *Lrp5* decreases the proliferation of MSCs and decreases bone mass. A gain of function mutation, *Lrp5G171V*, decreases osteoblast and osteocyte apoptosis, as well as osteoclastogenesis, resulting in high bone mass (Harada and Rodan, 2003; Johnson, 2004). Normally, the activation of LRP5/Frizzled by Wnt is balanced by the internalization of LRP5 through its binding to Dickkopf-1. The *G171V* mutation is thought to prevent this interaction, resulting in excessive binding of Wnt to LRP5 and high bone mass (Harada and Rodan, 2003).

Bone regeneration is also regulated locally by numerous cytokines and growth factors (Horowitz and Lorenzo, 1996; Lian et al., 1999; Harada and Rodan, 2003) that are sequestered in the bone matrix during skeletal development and released as the matrix is degraded by osteoclasts during bone remodeling (Pfeilshifter et al., 1986; 1987). Sequestration of growth factors released from resorbing bone matrix in newly formed matrix is one way that bone resorption might be coupled to bone regeneration. Skeletal cells make many of these growth factors, while others are made by extraskeletal cells and taken up from the blood. **TABLE 9.2** lists these factors according to their function in bone remodeling. BMPs induce osteogenic determination in MSCs (Rosen and Theis, 1992). TNF-α and IL-1 promote the formation of osteoclasts and thus stimulate bone resorption, whereas the remainder of the growth factors listed promote bone regeneration by stimulating

**TABLE 9.2** Growth Factors and Cytokines Involved in Local Control of Bone Remodeling, Summarized by Function

| Stimulate Osteogenic Competence of MSCs | Stimulate Proliferation and Differentiation of MSCs and Pre-osteoblasts into Osteoblasts |
|---|---|
| BMPs | IL-6, 11<br>LIF<br>Oncostatin-M<br>CNTF<br>BDNF<br>EGF<br>IGF-I<br>TGF-β<br>PDGF<br>FGF-1, 2 |
| Promote Osteoclast Differentiation Via Stromal Upregulation of M-CSF and RANKL | |
| IL-1<br>TNF-α | |

the proliferation of osteoblast precursors and their differentiation into osteoblasts via upregulation of the type I collagen gene and inhibiting transcription of the gene for MMP-3, which degrades collagen I (Trifitt, 1987; Canalis et al., 1988; Hauschka et al., 1988; Bonewald, 1996; Horowitz and Lorenzo, 1996; Filvaroff et al., 1999; Wronski, 2001). TGF-β also induces the apoptosis of osteoclasts (Bonewald, 1996).

## 3. Injury-Induced Regeneration of Bone

Flat bones, such as the bones of the skull, develop during embryogenesis by the direct differentiation of MSCs into osteoblasts, a process called intramembranous bone formation. Fractures of intramembranous bone are repaired in the same way, by differentiation of MSCs in the periosteum into osteoblasts. Long bones, however, exhibit endochondral development; i.e., a calcified cartilage template is formed first, which is then replaced by bone (Olsen, 1999). The chondrocytes of the template hypertrophy and die, releasing angiogenic signals that trigger the sprouting of capillaries in the periosteum, while osteoclasts excavate most of the calcified matrix (Alini et al., 1996; Carlevaro et al., 1997). The periosteal capillaries invade the matrix, accompanied by perivascular MSCs. Some of these MSCs differentiate into osteoblasts that replace the cartilage matrix with bone matrix, others form the endosteum and stroma of the bone marrow, and still others remain within the endosteum and stroma as MSCs. Fractured long bones exhibit characteristics of both intramembranous and endochondral bone formation.

### a. Events of Fracture Repair in an Endochondral Bone

**FIGURE 9.15** illustrates the repair of a fractured long bone. The phases of bone repair are similar to those of wounded skin, except that the outcome is regeneration instead of fibrosis. Regeneration is accomplished primarily by MSCs in the periosteum, with lesser contributions from the endosteum and marrow stroma (McKibben, 1978; Wornom and Buchman, 1992; Einhorn, 1998). Following fracture, blood vessels within and without the bone are torn, resulting in the formation of a fibrin clot (hematoma) in and around the break. Hypoxia results in osteocyte death for a limited distance on either side of the fracture. Platelets in the clot release PDGF and TGF-β, initiating an inflammatory phase in which the hematoma is invaded by neutrophils and macrophages (Einhorn, 1998). Some of the macrophages in the bone marrow become osteoclasts that degrade the matrix of the dead bone.

Within a few days after fracture, periosteal MSCs differentiate on both sides of the fracture to osteoblasts (hard callus) in a process of direct (intramembranous) ossification (Brighton and Hunt, 1991; Glowacki, 1998). The osteoblasts secrete a bone matrix rich in Type I collagen, and containing osteocalcin, the mineralization-associated glycoproteins osteonectin, osteopontin, and bone sialoprotein II (BSP-II), and numerous proteoglycans (Robey, 1996). Within the fracture space itself, the phases of repair appear to recapitulate the events of embryonic endochondral bone development through a cartilage template. MSCs in the periosteum, endosteum, and bone marrow proliferate to form a "soft callus." These MSCs condense and differentiate into chondrocytes that secrete cartilage-specific matrix composed of Type II and XI collagens, aggrecan, hyaluronic acid, and fibronectin. The chondrocytes undergo hypertrophy that is characterized by a switch to the production of Type X collagen and downregulation of the other collagen types. Subsequently, the cartilage matrix becomes calcified and the chondrocytes undergo apoptosis (Einhorn, 1998). Osteoclasts excavate the matrix of this calcified matrix template, and periosteal capillaries, induced by angiogenic factors produced by the hypertrophic chondrocytes, invade the matrix (Glowacki, 1998). The invading blood vessels are accompanied by MSCs that differentiate into osteoblasts, which replace the cartilage matrix with bone matrix. Presumably, the balance between bone matrix resorption and synthesis during fracture repair is maintained through the same systemic and local signals as in the maintenance regeneration of bone.

### b. Molecular Regulation of Cartilage Differentiation During Fracture Repair

**TABLE 9.3** summarizes the molecular elements and pathways known to regulate chondrogenesis during fracture repair. Soft callus formation and chondrocyte differentiation is controlled by the same transcription factors and signaling molecules involved in endochondral bone development and maintenance regeneration (Vortkamp et al., 1998). Commitment to chondrogenesis of the soft callus requires expression by MSCs of the transcription factor Sox9, which induces the expression of genes for cartilage markers such as type II, X, IX, and XI collagens and aggrecan (Bi et al., 1999). As the chondrocyte callus matures, *Ihh* transcripts are detected in chondrocytes and *Gli1* transcripts are expressed in a population of cells on the periphery of the callus that will reform the periosteum (Ferguson et al., 1998). In the embryonic development of endochondral bone, the products of these genes, and of the genes for PTH and PTHrP, are part of a feedback loop that controls the rate at which chondrocytes mature

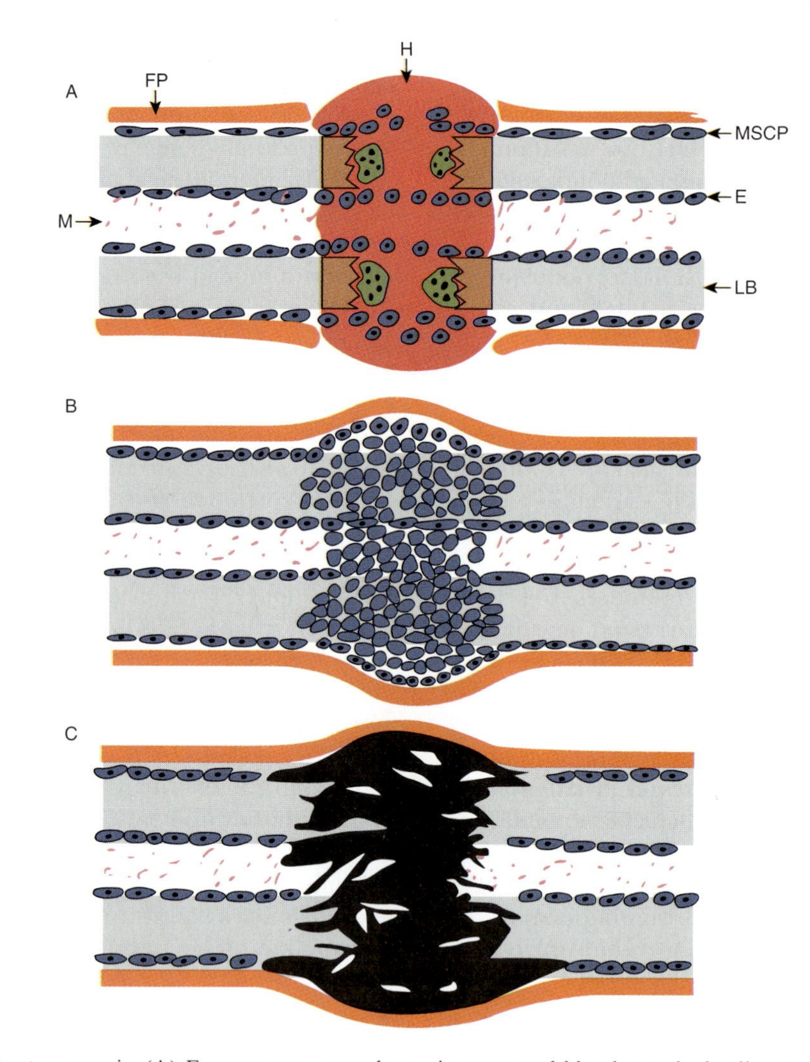


**FIGURE 9.15** Diagram of fracture repair. **(A)** Fracture tears muscle, periosteum and blood vessels, leading to the formation of a fibrin clot (hematoma, H) within and around the fracture space. Bone dies (brown) for a short distance adjacent to the fracture space and is degraded by osteoclasts (green cells). Living bone (LB) is depicted in gray. Mesenchymal stem cells (blue) are activated in the mesenchymal stem cell layer of the periosteum (MSCP) and endosteum (E) and migrate into and proliferate in, the hematoma. FP = fibrous layers of the periosteum. M = marrow (red dots). **(B)** Mesenchymal stem cells proliferate to form a soft callus that replaces the fibrin clot. These cells differentiate into chondrocytes (not shown) that form the template for ossification. **(C)** Blood vessels (not shown) invade the hypertrophied cartilage and osteoblasts derived from perivascular mesenchymal stem cells secrete new bone matrix (black). The new bone that overlaps the living bone externally on either side of the fracture gap is formed directly from periosteal MSCs without going through a cartilaginous phase. Eventually, the original organization into compact and trabecular bone is achieved.

**TABLE 9.3** Signaling Molecules Expressed in the Periosteum and Regenerating Tissue of a Fractured Endochondral Bone

| Tissue | Signals | Transcription Factors | Differentiaton Markers |
|---|---|---|---|
| Periosteum | BMP2,4,7 | Runx2 | |
| Soft Callus | BMP2,4,5,7<br>PDGF<br>FGF-2<br>IGF-1 | Sox9 | |
| Chondro Callus | Ihh<br>BMP2,4,6,7<br>TGF-β<br>FGF-1,2 | Runx2<br>Sox9 | Col II, IX, X, XI<br>Aggrecan |
| Forming bone | | Runx2 | Osteocalcin<br>Col I |

(Vortkamp et al., 1996; Pathi et al., 1999; Crowe et al., 1999; Zou et al., 1997; Lee et al., 1995; Amizuka et al., 1994; Long and Linsenmayer, 1998). During the replacement of the cartilage template with bone, transcripts of genes that encode signaling proteins and transcription factors active in osteoblast differentiation, such as BMPs, Runx2, and osteocalcin, are detected (Ferguson et al., 1998).

Less is known about the expression and function of other molecules during fracture repair that are involved in the early formative events of cartilage templates during ontogenesis. For example, it is not known whether the same upregulation of hyaluronidase and adhesion proteins (NCAM, fibronectin, Cyr61) seen in embryonic skeletal condensations (Knudson and Toole, 1985; Oberlender and Tuan, 1996; Gehris et al., 1997; Wong et al., 1997) is required for mesenchymal condensation within the soft callus. However, it is likely that these molecules play similar roles in fracture repair. It is less likely that molecules such as the Hoxa, Hoxd, and T-box transcription factors, sonic hedgehog (Shh), FGF-4, and FGF-8, and Lmx1, which are involved in axial patterning of the skeletal condensations of the limb bud (Mariani and Martin, 2003), play a role in condensation of the soft callus in an endochondral fracture, or if they do, it is not a patterning role. This is because condensation and chondrocytic differentiation of the soft callus are taking place within an already delineated space and therefore probably do not require patterning.

The same growth factors involved in the local control of bone remodeling are essential for fracture repair. Some of these growth factors are synthesized by the chondrocytes, MSCs, and osteoblasts of the regenerating bone, while others such as TGF-β and IGF I and II are released from the bone matrix by the degradative activity of osteoclasts (Pfeilschifter et al., 1986; 1987; Trippel, 1998).

Members of the TGF-β family are particularly important for chondrogenesis and osteogenesis in fractured bone. BMPs, which induce MSCs to commit to the chondrogenic/osteogenic lineage, are released from degrading bone matrix and they and their receptors are also strongly expressed by soft callus cells (Bostrom, 1998; Bostrom and Camacho, 1998; Reddi, 1998). BMPs were detectable by antibody staining in cells of the soft callus as early as three days after fracture in a rabbit mandible model (Jin and Yang, 1990; Yang and Jin, 1990). Antibody staining increased as cartilage differentiated, and by two weeks after fracture osteoblasts were stained. BMP-4 transcripts were expressed in mesenchymal cells of the periosteum and marrow during the early phase of rib fracture repair, then disappeared when chondrogenesis began (Nakase et al., 1994). Monoclonal antibodies to BMP-2, 4, and 7 showed an increasing intensity of staining for these BMPs in periosteal mesenchymal cells in the region of hard callus formation, in the proliferating mesenchymal cells of the early soft callus and chondroblasts differentiated from these cells. Less intense staining was seen in maturing and hypertrophic chondrocytes, but staining was again intense in osteoblasts replacing the cartilage with bone (Bostrom, 1998). A mutation in the mouse *short ear* gene, which encodes BMP-5, results in congenital bone defects and a reduced capacity to repair fractures, suggesting that this BMP plays an important role in both bone embryogenesis and regeneration (Kingsley et al., 1992). BMPs are not expressed in uninjured bone, but immunolocalization studies show that their receptors, BMPR-IA and IB, are expressed in the periosteal cells of uninjured bone and are upregulated in these cells after fracture, parallel with the upregulation of BMPs (Bostrom, 1998). Activin receptors are also expressed in proliferating and maturing chondrocytes in the fracture, but activin expression is weak, suggesting that activin receptors might act as receptors for BMPs (Bostrom, 1998).

Northern blotting and immunolocalization studies indicate that high levels of TGF-β and FGF-1 and 2 are expressed during chondrogenesis of the soft callus, but not in the region of hard callus formation (Bostrom and Asnis, 1998; Rosier et al., 1998; Canalis and Rydziel, 1996). FGF-2 and other members of the FGF family regulate the expression of Sox 9 (Murakami et al., 2000). TGF-β is present earlier in the hematoma and periosteum, but its source appears to be platelets and release from degrading bone matrix rather than synthesis by periosteal cells. PDGF, FGF-2, and IGF-I are expressed in the soft callus (Bolander, 1992; Trippel, 1998), suggesting that these growth factors are involved in fracture healing.

Transcriptional profiling of intact versus fractured rat femur by subtractive hybridization and microarray analysis has revealed that gene expression patterns change dramatically during fracture repair (Hadjiargyrou et al., 2002). Analysis of ~3,635 cDNAs indicated that 66% had homology to 588 known genes, 31% had homology to 821 expressed sequence tags (ESTs), and the remaining 3% were completely novel. Thirty-four percent of the cDNAs represented genes with unknown functions. The majority of these (663 genes) were grouped in two clusters marked by a sharp increase in activity at three days post-fracture that peaked at day 14 and then decreased. This pattern suggests that these genes are involved in the growth and differentiation of the fracture callus. The known genes were part of multiple families involved in the cell cycle, cell adhesion, ECM, cytoskeleton, inflammation, general metabolism, molecular processing, transcriptional activation, and cell signaling. Consistent with studies on other regenerating tissues, members of the Wnt signaling pathway (Wnt 5a, frizzled, and β-catenin) were identified in regenerating bone.

## REPAIR OF ARTICULAR CARTILAGE

Articular cartilage is typical hyaline cartilage with a matrix consisting primarily of hyaluronic acid, the proteoglycan aggrecan and type II collagen, with minor quantities of type IX and XI collagens (Wornam and Buchman, 1992; Reddi, 2003). Roughly 80% of the wet weight of cartilage is water, due to the high water-

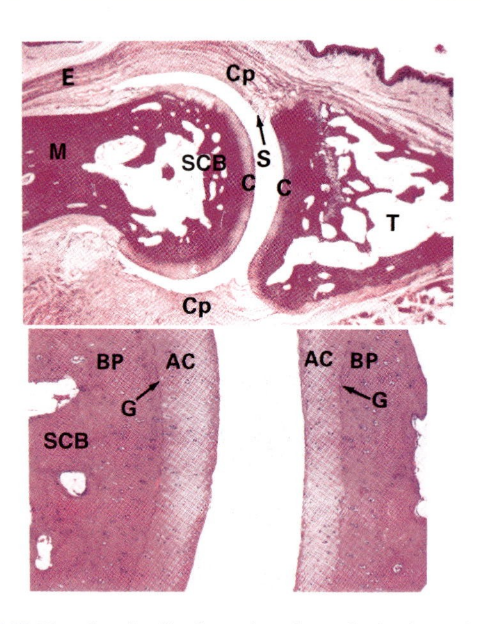


**FIGURE 9.16** **Top,** longitudinal section through the interphalangeal joint of a monkey finger (H&E stain). M = middle phalanx, T = terminal phalanx. C = articular (hyaline) cartilage of the two joint surfaces, SCB = subchondral bone. Cp = the fibrous capsule of the joint, S = the synovium, the specialized layer of connective tissue that lines the inner side of the capsule and secretes the synovial fluid. E = the extensor tendon, which inserts into the base of the terminal phalanx. **Bottom,** higher magnification view of the articular cartilage (AC). The articular cartilage is bonded to the subchondral bony end plate (BP). GP = a glycoprotein-rich material that appears to cement the articular cartilage too thhe bone. Several layers of small flattened chondrocytes constitute the surface of the hyaline cartilage. Beneath the surface layer are chondrocytes arranged in columns perpendicular to the surface. The articular cartilage is avascular. Reproduced with permission from Wheater et al., Wheater's Functional Histology (3rd ed). Copyright 1997, Elsevier.

binding capacity of HA in the matrix. The structure of hyaline cartilage matrix gives it its special physical properties of elasticity and simultaneous firmness and resistance to deformity, properties essential for the weight-bearing function of articular cartilage.

The structure of a joint and of articular cartilage is illustrated in FIGURE 9.16. Articular cartilage acts as a surface growth zone for the epiphysis of growing long bones. It is composed of two regions, a superficial zone of 3–4 rows of small flattened chondrocytes, and a middle zone of larger chondrocytes arranged in columns. Beneath the middle zone is a zone of calcified cartilage, which rests on endochondral bone of the epiphysis. During growth, the chondrocytes in about the third row of the superficial layer divide to provide cells that differentiate in sequence into the cell phenotypes of the middle zone, calcified zone and ultimately to endochondral bone (Wornam and Buchman, 1992).

Mitosis of superficial chondrocytes ceases in the adult, and thus adult articular cartilage must compensate for wear by production of new matrix rather than new cells. Articular cartilage is avascular, but because of the high water content of its matrix, it is easily nourished by diffusion of oxygen and nutrients from the synovial fluid of the joint capsule.

Osteoarthritis is a major disease of articular cartilage that occurs with age and/or injury. Two major retrogressive changes occur during the onset and progression of osteoarthritis (Stockwell, 1975). The first is calcification, which decreases the diffusion of nutrients and oxygen to the chondrocytes. Chondrocytes of the calcified matrix thus die, and the matrix is resorbed. The second change is termed cartilage fibrillation and involves a splitting of the matrix articular surface along the direction of orientation of the collagen fibers to expose these fibers, giving the surface a fuzzy appearance. This occurs at first in patches, which then enlarge. As the condition progresses, there is variable loss of cartilage and exposure of underlying bone, accompanied by increasing pain. The destruction of the cartilage appears due to the cleavage of aggrecan by the aggrecanase ADAMTS5 (a disintegrin and metalloproteinase with thrombospondin-like repeats). ADAMTS5 is the major aggrecanase expressed in normal or inflamed mouse joint cartilage (Stanton et al., 2005) and ADAMTS5 knockout mice are resistant to the cartilage erosion seen in wild-type mice following induction of osteoarthritis by surgically induced joint instability (Glasson et al., 2005).

The reparative ability of articular cartilage is low (Campbell, 1969; Wornom and Buchman, 1992). Injuries that affect only the cartilage (partial-thickness defects) do not repair, because the injury is isolated by the avascularity of the cartilage. There is no fibrin clot, no inflammatory response, and the chondrocytes at the edges of the defect do not re-enter the cell cycle. Injuries that penetrate into the bone (full-thickness defects) show better repair (FIGURE 9.17). Blood from the bone vasculature enters the wound to form a fibrin clot and a typical inflammatory response is mounted. The defect is repaired by MSCs from the bone and fibroblasts that form fibrocartilage, the equivalent of scar in skin wounds (Wornam and Buchman, 1992). However, the repair response in the absence of motion is poor. Repair is better if the joint is moved passively during the process, probably because the motion of the synovial fluid provides for better nutrient supply and waste removal than if the joint is immobilized (Salter et al., 1978). Under these circumstances, more typical hyaline cartilage can be formed (Salter, 1983), but the outcome of the response in humans is highly variable.

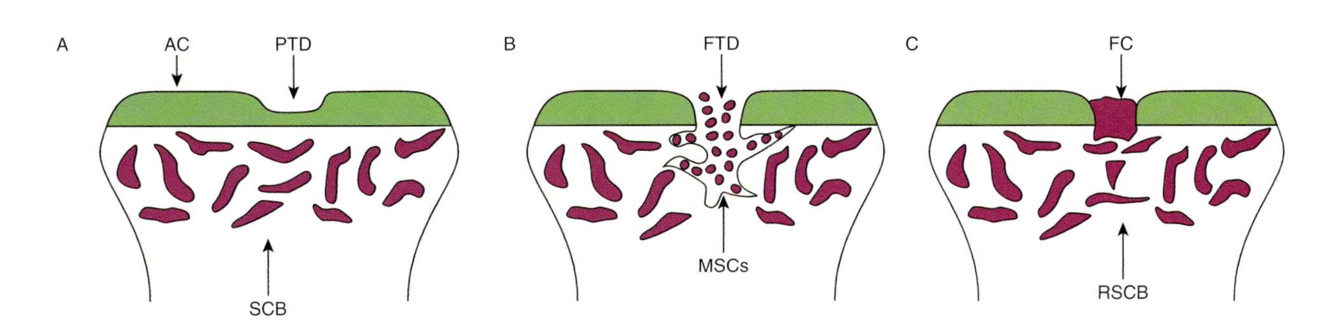


**FIGURE 9.17** Diagram illustrating healing of articular cartilage (AC). **(A)** Partial thickness defect (PTD) that does not penetrate into subchondral bone (SCB). The defect does not repair. **(B, C)** Full thickness defect (FTD) that penetrates into the subchondral bone. **(B)** Blood and mesenchymal stem cells (MSCs) fill the defect. **(C)** Within the cartilage portion of the defect the MSCs differentiate into fibrocartilage (FC). Below the fibrocartilage is regenerated subchondral bone (RSCB).

The nondividing, superficial chondrocytes of adult articular cartilage have the same appearance as mitotically active chondrocytes in growing epiphyses, suggesting that they might not have undergone irreversible structural changes that prevent them from re-entering the cell cycle (Ham and Cormack, 1979). Namba et al. (1998) have shown that partial-thickness incisions in the articular cartilage of fetal lambs are healed by the proliferation of chondrocytes in the lesion area. Clearly, fetal articular chondrocytes are still able to respond to injury by proliferating to fill the defect. Whether this response is possible because of a lower state of maturity of the immune system, as in fetal skin, or because of the intrinsic capacity of the chondrocytes, or a combination of these, is unknown.

The environment of adult articular chondrocytes might contain inhibitors that prevent them from undergoing mitosis after injury. If such inhibitors could be identified and neutralized, it might be possible to stimulate repair of articular cartilage without the involvement of injury to the underlying bone. Some of these inhibitors may have already been identified. The differentiation and morphogenesis of articular cartilage is dependent on the proper balance between anabolic and catabolic agents (Reddi, 2003). BMPs and cartilage-derived morphogenetic proteins (CDMPs) exert anabolic effects on articular cartilage synthesis. CDMP-1 (GDF-5) and CDMP-2 stimulate chondrogenesis *in vitro* and *in vivo* and are expressed in both normal and osteoarthritic cartilage. Members of the IL-17 family block the anabolic effects of BMPs and CDMPs. IL-17B is expressed in the midzone and deep chondrocytes of articular cartilage (Moseley et al., 2003). IL-17, IL-1, and TNF-α degrade osteoarthritic cartilage. The poor regeneration of damaged articular cartilage may thus be due in part to an imbalance between these anabolic and catabolic agents.

## REPAIR OF TENDON AND LIGAMENT

### 1. Structure of Tendons and Ligaments

Tendons and ligaments are regularly arranged, dense connective tissues with great tensile strength (**FIGURE 9.18**). Tendons attach muscles to bones, while ligaments attach bones to bones, stabilize joints, and help guide them through their normal range of motion. Both tissues are composed of flattened fibrocytes that secrete an ECM composed predominantly of parallel longitudinal bundles of collagen I and III fibers, but with a significant component of low-molecular-weight dermatan sulfate PG (Ham and Cormack, 1979; Amadio, 1992; Reddi, 2003). The PG component regulates the size and packing of the collagen fibrils and thus strength of the tendon or ligament, which is proportional to the diameter of the collagen fibrils. The collagen fibrils are wound into fascicles that spiral, particularly in tendons. Furthermore, the collagen fibrils have a crimped appearance in tendons at rest, which represents an inherent slackness that disappears with tensile stress (Amadio, 1992). During development, tendons and ligaments have a good blood supply, but in the adult, the capillary blood supply is minimal.

Some tendons that are close to bone (for example, the Achilles tendon) are enclosed in two sheaths of dense, irregular connective tissue. The outer sheath is attached to the structures that surround it, and the inner sheath is firmly attached to the tendon. Between the two is a space filled with an HA-rich lubricant similar to the synovial fluid of joints that allows the inner tendon sheath (and tendon) to glide within the outer sheath. Ligaments often have a vascular layer covering their surface that merges into the periosteum of the bone adjacent to the insertion sites of the ligament (Frank, 2004).

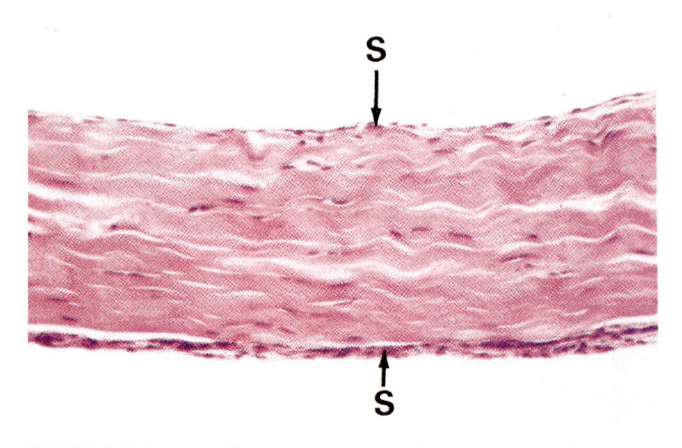


**FIGURE 9.18** Longitudinal section through a tendon (H&E stain). The collagen fibers in the tendon show characteristic crimping (waviness). The tendon is relatively acellular. S = tendon sheath lined with a synovium (S), which produces synovial fluid, allowing the tendon to glide within the sheath. Reproduced with permission from Wheater et al., Wheater's Functional Histology (3rd ed). Copyright 1997, Elsevier.

## 2. Repair of Tendons and Ligaments

Two types of injuries are common to tendons and ligaments: laceration and rupture. After an initial phase of inflammation, fibroblasts proliferate in the discontinuity and produce collagen fibers that initially are disorganized. Within a few weeks, however, they become aligned with the long axis of the tendon or ligament to form a scarlike structure (Amadio, 1992; Frank, 2004). As tension develops on the scar, it is remodeled so that the collagen fibrils align with the long axis of the tendon or ligament. This linear arrangement of collagen fibers is essential for tendon function, so the repair process essentially reproduces the original architecture of the tendon.

Human tendons that are ruptured or lacerated with damage to the synovial tendon sheaths have the ability to spontaneously repair without surgical intervention, with restoration of structure and function that is similar to that of normal adult tendons, provided that treatment is not delayed (Johnston, 1985; Cross et al., 1992; Eriksson et al., 1999; Chalmers, 2000; Papandrea et al., 2000; Ferretti et al., 2002). The repair is effected by fibroblasts from the inner sheath of the tendon, or, in tendons without sheaths, by fibroblasts that migrate in from the surrounding loose connective tissue. In tendons with sheaths, the inner sheath reforms in contact with the regenerating tendon and separates from the reforming outer sheath to restore gliding motion (Ham and Cormack, 1979; Johnston, 1985).

Ligaments are often torn in joint injuries, resulting in partial or complete discontinuities; they do not all heal in the same manner. Isolated injuries of the medial collateral ligament (MCL) heal well, without surgical intervention, but the cruciate ligaments heal poorly (Woo et al., 1999). Ligament repair in the rabbit MCL takes place in much the same way as in tendons, except that the repairing fibroblasts come from the ligament tissue itself, as well as from the surrounding connective tissues (Johnston, 1985). However, defects such as small collagen fibril diameter, failure of collagen crosslinks to mature, increased cellularity and incomplete resolution of matrix flaw, persist and the regenerated ligaments recover only 50% of normal failure loads (Frank, 2004).

## SUMMARY

Skeletal muscle and bone regenerate well after injury. Tendons and ligaments repair by scar tissue that is similar to the original but lacks its strength. Cartilage and menisci regenerate poorly or not at all.

The agents of skeletal muscle regeneration are satellite cells under the myofiber basement membrane having the phenotype $[\text{CXR } \beta\text{1-integrin CD34 c-met}]^{+}$ $[\text{CD45 Sca-1 Mac-1}]^{-}$. These cells are resident to myofibers and are self-renewing. While muscle harbors other cell types, including cells derived from the bone marrow, these populations do not contribute to muscle regeneration. SCs are normally held in a quiescent state by Notch signaling and do not express MRFs. Muscle injured by free-grafting degenerates, followed by a typical inflammatory response. During these phases, SCs detach from their basement membranes and proliferate within them, using the anaerobic pentose phosphate metabolic pathway. The proliferating SCs strongly upregulate Pax7 and MRFs. The MRFs are activated in the order MyoD, followed by myf5 and MRF4 and then by myogenin. Quiescent SCs are activated and their proliferation stimulated by HGF released from muscle ECM. PDGF, FGF-2, LIF and TGF-β have a modulating effect on SC proliferation. Subsequently, the SCs fuse and differentiate to form new myofibers. The application of tension and reinnervation of the regenerating muscle is essential for its complete structural and functional differentiation.

Bone is a dynamic tissue that is constantly remodeled through resorption of bone matrix by osteoclasts and regeneration by osteoblasts. Osteoclasts are multinucleate cells that differentiate from macrophages when stimulated by factors produced by stromal cells, including osteoblasts. The principal factors involved in the induction of osteoclast differentiation are RANK and c-Fms, receptors on the macrophage cell surface that bind the ligands RANKL and M-CSF, respectively. M-CSF is a soluble signal produced by stromal cells, whereas RANKL is a stromal cell surface molecule. Therefore, osteoclast differentiation requires contact between macrophages and stromal cells. Osteoblasts also make OPG, a decoy protein that competes with RANK for RANKL and inhibits osteoclast differentiation. Osteoclast differentiation is thus regulated by a balance between the concentrations of M-CSF and RANKL versus OPG. Osteoclasts adhere to the bone matrix and dissolve the hydroxyapatite and organic components via HCl and protease release, respectively. They then undergo apoptosis and their resorption space is occupied by osteoblasts, which rebuild the bone. Osteoblasts are postmitotic cells derived from MSCs in the periosteum, endosteum, and marrow. A wide variety of growth factors induce MSC proliferation. BMPs induce osteogenic determination in MSCs and still other growth factors promote osteoblast differentiation. Two transcription factors that play key roles in osteoblast differentiation are Runx2 and Osx.

Bone remodeling is under both systemic and local control pathways that either promote or inhibit the differentiation of osteoclasts and osteoblasts. Several systemic hormonal pathways are involved in bone remodeling. PTH, PTHrP, and 1,25-2OH vitamin $D_3$ increase bone resorption by stimulating stromal cells to express M-CSF and RANKL, thus increasing osteoclast differentiation. The adipocyte hormone leptin also increases osteoclast differentiation by RANKL upregulation in osteoblasts. Leptin binds to receptors in the hypothalamus, triggering the production of noradrenaline by sympathetic nerves that innervate bone. Noradrenaline binds to β2-ARs on osteoblasts, leading to upregulation of RANKL production. Conversely, sex steroids inhibit osteoclast differentiation, while intermittent PTH administration and high doses of vitamin $D_3$ stimulate osteoprogenitor cells to differentiate as osteoblasts. Local control is affected by a variety of cytokines released from resorbing matrix. TNF-α and IL-1 promote osteoclast differentiation, whereas LIF, IL-6, IL-11, oncostatin M, CNTF, and low concentrations of Wnt promote the proliferation of MSCs. BMPs induce osteogenic determination of MSCs and higher concentrations of Wnt promote the differentiation of MSCs into osteoblasts. There is cross-talk between systemic and local pathways. TSH inhibits osteoclast differentiation by negatively regulating TNF-α and inhibits MSC proliferation (and thus osteoblast differentiation) by suppressing the expression of the Wnt co-receptor LRP-5 on the MSC surface. Despite increasing knowledge of how bone is remodeled, however, we remain largely ignorant of the mechanisms that orchestrate the sequential appearance of osteoclasts and osteoblasts at bone remodeling sites.

When a long bone is fractured, a fibrin clot forms in the fracture space, followed by an inflammatory response. Rather than forming scar, however, the bone regenerates like a developing endochondral bone by the formation of a cartilage template. MSCs in the periosteum, endosteum, and bone marrow proliferate and differentiate into hypertrophic cartilage that is subsequently replaced by bone. The local molecular mediators of fracture repair appear to be identical to those involved in endochondral bone development and bone remodeling. BMPs determine MSCs to become chondrogenic. TGF-β and FGF-1 and 2, PDF and IGF-1 are all expressed in the soft callus and regulate chondrocyte differentiation. FGF-2 regulates expression of the transcription factor Sox-9, which activates the expression of type I, IX, and collagen genes and the gene for aggrecan protein. *Ihh* signaling pathway components are expressed in a population of cells on the periphery of the soft callus that will reform the periosteum, indicating that the same mechanism used to regulate the rate at which chondrocytes mature during the embryonic development of long bones is operative during fracture healing. As the cartilage template is replaced by perivascular MSC invasion, the expression of genes involved in osteoblast differentiation, such as Runx2 and osteocalcin, is detected. The expression of other genes, such as those that regulate MSC condensation during embryonic bone development, or genes that regulate skeletal patterning, have not been examined. Since fracture healing takes place within a defined space, patterning genes may not have to be activated.

In contrast to bone, articular cartilage does not repair at all if the injury does not penetrate the cartilage into the bone. Penetration into the bone evokes a fibrocartilaginous repair response by MSCs that migrate into the wound. Fetal articular cartilage can regenerate because its chondrocytes are still capable of division, whereas those of the adult are not. This has led to the idea that the lack of proliferation of adult chondrocytes might be due to inhibitors in their environment. BMPs and CDMPs stimulate chondrogenesis and are expressed in both normal and osteoarthritic cartilage. Their effects are blocked by members of the IL-17 family, which are not only expressed in articular cartilage, but also degrade osteoarthritic cartilage.

Human tendons have the ability to spontaneously repair without surgical intervention if treatment (immobilization) is not delayed. The repair is effected by fibroblasts of the tendon sheath. The fibroblasts lay down collagen fibers that initially are disorganized. Tension later aligns the collegen fibers with the long axis of the tendon to reproduce the original scarlike architecture of the tendon. Ligaments are structurally similar to tendons and regenerate in a similar way. The MCL regenerates well without surgical intervention, but the cruciate ligaments regenerate poorly. Even in cases of good regeneration, however, the regenerated ligaments recover only 50% of normal failure loads.

## REFERENCES


Abe E, Marians RC, Yu W, Wu X-B, Ando T, Li Y, Iqbal J, Eldeiry L, Rajendren G, Blair HC, Davies TF, Zaidi M (2003) TSH is a negative regulator of skeletal remodeling. Cell 115:151–162.

Akino K, Mineta T, Fukui M, Fujii T, Akita S (2003) Bone morphogenetic protein-2 regulates proliferation of human mesenchymal stem cells. Wound Rep Reg 11:354–360.

Alberts B, Bray D, Lewis J, Raff M, Roberts K, Watson JD (1994) Molecular Biology of the Cell, 3rd ed. New York, Garland Pub Co.

Alini M, Marriott A, Chen T, Abe S, Poole AR (1996) A novel angiogenic molecule produced at the time of chondrocyte hypertrophy during endochondral bone formation. Dev Biol 176:124–132.

Allen RE, Sheehan SM, Taylor RG, Kendall TK, Rice GM (1995) Hepatocyte growth factor activates quiescent skeletal muscle satellite cells in vitro. J Cell Physiol 165:307–312.

Allbrook D (1981) Skeletal muscle regeneration. Muscle Nerve 4:234–245.

Alliston T, Derynck R (2002) Interfering with bone remodeling. Nature 416:686–687.

Amadio PC (1992) Tendon and ligament. In: Cohen IK, Diegelmann RF, Lindblad WJ (eds.). Wound Healing: Biochemical & Clinical Aspects. Philadelphia, WB Saunders Co, pp 384–395.

Amizuka N, Warshawsky H, Henderson JE, Goltzman D, Karaplis AC (1994) Parathyroid hormone-related peptide-depleted mice show abnormal epiphyseal cartilage development and altered endochondral bone formation. J Cell Biol 126:1611–1623.

Amthor H, Christ B, Rashid-Doubell F, Kemp CF, Lang E, Patel K (2002) Follistatin regulates bone morphogenetic protein-7 (BMP-7) activity to stimulate embryonic muscle growth. Dev Biol 243:115–127.

Anderson JE, Liu L, Kardami E (1991) Distinctive patterns of basic fibroblast growth factor (bFGF) distribution in degenerating and regenerating areas of dystrophic (mdx) striated muscles. Dev Biol 147:96–109.

Armand A-S, Gaspera D, Launay T, Charbonnier F, Gallien C, Chanoine C (2003) Expression and neural control of follistatin versus myostatin genes during regeneration of mouse soleus. Dev Dynam 227:256–265.

Aubin JE, Liu F (1996) The osteoblast lineage. In: Bilezikian J, Raisz LG, Rodan GA (eds.). Principles of Bone Biology, San Diego, Academic Press, pp 51–68.

Baron R (1999) Anatomy and ultrastructure of bone. In: Favus MJ (ed.). Primer on the Metabolic Bone Diseases and Disorders of Mineral Metabolism, 4th ed. Philadelphia, Lippincott Williams and Wilkins, pp 3–10.

Beauchamp JR, Heslop L, Yu DSW, Tajbakhsh S, Kelly RG, Wernig A, Buckingham ME, Partridge TA, Aammit PS (2000) Expression of CD34 and Myf5 defines the majority of quiescent adult skeletal muscle satellite cells. J Cell Biol 151:1221–1233.

Bi W, Deng JM, Zhang Z, Behringer RR, de Crombrugghe B (1999) Sox9 is required for cartilage formation. Nat Genet 22:85–89.

Bischoff R (1986) A satellite cell mitogen from crushed adult muscle. Dev Biol 115:140–147.

Blaveri K, Heslop L, Yu DS, Rosenblatt D, Gross JG, Partridge A, Morgan JE (1999) Patterns of repair of dystrophic mouse muscle: Studies on isolated fibers. Dev Dynam 216:244–256.

Bolander ME (1992) Regulation of fracture repair by growth factors. Proc Soc Exp Biol Med 200:165–170.

Bonewald LF (1996) Transforming growth factor-β. In: Bilezikian JP, Raisz LG, Rhodan GA (eds.). Principles of Bone Biology. New York, Academic Press, pp 647–659.

Bostrom MPG (1998) Expression of bone morphogenetic proteins in fracture healing. Clin Orthopaed Rel Res 355S:S116–S123.

Bostrom MPG, Asnis P (1998) Transforming growth factor beta in fracture repair. Clin Orthopaed Rel Res 355S:S124–S131.

Bostrom MPG, Camacho NP (1998) Potential role of bone morphogenetic proteins in fracture healing. Clin Orthopaed Rel Res 355S:S274–S282.

Boyle WJ, Simonet S, Lacey DL (2003) Osteoclast differentiation and activation. Nature 423:337–342.

Brighton CT, Hunt RM (1991) Early histological and ultrastructural changes in medullary fracture callus. J Bone Joint Surg 73A: 832–847.

Bruder SP, Horowitz MC, Mosca JD, Haynesworth SE (1997) Monoclonal antibodies reactive with human osteogenic cell surface antigens. Bone 21:225–235.

Bruder SP, Kurth AA, Shea M, Hayes WC, Jaiswal N, Kadiyala S (1998) Bone regeneration by implantation of purified culture-expanded human mesenchymal stem cells. J Orthop Res 16: 155–162.

Burr DB, Robling AG, Turner CH (2002) Effects of biomechanical stress on bones n animals. Bone 30:781–786.

Burt-Pichat B, Lafage-Proust MH, Duboeuf F, Laroche N, Itzstein C, Vico L, Delmas PD, Chenu C (2005) Dramatic decrease of innervation density in bone after ovariectomy. Endocrinology 146:503–510.

Campbell CJ (1969) The healing of cartilage defects. Clin Orthop 64:45–63.

Canalis E, McCarthy T, Centrella M (1988) Growth factors and the regulation of bone remodeling. J Clin Invest 81:277–281.

Canalis E, Rydziel S (1996) Platelet-derived growth factor and the skeleton. In: Principles of Bone Biology, Bilezikian JP, Raisz LG, Rhodan GA (eds.). New York, Academic Press, pp 619–626.

Caplan AI (1991) Mesenchymal stem cells. J Orthop Res 9: 641–650.

Caplan AI, Fink DJ, Goto T, Linton AE, Young RG, Wakitani S, Goldberg VM, Haynesworth SE (1993) Mesenchymal stem cells and tissue repair. In: Jackson DW, Arnoczky S, Woo S, Frank C (eds.). The Anterior Cruciate Ligament: Current and Future Concepts. New York, Raven Press, pp 405–417.

Carlevaro MF, Albini A, Ribatti D, Gentili C, Benelli R, Cermelli S, Cancedda R, Cancedda FD (1997) Transferrin promotes endothelial cell migration and invasion: implication in cartilage neovascularization. J Cell Biol 136:1375–1384.

Carlson BM (1970) Organizational aspects of muscle regeneration. In: Research in Muscle Development and the Muscle Spindle. Excerpta Medica Internatl Congress Series 240:3–17.

Carlson BM (1974) Regeneration from short stumps of the rat gastrocnemius muscle. Experientia 30:275–276.

Carlson BM, Gutmann E (1972) Developent of contractile properties of minced muscle regenerates in the rat. Exp Neurol 36: 239–249.

Carlson BM, Gutmann E (1975) Regeneration in free grafts of normal and denervated muscles in the rat: Morphology and histochemistry. Anat Rec 1975; 183:47–62.

Carlson BM (1978) A review of muscle transplantation in mammals. Physiologia Bohemosovaca 27:387–400.

Carlson BM (1983) The biology of muscle transplantation. In: Freilinger G, Holle J, Carlson BM (eds.). Muscle Transplantation. New York, Springer-Verlag, pp 3–18.

Carlson BM (2003) Muscle regeneration in amphibians and mammals: passing the torch. Dev Dynam 226:167–181.

Carnac G, Fajas LL, l'Honore A, Sardet C, Lamb NJC, Fernandez A (2000) The retinoblastoma-like protein p130 is involved in the determination of reserve cells in differentiating myoblasts. Curr Biol 10:543–546.

Carrino DA, Oron V, Pechak D, Caplan AI (1988) Reinitiation of chondroitin sulphate proteoglycan synthesis in regenerating skeletal muscle. Development 103:641–656.

Chalmers J (2000) Treatment of Achilles tendon ruptures. J Orthopaed Surg 8:97–99.

Chenu C (2004) Role of innervation in the control of bone remodeling. J Musculoskel Neuron Interact 4:132–134.

Chen G, Birnbaum RA, Yablonka-Reuveni Z, Quinn LS (1994) Separation of mouse crushed muscle extract into distinct mitogenic activities by heparan affinity chromatography. J Cell Physiol 160:563–574.

Chen D, Ji X, Harris MA, Feng JQ, Karsenty G, Celeste AJ, Rosen V, Mundy GR, Harris SE (1998) Differential roles for bone morphogenetic protein (BMP) receptor type 1B and 1A in differentiation and specification of mesenchymal precursor cells to osteoblast and adipocyte lineages. J Cell Biol 141:295–305.

Cohn RD, Henry MD, Michele DE, Barresi R, Saito F, Moore SA, Flanagan JD, Skwarchuk MW, Robbins ME, Mendell JR, Williamson RA, Campbell KP (2002) Disruption of *Dag1* in differentiated skeletal muscle reveals a role for dystroglycan in muscle regeneraton. Cell 110:639–648.

Collins CA, Olsen I, Zammit PS, Heslop L, Petrie A, Partridge TA, Morgan JE (2005) Stem cell function, self-renewal and behavioral heterogeneity of cells from the adult muscle satellite cell niche. Cell 122:289–301.

Conboy IM, Rando TA (2002) The regulation of Notch signaling controls satellite cell activation and cell fate determination in postnatal myogenesis. Dev Cell 3:397–409.

Cornelison DDW, Wold BJ (1997) Single–cell analysis of regulatory gene expression in quiescent and activated mouse skeletal muscle satellite cells. Dev Biol 191:270–283.

Cornelison DDW, Filla MS, Stanley H, Rapraeger AC, Olwin BB (2001) Syndecan-3 and syndecan-4 specifically mark skeletal muscle satellite cells and are implicated in satellite cell maintenance and muscle regeneration. Dev Biol 239:79–94.

Cross MJ, Roger G, Anderson I, Kujawa P (1992) Regeneration of the semitendinosis and gracilis tendons following their transection for repair of the anterior cruciate ligament. Am J Sports Med 20:221–223.

Crowe R, Zikherman J, Niswander L (1999) Delta-1 negatively regulates the transition from prehypertrophic to hypertrophic chondrocytes during cartilage formation. Development 126:987–998.

d'Albis A, Couteaux R, Janmot C, Roulet A, Mira JC (1988) Regeneration after cardiotoxin injury of innervated and denervated slow and fact muscles of mammals. Myosin isoform analysis. Eur J Biochem 174:103–110.

DeBoer J, Wang HJ, van Blitterswijk C (2004) Effects of Wnt signaling on proliferation and differentiation of human mesenchymal stem cells. Tiss Eng 10:393–401.

DiMario J, Buffinger N, Yamada S, Strohman RC (1989) Fibroblast growth factor in the extra cellular matrix of dystrophic (mdx) mouse muscle. Science 244:688–690.

Dix DJ, Eisenberg BR (1991) Distribution of myosin mRNA during development and regeneration of skeletal muscle fibers. Dev Biol 143:422–426.

Ducy P, Schinke T, Karsenty G (2000) The osteoblast: A sophisticated fibroblast under central surveillance. Science 289:1501–1504.

Dusterhoft S, Pette D (1993) Satellite cells from slow rat muscle express slow myosin under appropriate culture conditions. Differentiation 53:25–33.

Duttaroy A, Bourbeau D, Wang E (1998) Apoptosis rate can be accelerated or decelerated by overexpression or reduction of the level of elongation factor EF-1α. Exp Cell Res 238:168–176.

Einhorn TA (1998) The cell and molecular biology of fracture healing. Clin Orthopaed Related Res 355S:7–21.

Elefteriou F, Ahn JD, Takeda S, Starbuck M, Yang X, Liu X, Kondo H, Richards WG, Bannon TW, Noda M, Clement K, Vaisse C, Karsenty G (2005) Leptin regulation of bone resorption by the sympathetic nervous system and CART. Nature 434:514–520.

Elmquist JK, Strewler GJ (2005) Do neural signals remodel bone? Nature 434:447–448.

Erben RG (2001) Vitamin D analogs and bone. J Musculoskel Neuron Interact 2:59–69.

Eriksson K, Larsson H, Wredmark T, Hamberg P (1999) Semitendinosus tendon regeneration after harvesting for ACL reconstruction: A prospective MRI study. Knee Surg Sports Traumatol Arthrocsopy 7:220–225.

Esser K, Gunning P, Hardeman E (1993) Nerve-dependent and -independent patterns of mRNA expression in regenerating skeletal muscle. Dev Biol 159:173–183.

Ferguson CM, Miclau T, Hu D, Alpern E, Helms JA (1998) Common molecular pathways in skeletal morphogenesis and repair. Ann NY Acad Sci 857:33–42.

Ferretti A, Conteduca F, Morelli F, Masi V (2002) Regeneration of the semitendinosus tendon after its use in anterior cruciate ligament reconstruction. Am Sports Med 30:204–561.

Filvaroff E, Erlebacher A, Ye J-Q, Gitelman SE, Lotz J, Heillman M, Derynck R (1999) Inhibition of TGF-β receptor signaling in osteoblasts leads to decreased bone remodeling and increased trabecular bone mass. Development 126:4267–4279.

Frank CB (2003) Ligament structure, physiology and function. J Musculoskel Neuron Interact 4:199–201.

Friedenstein AJ, Chailakhyan RK, Lalykina KS (1970) The development of fibroblast colonies in monolayer cultures of guinea pig bone marrow and spleen cells. Cell Tiss Kinet 3:393–402.

Friedenstein AJ, Chailakhyan RK, Gerasimov UV (1987) Bone marrow osteogenic stem cells: In vitro cultivation and transplantation in diffusion chambers. Cell Tiss Kinet 20:263–272.

Frost HM (1964) Laws of bone structure. Charles C Thomas, Springfield, IL.

Garrett KL, Anderson JE (1995) Colocalization of bFGF and the myogenic regulatory gene myogenenin in dystrophic mdx muscle precursors and young myotubes *in vivo*. Dev Biol 169: 596–608.

Garry DJ, Yang Q, Bassel-Duby R, Williams RS (1997) Persistent expression of MNF identifies myogenic stem cells in postnatal muscles. Dev Biol 188:280–294.

Gehris AL, Stringa E, Spina J, Desmond ME, Tuan RS, Bennett VD (1997) The region encoded by the alternatively spliced exon IIIA in mesenchymal fibronectin appears essential for chondrogenesis at the level of cellular condensation. Dev Biol 190: 191–205.

Glasson SS, Askew R, Sheppard B, Carito B, Blanchet T, Ma H-L, Flannery CR, Peluso D, Kanki K, Yang Z, Majumdar MK, Morris EA (2005) Deletion of active ADAMTS5 prevents cartilage degradation in a murine model of osteoarthritis. Nature 434: 644–648.

Glowacki J (1998) Angiogenesis in fracture repair. Clin Orthopaed Related Res 355S:S82–S89.

Gorza L, Sartore S, Triban C, Schiaffino S (1983) Embryonic-like myosin heavy chains in regenerating chicken muscle. Exp Cell Res 143:395–403.

Grigoriadis AE, et al. (1994) c-Fos: A key regulator of osteoclast-macrophage lineage determination and bone remodeling. Science 266:443–448.

Grim M, Bukovsky A (1986) How closely are rat skeletal muscle development and regeneration processes related? Biblthca Anat 29:154–172.

Gros J, Manceau M, Thome V, Marcelle C (2005) A common somatic origin for embryonic muscle progenitors and satellite cells. Nature 435:954–958.

Grounds MD (1991) Towards understanding skeletal muscle regeneration. Pathol Res Pract 187:1–22.

Grounds MD, Yablonka-Reuvini Z (1993) Molecular and cell biology of skeletal muscle regeneration. In: Partridge T (ed.). Molecular and Cell Biology of Muscular Dystrophy. London, Chapman and Hall, pp 210–256.

Gulati AK, Reddi AH, Zalewski AA (1983) Changes in the basement membrane zone components during skeletal muscle fiber degeneration and regeneration. J Cell Biol 97:957–962.

Hadjiargyrou M, Lombardo F, Zhao S, Ahrens W, Joo J, Ahn H, Jurman M, White DW, Rubin CT (2002) Transcriptional profiling of bone regeneration. J Biol Chem 277:30177–30182.

Halevy O, Piestun Y, Allouh MZ, Rosser BWC, Rinkevich Y, Reshef R, Rozenboim I, leklinski-Lee M, Yablonka-Reuveni Z (2004) Pattern of Pax7 expression during myogenesis in the post-hatch chicken establishes a model for satellite cell differentiation and renewal. Dev Dynam 231:489–502.

Ham AW, Cormack DH (1979) Histology, 8th ed. Philadelphia, JB Lippincott, pp 614–644.

Hansen-Smith FM, Carlson BM (1979) Cellular responses to free grafting of the extensor digitorum longus muscle of the rat. J Neurol Sci 41:149–173.

Harada S-I, Rodan G (2003) Control of osteoblast function and regulation of bone mass. Nature 423:349–354.

Hauschka PV, Chen TL, Maurakos AE (1988) Polypeptide growth factors in bone matrix. In: Cell and Molecular Biology of Vertebrate Hard Tissues (CIBA Foundation Symposium 136), Wiley, Chichester, pp 207–225.

Hinterberger TJ, Cameron JA (1990) Myoblasts and connective-tissue cells in regenerating amphibian limbs. Ontogenez 21:341–357.

Hinterberger TJ, Sassoon DA, Rhodes SJ, Konieczny SF (1991) Expression of the muscle regulatory factor MRF4 during somite and skeletal myofiber development. Dev Biol 147:144–156.

Hock JM (2001) Anabolic actions of PTH in the skeletons of animals. J Musculoskel Neuron Interact 2:33–47.

Hofbauer LC, Kuhne CA, Viereck V (2004) The OPG/RANKL/RANK system in metabolic bone diseases. J Musculoskel Neuron Interact 4:268–275.

Hoh JFY, Hughes S (1988) Myogenic and neurogenic regulation of myosin gene expression in cat jaw-closing muscles regenerating in fast and slow limb muscle beds. J Muscle Res Cell Motil 9: 59–72.

Horowitz MC, Lorenzo JA (1996) Local regulators of bone: Il-1, TNF, lymphotoxin, interferon-γ, Il-8, Il-10, Il-4, the LIF/Il-6 family, and additional cytokines. In: Principles of Bone Biology, Bilezikian JP, Raisz LG, Rhodan GA (eds.). New York, Academic Press, pp 687–700.

Jackson KA, Mi T, Goodell MA (1999) Hematopoietic potential of stem cells isolated from murine skeletal muscle. Proc Natl Acad Sci USA 96:14482–14486.

Jennische E, Ekberg S, Matejka GL (1993) Expression of hepatocyte growth factor in growing and regenerating rat skeletal muscle. Am J Physiol 265:122–128.

Jin Y, Yang LJ (1990) The relationship between bone morphogenetic protein and neoplastic bone diseases. Clin Orthopaed 259: 233–238.

Johnson SE, Allen RE (1995) Activation of skeletal muscle satellite cells and the role of fibroblast growth factor receptors. Exp Cell Res 219:449–453.

Johnson ML (2004) The high bone mass family—the role of Wnt/Lrp5 signaling in the regulation of bone mass. J Musculoskel Neuron Interact 4:135–138.

Johnston DE (1985) Tendons, skeletal muscles and ligaments in health and disease. In: Newton CD, Nunamaker DM (eds.). Textbook of Small Animal Orthopaedics, International Veterinary Information Service (www.ivis.org), Ithaca.

Jones PH (1982) *In vitro* comparison of embryonic myoblasts and myogenic cells isolated from regenerating adult rat skeletal muscle. Exp Cell Res 139:401–404.

Karsenty G (2003) The complexities of skeletal biology. Nature 423:316–318.

Khalyfa A, Carlson BM, Carlson JA, Wang E (1999) Toxin injury-dependent switched expression between EF-1a and its sister, S1, in rat skeletal muscle. Dev Dynam 216:267–273.

Kherif S, Lafuma C, Dehaupas M, Lachkar, Fournier J-G, Verdiere-Sahuque M, Fardeau M, Alameddine HS (1999) Expression of matrix metalloproteinases 2 and 9 in regenerating skeletal muscle: A study in experimentally injured and *mdx* muscles. Dev Biol 205:158–170.

Kingsley DM, Bland AE, Grubber JM, Marker PC, Russel LB, Copeland NG, Jenkins NA (1992) The mouse short ear skeletal morphogenesis locus is associated with defects in a bone morphogenetic member of the TGF beta superfamily. Cell 71:399–410.

Koishi K, Zhang M, McLennan IS, Harris AJ (1995) MyoD protein accumulates in satellite cells and is neurally regulated in regenerating myotubes and skeletal muscle fibers. Dev Dynam 202: 244–254.

Knudson CB, Toole BP (1985) Changes in the pericellular matrix during differentiation of limb bud mesoderm. Dev Biol 112: 308–318.

Kurek B, Nouri S, Kannourakis G, Murphy M, Austin L (1996) Leukemia inhibitory factor and interleukin-6 are produced by diseased and regenerating skeletal muscle. Muscle Nerve 19: 1291–1301.

Launay T, Armand AS, Charbonnier F, Mira JC, Donsez E, Gallien CL, Chanoine C (2001) Expression and neural control of myogenic regulatory factor genes during regeneration of mouse soleus. J Histochem Cytochem 49:887–899.

Lee K, Deeds JD, Serge GV (1995) Expression of parathyroid hormone-related peptide and its receptor messenger ribonucleic acids during fetal development in rats. Endocrinology 136:453–463.

Lee SJ, McPherron AC (2001) Regulation of myostatin activity and muscle growth. Proc Natl Acad Sci USA 98:9306–9311.

Lian JB, Stein GS, Canalis E, Robey PG, Boskey A (1999) Bone formation: Osteoblast lineage cells, growth factors, matrix proteins, and the mineralization process. In: Favus MJ (ed.). Primer on the Metabolic Bone Diseases and Disorders of Mineral Metabolism, 4th Edition. Philadelphia, Lippincott, Williams, and Wilkens, pp 14–29.

Long F, Linsenmayer TF (1998) Regulation of growth region cartilage proliferation and differentiation by perichondrium. Development 125:1067–1073.

Mackay AM, Beck SC, Murphy JM, Barry FP, Chichester C, Pittenger MF (1998) Chondrogenic differentiation of cultured human mesenchymal stem cells from marrow. Tiss Eng 4:415–428.

Mariani FV, Martin GR (2003) Deciphering skeletal patterning: Clues from the limb. Nature 423:319–325.

Marics I, Padilla F, Guillemot JF, Scaal M, Marcelle C (2002) FGFR4 signaling is a necessary step in limb muscle differentiation. Development 129:4559–4569.

Martin TJ (2004) Paracrine regulation of osteoclast formation and activity: Milestones in discovery. J Musculoskel Neuron Interact 4:243–253.

Matsuda RM, Spector DH, Strohman RC (1983) Regenerating adult chicken skeletal muscle and satellite cell cultures express embryonic patterns of myosin and tropomyosin isoforms. Dev Biol 100:478–488.

Mauro A (1961) Satellite cells of skeletal muscle fibers. J Biophys Biochem Cytol 9:493–495.

McKibben B (1978) The biology of fracture healing in long bones. J Bone Joint Surg 60B:150–162.

McKinney-Freeman SL, Jackson KA, Camargo FD, Ferrari G, Mavilio F, Goodell MA (2002) Muscle-derived hematopoietic stem cells are hematopoietic in origin. Proc Natl Acad Sci USA 99:1341–1346.

Megeny LA, Kablar B, Garrett K, Anderson JE, Rudnicki MA (1996) MyoD is required for myogenic stem cell function in adult skeletal muscle. Genes Dev 10:1173–1183.

Migdalska A, Molineaux G, Demuynck H, Evans S, Ruscetti F, Dexter TM (1991) Growth inhibitory effects of transforming growth factor β *in vivo*. Growth Factors 4:239–245.

Miyazawa K, Shimomura T, Kitamura N (1996) Activation of hepatocyte growth factor in injured tissues is mediated by hepatocyte growth factor activator. J Biol Chem 271:3615–3618.

Moore R, Walsh FS (1993) The cell adhesion molecule M-cadherin is specifically expressed in developing and regenerating, but not denervated muscle. Development 117:1409–1420.

Moseley TA, Haudenschild DR, Rose L, Reddi AH (2003) Interleukin-17 family and I-17 receptors. Cytokine Growth Factor Rev 14:155–174.

Mundlos S, Olsen BR (1997) Heritable diseases of the skeleton. Part I: Molecular insights into skeletal development-transcription factors and signaling pathways. FASEB J 11: 125–132.

Mundy GR (1999) Bone remodeling. In: Favus MJ (ed.). Primer on the Metabolic Bone Diseases and Disorders of Mineral Metabolism, 4th ed. Philadelphia, Lippincott, Williams and Wilkins, pp 30–38.

Nakase T, Takaoka K, Kimiaki H, Hirota S, Takemura T, Onoue H, Takebayashi K, Kitamura Y, Nomura S (1994) Alterations in the expression of osteonectin, osteopontin and osteocalcin mRNAs during the development of skeletal tissues in vivo. Bone Mineral 26:109–122.

Nakashima K, Zhou X, Kunkel G, Zhang Z, Deng JM, Behringer RR, deCrombrugghe B (2002) The novel zinc finger-containing transcription factor osterix is required for osteoblast differentiation and bone formation. Cell 108:17–29.

Namba RA, Meuli M, Sullivan KM, Le AX, Adzick NS (1998) Spontaneous repair of superficial defects in articular cartilage in a fetal lamb model. J Bone Joint Surg 80:4–10.

Nicolas N, Mira JC, Gallien CL, Chanoine C (2000) Neural and hormonal control of expression of myogenic regulatory factor genes during regeneration of Xenopus fast muscles: Myogenin and MRF4 mRNA accumulation are neurally regulated oppositely. Dev Dynam 221:112–122.

Oberlender SA, Tuan RS (1994) Expression and functional involvement of N-cadherin in embryonic limb chondrogenesis. Development 120:177–187.

Olsen BR (1999) Bone Morphogenesis and Embryologic Development. In: Favus MJ (ed.). Primer on the Metabolic Bone Diseases and Disorders of Mineral Metabolism, 4th ed. Philadelphia, Lippincott, Williams and Wilkins, pp 11–14.

Owen M (1987) Marrow stromal cells. J Cell Sci (Suppl) 10:63–76.

Owen M, Friedenstein AJ (1988) Stromal stem cells: Marrow-derived osteogenic precursors. Ciba Foundation Symposium 136: 420–460.

Papandrea P, Vulpiani MC, Ferretti A, Conteduca F (2000) Regeneration of the semitendinosus tendon harvested for anterior cruciate ligament reconstruction. Am J Sports Med 28:556–561.

Parker SB, Eichele, Zhang P, Rawls A, Sands AT, Bradley A, Olsen EN, Harper JW, Elledge SJ (1995) p53-independent expression of p21Cip1 in muscle and other terminally differentiating cells. Science 267:1024–1027.

Pastoret C, Sebille A (1995) ge-related differences in regeneration of dystrophic (Mdx) and normal muscle in the mouse. Muscle Nerve 18:1147–1154.

Pastoret C, Partridge TA (1998) Muscle regeneration. In: Ferretti P, Geraudie J (eds.). Cellular and Molecular Basis of Regeneration. New York, John Wiley and Sons, pp 309–334.

Pathi S, Rutenberg J, Johnson R, Vortkamp A (1999) Interaction of Ihh and BMP/Noggin signaling during cartilage differentiation. Dev Biol 209:239–253.

Pavlath GK, Thaloor D, Rando TA, Cheong M, English AW, Zheng B (1998) Heterogeneity among muscle precursor cells in adult skeletal muscles with differing regenerative capacities. Dev Dynam 212:495–508.

Pfeilschifter J, D'Souza S, Mundy GR (1986) Transforming growth factor beta is released from resorbing bone and stimulates osteoblast activity. J Bone Miner Res 1 (Suppl 1): 294.

Pfeilschifter J, Bonewald L, Mundy GR (1987) TGF-beta is released from bone with one or more binding proteins which regulate its activity. J Bone Miner Res 2 (Suppl 1): 249.

Pittinger MF, Mackay AM, Beck SC, Jaiswal RK, Douglas R, Mosca JD, Moorman MA, Simonetti W, Craig S, Marshak DR (1999) Multilineage potential of adult human mesenchymal stem cells. Science 284:143–147.

Pittinger MF, Marshak DR (2001) Adult mesenchymal stem cells. In: Marshak DR, Gardner RL, Gottlieb D (eds.). Stem Cell Biology. Cold Spring Harbor, Cold Spring Harbor Laboratory Press, pp 349–373.

Polesskaya A, Seale P, Rudnicki MA (2003) Wnt signaling induces the myogenic specification of resident $CD45^+$ adult stem cells during muscle regeneration. Cell 113:841–852.

Rapraeger AC (2000) Syndecan-regulated receptor signaling. J Cell Biol 149:995–998.

Reddi AH (1998) Initiation of fracture repair by bone morphogenetic proteins. Clin Orthopaed Rel Res 355S:S66–S72.

Reddi AH (2003) Cartilage morphogenetic proteins: Role in joint development, homeostasis, and regeneration. Ann Rheum Dis 62: ii73–ii78.

Relaix F, Rocancourt D, Mansouri A, Buckingham M (2005) A pax3/Pax-7 dependent population of skeletal muscle progenitor cells. Nature 435:948–953.

Robertson TA, Grounds MD, Papadimitriou JM (1992) Elucidation of aspects of murine skeletal muscle regeneration using local and whole body irradiation. J Anat 181:265–276.

Robey P (1996) Bone matrix proteoglycans and glycoproteins. In: Principles of Bone Biology. Bilezikian JP, Raisz LG, Rhodan GA (eds.). New York, Academic Press, pp 155–165.

Rodan GA, Martin TJ (2000) Therapeutic approaches to bone diseases. Science 289:1508–1514.

Rogers SL (1982) Muscle spindle formation and differentiation in regenerating rat muscle grafts. Dev Biol 94:265–283.

Rosen V, Thies RS (1992) The BMP proteins in bone formation and repair. Trends Genet 8:97–102.

Rosier RN, O'Keefe RJ, Hicks DG (1998) The potential role of transforming growth factor beta in fracture healing. Clin Orthopaed Rel Res 355S:S294–S300.

Saito T, Dennis JE, Lennon DP, Young RG, Caplan AI (1995) Myogenic expression of mesenchymal stem cells within myotubes of *mdx* mice *in vitro* and *in vivo*. Tissue Eng 1:327–343.

Salter RB (1983) Textbook of Disorders and Injuries of the Musculoskeletal System. Baltimore, The Williams & Wilkins Co.

Salter RB, Harris DJ, Clements ND (1978) The healing of bone and cartilage in transarticular fractures with continuous passive motion. Orthop Trans 2:77.

Schultz E (1996) Satellite cell proliferative compartments in growing skeletal muscles. Dev Biol 175:84–94.

Seale P, Rudnicki MA (2000) A new look at the origin, function, and "stem-cell" status of muscle satellite cells. Dev Biol 218:115–124.

Seale P, Sabourin LA, Girgis-Gabardo A, Mansouri A, Gruss P, Rudnicki MA (2000) Pax7 is required for the specification of myogenic satellite cells. Cell 102:777–786.

Seale P, Ishibashi J, Scime A, Rudnicki MA (2004) Pax7 is necessary and sufficient for the myogenic specification of $CD45^{+}$:$Sca1^{+}$ stem cells from injured muscle. PloS Biol 2:664–672.

Sherwood R, Christensen JL, Conboy IM, Conboy MJ, Rando TA, Weissman IL, Wagers AJ (2004) Isolation of adult mouse myogenic progenitors: functional heterogeneity of cells within and engrafting skeletal muscle. Cell 119:543–554.

Smith CK, Janney MJ, Allen RE (1994) Temporal expression of myogenic regulatory genes during activation, proliferation, and differentiation of rat skeletal muscle satellite cells. J Cell Physiol 159:379–385.

Song K, Wang Y, Sassoon D (1992) Expression of *Hox-7.1* in myoblasts inhibits terminal differentiation of and induces cell transformation. Nature 360:477–481.

Stanton H, Rogerson FM, East CJ, Golub SB, Lawlor KE, Meeker CT, Little CB, Last K, Farmer PJ, Campbell IK, Fourie AM, Fosang AJ (2005) ADAMT5 is the major aggrecanase in mouse cartilage *in vivo* and *in vitro*. Nature 434:648–652.

Stockwell RA (1975) Biology of Cartilage Cells. Cambridge, Cambridge University Press, pp 213–240.

Tajbakhsh S, Rocancourt D, Cossu G, Buckingham M (1997) Redefining the genetic hierarchies controlling skeletal myogenesis: *Pax-3* and *Myf-5* act upstream of *MyoD*. Cell 89:127–138.

Tajbakhsh S, Borelle U, Vivarelli E, Kelly R, Papkoff JJ, Duprez D, Buckingham M, Cossu G (1998) Differential activation of Myf5 and MyoD by different Wnts in explants of mouse paraxial mesoderm and the later activation of myogenesis in the absence of Myf5. Development 125:4155–4162.

Takayanagi H, Kim S, Matsuo K, Suzuki H, Suzuki T, Fato K, Yokochi T, Oda H, Nakamura K, Ida N, Wagner EF, Taniguchi T (2002) RANKL maintains bone homeostasis through c-Fos-dependent induction of *interferon*-β. Nature 416:744–749.

Takeda S, Elefteriou F, Levasseur R, Liu X, Zhao L, Parker K, Armstrong D, Ducy P, Karsenty G (2002) Leptin regulates bone formation via the sympathetic nervous system. Cell 111:305–317.

Tamai K, Semenov M, Kato Y, Spokony R, Liu C, Katsuyama Y, Hess F, Saint-Jeannet JP, He X (2000) LDL-receptor-related proteins in Wnt signal transduction. Nature 407:530–535.

Tatsumi R, Anderson E, Nevoret CJ, Halevy O, Allen RA (1998) HGF/SF is present in normal adult skeletal muscle and is capable of activating satellite cells. Dev Biol 194:114–128.

Tavassoli M, Crosby WH (1968) Transplantation of marrow to extramedullary sites. Science 161:548–556.

Teitlebaum SL (2000) Bone resorption by osteoclasts. Science 289: 1504–1508.

Triffit JT (1987) Initiation and enhancement of bone formation: A review. Acta Orthop Scand 58:673–684.

Urist MR (1965) Bone: Formation by autoinduction. Science 150:893–899.

Urist MR, McLean FC (1962) Osteogenic potency and new bone formation by induction in transplants to the anterior chamber of the eye. J Bone Joint Surg 34A:443–446.

Vortkamp A (2001) Interaction of growth factors regulating chondrocyte differentiation in the developing embryo. Osteoarth Cartilage 9 (Suppl A):S109–S117.

Vortkamp A, Lee K, Lanske B, Segre GV, Kronenberg H, Tabin CJ (1996) Regulation of rate of cartilage differentiation by Indian hedgehog and PTH-related protein. Science 273:613–622.

Vortkamp A, Pathi S, Peretti GM, Caruso EM, Zaleske DJ, Tabin CJ (1998) Recapitulation of signals regulating embryonic bone formation during postnatal growth and in fracture repair. Mech Dev 71:65–76.

Wagner KR, Max SR, Grollman M, Koski CL (1976) Glycolysis in skeletal muscle regeneration. Exp Neurol 52:40–48.

Wagner KR, Carlson BM, Max SR (1977) Developmental patterns of glycolytic enzymes in regenerating skeletal muscle after autogenous free grafting. J Neurol Sci 34:373–390.

Wehrli M, Dougan ST, Caldwell K, O'Keefe L, Schwartz S, Vaizel-Ohayon D, Schejter E, Tomlinson A, Di Nardo S (2000) Arrow encodes a D-receptor-related protein essential for wingless signaling. Nature 407:527–530.

Weis J, Kaussen M, Calvo S, Buonanno A (2000) Denervation induces a rapid nuclear accumulation of MRF4 in mature myofibers. Dev Dynam 218:438–451.

Wewer UM, Iba K, Durkin ME, Nielsen FC, Loechel F, Gilpin BJ, Kuang W, Engvall E, Albrechtsen R (1998) Tetranectin is a novel marker for myogenesis during embryonic development, muscle regeneration, and muscle cell differentiation *in vitro*. Dev Biol 200:247–259.

Whyte M, Mumm S (2004) Heritable disorders of the RANKL/OPG/RANK signaling pathway. J Musculoskel Neuron Interact 4:254–267.

Wolfman NM, Hattersley G, Cox K, Celeste AJ, Nelson R, Yamaji N, Dube JL, DiBiasio-Smith E, Nove J, Song JJ, Wozney JM, Rosen V (1997) Ectopic induction of tendon and ligament in rats by growth and differentiation factors 5, 6 and 7, members of the TGF-β family. J Clin Invest 100:321–330.

Wong M, Kireeva ML, Kolesnikova TV, Lau LF (1997) Cyr61, product of a growth factor-inducible immediate-early gene, regulates chondrogenesis in mouse limb bud mesenchymal cells. Dev Biol 192:492–508.

Woo SL-Y, Hildebrand K, Watanabe N, Fenwick JA, Papageorgiou CD, Wang JH-C (1999) Tissue engineering of ligament and tendon healing. Clin Orthopaed Related Res 367S:S312–S323.

Wornam IL, Buchman SR (1992) Bone and cartilaginous tissue. In: Cohen IK, Diegelmann RF, Lindblad W (eds.). Wound Healing: Biochemical And Clinical Aspects. Philadelphia, WB Saunders Co, pp 356–383.

Wronski T (2001) Skeletal effects of systemic treatment with basic fibroblast growth factor. J Musculoskel Neuron Interact 2:9–14.

Yablonka-Reuvini Z, Quinn LS, Nameroff M (1987) Isolation and clonal analysis of satellite cells from chicken pectoralis muscle. Dev Biol 119:252–259.

Yang LJ, Jin Y (1990) Immunohistochemical observations on bone morphogenetic protein in normal and abnormal conditions. Clin Orthop 257:249–256.

Zammit P, Beauchamp J (2001) The skeletal muscle satellite cell: Stem cell or son of stem cell? Differentiation 68:193–204.

Zammit PS, Heslop L, Hudon V, Rosenblatt JD, Tajbakhsh S, Buckingham ME, Beauchamp JR, Partridge TA (2002) Kinetics of myoblast proliferation show that resident satellite cells are competent to fully regenerate muscle. Exp Cell Res 281:39–49.

Zelzer E, Olsen BR (2003) The genetic basis for skeletal diseases. Nature 423:343–348.

Zhao P, Hoffman EP (2004) Embryonic myogenesis pathways in muscle regeneration. Dev Dynam 229:380–392.

Zou H, Wieser R, Massague J, Niswander L (1997) Distinct roles of type I bone morphogenetic protein receptors in the formation and differentiation of cartilage. Genes Dev 11:2191–2203.

# 10 Regenerative Medicine of Musculoskeletal Tissues

## INTRODUCTION

Disorders of the musculoskeletal system are responsible for an increasing number of medical and surgical treatments, particularly in developed countries. Many genetic disorders affect the muscles and skeleton, but these are relatively rare. Traumatic injuries to muscles, ligaments, and tendons, and articular cartilage and meniscus are increasingly common, as are degenerative diseases such as rheumatoid arthritis and osteoarthritis that are associated with aging of the population. These are the market drivers for anti-inflammatory drugs, orthopedic surgery, and the development of means to replace or regenerate failing joint tissues. Here, regenerative therapies for muscle damaged by muscular dystrophy and the construction of bioartificial muscle will be presented. The remainder of the chapter will focus on regenerative therapies for meniscus, articular cartilage, bone, tendon and ligament via chemical induction, cell transplants, and implants of bioartificial tissues. With a few exceptions, most of these therapies are still in experimental stages.

## REGENERATIVE THERAPIES FOR MUSCLE

### 1. Muscular Dystrophy

Muscular dystrophy is a diverse set of genetic myopathies characterized by progressive muscle weakness, atrophy, and replacement of myofibers by fat and scar tissue (Bushby, 2000; Burton and Davies, 2002). Several forms of the disease are caused by mutations in components of the dystrophin-glycoprotein complex (DGC), which links the cytoskeleton of the myofiber to the basement membrane (**FIGURE 10.1**). The proteins comprising the complex are the intercellular proteins dystrophin and the syntrophins, and the sarcolemmal proteins, the dystroglycans, sarcoglycans, and sarcospan (Cohn and Campbell, 2000). The DGC plays an important role in protecting the sarcolemma against muscle contraction-induced injury. Disruption of the DGC linkage causes sarcolemmal instability and structural damage that leads to increased calcium influx and myofiber death and necrosis, the major pathological finding in muscular dystrophies (Alderton and Steinhardt, 2000; Bushby, 2000; Cohn and Campbell, 2000).

There are three mouse models of muscular dystrophy, the dystrophin-deficient *mdx* mouse, the β or δ sarcoglycan-deficient (SGD) mouse, and the muscle creatine kinase-dystroglycan (MCK-DG) null mouse, in which the dystroglycan gene is inactivated by a *Cre-lox* construct driven by the MCK promoter (Cohn et al., 2002). All these models maintain muscle mass by regeneration from satellite cells during the early stage of the disease, but since the SCs also carry the genetic defect, the regenerated *mdx* and SGD mouse myofibers are dystrophic as well. Eventually the muscles exhaust their pool of SCs and become fatty and fibrotic (Cossu and Mavilio, 2000; Heslop et al., 2000). By contrast, MCK-DG myofibers are able to regenerate throughout life and the disease is mild (Cohn et al., 2002). This is because the null *Cre-lox* construct affects only the myofibers, leaving the SCs with normal dystroglycan expression, suggesting that muscular dystrophy in other mouse models and in humans might be treated by injections of normal SCs into the muscles.

Cell transplants have been tried many times over the last 25 years to cure muscular dystrophy. Syngeneic or allogeneic SCs or C2C12 cells were injected into the limb muscles of dystrophic mice in an attempt to improve their structure and function (Law, 1982; Partridge et al., 1989; Morgan et al., 1990; Saito et al., 1995; Blaveri et al., 1999). Wild-type SCs were reported to incorporate into the dystrophic muscle fibers, resulting in increases in cross-sectional area, total fiber number, membrane potentials, and twitch and tetanus tensions. Histological structure was temporarily improved and dystrophin levels were increased, particularly if the muscles were x-irradiated to suppress proliferation of host satellite cells. Initial clinical trials

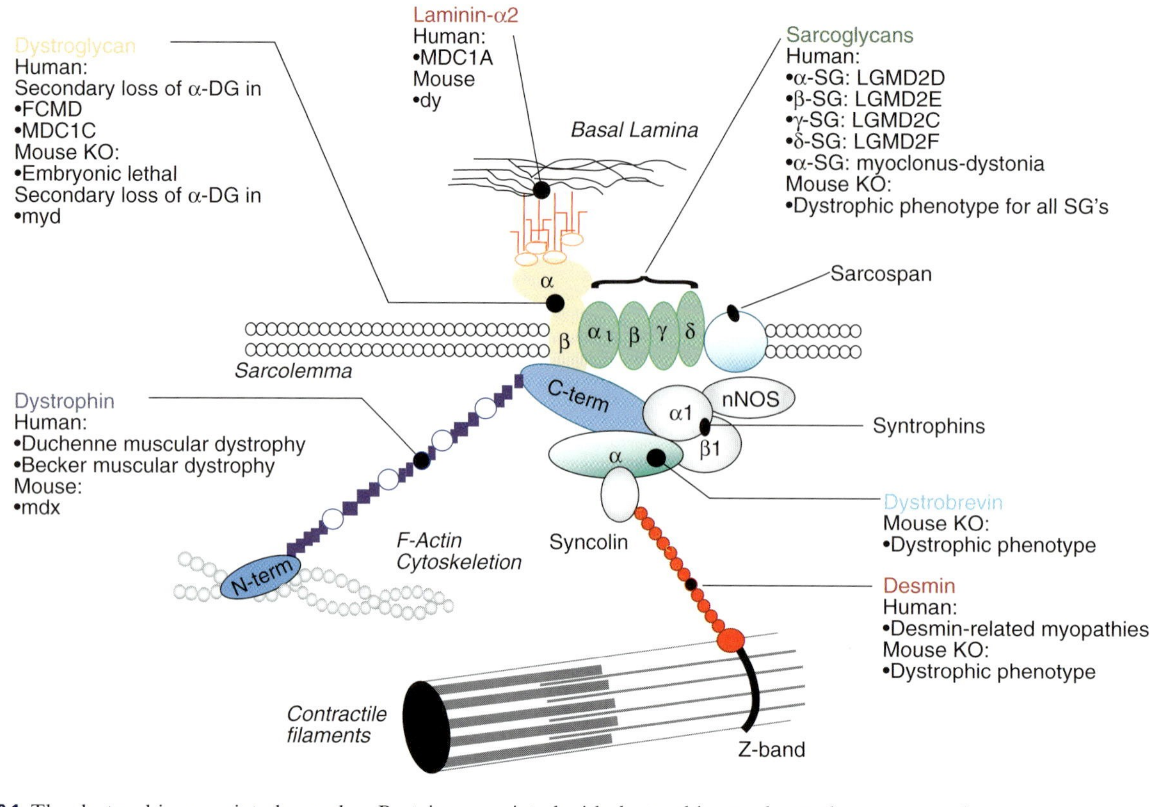


**FIGURE 10.1** The dystrophin-associated complex. Proteins associated with dystrophin are dystroglycans, sarcoglycans, sarcospan, and dystrobrevins. The complex serves to protect the sarcolemma from the local forces that develop during muscle contraction. Reproduced with permission from Burton and Davies, Muscular dystrophy—reason for optimism? Cell 108:5–8. Copyright 2002, Elsevier.

injecting allogeneic myoblasts into the muscles of non-immunosuppressed Duchenne muscular dystrophy patients indicated a transient restoration of dystrophin$^+$ fibers and improvement of muscle strength, which was then lost (Law, 1991; Huard et al., 1992a, b; Karpati et al., 1993; Mendell et al., 1995; Gussoni et al., 1992; Tremblay and Guerette, 1997; Beauchamp et al., 1997). The injected myoblasts exhibited low rates of spread and fusion, but the primary culprit in the loss of new myofibers was immunorejection, as shown by the demonstration of serum antibodies to dystrophin and dystrophin-associated proteins in some patients (Huard, 1992b) as well as the effective transplantation of myoblasts to nude, *scid*, and immunosuppressed mice (Huard et al., 1994a, b; Kinoshita et al., 1994). Better muscle regeneration and partial restoration of dystrophin were obtained in *mdx* mice by transplanting an expanded, high-proliferation subpopulation of SCs expressing desmin, CD34, Bcl-2, c-met, MNF, Flk-1, and Sca-1, and infected with a retroviral vector expressing interleukin-1 receptor antagonist protein (IL-1Ra), which inhibits the inflammatory cytokine IL-1 (Qu et al., 1998; Lee et al., 2000). These results suggest that selection of the SC subpopulation with the highest myogenic capability, as well as control of inflammation and immunorejection, are essential for success in alleviating human muscular dystrophy by SC transplant.

Most SC transplants for muscular dystrophy have used expanded populations of SCs. Recent evidence, however, suggests that expanding SCs *in vitro* is associated with a reduction in their regenerative capacity (Montarras et al., 2005). Grafting $10^4$ freshly isolated and purified [Pax7, CD34]$^+$ [CD45$^-$, Sca1$^-$] SCs from wild-type muscles into the muscles of *mdx* nude mice, where immunorejection is not a confounding factor, led to the restoration of dystrophin expression in an average number of 300 fibers. Grafting the same number of SCs expanded for three days in culture restored dystrophin expression in an average of only 88 fibers. Clonal assays showed that the cultured SCs had a lower proliferation potential and a tendency to differentiate more rapidly.

Regardless, allogeneic SC or myoblast transplants would require immunosuppression to prevent rejection. The use of autogeneic SCs transfected *ex vivo* with wild-type dystrophin or utrophin genes theoretically would overcome the immunorejection problem, but these genes are large and it is difficult to design a vector that can accommodate them. Furthermore, SCs become exhausted early in the life of dystrophic patients, due to the continual cycles of degeneration and regenera-

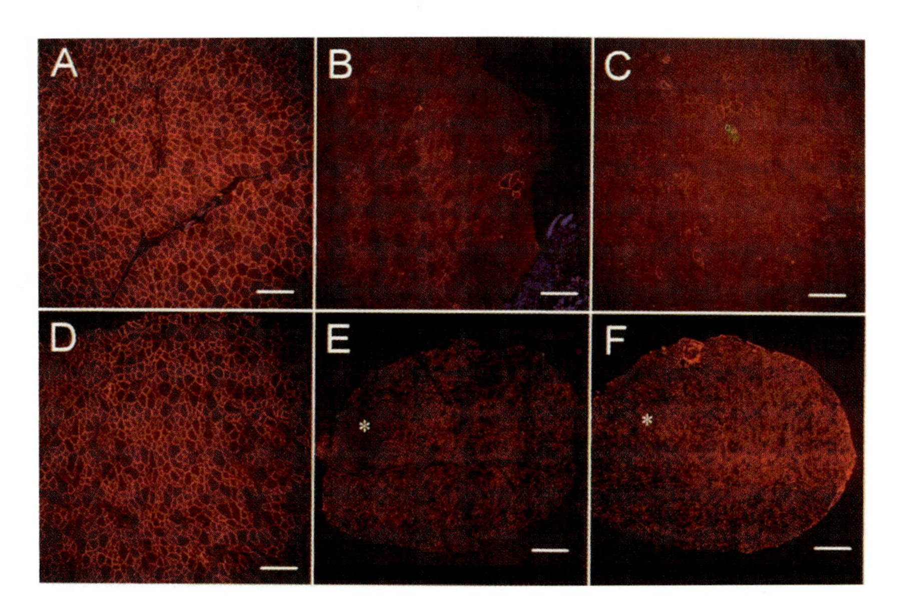


**FIGURE 10.2** Rescue of dystrophin in *mdx* mice by exon skipping. Transverse sections of tibialis anterior and extensor digitorum longus (asterisk) muscles immunostained with dystrophin antibody. Dystrophin appears red. **(A)** Normal C57BL6 muscle. Myofibers all stain for dystrophin. **(B)** Untreated *mdx* muscle showing loss of dystrophin. **(C-E)** *mdx* muscle 2, 4 and 13 weeks after intramuscular injection of vector showing recovery of dystrophin. **(F)** *mdx* muscle 4 weeks after intra-arterial injection of vector. Reproduced with permission from Goyenvalle et al., Rescue of dystrophic muscle through U7 snRNA-mediated exon skipping. Science 306:1796–1799. Copyright 2004, AAAS.

tion. The few SCs that can be recovered from muscle biopsies grow poorly *in vitro* and undergo rapid senescence (Cossu and Mavilio, 2000). A different genetic approach may be more promising. Most mutations in the dystrophin gene create a frameshift in the dystrophin mRNA that results in a lack of functional dystrophin. The *mdx* mouse carries a C to T mutation in exon 23 that creates a stop codon. Naturally occurring exon skipping gives rise to rare revertant fibers containing truncated, but functional, dystrophin proteins (Lu et al., 2000). Goyenvalle et al. (2004) have reported the achievement of long-term exon skipping in the muscle fibers of *mdx* mice by blocking translation of the mutated mRNA of exon 23 with an adenoviral construct containing antisense sequences linked to the U7 small nuclear RNA. When injected into the tibialis anterior muscle, the construct restored the histology and function of the muscle fibers to normal (**FIGURE 10.2**).

Another approach has been to transplant bone marrow cells into dystrophic muscle (Saito et al., 1995; Ferrari et al., 1998; Gussoni et al., 1999). The MSCs of bone marrow have the capacity to differentiate into muscle and are reported to be inherently nonimmunogenic because they do not constitutively express MHC class II antigens or the T-cell co-stimulatory molecule B7 (McIntosh and Bartholomew, 2000) and they inhibit the differentiation and activation of antigen-presenting dendritic cells (Zhang et al., 2004). Early attempts to use this approach were not very successful because of the very low starting number of MSCs in the transplants and still lower numbers of these cells that differentiate into muscle. Better results have been obtained by Dezawa et al. (2005). They expanded rat and human MSCs *in vitro* and then converted them to SCs by transfecting them with a gene encoding the Notch1 intracellular domain (NICD) and treating them with muscle stem cell conditioned medium. Clonal analysis of these treated cells indicated that 89% of the viable clones developed into multinucleated cells expressing skeletal muscle markers. When labeled with GFP and transplanted into normal or *mdx* mouse muscles damaged by cardiotoxin, conditioned medium-treated human MSCs were incorporated into a high percentage of regenerated myofibers. When these fibers were reinjured, many $GFP^+$ myofibers were again regenerated, indicating that some of the MSCs had been set aside as self-renewing satellite cells. Thus, transplantation of autologous MSCs induced by NICD and genetically manipulated to produce dystrophin, or transplantation of closely HLA-matched allogeneic NICD-MSCs may be useful in the treatment of muscular dystrophy.

## 2. Bioartificial Muscle

Bioartificial muscles theoretically could be used to replace whole muscles. Myooids have been made *in vitro* by co-culturing rat satellite cells and fibroblasts on strips of SYLGARD substrate with laminin-coated suture anchors pinned in place at either end 12 mm

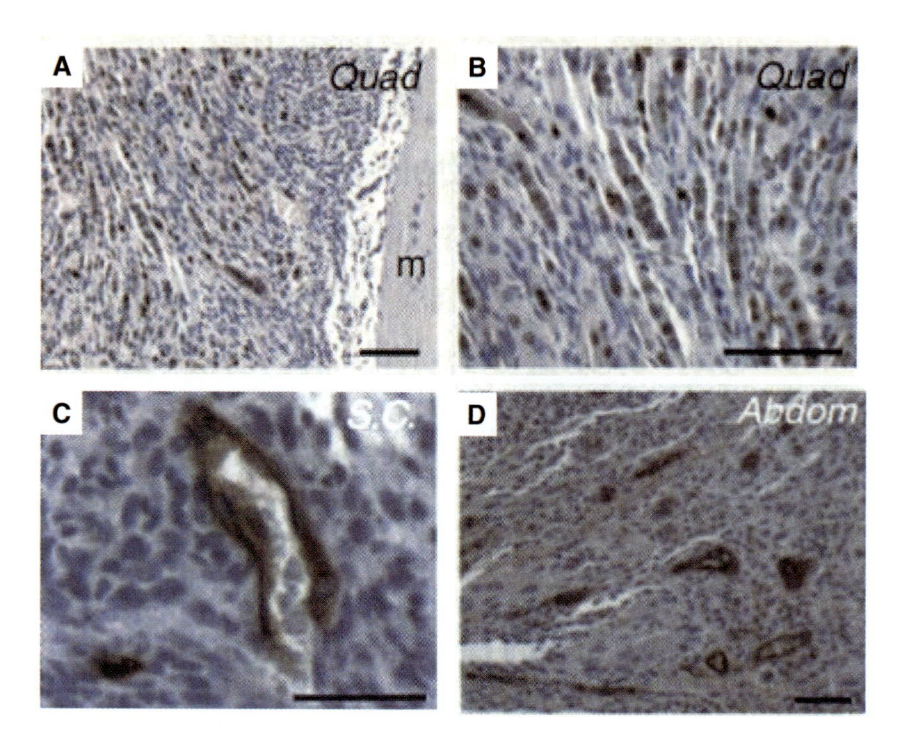


**FIGURE 10.3** *In vivo* analysis of bioartificial muscle constructs consisting of myoblasts, fibroblasts and endothelial cells implanted into immunodeficient hosts. **(A, B)** Constructs implanted into the quadriceps muscle and stained for myogenin, showing differentiation into aligned, multinucleated myotubes. **(C)** Construct implanted subcutaneously and stained with antibody to human CD31 to detect endothelial cells (brown). **(D)** Construct implanted into the abdominal muscles and stained with antibody to Von Willebrand factor to detect endothelial cells (brown). The muscle in **C and D** is clearly vascularized. Reproduced with permission from Levenberg et al., Engineering vascularized skeletal muscle tissue. Nature Biotech 23:879–884. Copyright 2005, Nature Publishing Group.

apart (Kosnik et al., 2001). As the SCs and fibroblasts proliferate to form a confluent layer, the SCs fuse to form myotubes that attach to the suture anchors and begin to contract spontaneously. The contractions delaminate the monolayer from the SYLGARD and it remodels itself into a cylinder of myogenic tissue surrounded by fibroblasts. As development of the myooid continues, septa comprised of fibroblasts and ECM form to create fascicles surrounded by perimysium. The isometric tetanic force generated by the myooids after electrical stimulation is about 1% of the control value for adult muscles.

Myooids grown in this way are very small in diameter (0.3–0.4 mm) because they must depend on diffusion for gas exchange and nutrients. Normal muscle is highly vascularized because of its tremendous energy needs, so the development of larger bioartificial muscles requires a way to vascularize the constructs *in vitro*. Levenberg et al. (2005) have grown a combination of myoblasts, embryonic fibroblasts, and endothelial cells on highly porous, three-dimensional sponges of 50% poly-(L-lactic acid) and 50% polylactic-glycolic acid. By one month in culture, the cells had formed muscle with stable endothelial vessels. Cultured muscle implanted into SCID mice subcutaneously or into the quadriceps muscle, or grafted in place of the anterior abdominal muscle segment continued to differentiate and was permeated with host vessels (**FIGURE 10.3**). Differentiation was poor when the muscle was implanted from cultures without endothelial cells. These results suggest that prevascularization prior to implantation greatly improves the survival and differentiation of bioartificial muscle. No measurements were made of the diameters of the cultured or *in vivo* muscles, but the authors reported that the myotubes *in vivo* were "relatively long and thick."

## REGENERATIVE THERAPIES FOR MENISCUS AND ARTICULAR CARTILAGE

### 1. Meniscus

The medial and lateral menisci are two C-shaped fibrocartilages that act as proprioceptive sensors for knee motion and as shock absorbers to cushion the articular cartilages of the femur and tibia (**FIGURE 10.4**). Meniscus tears are among the most frequently encountered knee injury and age-related problems in orthopedic practice. Meniscus injury or degeneration

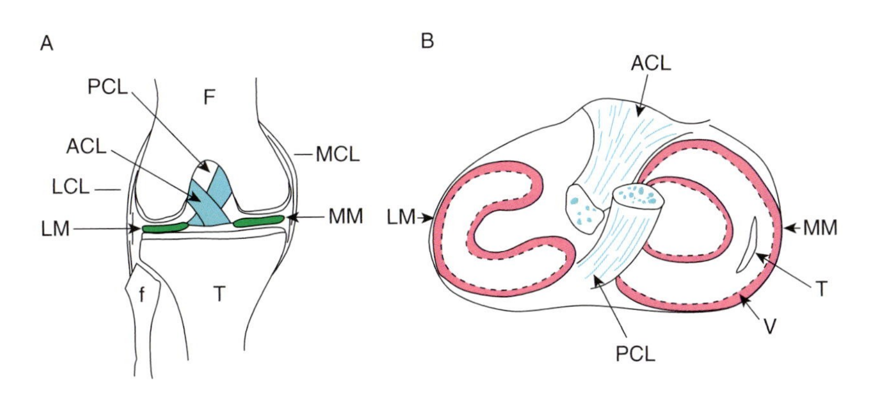


**FIGURE 10.4** Shape and spatial relationships of the menisci with respect to other structures in the human knee joint. **(A)** Front view of knee. **(B)** View of the tibial surface of the knee. F = femur, T = tibia, f = fibula, ACL = anterior cruciate ligament, PCL = posterior cruciate ligament, LCL = lateral collateral ligament, MCL = medial collateral ligament, MM = medial meniscus, LM = lateral meniscus, V (red) = vascularized rim of the menisci. T = tear in the non-vascular meniscal tissue.

leads to osteoarthritis of the articular cartilages of the knee joint. The menisci have some capacity for regeneration in their outer vascularized fibrocartilage zone, but little repair capacity in the nonvascular internal zone, which is composed of hyaline cartilage like that of the joint surface. The opinion of most orthopedic surgeons is that tears in any part of the meniscus do not heal in persons over the age of 60. Current treatments for meniscus tears are surgical repair, sometimes with the use of a fibrin clot or synovial flap, or partial or complete removal of the meniscus, which can accelerate osteoarthritic conditions. These treatments alleviate pain, but often do not stabilize the knee. Synthetic replacements have not produced good long-term results.

Allografts of cadaver meniscus have been more successful. Wirth et al. (2002) evaluated the results of lyophilized and deep-frozen allografts in 23 patients at 3 years and 14 years post-transplant. Significant gains in the Lysholm score were observed at three years and imaging showed that the deep-frozen grafts had undergone little deterioration. There was a slight drop in Lysholm score at 14 years, but imaging showed good preservation of the deep-frozen grafts. Lyophilized transplants did not fare as well and showed considerable deterioration. There is a shortage of meniscal donors, however, so ways have been sought to induce the regeneration of torn meniscus tissue using regeneration templates or to replace the meniscus entirely with bioartificial meniscus (Ibarra et al., 1997; Buma et al., 2004).

### a. Induction of Regeneration with Regeneration Templates

Acellular biomimetic scaffolds have been used in attempts to stimulate the regeneration of meniscus tissue. Tienen et al. (2003) implanted a polyurethane (ι-lactide/ε-caprolactone) scaffold into longitudinal lesions made in the avascular region of dog menisci. An access channel, 50% of the thickness of the meniscus, was made from the vascular perimeter to the lesion in order to encourage vascularization. The dogs regained their normal gait pattern by 10 days post-surgery. Six months post-surgery, histological studies indicated that the defect had filled primarily with fibrous tissue, although occasional islands of cartilage had formed.

A collagen/hyaluronic acid/chondroitin sulfate scaffold was reported to promote meniscus regeneration in dogs (Stone et al., 1990, 1992). A three-year phase I clinical trial of this scaffold was carried out in nine patients with meniscus defects of various sizes. Arthroscopy and MRI showed that the scaffold-induced regeneration of fibrocartilage that appeared similar to normal meniscus cartilage (Stone et al., 1997). At 36 months, the patients reported a drop in pain score of nearly 4-fold and an activity score that was near the pre-injury score.

Pig SIS was also reported to promote the regeneration of medial meniscus tissue in dogs (Cook et al., 1999). Four weeks after implanting the scaffold, lameness scores had decreased to pre-operative values. Histological studies, immunochemical staining for collagen II, and quantification of GAG content showed that the regenerated tissue closely resembled normal meniscal tissue.

### b. Bioartificial Meniscus

The development of bioartificial meniscus constructs in animals has been followed *in vitro* after subcutaneous implantation and after implantation into meniscal lesions (reviewed by Buma et al., 2004). The cell sources for these constructs have been the meniscus itself, articular chondrocytes, or MSCs of the bone marrow.

Mueller et al. (1999) reported that calf or canine meniscus cells seeded into a type II collagen-chondroitin-6-sulfate scaffold doubled their DNA content and increased their GAG content by 50% over three weeks of culture. The constructs did not contract during this period. Spherical chondrocyte-like cells and fibroblast-like cells were found throughout the matrix in approximately equal numbers. Iberra et al. (1997) seeded bovine meniscal chondrocytes into either a polyglycolic acid scaffold or calcium alginate gel and implanted the constructs subcutaneously into nude mice. They reported that over the course of 16 weeks, the cells formed a matrix similar to normal meniscal repair tissue.

Peretti et al. (2001) seeded chips of devitalized sheep meniscus with allogeneic articular chondrocytes and placed the chips into a bucket-handle lesion made in a larger piece of devitalized meniscus. After subcutaneous implantation of these constructs into nude mice, the chondrocytes synthesized a matrix that bonded the chips to the edges of the incision. No bond was made when the chips were unseeded. Similar chips were made with pig meniscus and chondrocytes and transplanted into longitudinal tears of the avascular meniscus (Peretti et al., 2004) (**FIGURE 10.5**). Good bonding of the chips to the margins of the lesions was reported in a number of cases (**FIGURE 10.6**), but no repair was noted in any of the untreated controls or controls treated with unseeded meniscus chips.

Walsh et al. (1999) embedded bone marrow-derived MSCs in a type I collagen sponge and found that this construct formed more fibrocartilage and hyaline cartilage than the sponge alone when grafted into medial meniscus defects in rabbits. However, the construct did not have the initial tensile strength required for secure anterior attachment to the intermeniscal ligament site.

These results suggest that acellular scaffolds can promote meniscal regeneration and that the construction of bioartificial meniscus is feasible. However, a number of questions remain. We do not know how well regenerated meniscus tissue or bioartificial meniscus will hold up under biomechanical load over the long term. Cell sources for bioartificial meniscus constructs remain problematic. The ideal cell source is meniscus cells from the patient, but it may not be easy to culture enough autogeneic human meniscus cells to make a

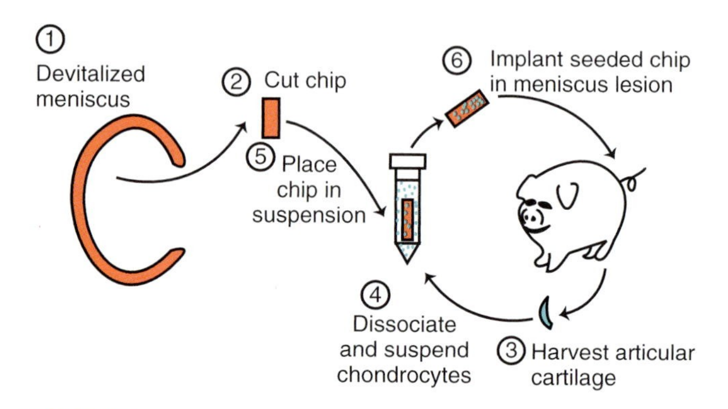


**FIGURE 10.5** Experimental protocol for testing the ability of autologous chondrocytes to regenerate meniscus tissue in pigs. **(1)** The inner portion of a pig devitalized meniscus was removed and **(2)** a flat chip was cut from it. **(3, 4)** Articular cartilage was harvested and a suspension of articular chondrocytes was prepared by enzymatic digestion of slices of articular cartilage. **(5)** The meniscal chips were then incubated with the chondrocyte suspension. **(6)** Once seeded, the chips were implanted in a curved (bucket-handle) lesion made in the non-vascular portion of a meniscus of the same pig that donated the chondrocytes.

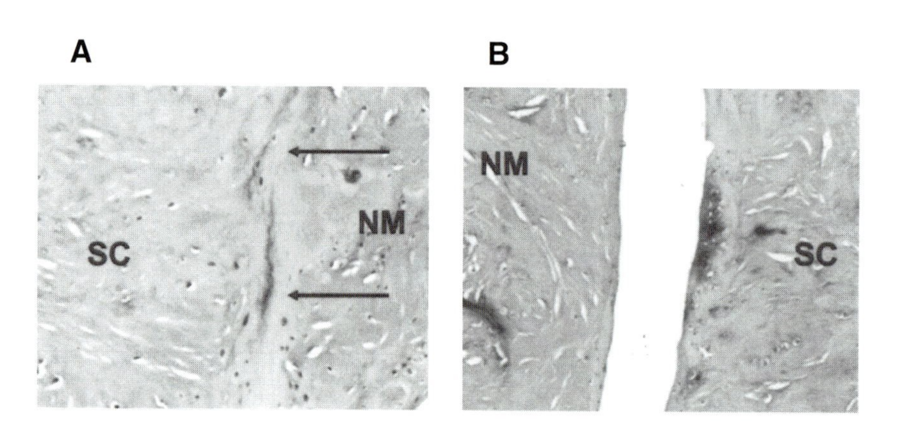


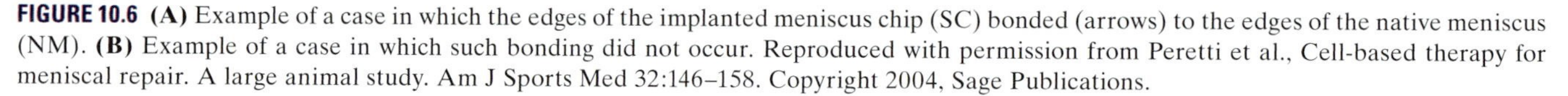

**FIGURE 10.6** **(A)** Example of a case in which the edges of the implanted meniscus chip (SC) bonded (arrows) to the edges of the native meniscus (NM). **(B)** Example of a case in which such bonding did not occur. Reproduced with permission from Peretti et al., Cell-based therapy for meniscal repair. A large animal study. Am J Sports Med 32:146–158. Copyright 2004, Sage Publications.

bioartificial meniscus. Finally, neither acellular scaffolds nor bioartificial meniscus prevents the appearance or progression of osteoarthritis. Advances in meniscal regeneration or bioartificial construction clearly will depend on choosing the right cell source and/or the right scaffold, coupled with an understanding of the factors that promote the original development of meniscus tissue.

## 2. Articular Cartilage

Defects in articular cartilage due to osteoarthritis or injury are extremely common. How a cartilage defect repairs depends on whether only the cartilage is injured, or the injury also involves the subchondral bone (see Chapter 6). Full-thickness osteochondral defects penetrate through the subchondral bone into the marrow cavity. These defects repair by the formation of a poorly functional fibrocartilaginous tissue derived from MSCs of the bone marrow (Shapiro et al., 1993). Partial defects that injure only the hyaline cartilage or the cartilage and subchondral bone without penetrating the marrow cavity grow larger in area and deeper with wear or disease progression and never repair spontaneously, because they have no access to bone marrow MSCs.

Many cases of severe damage to articular cartilage are treated simply by surgically removing the damaged tissue without any replacement. There are few even partially effective treatments for replacing damaged articular cartilage (O'Driscoll, 1998). Subchondral bone drilling, abrasion chondroplasty, and microfracture are procedures designed to recruit MSCs from the bone marrow into the lesion to regenerate the cartilage, but repair is no better than in a natural full osteochondral defect. Some success has been had with periosteal arthroplasty, in which periosteum is grafted over the lesion. Grafting plugs of autogeneic cartilage from nonweight-bearing regions of the knee into cartilage defects (mosaicplasty, a procedure analogous to hair follicle transplants), or grafting larger autogeneic or allogeneic osteochondral fragments has been reported to have a high success rate in alleviating symptoms (O'Driscoll, 1998). However, histological studies of autologous osteochondral grafts in sheep knees showed a lack of integration of the cartilage part of the graft with host tissue and degeneration of the cartilage three months after grafting (Tibesku et al., 2004), suggesting that autogeneic transplants are not an effective long-term treatement for cartilage defects.

Regenerative therapies being explored are the induction of cartilage regeneration *in situ* by growth factors and acellular regeneration templates, chondrocyte or MSC transplants, and implantation of bioartificial cartilage constructs (Hunziker, 1999; Grande et al., 1999; Tuan, 2004; Johnstone and Yoo, 1999; Otto and Rao, 2004).

### a. Induction of Cartilage Regeneration by Growth Factors and Regeneration Templates

Regeneration templates and growth factors known to have a role in chondrogenesis have been implanted into cartilage defects to encourage the regeneration of articular cartilage from local cells. In general, the results have been less satisfactory for partial defects than for full-thickness defects.

#### Growth Factors

TGF-β1 and 2 were reported to stimulate articular cartilage repair in the knee joints of mice rendered arthritic by intra-articular injection of zymosan (Glansbeek et al., 1998). These growth factor isoforms did not modify inflammation, but were reported to stimulate proteoglycan synthesis and restore the PG content of the cartilage. However, injection of BMP-2 did not stimulate repair at all.

FGF-2, as well as other growth factors, has been shown to promote the expansion and differentiation of cultured chondrocytes (reviewed by Reinholtz et al., 2004). Cuevas et al. (1998) reported that FGF-2 delivered to knee joints of rats by osmotic pump promoted repair of partial-thickness defects in articular cartilage by stimulating chondrocyte proliferation and ECM formation. However, Hunziker and Rosenberg (1996) found that FGF-2 in a fibrin carrier was unable to stimulate the regeneration of articular cartilage in rabbits and minipigs when delivered into partial-thickness lesions, which were treated with chondroitinase ABC to provide an adhesive surface for the migration of mesenchyme-like cells from the synovium. The fibrin clot degraded and was completely replaced by a loose connective tissue with a matrix consisting predominantly of collagen fibrils. No new cartilage was formed, even after 48 weeks. FGF-2 is released from the matrix after cutting pig articular cartilage and induces MMPs 1 and 3, as well as TIMP-1, which are likely involved in promoting the repair of the defect by fibrocellular scar tissue. The reasons for this disparity in results are unknown.

Better results have been obtained with full-thickness defects or partial-thickness defects that penetrate into the subchondral bone. Fujimoto et al. (1999) reported that FGF-2 in collagen sponge promoted the regeneration of full-thickness defects of rabbit articular cartilage. Tanaka (2004) carried out a dose-response study of the ability of FGF-2 in collagen gel to stimulate cartilage regeneration in full-thickness defects of rabbit knee condyles. Loading 100 ng of FGF-2/μl collagen gel promoted the regeneration of histologically well-organized hyaline cartilage. However, by 50 weeks after treatment, the regenerating cartilage became thinner and slight degenerative changes were noted. Furthermore, although initially the FGF-2 induced cartilage was superior histologically to the cartilage that regenerated in the control lesions, this difference diminished with time. Wakatani et al. (1998) reported that HGF promoted the regeneration of articular cartilage in full-thickness defects made in rabbit knee joints.

Similar results were obtained with BMP-2 and BMP-7 (Sellers et al., 1997; Hidaka et al., 2003; Cook et al., 2003). Full-thickness defects in the trochlear groove of the rabbit femur were filled with rhBMP-2 in collagen sponge (Sellers et al., 1997). Examination of the tissues at 24 weeks showed that the growth factor had accelerated the formation of new subchondral bone and improved the histological appearance of the overlying articular surface. The thickness of the cartilage repair was 70% that of the adjacent undamaged cartilage. On a histological grading scale, the rhBMP-2 treated defects showed much better filling, and the quality of cellular morphology and matrix staining was over threefold greater and twofold greater, respectively, than defects filled with collagen sponge alone or left unfilled (**FIGURE 10.7**). Integration of the repair cartilage with the undamaged cartilage at the edges of the defect, however, was not much better than in control animals, a finding that is common in attempts to repair articular cartilage. Cook et al. (2003) implanted the OP-1 Implant (Stryker Biotech) into full-thickness defects drilled in the medial femoral condyle of dogs. The OP-1 Implant consists of rhBMP-7 (rh osteogenic protein-1, OP-1) in Type I collagen. Repair of the cartilage was accelerated in the first 12 weeks after injury, but this advantage was lost by 52 weeks. Furthermore, the histological score for repair cartilage of rhBMP-7-treated defects was as much as 3.8 points higher than the control score at 12 weeks, but the difference declined to 0.7 points by 52 weeks.

The results of these experiments are very much like those obtained on the repair of skin wounds by the topical application of growth factors in that there is an acceleration of the repair process. At the end of the repair process, however, the treatments have produced little enhancement of repair over controls. Thus, a major issue is what happens at later time points to diminish the initial difference in repair? It is possible that a continual supply of growth factor is required to maintain the difference and that the supply of growth factor in single implants becomes exhausted before the repair is complete. Furthermore, given the complexity of the repair process, single growth factors are not likely to be optimal for repair. Combinations of growth factors may give better results, and their concentration and temporal sequence of deployment must also be considered.

Many people take over-the-counter formulations of glucosamine and chondroitin sulfate that are sold as remedies for osteoarthritis. These formulations may have some efficacy. A combination of glucosamine hydrochloride, low-molecular-weight chondroitin sulfate, and manganese ascorbate was reported to stimulate GAG synthesis and inhibit degradative enzyme activity in the existing articular cartilage of an osteoarthritic rabbit model (Lippiello et al., 2000). The progressive loss of chondrocytes, however, is not prevented.

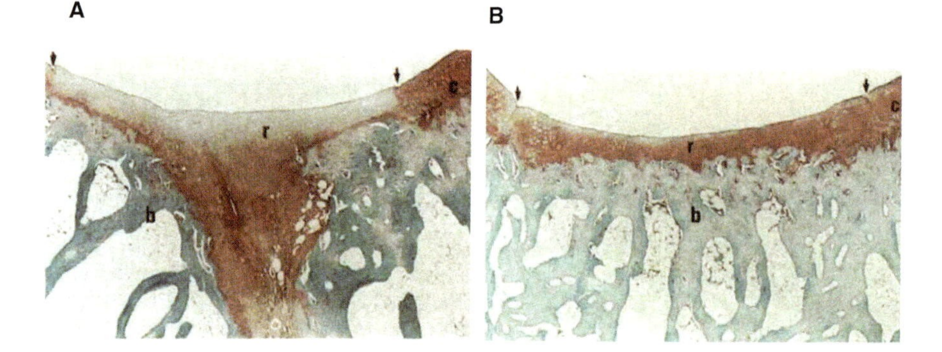


**FIGURE 10.7** **(A)** Repair of full-thickness cartilage defect in rabbit knee joint 24 weeks after implanting a collagen sponge in the defect. Safranin O and fast green stain. r = repair cartilage, which is not of high quality. c = normal articular cartilage, b = bone. Arrows mark the junction between the repair cartilage and normal cartilage. **(B)** Repair of full-thickness cartilage after implanting a collagen sponge impregnated with recombinant human BMP-2. The defect is filled with better repair cartilage that stains similar to the normal cartilage. Reproduced with permission from Sellers et al., the effect of recombinant human bone morphogenetic protein-2 (rhBMP-2) on the healing of full-thickness defects of articular cartilage. J Bone Joint Surg Am 79:1452–1463. Copyright 1997 by JBJS through the Copyright Clearance Center.

### Regeneration Templates

Acellular scaffolds have been partially successful in promoting regeneration in surgically created articular cartilage defects. Polyglycolic acid scaffolds were reported to promote cartilage regeneration in partial osteochondral defects made in the trochlear groove (Grande et al., 1999). Histological examination showed little repair in control (empty) defects, but implanted defects exhibited cartilage with a well-organized matrix, smooth articular surface, and good bonding to the native cartilage. The source of the cells that provided the repair was not identified. However, a 50:50 poly-DL-lactide-co-glycolide polymer incorporating TGF-β1 failed to improve the repair of articular cartilage in full-thickness osteochondral defects of the goat medial femoral condyle, although the repair of subchondral bone was enhanced (Athanasiou et al., 1997).

Demineralized cortical bone matrix implanted into full-thickness osteochondral defects drilled into the medial femoral condyle of rabbits was reported to induce cartilage regeneration (Gao et al., 2004). Histological evaluation 12 weeks after implanting the bone matrix revealed that the defects were repaired up to 95% of their depth by subchondral bone that graded into smooth-surfaced cartilage. This cartilage resembled normal articular cartilage histologically and was fully integrated with the articular cartilage around the graft. The demineralized bone matrix may harbor an array of cartilage-promoting growth factors and cytokines that are released over time after implantation and act combinatorially. While the most likely source of the cells involved in the repair was the bone marrow, contributions from the synovium or adjacent articular cartilage could not be ruled out.

Other regeneration templates that have been tested are hyaluronan-based polymers, chitosan, and SIS. The hyaluronan-based polymers stimulated the regeneration of subchondral bone and hyaline cartilage after implantation into full-thickness osteochondral defects made in the femoral condyles of rabbits (Solchaga et al., 2000). Chitosan, which has been used to accelerate the repair of skin wounds (Chapter 4), is marketed as a thermosensitive material (BST-Gel®, BioSyntech) to treat patients with severe articular cartilage damage. The material is a liquid at room temperature and a gel at body temperature. This formulation has been used to treat several patients, including some with osteoarthritis (probably extending through the subchondral bone) under Health Canada's Special Access Program for medical devices. The liquid chitosan is injected into the defect where it gels and serves as a scaffold to induce cartilage regeneration. These patients were reported to have experienced complete recovery following 12-week physiotherapy programs. SIS has been shown to promote the regeneration of rabbit ear cartilage (Pribitkin et al., 2004) and has been used successfully in a clinical setting to repair nasal septum defects (Ambro et al., 2003), although it is not yet known whether the repair includes the regeneration of septal cartilage.

### *b. Cell Transplants*

Experiments in animal models have treated partial-thickness lesions of articular cartilage with allografts of articular cartilage tissue (Meyers et al., 1989) or articular chondrocytes (Grande et al., 1987; Wakatani et al., 1989). These transplants formed tissue resembling normal hyaline cartilage. Isolated allogeneic chondrocytes are immunogenic, but allogeneic cartilage tissue is not, since it is avascular and the chondrocytes are hidden from the immune system by their production of matrix (Langer and Gross, 1974).

These experiments led to a human articular cartilage repair therapy (Brittberg et al., 1994, 1999) currently marketed by Genzyme for trauma-induced defects. A small piece of autologous cartilage is removed from a healthy site of the injured joint surface by arthroscopy. The chondrocytes are enzymatically disassociated and expanded *in vitro*. Clones exhibiting the cartilage phenotype are further cultured in suspension. The cells are then pelleted and implanted into the injured region, where they are held in place by a small flap of periosteal tissue (**FIGURE 10.8**). New hyaline cartilage is formed in the defect, and clinical results have been very satisfactory. Good clinical results have also been reported for transplants of chondrocytes in a type I collagen gel covered with a flap of periosteum (Ochi et al., 1998, 2001). The periosteal flap appears to make no cellular or secreted factor contributions, but serves only to keep transplanted chondrocytes in place, since chondrocytes in a collagen I gel and covered with freeze-thawed periosteum formed new hyaline cartilage that did not differ from that formed by chondrocytes covered with a living periosteal flap (Kajatani et al., 2004). How closely the replacement cartilage in these procedures matches native hyaline joint cartilage, and how long its integrity is maintained, are unknown.

Major issues regarding the use of cultured chondrocytes for transplantation are their dedifferentiation and capacity for redifferentiation after expansion *in vitro* and their capacity to adhere to the matrix of a cartilage defect. Chondrocytes dedifferentiate to a fibroblastic cell type as they divide in monolayer culture. Dedifferentiated human chondrocytes derived from the femoral lateral condyle have a surface expression profile very similar to that of MSCs and fat cells and also

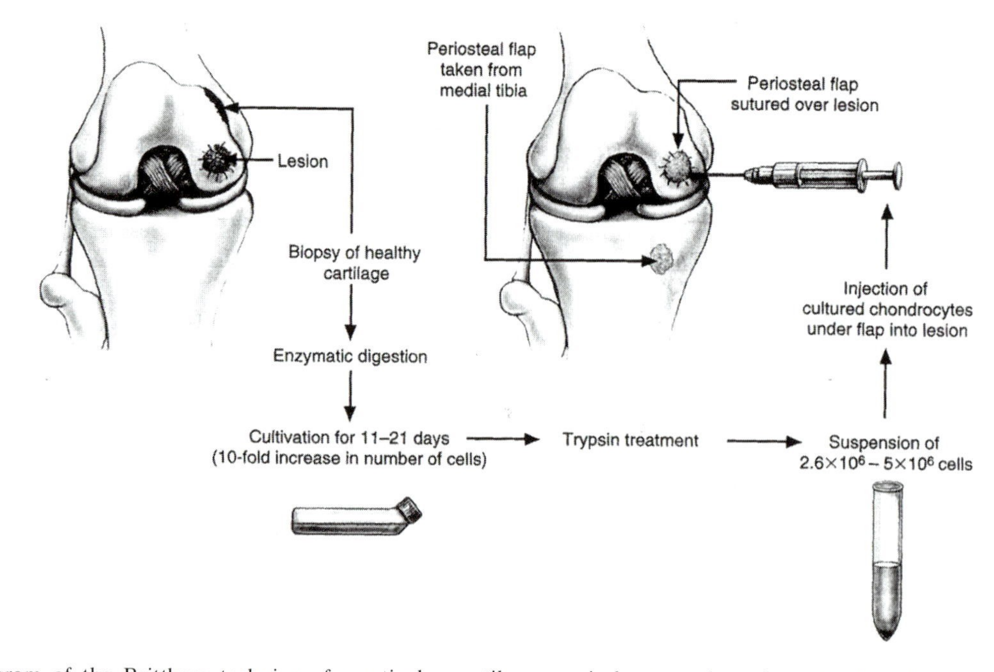


**FIGURE 10.8** Diagram of the Brittberg technique for articular cartilage repair by transplantation of cultured autologous chondrocytes. Reproduced with permission from Brittberg et al., Treatment of deep cartilage defects in the knee with autologous chondrocyte transplantation. New Eng J Med 331:899–895. Copyright 1994, Massachusetts Medical Society.

express a number of human ESC markers (de la Fuente et al., 2004). Under the appropriate signaling conditions, the dedifferentiated chondrocytes can redifferentiate into chondrogenic, osteogenic, adipogenic, myogenic, endothelial and neurogenic lineages when pelleted or embedded in a three-dimensional scaffold. However, culture *in vitro* has been reported to diminish the redifferentiation capacity of articular chondrocytes (Mandl et al., 2004).

Thus, culture conditions and manipulations to minimize dedifferentiation and/or maximize redifferentiation of chondrocytes are of high importance for producing high-quality chondrocytes for transplant. Low seeding density combined with no more than 2–3 passages was found to expand chondrocytes sufficiently to fill a 4-$cm^2$ defect with maximum retention of redifferentiation capacity, as measured by morphology and Type II collagen and sulfated GAG synthesis (Mandl et al., 2004; Veilleux et al., 2004). Factors in serum, which is usually included in chondrocyte expansion medium, also appear to decrease redifferentiation potential by decreasing the expression of Sox9. A serum-free expansion medium has been developed that induces dedifferentiated chondrocyte proliferation at levels equivalent to that in medium containing serum and maintains Sox9 expression up to at least six population doublings. Chondrocytes cultured in this medium differentiate into a tissue that has the cellular and molecular characteristics of hyaline cartilage (Mapeli et al., 2004). Consistent with these results, dedifferentiated human osteoarthritic chondrocytes from elderly patients, passaged eight times and transfected with *Sox9* in adenoviral or lentiviral vectors, redifferentiated into chondrocytes at a frequency significantly greater than controls, as measured by ratio of type I to type II collagen production, indicating that even osteoarthritic chondrocytes do not irreversibly lose the potential to regain chondrogenic features after extensive culture (Li et al., 2004).

Growth factors also enhance the differentiation of cultured chondrocytes (reviewed by Reinholtz et al., 2004). IGF-1, FGF-2, IL-4, and TGF-β increase chondrocyte expansion and differentiation, while PDGF decreases it. Furthermore, application of a combination of growth factors in sequence provides a greater stimulatory effect than application of single factors alone or together. FGF-2 applied to expanding monolayer cultures of chondrocytes enhanced their redifferentiation and their response to BMP-2 in three-dimensional cultures, but sequential treatment of chondrocytes on scaffolds with TGF-β and FGF-2 followed by IGF-1 yielded cartilage with the most enhanced biochemical and mechanical properties (Martin et al., 2001; Pei et al., 2002).

The capacity of dedifferentiated chondrocytes to adhere to the matrix of a cartilage defect is very important, not only because the ECM may have an influence of redifferentiation into chondrocytes, but also because such adherence dictates the ability of the tissue formed by transplanted cells to integrate with the tissue on the periphery of the defect, leaving no fissures between the two. Dennis et al. (2004) have described a "cell painting" procedure that fosters adherence of prechondrocytes to the defect surface. Protein G derivatized with the N-hydroxysuccinimide ester of palmitic acid (palmitized protein G, PPG) was incorporated onto the surfaces of dedifferentiated chondrocytes labeled with a fluorescent vital dye and the cells were incubated with antibodies to chondroitin-4-sulfate, keratan sulfate, and Type II collagen. The antibodies bind to the PPG with the antigen-binding domain extending outward from the cell surface. The antibody-coated cells exhibited a much higher capacity to bind *in vitro* to defects made in pieces of explanted articular cartilage. Results of applying this technique to filling cartilage defects *in vivo* have not yet been reported.

Hidaka et al. (2003) implanted allogeneic foal chondrocytes into full-thickness defects made in the lateral trochlear ridge of the femur in adult horses. The chondrocytes were transfected with a BMP-7 adenoviral vector for over expression of the growth factor. Although repair of the defect was accelerated over that of controls and the repair cartilage was markedly more hyaline, by 8 months after treatment no difference could be detected in the morphologic, biochemical, or biomechanical properties of controls and BMP-7 animals.

Xenografts of pig chondrocytes may be useful for repairing partial defects in human articular cartilage. Articular chondrocytes derived from pig femoral condyle and cultured for 14–21 days formed tissue with the characteristics of hyaline cartilage when implanted into articular cartilage defects of rabbit femoral condyles under a periosteal flap (Ramallal et al., 2004). Interestingly, the new tissue was composed of only 27.5% pig cells. Since the periosteal flap does not appear to contribute to new cartilage (Kajitani et al., 2004) this suggests that the remainder of the cells might be derived from host chondrocytes, perhaps stimulated to proliferate by the xenografted cells. Similar results were obtained in an *in vitro* model consisting of pig chondrocytes implanted into defects drilled into cores of human articular cartilage (Fuentes-Boquete et al., 2004).

Transplanted autologous mesenchymal stem cells were reported to form new articular cartilage in conjunction with MSCs from the host bone marrow, in full-thickness osteochondral defects in rabbit femoral condyle (Caplan et al., 1993; Wakitani et al., 1993) (**FIGURE 10.9**). The only deficiencies noted in the regenerated cartilage were thinning by 24 weeks post-implantation and incomplete integration with the surrounding articular cartilage. MSC transplantation has also been proposed as a way to repair partial-thickness

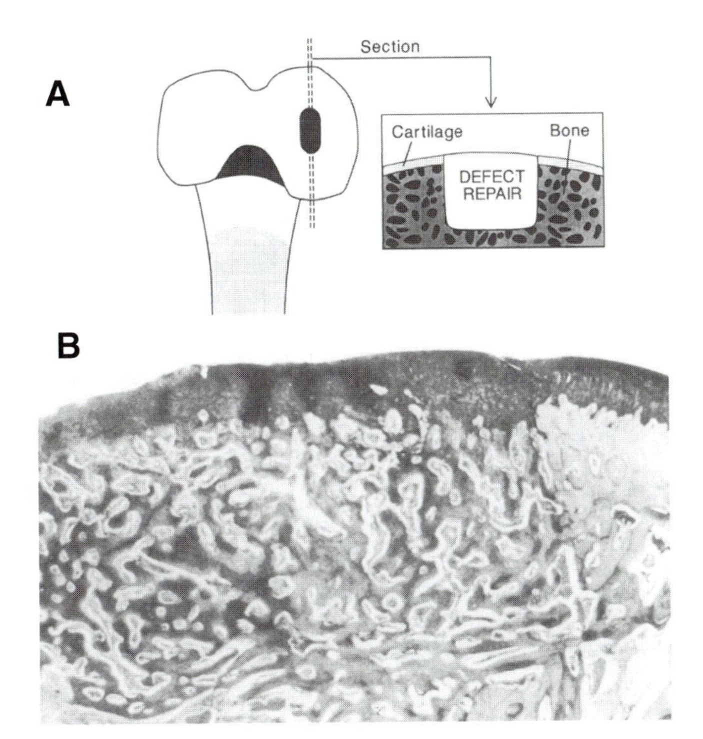


**FIGURE 10.9** **(A)** Diagram of the rabbit femoral condyle model for cartilage repair. **Left,** a full-thickness defect of specific size is drilled through the articular cartilage into the bone. **Right,** saggital section through the condyle showing the defect repair site. **(B)** Result of filling a condyle defect with mesenchymal stem cells. Relatively normal-appearing cartilage has regenerated in the defect and subchondral bone has also regenerated. Reproduced with permission from Caplan et al., Mesenchymal stem cells and tissue repair. In: Jackson DW, Arnoczky S, Woo S, Frank C (eds). The Anterior Cruciate Ligament: Current and Future Concepts. New York: Raven Press. Copyright 1993, Elsevier.

defects in articular cartilage (Johnstone and Yoo, 1999; Otto and Rao, 2004). Gelse et al. (2003) isolated MSCs from rat rib perichondrium. After expansion, the cells were transfected with adenoviral vectors carrying the genes for BMP-2 or IGF-1 to induce chondrogenesis. The cells were suspended in fibrin glue and used to fill partial-thickness defects in the articular cartilage of the patellar groove of the rat femur. Cells transfected with either vector produced hyaline repair cartilage with a PG-rich matrix containing type II collagen that restored the articular surface of most lesions. Untransfected cells formed fibrous tissue with a matrix containing type I collagen.

### c. Bioartificial Cartilage

A wide variety of polymers have been used as scaffolds to prepare bioartificial articular cartilage implants. These polymers can be molded to conform to the outline of a particular defect and are biodegradable.

Cells for the implant have been derived from cultured chondrocytes or bone marrow cells (Dennis et al., 2001).

Allogeneic chondrocytes (rat, calf, human) grown in a polyglactin mesh scaffold with stirring of the medium were reported to form good cartilage that when implanted into partial-thickness defects in rabbit articular cartilage repaired the defects (Freed and Vunjak-Novakovic, 1995). The mesh itself degrades, leaving only the cartilage cells and matrix. Three-dimensional scaffolds of polyglycolic or polyglactic acid, collagen gels, hybrid scaffolds of collagen plus polylactic acid or poly (DL-lactic-co-glycolic acid) (PLGA), and β-chitin from squid support the differentiation of cultured human or animal chondrocytes into tissue that histologically and biochemically resembles hyaline cartilage (Freed et al., 1993; Chen et al., 2004; Abe et al., 2004). Implants of allogeneic rabbit articular chondrocytes embedded in a type I collagen gel were able to repair full-thickness osteochondral defects (Wakatani et al., 1998). While the neocartilage resembled normal articular cartilage, there was no zonal differentiation.

Mesenchymal stem cells also form cartilage when seeded into a matrix composed of polylactide/alginate amalgam (Caterson et al., 2001). Stem cells isolated from rabbit muscle connective tissue via the freezing and secondary culture method described by Young et al. (1995) have been used to repair defects in the knee cartilage of rabbits (Grande et al., 1995). The cells were cultured in porous polyglycolic acid matrixes, and the constructs implanted into the defects. By 12 weeks post-implantation, the polyglycolic acid matrix had dissolved, and the implants had formed a well-integrated surface layer of cartilage in the defect that was approximately the same thickness as the normal articular cartilage. Similar results were obtained in full-thickness defects in the trochlear groove of rabbits after implantation of a polyglycolic acid scaffold seeded with periosteal cells (presumably MSCs) transfected with the BMP-7 or sonic hedgehog genes (Grande et al., 2003).

Bioartificial articular cartilage has also been made using rabbit articular chondrocytes and pig SIS as a scaffold (Peel et al., 1998). Chondrocytes were isolated enzymatically and seeded into the SIS without expansion. At 8 weeks in culture, the chondrocytes had formed a layer of toluidine blue-staining cartilage about 0.15 mm thick. The cultured cartilage contained large proteoglycans and type I cartilage and had a GAG content at eight weeks that was nearly 80% that of normal deep articular cartilage. When placed in full-thickness osteochondral defects of the femoral trochlear groove, the cartilage maintained its hyaline appearance, whereas defects receiving SIS alone repaired with fibrovascular tissue. Repair was not optimal, however. The hyaline cartilage was surrounded by variable amounts of fibrovascular tissue toward the joint surface and fibrocartilage tissue above or below the implant. Furthermore, the implant surface was concave.

Kisiday et al. (2002) developed a self-assembling peptide KLD-12 hydrogel with the sequence AcN-KLDLKLDLLDL-CNH$_2$. Chondrocytes seeded into the hydrogel were shown to maintain their morphology and synthesize a cartilage-like ECM rich in proteoglycans and type II collagen. Accumulation of the ECM over time was paralleled by increases in material stiffness, suggesting that this bioartificial cartilage was mechanically functional.

The long-term biochemical and structural integrity of implanted bioartificial cartilage has not been reported.

## REGENERATIVE THERAPIES FOR BONE

Mammalian endochondral or membranous bones are unable to regenerate across gaps that exceed a certain size, called the critical size gap. The critical size gap is different for each type of bone and in different species. In humans, the number of bone defects due to nonhealing fractures and conditions that necessitate bone removal is in the tens of thousands per year. The standard treatment for such defects is an autogeneic bone graft, which runs the risk of morbidity at both donor and host sites. Allogeneic bone grafts can also be used, but these are subject to immunorejection and a low risk of infectious disease.

Alternatives to bone grafting to regenerate bone across nonhealing defects include electrical stimulation, cell transplants, the use of osteogenic growth factors and acellular scaffolds, and implantation of bioartificial bone. Regardless of the approach used, bone regeneration has the same three requirements as bone development: osteoinductive signals; a matrix that traps the signal and provides an adhesive surface; and osteogenic cells that can adhere to the matrix and differentiate into osteoblasts in response to osteoinductive signals (Ripamonti, 2002). The ability to repair substantial bone defects may be closest to clinical reality than perhaps for any other musculoskeletal tissue.

### 1. Electromagnetic Field Stimulation

Areas of active bone growth and regeneration, or areas of bone deposition during physiological bone remodeling, are electronegative with respect to less active areas; thus, electric fields may be part of the normal process of bone development and regeneration.

Direct electrical currents were first used by Friedenberg et al. (1971) to successfully treat intractable bone fractures in humans. Nonhealing fractures have been reported to respond well to applications of either cathodal current or electromagnetic fields. Bone regeneration was accelerated in osteotomies that were made 3.5 to 4.5 cm distal to the head of the fibula in dogs (Bassett et al., 1974). A voltage field was induced in the fibula by inductively coupling pulsed electromagnetic fields of low frequency and strength directly to the bone across the skin. Not only did the stimulated osteotomies heal faster than control osetotomies, but the regenerated bone was more highly organized and stronger than control regenerated bone, even though the mass of callus formed was less than in controls. This method has been used to successfully treat tibial pseudoarthroses in young patients (Bassett et al., 1974; Lavine et al., 1972). Pseudoarthrosis is a rare, local bone dysplasia that has a very low probability of correction by conventional techniques. Increased bone deposition in rabbit heel bones with immobilization-induced osteoporosis has been reported after electrical stimulation, but the new bone is porous rather than compact (Kenner et al., 1975). Pulsed electromagnetic fields have also been used as a noninvasive postoperative treatment for lumbar vertebral fusion (Mark, 2000).

The effectiveness of pulsed electromagnetic fields (PEMF) over the years has been controversial. However, in a carefully controlled study on surgically created critical size gaps in the rat fibula, Ibiwoye et al. (2004) showed that PEMF treatment resulted in a 75% reduction in the loss of bone volume at the distal end of the fibula compared to controls and that in half the PEMF-treated animals bone growth was observed, whereas control osteotomies widened by 10%.

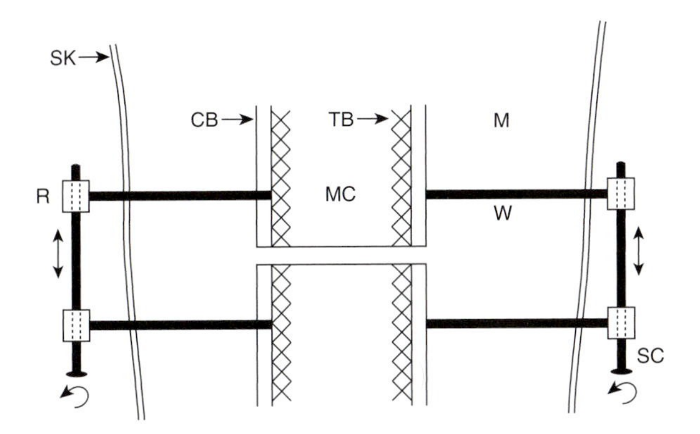


**FIGURE 10.10** Diagram of a frontal section of a human leg at femur level showing how the Ilizarov technique is used to lengthen bones. M = muscle, SK = skin, CB = cortical bone, TB = trabecular bone, MC = marrow cavity. The Ilizarov apparatus consists of a series of rings (R) with stiff wires (W) that attach to the bone. Rods are threaded through the rings. The cortical bone is cracked or cut and the rods are turned (arrows) so the rings distract the bone a few millimeters at a time.

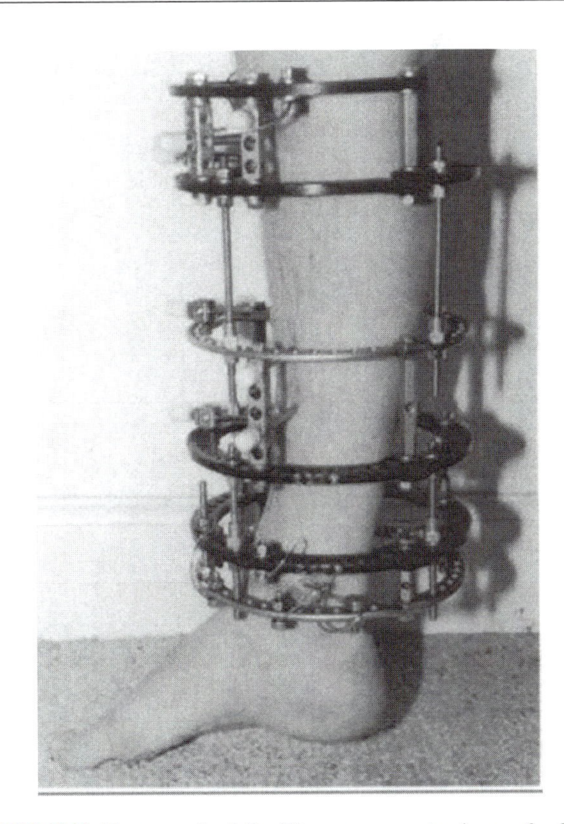

**FIGURE 10.11** Photograph of the Ilizarov apparatus in use for bone distraction. Reproduced with permission, courtesy of Dale Haines, http://www.ilizarov.org.uk.

## 2. Ilizarov Distraction Technique for Lengthening Bones

Ilizarov (1990) developed a procedure to lengthen abnormally short long bones in patients who have diseased bones or achondroplasia. The procedure involves cutting or cracking the cortical bone, taking care to maximally preserve bone marrow, blood vessels, and nerves, followed five to seven days later by the application of longitudinal tension (distraction) applied through a stable external fixation device (**FIGURES 10.10, 10.11**). The fixation device is lengthened each day in a series of small steps (no less than four) that separate the cut edges by one mm. MSCs are activated in the periosteum and marrow of the injury region, and the tension orients the differentiating chondrocytes into a growth platelike structure that regenerates endochondral bone in both proximal and distal directions. Once the desired length of the bone has been achieved, the regenerated part is allowed to ossify completely before engaging in full physical activity. The tension exerted by the fixation device undoubtedly also activates satellite cells, which would regenerate new

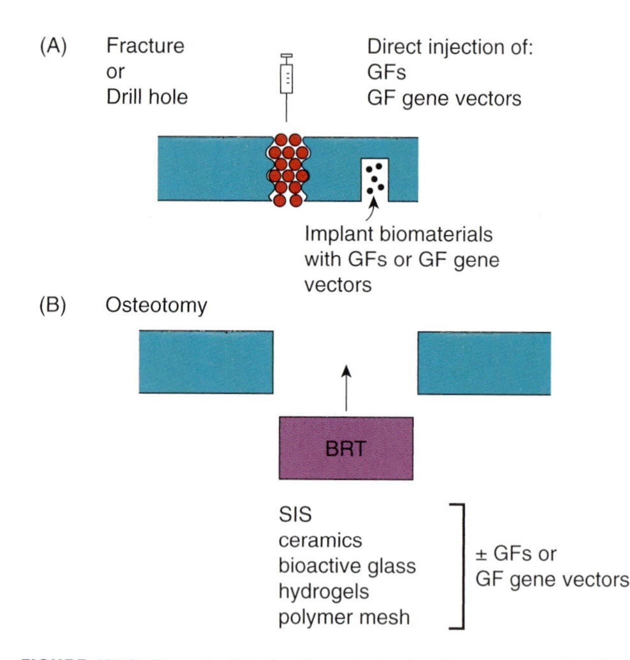


**FIGURE 10.12** Chemical induction strategies for regenerating bone. **(A)** The bone of an experimental animal is fractured or a hole is drilled in the bone. Growth factors or growth factor gene vectors are then injected directly into the fracture site, or biomaterials containing growth factors or growth factor gene vectors are implanted. **(B)** A gap is experimentally created in the bone by an osteotomy. A bone regeneration template (BRT), with or without (plus/minus sign) growth factors or growth factor gene vectors, is then inserted to replace the removed bone segment. Several types of materials have been used as regeneration templates.

myofibers to lengthen the muscles attached to the bone. Use of the limb and intensive physiotherapy during the lengthening procedure are essential to avoid contractures, subluxations, or even dislocations, since these can be caused by the resistance of myofascial tissues to elongation.

## 3. Induction of Bone Regeneration by Osteogenic Factors and Regeneration Templates

Osteogenic factors or genes encoding these factors can be injected directly into small lesion sites, such as fractures or holes drilled in cortical bone, to accelerate regeneration, based on the assumption the factors will initiate the entire cascade of regenerative events (**FIGURE 10.12**). Injection of rhBMP-2 or FGF-2 directly into the fracture site of a standard rat model accelerated the fracture healing process by two weeks (Yasko et al., 1992; Einhorn, 1999; Radomsky et al., 1998). Gelatin hydrogel containing 200 μg FGF-2 was successful in healing 10/10 ulnar non-union fractures in cynomolgus monkeys, whereas 4/10 fractures remained in a non-union state after 10 weeks. Mechanical strength, bone mineral content, and density were all increased in the FGF-2 treated fracture sites (Kawaguchi et al., 2001). The repair of cortical defects in the tibia of pigs was enhanced by application of a combination of PDGF-BB and IGF-I in a methylcellulose gel carrier, but not by either growth factor alone (Trippel, 1998). Percutaneous injection of the *BMP-6* gene in an adenoviral vector into rabbits with ulnar osteotomies accelerated the regeneration of bone over controls receiving vector alone (Bertone et al., 2004). Poly (DL-lactic-co-glycolic acid) (PLGA) porous microspheres incorporating the synthetic peptide TP508 (Chrysalin®, Chrysalis Biotechnology) were reported to induce a significantly higher degree of repair than controls receiving microspheres alone, when implanted into rabbit ulnar defects (Sheller et al., 2004). TP508 is a 23 amino acid peptide representing the natural amino acid sequence of the receptor-binding domain of human thrombin. These results are interesting in view of the role thrombin has been shown to play in lens and limb regeneration in newts (Chapters 3 and 14). This peptide is currently in Phase III trials for repair of bone fractures and is also in Phase II trials for the healing of diabetic ulcers.

To regenerate bone across gaps that do not greatly exceed the critical size gap, osteogenic factors or gene constructs have been incorporated into scaffolds that serve as osteoconductive regeneration templates for host MSCs (Einhorn, 1999; Boden, 1999; Winn et al., 1999; Rosa et al., 2002; Bourgeois et al., 2003) (**FIGURE 10.12**). The factors most commonly used to promote bone regeneration are the BMPs, which commit immigrating MSCs to osteoblastic differentiation (Reddi, 1998).

Scaffold materials fall into four general categories (Seeherman et al., 2002). These are (1) inorganics such as calcium phosphate ceramics and calcium-phosphate-based cements, (2) natural polymers, of which collagen is the most widely used, (3) synthetic polymers, of which the poly($\alpha$-hydroxy acids) are most commonly used, and (4) combinations of these. Each type of scaffold has its advantages and drawbacks.

Collagen-based scaffolds have frequently been used as templates to induce bone regeneration. Pig SIS was compared to particles of demineralized cortical bone for its ability to promote regeneration across an 11-mm defect made in the rat radius (Suckow et al., 1999). At 24 weeks, the amount of bone fill, evaluated by radiologic imaging, and the histological quality of the bone, was equivalent in the SIS and demineralized bone-treated groups. Bone regeneration was induced in femoral defects of rat and sheep and mandibular defects in dogs by BMPs in a collagen carrier (Rosen and Theis, 1992). Cook et al. (1994) found that BMP-7 (OP-1) in a collagen carrier induced bone regeneration across a critical size defect in rabbit ulna. Bone regeneration

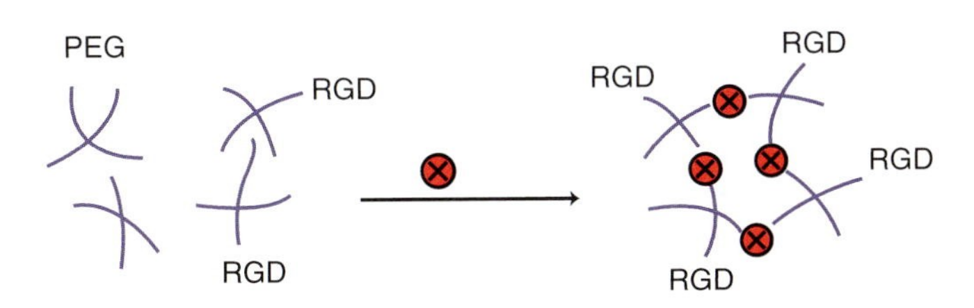


**FIGURE 10.13** Diagram illustrating preparation of a gel that degrades in response to MMPs secreted by adherent cells. First, a peptide containing an integrin-binding RGD ligand is reacted with four-armed end-functionalized polyethylene glycol (PEG) macromers. Second, the macromers are cross-linked by a peptide (X) that is a substrate for MMPs. When cells adhere to the gel via the RGD ligands, secreted MMPs degrade the gel at the sites of cross-linking. Growth factors can be mixed into the gel and will be released as the gel degrades.

across critical size defects in rabbit ulna and sheep femur was induced by implants of rhBMP-2 in a polyglycolic acid carrier or demineralized bone matrix, whereas no regeneration took place in controls with scaffolds alone (Bostrom et al., 1996). FGF-2 in a hyaluronan or gelatin hydrogel also promoted bone regeneration in calvarial defects of rats and fibular osteotomies of rabbits (Radomsky et al., 1998) and calvarial defects of monkeys (Tabata et al., 1999). In a study of 24 patients, Geesink et al. (1999) reported that BMP-7 in a collagen I carrier promoted the regeneration of bone across a gap created by a tibial osteotomy.

The type of carrier influences the release kinetics of bioactive factors (Winn et al., 1999). Sustained release of rhBMP-2 was exhibited by a type I collagen sponge and poly (D,L-lactide) carriers. By contrast, the protein was released from de-organified bovine bone matrix in an initial burst, but then appeared to bind irreversibly to the matrix. Lutolf et al. (2003) devised a polyethylene glycol scaffold that releases bioactive factors via the degradative activity of the immigrating cells (**FIGURE 10.13**). RGD adhesion peptides were coupled to the PEG chains, which were then cross-linked with an MMP cleavage site peptide to form a hydrogel. BMP-2 is incorporated into the scaffold as it gels. Incoming cells adhere to the RGD sites and release BMP-2 by their production of MMPs, which degrades the gel by acting on the cross-linking MMP cleavage peptide. This scaffold/BMP-2 combination promoted efficient and highly localized membrane bone regeneration in a critical-size defect in rat calvarium (**FIGURE 10.14**).

Bioactive factors can also be delivered to host osteogenic cells by gene-activated matrices (GAMS) (Goldstein et al., 1999; Bonadio, 2002; Ghivizzani, 2002). In this approach, viral or plasmid vectors containing genes for bioactive factors are incorporated into the scaffold and are taken up by host osteogenic cells as they migrate into the scaffold, where they are then used to synthesize the factor. Plasmids containing the hrPTH (1-34) and BMP-4 genes in a collagen sponge elicited the formation of new bone when the sponge was implanted into a segmental gap in the rat femur (Fang et al., 1996; Goldstein and Bonadio, 1998). The hrPTH 1-34 plasmid also elicited bone regeneration when administered in a collagen carrier to cylindrical defects (8 mm diameter × 8 mm deep) drilled into the femur or tibia of dogs (Goldstein and Bonadio, 1998).

Many ceramics such as hydroxyapatite mimic the mineral phase of bone and provide a good surface for osteoblast adhesion, but compromise biomechanical strength of regenerated bone because they are resistant to resorption and thus replacement by native bone matrix. Some calcium phosphate-based cements and ceramics, however, have been shown to be resorbable and to promote bone regeneration by host MSCs without the inclusion of growth factors (Cornell and Lane, 1998; Oreffo et al., 1998). Constanz et al. (1995) devised a paste consisting of monocalcium phosphate monohydrate, α-tricalcium phosphate and calcium carbonate, to which was added a solution of sodium phosphate. When injected into gaps in a bone, the paste rapidly hardened into a material resembling bone matrix that was invaded by host MSCs and/or osteoblasts, which secreted bone matrix. Sintered porous hydroxyapatites induced bone regeneration in calvarial defects of adult baboons (Ripamonti et al., 2001a). These scaffolds have a geometry that induces bone regeneration from host cells by adsorbing BMPs from the host environment onto their surfaces (Ripamonti et al., 2001b). The BMPs not only induce MSCs to differentiate as osteoblasts, but promote vascularization of the scaffold (Ripamonti, 2002). Such "smart" scaffolds that do not require the addition of exogenous growth factors are less expensive to use. Ripamonti (2002) reported that over 7,000 South African patients have been treated with these sintered hydroxyapatite scaffolds.

Still another bone-inducing scaffold material is resorbable glass particles (90–710 μm), composed of 45% $SiO_2$, 24.5% CaO, 24.5% $NaO_2$, and 6% $P_2O_5$. Bioactive glass granules have been mixed with autogeneic bone fragments to elevate the maxillary sinus (Cordioli et al., 2001). They have also been reported to promote excellent bone regeneration in 3-mm defects drilled into the cortical bone of the rat tibia (Oonishi et al., 1996; de Macedo et al., 2004). Host osteogenic cells migrate between the glass particles and into fissures in the particles. Bioactive glass particles have been used to regenerate bone in human periodontal disease (Yukna et al., 2001), but not other types of human bone defects. It would be interesting to know whether glass particles also adsorb growth factors from host surroundings and thus act as "smart" biomaterials.

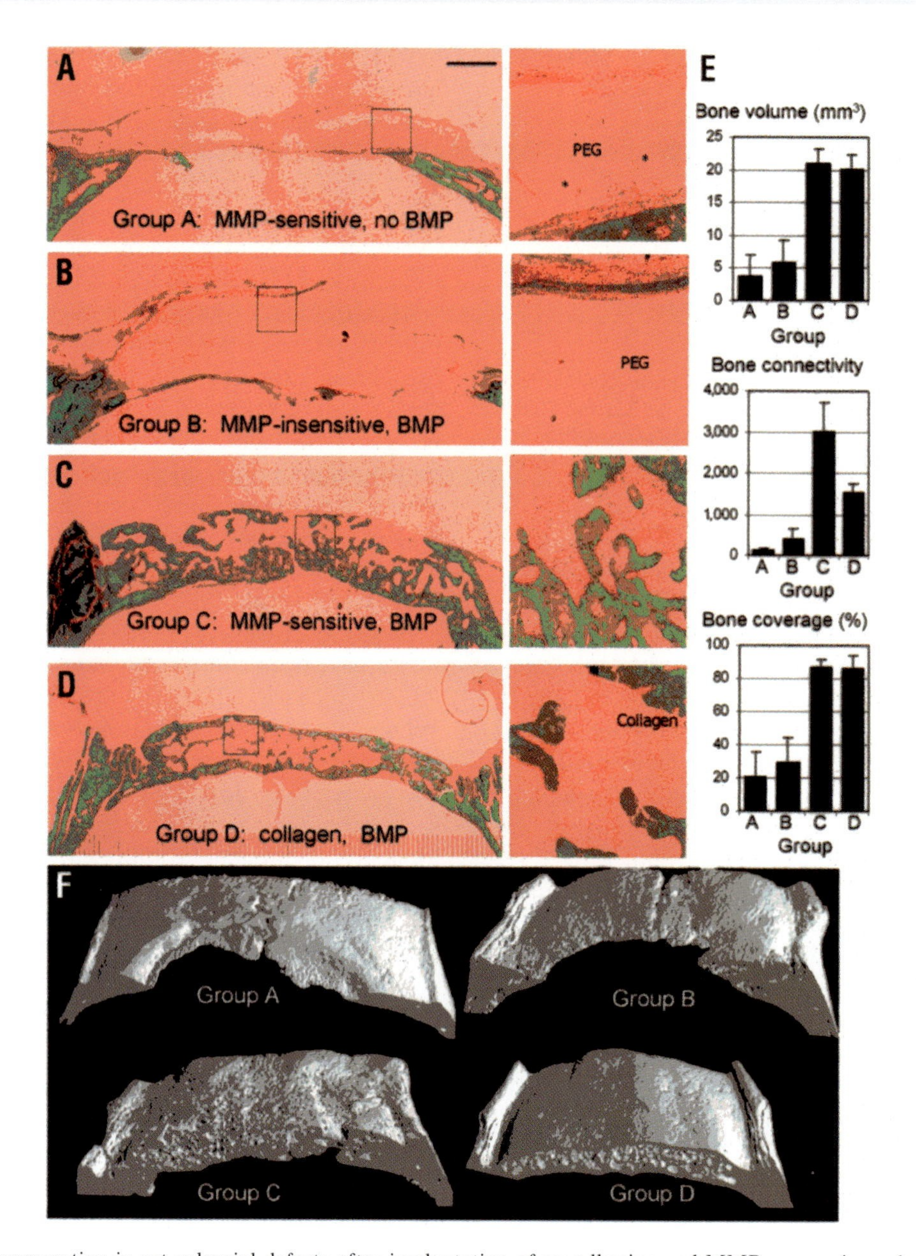


**FIGURE 10.14** Bone regeneration in rat calvarial defects after implantation of an adhesive and MMP-responsive gel containing recombinant human BMP-2 **(C)** compared to a gel that was MMP-sensitive but lacking BMP **(A)**, a gel containing BMP but lacking MMP-sensitivity **(B)** and a standard collagen sponge containing BMP **(D)**. **(F)** Microcomputed tomographic reconstructions show thicker bone regenerated in groups C and D. Bone volume, connectivity and coverage are quantified in **(E)**. The data show that the MMP-sensitive gel containing BMP was equivalent to the collagen sponge in terms of bone volume and coverage induced, but induced significantly better bone connectivity. Reproduced with permission from Lutolf et al., Repair of bone defects using synthetic mimetics of collagenous extracellular matrices. Nature Biotech 21:513–518. Copyright 2003, Nature Publishing Group.

Stevens et al. (2005) have described an innovative way to manufacture new bone for transplant by filling a sub-periosteal space in the tibia of rabbits with ~200 μl of alginate gel cross-linked by $CaCl_2$. Cells from the periosteum (presumably MSCs and/or preosteoblasts) migrated into the gel and formed compact bone in an intramembranous fashion, accompanied by neovascularization that attained normal levels by 12 weeks and dissolution of the alginate (**FIGURES 10.15–10.17**). The volume of bone formed was ~162 $mm^3$, which was calculated to be sufficient for a single bone fusion procedure in the same animal. The mechanical properties of the bone were comparable to those of normal bone. Significantly, the new bone was formed under the influence of only endogenous growth factors; the addition of exogenous TGF-β1 and FGF-2 did not improve the outcome. Interestingly, delivering liposome encapsulated Suramin (an inhibitor of vascular invasion) into

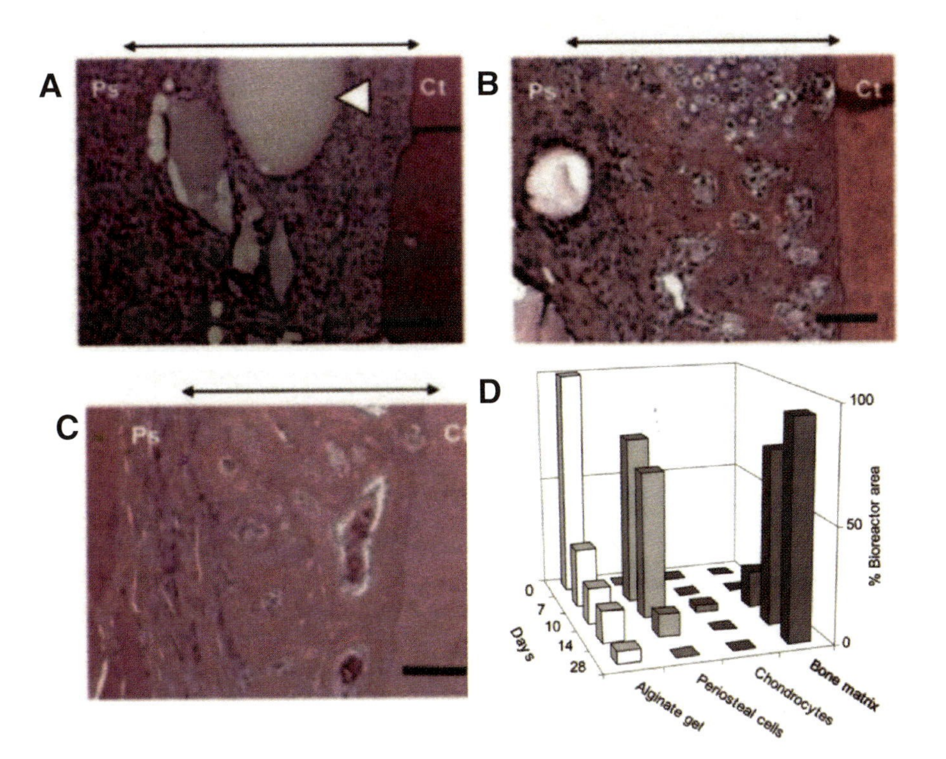


**FIGURE 10.15** Development of new bone in a space created between the cortical bone and periosteum of a rabbit tibia. **A-C,** H & E-stained longitudinal sections at 7, 10, and 14 days, respectively. Ps = periosteum, Ct = cortical bone, arrows indicate areaoccupied by the alginate gel. White arrowhead = unresorbed gel. Bar = 100 μm. Bone formation is evident at 10 days. **(D)** Percent of the development space occupied by alginate, migrating periosteal cells, chondrocytes and bone matrix over time, showing the course of new bone development and alginate resorption. Reproduced with permission from Stevens et al., *In vivo* engineering of organs: The bone bioreactor. Proc Nat'l Acad Sci USA 102:11450–11455. Copyright 2005, National Academy of Science, USA.

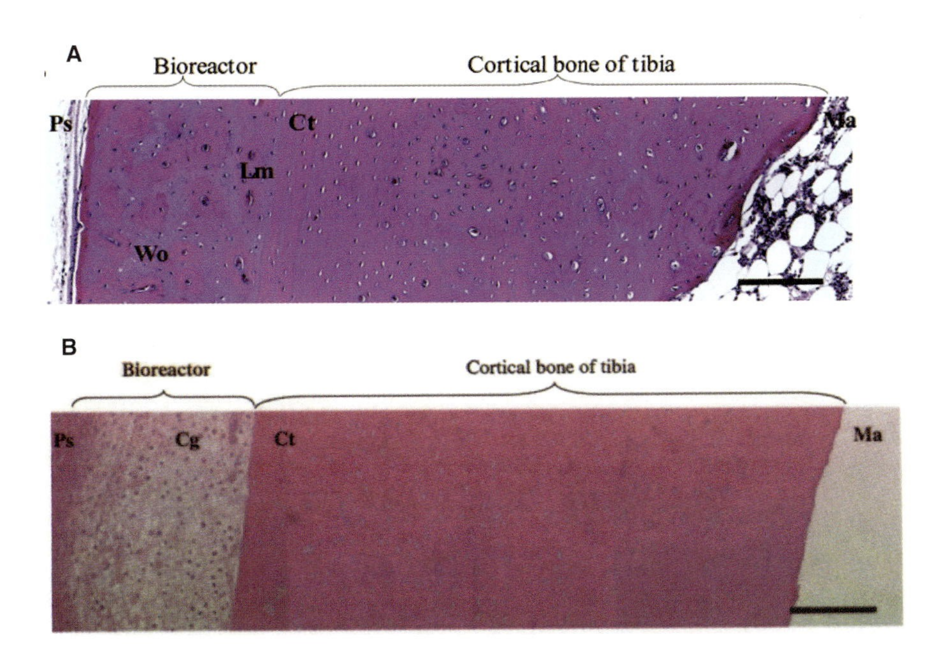


**FIGURE 10.16** **(A)** H&E-stained section of new bone formed after six weeks in the alginate-occupied space (bioreactor) between the cortical bone (Ct) and periosteum (Ps) of the rabbit tibia. Wo = woven (trabecular) bone, Lm = lamellar bone. **(B)** H&E-stained section of 10 day-old cartilage induced to differentiate in a space (bioreactor) between the cortical bone and periosteum of the rabbit tibia occupied by a hyaluronic acid gel containing Suramin. Ma = marrow space. Bar = 300 μm. Reproduced with permission from Stevens et al., *In vivo* engineering of organs: The bone bioreactor. Proc Nat'l Acad Sci USA 102:11450–11455. Copyright 2005, National Academy of Science, USA.

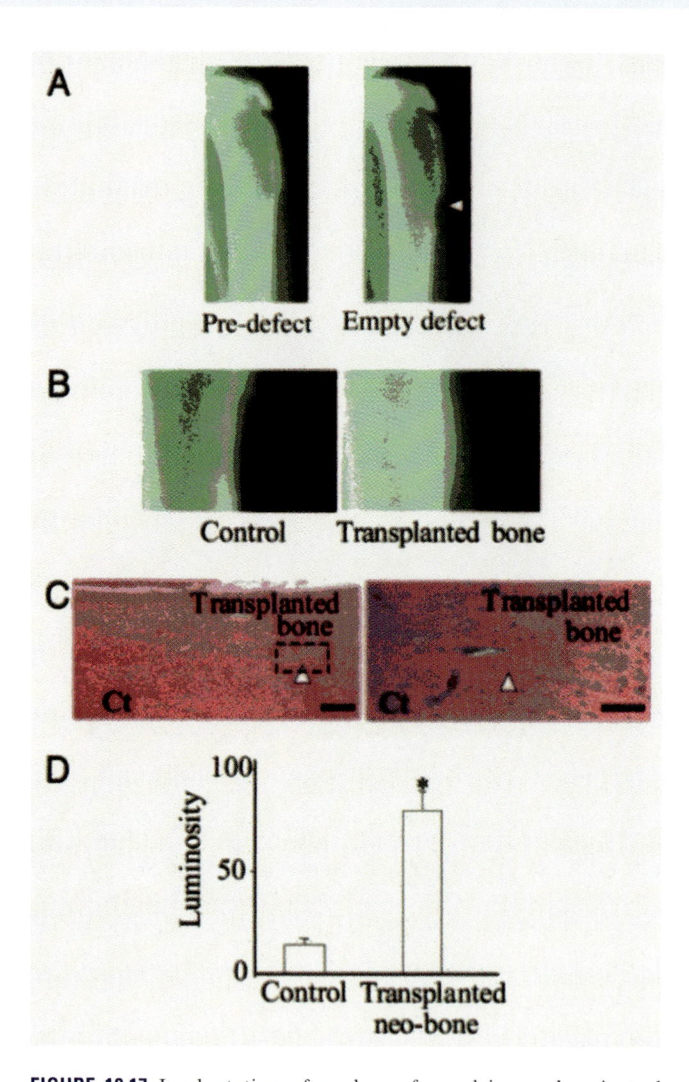


**FIGURE 10.17** Implantation of neobone formed in a subperiosteal space of one tibia to a defect created in the contralateral tibia. **(A)** Radiographs of normal tibia and empty defect (arrowhead). **(B)** Magnified view of control defect implanted with Gelfoam and defect implanted with neobone grown on the contralateral tibia six weeks after implant. The control defect remains unfilled, whereas the bone transplant fills the experimental defect. **(C)** H&E-stained sections at low magnification (left) and high magnification (right) showing the integration (arrowheads) of the transplanted neobone with the host cortical bone (Ct) six weeks after implant. **(D)** Radiographic densitometry of tibial defects at six weeks. Empty defect has a value of zero, normal cortical bone has a value of 100. Reproduced with permission from Stevens et al., *In vivo* engineering of organs: The bone bioreactor. Proc Nat'l Acad Sci USA 102:11450–11455. Copyright 2005, National Academy of Science, USA.

the bioreactor space resulted in neocartilage formation, suggesting the possibility of creating new cartilage for transplant by this method. Radiographical and histological evaluation six weeks after harvesting neobone and transplanting it to a C-shaped defect in the contralateral tibia revealed that the transplanted bone had remodeled and become integrated with the surrounding bone of the defect.

## 4. Cell Transplants and Bioartificial Bone

Whole bone marrow has been transplanted to replace defective MSCs in patients who have osteogenesis imperfecta, or "brittle bone" disease. In this disorder, a mutation in the type I collagen gene makes the bone matrix highly susceptible to fracture. Horwitz et al. (1999) administered chemotherapy to three infants with osteogenesis imperfecta to destroy HSCs and MSCs and then transfused them with whole bone marrow from HLA-identical siblings. Bone biopsies revealed that only a small percentage of the donor cells engrafted, but in the six months following transplant, the number of fractures suffered was reduced by 92%, suggesting that the transplanted MSCs proliferated and differentiated into enough normal osteoblasts to make a significant difference in bone matrix quality. Another possible treatment for this disease would be to isolate autogeneic MSCs from patients, replace the mutated gene by homologous recombination *in vitro*, then return the cells to the patient. Chamberlain et al. (2004) provided proof of principle for this approach. The mutant type I collagen gene in MSCs from individuals with osteogenesis imperfecta was disrupted by transduction with adenoviral vectors. The result was improved collagen processing, stability, and structure. Clones of transduced MSCs were cultured for two weeks in osteogenic differentiation medium, seeded in hydroxyapatite tricalcium phosphate matrices and implanted subcutaneously into SCID mice, where they were able to form bone.

Connolly (1995) summarized the results of injecting autologous bone marrow into non-union fractures of 100 patients. Eighty percent of the patients exhibited a bone-forming response, presumably by the injected cells, but possibly also due to factors secreted by these cells. The lack of bone-forming response in 20% of patients may have been due largely to advanced age. There is a decrease in the number of MSCs in marrow as a function of age (Chapter 13). Autologous marrow in a demineralized bone matrix carrier was also effective in promoting bone regeneration in a canine model of a gap non-union and spinal fusion to correct scoliosis (Tiedeman et al., 1991; Connolly, 1998).

Richards et al. (1999) injected MSCs in collagen gel directly into the distraction gap of rat femurs undergoing bone lengthening by the Ilizarov method to determine whether addition of cells would enhance the volume of bone produced. Collagen gel alone elicited larger bone volume, but volume enhancement by collagen gel with cells was greater. Lysis assays performed on the donor cells using host blood serum indicated that donor cell survival may have been an issue in this experiment, but even so, dying cells may have provided growth factors that stimulated bone regeneration by host cells.

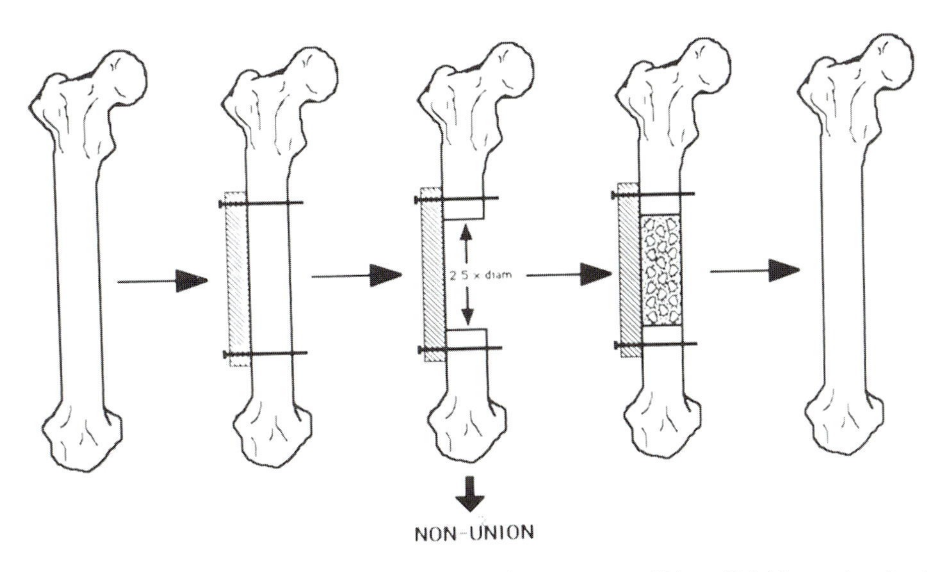


**FIGURE 10.18** Rat femur gap model for replacement of a segment of bone by a segment of bioartificial bone. A polyethylene plate is screwed into the femur for support and a defect equal to more than 2.5× the femur diameter is then created. Non-union is the result if the gap is left unfilled or filled with hydroxyapatite ceramic scaffold only. However, if the hydroxyapatite ceramic scaffold is loaded with MSCs, new bone is formed to fill the gap. Reproduced with permission from Caplan et al., Mesenchymal stem cells and tissue repair. In: Jackson DW, Arnoczky S, Woo S, Frank C (eds).The Anterior Cruciate Ligament: Current and Future Concepts. New York: Raven Press. Copyright 1993, Elsevier.

Although osteoblastic cells can be injected directly into nonhealing fractures to help regenerate bone, they cannot be used to replace bone in larger defects. Likewise, osteoconductive scaffolds, with or without osteoinductive factors, are unable to induce regeneration of bone across defects that greatly exceed the critical size gap, because of the distances host MSCs must migrate before differentiating into osteoblasts. Osteoconductive and inductive scaffolds seeded with mesenchymal stem cells prior to implantation are the solution to this problem. Culture-expanded MSCs seeded into a variety of scaffolds have been used to make such bioartificial bone (Caplan, 1993; Bruder et al., 1998a; Bruder and Caplan, 2000; Dennis et al., 2001). **FIGURE 10.18** illustrates a rat femur model for the use of bioartificial bone segments.

Kadiyala et al. (1997) seeded a hydroxyapatite/β-tricalcium phosphate cylinder throughout with cultured MSCs and implanted it into an 8-mm gap made in the adult rat femur. The MSCs differentiated into osteoblasts that by 8 weeks had secreted new bone matrix throughout the implant. Similar results were obtained with implants containing autologous MSCs that spanned 21-mm defects in dog femurs (Bruder et al., 1998b; Kraus et al., 1999). These studies were then expanded to show that human MSCs in a similar ceramic scaffold could make new bone to fill in an 8-mm gap in the femurs of athymic rats (Bruder et al., 1998c). In each of these studies, control implants (no cells) formed bone only at the interface between scaffold and native bone. Biomechanical tests showed that the new bone formed by the MSCs was stronger than the cell-free ceramic alone. However, because of slow scaffold resorption, the volume of space in the scaffold that was occupied by new bone matrix ranged only from 40–47%.

Some companies have reported the production of bioceramic "bone blanks" for seeding that in animal studies were resorbed over a period of three years. Millenium Biologix has received FDA approval for use of a resorbable, proprietary, multiphase calcium phosphate-based material called Skelite™ to repair defects in the spine, pelvis, and long bones (Flanagan, 2003). Nanoceramics (grain size less than 100 nm) may be more promising than conventional ceramics (grain size over 100 nm) in designing bone blanks (Webster et al., 2000). Surface occupancy of rat calvarial osteoblasts was less when cultured on nanophase alumina, titania, and hydroxyapatite than on the conventional forms, while proliferation and differentiation was greater.

Synthetic polymers or co-polymers, alone or in combination with calcium or hydroxyapatite, have also proven to be good scaffolds for making bioartificial bone. These polymers are resorbable and the apatite provides an adherent and osteoinductive surface for MSCs (Behravesh et al., 1999). Yaszemski et al. (1995) designed a composite of poly (propylene-fumerate), cross-linked through the fumarate double bond by a monomer, N-vinylpyrrolidine, and filled with NaCl and phosphate β-tricalcium. This composite is plastic when mixed fresh and can be pressed into bone defects. When hardened, it is strong enough to substitute temporarily

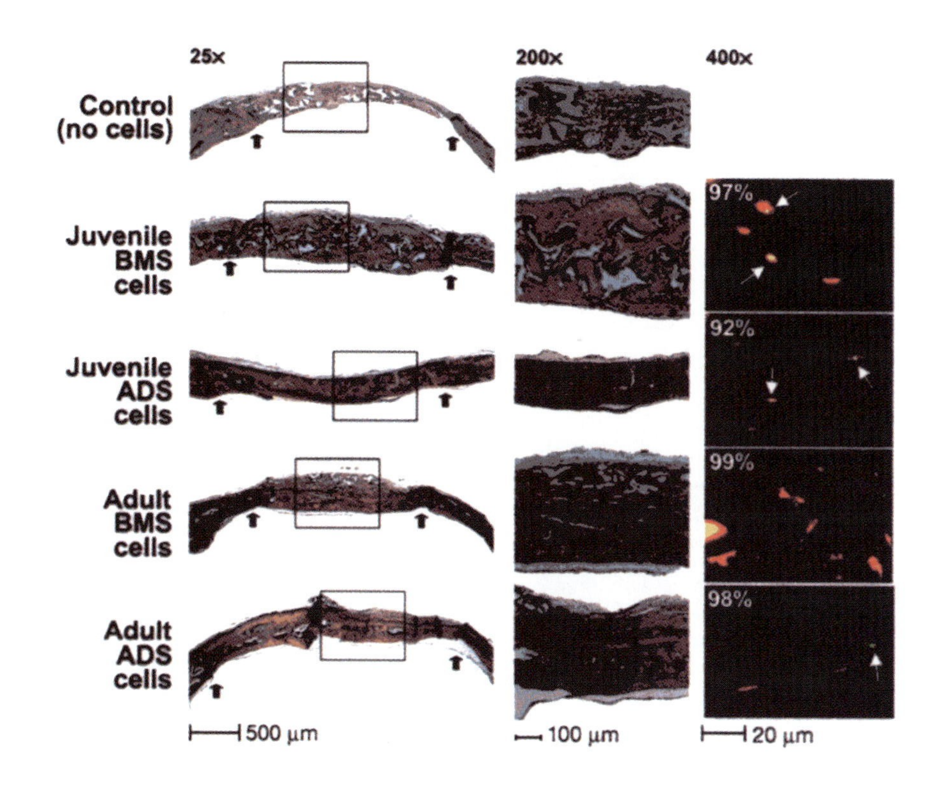


**FIGURE 10.19** Repair of 4 mm diameter circular calvarial defects in male mice after 12 weeks by implants of apatite-coated poly lactic-co-glycolic acid (PLGA) scaffolds seeded with adipose derived stromal cells from female mice. These cells are similar to MSCs (see Chapter 13). **Left,** low magnification of H&E-stained sections of controls implanted with acellular scaffold, and experimental animals implanted with scaffolds seeded with juvenile or adult bone marrow stromal cells (BMS) and juvenile or adult adipose-derived cells (ADS). Arrows indicate edges of defect. **Middle,** higher magnification of boxed areas. The defects were bridged by new bone in all cases. The bone formed by either juvenile or adult BMS or ADS cells was histologically similar, although juvenile BMS cells formed the thickest bone. **Right,** Fluorescent imaging of chromosomes in regenerated bone. Arrows point to male (host) chromosomes. Numbers indicate the percentage of cells in the regenerated bone derived from the female (donor) cells. Reproduced with permission from Cowan et al., Adipose-drived adult stromal cells heal critical-size mouse calvarial defects. Nature Biotech 22:560–567. Copyright 2004, Nature Publishing Group.

for bone, and it degrades as osteogenic cells invade it and regenerate new bone. Rat calvarial cells seeded onto the surface of a scaffold consisting of hydroxyapatite suspended in a 50:50 mixture of polylactide: glycolide were induced *in vitro*, in the presence of β-glycerophosphate, to form tissue resembling cancellous bone (Laurencin et al., 1996). Cultured adult female bone marrow stromal cells and calvarial-derived osteoblasts seeded into apatite-coated poly (lactic-co-glycolic acid) (PLGA) scaffolds, were able to form high-quality intramembranous bone and completely repair 5-mm circular defects in male mouse parietal bone in the absence of any exogenous growth factors (Cowan et al., 2004) (**FIGURE 10.19**). Ninety-nine percent of the osteoblasts comprising the new bone were donor-derived, as judged by XX vs. XY chromosomes. Interestingly, adipose-derived adult stromal (ADAS) cells in the same scaffold formed bone equally well. Ninety-eight percent of the osteoblasts in the new bone were derived from donor cells.

Growth factors can be added to bioartificial bone constructs to promote the differentiation of seeded cells. For example, *in vitro*, C3H10T1/2 cells suspended in a collagen gel formed bone in a three-dimensional poly-L-lactic acid scaffold in the presence of recombinant human BMP-2 (Kim et al., 1998). In a nude rat model, Gelfoam blocks impregnated with culture-expanded human MSCs and a solution containing 10 µg/ml each of BMP-2 and FGF-2 completely healed 4-mm defects in the parietal bone of nude rats. The quality of the bone formed was superior to that of MSCs in Gelfoam without the growth factors (Akita et al., 2004). The PEG scaffold developed by Lutolf et al. (2003) may also be a good candidate for seeding with MSCs to make bioartificial bone.

The ability exists to fabricate not only bone segments, but whole bioartificial bones. Kinoshita et al. (1996) reported the use of a poly (L-lactide) mesh to reconstruct the mandible. In experiments on dogs, a segment of mandible was removed and replaced with a PLLA

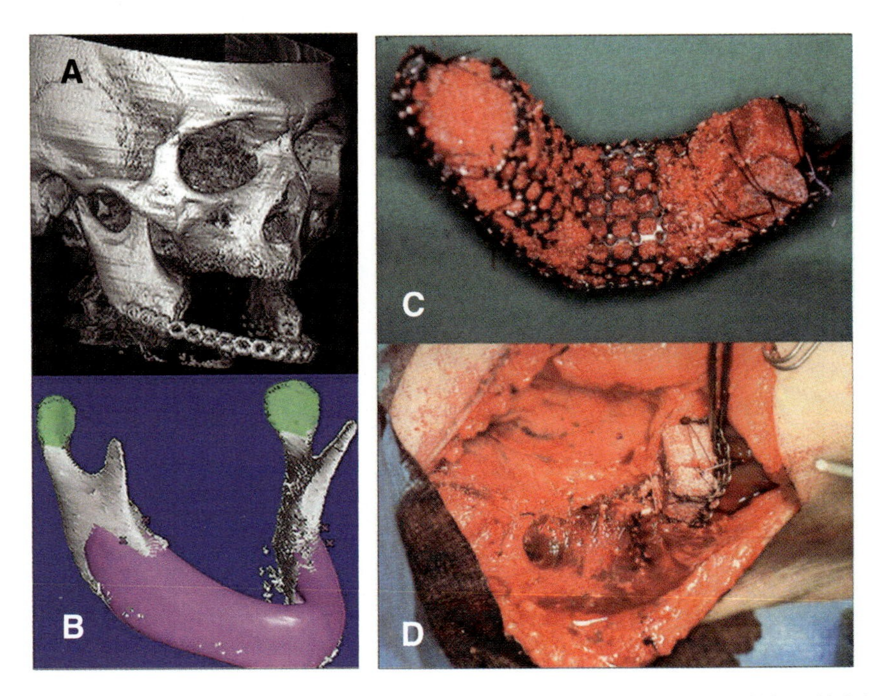


**FIGURE 10.20** **(A)** Three-dimensional CT scan of mandible defect. **(B)** Computer aided shape design of bioartificial mandible transplant. **(C)** Titanium mesh cage filled with bone mineral blocks infiltrated with recombinant human BMP7 and bone marrow mixture. **(D)** Implantation of the construct into the right latissmus dorsi muscle of patient. Reproduced with permission from Warnke et al., Growth and transplantation of a custom vascularized bone graft in a man. Lancet 364:766–770. Copyright 2004, Elsevier.

mesh cage containing autogeneic iliac particulate cancellous bone and marrow (PCBM). Radiograms taken six months after the operation revealed the regeneration of bone with the same height as the host bone, normal trabecular pattern, and a good alveolar ridge. Kinoshita et al. (1996) also reported two clinical cases of mandibular replacement using the same technique. In one case, the right side of the mandible was partially resected. The bioartificial replacement formed bone with an alveolar ridge that allowed the patient to wear a denture. In the other case, the whole front of the mandible from the left to right molar region had been lost, creating a cosmetic deformity that made speaking and chewing difficult. Three months after replacement with the bioartificial bone construct, the continuity of the mandible was restored, and the cosmetic appearance and speaking and chewing abilities of the patient were markedly improved.

A bioartificial mandible was implanted in another patient to replace the original, which had been removed nine years earlier because of oral cancer (Warnke et al., 2004). A three-dimensional CT scan of the patient's mouth was used to first generate a mandible blueprint. A three-dimensional titanium micromesh cage was then fashioned into the shape of the mandible and seeded with autologous bone marrow cells. Small blocks of bovine bone matrix and BMP-7 were also put into the cage to stimulate bone regeneration. The construct was then implanted into the latissmus dorsi muscle below the shoulder blade, where it remained for seven weeks and became vascularized (**FIGURES 10.20, 10.21**). The implant was then removed, along with some surrounding muscle and blood vessels, and transplanted to the skull, where its blood vessels were connected to blood vessels in the patient's neck. The construct continued to develop new bone (**FIGURE 10.22**), and the patient was able to chew solid food for the first time in nearly a decade.

Bioartificial phalanges of human dimensions have been constructed by wrapping calf periosteum around a co-polymer of polyglycolic and poly-L-lactic acids shaped like a phalange (Isogai et al., 1999). Separate co-polymer sheets were seeded with either calf articular chondrocytes from the shoulder or tenocytes from the flexor tendons. Chondrocyte-seeded polymer was sutured to (a) one end and (b) both ends of separate phalange constructs to create distal and middle phalanges with joint surfaces, respectively. The distal and middle phalanges were lined up in tandem and tenocyte-seeded polymer was sutured around them at their junction to create a tenocapsule. The constructs were then implanted subcutaneously into nude mice. Gross and histological examination at 20 weeks revealed that the constructs had the shape of phalanges with joints and were composed of bone, mature articular cartilage, and a tenocapsule.

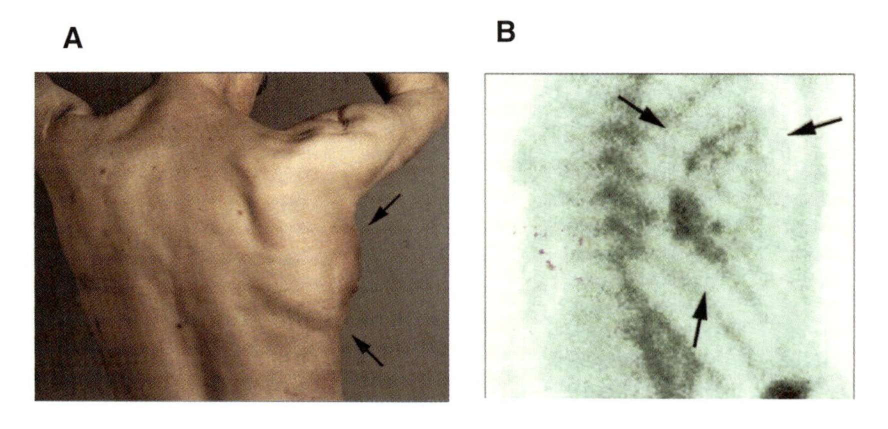


**FIGURE 10. 21** **(A)** Three weeks after implantation of the mandibular construct (arrows). **(B)** Scintigram after intravenous injection of technetium-99m-oxydronate tracer, done four weeks after implantation. The arrows outline the U-shaped mandibular construct. Reproduced with permission from Warnke et al., Growth and transplantation of a custom vascularized bone graft in a man. Lancet 364:766–770. Copyright 2004, Elsevier.

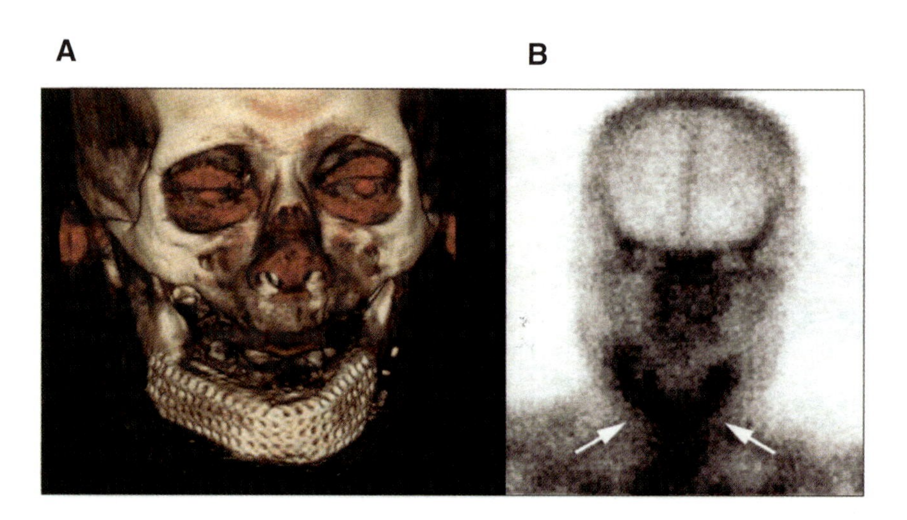


**FIGURE 10.22** **(A)** Three-dimensional CT scan after transplantation of the bioartificial mandible with enhancement of soft tissue (red). **(B)** Scintigram showing continued bone remodeling and mineralization (arrows). Reproduced with permission from Warnke et al., Growth and transplantation of a custom vascularized bone graft in a man. Lancet 364:766–770. Copyright 2004, Elsevier.

### 5. The Long Bones of Larval Salamanders Can Regenerate When Removed

The long bones of larval salamander limbs, which are mainly cartilaginous, can regenerate after their removal from the limb (Goss, 1958). This regeneration is accomplished by the formation of a mass of mesenchymal cells that accumulates in the space vacated by the bone, and subsequently differentiates into new cartilage and bone. Adult salamanders cannot regenerate missing bones or bone segments, even though they can regenerate the skeletons of their limbs after amputation. However, when the cartilaginous templates of individual bones are removed from the newly forming limb segments of adult newt limbs six weeks after amputation, they are regenerated as in larval limbs (Goss, 1958). The capacity for such regeneration declines rapidly starting at seven weeks after amputation as the skeletal elements begin to ossify. These observations suggest that only tissues in the early stages of differentiation are plastic enough to compensate for the loss of a skeletal element. The origin of the cells that replace the excised skeletal elements in these experiments has not been traced. This model system may be useful in understanding why mammalian long bones fail to regenerate across a critical size gap.

## REGENERATIVE THERAPIES FOR TENDON AND LIGAMENT

Tendons and ligaments heal by a process equivalent to scar formation in dermis, with elevation in levels of the same growth factors (Woo et al., 1999). Even with maximal maturation, the biomechanical properties of the repaired tissue never reach normal because the diameter and crosslinkages of the collagen fibrils are subnormal and proteoglycan profiles are altered. Ruptured tendons and ligaments are often reconstructed after injury with autografts or allografts, with the attendant problems of donor site morbidity and limited availability of tissue. Attempts have thus been made to improve the healing of ruptured tendon and ligament in animal models by injection of growth factors and cells into injury sites and the use of scaffolds and cells to make bioartificial tendons and ligaments (Pittenger et al., 2002; Carpenter and Hankenson, 2004). Since tendons and ligaments are different in molecular composition and the types of mechanical forces they experience, these differences have to be taken into account in designing scaffolds to regenerate or replace them.

## 1. Tendon

Tendons transduce the force of muscles to skeletal elements. Ruptured tendons are often repaired by suturing their ends together, but re-injuries are common, particularly during the early period of repair when the regenerating tissue is still weak (Kvist, 1994). Newer treatments designed to regenerate tendons or replace them with bioartificial constructs have met with some success compared with untreated injuries, but do not achieve a repair strength greater than that obtained by suturing the tendon ends together.

A number of growth factors enhance tendon repair *in vitro*. FGF-2 at a concentration of 10 ng/ml significantly accelerated the closure of wounds made in quiescent, confluent *in vitro* cultures of rat patellar tendon fibroblasts (Chan et al., 1997). The accelerated closure was the result of higher proliferation. PDGF-BB, IGF-1, and TGF-β1 are constitutively produced by cultured tendon fibroblasts and accelerate tendon repair by enhancing proliferation and synthesis of collagen and proteoglycans (Abrahamsson et al., 1991; Banes et al., 1995).

A single 10-μg injection of growth and differentiation factor (GDF)-6 (cartilage-derived morphogenetic protein-2) into the lesion site enhanced regeneration of the transected rabbit Achilles tendon *in vivo* (Forslund and Aspenburg, 2003). At 14 days post-injection, the treated tendons were larger than controls and their failure load and stiffness were increased by 35%, but they did not return to normal biomechanical strength. GDF 6, 7, and 8 lyophilized onto demineralized bone matrix particles induced tendon formation when implanted subcutaneously or intramuscularly into rats (Wolfman et al., 1997). Collagenase-induced lesions in the flexor digitorum superficialis tendon of horses showed a peak in the expression of TGF-β1 message at one week, but a 40% reduction in tissue levels of IGF-1 protein over the first two weeks. By four weeks, however, the level of IGF-1 protein exceeded that of normal tendon and remained elevated through eight weeks, suggesting that the administration of exogenous IGF-1 during the first two weeks after tendon injury might enhance tendon healing (Dahlgren et al., 2005).

Near-infrared light was reported to stimulate collagen production in surgically repaired transected Achilles tendons of rabbits (Reddy et al., 1998). Laser photostimulation resulted in a 26% increase in total collagen concentration. Differential extraction of the collagen showed a significant decrease in pepsin-soluble collagen and increases in neutral salt-soluble and insoluble collage in photostimulated tendon, suggesting a higher rate of remodeling and degree of mechanical integrity.

An Achilles tendon regeneration template consisting of collagen fibers promoted ingrowth of fibroblasts, but the regenerated tissue achieved only 36% of normal strength (Kato et al., 1991). In other studies, silicon tubes were used to bridge the ends of transected Achilles tendons in rabbits (Kato et al., 1991; Louie et al., 1997). A cable of new fibrous tissue was formed in the tubes that had the characteristic crimp of tendon fibers. Filling the tubes with a porous collagen and glycosaminoglycan (CG) matrix of the type used to regenerate dermis and peripheral nerve elicited the formation of denser tissue, but without the crimp, suggesting that the CG matrix inhibited the regeneration of tendon scar tissue (Louie et al., 1997). Biomechanical tests were not conducted on the regenerated tendon.

Cultured tendon fibroblasts and bone marrow MSCs in type I collagen gel have been used to enhance repair of 1-cm defects in the the rat and rabbit Achilles tendon (Caplan et al., 1993; Young et al., 1998; Butler and Awad, 1999). By 12 weeks post-transplant, the cross-sectional area, load-bearing properties, and maximum force generation of the treated tissue were twice that of the tissue in defects that had received only the collagen gel. Stiffness and strength were increased to 50–60% of uninjured tendons. Histological examination showed that the implanted cells differentiated into fibroblasts with good matrix organization. Awad et al. (2000) showed that the contraction kinetics of the gel were significantly influenced by the initial seeding density, with better alignment of cells and collagen fibers at higher densities.

## 2. Ligament

Ligaments stabilize joints. The ligaments of the knee (posterior and anterior cruciate ligaments, medial cruciate ligament), particularly the anterior cruciate ligament, are among the most often injured during sports activities. The gold standard for replacing a damaged ACL is an autograft of the patellar, gracilis, or semitendinosus tendon, but autografts are attended by donor site morbidity. Allografts and xenografts of cadaver and bovine ligaments have also been used, but there are problems with inflammation, immunorejection, and risk of disease transmission. Synthetics such as Dacron, Gore-Tex, polypropylene, polyesters, and carbon fibers have been used to replace ligaments, but have high rates of failure and often exhibit stress-shielding of adjacent tissues (reviewed by Vunjak-Novakovic et al., 2004). Furthermore, not all these materials encourage their replacement or augmentation by ingrowth of regenerating tissue. What is desired for regenerating or replacing a ligament is a scaffold that has high initial strength that will stabilize the joint, but one that allows cell ingrowth and transfers load bearing to the neoligament as it regenerates. An important factor in directing the differentiation of fibroblastic cells in scaffolds toward a ligamentous phenotype is mechanical stimulation.

PDGF-BB, TGF-β1, and FGF-2 applied to the injury region of knee ligaments in rat, rabbit, or canine models were reported to enhance the structural properties of the repair tissue over that of untreated injuries, although the improvement did not approach the values of uninjured controls (reviewed by Woo et al., 1999). Marked bone marrow cells injected into a transection site in the midregion of rat MCLs survived and by seven days had migrated into adjacent noninjured regions (Watanabe et al., 1998). Whether these cells made new ligament in the lesion, provided secreted factors that promoted regeneration, or did both, is unknown.

Pig SIS has been tested for its ability to enhance regeneration of transected ACL in goats and transected MCL in rabbits (Badylak et al., 1999; Mushal et al., 2004). SIS enhanced goat ACL regeneration over transected ligaments that received no treatment. The tissue regenerated into the SIS at one year was histologically comparable to that of a patellar tendon autograft and had approximately the same failure strength, which, however, was only 25% that of the normal ligament. SIS significantly improved MCL tissue regeneration in rabbits over transected nontreated controls, but the biomechanical strength of the regenerated tissue achieved only one-third that of the native ligament.

A collagen fiber scaffold has been developed for reconstruction of the ACL that consists of fibers aligned in parallel with tapered poly (methyl methylacrylate) plugs at both ends for insertion into the bone (Dunn et al., 1992). Braided poly-L-lactic acid prostheses have been used with some success to repair defects in rabbit ACL (Laitinen and Alitalo, 1993; cited in Bourke et al., 2004). They found that loose-braided poly-L-lactic acid fibers promoted the regeneration of new tissue within the scaffold better than tighter scaffolds that induced new tissue formation only on the outside of the scaffold. Poly desamino tyrosyl-tyrosine ethyl ester carbonate fibers are a new type of scaffolding that has been found to support fibroblast ingrowth when implanted subcutaneously in rats and to promote attachment of rabbit and human fibroblasts *in vitro* (Bourke et al., 2004). These fibers have also been fashioned into a scaffold that mimics a rabbit ACL and were shown to have strength values equal to or greater than those of normal ACL and higher than those of poly-L-lactic acid fibers.

Silk fibers treated to be nonimmunologic have been used to fashion a bioartificial ACL the size of a human ACL (Chen et al., 2003). Silk fibers have the biomechanical properties necessary to initially stabilize the joint, the geometry to allow ingrowth of cells to regenerate tissue, and a slow enough rate of degradation to gradually transfer biomechanical loads to the regenerating tissue as it matures. RGD adhesive peptides were coupled to silk fibers in a wire-rope configuration and the fibers were seeded with bone marrow stem cells or ACL fibroblasts. These constructs were incubated vertically in a bioreactor while applying linear and rotational displacement to the constructs. The inclusion of RGD peptides improved cell attachment, enhanced cell density, and resulted in significantly higher type I collagen content over 14 days of culture compared to silk fibers without RGD peptides. Silk fibers thus may have considerable potential for making bioartificial ligaments that approach the effectiveness of native ligaments.

## SUMMARY

Major efforts have been put into attempts to cure muscular dystrophy by wild-type myoblast or satellite cell transplantation. While these transplants have improved muscle structure and function in early clinical trials and in experimental animals, immunorejection is a major problem unless immunosuppression is used. Gene therapy with dystrophin or utrophin genes is difficult due to the large size of the genes. A promising genetic approach in mice is exon skipping by blocking translation of the mutant mRNA encoded by exon 23 with an antisense adenoviral construct, resulting in the production of truncated, but functional dystrophin protein. Injection of the antisense construct into the tibialis anterior muscle restored the histology and function of the muscle fibers to normal. Some progress has been made in constructing tiny bioartificial muscles, but constructing larger muscles will require solving the problem of vascularization; such efforts are underway.

Regeneration of meniscus and articular cartilage of the knees is a major goal of orthopedics, due to the high number of injuries and ostroarthritic degeneration suffered by these structures. Acellular regeneration templates of polyurethane, collagen/hyaluronic acid/ chondroitin sulfate (CHACS) or pig SIS promote the regeneration of fibrocartilage similar to meniscus tissue in dogs. The dogs regain normal gait. Human patients receiving CHACS implants also regenerated meniscus-like tissue and reported decreased pain, and increased activity scores after three years. Bioartificial meniscus tissue has been made by seeding canine, bovine, sheep, pig or rabbit meniscus cells or articular chondrocytes into scaffolds of collagen II/6-chondroitin sulfate, collagen I sponge, polyglycolic acid, alginate gel, or acellular meniscus matrix and then either implanting it subcutaneously into nude mice or into surgically created meniscus defects. Collectively, the results indicate that these bioartificial constructs all form meniscus-like fibrocartilage that is capable of bonding to host tissue.

There are a number of issues involved in the design of ways to regenerate or replace articular cartilage. First, autologous chondrocyte transplantation requires two surgeries. Second, whatever scaffold material is used must facilitate cell attachment, bond well to the native cartilage, and not shrink, while at the same time being biodegradable. Growth factors have done little to enhance the quality of repair tissue over controls in partial or full-thickness defects. Regeneration templates made from polyglycolic acid, demineralized cortical bone matrix, hyaluronan-based polymers, and a liquid thermosensitive chitosan preparation have produced better results. All have induced the regeneration of well-organized articular cartilage, presumably from bone marrow MSCs. Patients receiving liquid chitosan were reported to have experienced complete recovery from severe articular cartilage damage following 12-week physiotherapy programs.

Transplants of cultured chondrocytes are in clinical use to repair damaged human articular cartilage and are reported to give good results. Cultured chondrocytes dedifferentiate. Culturing dedifferentiated chondrocytes in a way that maintains maximum ability to redifferentiate into high-quality chondrocytes is a major concern for cells used as transplants or to construct bioartificial cartilage. Factors that promote redifferentiation of cartilage with the best biochemical and mechanical properties and ability to adhere to host cartilage appear to be low seeding density, few passages, maintenance of Sox-9 expression by serum-free medium, and the application of growth factors in the sequence TGF-β plus FGF-2 followed by IGF-I. The use of MSCs genetically manipulated to produce growth factors that promote chondrogenesis is a promising avenue for the regeneration of cartilage in partial-thickness defects.

Bioartificial articular cartilage has been made from bone marrow cells, stem cells isolated from connective tissue, and rabbit articular chondrocytes seeded into polylactide/alginate, polyglycolic acid, collagen I gel, and pig SIS matrices, respectively. Fibrocartilage or cartilage resembling articular cartilage formed after implantation into partial-thickness defects. Other experiments indicate that collagen, squid β-chitin, or hybrid scaffolds of collagen plus polylactic acid and polyglycolic acid support the differentiation of cultured human or animal chondrocytes into tissue that histologically and biochemically resembles hyaline cartilage. In most of these experiments, long-term study of the biochemical and structural integrity of the regenerated cartilage was not done.

Many experimental models use an osteochondral defect that penetrates into the bone marrow, whereas in humans defects occur mostly in the cartilage itself or penetrate only a short distance into the subchondral bone. We need to do more testing of regenerative protocols in such partial or shallow osteochondral defects. Finally, there is considerable debate about what cell type is best for bioartificial cartilage constructs. An argument can be made for MSCs and dedifferentiated chondrocytes, which are similar to one another and can respond to local cues that generate specific types of chondrocytes; furthermore, they can be genetically modified with growth factor genes that enhance their proliferation and differentiation.

Regenerative therapies for bone are probably closest to clinical reality. Pulsed electromagnetic fields, though controversial, have been reported for years to enhance the repair of non-union fractures. Bones can be lengthened by the Ilazarov distraction technique, which activates MCSs of the periosteum and marrow to form a growth platelike region. Newer therapies are being developed that involve inducing bone regeneration by osteogenic factors and acellular templates, as well as by cell transplantation and implantation of bioartificial bone.

Soluble factors involved in bone repair and transplants of osteogenic cells accelerate the repair of small lesion sites such as fractures. BMPs and FGF-2 accelerate fracture repair in rat fracture and rabbit osteotomies. The synthetic peptide TP508, a 23 amino acid sequence representing the receptor-binding domain of human thrombin, enhanced the repair of rabbit ulnar osteotomies. The injection of MSCs in collagen gel into Ilazarov distraction gaps of rat femurs or autologous bone marrow into non-union fractures of human patients enhanced repair in both cases. The induction of bone regeneration across larger gaps requires an osteoconductive bone regeneration template or an implant of bioartificial bone. Acellular bone regeneration templates incorporating BMPs induce regeneration of bone across gaps that do not exceed a critical size. Osteogenic cells migrate from host bone into the templates and regenerate bone over a distance restricted by how far the cells can migrate over a given period of time. Templates that have been shown to foster osteoblast migration and bone regeneration are pig SIS, polyglycolic and polylactic acids and co-polymers, collagen, hyaluronate or gelatin hydrogels, calcium phosphate ceramics and cements, and resorbable glass particles. BMPs or plasmids producing BMPs have been incorporated into collagen gels for release in the lesion site. MSCs or osteoblasts migrating into the scaffold pick up the plasmids and synthesize the BMPs. Patients have been successfully treated with BMP-7 in type I collagen and with "smart" ceramic scaffolds that absorb BMPs from their environment. One highly interesting template consists of BMP-2 incorporated into a hydrogel of RGD adhesion peptides coupled to polyethylene glycol chains cross-linked with a peptide that is an MMP cleavage site. Osteogenic cells adhere to the RDG sites and produce MMPs, which degrade the gel and release the BMP-2.

The replacement of large segments of bone can be achieved only by implants of bioartificial bone made by seeding osteoconductive templates with MSCs. Hydroxyapatite/calcium phosphate scaffolds loaded with MSCs make new bone, but the rate of resorption of these scaffolds is too slow to allow formation of the amount of new bone required for full strength. Synthetic polymers or co-polymers, alone or in combination with ceramics, have proven to form higher amounts of bone and repair circular defects in the skulls of rats and mice, particularly in the presence of BMP-2 and FGF-2.

Three cases of human bioartificial mandible construction have been reported in which poly (L-lactide) or titanium mesh was seeded with bone marrow cells and particles or blocks of bone and BMP-7. In two cases, this construct was implanted directly in place of the original mandible and in the other it was prevascularized by culture in the latissmus dorsi muscle before implantation. In all three cases, the patients regained chewing ability. The construction *in vitro* of bioartificial digits with a terminal and middle phalange from polyglycolic and poly-L-lactic acids and calf periosteum has also been reported. The digits were reported to form mature articular cartilage, bone, and a joint capsule after being implanted subcutaneously into nude mice. New bone and cartilage has also been grown by packing a subperiosteal space with alginate that allows the migration of MSCs into the alginate, where they differentiate into bone or cartlage.

Little progress has been made in enhancing tendon and ligament repair. The repair of wounds made in cultures of rat patellar tendon fibroblasts was accelerated by the growth factors PDGF-B, IGF-I, and TGF-β1. These growth factors enhance proliferation of tendon fibroblasts and their synthesis of collagen and proteoglycans. Growth and differentiation factor-6 enhanced the rate of regeneration of the lesioned rabbit Achilles tendon *in vivo*. Near-infrared light was reported to stimulate collagen production in surgically repaired Achilles tendons of rabbits. Acellular collagen regeneration templates induce some tendon regeneration; better regeneration is obtained when tendon fibroblasts and MSCs in type I collagen are implanted in the tendons. None of these treatments or constructs results in restoring tendon strength to any more than 60%. Ligament regeneration is also enhanced above what is observed in untreated injuries by PDGF-BB, TGF-β1, and FGF-2, but never approaches control levels. SIS, collagen, braided poly-L-lactic acid fibers, DTE carbonate fibers, and silk fibers have all shown promise as acellular regeneration templates or scaffolds for bioartificial ligament construction.

## REFERENCES

Abe M, Takahashi M, Tokura S, Tamura H, Nagano A (2004) Cartilage-scaffold composites produced by bioresorbable β-chitin sponge with cultured rabbit chondrocytes. Tiss Eng 10:585–594.

Abrahamsson SO, Lundborg G, Lohmander LS (1991) Recombinant human insulin-like growth factor-1 stimulates in vitro matrix synthesis and cell proliferation in rabbit flexor tendon. J Orthop Res 9:495–502.

Akita S, Fukui M, Nakgawa, Fujii T, Akino K (2004) Cranial bone defect heaing is accelerated by mesenchymal stem cells induced by administratuon of bone morphogenetic protein-2 and basic fibroblast growth factor. Wound Rep Reg 12:252–259.

Alderton JM, Steinhardt A (2000) How calcium influx through calcium leak channels is responsible for the elevated levels of calcium-dependent proteolysis in dystrophic myotubes. Trends Cardiovasc Med 10:268–272.

Ambro BT, Zimmerman J, Rosenthal M, Pribitkin EA (2003) Nasal septal perforation repair with porcine small intestinal submucosa. Arch Facial Plast Surg 5:528–529.

Asahara T, Kalka C, Isner JM (2000) Stem cell therapy and gene transfer for regeneration. Gene Ther 7:451–457.

Athanasiou K, Korvick D, Schenck R (1997) Biodegradable implants for the treatment of osteochondral defects in a goat model. Tiss Eng 4:363–373.

Awad HA, Butler DL, Harris MT, Ibrahim RE, Wu Y, Young RG, Kadiyala S, Boivin GP (2000) *In vitro* characterization of mesenchymal stem cell-seeded scaffolds for tendon repair: Effects of initial seeding density on contraction kinetics. Biomed Mater Res 51:233–240.

Badylak SF, Arnoczky S, Plouhar P, Haut R, Mendenhall V, Clarke R, Horvath C (1999) Naturally occurring extracellular matrix as a scaffold for musculoskeletal repair. Clin Orthopaed Rel Res 367S:S333–S343.

Banes AJ, Tsuzaki M, Hu P, Brigman B, Brown T, Almekinders L, Lawrence WT, Fischer T (1995) PDGF-B, IGF-1 and mechanical load stimulate DNA synthesis in avian tendon fibroblasts in vitro. J Biomech 28:1505–1513.

Bassett CAL, Pawluk RJ, Pilla AA (1974) Augmentation of bone repair by inductively coupled electromagnetic fields. Science 184: 575–577.

Beauchamp JR, Pagel CN, Partridge TA (1997) A dual-marker system for quantitative studies of myoblast transplantation in the mouse. Transplantation 63:1794–1797.

Behravesh E, Yasko AW, Engel PS, Mikos AG (1999) Synthetic biodegradable polymers for orthopaedic applications. Clin Orthopaed Rel Res 367S:S118–S129.

Bertone AL, Pittman DD, Bouxsein ML, Li J, Clancy B, Seeherman HJ (2004) Adenoviral-mediated transfer of human BMP-6 gene accelerates healing in a rabbit ulnar osteotomy model. J Orthopaed Res 22:1261–1270.

Blaveri K, Heslop L, Yu DS, Rosenblatt D, Gross JG, Partridge A, Morgan JE (1999) Patterns of repair of dystrophic mouse muscle: studies on isolated fibers. Dev Dynam 216:244–256.

Boden SD (1999) Bioactive factors for bone tissue engineering. Clin Orthopaed Rel Res 367S:S84–S94.

Bonadio J (2002) Genetic approaches to tissue repair. Ann NY Acad Sci 961:58–60.

Bostrom M, Lane JM, Tomin E, Browne M, Berberian W, Turek T, Smith J, Wozney J, Schildhauer T (1996) Use of bone morphogenetic protein-2 in the rabbit ulnar nonunion model. Clin Orthopaed Rel Res 327:272–282.

Bourgeois B, Laboux O, Obadia L, Gauthier O, Betti E, Aguado E, Daculsi G, Bouler J-M (2003) Calcium-deficient apatite: A first *in vivo* study concerning bone ingrowth. J Biomed Mater Res 65A:402–408.

Bourke SL, Kohn J, Dunn MG (2004) Preliminary development of a novel resorbable synthetic polymer fiber scaffold for anterior cruciate ligament reconstruction. Tiss Eng 10:43–52.

Brittberg M, Lindahl A, Nilsson AA, Ohlsson C, Isaksson O, Peterson L (1994) Treatment of deep cartilage defects in the knee with autologous chondrocyte transplantation. New Engl J Med 331:889–895.

Brittberg M (1999) Autologous chondrocyte transplantation. Clin Orthopaed Rel Res 367S:S147–S155.

Bruder SP, Jaiswal N, Ricalton NS, Kraus KH, Kadiyala S (1998a) Mesenchymal stem cells in osteobiology and applied bone regeneration. Clin Orthopaed Rel Res S355:S247–S256.

Bruder SP, Kraus KH, Goldberg VM, Kadiyala S (1998b) The effect of implants loaded with autologous mesenchymal stem cells on the healing of canine segmental bone defects. J Bone Joint Surg Am 80A:985–996.

Bruder SP, Kurth AA, Shea M, Hayes WC, Jaiswal N, Kadiyala (1998c) Bone regeneration by implantation of purified, culture-expanded human mesenchymal stem cells. J Orthoped Res 16:155–162.

Bruder SP, Caplan AI (2000) Bone regeneration through cellular engineering. In: Lanza RP, Langer R, Vacanti J (eds.). Principles of Tissue Engineering, 2nd ed. New York: Academic Press, pp 683–695.

Buma P, Ramrattan NN, van Tienen TG, Veth RPH (2004) Tissue engineering of the meniscus. Biomats 25:1523–1532.

Burton EA, Davies KE (2002) Muscular dystrophy—reason for optimism? Cell 108:5–8.

Bushby KM (2000) Genetics and the muscular dystrophies. Dev Med Child Neurol 42:780–784.

Butler D, Awad H (1999) Perspectives on cell and collagen composites for tendon repair. Clin Orthopaed Rel Res 367S:S324–S332.

Caplan AI, Fink DJ, Goto T, Linton AE, Young RG, Wakitani S, Goldberg VM, Haynesworth SE (1993) Mesenchymal stem cells and tissue repair. In: Jackson DW, Arnoczky S, Woo S, Frank C (eds.). The Anterior Cruciate Ligament: Current and Future Concepts. New York: Raven Press, pp 405–417.

Carpenter JE, Hankenson KD (2004) Animal models of tendon and ligament injuries for tissue engineering applications. Biomats 25:1715–1722.

Caterson EJ, Nesti LJ, Li WJ, Danielson KG, Albert TJ, Vaccaro AR, Tuan RS (2001) Three-dimensional cartilage formation by bone marrow-derived cells seerded in polyactide/alginate amalgam. J Biomed Mater Res 57:394–403.

Chamberlain JR, Schwarze U, Wang P-R, Hirata RK, Hankenson KD, Pace JM, Underwood RA, Song KM, Sussman M, Byers PH, Russell DW (2004) Gene targeting in stem cells from individuals with osteogenesis imperfecta. Science 303:1198–1201.

Chan BP, Chan KM, Maffulli N, Webb S, Lee K (1997) Effect of basic fibroblast growth factor: An in vitro study of tendon healing. Clin Orthopaed 342:239–247.

Chen G, Sato T, Ushida T, Ochiai N, Tateishi T (2004) Tissue engineering of cartilage using a hybrid scaffold of synthetic polymer and collagen. Tiss Eng 10:323–330.

Chen J, ltman GH, Karageorgiou V, Horan R, Collette A, Volloch V, Colabro T, Kaplan DL (2003) Human bone marrow stromal cell and ligament fibroblast responses on RGD-modified silk fibers. J Biomed Mater Res 67A:559–570.

Cohn RD, Henry MD, Michele DE, Barresi R, Saito F, Moore SA, Flanagan JD, Skwarchuk MW, Robbins ME, Mendell JR, Williamson RA, Campbell KP (2002) Disruption of *Dag1* in differentiated skeletal muscle reveals a role for dystroglycan in muscle regeneration. Cell 110:639–648.

Cohn RD, Campbell KP (2000) The molecular basis of muscular dystrophy. Muscle Nerve 23:1456–1471.

Connolly JF (1995) Injectable bone marrow preparations to stimulate osteogenic repair. Clin Orthop 313:8–18.

Constantz BR, Ison IC, Fulmer MT, Poser RD, Smith ST, VanWagoner M, Ross J, Goldstein SA, Jupiter JB, Rosenthal DI (1995) Skeletal repair by in situ formation of the moneral phase of bone. Science 267:1796–1799.

Cook JL, Tomlinson J, Kreeger JM, Cook CR (1999) Induction of meniscal regeneration in dogs using a novel biomaterial. Am J Sports Med 27:658–665.

Cook SD, Baffes GC, Wolfe MW, Sampath TK, Rueger DC, Whitecloud TS III (1994) The effect of recombinant human osteogenic protein-1 on healing of large segmental bone defects. J Bone Joint Surg 76A:827–838.

Cook SD, Patron LP, Salkeld SL, Rueger DC (2003) Repair of articular cartilage defects with osteogenic protein-1 (BMP-7) in dogs. J Bone & Joint Surg 85A (Suppl 3):116–123.

Cordioli G, Mazzocco C, Schepers E, Brugnolo E, Majzoub Z (2001) Maxillary sinus floor augmentation using bioactive glass granules and autogenous bone with simultaneous implant placement. Clinical and histological findings. Clin Oral Implants Res 12: 270–278.

Cornell CN, Lane JM (1998) Current understanding of osteoconduction in bone regeneration. Clin Orthopaed Rel Res 355S: S267–S273.

Cossu G, Mavilio F (2000) Myogenic stem cells for the therapy of primary myopathies: wishful thinking or therapeutic perspective? J Clin Invest 105:1669–1674.

Cowan CM, Shi Y-Y, Aalami OO, Chou Y-F, Mari C, Thomas R, Quarto N, Contag C, Wu B, Longaker MT (2004) Adipose-drived adult stromal cells heal critical-size mouse calvarial defects. Nature Biotech 22:560–567.

Cuevas P, Burgos J, Baird A (1988) Basic fibroblast growth factor (FGF) promotes cartilage repair in vivo. Biochem Biophys Res Commun 156:611–618.

Dahlgren LA, Mohammed HO, Nixon AJ (2005) Temporal expression of growth factors and matrix molecules in healing tendon lesions. J Orthopaed Res 23:84–92.

de la Fuente R, Abad JL, Garcia-Castro J, Fernandez-Miguel G, Petriz J, Rubio D, Vicario-Abejon C, Guillen P, Gonzalez MA, Bernad A (2004) Dedifferentiated adult articular chondrocytes: A population of human multipotent primitive cells. Exp Cell Res 297:313–328.

de Macedo N, da Silva Matuda F, de Macedo LGS, Gonzales MB, Ouchi SM, Carvalho YR (2004) Bone defect regeneration with bioactive glass implantation in rats. J Appl Oral Sci 12: #2, Apr/June.

Dennis JE, Solchaga LA, Caplan AI (2001) Mesenchymal stem cells for musculoskeletal tissue engineering. Landes Biosci 1:112–115.

Dennis JE, Cohen N, Goldberg VM, Caplan AI (2004) Targeted delivery of progenitor cells for cartilage repair. J Orthopaed Res 22:735–741.

Dezawa M, Ishikawa H, Itokazu Y, Yoshihara T, Hoshino M, Takeda S-I, Ide C, Nabeshima Y-I (2005) Bone marrow stromal cells generate muscle cells and repair muscle degeneration. Science 309:314–317.

Dunn MG, Tria AJ, Kato YP, Bechler JR, Ochner RA, Zawadsky JP, Silver FH (1992) Anterior cruciate ligament reconstruction using a composite collagenous prosthesis: A biomechanical and histologic study in rabbits. Am J Sports Med 20:507–515.

Einhorn TA (1999) Clinically applied models of bone regeneration in tissue engineering research. Clin Orthopaed Rel Res 367S: S59–S67.

Fang J, Zhu Y-Y, Smiley E, Bonadio J, Rouleau JP, Goldstein SA, McCauley LK, Davidson BL, Roessler BJ (1996) Stimulation of new bone formation by direct transfer of osteogenic plasmid genes. Proc Natl Acad Sci USA 93:5753–5758.

Ferrari G, Cusella-De Angelis G, Coletta M, Paolucci E, Stornaiuolo A, Cossu G, Mavilio F (1998) Muscle regeneration by bone marrow-derived myogenic progenitors. Science 279:1528–1530.

Flanagan N (2003) Engineering methods for tissue & cell repair. Gen Eng News 23:10.

Forslund C, Aspenberg P (2003) Improved healing of transected rabbit Achilles tendon after a single injection of cartilage-derived morphogenetic protein-2. Am J Sports Med 31:555–559.

Freed E, Marquis C, Nohira A, Emmanuel J, Mikos AG, Langer R (1993) Neocartilage formation *in vitro* and *in vivo* using cells cultured on synthetic biodegradable polymers. J Biomed Mats Res 27:11–23.

Freed LE, Vunjak-Novakovic G (1995) Tissue Engineering of Cartilage. In: Bronzino J (ed.). Biomedical Engineering Handbook. Boca Raton, FL: CRC Press, pp 1788–1806.

Friedenberg ZB, Roberts G, Didizian NH, Brighton CT (1971) Stimulation of fracture healing by direct current in the rabbit fibula. Bone Joint Surg 53A:1400–1408.

Fuentes-Boquete I, Lopez-Armada M, Maneiro E, Fernandez-Sueiro JL, Carames B, Galdo F, De Toro FJ, Blanco FJ (2004) Pig chondrocyte xenoimplants for human chondral defect repair: An in vitro model. Wound Rep Reg 12:444–452.

Fujimoto E, Ochi M, Kato Y, Mochizuki Y, Sumen Y, Ikuta Y (1999) Beneficial effect of basic fibroblast growth factor on the repair of full-thickness defects in rabbit articular cartilage. Arch Orthop Trauma Surg 119:139–150.

Gao J, Knaack D, Goldberg V, Caplan AI (2004) Osteochondral defect repair by demineralized cortical bone matrix. Clin Orthopaed Rel Res 427S:S62–S66.

Geesink RGT, Hoefnagels NM, Bilstra SK (1999) Osteogenic activity of OP-1 bone morphogenetic protein (BMP-7) in a human fibular defect. J Bone Joint Surg 81B:710–718.

Gelse K, von der Mark K, Aigner T, Park J, Schneider H (2003) Articular cartilage repair by gene therapy using growth factor-producing mesenchymal cells. Arth Rheum 48:430–441.

Ghivizzani SC (2002) Genetic approaches to the repair of connective tissues. Ann NY Acad Sci 961:65–67.

Glansbeek HL, van Beuningen HM, Vitters EL, van der Kraan PM, van den Berg WB (1998) Stimulation of articular cartilage repair in established arthritis by local administration of transforming growth factor-β into murine knee joints. Lab Invest 78:133–142.

Goldstein SA, Patil P, Moalli M (1999) Perspectives on tissue engineering of bone. Clin Orthopaed Rel Res 367:S419–S423.

Goldstein SA, Bonadio J (1998) Potential role for direct gene transfer in the enhancement of fracture healing. Clin Orthopaed Rel Res 355S:S154–S162.

Goss RJ (1958) Skeletal regeneration in amphibians. J Embryol Exp Morph 6:638–644.

Goyenvalle A, Vulin A, Fougerousse F, Leturcq F, Kaplan J-C, Garcia L, Danos O (2004) Rescue of dystrophic muscle through U7 snRNA-mediated exon skipping. Science 306:1796–1799.

Grande DA, Southerland SS, Manji R, Pate DW, Schwartz RE, Lucas PA (1995) Repair of articular cartilage defects using mesenchymal stem cells. Tiss Eng 1:345–353.

Grande DA, Halberstadt C, Naughton G, Schwartz R, Manji R (1997) J Biomed Mater Res 34:211–220.

Grande DA, Breitbart AS, Mason J, Paulino C, Laser J, Schwartz RE (1999) Cartilage tissue engineering: Current limitations and solutions. Clin Orthopaed Rel Res 367S:S176–S185.

Grande DA, Mason J, Light E, Dines D (2003) Stem cells as platforms for delivery of genes to enhance cartilage repair. J Bone & Joint Surg 85:111–116.

Gussoni E, Pavlath GK, Lanctot AM, Sharma KR, Miller RG, Steinman L, Blau HM (1992) Normal dystrophin transcripts detected in Duchenne muscular dystrophy patients after myoblast transplantation. Nature 356:435–439.

Heslop L, Morgan JE, Partridge TA (2000) Evidence for a myogenic stem cell that is exhausted in dystrophic muscle. J Cell Sci 113:2299–2308.

Hidaka C, Goodrich LR, Chen C-T, Warren RF, Crystal RG, Nixon AJ (2003) Acceleration of cartilage repair by genetically modified chondrocytes over expressing bone morphogenetic protein-7. J Orthopaed Res 21:573–583.

Horwitz EM, Prockop DJ, Fitzpatrick L, Koo WWK, Gordon P, Neel M, Sussman M, Orchard P, Marx JC, Pyeritz RE, Brenner MK (1999) Transplantability and therapeutic effects of bone marrow-derived mesenchymal cells in children with osteogenesis imperfecta. Nature Med 5:309–313.

Huard J, Bouchard J, Roy R, Malouin F, Dansereau G, Labrecque C, Albert N, Richards CL, Lemieux B, Tremblay JP (1992a) Human myoblast transplantation: Preliminary results of four cases. Muscle Nerve 15:550–560.

Huard J, Roy JP, Bouchard J, Malouin F, Richards CL, Tremblay JP (1992b) Human myoblast transplantation between immunohistocompatible donors and recipients produces immune reactions. Transpl Proc 24:3049–3051.

Huard J, Guerette B, Verrault, Tremblay G, Roy R, Lille S, Tremblay JP (1994) Human myoblast transplantation in immunodeficient and immunosuppressed mice: Evidence of rejection. Muscle Nerve 17:224–234.

Hunziker E (1999) Biologic repair of articular cartilage: Defect models in experimental animals and matrix requirements. Clin Orthopaed Rel Res 367S:S135–S146.

Hunziker EB, Rosenberg LC (1996) Repair of partial-thickness defects in articular cartilage: Cell recruitment from the synovial membrane. J Bone Joint Surg 78A:721–733.

Ibarra C, Jannetta C, Vacanti CA, Cao Y, Kim TH, Upton J, Vacanti JP (1997) Tissue engineered meniscus: A potential new alternative to allogeneic meniscus transplantation. Transpl Proceedings 29:986–988.

Ibiwoye MO, Powell KA, Grabiner MD, Patterson TE, Sakai Y, Zborowski M, Wolfman A, Midura RJ (2004) Bone mass is preserved in a critical-sized osteotomy by low energy pulsed magnetic fields as quantitated by in vivo micro-computed tomography. J Orthopaed Res 22:1086–1093.

Ilizarov GA (1990) Clinical application of the tension-stress effect for limb lengthening. Clin Orthopaed Rel Res 250:8–26.

Isogai N, Landis, Kim TH, Gerstenfeld LC, Upton J, Vacanti JP (1999) Formation of phalanges and small joints by tissue-engineering. J Bone Joint Surg 81A:306–316.

Johnstone B, Yoo J (1999) Autologous mesenchymal progenitor cells in articular cartilage repair. Clin Orthopaed Rel Res 367S: S156–S162.

Kadiyala S, Jaiswal N, Bruder SP (1997) Culture-expanded, bone marrow-derived mesenchymal stem cells can regenerate a critical-sized segmental bone defect. Tiss Eng 3:173–185.

Kajatani K, Ochi M, Uchio Y, Adachi N, Kawasaki K, Katsube K, Maniwa S, Furukawa S, Kataoka H (2004) Role of the periosteal flap in chondrocyte transplantation: An experimental study in rabbits. Tiss Eng 10:331–342.

Karpati G, Ajdukovic D, Arnold D, Gledhill RB, Guttmann R, Holland P, Koch PA, Shoubridge E, Spence D, Vanasse M (1993) Myoblast transfer in Duchenne Muscular Dystrophy. Ann Neurol 34:8–17.

Kato YP, Dunn MG, Zawadsky JP, Tria A, Silver FH (1991) Regeneration of Achilles tendon with a collagen tendon prosthesis. Results of a one-year implantation study. J Bone Joint Surg 73A:561–574.

Kawaguchi H, Nakamura K, Tanata Y, Ikada Y, Aoyama I, Anzai J, Nakamura T, Hiyama Y, Tamura M (2001) Acceleration of fracture healing in nonhuman primates by fibroblast growth factor-2. J Clin Endocrino Metab 86:875–880.

Kellner K, Tessmar J, Milz S, Angele P, Nerlich M, Schulz MB, Blunk T, Gopferich A (2004) PEGylation does not impair insulin efficacy in three-dimensional cartilage culture: An investigation toward biomimetic polymers. Tiss Eng 10:429–440.

Kenner GH, Gabrielson EW, Lovell JE, Marskall AE, Williams WS (1975) Electrical modification of disuse osteoporosis. Calcif Tiss Res 18:111–117.

Kim HD, Smith JG, Valentini RF (1998) Bone morphogenetic protein 2-coated porous poly-L-lactic acid scaffolds: Release kinetics and induction of pluripotent C3101/2 cells. Tiss Eng 4:35–51.

Kinoshita I, Vilquin JT, Guerette B, Asselin I, Roy R, Tremblay JP (1994) Very efficient myoblast allotransplantation in mice under F506 immunosuppression. Muscle Nerve 17:1407–1415.

Kinoshita Y, Kobayashi M, Fukuoka S, Yoyoka S, Ikada Y (1996) Functional reconstruction of the jaw bones using poly(L-lactide) mesh and autogeneic particulate cancellous bone and marrow. Tiss Eng 2:327–341.

Kisiday J, Jin M, Kurz B, Hung H, Semino C, Zhang S, Grodzinsky AJ (2002) Self-assembling peptide hydrogel fosters chondrocyte extracellular matrix production and cell division: Implications for cartilage tissue repair. Proc Natl Acad Sci USA 99:9996–10001.

Kosnik PE, Faulkner JA, Dennis RG (1001) Functional development of engineered skeletal muscle from adult and neonatal rats. Tiss Eng 7:573–584.

Kraus K, Kadiyala S, Wotton H, Kurth A, Shea M, Hannan M, Hayes WC, Kirkerhead CA, Bruder SP (1999) Critically sized osteo-periosteal femoral defects: A dog model. J Invest Surg 12: 115–124.

Kvist M (1994) Achilles tendon injuries in athletes. Sports Med 18:173–201.

Langer F, Gross AE (1974) Immunogenecity of allograft articular cartilage. J Bone Joint Surg 56:297–327.

Laurencin CT, Attawia MA, Elgendy HE, Herbert KM (1996) Tissue engineered bone-regeneration using degradable polymers: The formation of mineralized matrices. Bone 19S:93S–99S.

Lavine LS, Lustrin I, Shamos MH, Rinaldi RA, Liboff A (1972) Electric enhancement of bone healing. Science 178:1118–1121.

Law PK (1982) Beneficial effects of transplanting normal limb-bud mesenchyme into dystrophic muscles. Muscle Nerve 5:619–627.

Lee J, Qu-Petersen Z, Cao B, Kimura S, Jankowski R, Cummins J, Usas A, Gates C, Robbins P, Wernig A, Huard J (2000) Clonal isolation of muscle-derived cells capable of enhancing muscle regeneration and bone healing. J Cell Biol 150:1085–1099.

Levenberg S, Rouwkema J, Macdonald M, Garfein ES, Kohane DS, Darland DC, Marini R, can Blitterswijk CA, Mulligan RC, D'Amore PA, Langer R (2005) Engineering vascularized skeletal muscle. Nature Biotech 23:879–884.

Li Y, Tew SR, Russell AM, Gongalez R, Hardingham TE, Hawkins RE (2004) Transduction of passaged human articular chondrocytes with adenoviral, retroviral, and lentiviral vectors and the effects of enhanced expression of SOX9. Tiss Eng 10:575–584.

Lipiello L, Woodward J, Karpman R, Hammad TA (2000) In vivo chondroprotection and metabolic synergy ofglucosamine and chondroitin sulfate. Clin Orthopaed Rel Res 381:229–240.

Louie LK, Yannas IV, Hsu H-P, Spector M (1997) Healing of tendon defectsimplantedwithaporouscollagen-GAG matrix: Histological evaluation. Tiss Eng 3:187–195.

Lu QL, Morris GE, Wilton SD, Ly T, Artem'yeva OV, Strong P, Partridge TA (2000) J Cell Biol 148:985–995.

Lutolf MP, Weber F, Schmoekel HG, Schense JC, Kohler T, Muller R, Hubbell JA (2003) Repair of bone defects using synthetic mimetics of collagenous extracellular matrices. Nature Biotech 21:513–518.

Mapeli M, Randazzo N, Cancedda R, Dozin B (2004) Serum-free growth medium sustains commitment of human articular chondrocyte through maintenance of Sox9 expression. Tiss Eng 10:145–155.

Mandl EW, Van Der Veen SW, Verhaar JAN, Van Osch JVM (2004) Multiplication of human chondrocytes with low seeding densities accelerates cell yield without losing redifferentiation capacity. Tiss Eng 10:109–118.

Mark RA (2000) Spine fusion for discogenic low back pain: Outcomes in patients treated with or without pulsed electromagnetic field stimulation. Adv Therap 17:57–67.

Martin I, Jokob M, Schaefer D, Dick W, Spagnoli G, Heberer M (2001) Quantitative analysis of gene expression in human articular cartilage from normal and osteoarthritic joints. Osteoarthritis Cartilage 9:112–118.

Martin I, Suetterlin R, Baschong W, Heberer M, Vunjak-Novakovic G, Freed LE (2001) Enhanced cartilage tissue engineering by sequential exposure of chondrocytes to FGF-2 during 2D expansion and BMP-2 during 3D cultivation. J Cell Biochem 83:121–128.

McIntosh K, Bartholomew A (2000) Stromal cell modulation of the immune system-a potential role for mesenchymal stem cells. Graft 3:324–328.

Mendell R, Kissel JT, Amato AA, King W, Signore L, Prior TW, Sahenk Z, Benson S, Mcandrew PE, Rice R (1995) Myoblast transfer in the treatment of Duchenne Muscular Dystrophy. New Eng J Med 333:832–838.

Meyers MH, Akeson W, Convery FR (1989) Resurfacing of the knee with fresh osteochondral allograft. J Bone Joint Surg 71-A: 7-4-713.

Montarras D, Morgan J, Collins C, Relaix F, Zaffran S, Cumano A, Partridge T, Buckingham M (2005) Direct isolation of satellite cells for skeletal muscle regeneration. Science 309:2064–2067.

Morgan JE, Hoffman EP, Partridge TA (1990) Normal myogenic cells from newborn mice restore normal histology to degenerating muscles of the mdx mouse. J Cell Biol 111:2437–2449.

Mueller SM, Shortkroff S, Schneider TO, Breinan HA, Yannas IV, Spector M (1999) Meniscus cells seeded in type I and type II collagen-GAG matrices in vitro. Biomats 20:701–709.

Mushal V, Abramowitch SD, Gilbert TW, Tsuda E, Wang J H-C, Badylak SF, Woo S L-Y (2004) The use of porcine small intestinal submucosa to enhance the healing of the medial collateral ligament—a functional tissue engineering study in rabbits. Orthopaed Res 22:214–220.

Ochi M, Uchio Y, Matsusaki M, Wakatani S, Sumen Y (1998) Cartilage repair—a new surgical procedure of cultured chondrocyte transplantation. In: Chan KM, Fu F, Maffulli N, Rolf C, Kurosaka M, Liu S (eds.). Controversies in Orthopedic Sport Medicine. Hong Kong, Williams and Wilkins Asia-Pacific, pp 549–563.

Ochi M, Uchio Y, Tobita M, Kuriwaka M (2001) Current concepts in tissue engineering technique for repair of cartilage defect. Artif Organs 25:172–179.

O'Driscoll SW (1998) The healing and regeneration of articular cartilage. J Bone Joint Surg 80A:1795–1812.

Oonishi H, Kushitani S, Yasukawa E, Iwaki H, Hench LL, Wilson J, Tsuji E, Sugihara T (1996) Particulate bioglass compared with hydroxyapatite as a bone graft substitute. Clin Orthopaed 334: 316–325.

Oreffo ROC, Driessens FCM, Planell JA, Triffit JT (1998) Effects of novel calcium phosphate cements on human bone marrow fibroblastic cells. Tiss Eng 4:293–303.

Otto WR, Rao J (2004) Tomorrow's skeleton staff: Mesenchymal stem cells and the repair of bone and cartilage. Cell Prolif 37: 97–110.

Park Y, Lutolf MP, Hubbell A, Hunziker EB, Wong MY (2004) Bovine primary chondrocyte culture in synthetic matrix metalloproteinase-sensitive poly(ethylene glycol)-based hydrogels as a scaffold for cartilage repair. Tiss Eng 10:515–353.

Partridge TA, Morgan JE, Coulton GR, Hoffman EP, Kunkel LM (1989) Conversion of mdx myofibers from dystrophin-negative to -positive by injection of normal myoblasts. Nature 337:176–180.

Peel SAF, Chen H, Renlund R, Badylak SF, Kandel RA (1998) Formation of a SIS-cartilage composite graft *in vitro* and its use in the repair of articular cartilage defects. Tiss Eng 4:143–155.

Pei M, Seidel J, Vunjak-Novakovic G, Freed LE (2002) Growth factors for sequential cellular de- and re-differentiation in tissue engineering. Biochem Biophys Res Commun 294:149–154.

Peretti GM, Caruso EM, Randolph MA, Zaleske DJ (2001) Meniscal repair using engineered tissue. J Orthopaed Res 19:278–285.

Peretti GM, Gill TJ, Xu J-W, Randolph MA, Morse KR, Zaleske DJ (2004) Cell-based therapy for meniscal repair. A large animal study. Am J Sports Med 32:146–158.

Pittenger M, Vanguri P, Simonetti D, Young R (2002) Adult mesenchymal stem cells: Potential for muscle and tendon regeneration and use in gene therapy. Musculoskel Neuron Interact 2:309–320.

Pribitkin EA, Ambro BT, Bloeden E, O'Hara BJ (2004) Rabbit ear cartilage regeneration with a small intestinal submucosa graft. Laryngoscope 114:1–19.

Qu Z, Balkir L, van Deutekom JCT, Robbins PD, Pruchnic R (1998) Development of approaches to improve cell survival in myoblast transfer therapy. Cell Biol 142:1257–1267.

Radomsky ML, Thompson Y, Spiro RC, Poser JW (1998) Potential role of fibroblast growth factor in enhancement of fracture healing. Clin Orthopaed Rel Res 355S:S283–S293.

Ramallal M, Maneiro E, Lopez E, Fuentes-Boquete I, Lopez-Armada MJ, Fernandez-Sueiro JL, Galdo F, De Toro FJ, Blanco FJ (2004) Xeno-implantation of pig chondrocytes into rabbit to treat localized articular cartilage defects: An animal model. Wound Rep Reg 12:337–345.

Reddi AH (1998) Role of morphogenetic proteins in skeletal tissue engineering and regeneration. Nature Biotech 16:247–252.

Reddy GK, Stehno-Bittel L, Enwemeka C (1998) Laser photostimulation of collagen production in healing rabbit Achilles tendons. Lasers in Surg Med 22:281–287.

Reinholtz GG, Lu L, Saris DBF, Yaszemski MJ, O'Driscoll SW (2004) Animal models for cartilage reconstruction. Biomats 25: 1511–1521.

Richards M, Huibregtse BA, Caplan AI, Goulet JA, Goldstein SA (1999) Marrow-derived progenitor cell injections enhance new bone formation during distraction. J Orthopaed Res 900–908.

Ripamonti U (2002) Tissue engineering of bone by novel substrata instructing gene expression during de novo bone formation. Science in Africa, March, Merck Feature by Janice Limson.

Ripamonti U (2001a) Bone induction by BMPs/OPs and related family members in primates. J Bone Joint Surg 83A:S1-116–S1-127.

Ripamonti U (2001b) Induction of bone formation by recombinant human osteogenic protein-1 and sintered porous hydroxyapatite in adult primates. Plastic Reconstr Surg 107:977–988.

Rosa AL, Beloti MM, Van Noort R, Hatton PV, Devlin AJ (2002) Surface topography of hydroxyapatite affects ROS17/2.8 cells response. Pesqui Odontol Bras 16:209–215.

Rosen V, Thies RS (1992) The BMP proteins in bone formation and repair. Trends Genet 8:97–102.

Saito T, Dennis JE, Lennon DP, Young RG, Caplan AI (1995) Myogenic expression of mesenchymal stem cells within myotubes of *mdx* mice *in vitro* and *in vivo*. Tiss Eng 1:327–343.

Seeherman H, Wozney J, Li R (2002) Bone morphogenetic protein delivery systems. Spine 27(16S):S16–S23.

Sellers RS, Peluso D, Morris EA (1997) The effect of recombinant human bone morphogenetic protein-2 (rhBMP-2) on the healing of full-thickness defects of articular cartilage. J Bone Joint Surg Am 79:1452–1463.

Shapiro F, Koide S, Glimcher M (1993) Cell origination and differentiation in the repair of full-thickness defects of articular cartilage. J Bone Joint Surg Am 75:532–553.

Shea LD, Smiley E, Bonadio J, Mooney DJ (1999) DNA delivery from polymer matrices for tissue engineering. Nature Biotech 17:551–554.

Sheller MR, Crowther RS, Kinney JH, Yang, Di Jorio S, Breunig T, Carney DH, Ryaby JT (2004) Repair of rabbit segmental defects with the thrombin peptide, P508. J Orthopaed Res 22:1094–1099.

Solchaga LA, Yoo JU, Lundberg M, Dennis JE, Huinregtse BA, Goldberg VM, Caplan AI (2000) Hyaluronan-based polymers in the treatment of osteochondral defects. J Orthopaed Res 18: 773–780.

Stevens MM, Marini RP, Schaefer D, Aronson J, Langer R, Shastri VP (2005) *In vivo* engineering of organs: The bone bioreactor. Proc Nat'l Acad Sci USA 102:11450–11455.

Stone KR, Rodkey WG, Webber RJ, McKinney L, Steadman JR (1990) Future directions: Collagen-based prostheses for meniscal regeneration. Clin Orthop 252:129–135.

Stone KR, Rodkey WG, Webber RJ, McKinney L, Steadman JR (1992) Meniscal regeneration with copolymeric scaffolds. In vitro and in vivo studies evaluated clinically, histologically, and biochemically. Am J Sports Med 20:104–111.

Stone KR, Steadman JR, Rodkey WG, Li ST (1997) Regeneration of meniscal cartilage with use of a collagen scaffold. Analysis of preliminary data. J Bone Joint Surg 79A:1770–1777.

Suckow MA, Voytik-Harbin SL, Terril LA, Badylak SF (1999) Enhanced bone regeneration using porcine small intestinal submucosa. J Invest Surg 12:277–287.

Tabata Y, Yamamoto M, Ikada Y (1998) Biodegradable hydrogels for bone regeneration through growth factor release. Pure & Appl Chem 70:1277–1282.

Tanaka H, Mizokami H, Shiigi E, Murata H, Ogasa H, Mine T, Kawai S (2004) Effects of basic fibroblast growth factor on the repair of large osteochondral defects of articular cartilage in rabbits: Dose-response effects and long-term outcomes. Tiss Eng 10:633–641.

Tibesku CO, Szuwart T, Kleffner TO, Schlegel PM, Jahn UR, Van Aken H, Fuchs S (2004) Hyaline cartilage degenerates after autologous osteochondral transplantation. J Orthopaed Res 22:1210–1214.

Tiedeman J, Connolly J, Strates B, Lippiello L (1991) Treatment of nonunion by percutaneous injection of bone marrow and demineralized bone matrix: An experimental study in canines. Clin Orthop 268:294–302.

Tienen TG, Heijkants RGJC, De Groot JH, Pennings AJ, Veth RPH (2003) A porous polymer scaffold for meniscal lesion repair—a study in dogs. Biomats 24:2541–2548.

Tremblay J, Guerette B (1997) Myoblast transplantation: A brief review of the problems and of some solutions. Basic Appl Myol 7:221–230.

Trippel SB (1998) Potential role of insulinlike growth factors in fracture healing. Clin Orthopaed Rel Res 355S:S301–S313.

Tuan RS (2004) Biology of developmental and regenerative skeletogenesis. Clin Orthopaed Rel Res 427:S105–S117.

Veilleux NH, Yannas IV, Spector M (2004) Effect of passage number and collagen type on the proliferative, biosynthetic, and contractile activity of adult canine articular chondrocytes in type I and II collagen-glycosaminoglycan matrices *in vivo*. Tiss Eng 10:119–127.

Vunjak-Novakovic G, Altman G, Horan R, Kaplan DL (2004) Tissue engineering of ligaments. Ann Rev Biomed Eng 6:131–156.

Wakatani S, Kimura T, Hirooka A, Ochi T, Yoneda M, Yasui N, Owaki H, Ono K (1989) Repair of rabbit articular surfaces with allograft chondrocytes embedded in collagen gel. J Bone Joint Surg 71B:74–80.

Wakatani S, Imoto K, Kimura T, Ochi T, Matsumoto K, Nakamura T (1997) Hepatocyte growth factor facilitates cartilage repair: Full-thickness articular cartilage defect studied in rabbit knees. Acta Orthopaed Scand 68:474–480.

Wakatani S, Goto T, Young RG, Mansour JM, Goldberg VM, Caplan AI (1998) Repair of large full-thickness articular cartilage defects with allograft articular chondrocytes embedded in a collagen gel. Tiss Eng 4:429–444.

Walsh CJ, Goodman D, Caplan AI, Goldberg VM (1999) Meniscus regeneration in a rabbit partial meniscectomy model. Tiss Eng 5:327–337.

Warnke PH, Springer ING, Wiltfanf J, Acil Y, Eufinger H, Wehmoller M, Russo PAJ, Bolte H, Sherry E, Behrens E, Terheyden H (2004) Growth and transplantation of a custom vascularized bone graft in a man. Lancet 364:766–770.

Webster TJ, Ergun C, Doremus RH, Siegel RW, Bizios R (2000) Enhanced functions of osteoblasts on nanophase ceramics. Biomats 21:1803–1810.

Winn SR, Uludag H, Hollinger JO (1999) Carrier systems for bone morphogenetic proteins. Clin Orthopaed Rel Res 367S:S95–S106.

Wirth CJ, Peters G, Milachowski KA, Weismeier KG, Kohn D (2002) Long-term results of meniscal allograft transplantation. Am J Sports Med 30:174–181.

Wolfman NM, Hattersley G, Cox K, Celeste AJ, Nelson R, Yamaji N, Dube JL, DiBasio-Smith E, Nove J, Song JJ, Wozney JM, Rosen V (1997) Ectopic induction of tendon and ligament in rats by growth and differentiation factors 5, 6 and 7, members of the TGF-β gene family. Clin Invest 100:321–330.

Woo S, Hildebrand K, Nobuyoshi N, Fenwick JA, Papageorgiou CD, Wang J H-C (1999) Tissue engineering of ligament and tendon healing. Clin Orthopaed Rel Res 367S:S312–S323.

Yaszemski MJ, Payne RG, Hayes WC, Langer RS, Aufdemorte TB, Mikos AG (1995) The ingrowth of new bone tissue and initial mechanical properties of a degrading polymeric composite scaffold. Tiss Eng 1:41–52.

Yasko AW, Lane JM, Fellinger EJ (1992) The healing of segmental bone defects, induced by recombinant human bone morphogenetic protein (rhBMP-2). J Bone Joint Surg Am 74: 659–670.

Yukna RA, Evans GH, Aichelmann-Reidy MB, Mayer ET (2001) Clinical comparison of bioactive glass bone replacement graft material and expanded polytetrafluorethylene barrier membrane in treating human mandibular molar class II furcations. J Periodontol 72:125–133.

Young HE, Mancini ML, Wright RP, Smith JC, Black AC Jr, Reagan CR, Lucas PA (1995) Mesenchymal stem cells reside within the connective tissues of many organs. Dev Dynam 202:137–144.

Young RG, Butler DL, Weber W, Caplan AI, Gordon S, Fink DJ (1998) The use of mesenchymal stem cells in Achilles tendon repair. J Orthop Res 16:406–413.

Zhang W, Ge W, Li C, You S, Liao L, Han Q, Deng W, Zhao RC (2004) Effects of mesenchymal stem cells on differentiation, maturation, and function of human monocyte-derived dendritic cells. Stem Cells and Development 13:263–271.

# 11 Regeneration of Hematopoietic and Cardiovascular Tissues

## INTRODUCTION

Like the musculoskeletal system, the heart, blood vessels, and blood are derived from the mesodermal germ layer of the embryo. The blood cells and the endothelial cells of the vascular system arise during embryogenesis from common precursor cells called hemangioblasts that in the mouse embryo are already present in the posterior region of the mid primitive streak (Huber et al., 2004). The ingressing hemangioblasts express the VEGF receptor Flk-1 and the brachyury (T) gene. They form clusters of blood islands in the mesoderm of the yolk sac. The yolk sac is the earliest site of blood and vascular development, shifting later to the mesoderm of the so-called aorta-gonad-mesonephros (AGM) region (Moore and Metcalf, 1970; Dieterlen-Lievre et al., 1988; Medvinsky and Dzierzak, 1996; Robb and Choi, 2004). The central hemangioblasts of the blood islands become blood-forming HSCs, whereas the peripheral hemangioblasts become angioblasts (endothelial stem cells, EnSCs), that differentiate into capillary endothelial cells (**FIGURE 11.1**).

HSCs and EnSCs take up residence in the stroma of the marrow cavity as it is vascularized by invading capillaries. Either these stem cells could be circulating in the fetal blood and leave the capillaries for the stroma, or hemangioblasts in the walls of the invading blood vessels might give rise to them. Hemangioblasts are present in the peripheral blood of adult humans. These cells, which are $CD133^{+}$, have been expanded *in vitro* and transfected with an enhanced GFP construct (EGFP). Clones of single $EGFP^{+}$ cells were able to differentiate as either neutrophils or endothelial cells, indicating that they were true hemangioblasts. From the number of clones showing dual differentiation, it was estimated that 2% of the $EGFP^{+}$ cells were hemangioblasts (Loges et al., 2004).

The heart arises from a crescent-shaped field of mesoderm above the developing esophagus and respiratory system (Gilbert, 2003). This mesoderm forms a tube that loops to place the atrial portion anterior to the ventricular portion. In mammals, the atrial and ventricular portions of the looped tube become subdivided into two chambers each. The ventricular tube is bifurcated and extends as the aorta and pulmonary artery, whereas the bifurcated atrial tube extends as the superior and inferior vena cavae.

## REGENERATION OF HEMATOPOIETIC CELLS

### 1. Composition of Adult Blood

Blood consists of plasma and a myeloid component consisting of erythrocytes, platelets, granulocytes (basophils, eosinophils, and neutrophils), monocytes, and myeloid dendritic cells. Also circulating in the blood and lymph systems are cells of the immune system: B cells, T cells, natural killer cells, and lymphoid dendritic cells. Mature erythroid, myeloid, and lymphoid cells (with the exception of memory B and T cells), have half-lives of only days to weeks, and their numbers must be maintained by continual regeneration from multipotent HSCs residing in the bone marrow stroma of the ilium, vertebrae, and endochondral bones. Hematopoietic regeneration has been one of the most intensely studied developmental processes of the 20th century, and continues to be so today (Weissman, 2000).

### 2. Hematopoietic Stem Cells of the Bone Marrow

The existence of long-term (LT) self-renewing HSCs in the bone marrow has been demonstrated by bone marrow reconstitution experiments (**FIGURE 11.2**). In these experiments, mice and rats are lethally irradiated to eliminate the ability of bone marrow cells to divide, followed by injecting them with progressively diluted suspensions of labeled marrow cells. If the injected

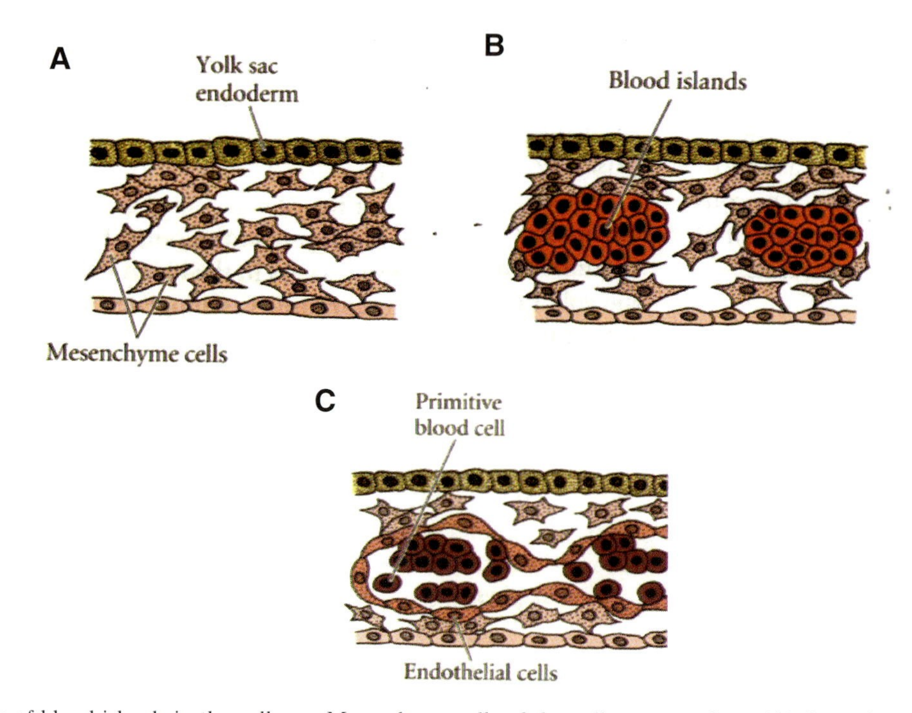


**FIGURE 11.1** Formation of blood islands in the yolk sac. Mesenchyme cells of the yolk sac mesoderm **(A)** form clusters of cells that are the blood islands **(B)**. **(C)** The cells on the periphery of the clusters differentiate into endothelial cells and the interior cells differentiate into hematopoietic stem cells. Reproduced with permission from Gilbert, Developmental Biology (7th ed). Copyright 2003, Sinauer Associates.

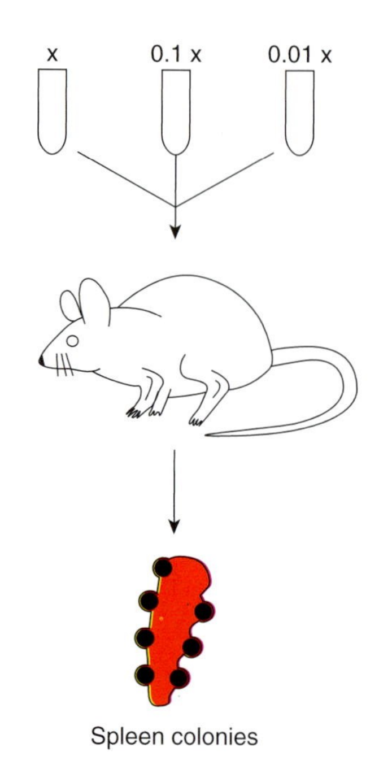


**FIGURE 11.2** Demonstration of long-term, self-renewing hematopietic stem cells by bone marrow reconstitution assays. Serially diluted suspensions of bone marrow cells (starting with concentration X) are injected into lethally X-irradiated mice or rats, which ablates the bone marrow. The injected cells reconstitute the bone marrow. LT-HSCs reveal themselves by their formation of blood cell colonies (erythroid, myeloid and lymphoid) on the spleen.

marrow contains LT-HSCs that give rise to all blood and immune cell lineages, this is revealed by the formation of labeled erythroid, myeloid, and lymphoid cell colonies in the spleen and labeled cells circulating in the blood (McCulloch, 2004; Ponting et al., 2004). Serial colony-forming assays have shown that a single transplanted LT-HSC can give rise to all blood and lymphoid cell types while self-renewing. The existence of separate short-term (ST) HSCs downstream from the LT-HSC (colony forming units, CFUs) has been demonstrated in the same way.

**FIGURE 11.3** illustrates current, alternative, and composite models of blood cell formation (Adolfsson et al., 2005). In the composite model, the ST-HSC gives rise to a common myeloid progenitor and a lymphoid-primed multipotent progenitor (LMPP). The LMPP gives rise to a common lymphoid progenitor and granulocyte/macrophage progenitors. The common myeloid progenitor gives rise to a megakaryocyte/erythrocyte progenitor as well as a granulocyte/macrophage progenitor. The common lymphoid progenitor gives rise to B cells and T cells. A variety of cytokines and growth factors produced by stromal cells have been shown to promote the proliferation and differentiation of the different lineages of blood and lymphoid cells that emerge from the initial progeny of the HSC. These include IL-1,3,6,7,8,11, and 12, LIF, stem cell factor (SCF), Flt ligand, and MCS-F (Charbord, 2004).

The bone marrow cell fraction that is $CD34^-$ $c\text{-}Kit^+$ $Sca\text{-}1^+$ $VEGFR2^+$ $Thy\text{-}1^{lo}$ $Lin^-$ is heavily enriched in

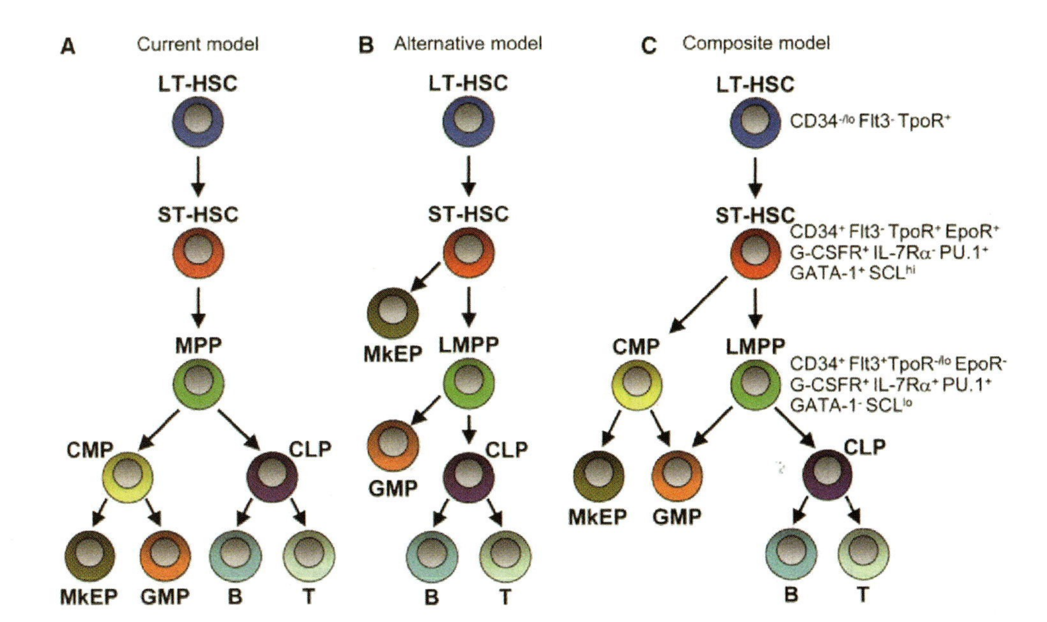


**FIGURE 11.3** Models for the generation of myeloid and lymphoid cell lineages from LT-HSCs. In the current model **(A)** the LT-HSC divides to give rise to a ST-HSC, which gives rise to a multipotent progenitor (MPP). The MPP begets a common lymphoid progenitor (CLP) and a common myeloid progenitor (CMP). The CLP gives rise to B and T cells, and the CMP gives rise to megakaryocyte/erythroid progenitors (MkEP) and granulocyte/macrophage progenitors (GMP). **(B)** In an alternative model, the ST-HSC gives rise to the MkEP, loses the ability to generate the MkEP and develops into a lymphoid-primed multipotent progenitor (LMPP). The LMPP begets the GMP, then loses the potential to form the GMP and generates the CLP. **(C)** Composite model in which the ST-HSC generates the CMP and LMPP, which both have the ability to generate the GMP. Genes expressed by the LT-HSC, ST-HSC and LMPP are shown. Reproduced with permission from Adolfsson et al., Identification of Flt3$^+$ lympho-myeloid stem cells lacking erythro-megakaryocytic potential: A revised road map for adult blood lineage commitment. Cell 121:295–306. Copyright 2005, Elsevier.

stem cells that can form all blood and lymphoid lineages; this phenotype is generally considered to be the LT-HSC (Spangrude et al., 1988; Ziegler et al., 1999; Melchers and Rolink, 2001). LT-HSCs also express two markers found on endothelial cells as well, the Trk Tie2, and PECAM-1 (CD31) Sato, 1995; Arai et al., 2004; Baumann et al., 2004). Another marker for LT-HSCs is the transcription factor Runx1 (North et al., 2002). A marker that appears to be selectively expressed on the surface of LT-HSCs is CD150, signaling lymphocyte activation molecule (SLAM) (Kiel et al., 2005; Wagers, 2005). Although its function is unknown, this protein is present on the surface of a large fraction of the cells comprising the enriched population of LT-HSCs and provides a simple way to isolate LT-HSCs and identify them in tissue sections. LT-HSCs cells are small (6-μm diameter), with dense chromatin, and are capable of differentiating *in vitro* into both lymphoid and myeloid lineages. They represent about one in every $10^5$ bone marrow cells (Berardi et al., 1995).

## 3. Regulation of LT-HSC Activation and Proliferation

The stromal cells of the bone marrow provide the microniche that regulates the survival, quiescence, activation, and proliferation of LT-HSCs, and thus the number of LT-HSCs and their progeny (**FIGURE 11.4**). LT-HSCs are in close proximity to the endosteal surface of the trabecular bone (Nillson et al., 2001; Calvi et al., 2003; Zhang et al., 2003). Staining for SLAM confirmed that LT-HSCs are enriched in this location and that they also are frequently associated with endothelial cells lining the marrow sinusoids (Kiel et al., 2005; Wagers, 2005). The HSCs are attached to a subset of spindle-shaped, N-cadherin$^+$ CD45$^-$ osteoblasts in the endosteum of the trabecular bone. N-cadherin, as well as β-catenin, which interacts with N-cadherin to form adherens junctions in addition to its transcriptional functions, was asymmetrically localized to the surface of the HSC adjacent to the osteoblasts. Hematopoietic differentiation appears to take place in a radial direction inward from this surface (Nilsson et al., 2001).

An important factor that enhances HSC survival in the stromal niche is VEGF. Mice null for VEGF exhibit reduced survival that is associated with significantly lower frequency of blood colony formation from bone marrow cells *in vitro* and reduced ability of bone marrow cells to rescue irradiated mice in a repopulation assay. Wild-type bone marrow cells also exhibit reduced blood colony formation when treated with the small molecule inhibitors of the VEGF receptors, ZD4190 and SU5416 (Gerber et al., 2002).

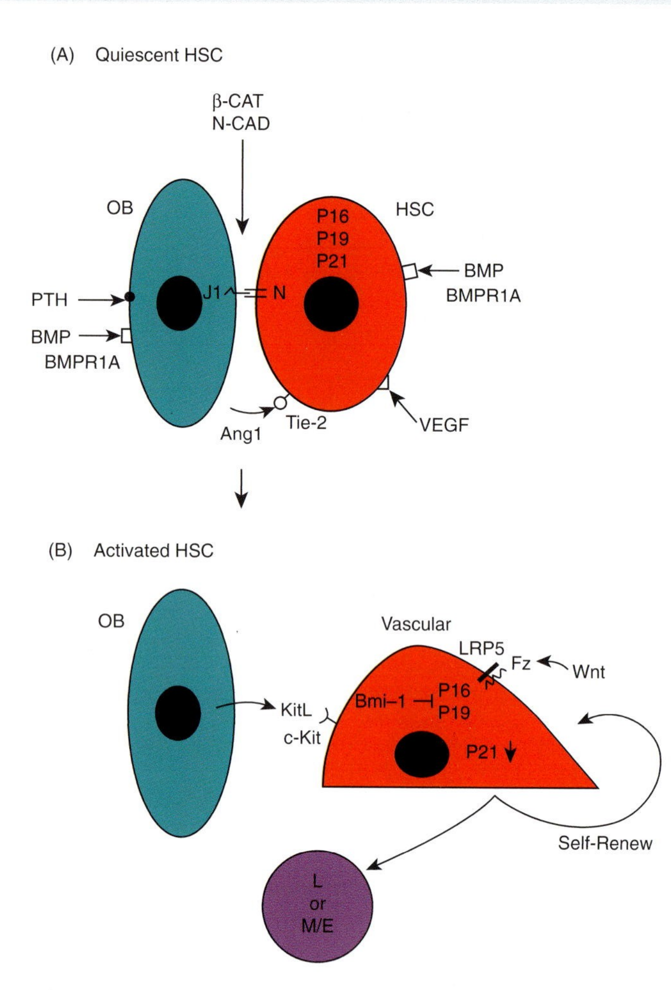


**FIGURE 11.4** Diagram summarizing niche factors that regulate HSC quiescence and activation. **(A)** HSCs are maintained in a quiescent state by several molecules associated with osteoblasts (OB) of the bone marrow. HSCs adhere to osteoblasts through N-cadherin (N-CAD) and β-catenin (β-CAT). Parathyroid hormone (PTH) binding to its receptor (dot) increases osteoblast number. The osteoblasts produce high levels of the Notch (N) ligand jagged-1 (J1), which maintains higher numbers of HSCs in a quiescent state. BMP counters this effect by binding to its receptor BMPR1A (square) to lower the number of osteoblasts. Ang-1 secreted by osteoblasts also maintains HSCs in the $G_0$ phase of the cell cycle by signaling through its receptor, Tie-2. The p16, 19, and 21 proteins suppress entry into the cell cycle. VEGF, which binds to its receptor VEGFR (triangle), is essential for HSC survival. **(B)** To divide, the HSC moves away from the osteoblast to a more vascular compartment. This is facilitated by KitL, which after binding to the c-kit receptor on HSCs, increases their motility. Wnt signals bind to LRP5/Frizzled to drive the HSC into the cell cycle, while Bmi-1 blocks the cell cycle inhibitors p16 and 19. The HSC divides to self renew and produce downstream lymphoid (L) or myeloid/erythroid (ME) cells.

HSC quiescence is maintained by a number of pathways. Osteoblasts in the endosteum appear to play a central role in regulating the number of HSCs in the hematopoietic niche through PTH and the Notch pathway. Mice transgenic for activated PTH/PTHrP receptors (PPRs) have markedly increased numbers of osteoblasts. Stimulation of these osteoblasts with PTH supports an increase in the number of HSCs *in vitro* and *in vivo* through the production of high levels of the Notch ligand jagged 1 (Calvi et al., 2003). Inhibition of Notch activation by γ-secretase inhibited the PTH-induced increase in HSCs *in vitro*. BMP signaling negatively regulates the HSC number. Mice with conditional inactivation of the BMP receptor type 1A (BMPR1A) exhibit an increased number of osteoblasts that is correlated with an increased number of HSCs (Zhang et al., 2003).

Another important interaction between osteoblasts and HSCs involved in maintaining HSC quiescence is mediated by Tie- 2 and its ligand, angiopoietin-1 (Ang-1), a 70 KD glycoprotein (Arai et al., 2004). Tie-2 is expressed by quiescent LT- HSCs, which are resistant to 5-FU, a molecule that kills cycling cells. After 5-FU treatment *in vivo*, increased numbers of Tie-2-expressing HSCs were found adhering exclusively to Ang-1-expressing osteoblasts of the endosteal bone surface. Bone marrow cells infected with a retroviral Ang-1 construct dramatically increased the number of cells in the $G_0$ phase of the cell cycle when injected into irradiated mice. Further evidence that Ang-1 signaling inhibits HSC division is that HSCs labeled *in vitro* with the fluorescent vital dye PKH26 lose fluorescence when cultured in the absence of Ang-1, indicating dilution of the dye by division, whereas cells cultured in the presence of Ang-1 retain their fluorescence.

Expression of the cdk inhibitor p21 is essential to maintain HSC quiescence (Cheng et al., 2000). The absence of p21 leads to exhaustion of the HSC pool. Mice null for p21 exhibit increased proliferation of HSCs; exposing these mice to 5-FU depletes the cycling HSCs by killing them. Serially transplanted bone marrow from p21-null donors to irradiated recipients results in hematopoietic failure. Two other genes that regulate HSC number by inhibiting the cell cycle are *p16* and *p19* (Park et al., 2003).

To divide, HSCs must physically move from the quiescent (osteoblast-adherent) part of the niche to a more proliferative vascular part of the niche. This movement is facilitated by proteases (Levesque et al., 2001) through the c-Kit/KitL signaling pathway. HSCs express c-Kit, which is the receptor for Kit ligand (KitL) secreted by stromal cells. Mutations in either c-Kit or KitL impair hematopoiesis (Huang et al., 1992). Bone marrow ablation in mice by 5-FU treatment induces the expression of stromal cell derived factor 1 (SDF-1) by stromal cells (Heissig et al., 2002). SDF-1 upregulates stromal MMP-9, which catalyzes the release of stromal KitL. Binding of KitL to c-Kit on the surface of HSCs increases their motility, allowing their recruitment into the proliferation-permissive vascular portion of the niche. MMP-9-null mice treated with 5-FU exhibit impaired release of stromal KitL and HSC motility, resulting in failure of hematopoietic recovery that can be reversed by exogenous KitL.

The Wnt signal transduction pathway is an important regulator of HSC activation and proliferation (Raya et al., 2003). HSCs exposed *in vitro* to purified Wnt3a protein, but not HSCs exposed to control medium lacking Wnt3a, were able to reconstitute the bone marrow of lethally irradiated mice (Willert et al., 2003). Overexpression of stable β-catenin expands the pool of HSCs and their progeny in long-term bone marrow cultures. HSCs transfected with an Lef-1/Tcf reporter construct and transplanted into irradiated mice activated the reporter, indicating that they respond to endogenous Wnt. Furthermore, ectopic exposure of HSCs *in vitro* to the soluble frizzled cysteine-rich domain (CRD), which competitively inhibits Wnt signaling, drastically reduced the ability of HSCs to proliferate. Ectopic expression of axin in HSCs, which increases β-catenin degradation, reduced their ability to rescue lethally irradiated mice in the bone marrow reconstitution assay. These treatments also reduced the expression of *Notch1* and *HoxB4*, genes that have also been implicated in the self-renewal of HSCs (Antonchuk et al., 2002).

Several other molecules are also important for LT-HSC activation and proliferation. Bmi-1, a protooncogene transcriptional repressor that regulates cell growth and differentiation genes in a variety of cell types, is essential for HSC self-renewal (Park et al., 2003; Lessard and Sauvageau, 2003). Bmi-1-null mice have an average number of HSCs that is 10 times less than normal, and the ability of Bmi-1-null bone marrow to reconstitute the hematopoietic system in irradiated hosts is significantly decreased. Microarray analysis revealed an assortment of genes affected by the lack of Bmi-1 (Park et al., 2003). The expression of the *p16* and *p19* genes was elevated. The ectopic expression of *p16* led to proliferative arrest and of *p19* to apoptosis in normal HSCs. These results indicate that Bmi-1 downregulates the inhibitory activity of these genes, allowing HSCs to enter the cell cycle.

Niche signals regulate specific sets of transcription factors that determine the patterns of gene activity defining each differentiation step. These transcription factors act through a variety of mechanisms that affect their binding to regulatory regions of genes (Orkin, 2001). In some cases, a single transcription factor acts as a dominant regulator. For example, the zinc-finger protein GATA–1 and the PU.1 protein induce blood cell lineages. The particular lineages induced may depend on the concentration of the transcription factor. The activation of genes for one lineage may be coupled to inactivation of the genes for another. For example, upregulation of cell surface antigens defining the blood cell lineages induced by GATA-1 and PU.1 is accompanied by downregulation of cell surface antigens characteristic of other lineages. In other cases, combinatorial positive and negative protein–DNA and protein–protein interactions among transcription factors and co-factors are probably the rule, even with dominant regulators. For example, GATA-1 must interact with a co-factor called FOG (Friend of GATA-1) for normal erythropoiesis and platelet formation.

## REGENERATION OF BLOOD VESSELS

### 1. Development and Structure of the Vascular System

The circulatory system is the first system to become functional during embryogenesis, because of the need for nutrition, oxygen, and waste elimination by cells throughout the developing embryo. Furthermore, as pointed out in Chapter 2, for the repair of ***any*** tissue —whether it be by fibrosis or regeneration of the original tissue architecture—the regeneration of blood vessels is crucial and ubiquitous.

The embryonic circulatory system develops via two processes, vasculogenesis and angiogenesis (Poole et al., 2001). Vasculogenesis refers to the initial formation of embryonic blood vessels from mesenchymal cells. Angiogenesis is the sprouting of new blood vessels from pre-existing ones and is involved in the elaboration of the initial circulatory system, as well as being the mechanism by which new blood vessels are regenerated after injury.

Vasculogenesis begins with the differentiation of hemangioblasts in the blood islands of the yolk sac into into angioblasts, the endothelial stem cells (EnSCs) (**FIGURE 11.5**). The EnSCs differentiate into endothelial cells (ECs), which express a number of endothelial-specific markers, including PECAM (CD31) CD34, CD146, VEGFR1 and 2, Tie-2, VE-cadherin, and von Willebrand factor (Rafii and Lyden, 2003). The ECs become organized into the primary capillary network, including the aorta and major veins of the circulatory system. They first coalesce to form cords, which are then reshaped into tubes. Pericytes, mesenchymal cells that express smooth muscle actin, become attached to the endothelial tubes and stabilize them. The capillary endothelium is thought to induce surrounding mesenchymal cells to become pericytes, but there is some evidence that pericytes might also be derived from EnSCs. When VEGF was added to embryoid bodies in the early stages of their differentiation (3–7 days post-plating), cells expressing CD31 formed cords and were also positive for α-smooth muscle actin and desmin, structural proteins that are markers for pericytes (Hagedorn et al., 2004). Endothelial cells and pericytes

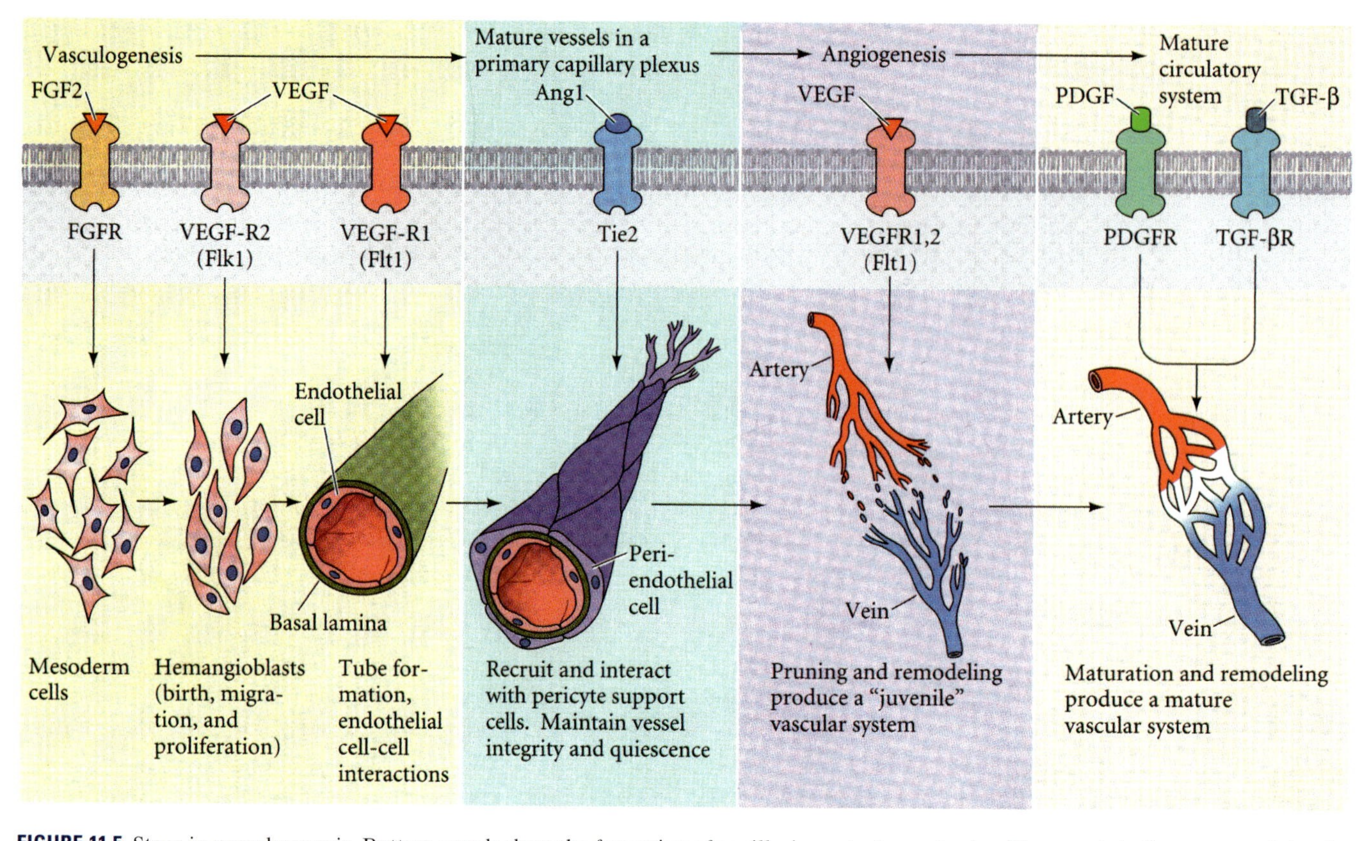


**FIGURE 11.5** Steps in vasculogenesis. Bottom panels show the formation of capillaries, arteries and veins. Top panels indicate some of the signaling molecules and their receptors that are important in promoting each step. Reproduced with permission from Gilbert, Developmental Biology (7[th] ed). Copyright 2003, Sinauer Associates.

interact throughout vasculogenesis. This interaction is mediated by Ang-1 produced by the pericytes, and the Tie-2 receptor, which is expressed on the surface of endothelial cells (Yancopoulos et al., 2001), as well as by VEGF and PDGF-B secreted by endothelial cells that bind to their receptors on pericyte cells (Lindhal et al., 1997). The lack of functional VEGF, Ang-1, PDGF-B, or their receptors results in defective vasculogenesis (Yancopoulos et al., 2001).

Upon completion of the initial phase of vasculogenesis, angiogenesis remodels and further develops the embryonic vascular system. Angiogenic remodeling involves the pruning and enlargement of blood vessels by fusion to form the mature vascular network. In addition, new capillaries are formed in previously avascular tissue by capillary growth. Mesenchymal cells differentiate into smooth muscle surrounding the endothelium. Veins and arteries become distinct from one another, through the expression by their endothelial cells of two different cell surface molecules belonging to the ephrin family. Arterial endothelial cells express the ligand Ephrin B2, whereas venous endothelial cells express EphB4, the receptor tyrosine kinase receptor for Ephrin B2 (Wang et al., 1998). Vasculogenesis occurs in EphB2 knockout mice, but angiogenesis does not. The Eph ligand–receptor interaction is believed to ensure that arterial capillaries connect end-to-end only with venous capillaries and that side-to-side fusion to make larger vessels occurs only between capillaries of the same type. The expression of EphB2 versus EphB4 appears to be under the control of the Notch signaling pathway (Lawson et al., 2001; Lawson and Weinstein, 2002). Activation of this pathway leads to EphB2 expression and artery formation through the transcription factor Gridlock, whereas its repression leads to low levels of Gridlock, expression of EphB4, and formation of veins. Notch expression, in turn, is under the control of VEGF (Lawson and Weinstein, 2002).

The mature vascular system consists of a network of arteries and veins whose diameters progressively narrow from the heart outward to small diameter arterioles and venules (Ham and Cormack, 1979). The arterioles and venules are connected by a vast capillary system. Mature capillaries consist of a layer of endothelial cells that synthesize a basement membrane on their external surface, in which pericytes are embedded, similar to the relationship between satellite cells and myofibers in skeletal muscle. The capillary endothe-

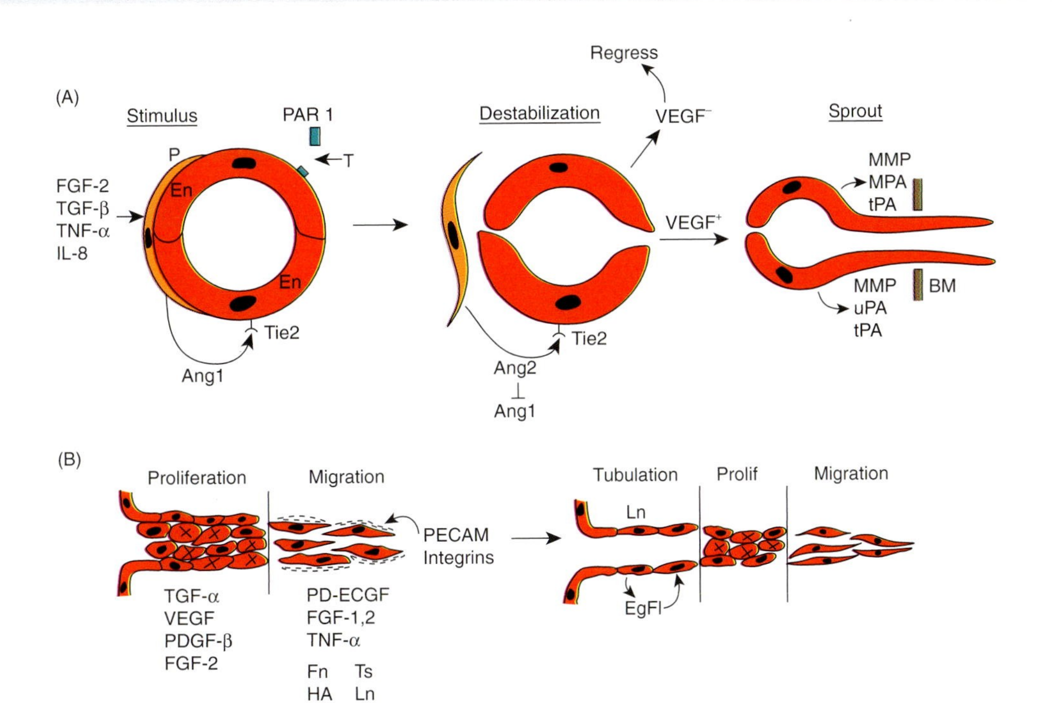


**FIGURE 11.6** Molecular factors regulating angiogenesis. **(A)** Initiation of sprouting. Endothelial cells (En) are stabilized by their interactions with pericytes (P), mediated by Ang1/Tie 2. Thrombin (T) cleaves protease-activated receptor (PAR-1), allowing it to activate endothelial cells. The endothelial cells respond to a combination of growth factors and cytokines released from platelets and ECM, and synthesized by macrophages, fibroblasts, endothelial cells and pericytes. Pericytes secrete Ang 2, which competes with Ang 1 for binding to Tie 2, allowing dissociation of pericytes from the capillary and destabilizing the endothelium. In the absence of VEGF, the endothelial cells secrete proteases (MMP, uPA, tPA) that degrade the basement membrane (BM) allowing sprouting of the endothelium away from the injured vessel. **(B)** Sprouting endothelial cells divide and form cords behind a distal zone of more flattened migrating cells that express PECAM and integrins on their surfaces. Cell division (x) and migration are each promoted by different sets of growth factors. Cell migration also requires several substrate molecules such as fibronectin (Fn), thrombospondin (Ts), hyaluronic acid (HA), and laminin (Ln). As the cords grow, the cells closest to the old vessel flatten and arrange themselves into tubules, which requires laminin and the secreted protein, EgFl.

lium is continuous throughout the arterial and venous systems. Arteries, arterioles, venules, and veins have three mural layers outside the endothelium, the intima, the media, and the adventitia. The intima and adventitia are elastic layers, whereas the media is a smooth muscle layer. These layers are less prominent and decrease in thickness with distance from the heart. In addition, the muscular media of veins and venules tends to be thinner than that of arteries and arterioles. Another major difference between arterioles and venules is that venules, like capillaries, are associated with pericytes.

## 2. Blood Vessel Regeneration in Wounded Tissue

### a. *Blood Vessels Regenerate from Two Sources*

New blood vessels originate from a major and a minor source. The minor source is the EnSCs from the bone marrow that differentiate into circulating endothelial cells that are incorporated into newly forming blood vessels. The major source is the endothelial cells of the injured vessels themselves (**FIGURE 11.6**). In an excisional wound, endothelial cells respond to angiogenic stimuli by migrating out of the vessel into the fibrin matrix to form cords of proliferating cells. As the cords grow on their distal ends, the more proximal cells undergo tubulogenesis into capillaries by flattening and configuring themselves into hollow cylinders, re-establishing tight junctions, and reconstituting the basement membrane. These capillaries then coalesce into new arterial and venous vessels of progressively larger diameter until the appropriate vascular configuration is achieved. Simultaneously, pericytes form vascular smooth muscle and fibroblasts form the elastic tissue of the intima, media, and adventitia. The pericytes that invest the new capillaries and venules are most likely derived from pericytes associated with the sprouting venules (Clark, 1996).

The presumption has been that the endothelium of blood vessels is a uniform population of differentiated cells that responds to angiogenic stimuli by migrating and proliferating. However, clonal proliferation studies *in vitro* on human umbilical vein endothelial cells (HUVECS) and human aortic endothelial cells (HAECS) have shown that the endothelial cell population of blood vessels is heterogeneous (Ingram et al., 2004). Single HUVECS and HAECS exhibited differential proliferation capacity. Approximately 28% of clonally plated endothelial cells that divided (52% of the cells plated) had a high proliferation capacity (i.e., formed colonies of >2,000 cells). These cells displayed higher telomerase activity than their low-proliferation counterparts. The data suggest that the endothelium of the umbilical cord and aorta contains a subpopulation of cells that is similar to EnSCs and that these cells may be the ones that proliferate to regenerate new blood vessels, although it remains to be demonstrated that the endothelium of other vessels has similar heterogeneity.

### a. Activation, Migration, and Proliferation of Endothelial Cells

Angiogenesis in adult wounds is regulated by a variety of soluble factors (Clark, 1996). The picture presented here is drawn from experiments on wounds, embryonic development, and other systems in which blood vessels are formed and regress, such as uterine tissue and tumors (**FIGURE 11.6**).

The initial step in the activation of endothelial cells after wounding may be the cleavage of protease-activated receptor 1 (PAR-1) by thrombin (Coughlin, 2000). PARs are unique in that they carry their own ligand in their extracellular domain. Cleavage of the receptor unmasks a new N-terminus that can bind with the intramembrane part of the receptor to effect transmembrane signaling via G proteins. Activation leads to a cascade of events regulated by a number of growth factors released from platelets and ECM fragments, and synthesized by macrophages, endothelial cells, pericytes, and fibroblasts. These include platelet-derived endothelial growth factor (PD-ECGF), TGF-α, TNF-α, PDGF, FGF-1 and 2, VEGF, and Ang 1 and 2 (Yancopoulos et al., 2001). It would be interesting to determine whether this activation may have features in common with the activation of PECs by thrombin in newt lens regeneration (Imokawa and Brockes, 2003, Chapter 3).

In uninjured endothelium, binding of Ang-1 (pericytes) to Tie-2 (endothelial cells) stabilizes vessel walls by promoting association of endothelial cells with pericytes (Davis et al., 1996; Suri-Jones et al., 1996; Vikkula et al., 1996), an interaction similar to that which stabilizes HSCs in a quiescent state by promoting their adhesion to osteoblasts (Arai et al., 2004). Activated endothelial cells lose their specialized junctions, loosening them from the vessel wall (Madri et al., 1996). This destabilization is due to the autocrine induction of Ang-2 in the pericytes. Ang-2 is antagonistic to Ang 1 and blocks the ability of mural Ang-1 to bind to endothelial Tie-2, destabilizing the vessel wall (Sato et al., 1995; Maisonpierre et al., 1997). In the presence of VEGF, which is synthesized in large quantities by the epidermis of healing skin wounds (Clark, 1996), this leads to sprouting; in the absence of VEGF, the blood vessels regress (Yancopoulos et al., 2001). Confluent monolayers of quiescent endothelial cells *in vitro* synthesize transcripts for 121 and 165 amino acid isoforms of VEGF. Wounding the monolayer leads to upregulation of a 189 amino acid isoform. Motility of the cells to close the wound is increased by 50% (Sammak et al., 1998). Expression of the 189 amino acid isoform can be induced by calcium, suggesting that a rise in calcium due to wounding may set off a cascade of gene activity that results in activation of the VEGF gene. One way in which VEGF stimulates angiogenesis is by elevating the activity of MMPs (Stetler-Stevenson, 1999; Coussens et al., 2002).

Activated endothelial cells express uPA, tPA, and collagenase, allowing them to break down the surrounding basement membrane and migrate into the fibrin matrix of the wound (Madri et al., 1996). Migration is dependent on the expression of membrane-tethered1-matrix metalloproteinase (MT1-MMP), which degrades the matrix (Hiraoka et al., 1998). Migrating endothelial cells express PECAM-1 in a diffuse pattern on their surfaces, suggesting that this adhesion molecule plays a role in their movement through the fibrin matrix (Madri et al., 1996). Simultaneously, expression of the RECK protein, which blocks angiogenesis, is downregulated (Oh et al., 2001), as is the expression of another angiogenesis inhibitor, endostatin (Wu et al., 2001). Endostatin is formed by MMP-catalyzed cleavage of the C-terminal 184 amino acids from collagen XVIII (Zatterstrom et al., 2000). It inhibits angiogenesis by antagonizing the binding of VEGF to its receptor, VEGFR2 (Kim et al., 2002), leading at high concentrations to $G_1$ arrest of endothelial cells by inhibiting cyclin $D_1$ (Hanai et al., 2002), and at lower concentrations to endothelial cell apoptosis and failure of normal vessel maturation (Berger et al., 2000; Bloch et al., 2000; Schmidt et al., 2004).

Other growth factors also play a role in endothelial cell activation, migration, and proliferation (Madri et al., 1996; Tomanek and Schatteman, 2000). Endothelial cell activation can be triggered *in vitro* by FGF-2, TGF-β1, IL-8, and TNF-α. Migration and proliferation of endothelial cells are modulated by a variety of growth factors, including PD-EGCF, FGF-1 and 2, TGF-α, TNF-α, TGF-β, VEGF, and PDGF (Tomanek and Schatteman, 2000). PD-ECGF and TNF-α have chemotactic activity for endothelial cells *in vitro* (Miyazono et al., 1987; Ishikawa et al., 1989). TGF-α binds to the EGF receptor on endothelial cells and EGF and TGF-α are equally potent in stimulating endothelial cell proliferation *in vitro*. However, TGF-α is 10 times more potent than EGF in inducing angiogenesis *in vivo* (Schreiber et al., 1980). TGF-β and FGF-1 and 2 may induce other cells such as fibroblasts to produce proteases, growth factors, or ECM molecules that directly stimulate angiogenesis (Folkman and Klagsbrun, 1987; Thompson et al., 1988; Risau, 1991). In addition to a role in endothelial cell-pericyte association, PDGF-B might stimulate mitosis of endothelial cells through an autocrine loop (Holmgren et al., 1991). Human recombinant IL-8 is strongly angiogenic to rat cornea capillaries *in vivo* and to human umbilical vein endothelial cells *in vitro*, activity that can be inhibited by antibodies to IL-8 (Koch et al., 1993).

Conditioned medium of mouse MSC cultures enhances the proliferation of endothelial cells and smooth muscle cells *in vitro* (Kinnaird et al., 2004). The conditioned medium contains VEGF, FGF-2, and monocyte chemoattractant protein-1 (MCP-1). The proliferation-enhancing effect of the medium was partly attenuated by antibodies to VEGF and FGF-2. VEGF secretion by human adipose stromal cells (ASCs) increased five-fold when the cells were cultured under hypoxic conditions (Rehman et al., 2004). Conditioned medium from these cultures significantly increased the proliferation, and decreased the apoptosis, of human endothelial cells *in vitro*.

### b. Tubulogenesis

Many *in vitro* and some *in vivo* studies indicate that ECM molecules synthesized by fibroblasts are important to the tubulogenesis of proliferating endothelial cells (FIGURE 11.6). Fibronectin, thrombospondin, HA, and Ln are essential for cell migration and normal endothelial tube formation, and endothelial cells express receptors for these molecules, including the β 1, 3, and 5 integrins (Whalen and Zetter, 1992; Tomanek and Schatteman, 2000). Laminin is particularly effective in causing endothelial cells to form hollow tubes. Two domains in the Ln molecule, the YISGR sequence in the B1 chain and the RGD-containing sequence in the A chain, appear to mediate this transformation. The RGD sequence binds to a cell surface integrin, whereas the YISGR sequence binds to a cell surface nonintegrin Ln-binding protein. ECM molecules convey regulatory information to cells by resisting tension generated within the cells by the actin cytoskeleton (Ingber and Folkman, 1989). The A chain mediates initial attachment to Ln, and the B chain mediates the cytoskeletal reorganization associated with tube formation. This reorganization presumably activates signal transduction pathways that induce patterns of gene activity leading to tubulogenesis (Juliano and Haskill, 1993). Basement membrane laminin-entactin complex promotes the stabilization of newly formed vessels (Nicosia et al., 1994). Other ECM molecules that may be involved in endothelial cell migration and tube formation are Tn and osteonectin, both of which promote reorganization of the actin cytoskeleton in cultured bovine arterial endothelial cells (Sage and Bornstein, 1991).

A secreted protein that plays a crucial role in tubulogenesis is Egfl7, an ECM-associated protein of ~30K molecular mass (Parker et al., 2004). Egfl7 is expressed at a high level in the angioblasts and ECs of all developing vessels of mouse, human and zebrafish embryos. It is undetectable in adult vessels, but is strongly upregulated in adult proliferating tissues, such as tumors, reproductive tissues and inflamed tissues, in which vascular growth and remodeling are taking place. When the Egfl7 transcript was knocked down in zebrafish embryos by morpholino antisense oligos, all primary arteries and veins formed in the correct patterns and EC markers were expressed normally. However, the vessels remained cords of cells that failed to undergo tubulogenesis (Parker et al., 2004).

### c. Cessation of Angiogenesis in Wounds

Late in the phase of granulation tissue formation in wounded dermis, angiogenesis ceases and the capillary network begins to regress via the apoptosis of endothelial cells (Clark, 1996). Over time, the density of the capillary network returns to normal. Cessation of angiogenesis and regression of blood vessels as granulation tissue is remodeled might involve the upregulation of Ang2 and the cessation of VEGF production as re-epithelialization comes to a close. A 140-kDa glycoprotein was identified in the conditioned medium of hamster cell cultures that suppressed new capillary formation in the rat cornea (Rastinejad et al., 1989). It is possible that this glycoprotein might be Ang 2, since Ang2 can either activate or antagonize Tie2 and is expressed in the endothelium of vessels undergoing regression (Yancopoulos et al., 2001).

### d. Endothelial Cells Derived from EnSCs Play a Minor Role in Blood Vessel Regeneration

EnSCs of the bone marrow have an antigenic phenotype of $[CD133\ VEGFR2]^{+}$ and lose expression of these markers when they differentiate into mature ECs (Rafii and Lyden, 2004). Minor numbers of circulating endothelial cells labeled with DiI or containing a

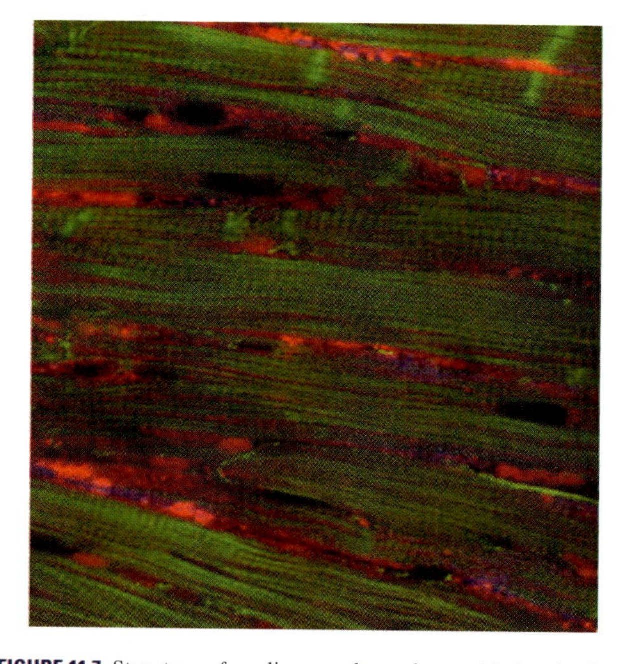

**FIGURE 11.7** Structure of cardiac muscle as observed in longitudinal section by confocal microscopy. Red staining shows the collagen network in the endomysium around the myofibers (green) and fibroblasts (blue). Reproduced with permission from Goldsmith et al., Organization of fibroblasts in the heart. Dev Dynam 230:787–794. Copyright 2004, Wiley-Liss.

constitutively expressing β-gal gene were shown to be incorporated into the endothelium of capillaries regenerating in response to limb ischemia created by removing a section of femoral artery in mouse and rabbit hosts (Asahara et al., 1997). Limb ischemia in mice and rabbits resulted in an elevated level of EnSCs in peripheral blood.

There is growing evidence that EnSCs and their more restricted progenitors are recruited to sites of tissue damage by angiogenic chemokines such as stromal cell-derived factor, VEGF, and GM-CSF that are upregulated by tissue ischemia. There is also evidence that co-recruitment of HSCs is necessary for the incorporation of EnSCs into neovessels, through the production of Ang-1 (Rafii and Lyden, 2003).

## REGENERATION OF CARDIAC MUSCLE

### 1. Structure of Cardiac Muscle

FIGURE 11.7 illustrates the organization of cardiac muscle. Mammalian cardiac muscle differs in structure from skeletal muscle in that it is not composed of syncytial multinucleate myofibers, but of mononucleate cardiomyocytes joined together in chains that form a branching pattern. Structures called intercalated discs join the individual cells. The discs have three functions: physical attachment of adjacent cells by desmosomes; connection of the actin filaments of adjacent cells (analogous to the function of Z lines of the sarcomeres of skeletal muscle); and synchronization of the contraction of the cells by their gap junctions, which allow the rapid conduction of action potentials from one cell to the next (Alberts et al., 1994). The heart muscle is called the myocardium. Some of the myocardial fibers are specialized as fast-conducting Purkinje fibers that function to propagate the electrical signal that synchronizes contraction. The myocardium is nourished by coronary arteries, capillaries, and veins that branch off the aorta and superior vena cava.

Internally, the mammalian myocardium is lined by a thin endothelium that is continuous with the endothelium of the vascular system. The myocardium is covered externally by a connective tissue sheath, the epicardium. Fibroblasts form a three-dimensional endomysium that surrounds groups of cardiomyocytes 2–5 cells thick (Weber, 1989; Bishop and Laurent, 1995; Goldsmith et al., 2004). Integrin receptors anchor the surface of the fibroblasts and cardiomyocytes to the collagen network of the ECM (Ross and Borg, 2001). Electron microscopy and staining for gap junction connexins revealed that the fibroblasts form gap junctions between themselves and with the cardiomyocytes as well (Goldsmith et al., 2004), suggesting that the fibroblasts have a role in electrically connecting the cardiomyocytes (Gudesius et al., 2003). This organization is thought to be important for maintaining the shape of the heart and for contraction. Myofibroblasts might exert force on the endomysial collagen network, contracting it in synchrony with cardiomyocyte contraction.

### 2. Mammalian Cardiac Muscle Does Not Regenerate, but Harbors Adult Stem Cells

Ventricular myocardial infarction is caused by the restriction of blood flow through the coronary arteries by thrombosis or plaque build-up, causing death of the cardiomyocytes in the ischemic area. Damaged cardiac muscle does not regenerate. While not occupying the most volume, fibroblasts are the most numerous cells in the heart, and they proliferate rapidly into the lesion to patch the infarcted area with noncontractile, collagenous scar tissue, distorting the normal epimysial network (FIGURE 11.8). The contractile force of the ventricles is diminished in proportion to the amount of cardiac muscle replaced by scar, and the remaining ventricular myocardium undergoes hypertrophy in an attempt to compensate for this diminished function.

Prior to being overtaken by scar tissue formation, however, the myocardium does initiate a regenerative response. Dividing cells with the characteristics of

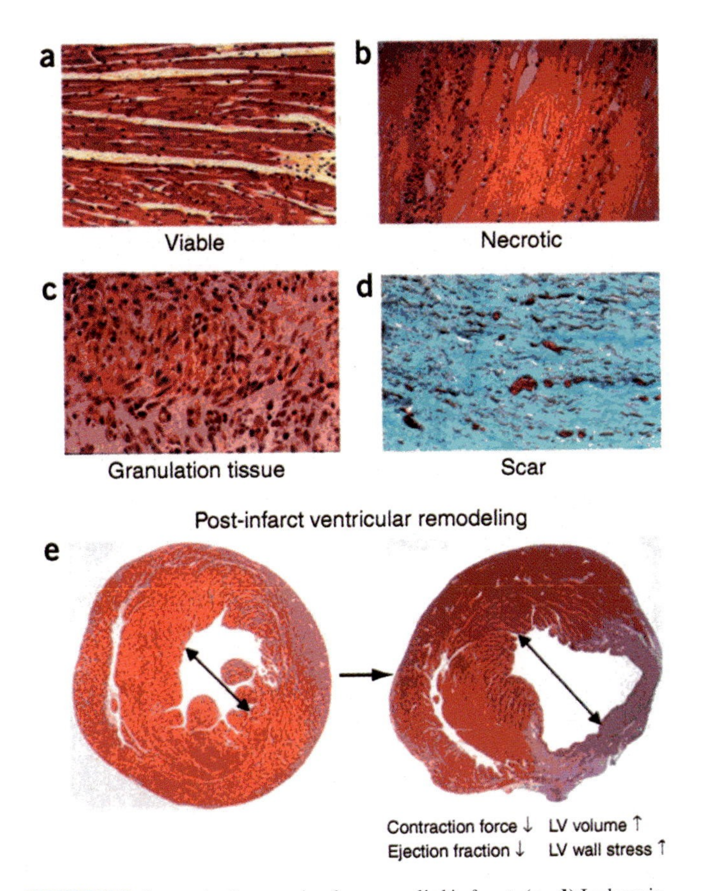


**FIGURE 11.8** Stages in the repair of myocardial infarct. **(a–d)** Ischemia in a region of the left ventricle causes necrosis of viable cardiomyocytes, which elicits an inflammatory response that promotes fibroblast proliferation and the formation of granulation tissue that subsequently remodels into scar. **(e)** The injured ventricular wall thins, with dilation of the ventricular cavity (arrows). This markedly increases mechanical stress on the ventricular wall and promotes progressive contractile dysfunction, as indicated by decreased force of contraction and diminished left ventricular ejection fraction. Reproduced from Laflamme and Murry, Regenerating the heart. Nature Biotech 23:845–856. Copyright 2005, Nature Publishing Group.

immature cardiomyocytes have been observed in the hearts of human patients who died 4–12 days after infarction (Beltrami et al., 2001; Anversa and Nadal-Ginard, 2002). Four percent ($2 \times 10^6$ cells) of the cardiomyocytes in the border zone of the infarcts were in mitosis, 84 times the value for control patients who had died of other causes and had no major risk factors for heart disease. This is clearly a regenerative response, comparable to the initiation of axon sprouting in the injured spinal cord.

What is the origin of the cells that divide after myocardial infarction? One possibility is that some cardiomyocytes undergo partial dedifferentiation and divide, like adult newt or fish cardiac muscle (see ahead). Mammalian cardiomyocytes can undergo partial dedifferentiation *in vitro*. Cultured adult rat cardiomyocytes come to resemble fetal cardiomyocytes in morphology and synthesis of the myosin heavy chain (Eppenberger et al., 1988). A high percentage of them (up to 25% ventricular and 63% atrial) synthesize DNA (Oberpriller et al., 1983; Rumyantsev, 1992; Claycomb, 1991, 1992), though only a very small percentage of (atrial) cells have been observed to divide. Entry into the cell cycle and dedifferentiation are linked in cardiomyocytes. Cardiomyocytes are prevented from entering the cell cycle by the Rb protein, which complexes with the E2F family of transcription factors (Flink et al., 1998). Transfection of the E2F-1 transcription factor into cultured ventricular cardiomyocytes induces DNA synthesis while simultaneously inhibiting the cardiac and skeletal α-actin promoters, sarcomeric actin synthesis, and serum response factor, which is crucial for the transcription of both sarcomeric α-actins in cardiac muscle (Kirshenbaum et al., 1996). The ability to partially dedifferentiate after injury, which is not shared by mammalian skeletal muscle, may be related to the fact that during embryonic development, cardiac myoblasts have parallel programs for proliferation and cytodifferentiation, rather than serial programs linked by a single switch (Claycomb, 1992). This unusual situation is associated with the fact that the heart is one of the first organs to form during development and must become functional while still growing.

A second possibility is that the dividing cells are stem cells recruited from the bone marrow. Quiani et al. (2002) have detected Y-chromosome$^+$ cells in the hearts of males who have received heart transplants from females. However, it has been shown that bone marrow stem cells have virtually no capacity to differentiate into cardiomyocytes *in vivo* (see Chapter 13). The Y$^+$ cells might be recipient endothelial cells incorporated into the donor cardiac vasculature.

The third possibility is that the heart contains cardiac stem cells that under normal conditions cycle at low frequency to maintain normal numbers of cardiomyocytes and respond to injury with more intense proliferation. Such cells have been thought not to exist, but three types of cardiac stem cells have recently been discovered (Parmacek and Epstein, 2005; Laflamme and Murry, 2005). The discovery of cardiac stem cells raises the possibility that in humans, they could be harvested, expanded, and used as replacements for damaged myocardium.

The first cardiac stem cell type is a Lin$^-$ [c-Kit Sca-1]$^+$ cell that in rats is distributed throughout the uninjured myocardium at a frequency of approximately one in every $1 \times 10^4$ cardiomyocytes (Beltrami et al., 2003). These cells are small, with a high nuclear/cytoplasmic ratio and were often found in clusters, suggesting clonal derivation (**FIGURE 11.9**). Many of the cells in the clusters either expressed the mitotic marker Ki67 or were

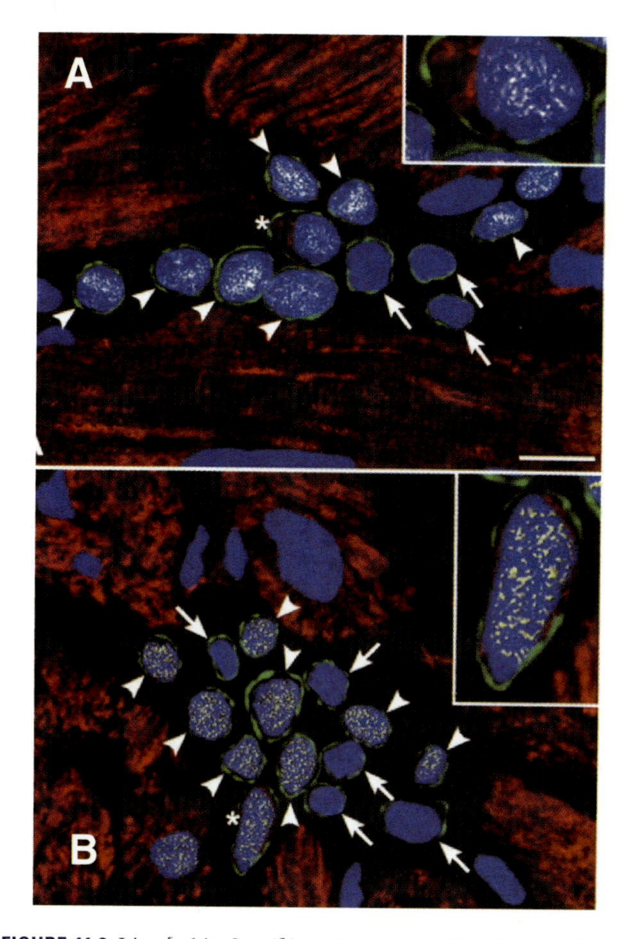


**FIGURE 11.9** Lin⁻ [c-kit, Sca-1]⁺ cardiac stem cells of the rat heart. Shown are two clusters of c-kit⁺ cells. **(A)** 11 cells, three of which are expressing c-kit only (green, arrows), 7 are expressing the cardiac transcription factor Nkx2.5 (white dots, arrowheads), and one expressing α-sarcomeric actin (red, asterisk). Inset, higher power view of the latter cell. **(B)** Fifteen c-kit⁺ cells. Five are expressing c-kit only (arrows), 8 are expressing the cardiac transcription factor MEF2C (arrowheads) and one is expressing MEF2 and α-sarcomeric actin (asterisk and see inset). Reproduced with permission from Beltrami et al., Adult cardiac stem cells are multipotent and support myocardial regeneration. Cell 114:763–776. Copyright 2003, Elsevier.

at early stages of cardiomyocyte differentiation. After isolation from the myocardium by FACS, they could differentiate *in vitro* into immature cardiomyocytes, smooth muscle cells, and endothelial cells. Similar [c-Kit Sca-1]⁺ cells that could self-renew and differentiate into the same cell types were found in human hearts (Messina et al., 2004), except that these cells also expressed CD34. Both mouse and human cells expanded *in vitro* were reported to regenerate cardiac muscle when injected into the borders of myocardial infarcts in rats or mice.

The second cardiac stem cell phenotype is Sca-1⁺ [c-Kit⁻ Lin⁻] (Oh et al., 2003). These cells express most cardiogenic transcription factors (GATA-4, MEF-2C, TEF-1) but not cardiac structural protein genes. They differentiated *in vitro* into cardiomyocytes expressing cardiac structural proteins when exposed to 5-azacytidine, and they homed to infarcted hearts and differentiated into cardiac muscle when injected intravenously.

The third stem cell type is derived from a population of embryonic cells that constitute a secondary embryonic heart field and express the gene *islet-1* in addition to other cardiac transcription factors. During development, these cells migrate to the anterior pharynx and enter the top and bottom of the looped heart tube, where they cease expression of *islet-1* as they differentiate into cardiomyocytes that contribute to the atria, the right ventricle, and perhaps the left ventricle as well (Cai et al., 2003). Laugwitz et al. (2005) identified small numbers of *islet-1* expressing cells in the outflow tract, atria, and right ventricle of mice, rats, and humans, consistent with their being leftover cells from the embryonic secondary heart field. These cells are [Sca-1 c-Kit]⁻. *Islet-1* expressing cells of the developing embryonic heart were marked by making ROSA26 mice transgenic for a tamoxifen-inducible fusion protein consisting of Cre and mutated estrogen receptors, as well as a *lac-Z* reporter gene prevented from expression by a floxed stop cassette. Tamoxifen was pulsed at an early stage of heart development to localize ERCre to the nucleus and allow expression of β-galactosidase (**FIGURE 11.10**). Marked cells isolated from young postnatal mice differentiated into beating cells with the molecular characteristics of cardiomyocytes when co-cultured with paraformaldehyde-killed neonatal cardiomyocytes in the presence of cardiomyocyte conditioned medium, indicating that they could differentiate without fusion into cardiomyocytes.

Given the presence of cardiac stem cells, why is the regeneration process not completed in the infarcted mammalian heart? The situation is likely to be similar to the abortive axon regeneration observed in the injured spinal cord. Fibroblast proliferation and scar tissue formation outrun and suppress the proliferation of the cardiac stem cells. The nature of this process might be profitably investigated in the MRL/MpJ mouse, which is able to regenerate myocardium (Leferovich et al., 2001). After cryogenic infarction in these mice, confocal microscopy demonstrated cells in which antibody staining for α-actinin and BrdU were co-localized, indicating that cells were proliferating and migrating into the infarct to regenerate cardiac muscle (**FIGURE 11.11**). These cells had a mitotic index of 10–20% as compared to 1–3% in controls, suggesting that what determines whether injury results in regeneration or scarring is the relative frequency of division of cardiomyocytes versus fibroblasts. Revascularization was also more evident at the injury site in the MRL

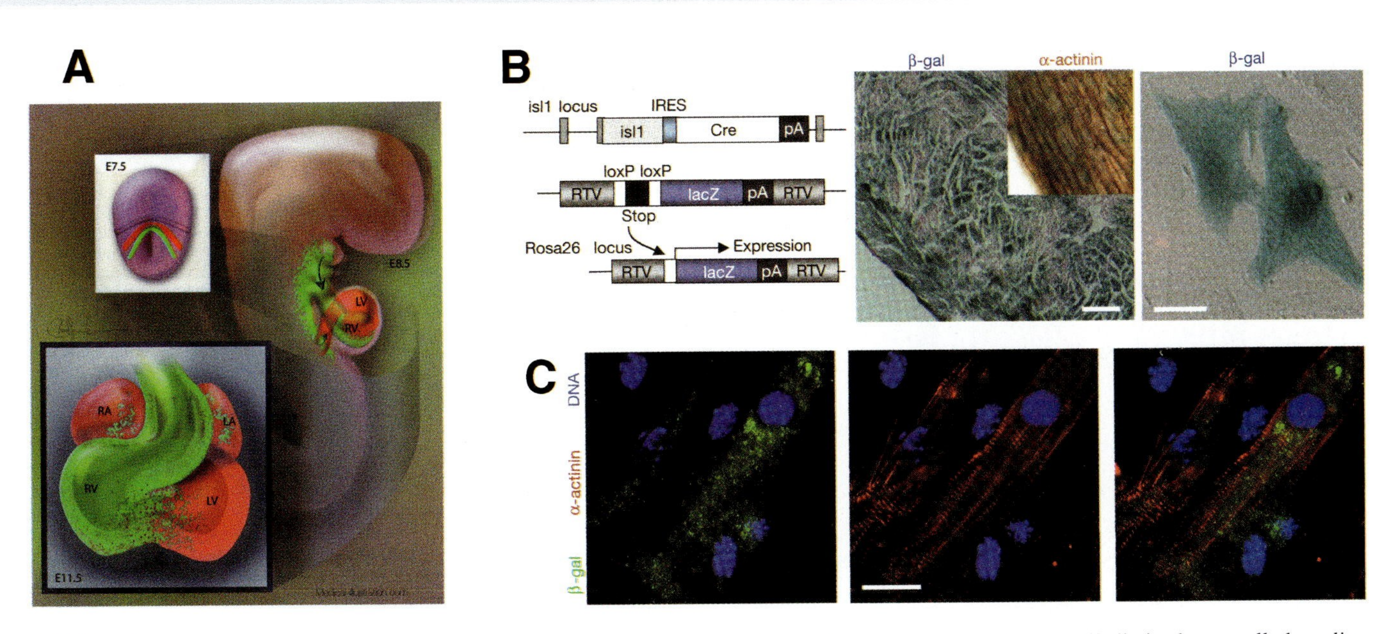


**FIGURE 11.10** [Sca-1, c-kit]⁻ *Islet-1* expressing cells from the secondary heart field. **(A)** At E7.5 heart-forming cells lie in the so-called cardiac crescent (red). Islet-1-expressing myocardial cells arise from a secondary crescent-shaped field next to the cardiac crescent (green). At E8.5, islet-1-expressing cells migrate to the anterior pharynx and invade the anterior and posterior tracts of the heart (arrows). By E11.5, they have given rise to cardiomyocytes of the right ventricle (RV), with scattered contributions to the right atrium, left atrium and left ventricle. **(B)** Method of marking cells expressing islet-1. Mice were made transgenic for *Cre* driven by the *isl1* promoter and a *lacZ* reporter gene whose expression is suppressed by a floxed stop cassette. Only islet-1-expressing cells will activate the reporter. Cells expressing β-gal in the right ventricle wall are shown in the middle; inset shows that these cells express α-actinin. Two isolated β-gal expressing cells are shown to the right. **(C)** Isolated cardiomyocytes immunostained for β-gal, α-actinin, and DNA. **(A)** Reproduced with permission from Parmacek and Epstein, Pursuing cardiac progenitors: regeneration redux. Cell 120:295–298. Copyright 2005, Elsevier. **(B, C)** Reproduced with permission from Laugwitz et al., Postnatal isl1⁺ cardioblasts enter fully differentiated cardiomyocyte lineages. Nature 433:647–653. Copyright 2005, Nature Publishing Group.

mice. By day 60, the mutant hearts were virtually scar-free and echocardiographic measurements indicated a return of heart function to normal. The lack of scarring in the infarcted hearts of these mice provides a potential opportunity to analyze the nature of the cell types resident to the heart, as well as what is different about the injury environment in these hearts compared to the injury environment in normal mice.

### 3. Amphibian and Zebrafish Cardiac Muscle Regenerates When Injured

Several amphibians and the zebrafish (and probably other teleost fish) can regenerate their myocardium. The hearts of these animals are therefore valuable models for the study of cardiac regeneration that may allow insights into how to regenerate mammalian cardiac muscle.

#### a. Amphibian

The amphibian heart is three-chambered, having two atria and one ventricle. The myocardium is thin, highly trabeculated, and avascular. It is nourished by incoming blood that flows into the many sinuses that permeate the muscle trabeculae. The myocardium is covered externally by a thin epicardium that has two layers, an external mesothelial layer and an internal fibroblastic connective tissue layer. The inner surface of the myocardium is lined by an endocardium of endothelial cells.

The regeneration of amphibian myocardium has been studied most extensively in adult newts. The cardiomyocytes of uninjured newt hearts do not divide at detectable levels. Newts can survive excision of 30–50% of the heart ventricle. Amputation of the ventricular apex, or mincing the tip of the ventricle and replacing the mince, results in the regeneration of myocardium from surviving cardiomyocytes, although restoration of the original ventricular morphology is limited (Oberpriller and Oberpriller, 1971, 1974, 1991; Bader and Oberpriller, 1978; Nag et al., 1979; Rumyantsev, 1991; Neff et al., 1996; Bettencourt-Diaz et al., 2003). **FIGURE 11.12** illustrates how myocardial regeneration takes place in the newt.

Histological, labeling, and ultrastructural studies of newt hearts at intervals after simple amputation of the

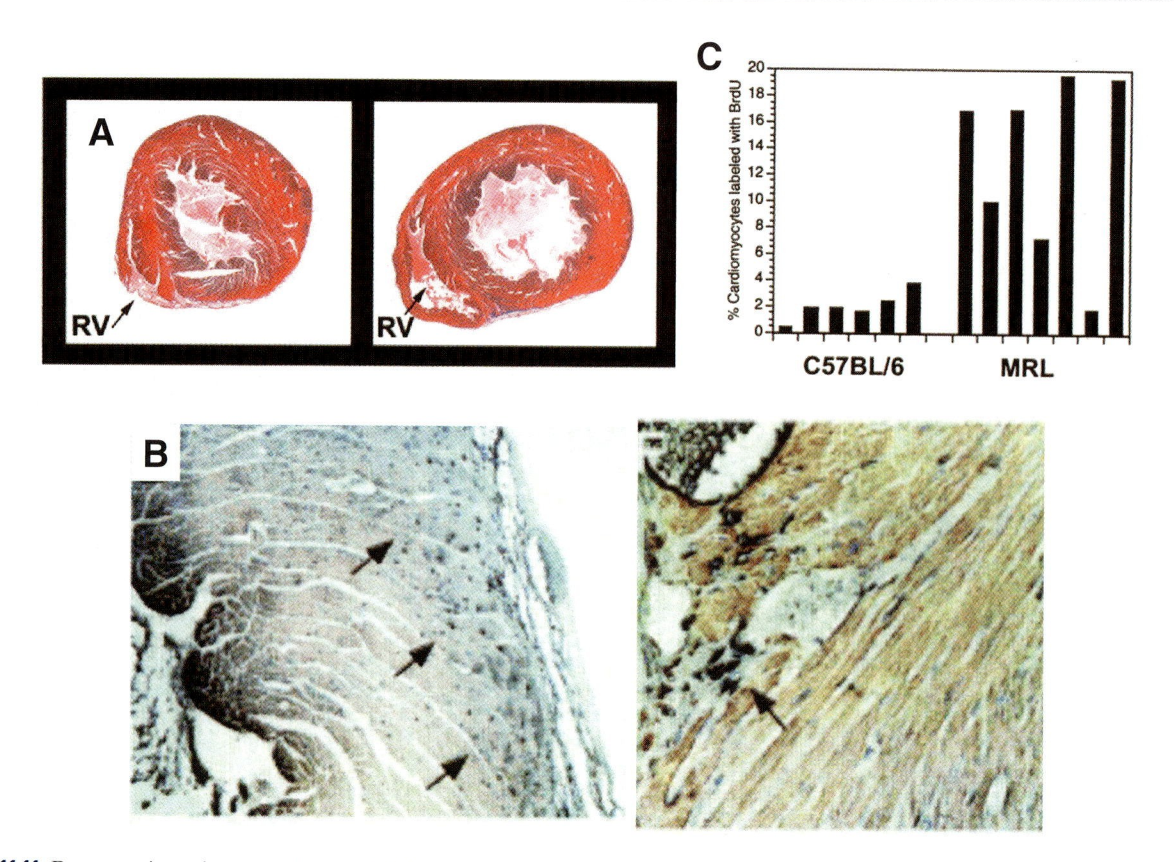


**FIGURE 11.11** Regeneration of myocardium in MRL mice. **(A)** Transverse sections of control and MRL hearts 60 days after cryoinjury. **Left,** control C57BL/6 heart showing scar tissue (arrow) in the injury area of the right ventricle (RV). **Right,** MRL heart with regenerated myocardium in the injury area. **(B) Left,** BrdU$^+$ nuclei (arrows) in cardiomyocytes near the injury site of the MRL right ventricle at 60 days post-injury. **Right,** BrdU$^+$ nuclei are also found in the injury area of the C57BL/6 right ventricle, but primarily in cells of the scar tissue. **(C)** BrdU labeling indices in cardiomyocytes of C57BL/6 and MRL right ventricles 60 days after cryoinjury. Each bar represents an individual animal. **(A)** courtesy of Dr. Ellen Heber-Katz; **(B, C)** Reproduced with permission from Leferovich et al., Heart regeneration in adult MRL mice. Proc Natl Acad Sci USA 98:9830–9835. Copyright 2001, National Academy of Science, USA.

ventricle suggested that uninjured cardiomyocytes at the edge of the wound divide after undergoing partial dedifferentiation (Oberpriller and Oberpriller, 1971, 1974; Nag et al., 1979, 1983; Rumyantsev, 1991). Compact myofibrils were disrupted in these cardiomyocytes, and myofibrillar patches with an embryonic type of organization were scattered throughout the cytoplasm (Nag et al., 1979). Intercalated discs are lost (Rumyantsev, 1991), resulting in the formation of mononucleate cells that re-enter the cell cycle. *In vitro* studies indicated that the dividing newt cardiomyocytes maintain their contractility after each mitosis, just like embryonic and fetal cardiomyocytes (Nag et al., 1979). $^3$H-thymidine labeling and counts of mitotic figures *in vivo* indicated that the percentage of cardiomyocytes undergoing DNA synthesis was 9–10% and the mitotic index was 0.9% (Oberpriller and Oberpriller, 1974). The outcome, however, was not perfect regeneration, but repair with scar tissue that included a limited number of new cardiomyocytes, suggesting that most of the labeling and mitosis was by fibroblasts.

Much higher DNA labeling (28%) and mitotic index (3.19%) was achieved at 20 days postoperation by mincing the amputated ventricle tip and replacing the mince on the wound surface (Bader and Oberpriller, 1978). In another study, 47% of the nuclei in the mince were labeled with $^3$H-thymidine at 45 days after operation (Oberpriller et al., 1988). In the 500-µm region of the ventricular wall adjacent to the mince, the labeling index was 9.25%. There was a low rate of binucleation and nuclear polyploidy in the minced ventricular myocardium, indicating mitosis with cytokinesis, rather than simple karyokinesis. The minced muscle rebuilt a continuous wall composed primarily of cardiac muscle that was continuous with the remaining ventricular myocardium. This reconstituted muscle, however, beat asynchronously with the remaining ventricle. In these studies, it was difficult to distinguish between dividing cardiomyocytes and other cell types such as endothelial cells and fibroblasts. However, the fact that during myocardial regeneration by minced tissue, the number of cardiomyocytes at 16 days postamputation decreased to

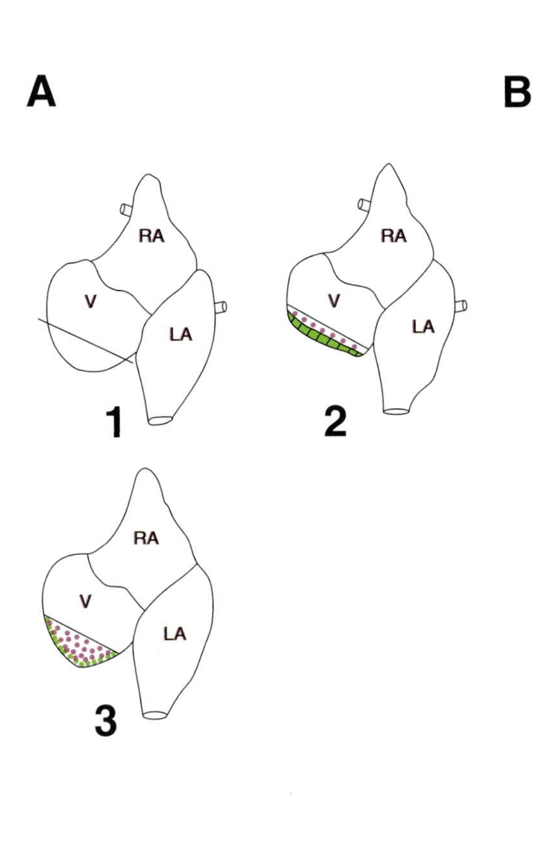


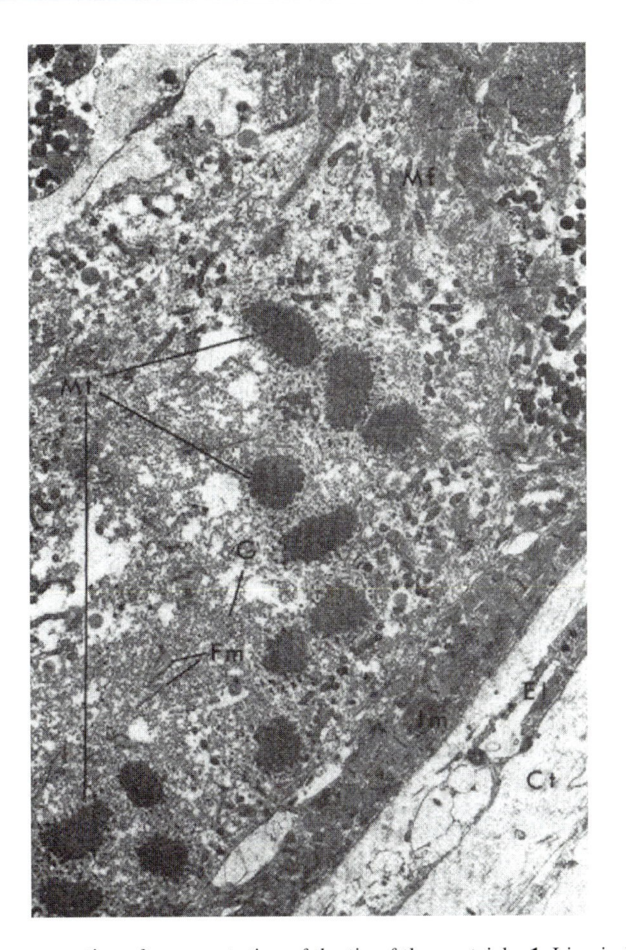


**FIGURE 11.12** Newt heart muscle regeneration. **(A)** Diagrams of heart regeneration after amputation of the tip of the ventricle. **1:** Line indicates amputation level. V = ventricle; RA = right atrium; LA = left atrium. **2:** The epicardium (green) regenerates first, followed by partial dedifferentiation of the myocardium (purple). **(3)** The dedifferentiated cells proliferate and redifferentiate into new cardiomyocytes. **(B)** Electron micrograph of a partially dedifferentiated dividing cardiomyocyte of a newt heart muscle explant after 14 days of culture *in vitro*. Mt = mitotic chromosomes. The cell contains patches of myofibrils (Mf), free myofilaments (Fm), and glycogen (G). Im = intact muscle cell; El = epithelial cell layer; Ct = loose connective tissue layer. **(B)** Reproduced with permission from Nag et al., DNA synthesis and mitosis in adult amphibian cardiac muscle cells in vitro. Science 205:1281–1282. Copyright 1979, AAAS.

21% of the starting number (probably by cell death) but then increased to 50% of the starting number at 50 days postamputation indicated that cardiomyocytes were indeed proliferating (Bader and Oberpriller, 1979).

The myocardium of the adult axolotl is also capable of regeneration after amputation of 10–15% of the ventricle (Flink, 2002). The epicardium regenerates first, under the fibrin clot. BrdU labeling studies showed that cells in both layers of the ventricular epicardium synthesized DNA at high frequency (up to 74% in the mesothelium). Ventricular cardiomyocytes labeled at 12.8%. The highest labeling (up to 50%) was in a 50μ wide zone subjacent to the regenerated epicardium, but labeling extended in a gradient as far as 750μ into the ventricular myocardium. BrdU-labeled cells also immunostained with MF20, an antibody to sarcomeric myosin, although the frequency of double-labeled cells compared to total BrdU-labeled cells was not reported. Patches of atrial cardiomyocytes (14%) and epicardium (64% mesothelium and 29% connective tissue layer) also incorporated BrdU, suggesting that amputation of ventricular tissue leads to the production of mitogenic factors that diffuse and activate entry into the cell cycle. Mitotic indices were not reported, so it is not known whether most of the cells that synthesize DNA also undergo mitosis.

The anuran myocardium, too, can mount a regenerative response to injury. Cardiomyocytes in the injured frog ventricular myocardium undergo DNA synthesis at a frequency of about 17% (Rumyantsev, 1977). An interesting natural analog of ventricular amputation is observed in the South American toad, *Bufo arenarum*. These toads lose 30% of their cardiomyocytes during winter hibernation, but the loss is fully made up during the summer (Aoki et al., 1989), though the mechanism by which this is achieved is unknown.

### b. Zebrafish

The zebrafish ventricle is organized into two layers of cardiomyocytes, a thin, vascularized external layer of compact cells, and an internal layer of cells organized into trabeculae (Poss et al., 2002). **FIGURE 11.13** illustrates regeneration of the fish ventricle. The ventricle regenerates from the edge of the wound after removal of 20–30% of its apex (Poss et al., 2002; Raya et al., 2003). The epicardium regenerates first, followed by the regeneration of new myocardium to cover the wound surface. The new myocardial tissue then expands to complete the restoration of the ventricle. Poss et al. (2002) assayed BrdU incorporation in regenerating ventricle under continuous and pulse labeling conditions. Continuous labeling showed that BrdU incorporation peaked at 32% of cells by 14 days after amputation. Incorporation decreased to 7% by 60 days, when regeneration was complete. Similar BrdU incorporation studies on fish transgenic for EGFP under the control of the *mlc2a* promoter showed a significant accumulation of EGFP/BrdU-positive cells with the morphological characteristics of cardiomyocytes in the myocardium next to the amputation plane (Raya et al., 2003). Cycling cells were located in both compact and trabecular layers at first, but by 14 days, most of the cycling cells

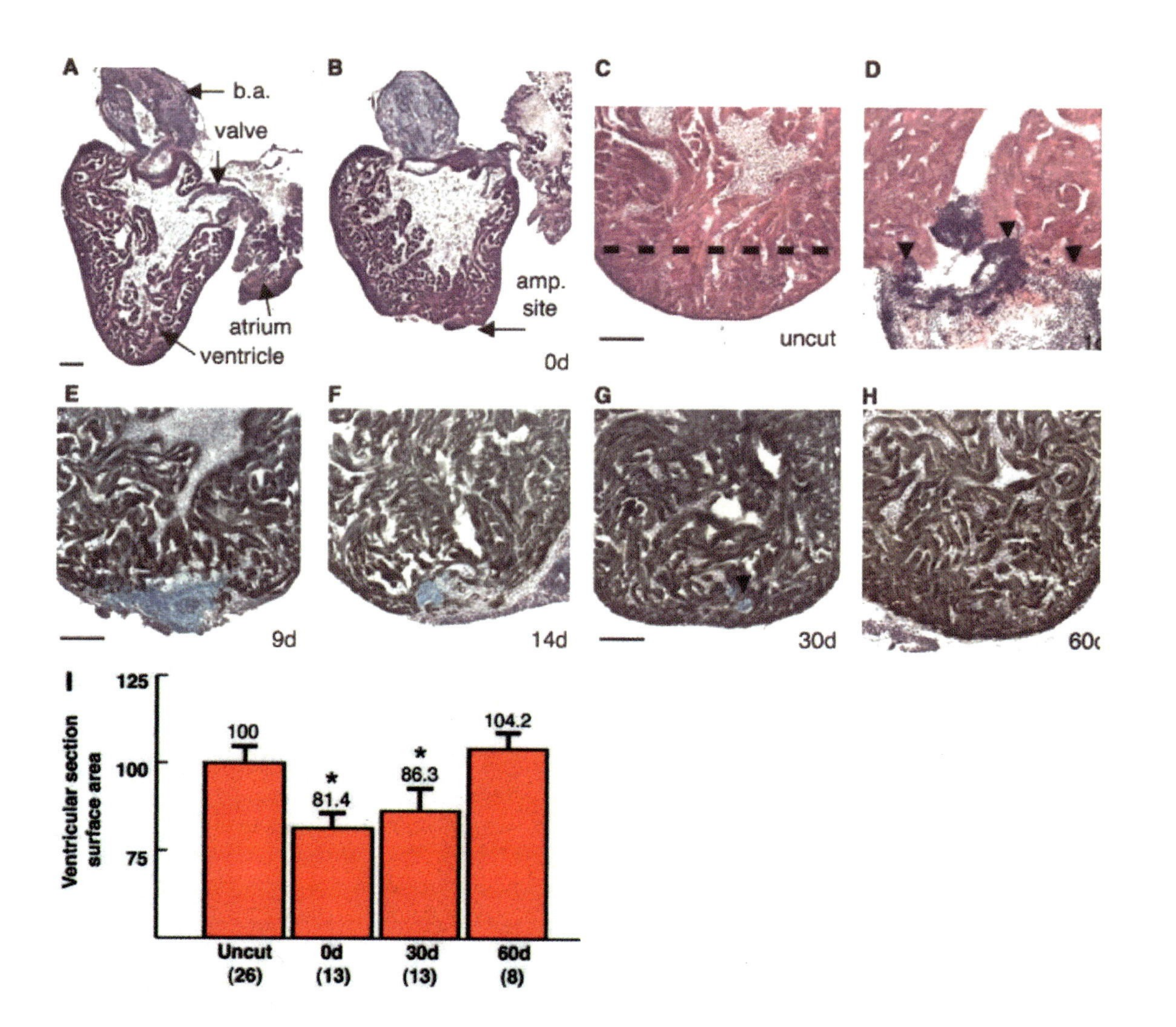


**FIGURE 11.13** Heart regeneration in the zebrafish. **(A)** Longitudinal section through an intact heart. ba, bulbus arteriosus. **(B)** Heart after amputation of 20% of ventricle. **(C)** Higher magnification of unamputated ventricular apex, showing the level of amputation. **(D)** One day post-amputation, showing plasma clot filled with erythrocytes (arrowheads). **(E–H)** The heart muscle is stained for myosin heavy chain (brown) and fibrin (blue). The fibrin disappears and is replaced by new cardiac muscle. **(F)** Area of ventricle as measured from largest section after removal of ~20% of the ventricle (0 days). By 60 days the area of the section has been restored to normal. Numbers in parentheses = number of hearts examined. Reproduced with permission from Poss et al., Heart regeneration in zebrafish. Science 298:2188–2190. Copyright 2002, AAAS.

were located in the compact muscle layer (Poss et al., 2002).

Pulse labeling experiments revealed an increase in the number of labeled cells after the pulse, with labeled cells being found adjacent to one another, and also showed that proliferation was diffuse during the regeneration of the initial ventricular wall, then occurred in a gradient that was highest in the cardiomyocytes subjacent to the regenerated epicardium (Poss et al., 2002). Furthermore, although low levels of expression of *nkx2.5* and *Tbx5* transcripts, early markers of the embryonic cardiomyocyte lineage are detectable in adult myocardium, these markers are not upregulated during regeneration (Raya et al., 2003). These observations suggest that the heart muscle regenerates by the division of adult cardiomyocytes. Whether or not partial dedifferentiation of the cardiomyocytes is involved in this division, as in newt cardiomyocytes, is not known. Finally, blocking cardiomyocyte division at the restrictive temperature in fish with a temperature-sensitive mutation in Mps1, a mitotic checkpoint kinase, inhibited regeneration and led to scarring (Poss et al., 2002), consistent with the conclusion of Leferovich et al. (2001) that the relative frequency of cardiomyocyte versus fibroblast division determines whether regeneration or scarring will take place.

Several genes were found to be strongly upregulated during fish heart regeneration (Raya et al., 2003). *MsxB* and *C* reached a peak of expression in the myocardium at two weeks postamputation. Expression of *notch1b* and its ligand *deltaC* was upregulated in the endocardium during the first week postamputation, suggesting that cardiomyocytes take on at least some stem cell attributes. Interestingly, these genes were not expressed in the developing embryonic heart, suggesting that the genetic program of regeneration does not recapitulate the embryonic program. However, *MsxB* and *C* genes were also upregulated after removal of 50% of the prospective atrium from 24-hr postfertilization embryos, indicating that embryonic cardiac cells respond to injury by initiating the adult program of gene activity.

### c. Might Stem Cells Contribute to the Regenerating Myocardium in Amphibians and Fish?

No cardiac stem cells have been detected in the amphibian or fish heart, so it has been assumed that the regeneration observed is due solely to the proliferation of partially dedifferentiated cardiomyocytes. However, with the discovery of stem cells in the mammalian heart, it might be instructive to analyze the uninjured amphibian and fish myocardium for cells with general stem cell markers to determine whether any putative stem cells might be present. Not all newt cardiomyocytes are equal in their ability to proliferate (Bettencourt-Diaz et al., 2003). Although the majority of cultured newt cardiomyocytes can enter S-phase, more than half of them stably arrest at $G_2$ or during cytokinesis. About a third of the cells complete mitosis, some of which enter successive cell divisions. These observations suggest that only a restricted subset of cardiomyocytes retains the proliferative potential to make a regenerative response to injury. What is different about these cells from other newt cardiomyocytes is not known. They may be more responsive to triggers of dedifferentiation, or they may be more stemlike in their phenotype. Another interesting question is whether thrombin activates a protein that triggers re-entry of cardiomyocytes into the cell cycle, as is the case for the regenerating lens and limb.

## SUMMARY

The blood, blood vessels, and the heart all arise from the mesodermal germ layer of the embryo. Blood cells are regenerated by HSCs in the bone marrow and blood vessels are regenerated both by endothelial cells in the walls of venules and circulating EnSCs from the bone marrow. Mammalian myocardium harbors cardiac stem cells that initiate a regenerative response after injury, but this response is quickly overtaken by fibroblast proliferation and scar formation. By contrast, amphibian and fish myocardium are able to regenerate new heart muscle.

HSCs are small cells with the surface phenotype [CD34 c-Kit Sca-1 VEGFR2]$^+$ Thy-1$^{lo}$ Lin$^-$ and express the transcription factor Runx-1. They are thought to divide asymmetrically to self-renew, while giving rise to a common erythroid/myeloid progenitor that gives rise to the blood cell lineages and a common lymphoid progenitor that gives rise to the cells of the immune system. Each step in the differentiation of these lineages is controlled by combinations of cytokines in the microniche that result in differential gene activity. The HSC is maintained in a quiescent state by interaction with a subset of N-cadherin$^+$ CD45$^-$ osteoblasts in the marrow stroma. This interaction is mediated by the Tie-2 receptor on HSCs and its ligand Ang-1 on osteoblasts. To divide, the HSC increases its motility through the binding of stromal-produced KitL to c-Kit on the HSC surface and moves to a proliferation-permissive vascular portion of the niche. Wnt3a and Notch signaling are necessary for proliferation and self-renewal, as is Bmi, a protein that represses the expression of the p16 and p19 genes, which suppress proliferation and promote apoptosis, respectively.

Blood vessels regenerate primarily from endothelial cells of injured vessel walls. In uninjured endothelium, the endothelial walls are stabilized by the interaction of Ang-1 on the surface of pericytes and Tie-2 on the surface of endothelial cells and by the cleavage product of collagen XVIII, endostatin, which arrests endothelial cells in $G_1$ by antagonizing the binding of VEGF to its receptor, VEGFR2. Injury activates the endothelial cells by downregulating endostatin and by thrombin cleavage of the extracellular domain of PAR-1. The shortened domain activates G proteins, leading to the autocrine induction of Ang-2. Ang-2 is antagonistic to Ang-1 and blocks the interaction between Ang-1 and Tie-2, destabilizing the vessel wall. Other growth factors that activate endothelial cells are FGF-2, TGF-β1, IL-8, and TNF-α. The endothelial cells that migrate may be a subpopulation in the vessel wall similar to EnSCs. Whatever their origin, the cells lose their intercellular junctions and express proteases that break down their basement membrane. In the presence of VEGF synthesized by the epidermis of the healing wound, the cells proliferate and migrate out as cords into the fibrin matrix of the wound. PD-EGCF and TNF-α are chemotactic for endothelial cells; TGF-α, TGF-β, FGF 1 and 2, and PDGF-B stimulate proliferation. The proliferating cells then rearrange themselves into endothelial tubes, a process that is mediated by Ln and a fibroblast-secreted protein, Egfl7, and ECM-associated proteins. Circulating EnScs have the phenotype $[CD133\ VEGFR2]^+$ and express the receptor for SDF-1. Injured tissue produces SDF-1, which recruits EnSCs to the site of injury. Here, they are incorporated into the new capillaries sprouting from the walls of injured vessels.

Mammalian cardiac muscle initiates a regenerative response to ischemic injury that is not sustained. Stem cells in the myocardium proliferate, but fibroblast proliferation is faster, suppressing the regenerative response and creating a scar. There are three distinct cardiac stem cell phenotypes that can differentiate *in vivo* and *in vitro* into cardiomyocytes: $[c\text{-Kit Sca-1}]^+$, $Sca\text{-}1^+$ $[c\text{-Kit Lin}]^-$, and $Isl\text{-}1^+$ $[Sca\text{-}1\ c\text{-Kit}]^-$. The relationship among these three subpopulations is not clear. They may regenerate different kinds of tissue, since the $Isl\text{-}1^+$ cells were found only in those parts of the heart that had an embryonic contribution from the secondary heart field. Heart muscle regenerates in the MRL/MpJ mouse after cryogenic infarction. Revascularization is greater at the infarct site. The frequency of mitosis in these injured hearts is 10–20%, compared to 1–3% in wild-type animals, suggesting that whether injury results in regeneration or scarring is a function of the relative frequency of mitosis of cardiomyocytes versus fibroblasts. This animal model offers the opportunity to investigate how stem cell populations and/or injury environments differ in regenerating versus nonregenerating mammalian heart tissue.

Two other useful research models of myocardial regeneration are the urodele and fish hearts. Both appear to regenerate by the partial dedifferentiation and proliferation of cardiomyocytes after amputation of the tip of the ventricle. In the adult newt, only about one-third of the cells that enter S phase complete mitosis, suggesting that a restricted subset of cardiomyocytes is capable of making a regenerative response. How these cells differ from other cardiomyocytes is unknown. Although low levels of *nkx2.5* and *Tbx5* transcripts, which are early markers of the embryonic cardiomyocyte lineage, are detectable in adult fish hearts, they are not upregulated during regeneration, suggesting that it is adult cardiomyocytes that regenerate the myocardium. On the other hand, *MsxB* and *C, Notch1b,* and *DeltaC* expression is upregulated during the first week after amputation of the fish ventricle, suggesting that cardiomyocytes express stem cell markers, perhaps through dedifferentiation.

## REFERENCES


Adolfsson J, Mansson R, Buza-Vidas N, Hultquist A, Liuba K, Jensen CT, Bryder D, Yang L, Borge O-J, Thoren LAM, Anderson K, Sitnicka E, Sasaki Y, Sigvardsson M, Jacobsen SEW (2005) Identification of $Flt3^+$ lympho-myeloid stem cells lacking erythro-megakaryocytic potential: A revised road map for adult blood lineage commitment. Cell 121:295–306.

Alberts B, Bray D, Lewis J, Raff M, Roberts K, Watson JD (1994) Molecular Biology of the Cell (3rd ed). New York: Garland Pub Co.

Antonchuk J, Sauvageau G, Humphries RK (2002) *HOXB4*-induced expansion of adult hematopoietic stem cells ex vivo. Cell 109: 39–45.

Anversa P, Nadal-Ginard B (2002) Myocyte renewal and ventricular remodeling. Nature 415:240–243.

Aoki A, Maldonado CA, Forsmann WG (1989) Seasonal changes of the endocrine heart. In: Functional Morphology of the Endocrine Heart, Forsmann W-G, Scheuermann DW, Alt J (eds.). New York: Steinkopff Verlag-Springer-Verlag, pp 61–68.

Arai F, Hirao A, Ohmura M, Sato H, Matsuoka S, Takubo K, Ito K, Koh GY, Suda T (2004) Tie2/angiopoietin-1 signaling regulates hematopoietic stem cell quiescence in the bone marrow niche. Cell 118:149–161.

Asahara T, Murohata T, Sullivan A, Silver M, van der Zee R, Li T, Witzenbichler B, Schatteman G, Isner JM (1997) Isolation of putative progenitor endothelial cells for angiogenesis. Science 275:964–967.

Bader D, Oberpriller JO (1978) Repair and reorganization of minced cardiac muscle in the adult newt (*Notopthalmus viridescens*). J Morph 155:349–358.

Baumann CI, Bailey AS, Li W, Ferkowicz MJ, Yoder MC, Fleming WH (2004) PECAM-1 is expressed on hematopoietic stem cells throughout ontogeny and identifies a population of erythroid progenitors. Blood 104:1010–1016.

Beltrami AP, Urbanek K, Kajstura J, Yan S-M, Finato N, Bussani R, Nadal-Ginard B, Silvestri F, Leri A, Beltrami A, Anversa P (2001) Evidence that human cardiac myocytes divide after myocardial infarction. New Eng J Med 344:1750–1756.

Beltrami AP, Barlucchi L, Torella D, Baker M, Limana F, Chimenti S, Kasahara H, Rota M, Musso E, Urbanek K, Leri A, Kajstura J, Nidal-Ginard B, Anversa P (2003) Adult cardiac stem cells are multipotent and support myocardial regeneration. Cell 114:763–776.

Berardi A, Wang A, Levine JD, Lopez P, Scadden D (1995) Functional isolation and characterization of human hematopoietic stem cells. Science 267:104–108.

Berger AC, Feldman AL, Gnant MF, Kruger EA, Sim BK, Hewitt S, Figg WD, Alexander HR, Libutti SK (2000) The angiogenesis inhibitor, endostatin, does not affect murine cutaneous wound healing. J Surg Res 91:26–31.

Bettencourt-Diaz M, Mittnacht S, Brockes JP (2003) Heterogeneous proliferative potential in regenerative adult newt cardiomyocytes. J Cell Sci 116:4001–4009.

Bishop JE, Laurent GJ (1995) Collagen turnover and its regulation in the normal and hypertrophying heart. Eur Heart J 16C:38–44.

Bloch W, Huggel K, Sasaki T, Grose R, Bugnon P, Addicks K, Timpl R, Werner S (2000) The angiogenesis inhibitor endostatin impairs blood vessel maturation during wound healing. FASEB J 14:2373–2376.

Borisov AB (1998) Cellular mechanisms of myocardial regeneration. In: Ferretti P and Geraudie J (eds.). Cellular and Molecular Basis of Regeneration. New York: John Wiley & Sons, pp 335–354.

Borisov AB (1999) Regeneration of skeletal and cardiac muscle in mammals: Do non-primate models resemble human pathology? Wound Rep Reg 7:26–35.

Breier G, Albrecht U, Sterrer J, Risau W (1992) Expression of vascular endothelial growth factor during embryonic angiogenesis and endothelial cell differentiation. Development 114:521–532.

Cai CL, Liang X, Shi Y, Chu PH, Pfaff SL, Chen J, Evans S (2003) Isl1 identifies a cardiac progenitor population that proliferates prior to differetiation and contributes a majority of cells to the heart. Dev Cell 5:877–889.

Calvi LM, Adams GB, Weibrecht KW, Weber JM, Olson DP, Knight MC, Martin RP, Schipani E, Diveti P, Bringhurst FR, Milner LA, Kronenberg HM, Scadden DT (2003) Osteoblastic cells cells regulate the haematopoietic stem cell niche. Nature 425:841–846.

Charbord P (2004) Stromal support of hematopoiesis. In: Stem Cells Handbook, Sell S (ed.). Towota NJ: Humana Press, pp 143–154.

Cheng T, Rodrigues N, Shen H, Yang Y-g, Dombkowski D, Sykes M, Scadden DT (2000) Hematopoietic stem cell quiescence maintained by $p21^{cip1/waf1}$. Science 287:1804–1808.

Clark RAF (1996) Wound repair: Overview and general considerations. In: Clark RAF (ed.). The Molecular and Cellular Biology of Wound Repair, New York: Plenum Press, pp 3–50.

Clauss M, Gerlach M, Gerlach H, Brett J, Wang F, Familletti PC, Pan E, Oleander JV, Connoly DT, Stern D (1990) Vascular permeability factor: A tumor-derived polypeptide that induces endothelial cell and monocyte procoagulant activity and promotes monocyte migration. J Exp Med 172:1535–1545.

Claycomb WC (1991) Proliferative potential of the mammalian ventricular cardiac muscle cell. In: Oberpriller JO, Oberpriller JC, Mauro A (eds.). The Developmental and Regenerative Potential of Cardiac Muscle. New York: Harwood, pp 351–363.

Claycomb WC (1992) Control of cardiac muscle cell division. Trends Cardiovasc Med 2:231–236.

Coughlin SR (2000) Thrombin signaling and protease-activated receptors. Nature 407:258–264.

Coussens LM, Fingleton B, Matrisian LM (2002) Matrix metalloproteinase inhibitors and cancer: Trials and tribulations. Science 295:2387–2392.

Davis S, Aldrich A, Jones P, Acheson A, Compton DL, Jain V, Ryan E, Bruno J, Radziejewski C, Maisonpierre PC, Yancopoulos GD (1996) Isolation of angiopoietin-1, a ligand for the TIE2 receptor, by secretion-trap expression cloning. Cell 87:1161–1169.

Dieterlen-Lievre F, Pardanaud L, Yassine F, Cormier F (1988) Early hematopoietic cells in the avian embryo. J Cell Sci S10:29–44.

Dumont DJ, Yamaguchi TP, Conlon RA, Rossant J, Breitman ML (1992) tek, a novel tyrosine kinase gene located on mouse chromosome 4, is expressed in endothelial cells and their presumptive precursors. Oncogene 7:1471–1480.

Eppenberger ME, Hauser J, Baechi T, Schaub M, Brunner UT, Dechesne CA, Eppenberger HM (1988) Immunocytochemical analysis of the regeneration of myofibrils in long-term cultures of adult cardiomyocytes of the rat. Dev Biol 130:1–15.

Flink IL, Oana S, Maitra N, Bahl JJ, Morkin E (1998) Changes in E2F complexes containing retinoblastoma protein family members and increased cyclin-dependent kinase inhibitor activities during terminal differentiation of cardiomyocytes. J Mol Cardiol 30: 563–578.

Flink IL (2002) Cell cycle reentry of ventricular and atrial cardiomyocytes and cells within the epicardium following amputation of the ventricular apex in the axolotl, *Ambystoma mexicanum*: confocal microscopic immunofluorescent image analysis of bromodeoxyuridine-labeled nuclei. Anat Embryol 205:235–244.

Folkman J, Klagsbrun M (1987) Angiogenic factors. Science 235:442–447.

Folkman J, D'Amore PA (1996) Blood vessel formation: What is its molecular basis? Cell 87:1153–1155.

Gerber H-P, Malik AK, Solar GP, Sherman D, Liang XH, Meng G, Hong K, Marsters JC, Ferrara N (2002) VEGF regulates haematopoietic stem cell survival by an internal autocrine loop mechanism. Nature 417:954–958.

Gilbert SF (2003) Developmental Biology, 7th ed. Sunderland: Sinauer Assoc Inc.

Goldsmith EC, Hoffman A, Morales MO, Potts JD, Price RL, McFadden A, Rice M, Borg TK (2004) Organization of fibroblasts in the heart. Dev Dynam 230:787–794.

Gudesius G, Miragoli M, Thomas ST, Rhor S (2003) Coupling of cardiac electrical activity over extended distances by fibroblasts of cardiac origin. Circ Res 93:421–428.

Hagedorn M, Balke M, Schmidt A, Bloch W, Kurz H, Javerzat S, Rousseau B, Wilting J, Bikfalvi A (2004) VEGF coordinates interaction of pericytes and endothelial cells during vasculogenesis and experimental angiogenesis. Dev Dynam 230:23–33.

Ham AW, Cormack DH (1979) Histology ($8^{th}$ ed). Philadelphia: JB Lippincott.

Hanai JI, Dhanabal M, Karumanchi SA, Albanese C, Waterman M, Chan B, Ramchandran R, Pestell R, Sukhatme VP (2002) Endostatin causes G1 arrest of endothelial cells through inhibition of cyclin 1. J Biol Chem 277:16464–16469.

Heissig B, Hattori K, Dias S, Friedrich M, Ferris B, Hackett NR, Crystal RG, Besmer P, Lyden D, Moore MAS, Werb Z, Rafii S (2002) Recruitment of stem and progenitor cells from the bone marrow niche requires MMP-9 mediated release of Kit-ligand. Cell 109:625–637.

Hiraoka N, Allen E, Apel IJ, Gyetko MR, Weiss SJ (1998) Matrix metalloproteinases regulate neovascularization by acting as pericellular fibrinolysins. Cell 95:365–377.

Holmgren L, Glaser A, Pfeifer-Ohlsson S, Ohlsson R (1991) Angiogenesis during human extraembryonic development involves the spatiotemporal control of PDGF ligand and receptor gene expression. Development 113:749–754.

Huang EJ, Nocka KH, Buck J, Besmer P (1992) Differential expression and processing of two cell associated forms of the kit-ligand: KL-1 and KL-2. Mol Biol Cell 3:349–362.

Huber TL, Kouskoff V, Fehling HJ, Palis J, Keller G (2004) Haemangioblast commitment is initiated in the primitive streak of the mouse embryo. Nature 432:625–630.

Ingber DE, Folkman J (1989) How does extracellular matrix control capillary morphogenesis? Cell 58:803–805.

Ingram DA, Mead LE, Moore DB, Woodard W, Fenoglio A, Yoder MC (2004) Vessel wall derived endothelial cells rapidly proliferate because they contain a complete hierarchy of endothelial progenitor cells. Blood First Edition Paper, published online: DOI 10.1182/blood-2004-08-3057.

Ishikawa F, Miyazoni K, Hellman U, Drexler H, Wernstedt C, Hagiwara K, Usuki K, Takaku F, Risau W, Heldin C-K (1989) Identification of angiogenic activity and the cloning and expression of platelet-derived endothelial growth factor. Nature 338: 557–562.

Jain RK (2003) Molecular regulation of blood vessel maturation. Nature Med 9:685–693.

Juliano and Haskill S (1993) Signal transduction from the extracellular matrix. J Cell Biol 120:577–585.

Keck PJ, Hauser SD, Krivi G, et al. (1989) Vascular permeability factor, an endothelial cell mitogen related to PDGF. Science 246:1309–1312.

Kiel MJ, Yilmaz OH, Iwashita T, Yilmaz OH, Terhorst C, Morrison SJ (2005) SLAM family receptors distinguish hematopoietic stem and progenitor cells and reveal endothelial niches for stem cells. Cell 121:1109–1121.

Kim YM, Hwang S, Kim YM, Pyun BJ, Kim TY, Lee ST, Gho YS, Kwon YG (2002) Endostatin blocks VEGF-mediated signaling via direct interaction with KDR/Flk-1. J Biol Chem 277:27872–27879.

Kinnaird T, Stabile E, Burnett MS, Shou M, Lee CW, Barr S, Fuchs S, Epstein SE (2004) Local delivery of marrow-derived stromal cells augments collateral perfusion through paracrine mechanisms. Circulation 109:1543–1549.

Kirshenbaum L, Abdellatif M, Chakraborty S, Schneider MD (1996) Human E2F-1 reactivates cell cycle progression in ventricular myocytes and represses cardiac gene transcription. Dev Biol 179:402–411.

Koch AE, Polverini PJ, Kunkel SL, Harlow LA, DiPietro LA, Elner VM, Elner SG, Strieter RM (1992) Interleukin-8 as a macrophage-derived mediator of angiogenesis. Science 1992; 258:1798–1801.

Laflamme MA, Murry CE (2005) Regenerating the heart. Nature Biotech 23:845–856.

Laugwitz K-L, Moretti A, Lam J, Gruber P, Chen Y, Woodard S, Lin L-Z, Cai C-L, Lu MM, Reth M, Platoshyn O, Yuan J X-J, Evans S, Chien KR (2005) Postnatal isl1$^+$ cardioblasts enter fully differentiated cardiomyocyte lineages. Nature 433:647–653.

Lawson ND, Scheer N, Pham VN, Kim C, Chitnis AB, Campos-Ortega JA, Weinstein BM (2001) Notch signaling is required for arterial-venous differentiation during embryonic vascular development. Development 128:3675–3683.

Lawson ND, Weinstein BM (2002) Arteries and veins: making a difference with zebrafish. Nature Rev Genet 3:674–682.

Leferovich J, Bedelbaeva K, Samulewicz S, Zhang X-M, Zwas D, Lankford EB, Heber-Katz E (2001) Heart regeneration in adult MRL mice. Proc Natl Acad Sci USA 98:9830–9835.

Lessard J, Sauvageau G (2003) *Bmi-1* determines the proliferative capacity of normal and leukaemic stem cells. Nature 423:255–260.

Leung DW, Cachianes G, Kuang W-J, Goeddel DV, Ferrara N (1989) Vascular endothelial growth factor is a secreted angiogenic mitogen. Science 246:1306–1309.

Levesque JP, Takamatsu Y, Nilsson SK, Haylock DN, Simmons PJ (2001) Vascular cell adhesion molecule-1 (CD106) is cleaved by neutrophil proteases in the bone marrow following hematopoietic progenitor mobilization by granulocyte colony-stimulating factor. Blood 98:1289–1297.

Lindahl P, Johansson BR, Leveen P, Betsholtz C (1997) Pericyte loss and microaneurysm formation in PDGF-B-deficient mice. Science 277:242–245.

Loges S, Fehse B, Brockmann MA, Lamszus K, Butzal M, Guckenbiehl M, Schuch G, Ergun S, Fischer U, Zander AR, Hossfeld DK, Fiedler W, Gehling UM (2004) Identification of the adult human hemangioblast. Stem Cells Dev 13:229–242.

Madri JA, Asankar S, Romanic AM (1996) Angiogenesis. In: Clark RAF (ed.). The Molecular and Cellular Biology of Wound Repair, 2nd ed. New York: Plenum Press, pp 355–372.

Maisonpierre PC, Suri C, Jones PF, Bartunkova S, Wiegand SJ, Radziejewski C, Compton D, McClain J, Aldrich TH, Papadopoulos N, Daly TJ, David S, Sato T, Yancopoulos G (1997) Angiopoietin-2, a natural antagonist for Tie2 that disrupts in vivo angiogenesis. Science 277:55–60.

McCulloch EA (2004) Normal and leukemic hematopoietic stem cells and lineages. In: Stem Cells Handbook, Sell S (ed.). Towota NJ: Humana Press, pp 119–132.

Medvinsky A, Dzierzak E (1996) Definitive hematopoiesis is autonomously initiated by the AGM region. Cell 86:897–906.

Melchers F, Rolink A (2001) Hematopoietic stem cells: Lymphopoiesis and the problem of commitment versus plasticity. In: Stem Cell Biology, Marshak DR, Gardner R, Gottleib D (eds.). Cold Spring Harbor: Cold Spring Laboratory Press, pp 307–327.

Messina E, De Angelis L, Frati G, Morrone S, Chimenti S, Fiordaliso F, Salio M, Battaglia M, Latronico MVG, Coletta M, Vivarelli E, Frati L, Cossu G, Giacomello A (2004) Isolation and Expansion of adult cardiac stem cells from human and murine heart. Circ Res 95:911–921.

Millauer B, Wizigmann-Voss S, Schnürch H, Martinez R, Moller NPH, Risau W, Ullrich A (1993) High affinity VEGF binding and developmental expression suggest that Flk-1 is a major regulator of vasculogenesis and angiogenesis. Cell 72:835–846.

Miyazono K, Okabe T, Urabe A, Takaku F, Heldin CH (1987) Purification and properties of an endothelial cell growth factor from human platelets. J Biol Chem 262:4098–4103.

Moore MAS, Metcalf D (1970) Ontogeny of the hematopoietic system: yolk sac origin of *in vivo* and *in vitro* colony forming cells in the developing mouse embryo. Br J Haem 18: 279–296.

Nag AC (1991) Reactivity of cardiac muscle cells under traumatic conditions. In: Oberpriller JO, Oberpriller JC, Mauro A (eds.). The Development and Regenerative Potential of Cardiac Muscle. New York: Harwood, pp 313–331.

Nag AC, Healy CJ, Cheng M (1979) DNA synthesis and mitosis in adult amphibian cardiac muscle cells in vitro. Science 205:1281–1282.

Neff AW, Dent AE, Armstrong JB (1996) Heart development and regeneration in urodeles. Int J Dev Biol 40:719–725.

Nicosia RF, Tuszynski GP (1994) Matrix-bound thrombospondin promotes angiogenesis *in vitro*. J Cell Biol 124:183–193.

Nillson SK, Johnston HM, Coverdale JA (2001) Spatial localization of transplanted hemopoietic stem cells: inferences for the localization of stem cell niches. Blood 97:2293–2299.

North TE, de Bruijn MFTR, Stacy T, Talebian L, Lind E, Robin C, Binder M, Dzierzak E, Speck N (2002) Runx1 expression marks long-term repopulating hematopoietic stem cells in the midgestation mouse embryo. Immunity 16:661–672.

Oberpriller JO, Oberpriller JC (1991) Cell division in adult newt cardiac myocytes. In: Oberpriller JO, Oberpriller JC, Mauro A (eds.). The developmental and regenerative potential of cardiac muscle. New York: Harwood, pp 293–312.

Oberpriller JO, Oberpriller JC (1974) Response of the adult newt ventricle to injury. J Exp Zool 187:249.

Oberpriller JO, Ferrans VJ, Carroll RJ (1983) DNA synthesis in rat atrial myocytes as a response to left ventricular infarction: an

autoradiographic study of enzymatically dissociated myocytes. J Mol Cell Cardiol 15:31–42.
Oh J, Takahashi R, Kondo S, Mizoguchi A, Adachi E, Sasahara RM, Nishimura S, Imamura Y, Kitayama H, Alexander DB, Ide C, Horan TP, Arakawa T, Yoshida H, Nishikawa S-I, Itoh Y, Seiki M, Itohara S, Takahashi C, Noda M (2001) The membrane-anchored MMP inhibitor RECK is a key regulator of extracellular matrix integrity and angiogenesis. Cell 107:789–800.
Oh H, Bradfute SB, Gallardo TD, Nakamura T, Gaussin V, Mishina Y, Pocius J, Michael LH, Behringer RR, Garry DJ, Entman ML, Schneider MD (2003) Cardiac progenitor cells from adult myocardium:homing, differentiation, and fusion after infarction. Proc Natl Acad Sci USA 100:12313–12318.
Park I-k, Qian D, Kiel M, Becker MW, Pihalja M, Weissman IL, Morrison SJ, Clarke MF (2003) Bmi-1 is required for maintenance of adult self-renewing haematopoietic stem cells. Nature 423:302–305.
Parker LH, Schmidt M, Jin S-W, Gray AM, Beis D, Pham T, Frantz G, Palmieri S, Hillan K, Stainier DYR, de Sauvage FJ, Ye W (2004) The endothelial-cell-derived secreted factor *Egfl7* regulates vascular tube formation. Nature 428:754–758.
Parmacek MS, Epstein A (2005) Pursuing cardiac progenitors: Regeneration redux. Cell 120:295–298.
Ponting I, Zhao Y, Anderson WF (2004) Hematopoietic stem cells. In: Stem Cells Handbook, Sell S (ed.). Towota N: Humana Press, pp 155–162.
Poole TJ, Finkelstein B, Cox C (2001) The role of FGF and VEGF in angioblast induction and migration during vascular development. Dev Dynam 220:1–17.
Poss, KD, Wilson LG, Keating MT (2002) Heart regeneration in zebrafish. Science 298:2188–2141.
Quiani F, Urbanek K, Beltrami AP, Finato N, Beltrami CA, Nadal-Ginard B, Kajstura J, Leri A, Anversa P (2002) Chimerism of the transplanted heart. New Eng J Med 346:5–15.
Rafii S, Lyden D (2003) Therapeutic stem and progenitor cell transplantation for organ vascularization and regeneration. Nature Med 9:702–712.
Rastinejad F, Polverini PJ, Bouck NP (1989) Regulation of the activity of a new inhibitor of angiogenesis by a cancer suppressor gene. Cell 56:345–355.
Raya A, Koth CM, Buscher D, Kawakami Y, Itoh T, Raya RM, Sternik G, Tsai H, Rodriguez-Esteban C, Izpisua-Belmonte JC (2003) Activation of Notch signaling pathway precedes heart regeneration in zebrafish. Proc Natl Acad Sci USA 100: 11889–11895.
Rehman J, Traktuev, Li J, Merfeld-Clauss S, Temm-Grove CJ, Bovenkerk JE, Pell CL, Johnstone BH, Considine RV, March KL (2004) Secretion of angiogenic and antiapoptotic factors by human adipose stromal cells. Circulation 109:1292–1298.
Risau W (1991) Vasculogenesis, angiogenesis and endothelial cell differentiation during embryonic development. In: Sherer G, Auerback R (eds.). The Development of the Vascular System. Basel: Karger, pp 58–68.
Robb L, Choi K (2004) Developmental origin of murine hematopoietic stem cells. In: Stem Cells Handbook, Sell S (ed.). Towota N: Humana Press, pp 133–142.
Ross RS, Borg TK (2001) Integrins and the myocardium. Circ Res 88:1112–1119.
Rumyantsev PP (1992) Reproduction of cardiac myoctyes developing in vivo and its relationship to processes of differentiation. In: Rumyantsev PP (ed.). Growth and Hyperplasia of Cardiac Muscle Cells. New York: Harwood, pp 70–159.
Sage H and Bornstein P (1991) Extracellular matrix proteins that modulate cell-matrix interactions. J Biol Chem 266:831–834.
Sammak PJ, Tran PO, Olson TA (1998) Vascular endothelial cell growth factor expression in endothelial cells is induced by mechanical wounding. J Minn Acad Sci 63:61–64.
Sato TN, Tozawa Y, Deitsch U, Wolburg-Bucholz K, Fujiwara Y, Gendron-Maguire M, Gridley T, Wolurg H, Risau W, Quin Y (1995) Distinct roles of the receptor tyrosine kinases Tie-1 and Tie-2 in blood vessel formation. Nature 376:70–74.
Schreiber AB, Winkler ME, Derynck R (1980) Transforming growth factor-alpha: A more potent angiogenic mediator than epidermal growth factor. Science 232:1250–1253.
Schmidt A, Wenzel D, Ferring I, Kazemi S, Sasaki T, Hescheler J, Timpl R, Addicks K, Fleischmann BK, Bloch W (2004) Influence of endostatin on embryonic vasculo-and angiogenesis. Dev Dynam 230:468–480.
Spangrude GJ, Heimfeld S, Weissman IL (1988) Purification and characterization of mouse hematopoietic cells. Science 241:58–62.
Stetler-Stevenson WG (1999) Matrix metalloproteinases in angiogenesis: a moving target for therapeutic intervention. J Clin Invest 103:1237–1241.
Suri-Jones F, Patan S, Bartunkova S, Maisonpierre PC, Davis S, Sato TN, Yancopoulos GD (1996) Requisite role of angiopoietin-1, a ligand for the TIE2 receptor, during embyonic angiogenesis. Cell 87:1171–1180.
Thompson JA, Anderson KD, DiPietro JM, Zweibel JA, Zametta M, Anderson WF, Maciag T (1988) Site-directed neovessel formation in vivo. Science 246:1349–1352.
Tomanek RJ and Schatteman GC (2000) Angiogenesis: New insights and therapeutic potential. Anat Rec (New Anat) 261: 126–135.
Vikkula M, Boon L, Carraway K, Calvert JT, Diamonti A, Goumnerov B, Pasyk KA, Marchuk DA, Warman M, Cantley LC, Mulliken JB, Olsen BR (1996) Vascular dysmorphogenesis caused by an activating mutation in the receptor tyrosine kinase TIE2. Cell 87:1181–1190.
Wagers AJ (2005) Stem cell grand SLAM. Cell 121:967–970.
Wang HU, Chen ZF, Anderson DJ (1998) Molecular distinction and angiogenic interaction between embryonic arteries and veins revealed by ephrin-B2 and its receptor Eph-B4. Cell 93:741–753.
Weber KT (1989) Cardiac interstitium in health and dusease: the fibrillar collagen network. J Am Coll Cardiol 13:1637–1652.
Weissman IL (2000) Stem cells: units of development, units of regeneration, and units in evolution. Cell 100:157–168.
Whalen GF, Zetter BR (1992) Angiogenesis. In: Cohen IK, Diegelmann RF, Lindblad WJ (eds.). Wound Healing: Biochemical and Clinical Aspects. Philadelphia: WB Saunders, pp 77–95.
Wu P, Yonekura H, Li H, Nozaki I, Tomono Y, Naito I, Ninomiya Y, Yamamoto H (2001) Hypoxia down-regulates endostatin production by human microvascular endothelial cells and pericytes. Biochem Biophys Res Comm 288:1149–1154.
Yancoupoulos GD, Davis S, Gale NW, Rudge JS, Wiegand SJ, Holash J (2001) Vascular-specific growth factors and blood vessel formation. Nature 407:242–248.
Zatterstrom UK, Felbor U, Fukai N, Olsen BR (2000) Collagen XVIII/endostatin structure and functional role in angiogenesis. Cell Struct Funct 25:97–101.
Zhang J, Niu C, Ye L, Huang H, He X, Tong W-G, Ross J, Haug J, Johnson T, Feng JQ, Harris S, Wiedemann LM, Mishina Y, Li L (2003) Identification of the haemopoietic stem cell niche and control of the niche size. Nature 425:836–841.
Ziegler B, Valtieri M, Porada GA, Maria R, Muller R, Masella B, Gabbianelli M, Casella I, Pelosi E, Bock T, Zanjani ED, Peschle C (1999) KDR receptor: A key marker defining hematopoietic stem cells. Science 285:1553–1558.

# 12 Regenerative Medicine of Hematopoietic and Cardiovascular Tissues

## INTRODUCTION

Cardiovascular disease is the number-one cause of mortality in the industrially developed world and a major cause of morbidity. Genetic and malignant diseases of the hematopoietic system are also relatively common, particularly in children. Regenerative medicine has had some success in treating hematopoietic diseases. Vascular grafts and heart transplants are currently the only replacement options for cardiovascular diseases, making these diseases intense targets for regenerative therapies. Cell transplants and construction of bioartificial tissues are the primary strategies that have been pursued, although there has been some work with soluble factors.

## THERAPIES FOR HEMATOPOIETIC DISORDERS

The most successful clinical regenerative medicine in existence today is bone marrow transplantation to regenerate the hematopoietic system after treatment for genetic or malignant hematopoietic disease. The diseased bone marrow is first treated via chemotherapy and/or irradiation to wipe out defective hematopoietic stem and progenitor cells. The hematopoietic system is then rescued (regenerated) by the infusion of allogeneic HSCs into the patient. HSC infusions are also given to patients with normal hematopoietic systems who have undergone chemotherapy and/or total body irradiation for solid tumors, since these treatments destroy the normal hematopoietic system as well. In this case, autologous HSCs for infusion can be sequestered prior to chemotherapy or irradiation.

Chemotherapy and irradiation to combat hematopoietic disease fall into two categories: ablative and nonablative. Ablative procedures use high doses of chemotherapy or irradiation in an attempt to kill off all the malignant or genetically defective hematopoietic cells, create marrow space for the transplanted HSCs, and immunosuppress the host so that the transplanted HSCs will not be rejected. Ablative therapy is the current clinical norm. However, it has very debilitating side effects and damages organs other than the bone marrow and blood. Thus, ablative therapy cannot be used on patients over 55 years of age or on younger patients who have other medical problems. Nonablative therapy is a newer approach that uses much milder, and thus less debilitating chemotherapy and/or irradiation, that aims not to eradicate the recipient hematopoietic system, but only to immunosuppress it enough to enable engraftment of transplanted HSCs.

The cells for transplant are obtained from bone marrow, peripheral blood after G-CSF stimulation, or, more recently, from umbilical cord blood. Autologous HSCs or hematopoietic progenitor cells are almost always obtained from peripheral blood. Bone marrow is the source for most allogeneic transplants, but peripheral blood is sometimes used as well. Umbilical cord blood is rich in HSCs and progenitor cells, which have certain advantages over marrow and peripheral blood cells, as will be explained ahead. The cells harvested for transplant include not only HSCs and progenitors, but also mature T cells. These T cells are a double–edged sword in that they can cause graft versus host disease (GVHD) but they also help eliminate abnormal hematopoietic cells of the host. This is an aspect of hematopoietic cell transplantation that clinical scientists are trying to harness as immunotherapy for hematopoietic disorders, particularly following nonablative chemotherapy or irradiation.

### 1. Myeloablative Therapy

Ablative therapy uses high doses of total body irradiation and antimitotic drugs to wipe out all dividing hematopoietic cells, where the defect is presumed to lie, leaving only quiescent HSCs of the bone marrow to gradually repopulate the system. These treatments cause debilitating anemia, mucositis, susceptibility to infection, and organ damage. Furthermore, relapse is frequent. Hematopoietic transplants are required after additional ablative chemotherapy following relapse, or immediately for patients who are considered a high risk for relapse.

### a. Transplants of Bone Marrow or Peripheral Blood

The standard source of HSCs for allogeneic transplants has been bone marrow aspirated from the iliac crest. Modern clinical hematopoietic regeneration began in 1968 with bone marrow transplants from HLA-identical siblings to treat SCID, Wiskott–Aldrich syndrome (an X-linked mutation involving T cells, B cells, and platelets), and advanced leukemia (Garovoy et al., 1997). Many recipients have survived for three decades or more. However, only 30% of patients can be expected to have an HLA-identical sibling or other relative. Today, the National Marrow Donation Program has HLA-typing information on over a million potential donors, making it possible for 50% of patients to find HLA matches from unrelated donors (Garovoy et al., 1997).

A good example of the use of allogeneic bone marrow transplants is in the treatment of leukemia patients who have relapsed after conventional chemotherapy, or who are at high risk for relapse. Patients are first put into remission by killing malignant cells with intensive chemotherapy and/or total body irradiation. HLA-matched allogeneic bone marrow cells are then infused into the blood at a concentration high enough to produce engraftment. Because of the high level of immunosuppression, patients are vulnerable to infections and also suffer from anemia. Gradually, over a period of about three months, the engrafted HSCs regenerate a new hematopoietic system. There may be mixed host/donor chimerism of hematopoietic cells or these cells may be derived totally from the donor. This chimerism leads to immunological tolerance of the transplanted HSCs and their progeny. Depending on the type of leukemia, the long-term cure rate can be as high as 80%, but averages around 60% (Garovoy et al., 1997).

Aside from infection, acute and chronic graft-versus-host disease (GVHD) is the major obstacle to recovery after allogeneic HSC transplantation, regardless of the source of the HSCs, and is more severe the greater the HLA-mismatch of the donor (Garovoy et al., 1997). GVHD is caused by donor T-lymphocyte attacks on the skin, intestine, and liver of the immunocompromised recipient, resulting in a clinical syndrome of skin rash, severe diarrhea, and jaundice. Acute GVHD is responsible for 10–30% of morbidity and mortality in the first 100 days following transplantation. Chronic GVHD affects 25–45% of bone marrow transplant patients

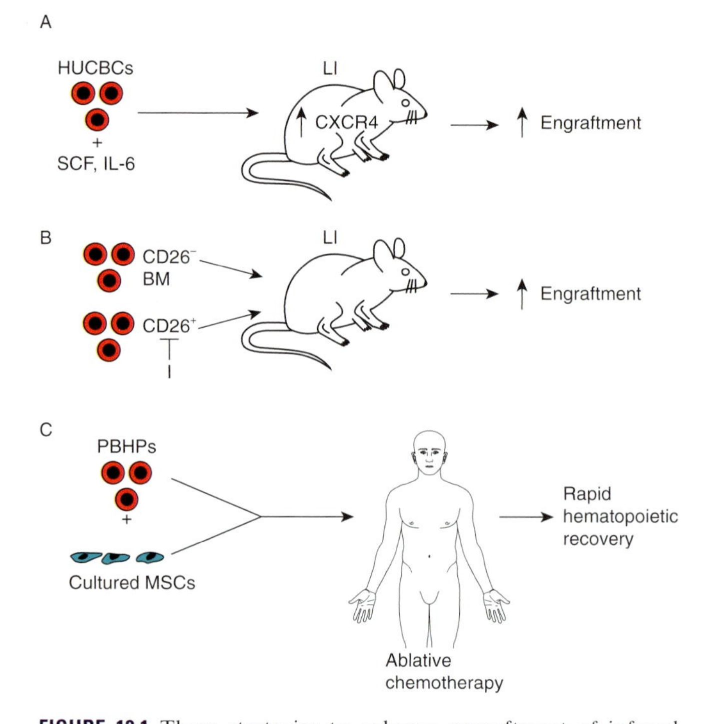


**FIGURE 12.1** Three strategies to enhance engraftment of infused hematopoietic stem cells. **(A)** Infusion of human umbilical cord blood cells (HUCBCs) along with stem cell factor (SCF) and interleukin 6 (IL-6) into lethally irradiated (LI) mice. The effect is to up-regulate the SCF receptor CXCR4 on the cells, promoting their engraftment. **(B)** Inhibition (I) of the CD26 antigen on HSCs or use of a $CD26^-$ HSC fraction provides better engraftment in lethally irradiated mice. **(C)** Infusion of peripheral blood hematopoietic progenitors (PBHPs) along with cultured MSCs into patients who have undergone ablative chemotherapy accelerates hematopoietic recovery because the MSCs provide factors for the survival and engraftment of the HSCs.

who survive more than 180 days. If the patient survives for 3–5 years, chronic GVHD usually resolves.

Standard procedures to combat or reduce the risk of GVHD are the administration of drugs such as cyclosporine A and/or prednisone to immunosuppress the donor T cells, or T cell depletion prior to infusion of the transplant. These methods, however, increase the risk of infection and the HSCs of T cell depleted, HLA-mismatched transplants are rejected at a high rate. Thus, ways have been sought to increase engraftment of donor HSCs and reduce the incidence and intensity of GVHD.

Several ways to increase donor HSC engraftment have been proposed (**FIGURE 12.1**). Bone marrow cells of mice that had received primary human umbilical cord cells along with SCF and IL-6 for 48 hr upregulated the cytokine receptor CXCR4 and exhibited a higher percentage of engraftment in secondary transplants (Peled et al., 1999). HSCs lacking expression of the cell surface antigen CD26 engrafted significantly better than HSCs that express this antigen (Christopherson et al., 2004). Transplantation of low-density $CD26^-$ $Sca\text{-}1^+lin^-$ bone marrow cells in mice resulted in a 2.6-fold increase in homing efficiency, and pretreatment of low-density $CD26^+$ cells with CD26 inhibitors resulted in a 1.5-fold increase. $CD26^-$ cells made a significantly greater contribution to peripheral blood leukocytes six months post-transplantation, contributed to significantly greater survival than $CD26^+$ cells at day 60, when delivered at concentrations below what is required for mouse survival, and made higher contributions in competitive repopulating assays and after secondary transplantation. These findings might possibly be translated to clinical hematopoietic transplantation.

HSCs are dependent on stromal cells for their survival and recipients of allogeneic bone marrow transplants regenerate their marrow stroma from autologous MSCs (Simmons et al., 1987). To increase engraftment of blood-forming cells, autologous MSCs cultured from bone marrow aspirates were injected with autologous hematopoietic progenitors from peripheral blood into breast cancer patients who had undergone ablative therapy (Koc et al., 2000). Clonogenic MSCs were detected in venous blood up to one hr after infusion in 13 of 21 patients (62%). There was rapid hematopoietic recovery, suggesting that augmenting transplants with MSCs facilitates engraftment of HSCs.

A potential way to reduce GVHD is to retain T cells in the transplant but "blind" them to the host antigens they normally attack. GVHD occurred less often and in milder form in bone marrow transplants to a strain of mouse whose antigen-presenting cells (APCs) were unable to present MHC class I peptides (Shlomchik et al., 1999). In humans, APCs might be neutralized by antibodies and replaced by APCs from the donor. Proof of principle that this is possible was shown by the ability of a monoclonal antibody to the β-integrin CD11c to bind to the surface of mouse dendritic cells in lymph nodes and spleen after injection (Shlomchik et al., 1999).

While reducing GVHD is a major goal, it has become clear that GVHD is also a weapon against leukemic cells, called the graft versus leukemia (GVL) effect. Minimizing GVHD in leukemia patients by an HLA-matched transplant from an identical twin or by transplanting T cell depleted bone marrow results in higher relapse rates, higher incidence of recurrent and life-threatening infections, and greater risk of transplant rejection. Patients with chronic GVHD have lower relapse rates. Furthermore, following relapse, sustained complete remission can be obtained by induction of GVHD through infusions of donor lymphocytes (Kolb et al., 1995; Porter et al., 1994). Clearly, while they cause GVHD, allogeneic T cells also actively attack recipient leukemia cells (Horowitz et al., 1990). In allogeneic bone marrow transplants, the issue is how to manage the control of GVHD while taking full advantage of the GVL effect of donor T lymphocytes. **FIGURE 12.2** summarizes the possible outcomes of host and donor T cell reactions, highlighting the two contradictory faces of GVHD.

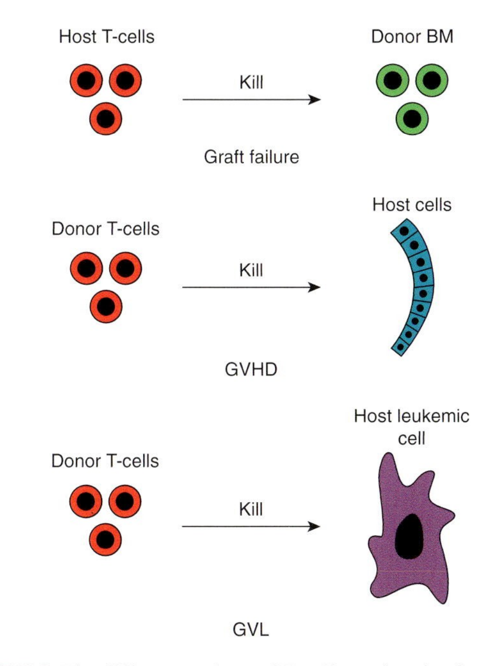


**FIGURE 12.2** The different actions of T-cells against foreign cells. **(A)** Host T-cells can kill donor bone marrow infusions, causing failure of engraftment. **(B)** Donor T-cells in a bone marrow transplant can kill host cells, causing graft vs host disease (GVHD). **(C)** Donor T-cells can also kill host leukemic cells, the graft vs leukemia (GVL) effect.

### b. Transplants of Umbilical Cord Blood

Until the late 1980s, human umbilical cord blood was routinely discarded as part of the "garbage" of birth. However, it has been known since the 1970s that cord blood contains HSCs (reviewed by Broxmeyer, 1993). The use of cord blood as a source of HSCs for transplant was first discussed in the early 1980s (Broxmeyer and Smith, 1994), and in 1988 the first transplant was performed in France to treat a six-year-old boy with Fanconi's anemia, using HLA-matched cord blood cells from his newborn sister (Gluckman et al., 1989). The recipient is alive and well today. The first unrelated donor cord blood transplant was performed in 1993 at Duke University, using a cryopreserved unit of blood from a public cord blood bank. It is now common for cord blood to be preserved at birth and placed in private or public cord blood banks, of which there are a number in the United States and Europe.

Cord blood has several advantages over bone marrow or peripheral blood as a source of HSCs. Lineage colony-forming cells are present at eight-fold higher frequency in cord blood than in adult bone marrow and can be collected at no risk to the mother (Broxmeyer et al., 1989, 1992; Lu et al., 1993). Cord blood can be cryopreserved indefinitely with high efficiency recovery of immature and mature progenitors and can be provided quickly to patients (Broxmeyer and Cooper, 1997). Cord blood cells can be expanded to a high degree before or after cryopreservation (Lu et al., 1995; Broxmeyer and Cooper, 1997). They can be transduced with foreign genes to a much higher level than G-CSF-stimulated peripheral blood and these genes are stably integrated and expressed even after cryopreservation (Lu et al., 1995; Pollok et al., 2001). Cord blood cells evoke GVHD less frequently and at lower intensity.

Cord blood transplants have been used to treat a wide variety of genetic and malignant hematopoietic diseases (Wagner et al., 1995; Kurtzberg et al., 1996; Gluckman et al., 1997; Thompson et al., 2000; Laughlin et al., 2004; Rocha et al., 2004; Sanz, 2004). Most transplants have been done on children because each unit of cord blood contains only about one-tenth the number of HSCs as bone marrow per unit volume, insufficient for transplant to adults. The results of these transplants are comparable to those of bone marrow transplants for most of the parameters examined, but there are differences for some parameters. The major drawback of a cord blood transplant is that recovery of the hematopoietic system is slower, because cord blood contains a higher proportion of immature progenitor cells than does bone marrow and has fewer progenitor cells per unit volume than bone marrow (Kurtzberg et al., 1996; Gluckman et al., 1997). The slower recovery is associated with higher rates of infection and early death. A major plus, however, is that the cytotoxic responses of cord blood lymphocytes are blunted compared with adult peripheral blood, for reasons that are not clearly understood (Risdon et al., 1994). Acute GVHD is less frequent and milder, and the probability of chronic GVHD is low. No correlation has been found between the incidence or extent of GVHD and the degree of HLA (1–3) mismatch, suggesting that cord blood transplants may allow wider discrepancies in HLA matching than bone marrow transplants. There has been some concern that the lesser frequency and intensity of GVHD may be associated with a reduction in the GVL effect, but the lack of significant differences in relapse rates between cord blood and bone marrow transplant patients suggests that this concern is not warranted (Barker et al., 2001; Sanz, 2004).

Comparison of the outcomes of bone marrow and cord blood transplants from unrelated donors in adults with leukemia indicated several differences associated with cell source. Laughlin et al. (2004) found that patients who received HLA-matched bone marrow transplants had the lowest rate of mortality (67%). The overall mortality of patients who received mismatched bone marrow transplants versus cord blood transplants was the same (78%). There was no significant difference in mortality between patients who received cord blood with one HLA mismatch versus two. The probability of leukemia-free survival for three years was 35% for HLA-matched marrow, 20% for mismatched marrow, and 26% for mismatched cord blood. As in studies with children, hematopoietic recovery was slower after cord blood transplants, but acute GVHD was less likely to occur; however, chronic GVHD was more likely to occur after cord blood infusion. Cord blood recipients were younger and more likely to have advanced leukemia than bone marrow recipients in this study. Furthermore, they received a dose of nucleated cells that was 10 times lower than recipients of bone marrow transplants. Thus, it would be of value to know whether the results of HLA-matched cord blood transplants would have produced even better results. In this study, data on HLA-matched cord blood transplants was too sparse to analyze.

Rocha et al. (2004) obtained somewhat different findings in a study comparing HLA-matched bone marrow transplants with HLA mismatched (1–3) cord blood transplants in adults with acute leukemia. As in the study of Laughlin et al. (2004), the recipients of cord blood were younger, had more advanced disease at the time of transplantation, and received one-tenth the number of nucleated cells. Slower hematopoietic reconstitution and lower risk for acute GVHD was confirmed for cord blood recipients. However, the incidence of chronic GVHD and overall survival was not significantly different between bone marrow and cord blood recipients (63% versus 55%). Persistent or recurrent leukemia caused about the same number of deaths in both groups. Deaths related to infection were significantly more common in the cord blood recipients, whereas deaths related to GVHD were significantly more common in bone marrow recipients.

The differences in the results of the Laughlin et al. (2004) and Rocha et al. (2004) studies may be due to the fact that the latter used data obtained only after 1998, when improvements in procedures would have provided better outcomes (Sanz, 2004). A single center study (Takahashi et al., 2004) comparing outcomes of HLA-matched donor bone marrow transplants versus HLA-unmatched (1–4) cord blood transplants for a variety of hematopoietic malignancies in adults, confirmed slower hematopoietic recovery in cord blood recipients and less need for GVHD steroid therapy. Unlike previous studies, however, this one found that cord blood recipients had significantly lower rates of transplant-related relapse and mortality and significantly higher rates of disease-free survival (**TABLE**

**TABLE 12.1** Comparison of cord blood and bone marrow transplants in patients with hematologic malignancies. The cumulative incidence of a number of events was tabulated for the patient populations. Cord blood performed better than bone marrow for all events. Data from Takahashi et al. (2004)

| Cumulative incidence of: | Bone marrow | Cord blood |
|---|---|---|
| GVHD grade III–V (100 days post-transplant) | 27% | 6% |
| Requiring steroid therapy (100 days post-transplant) | 48% | 16% |
| Terminating cyclosporine A or F506 (180 days post-transplant) | 16% | 40% |
| Transplant related mortality (5 years post-transplant) | 34% | <10% |
| Relapse (5 years post-transplant) | 34% | 18% |
| Disease-free survival (5 years post-transplant) | 30% | 74% |

**12.1**). Furthermore, the time from donor search to transplantation was significantly less for cord blood recipients (median of 2 months versus 11 months for bone marrow recipients).

From the data gathered so far, it seems likely that large-scale clinical studies will reveal that cord blood is a better source of cells than bone marrow or peripheral blood for hematopoietic regeneration. Clearly, it can be used as a source of cells for both children and adults and has some advantages over bone marrow, including fewer HLA-compatibility restrictions, more manageable GVHD, and cell expandability and cryopreservability. Millions of units of cord blood can be banked indefinitely and either used for autogeneic transplantation if the donor contracted a hematopoietic malignancy (a relatively low probability) or be made quickly available for allotransplant. Currently, the major problem with the use of cord blood in adults is the low number of stem and progenitor cells per unit volume. Collection units vary from 40–100 mL or more, but even the largest units contain significantly fewer HSCs than a typical bone marrow donation (Steinbrook, 2004). This problem should be easily overcome, however, either by concentrating HSCs from several collection units or by *ex vivo* expansion. It would be useful to have a quantitative assay to determine the repopulating capacity of human cord blood HSCs, such as the immune-deficient *bg/nu/xid* mouse assay described for human bone marrow HSCs (Kamel-Reid and Dick, 1988).

## 2. Non-Myeloablative Therapy

A major cause of mortality in ablative hematopoietic transplants is the side effects of the high-dose chemotherapy and irradiation used to eliminate the recipient hematopoietic system. Many patients are not eligible for allogeneic transplant because of medical problems that render them intolerant to the toxicity of these treatments. Nonablative (actually reduced ablative) approaches use lower-dose chemotherapy and irradiation that cause fewer and milder side effects that are well tolerated by patients. The goal is not to eliminate the recipient hematopoietic system, but to immunosuppress it to a degree that allows engraftment of donor HSCs.

A major principle of the nonablative approach is to use the full force of the GVL effect, rather than an ablative regimen, to eradicate malignant or genetically defective cells over time, including the addition after transplant of more donor T cells. Under these circumstances, donor HSCs will eventually produce a partial or (preferably) complete chimerism of the hematopoietic system that induces tolerance of the transplanted cells (Repka and Weisdorf, 2000; Little and Storb, 2000). GVHD will be a problem, but the incidence and intensity of GVHD should be no greater than with myeloablative transplants, and it can be treated the same way with immunosuppressive drugs. Furthermore, the much healthier condition of the recipient will enable higher tolerance of GVHD until it resolves.

**FIGURE 12.3** outlines the protocol for nonablative therapy. Prigozhina et al. (1997) described a nonablative method to induce tolerance to fully mismatched bone marrow cells and other tissues in mice. A short course (six daily exposures) of low-dose (200-cGy) total lymphoid irradiation was used to reduce the number of T cells. One day later, the mice were infused with a high concentration ($3 \times 10^7$ cells) of mismatched bone marrow, which activated any remaining host T cells specifically reactive to donor cells. One day after infusion, the mice were given a single intraperitoneal dose (200 mg/kg) of cyclophosphamide to eliminate these remaining T cells, thus paving the way for engraftment of donor bone marrow. The mice were then infused with a low number ($3 \times 10^6$) of T cell depleted donor bone marrow cells. This infusion converted recipients to stable mixed hematopoietic chimeras (19–53% donor cells). The recipient mice did not exhibit GVHD and were also able to accept grafts of skin, bone marrow stromal cells, and heart tissue that survived over the life of the animals. Mixed chimerism was also achieved by variations of this approach in DLA-matched dogs and SLA-matched swine (Storb et al., 1999; Fuchimoto et al., 2000).

Slavin et al. (1998) reported the results of nonablative conditioning prior to transplants of G-CSF-mobilized peripheral blood stem cells in 26 patients aged 1–61 years old (median, 31 years) suffering from a variety of hematopoietic malignant and genetic diseases. Immunosuppression over a period of 10 days prior to transplant was accomplished by administraton of fludarabine, busulfan, and anti-T-lymphocyte globu-

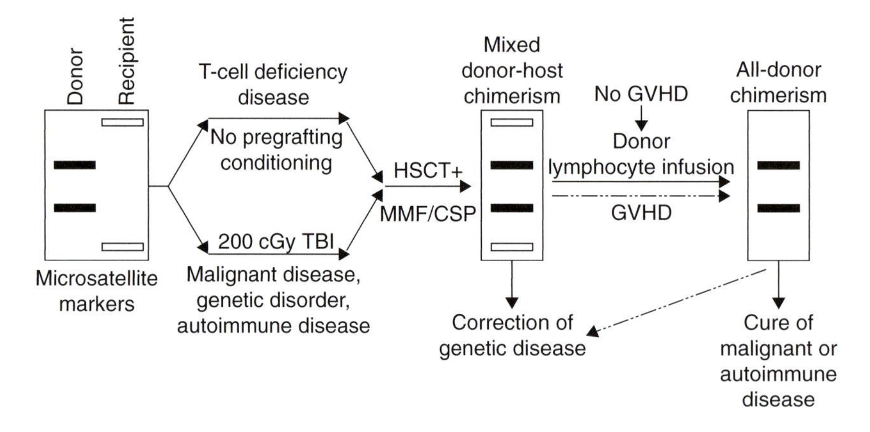


**FIGURE 12.3** Diagram of protocol for non-myeloablative induction of mixed donor-host chimerism by hematopoietic stem cell transplant (HSCT). Left box shows different microsatellite DNA markers. No pregraft immunosuppressive conditioning is needed for the treatment of T-cell deficiency disease. Low-level (200cGy) total body irradiation (TBI) or low-level chemotherapy is used as preconditioning for malignant disease, genetic disorder, or autoimmune disease. The HSCT is followed by a powerful post-grafting regimen using the antimetabolite mycophenolate mofitil (MMP) and the immunosuppressant cyclosporine (CSP). This procedure induces mixed donor-host chimerism, which may be sufficient to cure genetic disease. Malignancy and autoimmune disease may require complete replacement of host cells with donor cells (all-donor chimerism). If there is no GVHD, this can be achieved by an infusion of donor lymphocytes; spontaneous GVHD will also lead to all-donor chimerism. Reproduced with permission from Little and Storb, The future of allogeneic hematopoietic stem cell transplantation: minimizing pain, maximizing gain. J Clin Invest 105:1679–1681. Copyright 2000 JCI, through Copyright Clearance Center.

lin (ATG). Starting the day before transplant, cyclosporine A was given to combat GVHD. The donors were HLA-matched siblings. Partial chimerism was achieved in 9 patients and complete chimerism in 17 patients. The tolerance conferred by chimerism is useful in case of relapses, because it allows infusions of additional donor T cells to augment the GVL effect. Fourteen of the patients experienced no GHVD. The other 12 developed GVHD, and 4 of them died of severe GVHD, which occurred after early discontinuation of cyclosporine A. The other eight were successfully treated with prednisone.

Similar successes for both primary disease and relapses have been reported in other studies with low-dose chemotherapy regimens using a variety of immunosuppressive drugs and drug combinations (Carella et al., 2000), including indolent lymphoma (Khouri et al., 2001), SCID (Sarzotti et al., 2003), chronic lymphocytic leukemia (Khouri et al., 2004) acute myeloid leukemia (de Lima et al., 2004), and Hurler's syndrome (Staba et al., 2004). An antitumor effect of nonablative allogeneic peripheral blood transplants has also been reported in patients with metastatic renal cancer (Childs et al., 2000). This disease has an extremely poor prognosis with a median survival time of less than one year, but 9 of 19 patients were alive after 287 to 831 days after transplant. Ten patients (53%) showed partial (7 patients) or complete (3 patients) regression of metastatic disease.

Sorror et al. (2004) carried out a retrospective study that compared the morbidity and mortality of patients with ablative versus nonablative HSC transplants (**TABLE 12.2**). They found that even though nonablative patients were older, had higher pretransplant comorbidity scores, and had failed ablative procedures more often, they nevertheless experienced less high-grade toxicity, significantly less incidence of severe GVHD (77% versus 91%), and a lower nonrelapse mortality rate than the ablative patients. Both groups had similar one-year probabilities of chronic GVHD. Essentially the same results were found in a similar comparative study, except that nonablative transplantation was associated with a syndrome of acute GVHD that occurred in many patients after day 100 post-transplantation. Though the long-term results of nonablative hematopoietic transplant still need to be assessed, these procedures seem applicable to a wider range of patients

**TABLE 12.2** Kaplan-Meier estimates of two year overall survival of ablative vs non-ablative bone marrow transplant patients, stratified by CCI scores. CCI stands for comorbid condition index, the sum of a series of numerically weighted medical conditions that affect survival. High CCI scores have particularly negative effects on ablative patients, but survival of non-ablative patients with CCI scores of 0–2 is better than ablative patients with CCI scores of 1–2, and approaches that of ablative patients with CCI scores of 0 (50% vs. 63%). Data from Sorror et al. (2004)

| | % Survival |
|---|---|
| Ablative | |
| CCI = 0 | 63 |
| CCI = 1–2 | 22 |
| Non-Ablative | |
| CCI = 0–2 | 50 |
| CCI ≥ 3 | 9 |

because of their lower toxicity, their ability to generate stable chimerism and tolerance, their evocation of less severe GVHD, their association with fewer relapses and enhanced ability to deal with relapses, and survival rates equal to or better than ablative procedures.

## 3. The Future of Hematopoietic Cell Transplants

Although hematopoietic regenerative medicine via HSC transplant has been in existence for over 35 years, it has not yet been perfected to the point where it can be used in an uncomplicated way with highly efficient results, primarily because the transplanted stem cells are not just needed to reconstitute the hematopoietic system, but also to battle the underlying disease itself. The cells therefore need to be allogeneic, but allogeneic cells evoke GVHD. It seems possible, however, that improvements in outcome might be made by combining a number of protocols that would lead to high engraftment of HSCs, maximum reaction against defective host hematopoietic cells, and low incidence of GVHD. For example, we might combine nonablative chemotherapy with the use of allogeneic, HLA-matched umbilical cord blood HSCs that have been expanded *ex vivo* or concentrated from several collection units.

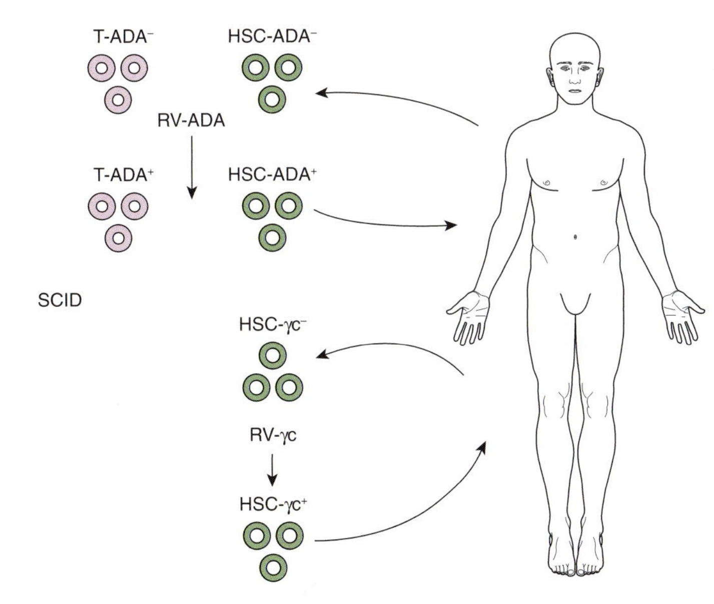


**FIGURE 12.4** Treatment for adenosine deaminase (ADA) deficiency and severe combined immunodeficiency disease (SCID). To treat ADA deficiency, $ADA^-$ T-cells ($T\text{-}ADA^-$) or HSCs ($HSC\text{-}ADA^-$) were harvested from the patient, transfected with a retroviral vector (RV) carrying the gene for ADA, and infused back into the patient. To treat SCID, $\gamma c^-$ HSCs ($HSC\text{-}\gamma c^-$) were harvested from the patient, transfected with a retroviral vector carrying the γc gene, and infused back into the patient.

## 4. Gene Therapy for Genetic Hematopoietic Diseases

The ability to insert a normal gene or genes into the genomes of hematopoietic stem cells *ex vivo* to compensate for defective alleles that cause hematopoietic diseases would be a simple and inexpensive way to cure these diseases. **FIGURE 12.4** illustrates how gene therapy is used for this purpose. Several gene therapy trials to cure severe combined immunodeficiency disease in children due to adenosine deaminase deficiency (ADA-SCID) or SCID-X1, an X-linked inherited disorder caused by γc cytokine deficiency, both of which lead to defective T and NK lymphocyte differentiation or function, have shown great promise.

Blaese et al. (1995) transfected peripheral T cells of two patients *ex vivo* with a retroviral construct containing the gene for ADA and infused the transfected cells back into the patients. Bordignon et al. (1995) also used retroviral vectors to transfer the ADA gene into peripheral T cells and bone marrow stem cells. In both cases, the immune system was normalized over two years of treatment and in the case of transfected bone marrow stem cells, peripheral T lymphocytes were progressively replaced by marrow-derived T cells expressing ADA. However, in both these trials, the patients were also maintained with injections of PEGylated ADA, which reverses much of the symptomology of the disease. Thus, the actual role of the transfected gene was difficult to assess.

In another study, however, in which ADA replacement enzyme therapy was not available, $CD34^+$ HSCs from two patients were transduced with a retroviral vector-ADA gene and infused back into the patients, thus allowing the effect of the gene therapy alone to be evaluated (Aiuti et al., 2002). The infusions were preceded by low-intensity, nonablative chemotherapy to provide a developmental advantage to the transduced cells. The patients responded well to the therapy, showing normal growth and development. There were sustained engraftment and differentiation of the transduced HSCs into multiple blood cell lineages and improved immune function.

SCID-X1 was treated in two patients with autologous $CD34^+$ HSCs transduced *ex vivo* with a γc Moloney retrovirus-derived vector. During a 10-month follow-up period, T and NK cells expressing the γc transgene were detected. Counts of B, T, and NK cells were comparable to age-matched normal children, as were the functions of these cells, including antigen-specific responses.

Both had normal growth and development without side effects.

As of 2004, a total of 18 SCID patients had been treated this way in Paris, Milan, and London. Seventeen were alive in February 2004 with fully reconstituted immune systems, but two of the French patients developed leukemia, which has been attributed to the insertion of the retrovirus near the cancer-promoting gene LMO-2 (Cavazzana-Calvo et al., 2004). Further trials have been put on hold, pending a review of the retroviral vector problem.

## THERAPIES FOR BLOOD VESSEL REGENERATION

Over 600,000 synthetic and biological vascular grafts are implanted yearly to replace blood vessels damaged by atherosclerosis (Nu and Allen, 1995). DeBakey established the use of Dacron tubes to replace damaged blood vessels in the late 1950s and 1960s. Dacron and Teflon are the vascular grafts of choice today for large vessels with high blood flow such as the aorta or iliac artery. Drawbacks to these grafts include decline in their patency rate to 50% after four years due to intimal hyperplasia, and risk of thrombosis. Dacron and Teflon are not suitable as grafts for smaller vessels with low blood flow (<6mm internal diameter) because of the substantial risk for thrombosis. Typically, surgeons use small-diameter autologous arteries or veins to replace small-diameter vessels such as the coronary arteries. The patency of these grafts also declines within a few years and suitable autografts are not always available. Thus, there is a need for better large-diameter vascular prostheses, as well as prostheses that can substitute for autologous small-diameter vessel grafts.

The ideal vascular replacement should have the typical three layers of endothelium, tunica media (smooth muscle), and adventitia (connective tissue) and be mechanically stable enough to be easily sutured. It should have the biochemical composition and distribution of collagens, elastin, and other matrix molecules that confer flexibility and burst strength comparable to the native vessel it will replace. The most common approaches to achieving this ideal have been to induce blood vessel regeneration with scaffolds of natural biomaterials or to construct implantable bioartificial blood vessels *in vitro*. The scaffolds in both cases typically are decellularized ECM, or cellular biological materials treated with cross-linking agents to reduce immunogenicity and enhance resistance to enzymatic degradation. Natural biomaterials offer many mechanical, chemical, and biological advantages over synthetic materials, which can induce a foreign body reaction and lead to bacterial infection (Schmidt and Baier, 2000).

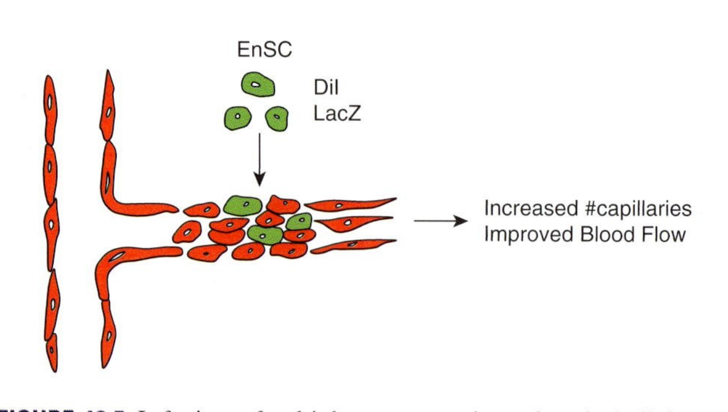


**FIGURE 12.5** Infusion of a high concentration of endothelial stem cells (EnSC) helps regenerate new capillaries in ischemic regions. Experiments using DiI or *Lac-Z* labeled EnSCs show that the EnSCs are incorporated into the new capillaries sprouting from existing vessels.

On the other hand, biodegradable synthetic polymers have shown promise in promoting vascular regeneration and in constructing bioartificial blood vessels, and transplants of endothelial stem cells have shown promise in regenerating vasculature.

### 1. Stem Cell Transplants

The current solution for vascular disease of veins and arteries is surgical removal, clearing, or implantation of a stent. As discussed in Chapter 11, circulating EnSCs derived from the bone marrow play a role in angiogenesis after vascular damage. The EnSCs complement existing endothelial cells in forming new blood vessels. There is also evidence that co-recruitment of HSCs is necessary for the incorporation of EnSCs into neovessels, through the production of Ang-1 (reviewed by Rafi and Lyden, 2003). Thus, infusions of endothelial stem cells at high concentrations have been explored as a way to enhance blood vessel regeneration (Lovell and Mathur, 2004, for review). **FIGURE 12.5** illustrates how infused EnSCs help to regenerate blood vessels.

Other studies have shown that the homing and incorporation of EnSCs after injury, as well as the proliferation of local endothelial cells, is mediated by growth factors and cytokines (**FIGURE 12.6**). There is growing evidence that these cells and their more restricted progenitors are recruited to sites of tissue damage by angiogenic chemokines such as VEGF-A and placental growth factor (PLGF) that are upregulated by tissue ischemia. The injection of labeled bone marrow cells into mice at birth resulted in increased tissue vascularity and incorporation of labeled cells into neovasculature (Young et al., 2002). Co-administration of VEGF with the bone marrow cells significantly increased this effect. The introduction of EnSCs into the limbs of animals made ischemic by resecting a segment of

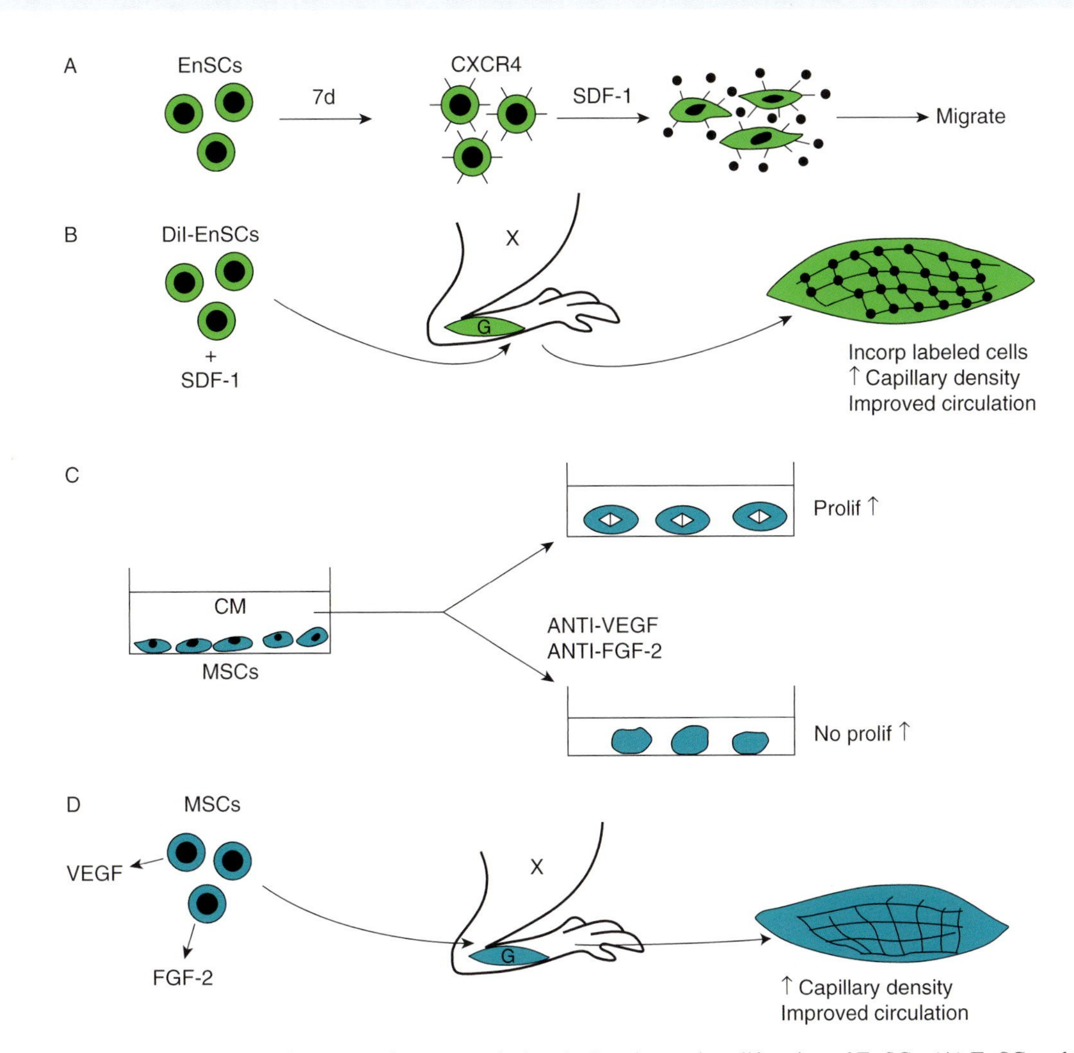


**FIGURE 12.6** SDF-1, VEGF and FGF-2 are important factors regulating the homing and proliferation of EnSCs. **(A)** EnSCs cultured for 7 days express the SDF-1 receptor CXCR4, enabling them to respond to SDF secreted at an injury site by migrating toward the source of the SDF (black dots). **(B)** DiI-labeled EnSCs plus SDF injected into the ischemic gastrocnemius muscle are incorporated into new capillaries; capillary density is increased and circulation is improved. X = site of ligation of femoral artery. **(C)** Conditioned medium (CM) of MSCs increases the proliferation of endothelial cells. The increase is abolished by antibodies to VEGF and FGF-2. **(D)** MSCs injected into the gastrocnemius muscle of an ischemic limb resulted in increased capillary density and improved circulation. The MSCs were not incorporated into new vessels, but increased the proliferation of endothelial cells via their secretion of VEGF and FGF-2.

femoral artery reduces the ischemia by improving collateral vessel formation. Limb ischemia in mice and rabbits resulted in an elevated level of EnSCs in peripheral blood. Ischemia not only resulted in the incorporation of DiI or *lac*-Z-labeled EnSCs into new limb vessels, improving blood flow, but also into neovessels in the corneas of animals that underwent microsurgery several days after artery occlusion (Takahashi et al., 1999). These effects were enhanced by exogenously administered GM-CSF. Similar results were obtained after intracardiac or intramuscular injection of human EnSCs into nude mice (Kalka et al., 2000; Iba et al., 2002).

An important chemoattractant in the homing process is stromal cell derived factor 1. Yamaguchi et al. (2003) showed that 66% of EnSCs expressed the receptor for SDF-1, CXCR4, after 7 days in culture and showed dose-dependent migration toward SDF-1 in a modified Boyden chamber assay. When SDF-1 was injected into ischemic gastrocnemius muscles of nude mice along with culture-expanded DiI-labeled human endothelial progenitor cells, labeled cells accumulated in the muscle, and blood flow and capillary density increased by 28 days after treatment.

Conditioned medium of mouse MSC cultures enhanced proliferation of endothelial cells and smooth muscle cells *in vitro* (Kinnaird et al., 2004). The conditioned medium was shown to contain VEGF, FGF-2, placental growth factor (PLGF), and monocyte chemoattractant protein-1 (MCP-1). The proliferation-enhancing effect of the medium was partly attenuated by antibodies to VEGF and FGF-2. Intramuscular injection of MSCs into adductor muscles rendered ischemic by ligation of the femoral artery improved circulation, reduced the incidence of auto-amputation and attenuated muscle atrophy and scarring. The MSCs were not incorporated into neovessels, but secreted VEGF and FGF-2 within the muscle. VEGF secretion by human adipose stromal cells (ADSCs) increased fivefold when the cells were cultured under under hypoxic conditions (Rehman et al., 2004). Conditioned medium from these cultures significantly increased the proliferation, and decreased the apoptosis, of human endothelial cells *in vitro*. Nude mice with ligated femoral arteries showed improved blood flow when GFP-labeled ADSCs were injected into the tibialis anterior muscle.

In another study, Walter et al. (2002) showed that treatment of rats with the statin simvastatin accelerated the re-endothelialization of balloon-injured segments of the carotid artery. Labeled bone marrow cells from *Tie2/lacZ* mice infused into nude rats, followed by carotid artery balloon injury, were incorporated into the re-endothelialized luminal surface of statin-treated animals at a frequency five times that of animals without statin treatment. Statin treatment increased circulating rat epithelial progenitor cells and increased the adhesiveness of cultured human epithelial progenitor cells to an umbilical vein endothelial cell monolayer by upregulation of $\alpha_5$, $\beta_1$, $\alpha_v$, and $\beta_5$ integrin subunits.

Tateishi-Yuyama et al. (2002) carried out a randomized, controlled human patient trial to assess the efficacy and safety of injecting autogeneic bone marrow mononuclear cells into the ischemic gastrocnemius muscle. One group of patients with unilateral ischemia received bone marrow cells in the gastrocnemius of the affected leg, whereas the other leg received saline. In another group of patients with bilateral ischemia, the gastrocnemius muscle of one leg received an injection of marrow cells, while the other leg received peripheral blood mononuclear cells. Injection of saline or peripheral blood mononuclear cells produced no improvement, but injection of bone marrow cells resulted in significant improvements in ankle-brachial index, transcutaneous oxygen pressure, rest pain, and pain-free walking time. The improvements were sustained at 24 weeks postinjection.

## 2. Acellular Regeneration Templates

Attempts have been made to replace segments of small-diameter blood vessels with small-bore blood vessel grafts whose matrix has been cross-linked by glutaraldehyde, polyepoxy compounds, or dye-mediated photooxidation. Of these, dye-mediated photooxidation seems to produce the best results. For example, cross-linked small-diameter sheep carotid artery segments implanted into pig carotid arteries exhibited patency rates of 100% after one year and patency rates of 60% after 150 days in a dog coronary artery bypass model. However, natural tissues need to be decellularized, because cellular remnants not only are a target for immune reactions, but they also can lead to calcification of the graft (Schmidt and Baier, 2000).

SIS and human arterial matrix have been investigated as acellular templates for vascular regeneration. Pig SIS has been tested as an acellular autoplastic, alloplastic, and xenoplastic vascular scaffold for inducing the regeneration of large- and medium-diameter blood vessels. The SIS is formed into a tube by wrapping it around a glass rod and suturing the edges. When these scaffolds were sutured into the femoral artery of dogs, they remained patent during a six-month observation period and promoted regeneration from the cut ends of the vessel (Badylak et al., 1989; Lantz et al., 1992, 1993; Sandusky et al., 1992). The grafts were completely endothelialized by 28 days postimplantation and were histologically similar to normal arteries at 90 days (Sandusky et al., 1995). SIS material was still present at two months, but in much diminished amounts. Follow-up periods of up to five years found no evidence of infection, intimal hyperplasia, or aneurismal dilation. The regenerated tissue in SIS grafts was greater than 10 times thicker than the implanted SIS and its mechanical stiffness and strength was higher than that of a normal artery (Hiles et al., 1995). The success of pig SIS in promoting blood vessel regeneration is attributed to its porosity (Hiles et al., 1993) and ability to provide an adhesive substrate for endothelial cells (Badylak et al., 1999).

In other studies, pig SIS was compared to autogeneic saphenous vein and Teflon as a substitute for a segment of carotid artery (Sandusky et al., 1992). Comparable patency was observed between SIS and saphenous vein grafts (83% versus 88%), whereas the patency rate of SIS grafts was significantly greater than for Teflon grafts (88% versus 25%). SIS grafts were found to be more resistant to infection than Teflon grafts (Badylak et al., 1994). Teflon grafts accumulated fibrin on their luminal surface, which formed fibrin thrombi with platelets and red cells. These grafts were not completely endothelialized even by 180 days (Sandusky et al., 1995). When pig SIS grafts of ~1-mm diameter were grafted into rat femoral artery, they were thrombogenic and occluded within the first hour postsurgery (Prevel et al., 1994). However, Huynh et al. (1999) used a modified pig SIS to construct a small-diameter (4-mm) blood vessel that functioned well immediately after transplant in a rabbit carotid bypass model. The SIS was impregnated with type I bovine collagen and cross-linked with

1-ethyl-3-(3-dimethylaminopropyl)carbodiimide hydrochloride (EDC), and the interior surface was coated with heparin-benzalkonium chloride complex (HBAC) to prevent thrombosis. After three months, the graft had been remodeled to the structure of a blood vessel by invading host endothelium, fibroblasts, and smooth muscle and showed vasomotor responsiveness to norepinephrine, serotonin, and bradykinin.

Experiments with piglets suggest that SIS might also be suitable as a blood vessel graft material for growing children. Autologous piglet SIS, prepared from short segments of jejunum, was used to make superior vena cava interposition grafts between the azygous vein and the cavo-atrial junction (Robotin-Johnson et al., 1998). By 90 days, the weight of the piglets had increased 630%, while the regenerated blood vessel increased in length by 147% and in circumference by 184%. Regenerated vessels remained patent and free of aneurysm over the course of the experiment. The histological appearance of the vessel tissue was similar to that of native vessels.

The LifeCell Corporation has developed an acellular vascular graft of human arterial matrix, along the lines of Alloderm®. This matrix is integrated into the ends of the host vessel and promotes the regeneration of endothelium, smooth muscle cells, and fibroblasts from those ends (Schmidt and Baier, 2000).

Synthetic materials have also been tested for their ability to support blood vessel regeneration. Iwai et al. (2004) created a patch for large vessels consisting of a poly (lactic-coglycolic acid)-bovine collagen sponge scaffold. The PLGA mesh was impregnated with a mixture of bovine collagen I and IV cross-linked with glutaraldehyde. Patches of 20 × 15 mm size, with or without being seeded with smooth muscle and endothelial cells grown from segments of the saphenous vein, were then implanted into the pulmonary trunk of dogs. The results over six months were the same whether the scaffolds were seeded with cells or not. The size of the patches did not change, there was no intimal thickening, and no thrombosis. The histological architecture and biochemical composition of the patch was the same in both groups, indicating that this scaffold was capable of inducing the regeneration of vascular tissue. Mechanical strength of the patch before and after implantation was greater than that of the native pulmonary artery.

## 3. Bioartificial Blood Vessels

Bioartificial blood vessels have been grown *in vitro*, using synthetic polymers and precultured autologous vascular cells, and used to completely replace or patch small and large blood vessels in animal studies.

Niklason et al. (1999) carried out a very thorough set of experiments to evaluate a small-caliber bioartificial artery construct. Their construct was prepared by seeding noncross-linked polyglycolic acid tubes with bovine aortic smooth muscle cells. The tubes were surface-hydrolyzed with sodium hydroxide to enhance adhesion of the cells to the polymer (Gao et al., 1998). They were cultured for eight weeks in bioreactors under pulsatile flow to mimic the hydrodynamic conditions of fetal development. An endothelial cell suspension was then injected into the lumen, allowed to adhere, and the constructs were cultured under luminal flow for an additional three days. Pre-seeding of scaffolds with smooth muscle cells and fibroblasts significantly increased the attachment of endothelial cells over constructs in which endothelial cells are seeded directly onto the scaffold wall (Gulbins et al., 2004).

Histological and biochemical evaluation of the constructs revealed the formation of an endothelium-lined smooth muscle wall containing collagen and smooth muscle actin that was similar to that of native artery (**FIGURE 12.7**). The bioartificial vessels exhibited burst strength greater than that of human saphenous vein and within the range suitable for arterial grafting. These properties were not attained when the constructs were not cultured under pulsatile flow, attesting to the importance of physical forces in vascular development (Jain, 2003). Bioartificial arteries made with autologous carotid artery cells were then implanted into the right saphenous artery of miniature pigs (**FIGURE 12.8**). Grafts grown under pulsatile flow remained open for the four-week duration of the experiment, whereas those grown in the absence of pulsatile flow thrombosed at three weeks.

Autologous endothelial stem cells of bone marrow or peripheral blood have been shown to form a nonthrombogenic endothelial surface when seeded onto small-diameter decellularized arteries or synthetic vascular grafts implanted into dogs or sheep. Such grafts remained patent for four months when grafted into the aorta or carotid artery and their contractility and relaxation responses were similar to those of native arteries (Noishiki et al., 1996; Kaushal et al., 2001). Opitz et al. (2004) have described a large-diameter bioartificial aorta made by seeding a poly-4-hydroxybutane scaffold with sheep autologous carotid artery smooth muscle cells and cultured under pulsatile flow. Two days after the addition of endothelial cells, the constructs were wrapped in sheep SIS to increase mechanical strength, and implanted into the descending aorta. They were evaluated histologically and biochemically for up to 24 weeks and for up to six months for graft performance. Up to three months after implantation, the grafts were fully patent, with no signs of occlusion, intimal thickening, or dilatation. Scanning electron microscopy

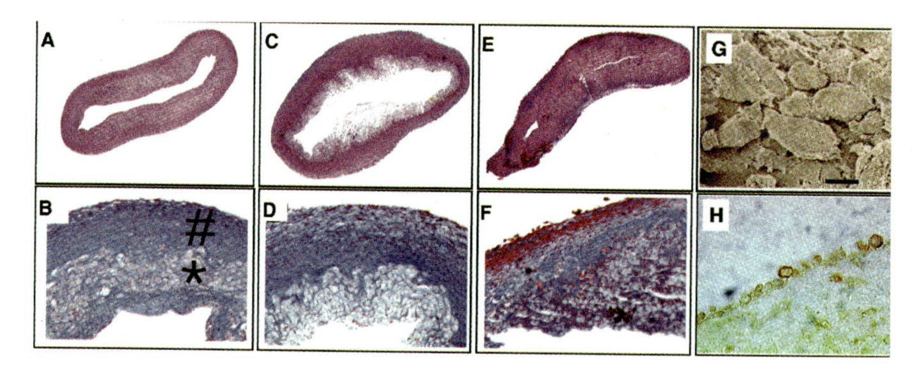


**FIGURE 12.7** Structure of bioartificial arteries cultured for 8 weeks. **A**, **C**, **E**, Verhoff's elastin stain (red). **B**, **D**, **F**, Masson's trichrome stain (collagen stains blue). # = dense cellular region. *remnants of polymer scaffold. **(A, B)** Pulsed vessel. **(C, D)** Non-pulsed vessel. **(E, F)** Pulsed vessel without medium supplementation. **(G)** Scanning EM of endothelial layer showing that the cells are less confluent and more rounded than those of a normal artery. **(H)** Endothelium stained for PECAM. Reproduced with permission from Niklason et al., Functional arteries grown in vitro. Science 284:489–493. Copyright 1999, AAAS.

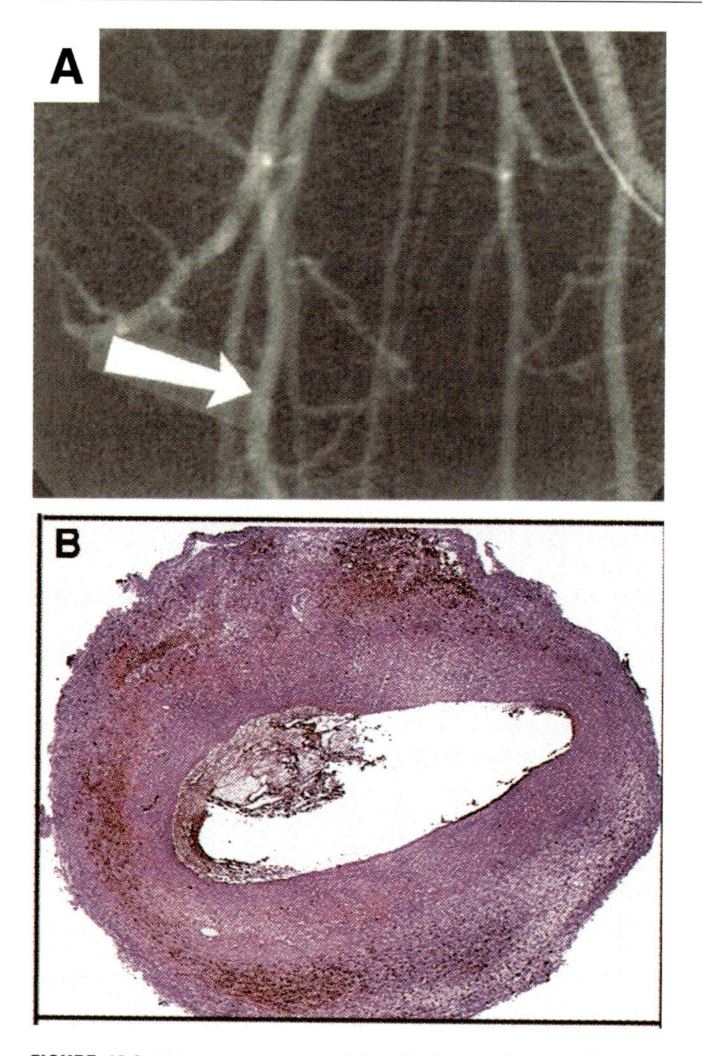


**FIGURE 12.8** **(A)** Arteriogram of hindlimb vessels of a Yucatan minipig that received a bioartificial xenograft vessel (arrow) of 3.5 cm in length in the right saphenous artery 24 days previously. **(B)** H&E stained section of the recovered xenograft. The graft was patent, but an inflammatory response is visible on the vessel wall. Reproduced with permission from Niklason et al., Functional arteries grown in vitro. Science 284:489–493. Copyright 1999, AAAS.

revealed a confluent luminal endothelium. By contrast, six-month grafts exhibited partial thrombus formation and significant dilatation. These abnormalities were associated with a significant deficiency in the number and distribution of elastic fibers compared to native artery. Elastin fibers are found throughout the wall of native aorta, whereas in the constructs, elastin fibers were located only in the luminal part of the wall. This work demonstrates the importance of elastin synthesis and distribution to achieving normal mechanical properties of bioartificial vessels.

Other experiments have focused on creating blood vessels *in vitro* from cells only, without any natural or synthetic scaffolds. L'Heureux et al. (1998) was one of the first to build a small-diameter human blood vessel in this way (**FIGURE 12.9**). Sheets of human endothelial and smooth muscle cells from umbilical veins and fibroblasts from dermis were grown in culture. The construct was made in a series of steps, starting with the production of an inner membrane (IM) by dehydrating a tubular fibroblast sheet. The IM was slipped over a perforated tubular mandrel (outside diameter 3 mm), wrapped with a sheet of smooth muscle, and cultured in a bioreactor under luminal flow for a week. A sheet of adventitial fibroblasts was then applied to the outside of the construct and it was cultured for another eight weeks before removing the mandrel and seeding the luminal surface with endothelial cells.

The mature construct strikingly resembled a human artery. Histological and immunochemical staining showed a three-layered vessel similar to a native artery, though the density of smooth muscle cells in the construct was lower. The ECM contained Type I, III, and IV collagens, as well as laminin, fibronectin, and chondroitin sulfates; the adventitia synthesized high amounts of elastic fibers. Ultrastructural analysis revealed a normal architectural organization of the collagen fibrils.

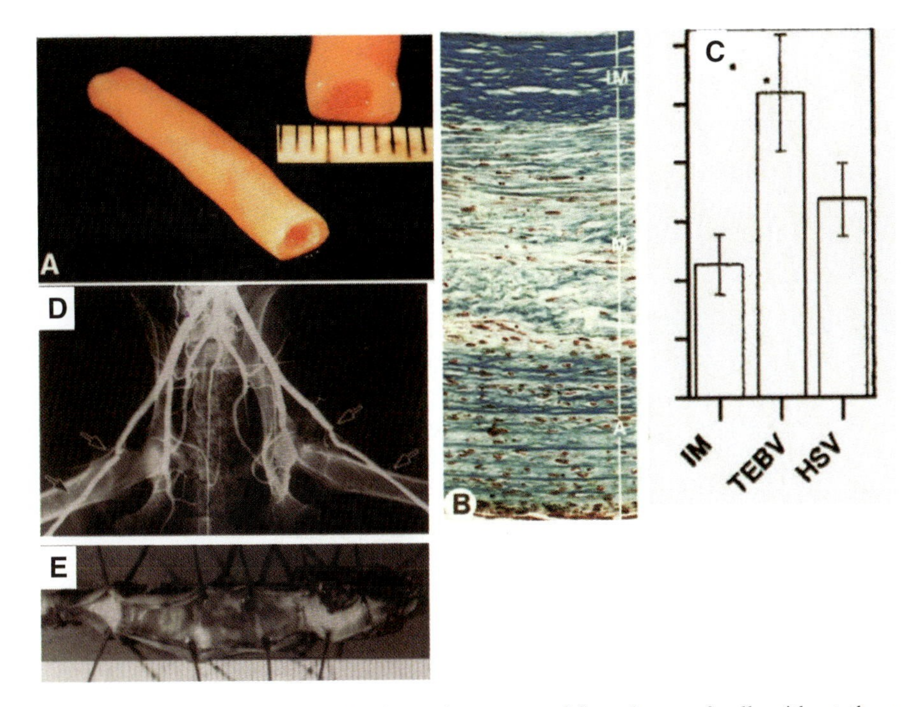


**FIGURE 12.9** Structure and properties of a bioartificial blood vessel constructed from layers of cells without the use of a polymer scaffold. **(A)** Gross appearance of the vessel at 9 weeks of culture. Ruler shows mm. **(B)** Histological organization of the vessel wall. Section stained with Masson's trichrome. The internal elastic lamina (IM) is larger than normal, but the remaining structure is similar to that of a normal muscular artery with a large media (M) and a surrounding adventitia (A). **(C)** Burst strength of the cultured IM layer, tissue-engineered blood vessel (TEBV) and human saphenous vein (HSV). Burst strength of the TEBV is considerably higher than that of human saphenous vein, which is considered to be the optimal graft for lower limb vascular reconstruction. **(D)** Angiogram of dog lower limbs 7 days after grafting two bioartificial blood vessels into the femoral arteries (arrows). The grafts are patent. **(E)** Luminal surface of recovered opened graft, showing that it is free of thrombus. Reproduced with permission from L'Heureux et al., A completely biological tissue-engineered human blood vessel. FASEB J 12:47–56. Copyright 1998, FASEB Journal, through Copyright Clearance Center.

The burst strength of the construct was significantly higher than that of human saphenous veins, most of which was due to the adventitia. The endothelial cells were positive for von Willebrand factor expression and ac-LD uptake, and did not adhere platelets of human blood. These human bioartificial blood vessels (5-cm length) were xenografted without endothelial cells (to avoid hyperacute rejection) into the femoral arteries of dogs. At the end of one week, the patency rate was only 50%, but the histological architecture of the graft was retained.

Vascular networks have been created *in vitro* from human umbilical cord endothelial cells transgenic for GFP and 10T1/2 fibroblasts, by seeding them into a three-dimensional fibronectin-type I collagen gel (Koike et al., 2004). The seeded collagen gels were implanted into mice with transparent windows and were observed to form stable vascular networks that became connected to the mouse arteriolar vasculature. The 10T1/2 cells differentiated into smooth muscle cells via heterotypic interactions with the endothelial cells, as evidenced by immunostaining for smooth muscle α-actin. These cells were lined with endothelial cells expressing the endothelial marker CD31. Local administration of the vasoconstrictor, endothelin-1, induced constriction of the bioartificial vessels. Endothelial cells seeded alone into collagen gels were unable to form stable vessels.

## THERAPIES FOR PROTECTION AND REGENERATION OF THE INFARCTED MYOCARDIUM

From the 1960s to the 1980s, Russian investigators pioneered attempts to induce regeneration of injured rat and rabbit myocardium. The myocardium was injured by induction of diptherial myocarditis or by electrodiathermocoagulation. The objective was to recruit circulating stem cells to the injury area and/or to stimulate local cardiomyocytes to divide. To do this, the injured heart was treated by the intraperitoneal injection of hydrolysates of skeletal and cardiac muscle, grafts of skeletal muscle, brain, and peripheral nerve to the infarct area, and intramuscular injections of cobalt-35 preparations (Polezhaev, 1972, 1979, 1980; Polezhaev and Tinyakov, 1980). A limited number of striated myocardial fibers were reported to have regenerated.

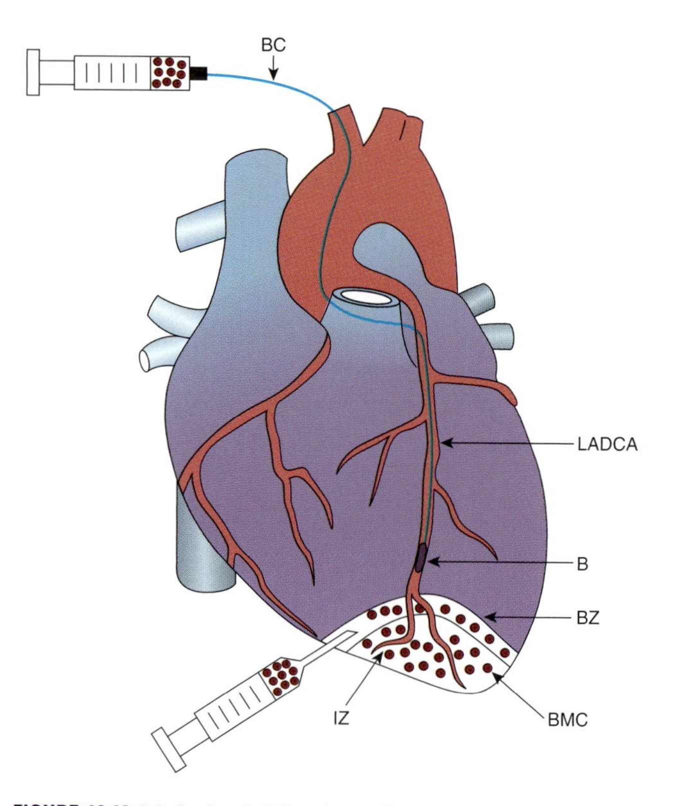


**FIGURE 12.10** Methods of delivering cells to the infarcted myocardium. **Top**, delivery of bone marrow cells (BMC) to the infarct area via a balloon catheter (BC) threaded into the coronary artery (in this case, the left anterior descending coronary artery, LADCA) feeding the infarct area. B = balloon inflated above the border zone (BZ) of the infarct zone (IZ). **Bottom**, direct injection of bone marrow cells into the border zone of the infarct.

Light and electron microscope observations suggested that cardiac myofibers dedifferentiated to provide a source of new cardiomyocytes. Polezhaev (1980) also suggested that new cardiomyocytes might arise from circulating stem cells situated near blood vessels, because cardiomyocytes with labeled nuclei were demonstrated in the scar tissue near blood vessels after $^3$H-thymidine injection.

These methods were rather crude and the rationale behind them was not always clear. With the rise of stem cell biology in the 1990s, cell therapy has been intensively explored as a means of rescuing the structure and function of the failing heart (Mathur and Martin, 2004; Lovell and Mathur, 2004; Laflamme and Murry, 2005, for reviews). The strategy for delivering various types of cells to myocardial infarcts is illustrated in **FIGURE 12.10**.

## 1. Transplants of Bone Marrow Cells

A number of studies have suggested that infused bone marrow cells improve left ventricular function after myocardial infarction in rats, rabbits, and pigs. Several mechanisms might account for this improvement. EnSCs in the bone marrow might participate in the formation of new cardiac microvessels regenerating from existing vasculature, or HSCs, MSCs, and EnSCs might undergo lineage conversion into new cardiomyocytes. Alternatively, or additionally, these cells might secrete factors that enhance angiogenesis from existing vasculature, promote proliferation and differentiation of resident stem cells into cardiac muscle, or inhibit scar formation. The weight of the evidence indicates that bone marrow cells are not converted into cardiomyocytes after transplantation (Murry et al., 2004; Balsam et al., 2004; Laflamme and Murry, 2005; Chapter 13). There is general consensus that the mechanism behind functional improvement after bone marrow infusion is the participation of EnSCs in the formation of new vessels and the promotion of angiogenesis from existing vessels, consistent with results observed with the peripheral vasculature, as well as paracrine action of injected cells on heart tissue that enhances cardiomyocyte survival and limits the extent of the infarct. This consensus is supported by evidence that transplantation of noncontractile cells, such as fibroblasts, into infarcted myocardium, also improves hemodynamic function (Li, 1999; Sakai et al., 1999; Hutcheson et al., 2000).

### a. Observations from Animal Experiments

Injection of allogeneic cardiac endothelial cells or autogeneic bone marrow or peripheral blood mononuclear cells into the ischemic zone of infarcted pig hearts resulted in significantly increased regional blood flow, number of visible collateral vessels, capillary density, and myocardial contractility, as well as reduction in the size of the ischemic area (Fuchs et al., 2001; Kamihata et al., 2001, 2002). Cardiac vessel density and left ventricular contractility were increased after injecting MSCs directly into the infarct area of rat hearts, and the rate of dilatation of the myocardium was decreased (Davani et al., 2003). Injection of the G-CSF-mobilized, EnSC-enriched fraction of human peripheral blood, or culture-expanded human EnSCs from peripheral blood, into the ischemic areas of nude rat and pig hearts after coronary artery ligation significantly increased myocardial vessel density and blood flow, decreased cardiomyocyte apoptosis and left ventricular scarring, and improved systolic function (Kocher et al., 2001; Kawamoto et al., 2001, 2003).

The effect of injected bone marrow or circulating stem cells on the vasculature of infarcted myocardium appears to be twofold: the enhancement of angiogenesis from existing myocardial blood vessels through secretion of angiogenic factors and promoting the incorporation of EnScs into the new capillaries (Perin et al., 2003). Marked bone marrow cells injected into infarcted mouse hearts were reported to improve hemodynamic function by as much as 40% (Orlic et al., 2001). Marked cells were found in the endothelium of new myocardial blood vessels (Tomita et al., 1999; Orlic et al., 2001; Jackson et al., 2001). Kocher et al. (2001) isolated hemangioblasts from adult human bone marrow, expanded them *in vitro*, and injected the cells into rat myocardial infarcts induced by coronary ligation. By two weeks, there was a significant increase in the microvascularity of the infarct, with human cells accounting for 20–25% of the increase in endothelial cells. Kobayashi et al. (2000) injected bone marrow cells labeled with the intracellular dye CFSE directly into the ischemic area of infarcted rat hearts and found that neovessel formation was significantly enhanced over controls at one week postinfarction, although this difference was not sustained over subsequent weeks.

Bone marrow cells injected into infarcted pig hearts upregulated expression of the angiogenesis-promoting factors FGF-2, VEGF, IL-1$\beta$, and TNF-$\alpha$ (Kobayashi et al., 2000; Kamihata et al., 2001, 2002). VEGF and MCP-1 increased in a time-dependent manner in the medium of pig bone marrow cell cultures and enhanced, in a dose-dependent manner, the proliferation of cultured pig aortic endothelial cells (Fuchs et al., 2001). Injection of VEGF into myocardial infarcts of sheep significantly increased the number of capillaries in the peri-infarct region (Chachques et al., 2004). Plasma levels of VEGF increased markedly in human patients with acute myocardial infarction, peaking on day 7 after onset. When cultured, $CD34^+$ cells from 7-day postinfarction peripheral blood produced more cell clusters and differentiated endothelial cells than cells from 1-day postinfarction blood of human patients (Shintani et al., 2001).

Bone marrow cells home to the ischemic tissue of infarcted mouse or rat myocardium after infusion into the circulation or mobilized from endogenous marrow. Ciulla et al. (2003) showed that bone marrow mononuclear cells labeled with PKH26 dye and infused into the femoral vein of isogenic rats 7 days after cryodamage to the anterior left ventricle wall, homed to the infarcted area. Askari et al. (2003) showed that SDF-1 is transiently expressed by myocardium after infarction, but not at high enough concentrations to lead to engraftment of endogenous bone marrow derived cells mobilized with filgrastim. However, transplantation of cardiac fibroblasts stably transfected with SDF-1 induced homing of endogenous $[CD117 \text{ (c-kit)}, CD34]^+$ cells (HSCs and EnSCs) to the myocardial infarct area of rats after filgrastim administration.

### b. Clinical Trials of Bone Marrow Transplants in Myocardial Infarct Patients

The improvements in cardiac function observed in the infarcted left ventricles of animals after injection of bone marrow cells led quickly to clinical trials of bone marrow cell transplants to enhance the myocardial circulation in infarcted human hearts. Mononuclear bone marrow cells were delivered onto the endothelial surface of the ischemic left ventricle of 14 patients in end-stage heart failure by cardiac catherization through the aortic valve, using electromechanical mapping (Perin et al., 2003). Follow-up examinations four to six months later showed symptomatic improvement over 7 control patients who received no bone marrow cells in terms of increased contractility (improved ejection fraction), reduction in end-systolic volume, and increased cardiac blood flow. In another uncontrolled study of 8 patients given bone marrow cells in the same way, three-month follow-ups showed improvement in symptoms, blood flow, and ventricular function (Tse et al., 2003).

These studies were performed on patients who received no other surgical intervention at the time of bone marrow cell delivery. In other clinical studies, the cells have been delivered to patients undergoing coronary artery bypass graft (CABG) or balloon angioplasty and insertion of a stent. Stamm et al. (2003) reported the results of multiple injections of the EnSC-enriched fraction of autologous bone marrow along the circumference of the infarct border after completing a CABG on six patients. When seen 9–16 months after surgery, all had resumed their daily activities with no ill effects of the transplant, and they reported significant improvement in exercise capacity. This study, however, did not have a control group that received stent therapy only.

Two nonrandomized studies have been done in which patients who received stents plus infusions of bone marrow or culture expanded peripheral blood stem cells through the left anterior coronary artery were matched to equal numbers of controls who received stents only (Strauer et al., 2002; Assmus et al., 2002). Significant improvements over controls were reported 4–5 months later in the transplant group for contractility, end-systolic left ventricular volumes, stroke volume, and reductions in infarct size and wall motion abnormalities. No adverse side effects of the transplants were observed.

Wollert et al. (2004) conducted the first randomized, double blind clinical trial involving 30 control patients who received only stents in the coronary artery supplying the infarct area and 30 test patients who received stents and autologous bone marrow cells in the infarct artery. The baseline left ventricular ejection fraction (measured by MRI) of both groups was 51%, as opposed to 67% in healthy adults. Six months later, the left ventricular ejection fraction had increased by 0.7% in the stent-only controls, whereas bone marrow recipients experienced a 6.7% improvement. This is a nearly 10-fold improvement over controls, and ~42% of the difference between patient baseline and normal adult values. While this appears impressive, Hescheler et al. (2004) have cautioned that these studies were done mainly on patients with moderately impaired left ventricular function, whereas the real test is what these cell transplants can do for patients with severe cardiac failure (35% ejection fraction or less). Furthermore, the full range of side effects has not been assessed for these transplants. In one study, an increase in stent restenosis was noted after infusion of G-CSF-mobilized peripheral blood stem cells, even though ejection fraction was increased by 7% (Kang et al., 2004). The reason for this effect is unclear, but may be the result of stem cells differentiating into smooth muscle or the aggregation of inflammatory cells within new capillaries that develop from the stem cells within the atherosclerotic plaque.

The improvements in cardiac function observed in human patients who received bone marrow infusions can be attributed to enhancement of vascularization and possibly also to inhibition of scar formation, although stimulation of resident cardiac stem cells cannot be ruled out. Regardless of the mechanisms involved, functional improvement has been minimal. Perin et al. (2003) have outlined a number of important questions that need to be answered in order to improve results. What type of stem cells should be used to ensure the best engraftment and delivery of angiogenic factors and incorporation into neovessels? Should these be whole bone marrow, mononuclear cells, purified HSCs, MSCs, or EnSCs, or some combination of purified fractions? What quantity and concentration of cells should be used? The answers to these questions depend on understanding the mechanisms by which stem cells home, engraft, survive, and differentiate. Is the functional and morphologic cardiac improvement achieved by increasing myocardial contractility, by limiting infarct expansion and scarring, or both? What is the lifespan of the transplanted cells? What is the long-term safety of this therapy; e.g., is there a danger of tumorigenesis?

## 2. Satellite Cell Transplants

Another approach to restoration of injured myocardium has been to inject satellite cells into the area of injury. Although Reinecke et al. (2002) demonstrated conclusively that adult rat SCs are not converted to cardiomyocytes when transplanted into the normal hearts of syngeneic rats (Chapter 13), the injection of SCs has been reported to improve myocardial function after infarct.

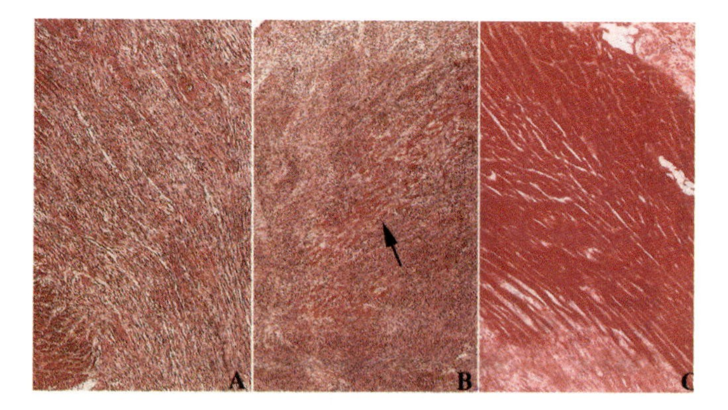


**FIGURE 12.11** H and E-stained sections of cryoinjured rabbit myocardium. **(A)** Infarct region showing loss of viable cardiomyocytes. **(B)** Three weeks after injecting skeletal myoblasts into the injury region. Arrow indicates diffusely dispersed myogenic tissue. **(C)** A different heart three weeks after injection of myoblasts into the injury region, showing highly organized myogenic tissue in the center of the cryo-induced scar. **B** and **C** are stained for myogenin. Reproduced from Atkins et al., Myocardial cell transplantation improves *in vivo* regional performance in infarcted rabbit myocardium. Cardiac Vasc Reg 1:43–53. Copyright 2000, Blackwell Publishing Co.

Koh et al. (1993) showed that C2C12 myoblasts injected into the uninjured myocardium of syngeneic mice withdrew from the cell cycle and formed typical myotubes. No overt cardiac arrythmias were noted and the donor cells survived for as long as three months. Data were not obtained as to whether the skeletal myotubes became electrically coupled to host cardiomyocytes or whether their contractile activity contributed to the function of the myocardium. Injection of SCs into cryo-injured or coronary artery-ligated hearts of rats, rabbits, and pigs was reported by several investigators to improve cardiac function, even though the myofibers differentiated from the SCs did not electrically couple to the native cardiomyocytes (Chiu et al., 1995; Taylor et al., 1998; Kao et al., 2000, 2001; Atkins et al., 2000a, b; Laflamme and Murry, 2005). **FIGURE 12.11** shows the results of one such experiment.

Satellite cell transplants have been used to treat myocardial failure in human patients. The first such transplant was carried out by Menasche (2000) in a Phase I trial on a 72-year-old patient suffering from severe congestive heart failure caused by extensive myocardial infarction. Satellite cells were isolated from a quadriceps muscle biopsy, expanded *in vitro* for two weeks and $800 \times 10^6$ cells (65% myoblasts) delivered into the myocardial scar via 30 injections with a small-gauge needle. Simultaneously, a double bypass was performed in viable but ischemic areas of the myocardium. Six months later, the patient's symptoms were dramatically improved. Echocardiogram showed evidence of new-onset contraction and fluoro-deoxyglucose PET scan showed increased metabolic activity of the infarct. The improvement was considered unlikely to be due to increased collateralization from the bypass region, because this region was far from the infarct.

Since then, the results of two larger-scale clinical trials have been reported. Menasche et al. (2003) treated 10 patients with severe left ventricular dysfunction (average left ventricular ejection fraction, 24%) by injection of cultured autologous SCs. Sixty-three percent of the cell-implanted infarcts were observed by echocardiography to have improved systolic contraction. The treatment increased the ejection fraction by 8%, which was not enough to improve above the 35% that defines severe dysfunction. In another study of five patients with average ejection fractions of 36%, Smits et al. (2003) injected cultured autologous satellite cells transendocardially into the myocardial infarct. At six months, the ejection fraction had improved by 9%, to an average of 45%, and MRI examination showed significant wall thickening in the target areas.

The improvements in cardiac function observed in these trials could have been due to augmentation of systole by the contraction of skeletal muscle fibers differentiated from the transplanted SCs. Other possibilities are that the transplanted cells make the scar region more pliable, or that SCs provide factors to the host myocardium that enhance survival of cardiomyocytes and inhibit the formation of scar tissue by fibroblasts. Regardless of mechanism, both studies revealed a potential problem that had not been apparent in animal studies. Several patients in both studies suffered episodes of ventricular arrhythmia. The cause of the arrhythmia is unknown, but might have been due to the lack of electrical coupling of donor skeletal muscle myofibers to host cardiac myofibers (Minami et al., 2003; Makkar et al., 2003).

## 3. Cardiomyocyte Transplants

The limited success of bone marrow or satellite cell transplants suggests that the best cell therapy for myocardial infarction is the transplantation of cardiomyocytes. Soonpa et al. (1994) established proof of concept that grafted cardiomyocytes could integrate into host heart tissue by injecting suspensions of fetal cardiomyocytes from *lac-Z* transgenic mice into the uninjured ventricular myocardium of syngeneic host mice (**FIGURE 12.12**). The injected cells continued to proliferate and

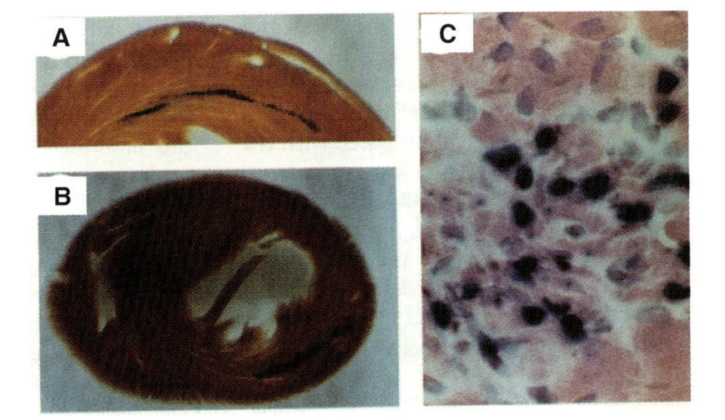


**FIGURE 12.12** Histological analysis of normal mouse myocardium that received an injection of cardiomyocytes from syngeneic fetal mice transgenic for *Lac-Z*. **(A, B)** Coronal sections of hearts showing β-gal expression by grafted cells. **(C)** High power view of grafted cardiomyocytes within the myocardium stained with H&E plus β-gal activity. Reproduced with permission from Soonpa et al., Formation of nascent intercalated discs between grafted fetal cardiomyocytes and host myocardium. Science 264:696–698. Copyright 1994, AAAS.

differentiated into mature cardiac muscle integrated with that of the host. Electron microscopic analysis indicated that the donor cells formed intercalated discs with the host myofibers, suggesting donor/host electrical coupling. No cardiac arrythmias were noted. Muller-Ehmsen et al. (2002) found that the survival of neonatal cardiomyocytes of male rats transplanted into syngeneic females decreased over time to ~15% at 12 weeks post-transplant. The loss of cells was not due to apoptosis, since the caspase inhibitor AcYVADcmk failed to improve survival. Similar results, with improvement of cardiac function, were obtained after transplanting fetal cardiomyocytes into the ventricular myocardium of dystrophic dogs (Koh et al., 1995) and the infarcted hearts of rats (Leor et al., 1996; Li et al., 1996, 1997; Yao et al., 2000) and pigs (Li et al., 2000). In these studies, the transplanted cells differentiated cross-striations, but the donor-derived cardiomyocytes were smaller than the host cardiomyocytes and were disoriented in their organization, suggesting that the transplanted cells achieved only the differentiation level of immature cardiomyocytes (Yao et al., 2000; Muller-Ehmsen et al., 2002).

$Lin^-$ c-$kit^+$ cardiac stem cells of adult rats are able to regenerate cardiac muscle when injected into the borders of myocardial infarcts (Beltrami et al., 2003) (**FIGURE 12.13**). Cardiomyocytes for transplant have also been generated from ESCs. Mouse ESC-derived cardiomyocytes *in vitro* were stably integrated into the ventricular myocardium of *mdx* dystrophic mice after transplantation, as shown by the presence of donor cells

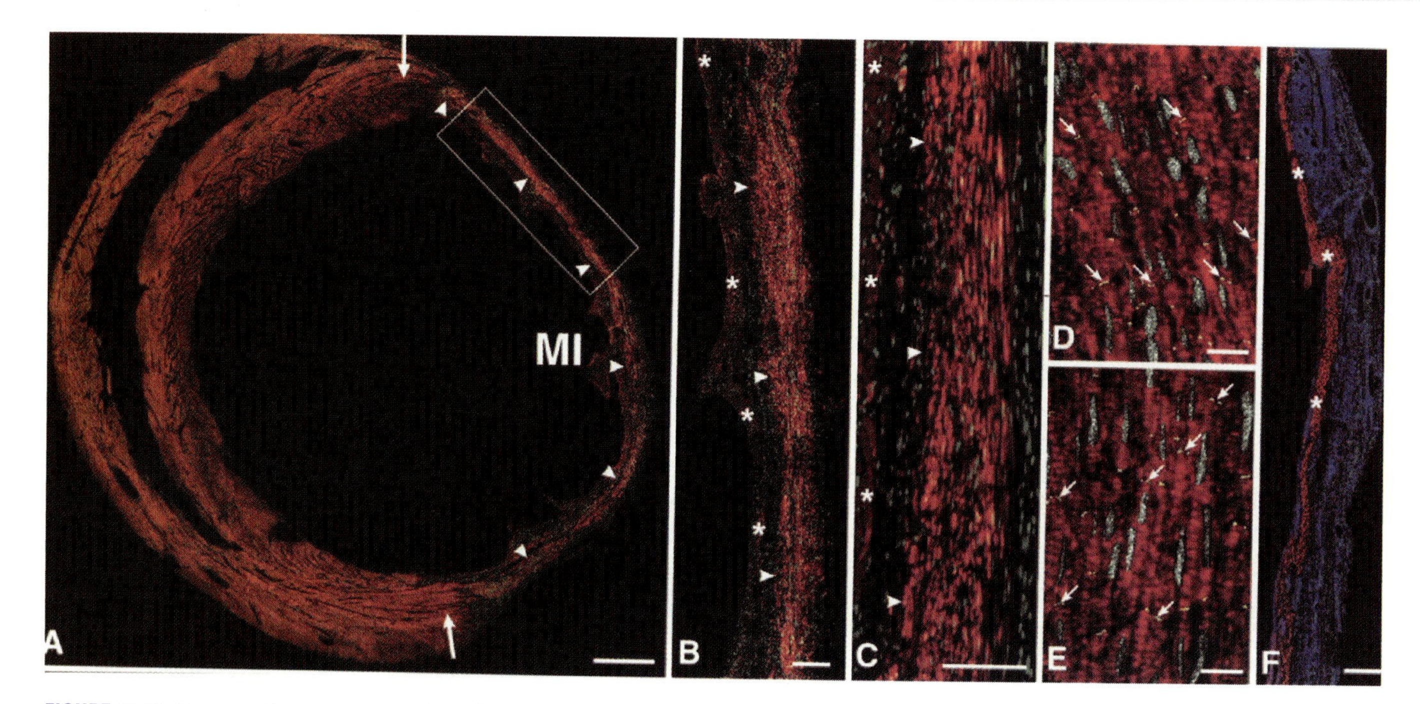


**FIGURE 12.13** Regenerating rat myocardium 20 days after injecting $Lin^-$ c-$kit^+$ cardiac stem cells into the border of a myocardial infarct. **(A)** Arrows indicate the sites of cell injection. MI = infarction area. Arrowheads indicate regenerated myocardium staining for cardiac myosin (red). (B, C) Higher magnification views of the boxed area in **A**, showing the regenerated cardiac myofiibers (arrowheads). Nuclei are green (propidium iodide stain). **(D, E)** Arrows indicate intercalated discs stained for connexin (yellow) in **D** and N-cadherin (yellow) in **E**. Nuclei are white-green in these sections (BrdU-PI). **(F)** Collagenous scar in non-treated infarct. Collagen is blue. Asterisks indicate spared endocardium. Reproduced with permission from Beltrami et al., Adult cardiac stem cells are multipotent and support myocardial regeneration. Cell 114:763–776. Copyright 2003, Elsevier.

positive for dystrophin (Klug et al., 1995, 1996). Johkura et al. (2003) reported that ESC-derived mouse cardiomyocytes transplanted into the retroperitoneum of adult nude mice became vascularized and differentiated into cardiac myocytes that expressed cardiac molecular markers and exhibited desmosomes, zona adherens, and gap junctions. Human ESC-derived cardiomyocytes proliferated and formed early sarcomeres when transplanted into the hearts of nude rats, whereas undifferentiated ESCs formed teratomas (Laflamme and Murry, 2005) (**FIGURE 12.14**), suggesting that cardiac muscle does not have the necessary factors to direct the differentiation of naïve ESCs.

The ability of ESC-derived human cardiomyocytes to function as pacemaker cells was tested by Kehat et al. (2004). They first demonstrated that human cardiomyocytes derived from embryoid bodies became mechanically and electrically coupled to neonatal rat ventricular cardiomyocytes in co-culture, beating in synchrony with the rat cells. ESC-derived human cardiomyocytes were then injected into the left ventricle of pigs in which a complete atrioventricular block had been induced by ablating the bundle of His. The transplanted cells restored normal electrical rhythm. Immunostaining with antihuman mitochondrial antibodies confirmed the presence of human cardiomyocytes in the hearts that were integrated with host cells. These cells reacted with α-actinin antibodies, but again were smaller, with morphology typical of embryonic-like cardiomyocytes.

Transplants of human fetal or neonatal cardiomyocytes for clinical purposes are neither practical nor ethical, and cardiomyocytes derived from human ESCs are subject to immunorejection. Bone marrow stem cells do not change fate to become cardiomyocytes *in vivo*, but MSCs can be made to differentiate into cardiomyocytes *in vitro* by treatment with 5-azacytidine (Makino et al., 1999). Cardiomyocytes derived in this way from autologous MSCs might then be used to regenerate functional myocardial tissue. Following this reasoning, Tomita et al. (1999) treated adult rat bone marrow cell cultures with 5-azaC and identified cells in the cultures with cardiac contractile proteins. BrdU-labeled cells were injected into cryoinjured ventricular myocardium. Control rats received bone marrow cells that were not treated. Both 5-azaC-treated and control cells appeared to differentiate into cardiomyocytes and to induce angiogenesis, but only the 5-azaC-treated

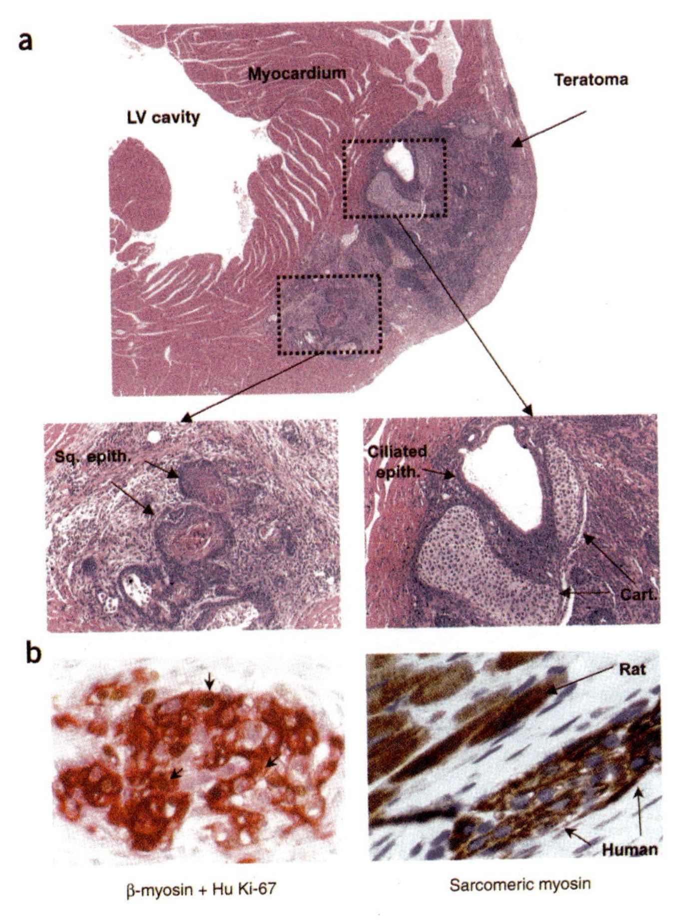


**FIGURE 12.14** Results of implanting embryonic stem cells (ESCs) and their derivatives into the myocardium of immunodeficient mice and rats. **(a)** Undifferentiated mouse ESCs formed teratomas after being implanted into immunodeficient mice. LV = left ventricle. Boxed areas are magnified to show the squamous epithelium, ciliated epithelium and cartilage that differentiated within the myocardium. **(b)** ESC-derived human cardiomyocytes transplanted into nude rats proliferated and differentiated into cardiac myofibers. Left panel shows implanted cells stained for β-myosin heavy chain and human Ki-67, which appears brown in proliferating cells (arrows). Right panel shows human cardiomyocytes at a stage of early sarcomere formation (arrows). The section is stained with a human-specific genomic probe and antibody to sarcomeric myosin (brown). Reproduced with permission from Laflamme and Murry, Regenerating the heart. Nature Biotech 23:845-856. Copyright 2005, Nature Publishing Group.

cells improved cardiac function. Bittira et al. (2002) and Saito et al. (2003) delivered labeled 5-azaC-treated MSCs to infarcted rat hearts before the cells had actually differentiated into cardiomyocytes. Again, the treated MSCs appeared to differentiate into cardiomyocytes within and around the scar, whereas untreated MSCs did not.

**FIGURE 12.15** summarizes the types of cell transplant therapies that have been used as therapy for myocardial infarct.

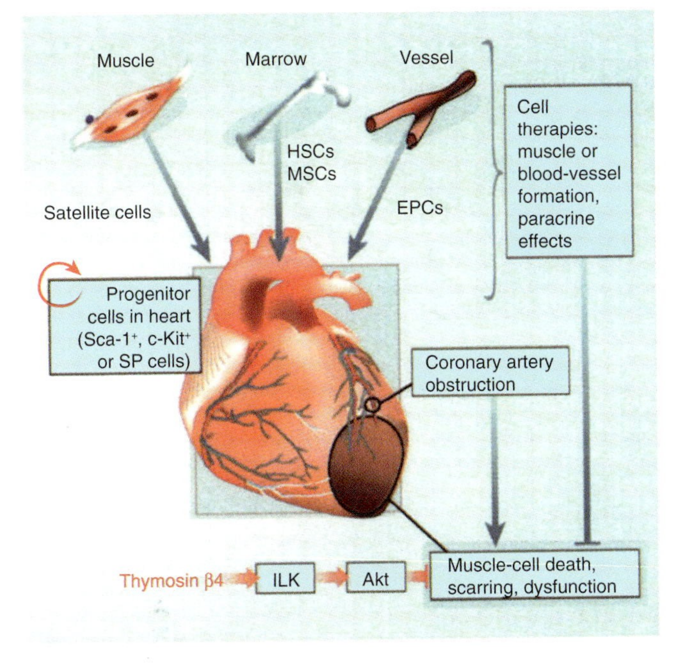


**FIGURE 12.15** Summary of the types of cell therapies that have been investigated for myocardial infarct and how they might work. Redrawn with permission after Schneider, Prometheus unbound, Nature 432:451–452. Copyright 2004, Nature Publishing Group.

## 4. Soluble Factors and Myocardial Regeneration

The evocation of angiogenesis in the peripheral and cardiac vascular system by factors secreted from transplanted stem cells has prompted the question of whether similar factors might promote the survival and regeneration of myocardium.

**FIGURE 12.16** illustrates an experiment in which fetal liver cells appeared to either stimulate or inhibit regenerative capacity of the myocardium in mice, depending on whether they were derived from wild-type or MRL/lpr donors. Fetal liver cells transplanted from male MRL/lpr mice, which can regenerate cardiac muscle, to irradiated B6 female mice reconstituted the hematopoietic system of the hosts through HSCs present in the fetal liver. The myocardium of the B6 mouse normally scars after cryoinjury, but after receiving MRL/lpr fetal liver cells it regenerated without scarring (Bedelbaeva et al., 2004). The regenerated cardiac myofibers contained many $Y^+$ nuclei and immunostained positively with antibodies to the cardiac markers MEF-2C and Nkx2.5 and cardiac troponin T, suggesting their origin from the transplant, although the identity of the cells was not certain. Cells from an adult liver stem cell line have also been reported to differentiate into cardiomyocytes *in vitro* and *in vivo* (Malouf et al., 2001; Muller-Borer et al., 2004). Interestingly, transplantation of

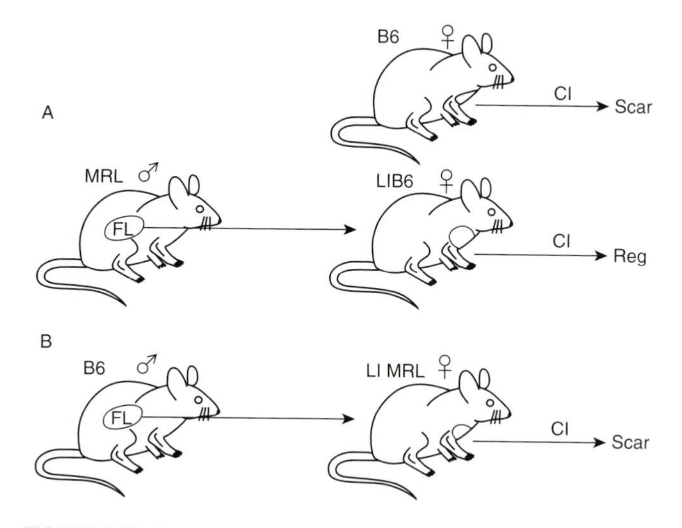


**FIGURE 12.16** Cross-injection of fetal liver cells (FL) between MRL and B6 male and female mice. The B6 myocardium normally scars after cryoinjury (CI), whereas the MRL myocardium regenerates. **(A)** Fetal liver cells from MRL male mice injected into lethally irradiated (LI) B6 female mice conferred regenerative ability on the B6 myocardium. **(B)** Conversely, fetal liver cells from B6 male mice injected into LI MRL female mice suppressed regeneration of the MRL myocardium.

fetal liver cells from B6 male mice into irradiated female MRL mice resulted in myocardial scarring after cryoinjury (Bedelbaeva et al., 2004). Whether or not donor $Y^+$ cells were incorporated into the scar tissue of the host myocardium was not reported. It is possible that in these experiments immune cells differentiating from HSCs of the B6 fetal liver induce host MRL cardiac fibroblasts to make a scar. Conversely, MRL fetal liver cells might supply immune cells to B6 hosts that induce a regenerative response by paracrine action, rather than differentiating into cardiomyocytes. Thus, as postulated for mammalian skin (Chapter 2) and amphibian limbs (Chapter 14), the difference between B6 and MRL mice in cardiac tissue repair may reside in differences in immune cells, particularly in the cytokines they produce.

In another series of experiments, Fraidenreich et al. (2004) showed that the epicardium produces short- and long-range signals essential for myocardial development, and by extension, for regeneration (**FIGURE 12.17**). Id transcription factors are not expressed in the myocardium, but are expressed in the epicardium. Knockout of any two of the three *Id1, 2*, or *3* genes in

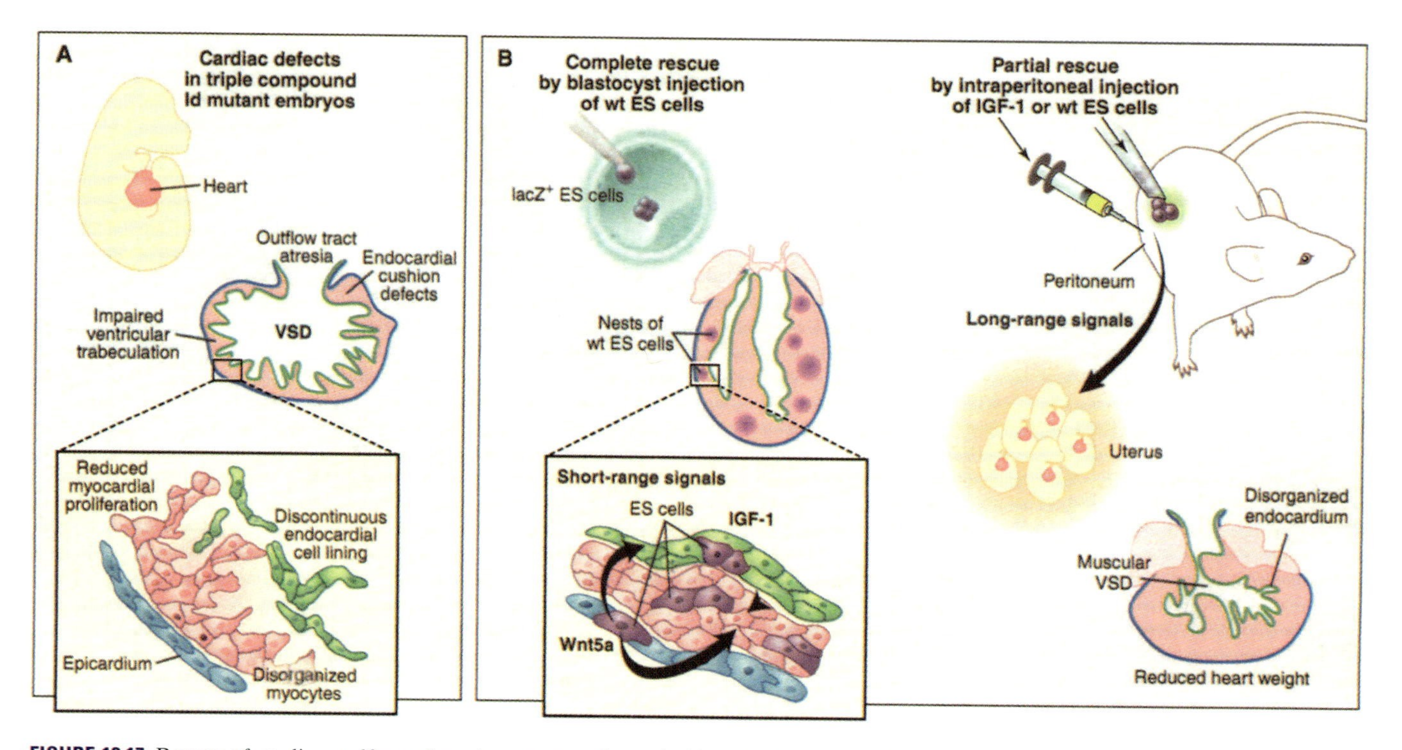


**FIGURE 12.17** Rescue of cardiac malformations in mutant embryos lacking the *Id* genes. **(A)** The mutant hearts have a thin myocardial wall, impaired ventricular trabeculation, endocardial cushion defects, outflow tract atresia and ventricular septal defects (VSD). Box illustrates reduced myocardial cell proliferation and a discontinuous endocardial cell lining. **(B)** Wild-type (wt) ESCs transgenic for *lacZ* effected complete rescue of the hearts when injected into mutant blastocysts (left), but only partial rescue when injected intraperitoneally into the mothers, suggesting that short-range and long-range *Id*-dependent signals are required for normal heart development. The short-range signal was identified as Wnt5a and the long-range signal as IGF-1. Reproduced with permission from Chien et al., ES cells to the rescue. Science 306:239–240. Copyright 2004, AAAS.

mice is lethal, due to a set of common cardiac defects in which cardiomyocyte proliferation is significantly reduced, myocardium is thinned, ventricular trabeculation is impaired, and ventricular septal defects appear (Fraidenreich et al., 2004). The epicardium, however, appears normal. These observations suggested that Id proteins are necessary for normal heart development via signaling from the epicardium to the myocardium. Fraidenreich et al. (2004) showed that culturing knockout cardiomyocytes in medium conditioned by wild-type epicardial cells rescued the proliferative defect, whereas conditioned medium from the knockout epicardium did not. Injection of knockout blastocysts with as few as 15 wild-type ESCs rescued cardiac development, including gene expression profiles. Furthermore, transplant of ESCs into the peritoneal cavity of knock-out-producing female mice prior to mating partially rescued cardiac development of the embryos, indicating that the cells were secreting a factor that could operate over distance. Microarray analysis was performed to identify factors expressed and secreted by the epicardium that were downregulated in *Id* knockout cardiomyocyte cultures or upregulated in wild-type cultures. IGF-1 was identified as the long-range factor, being able to mimic the partial rescue effect of maternally implanted ESCs, and Wnt5a as a factor that acts in a paracrine fashion on cardiomyocytes to bring about complete rescue. The expression of both is compromised in the epicardium of knockout mice.

Thymosin β4, which promotes skin wound repair (Chapter 4), also enhances the survival of cardiomyocytes in infarcted hearts (**FIGURE 12.18**). Bock-

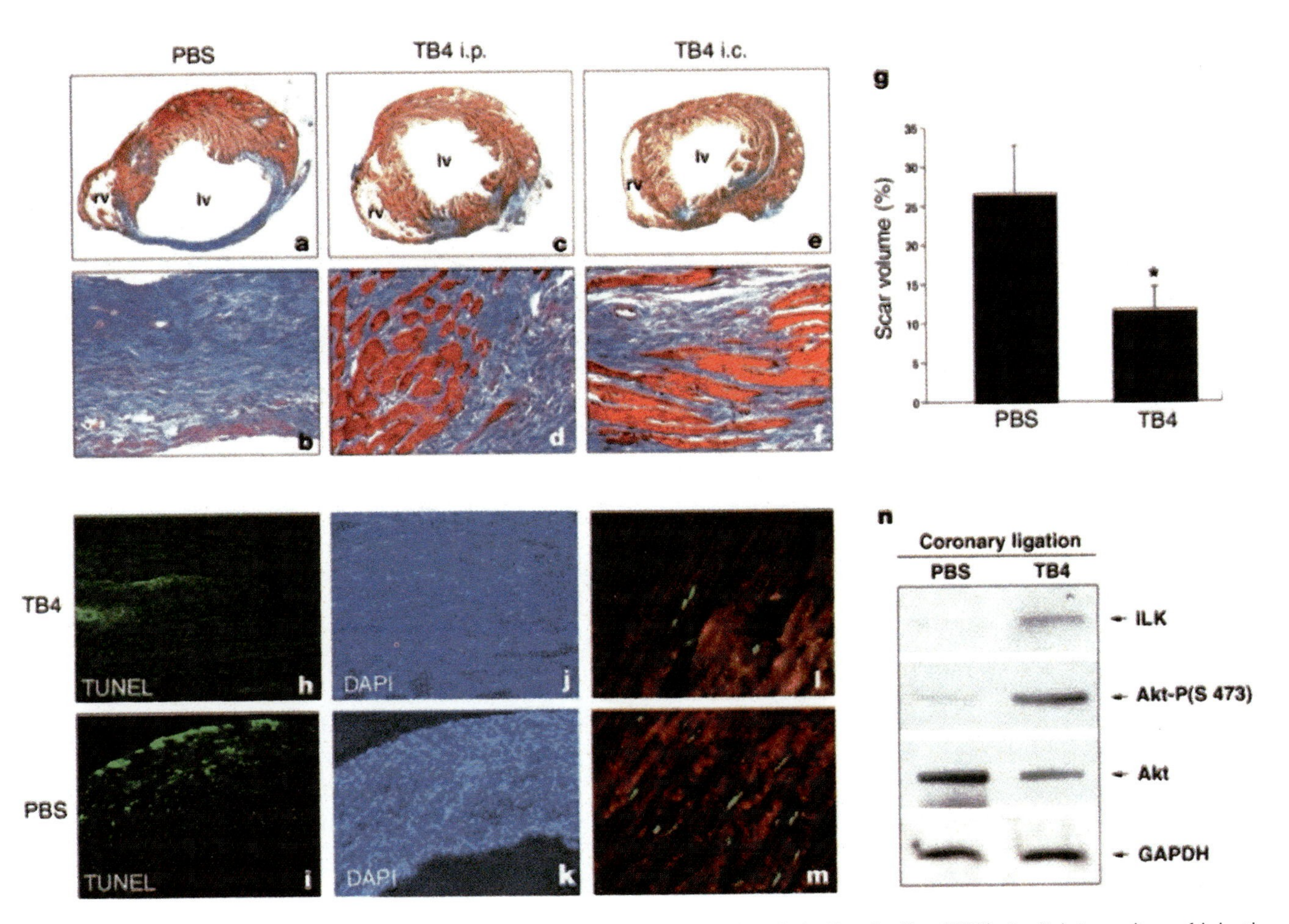


**FIGURE 12.18** Histological sections of infarcted mouse hearts treated with **(a, b)** phosphobuffered saline (PBS), **(c, d)** intraperitoneal injection of thymosin β4 (TB4 i.p.), **(e, f)** intracardiac injection of TB4. Collagen stains blue and cardiac myosin red. The TB4-treated infarct area contains many regenerated muscle fibers, whereas the PBS control infarct is completely fibrotic. lv = left ventricle, rv = right ventricle. **(g)** Quantitative analysis of scar volume shows that TB4 significantly reduces scar volume. **(h, i)** TUNEL staining reveals significantly less cardiomyocyte apoptosis (bright green) after TB4 treatment of the infarcted heart. **(l, m)** higher magnification of cardiac myofibers (stained red with anti-muscle actin antibody) showing fewer apoptotic cardiomyocytes (green) in infarcted hearts treated with TB4. **(j, k)** DAPI stain. **(n)** Western blot showing higher levels of phosphorylated (activated) AKT gene in TB4-treated infarcts. Reproduced with permission from Bock-Marquette et al., Thymosin β4 activates integrin-linked kinase and promotes cardiac cell migration, survival and cardiac repair. Nature 432:466–472. Copyright 2004, Nature Publishing Group.

**TABLE 12.3** Effect of transplanting MSCs transfected with *LacZ* (transfection control) or *Akt* constructs into mouse myocardial infarcts on infarct volume (V infarct, $mm^3$), left ventricular systolic performance (LVSP, mm Hg/gm), area of heart occupied by collagen (%), and cardiomyocyte diameter (μm), a measure of hypertrophy. Data from Mangi et al. (2003)

| | Sham | Saline | *Lac Z (#MSCs)* | | *AKT (#MSCs)* | |
|---|---|---|---|---|---|---|
| | | | $2.5 \times 10^5$ | $5 \times 10^6$ | $2.5 \times 10^5$ | $5 \times 10^6$ |
| V infarct | — | 140 | 115 | 85 | 60 | 15 |
| LVSP | 250 | 155 | 150 | 175 | 200 | 210 |
| Area of Collagen | 1 | 8 | — | 2 | — | 1 |
| Diameter of cardiomyocytes | 13 | 23 | — | 18 | — | 13 |

Marquette et al. (2004) reported that thymosin β4 expression is enriched in endothelial cells and migratory cardiomyocytes involved in the formation of cardiac valves, ventricular trabeculae, and outflow tract during embryonic development. The molecule binds to cultured endothelial cells and cardiomyocytes and enhances their migration, as well as enhancing the survival of cardiomyocytes of neonatal and two-week-old rats. Bock-Marquette (2004) found that thymosin β4 exerts its effects by activating the cell survival enzyme Akt (protein kinase B), which interferes with apoptotic pathways and renders heart tissue resistant to hypoxia (Fujio et al., 2000). This activation is dependent on an interaction of thymosin β4 with integrin-linked kinase (ILK). Administering thymosin β4 directly to the heart or intraperitoneally to mice immediately after coronary artery ligation more than doubled (from 28% to 58%) the ejection fraction of the infarcted hearts over controls, though this was still short of the 75% observed in sham-operated mice. This recovery was associated with significant reduction in scar and a marked decrease in cardiomyocyte apoptosis (Bock-Marquette et al., 2004).

These results are consistent with those of Mangi et al. (2003). They transfected rat MSCs with the *Akt* gene and injected $5 \times 10^6$ GFP-labeled Akt-MSCs from male donors into the infarcted myocardium of female hosts. The injected gene greatly increased survival of the transplanted cells (**TABLE 12.3**). Left ventricular function was completely restored and the infarct region was reduced to nearly zero, consistent with the notion that the injected cells exerted a powerful inhibitory effect on scar tissue (infarct) formation through paracrine factors.

Direct evidence for the paracrine action of MSCs in limiting the extent of myocardial infarct has been provided by conditioned medium experiments (Gnecchi et al., 2005) (**TABLE 12.4**). Exposure of hypoxic rat ventricular cardiomyocytes *in vitro* to conditioned medium derived from hypoxic cardiomyocytes transfected with the *Akt* gene raised the number of surviving cardiomyocytes by 40% compared to controls. Injection of concentrated conditioned medium from *Akt*-transfected hypoxic cardiomyocytes into five different sites of the infarct border zone 30 min after left coronary artery occlusion reduced the size of the infarct by 40% and apoptosis of cardiomyocytes by 69%.

**TABLE 12.4** Results of (**A**) injection into infarcted mouse heart myocardium of concentrated conditioned medium from cultures of MSCs grown for 12 hr under hypoxic conditions, and (**B**) exposure *in vitro* of adult rat cardiomyocytes subjected to hypoxia for 24 hr to concentrated conditioned medium from cultures of MSCs grown for 12 hrs under hypoxic conditions. The conditioned medium was derived from MSCs that were either untransfected (control), transfected with green fluorescent protein (GFP) or transfected with the AKT gene. Data from Gnecchi et al. (2005)

| | Baseline | *Conditioned Medium* Control | GFP | Akt |
|---|---|---|---|---|
| **A.** Intramyocardial Injection | | | | |
| Infarct size (% of left ventricle) | — | 35 | 30 | 13 |
| Apoptotic Index (%) | — | 10 | 8.5 | 3 |
| **B.** In vitro | | | | |
| # of cardiomyocytes per field | 153 | 70 | 100 | 130 |

The impaired contractile performance of the failing heart is associated with reduced levels of $Ca^{++}$ sensing and regulatory proteins, one of which is S100A1, a low-molecular-weight calcium-binding protein (Donato, 2003; Tsoporis et al., 2003). S100A1 overexpression strengthens the contractile force of the myocardium and inhibits the apoptosis of ventricular cardiomyocytes (Most et al., 2003a, b). Adenoviral constructs containing the S100A1 gene reversed the contractile dysfunction of cardiomyocytes obtained from infarcted rat hearts. These constructs also normalized S100A1 protein expression in the infarcted rat heart *in vivo* and improved ventricular function (Most et al., 2004).

## 5. Bioartificial Cardiac Muscle

The ability to replace a patch of myocardium damaged by ischemia with pieces of bioartificial myocardium may be the most immediately useful way to reverse heart failure; thus a number of groups are attempting to make bioartificial cardiac muscle by seeding cells into biodegradable scaffolds (reviewed by Zimmerman et al., 2004). Bioartificial myocardium, however, is even more challenging to construct than bioartificial tissues of the musculoskeletal or vascular system, because not only must it be integrated structurally and electrically into the host myocardium, it must be flexible and able to contract.

Cardiomyocytes from embryonic chicks, fetal rats, neonatal rats, and children undergoing heart repair (teratology of Fallot) have all been grown in two-dimensional or three-dimensional (planar biconcaval or ring configuration) scaffolds of solid Gelfoam, alginate or collagen I, or a liquid matrix composed of collagen I and Matrigel (Li et al., 1999; Carrier et al., 2000; Zimmerman et al., 2000; Kofidis et al., 2002; van Luyn et al., 2002). The cardiomyocytes differentiated well beyond their initial state, showed continuous rhythmic and synchronized beating for three months or more, and responded to electrical, $Ca^{++}$, and epinephrine stimulation with increased force generation. Differentiation and tissue architecture, as well as generation of contractile force, was enhanced by subjecting the scaffold to stretch (Fink et al., 2000; Zimmerman et al., 2001; Kofidis et al., 2002; Akhyari et al., 2002).

Bioartificial myocardium consisting of fetal cardiomyocytes in Gelfoam or alginate has been implanted both subcutaneously and into the scar of infarcted rat hearts (Li et al., 1999; Leor et al., 2000). In both cases, blood vessels grew into the implant, which became integrated into the host myocardial tissue, formed gap junctions, and contracted vigorously. Control rats showed dilatation of the left ventricle and deterioration in contractility, but implants into the myocardial scar attenuated dilatation and there was no decrease in contractility. The scaffold was almost completely biodegraded over the course of the experiments.

A major problem with grafts of bioartificial myocardium is the distortion of the ventricular wall geometry caused by the stiffness of the scaffold. To counter this problem, Eschengagen et al. (1997; Zimmerman et al., 2004) have formulated a liquid scaffold of collagen type I and Matrigel. Such a matrix, liquid at room temperature, but solidifying at body temperature, was seeded with EGFP-labeled rat ESCs and injected into the myocardial infarcts of rat hearts removed from the thoracic cavity (Kofidis et al., 2004). These hearts were then transplanted to host rats in a way that reperfused them through the coronary circulation, but without ejecting blood (the "nonworking heart" model). Despite the lack of ejection, the ventricles nevertheless contracted vigorously. Hearts injected with these constructs contracted nearly as well as transplanted, uninjured hearts and twice as well as infarcted hearts (no implant) and infarcted hearts that received ESCs alone. Hearts that received the scaffold alone contracted at a level intermediate between the latter two groups and the hearts that received the complete bioartificial tissue.

Another way to get around the deformation issue is to make myocardium by layering cell sheets. Shimizu et al. (2002) fabricated sheets of pulsatile cardiac tissue by growing neonatal rat cardiomyocytes on the temperature-responsive polymer poly (N-isopropylacrylamide) (PIPAAm). Confluent cell sheets detach from the polymer when the temperature is reduced and are layered on top of one another. Constructs made of four sheets establish intercellular communication through connexin 43. When these constructs were implanted subcutaneously into nude rats, they were observed macroscopically to beat. Histological examination showed typical myocardial tissue that was vascularized. The constructs survived through the 12 weeks of the experiment.

Krupnick et al. (2004) tested the efficacy of a scaffold consisting of a three-dimensional Matrigel/type I collagen hydrogel supported by a porous polylactide biodegradable mesh and reinforced with Gore-Tex. This material could be successfully stitched into the ventricular wall of rats. MSCs derived from fetal sheep liver were used to seed the scaffold, based on the idea that the MSCs would differentiate into cardiomyocytes. The construct was implanted into the left ventricle of an immunocompromised rat. The MSCs differentiated into spontaneously contracting cells that expressed cardiomyocyte-specific proteins, but due to diffusion limitations, only those cells on the periphery of the construct close to the host blood supply survived, whereas cells toward the center did not survive. The latter problem is similar to what others have experienced as well (Carrier et al., 1999).

Christman et al. (2004) reported that fibrin glue alone or in combination with skeletal myoblasts induced neovasculature formation in the hearts of myoinfarcted rats. Functional tests and histological evaluation showed that infarct wall thickness and myocardial function was preserved over a five-week period, compared to the deterioration observed in controls, suggesting that fibrin may be useful as a scaffold for making bioartificial myocardial tissue.

## SUMMARY

Hematopoietic diseases are among the greatest causes of mortality in children, and cardiovascular diseases are the number-one cause of mortality and morbidity in adults. Thus, it is not surprising that regenerative therapies for these diseases have been intensively pursued for decades.

Ablative chemotherapy and/or irradiation of the hematopoietic system and allogeneic transplantation of bone marrow or peripheral blood stem cells have been in clinical use since 1968 to treat hematopoietic diseases. Ablation destroys dividing cells and the hematopoietic system is slowly regenerated from a combination of host HSCs and donor HSCs. The introduction of donor T cells leads to mixed lymphocyte chimerism, which induces tolerance to the donor HSCs and their progeny. The main problems encountered in these procedures are infection, low engraftment of donor cells, and graft versus host disease. Sterile environment and antibiotics are used to deal with infection. One promising approach to increase engraftment of HSCs is to co-inject MSCs, which are essential to HSC survival and differentiation. Graft versus host disease is caused by attack of donor T cells on host organs and is managed with immunosuppressive drugs. However, donor T cells also attack malignant host cells, the graft versus leukemia effect. Thus, the problem now is how to balance this double-edged sword in favor of the GVL effect to combat disease. One approach is to use umbilical cord blood as a source of cells for transplant. GVHD is not as severe in cord blood transplants, and there does not appear to be any loss of GVL effect. Furthermore, larger HLA mismatches are tolerated and patient survival is just as good or better than for bone marrow or peripheral stem cell transplants.

Ablative procedures are very toxic and cannot be tolerated by patients who have additional medical problems. A less toxic approach, called nonablative therapy, uses lower doses of chemotherapy or irradiation to immunosuppress the recipient hematopoietic system only to a degree that allows engraftment of donor HSCs. Not all the diseased cells are wiped out by the lower-dose chemotherapy and irradiation, but mixed chimerism of T cells is greater in this approach, harnessing the full force of the GVL effect to kill the remaining ones. Though needing longer-term assessment, nonablative hematopoietic transplants appear to be associated with more stable chimerism and tolerance, less severe GVHD, fewer relapses, and survival rates equal to or better than ablative procedures. The possibility now exists of combining nonablative therapy with umbilical cord blood transplants using protocols that maximize engraftment, tolerance, and GVL effect, and minimize GVHD.

Gene therapy has been used to combat severe combined immunodeficiency disease in children due to ADA deficiency (ADA-SCID) and the X-linked disorder SCID-X1, which is caused by $\gamma c$ cytokine deficiency. Autologous peripheral T-cells were transfected with the ADA or $\gamma c$ genes in retroviral vectors and the transfected cells infused back into the patients. Eighteen patients had been successfully treated as of 2004, but two patients developed leukemia, putting the trials on hold. The development of leukemia was attributed to insertion of the retrovirus near a cancer-promoting gene. Nevertheless, as more is learned how to avoid these kinds of problems, gene therapy will become a powerful weapon against genetic hematopoietic diseases.

The ability to regenerate blood vessels and to create bioartificial blood vessels would be a major step forward in treating vascular disease. Transplants of endothelial stem cells have been used to augment EnSC incorporation into new capillaries, and acellular regeneration templates have been used to promote regeneration across gaps in large- and small-diameter blood vessels. Injection of EnSCs, and factors that increase the homing and proliferation of EnSCs, into ischemic muscles of mice resulted in enhanced angiogenesis and blood flow in the muscles. The effective factors were the chemoattractant SDF-1 and the growth factors VEGF and FGF-2. Injection of autogeneic bone marrow cells into the ischemic gastrocnemius muscles of patients resulted in significant improvements in ankle-brachial index, transcutaneous oxygen pressure, rest pain, and painfree walking time, suggesting that EnSCs in the injected bone marrow cells enabled the significant regeneration of new capillaries. The statin simvastatin accelerates re-endothelialization of injured carotid artery, suggesting that patients who take these cholesterol-lowering drugs may derive the additional benefit of an enhanced artery repair process.

Synthetic and biological acellular regeneration templates have been used to induce the regeneration of blood vessels or create bioartificial blood vessels. Poly(lactic-coglycolic acid) impregnated with a mixture of bovine collagens and IV and cross-linked with glutaraldehyde evoked the regeneration of dog pulmonary artery across $2 \times 15$ mm patch defects. The regenerated tissue was composed of all layers and there was no intimal thickening or thrombosis. Tubular pig SIS grafts induced the regeneration of short segments of the femoral and carotid arteries in dogs that was histologically similar to normal artery. Grafts of human arterial matrix have also been found to promote the regeneration of endothelium, smooth muscle, and fibroblasts from host vessels. Bioartificial blood vessels have been created by two methods. In the first method, smooth muscle cells were seeded into biodegradable synthetic scaffolds such as poly-4-hydroxybutane and polyglycolic acid and the constructs cultured under pulsatile flow for several weeks, followed by seeding endothelial cells into the lumen. In the second method, no scaffold was used. A sheet of smooth muscle was wrapped on a mandrel and cultured under flow. Fibroblasts were then added on the outside, the mandrel removed, and endothelial cells were added to the inside. The histological structure of these arteries was similar to that of normal arteries and had burst strengths comparable to, or higher than, human saphenous vein. Three-dimensional vascular networks have been created *in vitro* by seeding endothelial cells and fibroblasts into a 3D gel of Fn and collagen I. When the gel was implanted into mice the cells formed vessels with smooth muscle and endothelium and formed stable vascular networks connected to the host vasculature. Neither regeneration templates nor bioartificial vessels have yet been used in clinical studies.

Myocardial infarct in human patients has been treated with transplants of bone marrow cells, satellite cells, and cardiomyocytes. Numerous studies suggest that bone marrow cells infused into myocardial infarcts of rats, rabbits, and pigs improve cardiac function by upregulating paracrine factors such as FGF-2, VEGF, TNF-α, and IL-1β, thus enhancing angiogenesis from existing cardiac blood vessels and promoting the incorporation of mobilized EnSCs into new vessels. Other soluble factors necessary for the embryonic development of the myocardium, and secreted by the epicardium, that may be important for myocardial regeneration are IGF-1 and Wnt5a, and thymosin β4, which also promotes skin wound repair. Administration of thymosin β4 to mice with infarcted hearts doubled their ejection fraction over controls, bringing it to 77% of the sham control level. Thymosin β4 exerts its effects by activating protein kinase B (Akt) and reducing apoptosis of cardiomyocytes. Injection of MSCs transfected with *Akt* into the infarcted myocardium of mice completely restored left ventricular function and injection of conditioned medium from *Akt*-transfected hypoxic cardiomyocytes into the infarct border zone reduced the infarct area by 40% and cardiomyocyte apoptosis by nearly 70%. The infarcted heart has reduced levels of calcium-binding proteins. One of these proteins, S100A1, inhibits the apoptosis of cardiomyocytes and reversesthe contractile dysfunction of cardiomyocytes obtained from infarcted rat hearts.

Double blinded patient clinical trials injecting autologous bone marrow into the artery supplying the area of myocardial infarct resulted in a moderate (~7%) improvement in left ventricular ejection fraction that can be attributed to enhanced angiogenesis and possibly inhibition of scar tissue formation by paracrine factors secreted by the injected cells. Satellite cells also improved ejection fraction by 8–9% when injected transendocardially, perhaps by the same mechanisms, or by augmenting the contraction of cardiomyocytes; however, several patients suffered episodes of ventricular arrhythmia. The best cell therapy for patients would be human cardiomyocyte transplantation, because these cells would not only presumably supply paracrine factors that enhance angiogenesis and suppress scar formation, they would also integrate into the host myocardium. Autogeneic ESCs would be the best source of these cardiomyocytes (Chapter 15). Proof of principle that fetal or ESC-derived cardiomyocytes are stably integrated into the ventricular myocardium has been demonstrated for mice, rats, dogs, and pigs.

Bioartificial rat myocardium has been made by seeding Gelfoam, alginate, collagen I, or a mixture of collagen and Matrigel with cardiomyocytes or ESCs. The most successful construct to date has been a combination of ESCs in a liquid matrix of collagen I and Matrigel that solidifies at body temperature. This liquid mix was injected into the infarcts of hearts removed from the body, which were then transplanted to hosts where the matrix gelated. The myocardium of thesese hearts contracted twice as well as that of untreated infarcted hearts and nearly as well as transplanted uninjured hearts.

## REFERENCES

Aiuti A, Slavin S, Aker M, Ficara F, Deola S, Mortellaro A, Morecki S, Andolfi G, Tabucchi A, Carlucci F, Marinello E, Cattaneo F, Vai S, Servida P, Miniero R, Roncarolo MG, Bordignon C (2002) Correction of ADA-SCID by stem cell gene therapy combined with nonmyeloablative conditioning. Science 296:2410–2413.

Akhyari P, Fedak PWM, Weisel RD, Lee T-Y, Verma S, Mickle DAG, Li R-K (2002) Mechanical stretch regimen enhances the formation of bioengineered autologous cardiac muscle grafts. Circulation 106:I137–I142.

Askari A, Unzek S, Popovic Z, Goldman CK, Forudi F, Kiedrowski M, Rovner A, Ellis SG, Thomas JD, Diorleto PE, Topol EJ, Penn MS (2003) Effect of stromal-cell-derived factor-1 on stem-cell homing and tissue regenerartion in ischaemic cardiomyopathy. Lancet 362:697–703.

Assmus B, Schachinger, Teupe C, Britten M, Lehmann R, Dobert , Grunwald F, Aicher, Urbich C, Martin H, Hoelzer D, Dimmeler S, Zeiher AM (2002) Transplantation of progenitor cells and regeneration enhancement in acute myocardial infarction (TOPCARE-AMI). Circulation 106:3009–3017.

Atkins BZ, Hueman MT, Meuchel JM, Cottman MJ, Hutcheson KA, Taylor DA (2000a) Myocardial cell transplantation improves *in vivo* regional performance in infarcted rabbit myocardium. Cardiac Vasc Reg 1:43–53.

Atkins BZ, Mani S, Hutcheson K, Glower DD, Taylor DA (2000b) Transplanted autologous skeletal myoblasts improve myocardial performance after coronary artery ligation. Cardiac Vasc Reg 1:76–84.

Badylak SF, Lantz GC, Coffey A, Geddes LA (1989) Small intestinal submucosa as a large diameter vascular graft in the dog. J Surg Res 47:74–80.

Badylak SF, Coffey AC, Lantz GC, Tacker WA, Geddes A (1994) Comparison of the resistance to infection of intestinal submucosa arterial autografts versus polytetrafluoroethylene arterial prosteheses in a dog model. J Vasc Surg 19:465–472.

Badylak SF, Liang A, Record R, Tullius R, Hodde J (1999) Endothelial cell adherence to small intestinal submucosa: an acellular bioscaffold. Biomats 20:2257–2263.

Balsam LB, Wqagers AJ, Christensen JL, Kofidis T, Weissman IL, Robbis RC (2004) Haematopoietic stem cells adopt mature haematopoietic fates in ischaemic myocardium. Nature 428:668–673.

Barker JN, Davies SM, DeFor T, Ramsay NKC, Weisdorf DJ, Wagner J (2001) Survival after transplantation of unrelated donor umbilical cord blood is comparable to that of human leukocyte antigen-matched unrelated donor bone marrow: results of a matched-pair analysis. Blood 97:2957–2961.

Bedelbaeva K, Gourevitch D, Clark L, Chen P, Leferovich JM, Heber-Katz E (2004) The MRL mouse heart healing response shows donor dominance in allogeneic fetal liver chimeric mice. Cloning Stem Cells 6:352–363.

Beltrami AP, Barlucchi L, Torella D, Baker M, Limana F, Chimenti S, Kasahara H, Rota M, Musso E, Urbanek K, Leri A, Kajstura J, Nidal-Ginard B, Anversa P (2003) Adult cardiac stem cells are multipotent and support myocardial regeneration. Cell 114:763–776.

Bittira B, Wang J-S, Shum-Tim D, Chiu RC-J (2000) Marrow stromal cells as the autologous, adult stem cell source for cardiac myogenesis. Cardiac Vasc Reg 1:205–210.

Bittira B, Kuang J-Q, Al-Khaldi A, Shum-Tim D, Chiu RC-J (2002) In vitro preprogramming of marrow stromal cells for myocardial regeneration. Ann Thorac Surg 74:1154–1160.

Blaese RM, Culver KW, Miller AD, Carter CS, Fleisher T, Clerici M, Shearer G, Chamg L, Chiang Y, Tolstoshev P, Greenblat JJ, Rosenberg SA, Klein H, Berger M, Mullen CA, Ramsey WJ, Muul L, Morgan RA, Anderson WF (1995) T lymphocyte-directed gene therapy for ADA-SCI: Initial trial results after four years. Science 270:475–480.

Bock-Marquette I, Saxena A, White MD, DiMaio JM, Srivastava D (2004) Thymosin β4 activates integrin-linked kinase and promotes cardiac cell migration, survival and cardiac repair. Nature 432:466–472.

Bordignon C, Notarangelo LD, Nobili N, Ferrari G, Casorati G, Panina P, Mazzolari E, Maggioni D, Rossi C, Servida P, Ugazio AG, Mavilio F (1995) Gene therapy in peripheral blood lymphocytes and bone marrow for ADA-immunodeficient patients. Science 270:470–475.

Broxmeyer HE (1993) Cord blood stem cells. In: Anderson KC, Ness PM (eds.). Scientific Basis of Transfusion Medicine. Philadelphia: WB Saunders Co, pp 499–506.

Broxmeyer HE, Douglas GW, Hangoc G, Cooper S, Bard J, English D, Arny M, Thomas L, Boyse EA (1989) Human umbilical cord blood as a potential source of transplantable hematopoietic stem/progenitor cells. Proc Natl Acad Sci USA 86:3828–3832.

Broxmeyer H, Hangoc G, Cooper S, Ribeiro RC, Graves V, Yoder M, Wagner J, Vadhan-Raj S, Benninger L, Rubenstein P, Broun ER (1992) Growth characteristics and expansion of human umbilical cord blod and estimation of its potential for transplantation in adults. Proc Natl Acad Sci USA 89:4109–4113.

Broxmeyer HE, Cooper S (1997) High efficiency recovery of immature hematoppietic progenitor cells with extensive proliferative capacity from human cord blood cryopreserved for ten years. Clin Exp Immunol 107:45–53.

Broxmeyer HE, Smith FO (1994) Cord blood stem cell transplantation. In: Thomas ED, Blume KG, Forman SJ (eds.), Hematopoietic Cell Transplantation, 2nd ed. Malden: Blackwell Science Inc, pp 431–443.

Carella AM, Champlin, Slavin S, McSweeney P, Storb R (2000) Mini-allografts: ongoing trials in humans. Bone Marrow Transpl 25:345–350.

Carrier RL, Papadaki M, Rupnick M, Schoen FJ, Bursac N, Langer R, Freed LE, Vunjak-Novakovic G (1999) Cardiac tissue engineering: cell seeding, cultivation parameters, and tissue construct characterization. Biotechnol Bioeng 64:580–589.

Carrier RL, Papadaki M, Rupnick M, Schoen FJ, Bursac N, Langer R, Freed LE, Vunjak-Novakovic G (2000) Cardiac tissue engineering: Cell seeding, cultivation parameters, and tissue construct characterization. Biotech Bioeng 64:580–589.

Cavazzana-Calvo M, Hacein-Bey S, de Saint Basile G, Gross F, Yvon E, Nusbaum P, Selz F, Hue C, Certain S, Casanova J-L, Bousso P, Le Deist F, Fischer A (2000) Gene therapy of human severe combined immunodeficiency (SCID-X1 disease. Science 288:669–672.

Cavazzana-Calvo M, Thrasher A, Mavilio F (2004) The future of gene therapy. Nature 427:779–781.

Chachques JC, Duarte F, Cattadori B, Shafy A, Lila N, Chatellier G, Fabiani J-N, Carpientier AF (2004) Angiogenic growth factors and/or cellular therapy for myocardial regeneration: A comparative study. J Thor Cardiovasc Surg 128:245–253.

Chien KR, Moretti A, Laugwitz K-L (2004) ES cells to the rescue. Science 306:239–240.

Childs R, Chernoff A, Contentin N, Bahceci E, Schrump D, Leitman S, Read EJ, Tisdale J, Dunbar C, Linehan M, Young NS, Barrett AJ (2000) Regression of metastatic renal-cell carcinoma after nonmyeloablative allogeneic peripheral-blood stem-cell transplantation. New Eng J Med 343:750–758.

Chiu RCJ, Zibaitis A, Kao RL (1995) Cellular cardiomyoplasty: Myocardial regeneration with satellite cell implantation. Ann Thorac Surg 60:12–18.

Christman KL, Fok HH, Sievers RE, Fang Q, Lee RJ (2004) Fibrin glue alone and skeletal myoblasts in a fibrin scaffold preserve cardiac function after myocardial infarction. Tiss Eng 10:403–409.

Christopherson KW II, Hangoc G, Mantel CR, Broxmeyer HE (2004) Modulation of hematopoietic stem cell homing and engraftment by CD26. Science 305:1000–1003.

Ciulla MM, Lazzari L, Pacchiana R, Esposito A, Bosari S, Ferrero S, Gianelli U, Paliotti R, Busca G, Giorgetti A, Magrini F, Rebulla P (2003) Homing of peripherally injected bone marrow cells in the rat after experimental myocardial injury. Haematologica 88:614–621.

Davani S, Marandin A, Mersin N, Royer B, Kantelip B, Herve P, Etievent J-P, Kantelip J-P (2003) Mesenchymal progenitor cells differentiate into an endothelial phenotype, enhance vascular density, and improve heart function in a rat cellular cardiomyoplasty model. Circulation 208SII:II-253–II-258.

de Lima M, Anagnostopoulos A, Munsell M, Shahjahan M, Ueno N, Ippoliti C, Andersson BS, Gagewski J, Couriel D, Cortes J, Donato M, Neumann J, Champlin R, Giralt S (2004) Nonablative versus reduced-intensity conditioning regimens in the treatment of acute myeloid leukemia and high-risk myelodysplastic syndrome: Dose is relevant for long-term disease control after allogeneic hematopoietic stem cell transplantation. Blood 104:865–872.

Donato R (2003) Intracellular and extracellular roles of S100 proteins. Microsc Res Tech 60:540–551.

Eschengagen T, Fink C, Remmers U, Scholz H, Wattchow J, Weil J, Zimmerman W, Dohmen HH, Schafer H, Bishopric N, Wakatsuki T, Elson EL (1997) Three-dimensional reconstitution of embryonic cardiomyocytes in a collagen matrix; a new heart muscle model system. FASEB J 11:683–694.

Fink C, Ergun S, Kralisch D, Remmers U, Weil J, Eschenhagen T (2000) Chronic stretch of engineered heart tissue induces hypertrophy and functional improvement. FASEB J 14:669–679.

Fraidenreich D, Stillwell E, Romero E, Wilkes D, Manova K, Basson CT, Benezra R (2004) Rescue of cardiac defects in *Id* knockout embryos by injection of embryonic stem cells. Science 306:247–252.

Fuchimoto Y, Huang CA, Yamada K, Shimizu A, Kitamura H, Colvin RB, Ferrara V, Murphy MC, Sykes M, White-Scharf M, Neville DM, Sachs DH (2000) Mixed chimerism and tolerance without whole body irradiation in a large animal model. J Clin Invest 105:1779–1789.

Fuchs S, Baffour R, Zhou YF, Shou M, Pierre A, Tio FO, Weissman NJ, Leon MB, Epstein SE, Kornowski R (2001) Transendocardial delivery of autologous bone marrow enhances collateral perfusion and regional function in pigs with chronic experimental myocardial ischemia. J Am Coll Cardiol 37:1726–1732.

Fujio Y, Nguyen T, Wencker D, Kitsis RN, Walsh K (2000) Akt promotes survival of cardiomyocytes in vitro and protects against ischemia-reperfusion injury in mouse heart. Circulation 101:660–667.

Garovoy MR, Stock P, Keith F, Linker C (1997) Clinical transplantation. In: Stites DP, Terr AI, Parslow TG (eds.). Medical Immunology, 9th ed. Stamford: Appleton Lange, pp 802–826.

Gao J, Niklason L, Langer R (1998) Surface hydrolysis of poly(glycolic acid) meshes increases the seeding density of vascular smooth muscle cells. J Biomed Mater Res 42:417–424.

Gluckman E, Broxmeyer HE, Auerbach AD, Friedman HS, Douglas GW, Devergie A, Esperou H, Thierry D, Socie G, Lehn P, Cooper S, English D, Kurtzberg J, Bard J, Boyse EA (1989) Hematopoietic reconstitution in a patient with Fanconi anemia by means of umbilical-cord blood from an HLA-identical sibling. New Eng J Med 321:1174–1178.

Gluckman E, Rocha V, Boyer-Chammard, A, Locatelli F, Arcese W, Pasquini R, Ortega J, Souillet G, Ferreira E, Laporte J-P, Fernandez M, Chastang C (1997) Outcome of cord-blood transplantation from related and unrelated donors. New Eng J Med 337:373–381.

Gnecchi M, He H, Liang OD, Melo LG, Morello F, Mu H, Noiseux N, Zhang L, Pratt RE, Ingwall JS, Dzau VJ (2005) Paracrine action accounts for marked protection of ischemic heart by Akt-modified mesenchymal stem cells. Nature Med 11:367–368.

Gulbins H, Dauner M, Petzold R, Goldemund A, Anderson I, Doser M, Meiser B, Reichart B (2004) Development of an artificial vessel lined with human vascular cells. J Thor Cardiovasc Surg 128:372–377.

Hescheler J, Welz A, Fleischmann BK (2004) Searching for reputability: First randomized study on bone-marrow transplantation in the heart. Lancet 364:121–122.

Hiles MC, Badylak S, Geddes LA (1993) Porosity of porcine small-intestinal submucosa for use as a vascular graft. J Biomed Mats Res 27:139–144.

Hiles MC, Badylak SF, Lantz GC, Kokini K, Geddes LA, Morff RJ (1995) Mechanical properties of xenogeneic small-intestinal submucosa when used as an aortic graft in the dog. J Biomed Mats Res 29:883–891.

Hill JM, Zalos G, Halcox JPJ, Schenke WH, Waclawiw MA, Quyyumi AA, Finkel T (2003) Circulating endothelial progenitor cells, vascular function, and cardiovascular risk. New Eng J Med 348:593–600.

Horowitz MM, Gale RP, Sondel PM, Goldman JM, Kersey J, Kolb HJ, Rimm AA, Ringden O, Rozman C, Speck B (1990) Graft-versus-leukemia reactions after bone marrow transplantation. Blood 75:555–562.

Hutcheson KA, Atkins BZ, Hueman MT, Hopkins MB, Glower DD, Taylor DA (2000) Comparison of benefits on myocardial performance of cellular cardiomyoplasty with skeletal myoblasts and fibroblasts. Cell Transpl 9:359–368.

Huynh T, Abraham G, Murray J, Brockbank K, Hagen P-O, Sullivan S (1999) Remodeling of an acellular collagen graft into a physiologically responsive neovessel. Nature Biotech 17:1083–1086.

Iba O, Matsubara H, Nozawa Y, Fujiyama S, Amano K, Mori Y, Kojima H, Iwasaka T (2002) Angiogenesis by implantation of peripheral blood mononuclear cells and platelets into ischemic limbs. Circulation 106:2019–2025.

Iwai S, Sawa Y, Ichikawa H, Taketani S, Uchimura E, Chen G, Hara MH, Miyake J, Matsuda H (2004) Biodegradable polymer with collagen microsponge serves as a new bioengineered cardiovascular prosthesis. J Thor Cardiovasc Surg 128:472–479.

Jackson KA, Majika SM, Wang H, Pocius J, Hartley CJ, Majesky MW, Entman ML, Michael LH, Hirschi K, Goodell MA (2001) Regeneration of ischemic cardiac muscle and vascular endothelium by adult stem cells. J Clin Invest 107:1395–1402.

Jain RK (2003) Molecular regulation of vessel maturation. Nature Med 9:685–693.

Johkura K, Cui L, Suzuki A, Teng R, Kamiyoshi A, Okamura S, Kubota S, Zhao X, Asanuma K, Okouchi Y, Ogiwara N, Tagawa Y-I, Sasaki K (2003) Survival and function of mouse embryonic stem cell-derived cardiomyocytes in ectopic transplants. Cardiovasc Res 58:435–443.

Kalka C, Masuda H, Takahashi T, Kalka-Moll M, Silver M, Kearney M, Li T, Isner JM, Asahara T (2000) Transplantation of *ex vivo* expanded endothelial progenitor cells for therapeutic neovascularization. Proc Natl Acad Sci USA 97:3422–3427.

Kamahita H, Matsubara H, Nishiue T, Fujiyama S, Amano K, Iba O, Imada T, Iwasaka T (2002) Improvement of collateral perfusion and regional function by implantation of peripheral blood mononuclear cells into ischemic hibernating myocardium. Arterioscler Thromb Vasc Biol 22:1804–1810.

Kamahita H, Matsubara H, Nishiue T, Fujiyama S, Tsutsumi Y, Ozono R, Masaki H, Mori Y, Iba O, Tateishi E, Kosaki A, Shintani S, Murohara T, Imaizumi T, Iwasaka T (2001) Implantation of bone marrow mononuclear cells into ischemic myocardium enhances collateral perfusion and regional function via side supply of angioblasts, angiogenic ligands and cytokines. Circulation 104:1046–1056.

Kamel-Reid S, Dick JE (1988) Engraftment of immune-deficient mice with human hematopoietic cells. Science 242:1706–1709.

Kang H-J, Kim H-S, Zhang S-Y, Park K-W, Cho H-J, Koo B-K, Kim Y-J, Lee DS, Sohn D-W, Han B-H, Oh B-H, Lee M-M, Park Y-B (2004) Effects of intracoronary infusion of peripheral blood stem cells mobilized with granulocyte-colony stimulating factor on left ventricular systolic function and restenosis after coronary stenting in myocardial infarction: the MAGIC cell randomized clinical trial. Lancet 363:751–756.

Kang J-J, Zhang S-Y, Park K-W, Koo B-K, Kim Y-J, Lee DS, Sohn D-W, Han K-S, Oh B-H, Lee M-M, Park Y-B (2004) Effects of intracoronary infusion of peripheral blood stem-cells mobilized with granulocyte-colony stimulating factor on left ventricular systolic function and restenosis after coronary stenting in myocardial infarction: the MAGIC cell randomized clinical trial. Lancet 363:751–756.

Kao RL, Chin TK, Ganote CE, Hossler FE, Li C, Browder W (2000) Satellite cell transplantation to repair injured myocardium. Cardiac Vasc Reg 1:31–42.

Kao RL (2001) Autologous satellite cells for myocardial regeneration. J Reg Med (e-biomed) 2:1–8.

Kaushal S, Amiel GE, Guleserian KJ, Shapira OM, Perry T, Sutherland FW, Rabkin E, Moran AM, Schoen FJ, Atala A, Soler S, Bischoff J, Mayer JE (2001) Functional small-diameter neovessels created using endothelial progenitor cells expanded *ex vivo*. Nature Med 7:1035–1040.

Kawamoto A, Gwon H-C, Iwaguro H, Yamaguchi J-I, Uchida S, Masuda H, Silver M, Ma H, Kearney M, Isner JM, Asahara T (2001) Therapeutic potential of ex vivo expanded endothelial progenitor cells for myocardial ischemia. Circulation 103:634–637.

Kawamoto A, Gwon H-C, Iwaguro H, Yamaguchi J-I, Uchida S, Masuda H, Silver M, Ma H, Kearney M, Isner JM, Asahara T (2003) Therapeutic potential of ex vivo expanded endothelial progenitor cells for myocardial ischemia. Circulation 103:634–637.

Kawamoto A, Tkebuchava T, Yamaguchi J-I, Nishimura H, Yoon Y-S, Milliken C, Uchida S, Masuo O, Iwaguro H, Hong M, Hanley A, Silver M, Kearney M, Losordo DW, Isner JM, Asahara T (2003) Intramyocardial transplantation of autologous endothelial progenitor cells for therapeutic neovascularization of myocardial ischemia. Circulation 107:461–468.

Kawamoto A, Tkebuchava T, Yamaguchi J-I, Nishimura H, Yoon Y-S, Milliken C, Uchida S, Masuo O, Iwaguro H, Ma H, Hanley A, Silver M, Kearney M, Losordo DW, Isner JM, Asahara T (2003) Intramyocardial transplantation of autologous endothelial progenitor cells for therapeutic neovascularization of myocardial ischemia. Circulation 107:461–468.

Kehat I, Khimovich L, Caspi O, Gepstein A, Shofti R, Arbel G, Huber I, Satin ••, Itskovitz-Elder J, Gepstein L (2004) Electromechanical integration of cardiomyocytes derived from human embryonic stem cells. Nature Biotech 10:1282–1289.

Khouri IF, Saliba RM, Giralt SA, Lee MS, Okoroji GJ, Hagemeister FB, Korbling M, Younes A, Ippoliti C, Gajewski JL, McLaughlin P, Anderlini P, Donato ML, Cababillas FF, Champlin RE (2001) Nonablative allogeneic hematopoietic transplantation as adoptive immunotherapy for indolent lymphoma: Low incidence of toxicity, acute graft-versus-host disease, and treatment-related mortality. Blood 98:3595–3599.

Khouri I, Lee MS, Saliba RM, Andersson B, Anderlini P, Couriel D, Hosing C, Giralt S, Korbling M, McMannis J, Keating MJ, Champlin RE (2004) Nonablative allogeneic stem cell transplantation for chronic lymphocytic leukemia impact of rituximab on immunomodulation and survival. Exp Hematol 32:28–35.

Kinnaird T, Stabile E, Burnett MS, Shou M, Lee CW, Barr S, Fuchs S, Epstein SE (2004) Local delivery of marrow-derived stromal cells augments collateral perfusion through paracrine mechanisms. Circulation 109:1543–1549.

Klug M, Soonpaa MH, Field LJ (1995) DNA synthesis and multinucleation in embryonic stem cell-derived cardiomyocytes. Am J Physiol 269:H1913–H1921.

Klug M, Soonpa MH, Koh GY, Field LJ (1996) Genetically selected cardiomyocytes from differentiating embryonic stem cells form stable intracardiac grafts. J Clin Invest 98:1–9.

Kobayashi T, Hamano K, Li T-S, Katoh T, Kobayashi S, Matsuzaki M, Esato K (2000) Enhancement of angiogenesis by the implantation of self bone marrow cells in a rat ischemic heart model. J Surg Res 89:189–195.

Koc ON, Gerson SL, Cooper BW, Dyhouse SM, Haynesworth SE, Caplan AI, Lazarus HM (2000) Rapid hematopoietic recovery after coinfusion of autologous-blood stem cells and culture-expanded marrow mesenchymal stem cells in advanced breast cancer patients receiving high-dose chemotherapy. J Clin Oncol 18:307–316.

Kocher AA, Schuster MD, Szabolics MJ, Takuma S, Burkhoff D, Wang J, Homma S, Edwards NM, Itescu S (2001) Neovascularization of ischemic myocardium by human bone-marrow-derived angioblasts prevents cardiomyocyte apoptosis, reduces remodeling and improves cardiac function. Nature Med 7:430–436.

Kofidis T, Akhyari P, Boublik J, Theorodrou P, Martin U, Ruhparwar A, Fischer S, Eschenhagen T, Kubis HP, Kraft T, Leyh R, Haverich A (2002) In vitro engineering of heart muscle: Artificial myocardial tissue. J Thor Cardiovasc Surg 124:63–69.

Kofidis T, deBruin JL, Hoyt G, Lebl DR, Tanaka M, Yamane T, Chang C-P, Robbins RC (2004) Injectable bioartificial myocardial tissue for large-scale intramural cell transfer and functional recovery of injured heart muscle. J Thor Cardiovasc Surg 128:571–578.

Koh GY, Klug MG, Soonpaa MH, Field LJ (1993) Differentiation and long term survival of C2C12 myoblast grafts in heart. J Clin Invest 92:1548–1554.

Koh GY, Soonpaa MH, Klug M, Pride HP, Cooper BJ, Zipes DP, Field LJ (1995) Stable fetal cardiomyocyte grafts in the hearts of dystrophic mice and dogs. J. Clin Invest 96:2034–2042.

Koike N, Fukumura D, Gralla O, Au P, Schechner JS, Jain RK (2004) Creation of long-lasting blood vessels. Nature 428:138–139.

Kolb HJ, Mittermuller J, Clemm C, Holler E, Ledderose G, Brehm G, Heim M, Wilmanns W (1990) Donor leukocyte transfusions for treatment of recurrent chronic myelogenic leukemia in marrow transplant patients. Blood 76:2462–2465.

Kolb H, Schattenberg A, Goldman JM, Hertenstein B, Jacobsen N, Arcese W, Ljungman P, Ferrant A, Verdonck L, Niederwieser D, van Rhee F, Mittermueller J, de Witte T, Holler E, Ansari H (1995) Graft versus host leukemia effect of donor lymphocyte transfusions in marrow grafted patients. Blood 86:2041–2050.

Krupnick AS, Kreisel D, Riha M, Balsara R, Rosengard BR (2004) Myocardial tissue engineering and regeneration as a therapeutic alternative to transplantaton. In: Heber-Katz E (ed.). Regeneration: Stem cells and beyond. Curr Topics Microbiol Immunol 280:139–164.

Kurtzberg J, Laughlin M, Graham ML, Smith C, Olson JF, Halperin EC, Ciocci G, Carrier C, Stevens CE, Rubenstein P (1996) Placental blood as as source of hematopoietic stem cells for transplantation into unrelated recipients. New Eng J Med 335:157–166.

Laflamme MA, Murry CE (2005) Regenerating the heart. Nature Biotech 23:845–856.

Lantz GC, Badylak SF, Coffey A, Geddes LA, Blevins WE (1990) Small intestinal submucosa as a small diameter arterial autograft in the dog. J Invest Surg 3:217–227.

Lantz GC, Badylak SF, Coffey AC, Geddes LA, Sandusky GE (1992) Small intestinal submucosa as a superior vena cava graft in the dog. J Surg Res 53:175–181.

Lantz GC, Badylak SF, Hiles MC, Coffey AC, Geddes A, Kokini K, Sandusky GE, Morff RJ (1993) Small intestinal submucosa as a vascular graft: A review. J Invest Surg 6:297–310.

Laughlin M, Eapen M, Rubenstein P, Wagner JE, Zhang M-J, Champlin RE, Stevens C, Barker JN, Gale RP, Lazarus HM, Marks DI, van Rood JJ, Scaradavou A, Horowitz MM (2004) Outcomes after transplantation of cord blood or bone marrow from unrelated donors in adults with leukemia. New Eng J Med 351:2265–2275.

Leor J, Patterson M, Quinones MJ, Kedes LH, Kloner RA (1996) Transplantation of fetal myocardial tissue into the infarcted myocardium of rat. Circulation 94SII:332–336.

Leor J, Aboulafia-Etzion S, Dar A, Shapiro L, Barbash IM, Battler A, Granot Y, Cohen S (2000) Bioengineered cardiac grafts: New approach to repair the infarcted myocardium? Circulation 102: III56–III61.

L'Heureux N, Paquet S, Labbe R, Germain L, Auger FA (1998) A completely biological tissue-engineered human blood vessel. FASEB J 12:47–56.

Li R-K, Jia ZQ, Weisek RD, Mickle DA, Zhang J, Mohabeer MK, Rao V, Ivanov J (1996) Cardiomyocyte transplantation improves heart function. Ann Thor Surg 62:654–660.

Li R-K, Mickle DA, Wiesel RD, Zhang J, Rao V, Li G, Merante F, Jia ZQ (1997) Natural history of fetal rat cardiomyocytes transplanted into adult rat myocardial scar tissue. Circulation 96: II179–I186.

Li R-K (1999) Smooth muscle cell transplantation into myocardial scar tissue improves heart function. J Mol Cellul Cardiol 31: 513–522.

Li R-K, Jia Z-Q, Weisel RD, Mickle DAG, Choi A, Yau TM (1999) Survival and function of bioengineered cardiac grafts. Circulation 100:II63–II69.

Li R-K, Weisel RD, Mickle DA, Jia ZQ, Kim EJ, Sakai T, Tomita S, Schwartz L, Iwanochko M, Husain M, Vusimano RJ, Burns R, Yau TM (2000) Autologous porcine heart cell transplantation improved heart function after a myocardial infarction. J Thor Cardiovasc Surg 119:62–68.

Li ZH, Broxmeyer H, Lu L (1995) Cryopreserved cord blood myeloid progenitor cells can serve as targets for rwtroviral-mediated gene transduction and gene-transduced progenitors can be cryopreserved and recovered. Leukemia 9:S12–S16.

Little M-T, Storb R (2000) The future of allogeneic hematopoietic stem cell transplantation: Minimizing pain, maximizing gain. J Clin Invest 105:1679–1681.

Lovell MJ, Mathur A (2004) The role of stem cells for treatment of cardiovascular disease. Cell Prolif 37:67–87.

Lu L, Ge Y, Li Z-H, Freie B, Clapp DW, Broxmeyer HE (1995) $C34^+$ stem/progenitor cells purified from cryopreserved normal cord blood can be transduced with high efficiency by a retroviral vector and expanded ex vivo with stable integration and expression of Fanconi anemia complementation C gene. Cell Transpl 4:493–503.

Makkar RR, Lill M, Chen P-S (2003) Stem cell therapy for myocardial repair: Is it arrythmogenic? J Am Coll Cardiol 12:2070–2072.

Makino S, Fukuda K, Miyoshi S, Konishi F, Kodoma H, Pan J, Sano M, Takahashi T, Hori S, Abe H, Hata J, Umezawa A, Ogawa S (1999) Cardiomyocytes can be generated from marrow stromal cells *in vitro*. J Clin Invest 103:697–705.

Malouf NN, Coleman WB, Grisham JW, Lininger RA, Madden VJ, Sproul M, Anderson PAW (2001) Adult-derived stem cells from the liver become myocytes in the heart *in vivo*. Am J Pathol 158:1929–1935.

Mangi AA, Noiseux N, Kong D, He H, Rezvani M, Ingwall JS, Dzau VJ (2003) Mesenchymal stem cells modified with Akt prevent remodeling and restore performance of infarcted hearts. Nature Med 9:1195–1201.

Mathur A, Martin JF (2004) Stem cells and repair of the heart. Lancet 364:183–192.

Menasche PM (2000) Autologous skeletal myoblast transplantation for ischemic cardiomyopathy: First clinical case. Cardiac Vasc Reg 1:155–156.

Menasche P, Hagege AA, Vilquin J-T, Desnos M, Abergel E, Pouzet B, Bel A, Sarateanu S, Scorsin M, Schwartz K, Bruneval P, Benbunan M, Marolleau J-P, Duboc D (2003) Autologous skeletal myoblast transplantation for severe postinfarction left ventricular dysfunction. J Am Coll Cardiolgy 41:1078–1083.

Minami E, Reinecke H, Murry CE (2003) Skeletal muscle meets cardiac muscle: Friends or foes? J Am Coll Cardiol 41:1084–1086.

Most P, Remppis A, Pleger ST, Loffler E, Ehlermann P, Bernotat J, Kleuss C, Heierhorst J, Ruiz P, Witt H, Karczewski P, Mao L, Rockman HA, Duncan SJ, Katus HA, Koch W (2003a) Transgenic overexpression of the Ca2+ binding protein S100A1 in the heart leads to increased in vivo myocardial contractile performance. J Biol Chem 278:33809–33817.

Most P, Pleger ST, Volkers M, Heidt B, Boerries M, Weichenhan D, Loffler E, Janssen PML, Eckhart AD, Martini J, Williams ML, Katus HA, Remppis A, Koch WJ (2004) Cardiac adenoviral S100A1 gene delivery rescues failing myocardium. J Clin Invest 114:1550–1563.

Muller-Borer BJ, Cascio WE, Anderson PAW, Snowwaert JN, Frye JR, Desai N, Esch GL, Brackham JA, Bagnell CR, Coleman WB, Grisham JW, Malouf NN (2004) Adult-derived liver stem cells acquire a cardiomyocyte structural and functional phenotype *ex vivo*. Am J Pathol 165:135–145.

Muller-Ehmsen J, Whittaker P, Kloner RA, Dow JS, Sakoda T, Long TI, Laird PW, Kedes L (2002) Survival and development of neonatal rat cardiomyocytes transplanted into adult myocardium. J Mol Cell Cardiol 34:107–116.

Murasawa S, Llevadot J, Silver M, Isner FM, Losordo DW, Asahara T (2002) Constitutive human telomerase reverse transcriptase expression enhances regenerative properties of endothelial progenitor cells. Circulation 106:1133–1139.

Murry CE, Soonpa MH, Reinecke H, Nakajima H, Rubart M, Pasumarthi KBS, Virag JI, Barelmez SH, Poppa V, Bradford G, Dowell JD, Williams DA, Field LJ (2004) Haematopoietic stem celss do not transdifferentiate into cardiac myocytes in myocardial infarcts. Nature 428:664–668.

Niklason L, Gao J, Abbot WM, Hirschi KK, Houser S, Marini R, Langer R (1999) Functional arteries grown in vitro. Science 284: 489–493.

Noishiki Y, Tomizawa Y, Yamane Y, Matsumoto A (1996) Autocrine angiogenic vascular prosthesis with bone marrow transplantation. Nature Med 2:90–93.

Nu, DN, Allen RC (1995) Vascular grafts. In: Bronzino JD (ed.). The Biomedical Engineering Handbook. Boca Raton: CRC Press, pp 1871–1878.

Opitz F, Schenke-Layland K, Cohnert TU, Starcher B, Halbhuber KJ, Martin DP, Stock UA (2004) Tissue engineering of aortic tissue; dire consequence of suboptimal elastic fiber synthesis in vivo. Cardiovasc Res 63:719–730.

Orlic D, Kajstura J, Chimenti S, Jakoniuk I, Anderson SM, Li B, Pickel J, McKay R, Nadal-Ginard B, Bodine M, Leri A, Anversa P (2001) Bone marrow cells regenerate infarcted myocardium. Nature 410:701–705.

Peled A, Petit I, Kollet O, Magid M, Ponomaryov T, Byk T, Nagler A, Ben-Hur H, Many A, Schultz L, Lider O, Alon R, Zipori D, Lapidot T (1999) Dependence of human stem cell engraftment and repopulation of NOD/Scid Mice on CXCR4. Science 283: 845–848.

Perin EC, Geng Y-J, Willerson JT (2003) Adult stem cell therapy in perspective. Circulation 107:935–938.

Perin EC, Dohmann HFR, Orojevic R, Silva SA, Sousa ALS, Mesquita C, Rossi MID, Carvalho AC, Dutra HS, Dohmann HF, Silva GV, Belem L, Vivacqua R, Rangel FOD, Esporcatte R, Geng YJ, Vaughn WK, Assad JAR, Mesquita ET, Willerson JT (2003) Transendocardial, autologous bone marrow cell transplantation for severe, chronic ischemic heart failure. Circulation 107:2294–2302.

Polezhaev LV (1972) Loss and Restoration of Regenerative Capacity in Tissues and Organs of Animals. Cambridge: Harvard University Press.

Polezhaev LV (1979) Neoformation of muscle fibers and normalization of the myocardiac structure in diptherial myocarditis in rabbits under experimental conditions. Folia Biol (Krakow) 27:43–50.

Polezhaev LV (1980) Autoradiographic studies in experiments with stimulation of cardiac muscle regeneration in rats. Folia Biol (Krakow) 28:231–236.

Polezhaev LV, Tinyakov YG (1980) The use of electron microscopy in experiments on stimulation of cardiac muscle regeneration in rats. Folia Biol (Krakow) 28:225–230.

Pollok KE, van der Loo JCM, Cooper RJ, Hartwell JR, Miles KR, Breese R, Williams EP, Montel A, Seshadri R, Hanenberg H, Williams DA (2001) Differential transduction efficiency of SCID-repopulating cells derived fro umbilical cord blood and granulocyte colony-stimulating factor-mobilized peripheral blood. Human Gene Ther 12:2095–2108.

Porter DL, Roth MS, McGarigle C, Ferrara J, Antin JH (1994) Induction of graft-versus-host disease as immunotherapy for relapsed chronic myeoid leukemia. New Eng J Med 330:100–106.

Prevel CD, Eppley BL, McCarty M, Jackson JR, Voytik SL, Hiles MC, Badylak SF (1994) Experimental evaluation of small intestinal submucosa as a microvascular graft material. Microsurg 15:588–591.

Prigozhina TB, Gurevitch O, Zhu J, Slavin S (1997) Permanent and specific transplantation tolerance induced by a nonmyeloablative treatment to a wide variety of allogeneic tissues: I. Induction of tolerance by a short course of total lymphoid irradiation and selective elimination of the donor-specific host lymphocytes. Transplantation 63:1394–1399.

Rafi S, Lyden D (2003) Therapeutic stem and progenitor cell transplantation for organ vascularization and regeneration. Nature Med 9:702–712.

Rehman J, Traktuev D, Li J, Merfeld-Clauss S, Temm-Grove CJ, Bovenkerk JE, Pell CL, Johnstone BH, Considine RV, March KL (2004) Secretion of angiogenic and antiapoptotic factors by human adipose stromal cells. Circulation 109:1292–1298.

Reinecke H, Poppa V, Murry CE (2002) Skeletal muscle stem cells do not transdifferentiate into cardiomyocytes after cardiac grafting. J Mol Cell Cardiol 34:241–249.

Repka T, Weisdorf DJ (2000) Nonmyoablative HPC transplantation. Transfusion 40:758–760.

Risdon G, Gaddy J, Stehman FB, Broxmeyer HE (1994) Prolifrative and cytotoxic responses of human umbilical cord blood T lymphocytes following allogeneic stimulation. Cell Immunol 154:14–24.

Robotin-Johnson MC, Swanson PE, Johnson DC, Schuessler RB, Cox JL (1998) An experimental model of small intestinal submucosa as a growing vascular graft. J Thor Cardiovasc Surg 116:805–811.

Rocha V, Labopin M, Sanz G, Arcese W, Schwerdtfeger R, Bosi A, Jacobsen N, Ruutu T, de Lima M, Frassoni F, Gluckman E (2004) Transplants of umbilical-cord blood or bone marrow from unrelated donors in adults with acute leukemia. New Eng J Med 351:2276–2285.

Saito T, Kuang J-Q, Lin CCH, Chiu RC-J (2003) Transcoronary implantation of bone marrow stromal cells ameliorates cardiac function after myocardial infarction. J Thoracic Cardiovasc Surg 126:114–123.

Sakai T, Li RK, Weisel RD, Mickle DA, Kim EJ, Tomita S, Yau TM (1999) Autologous heart cell transplantation improves cardiac function after myocardial injury. Ann Thorac Surg 68:2074–2081.

Sandusky GE, Badylak SF, Morff RJ, Johnson WD, Lantz G (1992) Histologic findings after *in vivo* placement of small intestine submucosal vascular grafts and saphenous vein grafts in the carotid artery in dogs. Am J Pathol 140:317–324.

Sandusky GE, Lantz GC, Badylak S (1995) Healing comparison of small intestine submucosa and ePTFE grafts in the canine carotid artery. J Surg Res 58:415–420.

Sanz MA (2004) Cord-blood transplantation in patients with leukemia—a real alternative for adults. New Eng J Med 351:2328–2330.

Sarzotti M, Patel DD, Li X, Ozaki DA, Cao S, Langdon S, Parrott RE, Coyne K, Buckley RH (2003) J Immunol 170:2711–2718.

Scheubel RJ, Zorn H, Silber R-E, Kuss O, Morawietz H, Holtz J, Simm A (2003) Age-dependent depression in circulating endothelial progenitor cells in patients undergoing coronary artery bypass grafting. J Am Coll Cardiol 42:2073–2080.

Schmidt CE, Baier JM (2000) Acellular vascular tissues: Natural biomaterials for tissue repair and tissue engineering. Biomats 22:2215–2231.

Schneider MD (2004) Prometheus unbound. Nature 432:451–453.

Shintani S, Murohara T, Ikeda H, Ueno T, Honma T, Katoh A, Sasaki K-I, Shimada T, Oike Y, Imaizumi T (2001) Mobilization of endothelial cells in patients with acute myocardial infarction. Circulation 103:2776–2779.

Shlomchick WD, Couzens MS, Tang CB, McNiff J, Robert ME, Liu J, Schomlik MJ, Emerson SG (1999) Prevention of graft vs host disease by inactivation of host antigen-presenting cells. Science 285:412–415.

Shimizu T, Yamato M, Akutsu T, Setomaru T, Abe K, Kikuchi A, Umezu M, Okano T (2002) Fabrication of pulsatile cardiac tissue grafts using a novel 3-dimensional cell sheet manipulation twchnique and temperature-responsive cell culture surfaces. Circ Res 90:e40 (published online).

Simmons PJ, Przepiorka D, Thomas ED, Torok-Storb B (1987) Host origin of marrow stromal cells following allogeneic bone marrow transplantation. Nature 328:429–432.

Slavin S, Nagler A, Naparstek E, Kapelushnik Y, Aker M, Cividalli G, Varadi G, Kirschbaum M, Ackerstein A, Samuel S, Amar A, Brautbar C, Ben-Tal O, Eldor A, Or R (1998) Nonmyoablative stem cell transplantation and cell therapy as an alternative to conventional bone marrow transplantation with lethal cytoreduction for the treatment of malignant and nonmalignant hematologic diseases. Blood 91:756–763.

Smits PC, van Geuns R-J, Poldermans D, Bountioukos M, Onderwater EEM, Lee CH, Maat APWM, Serruys PW (2003) Catheter-based intramyocardial injection of autologous skeletal myoblasts as a primary treatment of ischemic heart failure. J Am Coll Cardiol 42:2063–2069.

Soonpaa MH, Koh GY, Klug MG, Field LJ (1994) Formation of nascent intercalated discs between grafted fetal cardiomyocytes and host myocardium. Science 264:696–698.

Sorror ML, Maris MB, Stirer B, Sandmaier BM, Diaconescu R, Flowers C, Maloney DG, Storb R (2004) Comparing morbidity and mortality of HLA-matched unrelated donor hematopoietic cell transplantation after nonmyeloablative and myeloablative conditioning: influence of pretransplantation comorbidities. Blood 104:961–968.

Staba S, Escolar M, Poe M, Kim Y, Martil PL, Szabolcs P, Allison-Thacker J, Wood S, Wenger DA, Rubenstein P, Hopwood JJ, Krivit W, Kurtzberg J (2004) Cord-blood transplants from unrelated donors in patients with Hurler's syndrome. New Eng J Med 350:1960–1969.

Stamm C, Westphal B, Kleine H-D, Petzsch M, Kittner C, Klinge H, Schumichen C (2003) Autologous bone-marrow stem-cell transplantation for myocardial infarction. Lancet 361:45–46.

Steinbrook R (2004) The cord-blood-bank controversies. New Eng J Med 351:2255–2257.

Storb R, Yu C, Zaucha JM, Deeg HJ, Georges G, Kiem H-P, Nash RA, McSweeney A, Wagner JL (1999) Stable mixed hematopoietic chimerism in dogs given donor antigen, CTLA4Ig, and 100cGy total body irradiation before and pharmacologic immunosuppression after marrow transplant. Blood 94:2523–2529.

Strauer BE, Brehm M, Zeus T, Kostering M, Hernandez A, Sorg RV, Kogler G, Wernet P (2002) Repair of infarcted myocardium by autologous intracoronary mononuclear bone marrow cell transplantation in humans. Circulation 106:1913–1918.

Takahashi T, Kalka C, Masuda H, Chen D, Silver M, Kearney M, Magner M, Isner JM, Asahara T (1999) Ischemia- and cytokine-induced mobilization of bone marrow-derived endothelial progenitor cells for neovascularization. Nature Med 5:434–438.

Takahashi S, Iseki T, Ooi J, Tomonari A, Kashiya Takasugi K, Shimohakamada Y, Yamada T, Uchimaru K, Tojo A, Shirafuji N, Kodo H, Tani K, Takahashi T, Yamaguchi T, Asano S (2004) Single-institute comparative analysis of unrelated bone marrow transplantation and cord blood transplantation for adult patients with hematologic malignancies. Transplantation 104: 3813–3820.

Tateishi-Yuyama E, Matsubara H, Murohara T, Ikeda U, Shintani S, Masaki H, Amano K, Kishimoto Y, Yoshimoto K, Akashi H, Shimada K, Iwasaka T, Imaizumi T (2002) Therapeutic angiogenesis for patients with limb ischemia by autologous transplantation of bone-marrow cells: A pilot study and a randomized controlled trial. Lancet 360:427–435.

Taylor DA, Atkins BZ, Hungspreugs P, Jones TR, Reedy MC, Hutcheson KA, Glower DD, Kraus WE (1998) Regenerating functional myocardium: Improved performance after skeletal myoblast transplantation. Nature Med 4:929–933.

Thompson BG, Robertson KA, Gowan D, Heilman D, Broxmeyer H, Emanuel D, Kotylo P, Brahmi Z, Smith FO (2000) Analysis of engraftment, graft-versus-host disease, and immune recovery following unrelated donor cord blood transplantation. Blood 96:2703–2711.

Tomita S, Li RK, Meisel RD, Mickle DA, Kim EJ, Tomita S, Jia ZQ, Yau TM (1999) Autologous transplantation of bone marrow cells improves damaged heart function. Circulation 100:II247–II256.

Tse H-F, Kwong Y-L, Chan JKF, Lo G, Ho C-L, Lau C-P (2003) Angiogenesis in ischemic myocardium by intramyocardial autologous bone marrow mononuclear cell implantation. Lancet 361: 47–49.

Tsoporis JN, Marks A, Zimmer DB, McMahon C, Parker RG (2003) The myocardial protein S100A1 plays a role in the maintenance of normal gene expression in the adult heart. Mol Cell Biochem 242:27–33.

Van Luyn MJA, Tio RA, van Seijen XJG Y, Plantinga JA, de Leij LFMH, DeJongste MJL, van Wachem PB (2002) Cardiac tissue engineering: Characteristics of in unison contracting two- and three-dimensional neonatal rat ventricle cell (co)-cultures. Biomats 23:4793–4801.

Wagner JE, Kernan NA, Steinbuch M, Broxmeyer HE, Gluckman E (1995) Allogeneic sibling umbilical-cord-blood transplantation in children with malignant and non-malignant disease. Lancet 346:214–219.

Wagner JE, Rosenthal J, Sweetman R, Shu XO, Davies SM, Ramsay NKC, McGlave PB, Sender L, Cairo MS (1995) Successful transplantation of HLA-matched and HLA-mismatched umbilical cord blood from unrelated donors: Analysis of engraftment and acute graft-versus-host-disease. Blood 88:795–802.

Walter D, Rittig K, Bahlmann FH, Kirchmair R, Silver M, Murayama T, Nishimura H, Losordo DW, Asahara T, Isner JM (2002) Statin therapy accelerates reendothelialization: A novel effect involving mobilization and incorporation of bone-marrow derived endothelial progenitor cells. Circulation 105:3017–3024.

Wollert KC, Meyer GP, Lotz J, Ringes-Lichtenberg S, Lippolt P, Breidenbach C, Fichtner S, Korte T, Hornig B, Messinger D, Arseniev L, Hertenstein B, Ganser A, Drexler H (2004) Intracoronary autologous bone-marrow cell transfer after myocardial infarction: The BOOST randomized controlled clinical trial. Lancet 364:141–148.

Yamaguchi J-I, Kusan KF, Masuo O, Kawamoto A, Silver M, Murasawa S, Bosch-Marcee M, Masuda H, Losordo DW, Isner JM, Asahara T (2003) Stromal cell-derived factor-1 effects on ex vivo expanded endothelial progenitor cell recruitment for ischemic neovascularization. Circulation 107:1322–1334.

Yao M, Hale S, Dow JS, Kloner RA (2000) A pilot study to assess the feasibility of transplanting fetal cardiac tissue into pericardium of infarcted rat heart. Cardiac Vasc Reg 1:221–227.

Young PP, Hofling AA, Sands MS (2002) VEGF increases engraftment of bone marrow-derived endothelial progenitor cells (EPCs) into vasculature of newborn murine recipients. Proc Natl Acad Sci USA 99:11951–11956.

Zimmerman W-H, Fink C, Krakisch D, Remmers U, Weil J, Eschenhagen T (2000) Three-dimensional engineered heart tissue from neonatal rat cardiac myocytes. Biotech Bioeng 68:106–114.

Zimmerman W-H, Schneiderbanger K, Schubert P, Didie M, Munzel F, Heubach JF, Kostin S, Neuhuber WL, Eschenhagen T (2002) Tissue engineering of a differentiated cardiac muscle construct. Circ Res 90:223–230.

Zimmerman W-H, Melnychenko I, Eschenhagen T (2004) Enginered heart tissue for regeneration of diseased hearts. Biomats 25: 1639–1647.

# 13 Regenerative Medicine: Developmental Potential of Adult Stem Cells

## INTRODUCTION

Cell transplant therapies of regenerative medicine are based on the idea that the transplanted cells will develop into new tissue that integrates into the surrounding tissue or secrete paracrine factors that promote the survival and proliferation of local cells within a lesion, reducing scarring, and perhaps inducing regeneration. Autogeneic or allogeneic adult stem cells expanded in culture have been used in many cases as a source of transplantable cells. Autogeneic ASCs are preferable, because both pre- and post-transplant immunosuppression are not required for cell survival. However, some ASCs are not easy to harvest without damaging the tissue in which they reside (e.g., NSCs and cardiac stem cells) and are difficult to expand *in vitro*, which would make them difficult to use in human patients, particularly in the case of autogeneic cells.

Over the past five years there have been numerous reports that ASCs have a much wider developmental potential (prospective potency) than they actually exhibit (prospective fate) in their normal microniche (Camargo et al., 2004). That is, when exposed to developmental signals they would not normally encounter, they can give rise to cell types other than the ones they would normally produce. If so, then an ASC that is easy to harvest and highly expandable *in vitro* would be able to provide enough cells for transplant to restore any tissue. Stem cells from bone marrow, umbilical cord blood (HSCs, MSCs, EnSCs), and fat meet these criteria. Bone marrow stem cells in particular have been reported capable of differentiating into a wide range of cells other than their normal prospective fate. Examples of bone marrow stem cells differentiating into kidney epithelium, β-cells, and cardiomyocytes have been mentioned in previous chapters.

The term that is commonly used for this phenomenon is transdifferentiation. Strictly speaking, transdifferentiation means that a differentiated cell shuts down its program of gene activity while simultaneously shifting to the program of another differentiated cell type, without the cell reverting to a more primitive state. Undifferentiated ASCs might be plastic enough to be redetermined along other lineage pathways, or they might require dedifferentiation to an earlier state of plasticity in order to be redetermined. In either case, such redetermination is called transdetermination. Here, I will use the term "lineage conversion" to indicate a change in prospective cell fate without implying what mechanism might be used.

This chapter reviews the evidence that ASCs can change their prospective fate under experimentally imposed conditions and explores the question of whether or not the normal developmental potential of ASCs wanes with age. The initial enthusiasm about the developmental potential of ASCs has been tempered by experiments showing that the frequency of their lineage conversion *in vivo*, when it occurs, is too low to regenerate new tissue, and by experiments revealing that many cases of what appear to be lineage conversions are artifacts. However, some experiments do suggest that cellular reprogramming might be achievable at higher frequencies by manipulations *in vitro*. The possibility that adult cells in general might be reprogrammed to become other cell types or to become pluripotent cells is a hot and fascinating topic and is considered in Chapter 15.

## ASSAYS TO TEST DEVELOPMENTAL PLASTICITY

**FIGURE 13.1** illustrates several types of assays to test the developmental potential of ASCs. The general procedure is to label the ASCs, followed by exposing the cells to foreign soluble or cell surface signals. Animals or cells can be made transgenic for markers such as β-galactosidase or GFP, or by incorporation of lipophilic dyes. Natural markers such as the Y chromosome, or species-specific DNA sequences and antigens are also used. BrdU is often given as a second label to follow division of the marked cells. Methods to expose ASCs to foreign signals include injecting the cells into irradiated or *scid* host mice (bone marrow reconstitution assay) or into early embryos (chimeric embryo assay), culturing them with other cell types *in vitro*, exposing them to medium conditioned by other cell types, or exposing them to sets of defined factors that specify differentiation pathways. The presence of cells containing the donor marker(s), but having the morphology,

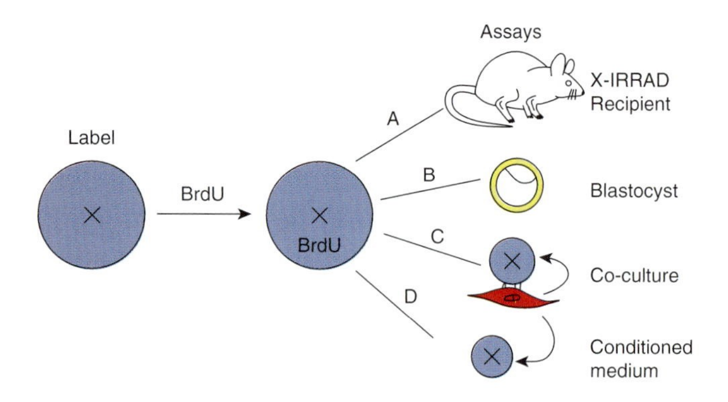


**FIGURE 13.1** Diagram showing assays for developmental potential of adult stem cells. Sex chromosomes or species-specific DNA sequences can be used as labels, or cells are labeled with dyes or transgenes (X). Sometimes labels will be combined, for example, the Y chromosome and a GFP transgene. BrdU is given to reveal DNA synthesis. The labeled cells are tested for their ability to differentiate into cell types other than those of their normal fate by *in vivo* (transplantation into x-irradiated recipients or into blastocysts) or *in vitro* (co-culture with differentiated cell types or conditioned medium) assays. Lineage conversion in co-culture could be by soluble signals (arrow) or by cell contact.

molecular phenotype, and function of differentiated cells from other lineages is taken as evidence that the cells have undergone lineage conversion.

Some ASCs have been tested for their developmental potential, while others have not (Rice et al., 2004). ASCs that have not been tested are epidermal and intestinal epithelial stem cells, auditory sensory cells, kidney epithelial cells, hepatocytes, retina stem cells, and cardiac stem cells. ASCs that have been tested are NSCs, unfractionated bone marrow, HSCs, SCs of muscle, stem cells residing in dermis, oval cells of the liver, MSCs of marrow, adipose stromal, dental pulp, and ESC-like cells derived from long-term cultures of bone marrow and connective tissues. Of these, some have been tested *in vivo*, some *in vitro*, and some in both.

## EXPERIMENTAL RESULTS

**TABLE 13.1** and **FIGURE 13.2** summarize the claims made for pluripotency that have been made for a wide variety of mammalian ASCs.

### 1. Neural Stem Cells

The developmental plasticity of NSCs has been tested *in vivo* and *in vitro*. Clonally derived NSCs from ROSA26 mice (transgenic for the *lacZ* gene) were tested *in vivo* in a bone marrow reconstitution assay using Balb/c mice as hosts (Bjornson et al., 1999). The results suggested that the NSCs were able to trans-

**TABLE 13.1** Normal Fate of Adult Stem Cells (ASCs) and Claims of Lineage Conversion (NSC = neural stem cell, SC = satellite cell, BM = bone marrow, HSC = hematopolietic stem cell, MSC = mesenchymal stem cell)

| | | Lineage Conversion | |
|---|---|---|---|
| ASC | Normal Fate | *In Vivo* | *In Vitro* |
| NSC | neurons, glia | myeloid, lymphoid, epidermal, notochord, mesonephros, somites, cardiac, digestive epith, liver | endothelial |
| Dermal neural precursors | Merkel cells | — | NSC, glia |
| SC | muscle | blood, cardiomyocytes | NSC |
| Adipose stem cells | adipocytes | osteoblasts | osteoblasts |
| Oval cells | hepatocytes | — | β-cells |
| Dental pulp stem cells | odontoblasts | — | neurons |
| Unfractionated BM | myeloid, lymphoid, cartilage, bone, endothelium | skeletal muscle, hepatocytes, neurons, kidney epith, cardiomyocytes, cheek epith | |
| HSC | myeloid, lymphoid | skeletal muscle, cardiomyocytes, neurons kidney epith, epidermal, lung epith, intestinal epith | neurons |
| MSC | cartilage, bone, fibroblasts, adipocytes | astrocytes, cardiomyocyte, Purkinje fibers, connective tissue, type 1 pneumocytes, kidney epith | neurons, glia, cardiomyocytes, skeletal muscle |

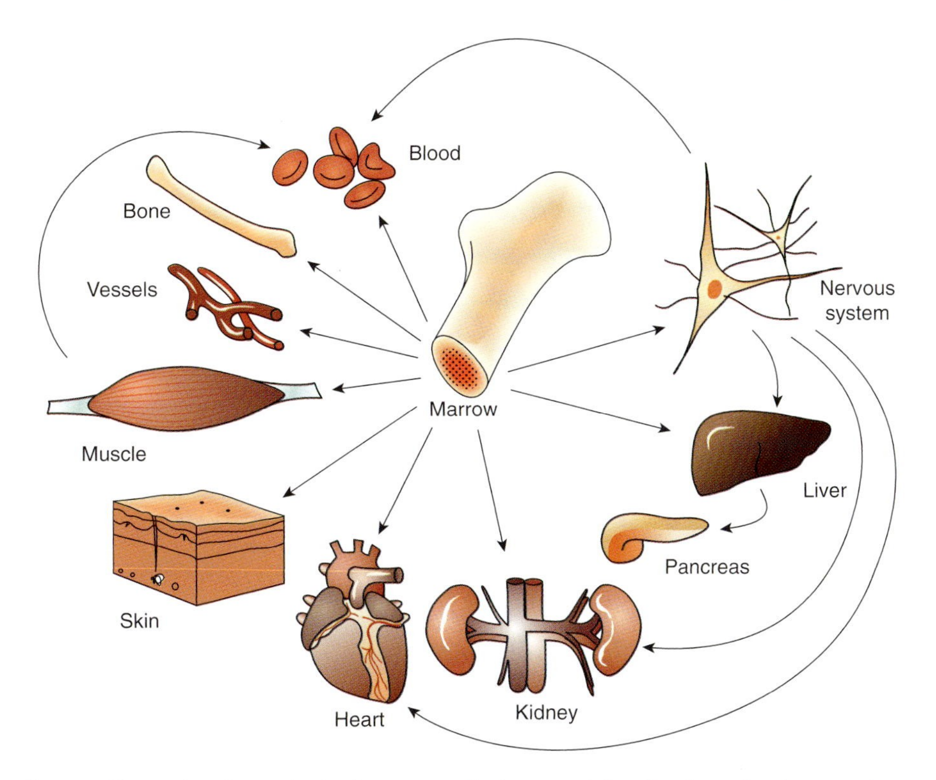


**FIGURE 13.2** Summary diagram illustrating tissues harboring stem cells that are claimed to have developmental potency greater than their normal fate. Stem cells of the bone marrow have been reported to have a particularly high degree of pluripotency. Redrawn with permission after Holden and Vogel, Plasticity: Time for a reappraisal? Science 296:2126–2129. Copyright 2002, AAAS.

differentiate into blood cells. Flow cytometry, using antibodies to CD3e, CD19, and CD11b, respectively, in combination with antibodies to the ROSA26-specific cell surface antigen H-2k$^b$, revealed the presence of T-lymphocytes, B-lymphocytes, and myeloid cells derived from β-galactosidase-expressing donor cells. In another study, NSCs derived from *lacZ*-labeled mouse neurospheres were also shown to form blood cells after injection into irradiated mice (Filip et al., 2004).

When injected into mouse blastocysts or into the prospective amniotic cavity of early chick embryos, clonally derived β-galactosidase-expressing neurospheres formed from ROSA26 NSCs participated in the formation of tissue and organ derivatives of all three germ layers in 25% of recipient chick embryos and 12% of recipient mouse embryos (Clarke et al., 2000). Nonneural tissues formed by donor cells were epidermis, notochord, mesonephros, somites, heart muscle, lung epitheium, stomach and intestinal epithelium, stomach and intestinal wall, and liver (**FIGURE 13.3**). The donor cells expressed the molecular markers typical for these tissues. The frequency of tissue chimerism in individual embryos ranged from 38% (heart) to 96% (intestinal epithelium). These results suggest that NSCs are pluripotent. However, injection into blastocysts of FACS-isolated mouse ESCs and NSC-derived neurospheres transgenic for *Sox2*-EGFP driven by a *Sox2* promoter fragment (P/Sox2-EGFP) gave different results (D'Amour and Gage, 2003). ESCs integrated efficiently into the blastocysts and participated in the formation of all tissues, whereas the NSCs did not.

NSCs have also been reported to change prospective fate *in vitro*. When co-cultured with human endothelial cells, clonally derived NSCs from mice transgenic for GFP differentiated at a frequency of 5–6% into cells that expressed the endothelial cell marker CD146 (Wurmser et al., 2004) (**FIGURE 13.4**). These cells were uninucleate, had no human chromosomes, and did not react to a human antiribonuclear protein antibody, indicating that this result was not due to fusion with the co-cultured endothelial cells. Clonal lines of GFP$^+$/CD146$^+$ cells expressed a wide variety of endothelial markers (VE cadherin, von Willebrand factor, etc.) over multiple passages and displayed the structural and functional features of endothelial cells, including the presence of Weible-Palade secretory vesicles and the ability to form capillaries when plated on Matrigel. Conditioned medium of human endothelial cell cultures was unable to induce the conversion of GFP$^+$ NSCs into endothelial cells, suggesting that the shift in fate requires cell contact. Evidence for this idea was obtained by culturing NSCs with endothelial cells fixed with paraformaldehyde. After five days of culture with fixed human endothelial cells, 2–4% of the NSCs

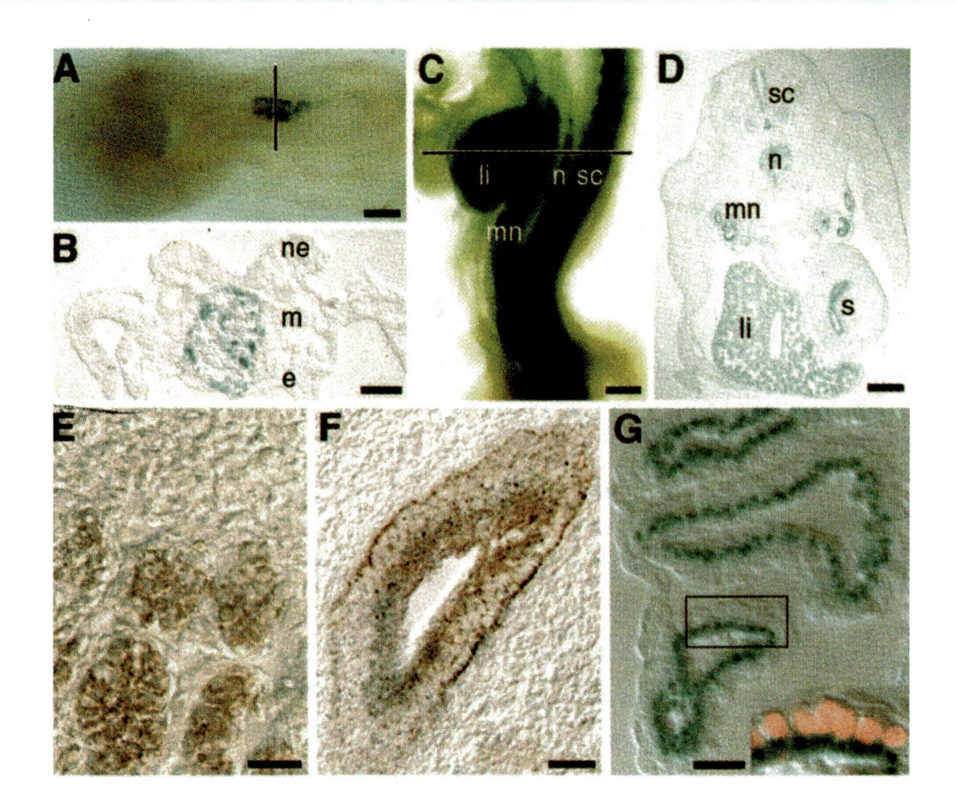


**FIGURE 13.3** Chimeric embryo assay for plasticity of mouse NSCs. ROSA 26 NSCs (expressing β-gal, blue) were injected into the amniotic cavity of early chick embryos. **(A)** Stage 8 chick embryo whole mount showing the transplanted cells. Line indicates plane of section in **(B)**. ne = neurectoderm, m = mesoderm, e = endoderm of host. **(C)** Trunk region of a highly chimeric stage 23 embryo. Li = liver, n = notochord, sc = spinal cord, mn = mesonephros. Line indicates plane of section in **(D)** showing same blue tissues. **(E, F)** mouse-derived cells stained with a mouse-specific antibody in the liver **(E)** and stomach **(F)** of the chick host. **(G)** Mesonephric tubules expressing cytoplasmic β-gal and nuclear Pax-2 (box, inset). Reproduced with permission from Clarke et al., Generalized potential of adult neural stem cells. Science 288:1660–1663. 

had differentiated into endothelial cells, as assayed by staining with a lectin that bound with much higher affinity to mouse NSC-derived endothelial cells than to human endothelial cells. A control experiment with fixed COS7 cells yielded no conversion to endothelial cells. Clonal GFP$^+$ NSCs injected into the telencephalon of embryonic day 14 mice differentiated primarily into glial cells, but 1.6% differentiated into endothelial cells with a single nucleus, suggesting that signals in brain tissue can effect conversion of NSCs at a low level.

Merkel cells are neural sensory receptors found in the dermis of the skin. In the rat, the number of these receptors is diminished by denervation but is restored by innervation (Nurse et al., 1984), suggesting that the skin might contain a precursor cell capable of differentiating into neural cell types. Such cells have been isolated by culturing dissociated dermal cells from juvenile mice in medium with EGF and FGF, which promoted the development of neurospheres from a subpopulation of the dermal cells (Toma et al., 2001). When plated on polylysine, in the absence of growth factors, individual neurosphere cells proliferated and differentiated into neuron-like cells that expressed neurofilament M, NSE, NeuN, and neuron-specific Tα1 α-tubulin at a frequency of 5–7%, as well as into glia-like cells expressing either GFAP (astrocytes), CNPase (oligodendrocytes), or both (Schwann cells) at a frequency of 7–11%. At serum concentrations of 3% and 20%, they also differentiated into adipocytes at frequencies of 1–25%. These stem cells are distinct from MSCs, because MSCs will not proliferate in the medium used to grow the dermal cells.

## 2. Satellite Cells

Satellite cells have been reported to be convertible to blood cells. FACS-purified SCs (c-kit$^+$/CD45$^-$) were reported to reconstitute the hematopoietic system after injection into lethally irradiated *mdx* or wild-type mice (Gussoni et al., 1999; Jackson et al., 1999; Howell et al., 2002). However, these results were subsequently shown to be due to contamination of the injected cells by CD45$^+$/Sca-1$^+$ hematopoietic cells normally resident in adult skeletal muscle (McKinney-Freeman et al., 2002).

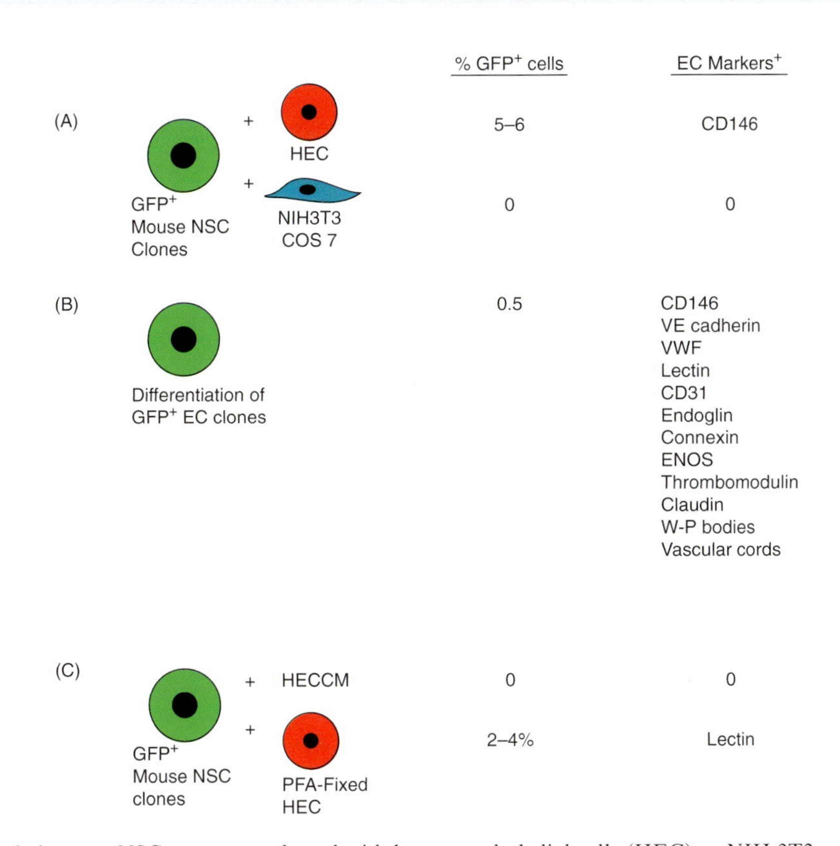


**FIGURE 13.4 (A)** GFP-labeled mouse NSCs were co-cultured with human endothelial cells (HEC) or NIH 3T3 cells or COS7 cells. A small percentage (5–6%) of the NSCs were converted to endothelial cells, as assayed by staining for the endothelial cell marker CD146. No conversion of 3T3 or Cos7 cells was observed. **(B)** A very small percentage (0.5%) of endothelial cell (EC) clones expressing GFP differentiated *in vitro* into endothelial cells expressing the wide variety of markers characteristic of these cells. **(C)** Experiment showing that contact with human endothelial cells, but not human endothelial cell conditioned medium (HECCM), converts NSCs to endothelial cells. Human endothelial cells that normally express an endothelial-specific lectin were fixed in culture with paraformaldehyde (PFA-fixed HEC) and co-cultured with GFP-labeled live mouse NSCs. A small percentage of cells (2–4%) subsequently expressed the endothelial lectin. Since the endothelial cells were dead, the signal they transmitted to the NSCs had to be through their cell surface (contact).

Muscle cell suspensions from male mice injected retro-orbitally into lethally irradiated female host mice resulted in reconstitution of a chimeric hematopoietic system in which the donor cells contributed 30–80%. However, immunofluorescent cell surface phenotyping indicated that the injected suspensions contained a substantial fraction (21.6%) of hematopoietic $CD45^+$ cells (Farace et al., 2004).

Rat SCs were reported to express markers characteristic of NSCs when cultured in neural differentiation medium (Alessandri et al., 2004). GFAP was expressed in 18% of the total cells seeded and β-tubulin III in 3%. However, SCs transplanted into lesioned spinal cords did not differentiate into either glia or neurons. These results suggested that skeletal muscle may contain a small population of NSCs that become confined to muscle during ontogenesis or are attracted there by muscle fibroblasts, which secrete stromal cell-derived factor 1 (SDF1), a chemoattractant for NSCs.

Several investigators have suggested, on the basis of histological observations, that SCs appeared to become converted to cardiomyocytes after transplantation into myocardial infarcts of rats or rabbits (Chiu et al., 1995; Taylor et al., 1998; Dorfman et al., 1998; Atkins et al., 1999; Kao et al., 2000). However, Reinecke et al. (2002) demonstrated conclusively that adult rat SCs are not converted into cardiomyocytes when transplanted into the normal hearts of syngeneic rats. Satellite cells labeled *in vitro* with BrdU never expressed cardiac-specific molecular markers after transplantation into the heart, but instead differentiated exclusively into slow-twitch skeletal muscle that expressed β-MHC. This muscle did not form electrically coupled junctions with host cardiac muscle, and atrophied by 12 weeks post-transplantation.

## 3. Liver Oval Cells

Oval stem cells were isolated and purified by FACS for the $Thy\text{-}1.1^+$ population (purity >95%), cultured for six months in serum-free medium, and then tested for their ability to transdifferentiate into insulin-producing pancreatic cells by growing them in medium supplemented with 10% fetal calf serum and a high (23 mM) glucose concentration (Yang et al., 2002). After two months of culture, the cells formed small, spheroidal, islet-like clusters that enlarged and became connected by ductal structures. These clusters expressed transcripts for a number of endocrine cell differentiation markers and hormones, such as the transcription factors PDX-1, Nkx2.2, Nkx6.1, PAX-4, and PAX 6, the cell surface marker Glut-2, and insulin I, insulin II, glucagons, and somatostatin. Immunocytochemistry showed that the cells responded to glucose stimulation by producing glucagon and insulin, and that the response was enhanced by nicotinamide. A preliminary study indicated that the hyperglycemia of streptozotocin-induced diabetic NOD/*scid* mice was normalized by implants of the transdifferentiated cells under the renal capsule. These results are perhaps not too surprising, given the closely related embryonic derivation of the liver and pancreas.

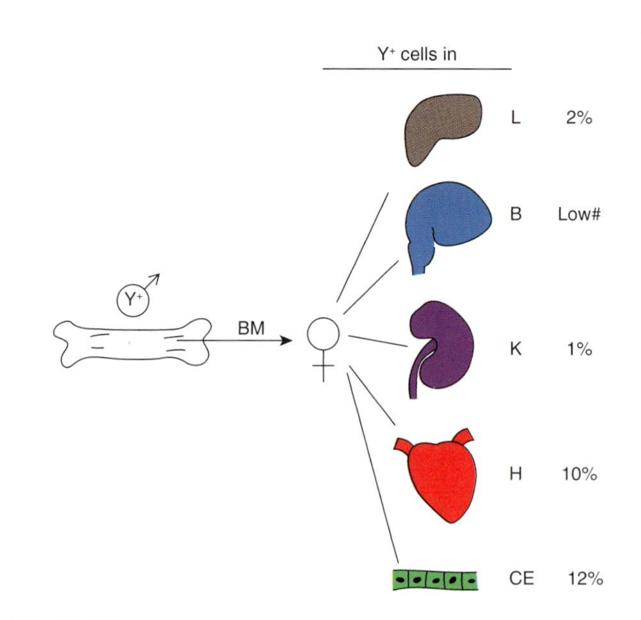


**FIGURE 13.5** Following bone marrow transplants from males to females, Y⁺ cells were reported to be present in several tissues of the recipient, suggesting the conversion of bone marrow cells to liver (L), neurons of the brain (B), kidney (K) epithelium, heart (H) and cheek epithelium (CE), at the indicated percentages.

## 4. Unfractionated Bone Marrow and HSCs

Bone marrow stem cells have by far been the subject of the most claims for ASC pluripotency.

### a. Unfractionated Bone Marrow

Unfractionated, marked bone marrow cells have been reported not only to reconstitute the hematopoietic system after injection into x-irradiated, *scid*, or PU.1-null mice, but also to differentiate into skeletal muscle after muscle injury (Ferrari, 1998; LaBarge and Blau, 2002), into hepatocytes after liver injury (Petersen et al., 1999; Thiese et al., 2000; Sell, 2001; Shafritz and Dabeva, 2002; Dabeva and Shafritz, 2003; Fausto, 2004; Alison et al., 2004), into neurons (Brazelton et al., 2000; Mezey et al., 2000), and into kidney epithelium after injury (Imasawa et al., 2001; Poulsom et al., 2003). The frequencies of transdifferentiation reported were low, 0.2–3.5% for muscle, 0.15–2% for hepatocytes, 0.2–2.3% for neurons, and occasional cells in kidney epithelium.

Observations on the tissues of deceased human patients who received cross-gender organ transplants have suggested that host bone marrow cells can be recruited and converted into out-of-lineage cell types (**FIGURE 13.5**). Studies of female patients with liver damage who received male bone marrow transplants identified $Y^+$ hepatocytes in the recipients at a frequency of 0.5–2% (Theise et al., 2000b; Alison et al., 2000). Analysis of the brains of female patients who died after receiving bone marrow transplants from male donors identified small numbers of $Y^+$ cells with neuronal markers (Mezey et al., 2003). In healthy female patients who had received male bone marrow transplants, a high percentage of cheek cells (12% in one patient), identified by their characteristic morphology and the production of a phenotype-specific protein, were $Y^+$ (Tran et al., 2003). Analysis of female kidneys transplanted into male recipients showed the presence of up to 1% $Y^+$ cells in renal tubule epithelium (Poulsom et al., 2003), and Quiani et al. (2002) reported that up to 10% of the cardiomyocytes in female human hearts transplanted to male recipients were $Y^+$. Laflamme et al. (2002) were unable to verify the latter observations and concluded that the recruited host cells are likely to be blood or endothelial cells.

Other studies also have failed to confirm the pluripotency of unfractionated bone marrow cells *in vivo*. Kanazawa and Verma (2003) were unable to detect any lineage conversion into hepatocytes of unfractionated bone marrow cells marked with EGFP or *LacZ* transgenes, or by the Y chromosome, after injection into mice with three different modes of liver injury, with or without x-irradiating the host animals. Sell (2001) has argued that oval cells of the liver may be derived from the bone marrow. To test this idea, ROSA-26 (*LacZ*$^+$) bone marrow cells were transplanted into wild-type mice (Wang et al., 2003), as shown in **FIGURE 13.6**. These mice were then treated with 3,5-diethoxycarbonyl-1, 4-dihydrocollidine (DDC) to evoke oval cell production and the oval cells were then transplanted into $Fah^{-/-}$ mice, where they repopulated the liver. The repopulated hepatocytes, however, did not express β-

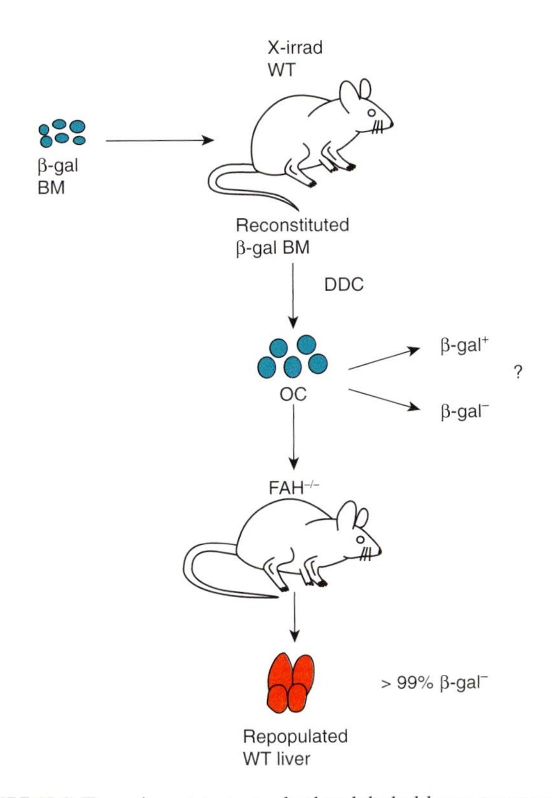


**FIGURE 13.6** Experiment to test whether labeled bone marrow cells (β-gal$^+$ BM) were transformed into oval cells of the liver. The labeled bone marrow was transplanted into lethally-irradiated wild-type (WT) mice, where it reconstituted the bone marrow of the recipients. The mice were then treated with DDC to activate oval cell (OC) production and the oval cells were injected into FAH$^{-/-}$ mice, where they repopulated the liver with wild-type hepatocytes. These hepatocytes were over 99% negative for β-gal staining, indicating that they were derived from the wild-type host's oval cells, not the transplanted bone marrow cells.

galactosidase, which they would have done had the oval cells of the original transplanted wild-type mice been derived from the bone marrow. Instead, the result indicates that the oval cells in the DDC-treated mice were generated intrahepatically. A similar conclusion was reached after grafting mutant DPP4$^-$ female rats with wild-type male bone marrow cells (Menthena et al., 2004). Although DPP4$^+$ cells were detected in the livers of all recipients one month after transplant, these cells disappeared after subjecting animals to three different liver injury protocols for activation and expansion of oval cells. GFP-labeled male bone marrow cells also failed to differentiate into pancreatic β-cells after injection into female mice that were treated with either streptozotocin or partial pancreatectomy (Lechner et al., 2004).

### *b.* HSCs

A Thy-1$^+$, Sca-1$^+$ population of bone marrow cells (presumably HSCs) isolated by immunomagnetic sorting were reported to form neurospheres when cultured *in vitro* in neuronal medium with FGF and EGF (Locatelli et al., 2003). Exposure of the neurosphere cells to RA resulted in their differentiation, at a frequency of 40%, into cells positive for the neuronal markers TuJ-1 and neurofilament (NF).

Several reports have suggested that HSCs (phenotype [Sca-1, c-kit, CD43, CD45]$^+$ [Lin CD34]$^-$) are pluripotent, as assayed in transplants to X-irradiated recipients. Marked (GFP, LacZ, Y chromosome) HSCs were reported to differentiate into skeletal muscle (Gussoni et al., 1999; Castro et al., 2002), neurons (Locatelli et al., 2003), brush border epithelial cells of kidney proximal tubules (Lin et al., 2003), and epithelium of the skin, liver, lung, and intestine (Krause et al., 2001; Fang et al., 2003). The reported frequencies of lineage conversion were low, no more than 4% (skeletal muscle). However, values of 20% were reported for lung alveolar epithelium (Krause et al., 2001). Kong et al. (2004) reported that HSCs from human umbilical cord blood differentiated into normal muscle cells when injected into mice with muscular dystrophies caused by mutations in the dysferlin gene. The number of human cells expressing the *dysf* gene, however, was less than 1%.

Several studies reported a low frequency of conversion (0.02–0.7%) of *LacZ* or *GFP* transgene-labeled HSCs into cardiac muscle after transplantation into myocardial infarcts of mice or rats (Orlic et al., 2001; Jackson et al., 2001; Eisenberg et al., 2003; Lanza et al., 2004; Zhang et al., 2004). The transgenes were driven by ubiquitous constitutive promoters. However, when the cardiac-specific α-MHC promoter was used to drive a β-gal reporter in transgenic mice, no β-gal expressing cells were detectable after injecting several subpopulations of HSCs from these mice into myocardial infarcts of recipient mice (Murry et al., 2004) (**FIGURE 13.7**). HSCs marked with an EGFP transgene driven by a ubiquitous constitutive promoter were injected into the myocardial infarcts of X-irradiated or unirradiated mice (Murry et al., 2004; Balsam et al., 2004). EGFP-expressing cells were abundant in the host hearts during the first week after transplant, but rapidly decreased in number thereafter, until only 0.02–0.03% co-expressed EGFP and cardiac α-myosin heavy chain. These were probably blood cells. In another study, bone marrow cells from male mice injected into female δ-sarcoglycan-null mice were incorporated into skeletal muscle and cardiac muscle. However, both skeletal and cardiac muscle fibers containing donor-derived nuclei failed to express sarcoglycan (Lapidos et al., 2004). These studies suggest that donor cells may initially be incorporated into foreign tissues, but that the vast majority of cells do not survive long-term and/or are unable to convert to the authentic phenotype of the host tissue.

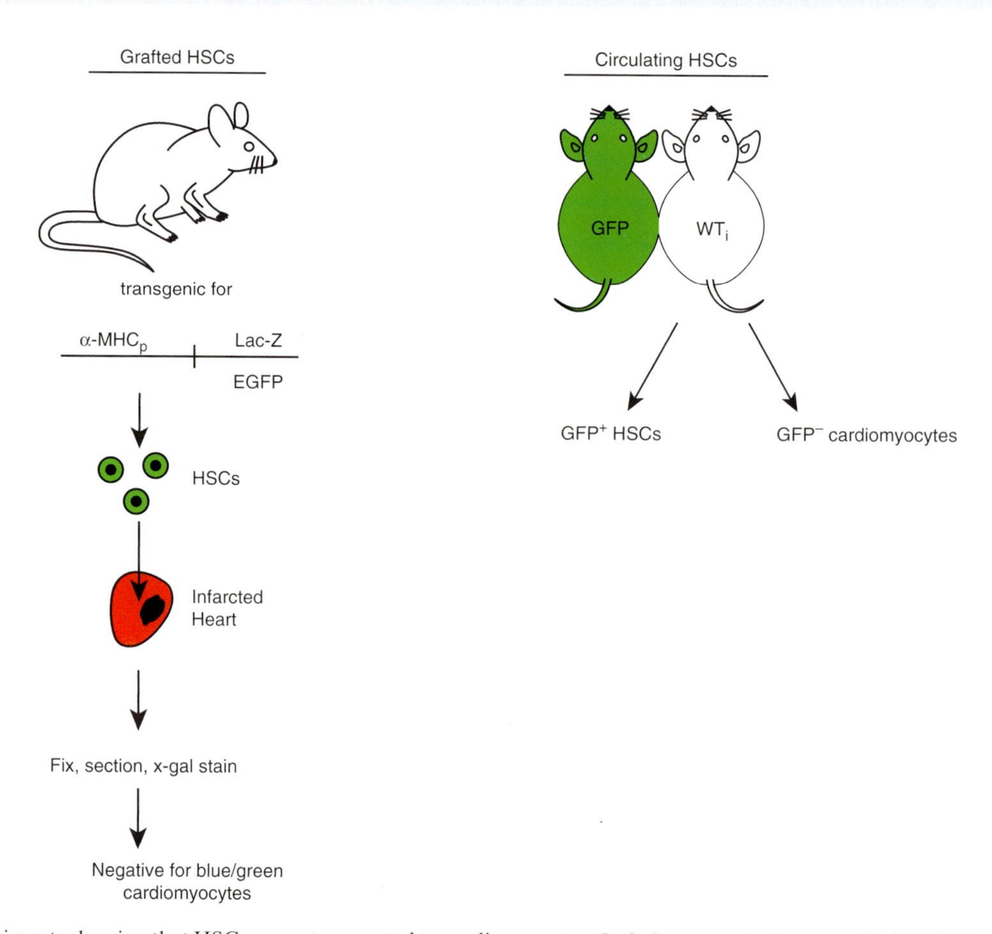


**FIGURE 13.7** Experiments showing that HSCs are not converted to cardiomyocytes. **Left**, hematopoietic stem cells (HSCs) from mice transgenic for the *lac-Z* or *GFP* gene driven by an α-myosin heavy chain promoter (α-MHCp) were injected into the infarct region of unlabeled mice. The injected hearts were fixed, sectioned, and stained for β-galactosidase activity. No labeled cardiomyocytes were found in the hearts, indicating that the injected HSCs were not converted to cardiomyocytes. **Right**, mice transgenic for GFP were parabiosed to wild-type mice (WTi) that were subsequently subjected to a myocardial infarct. Although $GFP^+$ hematopoietic stem cells were found circulating in the wild-type mice, no $GFP^+$ cardiomyocytes were found in the infarcted hearts, demonstrating that the injured myocardium does not recruit HSCs to repair damage.

In a review on heart regeneration, Laflamme and Murry (2005) identified two common artifacts that can mislead investigators into thinking that host bone marrow cells have undergone a lineage conversion in a transplanted organ or that transplanted donor bone marrow cells have undergone lineage conversion (**FIGURE 13.8**). These are the presence of host leukocytes in a transplanted organ, and an increase in autofluorescence of host cells after injury that be mistaken for a donor fluorescent marker.

## 5. MSCs of the Bone Marrow and Related MSCs

### a. *Bone Marrow MSCs*

Bone marrow MSCs appear able to change their fate in both *in vivo* and *in vitro* assays. Cultured rat marrow MSCs labeled with BrdU were reported to differentiate into cardiomyocytes after transplantation into the myocardium (Tomita et al., 1999; Wang et al., 2000; Bittira et al., 2000). Airey et al. (2004) marked human marrow MSCs with dsRed or GFP and injected them intraperitoneally into sheep fetuses *in utero*. Analysis of the hearts at a late stage of fetal development showed that the human cells engrafted in the heart and differentiated into Purkinje fibers. On average, over 43% of the total Purkinje fibers in random areas of both ventricles were of human origin, as opposed to only 0.01% of cardiomyocytes.

Marked MSCs infused into lethally irradiated mice homed to the bone marrow and were later found in the connective tissues of liver, lung, and thymus at a frequency of 0.2–2.3% (Pereira et al., 1995; Prockop, 1997). Marrow MSCs transgenic for *Lac-Z* were reported to differentiate into Type I pneumocytes when injected into mice whose lungs were injured by bleomy-

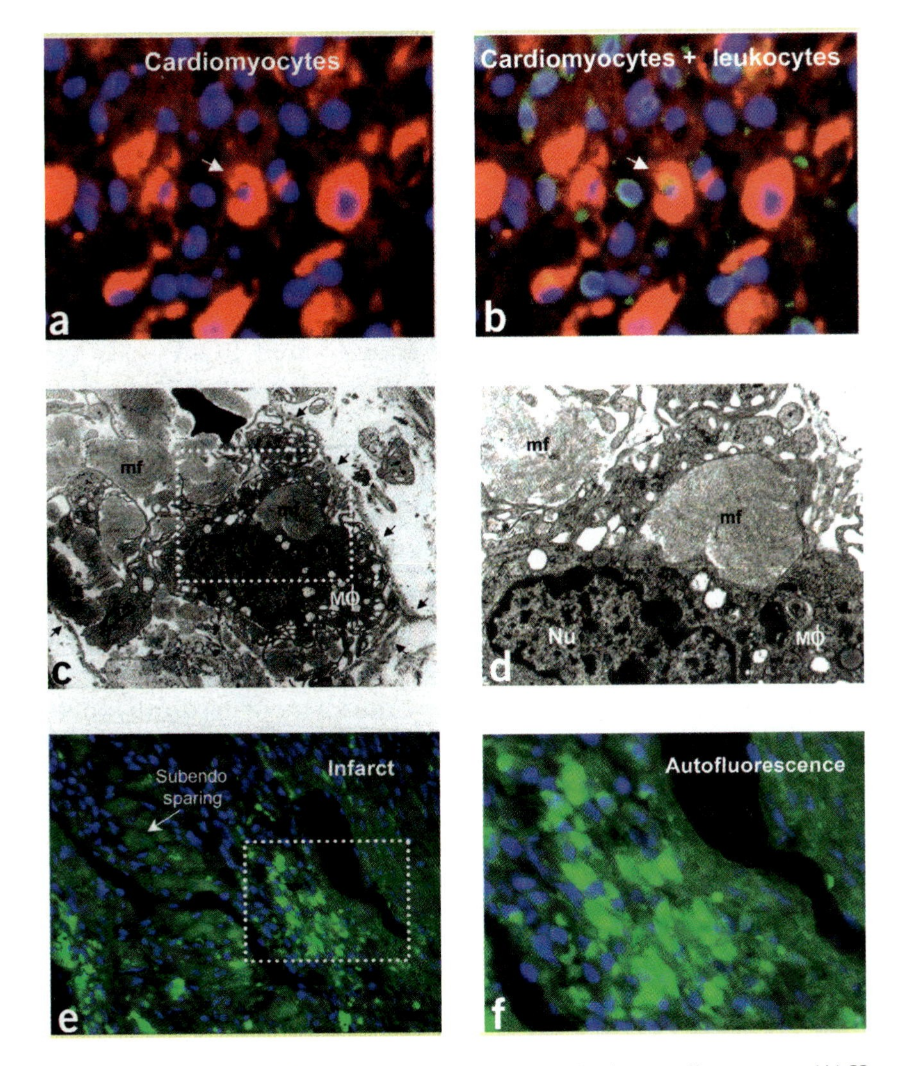


**FIGURE 13.8** Artifacts that can lead to the illusion of bone marrow stem cell conversion into cardiomyocytes. **(A)** Human heart transplant. The arrow appears to indicate a myosin-stained cardiomyocyte with a blue nucleus, but staining with markers for leukocytes **(B)** shows that the nucleus belongs to an intracytoplasmic leukocyte. **(C)** Electron micrograph of infarcted myocardium shows a necrotic cardiomyocyte with degenerating myofibrils (mf) being engulfed by a macrophage (Mϕ). Boxed region is shown at higher magnification in **(D)**, revealing a macrophage nucleus (Nu) in close proximity to myofibrils, showing how the macrophage could be mistaken for a myosin$^+$ marrow-derived cell. **(E, F)** Myocardial infarct in a mouse that had received no cells whatsoever, stained with a blue DNA dye to identify nuclei by fluorescence microscopy. The green color is due to autofluorescence. If GFP-expressing bone marrow cells had been introduced into this animal, autofluorescence of cardiomyocytes could give the mistaken impression that the labeled bone marrow cells had been converted to cardiomyocytes. Reproduced with permission from Laflamme and Murry, Regenerating the heart. Nature Biotech 23:845–856. Copyright 2005, Nature Publishing Group.

cin, or when cultured *in vitro* (Kotten et al., 2001). Male MSCs injected into female mice treated with cisplatin to induce renal tubule injury engrafted into the tubules, where they were reported to differentiate at low frequency into proximal tubule epithelial cells (Morigi et al., 2004). When injected into the brain ventricles of neonatal mice, MSCs migrated throughout the forebrain and cerebellum, where they appeared to differentiate into astrocytes (Azizi et al., 1998; Kopen et al., 1999).

Adult rat and human marrow MSCs were induced by β-mercaptoethanol or dimethyl sulfoxide/butylated hydroxyanisole treatment to differentiate at high frequency (up to 80%) *in vitro* into neural-like phenotypes expressing the neural markers NSE, NeuN, neurofilament-M, and tau (Woodbury et al., 2000). Individual clones of cells were self-renewing, giving rise to both neurons and MSCs. In another set of experiments, human or mouse MSCs cultured in neural differentiation medium containing all-trans RA and BDNF differentiated at low frequency into neural (0.5%) or glial (1%) phenotypes expressing NeuN or GFAP (Sanchez-Ramos et al., 2000). Human MSCs labeled with red or green fluorescent tracker dyes or mouse MSCs transgenic for *lac Z* expressed Neu-N at a frequency of 2–5% or GFAP at a frequency of 5–8% when co-cultured with mouse fetal midbrain cells. The morphology of these cells approximated that of immature neural and glial cells (Sanchez-Ramos et al., 2000). In another experiment, sequential culture of clonally derived MSCs with EGF and FGF-2, followed by culture in medium conditioned by postnatal hippocampal astrocytes, resulted in the differentiation of significant numbers of MSCs into cells positive for neurofilament and β-tubulin III (Joannides et al., 2003). The number of neuron-like cells was greater than with EGF and FGF-2 alone, suggesting that astrocytes have a positive influence on neuronal differentiation. Nearly 80% of mouse MSCs were reported to differentiate into neurons when cultured in the presence of RA and Shh together, or when incubated with prenatal cochlea explants or in medium conditioned by embryonic hindbrain/somite/otocyst (Kondo et al., 2005).

Treatment of mouse MSCs with 5-azacytidine *in vitro* induced them to differentiate as skeletal muscle cells (Wakatani et al., 1995). Cells of two MSC lines could differentiate as beating cells with a contractile protein profile of fetal ventricular cardiomyocytes, either spontaneously after long-term culture (Jiang et al., 2000) or after 5-azacytidine treatment (Makino et al., 1999). Both these lines, however, expressed the cardiac-specific transcription factor Nkx2.5, which would predispose them toward cardiomyocyte differentiation. Marrow MSCs transgenic for *Lac-Z* were also reported to differentiate into Type I pneumocytes when cultured *in vitro* (Kotton et al., 2001).

Murry et al. (2004) created chimeric embryoid bodies by co-culturing mouse ESCs with mouse MSCs transgenic for *Lac-Z* driven by the cardiac-specific α-MHC promoter. When the embryoid bodies were removed from suspension culture and plated, widespread areas of contractile activity developed. However, none of the cells that differentiated as cardiomyocytes expressed β-galactosidase, even though the presence of cells with the reporter gene was demonstrated by PCR, suggesting that the MSCs did not differentiate into cardiomyocytes. MSCs also failed to differentiate as cardiomyocytes (as measured by α-actinin expression) when co-cultured with neonatal rat ventricular cardiomyocytes separated by a semipermeable membrane (Xu et al., 2004). However, after a 7-day co-culture without separation, MSCs in contact with cardiomyocytes became positive for α-actinin and GATA-4, and formed gap junctions with native cardiomyocytes. These results suggest that exposure to soluble factors secreted by cardiomyocytes is insufficient to change the prospective fate of MSCs, but that cell–cell contact can do so, as in the conversion of NSCs to endothelial cells (Wurmser et al., 2004).

### b. Multipotent Adult Progenitor Cells of Bone Marrow

Multipotent cells related to the MSC, designated multipotent adult progenitor cells (MAPCs), have been isolated from human, rat, and mouse marrow cultures that have undergone 15 or more doublings (Reyes et al., 2001, 2002; Schwartz et al., 2002; Jiang et al., 2002). These cells are smaller (8–10 μm-diameter) than MSCs and have the antigenic phenotype [CD34, CD44, CD45, c-kit, MHC complex class I and II]$^-$. They express low levels of Flk-1, Sca-1, and Thy-1, and higher levels of CD13. The cells show similarities to ESCs in their expression of SSEA-I and the transcription factors Oct-4 and Rex-1, and in their LIF-dependence (murine ESCs) and independence (human ESCs) to maintain the undifferentiated state. They express high levels of telomerase, and their average telomere length remained unchanged for over 100 population doublings (Reyes et al., 2001; Jiang et al., 2002). About one in $10^3$ cultured cells was capable of generating a clone of MAPCs.

**FIGURE 13.9** illustrates the result of an *in vivo* assay of MAPC developmental potency. ROSA-26 mouse MAPCs were tested in the chimeric embryo assay (C57BL/6 hosts) after 55–65 population doublings. Chimerism was detected in 80% of the mice derived from blastocysts receiving 10–12 cells and in 33% of those receiving one cell. In both cases, the extent of chimerism in individual animals ranged between 0.1–45%. Examination of sections of whole chimeric mice and individual organs for β-galactosidase staining and phenotype-specific markers indicated that the MAPCs contributed to virtually every tissue of the chimeric mice: brain (including cortex, striatum, hippocampus, thalamus, and cererbellum), retina, lung, myocardium, skeletal muscle, liver, intestine, kidney, spleen, bone marrow, blood, and skin. However, no tests for fusion events were carried out and the long-term survival of the embryos was not assessed. Similar results were obtained after transplantation to *scid* recipients.

The progeny of single GFP-labeled MAPCs were induced to differentiate *in vitro* into endothelial cells, hepatocytes, and neural cells. Endothelial cell differentiation was induced with VEGF (Reyes et al., 2002). The cells first differentiated to angioblasts, identified by their expression of CD34, VE-cadherin and Flk1, then into more mature endothelial cells marked by expression of CD31, CD36, CD62, and the adhesion junction proteins ZO-1, β-catenin, and γ-catenin. These cells had the functional characteristics of endothelial

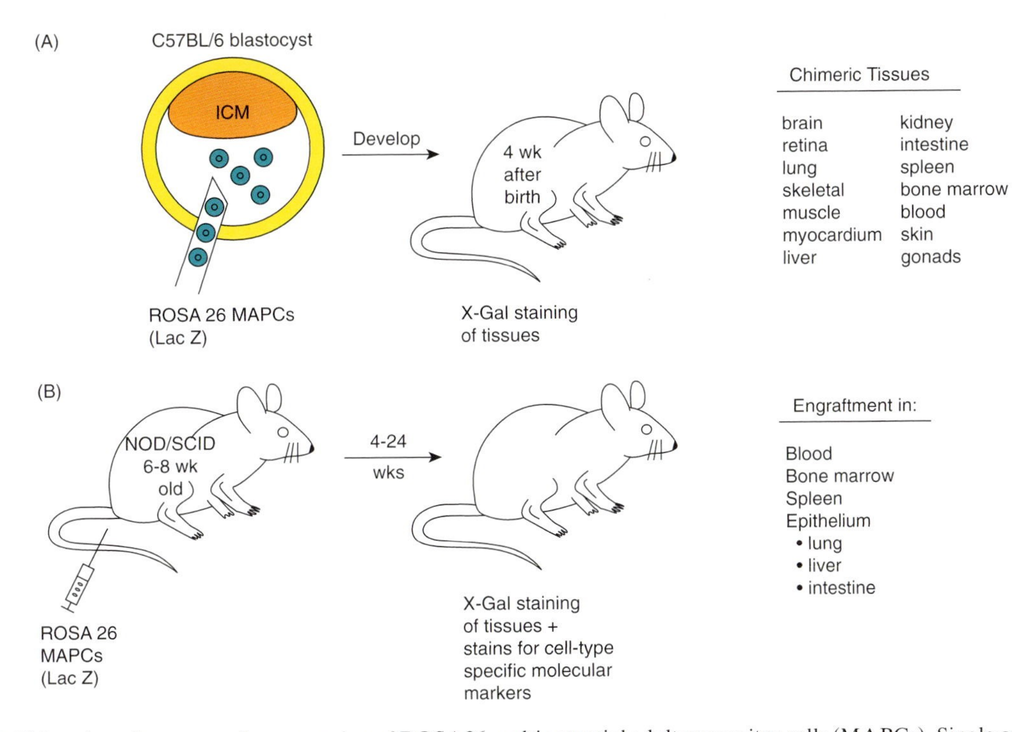


**FIGURE 13.9** **(A)** Chimeric embryo assay for conversion of ROSA26 multipotential adult progenitor cells (MAPCs). Single or multiple ROSA26 MAPCs were injected into the blastocysts of C57BL/6 mice and assayed for β-gal staining of tissues. β-gal staining was observed in a wide variety of tissues. **(B)** Conversion of MAPCs was also assayed by injecting them into NOD/SCID mice. β-gal staining was observed in tissues that normally undergo maintenance regeneration.

cells. They contained von Willebrand factor that was released by histamine treatment, took up LDL, responded to hypoxia by upregulating VEGF, Flk1 and the angiogenesis receptor tek, and upregulated HLA antigens and cell adhesion ligands in response to the inflammatory cytokine Il-1a. When plated on Matrigel, they formed vascular tubes. In parallel *in vivo* experiments, human MAPCs were injected into NOD/SCID mice undergoing extensive angiogenesis due to the presence of Lewis lung tumors or ear-hole punches. Thirty–35% of the tumor blood vessels and the blood vessels of the healed ear wound contained MAPC-derived endothelial cells that reacted to human β2-microglobulin and von Willebrand factor.

Hepatocyte differentiation was induced by culturing MAPCs on Matrigel in the presence of FGF-4 and HGF (Schwartz et al., 2002; Jiang et al., 2002). The cells differentiated first to a phenotype positive for hepatocyte nuclear factor-3β (HNF-3β), GATA 4, cytokeratin 19 (CK 19), transerythrin, and α-fetoprotein, then to a phenotype expressing CK18, HNF-4, and HNF-1α. These cells exhibited the morphological and functional characteristics of hepatocytes. They were epithelioid, secreted urea, and albumin, took up LDL, stored glycogen, and were phenobarbitol-inducible with regard to cytochrome 450.

MAPCs were induced to differentiate into neurons and glia by FGF-2, as assessed by PCR for neuroectodermal gene activity and reaction to antibodies to NF-200, galactocerebroside, and GFAP (Jiang et al., 2000). When the MAPCs were sequentially cultured with FGF-2, FGF-8, and BDNF, they differentiated into more mature dopaminergic, serotonergic, and GABA$^+$ neurons.

### c. Adipose Mesenchymal Stem Cells

Adipocytes (fat cells) are derived from embryonic mesenchyme. The connective tissue stroma of human subcutaneous or knee pad adipose tissue aspirate has been found to contain stem cells that are phenotypically similar to MSCs of the bone marrow (Zuk et al., 2002; Wickham et al., 2003). These cells, which are called adipose-derived stem (or stromal) cells (ADSCs), express nearly all of the same CD markers as marrow MSCs, though they differ in two of them, CD49d (α4 integrin), expressed by ADSCs but not marrow MSCs, and CD106 (vascular cell adhesion molecule, VCAM), expressed by marrow MSCs but not ADSCs (Zuk et al., 2002). Like marrow MSCs, clonal ADSCs are able to differentiate *in vitro* into mature adipocyte, osteoblast, chondrocyte, and muscle phenotypes when induced by sets of factors that promote the differentiation of these phenotypes (Zuk et al., 2002; Erikson et al., 2002; Mizuno et al., 2002; Hicok et al., 2004; Justesen et al., 2004). ADSCs were also reported to differentiate into immature neurons when exposed to neuronal differentiation factors (Zuk et al., 2002; Ashjian et al., 2003). ADSCs may be good candidates as gene delivery vehicles. When transfected with retroviral or lentiviral EGFP constructs and induced to differentiate into adipocytes or osteoblasts, the differentiated cells maintained expression of the EGFP (Morizono et al., 2003). The idea of using adipose tissue as a source of MSCs is appealing, because of the virtually unlimited supply and ease of harvest!

### d. Dental Pulp Cells

Dental pulp cells are a heterogeneous population of cells that includes progenitor odontoblasts and two types of stem cells that have similarities to MSCs of the bone marrow (Chapter 3). Human dental pulp cells labeled with the nuclear localizing dye bisbenzimide and cultured *in vitro* on polylysine with neural cells differentiated into cells with neural morphology at a frequency of 3.6% and expressed the neuron-specific markers PGP 9.5 and β-tubulin III (Nosrat et al., 2004). Although the cultures contained stem cells, it is not known whether it was these cells that actually became neural cells. These results are not surprising, given that the dental pulp is a derivative of the neural crest, which also gives rise to sensory neurons.

### e. Pluripotent Mesenchymal Cells from Connective Tissues

Young et al. (2004) and Young and Black (2004) have reported the isolation of pluripotent epiblastic-like stem cell PPELSCs from the connective tissue compartments of virtually every mammalian organ system, including those of humans, as well as from granulation tissue of skin wounds (Calcutt et al., 1993; Lucas et al., 1993) that can differentiate into all the derivatives specific to ectoderm, mesoderm, and endoderm. These investigators have been unable to confirm reports of lineage conversion of purified (or highly enriched) ASCs and argue that purported cases of such conversion are due to the contamination of the cell preparations with PPELSCs.

PPELSCs are small (6–8-μm diameter) with a high nuclear to cytoplasmic ratio, have extended capability for self-renewal, and are quiescent in serum-free defined medium lacking inhibitory factors such as LIF. Their molecular phenotype has some of the characteristics of ESCs, such as expression of telomerase, SSEA-1, 3, and 4, and Oct-4. PPELSCs can be stimulated to proliferate *in vitro* by PDGF-BB and to differentiate into cartilage, bone, adipocytes, fibroblasts, and skeletal myotubes by dexamethasone treatment, within the normal lineage range for MSCs. *In vivo*, clonally derived PPELSCs transfected with the *LacZ* gene were incorporated into myocardium, vasculature, and connective tissue (Young and Black, 2004), but β-galactosidase expressing cells were not tested for differentiation markers of cardiac cells.

Rat PPELSCs were induced to differentiate *in vitro* into three-dimensional pancreatic islet-like structures by plating them in a medium that promotes the differentiation of embryonic pancreatic cells into islet cells (Young and Black, 2004). The cells were reported to express molecules specific to the different kinds of islet cells: insulin, glucagon, and somatostatin. Under glucose stimulation, the induced β-cells secreted insulin at a level up to 49% of native islets. This insulin was shown by radioimmunoassy to be rat-specific insulin, not bovine insulin sequestered from the medium and then released.

### f. "Spore" Cells

A mesenchymal stem cell similar in morphology to the MAPC and PPELSC that can reconstitute site-specific tissue has been reported to reside in virtually every tissue of the body (Vacanti et al., 2001). Light, scanning, and transmission electron microscopy indicated that the cells are small, rounded entities, 3–5 μm in diameter, with a large nucleus and scant cytoplasm. They were named "spore cells," because of their resemblance to spores. BrdU incorporation studies indicate that the cells have a doubling time ranging from 12–36 hr, depending on the tissue from which it was derived. When cultured in a DMEM/F-12 medium, the cells reproduced the specific tissue from which they were derived and developed tissue-specific morphology after 7–10 days in culture.

### g. What Are the Multipotent/Pluripotent Mesenchymal Cells That Have Been Isolated from Bone Marrow, Fat, Dental Pulp, and Connective Tissue?

The nature and origin of the multipotent/pluripotent mesenchymal stem like cells that have been isolated from bone marrow, fat, dental pulp and connective tissue is not clear. One idea is that all of these mesenchymal stem like cells might be variants of ESC-like stem cells that persist throughout development to reside in all tissue compartments. The idea that such a cell exists in the bone marrow and circulates throughout the body was proposed some time ago (Owen and Friedenstein, 1988; Caplan, 1991). These pluripotent stem cells might colonize developing tissues through the vasculature and give rise to different kinds of ASCs, depending on local environmental cues, and some of the pluripotent cells might be preserved in the microvasculature of adult tissues. These cells might, in fact, be responsible for the rare lineage conversions observed. The idea of a universal, pluripotent stem cell is seductive, for such a cell could potentially be used to regenerate any tissue.

One cell type that is ubiquitously associated with arterioles, venules, and capillaries of all tissues is the pericyte. Pericytes are the source of smooth muscle cells in regenerating arterioles and venules and have long been considered by histologists to retain a high degree of mesenchymal developmental potential (Ham and Cormack, 1979). For example, ectopic bone can be induced in connective tissue by demineralized bone matrix (Urist, 1965) or BMPs (Lindholm and Urist, 1980; Urist et al., 1983) and is always associated with the ingrowth of blood vessels. This suggests that pericytes might be the source of the ectopic bone, which would be consistent with the fact that MAPCs, PPELSCs, and ADSCs can differentiate into chondrogenic and osteogenic phenotypes. It would be interesting and instructive to compare the molecular phenotype of pericytes to that of PPELSCs, DPSCs, ADSCs, and MSCs.

Alternatively, MAPCs, PPELSCs, DPSCs, and ADSCs might be the products of dedifferentiation to a wider state of developmental potential as a result of prolonged times in culture.

## MANY REPORTS OF LINEAGE CONVERSION ARE DUE TO FUSION WITH HOST CELLS

Many reports of stem cell pluripotency appear to be cases of fusion of the donor stem cells with differentiated host cells to form heterokaryons, in which the transcriptional profiles of the differentiated cell are adopted by the stem cell nucleus through intracellular reprogramming, giving the illusion of a prospective fate switch induced by extracellular signals (**FIGURE 13.10**). This type of reprogramming by heterokaryon formation between differentiated cell types has been demonstrated many times *in vitro* (Blau et al., 1985).

The first inkling that transdifferentiation could be confused with cell fusion was the demonstration that GFP-labeled mouse NSCs or bone marrow cells that apparently were converted into ESCs when co-cultured with ESCs had a 4N chromosome number and exhibited markers of both NSCs or bone marrow cells and ESCs (Terada et al., 2002; Ying et al., 2002). The frequency of fusion in the cultures was very low, around $10^{-6}$–$10^{-4}$.

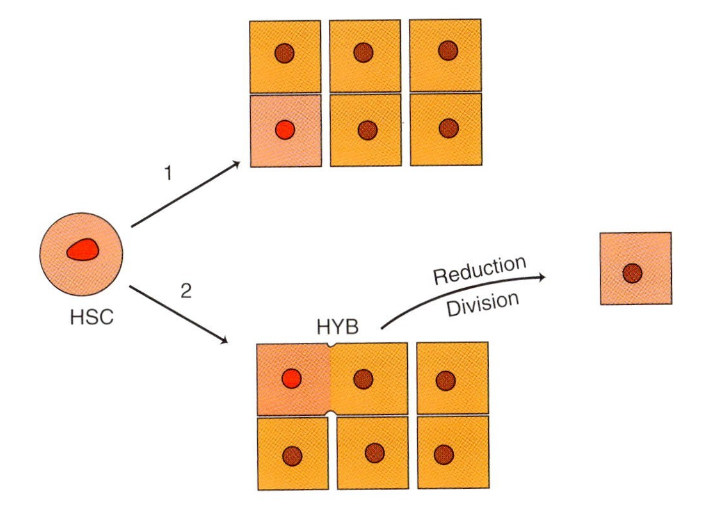


**FIGURE 13.10** How fusion of transplanted stem cells with host cells can give the appearance of lineage conversion by extracellular signals. **(A)** If a blood stem cell (HSC) were converted to a hepatocyte by extracellular signals in the liver environment (1), it and its daughters would contain only donor chromosomes and express only hepatocyte markers. However, as shown in (2) the reprogramming is done by fusion of the stem cells with hepatocytes, resulting in a hybrid cell (HYB) with two sets of chromosomes that express markers of both cell types. Reduction division may occur later to give the normal number of chromosomes.

NSCs were reported in several experiments to differentiate into muscle when co-cultured with myoblasts, cardiomyocytes, or ESCs. Fifty-seven percent of mouse NSCs transfected with the GFP gene and co-cultured with C2C12 myoblasts differentiated *in vitro* into skeletal myotubes expressing the muscle markers α-actinin-2 and myosin heavy chain (Rietze et al., 2001). GFP-labeled NSCs differentiated into cardiac muscle cells when co-cultured with cardiomyocytes from neonatal rats (Condorelli et al., 2001). Clarke et al. (2000) cultured ROSA26 NSCs with mouse embryoid bodies. Progeny of the NSCs expressed desmin and many cells fused to form muscle-like syncytia that were immunoreactive to myosin heavy chain. All of these results might be explained by fusion of the NSCs with the co-cultured muscle cells or muscle cells that differentiated from ESCs.

Fusion of injected stem cells has also been demonstrated *in vivo*. Marked HSCs injected at limiting dilution were reported to give rise to cells morphologically similar to hepatocytes that extensively reconstituted the liver of tyrosinemic ($FAH^-$) mice (Lagasse et al., 2000). However, a repeat of this experiment showed that the new liver cells were positive for molecular markers of both donor and host (Wang et al., 2003; Vassilopoulos et al., 2003; see Medvinsky and Smith, 2003), indicating that fusion took place at high frequency. Interestingly, donor cell markers were seen

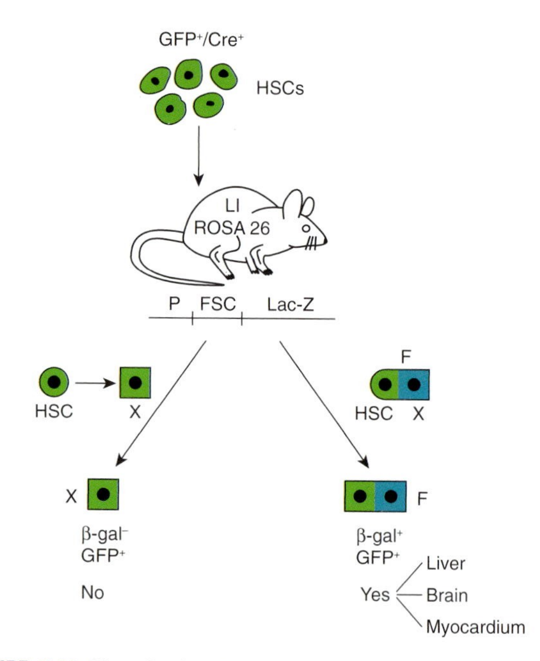


**FIGURE 13.11** Use of a Cre-lox marking protocol to detect fusion of transplanted stem cells with host cells. HSCs from mice transgenic for GFP and Cre were injected into lethally irradiated (LI) ROSA26 mice transgenic for a *Lac-Z* reporter gene separated from its promoter (P) by a floxed stop cassette (FSC). If lineage conversion of HSCs to cell X took place (left), cell X would express GFP but not β-gal. If fusion between HSCs and cell X took place (right) Cre would excise the FSC allowing expression of β-gal, along with GFP. This is what was observed at very low frequency in liver, brain and myocardium.

only after the host hematopoietic system was reconstituted. This could mean that it is not the HSCs themselves that are fusing with host hepatocytes (or other cell types), but differentiated products of the HSCs such as macrophages, which fuse with each other to form osteoclasts in bone remodeling or repair.

Alvarez-Dolado et al. (2003) lethally irradiated ROSA26 mice transgenic for a construct in which a floxed stop cassette was interposed between promoter and a *lac-Z* reporter gene, then engrafted them with HSCs from mice constitutively expressing Cre recombinase and GFP (**FIGURE 13.11**). Only cells derived by the fusion of donor HSCs with nonhematopoietic cells would express the *lacZ* reporter gene, because only in fused cells could Cre excise the floxed stop cassette of the ROSA26 cells, allowing them to express β-galactosidase. After 10 months, β-galactosidase and GFP-expressing cells were observed at very low frequency (0.0001–0.019%) in three tissues: brain, myocardium, and liver. These cells were shown to have two nuclei with different morphologies, bone marrow plus brain, myocardium, or liver. No $lacZ^{-}/GFP^{+}$ cells were observed anywhere except in the marrow and blood.

Weimann et al. (2003) showed in another study that bone marrow cells fuse at very low frequency with Purkinje neurons in the brain to form binucleate cells after injection into irradiated mice. These results indicate that transplanted HSCs are not converted to other cell types, but that they do fuse to other cell types at low frequency.

Other experiments are consistent with these results. Purified HSCs of ROSA26 mice failed to differentiate into neurons in irradiated C57B1/6 hosts when injected either before or after cortical stab injury (Castro et al., 2002). A few β-galactosidase$^{+}$ cells were observed, but these were associated with blood vessels and did not have the morphology of neurons. Likewise, GFP-labeled HSCs injected into lethally irradiated recipients reconstituted the hematopoietic system, but only one GFP$^{+}$ Purkinje neuron was found out of over 13 million cells examined in the brain. In addition, seven of 470,000 hepatocytes examined were GFP$^{+}$, but no GFP$^{+}$ cells were found in kidney, gut, skeletal muscle, cardiac muscle, or lung (Wagers et al., 2002). In another experiment, clonally derived NSCs injected into irradiated mice failed to reconstitute the bone marrow and did not regenerate any blood cells (Morshead et al., 2002).

The notion that fusion might account for all reports of lineage conversion has been challenged by Harris et al. (2004) in an experiment similar to that of Alvarez-Dolado et al. (2003). They injected male mouse bone marrow cells transgenic for an EGFP reporter gene, suppressed by a floxed stop cassette, into lethally irradiated female mice transgenic for *Cre* (**FIGURE 13.12**). Fusion events would be detectable by the expression of EGFP due to excision of the stop cassette by host cell Cre-recombinase. Injured muscle of host Cre mice injected with EGFP bone marrow exhibited EGFP expressing cells with two nuclei, showing that this system does detect fusion events. Y$^{+}$ cells expressing an epithelial cytokeratin gene were found at frequencies between 0.05% and 0.1% in the lung, liver, and wounded epidermis. RT-PCR analysis showed that none of the cells expressed EGFP, suggesting that lineage conversion, not fusion, had occurred. Since the bone marrow cells used in the experiments were not fractionated, and HSCs do not appear to undergo conversion, it is possible that MSCs were the cells that differentiated into epithelial cells.

## ADULT STEM CELL PLURIPOTENCY: FACT OR FICTION?

The overall conclusion to be drawn from the vast majority of experiments to test the developmental potency of ASCs is that they can at best be induced to

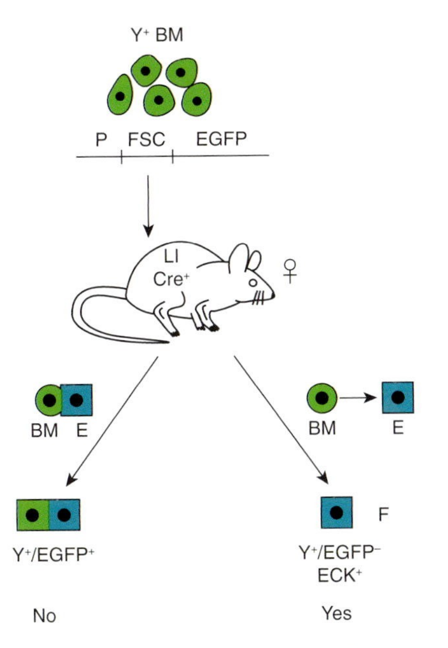


**FIGURE 13.12** A Cre-lox marking experiment that suggests lineage conversion of bone marrow cells to epithelial cells. Male bone marrow ($Y^+$ BM) transgenic for an enhanced green fluorescent protein (EGFP) reporter gene separated from its promoter (P) by a floxed stop cassette (FSC) was injected into lethally irradiated female mice transgenic for Cre. Fusion of the bone marrow cells with host epithelial cells (E) would be indicated by the presence of cells that are $Y^+$/EGFP$^+$. Lineage conversion of bone marrow cells would be indicated by the presence of $Y^+$/EGFP$^-$ cells expressing epithelial cytokeratin (ECK). This was the result, but at low frequency.

switch prospective fate at low frequency. In some cases, such as liver reconstitution under strong selection pressure, fusion of ASCs with host cells can be significant. Alvarez-Dolado et al. (2003) have pointed out that cell fusion is a normal developmental process that goes on not only in muscle development, but also in osteoclast formation by the fusion of monocytes (Boyle et al., 2003). Cardiomyocytes sometimes exhibit two nuclei under normal conditions, and hepatocytes are frequently polyploid (Ham and Cormack, 1979). Thus, the fusion of endogenous or exogenous stem cells to host differentiated cells may, at least in some tissues such as muscle and liver, be another mechanism by which injured tissue can regenerate itself. In fact, fusion of ESCs with somatic cells is being explored as a way to create pluripotent autogeneic ESCs for transplant by reprogramming the somatic cell nuclei (see Chapter 15).

Other cases of apparent pluripotency can be ascribed to contamination of injected cells with differentiated donor cell types, mistaken identity, fluorescence artifacts, and incorporation of ASCs into host tissues without their actually taking on the authentic phenotypes of these tissues. Furthermore, it appears that many cells that appear to have changed their prospective fate when assayed *in vivo* shortly after transplantation do not survive over longer periods of time. This means that the apparent higher frequency of lineage conversion in some *in vitro* studies should be viewed with skepticism until it has been shown in long-term cultures that it is real. Overall, the verdict is that unless the frequency can be increased significantly, lineage conversion of a single ASC type will not be a viable approach for cell therapies designed to replace damaged tissue.

However, there are some caveats to these conclusions. First, the type of assay used to assess developmental potency may affect the result. The developmental potential of most ASCs has been assessed through *in vivo* stem cell transplants to irradiated adult animals, or by *in vitro* assays. Of all ASCs that have been cultured for a relatively short period of time, only NSCs and MAPCs have been tested in the chimeric embryo assay. This is the only assay in which all the developmental signals that would reveal pluripotency are provided. In one case this assay has suggested that NSCs are pluripotent, but in another that they are not. The chimeric embryo assay should be used to test the developmental potential of other ASCs, not only because of the signaling environment, but because the embryos could be allowed to develop into neonates and beyond. This would allow us to determine the longevity and robustness of function of ASCs that may have changed their prospective fate, something that we currently do not know.

Second, we cannot ignore the possibility that MSC-like cells with many of the characteristics of ESCs (MAPCs, PPELSCs) may exist in the adult body or that cells such as pericytes or other cells can dedifferentiate to a more developmentally potent state. There is no theoretical reason why nuclear deprogramming and/or reprogramming leading to lineage conversion of ASCs or even differentiated cells should not be possible. Most somatic cells contain the whole genome, and both amphibian and mammalian adult somatic cell nuclei can be reprogrammed by enucleated egg cytoplasm to support the development of a complete organism, even though the percentage of success in such adult reproductive cloning is quite low (Gurdon and Laskey, 1975; Wilmut et al., 2002).

## EFFECTS OF AGE ON NUMBER AND DEVELOPMENTAL STATUS OF REGENERATION-COMPETENT CELLS

The ability of organisms to repair tissues by fibrosis or regeneration clearly declines with age (Nemoto and Finkel, 2004; Abbot, 2004). With aging, oxidative stress

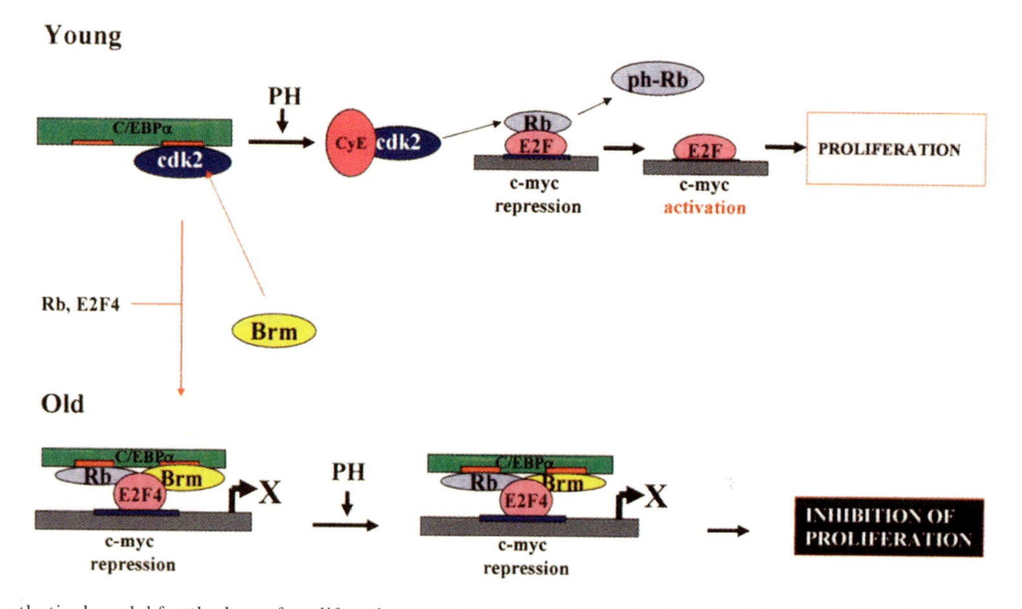


**FIGURE 13.13** Hypothetical model for the loss of proliferative response to partial hepatectomy in old animals. **Top**, in young animals cell cycling in uninjured liver is prevented by interactions of C/EBPα with cdks, Rb, E2Fs and Brm. Partial hepatectomy reduces the level of C/EBPα, allowing phosphorylation of Rb and its release from E2Fs, which activate transcription of genes such as c-myc that lead to hepatocyte proliferation. **Bottom**, in aged rats, C/EBPα level is not reduced after partial hepatectomy and hepatocyte proliferation is repressed by a C/EBPα/Rb/E2F4/Brm complex. Reproduced with permission from Iakova et al., Aging reduces proliferative capacities of liver by switching pathways of C/EBPα growth arrest. Cell 113:495–506. Copyright 2003, Elsevier.

takes place in the differentiated cells of tissues. There is an increasing imbalance between the intracellular levels of free radicals and cellular antioxidant systems that is due to both intrinsic changes and environmental insults. This imbalance is reflected in changes in cellular, tissue, and organ system structure and function. In mammals, these changes take place in all organs, but are most easily seen in the skin, where there are decreases in hair regeneration and the turnover rate of the epidermis, changes in proteoglycans that decrease thickness, elasticity, and hydration of the dermis, and increases in lipofuchsin deposits and benign lesions of various kinds, all of which are intensified by overexposure to the UV radiation from the sun (Jacobson and Flowers, 1996; Carrino et al., 2000). The rate of wound repair also declines with age and is associated with reduction in re-epithelialization (Cao et al., 2003), an increase in the number of infiltrating macrophages but a diminished capacity of macrophages for phagocytosis, delayed but increased T cell infiltration, and decrease in production of most cytokines, with the exception of MCP-1, which is increased (Swift et al., 2001). In addition, there is decreased wound contraction and increased collagen remodeling by MMPs with age (Fisher et al., 1996; Yao et al., 2001; Ballas and Davidson, 2001).

A major question is whether decline in the regenerative capacity of tissues is due to intrinsic cellular changes that decrease the numbers of regeneration-competent cells (RCCs) and/or their ability to proliferate and differentiate, or to a deteriorating general tissue environment that prevents their ability to respond to regenerative needs. The answer to this question is crucial if we are to develop a regenerative medicine based on the use of adult regeneration-competent cells. If the decline in regenerative capacity is due to declining numbers and intrinsic regenerative capacity of RCCs, their potential therapeutic effectiveness will be progressively compromised with age, unless they can be rejuvenated by genetic modification. By contrast, if the regenerative capacity of RCCs is maintained with age, then it may be possible to restore tissue structure and function in older individuals by supplying the cells with a cocktail of molecules representing a more "youthful," less stressful extracellular environment. Evidence obtained in experiments on tissues in young versus old rats, mice, and humans suggests that, at least in some cases, it is environmental deterioration that is responsible for loss of regenerative capacity.

The livers of old rats have a reduced proliferative response to partial hepatectomy. There is a significant delay in re-entering the cell cycle and a significant reduction in the number of dividing hepatocytes (Bucher, 1964; Fry et al., 1984). **FIGURE 13.13** shows why. The liver transcription factor C/EBPα mediates the growth arrest of hepatocytes in young rats by directly inhibiting the action of cyclin-dependent kinases and by forming a complex with Rb proteins, E2F transcription factors, and with a chromatin remodeling protein, Brm (Wang et al., 2001, 2002). Levels of C/EBPα are reduced after partial hepatectomy,

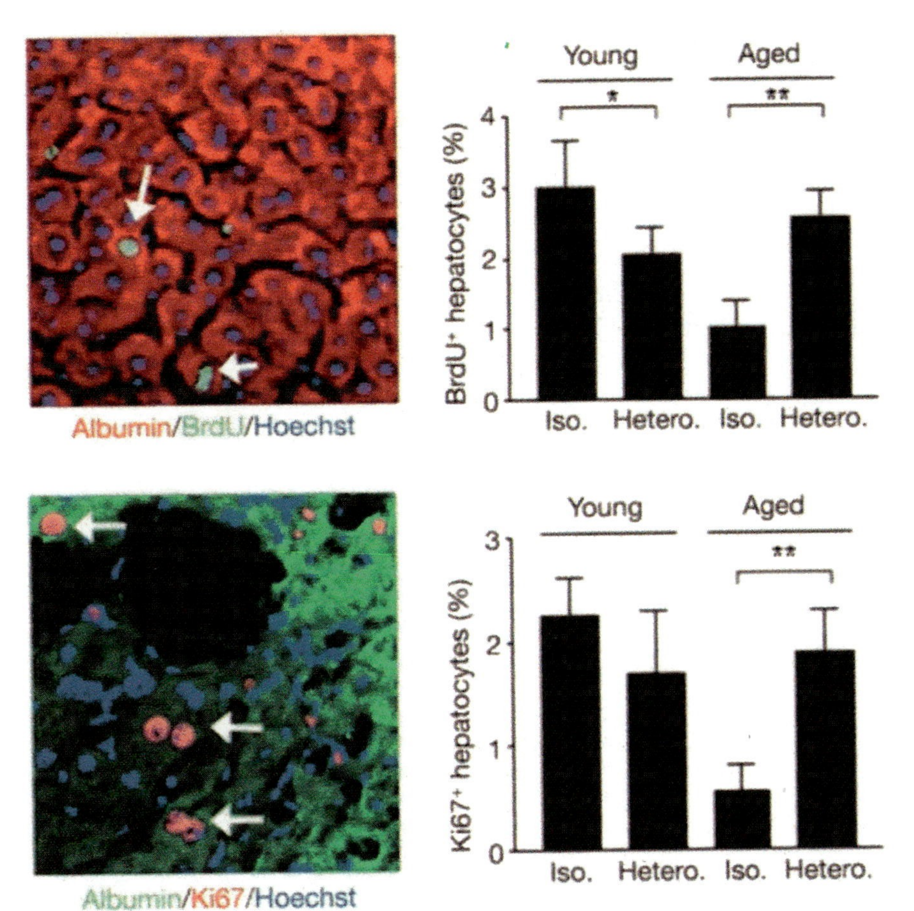


**FIGURE 13.14** Enhancement of maintenance liver regeneration in aged mice by parabiosis. **Left**, parabiotic animals were given BrdU and their regenerating liver cells stained for albumin, BrdU or Ki67 for proliferation, as well as with Hoechst nuclear dye. Arrows point to nuclei of cells co-staining for albumin and BrdU or Ki67. **Right**, bar graphs show significantly more proliferation in the livers of isochronous young parabiotic pairs than in old pairs, but pairing aged rats with young rats significantly increases proliferation of hepatocytes in the old rats. Reproduced with permission from Conboy et al., Rejuvenation of aged progenitor cells by exposure to a young systemic environment. Nature 3:760–764. Copyright 2005, Nature Publishing Group.

allowing this complex to dissociate. Cdks now phosphorylate Rb, releasing the E2Fs to activate target genes involved in proliferation. In aged rats, the level of C/EBPα is not reduced and the pathway of growth arrest switches to the repression of E2F target gene transcription by the formation of a C/EBPα-Rb-E2F4-Brm protein complex that continues to bind to the promoters of the target genes after partial hepatectomy, causing a significant loss of hepatocyte proliferation (Iakova et al., 2003).

The decline in the level of liver maintenance regeneration in aged mice can be reversed by parabiosing them to young mice (Conboy et al., 2005). Hepatocyte proliferation in the uninjured liver of the old member of a heterochronic pair was more than doubled, and approached the level in isochronic parabionts of young mice, while hepatocyte proliferation in the young member of a heterochronic pair was reduced by nearly a third (**FIGURE 13.14**). This enhancement and reduction were not due to circulating cells from either member of the pair, as shown by parabionts in which one member of the pair was transgenic for GFP. GFP-expressing cells were virtually nonexistent in the livers of the non-transgenic members of the pairs. Furthermore, formation of the inhibitory protein complex was diminished in the livers of the old mice and increased in the livers of the young mice of the pairs. It would be of interest to determine whether the same results would hold in an injured liver.

Muscle sarcopenia, characterized by loss of contractile protein from the myofiber, is one of the most obvious effects of aging in humans. There is a tendency for muscle to be replaced by fat and there is evidence to suggest that SCs become more adipocyte-like (Taylor-Jones et al., 2002). Muscle regeneration is diminished in aged mice and rats. The number of SCs is not diminished with age, but their ability to proliferate declines significantly due to a deficiency that results in the activation of only 25% of the number of SCs activated in young muscle (Conboy et al., 2003). The deficiency is due to the insufficient upregulation of the Notch ligand Delta and thus diminished activation of Notch. Blocking Notch activation in young muscle with an inhibitor caused a marked inhibition of regeneration (Conboy et al., 2003).

However, the poor regeneration of aged muscle can be reversed. Muscles of old rats transplanted into young recipients regenerate as well as young muscle transplanted into young rats, and the muscles of young rats transplanted into old recipients regenerate as poorly as old muscles transplanted into old rats (Carlson and Faulkner, 1989; Carlson et al., 2001). Activating Notch directly with an antibody to its extracellular domain in the muscles of old mice restores their regenerative ability to the level of young mice (Conboy et al., 2003). Furthermore, the number of activated SCs and bulk muscle regeneration was restored in the old mice of heterochronic parabionts to nearly the level of young mice, while the number of activated SCs and bulk muscle regeneration in young mice of the pairs was only slightly reduced (Conboy et al., 2005) (**FIGURE 13.15**).

These observations suggest that it is the loss of certain systemic factors that results in the decline of regenerative capacity in muscle and liver with age. Consistent with this hypothesis, the exposure of SCs from the muscles of old mice to the serum of young mice *in vitro* results in elevated expression of Delta, increased Notch activation, and enhanced proliferation (Conboy et al., 2005) (**FIGURE 13.15**). Furthermore, the muscles of old rats and humans are reported to be deficient in the production of mechano growth factor (MGH), a splice variant of IGF-I (Owino et al., 2001; Hameed et al., 2003). Strength training retards the onset of sarcopenia in humans (Clarke, 2004) and significantly increases MGH production, particularly in combination with growth hormone administration (Goldspink, 2004), again suggesting that systemic factors exert a significant effect on the ability of SCs in aging muscle to express an undiminished regenerative potential (Aagaard, 2004).

Regenerative capacity is also reduced in the skeletal, cardiovascular, and neural systems of old animals. In rats, the ability to repair fractures declines with age. The decline is associated with a deficit in the number of progenitor cells and their responsiveness to bioactive factors. There is a decrease in the number of MSCs in marrow as a function of age (Ergise, 1992; Oreffo et al., 1998) and the ability of marrow to form osteogenic colonies is less in older patients (Connolly, 1998). Fewer osteoprogenitor cells are mobilized from the marrow after fracture (Quarto et al., 1995), and an age-related decline in the bone-inductive response of nonosteogenic tissues to BMP-2 implants has been reported (Fleet et al., 1997). The mesodermal lineage-specific stem cells that have been isolated from mouse and human tissues of all ages show no age-related differences in their ability to differentiate into skeletal, smooth and cardiac muscle, adipocytes, various types of cartilage, bone, tendon, ligament, dermis, endothelium, and hematopoietic cells (Young et al., 2004). However, an age-related reduction in the chondro-osteogenic potential of MSCs was reported in first-harvest samples from the iliac crest of rabbits (Huibregtse et al., 2000). Cells derived from serial samples had improved chondro-osteogenic potential, suggesting that repeated sampling increased the recruitment of MSCs within the bone marrow. Interestingly, MSCs taken from the marrow of the tibia exhibited no age-related decline in chondro-osteogenic potential.

Stem cells of the hematopoietic and cardiovascular systems exhibit reduced capacities for regeneration with aging. The capacity for self-renewal of LT-HSCs does not appear to decline over the life of an animal, but the ability of the transit amplifying populations to proliferate in response to stress is diminished, perhaps by decreased telomerase activity (Globerson, 2001; Allsopp and Weissman, 2002).

Circulating endothelial cell progenitor cells are thought to continually repair the linings of blood vessels, preventing plaque formation and atherosclerosis. The number of circulating endothelial stem cells decreases with age in humans. An inverse relationship has been found between the number of circulating EnSCs and risk for cardiovascular disease (Hill et al., 2003). The number of circulating EnSCs was a better predictor for cardiovascular disease than the standard set of risk factors. The circulating EnSCs from high-risk subjects were shown to have much higher rates of senescence *in vitro* than those from low-risk subjects. Indeed, the loss of EnSCs with age has been directly associated with the loss of a protective effect against atherosclerosis (Rauscher et al., 2003). $ApoE^-$ mice have high cholesterol levels and become severely atherosclerotic. Injection of old $ApoE^-$ mice with bone marrow cells from young $ApoE^-$ mice, which have not yet developed atherosclerosis, prevented the progression of atherosclerosis in the old mice. The protective effect was not associated with any reduction in cholesterol levels. By contrast, bone marrow cells from old $ApoE^-$ mice failed to prevent atherosclerotic progression when injected into old $ApoE^-$ mice. Furthermore, the number of cells with vascular potential in the bone marrow of $ApoE^-$ mice declines with age. Together, these findings suggest that atherosclerosis develops as a consequence of an age-related decline in the number and quality of bone marrow stem cells with the capacity to differentiate as endothelial cells.

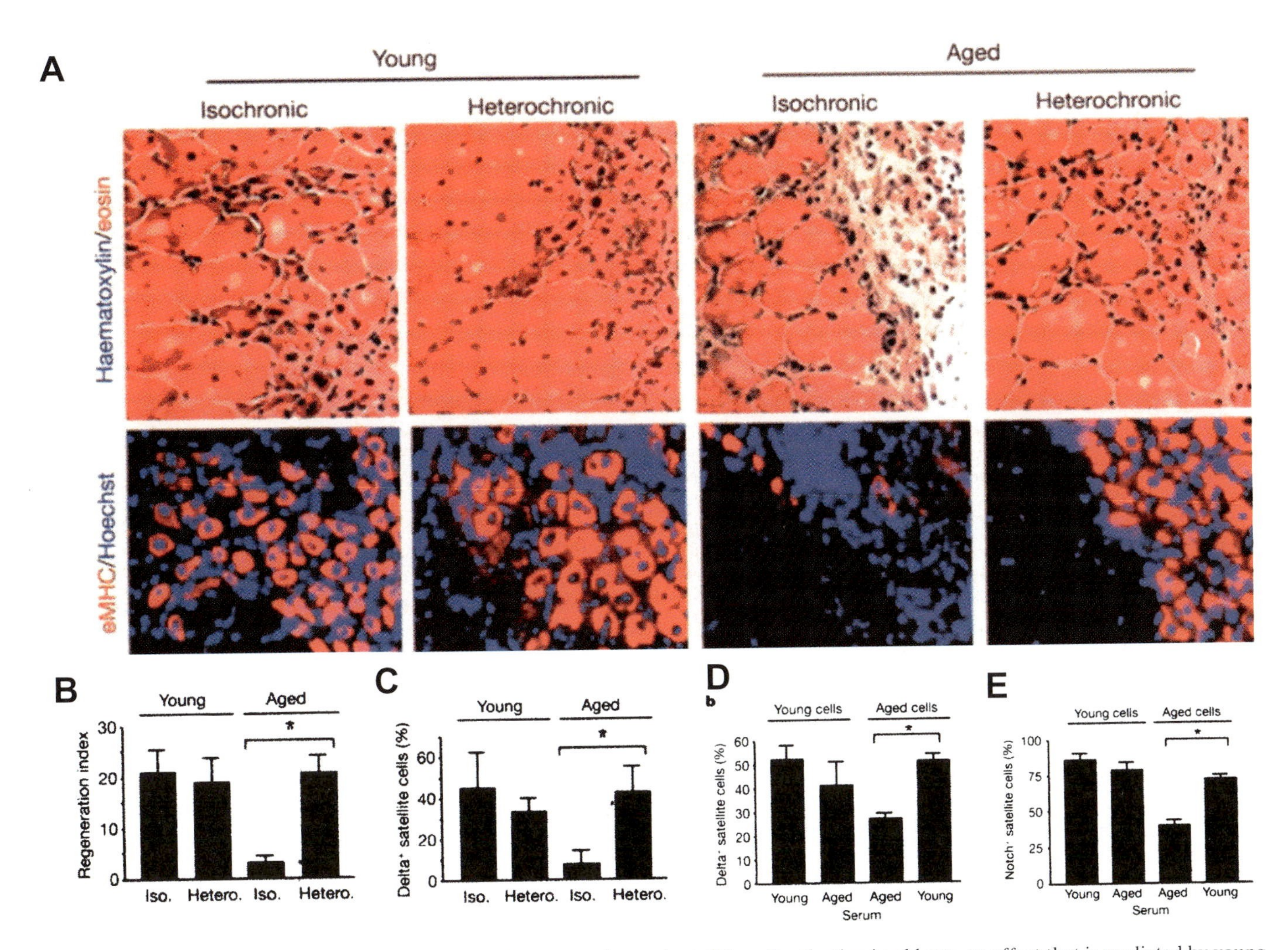


**FIGURE 13.15** Heterochronic parabiosis restores muscle regeneration and satellite cell activation in old rats, an effect that is mediated by young serum. **(A)** Young and aged mice were paired isochronically or heterochronically and the tibialis muscle injured locally by freezing. The muscles were analyzed histologically 5 days after injury. **Top**, hematoxylin and eosin stain. **Bottom**, immunostaining for embryonic myosin heavy chain (eMHC, red). Nuclei stained with Hoechst (blue). Regeneration of aged muscle is significantly improved by heterochronic parabiosis. **(B, C)** Bar graphs measuring regeneration index (number of embryonic MHC myofibers at injury site/total number of nuclei in the field) and percentage of Delta$^+$ (active) satellite cells. The data show that heterochronic parabiosis significantly improves the regeneration index and number of active SCs in aged muscle. **(D, E)** This effect is environmental, as shown by the significant increase in the number of Delta$^+$ and Notch$^+$ SCs when aged SCs were cultured for 24 hr in young serum. Reproduced with permission from Conboy et al., Rejuvenation of aged progenitor cells by exposure to a young systemic environment. Nature 433:760–764. Copyright 2005, Nature Publishing Group.

Aging has also been shown to decrease the potential for angiogenesis in the heart. Endothelial production of PDGF-BB mediated by cardiomyocyte-endothelial cell communication is disrupted in the aging mouse heart and contributes to impaired cardiac angiogenesis in old mice (Edelberg et al., 1998, 2002a). To determine whether EnSCs can restore PDGF-dependent angiogenesis in aged mice, a unique model was used (Edeberg et al., 2002b). In this model, neonatal mouse hearts are transplanted into the external ears of senescent syngeneic mice (**FIGURE 13.16**). These allografts are vascularized poorly due to the impairment in PDGF-BB expression by the endothelial cells of the host vessels; administration of exogenous PDGF-BB promotes vascularization of the graft. Senescent mice were given nonablative bone marrow transplants from young ROSA 26 (*lacZ*) mice prior to receiving grafts of neonatal hearts. The grafts were vascularized and endothelial precursors from the transplanted bone marrow were incorporated into the new blood vessels of the graft myocardium. In addition, the bone marrow-derived endothelial precursors restored the host angiogenic PDGF-B induction pathway and cardiac angiogenesis. Restoration of the pathway was blocked by anti-PDGF antibodies injected into the ear. Bone marrow transplanted from old ROSA 26 mice did not

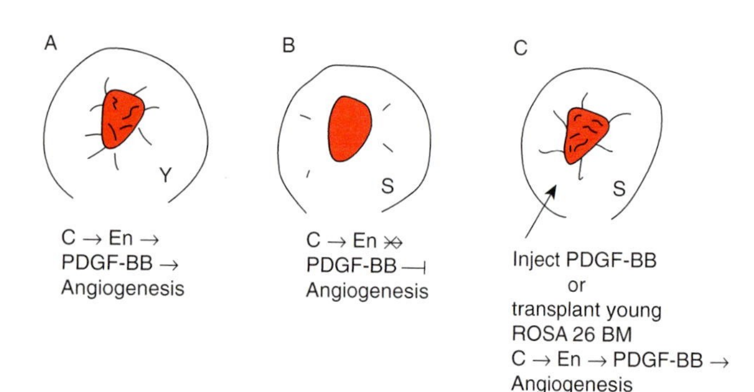


**FIGURE 13.16** Experimental model to show declining ability with age of endothelial cells to respond to signals from heart cells by producing PDGF-BB and promoting angiogenesis in the mouse heart. Neonatal hearts were transplanted under the skin of the ear. **(A)** When the heart was transplanted into the ear of a young (Y) mouse, endothelial cells responded to the cardiac signal (C) by producing PDGF-BB and participating in angiogenesis (lines). **(B)** After transplanting the heart into the ear of a senescent (S) mouse, endothelial cells did not respond (X) to the cardiac signal and cardiac angiogenesis was greatly diminished. **(C)** Heart transplanted into the ear of a senescent mouse, followed by injection of PDGF-BB or transplantion of young ROSA26 bone marrow (BM) cells. The injected PDGF substituted for the natural production of PDGF, whereas the bone marrow cells provided endothelial precursors that were able to respond to the cardiac signal, produce PDGF and restore angiogenesis.

restore the PDGF induction pathway. In a study of 50 patients who underwent bypass surgery, the numbers of EnSCs and levels of VEGF, IL-6, IL-8, IL-10 reached significantly higher levels 6 hr after operation. Preoperative numbers of EnSCs and VEGF concentration were lower with increasing age and were persistently lower in patients >69 yr old (Scheubel et al., 2003).

Thus, age is a limiting factor for mobilization of EnSCs, which would correlate with a higher incidence of atherosclerotic damage and the predictive value of EnSC number for atherosclerosis. "Old" EnScs might possibly be rejuvenated by gene therapy. For example, Murasawa et al. (2002) counteracted the reduction in telomere length that occurs with age by overexpression of the active subunit of the telomerase, TERT, resulting in an increase in the angiogenic capacity of EnSCs from healthy volunteers.

A decline in numbers of cardiac stem cells (CSCs) and in their ability to divide has been implicated in the decreased functional capacity of the heart with age (Torella et al., 2004). Gene products involved in growth arrest and senescence were found to increase with age in wild-type mice, while telomerase activity decreased. Cardiomyocytes die and are replaced by new ones derived from CSCs. With aging, the number of CSCs was attenuated by apoptosis until the loss of cardiomyocytes exceeded the gain of cardiomyocytes, leading ultimately to compromised function. Interestingly, these indicators of cellular senescence were not seen over the same time period in mice transgenic for IGF-1. IGF-1 activates the p13K-Akt pathway. While the amount of Akt protein did not differ between wild-type and transgenic animals, the latter showed increased phospho-Akt levels that were correlated with increased telomerase activity. Wild-type myocytes transfected with an adenoviral nuclear Akt construct exhibited a marked elevation in telomerase activity.

Microarray analysis of functional cells in the aging human frontal cortex indicates that after age 40, sets of genes that play central roles in synaptic plasticity, vesicular transport, mitochondrial function, stress response, antioxidation, and DNA repair are downregulated. This downregulation is associated with DNA damage in the promoters of these genes (Lu et al., 2004). In the mouse and rat brain, there is a decline in the production of new neurons by NSCs in the two regions that normally maintain new neuron production, the subventricular zone of the lateral ventricles and the hippocampus (Limke and Rao, 2003). The mechanism of the reduction is unknown, but could be due to decreased proliferation, or increased quiescence or apoptosis. However, there is no difference in the ability of NSCs from young versus old mice to form neurospheres *in vitro* (Tropepe et al., 1997), suggesting that environmental changes, not the intrinsic potential of the NSCs, are responsible for the decline in neurogenesis.

It will be instructive to perform experiments of the type devised to test the regenerative capacity of old liver and muscle on these other systems to determine whether their decline in regenerative capacity with age is due to the cellular environment or to intrinsic declines in the quality of their stem cells.

## SUMMARY

One of the visions of regenerative medicine is to use a pluripotent ASC that is easy to harvest and expand for cell therapies and bioartificial tissue constructs. *In vivo*, such cells would respond to signals at the site of tissue damage by changing their normal prospective fate to the phenotypes of the injured tissue, thus restoring the tissue. Alternatively, these cells could be manipulated *in vitro* to differentiate into the desired cell type for use. Either way, bioethical issues associated with ESCs as a therapeutic source of cells would be obviated. The term used here for this switch in cell fate is lineage conversion, but in most cases the actual mechanism of conversion (transdifferentiation, transdetermination) is unknown. Assays to detect lineage conversion include bone marrow reconstitution of x-irradiated or *scid* hosts, chimeric embryo formation, culturing the cells with other cell types *in vitro*, exposing the cells to conditioned medium, or exposing them to sets of defined differentiation factors.

In mammals, NSCs, SCs, adipocyte stem cells, liver oval cells, dental pulp cells, and bone marrow cells have been tested for pluripotency. SCs do not develop into other cell types *in vivo*. Not surprisingly, dental pulp cells, which are derived from the neural crest, can differentiate into neurons and liver oval cells, which are derived from the same endodermal evagination as the pancreas, can differentiate into β-cells. Adipocyte-derived stem cells are MSC-like cells that differentiate into the same phenotypes as marrow MSCs. NSCs, HSCs and marrow MSCs have been reported to be pluripotent.

In the bone marrow reconstitution assay, NSCs were reported to differentiate into blood cells. Chimeric embryo assays, however, gave conflicting results. In one set of experiments, NSCs appeared to differentiate into a wide range of cell types, but in another, they did not, in contrast to ESCs, which integrated efficiently into the blastocysts and participated in the formation of all tissues. When NSCs were co-cultured with endothelial cells, ~6% became endothelial. The conversion required cell contact, indicating that the crucial signals directing this conversion were on the surface of the endothelial cells.

Bone marrow stem cells have drawn the most attention as potential pluripotent ASCs. *In vivo*, unfractionated bone marrow cells, purified HSCs, and purified MSCs have been reported to differentiate into skeletal muscle, cardiac muscle, hepatocytes, neurons and glia, kidney tubule epithelium, epithelium of skin, alveoli and intestine, and connective tissue of liver, lung, and thymus. With the exception of alveolar epithelium and the Purkinje fibers of the heart, the frequency of conversion of bone marrow stem cells was usually less than 1%. HSCs and MSCs cultured in medium containing neural differentiation factors or in neural differentiation medium conditioned by astrocytes, differentiated into neurons at frequencies ranging between 0.5% and 80%. Marrow MSCs can differentiate into cardiomyocytes *in vitro*, but like the differentiation of NSCs to endothelial cells, this conversion requires contact between the MSCs and cardiomyocytes.

Many reports of low-frequency lineage conversion *in vivo* are the result of fusion of donor cells with host cells. Numerous marking experiments have demonstrated such fusion both *in vivo* and *in vitro*. Other cases of putative lineage conversion can be ascribed to contamination of the donor cell population with other differentiated cell types, incorporation of host leukocytes into donor tissues, autofluorescence artifacts, or incorporation of donor cells into host tissues, but without long-term survival or differentiation into authentic cell phenotypes of that tissue. Furthermore, donor cells that appear initially to have undergone lineage conversion often do not survive for extended periods of time. The bottom line is that the pluripotency of ASCs, if it exists, is too weak to be useful in regenerative medicine.

Pluripotent cells have been isolated from long-term cultures of bone marrow MSCs and MSCs from the connective tissue compartments of virtually all organ systems. These cells share some characteristics with ESCs and differentiate *in vivo* and *in vitro* into a wide variety of cell types at frequencies of from 5–90%. We do not know whether these cells exist as such in the bone marrow, are pericytes, or are the result of nuclear reprogramming by dedifferentiation resulting from extensive culturing.

Some regeneration-competent cells (hepatocytes, satellite cells, HSCs) seem to maintain their numbers and capacity for into old age, but cannot express this capacity because their environment deteriorates. The loss of hepatocyte proliferative capacity in old rats is reversed by parabiosis to young rats due to a reduction of a protein complex inhibitory to proliferation, while simultaneously the proliferative capacity of hepatocytes in the young rat liver is decreased due to an increase in the level of the inhibitory complex. The ability of skeletal muscle to regenerate in aged mice declines because insufficient upregulation of the Notch ligand, Delta after injury diminishes the number of SCs activated to 25% of the number activated in young muscle. Parabiosing old mice to young mice or activating Notch directly with an antibody to its extracellular domain restores the number of activated SCs and regenerative capacity.

These observations on restoration of liver regeneration in aged animals suggest that regenerative decline is due to the loss of certain systemic factors. Data consistent with this idea are the elevated expression of Delta, increased Notch activation, and enhanced proliferation of SCs from old mice cultured in the presence of serum from young mice, and a deficiency in production of mechano growth hormone (MGH) in both aging mice and humans.

There is a decrease with age in the number of osteoprogenitor cells produced after bone fracture that is associated with poorer fracture healing. Likewise, there is a reduction in circulating EnSCs that is associated with greater risk for, and development of, cardiovascular disease. Cardiomyocytes induce the production of PDGF-BB by the endothelial cells of heart vessels. Cardiac angiogenesis is impaired due to disruption of this interaction in old mice as well as in allografts of neonatal hearts grafted to the ears of old mice. Angiogenesis can be restored in these grafts by PDGF-

BB or by bone marrow transplants to the host from young mice. In the myocardium, cardiomyocytes die at a certain rate and are replaced by cardiac stem cells. The number of cardiac stem cells decreases with age until loss exceeds replacement, leading to compromised function. Mice transgenic for IGF-1 do not exhibit the decline in cardiac stem cells.

There is no difference in the ability of NSCs from young versus old mice to form neurospheres *in vitro*, but there is a decline in production of new neurons from NSCs in the subventricular zone and hippocampus. The decline may be due to downregulation of genes with age that play central roles in synaptic plasticity, vesicular transport, mitochondrial function, stress response, antioxidation, and DNA repair.

No experiments of the type performed on liver and skeletal muscle have been done on other tissues that directly test the regenerative quality of old ASCs by placing them in a young environment. Clearly, there is a need to more systematically test the intrinsic proliferation and differentiation potential of other aging tissues in this way and to determine how the composition of the stem cell microniche changes with age.

## REFERENCES


Aagaard P (2004) Making muscles "stronger": Exercise, nutrition, drugs. J Musculoskel Neuron Interact 4:165–174.

Abbot A (2004) Growing old gracefully. Nature 428:116–118.

Airey JA, Almeida-Porada G, Colletti EJ, Porada CD, Chamberlain J, Movsesian M, Sutko JL, Zanjani ED (2004). Human mesenchymal stem cells form Purkinje Fibers in Fetal sheep heart. Circulation 109:1401–1407.

Alessandri G, Pagano S, Bez A, Benetti A, Pozzi S, Iannolo G, Baronio M, Invernici G, Caruso A, Muneretto C, Bisleri G, Parati E (2004) Isolation and culture of human muscle-derived stem cells able to differentiate into myogenic and neurogenic cell lineages. Lancet 364:1872–1883.

Alison MR, Golding MH, Sarraf CE (1996) Pluripotential liver stem cells. Facultative stem cells located in the biliary tree. Cell Prolif 29:373–402.

Alison MR, Poulsom R, Jeffery R, Dhillon P, Quaglia A, Jacob J, Novelli M, Prentice G, Williamson J, Wright NA (2000) Hepatocytes from non-hepatic adult stem cells. Nature 406: 257.

Alison MR, Vig P, Russo F, Bigger BW, Amofah E, Themis M, Forbes S (2004) Hepatic stem cells: From inside and outside the liver? Cell Prolif 37:1–21.

Allsopp R, Weismann IL (2002) Replicative senescence of hematopoietic cells during serial transplantation: does telomere shortening play a role? Oncogene 21:3270–3273.

Alvarez-Dolado M, Pardal R, Garcia-Verdiugo JM, Fike R, Lee HO, Pfeffer K, Lois C, Morrison SJ, Alvarez-Buylla A (2003) Fusion of bone-marrow-derived cells with Purkinje neurons, cardiomyocytes and hepatocytes. Nature 425:968–973.

Ashjian PH, Elbarbary AS, Edmonds B, De Ugarte D, Zhu M, Zuk PA, Lorenz HP, Benhaim P, Hedrick MH (2003) In vitro differentiation of human processed lipoaspirate cells into early neural progenitors. Plastic Reconstr Surg 111:1922–1931.

Atkins BZ, Lewis CW, Kraus WE, Hutcheson K [illegible] Glower DD, Taylor DA (1999) Intracardiac transplantatio[illegible] myoblasts yields two populations of striated cells i[illegible] Surg 67:124–129.

Azizi SA, Stokes D, Augelli BJ, DiGirolamo C [illegible] Engraftment and migration of human bone [illegible] implanted into the brains of albino rats-sim[illegible] grafts. Proc Natl Acad Sci USA 95:3908–3[illegible]

Ballas CB, Davidson JM (2001) Delayed wo[illegible] is associated with increased collagen gel [illegible] tion by skin fibroblasts, not with differe[illegible] fibroblast cell populations. Wound Rep [illegible]

Balsam LB, Wagers AJ, Christensen JL, Kofidis [illegible] Robbis RC (2004) Haematopoietic stem cells adopt matu[illegible] matopoietic fates in ischaemic myocardium. Nature 428:668–673.

Bittira B, Wang J-S, Shum-Tim D, Chiu RC-J (2000) Marrow stromal cells as the autologous, adult stem cell source for cardiac myogenesis. Cardiac Vasc Reg 1:205–210.

Bjornson R, Rietze R, Reynolds BA, Magli MC, Vescovi AL (1999) Turning brain into blood: A hematopoietic fate adopted by adult neural stem cells *in vivo*. Science 283:534–537.

Blau H, Pavlath, Hardeman E, Chiu C-P, Silberstein L, Webster SG, Miller SC, Webster C (1985) Plasticity of the differentiated state. Science 230:758–766.

Boyle WJ, Simonet S, Lacey DL (2003) Osteoclast differentiation and activation. Nature 423:337–342.

Brazelton TR, Rossi FMV, Keshet G, Blau HM (2000) From marrow to brain: Expression of neuronal phenotypes in adult mice. Science 290:1775–1779.

Calcutt AF, Ossi P, Young HE, Southerland SS, Lucas PA (1993) Mesenchymal stem cells from wound tissue. Clin Res 41:336A.

Camargo FD, Chambers SM, Goodell MA (2004) Stem cell plasticity: From transdifferentiation to macrophage fusion. Cell Prolif 37:55–65.

Cao L, Li W, Kim S, Brodie SG, Deng C-X (2003) Senescence, aging, and malignant transformation mediated by p53 in mice lacking the Brca1 full-length isoform. Genes Dev 17:201–213.

Caplan AI (1991) Mesenchymal stem cells. J Orthop Res 9:641–650.

Carlson BM, Faulkner JA (1989) Muscle transplantation between young and old rats: Age of host determines recovery. Am J Physiol 256:C1262–C1266.

Carlson BM, Dedkov EI, Borisov AB, Faulkner JA (2001) Skeletal muscle regeneration in very old rats. J Gerontol A56:B224–B233.

Carrino DA, Sorrell JM, Caplan AI (2000) Age-related changes in the proteoglycans of human skin. Archiv Biochem Biophys 373:91–101.

Castro RF, Jackson KA, Goodell MA, Robertson CS, Liu H, Shine HD (2002) Failure of bone marrow cells to transdifferentiate into neural cells in vivo. Science 297:1299.

Chiu RC, Zibaitis A, Kao RL (1995) Cellular cardiomyoplasty: Myocardial regeneration with satellite cell implantation. Ann Thorac Surg 60:12–18.

Clarke D, Johansson C, Wilbertz J, Veress B, Nilsson E, Karlstrom H, Lendahl U, Frisen J (2000) Generalized potential of adult neural stem cells. Science 288:1660–1663.

Clarke MSF (2004) The effects of exercise on skeletal muscle in the aged. J Musculoskel Neuron Interact 4:175–178.

Conboy, IM, Conboy MJ, Smythe GM, Rando TA (2003) Notch-mediated restoration of regenerative potential to aged muscle. Science 302:1575–1577.

Conboy IM, Conboy MJ, Wagers AJ, Girma ER, Weissman IL, Rando TA (2005) Rejuvenation of aged progenitor cells by exposure to a young systemic environment. Nature 433:760–764.

Condorelli G, Borello U, De Angelis L, Latronico M, Sirabella D, Coletta M, Galli R, Balconi G, Follenzi A, Frati G, Cusella De Angelis MG, Gioglio L, Amuchastegui S, Adorini L, Naldini L, Vescovi A, Dejana E, Cossu G (2001) Cardiomyocytes induce endothelial cells to trans-differentiate into cardiac muscle: implications for myocardium regeneration. Proc Natl Acad Sci USA 98:10733–10738.

Connolly JF (1998) Clinical use of marrow osteoprogenitor cells to stimulate osteogenesis. Clin Orthopaed Rel Res 355S:S257–S266.

Dabeva MD, Shafritz DA (2003) Hepatic stem cells and liver repopulation. Sem Liver Disease 23:349–361.

D'Amour KA, Gage F (2003) Genetic and functional differences between multipotent neural and pluripotent embryonic stem cells. Proc Natl Acad Sci USA 100:11866–11872.

Dorfman J, Duong M, Zibaitis A, Pelletier MP, Shum-Tim D, Li C, Chiu RC (1998) Myocardial tissue engineering with autologous myoblast implantation. J Thorac Cardiovasc Surg 116:744–751.

Edelberg JM, Aird WC, Wu W, Rayburn H, Mamuya WS, Mercola M, Rosenberg RD (1998) PDGF mediates cardiac microvascular communication. J Clin Invest 102:837–843.

Edelberg JM, Lee SH, Kaur M, Tang L, Feirt NM, McCabe S, Bramwell O, Wong SC, Hong MK (2002a) Platelet-derived growth factor-ab limits the extent of myocardial infarction in a rat model: Feasibility of restoring impaired angiogenic capacity in the aging heart. Circulation 105:608–613.

Edelberg JM, Tang L, Hattori K, Lyden D, Rafii S (2002b) Young adult bone marrow-derived endothelial precursor cells restore aging-impaired cardiac angiogenic function. Circ Res 90:e89 (published online).

Eguchi G (1998) Transdifferentiation as a basis of eye lens regeneration. In: Ferretti P, Geraudie J (eds.). Cellular and Molecular Basis of Regeneration. New York: John Wiley & Sons, pp 207–229.

Eisenberg LM, Burns L, Eisenberg CA (2003) Hematopoietic cells from bone marrow have the potential to differentiate into cardiomyocytes *in vitro*. Anat Rec 274A:870–882.

Ergrise D, Martin D, Vienne A, Neve P, Schoutens A (1992) The number of fibroblastic colonies formed from bone marrow is decreased and the *in vitro* proliferation rate of trabecular bone cells increased in aged rats. Bone 13:355–361.

Erickson GR, Gimble JM, Franklin DM, Rice HE, Awad H, Guilak F (2002) Chondrogenic potential of adipose tissue-derived stromal cells *in vitro* and *in vivo*. Biochem Biophys Res Comm 290: 763–769.

Fang B, Shi M, Liao L, Yang S, Liu Y, Zhao RC (2003) Multiorgan engraftment and multilineage differentiation by human fetal bone marrow Flk1/CD31$^-$/CD34$^-$ progenitors. J Hematother Stem Cell Res 12:603–613.

Farace F, Prestoz L, Badaoui S, Guillier M, Haond C, Opolon P, Thomas J-L, Zalc B, Vainchenker W, Turhan AG (2004) Evaluation of hematopoietic potential generated by transplantation of muscle-derived stem cells in mice. Stem Cells Dev 13: 83–92.

Fausto N (2004) Liver regeneration and repair: hepatocytes progenitor cells and stem cells. Hepatology 39:1477–1487.

Ferrari G, Cusella-De Angelis G, Coletta M, Paolucci E, Stornaiuolo A, Cossu G, Mavilio F (1998) Muscle regeneration by bone marrow-derived myogenic progenitors. Science 279:1528–1530.

Filip S, Mokry J, Karbanova J, Vavrova J, English D (2004) Local environmental factors determine hematopoietic differentiation of neural stem cells. Stem Cells Dev 13:113–120.

Fisher GJ, Datta SC, Talwar HS, Wang Z-Q, Varani J, Kang SK, Voorhees JJ (1996) Molecular basis of sun-induced premature skin ageing and retinoid antagonism Nature 379:335–339.

Fleet JC, Cashman K, Cox K, Rosen V (1997) The effects of aging on the bone inductive activity of recombinant human bone morphogenetic protein-2. Endocrinology 137:4605–4610.

Globerson A (2001) Hematopoietic stem cell ageing. Norvartis Foundation Symp 235:85–96.

Goldspink G (2004) Age-related loss of skeletal muscle function; impairment of gene expression. J Musculoskel Neuron Interact 4:143–147.

Gurdon J, Laskey R, Reeves O (1975) The developmental capacity of nuclei transplanted from keratinized skin cells of adult frogs. J Embryol Exp Morph 34:93–112.

Gussoni E, Soneoka Y, Strickland CD, Buzney EA, Khan MK, Flint AF, Kunkel LM, Mulligan RC (1999) Dystrophin expression in the mdx mouse restored by stem cell transplantation. Nature 401: 390–394.

Ham AW, Cormack DH (1979) Histology, 8th ed Philadelphia: JB Lippincott, pp 614–644.

Harris RG, Herzog EL, Bruscia E, Grove JE, Van Arnam JS, Krause DS (2004) Lack of a fusion requirement for development of bone marrow-derived epithelia. Science 305:90–93.

Hicok K, Du Laney TV, Zhou YS, Halvorsen Y-D C, Hitt DC, Cooper LF, Gimble JM (2004) Human adipose-derived adult stem cells produce osteoid *in vivo*. Tiss Eng 10:371–380.

Hill JM, Zalos G, Halcox JPJ, Schenke WH, Waclawiw MA, Quyyumi AA, Finkel T (2003) Circulating endothelial progenitor cells, vascular function, and cardiovascular risk. New Eng J Med 348:593–600.

Holden C, Vogel G (2002) Plasticity: Time for a reappraisal? Science 296:2126–2129.

Howell J, Yoder M, Srour E (2002) Hematopoietic potential of murine skeletal muscle-derived CD45(−)Sca-1(+)c-kit(−) cells. Exp Hematol 30:915–924.

Huibregtse BA, Johnstone B, Goldberg VM, Caplan AI (2000) Effect of age and sampling site on the chondro-osteogenic potential of rabbit marrow-derived mesenchymal progenitor cells. J Orthopaedic Res 18:18–24.

Iakova P, Awad SS, Timchenko NA (2003) Aging reduces proliferative capacities of liver by switching pathways of C/EBP$\alpha$ growth arrest. Cell 113:495–506.

Imasawa T, Utsunomiya Y, Kawamura T, Zhong Y, Nagasawa R, Okabe M, Maruyama N, Hosoya T, Ohno T (2001) The potential of bone marrow-derived cells to differentiate to glomerular mesangial cells. J Am Soc Nephrol 12:1401–1409.

Jackson KA, Mi T, Goodell MA (1999) Hematopoietic potential of stem cells isolated from murine skeletal muscle. Proc Natl Acad Sci USA 96:14482–14486.

Jackson KA, Goodell MA (2003) Generation and stem cell repair of cardiac tissue. In: Sell S (ed.). Stem Cells Handbook. Towota: Humana Press, pp 259–266.

Jackson KA, Majika SM, Wang H, Pocius J, Hartley CJ, Majesky MW, Entman ML, Michael LH, Hirschi K, Goodell MA (2001) Regeneration of ischemic cardiac muscle and vascular endothelium by adult stem cells. J Clin Invest 107:1395–1402.

Jacobson RG, Flowers FP (1996) Skin changes with aging and disease. Wound Rep Reg 4:311–315.

Jiang H, Hidaka K, Morisaki H, Morisaki T (2000) MS-5 bone marrow stromal cell line differentiates into cardiac muscle cells after long-term culture. Cardiac Vasc Reg 1:274–282.

Jiang Y, Jahagirdar BN, Reinhardt R, Schwarts RE, Keene CD, Ortiz-Gonzalez XR, Reyes M, Lenvik T, Lund T, Blackstad M, Du J, Aldrich S, Lisberg A, Low WC, Largaespada DA, Verfaille CM (2002) Pluripotency of mesenchymal stem cells derived from adult marrow. Nature 418:41–49.

Joannides A, Gaughwin P, Scott M, Watt S, Compston A, Chandran S (2003) Postnatal astrocytes promote neural induction from adult human bone marrow-derived cells. J Hematother Stem Cell Res 12:681–688.

Joannides A, Gaughwin P, Schwiening C, Majed H, Sterling J, Compston A, Chandran S (2004) Efficient generation of neural precursors from adult human skin: astrocytes promote neurogenesis from skin-derived stem cells. Lancet 364:172–178.

Justesen J, Pedersen SB, Stenderup K, Kassem M (2004) Subcutaneous adipocytes can differentiate into bone-forming cells *in vitro* and *in vivo*. Tiss Eng 10:381–391.

Kanazawa Y, Verma IM (2003) Little evidence of bone marrow-derived hepatocytes in the replacement of injured liver. Proc Natl Acad Sci USA 100:11850–11853.

Kao RL, Chin TK, Ganote CE, Hossler FE, Li C, Browder W (2000) Satellite cell transplantation to repair injured myocardium. Cardiac Vasc Reg 1:31–42.

Kocher AA, Schuster MD, Szabolcs MJ, Takuma S, Burkhoff D, Wang J, Homma S, Edwards NM, Itescu S (2001) Neovascularization of ischemic myocardium by human bone marrow-derived angioblasts prevents cardiomyocyte apoptosis, reduces remodeling and improves cardiac function. Nature Med 7:430–436.

Kondo T, Johnson SA, Yoder MC, Romand R, Hashino E (2005) Sonic hedgehog and retinoic acid synergistically promote sensory fate specification from bone marrow-derived pluripotent stem cells. Proc Natl Acad Sci USA 102:4789–4794.

Kong KY, Ren J, Kraus M, Finklestein SP, Brown RH (2004) Human umbilical cord blood cells differentiate into muscle in *sjl* muscular dystrophy mice. Stem Cells 22:981–993.

Kopen GC, Prockop DJ, Phinney DG (1999) Marrow stromal cells migrate throughout forebrain and cerebellum where they differentiate into astrocytes after injection into neonatal mouse brains. Proc Natl Acad Sci USA 96:10711–10716.

Kotton DN, Ma BY, Cardoso WV, Sanderson EA, Summer RS, Williams MC, Fine A (2001) Bone-marrow derived cells as progenitors of lung epithelium. Development 128:5181–5188.

Krause DS, Theise ND, Collector MI, Hwang S, Gardner R, Neutzel S, Sharkis J (2001) Multi-organ, multilineage engraftment by a single bone marrow-derived stem cell. Cell 105:369–377.

Kumar A, Velloso C, Imokawa Y, Brockes JP (2000) Plasticity of retrovirus-labelled myotubes in the newt regeneration blastema. Dev Biol 218:125–136.

Labarge MA, Blau HM (2002) Biological progression from adult bone marrow to mononucleate stem cell to multinucleate muscle fiber in response to injury. Cell 111:589–601.

Laflamme MA, Murry CE (2005) Regenerating the heart. Nature Biotech 23:845–856.

Lagasse E, Connors, Al-Dhalimy M, Teitsma M, Dohse M, Osborne L, Wang X, Finegold M, Weissman I, Grompe M (2000) Purified hematopoietic stem cells can differentiate to hepatocytes in vivo. Nature Med 6:1229–1234.

Lanza R, Moore MA, Wakayama T, Perry AC, Shieh JH, Hendrikx J, Leri A, Chimenti S, Monsen A, Nurzynska D, West MD, Kajstura J, Anversa P (2004) Regeneration of the infarcted heart with stem cells derived by nuclear transplantation. Circ Res 94:820–827.

Lapidos KA, Chen YE, Earley JU, Heydemann A, Huber JM, Chien M, Ma A, McNally E (2004) Transplanted hematopoietic stem cells demonstrate impaired sarcoglycan expression after engraftment into cardiac and skeletal muscle. J Clin Invest 114: 1577–1585.

Lechner A, Yang Y-G, Blacken RA, Wang L, Nolan AL, Habener JF (2004) No evidence for significant transdifferentiation of bone marrow into pancreatic β-cells in vivo. Diabetes 53:616–623.

Limke TL, Rao MS (2003) Neural stem cell therapy in the aging brain: Pitfalls and possibilities. J Hematother Stem Cell Res 12:615–623.

Lin F, Igarashi P (2003) Searching for stem/progenitor cells in the adult mouse kidney. J Am Soc Nephrol 14:3290–3292.

Lin F, Cordes K, Li L, Hood L, Couser WG, Shankland SJ, Igarishi P (2003) Hematopoietic stem cells contribute to the regeneration of renal tubules after renal ischemia-reperfusion injury in mice. J Am Soc Nephrol 14:1188–1199.

Lindholm TS, Urist MR (1980) A quantitative analysis of new bone formation by induction in composite grafts of bone marrow and bone matrix. Clin Orthop 150:288–300.

Locatelli F, Corti S, Donadoni C, Guglieri M, Capra F, Strazzer S, Salani S, Del Bo R, Fortunato F, Bordoni A, Comi GP (2003) Neuronal differentiation of murine bone marrow Thy-1- and Sca-1-positive cells. J Hematother Stem Cell Res 12:727–734.

Lu T, Pan Y, Kao S-Y, Li C, Kohane I, Chan J, Yankner BA (2004) Gene regulation and DNA damage in the ageing human brain. Nature 429:883–891.

Lucas PA, Calcutt AF, Ossi P, Young HE, Southerland SS (1993) Mesenchymal stem cells from granulation tissue. J Cell Biochem 17E:122.

Makino S, Fukuda K, Miyoshi S, Konishi F, Kodoma H, Pan J, Sano M, Takahashi T, Hori S, Abe H, Hata J, Umezawa A, Ogawa S (1999) Cardiomyocytes can be generated from marrow stromal cells *in vitro*. J Clin Invest 103:697–705.

Martin K, Potten CS, Roberts SA, Kirkwood TBL (1998) Altered stem cell regeneration in irradiated intestinal crypts of senescent mice. J Cell Sci 111:2297–2303.

McKinney-Freeman SL, Jackson KA, Camargo FD, Ferrari, Mavilio F, Goodell MA (2002) Muscle-derived hematopoietic stem cells are hematopoietic in origin. Proc Natl Acad Sci USA 99:1341–1346.

Medvinsky A, Smith A (2003) Fusion brings down barriers. Nature 422:823–825.

Menthena A, Beb N, Oertel M, Grozdanov PN, Sandhu J, Shah S, Guha C, Shafritz DA, Dabeva MD (2004) Bone marrow progenitors are not the source of expanding oval cells in injured liver. Stem Cells 22:1049–1061.

Mezey E, Chandross KJ, Harta G, Maki RA, McKercher SR (2000) Turning blood into brain: Cells bearing neuronal antigens generated in vivo from bone marrow. Science 290:1779–1782.

Mezey E (2004) On bone marrow stem cells and openmindedness. Stem Cells Dev 13:147–152.

Mizuno H, Zuk PA, Zhu M, Lorenz HP, Benhaim P, Hedrick MH (2002) Myogenic differentiation by human processed lipoaspirate cells. Plastic Reconstr Surg 109:199–209.

Morigi M, Imberti B, Zoja C, Corna D, Tomasoni S, Abbate M, Rottoli D, Angioletti S, Benigni A, Perico Norberto, Alison M, Remuzzi G (2004) Mesenchymal stem cells are renotropic, helping to repair the kidney and improve function in acute renal repair. J Am Soc Nephrol 15:1794–1804.

Morizono K, De Ugarte DA, Zhu M, Zuk P, Barbary A, Ashjian P, Benhaim P, Chen ISY, Hedrick MH (2003) Multilineage cells from adipose tissue as gene delivery vehicles. Human Gene Ther 14:59–66.

Morshead CM, Benveniste P, Iscove NN, van der Kooy D (2002) Hematopoietic competence is a rare property of neural stem cells that may depend on genetic and epigenetic alterations. Nature Med 8:268–273.

Murry CE, Soonpa MH, Reinecke H, Nakajima H, Rubart M, Pasumarthi KBS, Virag JI, Barelmez SH, Poppa V, Bradford G, Dowell JD, Williams DA, Field LJ (2004) Haematopoietic stem cells do not transdifferentiate into cardiac myocytes in myocardial infarcts. Nature 428:664–668.

Nemoto S, Finkel T (2004) Ageing and the mystery at Arles. Nature 429:149–152.

Nosrat IV, Smith CA, Mullally P, Olson L, Nosrat CA (2004) Dental pulp cells provide neurotrophic support for dopaminergic neurons and differentiate into neurons *in vitro*; implications for tissue engineering and repair in the nervous system. Eur J Neurosci 19:2388–2401.

Nurse CA, Macintyre L, Diamond J (1984) Reinnervation of the rat touch dome restores the Merkel cell population reduced after denervation. Neurosci 13:563–571.

Oreffo ROC, Driessens FCM, Planell JA, Triffit JT (1998) Effects of novel calcium phosphate cements on human bone marrow fibroblastic cells. Tiss Eng 4:293–302.

Orlic D, Kajstura J, Chimenti S, Jakoniuk I, Anderson SM, Li, Pickel J, McKay R, Nadal-Ginard B, Bodine M, Leri A, Anversa P (2001) Bone marrow cells regenerate infarcted myocardium. Nature 410:701–705.

Owen M, Friedenstein AJ (1988) Stromal stem cells: marrow derived osteogenic precursors. Ciba Foundation Symposium 136:420–460.

Pereiera RF, Halford KW, O'Hara MD, Leeper DB, Sokolov B, Pollard MD, Bagasra O, Prockop DJ (1995) Cultured adherent cells from marrow can serve as long-lasting precursor cells for bone, cartilage, and lung in irradiated mice. Proc Natl Acad Sci USA 92:4857–4861.

Petersen BE, Goff J, Greenberger JS, Michalopoulos GK (1998) Hepatic oval cells express the hematopoietic stem cell marker Thy-1 in the rat. Hepatology 27:433–445.

Petersen BE, Bowen WC, Patrene KD, Mars M, Sullvan AK, Murase N, Boggs SS, Greenberger JS, Goff JP (1999) Bone marrow as a potential source of hepatic oval cells. Science 284:1168–1170.

Poulsom R, Alison MR, Cook T, Jeffery R, Ryan E, Forbes SJ, Hunt T, Wyles S, Wright NA (2003) Bone marrow stem cells contribute to healing of the kidney. J Am Soc Nephrol 14:S48–S54.

Prockop DJ (1997) Marrow stromal cells as stem cells for non-hematopoietic tissues. Science 276:71–74.

Quarto R, Thomas, Liang CT (1995) Bone progenitor cell deficits and age-related decline in bone repair capacity. Calcif Tiss Int 56:123–129.

Quiani F, Urbanek K, Beltrami AP, Finato N, Beltrami CA, Nadal-Ginard B, Kajstura J, Leri A, Anversa P (2002) Chimerism of the transplanted heart. New Eng J Med 346:5–15.

Rauscher FM, Goldschmidt-Clermont PJ, Davis BH, Wang T, Gregg D, Ramaswami P, Pippen AM, Annex BH, Dong C, Taylor DA (2003) Aging, progenitor cell exhaustion, and atherosclerosis. Circulation 108:457–463.

Reinecke H, Poppa V, Murry CE (2002) Skeletal muscle stem cells do not transdifferentiate into cardiomyocytes after cardiac grafting. J Mol Cell Cardiol 34:241–249.

Reyes M, Dudek A, Jahagirdar B, Koodie L, Marker PH, Verfaille CM (2002) Origin of endothelial progenitors in human postnatal bone marrow. J Clin Invest 109:337–346.

Rice CM, Scolding NJ (2004) Adult stem cells-reprogramming neurological repair? Lancet 364:193–199.

Sanchez-Ramos J, Song S, Cardozo-Pelaez F, Hazzi C, Stedeford T, Willing, Freeman TB, Saporta S, Janssen W, Patel N, Cooper DR, Sanberg PR (2000) Adult bone marrow cells differentiate into neural cells *in vitro*. Exp Neurol 164:247.

Schwartz RE, Reyes M, Koodie L, Jiang Y, Blackstad M, Lund T, Lenvik T, Johnson S, Hu W-S, Verfaille M (2002) Multipotent adult progenitor cells from bone marrow differentiate into functional hepatocyte-like cells. J Clin Invest 109:1291–1302.

Sell S (2001) Heterogeneity and plasticity of hepatocyte lineage cells. Hepatology (2001) 33:738–750.

Shafritz DA, Dabeva MD (2002) Liver stem cells and model systems for liver repopulation. J Hepatol 36:552–564.

Shatos MA, Mizumoto K, Mizumoto H, Kurimoto Y, Klassen H, Young MJ (2001) Multipotent stem cells from the brain and retina of green mice. (E-biomed) J Reg Med 2:13–15.

Steen TP (1968) Stability of chondrocyte differentiation and contribution of muscle to cartilage during limb regeneration in the axolotl (*Siredon mexicanum*). J Exp Zool 167:49–78.

Stewart CEH (2004) The physiology of stem cells: Potential for the elderly patient. J Musculoskel Neuron Interact 4:179–183.

Swift ME, Burns AL, Gray KL, Di Pietro LA (2001) Age-related alterations in the inflammatory response to dermal injury. J Invest Dermatol 117:1027–1035.

Taylor DA, Atkins BZ, Hungspreugs P, Jones TR, Reedy MC, Hutcheson KA, Glower DD, Kraus WE (1998) Regenerating functional myocardium: improved performance after skeletal myoblast transplantation. Nature Med 4:929–933.

Terada N, Hamazaki T, Oka M, Hoki M, Mastalerz DM, Nakano Y, Meyer EM, Morel L, Petersen BE, Scott EW (2002) Bone marrow cells adopt the phenotype of other cells by spontaneous cell fusion. Nature 416:542–545.

Theise ND, Nimmakayalu M, Gardner, Illei PB, Morgan G, Teperman L, Henegariu O, Krause D (2000) Liver from bone marrow in humans. Hepatology 32:11–16.

Thiese ND, Badve S, Saxena R, Henegariu O, Sell S, Crawford JM, Krause DS (2000) Derivation of hepatocyes from bone marrow cells in mice after radiation-induced myeloablation. Hepatology 31:235–240.

Toma G, Akhavan M, Fernandes KL, Barnabe-Heider F, Sadikot A, Kaplan DR, Miller FD (2001) Isolation of multipotent adult stem cells from the dermis of mammalian skin. Nature Cell Biol 3:778–784.

Tomita S, Li RK,Meisel RD, Mickle DA, Kim EJ, Tomita S, Jia ZQ, Yau TM (1999) Autologous transplantation of bone marrow cells improves damaged heart function. Circulation 100:II247–II256.

Torella D, Rota M, Nurzynska D, Musso E, Monsen A, Shiraishi I, Zias E, Walsh K, Rosenzweig A, Sussman MA, Urbanek K, Nadal-Ginard B, Kajstura J, Anversa P, Leri A (2004) Cardiac stem cell and myocyte aging, heart failure, and insulin-like growth factor-1 overexpression. Circ Res 94:514–524.

Tran SD, Pillemer SR, Dutra A, Barrett AJ, Brownstein MJ, Key S, Pak E, Leakan RA, Kingman A, Yamada KM, Baum BJ, Mezey E (2003) Differentiation of human bone marrow-derived cells into buccal epithelial cells in vivo: a molecular analytical study. Lancet 361:1084–1088.

Urist M (1965) Bone: Formation by autoinduction. Science 150:893–899.

Urist MR, Delange RJ, Finerman GAM (1983) Bone cell differentiation and growth factors. Science 220:680–686.

Vacanti MP, Roy A, Cortiella J, Bonassar L, Vacanti CA (2001) Identification and initial characterization of spore-like cells in adult mammals. J Cellular Biochem 80:455–460.

Vassilopoulos G, Wang P-R, Russell DW (2003) Transplanted bone marrow regenerates liver by cell fusion. Nature 422:901–904.

Wagers AJ, Sherwood RI, Christensen JL, Weissman IL (2002) Little evidence for developmental plasticity of adult hematopoietic stem cells. Science 297:2256–2259.

Wakitani S, Saito T, Caplan A (1995) Myogenic cells derived from rat bone marrow mesenchymal stem cells exposed to 5-azacytidine. Muscle Nerve 18:1417–1426.

Wang S, Shum-Tim D, Galipeau J, Chedrawy E, Eliopoulos N, Chiu RJ-C (2000) Marrow stromal cells for cellular cardiomyoplasty feasibility and potential clinica advantages. J Thoracic Surg 120: 999–1006.

Wang X, Willenbring H, Akkari Y, Yorimaru Y, Foster M, Al-Dhalimy M, Lagasse E, Finegold M, Olson S, Grompe M (2003) Cell fusion is the principal source of bone-marrow-derived hepatocytes. Nature 422:897–901.

Wang X, Foster M, Al-Dhalimy M, Lagasse E, Finegold M, Grompe M (2003) The origin and liver repopulating capacity of murine oval cells. Proc Natl Acad Sci 100:11881–11888.

Wickham MQ, Erickson GR, Gimble JM, Vail TP, Guilak F (2003) Multipotent stromal cells derived from the infrapatellar fat pad of the knee. Clin Orthopaed Rel Res 412:196–212.

Wilmut I, Beaujean N, de Sousa A, Dinnyes A, King T, Paterson LA, Wells DN, Young LE (2002) Somatic cell nuclear transfer. Nature 419:583–586.

Woodbury D, Schwarz EJ, Prockop DJ, Black IB (2000) Adult rat and human bone marrow stromal cells differentiate into neurons. J Neurosci Res 61:364–370.

Wurmser AE, Nakashima K, Summers RG, Toni N, D'Amour KA, Lie DC, Gage FH (2004) Cell fusion-independent differentiation of neural stem cells to the endothelial lineage. Nature 430:350–356.

Yang L, Li S, Hatch H, Ahrens K, Cornelius J, Petersen BE, Peck AB (2002) *In vitro* trans-differentiation of adult hepatic stem cells into pancreatic endocrine hormone-producing cells. Proc Natl Acad Sci USA 99:8078–8083.

Yao F, Visovatti S, Johnson CS, Chen M, Slama J, Wenger A, Eriksson E (2001) Age and growth factors in porcine full-thickness wound healing. Wound Rep Reg 9:371–377.

Ying Q-L, Nichols J, Evans EP, Smith AG (2002) Changing potency by spontaneous fusion. Nature 416:545–548.

Yoo JU, Johnstone B (1998) The role of osteochondral progenitor cells in fracture repair. Clin Orthopaed Related Res 355S:S73–S81.

Young HE, Duplaa C, Romero-Ramos M, Chesselet M-F, Vourch P, Yost MJ, Ericson K, Terracio L, Asahara T, Masuda H, Tamura-Inomiya S, Detmer K, Bray RA, Steele TA, Hixson D, el-Kalay M, Tobin BW, Russ RD, Horst MN, Floyd JA, Henson NL, Hawkins KC, Groom J, Parikh A, Blake L, Bland LJ, Thompson AJ, Kirinich A, Moreau C, Hudson J, Bowyer FP III, Lin TJ, Black AC (2004) Adult reserve stem cells and their potential for tissue engineering. Cell Biochem Biophys 40:1–80.

Young HE, Duplaa C, Yost MJ, Henson NL, Floyd JA, Detmer K, Thompson AJ, Powell SW, Gamblin TC, Kizziah K, Holland BJ, Boev A, Van De Water JM, Godbee DC, Jackson S, Rimando M, Edwards CR, Wu E, Cawley C, Edwards PD, MacGregor A, Bozof R, Thompson TM, Petro GJ Jr, Shelton HM, McCampbell BL, Mills JC, Flynt FL, Steele TA, Fearney M, Kirincich-Greathead A, Hardy W, Young PR, Amin AV, Wiliams RS, Horton MM, McGuinn S, Hawkins KC, Ericson K, Terracio L, Moreau C, Hixson D, Tobin BW, Hudson J, Bowyer FP III, Black AC (2004) Clonogenic analysis reveals reserve stem cells in postnatal mammals. II. Pluripotent epiblastic-like stem cells. Anat Rec Part A 277A:178–203.

Young HE, Black AC Jr (2004) Adult stem cells. Anat Rec Part A 276A:75–102.

Ying Q-L, Nichols J, Evans EP, Smith AG (2002) Changing potency by spontaneous fusion. Nature 416:545–548.

Zhang, Wang D, Estrov Z, Raj S, Willerson JT, Yeh ETH (2004) Both cell fusion and transdifferentiation account for the transformation of human peripheral blood CD34-positive cells into cardiomyocytes *in vivo*. Circulation 110:3803–3807.

Zuk PA, Zhu M, Ashjian P, De Ugarte DA, Huang JI, Mizuno H, Alfonso ZC, Fraser JK, Benhaim P, Hedrick MH (2002) Human adipose tissue is a source of multipotent stem cells. Mol Biol Cell 13:4279–4295.

# 14 Regeneration of Appendages

## INTRODUCTION

Several vertebrate species are able to epimorphically regenerate tissue of appendages or whole appendages such as fingertips, limbs, fins, tails, antlers, and ear tissue via the formation of a blastema of proliferating cells. For some of these structures, such as mammalian ear tissue and fingertips and antlers, the origin of the cells for regeneration is uncertain, but in others, such as fish fin regeneration and amphibian limb and tail regeneration, the evidence is that a blastema is formed by the dedifferentiation of mature cells local to the region of injury. Tail regeneration in amphibians and lizards was discussed in Chapter 5. Here, the focus is on amphibian limb regeneration and the regeneration of mammalian ear tissue, antlers, and digit tips. While not technically appendages, regeneration of the jaws in amphibians is discussed as well, since this regeneration is also achieved by dedifferentiation to form a regeneration blastema. This chapter is an updated version of two previous reviews by Nye et al. (2003) and Stocum (2004).

## AMPHIBIAN LIMB REGENERATION

### 1. Events of Limb Regeneration

Many species of larval and adult salamanders (urodeles), as well as early frog and toad (anuran) tadpoles, can regenerate their limbs. Spallanzani (1768) was the first to publish descriptions of amphibian limb regeneration, in adult newts. Newts, as well as the axolotl, *Ambystoma mexicanum* (**FIGURE 14.1**), have been favorite species for the study of limb regeneration. **FIGURE 14.2** illustrates the developmental stages of regenerating adult newt and larval salamander limbs (Goss, 1969; Iten and Bryant, 1973; Smith et al., 1974; Tank et al., 1976; Stocum, 1979; Young et al., 1983; Dent, 1964). The amputation surface is covered within a few hours by migrating epidermis. Dedifferentiated cells then accumulate under the wound epidermis to form a regeneration blastema, while at the same time the wound epidermis thickens apically to form an apical epidermal cap (AEC) several layers thick. The outer layers of the AEC are protective, whereas its basal layers appear to be anatomically and functionally equivalent to the apical ectodermal ridge (AER) of amniote embryonic limb buds (Christensen and Tassava, 2000). Capillaries and nerves begin regenerating into the forming blastema within a few days after amputation. Both the AEC and the regenerating nerves provide growth and trophic factors essential for the survival and proliferation of the blastema cells (Van Stone, 1955; Piatt, 1957; Thornton, 1968).

The nascent blastema grows rapidly to form a conical bud of undifferentiated cells that is morphologically similar to the embryonic limb bud. With further growth, the blastema undergoes patterned differentiation that replicates the amputated limb parts. Differentiation and morphogenesis take place in a proximal to distal and anterior to posterior sequence, except that in the proximodistal (PD) axis, the digits appear to begin differentiation prior to the carpals or tarsals. The limb buds and regeneration blastemas of urodeles exhibit a unique anterior to posterior sequence of differentation in the anteroposterior (AP) axis, whereas the blastemas and embryonic limb buds of anuran amphibians and the limb buds of amniotes follow a posterior to anterior sequence (Shubin and Alberch, 1986). Differentiation in the dorsoventral (DV) axis appears to be simultaneous across the axis. The remainder of the regenerative process consists of growth to match the size of the unamputated limb. Regardless of whether amputation is through the digits or the base of the humerus, the time required to regenerate the missing parts is the same, as observed originally by Spallanzani and confirmed many times since. Limb regeneration is two or more times faster in larvae than in adults (Voit et al., 1985).

In adult newts, the regenerant that results from one amputation is an exact replica of the original in over

90% of cases. Like the liver, the same limb can regenerate repeatedly. Serial amputation, however, results in decreasing morphogenetic fidelity. After four re-amputations through regenerants derived from the upper arm of adult newts, 81% of the regenerated limbs exhibited structural abnormalities such as interdigital webbing, reduction in number of skeletal elements, and even total inhibition of regeneration (Dearlove and Dresden, 1976).

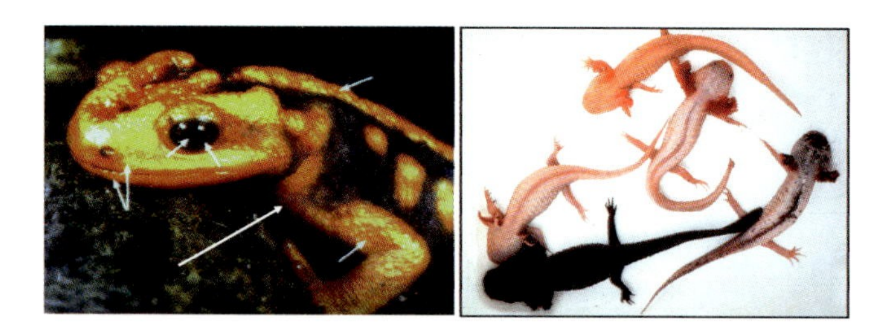

**FIGURE 14.1** The two most widely used amphibian species for studies of limb regeneration. **Left**, the adult newt. Arrows point to structures that can be regenerated (jaws, cornea, lens and neural retina, heart, spinal cord and limbs). Courtesy of Jeremy Brockes. **Right**, the axolotl. The dark animal is wild-type, the others are pigment mutants. Courtesy of the Indiana University Axolotl Colony.

## 2. Origin of the Blastema

Experiments in which *Ambystoma* salamander larvae were X-irradiated to prevent cell division in all but a shielded segment of the limb, showed that the cells of the regeneration blastema are derived from limb tissues local to the amputation plane (Butler and O'Brien, 1942; **FIGURE 14.3**). Light and electron microscopic studies suggested that blastema cells arise by the dedifferentiation of chondrocytes, muscle cells, dermal and muscle fibroblasts, and Schwann cells (Thornton, 1938, 1968; Hay, 1959; Hay and Fischman, 1961). Additional evidence for the origin of blastema cells by dedifferentiation was obtained by grafting marked limb tissues to host limbs and tracing the marked cells into the blastema (**FIGURE 14.4**). Triploid cartilage and dermis were grafted to diploid limbs of axolotls (Steen, 1968; Namenwirth, 1974; Dunis and Namenwirth, 1977; Muneoka et al., 1986). Mutant 1-nu myofibers were implanted into 2-nu *Xenopus* hindlimbs (Steen, 1973). Myotubes derived from a newt muscle satellite cell line and labeled *in vitro* with fluorescent rhodamine-conjugated lysinated dextran (Lo et al., 1993), the tracker dye PKH26, or a retrovirus expressing alkaline phospha-

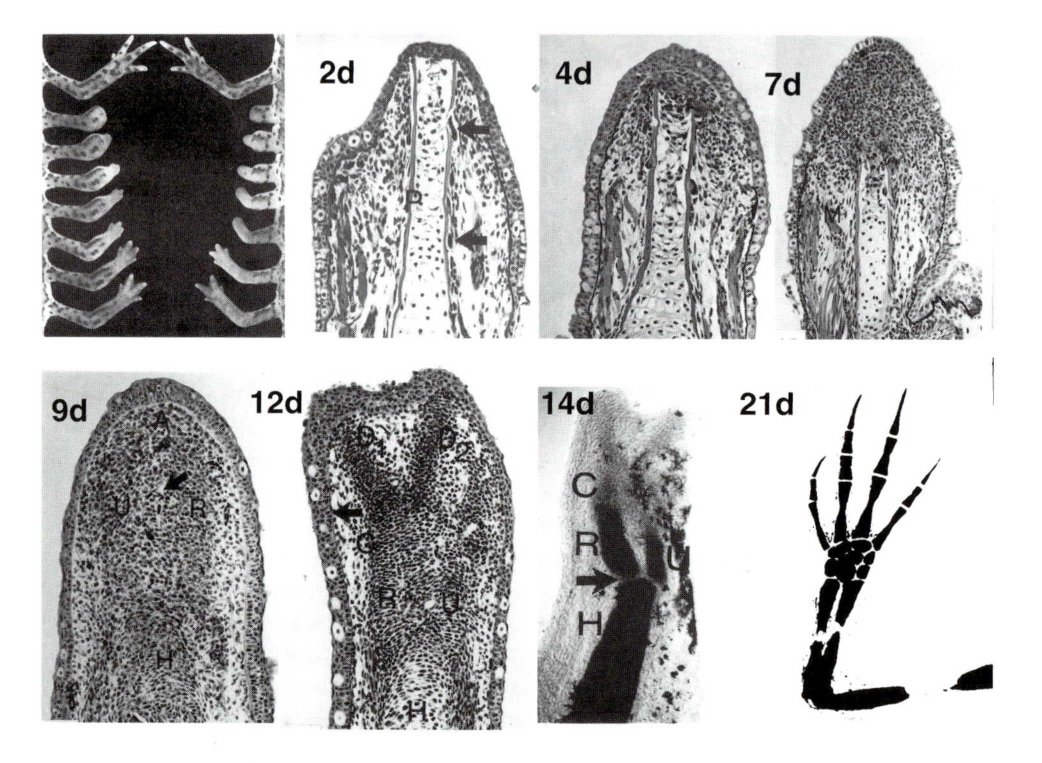


**FIGURE 14.2** Stages of limb regeneration. **Upper left panel** shows external views of stages of regeneration of an adult newt (*Notophthalmus viridescens*) forelimb amputated through the mid radius-ulna (left) and the mid-humerus (right). Stages go from top to bottom. Reproduced with permission from Goss, Principles of Regeneration. Copyright 1969, Elsevier. **Remaining panels**, show hematoxylin and light green stained longitudinal sections and methylene blue stained whole mounts of regenerating limbs of the larval salamander *Ambystoma maculatum* at different days post-amputation. P = periosteal bone shell, H = humerus, R = radius, U = ulna, C = carpal region, $D_1$ and $D_2$ = anterior digits 1 and 2. The arrows at 2d point to osteoclasts eroding the periosteal bone shell, at 9 days to a blood vessel, at 12 days to the re-forming basement membrane and at 14d to the elbow joint. Photos from D.L. Stocum.

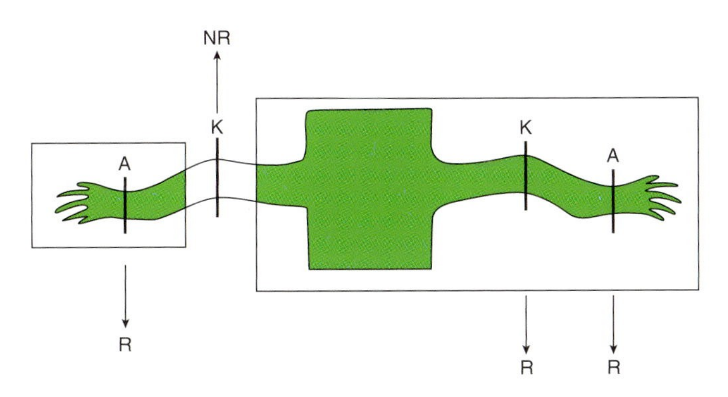


**FIGURE 14.3** Experiment showing the local origin of blastema cells in an amputated hindlimb. Larval salamanders were x-irradiated with limbs or parts of limbs shielded (boxes). Limbs amputated through unshielded regions failed to regenerate (NR), whereas amputation through shielded regions resulted in regeneration (R). A = ankle, K = knee.

tase (Kumar et al., 2000), were implanted into regenerating adult newt limbs. Marked, undifferentiated cells were detected in the blastema in each case, demonstrating that all cell types of the limb except epidermis (Riddiford, 1960; Namenwirth, 1974) undergo dedifferentiation to form the blastema. In addition, endogenous myofibers of the thin, transparent tails of newly hatched axolotls were labeled with rhodamine dextran *in vivo* and followed live after amputation. The labeled cells were shown directly to dedifferentiate and contribute to the blastema (Echeverri et al., 2001).

The dermis makes a disproportionate contribution to the blastema (Schmidt, 1968; Muneoka et al., 1986). In the axolotl, dermis represents ~19% of the limb tissue, while cartilage represents 6%. However, grafts of triploid dermis to diploid limbs contributed an average of 43% of the blastema cells and grafts of triploid carti-

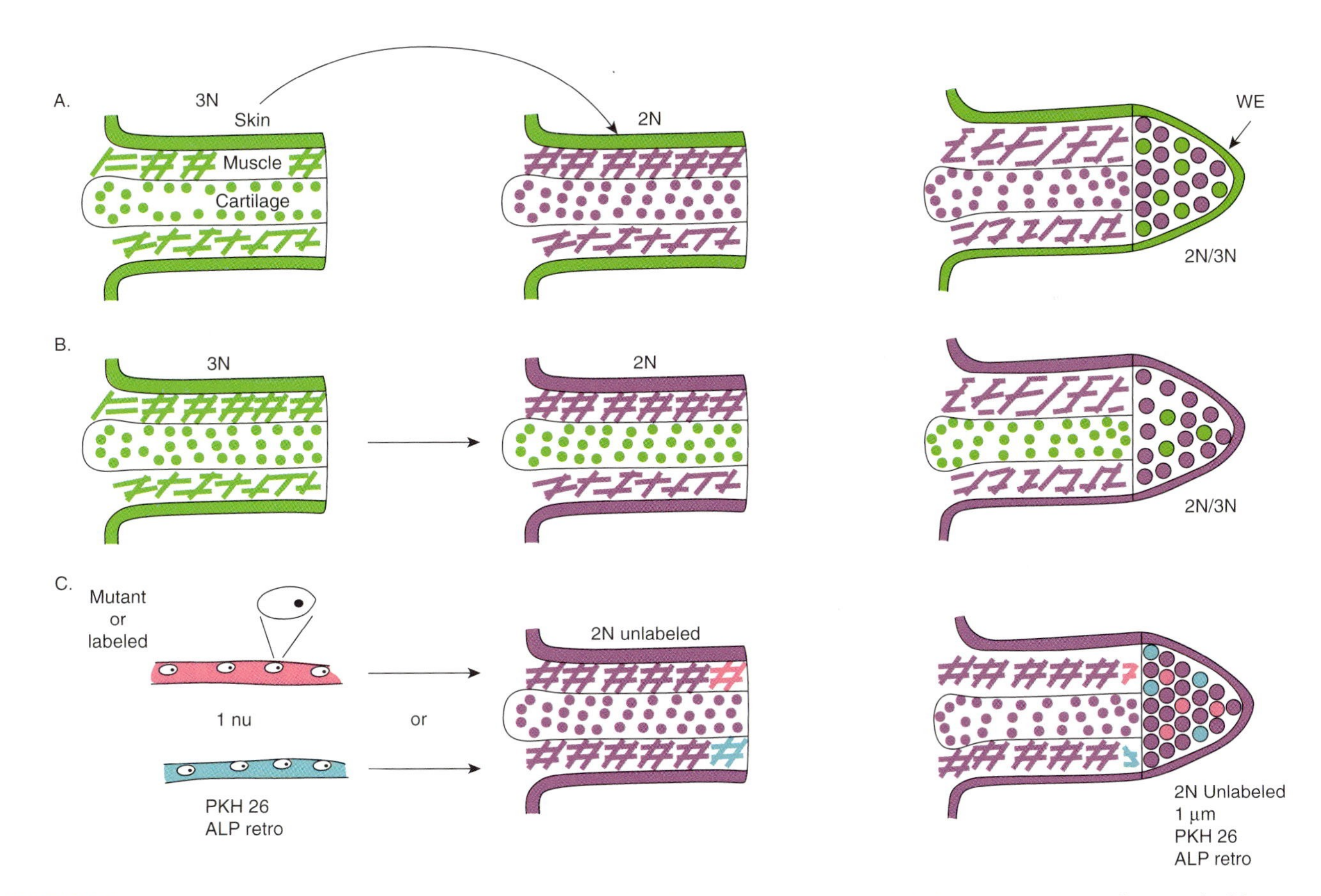


**FIGURE 14.4** Experiments demonstrating that dermal fibroblasts, chondrocytes and myofibers dedifferentiate and contribute to the blastema of a regenerating limb. Either skin **(A)** or cartilage **(B)** was grafted from a triploid axolotl (3N) in place of the same tissue in a diploid axolotl (2N). In each case, undifferentiated triploid cells were found among dedifferentiated cells of the host in the regeneration blastema. **(C)** Myofibers of 1-nucleolar mutant axolotls (pink) or derived from a newt blastema myoblast line *in vitro* and labeled with PKH26 dye or a retroviral alkaline phosphatase (ALP) (blue) were transplanted into the amputated limbs of unlabeled animals. The blastemas contained labeled undifferentiated cells among the host unlabeled cells.

lage only 2%. Thus, cells from the dermis are overrepresented in the blastema by more than a factor of two, and cells from cartilage are underrepresented by a factor of three. The percentage contributions from nondermal fibroblasts, muscle, and Schwann cells are unknown. Since limb muscle of amphibians contains SCs that regenerate injured muscle (Cameron et al., 1986) and the periosteum and bone marrow harbor MSCs that regenerate injured fractured bone, it is possible that these cells also contribute to the regeneration blastema, in addition to dedifferentiated cells. Proof of such a contribution requires the selective labeling of SCs and MSCs, but a way to do this has not yet been found.

A molecular biology of limb regeneration is emerging to complement the many studies that have been conducted on the tissue and cell level. Geraudie and Ferretti (1998), in a comprehensive review, listed all the molecules known to be synthesized in regenerating limbs, including soluble signaling molecules, ECM components, and transcription factors. Many of these will be mentioned in subsequent pages, along with more recent molecules first identified in embryonic limb buds and from screens of regenerating versus unamputated and regeneration-deficient limbs (e.g., King et al., 2003). A major deficiency is that a cell surface phenotype has not yet been established, so we do not know how comparable blastema cells are to mammalian stem cells such as MSCs. Several cytokeratin and intermediate filament proteins specific to blastema cells or cells of the wound epidermis have also been identified (Kintner and Brockes, 1984; Ferretti and Brockes, 1991; Ferretti and Ghosh, 1997; Castilla and Tassava, 1992; Onda and Tassava, 1991), but their functions are largely unknown.

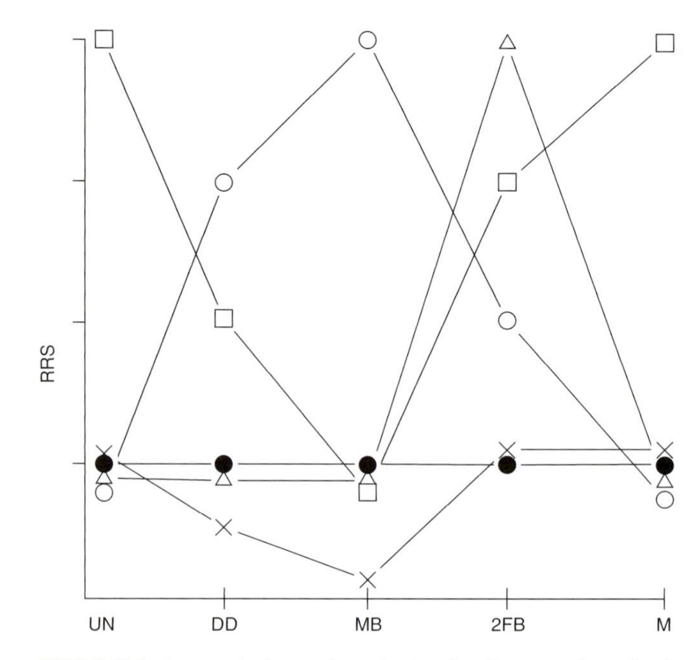


**FIGURE 14.5** Accumulation and synthesis of collagen and synthesis of chondroitin sulfate and hyaluronate at various stages of limb regeneration. Data from larval salamander and adult newt limbs. RRS on ordinate = relative rate of synthesis. Stages of regeneration on abscissa. UN = unamputated limb, DD = dedifferentiation, MB = medium bud, 2FB = two-fingerbuds, M = mature regenerant. X = type II collagen, closed circles = type I collagen, open triangles = chondroitin sulfate, open circles = hyaluronic acid, open squares = accumulation of collagen.

## 3. Mechanism of Blastema Formation

### a. *Matrix Degradation and Replacement*

The blastema is formed by degradation of the ECM local to the amputation surface, resulting in the loss of tissue organization (histolysis) and liberation of individual cells, followed by loss of phenotypic specialization and proliferation of the cells (Thornton, 1968). Regardless of the specialization of their parent cells, blastema cells have the morphological appearance of limb bud mesenchymal cells.

The ECM is degraded by acid hydrolases such as cathepsin D, acid phosphatase (Ju and Kim, 1998), β-glucuronidase, and carboxylester hydrolases (Schmidt, 1966), and by the MMPs 2 and 9 (gelatinases) and MMP3/10a and b (stromelysins), which are elevated during the period of blastema formation (Grillo et al., 1968; Dresden and Gross, 1970; Miyazaki et al., 1996; Ju and Kim, 1998; Yang et al., 1999; Park and Kim, 1999; Vinarsky et al., 2005). Differential display analysis between regenerating and nonregenerating newt limbs has also revealed a new collagenase that is active in degrading the ECM (Vinarsky et al., 2005). Acid hydrolases are undoubtedly released from dying cells and from osteoclasts, which are abundant in regenerating limbs (Stocum, 1979). Which cell types produce MMPs is unclear, but potential candidates are macrophages, wound epidermis, and osteoclasts, which produce MMPs after injury to other tissues (Chapters 2, 3, 9), as well as the blastema cells themselves. Regenerating axons are another potential source of MMPs. During embryogenesis, extending neurites secrete MMPs to cut a path through the ECM via their production of MMPs (Pittman and Buettner, 1989).

As the ECM is degraded, it is replaced by an embryonic limb bud-type ECM that helps maintain the blastema cells in an undifferentiated state (Stocum, 1995). FIGURE 14.5 illustrates the level of several molecules over the course of limb regeneration. Fibronectin, tenascin, and hyaluronate are present at high concentrations in embryonic limb buds and are markedly upregulated in the blastema, whereas the expression of laminin is extinguished. A new basement membrane does not form until the blastema begins to differentiate. Collagen II is also eliminated, leaving only the expression of collagen I, typical of mesenchymal cells of the limb bud. The expression of sulfated GAGs remains the same. Matrix degradation ceases once the blastema is formed. As the blastema differentiates, the ECM of the mature limb is reconstituted. Cessation of matrix degradation is likely to involve the upregulation of tissue inhibitors of metalloproteinases (TIMPS) as well as the downregulation of MMPs and acid hydrolases at any or all levels of gene regulation, from transcription to protein modification, but neither the regulation of protease activity nor expression of TIMPS have been systematically analyzed in regenerating limbs.

### b. Re-Entry of Cells into the Cell Cycle

Pulse labeling studies with $^3$H-thymidine have shown that the fraction of blastema cells that ultimately cycles is between 92–96% in larvae and over 90% in adults (Tomlinson et al., 1985; Goldhamer and Tassava, 1987; Tomlinson and Barger, 1987). The length of the cell cycle does not vary appreciably between larval and adult limbs, ranging from 40–53 hr (average = 46 hr) in both (Tassava and Olsen, 1985). The mitotic index in the regenerating limbs of *A. maculatum* is uniform until differentiation sets in (Stocum, 1980a), when cells in the proximal region of the blastema withdraw from the cell cycle, creating a distal to proximal gradient of mitosis (Litwiller, 1939; Smith and Crawley, 1977; Stocum, 1980a). Re-entry into the cell cycle and loss of phenotypic specialization has been studied in detail only in myofibers, which are structurally the most complex cells that contribute to the blastema. Not only must they re-enter the cell cycle, shut down the production of myogenic regulatory proteins, and dismantle the contractile protein apparatus, they must also fragment into mononucleate cells.

Nuclei of cultured newt and mouse myoblasts are induced to re-enter the cell cycle by either serum or several growth factors found in serum: PDGF, FGF-2, EGF, IGF-I, TGF-β, and TNF-α (Tanaka et al., 1997). These growth factors stimulate phosphorylation of the retinoblastoma (Rb) protein, releasing it from E2F transcription factors so they can activate genes essential for entry into S-phase. Serum also stimulates DNA synthesis in cultured newt myotubes, but not in cultured C2C12 mouse myotubes. However, the same individual growth factors that stimulate DNA synthesis in newt and mouse myoblasts fail to do so in newt myotubes (Tanaka et al., 1997). The data are summarized in **TABLE 14.1**. The results suggest that, as they differentiate from myoblasts, newt myotubes become refractory to stimulation by these growth factors, but express a receptor that can respond to another protein in serum that stimulates DNA synthesis. Mouse myotubes appear to lack this receptor, for they will only synthesize DNA in response to serum stimulation if they are part of a heterokaryon made by fusing C2C12 and newt myoblasts (Velloso et al., 2001). The stimulatory protein has so far escaped identification, but tests of serum fractions on cultured myoblasts and myotubes indicate that it requires thrombin for activation (**FIGURE 14.6**). Flurogenic assays have shown that thrombin activity increases during blastema formation (Tanaka et al., 1999). Although the thrombin-activated protein (TAP) is both necessary and sufficient to stimulate myonuclear entry into the cell cycle, it is not sufficient to drive them through mitosis; the nuclei arrest in the $G_2$ phase of the cycle. Thrombin itself might be activated by tissue

**TABLE 14.1** Effect of High Serum Concentration and Growth Factors (GFs) on DNA Synthesis in Cultured Myoblasts and Myotubes of Adult Newt and Mouse

| | *Myoblasts* | | *Myotubes* | |
|---|---|---|---|---|
| | Serum | GFs* | Serum | GFs* |
| Newt | + | + | + | – |
| Mouse (C2C12) | + | + | – | – |

*PDGF, FGF-2, EGF, IGF I, TGF-β, TNF-α

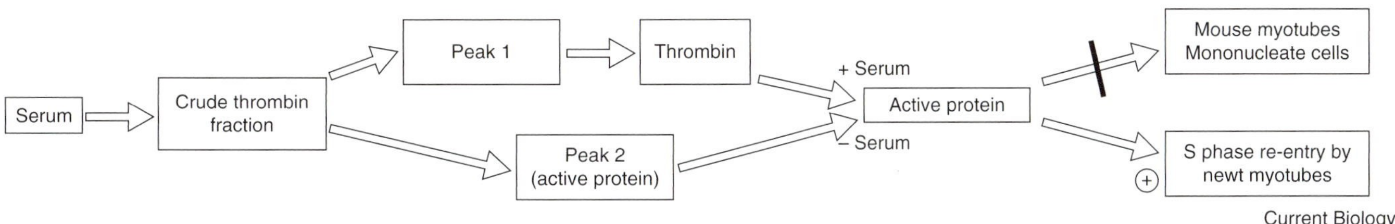


**FIGURE 14.6** Detection of a protein in serum that stimulates newt myonuclei to re-enter the S phase of the cell cycle. The crude thrombin fraction of serum was fractionated into two peaks. Further fractionation of peak 1 yielded pure thrombin, which produced an active protein when incubated with serum. Peak 2 contained the active protein itself. The active protein stimulated S phase re-entry in cultured newt myotubes, but not in mouse myotubes or myoblasts of either newt or mouse. Reproduced with permission from Stocum, Limb regeneration: Re-entering the cell cycle. Curr Biol 9:R644–R646. Copyright 1999, Elsevier.

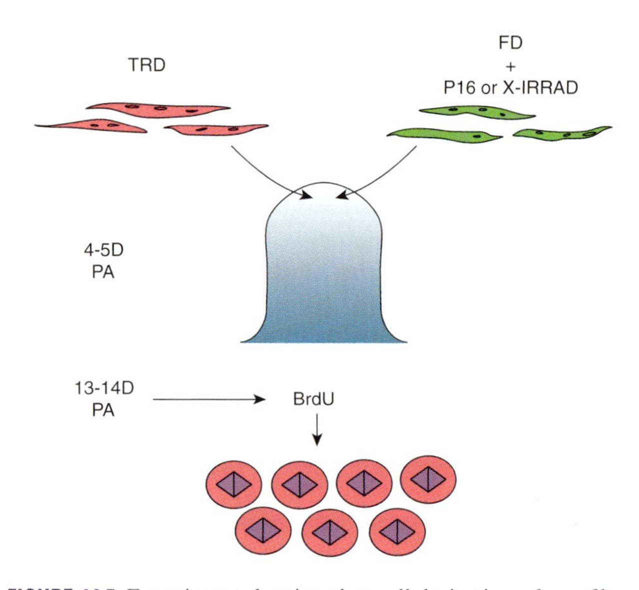


**FIGURE 14.7** Experiment showing that cellularization of myofibers is not coupled to cell cycle re-entry in regenerating newt limbs. Myofibers in which cell cycle re-entry was blocked by p16 transfection or X-irradiation were labeled with fluorescein dextran (FD) and co-implanted into the early blastema (4–5 days postamputation) with myofibers labeled with Texas Red-Dextran (TRD). BrdU was given two weeks post-amputation. Histological examination of limbs revealed that both sets of myofibers cellularized, but only the TRD-labeled mononucleate cells divided.

factor, as hypothesized for newt lens regeneration (Chapter 3), and the thrombin-activated protein revealed in limb regeneration experiments might also be the trigger for lens regeneration. The ubiquity of thrombin in wounds suggests that the same protein may be activated, and use the same TAP receptor, in all cases of injury-induced regeneration.

Re-entry into the cell cycle does not require fragmentation of the multinucleated myotubes into mononucleate cells, as shown by an experiment in which cultured newt myotubes were labeled with fluorescein dextran and X-irradiated or transfected with the p16 gene for cyclin-dependent kinase 4/6 inhibitor to block entry into S phase. These myotubes were then co-implanted with control myotubes labeled with Texas Red-Dextran into limb blastemas (**FIGURE 14.7**). Both the control and cell cycle-inhibited myotubes fragmented into mononucleate cells, but only the control myotubes synthesized DNA and divided, as shown by BrdU incorporation (Velloso et al., 2000). Completion of mitosis does appear to require mononucleate cell status. The mechanism of myotube fragmentation into single cells is not known, nor is it known whether the TAP is necessary to drive mononucleate cells such as chondrocytes and fibroblasts into the cell cycle as well, or whether this is a feature unique to myofibers.

### c. *Loss of Phenotypic Specialization*

We do not have a clear understanding of the mechanisms that initiate the loss of phenotypic specialization during limb regeneration. Degradation of the ECM by proteases would break contacts between ECM molecules and integrin receptors, leading to changes in cell shape and reorganization of the actin cytoskeleton. This reorganization in turn might activate signal transduction pathways that induce the upregulation of enzymes that dismantle the phenotype-specific internal structure of the cells. Two small molecules, one a tri-substituted purine called myoseverin and the other a disubstituted purine dubbed reversine, have been screened from combinatorial chemical libraries and found to cause cellularization of C2C12 mouse myofibers (Rosania et al., 2000; Chen et al., 2004). Myoseverin disrupts microtubules and upregulates growth factor, immunomodulatory, ECM-remodeling and stress-response genes, consistent with the activation of pathways involved in wound healing and regeneration, but does not activate a full program of myogenic dedifferentiation (Duckmanton et al., 2005). Reversine may act through protein kinase pathways. The mononucleate cells produced by reversine treatment behaved like MSCs; they were able to differentate *in vitro* into osteoblasts and adipocytes, as well as muscle cells. Myoseverin and reversine will be useful in analyzing the events of dedifferentiation and may have natural counterparts that can be isolated.

Several genes have been identified that might be involved in the initiation or maintenance of dedifferentiation. The transcription factor Msx1, which inhibits myogenesis (Song et al., 1992; Woloshin et al., 1995), is expressed in undifferentiated limb bud mesenchyme and regeneration blastemas (Hill et al., 1989; Robert et al., 1989; Crews et al., 1995; Simon et al., 1995). C2C12 mouse myotubes fragmented into mononucleate cells and dedifferentiated when transfected *in vitro* with *msx1* (Odelberg et al., 2000). Interestingly, these mononucleate cells also behaved like MSCs, differentiating *in vitro* into cells expressing chondrogenic, adipogenic, myogenic, and osteogenic molecular markers. These results suggest that *msx1* might be upregulated in regenerating amphibian limbs after myoseverin or reversine treatment. Results of other experiments, however, are conflicting. The *msx1* gene is not strongly upregulated until the conical blastema stage (Crews et al., 1995; Simon et al., 1995; Koshiba et al., 1998). The uptake of anti-*msx* morpholinos by newt muscle cells *in vitro* decreased their expression of *msx1* and prevented cellularization (Kumar et al., 2004). However, the *in vivo* electroporation of either the *msx1* gene or an anti-*msx1* morpholino into individual muscle fibers of the axolotl tail, followed by amputation, had no effect on the number of muscle fibers that dedifferentiated (Schnapp and Tanaka, 2005). Thus, although *msx-1* is associated with the dedifferentiated state in regenerating amphibian limbs and tails, its function in dedifferentiation is not yet clear.

Other genes that might play roles in dedifferentiation are *Nrad*, *radical fringe*, and *Notch*. Nrad is the orthologue of a diabetes-associated mammalian ras protein that is upregulated in the nuclei of newt limb myofibers within four hr after amputation (Shimizu-Nishikawa et al., 2001). The Notch ligand radical fringe is expressed in the dorsal ectoderm of amniote limb buds. The position of the AER in these limb buds is specified by the boundary between *rfrng*-expressing dorsal and nonexpressing ventral ectoderm (Laufer et al., 1997; Rodriguez-Esteban et al., 1997). In regenerating newt limbs, *rfrng* is expressed only in blastema cells (Cadinouche et al., 1999), suggesting that rfrng-activated Notch may be involved in initiating and/or maintaining dedifferentiation. The downstream targets of these proteins in regenerating limbs are unknown.

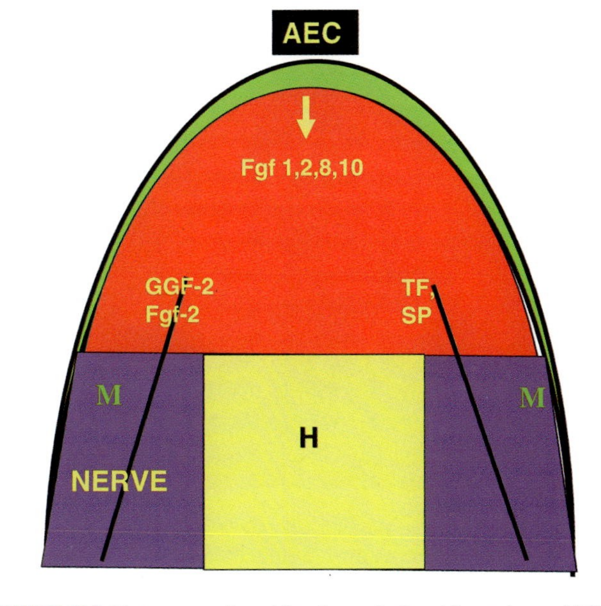


**FIGURE 14.8** Factors produced by the apical epidermal cap (AEC, green) and nerves that are vital to blastema cell (red) survival and proliferation. Fgf 1,2,6,10 = fibroblast growth factors. GGF-2 = glial growth factor-2, TF = transferrin, SP = substance P. H = humerus (yellow), M = muscle (purple).

## 4. Epidermal and Neural Requirements for Survival and Proliferation of Blastema Cells

Survival and proliferation of blastema stem cells requires the metabolic effects exerted by several endocrine hormones, primarily insulin, growth hormone, hydrocortisone, and thyroxine (Globus and Vethamany-Globus, 1985), but is also highly dependent on specific factors produced by the AEC (Thornton, 1968) and the nerves of the blastema (Singer, 1965). **FIGURE 14.8** illustrates some of the molecules expressed by the AEC and nerves that are potential survival and proliferation factors for blastema cells. To be defined as survival and proliferation factors, candidate molecules should meet four criteria (Brockes, 1984): (1) they should be secreted from the AEC or nerve endings into the blastema; (2) removal of the AEC or nerves should result in loss of the molecule from the blastema; (3) the molecule should be able to substitute for the AEC or nerves in maintaining mitosis and/or promoting regeneration to completion; and (4) selective neutralization of the molecule should abolish its mitogenic effect on blastema cells. Most of the candidate molecules meet at least one criterion, but only one, transferrin, so far meets them all. However, the majority of the candidates have been not been tested over the full range of criteria.

### a. AEC Factors

Blastema formation and growth require direct contact between the wound epidermis and underlying mesenchymal cells. Failure of regeneration is associated with the premature formation of a basement membrane under the wound epithelium (Stocum and Crawford, 1987). Preventing the formation of a wound epidermis by inserting the tips of amputated limbs into the coelom or folding a full-thickness skin flap over the amputation surface inhibits DNA synthesis by the initial blastema cells formed by histolysis and dedifferentiation and prevents blastema formation (Goss, 1956a; Mescher, 1976; Tassava and Garling, 1979; Loyd and Tassava, 1980). Removal of the wound epidermis at various stages of blastema growth and inserting the limb into the coelom or enclosing epidermis-free blastemas in fin tunnels leads to truncation of the regenerate skeletal pattern in the PD axis (Goss, 1956b; Stocum and Dearlove, 1972) (**FIGURE 14.9**). UV or X-irradiation of the wound epidermis also prevents blastema formation (Thornton, 1958; Lheureux, 1983; Lheureux and Carey, 1988). The early wound epidermis is highly active in RNA and protein synthesis (Hay, 1960), and blastema formation is retarded in limbs treated with the transcription inhibitor, actinomycin D (Carlson, 1969). The effect of AEC removal on proliferation in an already established blastema *in vivo* has not been quantitated, but the $^{3}$H-thymidine and mitotic indices of blastemas cultured in the absence of the AEC and in the presence of hormones and nerves, are reduced 3–4-fold (Globus et al., 1980; Smith and Globus, 1989). These observations show that the AEC plays a role in the outgrowth and patterning of the blastema that is equivalent to that of the AER

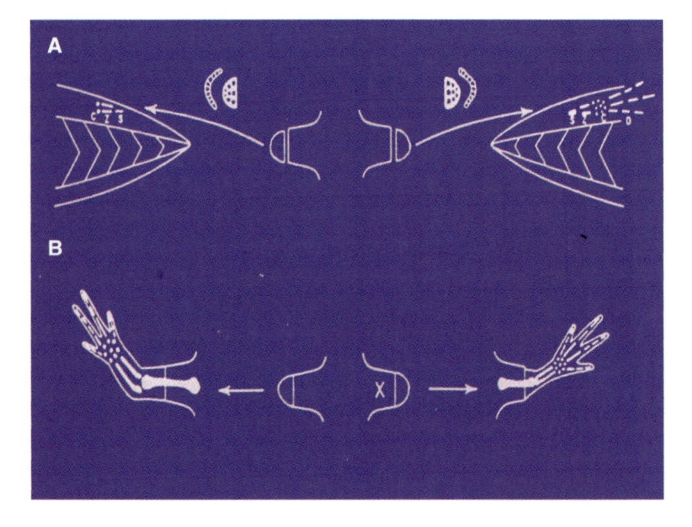


**FIGURE 14.9** Experiment showing the importance of the wound epidermis and nerves in promoting proliferation of blastema cells. **(A)** Epidermis-free conical blastemas derived from the distal level of the humerus were implanted into a tunnel bored in the connective tissue of the dorsal fin so that they were either completely enclosed (left), or with distal tips protruding to be covered with fin wound epidermis (right). The enclosed blastemas differentiated miniature sets of skeletal structures that ended with 2–3 basal carpals. The blastemas with tips re-covered by fin epidermis also developed miniature sets of skeletal structures, but these were complete in the PD axis. **(B)** Blastemas denervated at the conical stage by severing the nerves of the brachial plexus. The blastema cells cease mitosis and form a miniature regenerate that is complete in the PD axis.

in the amniote and anuran limb bud (Saunders, 1948; Tschumi, 1957).

In amniote limb bud development, FGF-4 and -8 are expressed in the AER and FGF-10 is expressed in the mesoderm (Martin, 1998). During early limb bud development, FGF-8 maintains mesenchymal FGF-10 expression and induces the expression of sonic hedgehog (*shh*) in the posterior mesenchyme. Shh then induces the expression of FGF-4 in the posterior region of the AER and these two molecules maintain each other's expression via a feedback loop. FGF-2 and -4 can substitute for the outgrowth-promoting effect of the AER when delivered to the mesenchyme in beads (Fallon et al., 1994; Niswander et al., 1993). Inactivation of both *Fgf-4* and *-8* genes in mouse limb buds inhibits their development by decreasing the number of cells recruited into the limb bud and causing abnormal cell death at a distance from the AER (Sun et al., 2002).

FGFs play an important role in the epithelial-mesenchymal interactions of limb regeneration as well. FGF-1 protein is expressed in the mesenchyme and AEC of medium bud stage blastemas (Boilly et al., 1991) and FGF-2 protein has been detected in the AEC of the mid-bud blastema of axolotls (Mullen et al., 1996). FGF-1 mRNA is transcribed at a low level in medium-bud stage axolotl and newt blastemas (Dungan et al., 2002). The receptor for FGF-1 and -2, FGFR1, is expressed on both AEC and blastema stem cells (Poulin et al., 1993; Poulin and Chiu, 1995), and $^{125}$I-labeled FGF-1 and -2 bind to sections or membrane preparations of AEC and blastemal mesenchyme (Boilly et al., 1991). FGF-1 elevates the mitotic index of epidermis-free blastema cells *in vitro* (Albert et al., 1987). FGF-1 injections or beads promote significant mitosis and blastema development up to a medium bud stage in amputated epidermis-free limbs inserted into the body cavity (Dungan et al., 2002). Blastemas release a mitogenic activity into culture medium that is depressed by one-third when the medium is treated with antibody to FGF-1 (Zenjari et al., 1997). FGF-2 has also been found to elevate the mitotic index of regenerating limbs covered with a full-thickness skin flap *in vivo* (Chew and Cameron, 1983). Specific inhibitors of the FGFR1 and 2 receptors inhibit the regeneration of limb buds in early *Xenopus* tadpoles (D'Jamoos et al., 1998). These observations suggest that the AEC promotes blastema cell survival and proliferation by its production of FGF-1 and -2.

*FGF-8* and *-10* transcripts are upregulated in both the AEC and the blastema mesenchyme of axolotls in patterns similar to those of the amniote limb bud (Christensen et al., 2001, 2002; Han et al., 2001). Expression of FGF-10 is much stronger in the mesenchyme cells than in the AEC (Christensen et al., 2002). FGF-8 mRNA is first detectable during dedifferentiation, increases to a peak at medium bud, but disappears by the palette stage (Han et al., 2001). Expression is confined to the basal layer of the AEC and a thin band of mesenchyme cells just under the AEC; it is strongest in the anterior AEC. Interestingly, this expression pattern differs from the expression pattern in the axolotl limb bud, where FGF-8 is expressed primarily in the mesenchyme (Han et al., 2001). In *Xenopus*, FGF-8 is also expressed in the AEC and FGF-10 in the mesenchyme of regenerating early tadpole limbs (Christen and Slack, 1997; Yokoyama et al., 2000). While these expression patterns are suggestive of AEC-mediated blastema cell survival and proliferation via FGF-8 and -10, more direct evidence for such a role is lacking.

The pathways that regulate the production of FGFs by the AEC are unknown. The genes for the transcription factors Msx2 and Dlx3, which are expressed during amniote limb bud development, are strongly expressed in the early wound epidermis of regenerating axolotl limbs (Mullen et al., 1996; Koshiba et al., 1998; Carlson et al., 1998). Expression of *dlx3* is first detected at the accumulation blastema-early bud stages, peaks at late bud, when it is also detected in the mesenchyme, and decreases to zero by late digit stages of regeneration (Mullen et al., 1996). The inhibitor of differentiation genes, *Id2*, *Id3*, and *HES1* are also upregulated in the

AEC of regenerating newt limbs (Shimizu-Nishikawa et al., 1999). The upregulation of *HES1* suggests the expression of Notch and Nrp-1 in the cells of the AEC. The role these genes play in the structure and function of the AEC is not clear.

### b. Neural Factors

Total denervation of an amputated urodele limb by cutting spinal nerves III, IV, and V does not prevent migration of wound epidermis, histolysis, or dedifferentiation, but mitosis of dedifferentiated cells does not take place and a blastema fails to form (Tassava and Mescher, 1975). Denervation depresses RNA and protein synthesis without changing the qualitative profile of the proteins synthesized (Dresden, 1969; Lebowitz and Singer, 1970; Morzlock and Stocum, 1972; Singer, 1978). Glycosaminoglycan (GAG) synthesis is strongly reduced (Smith et al., 1975; Mescher and Cox, 1988; Young et al., 1989). Hyaluronate synthesis is affected more than any other GAG, consistent with the fact that it is the major GAG synthesized during blastema formation and growth (Mescher and Munaim, 1986).

The nerve effect is quantitative; regeneration can be supported by either sensory or motor fibers, as long as the number of fibers is at or above a certain threshold (Singer, 1952, 1965). TUNEL staining indicates that the dedifferentiated cells in an amputated and denervated limb undergo apoptosis and are removed by macrophages (Mescher et al., 2000). The wound epidermis does not form an AEC in amputated and denervated limbs. However, it is not the lack of an AEC that prevents mitosis, as shown by denervating limbs in which a blastema (and thus an AEC) has already formed. The blastema cells of these limbs arrest in $G_1$ (Maden, 1979; Oudkhir et al., 1985). The mitotic index falls to zero (Goldhamer and Tassava, 1987), but the blastemas are able to undergo differentiation and morphogenesis into miniature regenerates because they have attained a critical mass of cells (Schotte and Butler, 1944; Maden, 1981; Singer and Craven, 1948) (**FIGURE 14.9**).

The limb nerves also express FGF-1 and -2, but otherwise their set of survival and proliferation promoting factors is different from that of the AEC (**FIGURE 14.8**). However, *FGF-1* transcript levels do not increase after amputation, and FGF-1 injections and bead implants do not rescue the regeneration of limbs denervated at early bud (Dungan et al., 2002). FGF-2 promotes the regeneration of denervated axolotl limbs to digit stages *in vivo* when delivered to early blastemas in bead implants (Mullen et al., 1996). Whether FGF-8 and -10 are present in limb nerves is not known. Denervation strongly downregulates the expression of FGF-8 transcripts in regenerating *Ambystoma* limbs, but only slightly affects the level of *FGF-10* mRNA, suggesting that nerve factors regulate *FGF-8* expression in the blastema (Christensen et al., 2001).

Regenerating limb nerves express three other factors that support blastema cell survival and proliferation. The first is glial growth factor 2 (GGF-2), a member of the neuregulin family. A newt neuregulin gene cloned from spinal cord is expressed in newt dorsal root ganglia (Wang et al., 2000). Recombinant human GGF-2 infused into denervated nerve-dependent axolotl limb blastemas maintained the $^3$H-thymidine labeling index at control levels and allowed regeneration to digit stages (Wang et al., 2000), similar to the rescue of denervated blastemas by implants of spinal ganglia (Tomlinson and Tassava, 1987). The second is substance P, a peptide neurotransmitter widely distributed throughout the mammalian central and peripheral nervous system that is mitogenic for mammalian connective tissue cells (Nilsson et al., 1985). Substance P is found in the cell bodies of adult newt sensory ganglia and in the AEC and in the mesenchyme of the blastema (Globus and Alles, 1990). The peptide is mitogenic for blastema cells *in vitro* at low concentrations and an antibody to substance P applied to spinal ganglia cultured transfilter to blastemas decreases the mitogenic effect of the ganglia in a dose-dependent fashion (Globus, 1988).

The third nerve factor is the iron-transport protein, transferrin. Transferrin is the only factor present in limb nerves that meets all the criteria for being a blastema cell survival/proliferation factor (Mescher, 1996). Iron is a co-factor for many enzymes essential for cell proliferation, including the rate-limiting enzyme for DNA replication, ribonucleotide reductase (Mescher and Munaim, 1988; Sussman, 1989). Transferrin is axonally transported and released distally in the nerves of regenerating axolotl limbs; nerve transection reduces the concentration of blastema transferrin by 50% (Kiffmeyer et al., 1991; Mescher and Kiffmeyer, 1992). Transferrin can substitute for the nerve or nerve extracts in maintaining blastema cell proliferation *in vitro* (Munaim and Mescher, 1986; Albert and Boilly, 1988) and *in vivo* (Mescher and Kiffmeyer, 1992). Chelation of ferric ions or immunoabsorption with antitransferrin antiserum abolishes the mitogenic effect of nerve extracts on blastema stem cells. The inhibitory effect is reversed by adding iron to the chelated extract or purified transferrin to the antiserum-treated extract (Munaim and Mescher, 1986; Mescher et al., 1997). The only credential missing from the transferrin portfolio is the demonstration *in vivo* that it can promote the regeneration of denervated limbs to completion.

The relationship between regenerating nerve fibers and blastema cells is a reciprocal one. The regeneration of nerve fibers into the blastema is dependent on factors produced by the blastema cells. Regeneration of axons from nerve cell bodies is promoted *in vitro* by co-culture of neurons with blastema tissue (Richmond and Pollack, 1983). Several known neurotrophic factors such as brain-derived neurotrophic factor (BDNF), neurotrophins 3 and 4 (NT3,4), glial-derived neurotrophic factor (GDNF), and hepatocyte growth factor/scatter factor (HGF/SF) can substitute for blastema tissue in promoting axon outgrowth *in vitro* (Tonge and Leclere, 2000). Several of these factors are the same ones produced by Schwann cells that promote neuron survival and axon outgrowth in regenerating peripheral nerves of mammals, raising the question of whether they might be produced by the subpopulation of blastema cells derived from Schwann cells. Regardless, axon outgrowth is significantly more vigorous with blastema tissue, suggesting that other, as yet unidentified, factors are produced by blastema cells that encourage neuron survival and axon outgrowth.

### c. Differential Roles of AEC and Nerves?

Are the AEC and nerves complementary to one another qualitiatively, with each providing essential factors to blastema cells that the other does not, or are they redundant? It might be argued that there is at least some redundancy, since FGF-1 supports regeneration to the medium bud stage in the absence of the AEC, and both FGF-2 and GGF-2 promote regeneration to digit stages. This question is difficult to answer definitively, since we have no data on the ability of each factor to support regeneration *in vivo* after knocking out all the others. However, other observations suggest that the AEC and nerves have different and synergistic functions.

The amphibian limb exhibits continual dependence on the AER/AEC for its outgrowth during ontogeny and during regeneration. Nerve dependence for regeneration arises only during the digital stages of limb development, when the limb bud is becoming heavily innervated. This dependence is associated with expression of an intermediate filament called 22/18 in a subpopulation of blastema cells derived from Schwann cells that has been shown to be the most sensitive to denervation (Fekete and Brockes, 1987, 1998; Brockes and Kintner, 1986). Ferretti and Brockes (1991) have suggested that blastema cells originating from Schwann cells might produce an inhibitor of mitosis when their own cycling is halted by denervation, thus halting the proliferation of blastema cells derived from other cell types. This idea is supported by the results of experiments in which pieces of peripheral nerves implanted into amputated adult newt limbs delayed regeneration in 80% of the cases (Irvin and Tassava, 1998). Implants of other tissues did not have this effect, implying that the nerve implants specifically raised the requirement for proliferation factors by increasing the number of nerve-dependent blastema cells derived from Schwann cells.

These observations, coupled with the fact that dedifferentiated Schwann cells in regenerating mammalian spinal nerves are dependent on axons for their proliferation, suggest that some nerve factors may differentially support the blastema cells derived from Schwann cells. More generally, the requirements of blastema cells for survival and proliferation factors might change qualitatively as the limb bud becomes heavily innervated, with nerves supplying factors that the AEC cannot, such as transferrin, substance P, and GGF-2, some of which are important not only to dedifferentiated Schwann cells, but to all blastema cells. In support of this notion, the epidermis-free mesenchyme cells of newt limb blastemas cultured *in vitro* transfilter to spinal ganglia arrest in $G_0$ and differentiate prematurely into cartilage (Globus et al., 1980; **FIGURE 14.10**), whereas the cells of denervated blastemas with AEC intact appear to arrest in $G_1$, S, or possibly $G_2$ (Maden, 1979; Tassava and Mescher, 1975), suggesting synergistic functions of AEC and nerve factors.

The limbs of aneurogenic embryos never become nerve-dependent for their regeneration (Yntema, 1959a, b), nor do their regeneration blastemas express 22/18, though they remain dependent on the AEC (Fekete and Brockes, 1988). Aneurogenic limbs acquire nerve-dependence when allowed to become innervated and exhibit a 22/18-reactive subpopulation of cells in their regeneration blastemas (Fekete and Brockes, 1987). Interestingly, when reciprocal combinations of skin and internal tissues were made from neurogenic and aneurogenic limbs, the combination of neurogenic skin and aneurogenic internal tissues failed to regenerate, whereas the combination of aneurogenic skin and neurogenic internal tissues regenerated normally (Steen and Thornton, 1963). This suggests that the wound epidermis undergoes a change in its ability to support blastema cells during the transition from nerve-independence to dependence. It would be of interest to investigate the types and quantities of survival and proliferation factor candidates in the AEC before, during, and after this transition period.

## 5. Developmental Plasticity of Blastema Cells in Regenerating Amphibian Limbs

The PECs of the newt eye, which are derived from the neural tube, clearly change their prospective fate to regenerate the lens, and the ependymal cells of the amphibian spinal cord appear able to differentiate into muscle and cartilage during tail regeneration (Chapters 3, 5). The plasticity of limb regeneration blastema cells

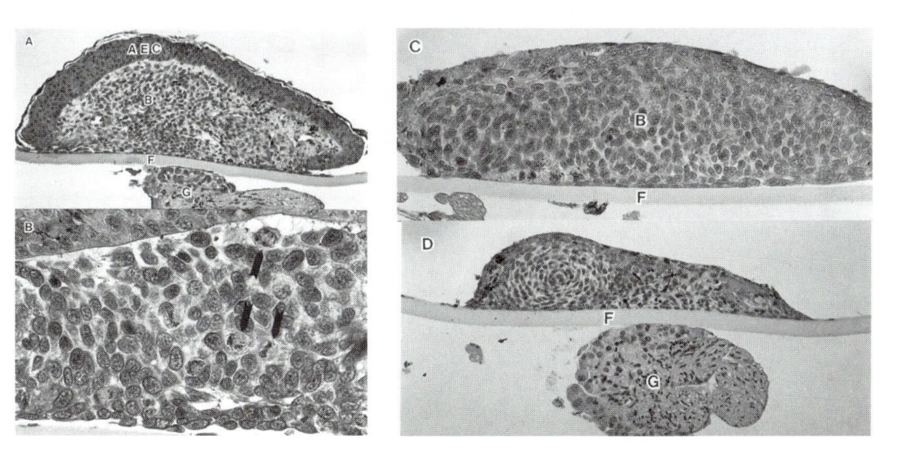


**FIGURE 14.10** Conical (medium bud) blastemas of adult newt limbs cultured *in vitro* transfilter to spinal ganglia in the presence **(A, B)** or absence **(C, D)** of the wound epidermis. F = Millipore filter, G = spinal ganglia. **(A)** The blastemal mesenchyme, B, remains undifferentiated under the AEC. **(B)** Higher magnification of **(A)**. Numerous mitotic figures are visible (arrowheads). **(C)** Epidermis-free blastemal mesenchyme early in culture; the mesenchyme cells are healthy but there are no mitotic figures. **(D)** Older culture of epidermis-free blastemal mesenchyme in which a whorl of cartilage is differentiating. Reproduced with permission from Globus et al., Effect of apical epidermal cap on mitotic cycle and cartilage differentiation in regeneration blastemata in the newt, *Notophthalmus viridescens*. Dev Biol 75:358–372. Copyright 1980, Elsevier.

has been tested in transplantation experiments. Triploidy was used as a marker to trace the fate of dermal fibroblasts and cartilage cells in regenerating axolotl limbs. The limb skin or cartilage of triploid animals was grafted in place of diploid skin and cartilage. The blastema cells derived from triploid dermal fibroblasts differentiated into both dermal fibroblasts and chondrocytes, and blastema cells derived from chondrocytes differentiated into chondrocytes and fibroblasts of the connective tissues of the perichondrium, joints, dermis, and skeletal muscle (Steen, 1968; Namenwirth, 1974; Dunis and Namenwirth, 1977; Holder, 1989). Blastema cells derived from cartilage implanted into X-irradiated amputated limbs were reported to differentiate into muscle cells, as well as into skeletal elements and connective tissues (Maden and Wallace, 1975; Desselle and Gontcharoff, 1978). Blastema cells derived from marked muscle cells were reported to differentiate into chondrocytes and other cell types after being implanted into amputated wild-type *Xenopus* or newt limbs (Steen, 1973; Casimir et al., 1988; Lo et al., 1993; Kumar et al., 2000). Thus, fibroblasts and myogenic cells appear able to become chondrocytes and chondrocytes can become fibroblasts and myogenic cells. Myogenic cells have not been observed to become fibroblasts, or vice versa, though fibroblasts could hypothetically become muscle cells by first becoming chondrocytes during one round of regeneration, which then could become muscle cells after a second round.

We do not know whether limb blastema stem cells have developmental potency beyond the range of cell types present in the limb. For example, might they be able to differentiate into lens or neural cells if transplanted into the lentectomized eye or injured spinal cord? Experiments done in the 1930s claimed to have shown that limb blastemas grafted into the lentectomized eye of newts differentiated into lens, but those claims could not be substantiated because no cell markers were employed to trace the origin of the regenerated lenses (Stocum, 1984, for review). Based on histological and molecular observations of cartilage and muscle in regenerating limbs, and pigmented epithelial cells in regenerating lens, it appears that lineage conversion by cells of these tissues involves dedifferentiation to an earlier state of development and transdetermination, but this has not been rigorously proven.

## 6. Spatial Organization of Tissue Patterns in the Regenerating Limb

### *a. The Blastema Is a Self-Organizing System*

The tissue pattern and morphogenesis of the limb regeneration blastema is determined from its inception, as shown by its ability to self-organize when grafted to ectopic locations. Blastemas or fractions of blastemas grafted to the dorsal fin develop autonomously (Stocum, 1968a, b, 1984, 1996, 2004) (**FIGURE 14.11**). **FIGURE 14.12** shows that grafting blastemas between forelimbs and hindlimbs, to different levels, or with axes disharmonious to the stump tissues is unable to alter their development. Forelimb blastemas grafted to hindlimbs always develop as forelimbs, and vice versa. When a

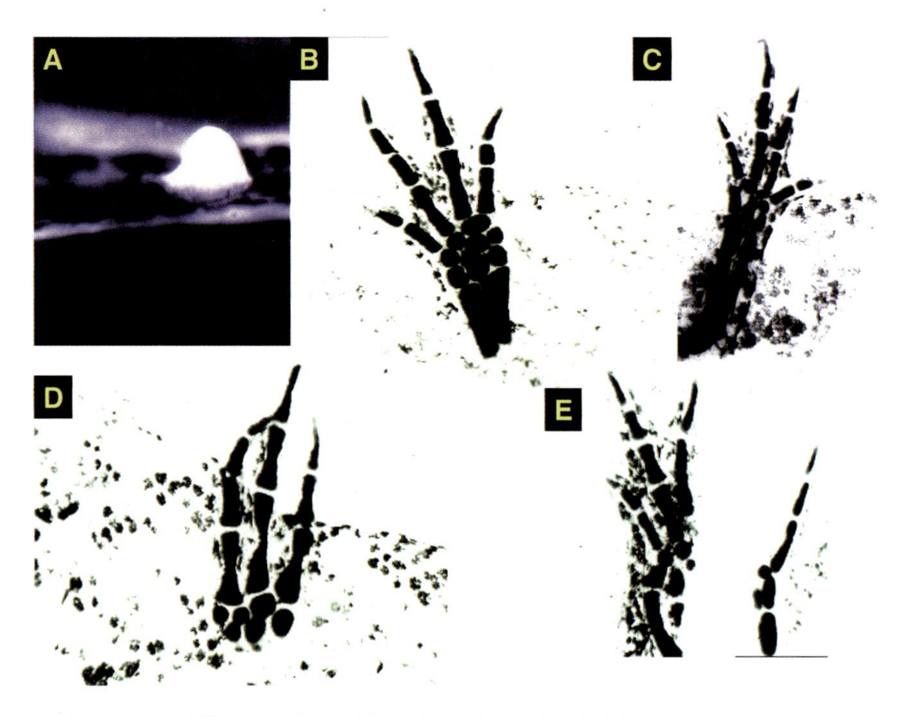


**FIGURE 14.11** Development of conical, undifferentiated blastemas derived from the distal humerus after grafting to a wound bed on the dorsal fin. **(A)** Four days after grafting. **(B–E)**, Whole mounts stained for cartilage. **(B)** Whole mount of fully developed regenerant. **(C)** Regenerant developed from a blastema plus a small segment of stump tissue. **(D)** Regenerant developed from a distal half blastema. **(E)** Regenerants developed from a blastema cut longitudinally into a $^{2}/_{3}$ fraction (left) and a $^{1}/_{3}$ fraction (right).

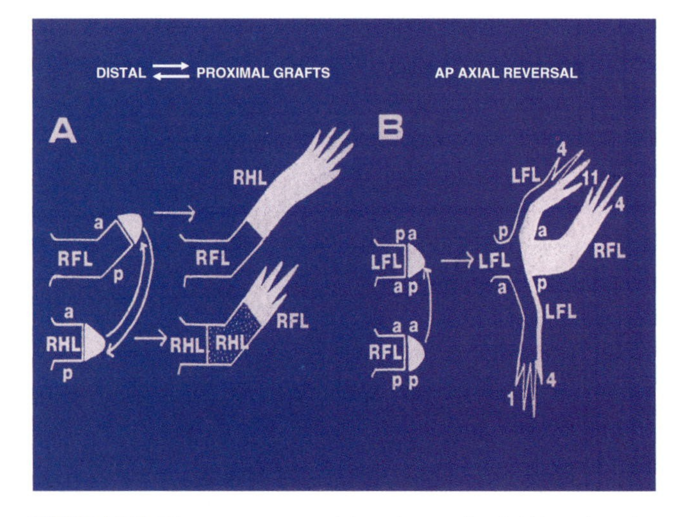


**FIGURE 14.12** Diagrams summarizing the result of **(A)** exchanging forelimb and hindlimb blastemas with simultaneous distal to proximal or proximal to distal shift, and **(B)** transposing a blastema from a right limb to the left limb stump with reversal of the AP axis. The blastema always develops in accord with its PD level and with the handedness of origin. White = graft, blue = host. a = anterior, p = posterior. RFL = right forelimb, LFL = left forelimb, RHL = right hindlimb, LHL = left hindlimb. Numbers 1 and 4 refer to the most anterior and posterior digits of the forelimb. After a distal to proximal graft, the intermediate structures (stippled) are filled in by blastema cells derived from the host stump level, but after a proximal to distal graft, there is no filling in of structures between the level of the graft and stump. When anterior and posterior tissues of graft and host are confronted after axial reversal, supernumerary limbs form, the cells of which are derived in variable percentages from graft and host.

distal blastema is grafted to a more proximal level, it differentiates according to its level of origin, not its new level. Likewise, when the AP or DV axis of the blastema is reversed with respect to the stump tissues by grafting the blastema to the contralateral limb stump, the blastema develops with the handedness of origin. In both cases, however, the juxtaposition of distal to proximal, anterior to posterior, or dorsal to ventral tissues stimulates intercalary regeneration between the confronted positions (FIGURES 14.13, 14.14). The cells of the stump dedifferentiate and fill in the structures between these positions (Stocum, 1975; Iten and Bryant, 1975; Pescitelli and Stocum, 1980). In the case of a distal to proximal blastema graft, the result is PD intercalation to restore a normal limb. Proximal blastemas grafted to a more distal level develop according to origin, producing a limb with serially repeated structures, but they do not elicit intercalary regeneration, showing that there is a preferred distal to proximal polarity to the recognition of a structural discontinuity and intercalary regeneration in the PD axis (Stocum and Melton, 1977). When the AP or DV axis of the blastema is reversed with respect to the adjacent tissues, confronting anterior and posterior, or dorsal and ventral half-circumferences of cells, supernumerary limbs are generated at the sites of confrontation by radial intercalary regeneration within the complete limb circumferences thus created, followed by proximodistal outgrowth (Carlson, 1974, 1975a, b; Bryant and Iten,

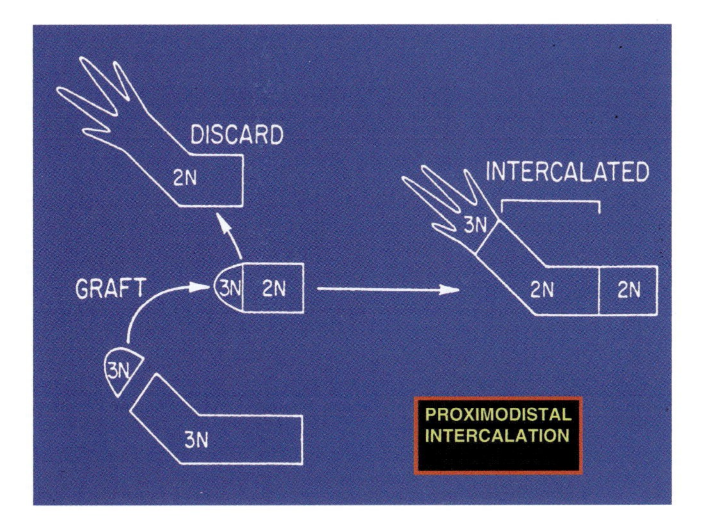


**FIGURE 14.13** PD intercalation after grafting a triploid (3N) wrist level blastema to the mid-humerus level of a diploid (2N) animal. The intercalated intermediate structures are composed of 2N cells, whereas the graft is 3N.

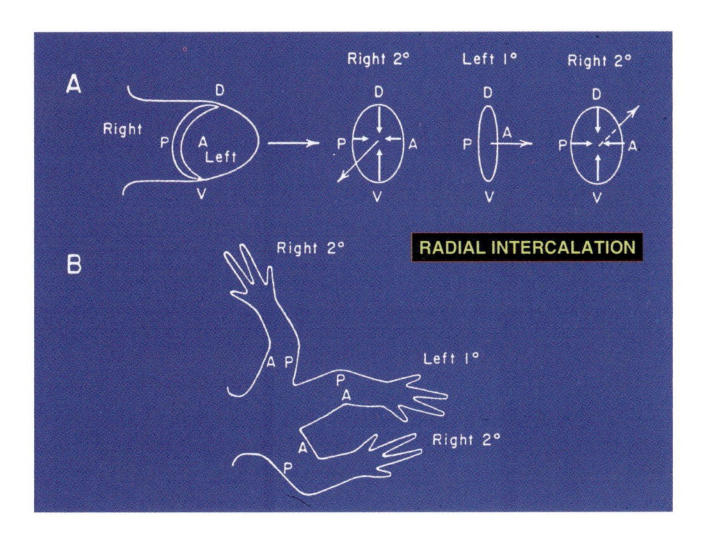


**FIGURE 14.14** Details of radial intercalation and supernumerary limb formation after transposing a left-handed blastema to a right-handed limb stump at the level of the proximal humerus, confronting anterior and posterior tissues on both sides of the grafted limb. **(A)** The operation justaposes two half limb circumferences to form a full circumference. The missing structures within the circumference are then radially intercalated to form the bases of two supernumerary blastemas. These grow out and develop with the handedness of the stump, while the primary blastema (graft) develops with the handedness of origin. **(B)** Visualization of the primary and supernumerary regenerants from the dorsal side of the animal.

1976; Cameron and Fallon, 1977; Tank, 1978; Holder and Tank, 1979; Stocum, 1980b). Both host and graft tissue contribute to the supernumerary limb to varying degrees (Stocum, 1982).

Supernumerary regenerants can also be created after rotating the dermis 180° around the PD axis, or after cross-transplantation of extensor and flexor muscles, but not by rotation or axial displacement of skeletal elements, nerves, or epidermis (Carlson, 1974, 1975a, b), showing that if the latter tissues carry positional identity, it is not asymmetrical enough to evoke intercalary regeneration. Supernumerary regenerants do not form after cross-transplanting muscles or rotating dermis unless the limb is amputated and blastema cells are formed (Carlson, 1974, 1975a). Positional identity is stable, since supernumeraries can be evoked by amputation two years after muscle cross-transplantation (Carlson, 1975b).

Collectively, the outcomes of spatially rearranging stump tissues or regeneration blastemas show that limb cells have positional identities that are remembered during the process of regeneration. The fibroblasts of the limb carry enough information for blastema self-organization. In axolotls, unirradiated triploid skin grafted to irradiated diploid limbs gave rise to a normally patterned skeleton and skin after amputation through the graft (Namenwirth, 1974). Irradiated limbs that received grafts of unirradiated tail skin, or irradiated tails that received grafts of unirradiated limb skin regenerated tails and limbs, respectively, demonstrating that dermal fibroblasts carry appendage-specific pattern information (Trampusch, 1958a, b). Freeze-thawing or X-irradiation of axolotl limb dermis abolishes its ability to elicit supernumerary regenerants after axial displacement, indicating that positional identity is carried by the cells themselves, not their ECM (Tank, 1981).

### b. A Boundary Model of Self-Organization Based on Intercalary Regeneration

The intercalary regeneration of the structures that normally lie between two disparate positions when they are artificially confronted suggests that the blastema self-organizes within a set of boundary positional identities that outline what is to be regenerated (Stocum, 1978, Nye et al., 2003; Stocum, 2004). FIGURE 14.15 illustrates a boundary model of regeneration for the PD axis of the blastema and in three dimensions. The limb can be viewed as a three-dimensional field of positional identities that constitute a "normal neighbor" map (Mittenthal, 1981). During initial dedifferentiation, before there is even a visible blastema, the essential outline of what is to be regenerated, consisting of positional identities representing circumferential, proximal, and distal boundaries, is present at the amputation surface. These boundaries are used as reference points to restore a full map within them by intercalary regeneration. The progeny of the blastema cells proliferated within the boundaries can intercalate PD positional identities distal to the amputation plane only (*rule of distal transformation*) and in any direction along the radii within the circumferential boundaries *(rule of*

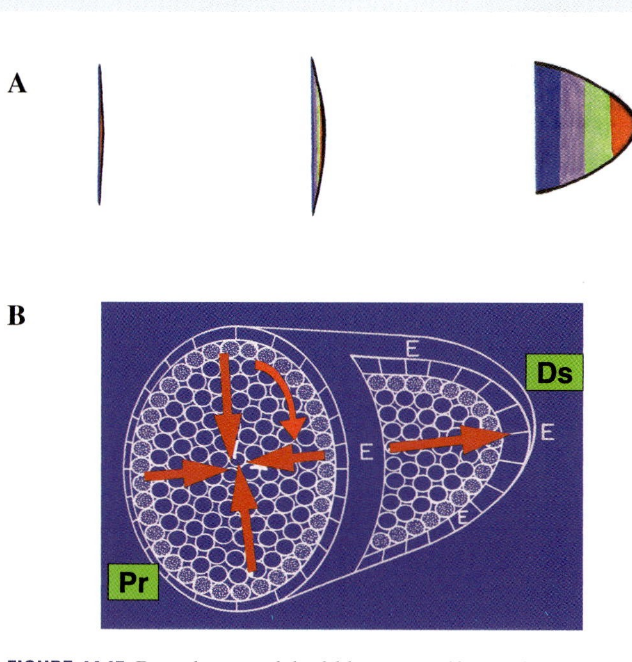


**FIGURE 14.15** Boundary model of blastema self-organization after amputation through the proximal humerus. **(A)** The PD axis viewed as a longitudinal section. **Left**, the initial population of blastema cells represents two levels, the level of amputation (blue) and the hand structures (red). These are the proximal and distal boundaries, respectively, of what is to be regenerated. **Middle**, intercalation of positional identities representing the radius/ulna and wrist between the proximal and distal boundaries. **Right**, the cone (medium bud) stage of regeneration with regenerating segments expanded (green = wrist, purple = radius/ulna). **(B)** Boundary model in three dimensions. Ds = distal, Pr = proximal, E = wound epidermis. Arrows at the base represent centripetal radial intercalation. Curved arrow represents circumferential intercalation. Arrow from Pr to Ds represents proximodistal intercalation.

*radial intercalation)*, or in any direction within any gap in the circumference (*rule of circumferential intercalation)*. Local, short-range interactions between cells with nonneighbor positional identities drive the intercalation of missing positional identities until a complete map is reconstituted (French et al., 1976; Maden, 1977; Mittenthal, 1981; Stocum, 1978a, 1980b, 1996, 2004). As the map is restored, the system becomes stable and differentiation in the spatial pattern specified by the map takes place.

The interactions between any two cells with nonneighboring positional identities follow the *rule of least (shortest) intercalation*, which states that given a choice of intercalating the greater or lesser number of positional identities that will result in a normal neighbor map, cells always choose the lesser number, or shorter route (French et al., 1976). The rule of least intercalation appears to be a biological analog of the principles of least action and least time in physics. The principle of least action states that an object going from one point to another takes the path over which the average kinetic energy minus the potential energy is least (Feynman et al., 1977a). The principle of least time (Fermat's principle) states that, out of all the possible paths light might take to go from point A to point B, the path taken is the one requiring the shortest time (Feynman et al., 1977b). These "least principles" appear to a feature of physical systems on many, if not all, levels of organization.

In the boundary model, the proximal boundary and the circumferential boundary positional identities are those of the PD level of amputation. However, we do not know the mechanism by which the distal boundary is established. Genes associated with both proximal and distal structures are expressed very early in blastema formation. *HoxA9* expression is associated with proximal structures, and *HoxA13* expression is associated with distal structures in regenerating axolotl limbs (Gardiner et al., 1995). *HoxA9* and *HoxA13* are both expressed from one day postamputation until redifferentiation begins in regenerating axolotl limbs. Each is expressed uniformly throughout the blastema until early bud. Thereafter, while *HoxA9* continues to be expressed throughout the bud, the expression of *HoxA13* becomes restricted to the prospective autopodium, suggesting that this gene is involved in the patterning of autopodial structures. *HoxA13* expression in distal blastemas is 30% higher at the medium bud stage than in proximal blastemas.

Two possible mechanisms for establishing the distal boundary suggest themselves (**FIGURE 14.16**). The first is that, as the blastema begins to form, some dedifferentiated cells assume a pattern of gene activity that defines the proximal boundary (*HoxA9*), whereas others randomly express *Hox A13* in addition, a pattern that characterizes the distal boundary. The distal boundary cells then sort out from the proximal boundary cells by virtue of their having a greater adhesive affinity for themselves and for the AEC than for proximal boundary cells. The second possible mechanism is that the pattern of distal gene expression (*HoxA9* + *HoxA13*) is initially expressed in every dedifferentiated cell. As the blastema grows, however, distal gene activity is maintained only in those blastema cells in contact with the wound epidermis, perhaps in response to FGF signals. No sorting is involved in this mechanism. In either case, distal cells would be confronted with proximal cells, evoking intercalary regeneration of intermediate positional identities. In addition to *Hoxa9* and *13*, there are undoubtedly other distal boundary genes that need to be identified.

This model allows for the very early specification of all the amputated segments of the limb. For example, if the forelimb were amputated at the level of the body

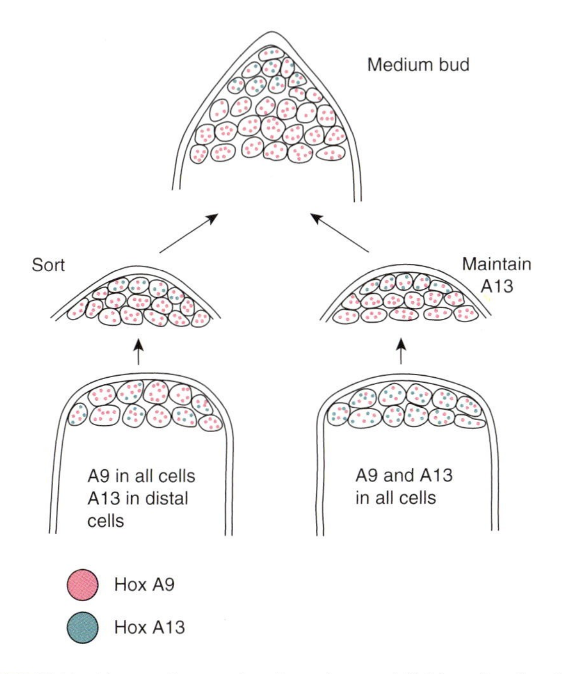


**FIGURE 14.16** Alternative mechanisms for establishing the distal and proximal boundaries. **Left**, the initial blastema cells adopt different patterns of gene activity that represent proximal (pink = Hox A9) and distal (blue = Hox A13). The distal cells have a higher affinity for the wound epidermis than the proximal cells and they sort out from one another. **Right**, all of the initial blastema cells express both proximal and distal genes, but only those cells that are in contact with the wound epidermis maintain expression of the distal boundary gene, Hox A13.

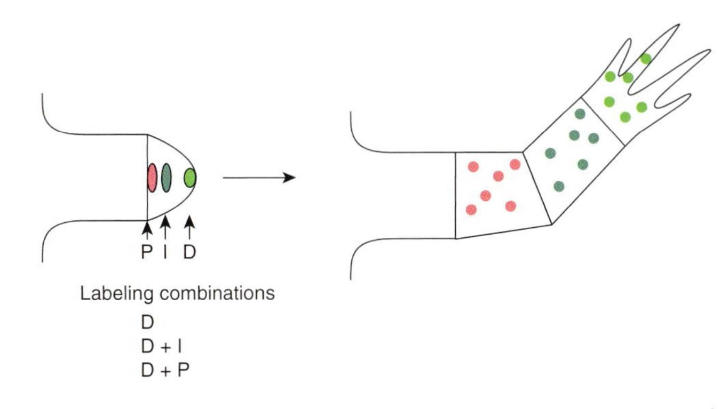


**FIGURE 14.17** Experiment showing the presence at the cone stage of regeneration of cells that will form the different parts of the regenerant. Amputation level is the mid-humerus. Binary combinations of distal (D, green), intermediate (I, blue), and proximal (P, pink) blastema cells were marked with dyes. The marked cells contributed to the regenerant according to their position in the blastema.

wall, removing the humerus, radius, ulna, and hand, positional identities for each of these segments could be intercalated very rapidly between proximal and distal boundary cells, even prior to the accumulation blastema stage. Intercalation of intrasegmental positional identities would follow and, along with proliferation, would expand each of the segments.

Mapping studies on both the embryonic chick limb bud and urodele limb regeneration blastema are consistent with a boundary model of self-organization. Experiments labeling cells at different distances from the AER with $^3$H-thymidine or cytoplasmic dyes showed that cells representing all three limb segments are present in the very early chick wing bud (Stark and Searls, 1973; Dudley et al., 2002). Echeverri and Tanaka (2005) marked distal, intermediate, and proximal cells of early blastemas derived from the upper arm with GFP or DsRed plasmids (**FIGURE 14.17**). The distal cells formed the hand, the intermediate cells the radius and ulna, and the proximal cells the humerus. Furthermore, marked cells derived from an early wrist blastema grafted into the base of an early humeral level blastema sorted out to participate in construction of the hand. This result, as well as the fact that wrist and elbow blastemas grafted to the base of a hindlimb blastema regenerating from mid-femur sort distally to articulate with the ankle and knee of the host hindlimb (Crawford and Stocum, 1988a; see ahead), suggest that the distal and proximal boundary qualities might be expressed in separate cells of the early blastema, which then sort out to juxtapose the boundaries.

### c. The Use of Retinoids to Investigate the Physical Basis of Positional Identity

Retinoic acid (RA, the acid derivative of vitamin A) administered systemically during the stage of initial blastema formation proximalizes, posteriorizes, and ventralizes the positional identities of blastema cells in a concentration-dependent way (Niazi and Saxena, 1978; Maden, 1982, 1997; Thoms and Stocum, 1984; Niazi et al., 1985; Kim and Stocum, 1986; Ludolph et al., 1990, 1993; Stocum and Mitashov, 1990; Stocum and Maden, 1990; Stocum, 1991; Monkmeyer et al., 1992; Niazi, 1996). At the maximum concentration of RA, the cells of a wrist-level blastema are proximalized to the level of the shoulder and a whole limb, complete with shoulder girdle, regenerates from the wrist (**FIGURE 14.18**). Three major RAR isoforms, α, β, and δ, have been identified in the regenerating newt limb (Maden, 1997; Hill et al., 1993; Ragsdale et al., 1989, 1992a, b). The α and β isoforms are expressed uniformly throughout the blastema (Giguere et al., 1989). Delta has two isoforms, of which δ2 has been identified by transfection experiments as the receptor that mediates proximalization (Pecorino et al., 1996) (**FIGURE 14.18**). It is not known whether δ2 also mediates posteriorization

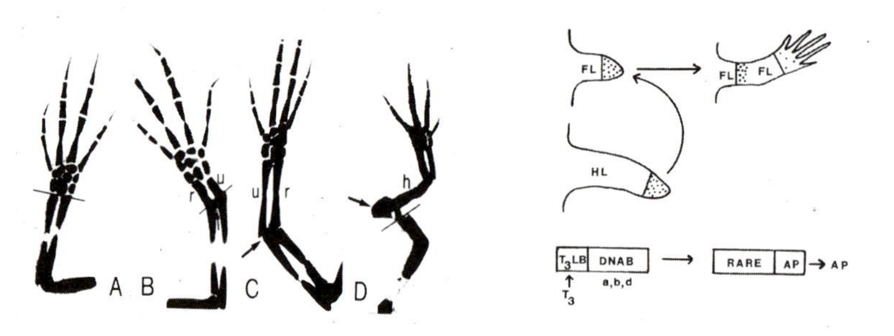


**FIGURE 14.18** Retinoic acid proximalizes positional identity. **Left**, effect of increasing doses of RA on axolotl limbs amputated through the wrist joint. Line indicates amputation plane, except for C, where it is indicated by the arrow. **A** = vehicle control, **B** = duplication of distal radius/ulna at 50µg/g body wt, **C** = duplication of a complete radius/ulna at 75µg/g body wt, **D** = duplication of a complete limb at 100µg/g body wt. Arrow in **D** indicates shoulder girdle. h = humerus, r = radius, u = ulna. **Right**, Experimental strategy to identify the retinoic acid receptor that mediates proximalization by RA in the adult newt. Chimeric receptors were made that consisted of the ligand binding domain of the T3 receptor ($T_3$LB) and the DNA binding domain (DNAB) of the α, β or δ RA receptors (a,b,d). These constructs, along with a reporter construct consisting of the alkaline phosphatase gene and a retinoic acid response element (RARE), was biolistically incorporated into a few cells of a hindlimb (HL) blastema (stipple) derived from the tarsus level. This blastema was grafted to the mid humerus level of a forelimb, and the reporter construct was activated by injecting thyroxine. The grafted blastema developed into a foot and intermediate forelimb structures were intercalated. Some of the modified cells remained as part of the hand, but others were proximalized, as indicated by the fact that they sorted out to a more proximal level as intercalary regeneration took place. Only the δ2 receptor was able to proximalize the cells.

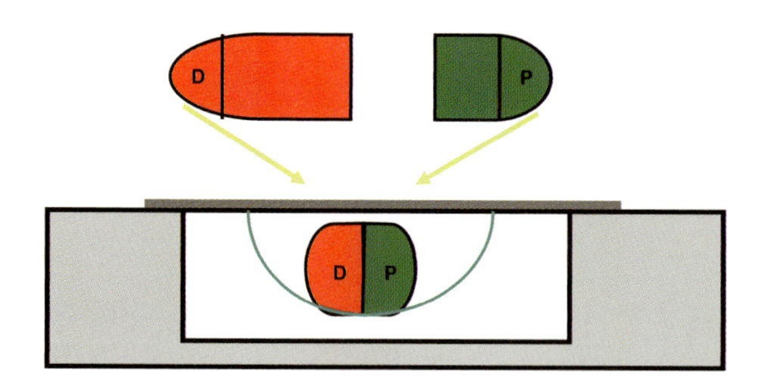


**FIGURE 14.19** Experimental strategy to test the adhesive behavior of blastemas derived from different PD levels. Distal (D, red) and proximal (P, green) newt limb blastemas were pressed together in culture medium and cultured in a drop of culture medium hanging from the underside of a glass coverslip.

and ventralization by RA, or whether these effects are mediated by other receptors.

Assays *in vitro* and *in vivo* showed that positional identity is encoded in the surface of blastema cells and that RA alters surface properties. In the *in vitro* assay, proximal and distal blastemas were pressed together at their bases and cultured in hanging drops (Nardi and Stocum, 1983) (**FIGURE 14.19**). The proximal blastemas surrounded the distal ones, a result that conforms to the differential adhesivity hypothesis of Steinberg (1978), suggesting that blastemas derived from distal levels are more adhesive than those derived from proximal levels (**FIGURE 14.20**). In the *in vivo* "affinophoresis" assay, axolotl blastemas derived from the wrist or elbow levels of the forelimb were grafted to the blastema-stump junction of a hindlimb host regenerating from the mid-thigh (Crawford and Stocum, 1988a; Egar, 1993). The grafted blastemas were displaced distally, i.e., sorted to their corresponding level on the host regenerate (ankle and knee, respectively), where they differentiated (**FIGURE 14.21**). Retinoic acid, in combination with the affinophoresis assay or an intercalation assay, was used to establish a direct correlation between positional identity and blastema cell affinity (Crawford and Stocum, 1988a, b). When the affinophoresis assay was performed using donor wrist and elbow blastemas proximalized by RA, distal displacement was abolished (Crawford and Stocum, 1988a). Similarly, grafting a RA-treated wrist-derived blastema to a more proximal limb stump abolished the intercalary regeneration that would normally have taken place between distal and proximal levels (Crawford and Stocum, 1988b).

These results argue that positional memory is a property of the cell surface and that RA alters gene activity affecting the qualitative and/or quantitative molecular composition of the cell surface. Consistent with this idea is the fact that tunicamycin, which inhibits biosynthesis of the oligosaccharide component of N-linked glycoproteins on the cell surface, blocks the proximalization of positional identity by RA (Johnson and Scadding, 1992). Furthermore, proximal and distal cells of chick limb buds sort out from one another *in vitro* (Wada et al., 1993; Wada and Ide, 1994; Ide et al., 1994, 1998; Tamura et al., 1997). This sorting is inhibited by antibodies to the EphA4 receptor and to N-cadherin, suggesting that the ephrin A family of ligands and receptor, as well as N-cadherin, play a role in effecting PD positional identity (Wada et al., 1998; Yajima et al., 1999). The ephrin A ligands are anchored

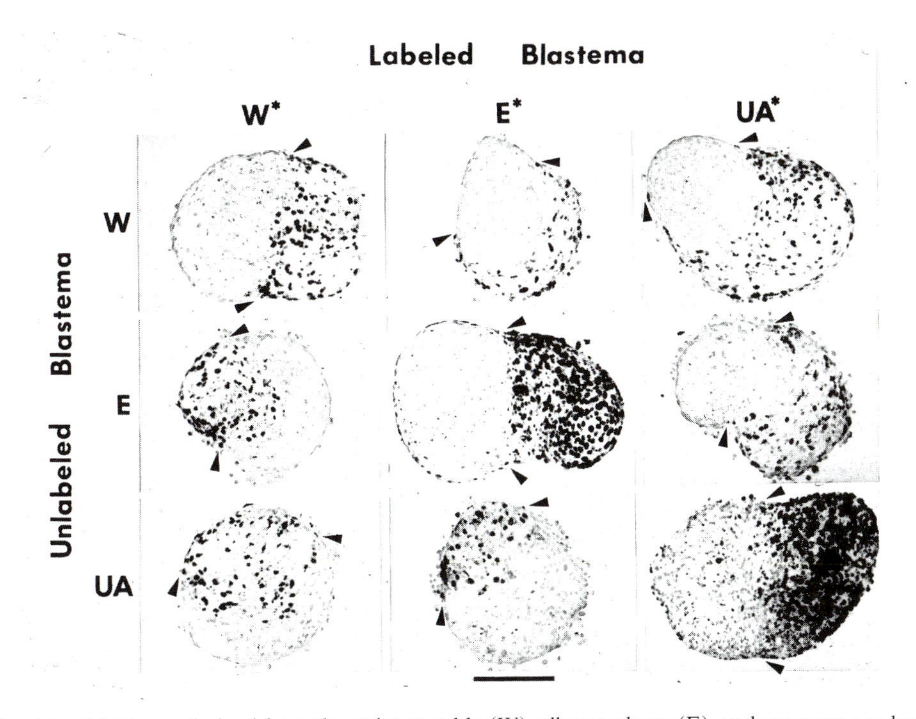


**FIGURE 14.20** Results of pairing blastemas derived from the wrist or ankle (W), elbow or knee (E), and upper arm or leg (UA) in each of nine possible binary pairs and culturing them in hanging drops. Asterisk denotes radiolabeled blastema. E*-E, UA*-E, UA*-UA and UA*-W are examples of cultured blastemas from hindlimbs. Dark spots lying within the unlabeled cell populations are either pigment granules (in the case of E*-E and UA*-UA), necrotic cells, or darkly staining peripheral nuclei (E*-E). Same level blastemas fused in a straight line, whereas in different-level pairs, the proximal mesenchyme tended to engulf the distal one. Bar = 0.5 mm. Reproduced with permission from Nardi and Stocum. Surface properties of regenerating limb cells: Evidence for gradation along the proximodistal axis. Differentiation 25:27–31. Copyright 1983, Blackwell Pub.

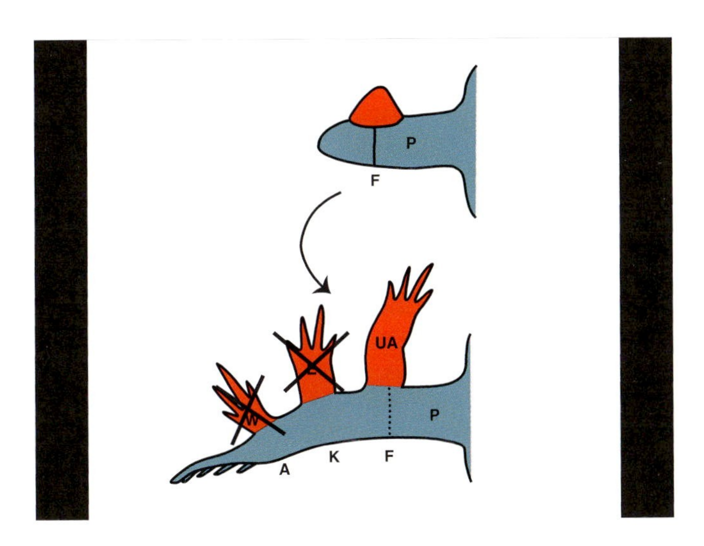


**FIGURE 14.21** *In vivo* affinophoresis assay. Conical blastemas derived from the wrist (W), elbow (E), or mid-upper arm (UA) levels of the axolotl forelimb were grafted to the blastema-stump junction (line and dotted line) on the dorsal side of a hindlimb regenerating from the mid-femur (F). P = posterior. Untreated forelimb blastemas sorted distally to their corresponding level (ankle = A, knee = K and mid-femur, F) along the PD axis of the hindlimb regenerant and developed according to origin. Retinoic acid treatment abolished the distal sorting behavior of the wrist and elbow blastemas (X).

to the cell surface by glycosylphosphatidylinositol (GPI). Treatment of chick limb buds with bacterial phosphatidylinositol-specific phospholipase C prevents the sorting of proximal and distal chick limb bud cells (Wada et al., 1998).

Ephrins and N-cadherin have not been analyzed with regard to their potential involvement in PD differential adhesivity of blastema cells in regenerating limbs. A screen of cDNA libraries from normal and RA-treated newt wrist-level blastemas has, however, identified a gene that is differentially regulated by RA (Morais da Silva et al., 2002). This gene, called proximodistal-1 (Prod1), encodes the homologue of the mammalian complement molecule, CD59, a GPI-anchored cell surface molecule. In the differential adhesion assay *in vitro*, treatment with antibodies to Prod1 or with PIPLC prevented proximal blastemas from surrounding distal ones, indicating that Prod1 plays a role in maintaining the gradient of PD adhesivity (**FIGURE 14.22**). Overexpression of Prod1 causes distal blastema cells to translocate to a more proximal position when transplanted into proximal blastemas (Echeverri and Tanaka, 2005), suggesting that Prod1 plays a major role in the expression of the adhesivity gradient exhibited by blastemas from different PD levels. Collectively, the data suggest that at the molecular level, blastemal

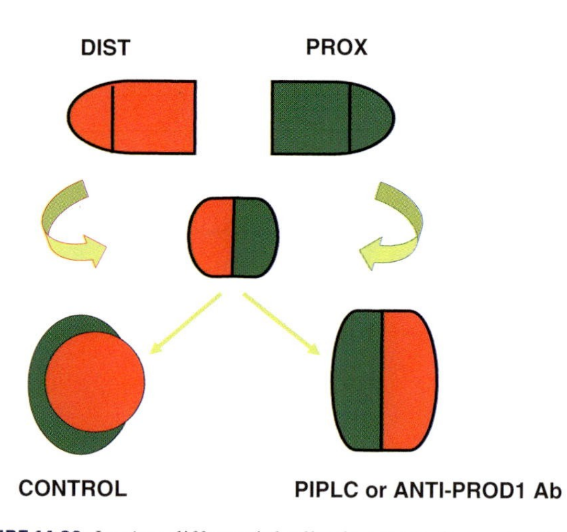


**FIGURE 14.22** *In vitro* differential adhesion assay demonstrating that Prod1 (CD59) is one of the surface molecules that determines the level of adhesivity associated with positional identity in regenerating newt limbs. Distal (green) blastemas engulf proximal (red) blastemas in control cultures. Treatment of the blastemas with PIPLC, which cleaves Prod1 from the surface, or with an antibody to Prod1, abolishes the engulfment behavior.

self-organization is orchestrated by short-range interactions between neighboring cells.

### d. Signaling Molecules and Transcription Factors Involved in Pattern Specification

The outgrowth and patterning of the embryonic limb bud is regulated by three interdependent signaling centers. The AER controls PD outgrowth of the limb bud and PD patterning, the zone of polarizing activity (ZPA) on the posterior edge of the mesenchyme controls AP patterning, and the dorsal ectoderm controls DV patterning (Tickle, 2003; Niswander, 2003). A variety of FGFs is required for PD outgrowth, and PD patterning is controlled by the spatial expression patterns of *HoxA-D* and *Meis 1* and *2* genes. The AP pattern of the limb bud seems to be determined even before the bud appears, except for the digit pattern, which is directed by *shh* and *BMP* genes expressed in the ZPA. *Wnt 7a* and *Lmx-1* are expressed in the dorsal mesenchyme, but are repressed in the ventral mesenchyme by *En-1*, to create DV pattern.

Patterning of the amphibian limb regeneration blastema appears to be regulated by many of the same transcription factors. In regenerating axolotl and newt limbs, *HoxD*, *HoxA*, and *Meis* genes are involved in PD patterning (Simon and Tabin, 1993; Gardiner and Bryant, 1996; Mercader et al., 2005). *Hox D10* is expressed at a 2–3-fold higher level in proximal versus distal blastemas (Simon and Tabin, 1993). *HoxA13* expression in distal blastemas is 30% higher at the medium bud stage than in proximal blastemas. Conversely, proximalization of distal blastemas by RA reduces the expression of *HoxA13* by 55% (Gardiner et al., 1995). *Meis 1* and *2* expression is preferentially associated with the stylopodial region of the regeneration blastema after amputation through the upper arm or leg (Mercader et al., 2005) (**FIGURE 14.23**). The expression of *HoxD10* and *Meis 1* and *2* is upregulated in distal blastemas proximalized by RA (Simon and Tabin, 1993; Mercader et al., 2005). Distal blastema cells are proximalized by overexpression of *Meis 2* and relocate more proximally (Mercader et al., 2005). Inhibition of *Meis* genes in one forelimb of a bilaterally amputated pair by microinjection of morpholinos into the blastema and then treatment of the animal with RA results in proximalization of the uninjected blastema to the level of the shoulder girdle, but allows the injected blastema to proximalize only to the level of the radius/ulna (Mercader et al., 2005). These observations indicate that *HoxD10* and *Meis* 2 are upregulated targets of RA proximalizing activity and involved in specifying the stylopodium, while *HoxA13* is a downregulated target of RA proximalizing activity that is involved in specifying the autopodium. These transcription factors may be the ones that directly or indirectly activate genes that control the degree of expression of surface molecules such as Prod1, ephrins, and N-cadherin that encode positional identity. As the blastema grows, FGFs from the AEC and the blastema cells may maintain proliferation of the cells until they are of sufficient mass to begin differentiation.

*Shh* plays a role in establishing the AP pattern of digits in the regeneration blastema (**FIGURE 14.24**). *Shh* is expressed on the posterior margin of medium bud stages of regenerating newt, axolotl, and *Xenopus* tadpole limbs and moves distally with outgrowth, becomes restricted to the posterodistal region as digits begin to condense, then ceases. Expression of *shh* is lost in amputated *Xenopus* hindlimbs as the tadpoles approach metamorphosis, coincident with the loss of regenerative ability in this frog (Imokawa and Yoshizato, 1997, 1998; Endo et al., 1997; Torok et al., 1999). When transfected into the anterior cells of axolotl blastemas, *shh* evokes digit duplications (Roy et al., 2000). Two patches of *Shh* expression are observed after reversing the AP axis grafting the blastema to the contralateral limb stump. One is the original patch expressed by the posterior tissue of the graft. The other is a new patch that arises in the posterior tissue of the host limb where it confronts anterior tissue of the graft (Imokawa and Yoshizato, 1997, 1998). A similar expression pattern is observed when the anteroposterior axis of young (stage 53) *Xenopus* limb bud tips is reversed (Endo et al.,

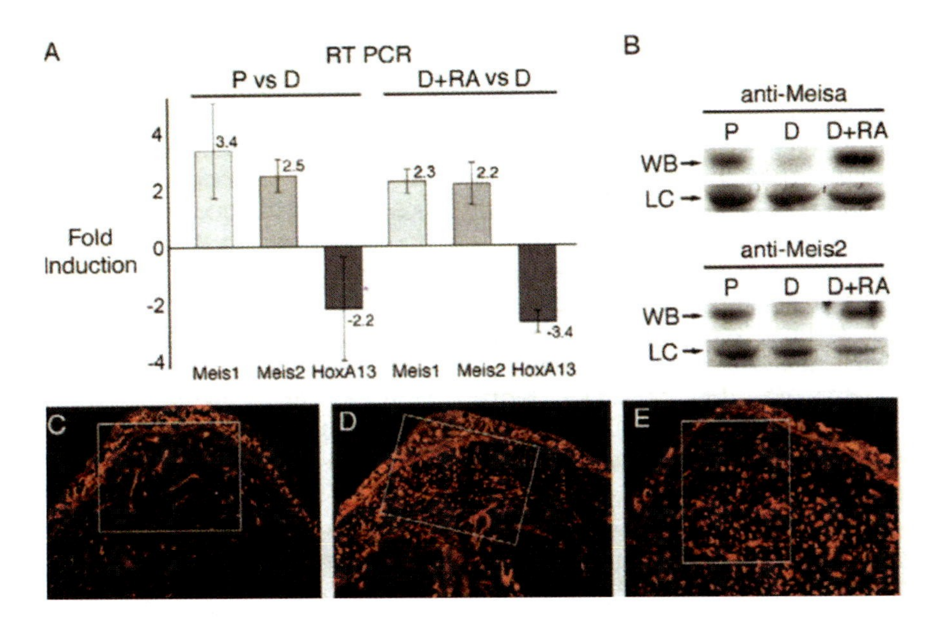


**FIGURE 14.23** Meis 1 and 2 expression in proximal (P) and distal (D) blastemas and its regulation by retinoic acid (RA). **(A)** Up-regulation of Meis 1 and 2 and down-regulation of HoxA13, in proximal vs distal blastemas and distal blastemas proximalized by RA vs. distal blastemas. **(B)** Western blots (WB) of proximal and distal blastemas and distal blastemas treated with RA (D + RA) using an antibody to Meis proteins (anti-Meisa) in general or to Meis 2 specifically (anti-Meis2). The anti-Meisa proteins were 4-fold more abundant in proximal vs distal blastemas and RA increased them by 6-fold in distal blastemas compared to proximal blastemas; anti-Meis2 protein was 3-fold more abundant in proximal vs distal blastemas and RA increased this protein 4.5 fold in distal blastemas compared to proximal blastemas. LC = loading control. **(C)** Distal blastema 4d post-amputation immunostained with Meis-a (red). **(D)** Distal blastema 3d after RA injection. **(E)** Distal blastema 6d after RA injection. Note lack of expression in **C** and increasing expression in **D** and **E**. Reproduced with permission from Mercader et al., Proximodistal identity during vertebrate limb regeneration is regulated by Meis homeodomain proteins. Development 132: 4131–4142. Copyright 2005, The Company of Biologists.

1997). The relatively late expression of the *shh* gene (at medium bud) suggests that it is not essential for AP patterning of most of the limb segments, but rather is involved in specification of digit number, as in amniote limb buds. Consistent with this idea, BMPs have the same effect of posteriorizing digit identity in *Xenopus* limb buds and regenerating axolotl limbs (Nye et al., unpublished results). Another gene that is differentially expressed along the AP axis of the developing and regenerating *Xenopus* limb is *XlSALL4*, a member of the *spalt* family that has human homologues important for AP limb development (Neff et al., 2005).

A hedgehog gene expressed in large areas of the early newt limb regeneration blastema is *Notophthalmus viridescens* banded hedgehog (*Nv-Bhh*) (Stark et al., 1998). *Bhh* is the equivalent of Indian hedgehog (*Ihh*) in other vertebrates (Zardoya et al., 1996). Ihh is known to be essential in signaling mechanisms that control cartilage proliferation, hypertrophic differentiation, and ossification in developing limbs (Vortkamp, 2001), but its very early appearance during regeneration implies that it has other, as yet unknown, roles.

Investigations of gene expression associated with DV axial polarity have so far been carried out only in

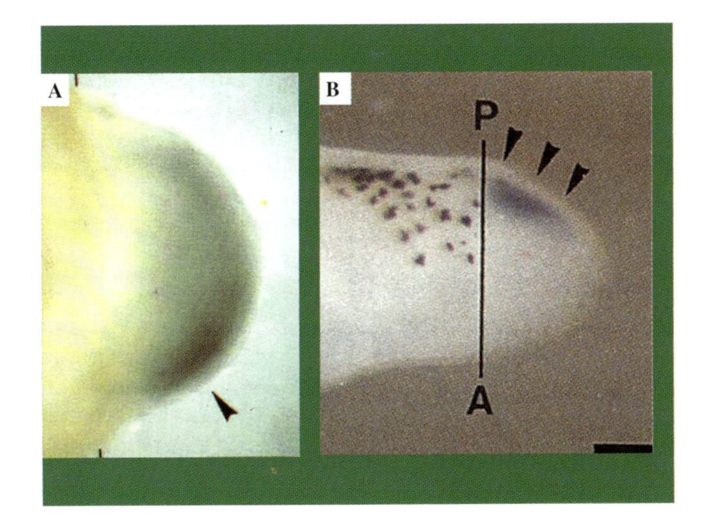


**FIGURE 14.24** Sonic hedgehog gene expression on the posterior edge of medium bud blastemas of **(A)** newt and **(B)** early *Xenopus* tadpoles. **(A)** reproduced with permission from Imokawa and Yoshizato, Expression of *Sonic hedgehog* gene in regenerating newt limb blastemas recapitulates that in developing limb buds. Proc Natl Acad Sci USA 94:9159–9164. Copyright 1997, National Academy of Science, USA. **(B)** Reproduced with permission from Endo et al., *Shh* expression in developing and regenerating limb buds of *Xenopus laevis*. Dev Dynam 209:227–232. Copyright 1997, Wiley-Liss.

regenerating *Xenopus* limbs. *Lmx-1* is expressed in the dorsal mesenchyme of early tadpole limb buds (stages 51–53) and in regeneration blastemas formed after amputation through the zeugopodium at these stages, but is not expressed in blastemas of stage 55 limbs, which regenerate poorly (Matsuda et al., 2001). The normal DV epidermal/mesodermal spatial relationship was altered by removing the epidermis from the zeugopodial mesoderm and reversing the DV axis of the mesoderm with respect to the limb stump (**FIGURE 14.25**). Fresh epidermis of normal DV polarity grew over the reversed segment, which was then allowed to regenerate. When this operation was performed at stage 52, *Lmx-1* expression and structural pattern of the regenerate conformed to the epidermal DV polarity, but when the operation was performed at stage 55, the original pattern of *Lmx-1* expression and DV polarity of the segment were maintained in the blastema (Matsuda et al., 2001). These results suggest that self-organization of DV polarity in the limb regeneration blastema is specified by the effect of the wound epidermis on the expression of *Lmx-1* up through stage 52 or 53; after that the polarity is fixed. In embryonic limb buds, *Wnt7a* and *engrailed* (*en1*) expression in the dorsal and ventral ectoderm, respectively, determine that *Lmx-1* will be expressed in the dorsal mesoderm only. The expression patterns of *Wnt-7a* and *En-1* in the regeneration blastema have not yet been reported.

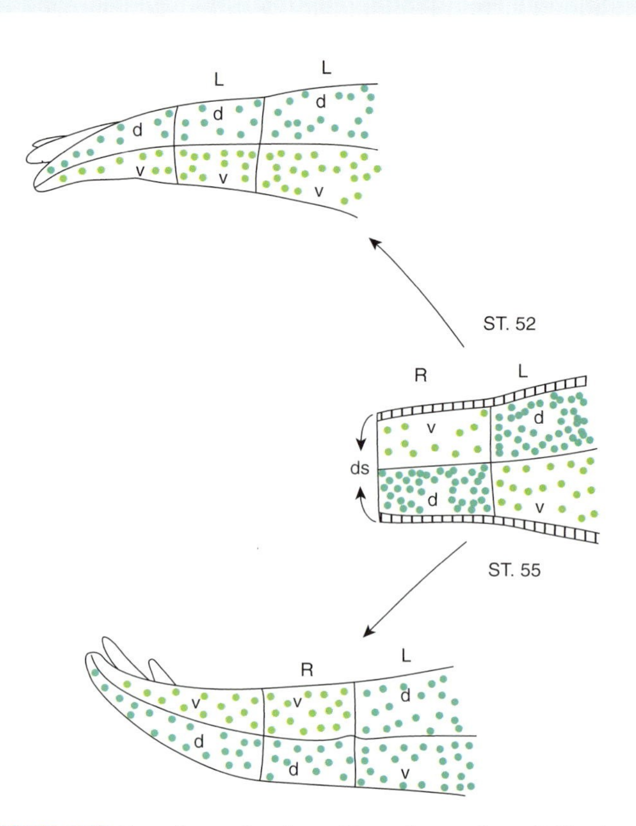


**FIGURE 14.25** Experiment showing epidermal control over DV polarity of regenerating limbs in *Xenopus* at early tadpole stages when the limb is still under development. The tibia-fibula mesodermal tissue of right (R) stage 52 or stage 55 limbs was stripped of epidermis and grafted in place of the tibia-fibula mesoderm of the left limb (L) with reversal of the DV axis. ds = distal, d = dorsal (blue), v = ventral (green). The host epidermis migrates over the grafted tissue (arrows). When the operation is done at stage 52, the mesoderm and regenerant conform to the polarity of the host epidermis, but the polarity cannot be reversed at stage 55.

## 7. Stimulation of Frog Limb Regeneration

The hindlimb buds of anuran tadpoles regenerate well during early stages of their differentiation, but lose the capacity for regeneration in a proximal to distal sequence in concert with progressive differentiation of the limb segments (Guyenot, 1927; Dent, 1962). The increasing deficiency of anuran limb regeneration with advancing developmental stage is correlated with changes in the cellular features of the blastema (Korneluk and Liversage, 1984; Wolfe et al., 2000). Studies on regeneration-deficient *Xenopus* limbs show that histolysis is minimal, that the morphology of the blastema is fibroblastic rather than mesenchymal, and that the amount of basement membrane and dermal tissue under the wound epidermis is increased, AEC thickness declines, and there is decreased ability to support the sprouting of nerves and blood vessels into the blastema (Wolfe et al., 2000). The fibroblastic cells proliferate and then differentiate into several cartilage nodules that fuse to form a symmetrical cartilage spike of varying length that is surrounded by connective tissue, but no muscle (**FIGURE 14.26**). The limbs of other frogs are unable to regenerate even a cartilage spike by late tadpole stages (Stocum, 1995).

The shift from regeneration-competence to deficiency in *Xenopus* limbs from early to late tadpole stages is not associated with changes in innervation requirements or loss of the ability of the wound epidermis to form an AEC. There is some evidence that it is the result of intrinsic cellular changes in the limbs, as shown by the fact that transplantation of regeneration-deficient limb buds to regeneration-competent limb buds, and vice versa, does not alter the regenerative capacity of the donor limb bud (Sessions and Bryant, 1988; Filoni et al., 1991). However, other evidence suggests that systemic changes associated with maturation of the immune system are responsible for the loss of regenerative capacity (Harty et al., 2003; Mescher and Neff, 2005). The inflammatory response in regeneration-competent early frog tadpoles and in larval and adult urodeles is nonexistent or minimal, whereas in adult *Xenopus* it is similar to that of mammals (Robert and Cohen, 1998). Antigen-presenting cells of tadpoles have only MHC class II proteins. MHC class I proteins

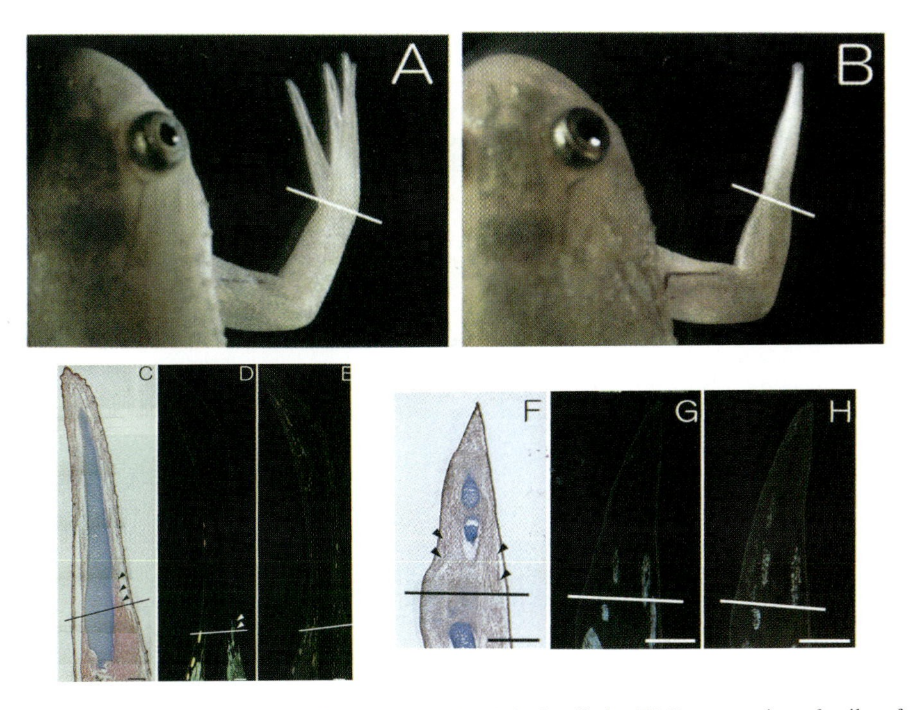


**FIGURE 14.26** Regeneration of *Xenopus* froglet limbs. **(A)** Unamputated right forelimb. **(B)** Regeneration of spike after amputation through the wrist. **(C)** Longitudinal section of spike stained with Mayer's hematoxylin, eosin and Alcian blue. The cartilage stains blue and the muscle red. Note the lack of muscle in the regenerant, except at its very base (arrowheads). Line indicates plane of amputation. **(D, E)** Spike immunostained for the muscle marker myosin heavy chain **(D)** and the satellite cell marker Pax-7 **(E)** (both green). Neither are present in the spike. Arrowheads point to a few myofibers at the base of the spike. **(F–H)** Sections of regenerants derived by amputation at the ankle level of a stage 52 hindlimb. Staining same as in **C–E**, except that **G** is stained for the muscle marker MF20. Both indicate the presence of muscle and satellite cells in the regenerant. Reproduced with permission from Satoh et al., Muscle formation in regenerating *Xenopus* froglet limb. Dev Dynam 233:337–346. Copyright 2005, Wiley-Liss.

are not expressed until stage 55 and then only at a low level until the end of metamorphosis. These differences in expression are correlated with the successive infiltration of three waves of lymphocytes, derived from different populations of precursor cells, into the developing thymus. The three waves of lymphocyte infiltration correspond to the development of the larval, premetamorphic, and froglet/adult immune systems, respectively (Du Pasquier et al., 1989; Du Pasquier and Flajnik, 1999; Rollins-Smith et al., 1997; Turpen, 1998). The differences in these three immune systems are reflected in their reaction to skin grafts. Tadpoles do not reject skin grafts from minor-histocompatability mismatches, whereas adult frogs do. Skin taken from a tadpole and cryopreserved while that same tadpole develops into an adult is rejected when grafted to the adult (Izutsu and Yoshizato, 1993). These differential responses undoubtedly involve changes in the complement of inflammatory cells and cytokines at wound sites (Harty et al., 2003; Mescher and Neff, 2005). Most likely, the adult inflammatory response prevents the tissue interactions required for regeneration in frog limbs by the early deposition of basement membrane and fibrous tissue.

Numerous attempts have been made to improve the regenerative capacity of limbs in late anuran tadpoles, froglets, or adult frogs by repeated trauma and irritation to increase histolysis, or augmentation of the nerve supply to increase blastema cell survival and proliferation (Singer, 1954; Polezhaev, 1972; Cecil and Tassava, 1986; Stocum, 1995). The general conclusion to be drawn from these experiments is that it is relatively easy to induce blastema formation in late-stage tadpoles, when regeneration competence has only recently disappeared. However, the induction of blastema formation in the limbs of froglets or adult frogs is much more difficult. Nevertheless, blastema formation followed by abnormal digit regeneration has been reported in froglets or adult frogs following treatment with retinoids (Sharma and Niazi, 1979; Niazi et al., 1979; Cecil and Tassava, 1986) (**FIGURE 14.27**). Regeneration-deficient limbs of *Xenopus* late tadpoles do not express FGF-8 and 10, HGF, or BMP4 after amputation, whereas these molecules are expressed in the amputated regeneration-competent limbs of early tadpoles (Yokoyama et al., 2000; Satoh et al., 2005a, b). These growth factors have been applied in beads or cells to regeneration-deficient limbs in attempts to improve their regeneration. Human

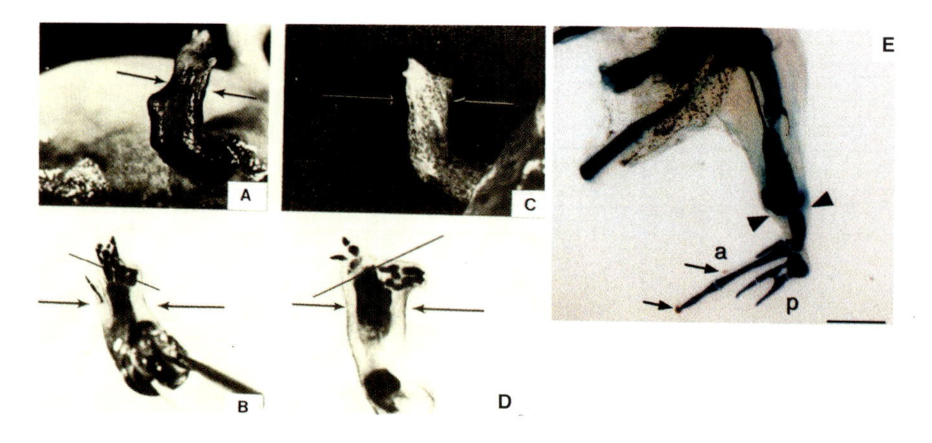


**FIGURE 14.27** Stimulation of regeneration in frogs. **(A–D)** Dimethyl sulfoxide **(A, B)** and RA-treated **(B, C)** adult *Rana catesbieana* forelimbs. Digit-like structures are visible in the *R. catesbieana* limbs grossly and after staining for cartilage in whole mounts. Arrows indicate amputation plane. Lines indicate where regenerates were removed for sectioning. **(E)** *Xenopus* froglet hindlimb treated with FGF-10 after amputation through the tibia/fibula. Whole mount stained for cartilage. Arrowheads indicate amputation plane. A partial foot has regenerated, but not more proximal structures. Arrows point to an anterior (a) digit with claw. p = posterior. **(A–D)** reproduced with permission from Cecil and Tassava, Forelimb regeneration in the postmetamorphic bullfrog: Stimulation by dimethyl sulfoxide and retinoic acid. J exp Zool 239:57–61. Copyright 1986, Wiley-Liss. **(E)** reproduced with permission from Yokoyama et al., FGF-10 stimulates limb regeneration ability in *Xenopus laevis*. Dev Biol 233:72–79. Copyright 2001, Elsevier.

FGF-8 provided only marginal improvement, but human FGF-10 restored the regeneration of digits while simultaneously inducing expression of FGF-8 in the AEC, indicating the dependence of FGF-8 expression on FGF-10 (Yokoyama et al., 2001) (**FIGURE 14.27**). However, more proximal elements were not restored in these tadpoles, nor was any attempt made to test the effect of FGF-10 on froglet or adult frog limbs.

The lack of muscle in the spike regenerates of *Xenopus* froglets is not due to a blastema environment that is nonsupportive of satellite cell differentiation, because satellite cells implanted into the blastema differentiate into muscle (Satoh et al., 2005a). Rather, the problem appears to be a lack of HGF signaling that activates migration of SCs from the stump into the blastema. The blastema lacks cells that express the SC marker Pax-7, but implanting HGF-expressing mouse P14 cells into the blastema results in the appearance of *Xenopus*-derived muscle cells in the regenerate, although the muscle appears in disorganized patches (**FIGURE 14.28**). This result suggests that muscle in the regenerating amphibian limb is derived primarily from SCs in the limb stump, although it cannot be ruled out that HGF might induce the formation of Pax-7 expressing cells from naïve blastema cells. Finally, the application of BMP4 to the regenerating cartilaginous spike induced the formation of a joint in the spike (Satoh et al., 2005b). Whether or not combinations of these proteins would be synergistic and result in further improvement of regeneration is unknown.

## 8. Comparative Analysis of Gene Activity in Regeneration-Competent versus Regeneration-Deficient Limbs

The amphibian limb is a valuable experimental model to understand how we might be able to regenerate mammalian tissues and complex structures that do not normally regenerate. Working on the assumption that mammals have more inherent ability to regenerate than they express and that the genes encoding proteins involved in regeneration are conserved across vast evolutionary distances, we can use strong regenerators such as the amphibians to define what those genes and their products are. As in fetal versus adult skin wounds, or intact versus fractured bone, molecular comparisons between regenerating and nonregenerating limbs can be made to reveal the patterns of gene activity that distinguish regeneration from scar tissue formation, not only in terms of factors that promote regeneration, but equally important, factors that inhibit regeneration and thus must be neutralized. Such comparisons can be made between the limbs of regeneration-competent versus -deficient species of amphibians (e.g., axolotl versus adult frog), between limbs of regeneration-competent and -deficient developmental stages of the same species (tadpoles versus metamorphosed frogs) or between regenerating amphibian and nonregenerating mammalian tissues (axolotl or newt versus mouse).

King et al. (2003) used suppression PCR-based subtractive hybridization to analyze the differences in

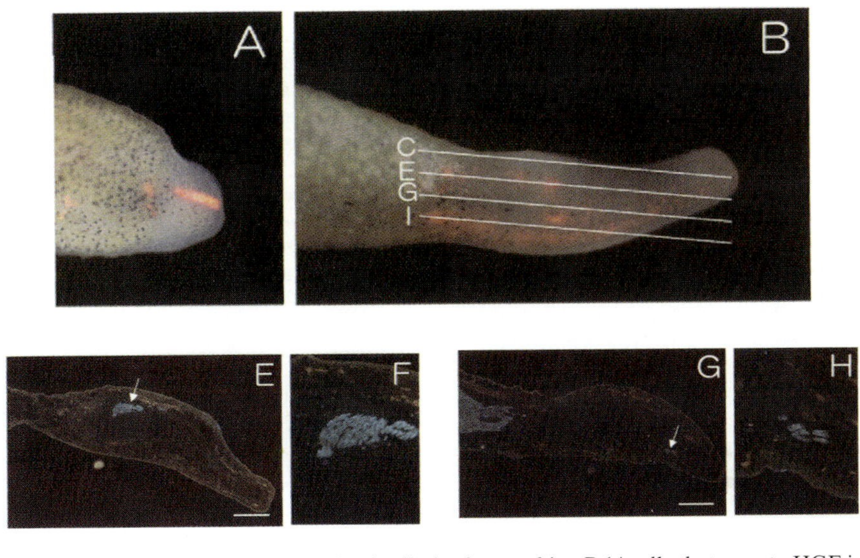


**FIGURE 14.28** Induction of muscle development in *Xenopus* froglet limb after grafting P-14 cells that secrete HGF into the wrist-derived blastema. The cells were labeled with the red fluorescent dye PKH26. **(A)** After implantation. **(B)** Detection of labeled cells throughout the developing spike. C-I indicate levels of longitudinal sections, two of which (**E** and **G**) are shown below. These sections are immunostained for the muscle marker MF20 and show the presence of muscle at both proximal and distal levels. **(F, H)** Higher magnification of the muscle in **E** and **G**. Reproduced with permission from Satoh et al., Muscle formation in regenerating *Xenopus* froglet limb. Dev Dynam 233:337–346. Copyright 2005, Wiley-Liss.

gene activity between stage-53 regeneration-competent and stage-59 regeneration-deficient *Xenopus* limbs. Reciprocal subtractions of cDNA libraries of these two stages were done by hybridizing each library with mRNA representing the other library to provide two subtracted libraries, one enriched in regeneration-permissive genes and the other in regeneration-inhibitory genes. Several thousand clones were analyzed from the subtracted libraries by random selection and sequencing or by screening subtracted cDNA clones on nylon membranes with probes from the other library. Over a hundred clones were identified from the array screens that were differentially expressed in regeneration-competent and -deficient blastemas. Many of these could be classified as transport-binding proteins, structural proteins, metabolic enzymes, transcription factors, etc., but over 60% of the clones in each library had no sequence match. Subtraction analysis has also been used to identify a number of genes that are common to both limb and lens regeneration in *Xenopus* (Wolfe et al., 2004) and a library of cDNA clones has been generated from the stage of maximal dedifferentiation of the regenerating adult newt forelimb (Vascotto et al., 2005). **TABLE 14.2** summarizes an abbreviated set of molecular markers and extracellular signals associated with regeneration-competent limb regeneration blastema cells, none of which is expressed in the pseudoblastema cells of regeneration-deficient limbs. In addition, the table summarizes features of regeneration competence at the cell and tissue levels.

To further understand the differences between regeneration competence and deficiency, the expression patterns of the genes identified and their proteins must be investigated, and their functions tested by knockdowns with RNAi or morpholinos (Schnapp and Tanaka, 2005), or by making frogs transgenic for genes associated with regeneration competence or deficiency (Amaya and Kroll, 1999). The ability to make transgenic newts (Ueda et al., 2005) and other transgenic urodeles will also be of great value in determining the function of genes that direct or inhibit the process of regeneration.

## REGENERATION OF AMPHIBIAN JAWS

The lower and upper jaws of the adult newt and larval salamander regenerate after amputation by formation of a blastema that reconstitutes the amputated parts (Goss and Stagg, 1958a; Graver, 1972; Ghosh et al., 1994; Ferretti and Ghosh, 1997). In the lower jaw, the blastema cells appear to be derived principally from dedifferentiating muscle (Goss and Stagg, 1958a), but in the upper jaw the remaining nasal cartilages also appear to contribute significantly to blastema formation. The blastema cells appear to be like those of the

**TABLE 14.2** Indicators of Cellular and Environmental Regeneration Competence

| Level | Positive Markers of Cellular Regeneration-Competence | Signals in a Regeneration-Competent Environment |
|---|---|---|
| Molecular | Msx2<br>Nrp-1/Msi-1<br>Tbx-3<br>Hoxa 9, 13<br>XlSALL4 | Wnt5a<br>BMP-4<br>Delta 1<br>FGF-2, 4, 8, 10<br>Shh<br>TGFβ5 |
| Cellular | Mesenchymal morphometry Proliferation | Fibronectin, hyaluronic acid |
| Tissue | Cell number and density<br>*Blastema shape, presence of apical epidermal cap, number and identity of regenerated skeletal elements | |

*Geometrically, a regeneration-competent blastema is a parabola in anteroposterior section, a hyperbola in dorsoventral section, and an ellipse in cross-section. The equation describing the volume of this conic is $1/2\pi abc$, where $a$ = the major axis of the ellipse, $b$ = the minor axis of the ellipse, and $c$ = the height of the parabola and hyperbola.

**Msx-2**: transcription factor implicated in maintaining mesenchymal cells in an undifferentiated state.

**Nrp-1/Musashi-1 (Msi-1)**: an RNA-binding protein that maintains Notch signaling. Msi-1 expression is correlated with stemness in *Xenopus* and axolotl (salamander) spinal cord and limb regeneration.

**XlSALL4**: encodes a zinc finger containing transcription factor of the SAL-type. Mutations of this gene in humans are correlated with limb abnormalities.

**Tbx-3, HOXA9, 13**: T-box and homeodomain transcription factors known to be involved early in the patterning of limb regeneration.

**Delta1:** signaling ligand for Notch, implicated in maintaining stemness.

**Shh:** Sonic hedgehog, a signaling molecule important in blastemal patterning.

**FGF-2, 4, 8, 10, BMP4**: signaling growth factors important for blastema cell survival and proliferation.

**Wnt5a:** implicated in activating stem cells.

**TGFβ5**: growth factor closely related to human TGF-β1; involved in immune regulation.

regenerating limb in their expression of cytokeratin and intermediate filament proteins, as well as tenascin and N-CAM (Ferretti and Ghosh, 1997). As the blastema forms, the wound epidermis that migrates over the amputation surface thickens, perhaps the equivalent of the AEC in regenerating limbs.

The highest density of blastema cells occurs on the cut ends of the jawbones. These cells regenerate the bones and also the dental lamina, teeth, and extra tooth buds (Chapter 3), while the intermandibular or intermaxillar blastema cells regenerate the muscle and connective tissues of the jaws. Lower jaw regeneration after transverse amputation is incomplete, in that the tongue and hyoid bone are not regenerated (Goss and Stagg, 1958a). Experiments in which quarters, halves, or whole mandibles were removed showed that the mandible of the larval salamander can regenerate in either an anterior or posterior direction, but that the adult newt jaw can regenerate only in an anterior direction (Graver, 1972).

Interestingly, removal of the intermandibular soft tissue results in the healing of oral and skin epidermis together to close the wound all round, followed by the formation of blastemas from muscle and connective tissue remaining at the lateral edges of the wound (Goss and Stagg, 1958b). These blastemas grow inward until they meet in the center to re-establish continuity of oral and skin epidermis (**FIGURE 14.29**). New salivary glands differentiate from the oral epidermis. The mesenchyme cells of the blastema regenerate not only new muscle and connective tissue, but also cartilage thought to represent the hyoid bone.

Regenerating jaws exhibit several differences from regenerating limbs (Ferretti and Ghosh, 1997). The cytokeratin NvKII is expressed in the wound epidermis of regenerating limbs, but is not expressed in the wound epidermis of the regenerating jaw. Unlike limbs, proximalization of jaw blastema cells by RA is not observed. In the maxilla, RA treatment produces truncated regenerates with a cleft lip and palate morphology (Ghosh et al., 1996). Finally, regeneration of the jaw does not appear to be dependent on innervation (Finch, 1969). Whether or not the jaw blastema cells are dependent on the wound epidermis for their survival and proliferation is not known.

## APPENDAGE REGENERATION IN MAMMALS

Several mammalian appendages have the ability to regenerate epimorphically. These interesting cases are regeneration from the edges of holes made in rabbit and MRL mouse ears, the annual regeneration of male deer antlers, and the regeneration of digit tips in fetal mice and humans. The digit tips of adult mice and humans

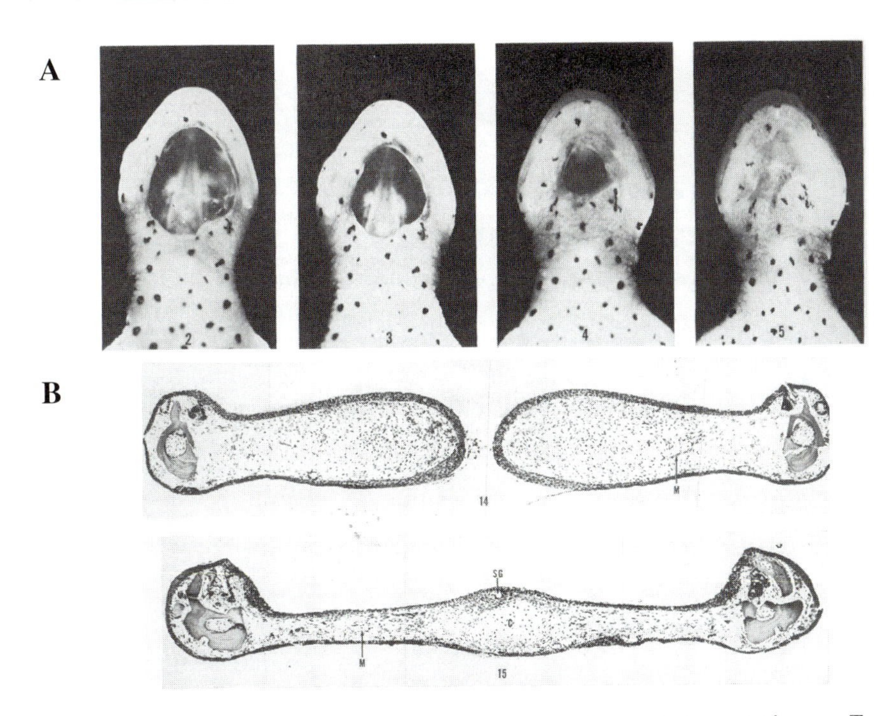


**FIGURE 14.29** Stages in the regeneration across an intermandibular gap in the newt *Notophthalmus viridescens*. **Top**, viewed from the ventral side of the animal. **Middle**, cross-section through the lower jaw four weeks after operation. Two large blastemas are converging toward the center of the wound, and differentiation has begun toward the periphery of the wound. **Bottom**, the blastemas have converged at 7 weeks after the operation and differentiation is occurring in the center. M = muscle, SG = salivary gland, C = cartilage (hyoid). Reproduced with permission from Goss and Stagg, Regeneration in lower jaws of newts after excision of the intermandibular regions. J exp Zool 137:1–12. Copyright 1958, Wiley-Liss.

are also able to regenerate, but this regeneration does not take place by blastema formation.

## 1. Regeneration of Ear Tissue

### a. *Rabbit Ear Tissue*

Rabbit ears consist of a sheet of stiff fibrocartilage covered by skin. They can regenerate the tissue removed by punching a 1–3-cm hole in the ear, in contrast to other wild-type mammalian ears that undergo simple wound healing around the circumference of the hole (Vorontsova and Liosner, 1960; Joseph and Dyson, 1966; Grimes and Goss, 1970). Rabbit ear regeneration takes place by the centripetal growth of tissue from the edges of the defect. Chondrogenesis begins at about three weeks in the regenerating tissue, and closure of the hole takes 6–8 weeks. Amputated toes transplanted into a rabbit ear hole do not regenerate, indicating that the regenerative environment of the ear does not promote the regeneration of toes (Williams-Boyce and Daniel, 1980).

Histological studies indicate that ear hole closure in the rabbit is by epimorphic regeneration (Goss and Grimes, 1975). A ring-shaped blastema of rapidly dividing mesenchymal cells forms at the edge of the wound and differentiates into new dermis and cartilage as the blastema grows to close the hole. Epidermal downgrowths penetrate between the cartilage and the dermis of the wound edge. Such downgrowths are not observed in the nonregenerating ear wounds of dogs and sheep, but are characteristic of the regenerating limbs of adult newts. The origin of the blastema cells in rabbit ear tissue is not clear. They might be derived from a population of stem cells, or by dedifferentiation of dermal fibroblasts and/or cartilage.

Regeneration of rabbit ear tissue requires both ear skin and cartilage. If belly skin is substituted for ear skin, a hole punched within the belly skin field fills in only about halfway. If the ear cartilage is removed before grafting the belly skin, or if the cartilage is removed and ear skin replaced, a hole punched in the cartilage-free region does not close (Goss and Grimes, 1972). Irradiated ears do not regenerate, but irradiated ears that receive unirradiated cartilage regenerate nearly as well as control ears. Unirradiated ears that receive irradiated cartilage exhibit nearly 80% closure, but the regenerated tissue does not undergo chondrogenesis (Grimes, 1974). Collectively, these data indicate that skin can sustain partial regeneration, but for complete regeneration, cartilage is required. This could mean that both cartilage and dermis normally contribute cells to the blastema, that cartilage exerts an inductive role on cells derived from skin, or some combination of these.

### b. Rodent Ear Tissue

The ear cartilage of adult rats does not regenerate. After making an incisional wound all the way through the ear, a fibrotic scar forms between the edges of the incised cartilage that stains intensely with antibody to type I collagen (Wagner et al., 2001). Similar incisions in the cartilage of neonatal rat ears, however, are repaired by the regeneration of chondrocytes that express type II collagen. Nuclear PCNA staining indicated that the regenerated chondrocytes were derived by the proliferation of neonatal chondrocytes in the lesion area.

Immunologists identify mice in experiments by ear punch-holes. In wild-type mice, such as C57B1/6, the punch holes heal like standard excisional skin wounds: they re-epithelialize, form scar tissue at the rim, and exhibit minimal closure. The MRL/lpr mouse displays lupus-like symptoms due to a mutation in the *fas* gene and has been used as an experimental model for study of this autoimmune disease. Interestingly, punch-holes in the ears of MRL/lpr mice close completely in less than four weeks (Heber-Katz et al., 2004). **FIGURE 14.30** illustrates this closure, which is very similar to that observed in the rabbit ear. Histological studies show that closure is accomplished by a blastema of proliferating cells at the wound edges. The blastema regenerates the normal structure of the ear tissue, including the supporting cartilage sheet (Desquenne-Clark et al., 1998; Kench et al., 1999; Heber-Katz et al., 2004). A similar result is observed with punch-holes in the ears of MRL/MpJ mice (Masinde et al., 2001). As in rabbit ear hole closure, the origin of the blastema is not certain, whether from a population of stem cells or by dedifferentiation of dermal fibroblasts and cartilage. Interestingly, dorsal skin wounds of MRL/MpJ mice repair by fibrosis (Rajnoch et al., 2003; Metcalfe and Ferguson, 2005), which raises the question of whether ear skin can regenerate in the absence of cartilage.

The capacity for mouse ear-tissue regeneration is quantitative and associated with a heritable multigenic trait. To identify the genetic loci underlying the regenerative ability of these mutant mice, a genome-wide scan using microsatellite DNA markers was performed on the F2 intercross progeny of regenerating and non-regenerating mouse strains, MRL/lpr × C57B1/6J (McBrearty et al., 1998), MRL/MP × SJL/J (Masinde et al., 2001), and MRL/MpJ and CAST/Ei, an inbred subspecies of *M. castaneus* (Heber-Katz et al., 2004). A normal quantitative distribution of regeneration was exhibited by the F2. Sixteen quantitative trait loci (QTL) associated with a high degree of regeneration have been identified on chromosomes 1, 2, 4, 6, 7, 8, 9, 11, 12, 13, 14, 15, and 17. The majority were inherited

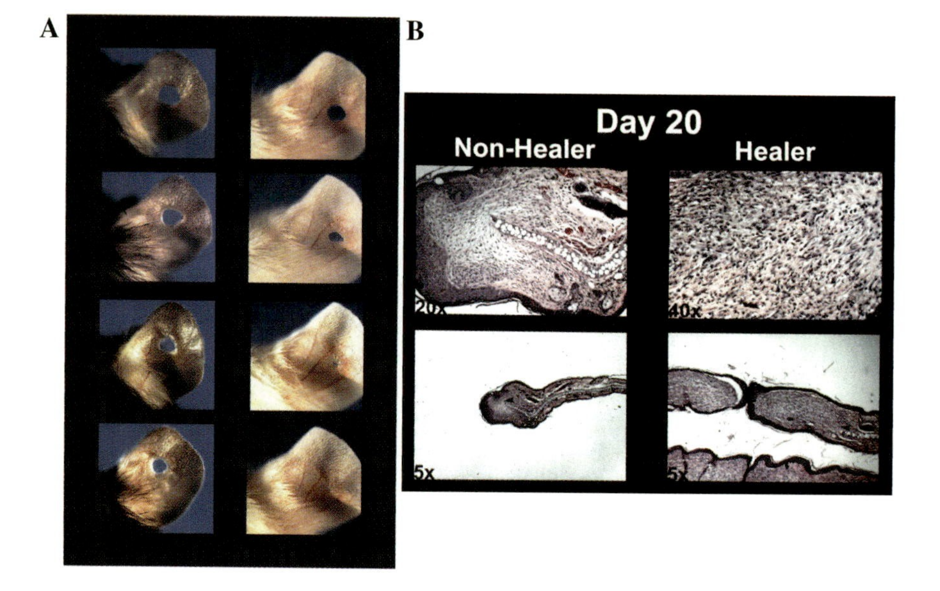


**FIGURE 14.30** Regeneration of ear tissue in the MRL mouse vs. wild-type mouse. **(A)** Ear hole punch in wild-type mouse on left undergoes only partial closure, whereas ear hole punch in MRL mouse closes completely. **(B)** Sections through wild-type (left) and MRL (right) ears show that the edge of the wild-type wound scars, whereas a blastema forms at the edge of the MRL wound and grows centripetally to close the wound. Courtesy of Dr. Ellen Heber-Katz.

from the regenerating MRL mice and a minority from the nonregenerating strains. These loci explain 70% of the variance in regeneration of the F2 mice (Masinde et al., 2001). Several of the QTLs appear to interact during regeneration (McBrearty et al., 1998; Masinde et al., 2001). Several genes known to be involved in limb development and regeneration have been identified at the QTLs: FGFR4; Gli3, the transcription factor activated by sonic hedgehog; the *Hoxc* cluster; latent TGF-β binding protein 1; and RARγ, the mammalian equivalent of the newt RARδ. Many other genes at these loci are involved in signal transduction pathways (McBrearty et al., 1998). None of the autoimmunity modifying loci co-localize with QTLs that control the healing phenotype.

Formation of a regeneration blastema in the amputated amphibian limb is dependent on the absence of a basement membrane between the AEC and the underlying blastema mesenchyme, allowing direct communication between the two. The basement membrane in the regenerating amphibian limb is prevented by MMPs from reforming prematurely. The status of the basement membrane in regenerating rabbit ears was not reported (Goss and Grimes, 1975), but studies in the MRL mouse indicate that, like in the regenerating amphibian limb, closure of ear holes is associated with lack of a basement membrane (Gourevich et al., 2003; Heber-Katz et al., 2004). A basement membrane is initially re-established, but disappears by day five post-wounding, whereas in wild-type mice, it remains. This difference is correlated with higher expression and activity of MMP-2 and 9 in the MRL ear, which in turn is correlated with the presence of a higher number of inflammatory cells. The higher activity of these enzymes may be due to the fact that the MRL ear also has lower amounts of TIMPs.

It would be interesting to know whether the blastema of the MRL mouse or rabbit ear is similar to that of the amphibian limb regeneration blastema in its dependence of wound epidermis and innervation for formation and outgrowth, and in the types of genes expressed.

## 2. Regeneration of Deer Antlers

Antlers are twin, branched outgrowths of bone extending from the foreheads of male deer. In temperate climates, they are used for display and fighting with rivals for females during the fall mating season. The antlers are shed in the spring, due to erosion of a narrow zone of bone at the base of the antler by osteoclasts, leaving a raw wound that is healed over by skin (Goss et al., 1992). New antlers are regenerated over a period of four months in the summer (Goss, 1970, 1974, 1983, 1995; Price et al., 1996). Antlers are the only example of a naturally repeated and complete epimorphic regeneration of an appendage in mammals, and as such are a valuable research model to help understand how we might regenerate mammalian appendages.

The first antlers of a male deer develop like limb buds, from bilateral blastemas that form on the pedicles under the influence of rising testosterone levels in maturing fawns. The pedicles are masses of trabecular bone laid down by the periosteum overlying the frontal protuberances, bumps of bone protruding from the frontal bone of the skull. The blastemas are formed from spindle-shaped mesenchymal cells in the lower layers of the pedicle periosteum. This has been shown by experiments in which surgical deletion of the periosteum results in an absence of antler formation, and grafting the pedicle periosteum to an ectopic position on the forehead or the foreleg results in ectopic antler formation (Goss et al., 1964; Hartwig, 1968a, b; Hartwig and Schrudde, 1974; Goss and Powel, 1985). The pedicle induces the epidermis of the overlying skin to become a velvet epidermis, so named because of its many fine hairs (Goss, 1972).

Antler regeneration takes place annually after the first antlers are cast. The longitudinal growth rate of the antler varies with the size of the antlers in the adult animal. In the large-antlered elk, for example, it is over two centimeters per day. Casting of the antler leaves an open wound that exposes the pedicle. The origin of the mesenchymal cells forming the blastemas of regenerating antlers is not as clear as it is for the development of the initial antlers. Histological studies suggest that the cells of the new antler blastema may be derived from the pedicle bone, the periosteum, and the dermis of skin surrounding the wound (Wislocki and Waldo, 1953; Goss, 1992, 1995). Unlike amphibian limb regeneration, there is no neural requirement for antler regeneration (Goss, 1985).

Whatever the source(s) of cells, histological and molecular studies indicate that antler regeneration is a process of modified endochodral bone formation (**FIGURE 14.31**). Several zones of differentiation can be discerned in the growing antler that resemble the distal half of a differentiating embryonic endochondral long bone (Price et al., 1996). The mesenchymal blastema forms a cap under the velvet epidermis at the growing tips of the antler branches. A dermis is interposed between the mesenchymal cells and the velvet epidermis. Proximal to the mesenchymal cells, chondrocytes are arranged in columns that differentiate in a proximal to distal sequence. The oldest chondrocytes toward the base of the antler are hypertrophied. More distally are maturing chondrocytes that grade off into chondrocyte progenitors just below the mesenchymal cells.

The hypertrophied chondrocytes calcify their matrix, which is subsequently degraded by osteoclasts and

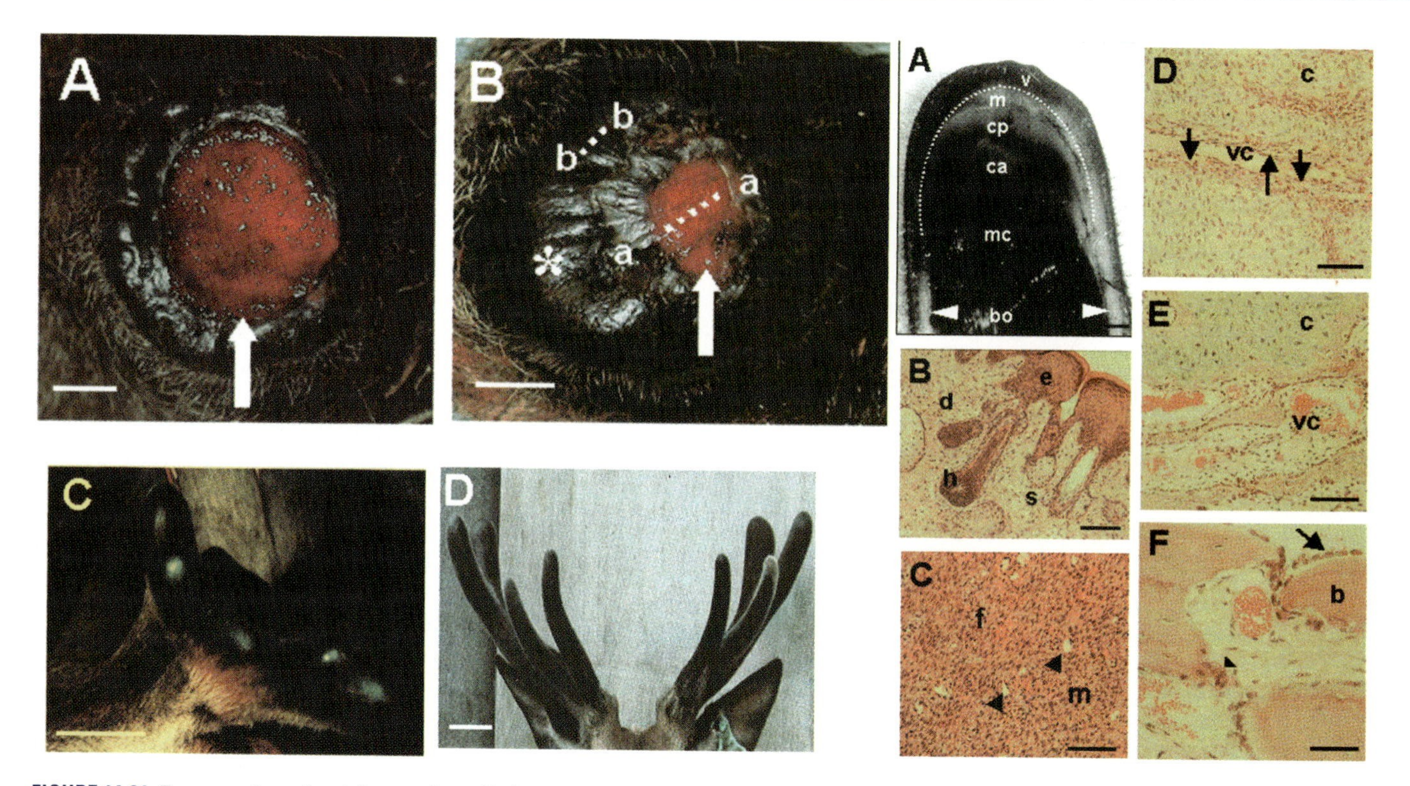


**FIGURE 14.31** Regeneration of red deer antlers. **(Left)** Anatomical views. **(A)**, 24 hr after casting the antler. **(B)**, Blastema 7 days after casting. White arrows indicate leading edge of the regenerating epidermis. Dotted lines a-a and b-b indicate regions from which sections were cut. Asterisk indicates the position of a future branch of the antler. **(C** and **D)**, 4 and 8 weeks of growth, respectively. **(Right)** Histology of the antler tip. **(A)**, longitudinal section showing zones of the antler blastema from least differentiated (top) to most differentiated (bottom). v = velvet epidermis, m = mesenchyme, cp = chondrocyte progenitor region, ca = non-mineralized cartilage, mc = mineralized cartilage, bo = bone. Arrowheads indicate intramembranous bone formation. **(B)**, velvet skin. e = epidermis, d = dermis, h = hair follicle, s = sebaceous gland. **(C)**, perichondrium. f = fibrous portion, m = mesenchymal portion. Arrowheads indicate blood vessels. **(D)**, non-mineralized cartilage. c = recently differentiated chondrocytes in columns separated by vascular channels (vc). Arrows indicate perivascular tissue. **(E)**, mineralized, hypertrophic cartilage (c) with larger vascular channels (vc). **(F)**, bone. Arrow indicates osteoblasts, arrowhead indicates an osteoclast. Reproduced with permission from Faucheaux et al., Recapitulation of the parathyroid hormone-related peptide-Indian hedgehog pathway in the regenerating deer antler. Dev Dynam 231:88–97. Copyright 2004, Wiley-Liss.

replaced by bone matrix secreted by osteoblasts. This process differs somewhat, however, from the regeneration of bone tissue in a fractured endochondral long bone. The latter ossifies by the invasion of blood vessels into the avascular cartilage template, accompanied by perivascular mesenchymal cells that become osteoblasts and bone marrow cells. By contrast, the cartilage template of the antler is penetrated from the outset by many blood vessels with perivascular osteoblasts. Thus, the osteoblasts for making bone matrix are an integral part of the template from the start. This mechanism of ossification may be related to the fact that antler bone does not contain marrow.

The molecular aspects of antler regeneration are similar to those of endochondral bone development or fracture healing. Type I collagen is the predominant collagen expressed in the mesenchyme cells at the tip of the growing antler (Price et al., 1996). Sulfated GAGs, collagen II, and aggrecan core proteoglycan protein are expressed by differentiating chondrocytes, and collagen X is expressed by hypertrophic chondrocytes (Moello et al., 1963; Frasier et al., 1975; Price et al., 1994, 1996). The osteoclasts that remodel the cartilage template of the antler first appear in association with perivascular tissues, indicative of their origin from monocytes, and have the same phenotype as those of appendicular endochondral bone (Faucheux et al., 2001).

Osteoclast formation in antlers is regulated by the PTHrP/PPR and RANKL/RANK pathways (Faucheux et al., 2002). PTHrP and PPR are expressed *in vivo* by osteoclasts differentiating in the perivascular stroma. Studies using micromass cultures of non-calcified antler cartilage showed that RANKL and M-CSF, which are produced by osteoblasts and promote osteoclast formation during the remodeling of endochondral bone

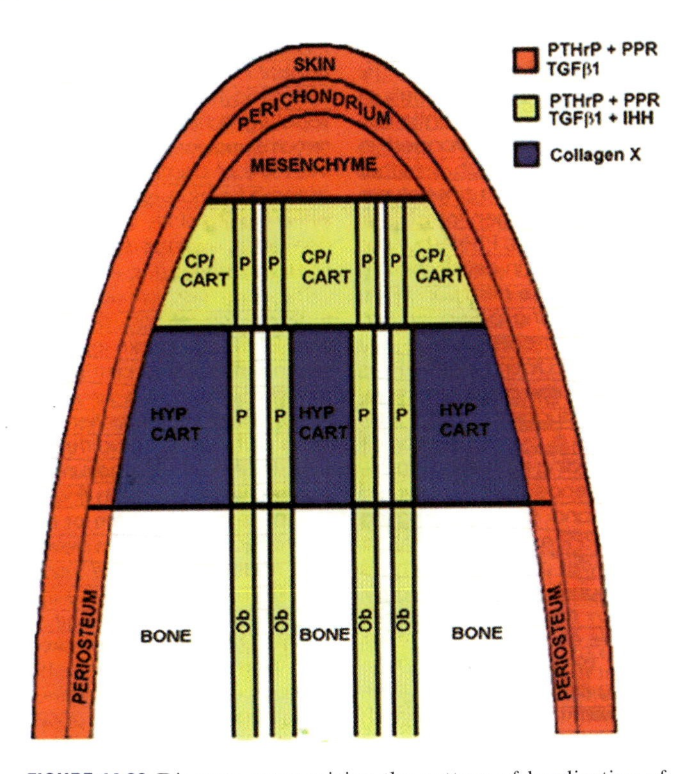


**FIGURE 14.32** Diagram summarizing the pattern of localization of collagen X, parathyroid related peptide (PTHrP), the parathyroid hormone (PTH)/PTHrP receptor (PPR), Indian hedgehog (IHH) and TGF-β1 in the growing antler tip. CP/CART = chondroprogenitors/cartilage, p = perivascular tissue, ob = osteoblasts. Reproduced with permission from Faucheaux et al., Recapitulation of the parathyroid hormone-related peptide-Indian hedgehg pathway in the regenerating deer antler. Dev Dynam 231:88–97. Copyright 2004, Wiley-Liss.

(Chapter 9), are constituitvely produced by the cultured chondrocytes (Faucheux et al., 2001). Exogenous RANKL and PTHrP both stimulated osteoclast differentiation in micromass cultures. The stimulatory effect of RANKL was less marked than that of PTHrP, and the addition of the decoy RANK ligand, osteoprotegerin, to the cultures did not completely block osteoclast formation in PTHrP-treated cultures (Faucheux et al., 2002).

Furthermore, chondrocyte differentiation in the growing antler appears to be regulated by the PTHrP/PPR pathway, through TGF-β and Ihh, as it is in developing embryonic endochondral bones (Faucheux et al., 2004) (**FIGURE 14.32**). In addition to being expressed in perivascular cells of cartilage, PTHrP, PPR, and TGF-β are all expressed in the undifferentiated mesenchyme, recently differentiated chondrocytes, skin, perichondrium, and osteoblasts, but not hypertrophied chondrocytes, of the growing antler. TGF-β1 stimulated PTHrP synthesis by cells from the mesenchyme, cartilage, and perichondrium (Faucheux et al., 2004), and PTHrP promoted proliferation and inhibited differentiation in micromass cultures (Faucheux and Price, 1999), suggesting that TGF-β1 regulates the synthesis of PTHrP and that PTHrP may maintain the proliferation of chondroprogenitor cells. Ihh was expressed only in recently differentiated chonodrocytes, osteoblasts, and perivascular cells. The co-localization of Ihh, TGF-β1, PTHrP, and PPR in these cells suggests a regulatory feedback loop for cartilage differentiation and ossification in which Ihh upregulates the expression of TGF-β1, which in turn upregulates PTHrP to keep feeding cartilage cells through the differentiation cycle (Faucheux et al., 2004).

Another molecule that plays an important role in the regulation of chondrocyte and osteoblast differentiation is retinoic acid. In addition to its role in establishing the ZPA of the developing limb bud, RA is required for the maturation of chondrocytes and osteogenesis. Inhibiting the binding of RA to RARs with RAR antagonists, or expression of a dominant negative RAR at later stages of limb bud development, results in impaired endochondral bone development due to failure of chondrocytes to hypertrophy and produce type X collagen (Yamaguchi et al., 1998; Koyama et al., 1999; De Luca et al., 2000). RA induces the differentiation *in vitro* of mouse and human osteoblasts (Oliva et al., 1993; Slootweg et al., 1996).

Local treatment of deer pedicles with RA prior to growth of the first antler resulted in increased antler size, suggesting that RA increases the proliferation of the mesenchymal cells derived from the periosteum (Kierdorf and Kierdorf, 1998; Kierdorf and Bartos, 1999). This suggestion is supported by the localization patterns of retinol, the retinol to RA-converting enzyme RALDH-2, and RARα and RXRβ, receptors for all-trans RA and 9-cis RA, respectively (Allan et al., 2002). Large amounts of retinol are found in antler tissues at all stages of differentiation. RALDH-2, RARα, and RXRβ are expressed in the skin, perichodrium, cartilage, perivascular cells, and osteoblasts at sites of bone formation, consistent with a critical role of RA in chondrogenesis and osteogenesis in the growing antler (Allen et al., 2002). In addition, all-trans RA increased the expression of alkaline phosphatase *in vitro* by antler mesenchymal cells, indicating that it plays a role in osteogenesis.

Eventually, the complete ossification of the antlers occludes their blood supply, leaving the antlers as masses of dead bone (Goss, 1970). The deer rub off the velvet epidermis and the antlers are ready for their function in mating contests. Once the mating season is over, the antlers are shed over a four-month period from December to March.

The annual growth cycle of antlers is controlled by endocrine and environmental factors (Goss, 1983), but how these factors interface with local factors that regulate antler development is not well understood. A common misconception is that testosterone stimulates antler regeneration (as opposed to formation of the first antlers). Testosterone is involved in the ossification of growing antlers, but it prevents the onset of regeneration when injected (Goss, 1970). What is clear is that the annual cycle of antler growth, in temperate climates, is regulated by photoperiod, superimposed on little-understood internal cycles. Changes in the length and frequency of the photoperiod alter the number of regeneration cycles but do not abolish them, as long as there is an inequity in the ratio of light to dark intervals (Goss, 1969a, b; 1970).

## 3. Regeneration of Mouse and Human Digit Tips

### a. Embryonic Mammalian Limb Buds Can Regenerate, But This Capacity Becomes Restricted to Digits as They Differentiate

A number of studies have shown that mouse and rat embryonic limb buds can regenerate early in their development but lose this capacity as the limb bud differentiates, much as in *Xenopus*. Nicholas (1926) amputated the 14-day forelimb bud of the rat at various levels *in utero*. The best response was evoked at the more distal amputation levels, where a knot of undifferentiated cells was formed at the amputation surface. Deuchar (1976) removed 11.5-day-old mouse embryos from the uterus, amputated one forelimb through its base, and cultured the embryos in roller tubes for 44 hours. The 11.5-day limb bud is undifferentiated and corresponds to stage 2/3 in the limb bud staging system constructed for mice by Wanek et al. (1989). Twenty-nine of 32 amputated embryos regenerated limb buds that were similar in shape and size to control buds, which still possessed an AER and were undifferentiated at the end of the experiment. Of the 29 regenerates, eight had a normal AER, and seven were reported to have an "incipient" AER. Regeneration *in vitro* of undifferentiated mouse and rat limb buds was also reported by Chan et al. (1991) and Lee and Chan (1991).

The importance of re-establishing the AER or its functional equivalent for limb bud regeneration is shown by the fact that early stage chick limb buds do not reconstitute an AER after amputation and do not regenerate. However, prospective stylopodial cells will restore the zeugopodium by intercalary regeneration when the distal tip of the early bud with AER is grafted to a more proximal level (Hampe, 1959) and cells from a later stage limb bud (stages 24–25) that have been specified as zeugopodium will change their fate to digits when grafted into a space made under the AER (Hayamizu et al., 1994). Furthermore, FGF-2 or -4 can substitute for the AER and evoke a regenerative response from amputated stage 24–25 chick limb buds (Taylor et al., 1994; Kostakopoulou et al., 1996, 1997). This response is accompanied by the upregulation of distally expressed genes such as *Msx1*, *Shh*, and *HoxD13* (Hayamizu et al., 1994; Kostakopoulou et al., 1996).

As in *Xenopus*, regeneration of the mouse limb bud becomes restricted to progressively more distal regions as the limb bud differentiates (Wanek et al., 1989; Reginelli et al., 1995). At stage 7/8 (embryonic day 12.5), one or two digits regenerated after amputation *in utero* of the mouse embryo footplate through the phalanges, but no regeneration took place after amputation through more proximal levels (Wanek et al., 1989). Eventually, regenerative capacity becomes restricted to the tips of the digits in the mouse stage 11 (14.5-day) embryo (Reginelli et al., 1995).

Newborn opposum limbs have been reported to regenerate after amputation through the zeugopodium (Mizell, 1968; Mizell and Isaacs, 1970). The amputation level, however, was not precise, and later work showed that the amputation plane was actually through the tarsals or phalanges (Fleming and Tassava, 1981). Thus the neonatal opposum can also regenerate digits, but not more proximal structures of the limb.

### b. Digit Tip Regeneration in Mammalian Embryos Is Regulated by Msx1 and BMP4

Msx1 has been implicated in the regulation of amniote limb bud outgrowth and amphibian limb regeneration. Msx1 is expressed in the apical mesenchyme of the limb bud and regeneration blastema, and its expression is dependent on FGFs from the AER and AEC, as well as being regulated by BMPs (Pizette et al., 2001), TGFβ (Ganan et al., 1996), and retinoic acid (Wang and Sassoon, 1995). Msx1 and BMP4 are also important regulators of digit tip regeneration in the stage 11 (14.5-day) mouse embryo. By this stage, regeneration will occur only from the apical portion of the developing digit tip that will become the terminal phalange, and it is only this region that continues to express Msx1 (Reginelli et al., 1995).

Studies of amputated stage 11 digit tips *in vivo* and in organ culture (Han et al., 2003) have shown that the apical epidermis is regenerated and that *Msx1* and *Bmp4* are co-expressed in the apical mesenchymal cells of the blastema surrounding the tip of the condensing cartilage of the terminal phalange (**FIGURE 14.33**). Regeneration of the digit did not take place if it was amputated proximal to the expression domain of these two genes. In addition, *Msx2* is expressed in the apical

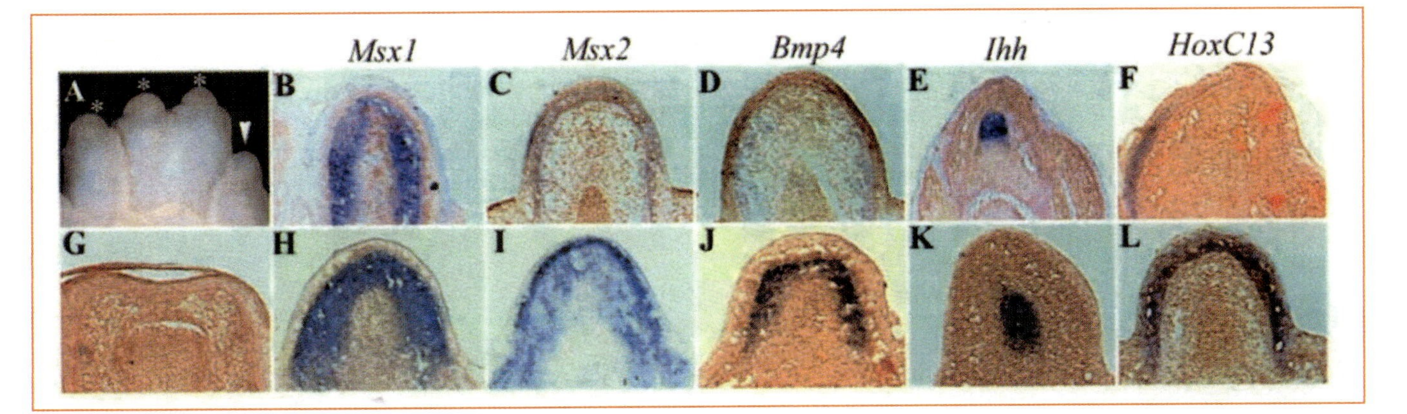


**FIGURE 14.33** Regeneration *in vivo* of mouse fetal digits amputated through the distal phalange. **(A)** E17.5 autopod, ventral view. **(B–F)** *In situ* hybridization of regenerating digit tips 4 days after amputating them at E.14.5. *Msx-1* is expressed in the mesenchyme surrounding the terminal phalanx, *Msx-2* in the apical epidermis, low levels of *Bmp-4* in the apical mesenchyme, *Ihh* in the re-forming terminal phalanx and *Hoxc13* in the nail organ. **(G)** Digits amputated proximal to the terminal phalanx do not regenerate. **(H–L)** *In situ* hybridization two days after amputation of the terminal phalanx. Higher levels of *Msx1, Msx2, Bmp4 and Ihh* are expressed. Reproduced with permission from Han et al., Digit regeneration is regulated by *Msx1* and BM4 in fetal mice. Development 130:5123–5132. Copyright 2003, The Company of Biologists.

epidermis and the immediately subjacent mesenchyme. The frequency of regeneration after amputation within the expression domain was about 90%, both *in vitro* and *in vivo*. Mutant mice with a targeted deletion of the *Msx1* gene regenerated at frequencies of 38% *in vivo* and 28% *in vitro*. Treatment of mutant cultures with BMP4 (1,000 ng/ml) rescued the frequency of regeneration back to 86%, but treatment with the BMP antagonist, Noggin (200 ng/ml), reduced the regeneration frequency to 8% and in wild-type cultures to 18% without affecting the expression of *Msx1*. The wild-type digit tips express *Ihh* in the chondrifying portion of the regenerating terminal phalange, and *HoxC13* in the epidermis of the prospective nail bed region. Interestingly, in those *Msx1* mutants that did regenerate, BMP4 was expressed, and the expression domain of *Msx2* was expanded to occupy the *Msx1* domain. These results suggest first that there is a degree of redundancy in the expression and function of Msx 1 and Msx2, and, second, that *BMP4* is the downstream target of Msx1 and Msx2.

Human fetal digit tips have also been successfully cultured *in vitro* (Zaaijer, 1958; Rajan and Hopkins, 1970; Zeltinger and Holbrook, 1997; Allan et al., 2005). Allan et al. (2005) showed that MSX1 is expressed beneath the developing nail field of the digit tips of human embryos ranging in estimated gestational age from 53–67 days. Four days after amputation of the digit tip, the epidermis, identified by the expression of keratins K14 and K19, had proliferated to nearly close the wound and there was robust proliferation of fibroblast-like cells. There was intense expression of MSX1 in the migrating epidermis, as well as in the mesenchyme subjacent to the presumptive nail bed. The proximal end of the cultured digit tip exhibited no proliferative response, and the skin retracted away from the end of the bone. The proliferative response at the distal amputation site decreased with estimated gestational age.

### c. Adult Mammals Retain the Ability to Regenerate Fingertips Distal to the Last Interphalangeal Joint

The ability of the terminal phalanges of adult mammalian digits to regenerate was first demonstrated in humans. Several cases of regeneration of amputated distal phalanges in children were reported in the 1970s (Douglas, 1972; Illingworth, 1974; Rosenthal et al., 1979). Since then, the terminal phalanges of fingers and toes have been found to regenerate in adults as well (Neumann, 1988) (**FIGURE 14.34**). Significantly, regeneration takes place only when the wound is left open, rather than following the common medical practice of sewing the skin shut over the amputation surface. Since epidermis obviously heals over the open wound, this fact suggests that an epithelial–mesenchymal interaction is required for fingertip regeneration in adult humans and other mammals, just as it is in amphibian limb regeneration.

The ability of fetal mice to regenerate terminal phalanges is retained by adult mice. Borgens (1982) amputated the terminal phalange of the middle toe of 4-week-old mice immediately proximal to the nail. The amputated digit tips were regenerated in four weeks' time and appeared normal, both in external appear-

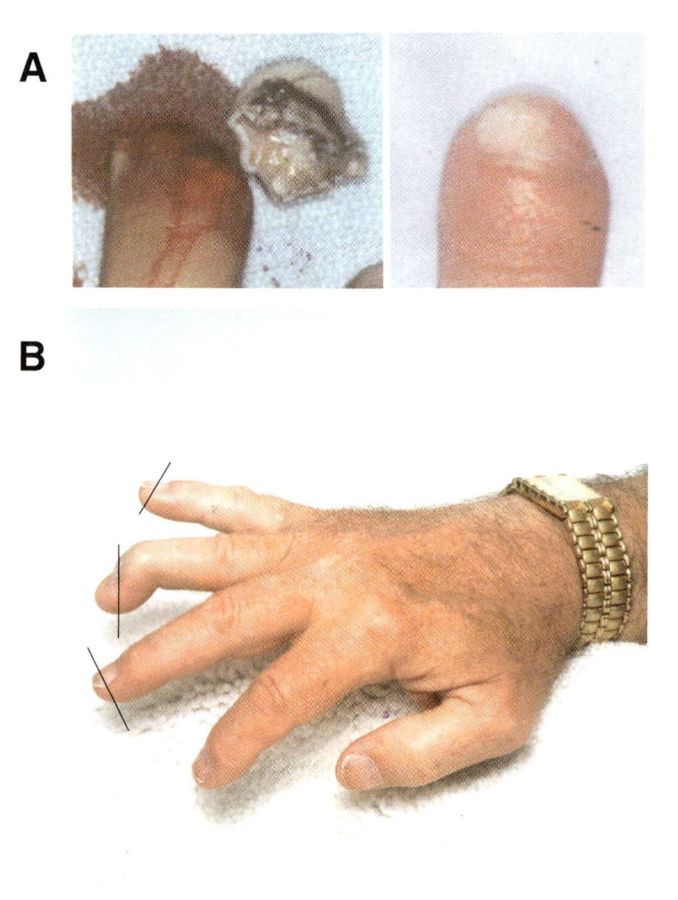


**FIGURE 14.34** Regeneration of fingertips in human children and adults. **(A)** Left, fingertip of a 7yr old girl amputated in a bicycle spoke accident, left unsutured and treated with dressing changes. Right, regenerated fingertip at 8 weeks. Courtesy of Dr. Christopher Allan **(B)** Regenerated tips of digits 3, 4 and 5 of a 76yr old man that were amputated in an industrial accident at age 29. Lines indicate level of amputation. Courtesy of Mr. Ben Bierly.

ance and in histological structure (**FIGURE 14.35**). No regeneration took place if the amputation plane was proximal to the joint, and this appears to be true for human fingertips as well. Other studies on mice by Neufeld and Zhao (1995) have confirmed these results.

Although fetal mouse digit tips regenerate the terminal phalangeal bone in an endochondral fashion, the adult bone appears to be regenerated by direct deposition of bone onto the remaining bone. There is no evidence of a classical mesenchymal cell blastema that differentiates a cartilage template like that of the regenerating digit tip in fetal mice (Borgens, 1982). Fibroblastic cells also appear at the amputation site and contribute to the regenerating digit tip, but their fate has not been determined. It might be presumed that they reform the dermis, periosteum, and loose connective tissues, although they might also contribute to the bone and fat pads. The nail matrix, nail bed, and nail plate regenerate from the epidermis.

Because the frequency of regeneration falls off rapidly proximal to the nail matrix (Neufeld and Zhao, 1995), it has been speculated that there may be a lower vascular supply proximal to this boundary that might inhibit new osteogenesis, or that the nail matrix (which is equivalent to the stem cell-containing matrix of a hair follicle) or the nail bed may be essential for bone regeneration. However, the vascular supply in mouse digits is greater proximal to the nail matrix than distally (Said et al., 2004). It is unlikely that the prospective nail matrix or bed contributes cells for the regeneration of any part of the terminal phalanx except the nail itself. There may, however, be an epithelial–mesenchymal interaction between the nail epithelium and the regenerating bone and other structures. Transplants of nail plate that included the nail matrix to an amputated proximal phalange resulted in bone growth toward the implant, rather than the usual capping off of the marrow cavity with bone (Mohammad et al., 1999). Han et al. (2003) reported that, in the unamputated terminal phalanges of neonatal mice, *Msx1* and *Bmp4* are expressed in the connective tissue subjacent to the nail matrix and bed, whereas *HoxC13* and *Msx2* are expressed within these tissues. Given that *Msx1* and *Bmp4* are expressed in the mesenchymal cells of the regenerating fetal digit tip, the connective tissue beneath the nail organ may be the source of cells for the regenerating terminal phalange.

## 4. Stimulation of Digit and Limb Regeneration in Mice and Rats

Attempts have been made to induce the regeneration of amputated proximal phalanges and limbs in mice and rats. Amputated proximal phalanges have been treated with trypsin, $CaCl_2$ (Scharf, 1961, 1963), and saturated solutions of NaCl (Neufeld, 1980) in attempts to induce histolysis and dedifferentiation. Repeated opening of the wound epidermis of amputated digits and treatment with saturated salt induced a small accumulation of semimesenchymal cells under the wound epithelium that incorporated $^3$H-thymidine and divided. The wound epithelium itself resembled that of the newt limb blastema. These accumulations were not followed beyond 14–16 days, but it is unlikely that the digits would have exhibited any substantial regeneration. Amputated rat limbs have been treated with muscle implants (Person et al., 1979), electrical current (Becker, 1972; Libbin et al., 1979; Smith, 1981), and maintaining animals on a vitamin A- and D-free diet or injecting them with a preparation of vitreous body (Polezhaev, 1972). The general conclusion from these experiments

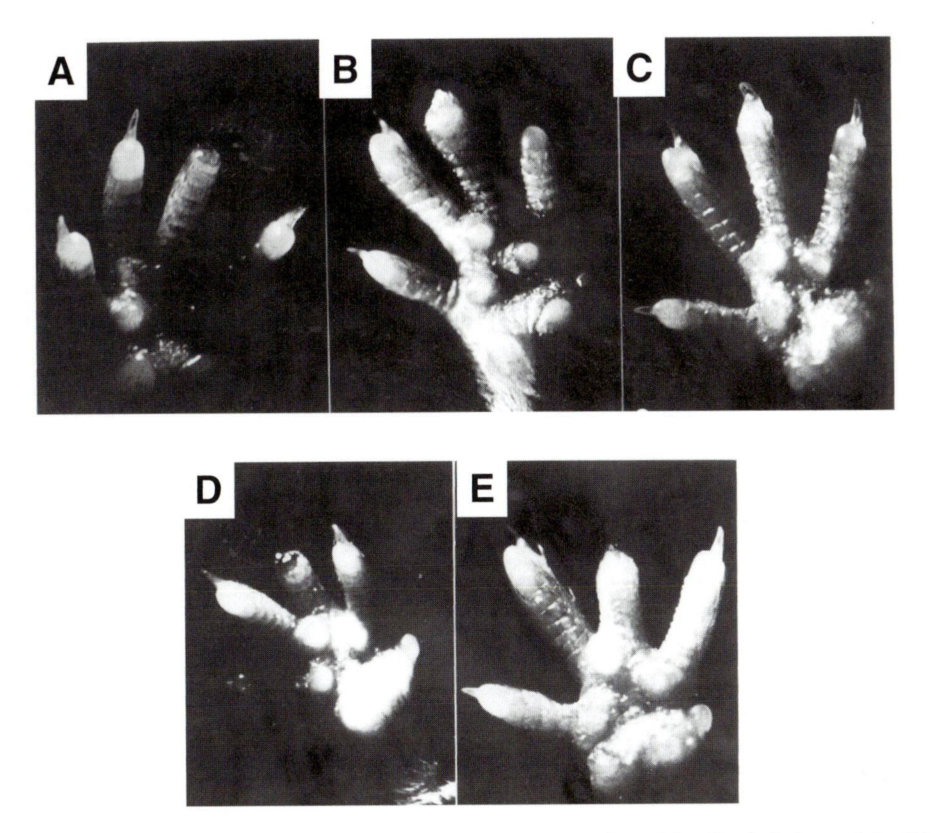


**FIGURE 14.35** Regeneration of digit tips in adult mice. **(A)** Immediately after amputation of the distal phalanx of the third toe of the foot. **(B)** Midway through the regenerative process. **(C)** Regeneration is complete, with formation of toe tip with nail. **(D)** Immediately after amputation of the third toe through the proximal phalanx. **(E)** Several weeks later, the toe has healed without regeneration. Courtesy of Dr. Richard Borgens. Reproduced with permission from Borgens, Mice regrow the tips of their foretoes. Science 217:747–750. Copyright 1982, AAAS.

is that various noxious treatments induced the accretion of extra cartilage from the amputated bone, but did not induce epimorphic regeneration.

## SUMMARY

Amphibians can replace their limbs and jaws by epimorphic regeneration. Deer, elk, and moose regenerate antlers, and the rabbit and MRL mouse regenerate ear tissue. Mice, rabbits, and humans regenerate the tips of the terminal phalanges. All of these structures provide research models by which we can further our understanding of how to expand the regenerative capacity of appendages in humans.

We know the most about mechanisms of epimorphic regeneration in the amphibian limb. Regeneration is achieved by the formation of a blastema derived by the dedifferentiation of dermis, cartilage, and muscle cells local to the amputation surface. Dedifferentiation is accomplished by the degradation of ECM by proteases, and the reversion of the liberated cells to a more embryonic phenotype. Re-entry of the blastema cells into the cell cycle is induced by an as-yet unidentified protein activated by thrombin, similar to the re-entry of PECs into the cell cycle during lens regeneration. We do not have a clear picture of the mechanism of dedifferentiation. Elements of the Notch signal transduction pathway are expressed in blastema cells, but other pathways may be involved as well. Dedifferentiation of mouse myofibers is induced *in vitro* by the trisubstituted purine myoseverin and the disubstituted purine reversine, suggesting that counterparts of these molecules might exist in the regenerating limb. The blastema cells require growth and trophic factors from the apical epidermal cap (AEC) and nerves for their survival and proliferation. Both the AEC and nerves provide FGFs for this purpose. In addition, the blastema cells are sustained by glial growth factor-2, substance P, and transferrin from nerves.

The blastema is a self-organizing entity from its inception. The developmental fate and handedness of the blastema cannot be altered by grafting it to an ectopic location, even under conditions that force it to repeat the earliest stages of blastema formation. The mechanism of self-organization may be through local cell interactions that intercalate missing limb structures within boundaries established in the early blastema.

The limb can be viewed as a three-dimensional "normal neighbor" map in which each cell knows its position relative to all other cells. When a limb is amputated, the cells that dedifferentiate inherit a positional identity of their position on the PD axis and the circumference of the limb. These positional identities constitute the PD and circumferential boundaries of what is to be regenerated. The mechanism of distal boundary formation is not yet understood, but may involve the random adoption by some dedifferentiated cells of distal positional identity followed by sorting out of these cells to a position directly adjacent to the AEC. Local cell interactions between the proximal and distal boundary cells induce proliferation and intercalation of any missing PD positional identities. Evidence from grafting experiments, labeling different populations of cells along the PD axis of the early blastema, and adhesion assays support this model. *In vitro* and *in vivo* adhesion assays, in conjunction with RA treatment, have shown that positional identity is encoded in the cell surface. One molecule that has been implicated in PD positional identity is Prod-1, a homologue of mammalian CD59 whose expression is differentially regulated by RA. Patterning genes that are activated by local cell interactions during self-organization are similar to those that have been identified in the developing embryonic limb bud. In the PD axis, *Hoxd10* and *Meis1* and *2* are involved in specification of the stylopodium and *Hoxa13* is involved in specification of the autopodium. In the AP axis, *Shh* plays a role in establishing digit number and identity, and *Lmx-1* in the development of dorsal tissue pattern.

Anuran amphibians lose their ability to regenerate as they progress from early to late tadpole stages. Blastema formation can be induced in limbs that have only recently lost regeneration competence by treatments that increase tissue breakdown, but induction of blastema formation is much more difficult in froglets and adult frogs. There is evidence that changes in the tissue environment, particularly the immune system, are responsible for the loss of regenerative capacity. Thus, one way to improve the regenerative response of adult frog limbs might be to suppress their immune system. Regenerative incompetence is associated with lack of expression of FGF 8 and 10, BMP4, and HGF, but other factors are involved as well. Molecular fingerprints have identified genes that are differentially expressed in regeneration-competent versus -deficient limbs. The identification of these genes, along with studies of their expression patterns and functional roles, promises to be of great value in determining how we might convert regenerative deficiency to regenerative competence.

The upper and lower jaws of urodele amphibians also regenerate by formation of a blastema of undifferentiated cells. Regeneration of the upper jaw is complete, including the regeneration of teeth and reserve tooth buds, but the regenerated lower jaw is missing the tongue and hyoid bone. Jaw regeneration differs from limb regeneration in that it is not dependent on nerves. Although the jaw blastema has a thickened apical wound epidermis, it is not known if jaw regeneration is dependent on this epidermis.

Appendage regeneration in mammals has been less well studied. Ear tissue of the rabbit and MRL mouse regenerates from a ring-shaped blastema after punching out a tissue button of skin and cartilage. The origin of this blastema is presumed to be local, but the cell types that contribute to it are not certain. In the MRL mouse, 16 quantitative trait loci have been identified that are associated with regeneration. It is not known whether there is a neural or wound epidermal requirement for ear tissue regeneration. Deer antlers are bony structures that regenerate from a blastema formed by MSCs from the periosteum and cells from dermis of the skin surrounding the wound created by the cast antler. The cells of the blastema proliferate and differentiate into a cartilage template. RA is a crucial stimulant of MSC proliferation, as well as chondrocyte and osteoblast differentiation. The rate at which chondrocytes differentiate is regulated by the Ihh/PhrP/PPR pathway. As it grows, bone formed by osteoblasts is remodeled by osteoclasts, whose differentiation is regulated by the PTHrp/PPR and the RANKL/RANK pathways.

Early mouse and chick limb buds can regenerate after amputation if the apical ectodermal ridge (AER) is replaced. In late fetal stages of the mouse, regenerative capacity becomes restricted to the terminal phalanges of the digits as the limb bud differentiates. Regeneration of the terminal phalange takes place by the formation of a mesenchymal blastema. Blastema formation and differentiation is promoted by Bmp4. *Msx1* expression is correlated with successful regeneration, and *Bmp4* appears to be its downstream target. Cultured human fetal digit tips also mount a regenerative response after amputation that includes intense expression of *MSX1*. The digit tips of adult mice and humans retain the capacity to regenerate amputated terminal phalanges, but this regeneration is achieved by the direct deposition of bone on pre-existing bone, rather than through a blastema. Attempts to induce the regeneration of adult digits after amputation through more proximal phalanges by irritants, trypsin, and electrical current have proven unsuccessful.

## REFERENCES

Albert P, Boilly B, Courty J, Barritault D (1987) Stimulation in cell culture of mesenchymal cells of newt limb blastemas by EDGFI Or II (basic or acidic FGF). Cell Diff 21:63–68.

Albert P, Boilly B (1988) Effect of transferrin on amphibian limb regeneration: A blastema cell culture study. Roux's Archiv Dev Biol 197:193–196.

Allan CH, Fleckman P, Gutierrez A, Caldwell T, Underwood R (2005) Expression of MSX1 and Keratins 14 and 19 after human fetal digit tip amputation in vitro. Wound Rep Reg 13:A24.

Allen SP, Maden M, Price JS (2002) A role for retinoic acid in regulating the regeneration of deer antlers. Dev Biol 251:409–423.

Amaya E, Kroll KL (1999) A method for generating transgenic frog embryos. Methods Mol Biol 97:393–414.

Boilly B, Cavanaugh KP, Thomas D, Hondermarck H, Bryant SV, Bradshaw RA (1991) Acidic fibroblast growth factor is present in regenerating limb blastemas of axolotls and bonds specifically to blastema tissues. Dev Biol 145:302–310.

Borgens RB (1982) Mice regrow the tips of their foretoes. Science 217:747–750.

Brockes JP (1984) Mitogenic growth factors and nerve dependence of limb regeneration. Science 225:1280–1287.

Brockes JP (1997) Amphibian limb regeneration: Rebuilding a complex structure. Science 276:81–87.

Brockes JP, Kintner CR (1986) Glial growth factor and nerve-dependent proliferation in the regeneration blastema of urodele amphibians. Cell 45:301–306.

Bryant SV, Iten LE (1976) Supernumerry limbs in amphibians: Experimental production in *Notophthalmus viridescens* and a new interpretation of their formation. Dev Biol 50:212–234.

Butler EG, O'Brien JP (1942) Effects of localized x-irradiation on regeneration of the urodele limb. Anat Rec 84:407–413.

Cadinouche MZA, Liversage RA, Muller W, Tsilfidis C (1999) Molecular cloning of the *Notophthalmus viridescens Radical Fringe* cDNA and characterization of its expression during forelimb development and adult forelimb regeneration. Dev Dyn 214:259–268.

Cameron JA, Hilgers AR, Hinterberger TJ (1986) Evidence that reserve cells are a source of regenerated adult newt muscle *in vitro*. Nature 321:607–610.

Cameron JA, Fallon JF (1977) Evidence for polarizing zone in the limb buds of *Xenopus laevis*. Dev Biol 55:320–330.

Carlson BM (1969) Inhibition of limb regeneration in the axolotl after treatment of the skin with Actinomycin D. Anat Rec 163: 389–402.

Carlson BM (1974) Morphogenetic interactions between rotated skin cuffs and underlying stump tissues in regenerating axolotl forelimbs. Dev Biol 39:263–285.

Carlson BM (1975) The effects of rotation and positional change of stump tissues upon morphogenesis of the regenerating axolotl limb. Dev Biol 47:269–291.

Carlson MRJ, Bryant SV, Gardiner DM (1998) Expression of *Msx-2* during development, regeneration, and wound healing in axolotl limbs. J Exp Zool 282:715–723.

Casimir CM, Gates P, Patient RK, Brockes JP (1988) Evidence for dedifferentiation and metaplasia in amphibian limb regeneration from inheritance of DNA methylation. Development 104:657–668.

Castilla M, Tassava RA (1992) Extraction of the WE3 antigen and comparison of reactivities of mAbs WE3 and WE4 in adult newt regenerate epithelium and body tissues. In: Taban CH, Boilly B (eds.). Keys for Regeneration, Monographs in Dev Biol. Basel: Karger, pp 116–130.

Cecil ML, Tassava RA (1986) Forelimb regeneration in the postmetamorphic bullfrog: Stimulation by dimethyl sulfoxide and retinoic acid. J Exp Zool 239:57–61.

Chan WY, Lee KKH, Tam PPL (1991) The regenerative capacity of forelimb buds after amputation in mouse embryos at the early organogenesis stage. J Exp Zool 260:74–83.

Chen S, Zhang Q, Wu X, Schultz PG, Ding S (2004) Differentiation of lineage-committed cells by a small molecule. J Am Chem Soc 126:410–411.

Chew K, Cameron JA (1983) Increase in mitotic activity of regenerating axolotl limbs by growth factor-impregnated implants. J Exp Zool 226:325–329.

Crews L, Gates PB, Brown R, Joliot A, Foley C, Brockes, J, Gann A (1995) Expression and activity of the newt Msx-1 gene in relation to limb regeneration. Proc Royal Soc London [Biol] 259:161–171.

Christen B, Slack JMW (1997) *FGF-8* is associated with anteroposterior patterning and limb regeneration in *Xenopus*. Dev Biol 192:455–466.

Christen B, Slack JMW (1998) All limbs are not the same. Nature 395:230–231.

Christensen RN, Tassava RA (2000) Apical epithelial cap morphology and fibronectin gene expression in regenerating axolotl limbs. Dev Dyn 217:216–224.

Christensen RN, Weinstein M, Tassava RA (2001) Fibroblast growth factors in regenerating limbs of *Ambystoma*: Cloning and semiquantitative RT-PCR expression studies. J Exp Zool 290:529–540.

Christensen RN, Weinstein M, Tassava RA (2002) Expression of fibroblast growth factors 4, 8, and 10 in limbs, flanks, and blastemas of *Ambystoma*. Dev Dyn 223:193–203.

Crawford K, Stocum DL (1988a) Retinoic acid coordinately proximalizes regenerate pattern and blastema differential affinity in axolotl limbs. Development 102:687–698.

Crawford K, Stocum DL (1988b) Retinoic acid proximalizes level-specific properties responsible for intercalary regeneration in axolotl limbs. Development 104:703–712.

Crews L, Gates PB, Brown R, Joliot A, Foley C, Brockes, J, Gann A (1995) Expression and activity of the newt Msx-1 gene in relation to limb regeneration. Proc Royal Soc London [Biol] 259:161–171.

Dearlove GD, Dresden M (1976) Regenerative abnormalities in *Notophthalmus viridescens* induced by repeated amputations. J Exp Zool 196:251–262.

De Luca F, Uyeda JA, Mericq V, Mancilla EE, Yanovski JA, Barnes KM, Zile MH, Baron J (2000) Retinoic acid is a potent regulator of growt plate chondrogenesis. Endocrinology 141:346–353.

Dent JN (1964) Limb regeneration in larvae and metamorphosing individuals of the South African clawed toad. J Morph 110: 61–77.

Desquenne-Clark L, Clark HK, Heber-Katz E (1998) A new murine model for mammalian wound repair and regeneration. Clin Immunol Immunopath 88:35–45.

Desselle J-C, Gontcharoff M (1978) Cytophotometric detection of the participation of cartilage grafts in regeneration of x-rayed urodele limbs. Biol Cellul 33:45–54.

Deuchar EM (1976) Regeneration of amputated limb buds in early rat embryos. J Embryoll Exp Morph 35:345–354.

D'Jamoos CA, McMahon G, Tsonis PA (1998) Fibroblast growth factor receptors regulate the ability for hindlimb regeneration in *Xenopus laevis*. Wound Rep Reg 6:388–397.

Douglas BS (1972) Conservative management of guillotine amputation of the finger in children. Aust Paediatr J 8:86–89.

Dresden MH (1969) Denervation effects on newt limb regeneration: DNA, RNA, and protein synthesis. Dev Biol 19:311–320.

Dresden MH, Gross J (1970) The collagenolytic enzyme of the regenerating limb of the newt *Triturus viridescens*. Dev Biol 22: 129–137.

Du Pasquier L, Flajnik M (1999) Origin and evolution of the vertebrate immune system. In: Paul WE (ed.). Fundamantal Immunology. Philadelphia: Lippincott-Raven, pp 605–650.

Du Pasquier L, Schwager J, Flajnik MF (1989) The immune system of *Xenopus*. Ann Rev Immunol 7:251–275.

Duckmanton A, Kumar A, Chang Y-T, Brockes J (2005) A single cell analysis of myogenic dedifferentiation induced by small molecules. Chemistry and Biology 12:1117–1126.

Dudley AT, Ros MA, Tabin CJ (2002) A re-examination of proximodistal patterning during vertebrate limb development. Nature 418:539–544.

Dungan KM, Wei TY, Nace JD, Poulin ML, Chiu I-M, Lang JC, Tassava RA (2002) Expression and biological effect of urodele fibroblast growth factor 1: Relationship to limb regeneration. J Exp Zool 292:540–554.

Dunis DA, Namenwirth M (1977) The role of grafted skin in the regeneration of X-irradiated axolotl limbs. Dev Biol 56:97–109.

Echeverri K, Clarke DW, Tanaka EM (2001) *In vivo* imaging indicates muscle fiber dedifferentiation is a major contributor to the regenerating tail blastema. Dev Biol 236:151–164.

Echeverri K, Tanaka EM (2005) Proximodistal patterning during limb regeneration. Dev Biol 279:391–401.

Egar MW (1993) Affinophoresis as a test of axolotl accessory limbs. In: Fallon JF, Goetinck PF, Kelley RO, Stocum DL (eds.). Limb Development and Regeneration, Part B. New York: Wiley-Liss, pp 203–211.

Endo T, Yokoyama H, Tamura K, Ide H (1997) *Shh* expression in in developing and regenerating limb buds of *Xenopus laevis*. Dev Dyn 209:227–232.

Fallon JF, Crosby GM (1977) Polarizing zone activity in limb buds of amniotes. In: Ede DA, Hinchliffe JR, Balls M (eds.). Vertebrate Limb and Somite Morphogenesis. Cambridge: Cambridge University Press, pp 55–70.

Fallon JF, Lopez A, Ros, MA, Savage MP, Olwin BB, Simandl BK (1994) Apical ectodermal ridge growth signal for chick limb development. Science 264:104–107.

Faucheux C, Nesbitt SA, Horton MA, Price JS (2001) Cells in regenerating deer antler cartilage provide a microenvironment that supports osteoclast differentiation. J Exp Biol 204:443–455.

Faucheux C, Horton MA, Price JS (2002) Nuclear localization of type I parathyroid hormone/parathyroid hormone-related protein receptors in deer antler osteoclasts: Evidence for parathyroid hormone-related protein and receptor activator of NF-κB-dependent effects on osteoclast formation in regenerating mammalian bone. J Bone Min Res 17:455–464.

Faucheux C, Nicholis BM, Allen S, Danks JA, Horton MA, Price JS (2004) Recapitulation of the parathyroid hormone-related peptide-indian hedgehog pathway in the regenerating deer antler. Dev Dynam 231:88–97.

Fekete D, Brockes JP (1987) A monoclonal antibody detects a difference in the cellular composition of developing and regenerating limbs of newts. Development 99:589–602.

Fekete DM, Brockes JP (1988) Evidence that the nerve controls molecular identity of progenitor cells for limb regeneration. Development 103:567–573.

Ferretti P, Brockes JP (1991) Cell origin and identity in limb regeneration and development. Glia 4:214–224.

Ferretti P, Ghosh S (1997) Expression of regeneration-associated cytoskeletal proteins reveals differences and similarities between regenerating organs. Dev Dyn 210:288–304.

Feynman R, Leighton R, Sands M (1977) The Feynman Lectures on Physics, Vol II. New York: Addison-Wesley Publishing Company, pp 19-4–19-14.

Feynman RP, Leighton RB, Sands M (1977) The Feynman Lectures on Physics, Vol I. New York: Addison-Wesley Publishing Company, pp 26-3–26-8.

Filoni S, Bernardini S, Cannata SM (1991) The influence of denervation on grafted hindlimb regeneration of larval *Xenopus laevis*. J Exp Zool 260:210–219.

Finch RA (1969) The influence of the nerve on lower jaw regeneration in the adult newt *Triturus viridescens*. J Morph 129:401–414.

Fleming M, Tassava RA (1981) Preamputation and postamputation histology of the neonatal opossum hindlimb: Implications for regeneration experiments. J Exp Zool 215:143–149.

French V, Bryant PJ, Bryant SV (1976) Pattern regulation in epimorphic fields. Science 193:969–981.

Ganan Y, Macias D, Duterque-oquillaud M, Ros MA, Hurle JM (1996) Role of TGF betas and BMPs as signals controlling the position of the digits and the areas of interdigital cell death in the developing chick limb autopod. Development 122:2349–2357.

Gardiner DM, Blumberg B, Komine Y, Bryant SV (1995) Regulation of *HoxA* expression in developing and regenerating axolotl limbs. Development 121:1731–1741.

Gardiner DM, Bryant SV (1996) Molecular mechanisms in the control of limb regeneration: The role of homeobox genes. Int J Dev Biol 40:797–805.

Geraudie J, Ferretti P (1998) Gene expression during amphibian limb regeneration. Int J Cytol 180:1–50.

Ghosh S, Thorogood P, Ferrretti P (1994) Regenerative capacity of upper and lower jaws in urodele amphibians. Int J Dev Biol 38: 479–490.

Ghosh S, Thorogood P, Ferretti P (1996) Regeneration of lower and upper jaws in urodeles is differentially affected by retinoic acid. Int J Dev Biol 40:1161–1170.

Giguere V, Ong S, Evans RM, Tabin CJ (1989) Spatial and temporal expression of the retinoic acid receptor in the regenerating amphibian limb. Nature 337:566–569.

Globus M (1988) A neuromitogenic role for substance P in urodele limb regeneration. In: Inoue S, Shirai T, Egar M, Aiyama S, Geraudie J, Nobunaga T, Sato NL (eds.). Regeneration and Development. Maebashi: Okada Printing & Publishing, pp 675–685.

Globus M, Alles P (1990) A search for immunoreactive substance P and other neural peptides in the limb regenerate of the newt *Notophthalmus viridescens*. J Exp Zool 254:165–176.

Globus M, Vethamany-Globus S, Lee YCI (1980) Effect of apical epidermal cap on mitotic cycle and cartilage differentiation in regeneration blastemata in the newt, *Notophthalmus viridescens*. Dev Biol 75:358–372.

Globus M, Vethamany-Globus S (1985) *In vitro* studies of controlling factors in newt limb regeneration. In: Sicard RE (ed.). Regulation of Vertebrate Limb Regeneration. New York: Oxford University Press, pp 106–127.

Goldhamer DJ, Tassava RA (1987) An analysis of proliferative activity in innervated and denervated forelimb regenerates of the newt *Notophthalmus viridescens*. Development 100:619–628.

Goss RJ (1956a) Regenerative inhibition following limb amputation and immediate insertion into the body cavity. Anat Rec 126: 15–27.

Goss RJ (1956b) The regenerative responses of amputated limbs to delayed insertion into the body cavity. Anat Rec 126:283–297.

Goss RJ (1969a) Photoperiodic control of antler cycles in deer I. Phase shift and frequency changes. J Exp Zool 170:311–324.

Goss RJ (1969b) Photoperiodic control of antler cycles in deer II. Alterations in amplitude. J Exp Zool 171:223–234.

Goss R (1969d) Principles of Regeneration. New York: Academic Press.

Goss RJ (1970) Problems of antlerogenesis. Clin Orthopaed 69:227–238.

Goss RJ (1985) Induction of deer antlers by transplanted periosteum I. Graft size and shape. J Exp Zool 235:359–373.

Goss RJ (1985) Tissue differentiation in regenerating antlers. Royal Soc New Zealand 22:229–238.

Goss RJ (1992) The mechanism of antler casting in the fallow deer. J Exp Zool 264:429–436.
Goss RJ (1995) Future directions in antler research. Anat Rec 241:291–302.
Goss RJ, Stagg MW (1958a) Regeneration of lower jaws in adult newts. J Morph 102:289–310.
Goss RJ, Stagg MW (1958b) Regeneration in lower jaws of newts after excision of the intermandibular regions. J Exp Zool 137: 1–12.
Goss RJ, Severinghaus CW, Free S (1964) Tissue relationships in the development of pedicles and antlers in the Virginia deer. J Mammol 45:61–68.
Goss R, Grimes LN (1975a) Tissue interactions in the regeneration of rabbit ear holes. Am Zool 12:151–157.
Goss RJ, Grimes LN (1975b) Epidermal downgrowths in in regenerating rabbit ear holes. J Morph 146:533–542.
Goss RJ, Van Praagh A, Brewer P (1992) The mechanism of antler casting in the fallow deer. J Exp Zool 264:429–436.
Gourevich D, Clark L, Chen P, Seitz A, Samulewicz S, Heber-Katz E (2003) Matrix metalloproteinase activity correlates with blastema formation in the regenerating MRL ear hole model. Dev Dynam 226:377–387.
Graver H (1972) The polarity of the dental lamina in the regenerating salamander jaw. J Embryol Exp Morph 30:635–646.
Grillo H, Lapiere CM, Dresden MH, Gross J (1968) Collagenolytic activity in regenerating forelimbs of the adult newt (*Triturus viridescens*). Dev Biol 17:571–583.
Grimes LN (1974) Selective x-irradiation of the cartilage at the regenerating margin of rabbit ear holes. J Exp Zool 206:237–241.
Grimes LN, Goss RJ (1970) Regeneration of holes in rabbit ears. Am Zool 10:537.
Guyenot E (1927) La perte du pouvoir régenerateur des Anoures, etudiée par les heterogreffess, et la notion de territories. Rev Suisse Zool 34:1–54.
Han M-J, An J-Y, Kim W-S (2001) Expression patterns of *Fgf-8* during development and limb regeneration of the axolotl. Dev Dyn 220:40–48.
Han M-J, Yang X, Farrington JE, Muneoka K (2003) Digit regeneration is regulated by *Msx1* and BM4 in fetal mice. Development 130:5123–5132.
Hartwig H (1968a) Durch periosteuerlagerung experimentelle erzegte, heterotope Stirnzapfenbildung beim Reh. Z Saugetierkunde 33:246–248.
Hartwig H (1968b) Verhinderung der Rosenstock und Stangenbildung beim Reh, Capreolus capreolus, durch Periostausschaltung. Zool Garten 35:252–255.
Hartwig H, Schrudde J (1974) Experimentelle Untersuchen zur Bildung der primaren Stirnauswuchse beim Reh (Capreolus capreolus). Z Jagdwiss 20:1–13.
Harty M, Neff AW, King MW, Mescher AL (2003) Regeneration or scarring: An immunologic perspective. Dev Dynam 226:268–279.
Hay ED (1959) Electron microscopic observations of muscle dedifferentiation in regenerating *Amblystoma* limbs. Dev Biol 3:26–59.
Hay ED, Fischman DA (1961) Origin of the blastema in in the regenerating newt *Triturus viridescens*. An autoradiographic study using tritiated thymidine to follow cell proliferation and migration. Dev Biol 3:26–59.
Hayamizu TF, Wanek N, Taylor G, Trevino C, Shi C, Anderson R, Gardiner DM, Muneoka K, Bryant SV (1994) Regeneration of *HoxD* expression domains during pattern regulation in chick wing buds. Dev Biol 161:504–512.
Heber-Katz E (2004) Spallanzani's mouse: A model of restoration and regeneration. In: Heber-Katz E (ed.). Regeneration: Stem Cells and Beyond, Curr Topics Microbiol Immunol 280:165–189.
Heber-Katz E, Chen P, Clark L, Zhang X-M, Troutman S, Blankenhorn EP (2004) Regeneration in MRL mice: Further genetic loci controlling the ear hole closure trait using MRL and *M. m. Castaneus* mice. Wound Rep Reg 12:384–392.
Hill RE, Jones PF, Rees AR, Sime CM, Justice MJ, Copeland NG, Jenkins NA, Graham E, Davidson DR (1989) A new family of mouse homeobox-containing genes: Molecular structure, chromosomal location, and developmental expression of Hox-7.1. Genes Dev 3:26–37.
Hill DS, Ragsdale CW, Brockes JP (1993) Isoform-specific immunological detection of newt retinoic acid receptor δ1 in normal and regenerating limbs. Development 117:937–945.
Holder N (1989) Organization of connective tissue patterns by dermal fibroblasts in the regenerating axolotl limb. Development 105:585–593.
Holder N, Tank PW (1979) Morphogenetic interactions occurring between blastemas and stumps after exchanging blastemas between normal and double-half forelimbs in the axolotl *Ambystoma mexicanum*. Dev Biol 74:302–314.
Ide H, Wada N, Uchiyama K (1994) Sorting out of cells from different parts and stages of the chick limb bud. Dev Biol 162: 71–76.
Ide H, Yokoyama H, Endo T, Omi M, Tamura K, Wada N (1998) Pattern formation in dissociated limb bud mesenchyme in vitro and in vivo. Wound Rep Reg 6:398–402.
Illingworth CM (1974) Trapped fingers and amputated fingertips in children. J Paediatr Surg 9:853–858.
Imokawa Y, Yoshizato K (1997) Expression of *Sonic hedgehog* gene in regenerating newt limb blastemas recapitulates that in developing limb buds. Proc Natl Acad Sci USA 94:9159–9164.
Imokawa Y, Yoshizato K (1998) Expression of *Sonic hedgehog* gene in regenerating newt limbs. Wound Rep Reg 6:366–470.
Irvin BC, Tassava RA (1998) Effects of peripheral nerve implants on the regeneration of partially and fully innervated urodele forelimbs. Wound Rep Reg 6:382–387.
Iten LE, Bryant SV (1973) Forelimb regeneration from different levels of amputation in the newt, *Notophthalmus viridescens*: Length, rate, and stages. W Roux Archiv 173:263–282.
Iten LE, Bryant SV (1975) The interaction between the blastema and stump in the establishment of the anterior-posterior and proximal-distal organization of the limb regenerate. Dev Biol 44: 119–147.
Izutsu Y, Yoshizato K (1993) Metamorphosis-dependent recognition of larval skin as non-self by inbred adult frogs (*Xenopus laevis*). J Exp Zool 266:163–167.
Johnson KJ, Scadding SR (1992) Effects of tunicamycin in retinoic acid induced respecification of positional vaues in regenerating limbs of the larval axolotl, *Ambystoma mexicanum*. Dev Dyn 193:185–192.
Joseph J, Dyson M (1966) Tissue replacement in the rabbit's ear. Brit J Surg 53:372–380.
Ju B-G, Kim W-S (1998) Upregulation of cathepsin D expression in the dedifferentiating salamander limb regenerate and enhancement of its expression by retinoic acid. Wound Rep Reg 6: S349–S358.
Kierdorf U, Bartos L (1999) Treatment of the growing pedicle with retinoic acid increased the size of first antlers in fallow deer (*Dama dama* L). Comp Biochem Physiol 124:7–9.
Kierdorf U, Kierdorf H (1998) Effects of retinoic acid on pedicle and first antler growth in a fallow buck (*Dama dama* L). Ann Anat 180:373–375.
Kiffmeyer WR, Tomusk EV, Mescher AL (1991) Axonal transport and release of transferring in nerves of regenerating amphibian limbs. Dev Biol 147:392–402.

Kim W-S, Stocum DL (1986a) Retinoic acid modifies positional memory in the anteroposterior axis of regenerating axolotl limbs. Dev Biol 114:170–179.

Kim W-S, Stocum DL (1986b) Effects of retinoic acid on regenerating normal and double half limbs of axolotls. Roux' Archiv Dev Biol 195:243–251.

King MW, Nguyen T, Calley J, Harty MW, Muzinich MC, Mescher AL, Chalfant C, N'Cho M, McLeaster K, McEntire J, Stocum DL, Smith RC, Neff AW (2003) Identification of genes expressed during *Xenopus laevis* limb regeneration by using subtractive hybridization. Dev Dynam 226:398–409.

Kintner CR, Brockes JP (1984) Monoclonal antibodies identify blastemal cells derived from dedifferentiating muscle in newt limb regeneration. Nature 308:67–69.

Korneluk RG, Liversage RA (1984) Tissue regeneration in the amputated forelimb of *Xenopus laevis* froglets. Canadian J Zool 62:2383–2391.

Koshiba K, Kuroiwa A, Yamamoto H, Tamura K, Ide H (1998) Expression of Msx genes in regenerating and developing limbs of axolotl. J Exp Zool 282:703–714.

Kostakopoulou K, Vogel A, Brickell P, Tickle C (1996) "Regeneration" of wing bud stumps of chick embryos and reactivation of Msx-1 and Shh expression in response to FGF-4 and ridge signals. Mech Dev 55:119–131.

Kostakopoulou K, Vargesson N, Carke JD, Brickell PM, Tickle C (1997) Local origin of cells in FGF-4-induced outgrowth of amputated chick wing bud stumps. Int J Dev Biol 41:747–750.

Koyama E, Golden E, Kirsch T, Adams SL, Chandraratna RAS, Michaille J-J, Pacifici M (1999) Retinoid signaling is required for chondrocyte maturation and endochondral bone formation during limb skeletogenesis. Dev Biol 208:375–391.

Kumar A, Velloso C, Imokawa Y, Brockes JP (2000) Plasticity of retrovirus-labelled myotubes in the newt regeneration blastema. Dev Biol 218:125–136.

Kumar A, Velloso C, Imokawa Y, Brockes JP (2004) The regenerative plasticity of isolated urodele myofibers and its dependence on MSX1. PloS Biol 2:E218.

Kurabuchi S, Inoue S (1983) Denervation effects on limb regeneration in postmetamorphic *Xenopus laevis*. Dev Growth Diff 25: 463–467.

Laufer E, Dahn R, Orozco OE, Yeo CY, Pisenti J, Henrique D, Abbot UK, Fallon JF, Tabin C (1997) Expression of *Radical fringe* in limb-bud ectoderm regulates apical ectodermal ridge formation. Nature 386:366–373.

Lebowitz P, Singer M (1970) Neurotrophic control of protein synthesis in the regenerating limb of the newt *Triturus*. Nature 225:824–827.

Lee KK, Chan WY (1991) A study on the regenerative potential of partially excised mouse embryonic fore-limb bud. Anat Embryol 184:153–157.

Lheureux E (1983) Replacement of irradiated epidermis by migration of non-irradiated epidermis in the newt limb: The necessity of healthy epidermis for regeneration. J Embryol Exp Morph 76:217–234.

Lheureux E, Carey F (1988) The irradiated epidermis inhibits newt limb regeneration by preventing blastema growth. Biol Struct Morph 1:49–57.

Li X, Gu W, Masinde G, Hamilton-Ulland M, Xu S, Mohan S, Baylink DJ (2001) Genetic control of the rate of wound healing in mice. Heredity 86:668–674.

Litwiller R (1939) Mitotic index and size in regenerating amphibian limbs. J Exp Zool 82:273–286.

Lo DC, Allen F, Brockes JP (1993) Reversal of muscle differentiation during urodele limb regeneration. Proc Natl Acad Sci (USA) 90:7230–7234.

Louis DS, Palmer AK, Burney RE (1980) Open treatment of digital tip injuries. J Amer Med Assoc 244:697–698.

Loyd RM, Tassava RA (1980) DNA synthesis and mitosis in adult newt limbs following amputation and insertion into the body cavity. J Exp Zool 214:61–69.

Ludolph D, Cameron JA, Stocum DL (1990) The effect of retinoic acid on positional memory in the dorsoventral axis of regenerating axolotl limbs. Dev Biol 140:41–52.

Maden M (1977) The regeneration of positional information in the amphibian limb. J Theor Biol 69:735–753.

Maden M (1979) Neurotrophic and X-ray blocks in the blastemal cell cycle. J Embryol Exp Morphol 50:169–173.

Maden M (1981) Morphallaxis in an epimorphic system: Size, growth control and pattern formation during amphibian limb regeneration. J Embryol Exp Morph 65:151–167.

Maden M (1982) Vitamin A and pattern formation in regenerating limbs. Nature 295:672–675.

Maden M (1983) The effects of vitamin A on limb regeneration in *Rana temporaria*. Dev Biol 98:409–416.

Maden M (1997) Retinoic acid and its receptors in limb regeneration. Sem Cell Dev Biol 8:445–453.

Maden M, Wallace H (1975) The origin of limb regenerates from cartilage grafts. Acta Embryol Exp 2:77–86.

Martin GR (1998) The roles of FGFs in the early development of vertebrate limbs. Genes Dev 12:1571–1586.

Masinde GL, Li X, Gu W, Davidson H, Mohan S, Baylink D (2001) Identification of wound healing/regeneration quantitative trait loci (QTL) at multiple time points that explain seventy percent of variance in (MRL/MpJ and SJL/J) mice $F_2$ population. Genome Res 11:2027–2033.

Matsuda H, Yokoyama H, Endo T, Tamura K, Ide H (2001) An epidermal signal regulates *Lmx-1* expression and dorsal-ventral pattern during *Xenopus* limb regeneration. Dev Biol 229:351–362.

McBrearty BA, Clark LD, Zhang XM, Blankenhorn EP, Heber-Katz E (1998) Genetic analysis of a mammalian wound-healing trait. Proc Natl Acad Sci USA 95:11792–11797.

Mercader N, Tanaka E, Torres M (2005) Proximodistal identity during vertebrate limb regeneration is regulated by Meis homeodomain proteins. Development 132:4131–4142.

Mescher AL (1976) Effects on adult newt limb regeneration of partial and complete skin flaps over the amputation surface. J Exp Zool 195:117–128.

Mescher AL (1996) The cellular basis of limb regeneration in urodeles. Int J Dev Biol 40:785–796.

Mescher AL, Cox CA (1988) Hyaluronate accumulation and nerve-dependent growth during regeneration of larval *Ambystoma* limbs. Differentiation 38:161–168.

Mescher AL, Munaim SI (1986) Changes in the extracellular matrix and glycosaminoglycan synthesis during the imnitiation of regeneration in adult newt forelimbs. Anat Rec 214:424–431.

Mescher AL, Munaim SI (1988) Transferrin and the growth-promoting effect of nerves. Int Rev Cytol 100:1–26.

Mescher AL, Kiffmeyer WR (1992) Axonal release of transferrin in peripheral nerves of axolotls during regeneration. In: Taban CH, Boilly B (eds.). Keys for Regeneration: Monographs in Dev Biol 23:100–109.

Mescher AL, Connell E, Hsu C, Patel C, Overton B (1997) Transferrin is necessary and sufficient for the neural effect on growth in amphibian limb regeneration blastemas. Dev Growth Diff 39:677–684.

Mescher AL, White GW, Brokaw JJ (2000) Apoptosis in regenerating and denervated, nonregenerating urodele forelimbs. Wound Rep Reg 8:110–116.

Mescher AL, Neff AW (2005) Regenerative capacity and the developing immune system. Adv Biochem Eng/Biotech 93:39–66.

Metcalfe AD, Ferguson MWJ (2005) Harnessing wound healing and regeneration for tissue engineering. Biochem Soc Trans 33 (Part 2):413–417.

Mittenthal JE (1981) The rule of normal neighbors: A hypothesis for morphogenetic pattern regulation. Dev Biol 88:15–26.

Miyazaki K, Uchiyawa K, Imokawa Y, Yoshizato K (1996) Cloning and characterization of of cDNAs for matrix metalloproteinases of regenerating newt limbs. Proc Natl Acad Sci USA 93:6819–6824.

Mizell M (1968) Limb regeneration: induction in the new-born opossum. Science 161:283–286.

Mizell M, Isaacs JJ (1970) Induced regeneration of hindlimbs in the newborn opossum. Am Zool 10:141–155.

Mohammad KS, Day FA, Neufeld DA (1999) Bone growth is induced by nail transplantation in amputated proximal phalanges. Calcif Tiss Int 65:408–410.

Monkmeyer J, Ludolph DC, Cameron JA, Stocum DL (1992) Retinoic acid-induced change in anteroposterior positional identity in regenerating axolotl limbs is dose-dependent. Dev Dynam 193:286–294.

Morais da Silva SM, Gates PB, Brockes JP (2002) The newt ortholog of CD59 is implicated in proximodistal identity during amphibian limb regeneration. Dev Cell 3:547–555.

Morzlock FV, Stocum DL (1972) Neural control of RNA synthesis in regenerating limbs of the adult newt *Triturus viridescens*. Roux's Archiv Dev Biol 171:170–180.

Mullen LM, Bryant SV, Torok MA, Blumberg B, Gardiner DM (1996) Nerve dependency of regeneration: The role of *Distal-less* and FGF signaling in amphibian limb regeneration. Development 122:3487–3497.

Muller TL, Ngo-Muller V, Reginelli A, Taylor G, Anderson R, Muneoka K (1999) Regeneration in higher vertebrates: Limb buds and digit tips. Sem Cell Dev Biol 10:405–413.

Munaim SI, Mescher AL (1986) Transferrin and the trophic effect of neural tissue on amphibian limb regeneration blastemas. Dev Biol 116:138–142.

Muneoka K, Fox WF, Bryant SV (1986) Cellular contribution from dermis and cartilage to the regenerating limb blastema in axolotls. Dev Biol 116:256–260.

Muneoka K, Holler-Dinsmore G, Bryant SV (1986) Intrinsic control of regenerative loss in *Xenopus laevis* limbs. J Exp Zool 240:47–54.

Namenwirth M (1974) The inheritance of cell differentiation during limb regeneration in the axolotl. Dev Biol 41:42–56.

Nardi JB, Stocum DL (1983) Surface properties of regenerating limb cells: Evidence for gradation along the proximodistal axis. Differentiation 25:27–31.

Neff AW, King MW, Harty MW, Nguyen T, Calley J, Smith RC, Mescher AL (2005) Expression of *Xenopus XlSALL4* during limb development and regeneration. Dev Dynam 233:356–367.

Neufeld D, Zhao W (1995) Bone regrowth following gigit-tip amputation in mice is equivalent in adults and neonates. Wound Rep Reg 3:461–466.

Niazi IA (1996) Background to work on retinoids and amphibian limb regeneration: Studies on anuran tadpoles—a retrospect. J Biosci 21:273–297.

Niazi IA, Saxena S (1978) Abnormal hindlimb regeneration in tadpoles of the toad *Bufo andersoni*, exposed to vitamin A. Folia Biol (Krakow) 26:3–11.

Niazi IA, Jangir OP, Sharma KK (1979) Forelimb regeneration at wrist level in adults of skipper frog *Rana cyanophylctis* (Schneider) and its improvement by vitamin A treatment. Indian J Exp Biol 17:435–437.

Niazi IA, Pescetelli MJ, Stocum DL (1985) Stage dependent effects of retinoic acid on regenerating urodele limbs. Roux's Archiv Dev Biol 194:355–363.

Nicholas JS (1926) Extirpation experiments upon the embryonic forelimb of the rat. Proc Soc Exp Biol Med 23:436–439.

Nilsson J, von Euler AM, Dalsgaard C-J (1985) Stimulation of connective tissue growth by substance P and substance K. Nature 315:61–63.

Niswander L, Tickle C, Vogel A, Booth I, Martin GR (1993) FGF-4 replaces the apical ectodermal ridge and directes outgrowth and patterning of the limb. Cell 75:579–587.

Niswander L (2003) Pattern formation: old models out on a limb. Nature Rev Genetics 4:133–143.

Nye HLD, Cameron JA, Chernoff EG, Stocum DL (2003) Regeneration of the urodele limb: A review. Dev Dynam 226:280–294.

Odelberg SJ, Kollhof A, Keating M (2001) Dedifferentiation of mammalian myotubes induced by msx-1. Cell 103:1099–1109.

Oliva A, Della Ragione F, Fratta M, Marrone G, Palumbo R, Zappia V (1993) Effect of retinoic acid on osteocalcin gene expression in human osteoblasts. Biochem Biophys Res Comm 191:908–914.

Onda H, Tassava RA (1991) Expression of the 9G1 antigen in the apical cap of axolotl regenerates requires nerves and mesenchyme. J Exp Zool 257:336–349.

Oudkhir M, Boilly B, Lheureux E, Lasalle B (1985) Influence of denervation on the regeneration of *Pleurodele* limbs: Cytophotometric study of nuclear DNA from blastema cells. Differentiation 29:116–120.

Park I-S, Kim W-S (1999) Modification of gelatinase activity correlates with the dedifferentiation profile of regenerating axolotl limbs. Mol Cells 9:119–126.

Pecorino LT, Entwistle, Brockes JP (1996) Activation of a single retinoic acid receptor isoform mediates proximodistal respecification. Curr Biol 65:63–69.

Pescitelli MJ, Stocum DL (1980) The origin of skeletal structures during intercalary regeneration of larval *Ambystoma* limbs. Dev Biol 79:255–275.

Piatt J (1957) Studies on the problem of nerve pattern. III. Innervation of the regenerated forelimb in *Amblystoma*. J Exp Zool 136:229–248.

Pittman RN, Buettner HM (1989) Degradation of extracellular matrix by neuronal proteases. Dev Neurosci 11:361–375.

Polezhaev LV (1972) Loss and Restoration of Regenerative Capacity in Tissues and Organs of Animals. Jerusalem: Keter Press, pp 153–199.

Poulin ML, Patrie KM, Botelho MJ, Tassava RA, Chiu I-M (1993) Heterogeneity in expression of fibroblast growth factor receptors during limb regeneration in newts (*Notophthalmus viridescens*). Development 119:353–361.

Poulin ML, Chiu I-M (1995) Re-programming of expression of the KGFR and *bek* variants of fibroblast growth factor receptor 2 during limb regeneration in newts (*Notopthalmus viridescens*). Dev Dyn 202:378–387.

Price JS, Oyajobi BO, Nalin AM, Frazer A, Russell BGG, Sandell LJ (1996) Chondrogenesis in the regenerating antler tip in red deer: Expression of collagen types I, IIA, IIB, and X demonstrated by in situ nucleic acid hybridization and immunocytochemistry. Dev Dynam 205:332–347.

Ragsdale CW, Petkovich M, Gates PB, Chambon P, Brockes JP (1989) Identification of a novel retinoic acid receptor in regenerating tissues of the newt. Nature 341:654–657.

Ragsdale CW, Gates PB, Brockes JP (1992a) Identification and expression pattern of a second isoform of the newt alpha retinoic acid receptor. Nuc Acid Res 20:5851.

Ragsdale CW, Gates PB, Hill D, Brockes JP (1992b) Delta retinoic acid receptor isoform δ1 is distinguished by its exceptional N-terminal sequence and abundance in the limb regeneration blastema. Mech Dev 40:99–112.

Rajan KT, Hopkins AM (1970) Human digits in organ culture. Nature 227:621–622.

Rajnoch C, Ferguson, Metcalfe AD, Herrick SE, Willis HS, Ferguson MW (2003) Regeneration of the ear after after wounding in different mouse strains is dependent on the severity of wound trauma. Dev Dynam 226:388–397.

Reginelli AD, Wang YQ, Sassoon D, Muneoka K (1995) Digit tip regeneration correlates with regions of Msx1 (Hox7) expression in fetal and newborn mice. Development 121:1065–1076.

Richmond MJ, Pollack ED (1983) Regulation of tadpole spinal nerve growth by the regenerating limb blastema in tissue culture. J Exp Zool 225:233–242.

Riddiford LM (1960) Autoradiographic studies of tritiated thymidine infused into the blastema of the early regenerate in the adult newt, *Triturus*. J Exp Zool 144:25–32.

Robert B, Sassoon D, Jacq C, Gehring W, Buckingham M (1989) Hox-7, a mouse homeobox gene with a novel pattern of expression during embryogenesis. EMBO J 8:91–100.

Robert J, Cohen N (1998) Evolution of immune surveillance and tumor immunity: studies in *Xenopus*. Immunol Rev 166:231–243.

Rodriguez-Esteban C, Schwabe JW, De La Pena J, Foys B, Eshelman B, Ispizua-Belmonte JC (1997) *Radical fringe* positions the apical ectodermal ridge at the dorsoventral boundary of the vertebrate limb. Nature 386:360–366.

Rollins-Smith LA, Flajnik MF, Blair PJ, Davis AT, Green WF (1997) Involvement of thyroid hormones in the expression of MHC class I antigens during ontogeny in *Xenopus*. Dev Immunol 5:133–144.

Rosania GR, Chang Y-T, Perez O, Sutherlin D, Dong H, Lockhart DJ, Schultz PG (2000) Myoseverin, a microtubule-binding molecule with novel cellular effects. Nature Biotech 18:304–308.

Rosenthal LJ, Reiner MA, Bleicher MA (1979) Nonoperative management of distal fingertip amputations in children. Pediatrics 64:1–3.

Roy S, Gardiner DM, Bryant SV (2000) Vaccinia as a tool for functional analysis in regenerating limbs: ectopic expression of *Shh*. Dev Biol 218:199–205.

Said S, Parke W, Neufeld DA (2004) Vascular supplies differ in regenerating and nonregenerating amputated rodent digits. Anat Rec 278A:443–449.

Samulewicz SJ, Clark L, Seitz A, Heber-Katz E (2002) Expression of Pref-1, a Delta-like protein, in healing mouse ears. Wound Rep Reg 10:215–221.

Satoh A, Ide H, Tamura (2005a) Muscle formation in regenerating *Xenopus* froglet limb. Dev Dynam 233:337–346.

Satoh A, Suzuki M, Amano T, Tamura K, Ide H (2005b) Joint development in *Xenopus laevis* and induction of segmentation in regenerating froglet limb (spike). Dev Dynam 233:1444–1453.

Saunders JW Jr (1948) The proximo-distal sequence of the origin of the parts of the chick wing and the role of the ectoderm. J Exp Zool 108:363–403.

Scadding SR, Maden M (1994) Retinoic acid gradients during limb regeneration. Dev Biol 162:608–617.

Schmidt AJ (1966) The Molecular Basis of Regeneration: Enzymes. Illinois Monographs Med Sci 6 (4). Urbana: University of Illinois Press.

Schmidt AJ (1968) Cellular Biology of Vertebrate Regeneration and Repair. Chicago: University of Chicago Press.

Schnapp E, Tanaka EM (2005) Quantitative evaluation of morpholino-mediated protein knockdown of GFP, MSX1, and PAX7 during tail regeneration in *Ambystoma mexicanum*. Dev Dynam 232:162–170.

Schotte OE, Butler EG (1944) Phases in regeneration of the urodele limb and their dependence on the nervous system. J Exp Zool 97:95–121.

Sharma K, Niazi IA (1979) Regeneration induced in the forelimbs by treatment with vitamin A in the froglets of *Rana breviceps*. Experientia 35:1571–1572.

Shimizu-Nishikawa K, Tazawa I, Uchiyama K, Yoshizato K (1999) Expression of helix-loop-helix type negative regulators of differentiation during limb regeneration in urodeles and anurans. Dev Growth Diff 41:731–743.

Shimizu-Nishikawa KS, Tsuji S, Yoshizato K (2001) Identification and characterization of newt rad (ras associated with diabetes), a gene specifically expressed in regenerating limb muscle. Dev Dynam 220:74–86.

Shubin NH, Alberch P (1986) A morphogenetic approach to the origin and basic organization of the tetrapod limb. Evol Biol 20:318–390.

Sessions SK, Bryant V (1988) Evidence that regenerative ability is an intrinsic property of limb cells in *Xenopus*. J Exp Zool 247: 39–44.

Simon H-G, Tabin CJ (1993) Analysis of *Hox-4.5* and *Hox-3.6* expression during newt limb regeneration: Differential regulation of paralogous *Hox* genes suggests different roles for members of different *Hox* clusters. Development 111:1397–1407.

Simon H-G, Nelson C, Goff D, Laufer E, Morgan BA, Tabin C (1995) Differential expression of myogenic regulatory genes and Msx-1 during dedifferentiation and redifferentiation of regenerating amphibian limbs. Dev Dyn 202:1–12.

Simon H-G, Kittappa R, Han PA, Tsilfildis C, Liversage RA, Oppenheimer S (1997) A novel family of T-box genes in urodele amphibian limb development and regeneration: candidate genes involved in vertebrate forelimb/hindlimb patterning. Development 124:1355–1366.

Singer M (1952) The influence of the nerve in regeneraton of the amphibian extremity. Quart Rev Biol 27:169–200.

Singer M (1954) Induction of regeneration of the forelimb of the postmetamorphic frog by augmentation of the nerve supply. J Exp Zool 126:419–472.

Singer M (1965) A theory of the trophic nervous control of amphibian limb regeneration, including a re-evaluation of quantitative nerve requirements. In: Kiortsis V, Trampusch HAL (eds.). Regeneration in Animals and Related Problems. Amsterdam: North Holland, pp 20–32.

Singer M (1978) On the nature of the neurotrophic phenomenon in urodele limb regeneration. Am Zool 18:829–841.

Singer M, Craven L (1948) The growth and morphogenesis of the regenerating forelimb of adult *Triturus* following denervation at various stages of development. J Exp Zool 108:279–308.

Slootweg MC, Salles JP, Ohlsson C, De Vries CP, Engelbregt CP, Netelenbos JC (1996) J Endocrinol 150:465–472.

Smith AR, Lewis JH, Crawly A, Wolpert L (1974) A quantitative study of blastemal growth and bone regression during limb regeneration in *Triturus cristatus*. J Embryol Exp Morph 32:375–390.

Smith AR, Crawley A (1977) The pattern of cell division during growth of the blastema of regenerating newt forelimbs. J Embryol Exp Morph 37:33–48.

Smith GN, Toole BP, Gross J (1974) Hyaluronidase activity and glycosaminoglycan synthesis in the amputated newt limb: Comparison of denervated nonregenerating limbs with regenerates. Dev Biol 43:221–232.

Smith MJ, Globus M (1989) Multiple interactions in juxtaposed monolayer of amphibian neuronal, epidermal, and mesodermal limb blastema cells. In Vitro Cell Dev Biol 25:849–856.

Song K, Wang Y, Sassoon D (1992) Expression of *Hox-7.1* in myoblasts inhibits terminal differentiation and induces cell transformation. Nature 360:477–481.

Spallanzani L (1768) Prodromo di un opera da imprimersi sopra la riproduzioni animali. Modena, Giovanni Montanari. English translation, M. Maty (1769): An Essay on Animal Reproduction. London: Becket and DeHondt.

Stark RJ, Searls RL (1973) A description of chick wing development and a model of limb morphogenesis. Dev Biol 33:138–153.

Stark DR, Gates PB, Brockes JP, Ferretti P (1998) *Hedgehog* family member is expressed throughout regenerating and developing limbs. Dev Dynam 212:352–363.

Steen TP (1968) Stability of chondrocyte differentiation and contribution of muscle to cartilage during limb regeneration in the axolotl (*Siredon mexicanum*). J Exp Zool 167:49–78.

Steen TP (1973) The role of muscle cells in *Xenopus* limb regeneration. Am Zool 13:1349–1350.

Steen TP, Thornton CS (1963) Tissue interaction in amputated aneurogenic limbs of *Ambystoma* larvae. J Exp Zool 154:207–221.

Steinberg MS (1978) Cell-cell recognition in multicellular assembly: Levels of specificity. In: Curtis ASG (ed.). Cell-cell recognition. Cambridge: Cambridge University Press, pp 25–49.

Stocum DL (1968a) The urodele limb regeneration blastema: A self-organizing system. I. Differentiation *in vitro*. Dev Biol 18:441–456.

Stocum DL (1968b) The urodele limb regeneration blastema: Self-organizing system. II. Morphgenesis and differentiation of autografted whole and fractional blastemas. Dev Biol 18:457–480.

Stocum DL (1975) Regulation after proximal or distal transposition of limb regeneration blastemas and determination of the proximal boundary of the regenerate. Dev Biol 45:112–136.

Stocum DL (1978) Organization of the morphogenetic field in regenerating amphibian limbs. Am Zool 18:883–896.

Stocum DL (1979) Stages of forelimb regeneration in *Ambystoma maculatum*. J Exp Zool 209:395–416.

Stocum DL (1980a) Patterns of mitotic index, cell density and growth in regenerating *Ambystoma maculatum* limbs. J Exp Zool 212:233–242.

Stocum DL (1980b) Autonomous development of reciprocally exchanged regeneration blastemas of normal forelimbs and symmetrical hindlimbs. J Exp Zool 212:361–371.

Stocum DL (1982) Determination of axial polarity in the urodele regeneration blastema. J Embryol Exp Morph 71:193–214.

Stocum DL (1984) The urodele limb regeneration blastema: Determination and organization of the morphogenetic field. Differentiation 27:13–28.

Stocum DL (1991) Limb regeneration: A call to arms (and legs). Cell 67:5–8.

Stocum DL (1995) Wound Repair, Regeneration, and Artificial Tissues. Austin: RG Landes Co.

Stocum DL (1996) A conceptual framework for analyzing axial patterning in regenerating urodele limbs. Int J Dev Biol 40:773–784.

Stocum DL (2004) Amphibian regeneration and stem cells. In: Regeneration: Stem Cells and Beyond. Heber-Katz E (ed.). Curr Top Microbiol Immunol 280:1–70.

Stocum DL, Dearlove GE (1972) Epidermal-mesodermal interaction during morphogenesis of the limb regeneration blastema in larval salamanders. J Exp Zool 181:49–62.

Stocum DL, Melton DA (1977) Self-organizational capacity of distally transplanted limb regeneration blastemas in larval salamanders. J Exp Zool 201:451–472.

Stocum DL, Crawford K (1987) Use of retinoids to analyze the cellular basis of positional memory in regenerating axolotl limbs. Biochem Cell Biol 65:750–761.

Stocum DL, Maden M (1990) Regenerating limbs. Methods in Enzymol 190:189–201.

Sun X, Mariani FW, Martin GR (2002) Functions of FGF signaling from the apical ectodermal ridge in limb development. Nature 418:501–508.

Sussman HH (1989) Iron and tumor cell growth. In: deSousa M, Brock JH (eds.). Iron in immunity, cancer and inflammation. New York: John Wiley and Sons, pp 261–282.

Tamura K, Yokouchi Y, Kuroiwa A, Ide H (1997) Retinoic acid changes the proximodistal developmental competence and affinity of distal cells in the developing chick limb bud. Dev Biol 188:224–234.

Tanaka EM, Gann F, Gates PB, Brockes JP (1997) Newt myotubes re-enter the cell cycle by phosphorylation of the retinoblastoma protein. J Cell Biol 136:155–165.

Tanaka EM, Dreschel N, Brockes JP (1999) Thrombin regulates S phase re-entry by cultured newt myoblasts. Curr Biol 9:792–799.

Tank PW (1978) The occurrence of supernumerary limbs following blastemal transplantation in the regenerating forelimb of the axolotl, *Ambystoma mexicanum*. Dev Biol 62:143–161.

Tank PW (1981) The ability of localized implants of whole or minced dermis to disrupt pattern formation in the regenerating forelimb of the axolotl. Am J Anat 162:315–326.

Tank PW, Carlson BM, Connelly TG (1976) A staging system for forelimb regeneration in the axolotl, *Ambystoma maculatum*. J Morph 150:117–128.

Tassava RA, Garling DJ (1979) Regenerative responses in larval axolotl limbs with skin grafts over the amputation surface. J Exp Zool 208:97–110.

Tassava RA, Mescher AL (1975) The roles of injury, nerves and the wound epidermis during the initiation of amphibian limb regeneration. Differentiation 4:23–24.

Tassava RA, Olsen CL (1985) Neurotrophic influence on cellular proliferation in urodele limb regeneration: *In vivo* experiments. In: Sicard RE (ed.). Regulation of Vertebrate Limb Regeneration. New York: Oxford University Press, pp 81–92.

Taylor GP, Anderson R, Regenelli AD, Muneoka K (1994) FGF-2 induces regeneration of the chick limb bud. Dev Biol 163:282–284.

Thoms SD, Stocum DL (1984) Retinoic acid-induced pattern duplication in regenerating urodele limbs. Dev Biol 103:319–328.

Thornton CS (1938) The histogenesis of muscle in the regenerating forelimb of larval *Ambystoma punctatum*. J Morph 62:17–46.

Thornton CS (1968) Amphibian limb regeneration. Adv Morph 7:205–249.

Tickle C (2003) Patterning systems—from one end to the other. Dev Cell 4:449–458.

Tomlinson BL, Goldhamer DJ, Barger PM, Tassava RA (1985) Punctuated cell cycling in the regeneration blastema of urodele amphibians: an hypothesis. Differentiation 28:195–199.

Tomlinson B, Barger PM (1987) A test of the punctuated-cycling hypothesis in *Ambystoma* forelimb regenerates: The roles of animal size, limb regeneration, and the aneurogenic condition. Diferentiation 35:6–15.

Tomlinson BL, Tassava RL (1987) Dorsal root ganglia stimulate regeneration of denervated urodele forelimbs: timing of graft implantation with respect to denervation. Development 99:173–186.

Tonge DA, Leclere PG (2000) Directed axonal growth towards axolotl limb blastemas *in vitro*. Neurosci 100:201–211.

Toole BP, Jackson GJ, Gross J (1972) Hyaluronate in morphogenesis: Inhibition of chondrogenesis *in vitro*. Proc Natl Acad Sci USA 69:1384–1386.

Torok MA, Gardiner M, Shubin N, Bryant SV (1998) Expression of *HoxD* genes in developing and regenerating axolotl limbs. Dev Biol 200:225–233.

Torok MA, Gardiner DM, Izpisua-Belmonte J-C, Bryant SV (1999) *Sonic hedgehog* (*shh*) expression in developing and regenerating axolotl limbs. J Exp Zool 284:197–206.

Trampusch HAL (1958a) The action of X-rays on the morphogenetic field. I. Heterotopic grafts on irradiated limb. Ned Akad Weten Proc K Series C 61:417–430.

Trampusch HAL (1958b) The action of X-rays in the morphogenetic field. II. Heterotopic skin on irradiated tails. Ned Akad Weten Proc K Series C 61:530–545.

Tschumi PA (1957) The growth of hindlimb bud of *Xenopus laevis* and its dependence upon the epidermis. J Anat 91:149–173.

Turpin JB (1998) Induction and early development of of the hematopoietic and immune systems in *Xenopus*. Dev Comp Immunol 22:265–278.

Ueda Y, Kondoh H, Mizuno (2005) Generation of transgenic newt *Cynops pyrrhogaster* for regeneration study. Genesis 41:87–98.

Van Stone JM (1955) The relationship between innervation and regenerative capacity in hind limbs of *Rana sylvatica*. J Morph 97:345–391.

Vascotto S, Beug S, Liversage RA, Tsilfildis C (2005) Identification of cDNAs associated with late dedifferentiation in adult newt forelimb regeneration. Dev Dynam 233:347–355.

Velloso CP, Kumar A, Tanaka EM, Brockes JP (2000) Generation of mononucleate cells from post-mitotic myotubes proceeds in the absence of cell cycle progression. Differentiation 66:239–246.

Velloso CP, Simon A, Brockes JP (2001) Mammalian postmitotic nuclei reenter the cell cycle after serum stimulation in newt/mouse hybrid myotubes. Curr Biol 11:855–858.

Vinarsky V, Atkinson DL, Stevenson T, Keating MT, Odelberg SJ (2005) Normal newt limb regeneration requires matrix metalloproteinase function. Dev Biol 279:86–98.

Voit EO, Anton HJ, Blecker J (1985) Regenerative growth curves. Math Biosci 73:253–269.

Vorontstova MA, Liosner LD (1960) Asexual Propagation and Regeneration. New York: Permagon Press.

Vortkamp A (2001) Interaction of growth factors regulating chondrocyte differentiation in the developing embryo. Osteoarth Cartilage 9 (Suppl A):S109–S117.

Wada N, Uchiyama K, Ide H (1993) Cell sorting out and chondrogenic aggregate formation in limb bud recombinant and in culture. Dev Growth Diff 35:421–430.

Wada N, Ide H (1994) Sorting out of limb bud cells in monolayer culture. Int J Dev Biol 38:351–356.

Wada N, Kimura I, Tanaka H, Ide H, Nohno T (1998) Glycosylphosphatidylinositol-anchored cell surface proteins regulate position-specific cell affinity in the limb bud. Dev Biol 202:244–252.

Wagner W, Reichl J, Wehrman M, Zenner H-P (2001) Neonatal rat cartilage has the capacity for tissue regeneraton. Wound Rep Reg 9:531–536.

Wanek N, Muneoka K, Holler-Dinsmore G, Bryant SV (1989) A staging system for mouse limb development. J Exp Zool 249:41–49.

Wanek N, Muneoka K, Bryant SV (1989) Evidence for regulation following amputation and tissue grafting in the developing mouse limb. J Exp Zool 249:55–61.

Wang L, Marchionni MA, Tassava RA (2000) Cloning and neuronal expression of a type III newt neuregulin and rescue of denervated nerve-dependent newt limb blastemas by rhGGF2. J Neurobiol 43:150–158.

Wang Y, Sassoon D (1995) ctoderm-mesenchyme and mesenchyme-mesenchyme interactions regulate *Msx-1* expression and cellular differentiation in the murine limb bud. Dev Biol 168:374–382.

Williams-Boyce PK, Daniel JC Jr (1980) Regeneration of rabbit ear tissue. J Exp Zool 212:243–253.

Wislocki G, Waldo CM (1953) Further observations on the histological changes associated with the shedding of the antlers of the white-tailed deer (Odocoileus vitginianus borealis). Anat Rec 117:353–376.

Wolfe D, Nye HLD, Cameron J (2000) Extent of ossification at the amputation plane is correlated with the decline of blastema formation and regeneration in *Xenopus laevis* hindlmbs. Dev Dynam 218:681–697.

Wolfe AD, Crimmins G, Cameron JA, Henry JJ (2004) Early regeneration genes: Building a molecular profile for shared expression in cornea-lens transdifferentiation and hndlimb regeneration in *Xenopus laevis*. Dev Dynam 230:615–629.

Woloshin P, Song K, Degnin A, Killary DJ, Goldhamer DJ, Sassoon D, Thayer MJ (1995) MSX1 inhibits MyoD expression in fibroblast X 10T1/2 cell hybrids. Cell 82:611–620.

Yajima H, Yonei-Tamura S, Watanabe N, Tamura K, Ide H (1999) Role of N-cadherin in the sorting-out of mesenchymal cells and in the positional identity along the proximodistal axis of the chick limb bud. Dev Dynam 216:274–284.

Yang EV, Gardiner DM, Bryant SV (1999) Expression of *Mmp-9* and related matrix metalloproteinase genes during axolotl limb regeneration. Dev Dyn 216:2–9.

Yannas IV, Colt J, Wai YC (1996) Wound contraction and scar synthesis during development of the amphibian *Rana catesbiana*. Wound Rep Reg 4:431–441.

Yannas IV (2001) Tissue and organ regeneration in adults. New York: Springer-Verlag Inc, pp 138–185.

Yntema CL (1959a) Regeneration of sparsely innervated and aneurogenic forelimbs of *Ambystoma* larvae. J Exp Zool 140:101–123.

Yntema CL (1959b) Blastema formation in sparsely innervated and aneurogenic forelimbs in *Amblystoma* larvae. J Exp Zool 142: 423–440.

Yokoyama H, Yonei-Tamura S, Endo T, Izpisua-Belmonte JC, Tamura K, Ide H (2000) Mesenchyme with *fgf10* expression is responsible for regenerative capacity in *Xenopus* limb buds. Dev Biol 219:18–29.

Yokoyama H, Ide H, Tamura K (2001) FGF-10 stimulates limb regeneration ability in *Xenopus laevis*. Dev Biol 233:72–79.

Young HE, Bailey CF, Dalley BK (1983) Gross morphological analysis of limb regeneration in postmetamorphic adult *Ambystoma*. Anat Rec 206:295–306.

Young HE, Dalley B, Markwald RR (1989) Effect of selected denervations on glycoconjugate composition and tissue morphology during the initiation phase of limb regeneration in adult *Ambystoma*. Anat Rec 223:230.

Zaaijer J (1958) The effect of epithelium on the in vitro development of embryonic limb-bones of human origin. Proc K Ned Akad Wet 61:255.

Zardoya R, Abouheif A, Meyer A (1996) Evolution and orthology of hedgehog genes. Trends Genet 12:496–497.

Zeltinger J, Holbrook KA (1997) A model system for long-term serum-free suspension organ culture of human fetal tissues: Experiments on digits and skin from multiple body regions. Cell Tiss Res 290:51–60.

Zenjari C, Boilly, Hondermarck H, Boilly-Marer Y (1997) Nerve-blastema interactions induce fibroblast growth factor-1 release during limb regeneration in *Pleurodeles waltl*. Dev Growth Diff 39:15–22.

# 15 Research Issues in Regenerative Medicine

## INTRODUCTION

Where are we in our quest to regenerate tissues damaged by injury or disease, in terms of the three strategies of regenerative medicine? We have had good success in replacing hematopoietic cells with bone marrow transplants, but despite a great deal of work, much less success in using adult stem cells to replace other kinds of tissues. The extent to which new tissue differentiates after transplanting site-specific stem cells to lesions of other tissues is unclear and needs to be assessed more accurately. Furthermore, the initial excitement over the possibility that bone marrow cells might be pluripotent and thus be "universal stem cells" has been blunted by the generally low rates of lineage conversion observed *in vivo*, although lineage conversion at higher rate *in vitro* remains a possibility, as does the existence of pluripotent cells akin to ESCs in the stromal compartments of organ systems. This means that our best bet currently for tissue replacement on a wide scale is the derivation of the required precursor or terminally differentiated cell types from ESCs. Neither have we experienced much success in devising functional bioartificial tissue constructs, except for the use of skin equivalents as living wound dressings. Research on the construction of bioartificial bone, cartilage, and tubular organs such as blood vessels, bladder, and ureter is promising, but we are not yet at the point where we can replace human tissue with confidence of success on a wide scale.

A major discovery emerging from work with ASC transplants is that they secrete paracrine factors that enhance the survival of host cells, reduce scarring, and perhaps stimulate regeneration from host cells. Thus, ASCs such as those in bone marrow can be viewed as bioreactors that can be used to deliver proteins and other factors to damaged tissues. This is an exciting development, because it means that the third strategy of regenerative medicine, the induction of regeneration *in situ* from the body's own tissues, would be feasible on a large scale. If we can identify these factors and how they work, we may be able to use them by themselves, or in conjunction with regeneration templates that mimic the natural ECM, to promote the regeneration of new tissue in an inexpensive, safe, and effective manner. The use of growth factors by themselves has had some success as a reparative therapy for damaged skin, neural tissues, and blood vessels, but less success as a therapy for damaged musculoskeletal tissues. Regeneration templates have been successful in promoting the repair of skin and small segments of other tissues, such as bone, blood vessels, intestine, and urinary conduit tissue, and growth factors added to the templates have improved their performance in some instances. Despite some temporary setbacks, gain of function gene therapy by infusion of genes for enzymes or growth factors, either by themselves or as part of regeneration templates, will become a powerful method for the correction of genetic diseases and for the induction of regeneration *in situ*. Likewise, the use of RNAi to eliminate specific gene products has enormous potential to suppress disease states caused by malfunctioning genes. These approaches should advance rapidly in the next few years.

There are many biological research issues and challenges yet to be met in developing a regenerative medicine, but none is insurmountable. The general challenge is that we do not yet have the depth of understanding of the biology of regeneration that is necessary for its translation into clinical therapies. The use of animal models that are strong regenerators of one or more tissues that mammals cannot regenerate—flatworms, amphibians, fish, crocodilians, sharks—has not been adequately exploited. There are also bioethical issues and challenges that must be addressed. Scientific advances are not made in a social vacuum, but within a set of societal beliefs and mores that often view these advances with some measure of distrust and fear. The interplay between the applications of scientific discoveries and social and political ideas within the existing set of social precepts has always acted as a selective force for the evolution of technology and culture, and will continue to do so. This chapter is devoted to an analysis of some of these biological and bioethical issues and challenges.

## BIOLOGICAL ISSUES AND CHALLENGES

Each regenerative medicine strategy faces its own specific biological issues and challenges, but a number of these issues and challenges are common to all three strategies. The primary challenges to developing cell transplantation and bioartificial tissue therapies are the provision of adequate cell sources, the ability to expand primary cultures of ASCs and ESCs, defining what factors are necessary to direct their differentiation into specific cell types, and immunorejection of allogeneic or xenogeneic cells. Additional challenges to bioartificial tissue construction are the design of scaffolds that mimic natural ECM, and how to provide adequate vascularization for the survival of tissues larger than a diffusion-limited size. Issues for the induction of regeneration *in situ* are the extent to which regeneration-competent cells are distributed throughout the body, determining what factors are required to activate regeneration-competent cells, what factors are required for their survival, proliferation, and differentiation into organized tissue, and what regeneration-inhibiting factors need to be neutralized. An issue facing all forms of regenerative therapy for damage that is due to degenerative disease is how to first halt the progress of the disease.

### 1. Cell Sources for Transplantation and Bioartificial Tissue Construction

Cells for transplantation or bioartificial tissue construction can be autogeneic, allogeneic, or xenogeneic with regard to species source. With regard to tissue source, they can be differentiated cells, ASC, or ESCs. Differentiated autogeneic cells would be the most desirable cells to use, because they are already fully functional. In either case, however, obtaining a sufficient number of differentiated cells for use requires either multiple donors or the expansion of the cells *in vitro*. Donors are in short supply and the expansion of most differentiated cells *in vitro* is difficult and requires dedifferentiation of the cells for proliferation. The issue then becomes whether or not the cells are fully functional when redifferentiated for transplant. Thus far, only cultured chondrocytes are in use clinically, for articular cartilage repair, with significant attention being paid to determining the culture conditions that will maximize the quality of chondrocyte redifferentiation (Chapter 10). Block et al. (1996) have devised a defined expansion medium containing HGF, EGF, and TGF-β1 that supports the dedifferentiation, proliferation, and redifferentiation of adult rat and human hepatocytes, and hepatocytes and β-cells have been expanded by immortalizing them (Chapter 8), but such expanded cells have not yet been used clinically. Fetal tissues are another potential source of allogeneic cells. However, fetal cells will not become a standard cell source because for bioethical reasons their derivation would have to be limited to spontaneously aborted fetuses, thus limiting the number of cells available for use.

Most of our attention is focused on ASCs and derivatives of ESCs as transplant sources (Parenteau and Young, 2002; Faustman et al., 2002). ESCs and ASCs each have advantages and disadvantages as sources for transplant (Vogel, 2001; Orkin and Morrison, 2002). The major advantages of ESCs for regenerative therapy are their ease of expansion, pluripotency and youthful robustness (Brivanlou et al., 2003). They can also be used to study the cellular and molecular aspects of early human development, and to screen drugs for their effects on early human cells in an effort to understand and prevent congenital defects. Furthermore, human ESCs made by somatic cell nuclear transfer (SCNT) can be used to study the genetic basis of diseases.

ESCs have several disadvantages. *In vivo*, they form benign teratomas of mixed cell type if allowed to differentiate on their own, requiring that their differentiation be directed *in vitro* so that all the cells go one way. They are also difficult to maintain in an undifferentiated state. The cells of ESC lines must be properly imprinted (one of the maternal or paternal alleles silenced, the other expressed). The use of ESCs in which imprinting has not yet been established or has been lost is associated with developmental abnormalities (Humpherys et al., 2001). There is also concern about chromosome abnormalities accumulating in cultures of ESCs, although the development of such abnormalities seems to depend on whether passaging is done by enzymatic or mechanical methods (Draper et al., 2004; Pera, 2004; Buzzard et al., 2004; Mitalipova et al., 2004). Finally, the derivatives of ESCs will be immunorejected if transplanted into a nonimmunoprivileged site. However, different hESC lines have distinct HLA profiles, suggesting that they would elicit different intensities of immune response (Carpenter et al., 2004).

ASCs have several advantages over ESCs. They are not immunorejected if autogeneic, but this advantage disappears if the cells are allogeneic or xenogeneic. ASCs do not form teratomas and the problem of directed differentiation is not as significant as it is with ESCs. They can survive and differentiate when transplanted, as long as the required signals are present in the injury environment. However, ASCs present another set of problems. Most are few in number, difficult to harvest, and difficult to expand *in vitro*, though there are exceptions such as marrow MSCs, adipose-derived MSCs, and umbilical cord blood cells. Another question concerns the effects of age on ASCs and their derivatives. Although the evidence suggests that regeneration-competent cells from at least some tissues of older animals, such as hepatocytes and satellite cells, remain fully competent to regenerate in a young environment (Chapter 13), for others the answer is unknown.

Opponents of ESC research have cited the ability of ASCs, particularly bone marrow cells, to be converted into out-of-lineage cell types, as well as their ability to cure a myriad of diseases and injuries, as reasons to abandon research on ESCs (see statement by David Prentice in Steinberg, 2000). However, these claims are misleading at best. The results of most experiments indicate that lineage conversion of ASCs *in vivo* is in most cases extremely infrequent, although MSCs may be an exception and other ASCs may be more amenable to reprogramming *in vitro*. Bone marrow and other connective tissues may harbor rare pluripotent cells that share some of the characteristics of ESCs (Jiang et al., 2002; Vacanti, 2003; Young and Black, 2004), or these cells might arise *in vitro* from pericytes or MSCs that dedifferentiate to a more primitive, pluripotent state. While it is true that hematopoietic stem cell transplants have been used successfully to regenerate the hematopoietic system in a wide variety of genetic and malignant blood diseases, reports of improvements in spinal cord injuries and neurodegenerative and cardiovascular disorders after transplant of ASCs such as olfactory ensheathing or bone marrow cells, while intriguing and worthy of further investigation, have produced modest results at best and have not established the level of rigorous proof required to substantiate the clinical efficacy of these treatments.

We should continue research on how to deprogram/reprogram ASCs and differentiated cells to create pluripotency. At the same time, it is unwise to place all our bets on one source of cells for regenerative therapies. It is highly likely that multiple cell sources will be required to treat different types of damage caused by injury or disease.

## 2. Expansion and Directed Differentiation of Regeneration-Competent Cells

### a. *Expansion and Directed Differentiation of ASCs*

Efforts are ongoing to determine what proliferation and differentiation factors will allow us to generate sufficient numbers of ASCs that will result in the differentiation of enough functional cells to repair lesions or provide survival, proliferation, and antiscarring factors in enough quantity to slow disease processes or regenerate new tissue from host cells. Much of the work has been focused on NSCs. FGF-2 and EGF are standard constituents of media promoting NSC proliferation *in vitro*, but Smith et al. (2003) have reported that the proliferation rate of NSCs is more dependent on species and anatomical region than on soluble factors and culture conditions. Thus, the region from which NSCs are harvested may be an important determiner of how the cells respond to culture conditions. Mouse neural progenitor cells can be selectively induced to differentiate into neurons while astrocyte differentiation is inhibited, when mixed with peptide amphiphile molecules plus the laminin pentapeptide sequence IKVAV (isoleucine-lysine-valine-alanine-valine) that self-assemble into a network of nanofibers (Silva et al., 2004).

Another way to generate large numbers of neuronal progenitor cells (or other ASCs) is to immortalize them by inhibiting checkpoints in the cell cycle, as has been done for hepatocytes and β-cells (Chapter 8). Roy et al. (2004) reported the immortalization of human neural progenitor cells by use of a retrovirus encoding hTER, the rate-limiting component of the telomerase enzyme complex. They established several cell lines from different regions of the spinal cord that give rise to specific neuronal and glial cell types. These lines exhibited no detectable replicative senescence or karyotypic abnormalities in over 168 doublings *in vitro*. The cells matured as neurons when grafted to the fetal rat brain and injured adult spinal cord, showing that immortalization does not affect their ability to differentiate and become postmitotic. As methods for sorting neuronal subtypes on the basis of surface phenotype become more sophisticated, it will be possible to select precursors for specific neuron types as well as other adult stem cells and immortalize each one for expansion and transplantation, or expand them with the proper mitogens.

Maintenance and injury-induced regeneration, where it occurs naturally, takes place within the context of a tightly regulated microniche of autocrine, paracrine, and juxtacrine signals and receptors generated by local stem cells or other regeneration-competent cells and surrounding cells. A better understanding of the composition and spatial organization of these microniches will be essential to achieving the expansion and self-renewal of these cells *in vitro* and how to direct their differentiation. The three-dimensional spatial organization of signals and receptors is an especially important factor in cell behavior to which more attention needs to be paid when culturing cells *in vitro*. Many ASCs are close to capillaries, which may provide essential signals to maintain stemness, but which also provide the oxygen and nutrients essential for stem cell survival and activity. Where the injury environment is not adequate to support regeneration in the presence of regeneration-competent cells, for example, in the spinal cord, dermis, pancreas, or myocardium, it will be imperative to determine what molecular additives will establish a microniche that supports regeneration and inhibits scarring. This "stem cell ecology" (Powell, 2005) also applies to the proliferation and directed differentiation of ESCs.

### b. Expansion and Directed Differentiation of ESCs

One of the difficulties in maintaining human ESCs in an undifferentiated state has been the necessity of growing them on mouse embryonic fibroblast feeder layers (Pera and Trounson, 2004). While mouse ESCs can be maintained in an undifferentiated state by providing LIF in a complex medium conditioned by embryonic feeder fibroblasts, primate ESCs tend to spontaneously differentiate into trophoblast and endoderm cells whether LIF is present or not. Thus, propagation of primate ESCs requires continual passaging to minimize the accumulation of differentiated cells in the cultures. Ezashi et al. (2005) have pointed out that the environment of the mammalian reproductive tract is hypoxic (1.5–5.3% $O_2$) but that general laboratory practice is to culture ESCs in air (21% $O_2$) plus 5% $CO_2$. They showed that hESCs cultured under 3–5% $O_2$ grew as well as those cultured in 21% $O_2$/5% $CO_2$. Furthermore, the frequency of differentiation was markedly reduced and the ability of the cells to form embryoid bodies was enhanced, making it easier to propagate the cells in an undifferentiated state.

Feeder layers limit the number of ESCs that can be produced. Human ESCs are difficult to separate from mouse feeder cells, which could result in xenoplastic contamination of transplants. Furthermore, there is a risk of transferring animal pathogens to humans by the use of cells derived from hESC lines grown on animal feeder cells in medium supplemented with animal-derived sera. Thus, protocols have been developed to grow hESCs on animal-derived ECM (Xu et al., 2001) or on human ECM in medium conditioned by mouse feeder cells (Amit et al., 2004). Newer protocols are being developed to grow hESCs in chemically defined medium without animal sera (Bodnar et al., 2004; Rippon and Bishop, 2004; Carpenter et al., 2004; Rosler et al., 2004). A major advance has been made by Klimanskaya et al. (2005), who describe a protocol to grow hESCs on ECM of mouse embryonic fibroblasts, in a defined medium. This medium, however, includes a serum replacement containing animal proteins. It would be desirable to replace these with human proteins. The optimum defined medium and substrate for growing hESCs will require identification of the factors produced by human feeder cells that maintain the cells in an undifferentiated state and promote their proliferation.

Protocols for differentiating ESCs to ASC-like cells or terminal phenotypes have been devised for a wide variety of cell types (Pera and Trounson, 2004). These protocols are based on knowledge (as yet incomplete) of the factors that constitute the developmental pathways leading to the production of these different cell types. Several protocols for differentiating mouse ESCs to neuron and glial precursors were mentioned in Chapter 6. Gottlieb was the first to devise a protocol to efficiently derive neurons from mouse embryoid bodies using RA (Bain et al., 1995; Gottlieb, 2000). RA and sonic hedgehog applied sequentially to mouse and human ESCs induced them to differentiate into posterior motor neurons (Wichterle et al., 2002; Li et al., 2005). The mouse ESC-derived motor neurons were able to populate the embryonic spinal cord, extend axons, and form synapses with target muscles. Protocols for differentiating human ESCs to neurospheres were developed by Zhang et al. (2001) and Reubinoff et al. (2001). Cells of the neurospheres differentiated to glutaminergic and GABA-producing neurons *in vitro* and migrated into many brain regions after injection into the brain ventricles of neonatal rats. Other protocols have enabled the differentiation of all major neuronal and glial subtypes from mouse ESCs (Barberi et al., 2003).

Protocols have been developed for the directed differentiation of nonneuronal murine cell types as well. In addition to the cell types mentioned in previous chapters, mouse ESCs have been differentiated to dendritic cells (Fairchild et al., 2000), T-cells (Schmitt et al., 2004), epidermis (Troy and Turksen, 2004), lens and neural retina (Hirano et al., 2004), cardiomyocytes, chondrocytes, osteoblasts, lung alveolar epithelium (Rippon and Bishop, 2004) and male gametes (Geijsen et al., 2004). Human ESCs have been differentiated into neural cells, cardiomyocytes, chondrocytes, osteoblasts, and lung alveolar epithelium (Rippon and Bishop, 2004), endothelial cells (Gerecht-Nir et al., 2004) and a broad range of hematopoietic cells (Zhan et al., 2004). Screens of combinatorial chemical libraries have identified several small molecules that induced the differentiation of neural cells and cardiomyocytes from hESCs (Ding and Schultz, 2004).

Identification of the factors that direct the differentiation of hESCs will be aided by the comparative analysis of transcriptome profiles between ESCs and adult stem cells. Ramalho-Santos et al. (2002) and Ivanova et al. (2002) compared the transcriptional profiles of mouse ESCs, NSCs, and HSCs by hybridizing DNA microarrays containing several thousand genes with mRNAs from each set of cells. Bioinformatic analysis was used to identify transcripts absent in differentiated cells but present in stem cells and to assign transcripts enriched in stem cells to functional categories. Both studies found a common set of over 200 genes enriched in all three sets of stem cells that appeared to define core "stemness," including sets of genes that encoded proteins for signaling pathways, cell cycle, stress protection, transcriptional regulation, and translational regulation. More than 50% of the commonly enriched genes were ESTs that represented unidentified genes. Other gene sets were distinct from one another. HSCs were found to be more similar to the main bone marrow population than to ESCs or NSCs, but NSCs were more similar to ESCs than to HSCs or differentiated cells.

Microarray analysis of transcription in mouse ESCs versus NSCs revealed a high degree of difference between them that correlated with the multipotent nature of the NSCs versus the pluripotent nature of the ESCs (D'Amour and Gage, 2003). Brandenberger et al. (2004) compared the transcriptional profiles of undifferentiated hESCs and embryoid bodies, and identified 532 genes that were significantly upregulated and 140 that were significantly downregulated in undifferentiated ESCs. Transcription factors known to be essential for pluripotency were upregulated, as were components of the FGF, WNT, NOD, and LIF pathways, suggesting the readiness of undifferentiated hESCs to differentiate when presented with the appropriate shift in balance of inhibitors and activators.

The goal is to be able to direct the differentiation of ESCs (or ASCs for that matter) to produce all the phenotypic subtleties inherent in the original development of the tissue. This requires the application of the right combination of factors in the appropriate concentrations at the appropriate times, parameters that are heavily influenced by three-dimensional organization. The directed differentiation of hESCs in three-dimensional bioartificial constructs has been demonstrated by mixing hESCs in 50:50 Matrigel:culture medium and seeding this mixture into $5 \times 4 \times 1\,mm^3$ pieces of 50:50 PLGA/PLLA (Levenberg et al., 2003). The culture medium was supplemented with RA, TGF-β1, Activin-A, or IGF-I. These factors induced differentiation of the hESCs into neural, cartilage, and liver cells, respectively. The differentiated cells exhibited three-dimensional organization resembling primitive tissue structures. All constructs except the RA-treated ones contained capillary-like networks of cells that expressed the endothelial markers CD34 and CD31. When transplanted into *scid* mice, the constructs continued to express human proteins and the capillary networks appeared to anastomose with the host vasculature.

### c. Available hESC Lines

There are currently ~15 hESC lines available for research that can be supported by federal funding. These are the lines that were established prior to August 9, 2001, when a presidential ban was placed on the generation of further lines using federal funds (Gearhart, 2004; Phimister and Drazen, 2004). Investigators report that these lines are difficult to obtain, are expensive, are difficult to maintain, and are poorly characterized. Seventeen new hESC lines (designated HUES 1–17) have now been established in the laboratory of Dr. Douglas Melton at Harvard University using private funds (Cowan et al., 2004). These lines were created from unused blastocysts and cleaved embryos from fertility clinics by protocols that follow a set of standards designed for ease of derivation, growth, molecular characterization, and directed differentiation of the cells (Brivanlou et al., 2003). The cells are available to investigators from Dr. Melton's laboratory under a Material Transfer Agreement.

## 3. Challenges for Bioartificial Tissue Construction

In addition to cell sources, the design of bioartificial tissues faces three other major technical challenges. The first is how closely we can mimic the three-dimensional biological composition and organization of the ECM in the biomaterials used as scaffolds for cells. The second is how to provide bioartificial tissues with vascularization to ensure cell survival, a problem faced by cell transplants and regeneration templates as well. Cell transplants and regeneration templates can regenerate only relatively small amounts of tissue *in vivo* unless the transplanted cells or cells of the host can migrate and expand extensively, which requires vascularization. The vision of tissue engineers is that bioartificial tissues will be designed and constructed *in vitro* that can replace large volumes of tissue or even replace organs. The volume of these constructs is even larger than that of cell transplants or regeneration templates, making rapid vascularization to nourish embedded cells even more crucial. The third challenge is to develop standardized procedures for producing and testing tissue-engineered products and components.

### a. Mimicking the ECM

The ECM is a complex, three-dimensional assembly of macromolecules synthesized by cells as an adjacent acellular basement membrane, or as an interstitial tissue matrix surrounding the cells. Interstitial ECM is composed of fibrous proteins (primarily collagens) embedded in a highly hydrated gel of GAGs and proteoglycans that is also a repository for signaling molecules such as growth factors, proteases, and their inhibitors (Voytik-Harbin, 2001). The natural matrix is "smart"; i.e., it releases the appropriate biological signaling information at the right times and places to promote and maintain cell adhesion, differentiation, and tissue organization. Thus, processed natural biomaterials such as cadaver dermis and pig SIS have been a logical choice for use as regeneration templates and scaffolds for bioartificial tissues. The processing removes cells, eliminating the immunorejection response, but it also changes the properties of the matrix, and may thus remove biological information essential for regeneration as well.

The use of synthetic biomaterials is advantageous because they can be manufactured in virtually unlimited quantities to specified standards, with additional shape-shifting features built in, such as liquidity and small volume at room temperature, changing to gelation, expansion, and space-filling at body temperature within a tissue gap. Duplicating the properties of natural ECM has long been a goal of biomaterials science. Hench and Polak (2002) describe the evolution of biomaterials from a first generation in the 1960s and 1970s that mimicked the physical properties of replaced tissue with minimal toxicity, to a second generation in the 1980s and 1990s that was bioactive and biodegradable as well. This second generation of biomaterials is the set from which most regeneration templates and bioartificial tissues described in previous chapters have been constructed (reviewed by Ratner, 1993; Peppas and Langer, 1994; Hubbell, 1995; Baldwin and Saltzman, 1996; Griffith, 2000; Griesler, 2001).

A third generation of biomaterials is under development that involves the tailoring of biodegradable polymers to elicit specific cellular responses by immobilizing cell-specific adhesive and signaling molecules on the material (Hench and Polak, 2002). Lutolf and Hubbell (2005) have comprehensively reviewed the current approaches to building biomimetic elements into synthetic polymers that can biomechanically organize cells in three dimensions and promote their differentiation into organized tissues. The focus is on micro- and nanofibrillar biomaterial gels, including self-assembling peptide and nonbiological amphiphiles, and nonfibrillar synthetic hydrophilic polymer hydrogels that have the physical and chemical properties of natural ECM. A number of biologically important signaling and enzyme-sensitive entities can be incorporated into these hydrogels, including recognition sequences for cell adhesion proteins, soluble growth factors, and protease-sensitive oligopeptide or protein elements. In particular, derivatized amino reactive polyethylene glycols (PEG) containing both peptide substrates for proteases and binding peptides for soluble factors or cell adhesion molecules appear to be promising for creating mimics of ECM-cell interactive processes (West and Hubbell, 1999; Mann et al., 2001; Gobin and West, 2002; Tessmar et al., 2004). Thus, research is being focused on ways to incorporate growth factors into the scaffold for delivery to the cells (Nimni, 1997). One such scaffold under development is composed of polyglycolic acid plus an active succinimidyl ester of polyethylene glycol, which will covalently link growth factors and other proteins to the polymer (Kellner et al., 2004).

There are significant technical hurdles yet to be overcome in the design of interactive synthetic biomimetic materials, particularly in making them mimic the specific microniche environments that can direct the differentiation of ASCs or derivatives of ESCs (Lutolf and Hubbell, 2005). These stem cells rely for their proliferation on a constellation of spatially organized autocrine and paracrine signals from, and justracrine contacts with, surrounding cells and matrix. The problem may turn out to be less daunting, however, if there are just a few factors required to initiate a cascade that is subsequently carried out largely by the stem cells themselves.

The development of new generations of biomaterials with interactive effects on cell behavior is being aided by high-throughput screening of biomaterial arrays (**FIGURE 15.1**). Anderson et al. (2004) have described a system by which to screen large numbers of polymers for their effects on cell behavior. The polymers were formed by the combination of 25 different acrylate, diacrylate, dimethylacrylate, and triacrylate monomers, mixed at a 70:30 ratio pairwise in all possible combinations and spotted onto a layer of poly(hydroxyethyl methacrylate (to inhibit cell growth between spots), that was on top of an epoxide-coated slide. The monomers polymerize and become firmly attached to the slide. Small numbers of human ESCs were then seeded onto the spots and incubated in the presence of RA for six days. The majority of the combinations supported cell attachment and spreading and also allowed their differentiation into cytokeratin positive epithelial cells. In a second screen, polymers that best supported cell growth revealed differences in their ability to support

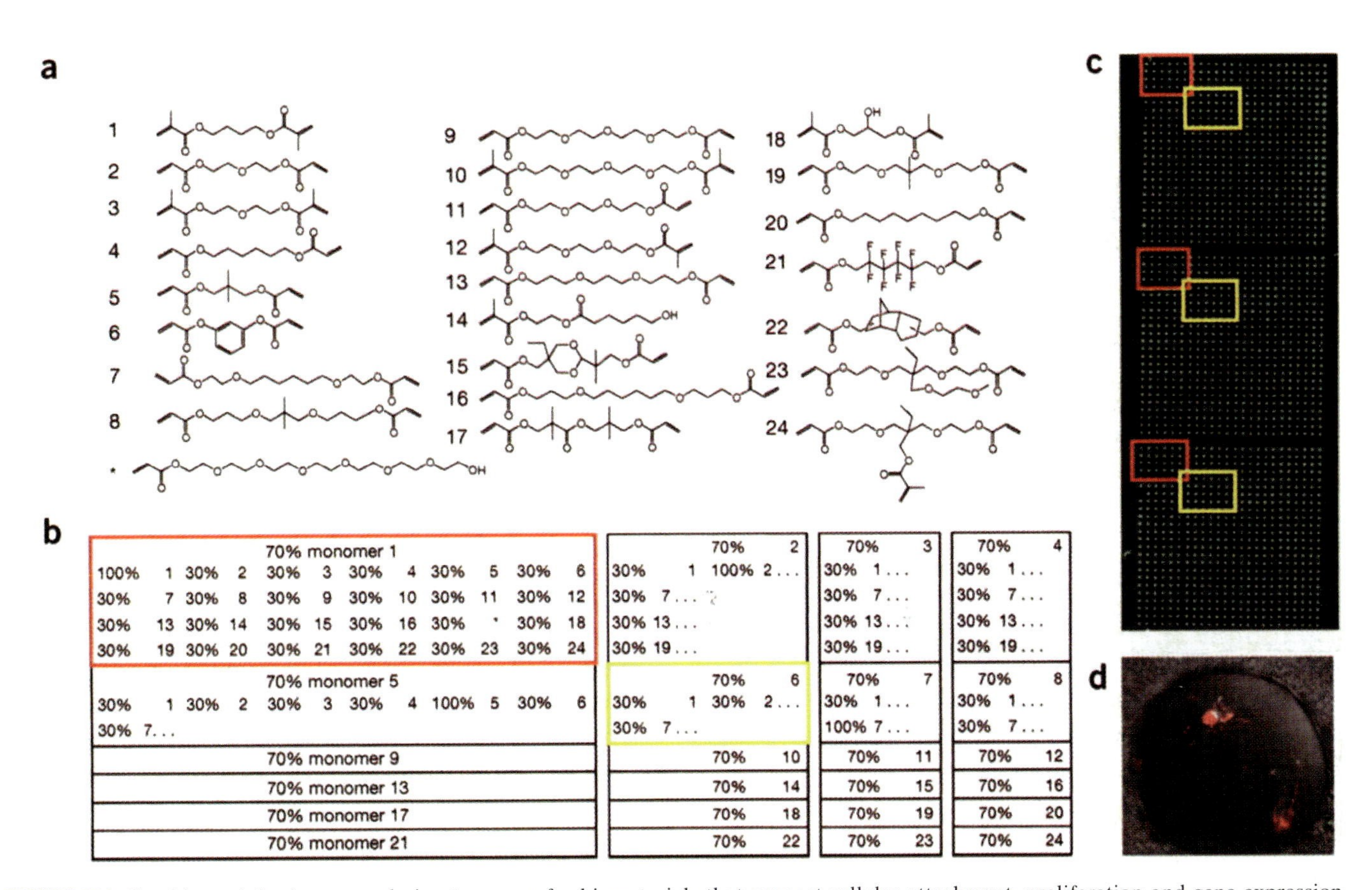


**FIGURE 15.1** Combinatorial microarray design to screen for biomaterials that support cellular attachment, proliferation and gene expression patterns. **(a)** 24 monomers used in making the biomaterial combinations. **(b)** Monomers were mixed at a 7:30 ratio pairwise in all possible combinations, with the exception of monomer 17. The 24 combinations of each 70% monomer with all the other 30% monomers were printed on an array as a 6 × 4 group of spots (red and yellow boxes). **(c)** Printed polymer array (three replicates) imaged by an array reader. Blocks composed of 70% monomer 1 and 70% monomer 6 are highlighted in red and yellow, respectively. Cells are added to the spots to screen for biomaterial effect. **(d)** Differential interference contrast light micrograph of polymer spot overlaid with a few fluorescent cells (red). Reproduced with permission from Anderson et al., Nanoliter-scale synthesis of arrayed biomaterials and application to human embryonic stem cells. Nature Biotech 22:863–866. Copyright 2004, Nature Publishing Group.

differentiation in the presence or absence of RA. Hubbell (2004) has pointed out that polymer biomaterials could also be used as tethering platforms to screen combinatorial libraries of molecules that are normally bound to the ECM for their effects on cell activity. If the activity of such molecules is dependent on their association with the ECM, they would not show an effect in a screen where they were presented to cells in a soluble form, but would reveal their effects if bound to a polymer (**FIGURE 15.2**).

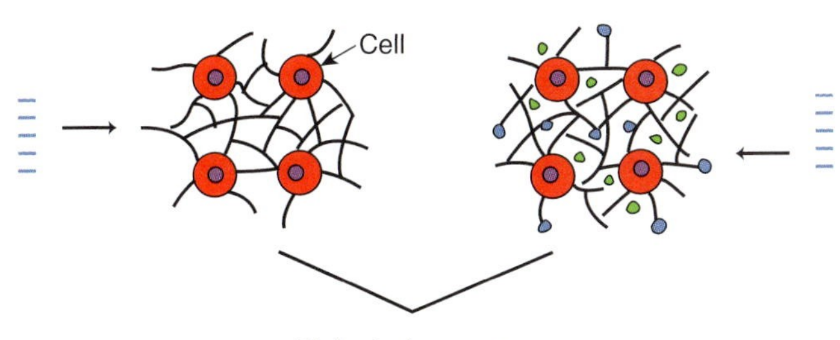


**FIGURE 15.2** Use of biomaterials as tethering platforms to screen combinatorial libraries of molecules whose activity is dependent on their association with the biomaterial. Stacks of blue lines are monomers. **Left**, a library of polymers is synthesized by co-polymerizing a number of monomers. These polymers (matrix) are screened for their effects on cells. **Right**, a library of small molecules that are either soluble (green dots) or adherent to the ECM (blue dots) can be screened for their effects on cells that adhere to the biomaterial. Reproduced with permission from Hubbell, Biomaterials science and high-throughput screening. Nature Biotech 22:828–829. Copyright 2004, Nature Publishing Group.

### *b. Vascularization*

A major technical obstacle in our ability to replace more than small amounts of damaged tissue by cell transplants (either through differentiation of the transplanted cells or their effects on host cells) or bioartificial tissue constructs is the massive amount of cell death that occurs due to hypoxia and lack of nutrients, particularly toward the center of a cell mass. Thus, another active area of research is to find ways to amplify and accelerate vascularization in cell transplants and bioartificial constructs.

Vascularization of cell transplants and implanted bioartificial tissues has been enhanced by the addition of endothelial cells or VEGF. For example, PGA scaffolds containing aortic smooth muscle cells, skeletal muscle cells, or aortic endothelial cells were implanted subcutaneously in rats and evaluated histologically at intervals up to four weeks (Holder et al., 1997). Scaffolds containing endothelial cells exhibited more vascularization than cellfree scaffolds or scaffolds seeded with smooth or skeletal muscle cells. Mouse hepatocytes seeded into small discs (~5-mm diameter, 1.5-mm thick) of PLGA containing VEGF demonstrated significantly greater survival that was associated with greater blood vessel density after subcutaneous transplantation compared to hepatocyte-seeded control scaffolds without VEGF (Smith et al., 2004). VEGF has also been covalently bound to one arm of a hydrogel made from four-armed PEG, with the other arms linked to protease-sensitive peptides or cell adhesive peptides (Zisch et al., 2003), or an engineered variant of VEGF has been covalently linked to fibrin for cell-induced release (Ehrbar et al., 2004). Human umbilical vein endothelial cells seeded in the gels showed active movement and VEGF-bound hydrogel or fibrin grafted to the chick chorioallantoic membrane or subcutaneously in rats elicited the invasion of host capillaries (**FIGURE 15.3**). The local concentration of VEGF is very important in determining the stability of the ingrowing vessels. Low concentrations favor stable vessels, whereas high concentrations lead to aberrantly formed and leaky vessels (Ozawa et al., 2004).

Another approach to the vascularization problem is the construction of three-dimensional scaffolds with microfluidic networks. The scaffold is built up from successive layers of biodegradable polymers imprinted with grooves to produce a three-dimensional structure with a network of channels. The construct can then be seeded with endothelial cells to line the channels. Alternatively, porous three-dimensional scaffolds that allow flowthrough of fluid could be seeded with microdots of the functional cell(s) of interest along with endothelial cells (**FIGURE 15.4**). The construct would be implanted immediately so that host vasculature would quickly connect with developing implant vasculature in these small tissue islands to establish a circulation, which would then serve to nourish the expansion/differentiation of the islands into continuous tissue. For cylindrical or tubular structures, cells could be seeded onto the outside or inside of the scaffold, where they would migrate inward or outward.

### c. Standardization of Products and Components

Scaffolds, cells, and/or any additional biomolecules added to the scaffolds will have to meet rigid standards for characterization, processing, and functionality in order to validate the performance of the end product (Picciolo and Stocum, 2001). Furthermore, developing standards for interactions among the components of a bioartificial tissue and with the host tissue will also be crucial to product validation. Test methods also need to be developed to assess the impact of product performance over the long term, something that has only rarely been done in animal experiments or clinical trials so far, and results need to be reported and thorougly disseminated.

## 4. The Challenge of Immunorejection

A primary concern in transplanting allogeneic or xenogeneic cells, either as suspensions, aggregates, or part of a bioartificial construct, is immunorejection. Although immunorejection can be held at bay with immunosuppressive drugs, we would like to evade it altogether. Strategies to avoid immune rejection of cells are (1) the creation of transgenic animals expressing regulatory molecules that inhibit the activation or synthesis of key proteins involved in rejection (reviewed by Cooper and Lanza, 2000), (2) mimicking immunoprivileged sites, such as the anterior chamber of the eye, by the expression of Fas ligand (reviewed by Bohana-Kashtan and Civin, 2004), (3) production of antibodies directed at T-cell receptors that recognize foreign cell surface MHC antigens (reviewed by Cooper and Lanza, 2000), (4) the production of genetically modified "stealth" cells that do not express surface MHC antigens (reviewed by Cooper and Lanza, 2000), and (5) the induction of peripheral tolerance to alloantigens by balancing the regulation and deletion of responder T-cells (reviewed by Lechler et al., 2003; Wood and Sakaguchi, 2003; Waldmann and Cobbold, 2004).

The most direct way of eliminating immune rejection, however, is through somatic cell nuclear transfer (SCNT), also known as therapeutic cloning (**FIGURE 15.5**). In this technique, the nucleus from a somatic cell of a prospective cell recipient is introduced into an enucleated egg and the egg is stimulated to divide. At the blastocyst stage, the inner cell mass is removed and the cells are used to create an autogeneic ESC line (aESC). Cells derived from this line will not be immunorejected when transplanted into the nuclear donor because they express the MHC antigens of the donor. Such aESC lines were first created for mice and were

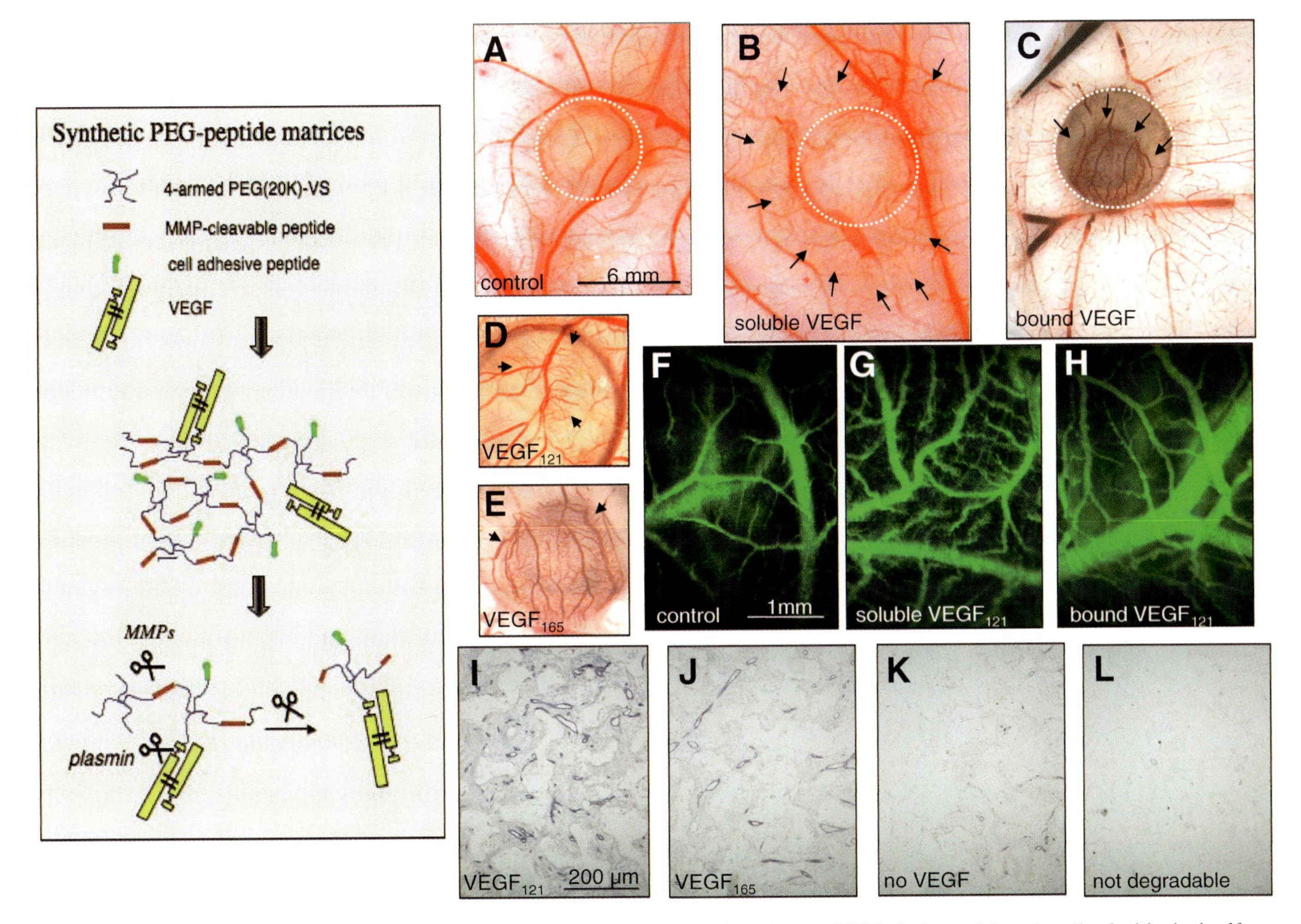


**FIGURE 15.3** **Left panel**, cell–mediated release of VEGF from hydrogels made from 4-armed PEG chains end-functionalized with vinyl sulfone (VS), conjugated with a cell adhesive peptide and VEGF, and cross-linked with a MMP-2 cleavage site peptide. VEGF could be released from this gel by either MMP-mediated degradation or plasmin-mediated cleavage of the matrix anchor. **Right panel**, **A–C**, chorioallantoic membrane (CAM) assay for angiogenesis. **(A)** Control hydrogel without VEGF does not induce angiogenesis. **(B)** Hydrogel with freely diffusible VEGF induces angiogenesis around the periphery of the gel. **(C)** Hydrogel with bound (conjugated) VEGF induces angiogenesis precisely at the area of graft-membrane contact, consistent with local cell-mediated release of VEGF from the polymer. Arrows indicate new vessels. **D**, **E**, higher magnification of vessels induced by conjugated VEGF gels. **F–H**, fluorescent images of CAM vasculature at the periphery of hydrogel grafts after perfusion with FITC. Only diffusible (soluble) FGF induced angiogenesis. **I–L**, sections of gels in porous polyurethane discs implanted subcutaneously in rats, stained with antibodies for endothelium (CD31). **I** and **J** are gels with conjugated VEGF, **K** is a gel with no VEGF, and **L** is a MMP-insensitive, non-degradable gel with conjugated VEGF. These data show that VEGF is released by the action of cells producing MMP-2. Courtesy of Dr. Andreas Zisch.

shown to be capable of differentiating into neurons and muscle *in vitro* (Munsie et al., 2000; Wakayama et al., 2001). Reproductive cloning using these aESCs as nuclear donors gave rise to fertile adult mice, demonstrating the pluripotency of their nuclei (Wakayama et al., 2001).

To be clinically useful, hESC lines must be derivable at relatively high efficiency from combinations of nuclei and eggs from different persons and genders. Cibelli et al. (2001) made the first attempt to derive autogeneic human ESCs. They performed SCNT from skin fibroblasts or cumulus cells to enucleated human eggs. Three eggs activated by cumulus cell SCNT developed to the four or six-cell stage. The development of a human blastocyst by SCNT, however, has not yet been achieved. Two reports (Hwang et al., 2004, 2005) had claimed to do so, but were later found to be fraudulent and were retracted.

Even assuming that ahESC lines can be created efficiently by SCNT, human eggs are likely to be in short supply. Thus, the possibility of using animal eggs as recipients of human somatic nuclei is being explored. Shen et al. (2003), at Shanghai Second Medical University, reported the first successful derivation of ESCs from blastocysts created by fusing enucleated rabbit eggs with somatic cell nuclei from human foreskin and facial skin [report published in the Chinese journal, Cell Research 13:251–263 (Dennis, 2003)]. The growth potential of these hybrid cells is not clear, and it is possible that they would become unstable due to species incompatibilities between the proteins produced by the human nucleus and those produced by the rabbit maternal mitocohondria.

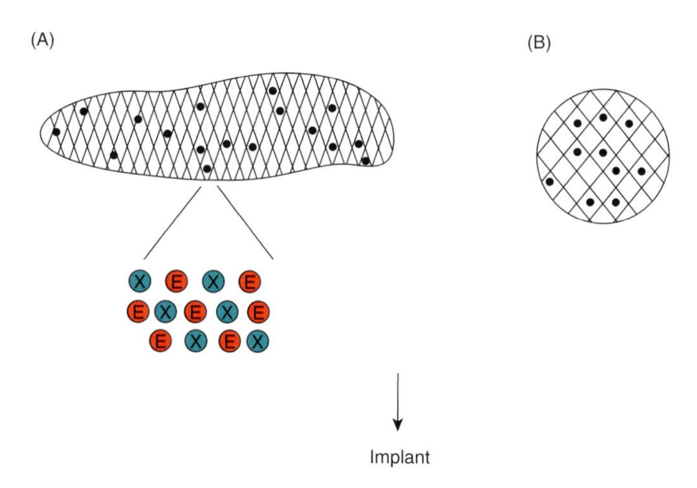


**FIGURE 15.4** A proposed method for vascularizing a bioartificial tissue or organ. **(A)** Surface view. Microdots of parenchymal cells (X)/endothelial cells (E) are seeded throughout a three-dimensional porous scaffold and the construct is implanted into the body, where the microdots will survive by diffusion. The endothelial cells will form capillaries and the construct will be invaded by capillaries from the host. As the vascular network grows, the parenchymal cells will be able to proliferate along with it. **(B)** Cross-section showing microdots throughout the scaffold.

Another approach to generate autogeneic hESCs, while getting around the problem of egg shortage, is to reprogram the nuclei of adult cells by fusing the cytoplasts of allogeneic or xenogeneic ESCs with the karyoplasts of somatic cells (Solter and Gearhart, 1999; Mantle et al., 2004). The resulting "cybrid" would theoretically reprogram the somatic cell nucleus to make the cell pluripotent. This approach is not straightforward, however. Do and Scholer (2004) reported that mouse ESC cytoplasts were unable to induce the expression of the *Oct4* gene in NSCs, which is required for pluripotency, but that ESC karyoplasts could induce *Oct4* expression by NSC nuclei. This result suggests that the ESC nucleus produces the same transcripts for proteins necessary to induce pluripotency that the egg produces, but that these proteins do not accumulate in the ESC cytoplasm as they do in the mature egg. Together with the results of the original experiments fusing NSCs and bone marrow cells with ESCs (Terada et al., 2002; Ying et al., 2002), these findings suggest that fusion of somatic cells with ESCs might be a way to reprogram somatic cell nuclei for the purpose of making autogeneic hESCs.

Tada et al. (2003) showed that hybrid species/cell mouse lymphocyte/ESC clones were pluripotent, as evidenced by their formation of teratomas with the morphology and transcriptional patterns of a variety of fully differentiated cells (FIGURE 15.6). Cowan et al. (2005) reported the reprogramming of human fibroblast nuclei after fusion with human ESCs to produce pluripotent hybrid cells (FIGURES 15.7–9). The hybrid

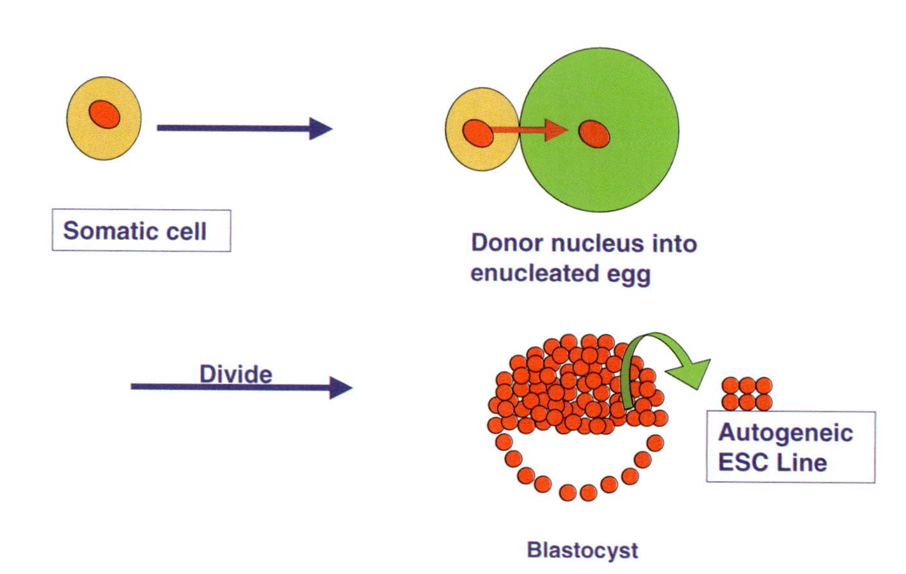


**FIGURE 15.5** Somatic cell nuclear transfer as a way of producing autogeneic ESCs whose derivatives will not be immunorejected. The somatic cell is fused with an enucleated egg and the diploid somatic nucleus enters the egg cytoplasm. The egg is induced to divide and allowed to develop to blastocyst stage, where the inner cell mass is removed and cultured to produce the autogeneic ES cell line.

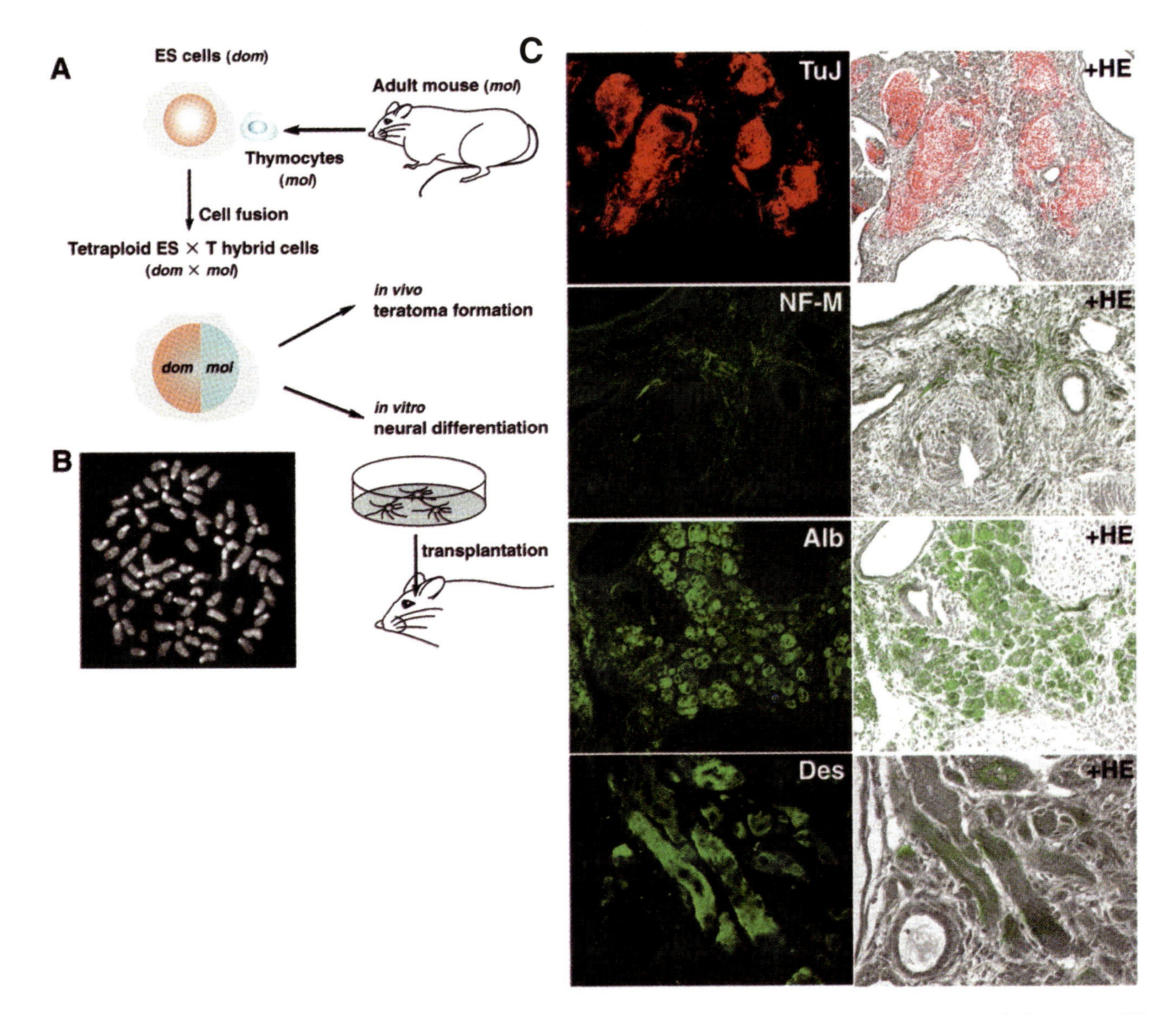


**FIGURE 15.6** Characterization of hybrid cells made by fusion of *mus domesticus (dom)* ESCs and adult *mus molossinus (mol)* thymocytes. **(A)** Experimental protocol. The tetraploid hybrid cells were injected subcutaneously into SCID mice or differentiated into $TH^+$ clones of neurons and transplanted into the striatum. **(B)** Metaphase spread of tetraploid chromosomes. **(C) Left panels**, antibody staining for expression of the ectodermal-specific markers Class III β-tubulin (TuJ) and neurofilament M (NF-M), the endodermal-specific marker albumin (Alb) and the mesodermal-specific marker desmin (Des). **Right panels**, sections counterstained with H&E. Reproduced with permission from Tada et al., Pluripotency of reprogrammed somatic genomes in embryonic stem cell hybrids. Dev Dynam 227:504–510. Copyright 2003, Wiley-Liss.

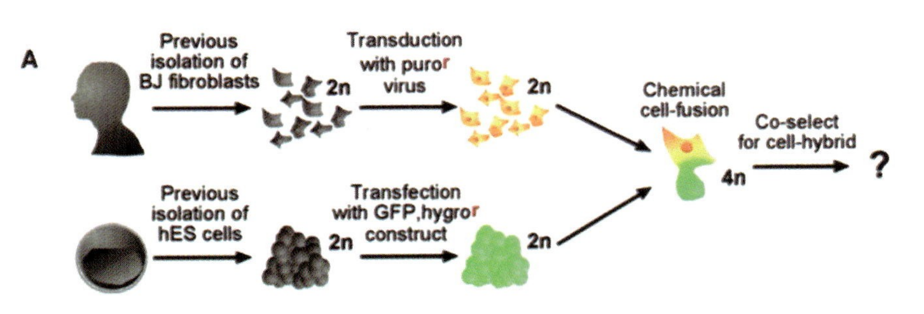


**FIGURE 15.7** Protocol for production of fibroblast/ESC hybrids. Fibroblasts were transfected with a retroviral puromycin construct and ESCs with GFP and hygromycin constructs. After polyethylene glycol-induced fusion of the cells, hybrids were selected by exposure of the cultures to puromycin and hygromycin. Reproduced with permission from Cowan et al., Nuclear reprogramming of somatic cells after fusion with human embryonic stem cells. Science 309:1369–1373. Copyright 2005, AAAS.

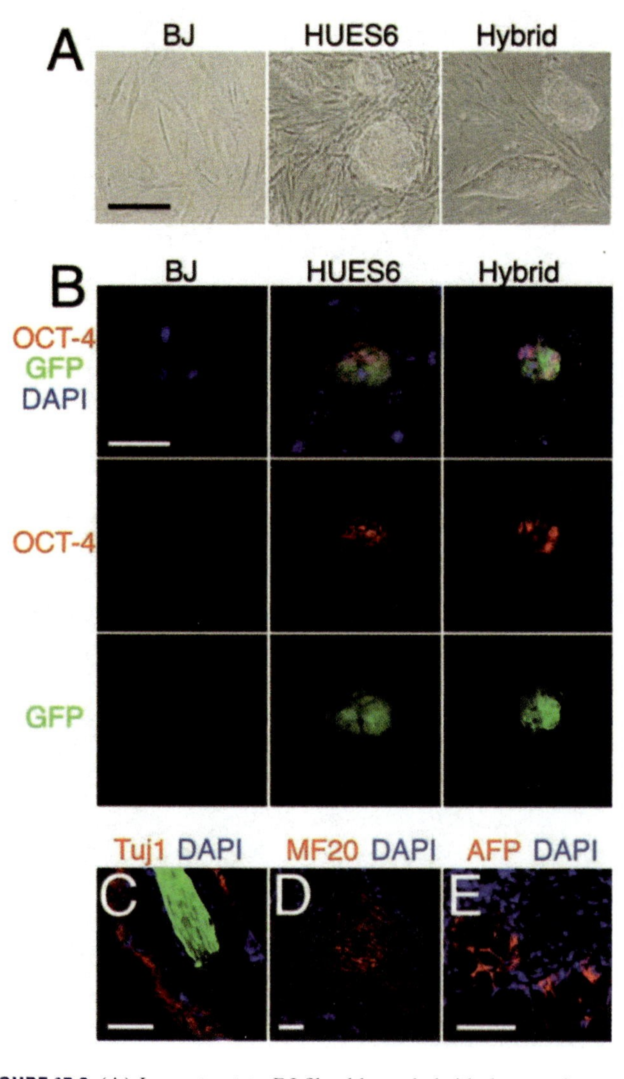


**FIGURE 15.8** **(A)** In contrast to BJ fibroblasts, hybrid, drug-resistant cells grew in compact, phase-bright cell clusters with a morphology identical to hES cells (HUES6). **(B)** Hybrid cells and HUES6 cells expressed GFP and OCT-4, whereas the GFP-negative BJ fibroblasts did not. Top panel is an overlay of the middle and bottom panels. DAPI = nuclear stain. **(C–E)** Immunostaining of teratomas derived from hybrid cells for phenotype-specific markers reveal differentiation into several cell types, including neurons (TuJ1) surrounding a hair follicle, skeletal muscle (MF-20) and intestinal endoderm (α fetal protein, AFP). Scale bars = 50μm. Reproduced with permission from Cowan et al., Nuclear reprogramming of somatic cells after fusion with human embryonic stem cells. Science 309:1369–1373. Copyright 2005, AAAS.

tetraploid cells had the growth characteristics of the ESCs, they formed teratomas with a variety of cell types and their transcriptional profile was overwhelmingly that of the ESCs. These experiments clearly show that it is possible to reprogram somatic cell nuclei to pluripotency. However, a hurdle still to be overcome is the presence of the ESC nucleus, which means that the hybrid cell is not fully autogeneic. Thus a way must be found to eliminate the ESC DNA, either before fusion without loss of reprogramming capacity, or after reprogramming has taken place.

## 5. Challenges for the Chemical Induction of Regeneration

### a. *Cell Sources in vivo: Ubiquity of Regeneration-Competent Cells*

A prerequisite for the induction of regeneration from the body's own tissues using regeneration templates of biological or artificial materials, combinations of signaling molecules, suppressors of scar formation, or some combination of these, is that regeneration-competent cells be widely distributed. To induce regeneration from the body's own tissues, we will need to know how ubiquitous regeneration-competent cells are, exactly what combinations of chemical and physical signals and neutralizers are necessary to stimulate regeneration and inhibit scarring in specific tissues, and the best ways to deliver these molecules or the genes that encode them.

As outlined in previous chapters, there is substantial evidence for the existence of ASCs in tissues that do not regenerate and the possibility exists that organ systems harbor rare pluripotent stem cells (Jiang et al., 2002; Vacanti et al., 2003; Young and Black, 2004). In addition, we could learn how to dedifferentiate or reprogram differentiated cells. The fact that cultured mouse muscle cells transfected with the *msx-1* gene or treated with protein extract from regenerating newt limbs undergo cellularization and dedifferentiation into cells with a developmental potential similar to that of MSCs (Odelberg et al., 2001; McGann et al., 2002) suggests that other mammalian cells could be induced to dedifferentiate *in vivo*, perhaps even to a pluripotent state. Libraries of small molecules generated by combinatorial chemical methods can be screened to identify molecules, such as reversine, that induce dedifferentiation (Ding and Schultz, 2004) (**FIGURE 15.10**).

Hakelien et al. (2002) reported that they were able to reprogram 293T fibroblasts *in vitro* with nuclear and cytoplasmic extract derived from stimulated human T-lymphocytes (Jurkat leukemia cell line). The treated fibroblasts exhibited nuclear uptake and assembly of transcription factors, activation of a chromatin remodeling complex, histone acetylation, and activation of genes specific for T-cells. They also responded to stimulation by upregulation of a T-cell-specific pathway (**TABLE 15.1**). In addition, 293T fibroblasts exposed to a neuronal cell precursor extract expressed neurofila-

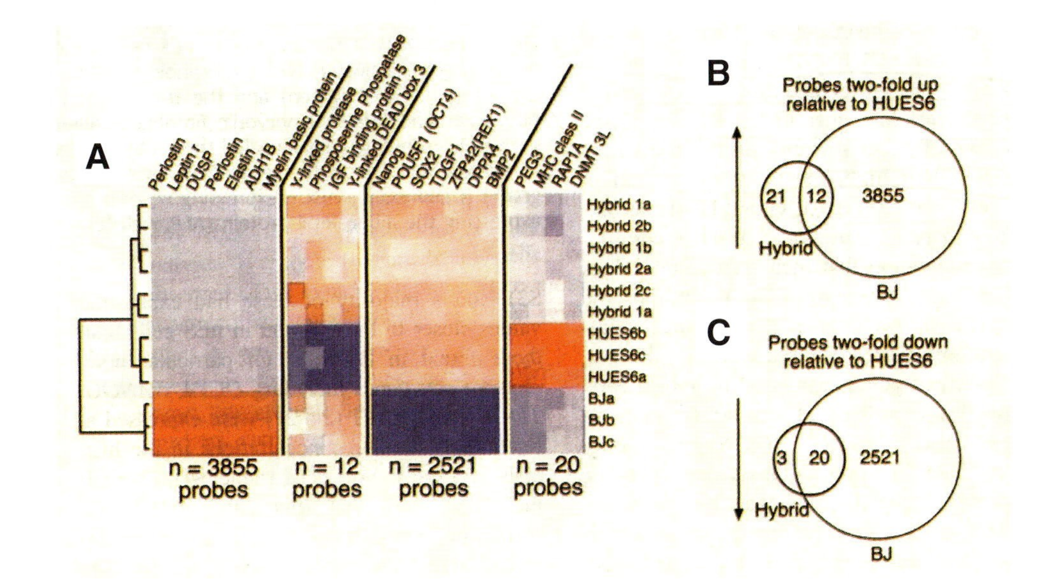


**FIGURE 15.9** Genome-wide transcription profiles demonstrating silencing of the fibroblast program and dominance of the ESC program in hybrid fibroblast/hESC cells. Genes with expression values 2X higher in BJ fibroblasts relative to ESCs (somatic-specific genes) or 2X lower in BJ fibroblasts relative to ESCs (ESC-specific genes) were identified and their expression analyzed in fibroblast/ESC hybrid cells. **(A)** Eisengram of representative genes from all three cell types with "on" genes represented by red and "off" genes represented by blue. **(B, C)** Venn diagrams showing the results of pair-wise comparisons from transcriptional profiles of the three cell types. Reproduced with permission from Cowan et al., Nuclear reprogramming of somatic cells after fusion with human embryonic stem cells. Science 309:1369–1373. Copyright 2005, AAAS.

**TABLE 15.1** Induction of a T-cell-specific signaling pathway in 293T fibroblasts reprogrammed by Jurkat cell extract. Jurkat cells, 293T fibroblasts treated with 293T extract and 293T fibroblasts treated with Jurkat extract were stimulated with anti-CD3 antibodies (T-cell-specific) and phorbolmyistylacetate (PMA, an activator of the protein kinase C pathway leading to T-cell activation). Stimulation results in expression of the high-affinity T-cell receptor IL-2R$\alpha$. Unstimulated or stimulated Jurkat and reprogrammed 293T fibroblasts express the low-affinity T-cell receptor IL-2R$\beta$. Data from Hakalien et al (2002).

| | % Cells Expressing | | | |
|---|---|---|---|---|
| | IL2-R$\beta$ | | IL-2R$\alpha$ | |
| | Unstimulated | Stimulated | Unstimulated | Stimulated |
| 293T cells + 293T extract | 0 | 0 | 0 | 0 |
| Jurkat cells | 90 | 90 | 0 | 90 |
| 293T cells + Jurkat extract | 75 | 75 | 0 | 75 |

ment protein 200 in polarized outgrowths resembling neurites. Changes in gene expression were stable over several weeks in culture. The reprogrammed fibroblasts did not have all the morphological features of T-cells or neurons, suggesting that their molecular profile is not completely reprogrammed. To what degree the T-cells are able to function as lymphocytes is as yet undetermined.

Finally, if we learn how hepatocytes manage to re-enter the cell cycle while maintaining all their differentiated functions, that knowledge might be applicable to inducing compensatory hyperplasia in other differentiated cell types as well.

### *b. Providing the Appropriate Microniche for Regeneration*

To enhance the regeneration of tissues that are normally able to regenerate, we can use molecular screens of culture media to identify the paracrine factors involved in maintaining or forming the regeneration microniche. Such screens, however, do not suffice to identify the factors required to initiate the regeneration of tissues that normally repair by fibrosis, because the microniche either does not contain regeneration-promoting factors or contains factors inhibitory to regeneration, or both. Here, we need to know what

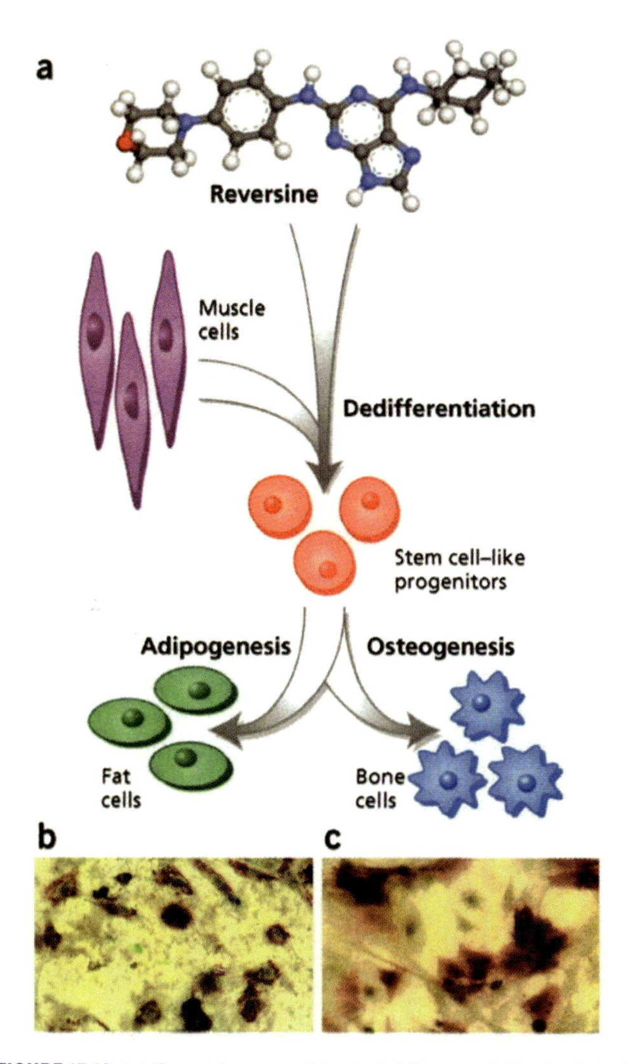


**FIGURE 15.10** **(a)** Screening a combinatorial library of di-substituted purines with C2C12 mouse muscle cells revealed reversine, a 2, 6-disubstituted purine, that caused the myofibers to cellularize and dedifferentiate to stem-like cells that could be induced to differentiate into adipocytes **(b)** or osteoblasts **(c)**. Reproduced with permission from Ding and Schultz, A role for chemistry in stem cell biology. Nature Biotech 22:833–840. Copyright 2004, Nature Publishing Group.

the molecular differences are between regenerative and fibrotic pathways if we are to be able to provide nonregenerating tissues with a regeneration-promoting microniche. These differences can be identified by differential screens of ECM and signaling molecules produced by regeneration-competent versus regeneration-deficient tissues. As outlined in previous chapters, comparison of regenerating versus nonregenerating tissues can be made between regenerating versus nonregenerating stages of development, regenerating versus nonregenerating species, and wild-type versus mutant tissues that differ in their ability to regenerate. The amphibians are particularly useful in this regard, because they can be used for each type of comparison: regeneration-competent versus regeneration-deficient stages of frog development, regeneration-competent adult salamander versus regeneration-deficient adult frog, and wild-type salamander (axolotl) limbs versus mutant *short-toe* limbs. Furthermore, the tissues of amphibians are much easier to manipulate surgically and chemically than those of mouse limbs and amphibians are much less expensive to maintain. Most importantly, amphibians regenerate a wide array of tissues that mammals cannot, so they are logical choices for the study of regeneration of those tissues.

Data analysis will be facilitated by bioinformatics and systems biology approaches. Systems biology is an analytical approach to the relationships among elements of a system, with the goal of understanding its emergent properties, whether this be a molecular system such as a transcription factor complex, a group of cells interacting for a specific function, such as the immune system, or an organ (Hood and Perlmutter, 2004). As they have stated, "In each case, one seeks to describe all of the elements of the system, define the biological networks that interrelate the elements of a system and characterize the flow of information that links these elements and their networks to an emergent biological process." In this way, we will be able to visualize the elements, interrelationships, and information flows that define the networks leading to fibrosis versus regeneration and determine the best places and times to target them with pharmaceutical agents that insure a regenerative outcome. The chemical induction of regeneration will be the least expensive and invasive method of regenerating a tissue, an advantage that cannot be overlooked.

### c. Gene Therapy as a Means for Chemical Induction of Regeneration

Over 600 clinical trials of gene therapy are underway in the United States for genetic diseases and cancers. Gene therapy may also be useful for the chemical induction of regeneration, particularly in cases where only a single protein is required to correct a deficiency. Genes can be delivered directly in plasmid or viral vectors to cells in the injury area, with or without electroporation (Aihara and Miyazaki, 1998), transfected into carrier cells such as fibroblasts or stem cells, which will then act as bioreactors to produce the protein(s) required, or incorporated into a biomaterial to be internalized by host cells involved in regeneration (Shea et al., 1999).

Three-dimensional cell-scaffold constructs promote efficient plasmid transfection of cells seeded into the constructs (Xie et al., 2001). Double promoter plasmids with the functional gene of interest and a reporter such as GFP can be used for rapid identification of transfected cells (Fu et al., 2002). Plasmids have been used for a number of tissue regeneration experiments. Plasmid DNA encoding human VEGF isoforms applied to the hydrogel polymer coating of an angioplasty balloon resulted in augmented collateral vessel development when delivered percutaneously to the ischemic iliac arteries of rabbits (Takeshita et al., 1996). A phase I clinical trial was performed in which VEGF plasmid DNA was injected directly into the ischemic myocardium of five male patients with angina who had failed conventional therapy (Losordo et al., 1998). All patients had significant reduction in angina associated with the development of increased collateral coronary circulation. Left ventricular ejection fraction was either unchanged (three patients) or improved (two patients) by a mean of 5%, approximately the same as was observed after transplant of bone marrow cells in other clinical trials (Chapter 12). The feasibility of using biodegradable polymers as carriers to deliver plasmid DNA constructs to cells at bone fracture or removal sites to stimulate the regeneration of new bone has been demonstrated (Fang et al., 1996; Niyibizi et al., 1998; Goldstein and Bonadio, 1998; Bonadio, 2000). A major drawback to the use of plasmid DNA is that the efficiency of uptake is low. Thus, high doses of DNA are required for tissue regeneration, and such doses can generate the formation of anti-DNA antibodies (Bonadio, 2002).

Viral vectors maximize the incorporation of genes into cells. Adenoviral vectors have frequently been used to deliver genes. Injection of adenoviral constructs containing the herpesvirus thymidine kinase gene or constitutively active form of the retinoblastoma protein into balloon catheter-injured pig or rat arteries decreased intimal hyperplasia (Ohno et al., 1994; Chang et al., 1995). Delivery of the telomerase gene in an adenoviral vector to the cirrhotic livers of telomerase-null mice restored telomerase activity, alleviated cirrhosis and promoted hepatocyte regeneration in models of liver injury (Rudolph et al., 2000). Adenoviral vectors, however, induce immune responses that decrease efficacy and increase toxicity (Bonadio, 2002).

Recombinant adeno-associated virus (rAAV) is viewed as one of the most promising viral vectors due to its lack of pathogenesis, nonimmunogenicity, and ability to incorporate into a wide variety of cells and tissues (Bonadio, 2002; Lu, 2004). Transfection of mouse liver with a rAAV vector containing the pancreatic transcription factor PDX-1 reprogrammed liver cells to produce insulin, though the liver cell type that was reprogrammed was unknown (Ferber et al., 2000). Lentivirus is being developed for clinical delivery of therapeutic neural genes. For example, genes for three key enzymes involved in dopamine production were reported by Oxford Biomedica to maintain dopamine production in the substantia nigra after injection of the genes in a lentiviral construct into the spinal cord of Parkinsonian rats, with substantial reduction in symptoms (Pearson, 2003).

A different form of gene therapy is RNAi (Novina and Sharpe, 2004). As described in Chapter 6, siRNAs have been delivered in lentiviral vectors to the spinal cord of a mouse ALS model to suppress mutant SOD genes. Soutschek et al. (2004) have described another way to deliver siRNAs that target the mRNA of apolipoprotein B by linking a cholesterol group to the sense strand of the siRNAs. The addition of the cholesterol group to the siRNAs allowed them to be efficiently delivered intravenously in mice. The siRNAs lowered the blood levels of both apolipoprotein B and cholesterol to those seen in mice with deletion of the apolipoprotein B gene and they did not change the levels of mRNAs for other genes, indicating that the effect was very specific to the apolipoprotein B mRNA. While encouraging, Rossi (2004) has pointed out a number of issues that must be addressed before this method is applicable to the same gene in humans, or to other genes as well. Treatment of high cholesterol in humans may have to be life-long, but the long-term effects of siRNA therapy are not known. Also not known is the benefit/risk ratio. The dosage of siRNA that would produce the desired effect is also problematic. Extrapolating from the mouse data on cholesterol reduction, humans would require regular infusions of gram quantities of siRNA-cholesterol conjugates, which would be very expensive and perhaps have undesirable side effects.

Nevertheless, RNAi promises to be one of the more potent tools to inhibit specific gene activity and to define the functional roles of genes in development. The power of RNAi to identify the functional roles of genes in regeneration has been demonstrated by expressing gene-specific dsRNA in bacteria and feeding it to planaria by mixing the bacteria with their food (blended liver and agarose) (Reddien et al., 2005). By this method, the effect of loss of function on planarian regeneration of a wide variety of genes was assessed. It is possible that similar studies could be performed on amphibians or fish by these methods.

## 6. The Challenge of Curing Disease

The technologies of regenerative medicine can be applied straightforwardly in cases of tissue injury due to trauma or ischemia. Their application to tissues damaged by degenerative diseases, however, is another matter. In most cases it will be imperative to cure the disease before regenerating the tissue. The need to cure the disease highlights the need for ESC research, because through SCNT or somatic cell-ESC fusion, nuclei from dysfunctional cells could be used to create ESCs for study of the genetic features by which diseases such as type 1 diabetes, Alzheimer's and ALS arise. How do these ESCs differ from wild-type ESCs? What are the pivotal molecular points at which the cells are turned to the dark side? Having found these points, it may be possible to devise individualized therapies to eliminate the disease process.

## BIOETHICAL ISSUES AND CHALLENGES

In Mary Shelly's 1818 novel *Frankenstein, or the Modern Prometheus*, the young medical student, Victor Frankenstein, suffers the terrible consequences of going beyond acceptable moral boundaries when he reanimates a corpse cobbled together from body parts procured from "the unhallowed damps of the grave." Many advances in science, and particularly in biology, have generated resistance based on moral and religious beliefs throughout the 19th and 20th centuries. The Darwinian idea of biological evolution through natural processes, though nearly 150 years old and supported by mountains of evidence, is still denied by creationists, based on religious ideas about complexity and the special place of man in the universe. Over the course of the 20th century, we developed a wide variety of technologies to combat infectious diseases, to perform sophisticated surgical operations, and to extend the lives of persons who would otherwise die. In many cases, these advances were both opposed and supported on moral grounds. For example, in the 1950s, there initially were deep divisions within the medical profession and the public over the feasibility and ethics of organ transplantation. Likewise, there were opposing views as to the ethics of inoculating children with the killed polio vaccine developed by Jonas Salk. In the last quarter of the 20th century, and during the first few years of the 21st century, we acquired the ability to manipulate animal and human reproductive processes. These manipulations led to the ability to create children by *in vitro* fertilization, as well as the cloning of mammals by SCNT and how to culture human ESCs *in vitro*. Each of these achievements has, in one way or another, given rise to Frankensteinian fears that the outcomes of the technologies involved go beyond acceptable moral boundaries.

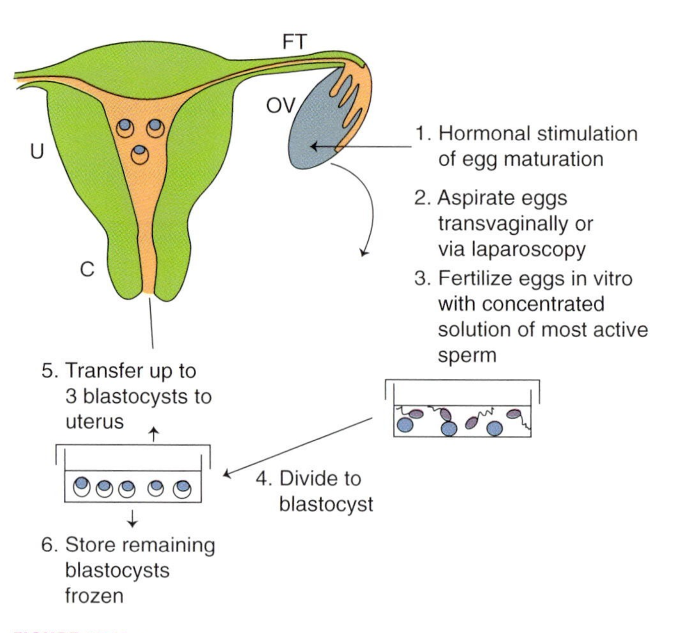


**FIGURE 15.11** *In vitro* fertilization technique. Maturation of eggs is stimulated by hormone injections (1) and the eggs are harvested by laparoscopy or trans-vaginal aspiration (2). The eggs are fertilized *in vitro* with concentrated sperm (3) and allowed to develop to the blastocyst stage (4). Up to three embryos are implanted in the uterus, while the rest are frozen.

### 1. In Vitro Fertilization, Reproductive Cloning, and Human ESCs

In the 1970s, Patrick Steptoe and Robert Edwards developed the assisted reproductive technology of *in vitro* fertilization. This technique involves the harvesting of several human eggs in a fertility clinic and fertilizing them *in vitro* (**FIGURE 15.11**). Up to three of the resulting blastocysts (~5 days postfertilization) are transferred to the uterus of the egg donor, where usually one (but sometimes more) will implant and result in the birth of an infant. The unused blastocysts are stored frozen for a specified length of time. Few of these extra blastocysts are ever used for further reproductive purposes. There are currently hundreds of thousands of blastocysts stored in fertility clinics across the globe that will eventually be discarded.

Reproductive cloning by SCNT has a long history dating to experiments done by embryologists in the early 1900s, but its modern application has its origins in research done on frogs by Briggs and King (1957) and Gurdon (Gurdon et al., 1975) in which nuclei from somatic cells at various stages of development were transplanted into enucleated eggs to test whether later stages of cell differentiation were due to loss of genes or the differential activity of a complete genome in different cell types. Transfer of nuclei from adult cells resulted in developmentally complete tadpoles in a small fraction of cases, indicating that the cells retained a complete genome and that the egg cytoplasm reprogrammed the transplanted nucleus to support the complete development of the egg. The same is true in mammals. Beginning with Dolly the sheep in 1996, a wide variety of farm animals and some endangered species have been cloned in this way (Wilmut et al., 2002). Hochedlinger and Jaenisch (2002) showed that the nuclei of even fully differentiated adult B and T cells of mice, in which genes are rearranged, could be reprogrammed and produce clones of mice by SCNT.

By the early 1980s, we had discovered how to propagate the pluripotent cells derived from the ICMs of mouse blastocysts (Evans and Kaufman, 1981). These embryonic stem cell cultures have been invaluable for understanding the general mechanisms of mammalian development. In 1998, Thompson et al. established human ESC lines from blastocysts left over from fertility treatments.

This achievement meant that hESCs could be used to study early human development, and screen drugs for toxic effects on human embryonic cells. While we do not yet have autogeneic hESCs, the ability to derive them will not be far off. We will then be able to study the genetics of disease states and differentiate an unlimited number of cells to regenerate damaged tissue without immunorejection. Our progress along these lines, however, has generated a bioethical debate on the moral status of the human embryo that has slowed research on hESCs.

## 2. The Controversy over Human ESC Research

To understand the controversy over human ESC research, it is helpful to look briefly at the initial reactions to organ transplants and *in vitro* fertilization. Solid organ transplants in humans were begun in the 1950s. At first, they were viewed as a "slippery slope" that could lead to a black market trade in which people would be murdered and their organs sold to the highest bidders. Worse, success rates were very low throughout the 1960s and moratoriums on some transplants, such as the heart, were imposed in the 1970s. However, organ transplants were the only hope for many people to extend their lives, and with the advent of immunosuppressive drugs such as cyclosporin A in the 1980s, success rates improved dramatically and resistance declined. Likewise, *in vitro* fertilization was at first condemned as unnatural and immoral, and raised fears of a black market for eggs. Since the birth in 1978 of the first child conceived *in vitro*, however, hundreds of thousands of normal children have been created in this way. *In vitro* fertilization has become an acceptable way for infertile couples to have children.

Opposition to these technologies evaporated because they successfully provided something of great benefit to their users without harming others. A key issue for both organ transplants and *in vitro* fertilization was the individual choice and consent of the individuals involved, based on their moral rights as persons. Organs may only be donated for transplant if the donor consents while living, or if the donor's family consents after his or her death. Likewise, individuals can make decisions through a living will about whether or not to have their life sustained after suffering irreversible coma or vegetative state, or a terminal illness. Couples who choose to have children by *in vitro* fertilization consent to the storage of excess embryos for a specified time, and these embryos cannot be used to make hESC lines without their consent.

But to make hESCs, the ability of the blastocyst to develop into a complete organism is destroyed, even though its cells live on. The fact that the embryo as an integrated whole is destroyed to create hESCs has engendered a debate about the moral status of the embryo that has made research on hESCs part of the politics of abortion. Three categories of views can be recognized in this debate; a detailed analysis of each category may be read in the report of the President's Council on Bioethics (2002).

On the one extreme are those who want all research on hESCs banned. No research on existing cell lines should be permitted, no new hESC lines should be derived from leftover blastocysts, and no derivation of ahESCs by SCNT should be allowed. Their argument is fundamentally as follows. The embryo has the moral right of existence, or personhood, from the moment of conception because the fertilized egg is a genetically distinct potential human life. Unlike an adult, or an agent acting under his or her orders, the embryo cannot consent to, or refuse anything, and therefore its moral right must be protected by society. Because the embryo has not yet exercised its right to be born, its parents cannot give consent to use it for research purposes. This argument is similar to the one used in support of maintaining persons in irreversible comas or vegetative states on life support when they have left no written instructions regarding what to do in such situations. Thus, destruction of a blastocyst, or allowing a person in a coma or vegetative state who has no living will, to die by withholding water and nourishment is considered by some to be murder by abortion or neglect. The making of new blastocysts by SCNT is considered particularly immoral, because it is abortion for a selfish end. Those who are totally opposed to any form of research on hESCs believe that if we open the door to this kind of research, it is only a matter of time before we embrace a totally utilitarian and dehumanizing view that destroys the dignity of human life and leads to crimes against any or all individuals deemed dispensable, undesirable, unable to function normally, or unable to speak for themselves. All of which has happened repeatedly throughout human history, the most recent examples being the atrocities of the Axis powers, the Stalinist Soviet Union, Maoist China, and the killing fields of Cambodia in the 20th century. The ultimate fear is that SCNT will inevitably lead to human reproductive cloning.

In the middle of the spectrum are people who do not object to the use of existing (or future) frozen blastocysts to derive new hESCs for research, but they do not want new blastocysts created by SCNT. They see the use of existing blastocysts as permissible, given the fact that they are there and if not used for reproductive purposes, will be destroyed anyway.

On the other end of the spectrum are those who believe that using existing or future blastocysts from fertility clinics for deriving new hESCs lines, or making new lines is acceptable if regulated according to a strict set of legal and ethical standards, much like those followed for organ donation. They are also in favor of deriving ahESCs by SCNT to study the genetic aspects of disease and as replacements for damaged tissues and organs.

Several arguments can be cited for this position (Young, 2000; Perry, 2000; Guenin, 2001; Brannigan, 2001; Reich, 2002; Reichardt, 2004). Collectively, they provide a strong argument that it would be immoral not to derive hESCs for research, including by SCNT, to alleviate human suffering. Two of the best analyses leading to this conclusion are those of May (2005) and Blackburn (2005) and should be consulted by anyone interested in the deliberations of the President's Council on Bioethics.

1. Abortion is medically defined as terminating the life of an embryo after implantation and placenta formation, which occurs in stages over the second and third weeks of development. Society accepts the practice of birth control through the use of methods that do not prevent fertilization, but prevent blastocyst implantation, such as the intrauterine device (IUD). Thus, destruction of an embryo that has not implanted is not abortion.
2. Since it is impossible to find enough surrogate mothers for all the excess embryos resulting from fertility treatments, the developmental potential of the blastocyst fails for lack of enablement (Guenin, 2001). These embryos cannot be kept indefinitely, and with the couple's permission are eventually destroyed after their fertility objectives are fulfilled. Since the cells of these embryos may save lives and alleviate suffering, we should use them for research aimed at that purpose.
3. Whereas genetic identity is set at fertilization, personhood is dependent on the attainment of developmental identity, which does not take place until gastrulation (~14 days postfertilization). Until then, individual cells of the two- or four-cell stage, or a fragmented blastocyst, are capable of making more than one individual (e.g., identical twins). Furthermore, although the blastocyst is certainly an early stage of human life, it is not a person. Personhood is attained only with the acquisition of sentience, the capacity to register sensory input from the environment, a capacity that is not acquired in the human embryo until at least eight weeks postfertilization. Interestingly, the point at which sentience is considered to be attained varies in different religions, but most place it at 40 days or more beyond conception (Reichardt, 2004).
4. A high percentage of human embryos conceived by sexual intercourse die during development. A woman has only a 20–30% chance of conceiving during a reproductive cycle and as many as 30–40% of successful conceptions miscarry early in development. Because of the high miscarriage rate, and because embryos derived by *in vitro* fertilization have a similar death rate, Grinnell (2003) has proposed that we learn how to identify the *in vitro* fertilization embryos that will die, and use only those embryos to make hESCs. This restriction would make blastocyst donation for the production of hESCs similar to the donation of organs from a brain-dead adult. In an actual experiment paralleling this idea, cells from morphologically abnormal partial frog blastulae, generated by SCNT from tadpole intestinal cells and destined to die, were able to proliferate and participate in the formation of axial muscle when transplanted into the blastocoel or above the lateral blastopore lip of gastrulae (Byrne et al., 2002) (**FIGURE 15.12**).
5. Brannigan (2001) has argued that even if the blastocyst were granted full moral status, research on hESCs is still permissible when the moral right of the embryo is in conflict with the moral rights of persons who suffer from conditions that could be helped by the use of hESCs. Whose moral right prevails must be decided by two factors, sentience and whether or not there are other options to relieve pain and suffering. These options would include the use of adult stem cells or the chemical induction of regeneration. Immunorejection of allogeneic cells would be controlled by immunosuppressive drugs, or ways would be discovered other than SCNT to

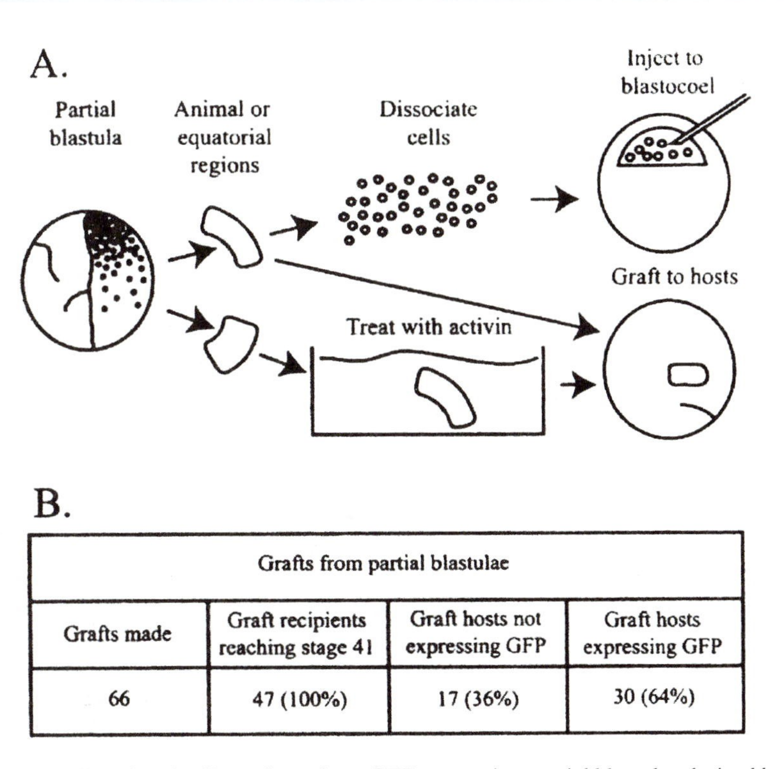


| Grafts from partial blastulae | | | |
|---|---|---|---|
| Grafts made | Graft recipients reaching stage 41 | Graft hosts not expressing GFP | Graft hosts expressing GFP |
| 66 | 47 (100%) | 17 (36%) | 30 (64%) |

**FIGURE 15.12** **(A)** Implants of *Xenopus* dissociated cells or tissue from GFP-expressing partial blastulae derived by SCNT to the blastocoel or the equatorial region of early gastrula hosts. The partial blastulae die, but the grafts are incorporated into the host tissues. **(B)** Experiment involving 66 grafts. Sixty-four percent of the grafts were incorporated into host tissues. Reproduced with permission from Byrne et al., From intestine to muscle: nuclear reprogramming through defective cloned embryos. Proc Natl Acad Sci USA 99:6059–6063. Copyright 2002, National Academy of Science, USA.

induce tolerance to allogeneic or xenogeneic cells. The blastocyst clearly is not sentient, and despite claims to the contrary, adult stem cells have so far not proven very useful in treating diseases and injuries other than hematopoietic disorders. Furthermore, we do not yet understand the mechanism of immune tolerance, and the use of immunosuppressive drugs introduces a wide range of side effects and a heightened incidence of malignant disease. The cost to society at large in terms of lost economic productivity and to patients and their families, in terms of physical and emotional suffering, is so enormous that if there is a moral conflict between the right of the embryo to exist and the needs of patients, the moral rights of suffering humanity should prevail.

6. Private companies will pursue ESC research (see Wilan et al., 2005) and could wind up owning this research to the exclusion of the NIH, which is the agency responsible for dissemination of research breakthroughs pertinent to the improvement of public health.

Different religions have a spectrum of official views that range from general support for hESC research (Judaism and Islam) to nearly total opposition (Catholicism). Within each religion, however, there is no uniform individual view (Indiana University Center for Bioethics Human Stem Cell Study Group, 2002). For example, the Catholic Church is willing to sanction research on ESCs derived from the germinal ridges of spontaneously aborted fetuses. Although most countries permit some research on hESCs, these different religious views are reflected in the different policies governments around the world have instituted on the derivation of blastocysts by SCNT. Some countries, such as the United Kingdom, Israel, Belgium, Singapore, and China permit the use of SCNT to create ahESCs for research purposes, while others such as Germany, Austria, and Italy have banned such research. Still others, such as the United States, lean strongly in this direction (Knowles, 2004), although within the United States, some states have provided funding for hESC research.

With regard to human reproductive cloning, there is uniform agreement among everyone—including scientists, religious leaders, and governments—that it should be banned. Everyone understands that this form of reproduction does not provide essential genetic recombination, that it is highly inefficient, and that the clones

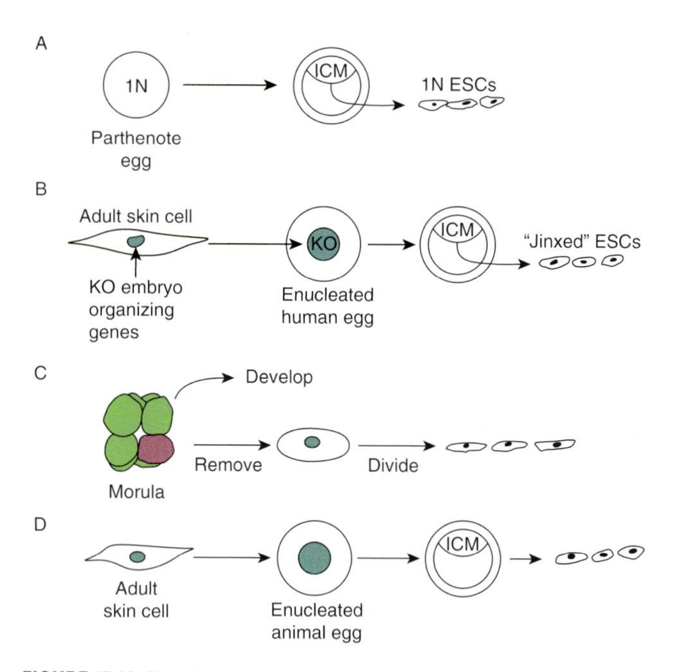


**FIGURE 15.13** Possible ways of making "non-embryos" that are either unable to develop **(A, B, C)** or use an animal egg **(D)**, to make hESC lines.

are susceptible to developmental abnormalities that reduce their ability to survive. This general agreement, however, will not prevent a rogue person from cloning a human being if he or she is so inclined. Therefore, banning research on ahESCs derived by SCNT in order to prevent human reproductive cloning makes no sense—they are quite different issues.

## 3. Resolving an Unresolvable Dilemma

We have already seen how fusion of somatic cells with existing ESCs might be used to create autogeneic pluripotent cells (Cowan et al., 2005). A number of other biological "fixes" have been proposed in an attempt to obviate the concerns of those who reject the creation of ESCs by SCNT (**FIGURE 15.13**). One idea is to create embryos that are "nonembryos." Two approaches have been proposed to do this. The first involves the creation of human parthenotes, which cannot develop into a human being, but can develop to a blastocyst stage (Maher, 2002; Holden and Vogel, 2004; Kiessling, 2005). An ESC line has been derived previously from parthenogenetic monkey blastocysts (Cibelli et al., 2002). The second approach is "alternate nuclear transfer," in which key genes involved in producing the signals required to form an organized embryo would be knocked out in donor somatic nuclei before transferring the nucleus to an enucleated egg, allowing enough development to create hESC lines, but not enough to produce a more advanced embryo (Holden and Vogel, 2004). A variant of this idea is to create a pluripotent cell, instead of an embryo, by transfecting either a somatic cell or an enucleated oocyte with pluripotency genes such as *Nanog* and fusing the cells (Vogel, 2005). Another idea is to extract a single blastomere from an eight-cell morula without injuring the remaining cells, a procedure that is already used for pre-implantation genetic diagnosis. Embryonic stem cells could potentially be derived from that single cell, and the remaining cells would continue development toward term (Pearson, 2004; Holden and Vogel, 2004). A third idea is to use animal eggs to create ahESC lines (Chen et al., 2003; see above).

None of these ideas has met with much enthusiasm, either for biological reasons and/or for bioethical ones. HESCs derived from parthenotes would be useful only to the egg donor and could possibly carry genetic abnormalities. Single blastomeres often die in culture, so this hurdle would have to be overcome in order to use single morula cells to generate hESCs. Furthermore, since morula cells have the potential to make a complete embryo, their use essentially destroys that potential, which would make this technique morally unacceptable to opponents of hESC research. A similar bioethical problem would be associated with alternate nuclear transfer, since inhibiting key genes required to block integrated development would be tantamount to abrogating the moral right of the embryo. While fusion of somatic cells and enucleated oocytes, one of which overexpresses pluripotency genes, would theoretically produce an ESC without going through an embryo stage, providing a supply of oocytes would be problematic.

In the final analysis, the different moral views on hESC research are not resolvable through bioethical argument and analysis, particularly in pluralistic societies. This Gordian knot can be cut only by an assessment of the utility of an idea or technology that weighs its potential benefits against its potential costs, including those costs that would have a dehumanizing effect. Guenin (2001) has argued that in the absence of interpersonally comparable utility measures, we cannot meaningfully sum utilities across a population. However, I would argue that history is replete with examples showing that the only measure that determines whether a controversial technology or idea survives and is adopted by society is whether or not it provides a high benefit to cost ratio across a population. In democratic societies, at least, that utility must always be measured within a framework of moral principles that includes a thorough understanding and assessment of the potential outcomes of the idea or technology, even though we cannot predict with certainty what the long-term effects of our decisions would be. Many ideas and technologies have been met with skepticism, if not outright derision or hostility, only to become reality because they proved themselves to be advantageous in some broad sector of human life. This process is aptly summarized in the truism that new ideas and technologies are first ridiculed, then violently opposed, and finally accepted, with opponents then claiming that it was all self-evident. Put in another way, if new technologies prove beneficial to enough people and do not harm the rest, they will eventually be accepted and become part of our social and moral fabric. Today's Frankenstein will be tomorrow's benefactor. The unacceptable will evolve into the status quo or generate an alternative route to the goal.

### 4. The Specter of Interspecies Grafting

In 1896, H.G. Wells published *The Island of Dr. Moreau*, the story of a vivisectionist who surgically altered animals to become more human-like. Today, it may be possible to do the same thing by embryological techniques. Experimental embryologists have long used interspecies grafts of cells that differ in some marker to follow the developmental fate of the grafted cells, giving valuable information about the origin of the different parts of the embryo and the mechanisms of their development. Human embryonic cells can also be grafted to the embryos of other species. The ability of human NSCs grafted into developing rat brains to migrate throughout the brain and differentiate into a variety of neurons and glia was mentioned in Chapter 6 as proof of concept that NSC-based cell therapy is feasible. Before NSC transplants for neurological disorders could undergo clinical trials in humans, transplants of human NSCs to higher-order animals would have to be conducted, preferably to nonhuman primates, to determine if there are any safety issues involved. Human NSCs have been grafted once into the ventricles of fetal bonnet monkeys (*Macaca radiata*). The cells migrated into the cortex and subventricular zone, where they differentiated into cortical neurons and glia, and contributed to the NSC population, respectively (Ourednik et al., 2001). No such chimeric animals have been brought to term, however.

The potential benefits of such experiments have to be very carefully weighed against the bioethical question of whether the cognitive capacity of the recipients might be altered in ways (presumably enhancements) that would make them more human and thus raise questions about their moral status as nonconsenting experimental subjects (Greene et al., 2005). The most likely scenario in which this could happen is after grafting large numbers of cells to the fetal nonhuman primate brain so that most of the cortex of the neonate is composed of human cells. One can even imagine an experiment in which the whole brain of a spontaneously aborted human embryo is grafted in place of the brain of an embryonic nonhuman primate.

Greene et al. (2005) have recommended support for the National Academy's recommendation that experiments proposing the grafting of human neural stem cells to the brains of adult nonhuman primates be subject to special review, and if permitted, that data on cognition and behavior of the recipients be collected and reported. Whether to allow the grafting of human NSCs to the embryos of nonhuman primates is an even more serious issue, because of the obvious moral issues surrounding the potential birth of such chimeras. Adding to the gravity of the issue, it is possible to imagine even more controversial experiments, such as bringing to term a chimera formed by mixing human and nonhuman primate blastomeres (or the blastomeres of mammals X and Y) and implanting the embryo into a surrogate mother. Like human reproductive cloning, it may be wiser to agree to ban the production of such chimeras.

## CONCLUDING REMARKS

Regenerative medicine is on the cusp of becoming a millennial revolution. Developing its potential will require the interdisciplinary efforts of biologists, chemists, experimental physicists, applied mathematicians, and computer scientists and engineers. The first wave of regenerative medicine is the transplantation of adult stem cells, which was begun in 1968 with the first bone marrow transplants, and is now being expanded to the transplantation of other kinds of adult stem cells and derivatives of embryonic stem cells, either as cell suspensions or as aggregates, or seeded into a biomimetic scaffold to make a bioartificial tissue. Cell therapy and bioartificial tissue construction is expensive because of the logistics of harvesting and expanding the cells, as well as the invasive procedures required. Nevertheless, as research advances, costs will come down. A large number of biotechnology companies have been formed to provide human cell-based regenerative medicine therapies for a variety of tissues (see Wilan et al., 2005).

Part of the first wave of regenerative medicine will be the chemical induction of regeneration using transplanted cells as bioreactors to provide paracrine survival and proliferation factors to host tissues. The second wave will be the chemical induction of regeneration by the delivery of regeneration-promoting proteins to a lesion site through noncellular means, or the insertion of genes that confer regeneration-competence into cells. These types of treatments will not only be relatively simple to administer clinically, but will also be much less expensive than cell transplantation therapies. To make the chemical induction of regeneration feasible, we must understand the biology of regeneration and how it differs from fibrosis to a much greater depth than is currently available. Only then will we know the appropriate places, times, and at what concentrations to intervene in the pathways of repair to choose regeneration over fibrosis.

Broad success will not be achieved by a singular focus on mammalian research models. We need to draw on nonmammalian models of regeneration as well. Furthermore, it is important to realize that one type of regeneration therapy will not fit all situations. Cell transplants, bioartificial tissues, and chemical induction of regeneration will likely each be required, depending on the type and extent of tissue damage. For example, it may not be possible to regenerate tissues beyond a critical size defect using a cell transplant, chemical cocktail, or regeneration template. Larger defects may require a regeneration template seeded with cells to make a bioartificial tissue.

How far we will be able to go in our quest to regenerate tissues and organs is anyone's guess. Twenty-five years ago, patients paralyzed as a result of spinal cord injuries were told that the spinal cord has no intrinsic capacity for regeneration and that there was no hope for recovery. Like other dogmas, this one, too, has crumbled under the weight of scientific advances. We know now that the mammalian spinal cord does have the intrinsic potential for regeneration and we have partly solved the problem of realizing that potential. Might we be able to induce the regeneration of myocardium, lung, kidney, or even appendages? These will be complex and difficult tasks, but we should not view them as impossible or even as improbable.

## SUMMARY

With the exception of hematopoietic stem cell transplants for malignant and genetic diseases of the hematopoietic system, we do not as yet have a broad-based-regenerative medicine. The further development of regenerative medicine requires a much deeper understanding of the biology of regeneration than we currently possess. Progress is being made, however, in laboratories around the world, and we can expect major conceptual and technical basic research breakthroughs that will lead to rapid advances in therapy.

The biological research challenges to be met are several. The challenge common to all the strategies of regenerative medicine is to identify the complete constellation of factors that constitutes a regenerative microniche for the many tissues that have the latent capacity for regeneration, but are induced to form scar tissue instead. Identification can be accomplished by comparative analyses of regeneration-competent versus regeneration-deficient tissues. Once the molecular elements of regeneration versus scarring are known, they must be related to one another to identify pathway networks that will reveal the most effective intervention strategies. In addition, each of the three major types of regenerative therapy has specific issues to be addressed. Therapies involving cell transplantation and construction of bioartificial tissues require reliable sources of expandable cells. These sources are differentiated cells that are expandable upon dedifferentiation, adult stem cells, and embryonic stem cells. Each has advantages and drawbacks. Dedifferentiated cells can be redifferentiated back to the original cell type, but if taken from the patient, will not be immunorejected. Most types of adult stem cells are difficult to harvest and to expand, but also will not be immunorejected. Embryonic stem cells have the best growth potential and are pluripotent, but require many more steps of directed differentiation than ASCs, and their derivatives are subject to immunorejection. A major question about all of the differentiated or redifferentiated products of expanded cell populations is whether they are equivalent to the original cells in their long-term structural integrity and function. In addition, there is a question about the effects of age on the proliferation potential and functional quality of the products of dedifferentiated cells and ASCs. A major problem in the propagation of ESCs in an undifferentiated state is their reliance on animal feeder cells and potential contamination with animal products. None of these problems is insurmountable. Advanced cell culture protocols are continually being developed that will ultimately produce usable cells of high quality. Methods to tolerize patients to allogeneic and even xenogeneic cells are being sought, but we already possess a way to bypass immunorejection of ESC derivatives by creating human autogeneic ESCs by somatic cell nuclear transfer and a new way of making autogeneic cells by fusing somatic cells to ESCs is being developed.

The successful construction and use of bioartificial tissues must overcome two problems. The first is to create a "smart" synthetic ECM that is interactive with cells; i.e., create a biomimetic microniche that makes the cells "at home." Some progress has been made in this direction with the design of polyethylene glycol scaffolds containing peptides that bind soluble molecules or adhesion proteins, or are substrates for proteases. The second is how to ensure the rapid vascularization of tissue constructs. Strategies to do this are the seeding of endothelial cells along with the functional cell types into scaffolds, and building up scaffolds in layers containing grooves that can form a microfluidic network when seeded with endothelial cells.

The chemical induction of regeneration *in situ* is the least invasive and expensive therapy, requiring only the delivery of a regeneration-promoting and/or fibrosis-inhibiting molecular cocktail to the site of injury in the form of an injection, a "smart" bandage, or a pill. This strategy relies on the widespread distribution of regeneration-competent cells, such as ASCs or pluripotent stem cells in connective tissue compartments, throughout the body. The challenge is to more accurately determine the extent of such cells. Alternatively, the challenge is to reprogram differentiated cells or make them dedifferentiate *in situ*, proliferate and redifferentiate into new tissue, or induce differentiated cells to undergo compensatory hyperplasia, as in the liver. Still another strategy in the chemical induction of regeneration is to use gene therapy to regain function in dysfunctional cells, or to use RNA interference to knock down mutant genes that are causing cell dysfunction.

A final challenge is the fact that regenerative therapies for many degenerative diseases, though they may slow the progress of the disease, will ultimately not be effective unless the disease is cured. ESCs created by nuclear transfer from the cells of patients with degenerative diseases would be a powerful tool with which to analyze the genetic basis for these diseases.

Our ability to create ESCs from the discarded embryos of fertility clinics and the potential to create autogeneic human ESCs by somatic cell nuclear transfer has opened an intense bioethical debate on the moral status of the embryo. Three viewpoints emerge from the debate: total opposition to any form of human ESC research; acceptance of hESC research except for the creation of hESCs by SCNT; and acceptance of hESC research including the creation of hESCs by SCNT. Those in total opposition to hESC research believe that the fertilized egg has the moral rights of personhood. Because the embryo is totally vulnerable, its moral rights must be protected by society. They believe that destruction of a blastocyst to make hESCs is tantamount to murder and that research on hESCs ultimately leads to a dehumanizing view of human life that strips it of its dignity and leads to crimes against humanity. A misleading view often projected by this group is that adult stem cells are as pluripotent as ESCs. Those on the opposite end of the spectrum argue that blastocysts are not sentient and therefore are not persons, and that the moral rights of sentient persons suffering from diseases that could be helped by hESC research should prevail over any moral right the blastocyst might have. A number of biological fixes have been proposed to try to resolve these disparate views, but the issue is fundamentally unresolvable. History shows that the only measure that determines whether a controversial technology or idea is accepted by society is whether or not it provides a high benefit-to-cost ratio across a broad spectrum of society and does no other harm to fundamental moral principles. If new technologies prove beneficial to enough people and do not harm the rest, they will eventually be accepted and become part of our social and moral fabric. Today's Frankenstein becomes tomorrow's benefactor.

Another bioethical controversy is whether or not nonhuman primates should be used as recipients to test the safety of human NSC transplants prior to clinical trials in humans and/or to learn whether the transplanted cells improve neural function in injured or diseased brains. It is possible that such transplants, particularly of large numbers of cells to cortical areas, could result in cognitive enhancements that "humanize" the recipients, changing their moral status.

Regenerative biology and medicine stands today on the verge of a millennial revolution. How far we will be able to go in our quest to regenerate tissues and organs is unknown. Just two decades ago, spinal cord patients were told to accept the fact that the spinal cord has no power of regeneration and that there was no hope for recovery. Like other dogmas, this one too has crumbled under the weight of advancing scientific evidence. Will it be possible to induce the regeneration of kidney tissue, lung tissue, myocardium, or even appendages? These will be complex and difficult tasks—biological Manhattan projects—but we should not view them as impossible or even improbable.

## REFERENCES


Abbot A, Cyranoski (2001) China plans hybrid embryonic stem cells. Nature 413:339.

Aihara H, Miyazaki J-I (1998) Gene transfer into muscle by electroporation in vivo. Nature Biotech 16:867–870.

Amit M, Shariki C, Margulets V, Itskovitz-Eldor J (2004) Feeder-layer and serum-free culture of human embryonic stem cells. Biol Reprod 70:837–845.

Anderson D, Levenberg S, Langer R (2004) Nanoliter-scale synthesis of arrayed biomaterials and application to human embryonic stem cells. Nature Biotech 22:863–866.

Bain G, Kitchens D, Yao M, Huettner JE, Gottleib DI (1995) Embryonic stem cells express neuronal properties in vitro. Dev Biol 168:342–357.

Baldwin SP, Saltzman WM (1996) Polymers for tissue engineering. Trends Polymer Sci 4:1–7.

Barberi T, Klivenyi P, Calingasan NY, Lee H, Kawamata H, Loonam K, Perrier A, Bruses J, Rubio ME, Topf N, Tabar V, Harrison NL, Beal MF, Moore MA, Studer L (2003) Neural subtype specification of fertilization and nuclear transfer embryonic stem cells and application in parkinsonian mice. Nature Biotech 21:1200–1207.

Bohana-Kashtan O, Civin CI (2004) Fas ligand as a tool for immunosuppression and generation of immune tolerance. Stem Cells 22:908–924.

Blackburn E (2005) Thoughts of a former Council member. Perspect Biol Med 48:172–180.

Block GD, Locker J, Bowen WC, Petersen BE, Katyal S, Strom SC, Riley T, Howard TA, Michalopoulos GK (1996) Population expansion, clonal growth, and specific differentiation patterns in primary cultures of hepatocytes induced by HGF/SF, EGF and TGFα in a chemically defined (HGM) medium. J Cell Biol 132: 1133–1149.

Block GD (2001) Enabling culture systems for homogenous, multiplex and composite tissue engineering with clonal expansion of differentiated, primary adult human cells. In: Holdridge GM (ed.). WTEC Panel Report on Tissue Engineering Research. Baltimore: International Technology Research Institute, pp 92–94.

Bodnar MS, Meneses JJ, Rodriguez RT, Firpo MT (2004) Propagation and maintenance of undifferentiated human embryonic stem cells. Stem Cells Dev 13:243–253.

Bohana-Kashtan O, Civin C (2004) Fas ligand as a tool for immunosuppression and generation of immune tolerance. Stem Cells 22:908–924.

Bonadio J (2000) Local gene delivery for tissue regeneration. J Reg Med (e-biomed) 1:25–29.

Brandenberger R, Wei H, Zhang S, Lei S, Murage J, Fisk G, Li Y, Xu C, Fang R, Giegler K, Rao MS, Mandalam R, Lebkowski J, Stanton LW (2004) Transcriptome characterization elucidates signaling networks that control human ES cell growth and differentiation. Nature Biotech 22:707–716.

Brannigan MC (2001) What if we assign moral status to human embryos? Genetic Eng News 12:6.

Briggs R, King T (1957) Changes in the nuclei of differentiating endoderm cells as revealed by nuclear transplantation. J Morph 100:269–311.

Brivanlou AH, Gage FH, Jaenisch R, Jessell T, Melton D, Rossant J (2003) Setting standards for human embryonic stem cells. Science 300:913–916.

Buzzard J, Gough NM, Crook JM, Colman A (2004) Karyotype of human ES cells during extended culture. Nature Biotech 22: 381–382.

Byrne JA, Simonsson S, Gurdon JB (2002) From intestine to muscle: Nuclear reprogramming through defective cloned embryos. Proc Natl Acad Sci USA 99:6059–6063.

Carpenter MK, Rosler ES, Fisk GJ, Brandenberger R, Ares X, Miura T, Lucero M, Rao MS (2004) Properties of four human embryonic stem cell lines maintained in a feeder-free culture system. Dev Dynam 229:243–258.

Chang MW, Barr E, Seltzer J, Jiang Y-Q, Nabel GJ, Nabel EG, Parmacek MS, Leiden JM (1995) Cytostatic gene therapy for vascular proliferative disorders with a constitutively active form of the retinoblastoma gene product. Science 267:518–522.

Cibelli JB, Grant A, Chapman B, Cunniff K, Worst T, Green HL, Walker S, Gutin PH, Vilner L, Tabar V, Dominko T, Kane J, ettstein PJ, Lanza RP, Studer L, Vrana KE, West MD (2002) Parthenogenetic stem cells in nonhuman primates. Science 295:819.

Cibelli JB, Kiessling AA, Cunniff K, Richards C, Lanza RP, West MD (2001) Somatic cell nuclear transfer in humans: Pronuclear and early embryonic development. J Reg Med (e-Biomed) 2:25–31.

Constans A (2002) Stem cell know-how. The Scientist 2 (Sept): 44–46.

Cooper DKC (2001) Immunomodulation: The immunology of allogeneic and xenogeneic hematopoietic cell transplantation and tolerance induction. In: Holdridge GM (ed.). WTEC Panel Report on Tissue Engineering Research. Baltimore: International Technology Research Institute, pp 82–91.

Cooper DKC, Lanza RP (2000) Xeno. New York: Oxford University Press.

Cowan CA, Klimanskaya I, McMahon J, Atienza J, Witmyer J, Zucker JP, Wang S, Morton CC, McMahon AP, Powers D, Melton DA (2004) Derivation of embryonic stem-cell lines from human blastocysts. New Eng J Med 350:1353–1356.

Cowan CA, Atienza J, Melton DA, Eggan K (2005) Nuclear reprogramming of somatic cells after fusion with human embryonic stem cells. Science 309:1369–1373.

D'Amour KA, Gage FH (2003) Genetic and functional differences between multipotent neural and pluripotent embryonic stem cells. Proc Natl Acad Sci USA 100:11866–11872.

Dennis C (2002) Stem cells rise in the east. Nature 419:334–336.

Dennis C (2003) Chinese fusion method promises fresh route to human stem cells. Nature 424:711.

Ding S, Schultz G (2004) A role for chemistry in stem cell biology. Nature Biotech 22:833–840.

Do JT, Scholer HR (2004) Nuclei of embryonic stem cells reprogram somatic cells. Stem Cells 22:941–949.

Draper JS, Smith K, Gokhale P, Moore HD, Maltby E, Johnson J, Meisner L, Zwaka TP, Thonson JA, Andrews PW (2004) Recurrent gain of chromosomes 17q and 12 in cultured human embryonic stem cells. Nature Biotech 22:53–54.

Ehrbar M, Djonov VG, Schnell C, Tschanz SA, Martiny-Baron G, Schenk U, Wood J, Burri PH, Hubbell JA, Zisch AH (2004) Cell-demanded liberation of $VEGF_{121}$ from fibrin implants induces local and controlled blood vessel growth. Circ Res 94:1124–1132.

Evans MJ, Kaufman M (1981) Establishment in culture of pluripotential cells from mouse embryos. Nature 292:154–156.

Ezashi T, Das P, Roberts R (2005) Low $O_2$ tensions and the prevention of differentiation of hES cells. Proc Natl Acad Sci USA 102:4783–4788.

Fairchild PJ, Brook FA, Gardner RL, Graca L, Strong V, Tone Y, Tone M, Nolan KF, Waldmann H (2000) Directed differentiation of dendritic cells from mouse embryonic stem cells. Curr Biol 10:1515–1518.

Fang J, Zhu Y-Y, Smiley E, Bonadio J, Rouleau JP, Goldstein SA, McCauley LK, Davidson BL, Roessler BJ (1996) Stimulation of new bone formation by direct transfer of osteogenic plasmid genes. Proc Natl Acad Sci USA 93:5753–5758.

Faustman DL, Pedersen RL, Kim SK, Lemischka IR, McKay RD (2002) Cells for repair. Ann NY Acad Sci 961:45–47.

Ferber S, Halkin A, Ber I, Einav Y, Goldberg I, Basshack I, Seijffers R, Kopolovic J, Kaiser N, Karasik A (2000) Pancreatic and duodenal homeobox gene 1 induces expression of insulin genes in liver and ameliorates streptozotocin-induced hyperglycemia. Nature Med 6:568–572.

Fu L, Buchholz D, Shi Y-B (2002) Novel double promoter approach for identification of transgenic animals: A tool for in vivo analysis of gene function and development of gene-based therapies. Mol Reprod Dev 62:470–476.

Gearhart J (2004) New human embryonic stem-cell lines—more is better. New Eng J Med 350:1275–1276.

Geijsen N, Horoschak M, Kim K, Gribnau J, Eggan K, Daley GQ (2004) Derivation of embryonic germ cells and male gametes from embryonic stem cells. Nature 427:148–154.

Gerecht-Nir S, Dazard J-E, Golan-Mashiach M, Osenberg S, Botvinnik A, Amariglio N, Domany E, Rechavi G, Givol D, Itskovitz-Eldor J (2005) Vascular gene expression and phenotypic correlation during differentiation of human embryonic stem cells. Dev Dynam 232:487–497.

Gobin AS, West JL (2002) Cell migration through defined, synthetic ECM analogs. FASEB J 16:751–753.

Goldstein SA, Bonadio J (1998) Potential role for direct gene transfer in the enhancement of fracture healing. Clin Orthopaed Related Res 355S:S154–S162.

Gottlieb DI (2000) An *in vitro* pathway from ES cells to neurons. J Reg Med (e-biomed) 1:7–9.

Greene M, Schill K, Takahashi S, Bateman-House A, Beauchamp T, Bok H, Cheney D, Coyle J, Deacon T, Dennett D, Donovan P, Flanagan O, Goldman S, Greely H, Martin L, Miller E, Mueller D, Siegel A, Solter D, Gearhart J, McKhann G, Faden R (2005) Moral issues of human—non-human primate neural grafting. Science 309:385–386.

Griesler HQ (2001) Biomolecules. WTEC Panel Report on Tissue Engineering Research. In: Holdridge GM (ed.). Baltimore, International Technology Research Institute, pp 31–47.

Griffith LG (2000) Polymeric biomaterials. Acta Mater 48:263–277.

Grinnell F (2003) Defining embryo death would permit important research. The Chronicle of Higher Education, May 16, B13.

Guenin LM (2001) Morals and primordials. Science 292:1659–1661.

Gurdon J, Laskey R, Reeves O (1975) The developmental capacity of nuclei transplanted from keratinized skin cells of adult frogs. J Embryol Exp Morph 34:93–112.

Hakelien A-M, Landsverk HB, Robl JM, Skalhegg BS, Collas P (2002) Reprogramming fibroblasts to express T-cell functions using cell extracts. Nature Biotech 20:460–466.

Hench LL, Polak JM (2002) Third-generation biomedical materials. Science 295:1014–1017.

Hirano M, Yamamoto A, Yoshimura N, Tokunaga T, Motohashi T, Ishizaki K, Yoshida H, Okazaki K, Yamazaki H, Hayashi S-I, Kunisada T (2003) Generation of structures formed by lens and retinal cells differentiating from embryonic stem cells. Dev Dynam 228:664–671.

Hochedlinger K, Jaenisch R (2002) Monoclonal mice generated by nuclear transfer from mature B and T cells. Nature 415:1035–1038.

Holden C, Vogel G (2004) A technical fix for an ethical bind? Science 306:2174–2176.

Holder WD, Gruber HE, Roland WD, Moore A, Culberson C, Loebsack AB, Burg KJL, Mooney DJ (1997) Increased vascularization and heterogeneity of vascular structures occurring in polyglycolide matrices containing aortic endothelial cells implanted in the rat. Tiss Eng 3:149–160.

Hood L, Perlmutter RM (2004) The impact of systems approaches on biological problems in drug discovery. Nature Biotech 22:1215–1217.

Hubbell JA (1995) Biomaterials in tissue engineering. Biotechnology 13:565–576.

Hubbell JA (2004) Biomaterials science and high-throughput screening. Nature Biotech 22:828–829.

Humpherys D, Eggan K, Akutsu H, Hochedlinger K, Rideout WM III, Biniszkiewicz D, Yanagimachi R, Jaenisch R (2001) Epigenetic instability in ES cells and cloned mice. Science 293: 95–97.

Hwang WS, Ryu YJ, Park JH, Park ES, Lee EG, Koo JM, Jeon HY, Lee BC, Kang SK, Kim SJ, Ahn C, Hwang JH, Park KY, Cibelli JB, Moon SY (2004) Evidence of a pluripotent human embryonic stem cell line derived from a cloned blastocyst. Science 303:1669–1674.

Hwang WS, Roh SI, Lee BC, Kang SK, Kwon DK, Kim SJ, Park SW, Kwon HS, Lee CK, Lee JB, Kim JM, Ahn C, Paek SH, Chang SS, Koo JJ, Yoon HS, Hwang JH, Hwang YY, Park YS, Oh SK, Kim HS, Park JH, Moon SY, Schatten G (2005) Patient-specific embryonic stem cells derived from human SCNT blastocysts. Science 308:1777–1783.

Indiana University Center for Bioethics Human Stem Cell Study Group (2002) Diverse Perspectives: Considerations About Embryonic Stem Cell Research, October, 2002. Available at http://www.bioethics.iu.edu/Diverse_Perspectives.pdf.

Ivanova NB, Dimos JT, Schaniel C, Hackney JA, Moore KA, Lemischka IR (2002) Stem cell molecular signature. Science 298: 601–604.

Jiang Y, Jahagirdar BN, Reinhardt R, Schwarts RE, Keene CD, Ortiz-Gonzalez XR, Reyes M, Lenvik T, Lund T, Blackstad M, Du J, Aldrich S, Lisberg A, Low WC, Largaespada DA, Verfaille CM (2002) Pluripotency of mesenchymal stem cells derived from adult marrow. Nature 418:41–49.

Keller GM (1995) *In vitro* differentiation of embryonic stem cells. Curr Opinion Cell Biol 7:862–869.

Kellner K, Tessmar J, Milz S, Angele P, Schulz MB, Nerlick M, Schulz MB, Blunk T, Gopferich A (2004) PEGylation does not impair insulin efficacy in 3-D cartilage culture: An investigation toward biomimetic polymers. Tiss Eng 10:429–441.

Kiessling AA (2005) Eggs alone. Nature 434:145.

Klimanskaya I, Chung Y, Meisner L, Johnson J, West MD, Lanza R (2005) Human embryonic stem cells derived without feeder cells. Lancet 365:1636–1641.

Klotzko AJ (2002) Take therapeutic cloning forward. The Scientist 16:11–12.

Knowles LP (2004) A regulatory patchwork—human ES cell research oversight. Nature Biotech 22:157–163.

Lechler RI, Garden OA, Turka LA (2003) The complementary roles of deletion and regulation in transplantation tolerance. Nature Reviews: Immunology 3:147–158.

Li X-J, Du Z-W, Zarnowska ED, Pankratz M, Hansen LO, Pearce RA, Zhang S-C (2005) Specification of motoneurons from human embryonic stem cells. Nature Biotech 23:215–221.

Losordo DW, Vale PR, Symes JF, Dunnington CH, Esakof DD, Maysky M, Ashare AB, Lathi, Isner JM (1998) Gene therapy for myocardial angiogenesis: Initia clinical results with direct myocardial injection of $phVEGF_{165}$ as sole therapy for myocardial ischemia. Circulation 98:2800–2804.

Lu Y (2004) Recombinant adeno-associated virus as delivery vehicle for gene therapy—a review. Stem Cells Dev 13:133–145.

Lutolf MP, Hubbell JA (2005) Synthetic biomaterials as instructive extracellular microenvironments for morphogenesis in tissue engineering. Nature Biotech 23:47–55.

Mann BK, Gobin AS, Tsai AT, Schmedlen R, West L (2001) Smooth muscle cell growth in photopolymerized hydrogels with cell adhesive and proteolytically degradable domains: Synthetic ECM analogs for tissue engineering. Biomats 22:3045–3051.

Mantle C, Guo Y, Kauffman C, English D, Broxmeyer HE (2004) Potential use of enucleated stem cell cytoplasts in investigations of stem cell fusion. Stem Cells Dev 13:165–167.

May WF (2005) The President's Council on Bioethics: My take on some of its deliberations. Perspect Biol Med 48:229–240.

Mitalipova MM, Rao RR, Hoyer DM, Johnson JA, Meisner LF, Jones KL, Dalton S, Stice SL (2005) Preserving the genetic integrity of human embryonic stem cells. Nature Biotech 23:19–20.

Munsie M, Michalska AE, O'Brien CM, Trounson AO, Pera MF, Mountford PS (2000) Isolation of pluripotent embryonic stem cells from reprogrammed adult mouse somatic cell nuclei. Curr Biol 10:989–992.

Nimni ME (1997) Polypeptide growth factors: Targeted delivery systems. Biomats 18:1201–1225.

Niyibizi C, Baltzer A, Lattermann C, Oyama M, Whalen JD, Robbins PD, Evans CH (1998) Potential role for gene therapy in the enhancement of fracture healing. Clin Orthopaed Related Res 355S:A148–S153.

Novina CD, Sharpe A (2004) The RNAi revolution. Nature 430: 161–164.

Ohno T, Gordon D, San H, Pomoili VJ, Imperiale MJ, Nabel GJ, Nabel EG (1994) Gene therapy for vascular smooth muscle cell proliferation after arterial injury. Science 265:781–784.

Orkin SH, Morrison S (2002) Stem-cell competition. Nature 418: 25–27.

Ourednik V, Ourednik J, Flax JD, Zawada WM, Hutt C, Yang C, Park KI, Kim SU, Sidman RL, Freed CR, Snyder EY (2001) Segregation of human neural stem cells in the developing primate forebrain. Science 293:1820–1824.

Ozawa CR, Banfi A, Glazer NL, Thurston G, Springer ML, Kraft PE, McDonald DM, Blau HM (2004) Microenvironmental VEGF concentration, not total dose, determines a threshold between normal and aberrant angiogenesis. J Clin Invest 113:516–527.

Parentau NL (2001) Cells. In: Holdridge GM (ed.). WTEC Panel Report on Tissue Engineering Research. Baltimore: International Technology Research Institute, pp 17–30.

Parentau NL, Young JH (2002) The use of cells in reparative medicine. Ann NY Acad Sci 961:27–39.

Pearson S (2003) Promise of successful gene therapy resurges. Gen Eng News 23, #20:1.

Pearson H (2004) Early embryos fuel hopes for shortcut to stem-cell creation. Nature 432:2–3.

Peppas NA, Langer R (1994) New challenges in biomaterials. Science 263:1715–1720.

Pera MA (2004) Unnatural selection of cultured human ES cells? Nature Biotech 22:42–43.

Pera MA, Trounson AO (2004) Human embryonic stem cells: Prospects for development. Development 132:5515–5525.

Perry D (2000) Patient's voices: The powerful sound in the stem cell debate. Science 287:1423.

Phimister EG, Drazen JM (2004) Two fillips for human embryonic stem cells. New Eng J Med 350:1351–1352.

Picciolo GL, Stocum DL (2001) ASTM lights the way for tissue engineered medical products standards. ASTM Standardization News 29:30–35.

Powell K (2005) It's the ecology, stupid! Nature 435:268–270.

President's Council on Bioethics (2002) Human Cloning and Human Dignity: An Ethical Inquiry. Chapter Six: The Ethics of Cloning-for-Biomedical Research. http://www.bioethics.gov/reports/cloningreport/research.html.

Ramalho-Santos M, Yoon S, Matsuzaki Y, Mulligan RC, Melton DA (2002) "Stemness": Transcriptional profiling of embryonic and adult stem cells. Science 298:597–600.

Ratner B (1992) New ideas in biomaterials science—a path to engineered biomaterials. J Biomed Mats Res 27:837–850.

Reddien PW, Bermange AL, Murfitt KJ, Jennings JR, Sanchez Alvarado A (2005) Identification of genes needed for regeneration, stem cell function, and tissue homeostasis by systematic gene perturbation in planaria. Dev Cell 8:635–649.

Reich JG (2002) The debate in Germany. Science 296:265.

Reichardt T (2004) Studies of faith. Nature 432:666–669.

Reubinoff BE, Itsykson P, Turetsky T, Pera MF, Reinhartz E, Itzik A, Ben-Hur T (2001) Neural progenitors from human embryonic stem cells. Nature Biotech 19:1134–1140.

Rippon HJ, Bishop AE (2004) Embryonic stem cells. Cell Prolif 37:23–34.

Rosler ES, Fisk G, Ares X, Irving J, Miura T, Rao MS, Carpenter MK (2004) Long-term culture of human embryonic stem cells in feeder-free conditions. Dev Dynam 229:259–274.

Rossi JJ (2004) A cholesterol connection in RNAi. Nature 432: 155–156.

Roy NS, Nakano T, Keyoung HM, Windrem M, Rashbaum WK, Alonso ML, Kang J, Peng W, Carpenter MK, Lin J, Nedergard M, Goldman SA (2004) Telomerase immortalization of neuronally restricted progenitor cells derived from the human fetal spinal cord. Nature Biotech 22:297–305.

Rudolph KL, Chang S, Millard M, Schreiber-Agus N, DePinho R (2000) Inhibition of experimental liver cirrhosis in mice by telomerase gene delivery. Science 287:1253–1256.

Schmitt T, de Pooter RF, Gronski M, Cho SK, Ohashi PS, Zuniga-Pflucker JC (2004) Induction of T cell development and establishment of T cell competence from embryonic stem cells differentiated *in vitro*. Nature Immunol 4:410–417.

Shea LD, Smiley E, Bonadio J, Mooney DJ (1999) DNA delivery from polymer matrices for tissue engineering. Nature Biotech 17:551–554.

Silva GA, Zeisler C, Niece KL, Benaish E, Harrington DA, Kessler JA, Stupp SI (2004) Selective differentiation of neural progenitor cells by high-epitope density nanofibers. Science 303:1352–1355.

Smith R, Bagga V, Fricker-gates RA (2003) Embryonic neural progenitor cells: The effects of species, region, and culture conditions on long-term proliferation and neuronal differentiation. J Hematother Stem Cell Res 12:713–725.

Smith MK, Peters MC, Richardson TP, Garbern JC, Mooney DJ (2004) Locally enhanced angiogenesis promotes transplanted cell survival. Tiss Eng 10:63–71.

Soltor D, Gearhart J (1999) Putting stem cells to work. Science 283:168–170.

Soutschek J, Akinc A, Bramlage B, Charisse K, Constien, Donoghue M, Elbashir S, Geick A, Hadwiger P, Harborth J, John M, Kesavan V, Lavine G, Pandey RK, Racie T, Rajeev KG, Rohl I, Toudjarska I, Wang G, Wuschko S, Bumcrot D, Koteliansky V, Limmer S, Manoharan M, Vornlocher H-P (2004) Therapeutic silencing of an endogenous gene by systematic administration of modified siRNAs. Nature 432:173–178.

Tada M, Morizane A, Kimura H, Ainscough JFX, Sasai Y, Nakatsuji N, Tada T (2003) Pluripotency of reprogrammed somatic genomes in embryonic stem cell hybrids. Dev Dynam 227:504–510.

Terada N, Hamazaki T, Oka M, Hoki M, Mastalerz DM, Nakano Y, Meyer EM, Morel L, Petersen BE, Scott EW (2002) Bone marrow cells adopt the phenotype of other cells by spontaneous cell fusion. Nature 416:542–545.

Tessmar J, Kellner K, Schultz MB, Blunk T, Gopferich A (2004) Toward the deveopment of biomimetic polymers by protein immobilization: PEGylation of insulin as a model reaction. Tiss Eng 10:441–453.

Takeshita S, Tsurumi Y, Couffinahl T, Asahara T, Bauters C, Symes J, Ferrara N, Isner JM (1996) Gene transfer of naked DNA encoding for three isoforms of vascular endothelial growth factor stimulates collateral development in vivo. Lab Invest 75:487–501.

Tessmar J, Kellner K, Schulz MB, Blunk T, Gopferich A (2004) Toward the development of biomimetic polymers by protein immobilization: PEGylation of insulin as a model reaction. Tiss Eng 10:441–453.

Thomson JA, Itskovitz-Eldor J, Shapiro SS, Waknitz MA, Swiergiel JJ, Marshall VS, Jones JM (1998) Embryonic stem cell lines derived from human blastocysts. Science 282:1145–1147.

Troy T-C, Turksen K (2005) Commitment of embryonic stem cells to an epidermal cell fate and differentiation in vitro. Dev Dynam 232:293–300.

Vacanti MP, Roy A, Cortiella J, Bonassar L, Vacanti CA (2001) Identification and initial characterization of spore-like cells in adult mammals. J Cellular Biochem 80:455–460.

Vogel G (2005) Embryo-free techniques gain momentum. Science 309:240–241.

Voytik-Harbin SL (2001) Three-dimensional extracellular matrix substrates for cell culture. Methods Cell Biol 63:561–581.

Wakayama T, Tabar V, Rodriguez I, Perry ACF, Studer L, Mombaerts P (2001) Differentiation of embryonic stem cell lines generated from adult somatic cells by nuclear transfer. Science 292:740–742.

Waldmann H, Cobbold S (2004) Exploiting tolerance processes in transplantation. Science 305:209–216.

West JL, Hubbell JA (1999) Polymeric biomaterials with degradation sites for proteases involved in cell migration. Macromolecules 32:241–244.

Wichterle H, Lieberam I, Porter JA, Jessell TM (2002) Directed differentiation of embryonic stem cells into motor neurons. Cell 110:385–397.

Wilan KH, Scott CT, Herrera S (2005) Chasing a cellular fountain of youth. Nature Biotech 23:807–815.

Wilmut I, Beaujean N, de Sousa PA, Dinnyes A, King TJ, Paterson LA, Wells DN, Young LE (2002) Somatic cell nuclear transfer. Nature 419:583–586.

Wood KJ, Sakaguchi S (2003) Regulatory T cells in transplantation tolerance. Nature Reviews Immunology 3:199–210.

Xie Y, Yang ST, Kniss DA (2001) Three-dimensional cell-scaffold constructs promote efficient gene transfection: Implications for cell-based gene therapy. Tiss Eng 7:585–598.

Xu C, Inokuma MS, Denham J, Golds K, Kundu P, Gold JD, Carpenter MK (2001) Feeder-free growth of undifferentiated human embryonic stem cells on defined matrices with conditioned medium. Nature Biotech 19:971–974.

Ying Q-L, Nichols J, Evans EP, Smith AG (2002) Changing potency by spontaneous fusion. Nature 416:545–548.

Young FE (2000) A time for restraint. Science 287:1424.

Young HE, Black AC Jr (2004) Adult stem cells. Anat Rec Part A 276A:75–102.

Zhan X, Dravid G, Ye Z, Hammond H, Shamblott M, Gearhart J, Cheng L (2004) Functional antigen-presenting lecocytes derived from human embryonic stem cells in vitro. The Lancet 364:163–171.

Zhang S-C, Wernig M, Duncan ID, Brustle O, Thomson JA (2001) *In vitro* differentiation of transplantable neural precursors from human embryonic stem cells. Nature Biotech 19:1129–1133.

Zisch AH, Lutolf MP, Ehrbar M, Raeber GP, Rizzi SC, Davies N, Schmokel H, Ezuidenhout D, Djonov V, Zilla P, Hubbell JA (2003) FASEB J 17:2260–2262.

# Index

Page numbers with "t" denote tables; those with "f" denote figures

**D**